The Oxford Handbook of
Transcranial Stimulation

Forthcoming and published titles in the Oxford Handbooks series:

The Oxford Handbook of Evolutionary Psychology
Dunbar and Barrett

The Oxford Handbook of Internet Psychology
Joinson, McKenna, Postmes, and Reips

The Oxford Handbook of Psycholinguistics
Gaskell

The Oxford Handbook of
Transcranial Stimulation

Edited by

Eric M. Wassermann
Brain Stimulation Unit,
National Institute of Neurological
Disorders and Stroke,
National Institutes for Health,
Bethesda MD
USA

Vincent Walsh
Institute of Cognitive
Neuroscience and Department
of Psychology,
University College London,
UK

Charles M. Epstein
Department of Neurology,
Emory University,
Atlanta GA,
USA

Tomáš Paus
Brain and Body Centre,
University of Nottingham, UK,
and Montreal Neurological
Institute, McGill University,
Canada

Ulf Ziemann
Department of Neurology,
Johann Wolfgang
Goethe-University,
Frankfurt am Main,
Germany

Sarah H. Lisanby
Brain Stimulation and
Therapeutic Modulation
Division, Department of
Psychiatry, Columbia University/
New York State Psychiatric
Institute, New York, USA

OXFORD
UNIVERSITY PRESS

UNIVERSITY PRESS

Great Clarendon Street, Oxford OX2 6DP

Oxford University Press is a department of the University of Oxford.
It furthers the University's objective of excellence in research, scholarship,
and education by publishing worldwide in

Oxford New York

Auckland Cape Town Dar es Salaam Hong Kong Karachi
Kuala Lumpur Madrid Melbourne Mexico City Nairobi
New Delhi Shanghai Taipei Toronto

With offices in

Argentina Austria Brazil Chile Czech Republic France Greece
Guatemala Hungary Italy Japan Poland Portugal Singapore
South Korea Switzerland Thailand Turkey Ukraine Vietnam

Oxford is a registered trade mark of Oxford University Press
in the UK and in certain other countries

Published in the United States
by Oxford University Press Inc., New York

© Oxford University Press, 2008

The moral rights of the author have been asserted

Database right Oxford University Press (maker)

First published 2008

British Library Cataloguing in Publication Data
Data available

Library of Congress Cataloguing in Publication Data

The Oxford handbook of transcranial stimulation/edited by Eric Wasserman,
Charles Epstien, Ulf Ziemann.
p.;cm. — (Oxford handbooks series)
Includes bibliographical references.
ISBN 978-0-19-856892-6 (alk. paper)
1. Magnetic brain stimulation—Handbooks, manuals, etc. 2. Mental illness—
Treatment—Handbooks, manuals, etc. I. Wasserman, Eric, Dr. II. Epstein,
Charles M. III. Ziemann, Ulf. IV. Title: Handbook of transcranial stimulation.
V. Series: Oxford handbooks.
[DNLM: 1. Transcranial Magnetic Stimulation—methods.
2. Evoked Potentials, Motor—physiology. 3. Mental Disorders—therapy.
WL 141 O98 2007]
RC386.6.M32094 2007
616.8—dc22
2007045277

ISBN 978-0-19-856892-6

10 9 8 7 6 5 4 3 2 1

Typeset in Minion
by Cepha Imaging Pvt Ltd, Bangalore, India
Printed in Great Britain
on acid-free paper by
Biddles Ltd., King's Lynn, Norfolk

Contents

Section III: The Motor-evoked Potential in Health and Disease 235
Eric M. Wassermann

Section IV: Transcranial Magnetic Stimulation in Perception and Cognition 409
Vincent Walsh

Acknowledgements

The Editors wish to thank Ms Devee Schoenberg for her expert help with many of the chapters.

List of Contributors

Linda S Aglio, Department of Anesthesia, Brigham & Women's Hospital, Boston MA, USA

Vahe E Amassian, Department of Physiology, SUNY Downstate Medical Center, Brooklyn, NY, USA

Andrea Antal, Department Clinical Neurophysiology, Georg-August-University, Göttingen, Germany

Julia Applebaum, Beersheva Mental Health Center, Beersheva, Israel

Robert Belmaker, Beersheva Mental Health Center, Beersheva, Israel

Reiner Benecke, Department of Neurology, University of Rostock, Germany

Alfredo Berardelli, Department of Neurological Sciences, University of Rome 'La Sapienza', Rome, Italy

Sven Bestmann, Sobell Department of Motor Neuroscience & Movement Disorders, Institute of Neurology, University College London, UK

Felix Blankenburg, UCL Institute of Cognitive Neuroscience & Department of Psychology, University College London, UK

Valentin Bohotin, Department of Neurology, Iasi UNniversity, Romania

Cathrin Buetefisch, Department of Neurology, Robert C. Byrd Health Science Center, Morgantown WV, USA

Stefano F Cappa, Department of Neuroscience, Vita-Salute University and S. Raffaele Scientific Institute, Milan, Italy

Robert Chen, Division of Neurology &Toronto Western Research Institute, University of Toronto, Canada

Bella Chudakov, Beersheva Mental Health Center, Beersheva, Israel

Joseph Classen, Department of Neurology, University of Würzburg, Germany

Leonardo Cohen, Human Cortical Physiology Section & Stroke Neurorehabilitation Clinic, NINDS, NIH, Bethesda MD, USA

Alan Cowey, Department of Experimental Psychology, University of Oxford, UK

Laila Craighero, Department of Biomedical Sciences, Section of Human Physiology, University of Ferrara, Italy

Zafiris J Daskalakis, Schizophrenia Program, Centre for Addiction & Mental Health, University of Toronto, Canada

Kent Davey, Center for Electromechanics, University of Texas, Austin TX, USA

Vedran Deletis, St. Luke's/Roosevelt Hospital, New York, USA

Mark Demitrack, Neuronetics Inc, Malvern PA, USA

Joseph Devlin, FMRIB Centre, Department of Clinical Neurology, University of Oxford, John Radcliffe Hospital, Oxford, UK

Vicenzo Di Lazzaro, Institute of Neurology, Catholic University, Rome, Italy

Jon Driver, UCL Institute of Cognitive Neuroscience & Department of Psychology, University College London, UK

Charles M Epstein, Department of Neurology, Emory University School of Medicine, Atlanta GA, USA

Luciano Fadiga, Department of Biomedical Sciences, Section of Human Physiology, University of Ferrara, Italy, and Italian Institute of Technology, Genoa, Italy

Marjorie Garvey, Neuroscience Research Center, National Rehabilitation Hospital, Washington DC, USA

Donald L Gilbert, Division of Neurology, Cincinnati Children's Hospital Medical Center, Cincinnati OH, USA

Benjamin Greenberg, Department of Psychiatry, Brown University Medical School, Providence RI, USA

Nimrod Grisaru, Beersheva Mental Health Center, Beersheva, Israel

Laverne D Gugino, Department of Anesthesia, Brigham & Women's Hospital, Boston MA, USA

Patrick Haggard, Institute of Cognitive Neuroscience, University College London, UK

Mark Hallett, Human Motor Control Section, National Institute of Neurological Disorders and Stroke, NIH, Bethesda MD, USA

Ritsuko Hanajima, Department of Neurology, Division of Neuroscience, Graduate School of Medicine, University of Tokyo, Japan

Ralph Hoffman, Department of Psychiatry. Yale University School of Medicine, New Haven CT, USA

Friedhelm Hummel, BrainImaging and NeuroStimulation Laboratory, Hamburg University Medical Center, Hamburg, Germany

Risto Ilmoniemi, Laboratory of Biomedical Engineering, Helsinki University of Technology, Finland

Alex Kaptsan, Beersheva Mental Health Center, Beersheva, Israel

Jari Karhu, Kuopio University Hospital, Kuopio, Finland

Nicholas Lang, Department Clinical Neurophysiology, Georg-August-University, Göttingen, Germany

Lucy Lee, Wellcome Department of Imaging Neuroscience, Institute of Neurology, University College London, UK

Jean-Pascal Lefaucheur, Service de Physiologie– Explorations Fonctionnelles, Hôpital Henri Mondor, Créteil, France

Andrea J Levinson, Mood and Anxiety Disorders Program, Centre for Addiction & Mental Health, University of Toronto, Canada

David Liebetanz, Department of Clinical Neurophysiology, Georg-August-University, Göttingen, Germany

Sarah H Lisanby, Brain Stimulation & Therapeutic Modulation Division, Department of Psychiatry, Columbia University/New York State Psychiatric Institute, New York, USA

Colleen Loo, School of Psychiatry, University of New South Wales, Sydney, Australia

Paul Maccabee, Department of Neurology, SUNY Downstate Medical Center, Brooklyn NY, USA

Alain Maertens de Noordhout, Department of Neurology, Headache Research Unit, Liège University, Belgium

Michel R Magistris, Unité d'ENMG et des Affections Neuromusculaires, Service de Neurologie, Hôpital Universitaire de Genève, Switzerland

Lotfi Merabet, Center of Noninvasive Brain Stimulation, Department of Neurology, Harvard Medical School, Boston MA, USA

Bertram Moeller, Schizophrenia Program, Centre for Addiction and Mental Health, University of Toronto, Canada

Michael Nitsche, Department Clinical Neurophysiology, Georg-August-University, Göttingen, Germany

Jacinta O'Shea, Department of Experimental Psychology, University of Oxford, UK

Alvaro Pascual-Leone, Center of Noninvasive Brain Stimulation, Department of Neurology, Harvard Medical School, Boston MA, USA

Walter Paulus, Department Clinical Neurophysiology, Georg-August-University, Göttingen, Germany

Tomáš Paus, Brain and Body Centre, University of Nottingham, UK, and Montreal Neurological Institute, McGill University, Canada

Martin Peller, Department of Neurology, Christian-Albrechts-University, Kiel, Germany

Marcella Rameriz, Department of Anesthesia, Brigham & Women's Hospital, Boston MA, USA

Marc E Richardson, Department of Anesthesia, Brigham & Women's Hospital, Boston MA, USA

Mark Riehl, Neuronetics Inc, Malvern PA, USA

Kai-M. Rösler, ENMG-Station & Neuromuskuläre Sprechstunde, Neurologische Klinik, Inselspital, Bern, Switzerland

Rafael Romero, Department of Anesthesia, Brigham & Women's Hospital, Boston MA, USA

Simone Rossi, Dipartimento di Neuroscienze, Sezione Neurologia, Università di Siena, Siena, Italy

Paolo Maria Rossini, Clinica Neurologica, Università Campus Biomedico, Rome, Italy

John C Rothwell, Sobell Department of Motor Neuroscience & Movement Disorders, Institute of Neurology, London, UK

Christian C Ruff, UCL Institute of Cognitive Neuroscience & Department of Psychology, University College London, UK

Elena Rusconi, Institute of Cognitive Neuroscience, University College London, and Inserme, U562 CEA, Saclay, NEUROSPIN, Gif-sur-Yvette, France

Matthew Rushworth, Department of Experimental Psychology, University of Oxford, UK

Francesco Sala, University Hospital, Verona, Italy

Friedhelm Sandbrink, Department of Neurology, Washington VAMC, Washington DC, USA

Jean Schoenen, Department of Neurology, Headache Research Unit, Liège University, Belgium

Simone Schutz-Bosbach, Institute of Cognitive Neuroscience, University College London, UK, and Max Planck Institute for Human Cognitive and Brain Sciences, Leipzig, Germany

Alona Shaldubina, Beersheva Mental Health Center, Beersheva, Israel

Hartwig R Siebner, Department of Neurology, Christian-Albrechts-University, Kiel, Germany

Martin Sommer, Department Clinical Neurophysiology, Georg-August-University, Göttingen, Germany

Arielle D Stanford, Brain Stimulation and Therapeutic Modulation Division, Department of Psychiatry, Columbia University/New York State Psychiatric Institute, New York, USA

Katja Stefan, Department of Neurology, of Rostock, Germany

Frithjof Tergau, Department Clinical Neurophysiology, Georg-August-University, Göttingen, Germany

Yoshikazu Ugawa, Department of Neurology, School of Medicine, Fukushima Medical University, Fukushima, Japan

Sedat Ulkatan, St. Luke's/Roosevelt Hospital, New York, USA

Carlo Umilta, Department of General Psychology, University of Padova, Italy

Stanislav R Vorel, Brain Stimulation & Therapeutic Modulation and Substance Abuse Divisions, Department of Psychiatry, Columbia University/New York State Psychiatric Institute, New York, USA

Vincent Walsh, Institute of Cognitive Neuroscience & Department of Psychology, University College London, UK

Eric M Wassermann, Brain Stimulation Unit, NINDS, National Institutes of Health, Bethesda MD, USA

Kate E Watkins, FMRIB Centre, Department of Clinical Neurology, University of Oxford, John Radcliffe Hospital, Oxford, UK

Alexander Wolters, Department of Neurology, University of Rostock, Germany

Ulf Ziemann, Department of Neurology, University of Frankfurt, Germany

Preface

It is now 27 years since Merton and Morton demonstrated electrical stimulation of the corticospinal tract through the scalp and skull, and 22 since Barker and colleagues did the same with magnetic pulses. Since then, this set of novel ways to activate the corticospinal tract has become a mainstream scientific tool and candidate treatment for brain disorders. The notion of affecting the brain with invisible emanations has had strong imaginative appeal since long before there were concepts of electricity or brain function and, unsurprisingly, work in this area has a high public profile, today.

Between 1985 and 1995, the term "transcranial magnetic stimulation" occurred in 361 papers listed in Medline. The rest of the 3,820 papers that can be retrieved today were published since then. The recent mushrooming of interest was driven primarily by the investigation of TMS as treatment, particularly for depression, but also by the diffusion of transcranial stimulation techniques out of motor physiology and into the study of cognition and perception where they have enabled significant advances, often in concert with functional brain mapping. These advances were supported or made possible by continued painstaking study of the human corticospinal system and the motor evoked potential (MEP). Work in this rich system has revealed important plastic capacities in cortical circuits and potent analogies between the effects of repetitive TMS and the synaptic phenomena of long-term potentiation and depression. Understanding the MEP has enabled investigators to use motor cortex TMS to demonstrate physiological states and disorders that affect the entire cortex and manifest in non-motor phenomena. The human corticospinal system also provided the ideal "preparation" for the rediscovery of the neuromodulatory effects of static electrical fields by clinicians and investigators of the human nervous system.

This book attempts to gather the divergent strands of this field in one place and provide an overview, as well as detailed information in each area. We aimed to make this book not merely complete and scholarly, but useful and accessible to current and prospective investigators in order to facilitate future advances. We thank the many contributors to this book, whose work has made this field so exciting.

SECTION I

Physics and Biophysics of TMS

Charles M. Epstein

Throughout history, whenever a new magnetic or electrical device has been discovered, physicians have immediately applied it to the human body in hopes of better understanding or of cures for devastating disease. The urge to use new techniques therapeutically may at times outrun a full knowledge of their scientific basis. But the principles of electromagnetism that underlie transcranial magnetic stimulation (TMS) were well known more than a century before its introduction by Barker and colleagues, and the failure to develop TMS sooner was due primarily to lack of the necessary high-power electronics. In contrast, direct-current transcranial stimulation was performed decades before but was seldom subjected to controlled trials. High-voltage transcranial pulse stimulation was also introduced prior to TMS, but found application in the operating room only afterwards.

Electromagnetic theory is both elegant and arcane. This section freely dispenses with most of it to concentrate on the portions that are most relevant – and, hopefully, most likely to be comprehended – for users of TMS. The basic circuitry of magnetic stimulators is much simpler than that of transistor radios. Understanding it helps a great deal in comprehending why, for example, only a limited range of TMS coils and waveforms are widely used, and the possible tradeoffs among different waveforms and coil shapes. It is our hope that such comprehension will, in turn, foster more informed and effective use in clinical and research applications.

CHAPTER 1

Electromagnetism

Charles M. Epstein

Electromagnetic induction

Magnetic stimulation follows the fundamental principles of electromagnetic induction: an electric current in the stimulation coil produces a magnetic field, and a *changing* magnetic field induces a flow of electric current in nearby conductors – including human tissue. On the assumption that readers who have mastered Maxwell's equations do not need to see them restated here, and that those who have not mastered them are unlikely to do so from this section, we will focus on those principles that are most relevant to the design and operation of magnetic stimulators.

A simple example of electromagnetic induction is shown in Figure 1.1: if two loops of wire are placed close together, a *changing primary current* in loop 1 produces a changing magnetic field, which generates an electric field, which in turn induces a *secondary current* of opposite direction in loop 2. The secondary currents induced in nearby conductors are commonly called *eddy currents*.

The strength of the electric field is measured by its *electromotive force*, or emf, which is measured in volts. Together, the two loops in Figure 1.1 make up a simple transformer. The directions of the magnetic and electric field lie at right angles to each other (see Figure 4.2). The magnitude of the electric field and the current produced by it are both proportional to the rate of change of the magnetic field:

$$E \sim dB/dt \qquad (1.1)$$

where E is the electric field, B is the magnetic field, and t is time.

The changing primary current also gives rise to an induced voltage in the primary loop itself. Lenz's law of inductance states that the emf induced in an electric circuit always acts to oppose the original changes in current and magnetic flux. Thus a coil of wire in isolation manifests *self*-inductance and develops a 'back emf'. This self-inductance acts to counter changes in the primary current. Inductors embody the following relationship:

$$V_L = LdI/dt \qquad (1.2)$$

where V is the back emf, L is the coil inductance in henrys, and I is the coil current.

Energy is stored in the magnetic field according to the inductance of the coil and the magnitude of the current:

$$J_L = 0.5LI^2 \qquad (1.3)$$

where J_L is the energy in joules.

At its maximum, the inductive energy in the stimulation coil is equal to the total power output of each pulse. Since a single transcranial magnetic stimulation (TMS) pulse may reach hundreds of joules, handling this large energy properly is crucial to stimulator design and places important constraints on pulse waveforms.

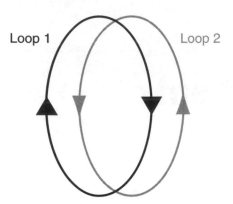

Fig. 1.1 Induction of electric current between adjacent wire loops.

A capacitor is a circuit element that consists essentially of two metal plates with a gap between them (left side of Figure 1.2). A voltage across the plates stores electric charge and energy according to

$$J_C = 0.5CV^2 \qquad (1.4)$$

where J_C is the energy in Joules, C is the capacitance, and V is the voltage across the two plates. Typically TMS capacitors are charged to thousands of volts.

The essential circuitry of a magnetic stimulator can be reduced to just three elements (Figure 1.2). These are the power capacitor, the inductance of the stimulation coil, and a switch which closes to connect them. Usually the switch is an electronic device called a thyristor or silicon-controlled rectifier (SCR). If the switch is closed and the stimulator fires, the simple circuit of Figure 1.2 is effectively reduced to the capacitor and the coil. At all times, the voltages across the capacitor and the coil are equal (though usually expressed as opposite in sign).

When a biphasic TMS pulse begins, all the energy of the circuit is stored in the charged capacitor. As the capacitor discharges and current flows, the energy of the capacitor is transferred to the coil (Figure 1.3). A simple way to look at energy flow through the inductor–capacitor system is to assume that when current is zero, all the energy is in the capacitor; when current is maximum, all the energy is in the coil. One-quarter of the way through a full cycle, the circuit voltage is zero, current is maximum, and all the energy has moved to the magnetic field of the inductor. Halfway through a cycle, the current is zero and all the energy is back in the capacitor but at a voltage that is the reverse of the starting value. Three-quarters of the way through one cycle, voltage is again zero and current is maximum in the opposite direction. At the end of one cycle, the current is zero and the remaining energy has again moved back to the capacitor at its original polarity. The duration of

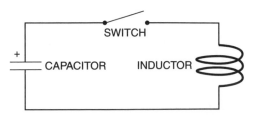

Fig. 1.2 Simplified TMS circuit. The capacitor is first charged to a high voltage, and then discharged into the inductor (the stimulation coil) when the switch is closed. Additional components are needed to stop the circuit from continuing to oscillate indefinitely after a single pulse. See Chapter 3 for more complete circuits.

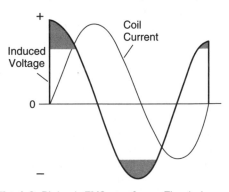

Fig. 1.3 Biphasic TMS waveforms. The darker curve corresponds to both the voltage across the stimulation coil and the induced voltage in the brain. The lighter curve tracks the current passing through the stimulation coil. Shaded areas show the times of highest induced voltage, when neuronal membranes are most likely to be depolarized.

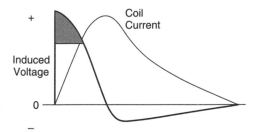

Fig. 1.4 Monophasic TMS waveform. The first portion of the induced voltage curve is the same as in Figure 1.3. Subsequently, however, the coil current is slowly dissipated over 300 μs rather than returning to the capacitor. Only in the initial, shaded segment of the curve is the induced voltage high enough to depolarize cell membranes. The magnitude of the magnetic field produced by the coil, *B*, is proportional to the coil current. Other key parameters including d*I*/d*t*, d*B*/d*t*, *E* (the magnitude of the induced electric field), the magnitude of the induced current density, and the voltage across the stimulation coil are all proportional to the induced voltage waveform.

a full cycle is the resonant period of the LC circuit, determined from

$$T = 2\pi\sqrt{LC} \qquad (1.5)$$

where cycle time T is typically 200–500 μs.

At the end of the cycle, with the capacitor recharged, the current could again start flowing from capacitor to coil, and continue oscillating back and forth indefinitely, except for two things:

(1) the energy would eventually be dissipated in circuit resistance, and

(2) this is the preferred time to turn the switch back off, while the energy has been recaptured in the capacitor and the current is zero.

If we tried to turn the switch off at any other time, the current would begin to fall very rapidly but the coil would continue to obey Equation 1.2. The rapid drop in current would drive d*I*/d*t* (and with it V_L) to a very large negative value, producing in all likelihood the explosive destruction of any circuit elements that stood in the way. With the large pulse energies used in TMS, the need to avoid catastrophic elevations of d*I*/d*t* practically confines pulse durations and waveforms to variations of the curves in Figure 1.3. In the special case of 'monophasic' TMS pulses, the first quarter of the waveform is identical to that in Figure 1.3, but the current is dissipated slowly rather than being allowed to recharge the capacitor (Figure 1.4). In theory, more complex pulse waveforms could be produced with additional switches, capacitors and inductors, but the bulk and weight of the system would increase considerably, with uncertain benefits.

Declaration of conflict of interest

Dr Epstein receives royalties from Neuronetics Inc. through Emory University in regard to iron-core TMS coils. He also receives consulting fees from Neuronetics.

CHAPTER 2

TMS waveform and current direction

Martin Sommer and Walter Paulus

Summary – Waveform and current direction determine the effectiveness of transcranial magnetic stimulation (TMS) in humans. For single-pulse TMS these factors have been shown to influence motor threshold, motor-evoked potential (MEP) latency and silent period duration detected by motor cortex stimulation and phosphene threshold tested by occipital stimulation. For paired TMS pulses they affect the range of inhibition and facilitation detected within the motor cortex, and for repetitive stimulation the extent and specificity of the induced motor cortex inhibition and facilitation. Thus, considering waveform and current direction may turn out to enhance clinical efficacy of repetitive TMS.

Introduction

The difference between biphasic and monophasic TMS is being increasingly noticed (Niehaus *et al.* 2000; Corthout *et al.* 2001; Kammer *et al.* 2001b) and taken into account also in the field of repetitive (r)TMS (Sommer *et al.* 2002b; Arai *et al.* 2005; Tings *et al.* 2005). The exact reason for differences, e.g. with regard to motor threshold, is still a matter of debate, as is the role of the different components of a biphasic pulse *in vivo* and *in vitro* (Maccabee *et al.* 1998; Kammer *et al.* 2001b).

Early TMS stimulators generated single pulses at a repetition rate of <0.25 Hz. Later, combining two (or more) single-pulse stimulators allowed the study of conditioning-test paradigms with interpulse intervals in the millisecond range. Later still, repetitive stimulators with pulse frequencies of up to 50 Hz made it possible to interfere with cortical processing and to lastingly modulate cortical excitability.

During this development little attention was paid to current direction and pulse configuration. The default pulse configuration of TMS stimulators actually changed over time, being monophasic in the early single-pulse stimulators and biphasic in most repetitive stimulators (Figure 2.1). The only area in which the current direction did play a role in the early days of TMS on was the motor threshold, which is lower with the current flowing in a posterior–anterior direction through the precentral gyrus. Hence, when using a round TMS coil, the appropriate coil surface has to be chosen according to the current direction (clockwise or counterclockwise) for an optimized stimulation of the desired hemisphere (Chiappa *et al.* 1991; Claus 2000).

As the induced magnetic field depends on the rate of change of the induced current (Jalinous 1995), the monophasic pulse has one sharp initial quarter cycle, whereas in the so-called biphasic pulse the second and third quarter cycles also contribute essentially to the induced effect (Bohning 2000; Di Lazzaro *et al.* 2001a). Studies in the visual system using biphasic pulses suggest that these later components may be even more important than the initial one (Corthout *et al.* 2001).

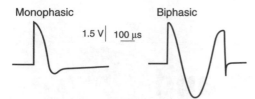

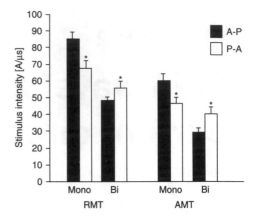

Fig. 2.1 Current induced in a probe coil of 1 cm diameter by different types of transcranial magnetic stimulators, recorded and stored by an oscilloscope. *Left*: Waveform induced by a MagPro stimulator in the 'monophasic' mode. *Right*: Waveform induced in the 'biphasic' mode. For all graphs the same Dantec MagPro stimulator and the same MC-B70 coil were used. Modified from Sommer *et al.* (2002b).

Fig. 2.2 Motor threshold with the target muscle at rest (RMT) or during voluntary target muscle contraction (AMT) of four different TMS types studied in 12 healthy subjects, mean ± SE. Biphasic (bi) or monophac (mono) stimuli with an anterior (P-A) or posterior (A-P) initial current direction (Dantec MagPro stimulator with a MC-B70 coil). *Significant *post hoc* differences between current directions for a particular waveform. Adapted from Sommer *et al.* (2006).

In our studies on rTMS effects (Sommer *et al.* 2001, 2002a) we began to observe differences between our findings and the published rTMS literature. We therefore hypothesized that pulse configuration and current direction might also play an important role in repetitive TMS. Subsequent studies confirmed this hypothesis. For this reason, we review here the relevant literature, following the development of TMS from single to paired pulses and to rTMS.

Single-pulse TMS

An influence of current direction on the motor threshold has been observed for many years. When stimulating the primary motor cortex of the left hemisphere using a round coil centered over the vertex, the current in the coil must flow counterclockwise when seen from above, in order to induce a clockwise current in the brain (i.e. a current flowing anteriorly in the brain) (Chiappa *et al.* 1991; Bohning 2000; Claus 2000). This could be related to the anatomical orientation of the pyramidal tract neurons and their axons (Brasil-Neto *et al.* 1992; Porter and Lemon 1993; Ziemann *et al.* 1996). For the sake of clarity, all current directions in this text refer to the current in the brain.

With regard to pulse configuration, biphasic single pulses yield a lower motor threshold than monophasic pulses (Niehaus *et al.* 2000; Kammer *et al.* 2001b; Sommer *et al.* 2006), suggesting that the second and third quarter-cycles

contribute essentially to the net effect of stimulation (see Figure 2.2). The mechanism for this is still unknown. Epidural recordings in vivo demonstrated a different *D*- and *I*-wave pattern for biphasic than for monophasic pulses, suggesting that either biphasic pulses stimulate interneurons at a different site than monophasic pulses, or that they activate a different subset of interneurons altogether (Di Lazzaro *et al.* 2004).

The visual cortex is sensitive to current orientation as well. The phosphene threshold is lower with latero-medial than with the opposite current orientation (Kammer *et al.* 2001a).

MEP latency depends on waveform and current direction as well. The pattern is similar to that of the motor threshold (Figure 2.3), with longer latencies for the monophasic posteriorly oriented pulses, whereas for the biphasic waveform anteriorly oriented pulses yield longer latencies.

Brasil-Neto *et al.* (1992) studied the amplitudes and the latency of MEPs induced in a small hand muscle and systematically varied the orientation of figure-of-eight coils over the

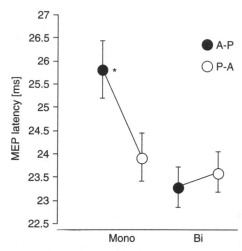

Fig. 2.3 Motor-evoked potential (MEP) latency (ms) of four different TMS types studied in 12 healthy subjects, mean ± SE. Asterisks indicate significant *post hoc* differences between current directions for a particular waveform. Biphasic (bi) or monophasic (mono) stimuli with an anterior (P-A) or posterior (A-P) initial current direction (Dantec MagPro stimulator with a MC-B70 coil). Adapted from Sommer *et al.* (in press).

primary motor cortex. They found an optimal MEP amplitude and the shortest latency with monophasic and biphasic pulses flowing anteriorly in the brain and approximately perpendicular to the presumed location of the central sulcus. Interestingly, for biphasic pulses at higher intensities, the orientational specificity was less marked, and another coil position yielding high amplitudes appeared about 180° opposite to the first. They concluded that the first and the second phase of the biphasic pulse became effective at higher intensities.

Another measure of single pulse TMS, the silent period duration, is sensitive to pulse configuration as well. Orth and Rothwell (2004) compared monophasic anteriorly and monophasic posteriorly oriented pulses with biphasic initially anteriorly oriented pulses. They found a longer silent period with monophasic posteriorly oriented pulses and concluded that inhibitory interneurons are best activated by posteriorly oriented pulses. A recent study by our laboratory essentially confirms these results (Sommer *et al.* in press).

Paired pulses

Few data have been published with regard to current direction or pulse configuration in paired-pulse protocols. The classical and best-studied conditioning-test paradigm to test intracortical inhibition and facilitation within the primary motor cortex has usually been investigated with monophasic pulses (Kujirai *et al.* 1993; Ziemann *et al.* 1996). Peinemann *et al.* (2001) have shown that when using biphasic pulses, the entire excitability curve is shifted towards facilitation, i.e. one detects less intracortical inhibition and more intracortical facilitation. This conclusion is impeded by the lack of a direct comparison between monophasic and biphasic pulses within the same group of subjects.

Chen *et al.* (2003) have studied whether orientation of a single pulse affects its transcallosally mediated inhibitory effect on a test pulse applied over the contralateral motor cortex. They found no such influence from current direction and concluded that the interneurons activating callosal fibers differ from those activating pyramidal tract cells, since the latter are sensitive to the current orientation.

The alternating use of mono- and biphasic pulses as conditioning or test pulse has so far not been possible, since pulses of different waveform or orientation cannot be applied through the same coil at an interval in the millisecond range. Using two different coils could be a feasible approach (Sommer *et al.* 2002c).

Repetitive TMS

Probably the most interesting data exist for rTMS, even though the proof of clinical relevance is still lacking. When repeating the 1 Hz rTMS studies of the NIH group (Chen *et al.* 1997; Muellbacher *et al.* 2000), we noticed that we did not find a significant post-rTMS inhibition of the corticospinal tract as reported there – not with short rTMS series of 80 pulses (Sommer *et al.* 2001), nor with long series of 900 pulses (Sommer *et al.* 2002a). A literature review yielded at least one other group that failed to find an MEP amplitude inhibition after several hundred pulses of 1 Hz frequency (Siebner *et al.* 1999; see Touge *et al.* 2001 for an overview).

The key difference seems to be the stimulator (and therefore the details of the biphasic pulse configurations) used. The NIH group (Chen *et al.* 1997; Muellbacher *et al.* 2000) used the Cadwell rapid-rate stimulator (Cadwell Laboratories Inc., Kennewick, WA, USA); we used a Magstim Rapid stimulator (the Magstim Company, Spring Gardens, UK); Siebner and colleagues applied a Dantec MagPro Stimulator (Dantec S.A., Skovlunde, Denmark) in the biphasic mode (H. R. Siebner, personal communication).

We therefore hypothesized that the reduction of corticospinal excitability after longer series of 1 Hz rTMS is specific for the configuration of the Cadwell stimulator. In an initial study we explored this hypothesis using two extreme pulse configurations available from the same stimulator in our laboratory (Dantec MagPro), monophasic and biphasic pulses, in the same group of 10 healthy subjects, with an established paradigm of 900 pulses at 1 Hz frequency applied over the hand area of the primary motor cortex. We chose a subthreshold intensity (90% resting motor threshold) to avoid potential oscillatory feedback from suprathreshold pulses (Narici *et al.* 1987; Salmelin and Hari 1994). Before and after rTMS, single pulses of the same configuration and type as used for rTMS were applied at 0.25 Hz to control for changes in corticospinal excitability, with an intensity adjusted to yield motor-evoked potentials of ~1 mV at baseline. Monophasic and biphasic pulses with an initial anterior–posterior orientation were studied in separate sessions. rTMS was interrupted every 180 pulses to interleave 10 suprathreshold pulses of baseline intensity to control for ongoing changes of corticospinal tract excitability. We indeed found a much stronger inhibition after monophasic than after biphasic rTMS (Figure 2.4).

In a similar study we tested whether effects of monophasic and biphasic rTMS also differ over the visual cortex (Antal *et al.* 2002). We tested the contrast sensitivity threshold before and after 1 Hz rTMS, applied 5 cm above the inion. Only monophasic, cranially oriented pulses reduced the contrast sensitivity threshold. Inverting the current direction resulted in less pronounced threshold reduction for either type of stimulation. A control for differences in rTMS intensity confirmed the stronger inhibition induced by monophasic rTMS.

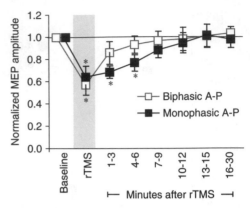

Fig. 2.4 Motor-evoked potential (MEP) amplitude before, during, and after rTMS (mean ± SE) in 10 healthy subjects. Asterisks indicate a significant difference from baseline (*t*-tests, $P < 0.05$). Bins at baseline and during rTMS did not differ significantly from each other and were pooled. After monophasic rTMS the induced corticospinal inhibition is much stronger than after biphasic rTMS. Adapted from Sommer *et al.* (2002b).

These results motivated us to screen faster rTMS frequencies for a potential role of pulse configuration. We therefore studied 80 pulses of suprathreshold 5 Hz stimuli over the hand area of the primary motor cortex and recorded from four contralateral hand or arm muscles to detect the facilitation of the MEP amplitudes and the spread of excitation as described earlier (Pascual-Leone *et al.* 1994). The induced MEP amplitude change depended both on pulse configuration and on current direction, with biphasic stimuli of either direction inducing a moderate facilitation, monophasic anteriorly directed pulses a strong facilitation, and monophasic posteriorly oriented pulses an initial MEP amplitude reduction waning off during the rTMS train (Figure 2.5). There were no lasting effects with this protocol. Voluntary target muscle contraction abolished the early inhibition with monophasic posteriorly oriented stimuli, but left the facilitation with monophasic anteriorly oriented stimuli unchanged (Tings *et al.* 2005).

We put forward three possible explanations for these findings, and we are currently undertaking further studies to test their validity. One explanation refers to the longer silent period

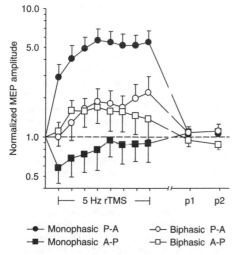

- Monophasic P-A
- Monophasic A-P
- Biphasic P-A
- Biphasic A-P

Fig. 2.5 Motor-evoked potential (MEP) amplitudes before, during and after 5 Hz rTMS of four different stimulus types indicated by different symbols. Mean ± SE. At baseline, 30 pulses were given at 0.1 Hz frequency. During rTMS, 80 pulses at 5 Hz were applied, and the resulting MEPs grouped in bins of 10 each. After rTMS, the baseline was repeated twice. *Significant difference from baseline at *post hoc* t-test ($P < 0.05$). Adapted from Tings *et al.* (2005).

observed with monophasic posteriorly oriented pulses (Orth and Rothwell 2004). Circuits responsible for long-latency intracortical inhibition could be similarly susceptible to the inhibitory effect of posteriorly oriented monophasic stimuli and would then cause the early inhibition seen with that type of stimulus. Alternatively, monophasic pulses might induce early D-waves as observed in some subjects studied with epidural recordings (Di Lazzaro *et al.* 2001b), resulting in a backpropagating excitation of the axon and therefore in a refractory state of the corticospinal neuron, blocking the excitation normally present with early I-waves. The net effect could be an inhibition (Di Lazzaro *et al.* 2002). Finally, we cannot rule out an orientation-sensitive membrane hyperpolarization or depolarization as a cause of the facilitation with anteriorly oriented pulses and the inhibition with posteriorly oriented, monophasic pulses.

Conclusion

Both pulse configuration and current direction affect the modulation of corticospinal excitability induced by rTMS. The effects during rTMS may differ from those outlasting rTMS. Monophasic, posteriorly oriented pulses seem to be particularly apt for inducing an inhibition of neuronal circuits. Further studies are needed to confirm the histological and physiological basis for these differences, and to clarify their clinical relevance. Investigators using rTMS should state the waveform and current direction under investigation.

Acknowledgements

The authors were supported by the Deutsche Forschungsgemeinschaft (DFG grant So-429/01, So-429/02, European Graduiertenkolleg 631 'Neuroplasticity: from Molecules to Systems'), and by an EU Marie Curie Training Site Fellowship (HPMT-2001-00413).

References

Antal A, Kincses TZ, Nitsche MA, Bartfai O, Demmer I, Sommer M, Paulus W (2002). Pulse configuration dependent effects of repetitive transcranial magnetic stimulation on visual perception. *Neuroreport 13*, 1–5.

Arai N, Okabe S, Furubayashi T, Terao Y, Yuasa K, Ugawa Y (2005). Comparison between short train, monophasic and biphasic repetitive transcranial magnetic stimulation (rTMS) of the human motor cortex. *Clinical Neurophysiology 116*, 605–613. Epub 2004 Nov 5.

Bohning DE (2000). Introduction and overview of TMS physics. In: George MS, Belmaker RH (eds), *Transcranial magnetic stimulation in neuropsychiatry*, pp. 13–44. Washington, DC and London: American Psychiatric Press, Inc.

Brasil-Neto JP, Cohen LG, Panizza M, Nilsson J, Roth BJ, Hallett M (1992). Optimal focal transcranial magnetic activation of the human motor cortex: effects of coil orientation, shape of the induced current pulse, and stimulus intensity. *Journal of Clinical Neurophysiology 9*, 132–136.

Chen R, Classen J, Gerloff C, Celnik P, Wassermann EM, Hallett M, Cohen LG (1997). Depression of motor cortex excitability by low-frequency transcranial magnetic stimulation. *Neurology 48*, 1398–1403.

Chen R, Yung D, Li JY (2003). Organization of ipsilateral excitatory and inhibitory pathways in the human motor cortex. *Journal of Neurophysiology 89*, 1256–1264. Epub 2002 Oct 30.

Chiappa KH, Cros D, Cohen D (1991). Magnetic stimulation: determination of coil current flow direction. *Neurology 41*, 1154–1155.

Claus D (2000). Motorisch evozierte Potentiale (MEP). In: Lowitsch K, Hopf HC, Buchner H, Claus D, Jörg J, Rappelsberger P, Tackmann W (eds), *Das EP-Buch*, pp. 173–232. Stuttgart and New York: Thieme.

Corthout E, Barker AT, Cowey A (2001). Transcranial magnetic stimulation: which part of the current waveform causes the stimulation? *Experimental Brain Research 141*, 128–132.

Di Lazzaro V, Oliviero A, Mazzone P, Insola A, Pilato F, Saturno E, Accurso A, Tonali P, Rothwell JC (2001a). Comparison of descending volleys evoked by monophasic and biphasic magnetic stimulation of the motor cortex in conscious humans. *Experimental Brain Research 141*, 121–127.

Di Lazzaro V, Oliviero A, Saturno E, Pilato F, Insola A, Mazzone P, Profice P, Tonali P, Rothwell JC (2001b). The effect on corticospinal volleys of reversing the direction of current induced in the motor cortex by transcranial magnetic stimulation. *Experimental Brain Research 138*, 268–273.

Di Lazzaro V, Oliviero A, Pilato F, Saturno E, Insola A, Mazzone P, Tonali PA, Rothwell JC (2002). Descending volleys evoked by transcranial magnetic stimulation of the brain in conscious humans: effects of coil shape. *Clinical Neurophysiology 113*, 114–119.

Di Lazzaro V, Oliviero A, Pilato F, Saturno E, Dileone M, Mazzone P, Insola A, Tonali PA, Rothwell JC (2004). The physiological basis of transcranial motor cortex stimulation in conscious humans. *Clinical Neurophysiology 115*, 255–266.

Jalinous R (1995). *Guide to the Magstim QuadroPulse*. Dyfed: The Magstim Company.

Kammer T, Beck S, Erb M, Grodd W (2001a). The influence of current direction on phosphene thresholds evoked by transcranial magnetic stimulation. *Clinical Neurophysiology 112*, 2015–2021.

Kammer T, Beck S, Thielscher A, Laubis-Herrmann U, Topka H (2001b). Motor threshold in humans: a transcranial magnetic stimulation study comparing different pulse waveforms, current directions and stimulator types. *Clinical Neurophysiology 112*, 250–258.

Kujirai T, Caramia MD, Rothwell JC, Day BL, Thompson PD, Ferbert A, Wroe S, Asselman P, Marsden CD (1993). Corticocortical inhibition in human motor cortex. *Journal of Physiology 471*, 501–519.

Maccabee PJ, Nagaranjan SS, Amassian VE, Durand DM, Szabo AZ, Ahad AB, Cracco RQ, Lai KS, Eberle LP (1998). Influence of pulse sequence, polarity and amplitude on magnetic stimulation of human and porcine peripheral nerve. *Journal of Physiology 513*, 571–585.

Muellbacher W, Ziemann U, Boroojerdi B, Hallett M (2000). Effects of low-frequency transcranial magnetic stimulation on motor excitability and basic motor behavior. *Clinical Neurophysiology 111*, 1002–1007.

Narici L, Romani GL, Salustri C, Pizella V, Modena I, Papanicolaou AC (1987). Neuromagnetic evidence of synchronized spontaneous activity in the brain following repetitive sensory stimulation. *International Journal of Neuroscience 32*, 831–836.

Niehaus L, Meyer BU, Weyh T (2000). Influence of pulse configuration and direction of coil current on excitatory effects of magnetic motor cortex and nerve stimulation. *Clinical Neurophysiology 111*, 75–80.

Orth M, Rothwell JC (2004). The cortical silent period: intrinsic variability and relation to the waveform of the transcranial magnetic stimulation pulse. *Clinical Neurophysiology 115*, 1076–1082.

Pascual-Leone A, Valls-Sole J, Wassermann EM, Hallett M (1994). Responses to rapid-rate transcranial magnetic stimulation of the human motor cortex. *Brain 117*, 847–858.

Peinemann A, Lehner C, Conrad B, Siebner HR (2001). Age-related decrease in paired-pulse intracortical inhibition in the human primary motor cortex. *Neuroscience Letters 313*, 33–36.

Porter R, Lemon RN (1993). Anatomical substrates for movement performance: cerebral cortex and the corticospinal tract. In: *Corticospinal function and voluntary movement*, Porter R and Lemon RN (eds), pp. 36–89. Oxford: Clarendon Press.

Salmelin R, Hari R (1994). Spatiotemporal characteristics of sensorimotor neuromagnetic rhythms related to thumb movement. *Neuroscience 60*, 537–550.

Siebner HR, Tormos JM, Ceballos-Baumann AO, Auer C, Catala MD, Conrad B, Pascual-Leone A (1999). Low-frequency repetitive transcranial magnetic stimulation of the motor cortex in writer's cramp. *Neurology 52*, 529–537.

Sommer M, Tergau F, Wischer S, Paulus W (2001). Paired-pulse repetitive transcranial magnetic stimulation of the human motor cortex. *Experimental Brain Research 139*, 465–472.

Sommer M, Kamm T, Tergau F, Ulm G, Paulus W (2002a). Repetitive paired-pulse transcranial magnetic stimulation affects corticospinal excitability and finger tapping in Parkinson's disease. *Clinical Neurophysiology 113*, 944–950.

Sommer M, Lang N, Tergau F, Paulus W (2002b). Neuronal tissue polarization induced by repetitive transcranial magnetic stimulation? *Neuroreport 13*, 809–811.

Sommer M, Ruge D, Tergau F, Beuche W, Altenmüller E, Paulus W (2002c). Intracortical excitability in the hand motor representation in hand dystonia and blepharospasm. *Movement Disorders 17*, 1017–1025.

Sommer M, Alfaro A, Rummel M, Speck S, Lang N, Tings T, Paulus W (2006). Half sine, monophasic and biphasic transcranial magnetic stimulation of the human motor cortex. *Clinical Neurophysiology 117*, 838–844.

Tings T, Lang N, Tergau F, Paulus W, Sommer M (2005). Orientation-specific fast rTMS maximizes corticospinal inhibition and facilitation. *Experimental Brain Research 124*, 323–333.

Touge T, Gerschlager W, Brown P, Rothwell J (2001). Are the after-effects of low-frequency rTMS on motor cortex excitability due to changes in the efficacy of cortical synapses? *Clinical Neurophysiology 112*, 2138–2145.

Ziemann U, Rothwell JC, Ridding MC (1996). Interaction between intracortical inhibition and facilitation in human motor cortex. *Journal of Physiology 496*, 873–881.

CHAPTER 3

TMS stimulator design

Mark Riehl

Introduction

Transcranial magnetic stimulators have recently progressed from basic implementations predicated on the technology of earlier single-pulse devices to integrated systems optimized for treatment of specific pathologies. Key factors in the design of such clinically targeted systems are reviewed here with discussion of underlying design principles, procedure-specific features, and clinical safety requirements.

TMS stimulator components

At the core of all TMS stimulators lies the circuitry that drives the treatment coil which in turn produces the magnetic stimulation pulse applied to the patient. This drive circuitry comprises a power source, an energy storage element (typically a capacitor) and a high-power switch which is precisely controlled by a processor that accepts control input from the equipment operator. These components together form the basic TMS stimulator which has been architecturally derived from simple single-pulse stimulator designs and is commonly used for TMS research (see Figure 1.2).

Magnetic pulse generation

The fundamental operating mechanism of a TMS stimulator is to create a changing magnetic field that can induce a current in adjacent conductive material (such as cortical tissue) as prescribed by Faraday's law. The most basic method of generating such a magnetic field pulse it to discharge a capacitor across a coil (Barker *et al.* 1985; Bohning 2000; Ruohonen 2003). The current through the coil changes according to the resonant frequency of the circuit. Since tissue stimulation is provoked by inducing a current of sufficient density in the tissue, which is proportional to the time rate of change of the magnetic flux density (i.e. dB/dt), it is important to design the stimulation circuit to produce the desired dB/dt. The peak magnitude of the magnetic field strength (i.e. the magnitude of the flux density) is often cited as the key specification of a TMS stimulator field, but dB/dt, the duration of the applied pulse, and the volume and location of the stimulated tissue are more fundamental to TMS effects.

Magnetic stimulators commonly drive a coil having an impedance (ratio of voltage to current) of <1 Ω. This very low characteristic impedance means that magnetic stimulators can be quite susceptible to energy losses and overheating from the internal resistance of the stimulating coils and other circuit components. Furthermore, the stimulated neural elements, which appear to be myelinated axons, behave as leaky capacitors, with time constants of a few hundred microseconds (Reilly 1989; Roth and Basser 1990). Stimulation with longer duration pulses is increasingly inefficient, because the initial charge on the axonal membrane has largely dissipated by the time the pulse is over. Therefore efficiency of stimulation is maximized, and circuit heating is minimized, by operation at the highest possible voltage and using pulses of the shortest possible duration. In practice, these design goals are in conflict, because higher-voltage circuit elements

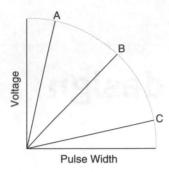

Fig. 3.1 Operating characteristics of magnetic stimulators. Lines A and C, respectively, represent maximally and minimally efficient designs. However, practical considerations generally require tradeoffs between maximum voltage and minimum pulse width, as represented by line B.

A high-voltage power supply is used to charge a capacitor, *C*, typically to 1–3 kV. In this configuration low-cost electrolytic capacitors may be used since the capacitor voltage is clamped by the diode and always remains positive. The coil is represented by the inductance *L*, which is typically in the range of 10–25 μH, and the resistance *R*, which is a combination of winding and cable resistance. Once the capacitor is charged, the trigger circuit gates the semiconductor switch on. The switching device is typically a thyristor such as a silicon-controlled rectifier (SCR) that has minimal losses and can handle peak current on the order of 10 000 A. Once the SCR is on, the capacitor voltage is applied to the coil and current begins to flow. The resonant frequency *F* of the circuit is given by

$$F = \frac{1}{\left(2\pi\sqrt{(LC)}\right)} \tag{3.1}$$

Values of *C* typically range from 100 to 500 μF. Pulse duration for monophasic pulses is usually on the order of 600 μs for designs using electrolytic capacitors. When the current peaks and begins reversing polarity, the diode begins conducting and the energy is dissipated through the circuit resistance. The capacitor must then be completely recharged for the next pulse. This type of circuit is inefficient in that less than half the energy originally stored in the capacitor is utilized to produce the magnetic pulse; the rest is dissipated as heat, and the capacitor must be

generally require longer switching times, are proportionally more expensive, and require more careful and elaborate design. As shown in Figure 3.1, different types of stimulators represent different tradeoffs among these limiting parameters.

Monophasic stimulators

Many basic magnetic stimulators and most early stimulators generate a monophasic coil current (i.e. of a single polarity). A simplified circuit diagram for generating a monophasic pulse is shown in Figure 3.2 (Reilly 1989).

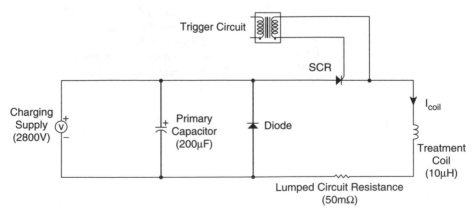

Fig. 3.2 Monophasic pulse circuit. SCR, silicon-controlled rectifier.

recharged from zero voltage which limits the minimum inter-pulse interval.

Electrolytic capacitors are attractive for energy storage because they are a great deal cheaper, smaller, and lighter than nonelectrolytics per coulomb of stored charge. The disadvantage of electrolytics is their relatively low voltage rating and high inductance, which force operation in the least efficient sector of Figure 3.1. However, for single-pulse TMS, systems based on electrolytics are the most compact and least expensive.

Figure 1.4 shows monophasic pulse waveforms for a circuit where $L = 5$ µH. $C = 200$ µF, $R = 50$ mΩ and the initial voltage is 2800 V. If a small pickup loop is placed in the tissue stimulation region of the coil magnetic field, a voltage, V_0, will be imposed across the loop that is nearly proportional to the current density that would be induced in conductive tissue at that location. Observing V_0 is therefore a simple means of estimating the induced current density waveform (J).

Biphasic stimulators

Stimulators that produce biphasic coil current waveforms are particularly useful for applications having short inter-pulse intervals requiring rapid recharging of the capacitor. A simplified circuit diagram for generating a biphasic (i.e. single sinusoidal cycle) pulses is shown in Figure 3.3.

The biphasic circuit is similar to the monophasic circuit except that the capacitor is not clamped, allowing the voltage to swing through both positive and negative values at the resonant frequency. This requires that the capacitor be nonpolarized (e.g. not electrolytic) and that a shunt diode be added the circuit to allow reverse current to bypass the SCR. A key advantage of this design is that the energy from half of the cycle is returned to the capacitor (minus resistive and load losses). At the completion of a full pulse there is a residual voltage on the capacitor of 50–80% of its original value, thus reducing the time required to recharge to full voltage for the next pulse. Not only does this reduce the minimum inter-pulse interval, but the total power required for a treatment session consisting of multiple pulse trains is reduced as well. Of course, good design practice requires management of parasitic resistance, inductance, and capacitance throughout the circuit to optimize performance and to eliminate artifacts and noise.

It is also important to consider that the charging power supply must be able to operate properly while the capacitor is being recharged by the return current from the inductor. This is frequently accomplished by isolating the power supply with a choke coil and clamping reverse voltage with a diode across the supply. Employing a second switching device such as an IGBT (insulated gate bipolar transistor) to isolate the charging supply is another common method. More advanced techniques employ current-limiting, constant-current, or constant-power designs. These approaches can optimize charging time as

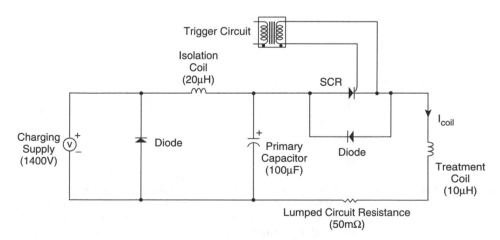

Fig. 3.3 Biphasic pulse circuit. SCR, silicon-controlled rectifier.

well as manage return current. Figure 3.4 shows biphasic pulse waveforms for a circuit where $L = 10\ \mu H$, $C = 100\ \mu F$, $R = 50\ m\Omega$, and the initial voltage is 1400 V.

The electrical characteristics of the coil must be considered along with either type of pulse circuit. As discussed previously, the circuit resonance frequency (and hence the TMS pulse width) is a function of the capacitance and the coil inductance. The coil resistance is also critical as it affects the damping rate and the level of heat generated in the coil winding during a pulse as

given by $P_{coil} \alpha (I_{coil}^2 * R_{coil})$, where P_{coil} is the average power dissipated in coil over a TMS pulse sequence. This relationship also emphasizes the fact that the current should be minimized in order to control coil heating; but to generate a magnetic field B of the desired depth and magnitude, consider that $B \alpha N * I_{coil}$, where N is the number of turns in the coil, so that if the current is reduced, the turns must be proportionately increased which raises the inductance and resistance. All of these factors must be traded off to arrive at a practical and optimized design.

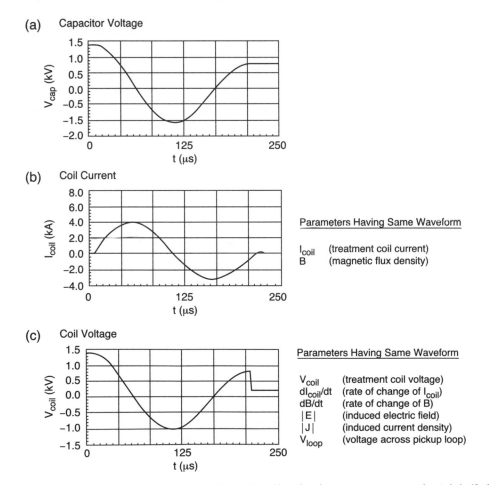

Fig. 3.4 Biphasic pulse waveforms. (a) Capacitor voltage V_{cap} showing recovery to approximately half of the starting voltage. (b) Resulting coil current I_c which is a decaying sine wave that varies as the first derivative of the voltage applied to the coil. The magnetic flux density B produced by the coil also exhibits this same wave shape. (c) Voltage applied to the coil V_{coil} showing a return to zero after completion of one full cosine cycle. This is the primary waveform used to determine TMS pulse width (~200 μS in this example). A number of other important parameters also exhibit this wave shape, including the induced current density in the target tissue and V_{loop}, the voltage observed across a pickup loop placed in the target region.

The TMS stimulator must also incorporate control logic to avoid problematic conditions. For example, inter-pulse intervals must allow for complete charging of the capacitor, and duty cycle limits for the semiconductor devices must not be exceeded.

TMS pulse sequence requirements

The clinical TMS stimulator must be capable of producing a broad range of pulse sequences in order to support the diverse emerging clinical indications and techniques. Let us consider the details of TMS pulse timing and terminology (Figure 3.5). In many TMS clinical applications, such as therapy for depression, pulses are delivered in rapid bursts or pulse trains (i.e. repetitive (r)TMS) lasting 2–10 s followed by a longer quiescent period of typically 10–30 s to prevent overheating and comply with safety guidelines. Within a pulse train the pulses are repeated with a frequency ranging from one to 50 pulses per second (pps), with most clinical protocols falling in the 5–10 pps range. If the pulse repetition rate is ≤1 s the technique is considered 'low frequency'.

The TMS stimulator is designed to control accurately all of the pulse timing as well as to maintain the proper drive level. Most current stimulators are optimized for operation from 5 to 10 pps but new techniques such as magnetic seizure therapy (MST) may require repetition rates up to 50 pps or higher. This places higher demands on the charging circuit in order to maintain the desired drive level.

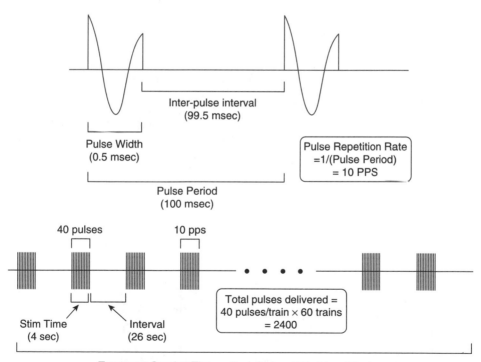

Fig. 3.5 *Upper:* Inter-pulse timing diagram: the timing of an induced electric field or induced current density pulse. Times shown are for illustration purposes and actual clinical protocols may vary significantly. Typical pulse widths are from 0.2 to 0.6 ms and repetition rates are often within the range of 5–10 pulses per second (pps). *Lower:* Pulse train timing. An example pulse sequence where the pulse repetition rate is 10 pps, stimulation time is 4 s, inter-train interval is 26 s, the total of pulses delivered is 2400, and the total treatment session time is 30 min.

Pulse characteristics

Previously we have considered how the TMS magnetic field pulse is created and combined with other pulses to form a useful therapeutic pulse sequence. It is perhaps beneficial to briefly examine the relationships between various pulsed parameters such as $|B|$, $|dB/dt|$, $|E|$ and $|J|$.

Although not treated with mathematical rigor here, these general relationships will provide a basis for discussing standardized calibration methods and consistent means of setting stimulator output levels on a particular patient.

Figure 3.6 illustrates a number of key relationships beginning with the pulsed magnetic field B. As discussed previously, the coil current

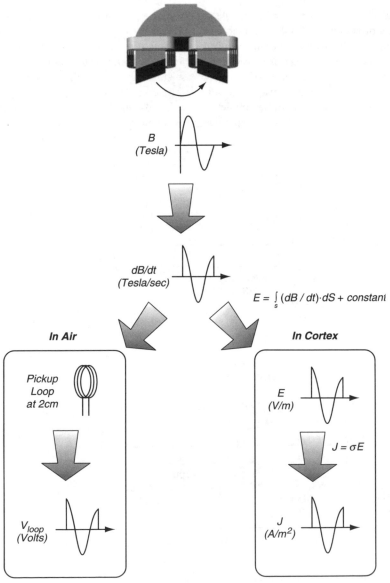

Fig. 3.6 Relationships between magnetic field waveforms and induced current density and induced voltage in a pickup loop. The voltage induced in a pickup loop in air can be a reasonable estimate of the induced current density in the cortex.

and the resulting B field are single-cycle sinusoids for a biphasic TMS system. If the coil is placed on a patient's head, Faraday's law informs us that the time rate of change of B as integrated over a surface normal to the B field yields the electric field E, which is induced in the cortex. The induced electric field in turn causes current to flow, yielding a current density, J. Note that E and J have the same basic wave shape as dB/dt. The magnitude of J is physiologically relevant since the capacity to stimulate a particular small target group of neurons is directly related to the magnitude of J at that target volume. The shape of the field and its orientation relative to the target tissue also play important roles in overall effectiveness of the stimulation, but setting J precisely is essential to consistent stimulation.

Now, instead of placing the treatment coil on the patient's head, consider placing a pickup loop near the treatment coil at the center of the target stimulation volume oriented so that the flux lines are orthogonal to the plane of the loop. The voltage induced in the loop is approximately proportional to dB/dt, and therefore also approximately proportional to E and J that would be induced if the head were in the treatment position. Therefore the voltage induced across a properly calibrated and placed pickup loop can be a reasonable and useful estimate of the induced current density in the cortex at the same spatial location. This is useful in defining a standard metric for setting the TMS stimulator output level which is examined later.

Setting stimulator output levels

Most TMS instruments allow the operator to adjust the output level (often inaccurately referred to as 'power level') by adjusting the voltage applied to the primary capacitor. The drive level corresponds to depth of penetration of the magnetic field into the patient's cortex. In a typical TMS treatment for depression, the drive level is calibrated to each individual patient by adjusting the instrument output with the coil positioned over a specific area of the motor strip and then observing the motor-evoked response. The motor threshold drive level (MT level) and corresponding field penetration may vary over a wide range across a patient population; so if MT is used to set stimulus intensity, then current density, penetration, and the volume of stimulated cortex will change widely as well.

It is desirable to limit stimulation only to the target neurons that produced the desired therapeutic result since stimulation of other neurons could produce unknown effects or could counter the primary therapeutic effect. For example, targeting only the near-surface cortical neurons in the left dorsolateral prefrontal cortex has been shown to be effective in many depression patients. Therefore producing a focused field set to the minimal effective level is a useful strategy. Often, the treatment level is set to a percentage multiple of the MT level and may be referred to as the '%MT'. %MT levels may range from 90% to 120%, or even higher.

Standard motor threshold units (SMTs)

It is beneficial to describe the stimulator output level in terms that can be measured directly (e.g. pickup loop voltage) or that correspond to a physiologically relevant parameter (e.g. E at the center of the stimulation volume) and which are independent of the stimulator. The output level setting has often been expressed as a percentage of the maximum stimulator output level which varies from instrument to instrument. An alternative approach is to define an SMT unit using a process as follows.

First, for a particular stimulator design, determine the average MT level setting for a statistically large group of patients (e.g. 300) using a consistent coil placement method and level setting algorithm. For this average stimulator setting, determine the voltage waveform induced in a standard pickup loop (e.g. 12 turns, 1.0 cm diameter) centered at a point 2.0 cm from the treatment origin and oriented so that the pickup loop is orthogonal to the direction of the B vector at that point. The treatment origin is the central point where the patient's head is tangential to the coil (in the case of a flat coil). For a nonplanar coil (i.e. being concave or having acutely angled pole faces) the treatment origin is the central surface point on a standard sphere (13 cm diameter) that is placed in a normal treatment position against the coil. The idea is to place the pickup loop at a point that would be

2 cm within the patient's head during typical treatment. The voltage waveform measured under these controlled conditions now corresponds to an average MT level setting and can be used to calculate B, dB/dt, E, and J for a specific coil design. Thus, for a particular coil design, using a standardized measurement method, a reference E can be determined at the 2.0 cm point. For example, it has been determined that this reference E is ~140 V/m at the 2.0 cm point for one particular coil design (Figure 3.7).

Note that E in air can be measured more directly over the coil by using a long rectangular loop oriented so that its long axis extends away from the coil down the central axis. The short leg of the loop is parallel to the direction of the electric field and positioned at the point where the field is to be measured (e.g. at the 2 cm point). The electric field at that point is given by the voltage across the loop divided by the dimension of the short axis. This approach is accurate only over the coil center. The small loop method does not yield the electric field as directly, but can be calibrated for a given magnet configuration to yield comparable results with easier setup. Now, 1.0 SMT unit can be defined as the instrument output level setting at which J is 140 V/m at a point 2.0 cm from the surface of a standard TMS head phantom (13 cm diameter cylinder with conductivity 1.0 S/m) placed in treatment position against the coil. This is a useful definition because stimulator output levels can be expressed simply

as a relative multiple of this normalized SMT. For example, 1.2 SMT would produce a 20% higher E and $|J|$. Also, the pickup loop voltage, which has a known relationship with SMT units, can be used as a metric to compare or calibrate SMT levels across different coils or stimulators.

Clinical procedure requirements and components

The clinical TMS system must facilitate the entire treatment procedure to produce a consistent therapy. In order to fashion TMS technology into a clinically practical and usable configuration, additional features must be considered beyond creating the appropriate magnetic field pulse. In particular, proper consideration of the unique procedural aspects of a particular TMS therapy (e.g. treating depression) is essential to a successful result. As indicated in Figure 3.8, patient positioning and comfort, accurate and repeatable coil positioning over the course of many treatment sessions, management and analysis of patient data to perceive trends and response, and incorporation of features to monitor the quality of stimulator operation are all fundamental aspects of a TMS stimulator system intended for routine clinical use by non-research personnel.

A typical session for treating depression may take 10–30 min, so patient comfort features

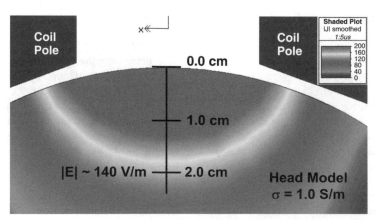

Fig. 3.7 Plot of $|J|$ in a standard head model at a stimulator output level corresponding to the average motor threshold drive level (MT level) of a large patient population. The value at 2.0 cm depth is 140 V/m. (Plate 1)

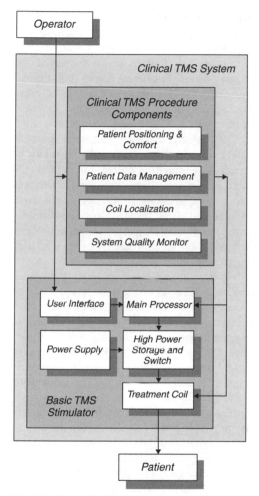

Fig. 3.8 Clinical TMS system components.

the MT position and navigating from that position to the treatment position. In research environments this is often accomplished by marking the patient's head or a swim cap to provide reference points for coil placement. Mechanical alignment approaches and image-guided graphical methods have proven useful, although requiring a three-dimensional image set may be limiting in routine clinical practice. Optical or ultrasonic coordinate measuring systems, such as those used in surgery, offer another accurate method of repeatable coil positioning, but cost may prohibit early adoption of this technology in a typical psychiatric practice. The positioning method and apparatus should be intuitive, repeatable, accurate to within ~1 cm, comfortable, and acceptable to the patient (i.e. minimal restraint, no visible markings on the patient, minimal hair disruption). The clinical system should also incorporate the capability of supporting the coil weight and maintaining its position throughout the procedure without excessively constraining the patient. Since the magnetic field diminishes rapidly with distance from the coil, ensuring stable contact with the patient's head aids in proper determination of MT level and consistent treatment levels. This may be achieved with mechanical means or preferably using a contact-sensing feature that provides real-time feedback to the operator regarding contact status which is available on some systems.

In clinical practice, TMS equipment may be used over many years and moved between locations. Such factors could affect calibration and performance, so incorporation of integral system diagnostics and a means of performing a quality test prior to each treatment should be considered.

Although delivering precise TMS therapy to the patient is the fundamental purpose of the TMS system, it is also important that the procedure be fully integrated into the clinical practice. Features that assist the clinician to recall treatment protocols, review patient treatment history, record observations, and to analyze trends are helpful. As TMS becomes more widely deployed it is likely that a clinical practice may utilize multiple stimulators, requiring that the TMS system also include centralized data management and synchronization capabilities.

are required to improve tolerability, and to minimize disruptions and undesired movement of the patient relative to the coil. A comfortable chair and headrest are very important, especially in patient populations who may not easily tolerate being still for extended periods. Also, since TMS pulses stimulate sensory neurons in the scalp as well as stimulating the cortex, discomfort can become a concern. Devices that reduce the superficial induced E field (Davey and Riehl 2006) have been used successfully and coil designs that minimize the superficial induced fields in the scalp may be helpful in improving tolerability.

The clinical system should also provide the clinician with a straightforward means of locating

Safety considerations

Selecting certain combinations of pulse sequence parameters can possibly lead to treatment that is excessive and which can pose some risks to the patient. The most important safety risk with repetitive TMS reported in the literature is risk of inducing seizure (Agnew and McCreery 1987; Wassermann 1998). Therefore the stimulator should be designed to present default sequences to the operator that have proven to be safe and effective. In addition, the stimulator should impose parametric limits to ensure low-risk operating conditions, both to the patient and to the equipment. TMS pulse parameter guidelines have been published by the 1998 NINDS Workshop that identify a conservative maximum safe stimulation time (in seconds) as a function of %MT, pulse repetition rate and pulse train stimulation time (Wassermann 1998). Informing the clinician when such guidelines are exceeded can minimize the opportunity for induced seizure or unnecessarily excessive treatment.

Generally, electrical and electronic medical equipment is designed to be compliant with widely recognized safety standards such as EN60601-1 (2004). Other similar standards (e.g. UL60601-1) include the same tests but are slightly customized for specific country requirements. Compliance with these standards requires the equipment manufacturer to certify that key safety parameters are within accepted limits. For a Class I device (i.e. where metal enclosures are protectively grounded) such as a TMS stimulator, the following are a few of the most critical and often cited safety requirements from this standard.

- Ground leakage current is not to exceed 300 μA.
- Enclosure leakage current not to exceed 100 μA. (with ground intact)
- Resistance between ground pin on power cord and metal enclosure is not to exceed 0.1 Ω.
- AC input circuitry must withstand 1500 V DC high-potential test for 60 s.
- Patient contact surfaces are not to exceed 41°C.

The last point is particularly relevant to TMS system design. Repetitive TMS requires that large current pulses (2–4 kA) be applied to the treatment coil which produces heat due to resistive and other losses. Efficient coil designs (e.g. those wound on ferromagnetic cores) can reduce this heating by as much as a factor of four, allowing operation with a much higher duty cycle for longer periods of treatment (Davey 2000; Epstein and Davey 2002). Less efficient coils must utilize cooling methods to dissipate the heat or interrupt operation when safety limits are reached.

The standard also addresses other practical safety considerations, such as flammability of materials, elimination of liquid ingress, avoidance of tip over conditions, and restricting access to high-voltage electronics. Single-point failures that could result in unsafe conditions must be eliminated in the design, and those related to shock hazard are tested when certifying to the standard.

There are other safety considerations not fully explored here. These include proper use of human factor analysis to minimize improper operation, biocompatibility of materials touching the patient, and addressing acoustic noise. The TMS coil produces an intense clicking sound when pulsed due to forces in the windings. The sound levels can be high enough to exceed 90 dB near the ear with the coil positioned over the dorsolateral prefrontal cortex. Foam earplugs are usually recommended to reduce exposure by 20–30 dB. Design risk analysis techniques (e.g. failure mode and effects analysis) are generally applied to mitigate these and other unique hazards that may arise from a particular stimulator design.

Conclusions

TMS stimulator design has progressed from basic monophasic designs to complete systems that facilitate a specific clinical treatment procedure. The fundamental pulse characteristics still generally fall into the single-cycle sinusoidal class, allowing straightforward methods of pulse measurement and stimulator calibration. The SMT unit is a useful universal standard for defining a stimulator output level that can be used with different coils and stimulators. The clinical TMS system must also incorporate patient positioning, patient comfort, and coil positioning features, as well as intuitive user controls and means of managing patient data to be a fully effective system. There are a number of specific safety considerations in the design of a TMS stimulator including heating, acoustic noise, and shock hazard. Nearly all safety concerns can be addressed through certification to an industry safety standard (e.g. EN60601-1).

Acknowledgments

The authors acknowledge Ken Ghiron PhD for providing magnetic field calculations and plots.

Declaration of conflict of interest

Mark Riehl is employed by Neuronetics Inc. which designs and conducts clinical trials on TMS systems.

References

Agnew WF, McCreery DB (1987) Considerations for safety in the use of extracranial stimulation for motor evoked potentials. *Neurosurgery 20*(1).

Barker AT, Jalinous R, Freeston IL (1985) Non-invasive magnetic stimulation of the human motor cortex. *Lancet 1*, 1106–1107.

Bohning DE (2000) Introduction and overview of TMS physics. In MS George and RH Belmaker (eds), *Transcranial magnetic stimulation in neuropsychiatry*, Chapter 2.

Davey, K (2000) Magnetic stimulation coil and circuit design. *IEEE Transactions on Biomedical Engineering 47*(11).

Davey KR, Riehl ME (2006) Suppressing the surface field during transcranial magnetic stimulation. *IEEE Transactions on Biomedical Engineering 53*(2).

Epstein CM, Davey KR (2002) Iron-core coils for transcranial magnetic stimulation. *Journal of Clinical Physiology 19*, 376–381.

Epstein, CM, Schwartzberg DG, Davey KR, Sudderth DB (1990) Localizing the site of magnetic brain stimulation in man. *Neurology 40*, 666–670.

Maccabee PJ, Eberle L, Amassian VE, Cracco RQ, Rudell A, Jayachandra M (1990) Spatial distribution of the electric field induced in volume by round and figure '8' coils: relevance to activation of sensory nerve fibers. *Electroencephalography and Clinical Neurophysiology 76*, 131–141.

Reilly JP (1989) Peripheral nerve stimulation by induced electric currents: exposure to time-varying magnetic fields. *Medical and Biological Engineering and Computing 7*, 101–110.

Reilly JP (1992) Electrical stimulation and electropathology. Cambridge university press, Chapter 9, 374–82.

Roth BJ, Basser PJ (1990) A model of the stimulation of a nerve fiber by electromagnetic induction. *IEEE Transactions on Biomedical Engineering 37*(6).

Ruohonen, J (2003) Background physics for magnetic stimulation. In W Paulus *et al.* (eds), Transcranial Magnetic Stimulation and Transcranial Direct Current Stimulation: *Clinical Neurophysiology*, 56(Suppl), Chap. 1.

Wassermann EM (1998) Risk and safety of repetitive transcranial magnetic stimulation: report and suggested guidelines. In International Workshop on the Safety of Repetitive Transcranial Magnetic Stimulation, June 5–7, 1996. *Electroencephalography and Clinical Neurophysiology 108*, 1–16.

CHAPTER 4

TMS stimulation coils

Charles M. Epstein

Circular TMS coils

The simplest TMS coil, and historically the first to be used, forms a simple circle. Typically the coil is 8–15 cm in outer diameter and includes 5–20 turns of wire. As shown in Figure 4.1, a changing current in the coil induces an antiparallel, circular current flow of opposite direction in the underlying brain.

The induced current tends to be maximum near the outer edge of the coil. In contrast, the magnetic field is maximum directly under the center of the coil. This discrepancy is an occasional source of confusion, and may lead to the erroneous assumption that the site of magnetic stimulation is beneath the coil center as well.

Because of their size, most circular TMS coils have good penetration to the cerebral cortex. They are commonly placed at the cranial vertex, where they can stimulate both hemispheres simultaneously. However, the effect on motor cortex is generally asymmetric, especially with monophasic pulse waveforms. Motor activation is substantially greater on the side in which the coil current flows from posterior to anterior across the central sulcus (Hess et al. 1990). Some commercial coils are labelled to show the direction of current flow, but the label does not necessarily clarify the situation. By long-standing and paradoxical convention, current flow is defined as positive to negative. Combined with the opposite directions of primary and secondary currents, this convention can render the definition of coil orientation quite confusing. Sometimes it is easiest to simply place the coil on a normal head and see which side twitches first.

TMS users familiar with this motor asymmetry have occasionally assumed that it represents unequal stimulation of the entire hemisphere. However, the induced magnetic and electric fields are quite symmetric in magnitude. The asymmetry of motor stimulation with a circular coil is probably a function of local cytoarchitecture within area 4 of the cerebral cortex. Extrapolating this finding to other brain functions and regions is likely to be inappropriate.

The main drawback of circular coils is their lack of focality. Not only does the circumference of the coil overlie a large area of brain, but the radius of the strongest field is seldom precisely known. If the coil is tilted so that only one edge contacts the scalp, the region of stimulation becomes smaller but efficiency of stimulation falls drastically.

Figure-8 TMS coils

If two round coils are placed side by side, so that the currents flow in the same direction at the junction point, the induced electric fields will add together and be maximum below the junction (Figure 4.2). This design, known as a 'figure-8' or 'butterfly' coil, allows focal stimulation at a limited and clearly definable location. Because of this greater focality, figure-8 coils are used more often than round coils in research and clinical applications.

The figure-8 coil junction can be extended along a line, producing a 'double-D' configuration and possibly augmentation of the central field. A flat figure-8 coil is a good compromise

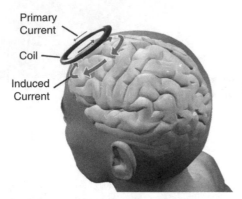

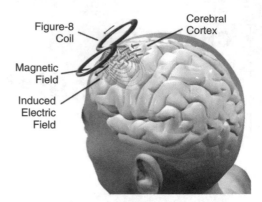

Fig. 4.1 Circular stimulation coil showing opposite directions of primary coil current and induced brain current. The size of the arrows does not reflect the size of the current, which is many orders of magnitude greater in the coil than in the brain.

Fig. 4.2 Magnetic and electric fields produced by a figure-8 coil over left prefrontal cortex. The two narrow black arrows show the current directions in the two side loops, which will sum at the coil junction. If the anterior–posterior coil orientation is defined by the plane of the coil junction, the magnetic field (small gray arrows) lies at right angles to that plane, extending from the center of one side loop to the center of the other. The induced electric field (large gray arrows) is in the plane of the coil orientation, and is largely constrained to lie parallel with the cortical surface.

between efficiency and focality. However, penetration of the induced electric field tends to be more limited than with a circular coil, because the two side loops are usually smaller. In most TMS applications the electric field induced by the side loops can be safely ignored. But with a double-cone figure-8 coil the side fields may approach 50% of the maximum field below the coil center. If stimulator output is increased above 150% of the subject's motor threshold, the field below the side loops will be in a range known to produce CNS effects.

With figure-8 coils, the isopotential lines of the induced electric field form an oval or rounded rectangle, whose long axis is parallel to the direction of current flow at the coil junction (Figure 4.3).

The 'size' of the cortical area stimulated depends on the specific intensity needed to produce a particular effect. In most situations this intensity is not exactly known, so that the area of stimulation in a given application cannot be precisely determined. An exception is the production of TMS motor maps (Chapter XXX): if the cortical representation of the muscle studied is a single point, then ideally the resulting map will be an oval whose shape reflects the electric field contour lines in Figure 4.3. The oval will become larger as stimulus intensity is increased, and its edges will be determined by the locations

at which the induced field falls to threshold for the intended neurophysiological effect.

Optimizing TMS coils

Calculations suggest that, for a circular coil, nearly optimal efficiency is obtained by designing the annular width and height at 60% and 20% of the radius, respectively (Mouchawar *et al.* 1991). A reasonable compromise between coil size and energy efficiency was achieved when the outer radius was twice the target depth. Round coils can also be bent into a spherical 'cap' shape to enclose the head and further focus the magnetic field, with some boost in the peak electric field (Kraus *et al.* 1993). Figure-8 coils folded into a similar configuration are commonly described as 'double-cone coils'.

Mathematical approaches for optimizing coil features such as penetration, focality, and efficiency have been proposed, but their application is not trivial (Roth *et al.* 1994b; Ruohonen *et al.* 1997).

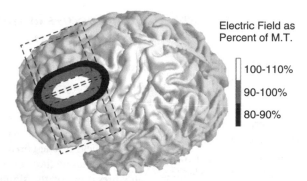

Electric Field as
Percent of M.T.

100-110%

90-100%

80-90%

Fig. 4.3 Isopotential lines of the electric field induced in cortex by a double-square iron-core coil over the left prefrontal region. The coil orientation is shown by dashed lines. Beneath the center of the coil, the peak electric field corresponds to 110% of resting motor threshold (MT).

Specialized TMS coils

Several more complex designs for multiloop coils have been proposed to increase focality or improve penetration to deep brain structures (Roth *et al.* 1994a; Ren *et al.* 1995). Hsu and Durand (2001) described a butterfly coil with two additional wing units and an extra bottom unit, both perpendicular to the plane of the butterfly coil. The wing units produce opposite fields to restrict the spread of induced currents, while the bottom unit enhances the induced fields at the excitation site. Although the extra loops represent additional mass and bulk, localization of the induced fields was improved by a factor of 3 compared with a standard butterfly coil. Efficiency of stimulation was mildly reduced.

Roth *et al.* (2002) designed a coil in the approximate shape of a hatband or tiara, which wraps over the top and sides of the cranium in a coronal plane. The multiple tangential elements of the 'Hesed' or H-coil were designed to reduce the field at the cortical surface while augmenting it at depth. The goal of reducing surface fields was met. However, at equal power, the induced electric field was not actually increased at depth compared to round or figure-8 coils. When driven from a Magstim Super Rapid Stimulator, a Hesed coil was able to activate motor cortex while held a distance of up to 5.5 cm from the head (Zangen 2005). Average motor threshold 4.5 cm from the scalp was 90%

output, comparable to 2.0 cm with the figure-8 coil. In this report, there was no correction for geometric forcing of the induced electric field towards zero in the center of the head. Even without such correction, the efficiency of the H-coil extrapolates to substantially worse than that of the conventional coil in a conventional position.

At the time of writing, the H-coil is the most promising device for directing stimulation towards deep brain structures. However, deep stimulation has yet to be demonstrated, and there are potential limitations in terms of high power requirement, mechanical stability, direction of induced current, and targeting of structures away from the mid-sagittal plane.

Many users of TMS have speculated that combinations of multiple coils should allow precise focusing of the induced electric field, through summation of many fields at different positions within the brain substance. Computer models suggest that this goal is theoretically achievable using dozens of small coils, whose individual efficiency is low (Ruohonen and Ilmoniemi 1998). The power requirement would be many times greater than that of any existing device.

Iron core TMS coils

The efficiency of energy transfer from TMS coils to tissue is extremely small, on the order of 0.0001%. This striking inefficiency is responsible for the high power requirement of magnetic stimulation, the bulky power supplies, and the tendency to overheating. Ferromagnetic materials are widely used to direct and augment magnetic flux for inductive applications, and should confer similar benefits for magnetic stimulation. However, achieving greater efficiency with a core made of iron or other ferromagnetic materials is not trivial. Essential features include:

• Very high saturation magnetization to permit operation at the range of 1.0–2.0 T needed for TMS

• Elimination of core eddy currents – that is, currents induced in the ferromagnetic core itself – which would impair efficiency and lead to overheating

• Practical coil and core geometries that can be achieved without the need for excess mass or esoteric shapes.

The coil configuration illustrated in Figure 3.6 represents one embodiment of a ferromagnetic core (Davey and Epstein 2000; Epstein and Davey 2002). The coil design is basically a figure-8 modified to a double square in order to accommodate ferromagnetic tape windings of equal width. The tape windings are only 1–2 ml thick, run parallel to the magnetic flux, and are insulated from each other to prevent eddy currents. The core extends beyond 180° in order to concentrate the field in a manner similar to a double-cone coil. This coil achieves clinical effects comparable to a conventional figure-8 device, while requiring only about 25% as much power and generating an even smaller fraction of heat. At the same time, it produces greater penetration than the conventional coil (although not comparable to the H-coil). The power supply is correspondingly smaller, lighter and cheaper, and can be built with lower voltage – and thus more reliable – components.

At present, the most efficient iron-core coils have a core mass of about 1.5 lb, a footprint of only 6.5 × 11 cm, and virtually no overheating even at intensive duty cycles. The disadvantages of iron-core coils include the additional cost and mass of the core, the existence of only figure-8 configurations, and the fact that a larger device would be required for monophasic than for biphasic current pulses.

Other factors in TMS coil design

Mechanical forces

The coil currents flowing in a magnetic field produce a mechanical force that pushes the windings apart (Mouchawar *et al.* 1991); this is known as the Lorentz force, and it *defines* the magnetic field B. The Lorentz force is responsible for the audible click when a magnetic stimulator fires. One of the simplest ways to make TMS coils more focal is to make them smaller. However, the Lorentz force increases as the coil becomes smaller, the magnetic flux density becomes tighter, and increasingly larger currents are required to activate the cerebral cortex. Cohen and Cuffin (1991) reported that for prototype figure-8 coils <2.5 cm the casings would fracture during stimulation. Devices smaller than this would appear to be dangerous and impractical, and also far less efficient than large coils for human TMS.

However, small coils are preferred for animal studies, since air-core coils larger than the target brain also have decreased efficiency (Weissman *et al.* 1992). Thus the tradeoff between coil size and mechanical integrity is likely to be most important for studies involving small animals.

Coil lead wires

The lead wire assembly can have substantial effects on stimulation coil characteristics. The coil cable contributes parallel capacitance, serial resistance, and serial inductance, all of which may potentially reduce effective coil output.

Serial inductance can significantly degrade system performance, especially where long lead wires may be desirable, such as in the vicinity of magnetic resonance imaging (MRI) systems. A 2 m pair of heavy-gauge, high-voltage wires may have an inductance of several microhenrys, which adds to the coil inductance and reduces the voltage appearing across the coil by a factor of

$$\frac{L_c}{(L_c + L_L)} \tag{4.1}$$

where L_C is the coil inductance and L_L is the lead inductance. An important design consideration for magnetic stimulators is to keep the coil inductance as low as feasible, which tends to accentuate the effect of excess lead inductance. A lead inductance of 1.1 µH into a stimulation coil inductance of 10 µH will reduce effective coil voltage by 10%. Since output power is proportional to V^2, this 10% decrement in coil voltage translates to an increased power requirement of 21%. Further, the increased total inductance of coil plus lead assembly increases pulse duration, which may produce an additional small loss of efficiency. Twisted pair leads will substantially reduce lead inductance, but at some cost in additional bulk and lead resistance. Coaxial cables, with an inductance around 0.2 µH/m, minimize power losses due to serial lead inductance. They should be preferred for most designs.

Lead and coil resistance also degrade performance, though not equally for all stimulator configurations. This is because the maximum induced electric field occurs at times when dI/dt is highest but coil current I is lowest (Figure 1.3). For the simplest monophasic stimulators, where the entire power capacitor charge is destined to be lost to resistance anyway, the lead wires need only be large enough to avoid local heating. For stimulators producing a biphasic pulse, where the most effective portion of the pulse occurs during the middle of the discharge cycle, the situation is different. Resistive losses in the first current peak will reduce the coil voltage in later portions of the pulse and also reduce the ultimate charge recovery, leading to increased power costs.

Parallel capacitance is unlikely to be a problem. Even using a coaxial cable, lead capacitance is in the range of 80 pF/m. This value is negligible in relation to the power capacitor, which is six to seven orders of magnitude larger.

Coil heating

Rapid TMS with conventional coils at hundreds of joules per pulse quickly results in coil heating. Cooling systems are generally required in coils intended for prolonged high-speed stimulation, although they add substantial weight and bulk. Water has a high specific heat and is an effective coolant, but water is also an effective conductor

at the high voltages used for TMS. Temperature fluctuations and other stresses associated with coil use may eventually crack the coil casing, bringing water into contact with high-voltage components and with the external environment. Oil has been proposed as a safer coolant but has a much lower specific heat, requiring more rapid circulation. Moving air appears to be the simplest and safest coolant for TMS coils, but air circulation is noisy and the systems required to deliver it remain bulky. Coils that are vulnerable to overheating should have an internal heat sensor that will turn power off if the temperature rises too high (41°C).

Electrical safety

To reduce the risk of lethal electrical shock the entire high-voltage power system, including the lead wires and stimulation coil, must be isolated from earth ground. This 'floating' high-voltage supply tends to produce excessive electrical noise from the stimulation pulse and the power supply, and therefore may need to be bypassed to ground with a large resistor and small high-voltage capacitor. The coil must be protected from electrical arcing to the subject, to the iron core if one is present, to system ground, and possibly to nearby control wiring or conductive coolants. At all points electrical insulation must be rated well above the peak coil voltage at 100% output. The coil windings must be insulated sufficiently to prevent internal short circuits, which could damage the system or ignite a fire. It is desirable for the coil connector to include an electrical interlock, which prevents operation unless a secure connection has been established.

Coil windings

TMS coil windings are made of multistranded insulated copper wire. Wire thickness must be sufficient to minimize resistive losses and consequent coil heating, but can be quite variable; the peak coil currents of different stimulator designs vary over a surprising range. Since resistive loss is proportional to the square of the current, and the cross-sectional wire area is proportional to the square of the diameter, the required wire diameters will also vary considerably. Silver is only about 5% more conductive than copper, so the benefit of silver wires would be negligible.

High-frequency alternating current tends to flow only around the outside of the wire bundle, not on the inside. This 'skin effect' reduces the effective cross-sectional area of the coil and lead wires, increasing the coil resistance, energy loss, and coil heating. *Litz wire* is a form of multi-stranded wire in which the strands are individually insulated from each other. Because the voltage difference between adjacent strands is small, the insulation layer can be quite thin. Substantial improvements in coil efficiency have been suggested with the use of Litz wire for the TMS coil and leads.

Flat rectangular wire has been used to allow tighter coil windings. To stabilize TMS coils against the Lorentz force, and to reduce the intensity of the acoustic artifact, the windings are commonly encased in a matrix of epoxy or plastic – a process referred to as 'potting'.

Sham TMS coils

The need for convincing placebo stimulation in research trials of TMS has led to the development of sham TMS coils, which are intended to prevent the subject and even the operator from knowing whether a given session involves real or sham stimulation. Ideally, sham stimulation should reproduce the external appearance of the coil and lead wires, the auditory click and mechanical tapping when it fires, and the complex sensations of scalp muscle contraction and electrical paresthesias that accompany real TMS. This goal has been difficult to reach.

The easiest form of sham stimulation is simply to tilt the coil on edge, outside the visual perception of the subject. Depending on the angle of tilt and the coil type, the resulting fields may be reduced by 24–73% compared to those produced by normal positioning (Lisanby 2001). This approach produces a sense of physical contact and duplicates the auditory click of normal operation. However, the contact sensation is different from that of the active coil surface, the operator is clearly not blinded, and there is concern that fields reaching more than 70% of real stimulation may have biological effects. We have used a sham consisting of an aluminum sheet mounted on the side of an iron core coil,

duplicating the 'footprint' of the normal coil surface. The induced fields in the sham configuration are only 30% of those in the active position, but are sufficient to make the aluminum vibrate with every pulse, producing a palpable impact along with identical auditory and contact sensations.

More recent sham systems are visually indistinguishable from active coils, and are placed on the head in the normal manner. The side loops of a figure-8 coil can be arranged so that the coil currents are opposite at the coil junction, drastically reducing the induced electric field. The most advanced sham coils are integrated with scalp electrodes, which deliver a small current and produce subjective scalp paresthesias at the moment the sham coil is triggered. Such devices are cumbersome, but allow the possibility of true placebo stimulation, in which the operator and the subject are both successfully blinded to treatment condition.

Distribution of the induced electric field

Depolarization of neurons is not due directly to the magnetic field, but rather to the electric field induced secondarily within the brain. In air, and within a few centimeters from the scalp surface, the induced field falls off exponentially at increasing distance from the coil (Figure 4.4). As illustrated in Figures 4.1 and 4.2, the electric field and the currents it produces are strongly constrained to an orientation parallel with the scalp and cortical surface. Radial fields and currents are smaller, and occur only as a result of local variations in tissue conductivity (Wagner 2004). Near the center of the head the induced electric field falls towards zero, regardless of coil configuration or position. This nonintuitive result is a simple consequence of geometry: radial currents are small, and in the center of the head all directions are radial.

Detailed measurement of induced fields can be made in model heads, and several mathematical models of TMS have been proposed (Tofts 1990; Davey *et al.* 2003; Miranda 2003; Wagner 2004). Such models can combine the two-dimensional features of Figures 4.3 and 4.4 into a three-dimensional analysis. With the caveat

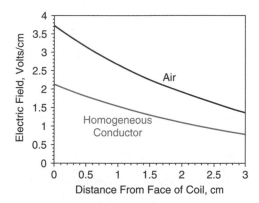

Fig. 4.4 Exponential decay of the induced electric field along a line perpendicular to the center of a figure-8 TMS coil, in air and in a model head containing a homogeneous conductor. Near the scalp surface, the ratio of field intensities in air and in the conductor appears to be consistent at around 0.58, regardless of coil type.

that the sites of maximum current intensity are not necessarily the sites of neuronal activation, the mathematical models can produce accurate estimates of TMS fields and currents. In general the models depend upon several simplifying assumptions, which do not apply to electromagnetic fields in general but are quite appropriate for TMS. The *quasistatic* assumption is that the induced electric field follows an identical time course in all parts of the brain. (We could not safely assume that a radio wave is received at the same instant at all points on earth; but we may do so for a TMS pulse in the human head.) The assumption of *negligible eddy currents* is that the electric currents induced in the brain are so much smaller than those in the TMS coil that they can be ignored in model calculations. (This is the converse of the statement that the coil couples to the brain with extremely low efficiency.) The assumption of *pure resistivity* means that the capacitance of cell membranes in human tissue is small enough that it too can be neglected.

For the most part, the fields and currents predicted by simple homogeneous head models do not differ greatly from those predicted by more complex models, even though the latter attempt to compensate for the different tissue layers and the division of brain into gray and white matter

(Davey *et al.* 2003; Wagner 2004). More realistic modelling of brain anatomy and conductivity may, however, allow better understanding of TMS effects.

Declaration of conflict of interest

Dr Epstein receives royalties from Neuronetics Inc. through Emory University in regard to iron-core TMS coils. He also receives consulting fees from Neuronetics.

References

Cohen D, Cuffin BN (1991). Developing a more focal magnetic stimulator. Part I: Some basic principles. *Journal of Clinical Neurophysiology 8*, 102–111.

Davey K, Epstein CM (2000). Magnetic stimulation coil and circuit design. *IEEE Transactions on Biomedical Engineering 47*, 1493–1499.

Davey K, Epstein CM, George M, Bohning DE (2003). Modeling the effects of electrical conductivity of the head on the induced electric field in the brain during magnetic stimulation. *Clinical Neurophysiology 114*, 2204–2209.

Epstein CM, Davey KR (2002). Iron-core coils for transcranial magnetic stimulation. *Journal of Clinical Neurophysiology 19*, 376–381.

Hess CW, Rosler KM, *et al.* (1990). Significance of the shape and size of the stimulating coil in magnetic stimulation of the human motor cortex. *Neuroscience Letters 100*, 347–352.

Hsu KH, Durand DM (2001). A 3-D differential coil design for localized magnetic stimulation. *IEEE Transactions on Biomedical Engineering 48*, 1162–1168.

Kraus KH, Gugino LD, *et al.* (1993). The use of a cap-shaped coil for transcranial magnetic stimulation of the motor cortex. *Journal of Clinical Neurophysiology 10*, 353–362.

Lisanby SHG, Luber D, Schroeder B, Sackeim CHA (2001). Sham TMS: intracerebral measurement of the induced electrical field and the induction of motor-evoked potentials. *Biological Psychiatry 49*, 460–463.

Miranda PC, Basser PJ (2003). The electric field induced in the brain by magnetic stimulation: a 3-D finite-element analysis of the effect of tissue heterogeneity and anisotropy. *IEEE Transactions on Biomedical Engineering 50*, 1075–1085.

Mouchawar GA, Nyenhuis JA, *et al.* (1991). Guidelines for energy-efficient coils: coils designed for magnetic stimulation of the heart. *Electroencephalography and Clinical Neurophysiology Supplement 43*, 255–267.

Ren C, Tarjan PP, *et al.* (1995). A novel electric design for electromagnetic stimulation – the Slinky coil. *IEEE Transactions on Biomedical Engineering 42*, 918–25.

Roth BJ, Maccabee PJ, *et al.* (1994a). In vitro evaluation of a 4-leaf coil design for magnetic stimulation of peripheral nerve. *Electroencephalography & Clinical Neurophysiology 93*, 68–74.

Roth BJ, Momen S, *et al.* (1994b). Algorithm for the design of magnetic stimulation coils. *Medical and Biological Engineering and Computing 32*, 214–216.

Roth Y, Zangen A, Hallett M (2002). A coil design for transcranial magnetic stimulation of deep brain regions. *Journal of Clinical Neurophysiology 19*, 361–370.

Ruohonen J, Ilmoniemi R (1998). Focusing and targeting of magnetic brain stimulation using multiple coils. *Medical and Biological Engineering and Computing 36*, 297–301.

Ruohonen J, Virtanen J, *et al.* (1997). Coil optimization for magnetic brain stimulation. *Annals of Biomedical Engineering 25*, 840–849.

Tofts PS (1990). The distribution of induced currents in magnetic stimulation of the nervous system. *Physics in Medicine and Biology 35*, 1119–1128.

Wagner TAZ, Grodzinsky MAJ, Pascual-Leone, A (2004). Three-dimensional head model simulation of transcranial magnetic stimulation. *IEEE Transactions on Biomedical Engineering 51*, 1586–1598.

Weissman JD, Epstein CM, *et al.* (1992). Magnetic brain stimulation and brain size: relevance to animal studies. *Electroencephalography and Clinical Neurophysiology 85*, 215–219.

Zangen ARY, Voller B, Hallett M (2005). Transcranial magnetic stimulation of deep brain regions: evidence for efficacy of the H-coil. *Clinical Neurophysiology 116*, 775–779.

CHAPTER 5

Magnetic field stimulation: the brain as a conductor

Kent Davey

Introduction

The brain is an inhomogeneous conductor consisting of white matter, gray matter, and cerebral spinal fluid with conductivities 0.48, 0.7, and 1.79 S/m, respectively (Polk and Postow 1996; Baumann *et al.* 1997). The skull is essentially a zero current density region since its conductivity is about 100 times smaller, 32–80 mS/m (Hoekema *et al.* 2003). Analysis shows that for the purposes of magnetic stimulation, the brain can be treated as a homogeneous conductor; differences in the computations of the induced electric field for the homogeneous and inhomogeneous models are insignificant.

Currents in the brain are induced by a changing magnetic field, but they are too small to influence that field. The induction is a one-way coupling, and thus the problem is not a true eddy current problem. Faraday's and Ampère's laws are easily applied to predict the induced current subject to the condition that the normal component of current density goes to zero at the scalp interface. Iron core stimulators constructed as tape cores are more efficient than air core stimulators. With both air and iron core stimulators, the field will always be higher on the scalp than within the white and gray matter. The higher induced surface fields cause pain in

some patients. This effect can be mitigated by shields and stimulator topologies that spread out the field, but it can never be eliminated. No inversion can ever be realized wherein the induced field is larger at depth than at all places on the scalp.

A properly designed brain stimulation system starts with the target stimulation depth, and it should incorporate the neural strength–duration response characteristics. Higher-frequency pulses require stronger electric fields. At the heart of the process is the transfer of charge across the nerve membrane commensurate to raise its intracellular potential about 30–40 mV. Think of this membrane as a capacitance that behaves more like a short circuit at high frequency. A nerve's chronaxie and rheobase values can be used to dictate the electric field required for stimulation as a function of frequency. The system's parameters can then be chosen to minimize stimulator energy and size.

Changes to TMS stimulators are not likely to come from the superconducting community or the ultracapacitor and supercapacitor community. Because of the large air gaps involved, 3% grain-oriented steels and vanadium cores are about as suitable for standard C cores as one might expect. The malleability offered by modern powdered cores might, however, offer interesting penetration and flexibility options.

Background information

An important question for researchers in this arena is determining exactly where in the brain TMS induces electrical activity, and whether this shifts as a function of differences in conductivity and organization of gray matter, white matter, and cerebrospinal fluid (CSF) (Wassermann et al. 1996; Bohning et al. 1997, 2001). Several effective homogeneous models of the TMS magnetic field have been suggested (Ueno et al. 1988, 1990; Roth and Basser 1990). Liu and Ueno (1998) proposed that when current flows from a lower conductivity region to a higher one, the interface acts as a virtual cathode. An analogy is then drawn to infer the similarities between conventional electric stimulation and magnetic stimulation.

I would support that inference and underscore the fact that at the point that positive ions are driven into a nerve cell, its intracellular potential will rise, and if the rise is sufficient, an action potential results. The nerve cell cannot distinguish whether the rise occurred because of a rapidly changing magnetic field or an imposed electric field. The inner skull boundary condition ensures that the normal component of current density is essentially zero on that boundary.

The cortex is characterized by neural bends and terminations, both of which activate on the electric field, not its gradient. Because of the small conductivity of the cortex, the induced B field is considerably smaller than the source field. For air core stimulators, the magnetic field is dictated entirely by the source current $\vec{J}_s$. With time harmonic stimulation at frequency ω, the electric field is determined by combining Ampère's and Faraday's laws:

$$\nabla \times \nabla \times \vec{E} = j\omega\mu_0\vec{J}_s \qquad (5.1)$$

The electric field boundary condition $\hat{n} \cdot \vec{E} = 0$ must be imposed to ensure that no normal component current exists at the skull interface.

Figure 3.2 shows a typical stimulation circuit in which low-voltage AC is transformed to a higher voltage and then rectified. This higher-voltage DC charges a capacitor which is fired via a thyristor switch into the stimulator core. This circuit goes through one complete resonance cycle before the diode thyristor shuts down and further current flow is prohibited by the diode. During the firing cycle, the circuit can be treated as a simple resistance–inductance–capacitance (RLC) resonance circuit. The current is

$$I(t) = \frac{V}{\omega L} e^{-(\alpha t)} \sin(\omega t),$$

where

$$\alpha = \frac{R}{2L} \qquad (5.2)$$

$$\omega = \sqrt{\frac{1}{LC} - \alpha^2}$$

This is the equation for a damped sinusoid. A typical trace is shown in Figure 3.4b.

Of particular interest is the time and value of the current peak, which can be calculated using typical values of $V = 1.5$ kV, $C = 15\ \mu$F, $L = 11\ \mu$H, and $R = 0.2\ \Omega$:

$$t_p = \frac{\tan^{-1}\left(\frac{\omega}{\alpha}\right)}{\omega} \qquad (5.3)$$

Table 5.1 Neural magnetic stimulation response parameters

	Strength duration curve parameters			
	Rheobase (β)		Chronaxie (γ)	
	Median (V/m)	SD (V/m)	Mean (μs)	SD (μs)
Sensory	6.75	2.06	329	78.4
Motor	16	6.1	203	78.5

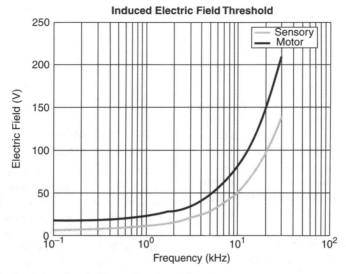

Fig. 5.1 Neural stimulation threshold as a function of frequency.

$$I_p = V\, e^{-\left(\frac{\tan^{-1}\left(\frac{\omega}{\alpha}\right)}{\omega/\alpha}\right)}\sqrt{\frac{C}{L}} \qquad (5.4)$$

Neural response

Motor and sensory thresholds for time-varying magnetic fields are related to the rheobase and chronaxie strength through strength–duration curves. For magnetic stimulation, Geddes had reported the rheobase and chronaxie results summarized in Table 5.1 (Bourland *et al.* 1996). Duration was defined as 'onset to zero', or onehalf-cycle. In terms of the stimulus frequency f, and the table parameters β and γ, the electric field is

$$E = \beta \cdot (1 + 2\gamma f) \qquad (5.5)$$

Figure 5.1 shows the required induced electric field as a function of frequency.

Conductivity considerations

The intent of this section is to give the reader a feel for the shape of the electric field induced in the brain, and to defend the thesis that attention to the exact conductivity distribution within the brain is unwarranted.

Homogeneous model

Consider first the homogeneous model shown in Figure 5.2. The brain is depicted as a homogeneous sphere with radius 7.5 cm. The outer surface corresponds to the skull. A quarter of the problem is worked due to symmetry. The arrows depict how the E field curls around the flux face of the magnet. When the conductivity is halved to 0.37 S/m, the resulting E field differs by a maximum of 0.17%. This was computed by

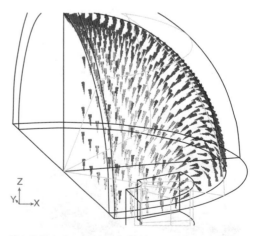

Fig. 5.2 Induced electric field arrow plot modelling the brain with a homogeneous conductivity of 0.75 S/m. (Plate 2)

breaking the quarter brain region into 778 sub-volumes and computing the E field at the center of those volumes.

Inhomogeneous model 1: concentric spheres

Consider positioning the gray and white matter as a number of concentric spherical bands as shown in Figure 5.3. The band pattern has been intentionally altered so that the two eighth sections would themselves exhibit a contrast.

The mean absolute value of the difference between this model and a pure homogeneous model is 6.8%.

Inhomogeneous model 2: concentric wedges with CSF

A final test might suffice to underscore the point being made that paying attention to the distribution of conductivity in the brain is unwarranted. Baumann *et al.* (1997) have shown that the electrical conductivity of CSF is 1.45 S/m at room temperature (25°C) and 1.79 S/m at body temperature (37°C) across the frequency range 10 Hz–10 kHz. Using the latter value, consider analyzing another wedge-shaped model of the brain, this time with CSF distributed in equal volume with white and gray matter as suggested in Figure 5.4.

In this extreme case, the volume of CSF is assumed to be equal to that of gray and white matter. The difference between the E fields increases to a mean absolute value of 15.2% from the homogeneous model because of the higher conductivity of the CSF. The mean of the difference is only 2.6%. The fact that such an extreme distribution of the three fields returns

such a small difference supports the claim that efforts to model the brain's composition accurately are unwarranted. If it is necessary to predict the exact site of stimulation, Wagner *et al.* (2004) make a case for building individualized models with precise locations of the CSF, white matter, and gray matter.

Conclusions about conductivity effects

As long as the tissue conductivity differences are small, two homogeneous models will deliver the same induced E field regardless of the conductivity distribution. The word 'small' applies when the magnetic field generated by the induced currents is insignificant compared with the stimulation field. The total integrated E field around a loop is fixed by the primary B field. The models analyzed contained well-defined borders between white and gray matter. Rather extreme distributions of matter in the brain were modelled to confirm that the induced fields have a mean variance from the homogeneous field of about 7% for white and gray matter, and 15% when CSF is added. The boundary condition that the normal current density must be continuous dictates the maximum departure from the homogeneous electric field case of half the ratio of the conductivities of the media involved.

Suppressing the surface field

The use of strong electric fields to treat many neurological disorders is well established. Both in the treatment of incontinence and clinical depression, the electric field should be sufficiently strong to

Fig. 5.3 Concentric sphere distribution of white and gray matter.

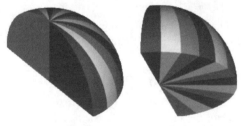

Fig. 5.4 Combination of white matter, gray matter, and cerebral spinal fluid in a wedge-shaped model of the brain.

initiate an action potential. A changing magnetic field induces an electric field within a conducting medium.

The most basic magnetic stimulator is a simple coil of wire, such as those shown in Figures 4.1 and 4.2. Iron core stimulators (Figure 3.6) require a specially constructed tape-wound core to suppress eddy currents. These tape-wound cores require about 25% of the energy of air core stimulators to realize the same induced electric field.

Magnetic-induced electric fields are effective for clinical depression treatment, but their use is sometimes accompanied by discomfort. Both the magnetic field and the induced electric field fall off exponentially with distance from the core in the near-field region. Neurons in the scalp are subjected to the strongest field. Other areas which register problems occur within neurons passing through an opening in bone. This chapter focuses on the first problem. It is an oxymoron to think about no surface field magnetic stimulation. However, there are means by which the ratio of target-depth electric field to surface field can be reduced considerably.

The electric field results displayed are computed using a boundary element solver (Yildir *et al*. 1993; Zheng *et al*. 1996). For simplicity and speed, the results are computed in two dimensions, through a midplane cut through the stimulation core (Figure 5.5).

Methods considered

Surface shield

A surface conductor like that shown in Figure 5.5b has the ability to suppress the local electric field and drive it to zero under the conductor if it is in electrical contact with the scalp. The integral form of Faraday's law

$$\oint \vec{E} \cdot \vec{dl} = -\frac{d}{dt} \int \vec{B} \cdot \hat{n} \, dS \qquad (5.6)$$

makes it clear that this technique will both move and increase the magnitude of the peak field. A better approach is to insulate the conductor from the surface.

Reverse excited secondary coils

A secondary coil can be inserted within the primary coil, between the primary coil and the scalp. If the secondary coil has a reverse excitation from the primary winding, it will lower the surface field. Because the secondary coil is an air core winding, and the winding spread is small, the field penetration into the brain is reduced.

Stretched core

Figure 5.6 shows a magnetic core stimulator with the angle opened considerably. If an inversion of the excitation current is allowed, spanning an angle β, the electric field is suppressed along the vertical midline.

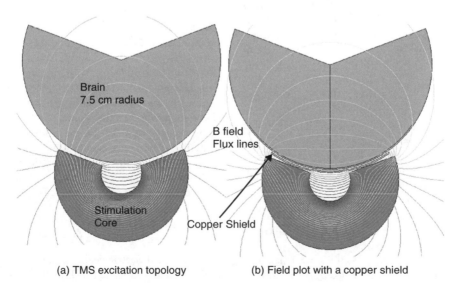

(a) TMS excitation topology   (b) Field plot with a copper shield

Fig. 5.5 Transcranial magnetic stimulation *B* field plot without (a) and with (b) a passive copper shield.

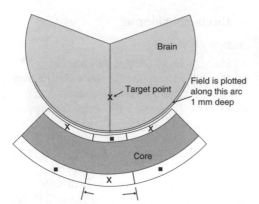

Fig. 5.6 Opening the angle of the core yields deeper field penetration.

Stretched core, no reversed excitation winding

Figure 5.12 shows a simpler core without reversed excitation. The core angle is opened further, from 90 to 140°. At each opening, compute the correct excitation under saturation so that the induced E field 3 cm down remains at 1 V/cm.

A comparison of the three methods

Three methods have been suggested for mitigating the surface field: (a) using a passive conducting shield; (b) actively exciting a smaller coil with opposite field polarity; and (c) opening up the excitation core angle. Of the three methods, the third is favored. The reverse field excitation has the drawback that it requires either another power supply or additional load to the existing supply. If that reverse field excitation fails or suffers a phase lag, the patient will suddenly experience more pain. The passive shield is rated second because there is little probability that it will fail. However, it suffers the problem that an additional load is imposed on the stimulator supply as well. The more serious difficulty is the heat dissipation within the shield. Continuous excitation will register a possibly dangerous rise of the shield temperature. The open angle core poses the least additional burden on the stimulator power supply, and it does not suffer the safety problems of the reverse excitation or thermal shield.

One option to suppress the surface field is to superimpose a secondary electric field with electrodes. Surface electric fields are placed normal to the induced surface field. Every other electrode must be excited with independent potential sources, and chosen to inject a current opposite to the induced current. The skull insures that the injected current does not penetrate into the white and gray matter. Implementation problems will probably preclude developing this system.

Optimization of magnetic stimulators

What constitutes an optimized system? Among the items that might be optimized are the following:

- Capacitor size
- Stimulation coil size
- Voltage
- Energy.

Energy involves both the capacitor and the voltage. The number of turns N increases the resistance and lowers the peak current in Equation 5.4.

Air core

An air core optimization is simplest. Many finite element and boundary element programs are suitable for analyzing this type of problem. Since the air core represents a linear analysis, a three-dimensional boundary element analysis (Zheng 1997) is employed to predict the electric field as a function of depth for various core sizes.

Analytic optimization

Consider an air core in which energy is to be minimized and the core shape is known. If the shape is known, then the problem can be solved using a numerical solver for the induced field E_0 at desired depth, at current I_0 and radian frequency ω_0. The actual induced electric field will scale from this value by the number of turns N, the actual peak current I_p, and the frequency ω:

$$E = NE_0 \left(\frac{I_p}{I_0} \right) \cdot \left(\frac{\omega}{\omega_0} \right) \tag{5.7}$$

The induced electric field is required to satisfy the requirement dictated in Equation 5.5; this can be interpreted as a requirement on voltage V:

$$V = NI_0 \left(\frac{\omega_0}{\omega}\right)\left(\frac{\beta(1+\gamma\omega/\pi)}{E_0}\right) e^{-\left(\frac{\tan^{-1}\left(\frac{\omega}{\alpha}\right)}{\omega/\alpha}\right)} \sqrt{\frac{L}{C}}$$

(5.8)

$$W = \frac{1}{2}L_0 I_0^{\ 2}\left(\frac{\omega_0}{\omega}\right)^2 \left(\frac{\beta(1+\gamma\omega/\pi)}{E_0}\right)^2 e^{-2\left(\frac{\tan^{-1}\left(\frac{\omega}{\alpha}\right)}{\omega/\alpha}\right)}$$

(5.10)

Let L_0 represent the inductance of the core with one turn. The resistance is somewhat complicated because it must account for heat lost in the thyristor and the wire. As will be seen shortly, it must also account for the eddy and hysteresis loss in the core if it is magnetizable. For the moment, consider only the loss from the wire, and consider the core to be filled with wire so that additional turns are added at the expense of a smaller cross-section. In this approximation, inductance and resistance will scale as N^2:

$$L = N^2 L_0$$
$$R = N^2 R_0$$

(5.9)

The energy can be written in terms of the two remaining unknowns C and N as

Consider the one-turn circular air core stimulation coil. The inductance is 0.004 µH for an ID = 3.214 cm, OD = 10.66 cm, and height = 5.9 cm. The core induces an electric field of 4.273 mV/m with 1 A of excitation with characteristic frequency 5.208 kHz. Using these parameters in Equation 5.10 yields the energy requirement shown in Figure 5.7 for a spread of capacitance values and number of turns. The equations clearly suggest the use of a small number of turns and a large capacitor. As will be seen shortly, when more realistic relationships are employed to relate resistance and inductance to the number of turns by incorporating parasitic lead inductance and resistance loss from the thyristor and core, this trend will change.

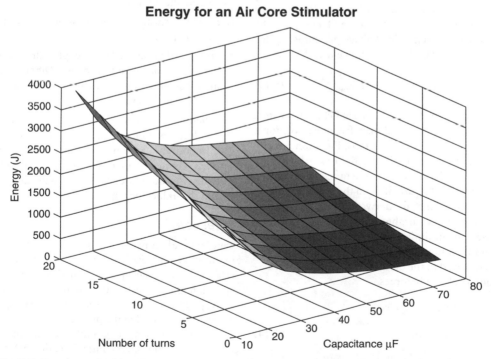

Energy for an Air Core Stimulator

Fig. 5.7 Energy required for stimulation at a depth of 1 cm for a spread of capacitance and turns options.

Numerical optimization

When the problem is considered as a four-parameter optimization in the variables C, V, N and core size x it can no longer be solved analytically. A numerical approach allows the parameters, such as resistance, to be treated more realistically, with the inclusion of proper bounds on voltage. Assume the core size to be a scale parameter x, scaling all the dimensions equally from the core origin. If ζ_0 represents the length of the core winding with one turn, then the length ζ of the winding with N turns, scaled by a value x, is

$$\zeta = N x \zeta_0 \qquad (5.11)$$

The combined resistance of both the parasitic core resistance and the diode R_0 with one turn is about 20 mΩ. A reasonable approximation to the resistance to be used in Equation 5.2 is

$$R = R_0 \left(0.9 + 0.1 \frac{\zeta}{\zeta_0} \right) \qquad (5.12)$$

The leads have a parasitic inductance $L_{\text{Parasitic}}$ equal to about 3 μH. Allow the core to vary through a spread of sizes and compute the inductance as the flux linkage per amperage for each size $L_0(x)$. The inductance with N turns is

$$L = N^2 L_0(x) + L_{\text{Parasitic}} \qquad (5.13)$$

Compute the induced electric field $E_0(x)$ with a current of I_0 A at radian frequency ω_0 for a spread of stimulation depths. Equation 5.7 dictates the induced electric field as delivered by the stimulator. If the inductance and induced electric field E_0 are fitted to the core size using a smooth spline, its derivative can be approximated and a variable metric procedure can be used to minimize an optimization index. If a combination of energy and stimulator core size is involved in the design objective, the optimization problem becomes

$$\text{Min } \Im = \frac{1}{2} C V^2 x$$

Subject to $\qquad\qquad\qquad (5.14)$

$$N E_0 \left(\frac{I_p}{I_0} \right) \cdot \left(\frac{\omega}{\omega_0} \right) = \beta \left(1 + \frac{\gamma \omega}{\pi} \right)$$

The peak stimulator current I_p is determined from Equation 5.4. This index is one of many options open to the designer. One of the applications motivating this research was using these stimulators in the field for alertness assistance. In such mobile contexts, minimizing size and energy consumption is warranted.

Results of the air core numerical optimization

Stimulation depth is a key parameter in the optimization. Figure 5.8 shows how the system energy changes with target stimulation depth. Here a core shell with ID = 1.836 cm, OD = 6.096 cm, and height = 3.38 cm is scaled in all dimensions by a scale factor which varied from 1 to 1.75. The capacitance was allowed to vary from 5 to 75 μF, the voltage from 500 V to 3 kV, and the number of turns from 2 to 18. The problem has many local minima. A Monte Carlo method is employed to randomly vary the starting guess to increase the probability that the global minimum is found.

Table 5.2 shows the results of the optimization for each of the parameters. Among the key lessons are the following:

1. Smaller cores are desired for the lower stimulation depths.

2. Deeper stimulation target depths drive up both the capacitance and voltage. The voltage increases more slowly since it affects the optimization by its square.

3. When parasitic losses such as the switching and lead resistance are considered, the optimization always favors a higher number of turns. The neural response shown in Figure 5.1 is driving the frequency down with depth, and the inductance up.

Steel core

Tape wound cores substantially reduce the required system size and energy requirements (Davey et al. 1989; Epstein and Davey 2002), although their construction is more difficult (Davey and Epstein 2000). This advantage comes at the price that the problem is nonlinear. The nonlinear element complicates the optimization in two respects. First, Equation 5.2 no longer describes the current. The magnitude will be dictated by the degree of saturation.

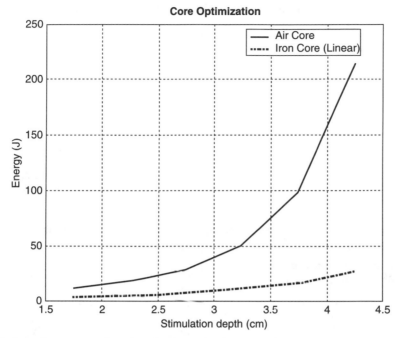

Fig. 5.8 Optimized energy as a function of various stimulation depths.

Second, the frequency is no longer a simple index. A core in saturation is characterized by a higher frequency and a lower amplitude. A Fourier decomposition must be performed to determine the fundamental frequency amplitude and at least the first harmonic.

Linear

Much is to be learned by examining what should be expected from a steel core. The gap is very large. Treating the core as linear with a relative

permeability of 1000 is a reasonable approximation. The energy drops with steel core in this approximation. Since the inductance is so high, the optimization parameters take a different posture. Important trends with iron cores are the following:

1. Deeper stimulation target depths require larger cores, as with air cores.

2. Because of the high inductance, low capacitance is desirable.

Table 5.2 Optimization results for an air core stimulator

Stimulation depth (cm)	Scale parameter	Capacitance (µF)	Voltage (kV)	No. of turns	Frequency (kHz)	Stimulation current (kA)
1.75	1.1621	5	2.1368	18	26.3319	30.6324
2.25	1.3918	5	2.6897	18	24.9194	36.3856
2.75	1.6752	6.1056	3	18	21.2232	41.8804
3.25	1.75	10.9834	3	18	15.5839	54.3823
3.75	1.75	21.7783	3	18	11.0567	74.5978
4.25	1.75	47.6275	3	18	7.4599	105.7719

Table 5.3 Optimization parameters for a linear iron core stimulator using $\mu_r = 1000$

Stimulation depth (cm)	Scale parameter	Capacitance (µF)	Voltage (kV)	No. of turns	Frequency (kHz)	Stimulation current (kA)
1.75	1.0022	5	1.0616	17.9702	13.7545	8.0951
2.25	1.2508	5	1.2975	15.2522	14.4552	8.8111
2.75	1.3283	5	1.6264	14.5454	14.4532	10.5291
3.25	1.3579	5	2.0476	14.334	14.4539	13.0631
3.75	1.3738	5	2.575	14.2388	14.4533	16.3166
4.25	1.3913	5.9533	3	13.7745	13.558	20.4981

3. As with air cores, voltage must increase with target stimulation depth.

4. Deeper target depths are commensurate with lower stimulation frequencies, and a lower frequency.

The required stimulation current increases nearly linearly with depth (4.25/1.75 = 2.43; 20.49/8.09 = 2.53). By contrast the required air core amp-turn excitation increases by 105.7/30.6 = 3.45. The iron core field does not fall off as rapidly with distance (Table 5.3).

Saturable cores

The analysis becomes nonlinear with real magnetizable cores. This complicates the analysis details, but the approach is unchanged. Discussion of this can be found in Davey and Riehl (2005). Table 5.4 shows the results of the nonlinear analysis allowing the capacitance to vary from 5 to 35 µF, the number of turns from 1 to 18, the voltage from 400 V to 1.5 kV, and

the core scale parameter from 1 to 1.75, using a parasitic inductance of 4.5 µH. Please note that the saturable core requires less energy; the linear core was modelled with a permeability of 1000; the saturable M-19 cores have a low field permeability of nearly 6000.

Cortical stimulation

The results quoted are consistent with the strength duration information of Table 5.1 and dependent on the optimization criteria targeted which could in general be different from Equation 5.14. The presence of a myelin sheath will increase both β and γ in Table 5.1. Geddes (1999) shows some of the variation of γ. More reasonable values of β and γ for the cortex are suggested by this team to be 32 V/m and 406 µs, and yield the optimization results in Table 5.5. Doubling these parameters from their former values in Table 5.1 increases the energy by more than an order of magnitude. Until data analogous to Table 5.1 are available for the cortex,

Table 5.4 Results for a nonlinear core analysis

Target depth (cm)	Scale parameter	Capacitance (µF)	Voltage (kV)	No. of turns	Frequency (kHz)	Stimulation current (kA)	Energy (J)
1.75	1.7167	15.8764	0.4206	13.655	6.2797	1.4495	1.4041
2.25	1.75	35	0.4	13.984	4.0998	2.0442	2.8002
2.75	1.5021	34.9972	0.4002	18	3.54	2.3122	2.8027
3.25	1.75	34.9537	0.5752	18	3.2554	3.3454	5.7818
3.75	1.75	34.9579	0.6896	18	3.2552	4.0112	8.3118
4.25	1.75	34.9994	0.8136	18	3.2533	4.7351	11.583

Table 5.5 Results for the nonlinear core analysis after doubling the rheobase and chronaxie values

Target depth (cm)	Scale parameter	Capacitance (μF)	Voltage (kV)	No. of turns	Frequency (kHz)	Stimulation current (kA)	Energy (J)
1.75	1.7353	34.9637	1.0384	17.867	3.2801	5.685	18.851
2.25	1.747	27.5194	1.3949	18	3.6591	6.8373	26.772
2.75	1.75	27.176	1.75	18	3.7858	8.3957	41.613
3.25	1.75	62.4074	1.75	18	2.772	13.5831	95.561
3.75	1.7501	112.4891	1.75	18	1.794	27.2075	172.25
4.25	1.7502	207.656	1.75	18	1.342	45.9637	317.97

the higher values of 32 V/m and 406 μs for β and γ appear reasonable, since they are consistent with excitation levels in the laboratory.

Conclusions of optimization considerations

The frequency, system voltage, capacitance, core stimulator size, and number of turns should be treated as unknowns in a TMS stimulation design. Based on the neural magnetic stimulation response parameters, and the electric field as computed through a boundary element solver, the ideal parameters for the system can be derived. A trust region technique is used to solve the four-parameter optimization problem. The result is target depth dependent, and is certainly dependent on the shape of the stimulation coil. Deeper targets are commensurate with lower excitation frequency and higher amp-turn products. Rheobase and chronaxie values of 32 V/m and 406 μs appear to be consistent with laboratory data.

Possible topological changes to consider in the future

Vanadium and 3% grain-oriented steel cores are already near the top of what a researcher might desire for C-shaped cores. AC superconductors are not likely to help soon, since their magnetic field limitations in frequency and field are below that of contemporary room temperature devices. Operation at 5 kHz would pose serious problems to the superconductor's stability. Both super- and ultracapacitors are characterized by low voltage.

There is, however, at least one topological change that might prove interesting.

Consider a bowl of powdered iron with a doughnut-shaped cavity as suggested in Figure 5.9. The cavity will house a toroid excitation coil. Tape-wound cores are typically C-shaped with the winding in the center of the C. Among the advantages of the core proposed in Figure 5.9 is a much improved magnetic circuit, with lower magnetic reluctance than that possible with the C core. Iron powder also has a higher field saturation than 3% grain-oriented silicon steel (2.15 versus 1.85 T) albeit with higher hysteresis loss. The field is able to spread out through 360° to return around the excitation coil as shown in Figure 5.10. The 7.5 cm brain is modelled with conductivity 0.48 S/m. Note how the magnetic field concentrates in the

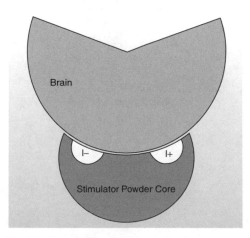

Fig. 5.9 Proposed powder iron stimulator core.

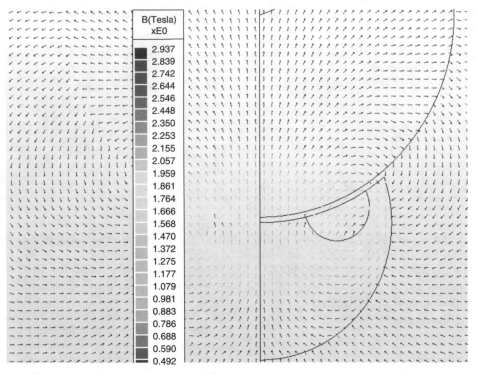

Fig. 5.10 Magnetic field for a 4 inch outer diameter stimulator excited with 20 000 AT at 5740 Hz. (Plate 3)

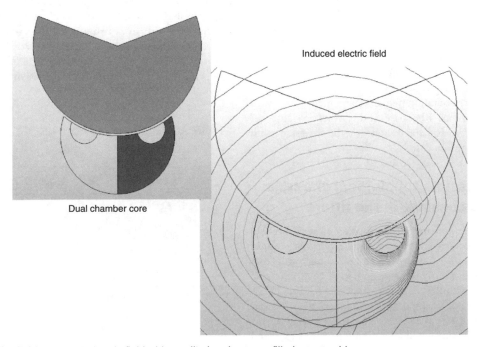

Fig. 5.11 Induced electric field with a split chamber core, filled on one side.

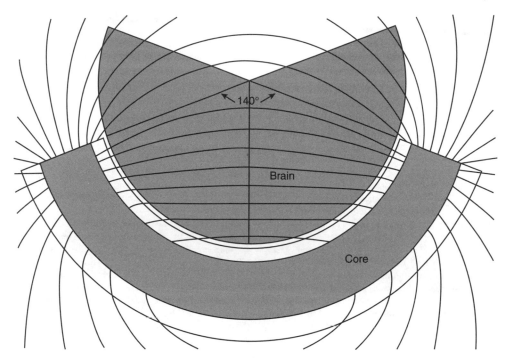

Fig. 5.12 Core opened to 140°, 3790 AT excitation at 5280 Hz.

center of the core but is quite small on the perimeter. Since the stimulator is bowl-shaped, the analysis is performed with an axysymmetric computation. Since cranial threshold occurs near 1 V/cm, it is calculated that for the excitation chosen, stimulation would occur down to 3.5 cm. By contrast, a corresponding C core stimulates down to only 3 cm under the same excitation.

The excitation region generated by the rotational symmetric geometry of Figure 5.9 is disk-shaped. Suppose an asymmetric excitation is desired. Assume the powder core region is built with partitions. The simplest partition would be a dual-chamber geometry as suggested in Figure 5.11. Note how the induced electric field focuses on the right chamber when the left chamber is not filled with iron powder. It should be clear that splitting the bowl into four chambers and filling one of the four will further focus the volume of neurons being excited.

In practice, the powder would be housed in a plastic reinforced bag and molded around the excitation coil. An inherent advantage would be the ability to shape the surrounding structure to the contour of the patient's head in a customized fashion.

References

Baumann SB, Wozny DR, Kelly SK, Meno FM (1997) The electrical conductivity of human cerebrospinal fluid at body temperature. *IEEE Transactions on Biomedical Engineering 44*, 220–223.

Bohning DE, Epstein CM, Vincent DJ, George MS (1997) Deconvolution of transcranial magnetic stimulation (TMS) maps. *Neuroimage 5*, S520.

Bohning DE, He L, George MS, Epstein CM (2001) Deconvolution of transcranial magnetic stimulation (TMS) maps. *Journal of Neural Transmission 108*, 35–52.

Bourland JD, Nyenhuis JA, Noe WA, Schaefer JD, Foster KS, Geddes LA (1996) Motor and sensory strength–duration curves for MRI gradient fields. *Proceedings of the International Society of Magnetic Resonance in Medicine*, 4th Scientific Meeting and Exhibit, New York, NY, p. 1724.

Davey K, Epstein CM (2000) Magnetic stimulation coil and circuit design. *IEEE Transactions on Biomedical Engineering 47*, 1493–1499.

Davey K, and Riehl M (2005) Designing transcranial magnetic stimulation systems. *IEEE Transactions on Magnetics 41*, 1142–1148.

Davey KR, Cheng CH, Epstein CM (1989) An alloy–CORE electromagnet for transcranial brain stimulation. *Journal of Clinical Neurophysiology 6*, 365.

Epstein CM, Davey KR (2002) Iron-core coils for transcranial magnetic stimulation. *Journal of Clinical Neurophysiology 19*, 376–381.

Geddes LA (1999) Chronaxie. *Australasian Physical and Engineering Sciences in Medicine 22*, 17.

Hoekema R, Wieneke GH, Leijten FS, van Veelen CW, van Rijen PC, Huiskamp GJ, Ansems J, van Huffelen AC (2003) Measurement of the conductivity of skull, temporarily removed during epilepsy surgery. *Brain Topography 16*, 29–38.

Liu R, Ueno S (1998) Stimulation of the influence of tissue inhomogeneity on nerve excitation elicited by magnetic stimulation. *Proceedings of the 20th Annual International Conference of the IEEE/Engineering in Medicine and Biology Society*, pp. 2998–3000.

Polk C, Postow E (1996) *Biological effects of electromagnetic fields*, 2nd edn, p. 67. Boca Raton: CRC Press.

Roth BJ, Basser PJ (1990) A model of the stimulation of a nerve fiber by electromagnetic induction. *IEEE Transactions on Biomedical Engineering 37*, 588–597.

Ueno S, Tashiro T, Harada K (1988) Localized stimulation of neural tissues in the brain by means of a paired configuration of time-varying magnetic fields. *Journal of Applied Physics 64*, 5862–5864.

Ueno S, Matsuda T, Fujiki M (1990) Functional mapping of the human motor cortex obtained by focal and vectorial magnetic stimulation of the brain. *IEEE Transactions on Magnetics 26*, 1539–1544.

Wagner TA, Zahn M, Grodzinsky AJ, Pascual-Leone A (2004) Three-dimensional head model simulation of transcranial magnetic stimulation. *IEEE Transactions on Biomedical Engineering 5*, 1586–1598.

Wassermann EM, Wang B, Zeffiro TA, Sadato N, Pascual-Leone A, Toro C, Hallett M (1996) Locating the motor cortex on the MRI with transcranial magnetic stimulation and PET. *NeuroImage 3*, 1–9.

Yildir YB, Prasad KM, Zheng D (1993) Computer aided design in electromagnetic boundary element method and applications. *Control and Dynamic Systems 59*, 167–223.

Zheng D (1997) Three dimensional eddy current analysis by the boundary element method. *IEEE Transactions on Magnetics 33*, 1354–1357.

Zheng D, Davey KR, Zowarka R, Pratap S (1996) Pushing the limits of 2-D boundary element eddy current codes – connectivity. *International Journal of Numerical Modeling: Electronic Networks, Devices and Fields 9*, 115–124.

CHAPTER 6

Lessons learned from magnetic stimulation of physical models and peripheral nerve *in vitro*

Paul J. Maccabee and Vahe E. Amassian

Introduction

Following the introduction of modern magnetic coil (MC) stimulators by Barker and colleagues (Polson *et al.* 1982), the relationship between the induced topographic electric field and specific sites and regions of nerve activation in the brain are still not precisely known. Nevertheless, there is much more specific information available concerning peripheral nerve and nerve root stimulation. Important parameters include MC design and orientation in relation to nerve fiber trajectory, induced electric field distribution, and induced pulse profile. The study of these variables provides a conceptual framework which contributes to our understanding of some of the mechanisms involved in brain stimulation. The derived principles also guide the rational creation of a test to predictably stimulate the proximal and distal cauda equina, an important adjunct useful for detection of demyelinating peripheral neuropathy.

Electric field (EF) path

In general, EFs induced by magnetic fields pulsed into adjacent volume conductors are typically measured with coaxial cable electrodes, and are expressed as the voltage drop between the bared distal tip and shield in units of V/m. The EF vector is oriented in a plane which is typically predominantly parallel to the inner surface of the volume conductor in contact with the MC (Toft 1990; Cohen and Cuffin 1991). In homogeneous media, a round MC held flat (i.e. tangential) to the surface of a volume conductor induces an annular electric field opposite in direction to that in the MC (Figure 6.1). The EF is maximal beneath the windings and falls off towards the center of the coil where the magnetic field is maximal. The EF remains very prominent moving outward from the windings. Holding a round MC flat over the top of the cranial vault centered at the vertex (vertex–tangential orientation) typically elicits responses in hand or arm muscles. When a round MC is held orthogonal to a volume conductor, the induced EF consists of two loops whose magnitude is predominantly parallel to the inner surface of the volume conductor, each loop running together directly under the contacting edge of the MC. Similarly, a tangentially orientated figure-8 MC also induces two electric field loops which superimpose maximally under the region of the long

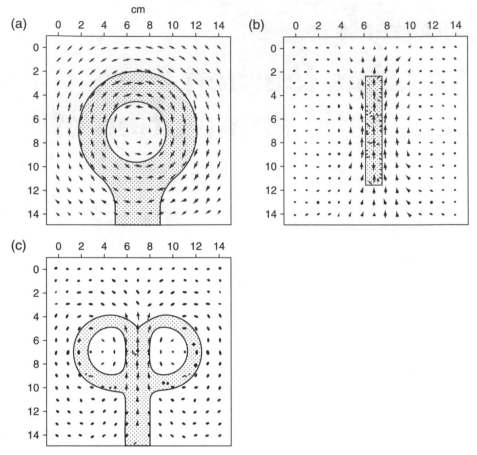

Fig. 6.1 Current arrow diagrams computed in the plane tangential to the bottom surface of a cylindrical tank filled with isotonic saline. Electric fields induced by a tangentially oriented round magnetic coil (MC) (a), an orthogonally oriented round MC (b), and a tangentially oriented figure-8 MC (c). Distance from inner surface of tank to recording electrodes was 0.5 cm. MC stimulator output 20% of maximum. Magnitude of electric field proportional to arrow length. In all three plots, maximal values are normalized. Reproduced from Maccabee et al. (1991b).

axis of the junction (Ueno et al. 1988). The initially orthogonally orientated round MC, when progressively tilted, provided the first example of localized brain stimulation, giving rise to preferential movements of individual digits (Amassian et al. 1989). However, an orthogonal orientation is generally not useful owing to insufficient flux conveyed to the volume conductor. By contrast, the figure-8 MC provides relatively exquisite directional and focal sensitivity, as for example, in brain mapping tasks.

In an inhomogeneous media volume conductor, such as a segment of human cervical–thoracic vertebral spine submerged in saline and located eccentrically within a large cylindrical tank,

the path of the EF within the spinal canal and across the intervertebral neuroforamina are similar to that in the homogeneous volume conductor (Maccabee et al. 1991a,b; Epstein et al. 1991). Notably, near and within a single neuroforamen, the EF (and its spatial derivative, see below) is markedly increased compared to that within the central long axis of the vertebral canal (Figure 6.2). By contrast, the EF derivative induced in the long axis of the spinal canal is relatively insignificant. Thus, in human subjects, it is easy to excite segmental nerve roots near and at neuroforamina (Ugawa et al. 1989). However, induced current easily spreads laterally (e.g. proximal brachial plexus) unless the

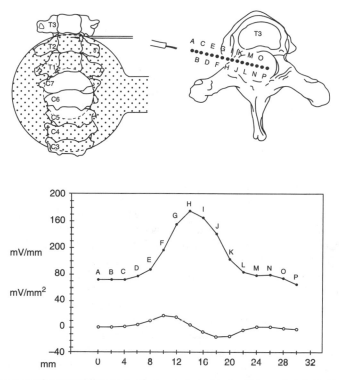

Fig. 6.2 The trans-neuroforaminal electric field recorded proximal, within and distal to the intervertebral foramen between T2 and T3. A 4 mm linear coaxial cable probe was used. Each data point designated by an alphabetic letter, obtained at 2 mm intervals, corresponds to the midpoint of the probe. The lower trace indicates the first spatial derivative (mV/mm^2) of the electric field. The magnetic coil (MC) was oriented symmetrical–tangentially. Reproduced from Maccabee *et al.* (1991b).

induced current is directed inwards. There are no known reports of direct spinal cord stimulation using an MC.

Similar findings are also recorded from a segment of lumbosacral spine immersed within a large cylindrical volume conductor. In the long axis of the sacral canal, a figure-8 coil induces the maximal field in the long axis of the sacral canal when the MC junction is oriented tangentially (i.e. long axis of the junction over the long axis of the sacrum). By contrast, the lumbar nerves transiting L4–L5 and L5–S1 neuroforamina are best oriented to channel horizontally induced current conferred by a roughly horizontal MC junction (Maccabee *et al.* 1996).

It is likely that the localized rise in the EF and its spatial derivative at a neuroforamen is an important, but probably not exclusive, mechanism involved in excitation of cervical, thoracic, and lumbar segmental nerve roots (see below).

Radial electric field

Using any MC design or orientation, there can be no induced radial electric field in a perfectly spherical volume conductor (Cohen and Cuffin 1991). A radial EF may be present in a non-spherical volume conductor such as the skull, but does not achieve more than one-sixth the magnitude of the EF induced parallel to the inner table of the skull. In a homogeneous cylindrical volume conductor such as a phantom arm, radial electric fields are also insignificant (Roth *et al.* 1990).

Electric field in-depth

It might be expected that an orientation of the EF vector predominantly parallel to the inner surface of the volume conductor would easily excite tangentially oriented superficial cortical fibers. However, the largest cortical neurons, the Betz cells and their fibers, are not readily excited directly with a vertex–tangential round MC orientation, but are directly excited with an induced lateral–medial EF (see below). Moreover, in addition to a perfect sphere, recordings in various other volume conductors also reveal a significant decrease of the EF with increasing distance from the inner surface of the volume conductor (Cohen *et al.* 1990; Epstein *et al.* 1990; Maccabee *et al.* 1990; Eaton 1992). Further calculations indicate the impossibility of selective three-dimensional targeting. At frequencies employed for TMS 'using any superposition of simultaneous external current sources' it is not possible 'to produce a three-dimensional local maximum of the electric field strength inside the brain' (Heller and van Hulsteyn 1992).

An ingenious new design, the Hesed (H)-coil, appears to confer a greater intensity of stimulation in deep brain regions without inducing a much greater stimulation of superficial cortical regions (Roth *et al.* 2002; Zangen *et al.* 2005). Compared to the H-coil, a standard figure-8 is more focal, but demonstrates 'more significant reduction . . . in the percentage of a deep region field relative to field at the surface.' The H-coil may therefore exhibit advantages in research and therapeutic applications.

Spatial derivative of the induced EF

According to classic cable theory and *in vitro* experiments, excitation of a straight nerve in a homogeneous media volume conductor occurs at the negative-going first spatial derivative of the induced electric field parallel to the nerve (Figure 6.3, left traces) (Katz 1939; Rattay 1986; Durand *et al.*, 1989; Reilly 1989; Roth and Basser 1990; Nilsson *et al.* 1992; Maccabee *et al.* 1993). The spatial derivative locus occurs where the uniform EF curves away from the linear axon. In clinical situations, exclusive primary excitation by the spatial derivative may be unlikely owing to other, more potent mechanisms (bend, neuroforamen, secondary sources, see below). An exception may be large coil excitation of distal peripheral limb nerves. Experiments in the visual system also show that excitation does not occur at the spatial derivative sites, but

rather closer to the peak of the induced EF (Amassian *et al.* 1994).

Low-threshold sites

Straddling a nerve *in vitro* with solid non-conducting lucite cylinders creates a localized spatial narrowing and marked increase in the induced electric field there, resulting in a lower threshold of excitation (Figure 6.3, right traces). This model is reminiscent of a nerve transiting a model neuroforamen (cf. Figure 6.2 above).

In vitro, when a nerve is bent and induced current is directed along the nerve towards the bend, the threshold of activation is markedly reduced there. Moreover, the angle of the bend from 0 to ≥90° grades the decrease in threshold (Figure 6.4). This mechanism may also account for segmental cervical nerve root excitation just distal to the foramen (Mills *et al.* 1993). Low-threshold excitation also occurs when current is directed along the nerve towards a cut nerve ending. These observations support theoretical notions that an effective gradient occurs if an axon either alters its spatial orientation within a focally uniform field, or abruptly terminates (Reilly 1989). In our view effective excitation may occur when an axon bends out of a uniform predominantly linear induced electric field or when the induced electric field bends away from the linear axon (Figure 7 in Amassian *et al.* 1992).

Low-threshold excitation of nerve may also arise at inhomogeneities. Nerve excitation may arise from separated charges accumulating at inhomogeneities within the nerve bundle or at the surface of bone adjacent to the axon, referred to as 'secondary source' excitation (Nagarajan and Durand 1996). In this terminology, 'primary source' refers to excitation mediated by the negative-going first spatial derivative of the induced electric field described above when exciting a straight nerve in a homogeneous media volume conductor. Possible sites of secondary source excitation include nerve root excitation at neuroforamina along the entire spine in addition to other mechanisms cited above.

Note that in the neuroforamen models (Figure 6.2 and Figure 6.3, right traces), excitation still occurs near or at the peak of the measured spatial derivative sites along the axis of the

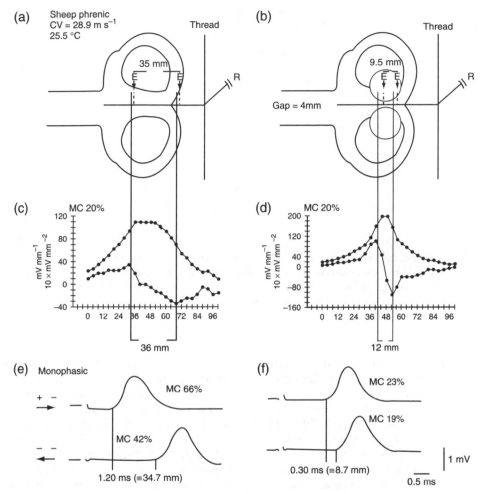

Fig. 6.3 Magnetic coil (MC) stimulation of sheep phrenic nerve immersed in homogeneous and inhomogeneous media volume conductors. (a) Accurate dimensional relationship of the figure-8 MC to the nerve trajectory within the trough. E indicates sites of excitation by monopolar MC pulses in both directions, obtained by matching latencies of MC-induced responses with those elicited by direct electric stimulation. (b) Accurate dimensional relationship of nerve trajectory in trough and lucite cylinders astride the nerve. (c, d) The measured electric field (mV/mm) and its first spatial derivative (mV/mm^2) (note 10-fold increase in amplification) corresponding to experimental set-ups illustrated at left and right, respectively. (e, f) Nerve responses elicited by monophasic current pulses. The electric fields were measured and the nerve responses were elicited in the same Ringer solution, at ambient room temperature. MC immediately beneath trough. Reproduced from Maccabee *et al.* (1993).

nerve because the EF measurement is the sum of the primary and secondary sources.

Excitation of sacral motor nerve roots is a special case, however, as the ventral sacral neuroforamina are oriented anterior–posterior, and therefore are unlikely to channel induced current there. Possibly, sacral nerve rootlets within the sacral canal are excited by secondary sources arising from the inner surfaces of the anterior and posterior sacral plates. Theoretically, excitation of sacral rootlets at distal cauda equina could also occur at the sudden change in conductivity where the rootlets emerge from the thecal sac or where the roughly vertical rootlets

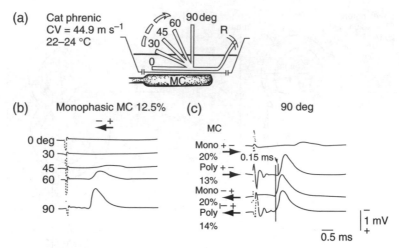

Fig. 6.4 Magnetic coil (MC) stimulation of straight and bent nerve (cat phrenic). (a) Various bend angles used. (b) Nerve responses are elicited by monophasic current, which is outward at the bend. (c) Nerve responses are elicited by monophasic and polyphasic current of both polarities, with a 90° bend in the nerve. Reproduced from Maccabee *et al.* (1993).

within the central sacrum bend anteriorly to enter the ventral foramina (MacDonell *et al.* 1992; Maccabee *et al.* 1996).

A common property of low-threshold sites is their optimal excitation at the peak of the induced EF. An extremely important principle is the further observation that excitation at low-threshold sites may commonly take place when the MC is at a distance from the excitation site, as long as the EF is of sufficient magnitude and, if appropriate, direction (see Figure 9 in Maccabee *et al.* 1993). In humans, there is a lack of latency shift of either evoked compound motor unit action potentials or motor units at threshold upon moving a round or figure-8 MC along the rostral–caudal axis of the vertebral column (Ugawa *et al.* 1989; Britton *et al.* 1990; Chokroverty *et al.* 1990; Maccabee *et al.* 1991). Excitation at a distance is well known in brain mapping, in which it is necessary to identifiy 'hot spots' at lowest threshold. It is likely that MC stimulation at intensities above threshold accesses sites and regions somewhat distant from the coil.

In the brain, the nonlinear trajectory of the cerebral cortical axon suggests that excitation may occur at bends in this trajectory (Amassian *et al.* 1992). This hypothesis was tested using a model system consisting of a peripheral nerve immersed in isotonic saline within a model human skull. Bends were made in the nerve trajectory to resemble those of corticospinal tract fibers traveling from motor cortex to the internal capsule. Pulsed magnetic fields applied to this model skull excited the nerve fibers at these bends and exactly mimicked the observation in the human, if induced current by a figure-8 coil junction was directed lateral to medial (hand muscle recordings). Alterations of induced EF distribution and density occur in the presence of normal tissue inhomgeneity as well as in pathological situations such as encephalomalacia following stroke (Kobayashi *et al.* 1997; Wagner *et al.* 2006).

Induced pulse profile

Typically, an induced monophasic pulse consists of a quarter-cycle first phase followed by a very small reversed second phase. By contrast, the polyphasic pulse consists of a quarter-cycle first phase followed by a somewhat lower-amplitude reversed second phase consisting of two quarter cycles. A biphasic pulse consists of four quarter cycles. Commercial stimulators show a range of quarter-cycle durations, extending from 70 to >100 μs. The optimal duration of a quarter cycle is not clearly defined (Panizza *et al.* 1992).

In vitro studies provide insight into the different properties of monophasic vs polyphasic pulses (Maccabee *et al.* 1998). A predominantly monophasic induced pulse excites a straight nerve in homogeneous media at one or the other spatial derivative sites, depending upon the direction of induced current. By contrast, a polyphasic pulse excites a straight nerve at both spatial derivative sites, by the negative-going first phase (quarter cycle I) at one location, and using our pulse duration, by the negative-going reversed second phase (quarter cycles II and II) 150 μs later at the other location (Figure 6.4C). Thus, both *in vitro* and clinically, a polyphasic pulse is not sensitive to the direction of induced current compared to a monophasic pulse. This property is well known in brain stimulation using a vertex–tangential round MC centered over the vertex (induced monophasic posterior-to-anterior current optimal) or a figure-8 coil with junction axis roughly placed over the long axis of the motor strip (induced monophasic lateral–medial current optimal for hand muscle stimulation).

Another important difference is the increased stimulation efficacy of the polyphasic pulse. Previously, MC pulse stimulation of peripheral limb nerves in normal subjects revealed that larger responses were elicited by polyphasic compared to monophasic pulses (McRobbie and Foster 1984; Claus *et al.* 1990). *In vitro*, at a derivative site on a straight nerve, at a bend in the nerve, and at a cut nerve ending, the polyphasic pulse elicits the largest amplitude response when the hyperpolarizing first phase (quarter cycle I) is followed by a depolarizing second phase (quarter cycles II and III). A lower-amplitude response is elicited by the sequence of a depolarizing quarter cycle I followed by a tiny hyperpolarization; an ever smaller response by the sequence of a depolarizing quarter cycle I followed by a larger hyperpolarization; the smallest response is elicited by a depolarizing quarter cycle followed by an even larger hyperpolarization. Simulation studies demonstrated that the increased efficacy of the optimal pulse configuration is attributed to the greater duration of the second phase (quarter cycles II and III) which confers a greater outward charge transfer.

In vitro, the foramen model, consisting of solid lucite cylinders straddling the nerve, reveals that the largest responses may be elicited by polyphasic pulses of either sequence (i.e. depolarization–hyperpolarization and vice versa). MC root stimulation of cervical, thoracic, lumbar, and sacral levels mimics this specific model. Possibly, this indicates secondary sources of excitation.

Studies of brain stimulation agree with the *in vitro* conclusion that the polyphasic pulse is more powerful than the monophasic pulse (Niehaus *et al.* 2000; Al-Mutawaly *et al.* 2003; Arai *et al.* 2005; Sommer *et al.* 2006). Notably, many commercially available repetitive stimulators induce a cosine-shaped half cycle, preceded and followed by quarter cycles of opposite polarity. The shape of the induced pulse may be a relevant clinical response factor in repetitive TMS.

Stimulation of proximal cauda equina

On the basis of experiments with physical models (Figure 3 in Maccabee *et al.* 1991) it was hypothesized that monophasic current induced in a caudal-to-rostral direction into the spinal-fluid-filled thecal sac at the proximal cauda equina would be displaced along the surface of the conus medullaris, preferentially exciting motor roots at the root exit zone. This hypothesis was confirmed using a large figure-8 MC, with its junction along the long axis of the vertebral spine, and the center of the junction optimally located approximately over the cauda equina (Maccabee *et al.* 1996). As anticipated, the reverse direction of induced monophasic current did not excite the proximal cauda equina at the same intensity of stimulation (Figure 6.5). However, an appropriately directed polyphasic pulse (i.e. depolarization by quarter cycles II and III) significantly elicits a larger-amplitude response.

Clinical studies of cauda equina stimulation

Employing the principles described above, it is possible to evaluate segmental conduction time of the entire peripheral motor axis in lower limbs. A study was recently completed in normal subjects and patients with diverse peripheral

Subject 2, Normal male
Stim: Proximal Cauda Equina
Distal Divergence position 16 cm
Rec: Lt. FHB

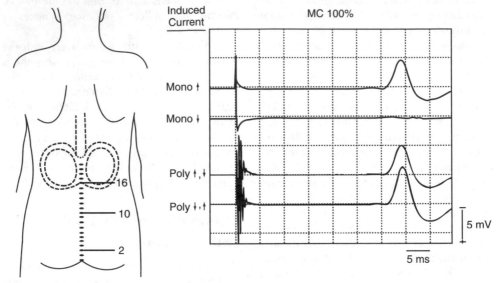

Fig. 6.5 Proximal cauda equina Compound muscle action potentials (CMAPs) in left flexor hallucis brevis (FHB) elicited by the mid-size twin magnetic coil (MC) held vertically at the optimal level (distal divergence at position 16). *Top trace*: monophasic current is induced in a caudal-to-cranial direction. *Second trace from top*: reversal of monophasic current in this typical thin normal subject elicited no response, despite the same 100% MC output. *Third trace from top*: a slightly larger amplitude CMAP is elicited by a polyphasic pulse (phase 1 directed cranially, i.e. depolarizing at proximal cauda equina; phase 2 directed caudally, i.e. hyperpolarizing at proximal cauda equina). *Lower trace*: a substantially larger amplitude response is elicited by the reverse polyphasic pulse yielding a hyperpolarizing depolarizing sequence at proximal cauda equina. Reproduced from Maccabee et al. (1996).

demyelinating and nondemyelinating neuropathies (Maccabee *et al.* 2006). Findings include:

1. Cauda equina conduction time (CECT) may be prolonged into the demyelinating range in both acute and chronic inflammatory myelinating polyneuropathy (AIDP, CIDP).

2. The segmental conduction time from distal cauda equina to popliteal fossa is frequently demyelinated in CIDP, rarely demyelinated in AIDP, and so far never demyelinated in motor neuron disease.

3. In an IVIG responsive patient with CIDP seen 9 days after onset of symptoms, demyelination was demonstrated in the distal cauda equina to the popliteal fossa segment, but not in the CECT segment until 6 months later.

4. In anti-myelin associated glycoprotein (anti-MAG) polyneuropathy, in addition to typical profound prolongation of terminal motor latency, there is also profound demyelinative slowing in the distal cauda equina to popliteal fossa segment, and lesser demyelinative slowing in the CECT segment.

These observations are uniquely informative in peripheral neuropathy.

Comment

The forgoing emphasizes the complex relationship between coil design and induced EF, peak of the induced EF, the spatial derivative, the bend, the neuroforamen and nerve models, the notion of secondary sources and tissue inhomogeneities, the induced pulse profile, and the fundamental physiology and neuroanatomy. Complexity will substantially increase in the presence of selected injury to nerves, brain, and adjacent bony structures.

References

Amassian VE, Cracco RQ, Maccabee PJ (1989). Focal stimulation of human cerebral cortex with the magnetic coil: a comparison with electric stimulation. *Electroencephalography ond Clinical Neurophysiology 74*, 401–416.

Amassian VE, Eberle LP, Maccabee PJ, Cracco RQ (1992). Modelling magnetic coil excitation of human cerebral cortex with a peripheral nerve immersed in a brain-shaped volume conductor. The significance of fiber bending in excitation. *Electroencephalography and Clinical Neurophysiology 85*, 291–301.

Amassian VE, Maccabee PJ, Cracco RQ, et al. (1994). The polarity of the induced electric-field influences magnetic coil inhibition of human visual cortex – implications for the site of excitation. *Electroencephalography and Clinical Neurophysiology 93*, 21–26.

Al-Mutawaly N, de Bruin H, Hasey G (2003). The effects of pulse configuration on magnetic stimulation. *Journal of Clinical Neurophysiology 20*, 361–370.

Arai N, Okabe S, Furubayashi T, Terao Y, Yuasa K, Ugawa Y (2005). Comparison between short train, monophasic and biphasic repetitive transcranial magnetic stimulation (rTMS) of the human motor cortex. *Clinical Neurophysiology 116*, 605–613.

Britton TC, Meer BU, Herdman J, Benecke R (1990). Clinical use of the magnetic stimulator in the investigation of peripheral conduction time. *Muscle and Nerve 13*, 396–406.

Chokroverty S, Spire JP, DiLullo J, Moody E, Maselli R (1990). Magnetic stimulation of the human peripheral nervous system. In: S Chokroverty (ed.), *Magnetic stimulation in clinical neurophysiology*, pp. 249–273. Boston, MA: Butterworth.

Claus D, Murray NMF, Spitzer A, Flugel D (1990). The influence of stimulus type on the magnetic excitation of nerve structures. *Electroencephalography and Clinical Neurophysiology 75*, 342–349.

Cohen D, Cuffin BN (1991). Developing a more focal magnetic simulator. I. Some basic principles. *Journal of Clinical Neurophysiology 8*, 102–111.

Cohen LG, Roth BJ, Nilsson J, et al. (1990). Effects of coil design on delivery of focal magnetic stimulation. *Technical Considerations 75*, 350–357.

Durand D, Ferguson AS, Dalbasti T (1989). Induced electric fields by magnetic stimulation in non-homogeneous conducting media. *Proceedings of the IEEE Engineering, Medicine and Biology Society 11*, 1252–1253.

Eaton H (1992). Electric field induced in a spherical volume conductor from arbitrary coils: application to magnetic stimulation and MEG. *Medical and Biological Engineering and Computing 30*, 433–440.

Epstein CM, Schwartzberg DG, Davey KR, Sudderth DB (1990). Localizing the site of magnetic brain stimulation in humans. *Neurology 40*, 666–670.

Epstein CM, Fernandez-Beer E, Weissman JD, Matsuura S (1991). Cervical magnetic stimulation: the role of the neural foramen. *Neurology 41*, 677–680.

Heller L, van Hulsteyn DB (1992). Brain stimulation using electromagnetic sources. Theoretical aspects. *Journal of Biophysics 63*, 129–138.

Katz B (1939). *Electric excitation of nerve*. London: Oxford University Press.

Kobayashi M, Ueno S, Kurokawa T (1997). Importance of soft tissue inhomogeneity in magnetic peripheral nerve stimulation. *Electroencephalography and Clinical Neurophysiology 105*, 406–413.

Maccabee PJ, Eberle L, Amassian VE, Cracco RQ, Rudell A, Jayachandra M (1990). Spatial distribution of the electric field induced in volume by round and figure 8 magnetic coils: relevance to activation of sensory nerve fibers. *Electroencephalography and Clinical Neurophysiology 76*, 131–141.

Maccabee PJ, Amassian VE, Cracco RQ, Eberle LP, Rudell AP (1991a). Mechanisms of peripheral nervous system stimulation using the magnetic coil. *Electroencephalography and Clinical Neurophysiology Suppl 43*, 344–361.

Maccabee PJ, Amassian VE, Eberle LP, et al. (1991b). Measurement of the electric field induced into inhomogenous volume conductors by magnetic coils: application to human spinal neurogeometry. *Electroencephalography and Clinical Neurophysiology 81*, 224–237.

Maccabee PJ, Amassian VE, Eberle LP, Cracco RQ (1993). Magnetic coil stimulation of straight and bent amphibian and mammalian peripheral nerve in vitro: locus of excitation. *Journal of Physiology 460*, 201–219.

Maccabee PJ, Lipitz ME, Desudchit T, et al. (1996). A new method using neuromagnetic stimulation to measure conduction time within the cauda equina. *Electroencephalography and Clinical Neurophysiology 101*, 153–166.

Maccabee PJ, Nagarajan SS, Amassian VE, Durand DM, et al. (1998). Influence of pulse sequence, polarity and amplitude on magnetic stimulation of human porcine and peripheral nerve. *Journal of Physiology 513.2*, 571–585.

Maccabee PJ, Stein IAG, Eberle LP, Willer JA, Lipitz MA, Amassian VE (2006). Detection of demyelinating neuropathy using cauda equina nerve conduction studies. *Annals of Neurology, 60* (Suppl 10), S93.

MacDonell RAL, Cros D, Shahani BT (1992). Lumbosacral nerve root stimulation comparing electrical with surface magnetic coil techniques. *Muscle and Nerve 15*, 885–890.

McRobbie D, Foster MA (1984). Thresholds for biological affects of time-varying magnetic fields. *Clinical Physics and Physiological Measurement 5*, 67–78.

Mills KR, McLeod C, Sheffy J, Loh L (1993). The optimal current direction for excitation of human cervical motor roots with a double magnetic stimulator. *Electroencephalography and Clinical Neurophysiology 89*, 138–144.

Nagarajan SS, Durand DM (1996). A generalized cable equation for magnetic stimulation of axons. *IEEE Transactions on Biomedical Engineering 43*, 304–312.

Niehaus L, Meyer BU, Weyh T (2000). Influence of pulse configuration and direction of coil current on excitatory effects of magnetic motor cortex and nerve stimulation. *Clinical Neurophysiology 111*, 75–80.

Nilsson J, Panizza M, Roth BJ, *et al.* (1992). Determining the site of stimulation during magnetic stimulation of a peripheral nerve. *Electroencephalography and Clinical Neurophysiology 85*, 253–264.

Panizza M, Nilsson J, Roth BJ, Basser PJ, Hallett M (1992). Relevance of stimulus duration for activation of motor and sensory fibers: implications for the study of H-reflexes and magnetic stimulation. *Electroencephalography and Clinical Neurophysiology 85*, 22–29.

Polson MJR, Barker AT, Freeston IL (1982). Stimulation of nerve trunks with time-varying magnetic fields. *Medical and Biological Engineering and Computing 20*, 243–244.

Rattay F (1986). Analysis of models for the external stimulation of axons. *IEEE Transactions in Biomedical Engineering 33*, 974–977.

Reilly JP (1989). Peripheral nerve stimulation by induced electric currents: exposure to time-varying magnetic fields. *Medical and Biological Engineering and Computing 27*, 101–110.

Roth BJ, Basser P (1990). Model of the stimulation of a nerve fiber by electromagnetic induction. *IEEE Transactions on Biomedical Engineering 37*, 588–597.

Roth BJ, Cohen LG, Hallett M, Friauf W, Basser PJ (1990). A theoretical calculation of the electrical field induced by magnetic stimulation of a peripheral nerve. *Muscle and Nerve 13*, 734–741.

Roth Y, Zangen A, Hallett M (2002). A coil design for transcranial magnetic stimulation of deep brain regions. *Journal of Clinical Neurophysiology 19*, 361–370.

Sommer M, Alfaro A, Rummel M, Speck S, Lang N, Tings T, Paulus W (2006). Half sine, monophasic and biphasic transcranial magnetic stimulation of human motorcortex. *Clinical Neurophysiology 117*, 838–844.

Toft PS (1990). The distribution of induced currents in magnetic stimulation of nervous system. *Physics in Medicine and Biology 35*, 1119–1128.

Ueno S, Tashiro T, Harada K (1988). Localized stimulation of neural tissues by means of a paired configuration of time-varying magnetic fields. *Journal of Applied Physiology 64*, 5862–5864.

Ugawa Y, Rothwell JC, Day BL, Thompson PD, Marsden CD (1989). Magnetic stimulation over the spinal enlargements. *Journal of Neurology, Neurosurgery and Psychiatry 52*, 1025–1032.

Wagner T, Fregni F, Eden U, *et al.* (2006). Transcranial magnetic stimulation and stroke: a computer-based human model study. *NeuroImage 30*, 857–870.

Zangen A, Roth Y, Voller B, Hallett M (2005). Transcranial magnetic stimulation of deep brain regions: evidence for efficacy of the H-coil. *Clinical Neurophysiology 116*, 775–779.

CHAPTER 7

Direct current brain polarization

Eric M. Wassermann

Introduction

The transcranial application of weak direct current (DC) to the brain is an effective neuromodulation technique with conceptual roots in antiquity and more than a century of experimental and therapeutic use (Lolas 1977; Priori, 2003). Ignored for decades despite interesting findings, focal DC brain polarization is now undergoing renewed interest, in part because of the wide acceptance of TMS as a research tool and candidate treatment for brain disorders.

The effects of static electrical fields on cortical neurons *in vivo* have been known almost since the advent of intracellular recording. In mammals, weak polarizing currents applied to the brain surface can produce lasting changes in cortical-evoked potentials and the activity of individual cortical neurons (Bindman *et al.* 1962; Creutzfeldt *et al.* 1962; Hern *et al.* 1962; Purpura and McMurtry 1964). These effects are highly selective for neurons oriented longitudinally in the plane of the electric field. For instance, when the active electrode is applied to the cortical surface, the effect is selective for radially oriented cortical output (pyramidal) neurons (Hern *et al.* 1962). The spatial selectivity of the DC effect can also be demonstrated in isolated neuronal preparations *in vitro* (Terzuolo and Bullock 1956) and other brain areas, e.g. cerebellum (Chan and Nicholson 1986), and appears to be mediated by effects on voltage-sensitive cation channels (Chan *et al.* 1988; Lopez *et al.* 1991). In virtually all cases described, surface-anodal polarization of the cortex (anode near the dendritic poles of radially oriented neurons) increases the firing rates of spontaneously active cells, but does not cause spontaneous firing. Surface-cathodal polarization has the opposite effect, down-modulating firing.

Modern human experimentation with DC current has generally employed current-controlled apparatus to deliver currents in the 0.5–2 mA range through large (25–35 cm^2) moistened sponge electrodes applied to the head. Current densities at the electrode face have been between 20 and 80 µA/cm^2. Physical models suggest that approximately half the current is shunted through the scalp (Miranda *et al.* 2006) and another significant fraction (perhaps as much as 7/8!) through the cerebrospinal fluid (Nathan *et al.* 1993). With a current of 2.0 mA delivered to the scalp, current density in nearby brain tissue has been estimated at 1 µA/cm^2, yielding a field strength of 0.22 mV/mm. There are no specific estimates of the focality of the technique, but common sense and modelling (Miranda *et al.* 2006) indicate that a large area of brain under the electrode is polarized. The physiological action of the current is presumably near the surface, since the electric field diffuses rapidly in the head. Recent experimenters (see below) have tended to place both electrodes on the head. However, at least in theory, the reference or 'indifferent' electrode can be placed anywhere on the body, thereby ensuring that it exerts no physiological effects of its own.

DC polarization often produces a tingling sensation in the scalp under the electrodes, presumably due to electrolysis on the skin. However, this seems to be most noticeable at the beginning of a session and terminating the current after a brief period may provide an acceptable sham. If the current is switched on or off there can be prominent retinal phosphenes, particularly with electrodes located near the eyes. This can be avoided by ramping the current up and down. To date, no adverse events more serious than local skin irritation have been reported with exposures of up to 60 min, and safety concerns, principally those related to intracranial electrolysis and generation of reactive ionic species, are theoretical at this point. The histotoxic effect of unbalanced waveforms delivered through small metal electrodes applied directly to the pial surface (McCreery and Agnew 1990) is not expected here.

Human motor cortex and motor effects

Priori et al. (1998) were the first to show that a weak current (0.5 mA through a 25 cm^2 electrode) could modulate the excitability of the human motor cortex, as measured by the amplitude of the motor-evoked potential (MEP) from TMS. Contrary to all previous and subsequent experience including their own (Ardolino et al. 2005), they reported that surface anodal current decreased, and cathodal current increased, local excitability. The effect on the MEP was described and explored further by Nitsche et al. (Nitsche and Paulus 2000, 2001; Nitsche et al. 2003b) who obtained their best results with the reference electrode over the orbit contralateral to the treated motor cortex (Nitsche and Paulus 2000). This arrangement has the disadvantage of having the reference electrode over brain and theoretically capable of producing its own effects. In subsequent experiments (Nitsche et al. 2003a), they showed that the acute and lasting facilitatory effects of surface-anodal polarization on MEP amplitude were blocked by the Na$^+$ channel blockers carbamazepine and flunarazine, while the lasting

effect was prevented with dextromethorphan, which antagonizes N-methyl-D-aspartate glutamate receptors, in addition to other effects. This was interpreted as implying involvement of 'plastic' mechanisms in the cortex, presumably involving changes in the efficacy of excitatory synapses. However, consistent with the in vitro and animal data, the effect of motor cortex stimulation is present when the MEP is evoked with an electric pulse that stimulates the corticospinal fibers below the cortex (Ardolino et al. 2005), suggesting that the change is not mediated by such a cortical mechanism and relies on the type of change in membrane dynamics indicated by the animal and in vivo data reviewed above. This category of mechanism is compatible with blockade by Na$^+$ channel blockers (Chan et al. 1988; Lopez et al. 1991).

Another interesting and impressive effect of DC polarization is its ability to precondition the cortex to respond in opposite ways to treatment with 1 Hz rapid (r)TMS (Siebner et al. 2004). Pretreatment with surface-anodal current enhances the depressant effect of 1 Hz rTMS on the MEP, whereas, after cathodal pretreatment, 1 Hz rTMS causes MEPs to increase in amplitude. This phenomenon is presumably an example of 'metaplasticity' (Abraham and Bear 1996) or homeostatic plasticity (Turrigiano 1999), a set of mechanisms whereby the frequency-dependent synaptic response to stimulation is modulated by previous experience.

Surface-anodal polarization of the motor cortex may also have functional consequences and potential clinical utility in motor disorders. For instance, it appears to enhance aspects of dexterity in the nondominant hand when applied contralaterally (Boggio et al. 2006a). No improvement in performance was found for the dominant hand; however, the test of hand function used by these authors was designed for patients with motor impairments. Others have used the same outcome measure to demonstrate temporary functional improvement in the affected hand in a challenge trial in chronic stroke patients (Hummel et al. 2005). Similar effects on hand function have also been found in Parkinson's disease (Fregni et al. 2006b).

The one study hitherto combining functional imaging with DC polarization of the motor cortex (Lang et al. 2005) used an exploratory design to demonstrate widespread and persistent changes after 10 min of either surface-cathodal or -anodal polarization. Therefore, caution may be advisable in interpreting some DC polarization results.

Sensory and perceptual effects

In a manner analogous to its effects in the motor cortex, DC polarization can also modulate the excitability of visual areas. Surface-anodal polarization of the occipital cortex enhances contrast sensitivity, while cathodal current does the opposite (Antal *et al.* 2001). Effects on visual cortex can also be measured by the threshold for the production of phosphenes with TMS (Antal *et al.* 2003a,b), which decreases with anodal current, or the amplitude of the visual-evoked potential (Antal *et al.* 2004), which increases. Cathodal current delivered to the somatosensory cortex reduced a tactile discrimination threshold (Rogalewski *et al.* 2004) and consistent findings have been adduced for the cortical components of the somatosensory-evoked potential (Matsunaga *et al.* 2004).

As with motor processes, DC polarization can produce potentially useful clinical effects. Anodal polarization of the motor cortex appears to have an analgesic effect in spinal cord injury patients with neurogenic pain (Fregni *et al.* 2006a).

Cognitive effects

As might be expected, DC can enhance cognitive processes occurring in the treated area. In one of the earliest demonstrations of this principle, anodal polarization of the motor cortex was used to speed the implicit acquisition of a repeated sequence of key presses, known as the serial reaction time task (Nitsche *et al.* 2003c). Interestingly, application of the same current to premotor and prefrontal areas, regions thought to participate in implicit learning, had no such effect.

In other studies, however, prefrontal cortex polarization has produced an array of interesting and potentially useful effects. Several studies have demonstrated effects on forms of learning and memory. In one instance, a nodal polarization of the left prefrontal cortex accelerated the acquisition of implicit knowledge about the probabilistic relationship between sets of cues and outcomes (Kincses *et al.* 2004). Essentially the same treatment improved response accuracy on a '3-Back' delayed match to sample task in healthy subjects (Fregni *et al.* 2005) and patients with Parkinson's disease (Boggio *et al.* 2006b). In a particularly interesting study (Marshall *et al.* 2004), anodal current applied to both lateral frontal areas during slow-wave sleep, but not while awake, enhanced retention of word pairs and mirror-tracing skill acquired previously. Unlike most other recent studies, the current was applied through small (8 mm diameter) electrodes, producing a relatively high current density (26 µA/cm^2).

We performed a large ($n = 103$), single-blind trial designed primarily to establish safety (Iyer *et al.* 2005). Anodal polarization of the left prefrontal area at 40 µA/cm^2 produced a trend-level increase in the ability to generate lists of words beginning with specified letters (Stuss *et al.* 1998), relative to sham and cathodal polarization. The same effect reached statistical significance when the current was doubled. Cathodal current produced no significant changes and there were no effects of either polarity on the electroencephalogram or a variety of other tasks, such as response inhibition and reaction time, that were included as safety tests and controls for nonspecific effects.

Effects on mood and affect

The earliest clinical application of DC polarization was in the field of mood disorders and there seems to have been considerable interest up until the 1970s (Lolas 1977); however, this apparently faded with the ascendancy of pharmacological treatment. In one notable study (Lippold and Redfearn 1964), 32 subjects (most judged to be subclinically depressed), recruited from the staff of a hospital clinic, were treated with currents of up to 0.5 mA, delivered through two 0.5 inch diameter electrodes placed above the orbits and referenced to an electrode at one knee, for periods of as long as 2 h while they went about their work. Currents were reduced in some subjects during the study because of apparent adverse effects of cathodal polarization on mood. Clinician raters, blind to the polarity, observed and questioned the subjects periodically. They were able to guess the polarity correctly in 26 of the 32 subjects, based on these observations. As predicted, when the frontal electrodes were positively charged (anodal), affect tended to become elated and talkative. When the current was reversed, affect became withdrawn and depressed. In a subsequent randomized, double-blind trial in 24 clinically depressed patients (Costain *et al.* 1964), anodal current and sham were delivered for 12 days each in a crossover design. While patients improved on both treatments, active current was associated with a significantly greater improvement in nurse and physician ratings.

Conclusion

Transcranial brain polarization is an old technique that is currently undergoing a revival and yielding interesting data. Due to its lack of temporal and spatial resolution, it does not appear particularly useful for exploring neurophysiological mechanisms. Rather, it is the possibility of safe, noninvasive, inexpensive, and wearable neuromodulatory treatments for brain disorders and, perhaps, the enhancement of normal function that is most attractive.

References

Abraham WC, Bear MF (1996). Metaplasticity: the plasticity of synaptic plasticity. *Trends in Neurosciences* 19, 126–130.

Antal A, Nitsche MA, Paulus W (2001). External modulation of visual perception in humans. *Neuroreport 12*, 3553–3555.

Antal A, Kincses TZ, Nitsche MA, Paulus W (2003a). Manipulation of phosphene thresholds by transcranial direct current stimulation in man. *Experimental Brain Research 150*, 375–378.

Antal A, Kincses TZ, Nitsche MA, Paulus W (2003b). Modulation of moving phosphene thresholds by transcranial direct current stimulation of V1 in human. *Neuropsychologia 41*, 1802–1807.

Antal A, Kincses TZ, Nitsche MA, Bartfai O, Paulus W (2004). Excitability changes induced in the human primary visual cortex by transcranial direct current stimulation: direct electrophysiological evidence. *Investigative Ophthalmological and Visual Science 45*, 702–707.

Ardolino G, Bossi B, Barbieri S, Priori A (2005). Non-synaptic mechanisms underlie the after-effects of cathodal transcutaneous direct current stimulation of the human brain. *Journal of Physiology 568*, 653–663.

Bindman LJ, Lippold OC, Redfearn JW (1962). Long-lasting changes in the level of the electrical activity of the cerebral cortex produced bypolarizing currents. *Nature 196*, 584–585.

Boggio PS, Castro LO, Savagim EA, Braite R, Cruz VC, Rocha RR, *et al.* (2006a). Enhancement of non-dominant hand motor function by anodal transcranial direct current stimulation. *Neuroscience Letters 404*, 232–236.

Boggio PS, Ferrucci R, Rigonatti SP, Covre P, Nitsche M, Pascual-Leone A, *et al.* (2006b). Effects of transcranial direct current stimulation on working memory in patients with Parkinson's disease. *Journal of Neurological Science.*

Chan CY, Nicholson C (1986). Modulation by applied electric fields of Purkinje and stellate cell activity in the isolated turtle cerebellum. *Journal of Physiology 371*, 89–114.

Chan CY, Hounsgaard J, Nicholson C (1988). Effects of electric fields on transmembrane potential and excitability of turtle cerebellar Purkinje cells in vitro. *Journal of Physiology 402*, 751–771.

Costain R, Redfearn JW, Lippold OC (1964). A controlled trial of the therapeutic effect of polarization of the brain in depressive illness. *British Journal of Psychiatry 110*, 786–799.

Creutzfeldt OD, Fromm GH, Kapp H (1962). Influence of transcortical d-c currents on cortical neuronal activity. *Experimental Neurology 5*, 436–452.

Fregni F, Boggio PS, Nitsche M, Bermpohl F, Antal A, Feredoes E, *et al.* (2005). Anodal transcranial direct current stimulation of prefrontal cortex enhances working memory. *Experimental Brain Research 166*, 23–30.

Fregni F, Boggio PS, Lima MC, Ferreira MJ, Wagner T, Rigonatti SP, *et al.* (2006a). A sham-controlled, phase II trial of transcranial direct current stimulation for the treatment of central pain in traumatic spinal cord injury. *Pain 122*, 197–209.

Fregni F, Boggio PS, Santos MC, Lima M, Vieira AL, Rigonatti SP, *et al.* (2006b). Noninvasive cortical stimulation with transcranial direct current stimulation in Parkinson's disease. *Movement Disorders 21*, 1693–1702.

Hern JEC, Landgren S, Philips CG, Porter R (1962). Selective excitation of corticofugal neurones by surface-anodal stimulation of the baboon's motor cortex. *Journal of Physiology (London) 168*, 890–910.

Hummel F, Celnik P, Giraux P, Floel A, Wu WH, Gerloff C, *et al.* (2005). Effects of non-invasive cortical stimulation on skilled motor function in chronic stroke. *Brain 128*, 490–499.

Iyer MB, Mattu U, Grafman J, Lomarev MP, Sato S, Wassermann EM (2005). Safety and cognitive effect of frontal DC brain polarization in healthy individuals. *Neurology 64*, 872–876.

Kincses TZ, Antal A, Nitsche MA, Bartfai O, Paulus W (2004). Facilitation of probabilistic classification learning by transcranial direct current stimulation of the prefrontal cortex in the human. *Neuropsychologia 42*, 113–117.

Lang N, Siebner HR, Ward NS, Lee L, Nitsche MA, Paulus W, et al. (2005). How does transcranial DC stimulation of the primary motor cortex alter regional neuronal activity in the human brain? *European Journal of Neuroscience* 22, 495–504.

Lippold OC, Redfearn JW (1964). Mental changes resulting from the passage of small direct currents through the human brain. *British Journal of Psychiatry* 110, 768–772.

Lolas F (1977). Brain polarization: behavioral and therapeutic effects. *Biological Psychiatry* 12, 37–47.

Lopez L, Chan CY, Okada YC, Nicholson C (1991). Multimodal characterization of population responses evoked by applied electric field in vitro: extracellular potential, magnetic evoked field, transmembrane potential, and current-source density analysis. *Journal of Neuroscience* 11, 1998–2010.

McCreery DB, Agnew WF (1990). Mechanisms of stimulation-induced damage and their relation to guidelines for safe stimulation. In: WF Agnew, DB McCreery (eds), *Neural prostheses: fundamental studies*, pp. 297–317. Englewood Cliffs, NJ: Prentice Hall.

Marshall L, Molle M, Hallschmid M, Born J (2004). Transcranial direct current stimulation during sleep improves declarative memory. *Journal of Neuroscience* 24, 9985–9992.

Matsunaga K, Nitsche MA, Tsuji S, Rothwell JC (2004). Effect of transcranial DC sensorimotor cortex stimulation on somatosensory evoked potentials in humans. *Clinical Neurophysiology* 115, 456–460.

Miranda PC, Lomarev M, Hallett M (2006). Modeling the current distribution during transcranial direct current stimulation. *Clinical Neurophysiology* 117, 1623–1629.

Nathan SS, Sinha SR, Gordon B, Lesser RP, Thakor NV (1993). Determination of current density distributions generated by electrical stimulation of the human cerebral cortex. *Electroencephalography and Clinical Neurophysiology* 86, 183–192.

Nitsche MA, Paulus W (2000). Excitability changes induced in the human motor cortex by weak transcranial direct current stimulation. *Journal of Physiology* 527(Pt 3), 633–639.

Nitsche MA, Paulus W (2001). Sustained excitability elevations induced by transcranial DC motor cortex stimulation in humans. *Neurology* 57, 1899–1901.

Nitsche MA, Fricke K, Henschke U, Schlitterlau A, Liebetanz D, Lang N, et al. (2003a). Pharmacological modulation of cortical excitability shifts induced by transcranial direct current stimulation in humans. *Journal of Physiology* 553, 293–301.

Nitsche MA, Liebetanz D, Antal A, Lang N, Tergau F, Paulus W (2003b). Modulation of cortical excitability by weak direct current stimulation–technical, safety and functional aspects. *Clinical Neurophysiology* 56(Suppl), 255–276.

Nitsche MA, Schauenburg A, Lang N, Liebetanz D, Exner C, Paulus W, et al. (2003c). Facilitation of implicit motor learning by weak transcranial direct current stimulation of the primary motor cortex in the human. *Journal of Cognitive Neuroscience* 15, 619–626.

Priori A (2003). Brain polarization in humans: a reappraisal of an old tool for prolonged non-invasive modulation of brain excitability. *Clinical Neurophysiology* 114, 589–595.

Priori A, Berardelli A, Rona S, Accornero N, Manfredi M (1998). Polarization of the human motor cortex through the scalp. *Neuroreport* 9, 2257–2260.

Purpura DP, McMurtry JG (1964). Intracellular activities and evoked potential changes during polarization of motor cortex. *Journal of Neurophysiology* 18, 166–185.

Rogalewski A, Breitenstein C, Nitsche MA, Paulus W, Knecht S (2004). Transcranial direct current stimulation disrupts tactile perception. *European Journal of Neuroscience* 20, 313–316.

Siebner HR, Lang N, Rizzo V, Nitsche MA, Paulus W, Lemon RN, et al. (2004). Preconditioning of low-frequency repetitive transcranial magnetic stimulation with transcranial direct current stimulation: evidence for homeostatic plasticity in the human motor cortex. *Journal of Neuroscience* 24, 3379–3385.

Stuss DT, Alexander MP, Hamer L, Palumbo C, Dempster R, Binns M, et al. (1998). The effects of focal anterior and posterior brain lesions on verbal fluency. *Journal of the International Neuropsychological Society* 4, 265–278.

Terzuolo CA, Bullock TH (1956). The measurement of imposed voltage gradient adequate to modulate neuronal firing. *Proceedings of the National Academy of Sciences of the USA* 42, 687–693.

Turrigiano GG (1999). Homeostatic plasticity in neuronal networks: the more things change, the more they stay the same. *Trends in Neurosciences* 22, 221–227.

CHAPTER 8

Transcranial electrical stimulation and intraoperative neurophysiology of the corticospinal tract

Vedran Deletis, Francesco Sala, and Sedat Ulkatan

Introduction

The revolutionary discovery of transcranial electrical stimulation (TES) by Merton and Morton became a well-recognized method for corticospinal tract (CT) activation and has been used in the clinical setting for the last 25 years (Merton and Morton 1980). Because of the painless method of brain stimulation and other advantages after introducing TMS of the brain, TES is mostly used during intraoperative monitoring of the functional integrity of the CT, in comatose patients and as a research tool in clinical neurophysiology. In this chapter, we will describe mainly the use of TES during surgery in anesthetized patients and highlight the underlying physiology of the motor-evoked potentials (MEPs) elicited by this method.

Stimulation

Electrode montages over the scalp for eliciting MEPs, using single- and multipulse stimulation techniques

Muscle MEPs elicited by multipulse stimulation techniques

A train of multipulse stimuli (three to nine) with interstimulus interval within the train of stimuli

of 4 ms and a train repetition rate of 1–2 Hz should be used when eliciting muscle MEPs.

In order to obtain optimal quality of MEP recordings, the position of the stimulating electrodes over the scalp and their combinations (montage) should be precisely determined. We prefer to use corkscrew electrodes (CS electrodes; Viasis, Neurocare, Madison, WI, USA) inserted subcutaneously over the scalp for stimulation because of their low impedance (Journee *et al.* 2004), secure placement, and minimum possibility of accidental detachment. We prefer as well placement of all six electrodes (according to 10/20 International EEG system) as illustrated in Figure 8.1. Using different montages of stimulating electrodes in most instances gives us flexibility to optimize eliciting MEPs without vigorous muscle twitching, which can disturb surgery. Selective muscle MEPs, even in the hemi-body muscles, can be obtained using montage C3/Cz or C4/Cz. From a practical point of view the more lateral stimulating montage is not always optimal because it can induce more vigorous muscle twitching and to some extent body movements. In most cases C1/C2 is a better electrode montage for eliciting muscle MEPs in all limbs or hemi-body muscles. It is an interesting observation that in a certain number of patients only the montage

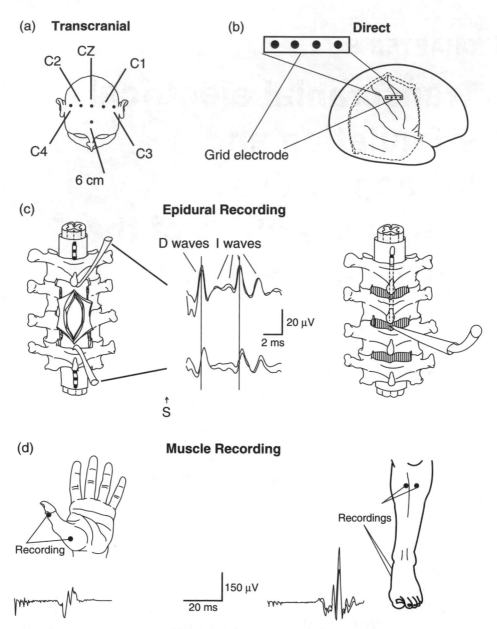

Fig. 8.1 Schematic drawing of intraoperative methodology for eliciting and recording motor evoked potentials from the spinal cord and limb muscles. (a, b) Illustration of electrode positions for transcranial (a) and direct (b) electrical stimulation of the motor cortex according to the International 10–20 EEG system. (c) Schematic diagram of the positions of the catheter electrodes (each with three recording surfaces) placed cranial to the tumor (control electrode) and caudal to the tumor to monitor the incoming signal passing through the site of surgery. In the center are D- and I-wave records rostral and caudal to the tumor site. Note the peak latency difference between cranial and caudal recordings of the D- and I-waves which are marked with vertical lines. (d) Recording of muscle motor-evoked potentials from the thenar, tibialis anterior and abductor hallucis muscles after eliciting them with a short train of electrical pulses applied transcranially. Modified from Deletis et al. (2001).

Cz vs 6 cm in front of Cz can elicit muscle MEPs, especially from lower extremities, without disturbing body jerks.

In an anesthetized patient, a multipulse stimulating technique should be used to successfully elicit muscle MEPs, either in the limb muscles via CTs or muscles innervated by cranial motor nerves via corticobulbar (corticonuclear) pathways. In both instances it is necessary to use a multipulse stimulating technique (a) because I-waves are fewer and/or less pronounced, or even non-existent, and (b) because of the low excitability of the alpha motor neuronal pools (see 'Relationship between descending drive from the electrically activated CT and generation of the muscle MEPs').

D- and I-waves elicited by single-pulse stimulating technique

A *single stimulus* applied over the scalp is sufficient when eliciting D- and I-waves and recording them over the spinal cord. During *moderate* depth of anesthesia, cortical synapses from interneurons oriented vertically to the CTs are shut down, therefore electrical stimulation of the motor cortex can only activate fast neurons of the CTs directly below the cell bodies generating D-waves. Besides direct activation of the CT (D-wave) during *light* anesthesia, indirect (transynaptical) activation can also be achieved to generate I-waves (Patton and Amassian 1954). Due to the high sensitivity of the trans-synaptically activated CTs to the nonsurgically induced influences, only monitoring D-wave (nonsynaptically activated CT) is appropriate.

Recordings

Recording of MEPs (from the epi/subdural space and from the limb muscles)

In order to record conducted responses in the CT (D- and I-waves) after TES, a recording electrode(s) has to be placed in the epi- or subdural space of the spinal cord. If the spinal cord is surgically exposed, placement of a specially designed epidural/subdural catheter electrode has to be performed by the surgeon.

Otherwise this can be done percutaneously using a Touhy needle when spinal cord is not surgically exposed (Figure 8.2). To record muscle MEPs from the limb muscle, either surface or needle electrodes are suitable. Exceptions are recordings of corticobulbar responses from muscles innervated by cranial motor nerves. We prefer to use a small diameter wire electrode (76 µm in diameter, Teflon-coated wires striped at the tip 2 mm, formed as a hook) inserted in the muscle via a 27 gauge needle. This is an important methodological detail because it prevents far field recordings from an adjacent muscle (e.g. as a needle electrode with a large surface in the orbicularis oris muscle can record activity from tongue muscles). Using wire electrodes, as mentioned earlier, can prevent serious false-positive or false-negative data because a strong TES activates corticobulbar pathways for nearly all motor cranial nerves innervating muscles. Furthermore, a hook wire electrode prevents displacement and detachment of electrodes during muscle contraction (Figure 8.3). For details see 'Posterior fossa surgery'.

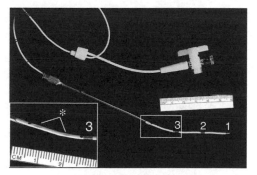

Fig. 8.2 Semi-rigid catheter electrode for recording motor-evoked potentials (D-wave) from the spinal cord, either epi- or subdurally; the electrode has passed through a 14 gauge Touhy needle for percutaneous placement epidurally. To the left (inset) are two openings marked with asterisks for flushing the three cylindrical recording contacts (1, 2, and 3) through the injection site (top, right). The same electrode can be placed in open surgical field after laminectomy without the need for a Touhy needle. Modified from Deletis (2002).

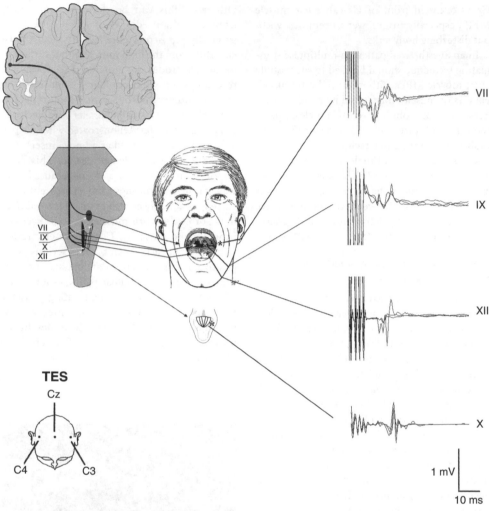

Fig. 8.3 Schematic of intraoperative eliciting and recordings of the corticolbulbar motor-evoked potentials (MEPs from muscle innervating by motor cranial nerves). *Lower left:* Schematic of positioning stimulating electrode over the scalp. *Upper left:* Schematic of corticobulbar pathways innervating motor cranial nerves nuclei (n. VII, IX, X, XII). *Middle:* positioning of recording electrodes inserted in m. orbicularis oris (n. VII), pharyngeal muscle (n. IX), tongue muscles (nerve XII) and vocal muscles (n. X) for monitoring corticobulbar MEPs. *To the right:* typical examples of corticobulbar MEPs recorded from cranial motor nerves innervated muscles. TES, transcranial electrical stimulation. NOTE: asterisk * depicts the insertion point for the recording electrode.

Relationship between electrode montage and current penetration within the brain

The choices of electrode montage for stimulation over the scalp together with the intensity of stimulation are two main factors that determine the depth of the current penetrating the brain. Figure 8.4 (left) shows the difference between stimulating montage C1/C2 vs C3/C4 and the activation of the CTs at different depths. Difference in the latencies of the two D-waves is as much as 1.9 ms, indicating activation of the CT at the level of the brain stem when using lateral montage (Rothwell *et al.* 1984). To the right

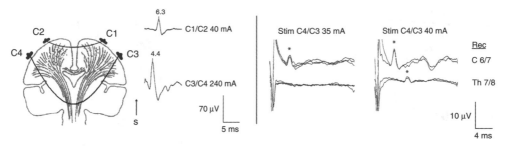

Fig. 8.4 *To the right:* Transcranial electrical stimulation over the C4 anode/C3 cathode with recordings of the D-wave over the C6/C7 segment (above), and the T7/T8 segment of the spinal cord (below). Stimulus intensity was 35 and 40 mA, respectively. Stronger stimuli elicit the D-wave over the thoracic spinal cord whereas a weaker stimulus (35 mA) elicits only the D-wave over the cervical spinal cord. *To the left:* Difference in amplitude and latencies of the D-wave records epidurally over the upper thoracic spinal cord in a patient undergoing surgery for a spinal cord tumor. Note the 1.9 ms difference between latencies of the D-wave when elicited with low intensity of current and stimulating montage C1/C2 vs high intensity of current and montage C3/C4. Note the higher amplitude of the D-wave when more axons of the corticospinal tract are recruited and current penetrates deep in the brain (C3/C4 montage and 240 mA stimulating current). Modified from Deletis (2002).

of Figure 8.4 is an example of how the low intensity of stimulation can selectively activate CT fibers only for the upper extremity without activating the CT tract for the lower extremity. This is shown by the presence of the D-wave recorded at the cervical spinal cord and its absence from the recording electrode positioned at the lower thoracic spinal cord. Increasing intensity of stimulation activates the CT tract for the lower extremity shown in the appearance of the D-wave at the thoracic spinal cord when using the higher stimulus intensity. For stimulation during brain surgery, it is highly advisable to use a grid electrode placed on to the exposed brain motor cortex in order to minimize a deep penetration of the stimulating current within the brain and possibly bypassing the site of surgery (by stimulating CT distally to the surgical site).

Relationship between descending drive from the electrically activated CT and generation of the muscle MEPs

One of the preconditions in the generation of the muscle MEPs intraoperatively after TES is to bring alpha motor neurons from the resting state to the firing level. This can be achieved either by a combination of the D- and I-waves or by multiple D-waves generated by TES with multipulse stimuli. The efficacy of multipulse stimulation not only comes from the generation of the multiple D-waves, easily bringing the alpha motor neuronal pool to the firing level, but from a second, important, phenomenon. TES with multipulses packed in the form of a short train can facilitate the generation of I-waves not being present after application of one or two TESs (Figure 8.5). The phenomenon of 'facilitation of I-waves' can be seen in the early publication of Philips and Porter, but was not recognized by those authors (Philips and Porter 1964). Clearly visible in Figure 8.5 is their recording from the baboon's spinal cord (to the right) as well as our recordings done from the patient (to the left), showing that three stimuli applied over the motor cortex can generate four descending volleys, three D-waves and one I-wave (Deletis *et al.* 2001). Furthermore, it is obvious from their intracellular recordings of the alpha motor neuron that the excitatory postsynaptic potential showed a significant elevation in amplitude after a third stimulus reached the alpha motor neuron. In monopolar recordings, these I-waves

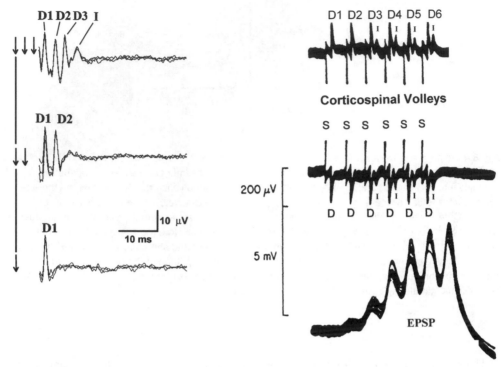

Fig. 8.5 Facilitation of the I-waves. *To the left:* TES with one and two stimuli elicit one and two D-waves respectively, while three stimuli elicited four waves (three D-waves and one I-wave). Recordings were done from the patient's epidural space at the upper thoracic spinal cord. *To the right, upper:* Phillips and Porter's (1964) recordings in baboon of conducted CT responses (D- and I-waves); *lower:* excitatory postsynaptic potentials (EPSPs) recorded intracellularly from a cervical motor neuron, following repetitive anodal stimulation of motor cortex. The D-wave responses are uniform during the train; I-waves are progressively recruited especially after the third stimulus. In the upper right we inverted polarity of recorded D- and I-waves (negative up) in order to highlight the similarity between recording in a baboon and a man (left), phenomenon of I-wave facilitation. S, stimulus artifact; D, D-wave; I, I-wave. Modified from Philips and Porter (1964) and Deletis *et al.* (2001).

appeared as low amplitude, but it is in fact many times larger. The reason for low amplitude is in their relative desynchronization, easily demonstrated by 'killed end recordings' which show much greater amplitude than in monopolar recordings (Amassian and Deletis 1999).

Due to the refractory period of the fast neurons of the CT, descending volleys cannot be conveyed at fast frequencies and a fully recovered D-wave. It has been shown that for complete recovery, the time period of the D-wave at a moderate intensity of stimulation, interstimulus interval (ISI) is 4 ms (250 Hz) and after strong stimulation is 2 ms (500 Hz) (Novak *et al.* 2004). This predetermines optimal ISI in the

train of TES for eliciting muscle MEPs when C1/C2 stimulation montage is used (Figure 8.6).

The D-wave collision technique with intraoperatively CT mapping of the spinal cord

This newly developed technique has allowed us to map intraoperatively and find the anatomical position of the CT within the surgically exposed spinal cord. It involves simultaneous TES of the motor cortex with concurrent stimulation of the CT from the surgically exposed spinal cord (Figure 8.7).

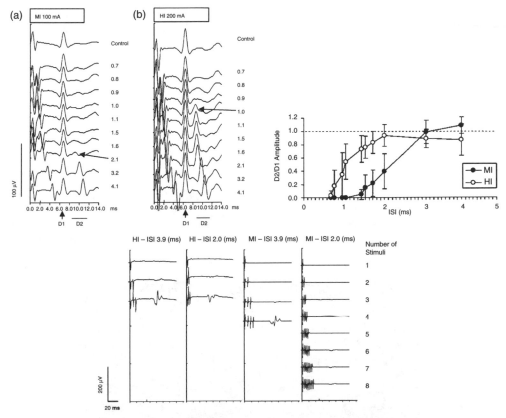

Fig. 8.6 Refractory period of human corticospinal tract determined by paired medium intensity (MI) and high intensity (HI) of transcranial electrical stimuli. *Left, upper:* Epidural recordings of the same patient illustrate D-wave recovery with an increase of interstimulus interval (ISI) at MI (a) and HI (b). *Right, upper:* Recovery of D-wave amplitude (mean D2/D1, error bars representing a standard deviation as a function of ISI (ranging from 0.7 to 4.1 ms). Stimuli of HI (open circles) show a faster recovery than stimuli of MI (filled circles). *Lower:* muscle MEPs were recorded from right m. tibialis anterior after TES to C1 (anode) and C2 (cathode). Number of stimuli in trains of short (2.0 ms) and long (4.1 ms) ISI was increased until a muscle MEP response was elicited. Note that no muscle MEPs were generated when MI stimuli were applied at a short ISI. Modified from Novak *et al.* (2004).

In 18 patients undergoing surgery for thoracic intramedullary spinal cord tumor (ISCTs), we elicited D-waves by TES and recorded them cranially and caudally to the spinal cord tumor. Simultaneous with TES, we stimulated the surgically exposed spinal cord (caudal to the tumor site) with a miniature bipolar hand-held probe (#5522.010, Inomed GmbH, Germany). The tips of the probe were 1 mm apart delivering constant current stimuli up to 2.5 mA in intensity, 0.5 ms in duration and repetition rate of

1 Hz. Whenever the stimulating probe was in a close proximity to the CT, the D-wave elicited by TES collided with the 'anti-D-wave' elicited by the spinal cord stimulation. This collision resulted in diminished amplitude of the D-wave recorded cranially to the lesion after collision (Deletis and Camargo 2001). This technique helps the surgeon to localize CT with spinal cord and modify resection of spinal cord tumors (or other pathology) by 'visualizing' CT tract position within the spinal cord.

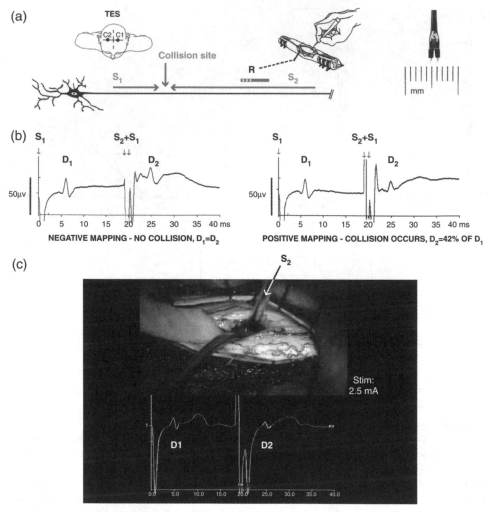

Fig. 8.7 Mapping of the corticospinal tract (CT) by D-wave collision technique (see text for explanation). (a) S1, transcranial electrical stimulation (TES); S2, spinal cord electrical stimulation; D1, control D-wave (TES only); D2, D-wave after combined stimulation of the brain and spinal cord; R, cranial electrode for recording the D-wave in the spinal epidural space. *To the right*: A tip of the hand-held stimulating probe with a scale in millimeters. (b) *To the left*: Negative mapping results (D1 = D2). To the right: positive mapping results (D2-wave amplitude significantly diminished after collision). (c) Intraoperative mapping of the CT within spinal cord in 44-year-old patient with intramedullary arteriovenous malformation at T3–T5 level. Stimulating probe delivering 2.5 mA current pulse in close proximity with CT, revealed by decrement of the D2-wave in comparison with D1-wave (control). Modified from Deletis and de Camargo (2001).

Intraoperative use of MEPs

Intraoperative MEPs have been used for almost a decade in the clinical setting. It was essential to develop the TES multipulse method that allowed muscle MEP recordings under general anesthesia (Taniguchi *et al.* 1993; Pechstein *et al.* 1996). MEPs are currently used during spinal and intracranial surgical procedures to monitor the functional integrity of motor pathways and prevent irreversible damage. However, the monitoring strategy and the interpretation of changes in the evoked responses vary between the intracranial and the spinal surgeries.

Spine and spinal cord surgery

For many years, before the advent of MEPs, only somatosensory-evoked potentials (SSEPs) were monitored during spinal procedures. However, SSEPs are not aimed to reflect the functional integrity of motor pathways and the assumption that they could do so has resulted in a number of so-called false negative results, meaning postoperative motor deficit in spite of unchanged intraoperative SSEPs. Theoretically, SSEPs may indirectly provide information on motor tract integrity for those procedures (like scoliosis surgery) where motor and sensory pathways are expected to be injured simultaneously due to distracting maneuvers of the spinal cord. Even so, Nuwer et al. (1995) reported a number of false-negative results in a large series of scoliosis surgeries. Nowadays, therefore, some agreement exists on the need for a combined SSEP–MEP monitoring approach also for orthopedic procedures. Interestingly, the use of D-wave monitoring for scoliosis surgery has been recently questioned by Ulkatan et al. (2006) due to the observation that misleading D-wave recordings (false results) can occur after correction of scoliosis due to the new anatomical relationship between de-rotated spinal cord and epidural recording catheter.

For other procedures like spinal cord tumor surgery or endovascular procedures for the embolization of spinal cord arteriovenous malformations, the combined used of SSEPs and MEPs is almost mandatory. During surgeries involving directly the spinal cord, there is a danger of selective injury to either the somatosensory or the motor pathways, and there is no scientific justification for a selective use of either SSEPs or MEPs. In ISCT surgery, for example, a selective injury to the CT without affecting the dorsal column may well occur and this is true also when the mechanism of spinal cord injury is purely vascular, as it could be during the embolization of a spinal arteriovenous malformation. The possibility of an anterolateral corticospinal tract insult occurring independently from an injury to the dorsal column – which would be reflected by the SSEPs – is possibly due to the different location and vascular supply of these structures. Dorsal columns, monitored by SSEPs, are in the vascular territory of the posterior spinal arteries, whereas anterior and lateral CTs are in the territory of the anterior spinal artery.

Overall, the most reliable monitoring protocol for a surgical procedure involving the spinal cord is based on the combined use of SSEPs, muscle MEPs and epidural MEPs (D-wave).

Muscle MEPs are generated by CT as well as by polysynaptic pathway, and are therefore sensitive to anesthesia and can be entirely blocked by muscle relaxants. Moreover, they can induce some muscle twitching that can be disturbing for the surgeon. However, they can provide specific information on different muscle groups from the left and right side and from upper and lower extremities. Finally, they are recordable also in patients with spinal cord lesion encompassing the most caudal spinal cord levels (conus).

D-wave recordings are very robust even under general anesthesia and are not impaired by the use of muscle relaxants. These recordings provide specific and semiquantitative information on the functional integrity of the fast conducting fibers of the CT and they are essential in predicting the prognosis and establishing reliable warning criteria. However, D-wave recordings still do not provide specific information on each of the lateral CT and cannot differentiate between muscle groups. Furthermore, D-wave cannot be recorded below the level of T12 because there are not enough CT fibers to record from. Previous scarring can sometimes impair electrode placement. Finally, in about one-third of the cases the D-wave is not recordable because of a desynchronization phenomenon (Deletis 2002).

With these limitations in mind, criteria to interpret MEP change during ISCT surgery have been established on the basis of a few hundreds of monitored cases (Kothbauer et al. 1997, 1998; Morota et al. 1997; Sala et al. 2006) and, up to now, no exceptions have been reported. To achieve a good long-term postoperative outcome, it is imperative to maintain the D-wave amplitude above 50% of its initial value. When this criterion is satisfied, the patient will either have an unchanged motor status at the end of the procedure if muscle MEPs have been preserved or will have a so-called transient para(mono)plegia if muscle MEPs have been lost uni- or bilaterally. The explanation we offered for the transient paraplegia phenomenon (i.e. transient postoperative motor deficit, occasionally as severe as a true plegia, that invariably recovers over a period of days, weeks or months after surgery) is as follows: the D-wave is generated exclusively by fast neurons of the CTs, while mMEPs are generated also by CT and other descending tracts within the spinal cord. An injury to these non-CT tracts can be functionally compensated postoperatively by the CTs, but not vice versa.

Applying these criteria, we have recently shown that MEP monitoring significantly improves long-term outcome after ISCT surgery (Sala *et al.* 2006). Interestingly, in the same study, short-term evaluation, early after surgery, did not show an advantage for the monitored vs nonmonitored group, and this was likely due to the transient paraplegia phenomenon that would mask the beneficial role of monitoring, when patients are evaluated in the early postoperative stage. It is important to notice that for spinal cord monitoring, unlike the amplitude criteria used for the D-wave, yes/no criteria are used for muscle MEPs. In other words, only presence/absence responses are considered but not changes in the morphology (biphasic versus polyphasic), amplitude or latency due to the extreme variability of these parameters even in the neurologically intact patient.

Posterior fossa surgery

Transcranial MEPs are also currently used during intracranial neurosurgical procedures, involving either the posterior fossa or the supratentorial compartment. Recently, current concepts in muscle MEP monitoring have been extended to the monitoring of the motor cranial nerves. So-called 'corticobulbar MEPs' can be used to monitor the functional integrity of corticobulbar pathways from the cortex through the cranial motor nuclei and to the muscle innervated by cranial nerves. Methodological aspects have appeared in the literature only recently and mostly with regard to the VII cranial nerve monitoring (Sala *et al.* 2004; Dong *et al.* 2005). A C3/Cz and C4/Cz montage is usually preferred to elicit right- and left-side muscle MEPs respectively. A train of four stimuli, ISI

of 2 ms at 1–2 Hz and intensity between 60 and 100 mA are generally used. For recordings, duration of individual stimuli of 0.5 ms, wire electrodes are inserted in the orbicularis oculi and oris (VII), posterior pharyngeal wall (IX–X), and tongue (XII). Reproducible muscle MEPs are continuously monitored. Nevertheless, this technique has not yet been standardized and some limitations still exist.

First, the use of a lateral montage (C3 or C4) as anodal stimulating electrode increases the risk that a strong TES may activate the cranial nerve directly rather than via the corticobulbar pathway. Therefore, the stimulation intensity should be kept as low as possible. In order to decrease the risk of a peripheral activation, once a corticobulbar MEP is recorded we repeat the stimulation keeping the same parameters except for the number of stimuli, which is reduced to one. This is based on the fact that cranial nerves can be activated by single pulses under anesthesia whereas bulbar motor neurons should require temporal summation of descending drive through corticobulbar pathways, as it occurs after multipulse stimulation. Therefore, if a response is still recorded after a single stimulus, this is likely to be the result of a direct activation of the peripheral nerve and cannot be considered reliable for monitoring. The longer latency of corticobulbar MEPs, as compared to that obtained after stimulation of intracranial portion of the motor cranial nerve, also distinguishes corticobulbar-activated response.

Second, the amplitude of corticobulbar MEPs recorded after TES versus that of CMAP recorded after direct intracranial supramaximal motor cranial nerve stimulation is much smaller, suggesting that after TES only a subset of motor neurons of cranial nerve nuclei are activated. If so, correlation between intraoperative corticobulbar response and postoperative outcome may not be totally accurate because of the possibility of false-positive and false-negative responses (Dong *et al.* 2005).

Brain surgery

Only few data have been reported on the use of muscle MEP monitoring after TES in brain surgery (Cedzich *et al.* 1996; Kombos *et al.* 2000; Neuloh *et al.* 2004). MEP monitoring can be used in combination with direct cortical and subcortical mapping, to increase the safety of surgery for tumors involving cortical and subcortical motor areas such as the central region, supplementary motor area and the insula. A C1/C2 or C2/C1 electrode montage is usually used to elicit contralateral limb muscle MEPs; Cz/6 cm (where 6 cm is a point 6 cm in front of Cz) can be used to elicit muscle MEPs from lower extremities. With these montages (C1/C2 or Cz/C6 cm) the risk that the current may activate the CTs deep in the brain is reduced; however, avoiding high current intensity further diminishes this risk and, when feasible, continuous MEP monitoring directly by stimulation through the cortical strip electrode is advisable because much less current is needed. Otherwise, stimulation parameters for TES in brain surgery are similar to those used for spinal cord monitoring.

Due to the considerably lower use of MEP monitoring in brain surgery as compared to spine and spinal cord surgery, criteria for MEP interpretation are less defined and standardized. The 'take home message', however, is that absolute criteria (presence/absence), like those applied in ISCT surgery, cannot be applied to brain surgery. Significant decreases in MEP amplitude (by 50–80% of baseline values) as well as 10–15% prolongation in latency have been described as warning signs that correlate with some degree of postoperative motor outcome, and muscle MEP loss consistently correlated with permanent paresis (Kombos et al. 2000; Neuloh et al. 2004). As reported by Neuloh et al. (2004), even an irreversible MEP deterioration or a transient loss may correlate with a permanent new motor deficit in a small percentage of patients.

Report on the use of D-wave monitoring in brain surgery is only anecdotal because most authors refrain from an invasive percutaneous placement of the epidural recording electrode in patients where there is no direct surgical access to the spinal cord. However, these recordings are extremely interesting (Yamamoto et al. 2004; Fujiki et al. 2006) because they provide unique data on the specific behavior of the CT during these procedures. The most intriguing result from the Yamamoto study is that hemi-body postoperative persistent motor deficit remained only in patients who had a decrease of D-wave amplitude

of more than 30% over baseline values, when monohemispheric stimulation was performed by using grid electrode. A decrease of less than 30% would correlate with either no deficit or only transient motor deficit. This empirical threshold of 30% corresponds to the 50% threshold we observed in ISCT surgery when the D-wave was elicited by stimulation of right and left CT (by bihemispheric stimulation) and correlates with the fact that muscle MEP warning criteria are also different in brain and spinal cord surgery.

Overall, it seems that different mechanisms may be involved in the pathophysiology of postoperative paresis in brain and spinal cord surgeries so that different MEP monitoring criteria should be used to warn the surgeon in time, avoid irreversible damage, and accurately predict the prognosis.

References

Amassian VE, Deletis V (1999). Relationships between animal and human corticospinal responses. *Electroencephalography and Clinical Neurophysiology* 51(Suppl), 79–92.

Cedzich C, Taniguchi M, Schafer S, Schramm J (1996). Somatosensory evoked potential phase reversal and direct motor cortex stimulation during surgery in and around the central region. *Neurosurgery 38*, 962–970.

Deletis V (2002). Intraoperative neurophysiology and methodology used to monitor the functional integrity of the motor system. In: V Deletis and J Shils (eds), *Neurophysiology in neurosurgery. A modern intraoperative approach*, pp 25–51. New York: Academic Press.

Deletis V, de Camargo AB (2001). Interventional neurophysiological mapping during spinal cord procedures. *Stereotactic and Functional Neurosurgery 77*, 25–28.

Deletis V, Rodi Z, Amassian V (2001). Neurophysiological mechanisms underlying motor evoked potentials (mep's) elicited by a train of electrical stimuli. Part 2. Relationship between epidurally and muscle recorded MEPs in man. *Clinical Neurophysiology 112*, 445–445.

Dong CC, Macdonald DB, Akagami R, Westerberg B, Alkhani A, Kanaan I, Hassounah M (2005). Intraoperative facial motor evoked potential monitoring with transcranial electrical stimulation during skull base surgery. *Clinical Neurophysiology 116*, 588–596.

Fujiki M, Furukawa Y, Kamida T, Anan M, Inoue R, Abe T, Kobayashi H (2006). Intraoperative corticomuscular potentials for evaluation of motor function: a comparison with corticospinal D and I waves. *Journal of Neurosurgery 104*, 85–92.

Journee HL, Polak HE, de Kleuver M (2004). Influence of electrode impedance on threshold voltage for transcranial electrical stimulation in motor evoked potential monitoring. *Medical and Biological Engineering and Computing 42*, 557–561.

Kombos T, Suess O, Funk T, Kern BC, Brock M (2000). Intra-operative mapping of the motor cortex during surgery in and around the motor cortex. *Acta Neurochirurgica 142*, 263–268.

Kothbauer, K, Deletis V. Epstein FJ (1997). Intraoperative spinal cord monitoring for intramedullary surgery: an essential adjunct. *Pediatric Neurosurgery 26*, 247–254.

Kothbauer K, Deletis V, Epstein FJ (1998). Motor evoked potential monitoring for intramedullary spinal cord tumor surgery: correlation of clinical and neurophysiological data in a series of 100 consecutive procedures. *Neurosurgery Focus 4*, 1–9.

Merton PA, Morton HB (1980). Electrical stimulation of human motor and visual cortex through the scalp. *Journal of Physiology 305*, 9–10P.

Morota N, Deletis V, Shlomi C, Kofler M, Cohen H, Epstein F (1997). The role of motor evoked potentials (meps) during surgery of intramedullary spinal cord tumors. *Neurosurgery 41*, 1327–1366.

Neuloh G, Pechstein U, Cedzich C, Schramm J (2004). Motor evoked potential monitoring with supratentorial surgery. *Neurosurgery 54*, 1061–1070; discussion 1070–1072.

Novak K, de Camargo AB, Neuwirth M, Kothbauer K, Amassian V, Deletis V (2004). The refractory period of fast conducting corticospinal axons in man and its implication for intraoperative monitoring of motor evoked potentials. *Clinical Neurophysiology 115*, 1931–1941.

Nuwer MR, Dawson EG, Carlson LG, *et al.* (1995). Somatosensory evoked potential spinal cord monitoring reduces neurologic deficits after scoliosis surgery: results of a large multicenter survey. *Electroencephalography and Clinical Neurophysiology 96*, 6–11.

Patton HD, Amassian VE (1954). Single and multiple unit analysis of cortical state of pyramidal tract activation. *Journal of Neurophysiology 17*, 345–363.

Pechstein, U, Cedzich C, Nadstawek J, Schramm, J (1996). Transcranial high-frequency repetitive electrical stimulation for recording myogenic motor evoked potentials with the patient under general anesthesia. *Neurosurgery 39*, 335–343; discussion 343–344.

Philips CG, Porter R (1964). The pyramidal projection to motoneurons of some muscle groups of the baboon's forelimbs. *Progress in Brain Research 12*, 222–245.

Rothwell J, Burke D, Hicks R, Stephen J, Woodforth I, Crawford M (1994). Transcranial electrical stimulation of the motor cortex in man: further evidence for the site of activation. *Journal of Physiology (London) 481*, 243–250.

Sala F, Lanteri P, Bricolo A (2004). Intraoperative neurophysiological monitoring of motor evoked potentials during brain stem and spinal cord surgery. *Advanced and Technical Standards in Neurosurgery 29*, 133–169.

Sala F, Palandri G, Lanteri P, Deletis V, Facioli F, Bricolo A (2006). Motor evoked potential monitoring improves outcome after surgery for intramedullary spinal cord tumor: a historical control study. *Neurosurgery 58*, 1129–1143.

Taniguchi M, Cedzich C, Schramm J (1993). Modification of cortical stimulation for motor evoked potentials under general anesthesia: technical description. *Neurosurgery 32*, 219–226.

Ulkatan S, Neuwirth M, Bitan F, Minardi C, Kokoszka A, Deletis V (2006). Monitoring of scoliosis surgery with epidurally recorded motor evoked potentials (D wave) revealed false results. *Clinical Neurophysiology 117*, 2093–2101.

Yamamoto T, Katayama Y, Nagaoka T, Kobayashi K, Fukaya C (2004). Intraoperative monitoring of the corticospinal motor evoked potential (D-wave): clinical index for postoperative motor function and functional recovery. *Neurologia Medicochirurgica 44*, 170–180; discussion 181–182.

TMS Measures of Motor Cortical and Corticospinal Excitability: Physiology, Function, and Plasticity

Ulf Ziemann

The first three chapters in this section describe the TMS measures that are currently available to probe motor cortical and corticospinal excitability. These chapters will provide a detailed account on the techniques, physiology, and basic and clinical applications; and they are fundamentally important because they set the stage for a broad readership to understand what can be measured with TMS and how and why. The next three chapters contain more advanced physiology of the TMS measures by means of electrophysiological and pharmacological profiling. This is an extremely dynamic and exciting field, which leads to a fundamental understanding of what is excited and measured by TMS, even down to the level of particular neuronal cortical circuits. This opens up fascinating avenues to use TMS measures for studying, beyond motor cortical excitability, motor function and mechanisms of plasticity. Chapter 15 is directly at this intersection. This chapter will focus on explaining the link between TMS measures and voluntary motor function. Finally, the last three chapters show how TMS measures can be used to study plasticity of human cortex. This is of utmost interest for all of those who want to understand how the brain changes in response to experimental manipulation, and how this may be modulated, for instance to purposefully enhance plasticity in the setting of rehabilitation.

CHAPTER 9

The size of motor-evoked potentials: influencing parameters and quantification

Kai M. Rösler and Michel R. Magistris

Summary – We discuss parameters influencing the size of motor-evoked potentials (MEPs) in normal and pathological conditions, and the methods that minimize these influences to allow a meaningful quantification of the MEPs. A better understanding of the parameters influencing the size of MEPs will provide new insights into the function of our central nervous motor system.

Introduction

Transcranial magnetic stimulation of the brain induces muscle responses termed motor-evoked potentials (MEPs). MEPs are widely used to study the physiology of corticospinal conduction in healthy subjects and in patients with diseases of the central nervous system. A variety of parameters of MEPs can be studied, including the latency yielding the *central motor conduction time* (CMCT), the *size of the MEP* (amplitude, duration and area), and others (such as stimulation thresholds, silent period, facilitation, etc.). In this chapter we will focus on the size of MEPs. We will discuss some mechanisms influencing it and the methods that minimize these influences.

If a peripheral nerve is stimulated, the resulting compound motor action potential (CMAP) reflects the number of activated motor units (McComas 1995), which is roughly proportional to the number of activated motor axons. Theoretically, the size of an MEP should also relate to the number of activated corticospinal motor neurons. However, this relation is obscured by some particular characteristics of MEPs, making the interpretation of MEP size measurements difficult. Three basic physiological mechanisms may influence the size of MEPs. These are:

- the number of recruited motor neurons in the spinal cord
- the number of motor neurons discharging more than once to the stimulus
- the synchronization of the TMS-induced motor neuron discharges.

In studies of corticomuscular conduction in healthy subjects and patients, it is often assumed that the size of an MEP reflects the number of activated motor neurons. However, it is desirable to know the contribution to MEP size of any of the influential mechanisms mentioned.

Characteristics of MEP size

Stimulus intensity

If electrical stimuli of increasing intensity are applied to a peripheral nerve, the size of the induced muscle response increases. When the stimulus is strong enough, the CMAP size saturates, i.e. the size will not increase if the stimulus intensity is further increased. These characteristics of peripheral nerve stimulation are easily explained by increasing recruitment of more and more nerve fibers by the increasing stimulus. Also, MEPs increase with increasing stimulus intensity, suggesting that stronger stimuli may also recruit more corticospinal and spinal motor neurons, or both. However, even with very strong brain stimuli, the MEP will usually not reach the same size as the CMAP to maximal peripheral stimulation (Figure 9.1), suggesting that either (i) factors other than stimulus intensity may influence the MEP size, or (ii) factors other than the number of excited cells may play a role in determining the size of an MEP. MEPs do eventually saturate with increasing stimulation intensity, but instead of reaching a maximal amplitude, the MEP size varies from one stimulus to the next (see below).

The stimulus–response relationship varies considerably between subjects. If the TMS intensity is standardized to the individual MEP threshold (usually defined as the minimum TMS intensity that is sufficient to elicit MEPs of liminal size in 50% of the trials; Rossini *et al.* 1999),

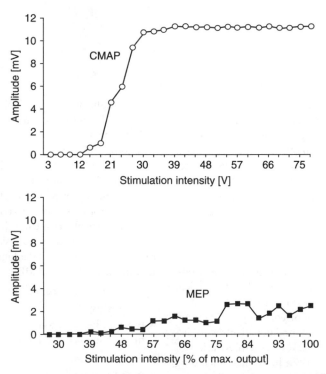

Fig. 9.1 Compound muscle action potentials (CMAP, *upper panel*) and motor-evoked potentials (MEP, *lower panel*) are obtained from abductor digiti minimi (ADM) in a normal subject in response to stimuli of increasing intensities. Electrical stimuli are applied to the ulnar nerve at the wrist; they are increased by steps of 3 V; each point of the curve corresponds to the response of a single stimulus. Transcranial magnetic stimuli are performed with a MagStim 200, ADM being at rest; they are increased by steps of 3% (of maximal stimulator output); each point of the curve corresponds to the amplitude averaged from three responses. Thresholds of both curves are arbitrarily aligned.

e.g. at 20% above threshold, not the same MEP size is reached in every subject, and considerable interindividual differences between 6% and 100% of MEP amplitude at maximum intensity have been observed (van der Kamp *et al.* 1996). Similar observations were made by Rösler *et al.* (2002). The stimulus–response relationship also varies between muscles in the same individual, e.g. in small hand muscles (Ziemann *et al.* 2004). Likewise, the ratio of the maximal MEP and the peripherally evoked maximal CMAP differs markedly and unpredictably between normal subjects (Magistris *et al.* 1998; Rösler *et al.* 2002). The MEP of abductor digiti minimi (ADM) is normally greater than 15% of the maximal CMAP evoked by ulnar nerve stimulation at the wrist (Hess *et al.* 1987b). To account for the dispersion that occurs along the peripheral nerve, one may prefer to compare the MEP with a CMAP evoked by more proximal stimuli (i.e. the response evoked by supramaximal stimulation of the brachial plexus at Erb's point; Roth and Magistris 1987), and the normal amplitude ratio of the ADM is then >33% (Magistris *et al.* 1998). The ratio between MEP and peripheral CMAP is even smaller in foot muscles. For abductor hallucis, a ratio of >21% is still considered normal, if the peripheral CMAP is obtained by proximal stimulation of the sciatic nerve in the gluteal region (Bühler *et al.* 2001).

Role of the coil and coil positioning

The size of an MEP depends on the localization of the magnetic stimulus and on the direction of the induced electrical field. Early TMS studies noted that, when viewed from above, clockwise current orientation within a circular stimulating coil centered over the vertex led to preferential stimulation of the right hemisphere, while counterclockwise current flow preferentially stimulated the left hemisphere (Hess *et al.* 1987a). The circular coil delivers a relatively large and diffuse field over the brain, so that slight movements of the coil over the head do not critically influence the MEP size (Gugino *et al.* 2001). To allow for more 'focal' stimulation, figure-8 coils were designed (Rösler *et al.* 1989). With this type of coil, the maximum field strength is produced at the junction of the two

singular coil elements, and stimulation is maximal beneath this point. Moving the coil over the head will influence the size of the MEP considerably. Figure-8 coils may therefore be employed to map the cortical motor output by correlating the size of the evoked MEP with the location of stimulation over the scalp (Pascual-Leone and Torres 1993; Wilson *et al.* 1993). On the other hand, this type of focal stimulation makes it important to place the coil appropriately over the 'hot spot' when maximal responses are desired.

The direction of the induced magnetic field is perpendicular to the long axis of the figure-8 coil. The orientation of the magnetic field over the brain is important, since large differences have been demonstrated to occur with different coil orientations (Rösler *et al.* 1989). For hand and facial muscles, the optimum field orientation is perpendicular to the central sulcus (Brasil-Neto *et al.* 1992; Mills *et al.* 1992; Dubach *et al.* 2004), whereas for leg muscles an orientation perpendicular to the interhemispheric cleavage is preferable (Rösler *et al.* 1989).

Facilitation

The MEP is influenced by the excitability of the corticospinal pathway, which is variable and can be facilitated by a number of mechanisms (Hess *et al.* 1986; Andersen *et al.* 1999). In particular, voluntary 'background' contraction of the target muscle 'facilitates' the MEP by reducing its threshold, shortening its latency and – most notably – increasing its size. Facilitation by voluntary background contraction is observed in all target muscles, and with both electrical and magnetic transcranial stimulation. The relationship between response amplitude and voluntary contraction has been determined for a number of muscles. In the hand muscles, facilitation is observed with very small force levels, and becomes maximal and saturates with a contraction force of about 5–10% of the maximum (Figure 9.3; Hess *et al.* 1987a; Rösler *et al.* 2002). In other muscles of forearm and arm, and also in muscles of the lower extremity, stronger contractions are required and the relation between force and facilitation may be more linear (Ravnborg *et al.* 1991; Kischka *et al.* 1993; Bühler *et al.* 2001). Facilitation is enhanced if

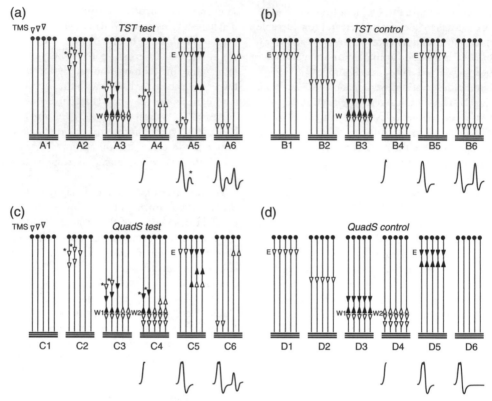

Fig. 9.2 (a) Triple stimulation technique (TST) principle. The motor tract is simplified to five spinal motor neurons (MNs); horizontal lines represent the muscle fibers of the five motor units. Black arrows represent action potentials that collide; arrows with an asterisk mark repetitive MN discharges (repMNDs). (a) In TST$_{test}$, (A1) a submaximal transcranial stimulus excites three spinal MNs out of five (open arrows). (A2) On these three neurons, TMS-induced action potentials descend. Desynchronization of the three action potentials has occurred (possibly at spinal MN level); two out of three spinal MNs are excited twice causing repMNDs. (A3) After a delay, a maximal stimulus is applied at the wrist (W). (A4) It gives rise to a first main negative deflection of the recording trace. The antidromic action potentials collide with the first descending action potentials on MNs 1, 2, and 3. The repMND on MNs 1 and 2 do not collide and give rise to a small negative deflection of the recording trace (asterisk). The action potential on neurons 4 and 5 continues to ascend. (A5) After an appropriate delay, a maximal stimulus is applied at Erb's point (E). On MNs 4 and 5, the descending action potentials collide with the ascending action potentials. (A6) A synchronized response from the three MNs that were initially excited by the transcranial stimulus is recorded as the second main negative deflection of the TST$_{test}$ trace. (b) In TST$_{control}$, (B1) a maximal stimulus is applied at Erb's point. (B2) On five out of five neurons action potentials descend. (B3) After a delay, a maximal stimulus is applied at the wrist. (B4) The orthodromic action potentials are recorded as the first negative deflection of the TST control trace, the antidromic action potentials collide with the action potentials of the first stimulus at Erb's point. (B5) After a delay, a maximal stimulus is applied at the Erb's point. (B6) A synchronized response from the five MNs is recorded as the second negative deflection of the TST control trace. The test response is quantified as the ratio of TST$_{test}$ to TST$_{control}$ curves (3/5 = 60% in this example). (c) Quadruple stimulation technique (QuadS) principle. In QuadS$_{test}$, (C1) a submaximal transcranial stimulus excites three spinal MNs out of five. (C2) On these three neurons, TMS-induced action potentials descend. Two out of three spinal MNs are excited twice (neurons 1 and 2). (C3) After a delay, a first maximal stimulus is applied at the wrist (W1). (C4) It gives rise to a first negative deflection of the recording trace. The antidromic action

stimuli are applied at the start of the contraction (e.g. during a ballistic contraction) rather than during isometric contraction (Mills and Kimiskidis 1996). Facilitation also occurs with contraction of muscles remote from the target muscle, on the same or on the opposite limb (Hess *et al.* 1986), or in the face (Andersen *et al.* 1999). Some facilitation of lower limb muscles also occurs in response to a contraction of the hands (Péréon *et al.* 1995), whereas contraction of lower limb muscles causes only a little facilitation on hand muscles. The increase in the amplitude of the MEP that occurs in response to voluntary contraction is probably mainly caused by an increasing number of spinal motor neurons brought to fire by the transcranial stimulus, since desynchronization of the motor neuron discharges appears to be unchanged by voluntary contraction (Rösler *et al.* 2002), and repetitive discharges of spinal motor neurons occur only with high stimulation intensities, higher than those necessary to observe facilitation (Z'Graggen *et al.* 2005). Facilitation may relate either to a larger corticospinal volley reaching the spinal motor neuron pool, or to the excitation of a larger proportion of the spinal motor neurons in response to the descending corticospinal volley, or to both. When stimuli are given at the base of the skull, and excite the motor pathway in the region of the cervico-medullary junction, the response is much larger in contracting than in relaxed muscles (Ugawa *et al.* 1994). This demonstrates the role played by the increased excitability of the spinal motor neurons. During strong voluntary contractions, however, additional facilitation is observed in response to magnetic cortical stimulation, with a corticospinal volley of increased duration (Di Lazzaro *et al.* 1998).

Along with the increase in MEP size, the latency of the transcranial MEP shortens during voluntary contraction (Day *et al.* 1987b; Hess *et al.* 1987a). The reduced latency suggests that the spinal motor neurons fire in response to an earlier I-wave or to the D-wave during voluntary contraction. A shortening of latency does not occur in response to the single volley evoked by brain-stem stimulation (Berardelli *et al.* 1991).

A variety of other maneuvers increase the size of the MEP. These are: conditioning transcranial stimulation (use of paired stimuli); tendon, cutaneous nerve or muscle afferent input (Claus *et al.* 1988; Kossev *et al.* 1999; Siggelkow *et al.* 1999; Kofler *et al.* 2001); muscle stretch; thinking about a movement or contraction of the target muscle (Izumi *et al.* 1995; Kiers *et al.* 1997); speech (when speaking aloud); and a variety of cognitive maneuvers (Pascual-Leone *et al.* 1992, 1994). In our experience none of these facilitation maneuvers is as effective as background contraction of the target muscle (unpublished observations). Some of these maneuvers have been used to minimize the trial-to-trial variability of the size of MEPs (Nielsen 1996).

In contrast to the MEP facilitation discussed above, a number of stimulation paradigms cause inhibition and reduce the size of MEPs. These are: paired stimuli given over the same

potentials collide with the first descending action potentials on MNs 1, 2, and 3. After 3 ms a second supramaximal stimulus is applied at the wrist (W2). The orthodromic action potentials give rise to a second negative deflection of the recording trace, merging with the first one. The antidromic action potentials evoked by this stimulus collide with the repMNDs on MNs 1 and 2. The action potential on neurons 3, 4, and 5 continues to ascend. (C5) After an appropriate delay, a maximal stimulus is applied at Erb's point. On MNs 3, 4, and 5, the descending action potentials collide with the ascending action potentials. (C6) A synchronized response from the two MNs that conducted repMND is recorded as the second deflection of the QuadS$_{test}$ trace. (d) In QuadS$_{control}$, (D1) a maximal stimulus is applied at Erb's point. (D2) On 5/5 neurons, action potentials descend. (D3) After a delay, a maximal stimulus is applied to the wrist. (D4) It is recorded as the first deflection of the TST control trace. After a delay of 3 ms a second supramaximal stimulus is applied at the wrist. (D5) After an appropriate delay, a maximal stimulus is applied at Erb's point. The antidromic action potentials evoked by this stimulus collide with the antidromic action potentials from the second stimulus applied at the wrist. (D6) As a consequence no response is recorded in the control trace. The test response is quantified as the ratio of QuadS$_{test}$ to TST$_{test}$ curves (2/3 = 66% in this example).

hemisphere with interstimulus intervals different from those causing facilitation (Inghilleri *et al.* 1993; Kujirai *et al.* 1993; Uncini *et al.* 1993), reviewed by (Mills 1999), cf. Chapter 11 in this book, or TMS given on the opposite hemisphere before a test stimulus (the so-called 'transcallosal inhibition'; Ferbert *et al.* 1992); and cerebellar stimulation (Ugawa *et al.* 1991). A reduction of MEP size is also observed after muscular fatigue (Taylor and Gandevia 2001).

Basic physiological mechanisms influencing MEP size

As mentioned above, the MEPs will usually not reach the same size as the CMAP to maximal peripheral stimulation (Figure 9.1), and they vary from stimulus to stimulus, suggesting that factors other than stimulus intensity or number of excited motor neurons may influence its size. Two basic causative mechanisms have been identified: (i) desynchronization of TMS-evoked motor neuron discharges, and (ii) repetitive discharges of motor neurons in response to a single brain stimulus (Magistris *et al.* 1998; Z'Graggen *et al.* 2005). If these two mechanisms can be dealt with, the size characteristics of MEPs may be better understood. We have developed a method that eliminates the influences of desynchronization and repetitive discharges on MEPs, allowing some new insights into the basic physiology of MEPs. In the following, we will first describe this technique, and thereafter the new insights gained with it.

The triple stimulation technique (TST)

The TST was developed to eliminate the effects of phase cancellation from the MEPs, to allow for a better quantification (see below). It was described in detail in several recent publications (Magistris *et al.* 1998, 1999; Bühler *et al.* 2001; Humm *et al.* 2004b). In short, the TST is a collision method using a sequence of three stimuli to the brain, the ulnar nerve at the wrist, and the brachial plexus at Erb's point (Magistris *et al.* 1998, 1999). The TST test response is calibrated by a TST control response, for which the brain stimulus is replaced by a proximal nerve stimulus (succession of stimuli: Erb – ulnar nerve at the wrist – Erb). The original TST protocol was

adapted for use with a variety of muscles, and now includes three small hand muscles and one foot muscle (Bühler *et al.* 2001; Humm *et al.* 2004b). The TST eliminates influences on MEP size caused by the desynchronization of TMS-induced motor neuron discharges, and it eliminates influences of repetitive motor neuron discharges from the response. Therefore, it allows a quantification of the percentage of the motor neuron pool of the target muscle that is driven to discharge by TMS (see Figure 9.2a,b for a summary of the principle of the technique). The delays between stimuli are calculated as follows:

delay I = minimal MEP latency – $CMAP_{wrist}$ latency,

delay II = $CMAP_{Erb}$ latency – $CMAP_{wrist}$ latency.

For the TST control recording, the brain stimulus is replaced by a maximal electrical stimulus to the brachial plexus at Erb's point (Figure 9.2b), with appropriate adjustments of the delays (delay I = delay II = $CMAP_{Erb}$ latency – $CMAP_{wrist}$ latency). The size ratio of the TST test and TST control curves is termed 'TST amplitude ratio' or 'TST area ratio'; it is an estimate of the proportion of the motor neuron pool of the target muscle that was driven to discharge by the transcranial stimulus.

It can easily be seen in Figure 9.2 that the TST results in a 'resynchronization' of the action potentials evoked by the transcranial stimulus. Moreover, the influence of multiple discharges of spinal neurons is avoided. Indeed, the collision affects only the first descending action potential descending on a given peripheral axon, whereas any following action potentials (i.e. the repetitive discharges) escape collision. The latter are recorded between the two main deflections of the TST curve (indicated by an asterisk in Figure 9.2a, A5).

Desynchronization: how motor units summate within the MEP

The TST demonstrated that, with TMS, nearly 100% of all spinal motor neurons supplying a target muscle can be excited in healthy subjects, yet that conventional TMS with equal stimulation intensity yields MEPs which are smaller than CMAPs to supramaximal peripheral nerve stimulation (Figure 9.3). Therefore, the main

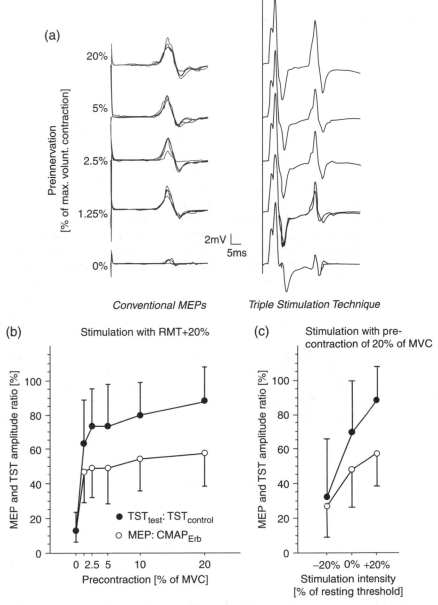

Fig. 9.3 Motor-evoked potential (MEP) sizes are always smaller than triple stimulation technique (TST) response sizes. (a) MEP and TST recordings in one subject, with suprathreshold stimulation (= threshold + 20%). The responses for five different levels of preinnervation (0–20% of maximal voluntary contraction) are shown. The TST responses are composed of a first negative deflection, which stems from the wrist stimulus and is not measured, and a second negative deflection which stems from Erb stimulation and quantifies the number of spinal motor neurons brought to discharge by the TMS. (b) Comparison of MEP and TST amplitude ratios of 10 subjects at six precontraction force levels and with a stimulus intensity of threshold + 20%. Averages of TST amplitude ratios and MEP amplitude ratios plus 1 SD, are given. (c) Stimulation was applied during a fixed precontraction level of 20% of maximal voluntary contraction force. The relation of response amplitude as a function of the stimulation intensity is shown. Modified from Rösler *et al.* (2002).

cause of the MEP size reduction is desynchronization of the TMS induced motor neuron discharges (Magistris *et al.* 1998). Desynchronization
induces phase cancellation and reduces the MEP
size (phase cancellation: the negative phases of
individual motor unit potentials are cancelled
by positive phases of others). Theoretically,
desynchronization should affect amplitudes more
than areas, and this is indeed what is observed
(Rösler *et al.* 2002; Z'Graggen *et al.* 2005).

We have used the TST (Figures 9.2 and 9.3) to
eliminate the effects of desynchronization from
the responses, using a variety of TMS intensities,
and different facilitation maneuvers. Comparison of the TST results with those of the
conventional MEPs allows an estimate of the
degree of size reduction caused by the discharge
desynchronization. For the ADM, four previous
studies suggested an average MEP amplitude
drop in healthy subjects and patients of about
one-third of the 'true' size (Magistris *et al.* 1998,
1999; Rösler *et al.* 2000, 2002; Table 9.1).
However, this effect varied markedly between
subjects (see below). For the abductor hallucis
(AH) muscle, the average amplitude drop is
greater than for ADM (~55%), due to the considerably greater degree of discharge desynchronization (Bühler *et al.* 2001; Table 9.1).

This effect of desynchronization appears to be
the same with submaximal and maximal stimulation, and does not depend on the amount of
facilitatory muscle contraction (Figure 9.3;
Rösler *et al.* 2002). Also, short-interval intracortical inhibition measured using the protocol
described by Kujirai *et al.* (1993) does not affect
the degree of desynchronization (Mall *et al.*
2001). However, the size-reducing effect of the
discharge desynchronization differs greatly (and
unpredictably) between subjects (Magistris *et al.*
1998; Rösler *et al.* 2002). The MEP vs TST
amplitude relation ranges between 0.37 and
0.75 in our subjects, i.e. MEP desynchronization
reduces the MEP size by 25–60% or more. This
degree of motor neuron discharge desynchronization after TMS may have the characteristics
of an individual trait. However, so far no association has been found with anthropometric data
such as age, gender, or height of the subjects, or
with electrophysiological data such as MEP
duration, latency, CMCT, and motor threshold.

It is unclear whether motor neuron discharge
desynchronization has a functional significance
for the individual. Desynchronization of motor
action potentials observed in peripheral neuropathies has no detrimental effect on motor
function (as long as it is not associated with

Table 9.1 Normal means and normal limits for examination of two muscles using the triple stimulation technique (TST)

	ADM[a]		AH[b]	
	Mean (SD)	Normal limit	Mean (SD)	Normal limit
TST amplitude ratio	99.1% (2.14)	≥93%	95.0% (4.06)	≥88%
TST area ratio	98.5% (2.48)	≥92%	96.1% (8.30)	≥84%
MEP amplitude ratio	66.1% (12.99)	≥33%	37.2% (9.72)	≥21%
MEP area ratio	96.8% (17.95)	≥52%	99.7% (38.45)	≥43%
Mean MEP amplitude loss (vs TST)	34.3%[c]		57.8% (9.12)	
Mean MEP area loss (vs TST)	16.2%[c]		58.9% (11.36)	
TST amplitude variability (CV)	2.6%[a,d]			
MEP amplitude variability (CV)	8.1%[a,d]			

[a]Abductor digiti minimi; Magistris *et al.* (1998).
[b]Abductor hallucis; Bühler *et al.* (2000).
[c]Rösler *et al.* (2002).
[d]Variability with maximal TMS, exciting nearly 100% of all spinal motor neurons.
CV, coefficient of variation.

conduction block). We suspect the same to be true for the motor neuron discharge desynchronization observed after TMS.

Repetitive motor neuron discharges

Another influence on the size of the MEP is that of repetitive motor neuron discharges. After a single brain stimulus, spinal motor neurons may discharge more than once (Day *et al.* 1987a; Hess *et al.* 1987a). Repetitive discharges can be directly seen in TMS recordings from a partially denervated muscle, where only one or two motor units remain active in the recording area of the needle electrode (Magistris *et al.* 1998; Figure 9.4). Repetitive discharges are an indirect consequence of the nature of the descending corticospinal volley after TMS, which contains a succession of a D-wave and multiple I-waves converging upon the spinal motor neuron (Patton and Amassian 1954; Amassian *et al.* 1987), eventually causing it to discharge repetitively (Day *et al.* 1989). They occur after cortex stimulation, but not after stimulation at the level of the brainstem, pointing to the importance of D- and I-waves driving motor neurons to discharge repetitively (Day *et al.* 1987a; Berardelli *et al.* 1991). Repetitive discharges may increase the size of the MEP, but it is difficult to estimate the amount of this increase. Twitch force measurements have indicated that with increasing stimulation intensity and increasing levels of background voluntary muscle contraction, the size (amplitude and area) of the MEP will eventually reach a plateau, while the twitch force will continue to increase (Kiers *et al.* 1995). This discrepancy was related to the number of motor neurons firing repetitively, and an important conclusion of these authors was that 'the MEP amplitude, and to a lesser extent area, do not accurately reflect the net motor output' (Kiers *et al.* 1995).

To estimate repetitive motor neuron discharges, a single collision method can be used. A transcranial magnetic stimulus is followed by a supramaximal distal peripheral nerve stimulus (Hess *et al.* 1987a; Berardelli *et al.* 1991; Naka and Mills 2000). If appropriately timed, the antidromic action potentials ascending from the distal stimulus collide with the first descending action potential on each motor neuron, such

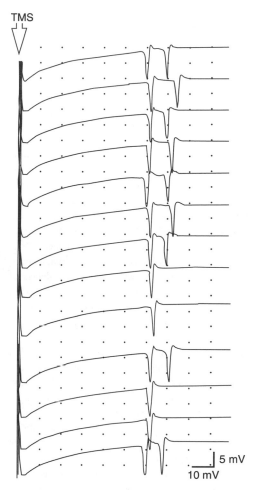

Fig. 9.4 Single motor unit potential evoked by serial TMS. Recording via concentric needle electrode from m. extensor digitorum brevis pedis in a 70-year-old woman suffering from a severe axonal S1 root disorder which has left only one motor unit at the needle recording site. Repetitive (i.e. double) discharges are observed after nine of 13 stimuli. The shortest interdischarge interval is just less than 5 ms. There is a marked latency variability ranging from 57.1 to 61.3 ms.

that only repetitive motor neuron discharges are recorded. The single collision method has disadvantages that impair a precise quantification of repetitive motor neuron discharges. First, as mentioned above, action potentials evoked by brain stimuli are not synchronous, which induces phase cancellation reducing the size of the compound response in an unpredictable manner.

Second, the recorded muscle response may contain indirect components such as H-reflex or F-wave (Mazzocchio et al. 1995; Naka and Mills 2000). Finally, if repetitive discharges follow the first discharge with a very short interval, they may escape the collision and thus quantification (Naka and Mills 2000).

We have used a different approach to quantify repetitive motor neuron discharges, combining the TST with the single collision method described above. To quantify repetitive motor neuron discharges, a fourth (and if necessary a fifth) stimulus can be given to the peripheral nerve at the wrist, using an external electrical stimulator (Quadruple stimulation technique, QuadS; and Quintuple stimulation technique, QuintS). The sequence of stimuli is then: brain – wrist 1 – wrist 2 (– wrist 3) – Erb. The delays of brain – wrist 1 – Erb are the same as those used for the TST_{test}. An appropriate interstimulus interval between wrist 1 and wrist 2 (and wrist 3) is 3 ms (Z'Graggen et al. 2005). The antidromic action potentials evoked by the wrist 2 stimulus collide with action potentials arising from second spinal motor neuron discharges, which may or may not follow the first action potential produced by the first spinal motor neuron discharges (Figure 9.2C). Under these circumstances, the second main negative deflection of the QuadS recording trace (evoked by Erb stimulus) is a 'resynchronized' compound response corresponding to the sum of second motor neuron discharges (Figure 9.2C). By comparing it to the TST_{test} curve (Figure 9.2A), the percentage of motor neurons discharging twice in response to the brain stimulus was calculated.

Using this technique, a number of interesting observations could be made concerning repetitive motor neuron discharges. First, repetitive discharges have a relatively high threshold. Notable amounts of repetitive motor neuron discharges occurred only when some ≥75% of the motor neurons were recruited by TMS (Z'Graggen et al. 2005). Second, repetitive discharges vary considerably between subjects. In some subjects, they will appear easily with relatively high stimulus intensities, in others, facilitation maneuvers and very intense stimuli are needed to produce repetitive discharges. Triple repetitive discharges were never observed with stimulation intensities of 150% of resting motor

threshold and 20% of maximal voluntary muscle force to facilitate the responses, again pointing to the high threshold of repetitive discharges (Z'Graggen et al. 2005). Third, repetitive stimuli make up for much of the difference between MEP amplitude and area ratios, since they affect area more than amplitude (Z'Graggen et al. 2005).

MEP size variability

A characteristic of MEPs is their variability in size and shape from one stimulus to the next, even if the stimulus parameters are kept constant (Kiers et al. 1993; Woodforth et al. 1996; Ellaway et al. 1998). All of the above-mentioned mechanisms may contribute to this variability: varying numbers of excited motor neurons, varying numbers of repetitive discharges, and varying synchronization. It is not yet clear how much these mechanisms interact to produce the observed variability. Varying numbers of excited motor neurons may be caused by subtle excitability changes of the involved neurons, which in turn may be the consequence of small variations of facilitation by voluntary contraction or cognitive events. Variation may also stem from inadvertent movements of the coil during stimulation, even though previous studies have shown that the contribution of coil movements does not account for the observed MEP variability (Ellaway et al. 1998; Gugino et al. 2001).

As demonstrated by the recording in Figure 9.4, there is a considerable variability of the corticomuscular latency of single motor units. Moreover, in situations where two motor units could be studied, the units appeared to vary independently of each other (Magistris et al. 1998), suggesting that synchronization of motor unit discharges varies importantly from stimulus to stimulus. Such varying synchronization induces different degrees of phase cancellation and therefore produces size variation of the MEP. Hence, if the TMS response is 'resynchronized' by the TST, the variability is expected to be reduced. Indeed, this is what was observed previously (Table 9.1; Magistris et al. 1998).

The variability of repetitive motor neuron discharges after TMS has never been quantified, but appeared quite substantial in a recent study (Z'Graggen et al. 2005). The example in

Figure 9.4 shows appearance of double discharges in nine of 13 equal brain stimuli.

To date, it is unclear which physiological mechanisms are involved in causing the variability of MEPs, be it by varying desynchronization and/or by a varying amount of repetitive discharges. Using cervical epidural recordings in patients undergoing spinal surgery, the variability of I-waves was found to be considerable (Burke et al. 1995), and was therefore held responsible for at least some of the observed MEP size variability (Ellaway et al. 1998). On the other hand, Woodforth et al. (1996) reported that the variability of the epidurally recorded descending volleys after transcranial electrical stimulation was much smaller than that of the resulting MEP in the anterior tibial muscle, suggesting a substantial contribution from subcortical processes to MEP variability. We have observed that desynchronization was similar for electrical and magnetic stimulation, also pointing to a subcortical mechanism. This is in agreement with previous observations on the influence of spinal motor neuron excitability on corticomuscular latencies to transcranial stimulation (Mills et al. 1991). However, additional contributing mechanisms, such as changes in excitability of cortical motor neurons, variations of the site of excitation within the brain, or the possibility of different conducting pathways cannot be ruled out.

Pathological conditions

Pathological conditions may modify the above-discussed parameters and influence the size of the MEPs by lesions of motor neurons or of their axons, central conduction velocity slowing, or conduction block. Furthermore, changes in excitability of the cortical and spinal motor neurons may occur in a number of conditions.

Lesions of motor neurons or of their axons and conduction block, whether of central or peripheral origin, cause a decrease in MEP size. Because of the normally smaller size of the MEP compared to the maximal CMAP, and the large trial-to-trial variability of the MEP, standard MEPs are not sufficiently sensitive to detect and to quantify the proportion of central neuronal loss. Therefore, the size parameter was not considered in many patient studies. With the introduction of the TST, it has become clear that

conduction deficits (reduced TST_{test}:$TST_{control}$ ratio) occur frequently (more frequently than conduction slowing) in patients with diseases affecting the central motor pathways, and that this conduction deficit is related to the reduction of muscle force experienced by the patient (Magistris et al. 1999; Rösler et al. 2000; Bühler et al. 2001; Humm et al. 2004a).

References

Amassian VE, Stewart M, Quirk GJ, Rosenthal JL (1987). Physiological basis of motor effects of a transient stimulus to cerebral cortex. Neurosurgery 20, 74–93.

Andersen B, Rösler KM, Lauritzen M (1999). Non-specific facilitation of responses to transcranial magnetic stimulation. Muscle and Nerve 22, 857–863.

Berardelli A, Inghilleri M, Rothwell JC, Cruccu G, Manfredi M (1991). Multiple firing of motoneurones is produced by cortical stimulation but not by direct activation of descending motor tracts. Electroencephalography and Clinical Neurophysiology 81, 240–242.

Brasil-Neto JP, Cohen LG, Panizza M, Nilsson J, Roth BJ, Hallett M (1992). Optimal focal transcranial magnetic activation of the human motor cortex, effects of coil orientation, shape of the induced current pulse, and stimulus intensity. Journal of Clinical Neurophysiology 9, 132–136.

Bühler R, Magistris MR, Truffert A, Hess CW, Rösler KM (2001). The triple stimulation technique to study central motor conduction to the lower limbs. Clinical Neurophysiology 112, 938–949.

Burke D, Hicks R, Gandevia SC, Stephen J, Woodforth I, Crawford M (1993). Direct comparison of corticospinal volleys in human subjects to transcranial magnetic and electrical stimulation. Journal of Physiology 470, 383–393.

Burke D, Hicks R, Stephen J, Woodforth I, Crawford M (1995). Trial-to-trial variability of corticospinal volleys in human subjects. Electroencephalography and Clinical Neurophysiology 97, 231–237.

Claus D, Mills KR, Murray NM (1988). Facilitation of muscle responses to magnetic brain stimulation by mechanical stimuli in man. Experimental Brain Research 71, 273–278.

Day BL, Rothwell JC, Thompson PD, Dick JPR, Cowan JMA, Berardelli A, Marsden CD (1987a). Motor cortex stimulation in intact man. (2) Multiple descending volleys. Brain 110, 1191–1209.

Day BL, Thompson PD, Dick JPR, Nakashima K, Marsden CD (1987b). Different sites of action of electrical and magnetic stimulation of the human brain. Neuroscience Letters 75, 101–106.

Day BL, Dressler D, Maertens de Noordhout A, et al. (1989). Electric and magnetic stimulation of human motor cortex: surface EMG and single motor unit responses. Journal of Physiology (London) 412, 449–473.

Di Lazzaro V, Restuccia D, Oliviero A, et al. (1998). Effects of voluntary contraction on descending volleys evoked by transcranial stimulation in conscious humans. Journal of Physiology 508, 625–633.

Dubach P, Guggisberg AG, Rösler KM, Hess CW, Mathis J (2004). Significance of coil orientation for motor evoked potentials from nasalis muscle elicited by transcranial magnetic stimulation. *Clinical Neurophysiology* 115, 862–870.

Ellaway PH, Davey NJ, Maskill DW, Rawlinson SR, Lewis HS, Anissimova NP (1998). Variability in the amplitude of skeletal muscle responses to magnetic stimulation of the motor cortex in man. *Electroencephalography and Clinical Neurophysiology* 109, 104–113.

Ferbert A, Priori A, Rothwell JC, Day BC, Colebatch JG, Marsden CD (1992). Interhemispheric inhibition of the human motor cortex. *Journal of Physiology* 453, 525–546.

Gugino LD, Romero JR, Aglio L, *et al.* (2001). Transcranial magnetic stimulation coregistered with MRI: a comparison of a guided versus blind stimulation technique and its effect on evoked compound muscle action potentials. *Clinical Neurophysiology* 112, 1781–1792.

Hess CW, Mills KR, Murray NMF (1986). Magnetic stimulation of the human brain: facilitation of motor responses by voluntary contraction of ipsilateral and contralateral muscles with additional observations on an amputee. *Neuroscience Letters* 71, 235–240.

Hess CW, Mills KR, Murray NMF (1987a). Responses in small hand muscles from magnetic stimulation of the human brain. *Journal of Physiology* 388, 397–419.

Hess CW, Mills KR, Murray NMF, Schriefer TN (1987b). Magnetic brain stimulation: central motor conduction studies in multiple sclerosis. *Annals of Neurology* 22, 744–752.

Humm AM, Beer S, Kool J, Magistris MR, Rösler KM (2004a). Quantification of Uhthoff's phenomenon in multiple sclerosis: a magnetic stimulation study. *Clinical Neurophysiology* 115, 2493–2501.

Humm AM, Z'Graggen WJ, von Hornstein NE, Magistris MR, Rösler KM (2004b). Assessment of central motor conduction to intrinsic hand muscles using the triple stimulation technique: normal values and repeatability. *Clinical Neurophysiology* 115, 2558–2566.

Inghilleri M, Berardelli A, Cruccu G, Manfredi M (1993). Silent period evoked by transcranial stimulation of the human cortex and cervicomedullary junction. *Journal of Physiology* 466, 521–534.

Izumi S-I, Findley TW, Ikai T, Andrews J, Daum M, Chino N (1995). Facilitatory effect of thinking about movement on motor-evoked potentials to transcranial magnetic stimulation of the brain. *American Journal of Physical and Medical Rehabilitation* 74, 207–213.

Kiers L, Cros D, Chiappa KH, Fang J (1993). Variability of motor potentials evoked by transcranial magnetic stimulation. *Electroencephalography and Clinical Neurophysiology* 89, 415–423.

Kiers L, Clouston P, Chiappa KH, Cros D (1995). Assessment of cortical motor output: compound muscle action potential versus twitch force recording. *Electroencephalography and Clinical Neurophysiology* 97, 131–139.

Kiers L, Fernando B, Tomkins D (1997). Facilitatory effect of thinking about movement on magnetic motor-evoked potentials. *Electroencephalography and Clinical Neurophysiology* 105, 262–268.

Kischka U, Fajfr R, Fellenberg T, Hess CW (1993). Facilitation of motor evoked potentials from magnetic brain stimulation in man: a comparative study of different target muscles. *Journal of Clinical Neurophysiology* 10, 505–512.

Kofler M, Fuhr P, Leis AA, Glocker FX, Kronenberg MF, Wissel J, Stetkarova I (2001). Modulation of upper extremity motor evoked potentials by cutaneous afferents in humans. *Clinical Neurophysiology* 112, 1053–1063.

Kossev A, Siggelkow S, Schubert M, Wohlfarth K, Dengler R (1999). Muscle vibration: different effects on transcranial magnetic and electrical stimulation. *Muscle and Nerve* 22, 946–948.

Kujirai T, Caramia MD, Rothwell JC, *et al.* (1993). Corticocortical inhibition in human motor cortex. *Journal of Physiology* 471, 501–519.

McComas AJ (1995). Motor-unit estimation: the beginning. *Journal of Clinical Neurophysiology* 12, 560–564.

Magistris MR, Rösler KM, Truffert A, Myers JP (1998). Transcranial stimulation excites virtually all motor neurones supplying the target muscle. A demonstration and a method improving the study of motor evoked potentials. *Brain* 121, 437–450.

Magistris MR, Rösler KM, Truffert A, Landis T, Hess CW (1999). A clinical study of motor evoked potentials using a triple stimulation technique. *Brain* 122, 265–279.

Mall V, Glocker F, Fietzek U, *et al.* (2001). Inhibitory conditioning stimulus in transcranial magnetic stimulation reduces the number of excited spinal motor neurons. *Clinical Neurophysiology* 112, 1810–1813.

Mazzocchio R, Rothwell JC, Rossi A (1995). Distribution of Ia effects onto human hand muscle motoneurones as revealed using an H reflex technique. *Journal of Physiology* 489(Pt 1), 263–273.

Mills KR (1999). Magnetic brain stimulation: a review after 10 years experience. *Electroencephalography and Clinical Neurophysiology* 49(Suppl), 239–244.

Mills KR, Kimiskidis V (1996). Motor cortex excitability during ballistic forearm and finger movements. *Muscle and Nerve* 19, 468–473.

Mills KR, Boniface SJ, Schubert M (1991). The firing probability of single motor units following transcranial magnetic stimulation in healthy subjects and patients with neurological disease. *Electroencephalogy and Clinical Neurophysiology* 43(Suppl), 100–110.

Mills KR, Boniface SJ, Schubert M (1992). Magnetic brain stimulation with a double coil: the importance of coil orientation. *Electroencephalogy and Clinical Neurophysiology* 85, 17–21.

Naka D, Mills KR (2000). Further evidence for corticomotor hyperexcitability in amyotrophic lateral sclerosis. *Muscle and Nerve* 23, 1044–1150.

Nielsen JF (1996). Improvement of amplitude variability of motor evoked potentials in multiple sclerosis patients and in healthy subjects. *Electroencephalography and Clinical Neurophysiology* 101, 404–411.

Pascual-Leone A, Torres F (1993). Plasticity of the sensorimotor cortex representation of the reading finger in Braille readers. *Brain* 116, 39–52.

Pascual-Leone A, Cohen LG, Hallett M (1992). Cortical map plasticity in humans (letter to the editor). *Trends in Neurological Sciences 15*, 13–14.

Pascual-Leone A, Grafman J, Hallett M (1994). Modulation of cortical motor output maps during development of implicit and explicit knowledge. *Science 263*, 1287–1289.

Patton HD, Amassian VE (1954). Single- and multiple-unit analysis of cortical stage of pyramidal tract activation. *Journal of Neurophysiology 17*, 345–363.

Péréon Y, Genet R, Guihéneuc P (1995). Facilitation of motor evoked potentials: timing of Jendrassik maneuver effects. *Muscle and Nerve 18*, 1427–1432.

Ravnborg M, Blinkenberg M, Dahl K (1991). Standardization of facilitation of compound muscle action potentials evoked by magnetic stimulation of the cortex. Results in healthy volunteers and in patients with multiple sclerosis. *Electroencephalography and Clinical Neurophysiology 81*, 195–201.

Rösler KM, Hess CW, Heckmann R, Ludin HP (1989). Significance of shape and size of the stimulating coil in magnetic stimulation of the human motor cortex. *Neuroscience Letters 100*, 347–352.

Rösler KM, Truffert A, Hess CW, Magistris MR (2000). Quantification of upper motor neuron loss in amyotrophic lateral sclerosis. *Clinical Neurophysiology 111*, 2208–2218.

Rösler KM, Petrow E, Mathis J, Arányi Z, Hess CW, Magistris MR (2002). Effect of discharge desynchronization on the size of motor evoked potentials: an analysis. *Clinical Neurophysiology 113*, 1680–1687.

Rossini PM, Berardelli A, Deuschl G, *et al.* (1999). Applications of magnetic cortical stimulation. *Electroencephalogy and Clinical Neurophysiology 52*(Suppl), 171–185.

Roth G, Magistris MR (1987). Detection of conduction block by monopolar percutaneous stimulation of the brachial plexus. *Electromyography and Clinical Neurophysiology 27*, 45–53.

Siggelkow S, Kossev A, Schubert M, Kappels IIIl, Wolf W, Dengler R (1999). Modulation of motor evoked potentials by muscle vibration: the role of vibration frequency. *Muscle and Nerve 22*, 1544–1548.

Taylor JL, Gandevia SC (2001). Transcranial magnetic stimulation and human muscle fatigue. *Muscle and Nerve 24*, 18–29.

Ugawa Y, Day BL, Rothwell JC, Thompson PD, Merton PA, Marsden CD (1991). Modulation of motor cortical excitability by electrical stimulation over the cerebellum in man. *Journal of Physiology 441*, 57–72.

Ugawa Y, Uesaka Y, Terao Y, Hanajima R, Kanazawa I (1994). Magnetic stimulation of corticospinal pathways at the foramen magnum level in humans. *Annals of Neurology 36*, 618–624.

Uncini A, Treviso M, Di Muzio A, Simone P, Pullman S (1993). Physiological basis of voluntary activity inhibition induced by transcranial cortical stimulation. *Electroencephalography and Clinical Neurophysiology 89*, 211–220.

van der Kamp W, Zwinderman AH, Ferrari MD, van Dijk JG (1996). Cortical excitability and response variability of transcranial magnetic stimulation. *Journal of Clinical Neurophysiology 13*, 164–171.

Wilson SA, Thickbroom GW, Mastaglia FL (1993). Transcranial magnetic stimulation mapping of the motor cortex in normal subjects. The representation of two intrinsic hand muscles. *Journal of Neurological Science 118*, 134–144.

Woodforth IJ, Hicks RG, Crawford MR, Stephen JP, Burke DJ (1996). Variability of motor-evoked potentials recorded during nitrous oxide anesthesia from the tibialis anterior muscle after transcranial electrical stimulation. *Anesthesia and Analgesia 82*, 744–749.

Z'Graggen WJ, Humm AM, Durisch N, Magistris MR, Rösler KM (2005). Repetitive spinal motor neuron discharges following single transcranial magnetic stimuli: a quantitative study. *Clinical Neurophysiology 116*, 1628–1637.

Ziemann U, Ilić TV, Alle H, Meintzschel F (2004). Cortico-motoneuronal excitation of three hand muscles determined by a novel penta-stimulation technique. *Brain 127*, 1887–1898.

CHAPTER 10

The cortical silent period

Alexander Wolters, Ulf Ziemann, and Reiner Benecke

Summary – The cortical silent period (cSP) refers to an interruption of voluntary muscle contraction by transcranial stimulation of the contralateral motor cortex. This chapter summarizes the physiology of the cSP, provides guidelines how the cSP should be recorded and analysed, and refers to useful clinical applications. Evidence will be provided that spinal inhibitory mechanisms contribute to the early part of the cSP up to its first 50 ms, but that its later part is generated exclusively by long-lasting inhibition originating within the motor cortex. This makes the cSP an attractive probe to assess motor cortical inhibition in health and disease.

Introduction

Marsden *et al.* (1983) were the first to show that transcranial electrical stimulation (TES) of the motor cortex in humans produces not only a muscle twitch in voluntarily activated contralateral limb muscles but also a subsequent period of electromyographic (EMG) suppression, lasting up to 100–300 ms. They proposed that the mechanisms of this cortically induced silent period are different from the ones responsible for the much shorter (40–50 ms) spinal silent period elicited by supramaximal electrical peripheral nerve stimulation (Merton 1951). TMS, invented soon after by Barker *et al.* (1985), raised the opportunity to study excitatory and inhibitory motor pathways with a noninvasive and painless method. Since the first report of an inhibitory motor phenomenon elicited by TMS (Calancie *et al.* 1987), this so-called cortical silent period (cSP) has meanwhile been studied extensively under physiological and pathological conditions. Investigation of the cSP can be easily performed: a single suprathreshold TMS pulse is applied to the motor cortical representation of a tonically preactivated target muscle, thereby producing a period of EMG silence in contralateral small hand muscles (Figure 10.1).

Site of origin of the cortical silent period

Theoretically, various mechanisms at the spinal and supraspinal level can account for the cSP. Possible spinal (or segmental) mechanisms include the following.

1. A suprathreshold TMS pulse applied to the motor cortex leads to activation of a potentially extensive portion of spinal motor neurons (MNs) (cf. Chapter 9, this volume). This produces recurrent spinal inhibition by activation of inhibitory Renshaw cells.

2. Fibers descending from the motor cortex can activate spinal inhibitory Ia-interneurons, which in turn produce inhibitory postsynaptic potentials in spinal MNs, but this effect is limited to a duration of <100 ms (Person and Kozhina 1978).

3. Refractoriness of spinal MNs may also contribute to the cSP, although this accounts for

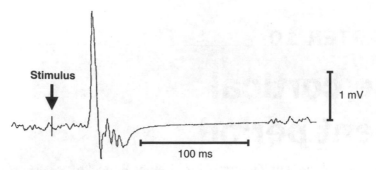

Fig. 10.1 Single-trial electromyogram recording of a physiological cortical silent period (cSP), elicited by focal TMS of the motor cortical representation of the contralateral first dorsal interosseous muscle during tonic contraction with 20% of maximum voluntary strength. Stimulus intensity, 1.2 times of the resting motor threshold; arrow, time of the TMS pulse.

only 5–8 ms after peripheral excitation (Pierrot-Deseilligny *et al.* 1976), and therefore cannot explain an inhibitory phenomenon lasting up to 300 ms. The duration of refractoriness of the spinal MN pool after excitation by TMS is not exactly known, but it is unlikely that it lasts longer than some 5 ms because repetitive discharges of spinal MNs elicited by a single TMS pulse may occur with an interspike interval of 5 ms (cf. Chapter 9, this volume).

4. Suprathreshold TMS produces a brisk muscle twitch (the motor-evoked potential (MEP), cf. Chapter 9, this volume) that interrupts the activity of muscle spindles (reduction of activity in Ia-afferents) and activates Golgi tendon organs (activation of Ib-afferent volleys), resulting in less activation and more inhibition of spinal MNs, respectively.

This inhibitory proprioceptive input to the spinal MN pool seems to play only a minor role in generating the cSP since (i) no change in its duration occurs when the joint is held in a fixed position (Roick *et al.* 1993), (ii) a cSP can also be observed in a state of complete peripheral deafferentation (Fuhr *et al.* 1991), and (iii) a cSP can be obtained without a preceding muscle twitch (see below). In summary, spinal inhibitory mechanisms may contribute to the cSP, but due to their short duration this is limited to the early part of the cSP, i.e. its first ~50 ms.

Spinal MN excitability during the cSP can be tested by the Hoffmann (H)-reflex. The H-reflex recruits predominantly small spinal MNs which are also preferentially activated by moderate voluntary muscle activation. Hence, H-reflex and cSP testing during moderate voluntary muscle activation address at least a partially overlapping pool of spinal MNs. In the case of a significant contribution of spinal inhibitory mechanisms to the cSP, H-reflex amplitudes should be reduced during the cSP. In line with this expectation, a clearly decreased or even abolished H-reflex amplitude occurred in the flexor carpi radialis muscle up to 50 ms after TMS to the hand area of the contralateral motor cortex (Fuhr *et al.* 1991). Similar findings were obtained in the soleus muscle (Ziemann *et al.* 1993), although one other report did not find a reduction of H-reflex amplitude during the cSP in this muscle (Roick *et al.* 1993).

Several authors used a cortical paired-pulse protocol to directly test the hypothesis of a motor cortical contribution to the cSP (e.g. Day *et al.* 1989; Inghilleri *et al.* 1993; Roick *et al.* 1993). The rationale for these paired-pulse studies is that the first stimulus produces the cSP and the second stimulus probes corticospinal excitability during the course of the cSP. Roick *et al.* (1993) employed a paired TMS protocol with strongly suprathreshold TMS shocks delivered at various interstimulus intervals (ISIs). ISIs of 20–30 ms revealed MEP facilitation, whereas at ISIs between 50 and 150 ms (corresponding to the time course of the cSP induced by the conditioning stimulus) the corticospinal excitability was unchanged. In contrast, if the second stimulus was of weaker intensity, then MEPs were inhibited at ISIs between 50 and 150 ms. High-intensity TMS is capable of activating cortico-cospinal neurons directly whereas low-intensity TMS leads predominantly to an indirect (trans-synaptic) activation (Edgley *et al.* 1990). In line with this, cervical epidural recordings of the descending corticospinal volley induced by TMS demonstrated that indirect (I)-waves elicited by a second TMS pulse given 100–200 ms after a conditioning first TMS pulse (i.e. during the cSP) were inhibited (Chen *et al.* 1999). Together, these findings strongly suggest that cortical inhi-bition plays a major role in the generation of the cSP.

There exist more pieces of evidence to suggest that the cSP is generated predominantly within the motor cortex: in animal studies direct corti-cal stimulation can induce a long-lasting post-synaptic inhibition of 100–300 ms (Krnjevic *et al.* 1966). The cSP induced by TMS in humans is significantly longer even at moderate stimula-tion intensities than the one generated by TES (Inghilleri *et al.* 1993; Brasil-Neto *et al.* 1995). TES activates preferentially the subcortical axon of the pyramidal neuron (D-wave excitation) (Edgley *et al.* 1990). By contrast, at moderate stimulation intensity, TMS predominantly excites axons of excitatory intracortical interneu-rons, which in turn synapse onto pyramidal neu-rons (I-wave excitation) (Amassian *et al.* 1987; Day *et al.* 1989; Di Lazzaro *et al.* 2004) (cf. also Chapter 14, this volume). Similarly, the cSP elicited by TMS is much longer than the silent period elicited by subcortical electrical stimula-tion through electrodes implanted into the globus pallidus internus for treatment of dys-tonic patients (Kühn *et al.* 2004a).

The cSP induced by TMS can be elicited at low stimulation intensities in the absence of a preceding MEP (Davey *et al.* 1994; Classen and Benecke 1995). Inhibition in the absence of corticospinal excitation underlines the cortical origin of the inhibition. This was further sup-ported in a study on single motor units that demonstrated long-lasting silent periods, which in some single motor units could last up to sev-eral hundred milliseconds, even at low TMS intensity and in the absence of preceding excita-tion (Classen and Benecke 1995), a phenomenon that never occurs in single motor units tested

during spinal inhibition elicited by peripheral nerve stimulation (Kudina and Pantseva 1988). Finally, another piece of evidence for a cortical origin of the cSP arises from the finding of an abolished cSP in stroke patients suffering from isolated ischemic lesions within the primary motor cortex (Schnitzler and Benecke 1994). Despite the lack of cSP, the MEP and the spinal silent period were preserved, indicating a strong contribution of inhibitory motor cortical mech-anisms even to the early part of cSP.

In summary, it is now generally agreed that spinal inhibitory mechanisms may contribute to the early part of the cSP up to its first 50 ms, but its later part is generated exclusively by inhibi-tion that originates within the motor cortex. Therefore, the cSP can be considered as a probe of motor cortical inhibition.

Physiological properties of the cSP

Under physiological conditions, the cSP threshold is usually slightly lower than the MEP threshold (Davey *et al.* 1994; Classen and Benecke 1995). The cSP duration is largely a linear function of stimulus intensity (Inghilleri *et al.* 1993; Roick *et al.* 1993; Orth and Rothwell 2004), but see (Kimiskidis *et al.* 2005) (Figure 10.2), while MEP size plateaus at high stimulus inten-sity (cf. Chapter 9, this volume). A subthreshold conditioning pulse reduced MEP size signifi-cantly more than cSP duration (Trompetto *et al.* 2001). These findings support the idea that MEP and cSP are generated by different mechanisms (Hallett 1995). The optimal point to elicit a cSP of maximal duration in hand muscles is located in the hand area of the contralateral motor cor-tex, but there are conflicting reports on its exact localization. Two studies demonstrated an iden-tical optimal coil position to elicit cSP and MEP (Roick *et al.* 1993; Wilson *et al.* 1993b). One other study reported that the optimal point to elicit the cSP was ~2 cm more lateral than the optimal site for eliciting the MEP (Wassermann *et al.* 1993). One possible explanation for this discrepancy is the smaller size of the focal coil used in the latter study, which may be more sen-sitive towards detecting discrete differences in the topography of cortical representations.

Contraction strength of the target muscle does not influence cSP duration significantly (Haug *et al.* 1992; Inghilleri *et al.* 1993; Roick *et al.* 1993; Taylor *et al.* 1997; Wu *et al.* 2002) (Figure 10.2). The cSP duration is longest in small hand muscles (up to 300 ms), shorter in leg muscles (up to 100 ms), proximal arm muscles (Roick *et al.* 1993), axial muscles (Ferbert *et al.* 1992; Lefaucheur and Lofaso 2002; Lefaucheur 2005a), facial muscles (Paradiso *et al.* 2005; Werhahn *et al.* 1995) and the tongue (Katayama *et al.* 2001). It is likely that the cSP in facial muscles is exclusively of cortical origin because the R1 component of the blink reflex is preserved throughout the entire facial cSP (Leis *et al.* 1993). The motor cortical representation of the cSP is completely lateralized to the contralateral motor cortex for distal limb muscles, but shows a more bilateral distribution for axial muscles. One study found that the cSP duration in small hand muscles is shorter in the non-dominant compared with the dominant motor cortex (Priori *et al.* 1999) but this asymmetry was not confirmed by another study (Cicinelli *et al.* 1997). In any case, the interhemispheric difference of cSP duration is low (usually <10 ms) in healthy subjects, and therefore interhemispheric comparison of cSP duration is attractive

for clinical applications to detect lateralized pathology. The intersession variability of the cSP duration in a given subject is also low, typically <10% (Kukowski and Haug 1992; Orth and Rothwell 2004). Therefore, measurement of cSP duration is also suitable for longitudinal studies in patients, or before and after experimental manipulation. The cSP duration shows a high interindividual variability, ranging from 20% to 35% (Orth and Rothwell 2004). Thus, interpretation of cSP duration in pathological conditions is meaningful only in group comparisons rather than at the level of the individual patient. There exist no systematic studies that have investigated the effects of gender and age on the duration of the cSP, but clinical observations provided no evidence for a relevant influence of these two factors. In children, a reproducible cSP can be elicited from the age of 5 years onwards (Heinen *et al.* 1998; Garvey and Gilbert 2004).

The direction of current flow induced by TMS in the motor cortex determines which neuronal elements are preferentially being activated. An induced current in posterior–anterior direction recruits predominantly interneurons that are involved in the generation of I1 waves, whereas an induced current in anterior–posterior direction recruits preferentially

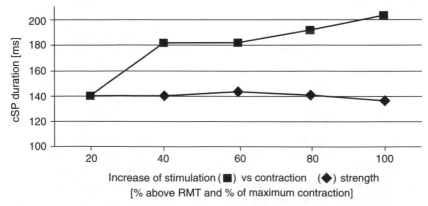

Increase of stimulation (■) vs contraction (◆) strength
[% above RMT and % of maximum contraction]

Fig. 10.2 Modulation of cortical silent period (cSP) duration in the first dorsal interosseous muscle by stimulus intensity and level of tonic voluntary muscle contraction in one representative subject. Squares: stimulus intensity is increased in steps of 0.2 × motor-evoked potential (MEP) threshold from 1.2 to 2.0 × MEP threshold while the level of contraction is kept constant at 20% of maximum. Diamonds: level of voluntary contraction is increased in steps of 20% of maximum strength from 20 to 100% of maximum strength while stimulus intensity is kept constant at 1.2 × MEP threshold. Note that stimulus intensity but not level of contraction affects the cSP duration. RMT, resting motor threshold.

interneurons that are involved in the generation of I3 waves (Day *et al.* 1989; Sakai *et al.* 1997) (cf. Chapter 14, this volume). Duration of cSP is ~30 ms longer when preferentially I3 waves are recruited (Orth and Rothwell 2004; Trompetto *et al.* 2001). This is another piece of evidence for a mainly intracortical origin of the cSP.

It is currently thought that the cSP is mediated by gamma-aminobutyric acid (GABA)$_B$ receptors (GABA$_B$R). The silent period in single motor units of small hand muscles can last up to 1000 ms (Classen and Benecke 1995). Corresponding to this duration, a GABA$_B$R-mediated inhibitory postsynaptic potential, lasting up to 1200 ms, has been demonstrated in pyramidal cells of human neocortical slices (McCormick 1989). In line with this, high doses of intrathecally applied baclofen, a specific GABA$_B$R agonist, resulted in significant cSP lengthening in a single patient with generalized dystonia (Siebner *et al.* 1998).

In addition to this general view that cSP duration reflects long-lasting GABA$_B$R-mediated inhibition in motor cortex, another view is that the cSP simply reflects the disruption of access of voluntary motor drive to the primary motor cortex. Support for this notion comes from the finding that, while active MEP threshold is usually clearly lower than resting MEP threshold (Ziemann *et al.* 1996; Chen *et al.* 1998), this difference disappears during the cSP (Tergau *et al.* 1999). Also, the modulation of cSP duration by motor instruction and motor attention (Mathis *et al.* 1998, 1999; Hess *et al.* 1999) is in line with view.

Modulation of cSP duration

Conditioning electrical stimulation of cutaneous nerves shortens the cSP (Hess *et al.* 1999; Classen *et al.* 2000). This effect shows a topographic gradient and is most pronounced if a cutaneous digital nerve adjacent to the target muscle is being stimulated (Classen *et al.* 2000), suggesting that the cSP is involved in sensorimotor integration that is somatotopically organized.

A conditioning TMS pulse also affects cSP duration. The interaction between the two TMS pulses is complex and depends on stimulus intensity and interstimulus interval (Shimizu *et al.* 1999; Wu *et al.* 2000). With stimulus intensity above MEP threshold and interstimulus intervals of 15–30 ms, consistent lengthening of cSP duration occurs in the absence of modulation of MEP amplitude (Shimizu *et al.* 1999; Wu *et al.* 2000). This effect is most likely attributed to temporal summation of inhibitory postsynaptic potentials at the level of motor cortex. At longer interstimulus intervals of 60–110 ms the predominant effect is shortening of the cSP (Wu *et al.* 2000). The physiology of this effect is more difficult to interpret because the MEP amplitude is inhibited at the same time. It is likely that spinal (less activation of spinal inhibition) and cortical mechanisms (decrease of excitability of cortical inhibitory interneurons) are involved (Wu *et al.* 2000). Conditioning TMS of one motor cortex modulates cSP duration elicited in the other motor cortex. The predominant effect is cSP shortening at interstimulus intervals of 10–20 ms in the absence of significant modulation of MEP amplitude (Schnitzler *et al.* 1996), indicating an interhemispheric interaction between the motor cortices on inhibitory interneurons. It is likely that testing these various techniques of modulating cSP duration by conditioning stimulation will be useful in exploring pathophysiology of neurological disease where the level of motor cortical inhibition and its control by afferent input are abnormal, such as in particular forms of epilepsy.

There is extensive evidence that repetitive (r) TMS is capable of modulating motor cortical excitability when measured by change in MEP amplitude (for review, Ziemann 2004). Considerably less is known about rTMS-induced change of cSP duration. Higher-frequency rTMS (2–15 Hz) at stimulus intensities above MEP threshold leads to a lengthening of the cSP duration during the train (Berardelli *et al.* 1999; Romeo *et al.* 2000), but this effect does not outlast the train by more than a few seconds. This lengthening of the cSP duration is probably produced by temporal summation of inhibitory postsynaptic potentials. A 30 min train of low-frequency rTMS (0.3 Hz) of suprathreshold intensity (115% resting MEP threshold) lengthened the cSP, and this effect lasted for ≥90 min (Cincotta *et al.* 2003). It was specific for cSP duration whereas other measures of motor cortical excitability, in particular MEP threshold and MEP intensity curve, did not change. This raises the possibility that low-frequency rTMS may be of therapeutic value in pathological conditions with abnormally short cSP duration, indicating a deficiency of cortical inhibitory mechanisms, such as in epilepsy (see below). Finally, cSP duration also lengthens by paired associative stimulation (90 pairs of electrical stimulation to a peripheral nerve at the wrist followed 25 ms later by focal TMS to the hand area of the contralateral motor cortex; stimulation frequency, 0.1 Hz), and this effect outlasts the interventional stimulation protocol (Stefan et al. 2000).

Practical aspects of performing cSP measurements

CSP measurements are easy to obtain but require a standardized protocol to allow useful interpretation. The cSP duration does not vary with the level of voluntary tonic contraction of the target muscle (Haug et al. 1992; Kukowski and Haug 1992; Triggs et al. 1992; Inghilleri et al. 1993; Roick et al. 1993), or slightly shortens (Cantello et al. 1992; Wilson et al. 1993a; Mathis et al. 1998) (Fig. 10.2), but may be significantly affected by motor instruction and motor attention (Mathis et al. 1998, 1999; Hess et al. 1999). Therefore, it is important to ask the subject to maintain a certain level of tonic contraction until well after the end of the trial. Since intertrial variability of cSP duration is minimal with maximum voluntary contraction (Mathis et al. 1998), a high level of contraction is recommended. The cSP duration varies with stimulus intensity (cf. Fig. 10.2). Therefore, it is important to define stimulus intensity appropriately. Most previous studies related stimulus intensity to resting or active MEP threshold. However, this cannot be recommended any longer because MEP threshold and cSP threshold are usually related (Kimiskidis et al. 2005) but may vary independently as a consequence of experimental manipulation, e.g. drug exposure (Kimiskidis et al. 2005) or disease (Chistyakov et al. 2001). Therefore, stimulus intensity should be related to either maximum stimulator output (Kimiskidis et al. 2005) or cSP threshold (Chistyakov et al. 2001). At least four repeats should be obtained at a given stimulus intensity to estimate the mean cSP duration within 10% of the true mean (Kimiskidis et al.

2005). In addition, it is recommended to obtain cSP duration–stimulus intensity curves. The data can be modelled by a linear (Orth and Rothwell 2004) or a sigmoid function (Kimiskidis et al. 2005), and provide comprehensive information on cSP threshold, slope and maximum cSP duration. Analysis of cSP duration should be performed by automated statistical analysis to eliminate subjective observer bias. This could be done by rectification of the EMG signal of the single trials and subsequent averaging of all repeats of the same intensity condition. CSP offset is defined as the time when the voluntary EMG has recovered to a certain level of the prestimulus EMG. CSP onset is the time when voluntary EMG drops consistently below this level. Several protocols for automated statistical analysis of cSP onset and offset have been published (Nilsson et al. 1997; Garvey et al. 2001; Daskalakis et al. 2003).

cSP duration in pathological conditions

Exploration of cSP duration in pathological conditions can contribute to a better understanding of the underlying pathophysiology. Results of cSP measurements in ischemic stroke patients are somewhat inconsistent across the published literature, which is most likely attributable to differences in cSP stimulation protocols, lesion localization and size, patient demographics, and delay of cSP measurement after the stroke event (for review see van Kuijk et al. 2005). Most of the studies in stroke patients show a lengthening of the cSP in the affected hand compared with the nonaffected hand (Kukowski and Haug 1992; Haug and Kukowski 1994; Braune and Fritz 1995, 1996; Faig and Busse 1996; Classen et al. 1997; Ahonen et al. 1998; Liepert et al. 2000, 2005a,b; Battaglia et al. 2006). Some of these studies indicate a greater sensitivity of cSP abnormalities than MEP abnormalities, and sometimes abnormal cSP lengthening occurs even in the presence of normal hand function (Haug and Kukowski 1994; Braune and Fritz 1995; Ahonen et al. 1998).

The relation between lesion localization and alteration of cSP duration was analyzed carefully in three studies (Schnitzler and Benecke 1994; von Giesen et al. 1994; Liepert et al. 2005a).

Lesions outside the motor cortex typically result in a lengthening of the cSP, whereas lesions affecting (exclusively) the motor cortex lead to a shortening or even an abolition of the cSP. The lengthening is most likely attributable to a reduction in the afferent excitatory cortico-cortical and thalamo-cortical drive to motor cortex, resulting in a shift of excitability towards inhibition, whereas cSP shortening indicates a selective vulnerability of motor cortical inhibitory interneurons in the motor cortex by ischemic stress.

What are the functional consequences of altered cSP duration? Extreme lengthening of the cSP (up to several seconds) in hemiparetic patients with subcortical ischemic stroke and preserved MEP responses is associated with motor neglect of the affected limb (Classen et al. 1997). Clinical improvement of this motor neglect is paralleled by a shortening of the cSP, suggesting that exaggerated inhibition of the affected motor cortex contributes to the generation of hemineglect (Classen et al. 1997). In a single patient with focal motor epilepsy and subsequent Todd's paresis, a clearly prolonged cSP appeared in the motor cortex contralateral to the paretic arm (Classen et al. 1995). It was speculated that this increase in cortical inhibition contributed to the Todd's paresis. A shortened cSP duration may be associated with the development of spasticity in the affected limb (Cruz Martinez et al. 1998). This notion is supported by the finding that acute stroke patients who exhibit a significant shortening of the cSP with increasing levels of voluntary contraction of the target muscle present later in the chronic phase after stroke with poorer functional outcome and higher spasticity scores when compared with stroke patients without a significant effect of contraction strength on cSP duration (Catano et al. 1997). Finally, out of a cohort of 84 patients with a first-time ischemic stroke, six patients showed a significantly shortened cSP in the affected motor cortex. This electrophysiological abnormality was clinically associated with focal motor seizures in five of these six patients, whereas none of the other 76 patients with normal or prolonged cSP durations developed seizures (Kessler et al. 2002) (Figure 10.3). This finding strongly suggests that a shortened cSP duration in stroke reflects deficient inhibitory mechanisms in the affected motor cortex associated with an increased risk for post-stroke seizures.

Measurement of cSP duration is of interest in patients with epilepsy in order to assess potentially altered motor cortical inhibition. One patient with focal motor epilepsy showed a dramatic shortening of the cSP in a hand muscle contralateral to the epileptic motor cortex at the time when this motor cortex showed high electroencephalogram (EEG) spiking activity (Inghilleri et al. 1998). This may indicate a breakdown of cortical inhibitory mechanisms, that in turn leads to epileptic EEG activity. In patients with focal epilepsy several studies showed consistently that cSP duration is shortened in the epileptic hemisphere (Cicinelli et al. 2000; Hamer et al. 2005). Therefore, the cSP intensity curve may have a lateralizing value in focal epilepsy, with shorter cSP in the epileptic hemisphere. On the other hand, a different study in patients with focal motor epilepsy showed that the cSP was lengthened bilaterally, but less prominently in the epileptic hemisphere (Cincotta et al. 1998). This cSP lengthening may indicate increased interictal cortical inhibition to ward off further epileptic activity. A lengthened cSP was also found in patients with idiopathic generalized epilepsy (Macdonell et al. 2001; Tataroglu et al. 2004). In summary, the cSP findings in patients with epilepsy are variable, but it appears that the time from the last seizure, the site of a cortical lesion, the interictal EEG spiking activity, the type of epilepsy, and antiepileptic medication are determinants of cSP duration.

The cSP duration was also extensively studied in movement disorders because alterations of the balance between motor cortical excitation and inhibition are thought to play a crucial role in the various forms of motor abnormalities in these disorders. In Parkinson's disease, cSP duration is reduced, predominantly on the clinically more affected side (Cantello et al. 1991; Haug et al. 1992; Priori et al. 1994a; Nakashima et al. 1995; Berardelli et al. 1996; Valzania et al. 1997; Dioszeghy et al. 1999). This finding was attributed to a decreased facilitation of motor cortical inhibitory interneurons by the thalamo-cortical projection (Berardelli et al. 1996). This is supported by stereotactic pallidotomy or subthalamic nucleus stimulation in patients with

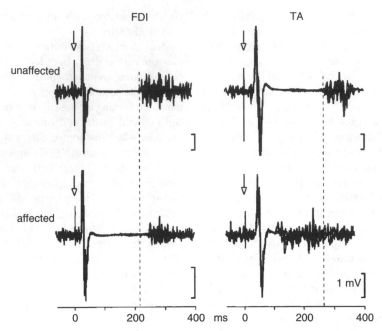

Fig. 10.3 Cortical silent period (cSP) recordings (three superimposed trials in each diagram) in both first dorsal interosseous (FDI) and tibial anterior (TA) muscles in a patient who had small cerebral ischemic strokes in the territory of the left middle cerebral artery including the precentral gyrus. Dotted vertical lines indicate the end of cSPs in the unaffected limbs. Open arrows: magnetic stimulus. Vertical bars indicate 1 mV. cSP measurements were performed in the acute phase after the stroke. Note that the cSP of the clinically dominantly affected right TA was markedly shortened. This patient developed several focal motor seizures of the right leg 2 months later. CSP shortening of the affected right TA reflects reduced inhibition in the affected motor cortex which may be involved in the development of seizures. From Kessler *et al.* (2002), with permission.

Parkinson's disease. Both forms of treatment increase thalamofugal firing, and lengthen the cSP (Young *et al.* 1997; Däuper *et al.* 2002), although this finding is not unanimously reported (Cunic *et al.* 2002). Also, administration of dopaminergic medication can normalize cSP duration (for review see Lefaucheur 2005b). In patients with atypical parkinsonian syndromes, cSP is also shortened in corticobasal-ganglionic degeneration (Kühn *et al.* 2004b) but prolonged in multiple system atrophy and progressive supranuclear palsy (Kühn *et al.* 2004b). In Huntington's disease, the cSP is prolonged and this abnormality correlates with the severity of chorea (Priori *et al.* 1994b; Tegenthoff *et al.* 1996; Modugno *et al.* 2001). In patients with task-specific dystonia (writer's cramp) the cSP is abnormally short, in particular during specific motor tasks such as the pincer grip (Tinazzi *et al.* 2005) or dystonic contraction (Filipovic *et al.* 1997). In patients with cranial dystonia, cSP duration is shortened in facial muscles (Curra *et al.* 2000).

Abnormal cortical inhibition has also been implicated in various psychiatric disorders. The cSP is abnormally short or even lacking in adult patients with Tourette's disorder associated with tics in the EMG target muscle (Ziemann *et al.* 1997) and in children with tic disorder independent of tic location (Moll *et al.* 1999, 2001), whereas cSP duration is normal in children with attention-deficit hyperactivity disorder without tics (Moll *et al.* 2001). This suggests uncontrolled access of voluntary drive to the corticospinal system in tic disorders. CSP duration is also shortened in drug-naïve schizophrenics

(Daskalakis *et al.* 2002), a finding that is in line with other evidence of reduced cortical inhibition in this disease. Drug-free patients with unipolar major depression also exhibit an abnormally short cSP (Bajbouj *et al.* 2006).

References

Ahonen JP, Jehkonen M, Dastidar P, Molnar G, Hakkinen V (1998). Cortical silent period evoked by transcranial magnetic stimulation in ischemic stroke. *Electroencephalography and Clinical Neurophysiology* 109, 224–229.

Amassian VE, Stewart M, Quirk GJ, Rosenthal JL (1987). Physiological basis of motor effects of a transient stimulus to cerebral cortex. *Neurosurgery 20*, 74–93.

Bajbouj M, Lisanby SH, Lang UE, Danker-Hopfe H, Heuser I, Neu P (2006). Evidence for impaired cortical inhibition in patients with unipolar major depression. *Biological Psychiatry 59*, 395–400.

Barker AT, Jalinous R, Freeston IL (1985). Non-invasive magnetic stimulation of the human motor cortex [letter]. *Lancet* i (8437), 1106–1107.

Battaglia F, Quartarone A, Ghilardi MF, et al. (2006). Unilateral cerebellar stroke disrupts movement preparation and motor imagery. *Clinical Neurophysiology 117,* 1009–1016.

Berardelli A, Rona S, Inghilleri M, Manfredi M (1996). Cortical inhibition in Parkinson's disease. A study with paired magnetic stimulation. *Brain 119,* 71–77.

Berardelli A, Inghilleri M, Gilio F, et al. (1999). Effects of repetitive cortical stimulation on the silent period evoked by magnetic stimulation. *Experimental Brain Research 125*, 82–86.

Brasil-Neto JP, Cammarota A, Valls-Sole J, Pascual-Leone A, Hallett M, Cohen LG (1995). Role of intracortical mechanisms in the late part of the silent period to transcranial stimulation of the human motor cortex. *Acta Neurologica Scandinavica 92*, 383–386.

Braune HJ, Fritz C (1995). Transcranial magnetic stimulation-evoked inhibition of voluntary muscle activity (silent period) is impaired in patients with ischemic hemispheric lesion. *Stroke 26*, 550–553.

Braune HJ, Fritz C (1996). Asymmetry of silent period evoked by transcranial magnetic stimulation in stroke patients. *Acta Neurologica Scandinavica 93*, 168–174.

Calancie B, Nordin M, Wallin U, Hagbarth KE (1987). Motor-unit responses in human wrist flexor and extensor muscles to transcranial cortical stimuli. *Journal of Neurophysiology 58*, 1168–1185.

Cantello R, Gianelli M, Bettucci D, Civardi C, De Angelis MS, Mutani R (1991). Parkinson's disease rigidity: magnetic motor evoked potentials in a small hand muscle. *Neurology 41*, 1449–1456.

Cantello R, Gianelli M, Civardi C, Mutani R (1992). Magnetic brain stimulation: the silent period after the motor evoked potential. *Neurology 42*, 1951–1959.

Catano A, Houa M, Noel P (1997). Magnetic transcranial stimulation: clinical interest of the silent period in acute and chronic stages of stroke. *Electroencephalography and Clinical Neurophysiology 105*, 290–296.

Chen R, Tam A, Butefisch C, Ziemann U, Rothwell JC, Cohen LG (1998). Intracortical inhibition and facilitation in different representations of the human motor cortex. *Journal of Neurophysiology 80*, 2870–2881.

Chen R, Lozano AM, Ashby P (1999). Mechanism of the silent period following transcranial magnetic stimulation. *Experimental Brain Research 128*, 539–542.

Chistyakov AV, Soustiel JF, Hafner H, Trubnik M, Levy G, Feinsod M (2001). Excitatory and inhibitory corticospinal responses to transcranial magnetic stimulation in patients with minor to moderate head injury. *Journal of Neurology, Neurosurgery and Psychiatry 70*, 580–587.

Cicinelli P, Traversa R, Bassi A, Scivoletto G, Rossini PM (1997). Interhemispheric differences of hand muscle representation in human motor cortex. *Muscle and Nerve 20*, 535–542.

Cicinelli P, Mattia D, Spanedda F, et al. (2000). Transcranial magnetic stimulation reveals an interhemispheric asymmetry of cortical inhibition in focal epilepsy. *Neuroreport 11*, 701–707.

Cincotta M, Borgheresi A, Lori S, Fabbri M, Zaccara G (1998). Interictal inhibitory mechanisms in patients with cryptogenic motor cortex epilepsy: a study of the silent period following transcranial magnetic stimulation. *Electroencephalography and Clinical Neurophysiology 107*, 1–7.

Cincotta M, Borgheresi A, Gambetti C, et al. (2003). Suprathreshold 0.3 Hz repetitive TMS prolongs the cortical silent period: potential implications for therapeutic trials in epilepsy. *Clinical Neurophysiology 114*, 1827–1833.

Classen J, Benecke R (1995). Inhibitory phenomena in individual motor units induced by transcranial magnetic stimulation. *Electroencephalography and Clinical Neurophysiology 97*, 264–274.

Classen J, Witte OW, Schlaug G, Seitz RJ, Holthausen H, Benecke R (1995). Epileptic seizures triggered directly by focal transcranial magnetic stimulation. *Electroencephalography and Clinical Neurophysiology 94*, 19–25.

Classen J, Schnitzler A, Binkofski F, et al. (1997). The motor syndrome associated with exaggerated inhibition within the primary motor cortex of patients with hemiparetic stroke. *Brain 120*, 605–619.

Classen J, Steinfelder B, Liepert J, et al. (2000). Cutaneo-motor integration in humans is somatotopically organized at various levels of the nervous system and is task dependent. *Experimental Brain Research 130*, 48–59.

Cruz Martinez A, Munoz J, Palacios F (1998). The muscle inhibitory period by transcranial magnetic stimulation. Study in stroke patients. *Electromyography and Clinical Neurophysiology 38*, 189–192.

Cunic D, Roshan L, Khan FI, Lozano AM, Lang AE, Chen R (2002). Effects of subthalamic nucleus stimulation on motor cortex excitability in Parkinson's disease. *Neurology 58*, 1665–1672.

Curra A, Romaniello A, Berardelli A, Cruccu G, Manfredi M (2000). Shortened cortical silent period in facial muscles of patients with cranial dystonia. *Neurology 54*, 130–135.

Daskalakis ZJ, Christensen BK, Chen R, Fitzgerald PB, Zipursky RB, Kapur S (2002). Evidence for impaired cortical inhibition in schizophrenia using transcranial magnetic stimulation. *Archives of General Psychiatry 59*, 347–354.

Daskalakis ZJ, Molnar GF, Christensen BK, Sailer A, Fitzgerald PB, Chen R (2003). An automated method to determine the transcranial magnetic stimulation-induced contralateral silent period. *Clinical Neurophysiology 114*, 938–944.

Däuper J, Peschel T, Schrader C, et al. (2002). Effects of subthalamic nucleus (STN) stimulation on motor cortex excitability. *Neurology 59*, 700–706.

Davey NJ, Romaiguere P, Maskill DW, Ellaway PH (1994). Suppression of voluntary motor activity revealed using transcranial magnetic stimulation of the motor cortex in man. *Journal of Physiology (London) 477*, 223–235.

Day BL, Dressler D, Maertens de Noordhout A, et al. (1989). Electric and magnetic stimulation of human motor cortex: surface EMG and single motor unit responses. *Journal of Physiology (London) 412*, 449–473.

Di Lazzaro V, Oliviero A, Pilato F, et al. (2004). The physiological basis of transcranial motor cortex stimulation in conscious humans. *Clinical Neurophysiology 115*, 255–266.

Dioszeghy P, Hidasi E, Mechler F (1999). Study of central motor functions using magnetic stimulation in Parkinson's disease. *Electromyography and Clinical Neurophysiology 39*, 101–105.

Edgley SA, Eyre JA, Lemon RN, Miller S (1990). Excitation of the corticospinal tract by electromagnetic and electrical stimulation of the scalp in the macaque monkey. *Journal of Physiology (London) 425*, 301–320.

Faig J, Busse O (1996). Silent period evoked by transcranial magnetic stimulation in unilateral thalamic infarcts. *Journal of the Neurological Sciences 142*, 85–92.

Ferbert A, Caramia D, Priori A, Bertolasi L, Rothwell JC (1992). Cortical projection to erector spinae muscles in man as assessed by focal transcranial magnetic stimulation. *Electroencephalography and Clinical Neurophysiology 85*, 382–387.

Filipovic SR, Ljubisavljevic M, Svetel M, Milanovic S, Kacar A, Kostic VS (1997). Impairment of cortical inhibition in writer's cramp as revealed by changes in electromyographic silent period after transcranial magnetic stimulation. *Neuroscience Letters 222*, 167–170.

Fuhr P, Agostino R, Hallett M (1991). Spinal motor neuron excitability during the silent period after cortical stimulation. *Electroencephalography and Clinical Neurophysiology 81*, 257–262.

Garvey MA, Gilbert DL (2004). Transcranial magnetic stimulation in children. *European Journal of Paediatric Neurology 8*, 7–19.

Garvey MA, Ziemann U, Becker DA, Barker CA, Bartko JJ (2001). New graphical method to measure silent periods evoked by transcranial magnetic stimulation. *Clinical Neurophysiology 112*, 1451–1460.

Hallett M (1995). Transcranial magnetic stimulation. Negative effects. *Advances in Neurology 67*, 107–113.

Hamer HM, Reis J, Mueller HH, et al. (2005). Motor cortex excitability in focal epilepsies not including the primary motor area – a TMS study. *Brain 128*, 811–818.

Haug BA, Kukowski B (1994). Latency and duration of the muscle silent period following transcranial magnetic stimulation in multiple sclerosis, cerebral ischemia, and other upper motoneuron lesions. *Neurology 44*, 936–940.

Haug BA, Schonle PW, Knobloch C, Kohne M (1992). Silent period measurement revives as a valuable diagnostic tool with transcranial magnetic stimulation. *Electroencephalography and Clinical Neurophysiology 85*, 158–160.

Heinen F, Glocker FX, Fietzek U, Meyer B-U, Lucking C-H, Korinthenberg R (1998). Absence of transcallosal inhibition following focal magnetic stimulation in preschool children. *Annals of Neurology 43*, 608–612.

Hess A, Kunesch E, Classen J, Hoeppner J, Stefan K, Benecke R (1999). Task-dependent modulation of inhibitory actions within the primary motor cortex. *Experimental Brain Research 124*, 321–330.

Inghilleri M, Berardelli A, Cruccu G, Manfredi M (1993). Silent period evoked by transcranial stimulation of the human cortex and cervicomedullary junction. *Journal of Physiology (London) 466*, 521–534.

Inghilleri M, Mattia D, Berardelli A, Manfredi M (1998). Asymmetry of cortical excitability revealed by transcranial stimulation in a patient with focal motor epilepsy and cortical myoclonus. *Electroencephalography and Clinical Neurophysiology 109*, 70–72.

Katayama T, Aizawa H, Kuroda K, et al. (2001). Cortical silent period in the tongue induced by transcranial magnetic stimulation. *Journal of the Neurological Sciences 193*, 37–41.

Kessler KR, Schnitzler A, Classen J, Benecke R (2002). Reduced inhibition within primary motor cortex in patients with poststroke focal motor seizures. *Neurology 59*, 1028–1033.

Kimiskidis VK, Papagiannopoulos S, Sotirakoglou K, Kazis DA, Kazis A, Mills KR (2005). Silent period to transcranial magnetic stimulation: construction and properties of stimulus-response curves in healthy volunteers. *Experimental Brain Research 163*, 21–31.

Krnjevic K, Randic M, Straughan DW (1966). Nature of a cortical inhibitory process. *Journal of Physiology 184*, 49–77.

Kudina LP, Pantseva RE (1988). Recurrent inhibition of firing motoneurons in man. *Electroencephalography and Clinical Neurophysiology 69*, 179–185.

Kühn AA, Brandt SA, Kupsch A, et al. (2004a). Comparison of motor effects following subcortical electrical stimulation through electrodes in the globus pallidus internus and cortical transcranial magnetic stimulation. *Experimental Brain Research 155*, 48–55.

Kühn AA, Grosse P, Holtz K, Brown P, Meyer BU, Kupsch A (2004b). Patterns of abnormal motor cortex excitability in atypical parkinsonian syndromes. *Clinical Neurophysiology 115*, 1786–1795.

Kukowski B, Haug B (1992). Quantitative evaluation of the silent period, evoked by transcranial magnetic stimulation during sustained muscle contraction, in normal man and in patients with stroke. *Electromyography and Clinical Neurophysiology 32*, 373–378.

Lefaucheur JP (2005a). Excitability of the motor cortical representation of the external anal sphincter. *Experimental Brain Research 160*, 268–272.

Lefaucheur JP (2005b). Motor cortex dysfunction revealed by cortical excitability studies in Parkinson's disease: influence of antiparkinsonian treatment and cortical stimulation. *Clinical Neurophysiology 116*, 244–253.

Lefaucheur JP, Lofaso F (2002). Diaphragmatic silent period to transcranial magnetic cortical stimulation for assessing cortical motor control of the diaphragm. *Experimental Brain Research 146*, 404–409.

Leis AA, Kofler M, Stokic DS, Grubwieser GJ, Delapasse JS (1993). Effect of the inhibitory phenomenon following magnetic stimulation of cortex on brainstem motor neuron excitability and on the cortical control of brainstem reflexes. *Muscle and Nerve 16*, 1351–1358.

Liepert J, Storch P, Fritsch A, Weiller C (2000). Motor cortex disinhibition in acute stroke. *Clinical Neurophysiology 111*, 671–676.

Liepert J, Restemeyer C, Kucinski T, Zittel S, Weiller C (2005a). Motor strokes: the lesion location determines motor excitability changes. *Stroke 36*, 2648–2653.

Liepert J, Restemeyer C, Munchau A, Weiller C (2005b). Motor cortex excitability after thalamic infarction. *Clinical Neurophysiology 116*, 1621–1627.

McCormick DA (1989). GABA as an inhibitory neurotransmitter in human cerebral cortex. *Journal of Neurophysiology 62*, 1018–1027.

Macdonell RA, King MA, Newton MR, Curatolo JM, Reutens DC, Berkovic SF (2001). Prolonged cortical silent period after transcranial magnetic stimulation in generalized epilepsy. *Neurology 57*, 706–708.

Marsden CD, Merton PA, Morton HB (1983). Direct electrical stimulation of corticospinal pathways through the intact scalp in human subjects. *Advances in Neurology 39*, 387–391.

Mathis J, de Quervain D, Hess CW (1998). Dependence of the transcranially induced silent period on the 'instruction set' and the individual reaction time. *Electromyography and Clinical Neurophysiology 109*, 426–435.

Mathis J, de Quervain D, Hess CW (1999). Task-dependent effects on motor-evoked potentials and on the following silent period. *Journal of Clinical Neurophysiology 16*, 556–565.

Merton PA (1951). The silent period in a muscle of the human hand. *Journal of Physiology 114*, 183–198.

Modugno N, Curra A, Giovannelli M, et al. (2001). The prolonged cortical silent period in patients with Huntington's disease. *Clinical Neurophysiology 112*, 1470–1474.

Moll GH, Wischer S, Heinrich H, Tergau F, Paulus W, Rothenberger A (1999). Deficient motor control in children with tic disorder: evidence from transcranial magnetic stimulation. *Neuroscience Letters*, *272*, 37–40.

Moll GH, Heinrich H, Trott GE, Wirth S, Bock N, Rothenberger A (2001). Children with comorbid attention-deficit-hyperactivity disorder and tic disorder: evidence for additive inhibitory deficits within the motor system. *Annals of Neurology 49*, 393–396.

Nakashima K, Wang Y, Shimoda M, Sakuma K, Takahashi K (1995). Shortened silent period produced by magnetic cortical stimulation in patients with Parkinson's disease. *Journal of the Neurological Sciences 130*, 209–214.

Nilsson J, Panizza M, Arieti P (1997). Computer-aided determination of the silent period. *Journal of Clinical Neurophysiology 14*, 136–143.

Orth M, Rothwell JC (2004). The cortical silent period: intrinsic variability and relation to the waveform of the transcranial magnetic stimulation pulse. *Clinical Neurophysiology 115*, 1076–1082.

Paradiso GO, Cunic DI, Gunraj CA, Chen R (2005). Representation of facial muscles in human motor cortex. *Journal of Physiology 567*, 323–336.

Person RS, Kozhina GV (1978). Study of orthodromic and antidromic effects of nerve stimulation on single motoneurons of human hand muscles. *Electromyography and Clinical Neurophysiology 18*, 437–456.

Pierrot-Deseilligny E, Bussel B, Held JP, Katz R (1976). Excitability of human motoneurons after discharge in a conditioning reflex. *Electromyography and Clinical Neurophysiology 40*, 279–287.

Priori A, Berardelli A, Inghilleri M, Accornero N, Manfredi M (1994a). Motor cortical inhibition and the dopaminergic system. Pharmacological changes in the silent period after transcranial brain stimulation in normal subjects, patients with Parkinson's disease and drug-induced parkinsonism. *Brain 117*, 317–323.

Priori A, Berardelli A, Inghilleri M, Polidori L, Manfredi M (1994b). Electromyographic silent period after transcranial brain stimulation in Huntington's disease. *Movement Disorders 9*, 178–182.

Priori A, Oliviero A, Donati E, Callea L, Bertolasi L, Rothwell JC (1999). Human handedness and asymmetry of the motor cortical silent period. *Experimental Brain Research 128*, 390–396.

Roick H, von Giesen HJ, Benecke R (1993). On the origin of the postexcitatory inhibition seen after transcranial magnetic brain stimulation in awake human subjects. *Experimental Brain Research 94*, 489–498.

Romeo S, Gileo F, Pedace F, et al. (2000). Changes in the cortical silent period after repetitive magnetic stimulation of cortical motor areas. *Experimental Brain Research 135*, 504–510.

Sakai K, Ugawa Y, Terao Y, Hanajima R, Furabayashi T, Kanazawa I (1997). Preferential activation of different I was by transcranial magnetic stimulation with a figure-of-eight shaped coil. *Experimental Brain Research 113*, 24–32.

Schnitzler A, Benecke R (1994). The silent period after transcranial magnetic stimulation is of exclusive cortical origin: evidence from isolated cortical ischemic lesions in man. *Neuroscience Letters 180*, 41–45.

Schnitzler A, Kessler KR, Benecke R (1996). Transcallosally mediated inhibition of interneurons within human primary motor cortex. *Experimental Brain Research 112*, 381–391.

Shimizu T, Oliveri M, Filippi MM, Palmieri MG, Pasqualetti P, Rossini PM (1999). Effect of paired transcranial magnetic stimulation on the cortical silent period. *Brain Research 834*, 74–82.

Siebner HR, Dressnandt J, Auer C, Conrad B (1998). Continuous intrathecal baclofen infusions induced a marked increase of the transcranially evoked silent period in a patient with generalized dystonia. *Muscle and Nerve 21*, 1209–1212.

Stefan K, Kunesch E, Cohen LG, Benecke R, Classen J (2000). Induction of plasticity in the human motor cortex by paired associative stimulation. *Brain 123*, 572–584.

Tataroglu C, Ozkiziltan S, Baklan B (2004). Motor cortical thresholds and cortical silent periods in epilepsy. *Seizure 13*, 481–485.

Taylor JL, Allen GM, Butler JE, Gandevia SC (1997). Effect of contraction strength on responses in biceps brachii and adductor pollicis to transcranial magnetic stimulation. *Experimental Brain Research 117*, 472–478.

Tegenthoff M, Vorgerd M, Juskowiak F, Roos V, Malin JP (1996). Postexcitatory inhibition after transcranial magnetic single and double brain stimulation in Huntington's disease. *Electromyography and Clinical Neurophysiology 101*, 298–303.

Tergau F, Becher V, Canelo M, *et al.* (1999). Complete suppression of voluntary motor drive during the silent period after transcranial magnetic stimulation. *Experimental Brain Research 124*, 447–454.

Tinazzi M, Farina S, Edwards M, *et al.* (2005). Task-specific impairment of motor cortical excitation and inhibition in patients with writer's cramp. *Neuroscience Letters 378*, 55–58.

Triggs WJ, Macdonell RA, Cros D, Chiappa KH, Shahani BT, Day BJ (1992). Motor inhibition and excitation are independent effects of magnetic cortical stimulation. *Annals of Neurology 32*, 345–351.

Trompetto C, Buccolieri A, Marinelli L, Abbruzzese G (2001). Differential modulation of motor evoked potential and silent period by activation of intracortical inhibitory circuits. *Clinical Neurophysiology 112*, 1822–1827.

Valzania F, Strafella A, Quatrale R, *et al.* (1997). Motor evoked responses to paired cortical magnetic stimulation in Parkinson's disease. *Electroencephalography and Clinical Neurophysiology 105*, 37–43.

van Kuijk AA, Pasman JW, Geurts AC, Hendricks HT (2005). How salient is the silent period? The role of the silent period in the prognosis of upper extremity motor recovery after severe stroke. *Journal of Clinical Neurophysiology 22*, 10–24.

von Giesen HJ, Roick H, Benecke R (1994). Inhibitory actions of motor cortex following unilateral brain lesions as studied by magnetic brain stimulation. *Experimental Brain Research 99*, 84–96.

Wassermann EM, Pascual-Leone A, Valls-Sole J, Toro C, Cohen LG, Hallett M (1993). Topography of the inhibitory and excitatory responses to transcranial magnetic stimulation in a hand muscle. *Electromyography and Clinical Neurophysiology 89*, 424–33.

Werhahn KJ, Classen J, Benecke R (1995). The silent period induced by transcranial magnetic stimulation in muscles supplied by cranial nerves: normal data and changes in patients. *Journal of Neurology, Neurosurgery and Psychiatry 59*, 586–596.

Wilson SA, Lockwood RJ, Thickbroom GW, Mastaglia FL (1993a). The muscle silent period following transcranial magnetic cortical stimulation. *Journal of the Neurological Sciences 114*, 216–222.

Wilson SA, Thickbroom GW, Mastaglia FL (1993b). Topography of excitatory and inhibitory muscle responses evoked by transcranial magnetic stimulation in the human motor cortex. *Neuroscience Letters 154*, 52–56.

Wu L, Goto Y, Taniwaki T, Kinukawa N, Tobimatsu S (2002). Different patterns of excitation and inhibition of the small hand and forearm muscles from magnetic brain stimulation in humans. *Clinical Neurophysiology 113*, 1286–1294.

Wu T, Sommer M, Tergau F, Paulus W (2000). Modification of the silent period by double transcranial magnetic stimulation. *Clinical Neurophysiology 111*, 1868–1872.

Young MS, Triggs WJ, Bowers D, Greer M, Friedman WA (1997). Stereotactic pallidotomy lengthens the transcranial magnetic cortical stimulation silent period in Parkinson's disease. *Neurology 49*, 1278–1283.

Ziemann U (2004). TMS induced plasticity in human cortex. *Reviews in the Neurosciences 15*, 252–266.

Ziemann U, Netz J, Szelenyi A, Hömberg V (1993). Spinal and supraspinal mechanisms contribute to the silent period in the contracting soleus muscle after transcranial magnetic stimulation of human motor cortex. *Neuroscience Letters 156*, 167–171.

Ziemann U, Lönnecker S, Steinhoff BJ, Paulus W (1996). Effects of antiepileptic drugs on motor cortex excitability in humans: a transcranial magnetic stimulation study. *Annals of Neurology 40*, 367–378.

Ziemann U, Paulus W, Rothenberger A (1997). Decreased motor inhibition in Tourette disorder: Evidence from transcranial magnetic stimulation. *American Journal of Psychiatry 154*, 1277–1284.

CHAPTER 11

Paired-pulse measures

Ritsuko Hanajima and Yoshikazu Ugawa

Summary – Paired-pulse TMS techniques include several protocols to study modulation of human motor cortical excitability by local circuits or afferent input from other areas of the brain. These modulating inputs elicit inhibitory or facilitatory effects on the motor cortex through intracortical, intrahemispheric, or interhemispheric connections. A conditioning stimulus is given somewhere to the brain prior to a test stimulus that is delivered over the motor cortex. Changes in motor cortical excitability produced by the conditioning pulse are estimated by changes in the size of the conditioned motor-evoked potential (MEP), compared to the test MEP elicited by the test stimulus alone. In this chapter, physiology and applications of the currently available paired-pulse protocols will be reviewed.

Short-interval intracortical inhibition (SICI)

Short-interval intracortical inhibition (SICI) was first reported by Kujirai *et al.* (1993) (Figure 11.1). The conditioning and test magnetic stimuli are applied over the hand area of the motor cortex through the same coil. The intensity of the conditioning stimulus is set below motor-evoked potential (MEP) threshold, and the intensity of the suprathreshold test stimulus is adjusted to elicit control MEPs with a peak-to-peak amplitude of ~1 mV. The subthreshold conditioning stimulus suppresses the control MEP at interstimulus intervals (ISIs) of 1–5 ms. The same subthreshold conditioning stimulus does not reduce the amplitude of MEPs elicited by transcranial electrical stimulation (TES), which activates axons of corticospinal neurons predominantly directly. The conditioning stimulus does not inhibit the size of the spinal H-reflex either. Based on these results, the reduction of the test MEP elicited by TMS is considered to reflect inhibition at the level of the primary motor cortex (Kujirai *et al.* 1993). The amount of this inhibition is sensitive to several factors of experimental manipulation. One of the important factors is the state of the target muscle. During a minimal voluntary tonic contraction of the target muscle, SICI is significantly reduced (Ridding *et al.* 1995c). Patients may unintentionally contract their target muscle more often than normal subjects. This may result in false underestimation of the amount of inhibition in the patient group even though their inhibitory system in the motor cortex is intact. Therefore, it is important that complete voluntary relaxation of the target muscle is carefully monitored during SICI measurements. This can be obtained by continuous visual and auditory feedback of the electromyogram (EMG) raw signal at high gain.

The magnitude of SICI depends critically on the intensity of both the conditioning and test stimulus. In particular, the effects of the intensity of the conditioning stimulus have been studied extensively. The intensity was originally

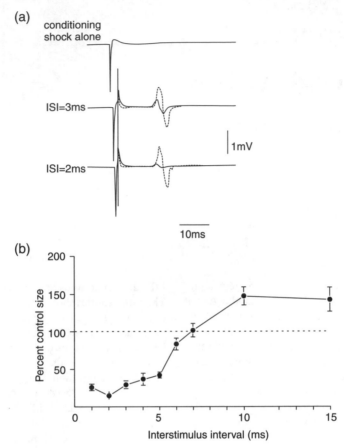

Fig. 11.1 Paired-pulse stimulation to test short-interval intracortical inhibition (SICI) and intracortical facilitation (ICF) (Kujirai et al. 1993). (a) Motor-evoked potentials (MEPs) recorded from the voluntarily relaxed first dorsal interosseous muscle of one representative healthy subject. The top trace shows MEPs to the conditioning stimulus alone (no response), and the middle and the bottom traces are MEPs to paired stimuli at interstimulus intervals (ISIs) of 3 ms (middle trace) and 2 ms (bottom trace). Averaged MEPs to paired stimulation (solid lines) are superimposed with MEPs to test stimulus alone (dotted lines). MEP sizes to paired stimuli are smaller than those to a test stimulus given alone. (b) The average time course of the paired stimulation effect obtained from 10 normal subjects. The ordinate indicates the percentage of the conditioned MEP size to the control MEP, and the abscissa the interstimulus interval (ISI). At ISIs of 1–5 ms, significant inhibition was obtained. The inhibition is followed by facilitation at ISIs of 10 and 15 ms. Reproduced from Figure 1 in Kujirai et al. (1993), with permission from the Blackwell Publishing Company.

related to the resting motor threshold (RMT), but was more recently related to the active motor threshold (AMT) in order to ensure that the conditioning stimulus is subthreshold for activation of the corticospinal projection (Di Lazzaro et al. 1998). The SICI intensity curve is typically U-shaped with maximum inhibition occurring at intensities of the conditioning stimulus of ~70–80% RMT (Kujirai et al. 1993; Kossev et al. 2003). The SICI threshold is highly correlated with the AMT (Orth et al. 2003), and is in the order of ~70% AMT (Ziemann et al. 1996b; Orth et al. 2003). When the intensity of the conditioning stimulus is related to AMT, maximum SICI occurs usually at intensities of ~90–100% AMT (Orth et al. 2003). This is approximately equivalent to 70–80% RMT, and therefore consistent with the results of Kujirai *et al.* (1993). The U-shape of the SICI intensity curve implies that, with intensities of the conditioning stimulus at or above AMT, higher-threshold excitatory inputs to the corticospinal neurons are activated in addition to the lower-threshold inhibitory inputs responsible for the generation of the SICI (Ilić *et al.* 2002). Therefore, SICI can be tested selectively if the intensity of the conditioning stimulus is below AMT, but a net effect of superimposed inhibition and short-interval intracortical facilitation (SICF, see below) is obtained if higher intensities of the conditioning stimulus are applied. Despite clear differences in MEP threshold and MEP intensity curves, the absolute intensities required to produce SICI are similar between target muscles of different body representations, suggesting that the intracortical mechanisms for inhibition are not related to the strength of the corticospinal projections (Chen *et al.* 1998).

In order to analyze the mechanisms underlying this form of motor cortical inhibition, SICI was studied invasively by direct stimulation of the motor cortex or epidural recording of the descending corticospinal volley from the spinal cord. In a patient with epilepsy, in whom a grid of 64 subdural electrodes was implanted over the left frontotemporal cortex for spike detection, focal paired electrical stimuli applied to the motor-cortex-induced inhibition of the test response at ISIs of 1–5 ms only if the conditioning stimulus was delivered within 1–2 cm of the site of the test stimulus (Ashby *et al.* 1999). This suggests that SICI is generated in local circuits in close proximity to the corticospinal neurons of the target muscle. Di Lazzaro *et al.* (1998) recorded the descending corticospinal volleys elicited by single- and paired-pulse TMS of the motor cortex, using epidural electrodes implanted over the cervical spinal cord for intractable pain relief. The intensity of the conditioning stimulus was set below AMT, and did not elicit any descending corticospinal volleys. The test stimulus evoked several so-called I-waves, i.e. corticospinal discharges that are produced by indirect activation of corticospinal neurons by excitation of presynaptic excitatory interneurons (cf. Chapter 14, this volume). The I-waves are numbered according to their order

of appearance. At ISIs of 1–5 ms, the conditioning stimulus reduces the amplitude of the later I-waves (I2-, I3-, I4-waves). In contrast, the first I-wave (I1-wave), which appears ~1.5 ms after the D-wave (elicited by direct activation of the corticospinal axon), remains unaffected by the conditioning stimulus. A similar difference between I1- and I3-waves with respect to their susceptibility to inhibitory modulation was demonstrated noninvasively by EMG recordings in normal subjects (Hanajima *et al.* 1998b). When subjects maintain a slight voluntary contraction of the target muscle, small MEPs elicited by anteriorly or medially directed currents in the motor cortex are produced predominantly by activation of I1-waves or D-waves (Sakai *et al.* 1997) (cf. Chapter 14, this volume). In contrast, small MEPs elicited by posteriorly or laterally directed currents are produced predominantly by I3-waves (Sakai *et al.* 1997). At ISIs of 1–5 ms, MEPs produced by I1-waves are not inhibited by the conditioning stimulus. In contrast, MEPs produced by I3-waves are suppressed by the conditioning stimulus, and this inhibition lasts ≥20 ms (Hanajima *et al.* 1998b). The same results are obtained when single motor units are recorded by needle EMG (Figure 11.2). In summary, these findings strongly support the notion that SICI is generated by synaptic inhibitory mechanisms at the level of local interneurons in the primary motor cortex.

Accordingly, in early research the SICI was considered to reflect inhibition mediated through the gamma-aminobutyric acid A receptor (GABA$_A$R) (Kujirai *et al.* 1993). The duration of the inhibition of I3-waves of ~20 ms (Hanajima *et al.* 1998b) is consistent with the duration of GABA$_A$R-mediated inhibitory postsynaptic potentials in animal preparations (Krnjevic *et al.* 1964, 1965; Connors *et al.* 1988). Pharmacological experiments support this view that SICI reflects GABA$_A$R-mediated inhibition (cf. Chapter 13, this volume). In essence, benzodiazepines, which act as positive allosteric modulators at the GABA$_A$R, consistently increase SICI (Ziemann *et al.* 1996a; Di Lazzaro *et al.* 2000, 2005a,b; Ilić *et al.* 2002).

Recently, it was demonstrated that the SICI at ISIs of 1–5 ms consists of at least two phases of inhibition, which are physiologically distinct.

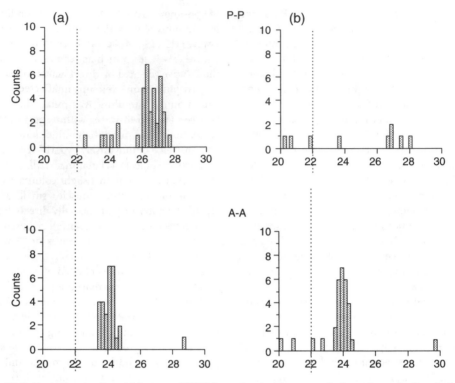

Fig. 11.2 Post-stimulus time histograms (PSTHs) constructed from one single motor unit after single or paired-pulse TMS of the contralateral motor cortex (Hanajima *et al.* 1998b). Control PSTHs are shown on the left (a) and conditioned PSTHs at an ISI of 4 ms are on the right (b). Dotted vertical lines indicate the latency of the D-wave. The upper two PSTHs are recorded with both the conditioning and test stimulus directed antero-posteriorly in the brain (P-P). Under this condition, the test stimulus elicited a single I3-peak (approximately delayed by 4 ms with respect to the D-wave). The I3-peak was strongly inhibited by the conditioning stimulus. The lower two PSTHs are obtained with both stimuli directed postero-anteriorly in the brain (A-A). Under this condition, the test stimulus elicited a single I1-peak (approximately delayed by 1.5 ms with respect to the D-wave). The I1-peak was not affected by the conditioning stimulus. Reproduced from Figure 1A and C in Hanajima *et al.* (1998b), with permission from the Blackwell Publishing Company.

Fisher *et al.* (2002), using a computer-assisted threshold-tracking method, showed that maximal inhibition occurred at discrete ISIs of 1.0 and 2.5 ms. Inhibition occurs at significantly lower conditioning stimulus intensities at the ISI of 1.0 ms than at the ISI of 2.5 ms. Voluntary activity reduces inhibition at both ISIs, but has a much greater effect on the inhibition at the ISI of 2.5 ms (Fisher *et al.* 2002). Roshan *et al.* (2003) showed that the magnitudes of SICI at ISIs of 1 and 2.5 ms do not correlate. This was interpreted as convergent evidence that there are at least two distinct phases of inhibition that constitute SICI.

In addition, at ISIs of 3–5 ms, SICI is produced specifically by inhibition of I3-waves, but not by inhibition of I1-waves. In contrast, at the ISI of 1 ms, SICI is produced by moderate inhibition of both I3-waves and I1-waves, and even D-waves are slightly suppressed (Hanajima *et al.* 2003). This led to the conclusion that the SICI at an ISI of 1 ms is, at least in part, caused by axonal refractoriness rather than by synaptic inhibition (Fisher *et al.* 2002; Hanajima *et al.* 2003).

Most SICI studies are performed in intrinsic hand muscles, but SICI can be obtained in any other muscles, although sometimes to a lesser extent. SICI is present in proximal arm muscles at ISIs of 1–3 ms, but its amount is significantly less than in intrinsic hand muscles (Abbruzzese et al. 1999; Shimizu et al. 1999). This may indicate a difference in the functional organization between proximal and distal arm representations in human motor cortex, with a requirement for more deliberate inhibitory control of finely tuned individuated finger movements. SICI can also be obtained in leg muscles (Stokic et al. 1997), and in muscles supplied by cranial nerves (Hanajima et al. 1998a; Kobayashi et al. 2001; Muellbacher et al. 2001; Paradiso et al. 2005).

SICI has been studied in various neurological and psychiatric disorders. In many instances, an abnormally reduced SICI was found, suggesting that alteration in SICI may be a sensitive but relatively unspecific indicator of deficient motor cortical inhibition (for review, Ziemann 1999). This chapter will focus on SICI studies in movement disorders in order to give a flavor of the usefulness of this technique in clinical neurophysiology. The SICI is abnormally reduced in Parkinson's disease (Ridding et al. 1995a; Hanajima et al. 1996; Strafella et al. 2000; Pierantozzi et al. 2001, 2002; Cunic et al. 2002; Bares et al. 2003; Buhmann et al. 2004). Hanajima et al. (1996) reported a direct correlation between reduction of SICI and reduced blood flow in the basal ganglia, supporting the notion that the excitability of motor cortical inhibitory interneurons is reduced by altered afferent signalling from the basal ganglia to motor cortex. This reduction in SICI tends to be normalized by dopaminergic treatment (Ridding et al. 1995a; Strafella et al. 2000; Pierantozzi et al. 2001, 2002), or deep brain stimulation (Cunic et al. 2002; Pierantozzi et al. 2002). However, an abnormally reduced SICI was not unanimously found in all studies (Berardelli et al. 1996; Däuper et al. 2002; MacKinnon et al. 2005). Patients with movement disorders may have difficulties in fully relaxing the target muscle. Therefore, Berardelli et al. (1996) studied SICI during a slight voluntary contraction of the target muscle and argued that the reduced SICI in patients with Parkinson's disease found at rest but not during voluntary muscle activation might have been caused by unintentional and subliminal activation of corticospinal neurons in the rest condition. Thus, it might be advisable to study SICI in

movement disorders during slight voluntary contraction of the target muscle. Ridding et al. (1995a) discussed the possibility that the reduction in SICI might be caused by hyperexcitability of excitatory neural elements in the motor cortex rather than by hypoexcitability of inhibition. This was addressed experimentally in a recent study (MacKinnon et al. 2005). The measurement of SICI intensity curves showed that the SICI threshold and the magnitude of maximal SICI are normal in Parkinson's disease, but that SICI is reduced if intensities of the conditioning stimulus above AMT are used. This strongly favours the idea that the SICI per se is not deficient in Parkinson's disease, but rather that the short-interval intracortical facilitation (SICF, see below) is hyperexcitable (MacKinnon et al. 2005). A reduced SICI was also obtained in various atypical parkinsonian syndromes, including predominantly parkinsonian multiple system atrophy (Marchese et al. 2000; Kühn et al. 2004), vascular parkinsonism (Marchese et al. 2000), progressive supranuclear palsy (Kühn et al. 2004), and corticobasal-ganglionic degeneration (Hanajima et al. 1996; Kühn et al. 2004), whereas SICI is normal in predominantly cerebellar multiple system atrophy (Marchese et al. 2000).

SICI is reduced in various forms of dystonia, such as generalized dystonia (Gilio et al. 2000), writer's cramp (Ridding et al. 1995b; Siebner et al. 1999; Sommer et al. 2002; Stinear and Byblow 2004a) (Figure 11.3), spasmodic torticollis (Hanajima et al. 1998a), or blepharospasm (Sommer et al. 2002). Stinear and Byblow (2004a) obtained SICI intensity curves and demonstrated that patients with focal hand dystonia have a higher SICI threshold when the intensity of the conditioning stimulus was expressed as a ratio of the AMT. This indicates a true hypoexcitability of those inhibitory neural elements that are responsible for the generation of the SICI, and highlights again the advantage of obtaining SICI intensity curves rather than testing SICI at a single intensity of the conditioning stimulus. The deficient SICI in dystonia may be normalized after treatment with botulinum toxin (Gilio et al. 2000), or low-frequency 1 Hz repetitive TMS of the motor cortex (Siebner et al. 1999). The occurrence of abnormalities in SICI in dystonic patients may be task dependent (Gilio et al. 2003; Stinear and

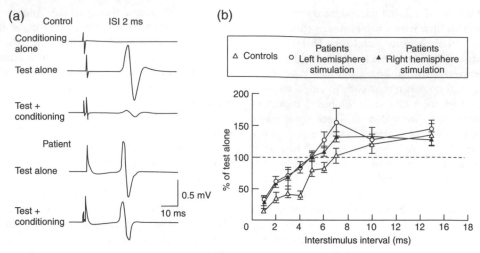

Fig. 11.3 Short-interval intracortical inhibition (SICI) in patients with focal hand dystonia (Ridding *et al.* 1995b). (a) The upper three traces show MEPs from the first dorsal interosseous muscle in a control subject and the lower two traces are those in a patient with focal hand dystonia. At an interstimulus interval (ISI) of 2 ms, the conditioned response (test + conditioning) is clearly smaller than the control response (test alone) in the normal subject. By contrast, in the patient, the conditioned response is less suppressed. (b) Time courses of the effects of the conditioning stimulus on the test response in normal subjects and patients with focal hand dystonia. Compared with normal subjects (open triangles), the suppression was reduced in patients with focal hand dystonia in the clinically affected right hand (open circles) and in the clinically nonaffected left hand (filled triangles). Reproduced from Figure 1 in Ridding *et al.* 1995b, with permission from the BMJ Publishing Group.

Byblow 2004b; Bütefisch *et al.* 2005). For instance, healthy subjects show an increase in SICI in a motor representation that is voluntarily relaxed during voluntary activation of another motor representation (selective activation) while dystonic patients fail to show this task-dependent increase of SICI (Bütefisch *et al.* 2005). This abnormality may contribute to the impairment of these patients in selective activation of a target muscle.

Finally, somewhat inconsistent results on SICI have been described in Huntington's disease. Whereas SICI was found to be reduced in one study (Abbruzzese *et al.* 1997), it was found to be normal in other studies (Hanajima *et al.* 1996, 1999; Priori *et al.* 2000).

In summary, exploration of motor cortical excitability in movement disorders by paired-pulse TMS protocols at short interstimulus intervals is useful. It is possible to identify specific abnormalities in the balance between inhibitory (SICI) and facilitatory (SICF) processes, even if the pathology lies in abnormal

afferent signalling to the motor cortex rather than in the motor cortex itself. SICI measurements can be used to monitor treatment effects, and they can enhance our understanding of the pathophysiology that underlies movement disorders.

Intracortical facilitation (ICF)

Kujirai *et al.* (1993) reported that a test MEP was facilitated by a subthreshold conditioning stimulus at ISIs of 10–15 ms, using the same protocol as the measurements for SICI (Figure 11.1B). This facilitation is not associated with facilitation of the H-reflex, and it was therefore concluded that it occurs at the level of the motor cortex (intracortical facilitation: ICF) (Ziemann *et al.* 1996b). The ICF threshold is slightly higher than the SICI threshold (Ziemann *et al.* 1996b). The ICF is maximal when the conditioning stimulus induces an anteriorly directed current in the brain, whereas the magnitude of SICI is much less affected by current direction of the conditioning stimulus (Ziemann *et al.* 1996b). Based on these findings, it was proposed that distinct cortical neuronal circuits are involved in the generation of ICF and SICI. ICF is a net facilitation consisting of prevailing facilitation and weaker inhibition. It is currently thought that the facilitation is mediated by glutamatergic *N*-methyl-D-aspartate receptors (NMDAR). This is supported by pharmacological studies that show a decrease in ICF by NMDAR antagonists (Ziemann *et al.* 1998a; Schwenkreis *et al.* 1999) (cf. Chapter 13, this volume). The inhibition probably comes from the tail of the GABA$_A$R-mediated SICI that has a duration of ~20 ms (Hanajima *et al.* 1998b). Accordingly, GABA$_A$R agonists also decrease ICF (Ziemann *et al.* 1996a). In summary, ICF has been much less extensively studied compared to SICI but the available evidence points to the idea that ICF tests mainly excitability of NMDAR-dependent excitatory interneuronal circuits in motor cortex.

Short-interval intracortical facilitation (SICF)

Paired magnetic stimuli of motor threshold intensity (Tokimura *et al.* 1996; Ilić *et al.* 2002), or a suprathreshold conditioning stimulus followed by a subthreshold test stimulus (Ziemann *et al.* 1998b; Chen and Garg 2000; Ilić *et al.* 2002) result in MEP facilitation at discrete ISIs of 1.0–1.5 ms, 2.5–3.0 ms, and at ~4.5 ms (Figure 11.4). Paired TES or substitution of the magnetic test stimulus by a TES test pulse does not produce MEP facilitation at these intervals (Tokimura *et al.* 1996; Ziemann *et al.* 1998b; Chen and Garg 2000). This strongly suggests that this facilitation originates at the level of the motor cortex (short-interval intracortical facilitation: SICF). The effective ISIs for producing SICF show the same periodicity as subsequent I-waves in the corticospinal volley (interpeak intervals of ~1.5 ms), pointing to the idea that

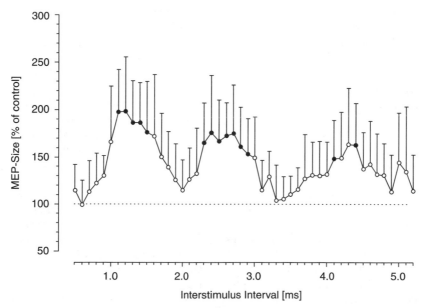

Fig. 11.4 Time course of the short-interval intracortical facilitation (Ziemann *et al.* 1998b). The mean (+SD) percentage of the conditioned MEP to the control MEP is plotted against the interstimulus interval. MEPs are significantly facilitated at interstimulus intervals of 1.1–1.5, 2.3–2.9, and 4.1–4.4 ms (filled circles). Reproduced from Figure 1B in Ziemann *et al.* (1998b), with permission from the Blackwell Publishing Company.

activation of the neural elements involved in I-wave generation is responsible for SICF.

Epidural recordings of the descending corticospinal volley revealed that I2- and I3-waves are enhanced by paired TMS at ISIs of 1–1.4 ms, whereas the I1-wave remains unaffected. This strongly supports further the notion of an intracortical origin of this facilitation (Di Lazzaro et al. 1999b). Single motor unit recordings showed consistently that facilitation occurs predominantly at the I3-peak elicited by the conditioning stimulus in the peristimulus time histograms, and sometimes even at the I2-peak, but never at the I1-peak (Hanajima et al. 2002; Ilić et al. 2002).

These characteristics of the SICF, in particular its occurrence at discrete interstimulus intervals, are best explained by the assumption that this facilitation originates mainly nonsynaptically by means of the subthreshold test pulse directly activating the axon initial segments of those excitatory motor cortical interneurons which were made hyperexcitable by excitatory postsynaptic potentials elicited by the conditioning pulse (Hanajima et al. 2002; Ilić et al. 2002). The clinical utility of testing SICF is not yet clear but it might offer a way to measure the excitability of those interneurons that are involved in the generation of I-waves noninvasively.

Long-interval intracortical inhibition (LICI)

Paired-pulse TMS of the motor cortex with two suprathreshold stimuli given through the same stimulating coil results in inhibition of the test MEP at long ISIs of 50–200 ms (Claus et al. 1992; Valls-Sole et al. 1992). The interaction between the two stimuli depends on the ISI and on the stimulus intensity. At clearly suprathreshold stimulus intensities, MEPs to the second stimulus are facilitated at ISIs of 25–50 ms, and inhibited at ISIs of 60–150 ms (Valls-Sole et al. 1992). It was suggested that the inhibition at later intervals results from both spinal and supraspinal inhibitory mechanisms. Epidural recordings of the descending corticospinal volley showed that the later I-waves are inhibited at ISIs of 50–200 ms, whereas the D-wave is not affected (Nakamura et al. 1995, 1997). This strongly

suggests that the inhibition at long intervals is a predominantly intracortical phenomenon. This long-interval intracortical inhibition (LICI) seems to resemble the cortical silent period (cf. Chapter 10, this volume) although the two phenomena are not identical. Another epidural recording study of the descending corticospinal volley in patents with intractable pain also reported that I2- and later I-waves are suppressed at ISIs of 100 or 150 ms (Di Lazzaro et al. 2002). However, at an ISI of 50 ms, the I-waves are enhanced even though the MEP recorded by surface EMG is inhibited. Therefore, it can be concluded that, as for the cortical silent period (cf. Chapter 10, this volume), MEP inhibition at 50 ms is generated by a subcortical (probably spinal) mechanism that overrules supraspinal facilitation (Di Lazzaro et al. 2002).

It is very likely that LICI reflects cortical inhibition mediated through the $GABA_B R$ because baclofen, a specific $GABA_B R$ agonist, enhances LICI (McDonnell et al. 2006) (cf. Chapter 13, this volume). The different pharmacological profiles of SICI ($GABA_A$R-mediated) and LICI ($GABA_B$R-mediated) suggest that it is now possible to study distinct GABAergic inhibitory circuits in human cortex noninvasively by means of different paired-pulse TMS protocols. In addition, the interactions between these circuits can be tested by triple-pulse TMS protocols (cf. Chapter 12, this volume). These revealed that SICI is reduced in the presence of LICI, most likely through presynaptic $GABA_B$R-mediated inhibition of those inhibitory interneurons involved in SICI (Sanger et al. 2001).

LICI was tested in several studies in patients with various movement disorders. In a patient with vascular parkinsonism, LICI was pronounced with suppression of all waves of the descending corticospinal volley except the D-wave (Di Lazzaro et al. 2002). This supports previous studies that demonstrated significantly enhanced LICI in patients with Parkinson's disease, despite abnormal shortening of the cortical silent period (Berardelli et al. 1996; Valzania et al. 1997). It was suggested that the enhancement of LICI is a correlate of bradykinesia in these patients (Berardelli et al. 1996). In patients with paroxysmal kinesigenic dyskinesias, deficiency of LICI was the only abnormality among a broad array of TMS measures of motor cortical excitability (Kang *et al.* 2006). The duration of LICI was abnormally prolonged in patients with sporadic pure cerebellar ataxia (Tamburin *et al.* 2004).

Interhemispheric facilitation, interhemispheric inhibition, and ipsilateral silent period

Using two magnetic stimulating coils, it is possible to study the effects of a conditioning stimulus applied to the motor cortex of one hemisphere on the MEP elicited by a test stimulus to the contralateral motor cortex (Ferbert *et al.* 1992). The main effect is an inhibition of the test MEP if the conditioning stimulus precedes the test stimulus by some 7 ms or more (Ferbert *et al.* 1992). Cracco *et al.* (1989) showed that motor cortical stimulation evoked an electroencephalogram potential over the contralateral motor cortex with an onset latency of 8.8–12.2 ms. In patients with cortical myoclonus (Shibasaki *et al.* 1978; Wilkins *et al.* 1984; Brown *et al.* 1991), differences in the onset latency of jerks between homologous muscles on both sides were in the order of 10 ms. Therefore, the interval for the interhemispheric inhibition in the Ferbert protocol is consistent with the transcallosal latency reported in those papers. The inhibitory effect of the conditioning stimulus on test MEPs in an intrinsic hand muscle is topographically specific and becomes maximal when the conditioning coil is located over the hand area of the motor cortex (Ferbert *et al.* 1992). This inhibition does not occur if the test MEP is elicited by TES, or if H-reflexes are tested ipsilateral to conditioning stimulus (Ferbert *et al.* 1992). Therefore, it was proposed that the inhibition was produced at a cortical level and mediated by transcallosal motor fibers (interhemispheric inhibition: IHI). In epidural recordings of the descending corticospinal volley, I2- and I3-waves of the test response are inhibited by the conditioning stimulus over the opposite motor cortex, whereas D- and I1-waves remain unaffected (Di Lazzaro *et al.* 1999a). In addition, IHI is normal in stroke patients with a subcortical ischemic lesion of the corticospinal tract below the centrum semiovale sparing the transcallosal motor fibers (Boroojerdi

et al. 1996). These findings further support the notion that IHI occurs at the level of motor cortex, and is not mediated by an ipsilateral corticospinal projection but via a transcallosal route. It is thought that the mechanisms underlying IHI are similar to those involved in the suppression of tonic voluntary contraction elicited by TMS of the ipsilateral motor cortex. This phenomenon is referred to as the ipsilateral silent period (iSP) (Ferbert *et al.* 1992; Meyer *et al.* 1995). In patients with lesions, in particular in the posterior half of the trunk of the corpus callosum, or agenesis of the corpus callosum, the iSP is reduced or absent (Meyer *et al.* 1995, 1998). The IHI shows a maturational profile and can be obtained only in children after the age of 5 years, which further supports the notion of its mediation by transcallosal fibers that show a similar maturational profile of myelination (Heinen *et al.* 1998).

Magnitude of IHI, and iSP duration and iSP area increase with stimulus intensity (Chen *et al.* 2003). However, several other observations suggest that IHI and iSP are not mediated by the same mechanism. ISP shows directional preference (i.e. dependence on the current direction induced by the conditioning stimulus), whereas IHI does not do this to the same extent (Chen *et al.* 2003). Although IHI measures only the depth of inhibition, whereas iSP area measures both the depth and duration of inhibition, if these measures are mediated by the same mechanism, they can be expected to correlate. For IHI tested at an ISI of 8 ms, no correlation was found between iSP and IHI at any of the intensities and current directions tested (Chen *et al.* 2003). On the other hand, IHI tested at an ISI of 40 ms both at rest and during muscle activation significantly correlated with iSP duration for some of the stimulus intensities and current directions tested (Chen *et al.* 2003). These observations suggest that similar circuits may mediate IHI at 40 ms and iSP. In summary, IHI and iSP should be considered complementary rather than equivalent measures of interhemispheric inhibition.

At shorter ISIs of 4–5 ms, a weak interhemispheric facilitation of the test MEP can be obtained (Hanajima *et al.* 2001). This facilitation was observed consistently when the conditioning and test stimuli are applied strictly over the motor cortical representations of the target muscles in both hemispheres and their intensities are adjusted appropriately (Ugawa *et al.* 1993). Using the methods to elicit preferentially I1- or I3-waves with differently directed currents in the brain (Sakai *et al.* 1997), MEPs elicited predominantly by I3-waves are facilitated by a medially directed conditioning stimulus or anodal TES over the opposite motor cortex, whereas MEPs elicited predominantly by D-waves or I1-waves are not facilitated by the same conditioning stimulus (Hanajima *et al.* 2001). This suggests that the interhemispheric facilitation (IHF) also originates at the cortical level. Its timing is still compatible with an effect mediated by the corpus callosum because the facilitation affects predominantly I3-waves that take some 4.5 ms to reach the corticospinal neurons (Hanajima *et al.* 2001). The facilitation is visible only with low-intensity conditioning pulses just above AMT, but vanishes with higher-intensity conditioning stimuli while IHI becomes more prominent (Hanajima *et al.* 2001). This is in agreement with the notion of strong surround inhibition, which may render it difficult to obtain IHF in humans using transcranial stimulation. A similar problem was encountered in animal experiments (Asanuma and Okamoto 1959, 1962).

Testing of iSP duration has drawn considerable attention in neurologists who are interested in the neurophysiological assessment of neurodegenerative movement disorders. The iSP is normal in Parkinson's disease and multiple system atrophy, whereas it is abnormal (delayed iSP onset, prolonged iSP duration, or loss of iSP) in patients with progressive supranuclear palsy and corticobasal-ganglionic degeneration (Trompetto *et al.* 2003; Kühn *et al.* 2004; Wolters *et al.* 2004). These abnormalities in functional interhemispheric connectivity are associated with atrophy of the trunk of the corpus callosum in morphometric imaging (Trompetto *et al.* 2003; Wolters *et al.* 2004).

Recently, Mochizuki *et al.* (2004) reported interhemispheric interactions between the right dorsal premotor cortex and the left primary motor cortex. A conditioning stimulus over the dorsal premotor cortex at an intensity of either 90 or 110% RMT inhibits test MEPs in hand muscles elicited by stimulation of the contralateral primary motor cortex at ISIs of 8–10 ms.

This effect is different from IHI between the two primary motor cortices because it has a lower threshold and is not affected by voluntary contraction of the target muscle contralateral to the conditioning stimulus whereas this maneuver enhances IHI (Mochizuki *et al.* 2004). These findings may reflect the existence of commissural fibers between dorsal premotor cortex and contralateral primary motor cortex. It is possible that these play a role in bimanual coordination. The fascinating conclusion that emerges from these studies on interhemispheric interactions is that it is now possible by means of TMS protocols to chart long-range functional interhemispheric connectivity of remote areas of the human brain.

Modulation of motor cortical excitability by cerebellar stimulation

Ugawa *et al.* (1991) were the first to report that electrical stimulation over the back of the head at the level of inion can activate the cerebellum and induces a cerebellar modulating effect on excitability of the motor cortex. MEPs elicited by a test TMS stimulus over the primary motor cortex are inhibited by electrical stimulation of the contralateral cerebellar hemisphere at ISIs of 5–6 ms (Ugawa *et al.* 1991) (Figure 11.5). The same inhibition can be evoked by conditioning magnetic stimulation over the back of the head with a double-cone coil (Ugawa *et al.* 1995). Test MEPs elicited by TES of the motor cortex are not inhibited by cerebellar stimulation. This suggests that this form of inhibition (cerebellar inhibition, CBI) occurs at the level of the motor cortex.

Patients with ataxia of various etiologies were studied with this protocol (Ugawa *et al.* 1997). In patients with degenerative late-onset ataxia, reduction of CBI at ISIs of 5–7 ms correlates with the severity of ataxia. In patients with a lesion in the cerebellum or cerebello-thalamo-cortical pathway, CBI is reduced or absent. By contrast, in patients with a lesion in the afferent pathway to the cerebellum, CBI is normal despite the presence of ataxia clinically. These findings support the idea that CBI of motor cortex originates from activation of Purkinje cells,

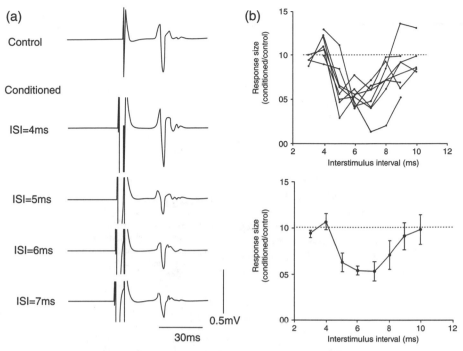

Fig. 11.5 Cerebellar inhibition (CBI) by electrical stimulation over the back of the head in normal subjects (Ugawa *et al.* 1995). (a) Motor-evoked potentials (MEPs) recorded from the voluntarily relaxed first dorsal interosseous muscle of one representative healthy subject. The top trace shows MEPs to the test stimulus alone, and the other traces are conditioned MEPs at interstimulus intervals (ISIs) of 4, 5, 6, and 7 ms. MEP sizes conditioned by cerebellar stimuli are smaller than those to a test stimulus given alone at ISIs of 5–7 ms. (b) The time course of CBI obtained from normal subjects (individual data, *upper panel*; average data and standard errors, *lower panel*). The ordinate indicates the ratio of the conditioned MEP size over the control MEP, and the abscissa the interstimulus interval (ISI). At ISIs of 5–7 ms, significant inhibition was obtained. Reproduced from Figure 1 in Ugawa *et al.* (1995), with permission from the Blackwell Publishing Company.

which inhibit a tonic facilitation of the motor cortex through the cerebello-thalamo-cortical pathway (Di Lazzaro *et al.* 1994; Ugawa *et al.* 1994, 1997). Another line of evidence supporting the notion that CBI is mediated by the cerebello-thalamo-cortical projection comes from patients with essential tremor who are treated with deep brain stimulation of the cerebellar thalamus. Deep brain stimulation of the ventralis intermedius (Vim) nucleus of the thalamus results in a specific enhancement of the inhibition from the cerebellum to the contralateral motor cortex without affecting excitability in the motor cortex *per se* (Molnar *et al.* 2004).

Stimulation over the back of the head may provoke other unwanted effects in addition to activation of inhibitory cerebellar output. It may result in activation of cervical spinal nerves at the C6/7 level in the brachial plexus that also induces suppression of the test MEP, but at slightly later ISIs of 7 or 8 ms, and lasting for some 5 ms (Werhahn *et al.* 1996). In patients with agenesis or large lesions of one cerebellar hemisphere, stimulation over the back of the head suppressed responses to motor cortical stimulation at ISIs of 8–10 ms (Meyer *et al.* 1994). These findings suggest that the suppression of the test MEP at these later intervals does not require cerebellar activation, and is probably mediated by activation of cervical spinal nerves or roots, antidromic activation of the corticospinal tract, or orthodromic activation of somatosensory tracts (Meyer *et al.* 1994). In order to escape this potential confounder, it is advisable to study CBI at ISIs of 5–6 ms.

In addition, a weak facilitatory effect of cerebellar electrical stimulation can be obtained on test MEPs elicited from the contralateral motor cortex by predominant activation of I3-waves (Iwata *et al.* 2004; Iwata and Ugawa 2005). This facilitation occurs at earlier intervals (ISI = 3 ms) than the cerebellar inhibition and only when high-voltage electrical stimuli are given with the anode on the ipsilateral mastoid process and a cathode over the contralateral process. Reversal of the polarity of the conditioning stimulus abolishes the facilitation. Based on the effective polarity of the conditioning stimulus and the time course of the facilitation, it was suggested that this effect is due to motor cortical facilitation (cerebellar facilitation, CBF) elicited by activation of the excitatory dentato-thalamo-cortical pathway at the deep cerebellar nuclei or superior cerebellar peduncle.

Modulation of motor cortical excitability by stimulation of premotor and supplementary motor cortex

Functional connectivity between the premotor cortex or supplementary motor cortex and the primary motor cortex can be studied by using two very small figure-8 stimulating coils (Civardi *et al.* 2001). Test MEPs elicited by TMS of the hand area of the primary motor cortex can be modified by conditioning stimuli delivered either 3–5 cm anterior to primary motor cortex (i.e. the premotor cortex) or in the midline 6 cm anterior to the vertex (i.e. the supplementary motor cortex). The test MEP is consistently inhibited if the conditioning stimulus is given 6 ms prior to the test stimulus and its intensity is set to 90% AMT (Civardi *et al.* 2001). By contrast, increasing the intensity of the conditioning stimulus to 120% AMT produces facilitation of the test MEP (Civardi *et al.* 2001). Conditioning stimulation over the premotor cortex does not affect test MEPs elicited by TES, or the H-reflex amplitude. Therefore, modulation of magnetic test MEPs is thought to occur at the level of motor cortex, and to be mediated by inputs from the premotor and supplementary cortex to the primary motor cortex. This protocol opens up a window to study functional connectivity between primary motor cortex and higher-order motor areas. This will be of substantial interest for the exploration of normal motor behavior, for instance preparation for movement, but also for studying patients with movement disorders where alterations in functional connectivity between these areas have long been suspected.

References

Abbruzzese G, Buccolieri A, Marchese R, Trompetto C, Mandich P, Schieppati M (1997). Intracortical inhibition and facilitation are abnormal in Huntington's disease: a paired magnetic stimulation study. *Neuroscience Letters* 228, 87–90.

Abbruzzese G, Assini A, Buccolieri A, Schieppati M, Trompetto C (1999). Comparison of intracortical inhibition and facilitation in distal and proximal arm muscles in humans. *Journal of Physiology* 514, 895–903.

Asanuma H Okamoto K (1959). Unitary study on evoked activity of callosal neurons and its effect on pyramidal tract cell activity on cats. *Japanese Journal of Physiology* 9, 437–483,

Asanuma H Okuda O (1962). Effects of transcallosal volleys on pyramidal tract cell activity of cat. *Journal of Neurophysiology* 25, 198–208

Ashby P, Reynolds C, Wennberg R, Lozano AM, Rothwell J (1999). On the focal nature of inhibition and facilitation in the human motor cortex. *Clinical Neurophysiology* 110, 550–555.

Bares M, Kanovsky P, Klajblova H, Rektor I (2003). Intracortical inhibition and facilitation are impaired in patients with early Parkinson's disease: a paired TMS study. *European Journal of Neurology* 10, 385–389.

Berardelli A, Rona S, Inghilleri M, Manfredi M. (1996). Cortical inhibition in Parkinson's disease. A study with paired magnetic stimulation. *Brain* 119, 71–77.

Boroojerdi B, Diefenbach K, Ferbert A (1996). Transcallosal inhibition in cortical and subcortical cerebral vascular lesions. *Journal of the Neurological Sciences* 144, 160–170.

Brown P, Day BL, Rothwell JC, Thompson PD, Marsden CD (1991). Intrahemispheric and interhemispheric spread of cerebral cortical myoclonic activity and its relevance to epilepsy. *Brain* 114, 2333–2351.

Buhmann C, Gorsler A, Bäumer T, *et al.* (2004). Abnormal excitability of premotor-motor connections in de novo Parkinson's disease. *Brain* 127, 2732–2746.

Bütefisch CM, Boroojerdi B, Chen R, Battaglia F, Hallett M (2005). Task-dependent intracortical inhibition is impaired in focal hand dystonia. *Movement Disorders* 20, 545–551.

Chen R, Garg R (2000). Facilitatory I wave interaction in proximal arm and lower limb muscle representations of the human motor cortex. *Journal of Neurophysiology* 83, 1426–1434.

Chen R, Tam A, Bütefisch C, et al. (1998). Intracortical inhibition and facilitation in different representations of the human motor cortex. *Journal of Neurophysiology 80*, 2870–2881.

Chen R, Yung D, Li JY (2003). Organization of ipsilateral excitatory and inhibitory pathways in the human motor cortex. *Journal of Neurophysiology 89*, 1256–1264.

Civardi C, Cantello R, Asselman P, Rothwell JC (2001). Transcranial magnetic stimulation can be used to test connections to primary motor areas from frontal and medial cortex in humans. *NeuroImage 14*, 1444–1453.

Claus D, Weis M, Jahnke U, Plewe A, Brunhölzl C (1992). Corticospinal conduction studied with magnetic double stimulation in the intact human. *Journal of the Neurological Sciences 111*, 180–188.

Connors BW, Malenka RC, Silva LR (1988). Two inhibitory postsynaptic potentials, and GABAA and GABAB receptor- mediated responses in neocortex of rat and cat. *Journal of Physiology 406*, 443–468.

Cracco RQ, Amassian VE, Maccabee PJ, Cracco JB (1989). Comparison of human transcallosal responses evoked by magnetic coil and electrical stimulation. *Electroencephalography and Clinical Neurophysiology 74*, 417–424.

Cunic D, Roshan L, Khan FI, Lozano AM, Lang AE, Chen R (2002). Effects of subthalamic nucleus stimulation on motor cortex excitability in Parkinson's disease. *Neurology 58*, 1665–1672.

Däuper J, Peschel T, Schrader C, et al. (2002). Effects of subthalamic nucleus (STN). stimulation on motor cortex excitability. *Neurology 59*, 700–706.

Di Lazzaro V, Molinari M, Restuccia D, Leggio MG, Nardone R, Fogli D, Tonali P (1994). Cerebro-cerebellar interactions in man: neurophysiological studies in patients with focal cerebellar lesions. *Electroencephalography and Clinical Neurophysiology 93*, 27–34.

Di Lazzaro V, Restuccia D, Oliviero A, et al. (1998). Magnetic transcranial stimulation at intensities below active motor threshold activates intracortical inhibitory circuits. *Experimental Brain Research 119*, 265–268.

Di Lazzaro V, Oliviero A, Profice P, et al. (1999a). Direct demonstration of interhemispheric inhibition of the human motor cortex produced by transcranial magnetic stimulation. *Experimental Brain Research 124*, 520–524.

Di Lazzaro V, Rothwell JC, Oliviero A, et al. (1999b). Intracortical origin of the short latency facilitation produced by pairs of threshold magnetic stimuli applied to human motor cortex. *Experimental Brain Research 129*, 494–499.

Di Lazzaro V, Oliviero A, Meglio M, et al. (2000). Direct demonstration of the effect of lorazepam on the excitability of the human motor cortex. *Clinical Neurophysiology 111*, 794–799.

Di Lazzaro V, Oliviero A, Mazzone P, et al. (2002). Direct demonstration of long latency cortico-cortical inhibition in normal subjects and in a patient with vascular parkinsonism. *Clinical Neurophysiology 113*, 1673–1679.

Di Lazzaro V, Oliviero A, Saturno E, et al. (2005a). Effects of lorazepam on short latency afferent inhibition and short latency intracortical inhibition in humans. *Journal of Physiology 564*, 661–668.

Di Lazzaro V, Pilato F, Dileone M, Tonali PA, Ziemann U (2005b). Dissociated effects of diazepam and lorazepam on short latency afferent inhibition. *Journal of Physiology 569*, 315–323.

Ferbert A, Priori A, Rothwell JC, Day BL, Colebatch JG, Marsden CD (1992). Interhemispheric inhibition of the human motor cortex. *Journal of Physiology 453*, 525–546.

Fisher RJ, Nakamura Y, Bestmann S, Rothwell JC, Bostock H (2002). Two phases of intracortical inhibition revealed by transcranial magnetic threshold tracking. *Experimental Brain Research 143*, 240–248.

Gilio F, Curra A, Lorenzano C, Modugno N, Manfredi M, Berardelli A (2000). Effects of botulinum toxin type A on intracortical inhibition in patients with dystonia. *Annals of Neurology 48*, 20–26.

Gilio F, Curra A, Inghilleri M, Lorenzano C, Suppa A, Manfredi M, Berardelli A (2003). Abnormalities of motor cortex excitability preceding movement in patients with dystonia. *Brain 126*, 1745–1754.

Hanajima R, Ugawa Y, Terao Y, Ogata K, Kanazawa I (1996). Ipsilateral cortico-cortical inhibition of the motor cortex in various neurological disorders. *Journal of the Neurological Sciences 140*, 109–116.

Hanajima R, Ugawa Y, Terao Y et al. (1998a). Cortico-cortical inhibition of the motor cortical area projecting to sternocleidomastoid muscle in normals and patients with spasmodic torticollis or essential tremor. *Electroencephalography and Clinical Neurophysiology 109*, 391–396.

Hanajima R, Ugawa Y, Terao Y et al. (1998b). Paired-pulse magnetic stimulation of the human motor cortex: differences among I waves. *Journal of Physiology 509*, 607–618.

Hanajima R, Ugawa Y, Terao Y, et al. (1999). Intracortical inhibition of the motor cortex is normal in chorea. *Journal of Neurology, Neurosurgery and Psychiatry 66*, 783–786.

Hanajima R, Ugawa Y, Machii K et al. (2001). Interhemispheric facilitation of the hand motor area in humans. *Journal of Physiology 531*, 849–859.

Hanajima R, Ugawa Y, Terao Y, et al. (2002). Mechanisms of intracortical I-wave facilitation elicited with paired-pulse magnetic stimulation in humans. *Journal of Physiology 538*, 253–261.

Hanajima R, Furubayashi T, Iwata NK, et al. (2003). Further evidence to support different mechanisms underlying intracortical inhibition of the motor cortex. *Experimental Brain Research 151*, 427–434.

Heinen F, Glocker FX, Fietzek U, Meyer BU, Lücking C-H, Korinthenberg R (1998). Absence of transcallosal inhibition following focal magnetic stimulation in preschool children. *Annals of Neurology 43*, 608–612.

Ilić TV, Meintzschel F, Cleff U, Ruge D, Kessler KR, Ziemann U (2002). Short-interval paired-pulse inhibition and facilitation of human motor cortex: the dimension of stimulus intensity. *Journal of Physiology 545*, 153–167.

Iwata NK, Ugawa Y (2005). The effects of cerebellar stimulation on the motor cortical excitability in neurological disorders: a review. *Cerebellum 4*, 218–223.

Iwata NK, Hanajima R, Furubayashi T, *et al.* (2004). Facilitatory effect on the motor cortex by electrical stimulation over the cerebellum in humans. *Experimental Brain Research 159*, 418–424.

Kang SY, Sohn YH, Kim HS, Lyoo CH, Lee MS (2006). Corticospinal disinhibition in paroxysmal kinesigenic dyskinesias. *Clinical Neurophysiology 117*, 57–60.

Kobayashi M, Theoret H, Mottaghy FM, Gangitano M, Pascual-Leone A (2001). Intracortical inhibition and facilitation in human facial motor area: difference between upper and lower facial area. *Clinical Neurophysiology 112*, 1604–1611.

Kossev AR, Siggelkow S, Dengler R, Rollnik JD (2003). Intracortical inhibition and facilitation in paired-pulse transcranial magnetic stimulation: effect of conditioning stimulus intensity on sizes and latencies of motor evoked potentials. *Clinical Neurophysiology 20*, 54–58.

Krnjevic K, Randic M, Straughan DW (1964). Cortical inhibition. *Nature 201*, 1294–1296.

Krnjevic K, Randic M, Straughan DW (1966). Pharmacology of cortical inhibition. *Journal of Physiology 184*, 78–105.

Kühn AA, Grosse P, Holtz K, Brown P, Meyer BU, Kupsch A (2004). Patterns of abnormal motor cortex excitability in atypical parkinsonian syndromes. *Clinical Neurophysiology 115*, 1786–1795.

Kujirai T, Caramia MD, Rothwell JC *et al.* (1993). Corticocortical inhibition in human motor cortex. *Journal of Physiology 471*, 501–519.

McDonnell MN, Orekhov Y, Ziemann U (2006). The role of GABA(B). receptors in intracortical inhibition in the human motor cortex. *Experimental Brain Research*, in press.

MacKinnon CD, Gilley EA, Weis-McNulty A, Simuni T (2005). Pathways mediating abnormal intracortical inhibition in Parkinson's disease. *Annals of Neurology 58*, 516–524.

Marchese R, Trompetto C, Buccolieri A, Abbruzzese G (2000). Abnormalities of motor cortical excitability are not correlated with clinical features in atypical Parkinsonism. *Movement Disorders 15*, 1210–1214.

Meyer BU, Röricht S, Machetanz J (1994). Reduction of corticospinal excitability by magnetic stimulation over the cerebellum in patients with large defects of one cerebellar hemisphere. *Electroencephalography and Clinical Neurophysiology 93*, 372–379.

Meyer BU, Röricht S, Gräfin von Einsiedel H, Kruggel F, Weindl A (1995). Inhibitory and excitatory interhemispheric transfers between motor cortical areas in normal humans and patients with abnormalities of the corpus callosum. *Brain 118*, 429–440.

Meyer BU, Röricht S, Woiciechowsky C. (1998). Topography of fibers in the human corpus callosum mediating interhemispheric inhibition between the motor cortices. *Annals of Neurology 433*, 60–69.

Mochizuki H, Huang YZ, Rothwell JC (2004). Interhemispheric interaction between human dorsal premotor and contralateral primary motor cortex. *Journal of Physiology 561*, 331–338.

Molnar GF, Sailer A, Gunraj CA, Lang AE, Lozano AM, Chen R (2004). Thalamic deep brain stimulation activates the cerebellothalamocortical pathway. *Neurology 63*, 907–909.

Muellbacher W, Boroojerdi B, Ziemann U, Hallett M (2001). Analogous corticocortical inhibition and facilitation in ipsilateral and contralateral human motor cortex representations of the tongue. *Journal of Clinical Neurophysiology 18*, 550–558.

Nakamura H, Kitagawa H, Kawaguchi Y, Tsuji H, Takano H, Nakatoh S (1995). Intracortical facilitation and inhibition after paired magnetic stimulation in humans under anesthesia. *Neuroscience Letters 199*, 155–157.

Nakamura H, Kitagawa H, Kawaguchi Y, Tsuji H (1997). Intracortical facilitation and inhibition after transcranial magnetic stimulation in conscious humans. *Journal of Physiology 498*, 817–823.

Orth M, Snijders AH, Rothwell JC (2003). The variability of intracortical inhibition and facilitation. *Clinical Neurophysiology 114*, 2362–2369.

Paradiso GO, Cunic DI, Gunraj CA, Chen R (2005). Representation of facial muscles in human motor cortex. *Journal of Physiology 567*, 323–336.

Pierantozzi M, Palmieri MG, Marciani MG, Bernardi G, Giacomini P, Stanzione P (2001). Effect of apomorphine on cortical inhibition in Parkinson's disease patients: a transcranial magnetic stimulation study. *Experimental Brain Research 141*, 52–62.

Pierantozzi M, Palmieri MG, Mazzone P, *et al.* (2002). Deep brain stimulation of both subthalamic nucleus and internal globus pallidus restores intracortical inhibition in Parkinson's disease paralleling apomorphine effects: a paired magnetic stimulation study. *Clinical Neurophysiology 113*, 108–113.

Priori A, Polidori L, Rona S, Manfredi M, Berardelli A (2000). Spinal and cortical inhibition in Huntington's chorea. *Movement Disorders 15*, 938–946.

Ridding MC, Inzelberg R, Rothwell JC (1995a). Changes in excitability of motor cortical circuitry in patients with Parkinson's disease. *Annals of Neurology 37*, 181–188.

Ridding MC, Sheean G, Rothwell JC, Inzelberg R, Kujirai T. (1995b). Changes in the balance between motor cortical excitation and inhibition in focal, task specific dystonia. *Journal of Neurology, Neurosurgery and Psychiatry 59*, 493–498.

Ridding MC, Taylor JL, Rothwell JC (1995c). The effect of voluntary contraction on cortico-cortical inhibition in human motor cortex. *Journal of Physiology 487*, 541–548.

Roshan L, Paradiso GO, Chen R (2003). Two phases of short-interval intracortical inhibition. *Experimental Brain Research 151*, 330–337.

Sakai K, Ugawa Y, Terao Y, Hanajima R, Furubayashi T, Kanazawa I (1997). Preferential activation of different I waves by transcranial magnetic stimulation with a figure-of-eight-shaped coil. *Experimental Brain Research 113*, 24–32.

Sanger TD, Garg RR, Chen R (2001). Interactions between two different inhibitory systems in the human motor cortex. *Journal of Physiology 530*, 307–317.

Schwenkreis P, Witscher K, Janssen F (1999). Influence of the N-methyl-D-aspartate antagonist memantine on human motor cortex excitability. *Neuroscience Letters 270*, 137–140.

Shibasaki H, Yamasaki H, Yamashita Y, Kuroiwa Y (1978). Electroencephalographic studies of myoclonus. Myoclonus-related cortical spikes and high amplitude somatosensory evoked potentials. *Brain 101*, 447–460.

Shimizu T, Filippi MM, Palmieri MG, *et al.* (1999). Modulation of intracortical excitability for different muscles in the upper extremity: paired magnetic stimulation study with focal versus non-focal coils. *Clinical Neurophysiology 110*, 575–581.

Siebner HR, Tormos JM, Ceballos-Baumann AO, Auer C, Catala MD, Conrad B, Pascual-Leone A (1999). Low-frequency repetitive transcranial magnetic stimulation of the motor cortex in writer's cramp. *Neurology 52*, 529–537.

Sommer M, Ruge D, Tergau F, Beuche W, Altenmuller E, Paulus W (2002). Intracortical excitability in the hand motor representation in hand dystonia and blepharospasm. *Movement Disorders 17*, 1017–1025.

Stinear CM, Byblow WD (2004a). Elevated threshold for intracortical inhibition in focal hand dystonia. *Movement Disorders 19*, 1312–1317.

Stinear CM, Byblow WD (2004b). Impaired modulation of intracortical inhibition in focal hand dystonia. *Cerebral Cortex 14*, 555–561.

Strafella AP, Valzania F, Nassetti SA, Tropeani A, Bisulli A, Santangelo M, Tassinari CA (2000). Effects of chronic levodopa and pergolide treatment on cortical excitability in patients with Parkinson's disease: a transcranial magnetic stimulation study. *Clinical Neurophysiology 111*, 1198–1202.

Stokic DS, McKay WB, Scott L, Sherwood AM, Dimitrijevic MR (1997). Intracortical inhibition of lower limb motor-evoked potentials after paired transcranial magnetic stimulation. *Experimental Brain Research 117*, 437–443.

Tamburin S, Fiaschi A, Marani S, Andreoli A, Manganotti P, Zanette G (2004). Enhanced intracortical inhibition in cerebellar patients. *Journal of the Neurological Sciences 217*, 205–10.

Tokimura H, Ridding MC, Tokimura Y, Amassian VE, Rothwell JC (1996). Short latency facilitation between pairs of threshold magnetic stimuli applied to human motor cortex. *Electroencephalography and Clinical Neurophysiology 101*, 263–272.

Trompetto C, Buccolieri A, Marchese R, Marinelli L, Michelozzi G, Abbruzzese G (2003). Impairment of transcallosal inhibition in patients with corticobasal degeneration. *Clinical Neurophysiology 114*, 2181–2187.

Ugawa Y, Day BL, Rothwell JC, Thompson PD, Merton PA, Marsden CD (1991). Modulation of motor cortical excitability by electrical stimulation over the cerebellum in man. *Journal of Physiology 441*, 57–72.

Ugawa Y, Hanajima R, Kanazawa I (1993). Interhemispheric facilitation of the hand area of the human motor cortex. *Neuroscience Letters 160*, 153–155

Ugawa Y, Hanajima R, Kanazawa I (1994). Motor cortex inhibition in patients with ataxia. *Electroencephalography and Clinical Neurophysiology 93*, 225–229.

Ugawa Y, Uesaka Y, Terao Y, Hanajima R, Kanazawa I (1995). Magnetic stimulation over the cerebellum in humans. *Annals of Neurology, 37*, 703–713.

Ugawa Y, Terao Y, Hanajima R, *et al.* (1997). Magnetic stimulation over the cerebellum in patients with ataxia. *Electroencephalography and Clinical Neurophysiology, 104*, 453–458.

Valls-Sole J, Pascual-Leone A, Wassermann EM, Hallett M (1992). Human motor evoked responses to paired transcranial magnetic stimuli. *Electroencephalography and Clinical Neurophysiology 85*, 355–364.

Valzania F, Strafella AP, Quatrale R, *et al.* (1997). Motor evoked responses to paired cortical magnetic stimulation in Parkinson's disease. *Electroencephalography and Clinical Neurophysiology 105*, 27–43.

Werhahn KJ, Taylor J, Ridding M, Meyer BU, Rothwell JC (1996). Effect of transcranial magnetic stimulation over the cerebellum on the excitability of human motor cortex. *Electroencephalography and Clinical Neurophysiology 101*, 58–66.

Wilkins DE Hallett M, Berardelli A, Walshe T, Alvarez N (1984) Physiologic analysis of the myoclonus of Alzheimer's disease. *Neurology 34*, 898–903.

Wolters A, Classen J, Kunesch E, Grossmann A, Benecke R (2004). Measurements of transcallosally mediated cortical inhibition for differentiating parkinsonian syndromes. *Movement Disorders 19*, 518–528.

Ziemann U (1999). Intracortical inhibition and facilitation in the conventional paired TMS paradigm. *Electroencephalography and Clinical Neurophysiology, 51*(Suppl), 127–136.

Ziemann U, Lönnecker S, Steinhoff BJ, Paulus W (1996a). The effect of lorazepam on the motor cortical excitability in man. *Experimental Brain Research 109*, 127–135.

Ziemann U, Rothwell JC, Ridding MC (1996b). Interaction between intracortical inhibition and facilitation in human motor cortex. *Journal of Physiology 496*, 873–881.

Ziemann U, Chen R, Cohen LG, Hallett M (1998a). Dextromethorphan decreases the excitability of the human motor cortex. *Neurology 51*, 1320–1324.

Ziemann U, Tergau F, Wassermann EM, Wischer S, Hildebrandt J, Paulus W (1998b). Demonstration of facilitatory I wave interaction in the human motor cortex by paired transcranial magnetic stimulation. *Journal of Physiology 511*, 181–190.

CHAPTER 12

Evaluating the interaction between cortical inhibitory and excitatory circuits measured by TMS

Zafiris J. Daskalakis and Robert Chen

Introduction

Transcranial magnetic stimulation was first introduced in the late 1980s, making noninvasive repeated cortical stimulation in humans possible. Since then numerous studies have used TMS as an investigational tool to elucidate cortical physiology and to probe cognitive processes. Recent studies have demonstrated that when TMS paradigms are combined, the changes induced can be used to delineate neuronal interactions that occur in the human motor cortex. This chapter will begin by briefly introducing the various TMS paradigms used to evaluate various inhibitory and excitatory circuits in the cortex, and then discuss studies which have evaluated the interaction of these circuits and information gathered on cortical neuronal connectivity.

TMS measures of intracortical inhibition

Intracortical inhibition refers to inhibitory circuits that arise within the same cortical area as the cortical targets for inhibition (cortical output neurons for motor cortex). Thus, the conditioning and test stimuli are applied to the same site. The ability to measure cortical inhibition is based on the principal that TMS can stimulate cortical inhibitory and excitatory interneurons in addition to corticospinal output neurons (Rothwell 1997; Hallett 2000). TMS paradigms that demonstrate intracortical inhibition include short-interval cortical inhibition (SICI) (Kujirai *et al.* 1993) (cf. Chapter 11, this volume), cortical silent period (cSP) (Cantello *et al.* 1992) (cf. Chapter 10, this volume) and long-interval cortical inhibition (LICI) (Valls-Sole *et al.* 1992) (cf. Chapter 11, this volume).

SICI involves a paired-pulse TMS paradigm with a subthreshold conditioning stimulus (CS) followed by a suprathreshold test stimulus (TS). A suprathreshold magnetic pulse activates cortical pyramidal neurons indirectly, via excitatory interneurons, leading to corticospinal output that can be measured peripherally as a motor-evoked potential (MEP) (cf. Chapter 9, this volume). In contrast, a low-intensity subthreshold pulse only excites cortical interneurons, and therefore does not result in an MEP. By combining a subthreshold pulse with a suprathreshold pulse, one can assess the inhibitory effects of interneurons on cortical output (Kujirai et al. 1993). When a subthreshold pulse precedes the test pulse by 1–5 ms, the activity of inhibitory interneurons can be measured and the MEP response is typically inhibited by 50–90%.

CSP experiments involve motor cortical stimulation superimposed on background EMG activity. At high stimulus intensities a cessation of all electromyographic activity occurs, producing a silent period. The first part of the silent period is, in part, due to spinal inhibition (Fuhr et al. 1991), but the second part of the silent period (>50 ms) is due to reduced cortical excitability (Cantello et al. 1992; Nakamura et al. 1997; Chen et al. 1999b).

LICI is also a paired-pulse TMS inhibitory paradigm. Unlike SICI, however, a suprathreshold CS is followed by a suprathreshold TS to produce inhibition at long interstimulus intervals (ISIs) of 50–200 ms (Valls-Sole et al. 1992). Preliminary evidence suggests that LICI and cSP are likely mediated by similar mechanisms related to gamma-aminobutyric acid B receptor ($GABA_BR$) mediated inhibitory neurotransmission. This is based on the finding that both measures of cortical inhibition are of similar duration (Wassermann et al. 1996) and both are facilitated by tiagabine, a GABA receptor uptake inhibitor (Werhahn et al. 1999). In contrast, tiagabine decreased SICI (Werhahn et al. 1999) that is thought to be mediated largely by $GABA_AR$ inhibitory activity (Ziemann et al. 1996) (cf. Chapter 13, this volume).

Cortical inhibition from stimulation of other brain areas

Interhemispheric inhibition (IHI)

Stimulation of the motor cortex ipsilateral to the target muscle by a CS inhibits the size of the MEP produced by a TS of the contralateral motor cortex given 6–15 ms later by ~50–75% (Ferbert et al. 1992). IHI also occurs at longer ISIs up to ~50 ms, and IHIs at short and long intervals are likely mediated by different mechanisms (Chen et al. 2003). IHI is likely mediated predominately by transcallosal inhibition

(Ferbert et al. 1992; Di Lazzaro et al. 1999), although subcortical circuits may also be involved (Gerloff et al. 1998) (cf. Chapter 11, this volume).

Cerebellar inhibition (CBI)

The cerebellum plays a major role in the planning, initiation, and organization of movement (Allen and Tsukahara 1974). These effects are mediated, in part, through its influence on the motor cortex and corticospinal outputs. Purkinje cells, the output neurons of the cerebellar cortex, have inhibitory connections with the deep cerebellar nuclei (DCN), which have a disynaptic excitatory pathway through the ventral thalamus to the motor cortex (Allen and Tsukahara 1974). Inhibitory Purkinje cell output results in a reduction of excitatory output from DCN to the motor cortex via the thalamus that leads to modification of motor control. Abnormalities of these pathways may result in ataxia or dysmetria of movement and possibly psychosis, considered by some to be a dysmetria of thought (Andreasen et al. 1999). Consequently, understanding the cerebellar cortical connectivity is imperative if we are to understand the mechanisms that play a role in the pathophysiology of these complex disorders.

Activity in the cerebello-thalamo-cortical pathway may be demonstrated noninvasively in humans. Electrical (Ugawa et al. 1991) or magnetic (Ugawa et al. 1995; Pinto and Chen 2001) stimulation of the cerebellum 5–7 ms prior to magnetic stimulation of the motor cortex inhibits the MEP produced by motor cortical stimulation. We will refer to this inhibition as cerebellar inhibition (CBI). Several lines of evidence suggest that CBI occurs at the level of the cerebral cortex (Granit and Phillips 1957; Ito et al. 1970; Uno et al. 1970; Phillips and Porter 1977; Ugawa et al. 1991, 1997). CBI can also be considered an index of cerebello-thalamo-cortical connectivity (cf. Chapter 11, this volume).

Motor cortical inhibition from afferent input

Inhibition of the motor cortex can also be induced through stimulation of peripheral nerves. For example, stimulation of the median nerve at the level of the wrist between 20 and 600 ms prior to TMS over the hand area of the contralateral motor cortex can produce a significant attenuation of the TMS-induced MEP (Chen *et al.* 1999a). At an ISI of 200 ms this form of inhibition has been shown to be mostly cortical in origin as spinal excitability remains largely unchanged (Chen *et al.* 1999a; Classen *et al.* 2000). This form of cortical inhibition following median nerve stimulation has been referred to as long-latency afferent inhibition (LAI) (Sailer *et al.* 2002, 2003).

Interactions between TMS inhibitory circuits

While there is considerable information on each of the inhibitory circuits assessed by TMS (cf. Chapter 11, this volume), few studies examined how they interact with each other. One way to investigate whether experimental phenomena (e.g. SICI, LICI, and IHI) share common mechanisms of action is to assess whether their profiles of response are similar or dissimilar under conditions of controlled perturbations. In experiments involving inhibitory interactions, this is achieved in two ways: first, by a controlled manipulation of TS intensities on measures such as SICI, LICI and IHI; second, by examining the impact of one inhibitory phenomenon on the other. This is accomplished by examining the interactions between SICI, LICI, IHI, CBI, LAI, and intracortical facilitation (ICF) (cf. Chapter 11, this volume) by activating two circuits at the same time. ICF is studied because it may interact with different inhibitory circuits (Daskalakis *et al.* 2004).

Simultaneous activation of two cortical inhibitory circuits

Since the timing or location of the CS for cortical inhibitory paradigms such as SICI, LICI, IHI, CBI, and LAI varies, there are opportunities to deliver two CS before the TS. Combining two CS and one TS (CS_1–CS_2–TS) provides valuable information regarding the effects of one inhibitory CS on the other. For example, the maximal inhibition of LICI occurs at an ISI of ~100 ms (Valls-Sole *et al.* 1992), whereas

maximal inhibition for IHI occurs at an ISI of ~10 ms (Ferbert *et al.* 1992). As such, when an LICI CS (CS_1) precedes an IHI CS (CS_2) the effects of LICI on IHI can be evaluated. Clarifying such interactions can provide information on how cortical circuits are related to each other and help to elucidate the neurophysiological underpinnings of these inhibitory/facilitatory paradigms. It may also help to interpret abnormalities reported in diseases and may be a new way of investigating the pathophysiology of neurological and psychiatric disorders.

In the following sections we will review several studies using the aforementioned approach to evaluate the interaction between these inhibitory and facilitatory paradigms (Sanger *et al.* 2001; Daskalakis *et al.* 2002b, 2004; Sailer *et al.* 2002; Kukaswadia *et al.* 2005). We will conclude by interpreting the results of these studies and discuss their meaning in terms of cortical physiology and connectivity.

Interactions between LICI and SICI

Sanger *et al.* (2001) explored the relationship between LICI and SICI/ICF. The first experiment evaluated how different test MEP amplitudes change LICI, SICI and ICF. It was demonstrated that LICI decreased with increased TS intensities (adjusted to result in test MEP amplitudes of ~0.2, 1, and 4 mV), with maximal inhibition at a test MEP amplitude of 0.2 mV and minimal inhibition with a test MEP of 4 mV. By contrast, SICI increased with increased TS intensities with maximal inhibition similar between test MEP amplitudes of 1 and 4 mV and minimal inhibition with a target test MEP of 0.2 mV. ICF was similar between test MEP amplitudes of 0.2 and 1 mV but decreased significantly at a test MEP amplitude of 4 mV. Since the responses to changes in test MEP amplitude are different for LICI and SICI, the inhibitory mechanisms mediating LICI and SICI are likely distinct.

The second experiment evaluated the interactions between LICI and SICI/ICF by activating these circuits either separately or together. One difficulty with determining whether an inhibitory circuit is increased or decreased in the presence of a second inhibitory circuit is that the degree of inhibition is influenced by the TS intensity and test MEP amplitude. Normally, the relationship between TS intensity and test MEP amplitude is straightforward with higher TS intensity leading to higher test MEP amplitude, at least in the range of MEP amplitudes used in these studies. However, with two inhibitory circuits activated simultaneously, the first circuit would decrease the test MEP amplitude upon which the second inhibitory circuit would act. In order to obtain the same test MEP amplitude with two circuits activated together compared to the second inhibitory circuit alone, the TS intensity needs to be increased to compensate for the inhibition from the first circuit. Since it is not possible to match both the TS intensity and MEP amplitude simultaneously, the experiments were designed to compare the second inhibition in the presence of the first inhibition to the second inhibition alone matched for test MEP amplitude and TS intensity in separate blocks of trials.

Therefore, two different TS intensities were used in order that the degree of SICI in the presence of LICI can be compared to SICI alone matched for both TS intensity and test MEP amplitude. It was demonstrated that in the presence of such a LICI, SICI is significantly reduced whether matched for TS intensity or MEP amplitude (Figure 12.1). These findings suggest that LICI inhibits SICI. This was supported by further experiments showing that progressively stronger LICI (elicited by stronger CS for LICI) results in a greater reduction in SICI. Moreover, a weak CS for LICI that does not suppress MEP amplitude also leads to a reduction of SICI, suggesting that LICI-mediated inhibition of SICI is due to a different mechanism from LICI-mediated MEP inhibition. ICF shows a nonsignificant increase in the presence of LICI.

A potential caveat to this approach of activating different inhibitory circuits simultaneously is the phenomenon of occlusion or saturation. This may occur if the same or overlapping populations of inhibitory interneurons are involved in two inhibitory paradigms (e.g. SICI and LICI). In the presence of one form of inhibition, fewer inhibitory interneurons would be available to be activated by the second inhibitory mechanism, leading to an apparent reduction in

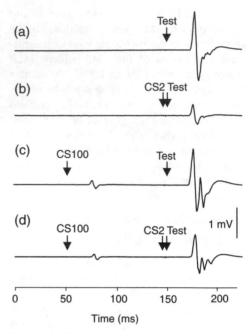

Fig. 12.1 Effects of long-interval cortical inhibition (LICI) on short-interval cortical inhibition (SICI) in a single subject. Traces represent the averaged waveform for a single subject.
(a) Response to test stimulus (TS) 1 mV alone.
(b) SICI alone: a conditioning stimulus (CS2) inhibited the test response compared with (a). The TS was the same as in (a). (c) LICI alone: a conditioning stimulus (CS100) using a TS that evokes a 1 mV motor-evoked potential (MEP) if preceded by a CS100 stimulus (i.e. TS 1 mV$_{CS100}$). The test MEP amplitude here is matched with that in (a). (d) Combined LICI and SICI: in the presence of CS100, the CS2 pulse caused little inhibition compared to that shown in (c). Adapted from Sanger *et al.* (2001).

the efficacy of the second inhibitory mechanism in the presence of the first one. This phenomenon is unlikely, given several observations. First, the different effects of TS intensities on SICI and LICI argue against the idea that the same or overlapping neuronal population mediates SICI and IHI. Second, the demonstration that weak LICI that does not suppress MEP amplitude also reduces SICI is against the occlusion model. Moreover, the MEP amplitude produced by the CS100 (100 ms before TS, eliciting LICI)–CS2 (2 ms before TS, eliciting SICI)–TS combination was higher than the MEP amplitude produced by the CS2–TS combination. This MEP facilitation due to the CS100 pulse cannot be explained by the occlusion model. That is, if the same population of inhibitory interneurons is shared by SICI and LICI, and CS100 (eliciting LICI) saturates all inhibitory interneurons, then lack of inhibition but not facilitation would be expected as the maximum occlusion effect.

From these experiments several conclusions were derived: first, that the inhibitory mechanisms mediating LICI are distinct from those mediating SICI; second, cortical neurons activated at low intensities are more susceptible to LICI whereas those activated at higher intensities are more inhibited by SICI. These findings are consistent with several additional lines of evidence suggesting that SICI and LICI are mediated by distinct inhibitory neurotransmitter systems. SICI is typically short lasting (1–20 ms) (Roshan et al. 2003) and is activated through subthreshold CS, whereas LICI is typically long lasting (50–200 ms) (Valls-Sole et al. 1992) and is typically activated though suprathreshold CS.

The third finding was that LICI inhibits SICI. This finding reinforces the association of these forms of cortical inhibition with different GABAergic receptor subtypes. Previous studies demonstrated that gamma-aminobutyric acid A receptor ($GABA_A R$) mediated inhibition is inhibited by $GABA_B$ activity (Davies et al. 1990). $GABA_B Rs$ are pre- and postsynaptic and the presynaptic receptors act to inhibit GABA release. The finding that LICI (likely mediated by $GABA_B Rs$, cf. Chapter 13, this volume) inhibits SICI is consistent with this interaction and provides a method to demonstrate this interaction in the motor cortex in human subjects.

This interaction may explain why some agents that increase inhibition in the cortex are also associated with a heightened risk of seizure. For example, baclofen, a $GABA_B R$ agonist used to treat several neurological disorders including spasticity and dystonia, has also been shown to lengthen the SP, a measure of cortical inhibition (Siebner et al. 1998). Treatment with baclofen has been associated with increased risk of seizures (Schuele et al. 2005). As seizures may be associated with reduced SICI (Delvaux et al. 2001) as well as reduced $GABA_A R$-mediated inhibition (Roberts 1986), it is possible that the

mechanism through which baclofen predisposes to seizures is through $GABA_B$-mediated attenuation of $GABA_A$ inhibitory neurotransmission (McDonnell et al. 2006). In addition, other centrally acting drugs may also act through this mechanism. For example, the mode of action of clozapine, an atypical antipsychotic, remains elusive. Clozapine is associated with significant sedation yet also a heightened risk of seizures (Iqbal et al. 2003). We recently demonstrated that clozapine treatment in patients with schizophrenia is associated with prolonged duration of the SP (Daskalakis et al. 2005), suggesting that (i) clozapine is associated with enhanced $GABA_B$ neurotransmission and (ii) the heightened risk of seizures associated with clozapine treatment is perhaps mediated through $GABA_B$ reduction of $GABA_A$ inhibitory activity. Whether or not facilitation of $GABA_B$-mediated inhibition is a mechanism through which clozapine exerts its unique antipsychotic effects (Kane et al. 1988), however, has yet to be determined.

Interactions between SICI, LICI, and LAI

Sailer et al. (2002) investigated how LAI interacts with different TMS-induced inhibitory processes. LAI is elicited by electrical stimulation of the median nerve at the wrist ~200 ms prior to suprathreshold TMS pulse of the contralateral motor cortex and results in a significant attenuation of the MEP. Previous studies demonstrated that deafferentation with ischemic nerve block or following extremity amputation results in reduced motor cortical inhibition (Sanes et al. 1990; Brasil-Neto et al. 1992; Chen et al. 1998; Ziemann et al. 1998), suggesting that changes in sensory input can alter cortical inhibitory circuits.

With increasing TS intensities, LICI and LAI are reduced but SICI and ICF were increased. When LAI and LICI were delivered together, their combined inhibitory effect is significantly reduced compared to the individual inhibitory effects, suggesting that LICI inhibits LAI or that LAI inhibits LICI. Since the interaction between LAI and LICI is related to the strength of baseline LAI but not LICI, it is likely that LAI inhibits LICI rather than LICI inhibiting LAI (Sailer et al. 2002).

When LAI was combined with SICI, it was found that SICI is not reduced in the presence of LAI and their effects are additive. This finding suggests that LAI and SICI exert cortical inhibition via independent mechanisms. This is consistent with previous studies demonstrating that sensory inputs to the motor cortex terminate primarily in superficial cortical layers (layer II), the same cortical region where the slow inhibitory postsynaptic potentials (IPSPs) that may mediate LICI are found. This anatomical distribution is different from that of fast IPSPs that may mediate SICI which are broadly distributed throughout all six cortical layers, and explains, in part, why there was no interaction between SICI and LAI.

Collectively, these results suggest that the inhibitory mechanisms mediating LAI are distinct from those mediating LICI or SICI, although LAI interacts with LICI to reduce its inhibitory effect.

Interactions between SICI, LICI, and IHI

While the aforementioned study demonstrated that LICI and SICI are likely mediated by distinct inhibitory systems, their interactions with several other processes that inhibit the human motor cortex had yet to be elucidated. Therefore, further experiments were conducted in an effort to evaluate the interactions between LICI, SICI, and IHI. It was anticipated that studies involving IHI would provide useful information regarding how homologous cortices influence one another. We evaluated the response of LICI, SICI, ICF, and IHI to three different test MEP amplitudes (Figure 12.2) (Daskalakis *et al.* 2002b). Similar to previous studies (Sanger *et al.* 2001), LICI decreases and SICI increases with higher test MEP amplitudes, with maximum SICI between test MEP amplitudes of 1 and 4 mV. There is little change in IHI between test MEP amplitudes of 0.2 and 1 mV but it is significantly reduced at a test MEP amplitude of 4 mV, similar to that demonstrated with LICI.

The second experiment evaluated the effects of a preceding IHI produced by a suprathreshold CS delivered to the opposite motor cortex on SICI and ICF. Again, two different TS intensities

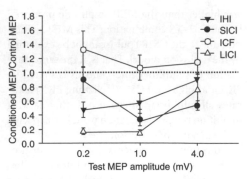

Fig. 12.2 Effects of increasing test stimulus (TS) intensity on cortical inhibition and facilitation. Data from 11 subjects. Each measure is expressed as a ratio (mean ± SE) of the conditioned motor-evoked potential (MEP) amplitude to the unconditioned MEP amplitude. Values <1 indicate inhibition; values >1 indicate facilitation. With increasing TS intensity, short-interval cortical inhibition (SICI) increased whereas long-interval cortical inhibition (LICI) and interhemispheric inhibition (IHI) decreased. Intracortical facilitation (ICF) showed no significant change. Adapted from Daskalakis *et al.* (2002b).

were used to allow comparison of SICI and ICF in the presence of IHI with SICI and ICF alone matched for test MEP amplitude and TS intensity in separate blocks of trials. IHI significantly reduces SICI, suggesting that IHI inhibits SICI (Daskalakis *et al.* 2002b), similar to the effects of LICI on SICI (Sanger *et al.* 2001) (Figure 12.3).

Several observations suggest that occlusion is unlikely to explain the interaction between IHI and SICI. First, SICI and IHI behave differently to changes in test MEP amplitude, suggesting that different neuronal populations mediate SICI and IHI. Second, the demonstration of MEP facilitation with the CCS10 (contralateral conditioning stimulus delivered 10 ms before TS, eliciting IHI)–CS2 (eliciting SICI)–TS pulse combination compared to the CS2–TS pulse combination in some subjects, cannot be explained by the occlusion model. Third, occlusion model predicts that subjects with greater IHI and SICI will have larger reduction of SICI in the presence of IHI. Although the change in SICI in the presence of IHI correlated with IHI, there was no correlation with SICI.

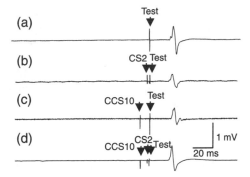

Fig. 12.3 Effects of interhemispheric inhibition (IHI) on short-interval cortical inhibition (SICI) in a single subject. These traces represent the averaged waveform form a single subject. In all traces the test stimulus (TS) intensity was adjusted to produce 1 mV motor-evoked potentials (MEPs) when preceded by a CCS10 (conditioning stimulus delivered to the opposite motor cortex 10 ms earlier). (a) Response to TS 1 mV$_{CCS10}$ alone. (b) SICI alone: the conditioning stimulus (CS2) inhibited the test MEP compared to (a). (c) IHI alone: the contralateral conditioning stimulus (CCS10) also inhibited the test response compared to (a). (d). Combined IHI and SICI: when the CCS10 preceded the CS2, CS2 led to facilitation rather than inhibition of the test MEP compared to (c). Adapted from Daskalakis et al. (2002b).

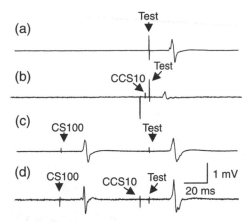

Fig. 12.4 Effects of long-interval cortical inhibition (LICI) on interhemispheric inhibition (IHI) in a single subject. Traces represent the averaged waveform for a single subject. (a) Response to test stimulus (TS) 1 mV alone. (b) IHI alone: a contralateral conditioning stimulus (CCS10) inhibited the test response compared to (a). The TS was the same as in (a). (c) LICI alone: a conditioning stimulus (CS100) using a TS that evokes a 1 mV motor-evoked potential (MEP) if preceded by a CS100 stimulus. The test MEP amplitude here is matched with that in (a). (d) Combined LICI and IHI: in the presence of CS100, the CCS10 pulse caused no inhibition but a slight MEP facilitation compared to that shown in (c). Adapted from Daskalakis et al. (2002b).

The third experiment evaluated the effects LICI on IHI and it was demonstrated that IHI is reduced in the presence of LICI (Figure 12.4). If LICI and IHI are mediated by similar inhibitory mechanisms, the neurons mediating LICI and IHI may inhibit one another. Single-cell recordings demonstrated a propensity for GABA$_B$ receptors to cause auto-inhibition (Rohrbacher et al. 1997). Therefore, cortical interneurons activated by LICI may result in a decreased inhibitory effect of IHI, perhaps through auto-inhibition through GABA$_B$ presynaptic auto-receptors.

These results also provide additional evidence that IHI occurs at the cortical level (Ferbert et al. 1992; Di Lazzaro et al. 1999), although it may be partially mediated by subcortical effects (Gerloff et al. 1998). Since both SICI (Kujirai et al. 1993; Nakamura et al. 1997; Di Lazzaro et al. 1998) and LICI (Nakamura et al. 1997; Chen et al. 1999b)

are cortically mediated phenomena, our finding that contralateral motor cortex stimulation significantly influences SICI and LICI suggests that IHI occurs predominately at a cortical level.

From this study, it was suggested that LICI and IHI may share common mechanisms for several reasons. First, the size of the test MEP has similar effects on LICI and IHI with both LICI and IHI decreasing with increasing test MEP size. Second, subthreshold conditioning stimuli are required to activate SICI inhibitory pathways, whereas suprathreshold conditioning stimuli are required to activate LICI and IHI inhibitory pathways, suggesting that the activation thresholds for neurons mediating LICI and IHI are higher than the activation thresholds for those mediating SICI. The relationship between IHI and LICI was further examined in the study of Kukaswadia et al. (2005) described below.

Interactions between LAI and IHI

Kukaswadia *et al.* (2005) examined the interactions between LAI and IHI. Since IHI at short ISIs (~10 ms, IHI10) are likely mediated by different mechanisms from those mediating IHI at longer ISIs (~40 ms, IHI40), both IHI10 and IHI40 were tested. If a similar population of inhibitory neurons mediates IHI and LICI, the interaction between LAI and IHI should be similar to the interaction between LICI and IHI (Daskalakis *et al.* 2002b).

With increasing test MEP amplitude, LAI, IHI10 and IHI40 all decrease. When testing the interactions between LAI and IHI, LAI results in a more marked decrease in IHI40 than at IHI10. The strength of LAI correlates with the degree of reduction of IHI40 but not IHI10. Moreover, when the strength of LAI is decreased, the inhibitory effect on IHI10 disappears whereas the inhibitory effect on IHI40 is still pronounced. It was concluded that LAI likely inhibits IHI40 whereas LAI and IHI10 probably do not directly inhibit each other. Therefore, the inhibitory mechanisms mediating LICI are closer to those mediating IHI40 than IHI10, as the interactions between LAI and IHI40 are similar to those between LAI and LICI (Sailer *et al.* 2002), whereas the interaction between LAI and IHI10 is not.

Physiological and pharmacological studies have suggested that LICI may be mediated by GABA$_B$Rs, whereas SICI probably involves GABA$_A$Rs (Roick *et al.* 1993; Siebner *et al.* 1998; Werhahn *et al.* 1999) (cf. Chapter 13, this volume). Since LICI and IHI40 have similar properties, IHI40 may also be related to GABA$_B$ activity. This is consistent with the finding that lorazepam increased SICI but did not change IHI, suggesting that IHI is not related to GABA$_A$ activity (Ziemann *et al.* 1996).

Interactions between SICI, LICI and CBI

The cerebellum plays a major role in the planning, initiation, and organization of movement (Allen and Tsukahara 1974). These effects are mediated, in part, through its influence on the motor cortex. Consequently, it is important to understand the cerebellar cortical connectivity. Therefore the connectivity between the cerebellum and the contralateral motor cortex in humans was investigated by examining how CBI interacts with cortical inhibitory and excitatory circuits (Daskalakis *et al.* 2004).

Increasing amplitudes of the test MEP result in less LICI, CBI, and ICF but greater SICI, suggesting that CBI and LICI may share similar inhibitory mechanisms. In the presence of CBI, SICI is reduced and ICF is increased (Figure 12.5). These changes are more pronounced in subjects with stronger CBI, suggesting that, similar to IHI, cerebellar stimulation reduces local inhibitory mechanisms in motor cortex, in addition to its influence on motor output neurons.

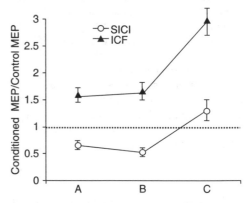

Fig. 12.5 Effects of cerebellar inhibition (CBI) on short-interval cortical inhibition (SICI) and intracortical facilitation (ICF). Data from 11 subjects. Both inhibition and facilitation are expressed as a ratio (mean ± SE) of the conditioned motor-evoked potential (MEP) amplitude to the unconditioned MEP amplitude. Values >1 represent facilitation; values <1 represent inhibition. Points above A represent SICI and ICF using a test stimulus (TS) that evokes a 0.5 mV MEP and points above B represent SICI and ICF with a TS that evokes a 0.5 mV MEP if preceded by a CCS5 stimulus (stimulation to the contralateral cerebellum 5 ms earlier, TS 0.5 mV$_{CCS5}$). Points above C demonstrate the combined approach in which a CS2 or CS10 are preceded by CCS5. Here the TS was 0.5 mV$_{CCS5}$. There was significantly less SICI and more ICF in the presence of CBI (C) compared with SICI and ICF in the absence of CBI (A and B). Adapted from Daskalakis *et al.* (2004).

Projections from the contralateral cerebellum and contralateral motor cortex were both shown to interact with inhibitory circuits in the motor cortex. Thus, projections of these distant CNS sites inhibit activity of both pyramidal and nonpyramidal neurons in the motor cortex. Previous studies suggests that cerebellar projections to the motor cortex via the thalamus terminate on both excitatory and inhibitory neurons (Noda et al. 1984; Ando et al. 1995; Holdefer et al. 2000). If TMS of the cerebellum activates the inhibitory Purkinje cells, the excitatory drive from the deep cerebellar nucleus (dentate and interpositus nuclei) to the motor cortex via the ventrolateral nucleus of the thalamus would be reduced. However, if the cerebello-thalamo-cortical pathway terminates on inhibitory neurons, as has been previously suggested (Na et al. 1997), it is anticipated that cerebellar stimulation would reduce local inhibitory mechanisms in motor cortex (Figure 12.6). This is consistent with the results of this experiment, which showed that cerebellar stimulation inhibits SICI. Moreover, in subjects with greater cerebellar inhibition, there is a more prominent reduction in SICI, lending further support to the hypothesis that cerebellar stimulation results in reduced local inhibitory activity in contralateral motor cortex.

In this study we also found that ICF is increased when the contralateral cerebellum is stimulated. Mediational statistics (for review see Baron and Kenny 1986) suggest that CBI-induced changes in ICF are mediated through changes in SICI. That is, CBI-induced change in ICF is mediated through the CBI-induced change in SICI. This finding suggests that ICF is not exclusively mediated by excitatory interneurons (Nakamura et al. 1997), but is related to a net balance between inhibition and excitation (cf. Chapters 11 and 13, this volume). In the presence of reduced inhibition (decreased SICI in the presence of CBI), excitatory circuits predominate, resulting in increased ICF.

The interaction between CBI and LICI leads to a significant reduction in CBI (Figure 12.7). The suprathreshold conditioning pulse for LICI can cause reduction of CBI through several possible pathways. One explanation is that the suprathreshold conditioning stimulus of LICI may disrupt the cerebello-thalamo-cortical

inhibitory pathway at the level of the motor thalamus. For example, cortical stimulation results in activation of reticular nuclei neurons and thalamic inhibitory neurons, which, in turn, inhibit the cerebello-thalamo-cortical pathway (Ando et al. 1995; Zhang and Jones 2004). Another possibility is that motor cortex stimulation from the CS100 pulse that elicits LICI influences the cerebellar cortex through activation of the mossy fiber system that arises from the pontine nuclei via the cortico-ponto-cerebellar pathway (Kelly and Strick 2003). Activation of this pathway may inhibit Purkinje cells through activation of inhibitory neurons (Golgi and basket cells) in the cerebellar cortex. Delivery of the cerebellar CS in the presence of inhibited Purkinje cells would produce the reduction of CBI seen in this experiment. Yet another possibility is that motor cortex stimulation results in decreased Purkinje cell inhibitory output through activation of the inferior olive (Schwarz and Welsh 2001). Here, the collaterals of the climbing fiber system from the inferior olive also innervate the Golgi (Schulman and Bloom 1981) and basket cells (Lemkey-Johnston and Larramendi 1968), leading to suppression of Purkinje cells.

Collectively, these data suggest that the cerebellum and contralateral motor cortex form reciprocal connections, with the cerebellum being able to influence both pyramidal and nonpyramidal neurons (interneurons) in the motor cortex while the motor cortex may, in turn, modulate cerebellar output.

Summary of the interactions between cortical inhibitory and excitatory circuits

The interactions between different cortical inhibitory and excitatory systems are summarized in Table 12.1. Several novel findings were elucidated in this series of experiments. First, these studies provided evidence that at least two distinct systems (SICI and LICI) coordinate inhibition in the motor cortex and that these two systems interact with one another, with the LICI acting to inhibit SICI. Second, additional evidence links these inhibitory paradigms to at least two types of GABAergic receptor-mediated

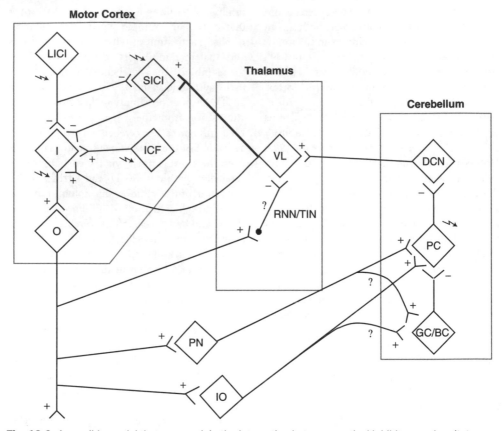

Fig. 12.6 A possible model that may explain the interaction between cortical inhibitory and excitatory circuits, and the cerebellum. Each diamond schematically represents a population of neurons mediating either inhibitory or facilitatory processes (i.e. short-interval cortical inhibition (SICI), long-interval cortical inhibition (LICI), intracortical facilitation (ICF)) or an anatomic location (i.e. deep cerebellar nuclei (DCN), ventrolateral nucleus of the thalamus (VL), inferior olive (IOP), cerebellar Purkinje cell (PC)). The diamond-labelled I represents cells leading to descending I-waves and O represents corticospinal output neurons. LICI is shown as inhibiting SICI based on the result of a previous study (Sanger *et al.* 2001). 'Bolts' represent the presumed site of TMS stimulation. The question-marks indicate pathways that may explain some of our experimental findings, but whether they are involved remains speculative. Lines in bold represent connections confirmed by these experimental findings. Our finding of reduced SICI in the presence of cerebellar inhibition (CBI) can be explained by activation of the PC leading to suppression of excitatory output from DCN). This results in suppression of excitatory output from the VL, leading to decreased excitatory drive to output neurons (causing decreased MEP amplitude) as well as inhibitory (SICI) interneurons (bold line). TMS-induced activation of corticospinal output neurons by the conditioning pulse for LICI may activate thalamic inhibitory neurons (TIN) or reticular nuclei neurons (RNN) which, in turn, inhibit thalamocortical neurons; this may account for the decreased CBI in the presence of LICI. Alternatively, activation of the mossy fibers that come from the pontine nuclei (PN) via the ponto-cerebellar pathway may inhibit Purkinje cells through activation of inhibitory Golgi cells (GC) and basket cells (BC). Another possibility is that cortical projection activates the inferior olive (IO) and the collaterals of the climbing fibers also innervate the inhibitory GC and BC that may also lead to decreased PC output. It must be remembered that that although these pathways exist, their involvement in these experimental paradigms remains speculative. Adapted from Daskalakis *et al.* (2002b).

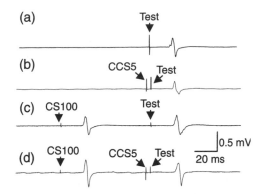

(a)

(b)

(c)

(d)

Fig. 12.7 Effects of long-interval cortical inhibition (LICI) on cerebellar inhibition (CBI) in a single subject. Traces represent the averaged waveform for a single subject. (a) Response to TS 0.5 mV alone. (b) CBI alone: a cerebellar conditioning stimulus (CCS5) inhibited the test response compared with (a). The TS was the same as in (a). (c) LICI alone: a conditioning stimulus (CS100) using a test stimulus (TS) that evokes a 0.5 mV motor-evoked potential (MEP) if preceded by a CS100 stimulus. The test MEP amplitude here is matched with that in (a). (d) Combined LICI and CBI: using both CS100 and CCS5 conditioning stimuli caused MEP facilitation compared to that shown in (b) and (c). Adapted from Daskalakis *et al.* (2002b).

inhibitory system, GABA$_A$ and GABA$_B$. Third, it was shown that distant brain areas such as the contralateral motor cortex and cerebellum form reciprocal connections with intracortical inhibitory systems such as SICI and LICI in the motor cortex. Fourth, we have confirmed that ICF is influenced by SICI. Finally, these studies suggest that LAI is mediated through inhibitory circuits that are different from SICI or LICI, but that LAI inhibits LICI.

Figure 12.8 shows a schematic representation of how the different systems may interact. It is hypothesized that SICI inhibits a set of neurons producing descending I-waves (indirect waves, cf. Chapter 14, this volume) via GABA$_A$Rs leading to MEP inhibition. LICI and IHI40 are mediated through a set of common inhibitory neurons and cause MEP inhibition via postsynaptic GABA$_B$Rs and inhibit SICI via presynaptic GABA$_B$Rs. LAI and IHI10 inhibit the I-wave-generating neurons via a different pathway, whereas CBI reduces MEP size and SICI due to reduction of facilitation. It should be noted that the connections shown in Figure 12.8 are preliminary and may need to be altered and expanded as soon as data from future studies become available.

Table 12.1 Summary of the known interactions between different cortical inhibitory and excitatory circuits

	First circuit			
	LICI	IHI10	LAI	CBI
Second circuit				
SICI	a	a	c	a
ICF	c	c	b	b
LICI	N/A	a	a	c
IHI10	a	N/A	c	?
IHI40	?	?	a	?
LAI	c	?	N/A	?
CBI	a	?	?	N/A

The effects of the first circuit on the second circuit are shown.
[a]Inhibition; [b]facilitation; [c]no interaction; ?, not studied; N/A, not applicable.
CBI, cerebellar inhibition; ICF, intracortical facilitation; IHI10, interhemispheric inhibition at interstimulus interval of 10 ms; IHI40, interhemispheric inhibition at interstimulus interval of 40 ms; LICI, long-interval intracortical inhibition at interstimulus interval of 100 ms; LAI, long latency afferent inhibition at interstimulus interval of 200 ms; SICI, short-interval intracortical inhibition at interstimulus interval of 2 ms.

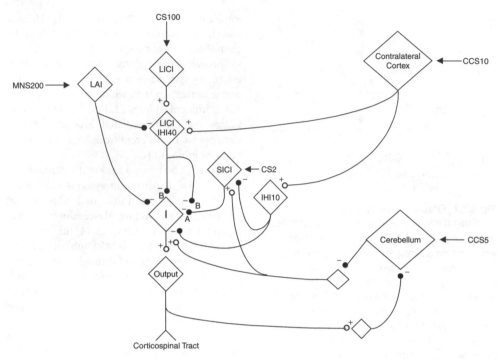

Fig. 12.8 Each diamond represents a population of neurons mediating an inhibitory or excitatory process. In this model it is suggested that at least two distinct systems mediate inhibition in the cortex. These are inhibitory systems related to long-interval cortical inhibition (LICI), likely mediated by gamma-aminobutyric acid (GABA)$_B$ receptors, and short-interval cortical inhibition (SICI), likely mediated by GABA$_A$ receptors. In addition, cortical inhibition may occur with cerebellum (CCS5) and contralateral motor cortex stimulation (interhemispheric inhibition (IHI) 10 and IHI40). LICI may be related to IHI40. Cortical inhibition associated with median nerve stimulation with an interstimulus interval of 200 ms (MNS200) (i.e. long-latency afferent inhibition (LAI)) inhibits LICI but does not interact with SICI. I represents a group of neurons generating I-waves. Output neurons represent descending corticospinal tract neurons.

Understanding disease states and mechanisms of therapeutic interventions

The study of the interactions among cortical inhibitory and excitatory circuits may help to elucidate pathophysiology of neurological and psychiatric diseases. For example, patients with schizophrenia demonstrated reduced inhibition in several different circuits (e.g. SICI, cSP, IHI) that may be normalized following treatment with antipsychotic medications (Daskalakis et al. 2002a). What remains unclear, however, is whether such inhibitory deficiencies are primary (due to reduction in the number or function of inhibitory interneurons) or secondary (e.g. due to excessive inhibition from another inhibitory system). Examining the interactions, such as the inhibitory effects of LICI on SICI, may help to investigate mechanisms through which medications may exert their therapeutic effects. Atypical antipsychotic medications (e.g. olanzapine), for example, have been shown to increase inhibition in the cortex in patients with schizophrenia (Daskalakis et al. 2002a, 2005; Fitzgerald et al. 2002), although in healthy human subjects typical neuroleptics (e.g. haloperidol) may have opposite effects (Ziemann et al. 1997). Future studies attempting to evaluate the effects of antipsychotic medications on the interaction between SICI and LICI, or between other forms of inhibitory processes, may yield important information on the dynamic relationship between these distinct inhibitory and excitatory circuits in such disease states.

In conclusion, the series of experiments was useful in evaluating the interactions between various cortical inhibitory and excitatory circuits and also for elucidating the underlying connectivity mediating such inhibition and facilitation in the human motor cortex. Future work using such an approach may also prove to be helpful in understanding the pathophysiology of diseases and the mechanisms through which various neuropharmacological agents exert their therapeutic effects.

References

Allen GI, Tsukahara N (1974). Cerebrocerebellar communication systems. *Physiology Review 54*, 957–1006.

Ando N, Izawa Y, Shinoda Y (1995). Relative contributions of thalamic reticular nucleus neurons and intrinsic interneurons to inhibition of thalamic neurons projecting to the motor cortex. *Journal of Neurophysiology 73*, 2470–2485.

Andreasen NC, Nopoulos P, O'Leary DS, Miller DD, Wassink T, Flaum M (1999). Defining the phenotype of schizophrenia: cognitive dysmetria and its neural mechanisms. *Biological Psychiatry 46*, 908–920.

Baron RM, Kenny DA (1986). The moderator-mediator variable distinction in social psychological research: conceptual, strategic, and statistical considerations. *Journal of Personal and Social Psychology 51*, 1173–1182.

Brasil-Neto JP, Cohen LG, Pascual-Leone A, Jabir FK, Wall RT, Hallett M (1992). Rapid reversible modulation of human motor outputs after transient deafferentation of the forearm: a study with transcranial magnetic stimulation. *Neurology 42*, 1302–1306.

Cantello R, Gianelli M, Civardi C, Mutani R (1992). Magnetic brain stimulation: the silent period after the motor evoked potential. *Neurology 42*, 1951–1959.

Chen R, Corwell B, Yaseen Z, Hallett M, Cohen LG (1998). Mechanisms of cortical reorganization in lower-limb amputees. *Journal of Neuroscience 18*, 3443–3450.

Chen R, Corwell B, Hallett M (1999a). Modulation of motor cortex excitability by median nerve and digit stimulation. *Experimental Brain Research 129*, 77–86.

Chen R, Lozano AM, Ashby P (1999b). Mechanism of the silent period following transcranial magnetic stimulation. Evidence from epidural recordings. *Experimental Brain Research 128*, 539–542.

Chen R, Yung D, Li JY (2003). Organization of ipsilateral excitatory and inhibitory pathways in the human motor cortex. *Journal of Neurophysiology 89*, 1256–1264.

Classen J, Steinfelder B Liepert J, *et al.* (2000). Cutaneomotor integration in humans is somatotopically organized at various levels of the nervous system and is task dependent. *Experimental Brain Research 130*, 48–59.

Daskalakis ZJ, Christensen BK, Chen R, Fitzgerald PB, Zipursky RB, Kapur S (2002a). Evidence for impaired cortical inhibition in schizophrenia using transcranial magnetic stimulation. *Archives of General Psychiatry 59*, 347–354.

Daskalakis ZJ, Christensen BK, Fitzgerald PB, Roshan L, Chen R (2002b). The mechanisms of interhemispheric inhibition in the human motor cortex. *Journal of Physiology 543*, 317–326.

Daskalakis ZJ, Paradiso GO, Christensen BK, Fitzgerald PB, Gunraj C, Chen R (2004). Exploring the connectivity between the cerebellum and motor cortex in humans. *Journal of Physiology 557*, 689–700.

Daskalis ZJ, Christensen BK, Fitzgerald PB, Möller BM, Fountain SI, Chen R (2007). Increased cortical inhibition in persons with schizophrenia treated with clozapine. *Journal of Psychopharmacology,* in press.

Davies CH, Davies SN, Collingridge GL (1990). Paired-pulse depression of monosynaptic GABA-mediated inhibitory postsynaptic responses in rat hippocampus. *Journal of Physiology 424*, 513–531.

Delvaux V, Alagona G, Gerard P, De Pasqua V, Delwaide PJ, Maertens de Noordhout A (2001). Reduced excitability of the motor cortex in untreated patients with de novo idiopathic 'grand mal' seizures. *Journal of Neurology, Neurosurgery and Psychiatry 71*, 772–776.

Di Lazzaro V, Restuccia D, Oliviero A, *et al.* (1998). Magnetic transcranial stimulation at intensities below active motor threshold activates intracortical inhibitory circuits. *Experimental Brain Research 119*, 265–268.

Di Lazzaro V, Oliviero A, Profice P, *et al.* (1999). Direct demonstration of interhemispheric inhibition of the human motor cortex produced by transcranial magnetic stimulation. *Experimental Brain Research 124*, 520–524.

Ferbert A, Priori A, Rothwell JC, Day BL, Colebatch JG, Marsden CD (1992). Interhemispheric inhibition of the human motor cortex. *Journal of Physiology 453*, 525–546.

Fitzgerald PB, Brown TL, Daskalakis ZJ, Kulkarni J (2002). A transcranial magnetic stimulation study of the effects of olanzapine and risperidone on motor cortical excitability in patients with schizophrenia. *Psychopharmacology (Berlin) 162*, 74–81.

Fuhr P, Agostino R, Hallett M (1991). Spinal motor neuron excitability during the silent period after cortical stimulation. *Electroencephalography and Clinical Neurophysiology 81*, 257–262.

Gerloff C, Cohen LG, Floeter MK, Chen R, Corwell B, Hallett M (1998). Inhibitory influence of the ipsilateral motor cortex on responses to stimulation of the human cortex and pyramidal tract. *Journal of Physiology 510*, 249–259.

Granit R, Phillips CG (1957). Effects on Purkinje cells of surface stimulation of the cerebellum. *Journal of Physiology 135*, 73–92.

Hallett M (2000). Transcranial magnetic stimulation and the human brain. *Nature 406*, 147–150.

Holdefer RN, Miller LE, Chen LL, Houk JC (2000). Functional connectivity between cerebellum and primary motor cortex in the awake monkey. *Journal of Neurophysiology 84*, 585–590.

Iqbal MM, Rahman A, Husain Z, Mahmud SZ, Ryan WG, Feldman JM (2003). Clozapine: a clinical review of adverse effects and management. *Annals of Clinical Psychiatry 15*, 33–48.

Ito M, Yoshida M, Obata K, Kawai N, Udo M (1970). Inhibitory control of intracerebellar nuclei by the purkinje cell axons. *Experimental Brain Research 10*, 64–80.

Kane J, Honigfeld G, Singer J, Meltzer H (1988). Clozapine for the treatment-resistant schizophrenic. A double-blind comparison with chlorpromazine. *Archives of General Psychiatry 45*, 789–796.

Kelly RM and Strick PL (2003). Cerebellar loops with motor cortex and prefrontal cortex of a nonhuman primate. *Journal of Neuroscience 23*, 8432–8444.

Kujirai T, Caramia MD, Rothwell JC, *et al.* (1993). Corticocortical inhibition in human motor cortex. *Journal of Physiology (London) 471*, 501–519.

Kukaswadia S, Wagle-Shukla A, Morgante F, Gunraj C and Chen R (2005). Interactions between long latency afferent inhibition and interhemispheric inhibitions in the human motor cortex. *Journal of Physiology 563*, 915–924.

Lemkey-Johnston N, Larramendi LM (1968). Types and distribution of synapses upon basket and stellate cells of the mouse cerebellum: an electron microscopic study. *Journal of Comparative Neurology 134*, 73–112.

McDonnell MN, Orekhov Y and Ziemann U (2006). The role of GABA(B) receptors in intracortical inhibition in the human motor cortex. *Experimental Brain Research 173*, 86–93.

Na J, Kakei S and Shinoda Y (1997). Cerebellar input to corticothalamic neurons in layers V and VI in the motor cortex. *Neuroscience Research 28*, 77–91.

Nakamura H, Kitagawa H, Kawaguchi Y and Tsuji H (1997). Intracortical facilitation and inhibition after transcranial magnetic stimulation in conscious humans. *Journal of Physiology 498*, 817–823.

Noda T, Yamamoto T (1984). Response properties and morphological identification of neurons in the cat motor cortex. *Brain Research 306*, 197–206.

Phillips CG and Porter R (1977). *Corticospinal neurons. Their role in movement*. London: Academic Press.

Pinto AD and Chen R (2001). Suppression of the motor cortex by magnetic stimulation of the cerebellum. *Experimental Brain Research 140*, 505–510.

Roberts E (1986). Failure of GABAergic inhibition: a key to local and global seizures. *Advances in Neurology 44*, 319–341.

Rohrbacher J, Jarolimek W, Lewen A, Misgeld U (1997). GABAB receptor-mediated inhibition of spontaneous inhibitory synaptic currents in rat midbrain culture. *Journal of Physiology 500*, 739–749.

Roick H, von Giesen HJ, Benecke R (1993). On the origin of the postexcitatory inhibition seen after transcranial magnetic brain stimulation in awake human subjects. *Experimental Brain Research 94*, 489–98.

Roshan L, Paradiso GO, Chen R (2003). Two phases of short-interval intracortical inhibition. *Experimental Brain Research 151*, 330–337.

Rothwell JC (1997). Techniques and mechanisms of action of transcranial stimulation of the human motor cortex. *Journal of Neuroscience Methods 74*, 113–22.

Sailer A, Molnar GF, Cunic DI, Chen R (2002). Effects of peripheral sensory input on cortical inhibition in humans. *Journal of Physiology (London) 544*, 617–629.

Sailer A, Molnar GF, Paradiso G, Gunraj CA, Lang AE, Chen R (2003). Short and long latency afferent inhibition in Parkinson's disease. *Brain 126*, 1883–1894.

Sanes JN, Suner S, Donoghue JP (1990). Dynamic organization of primary motor cortex output to target muscles in adult rats. I. Long-term patterns of reorganization following motor or mixed peripheral nerve lesions. *Experimental Brain Research 79*, 479–491.

Sanger TD, Garg RR, Chen R (2001). Interactions between two different inhibitory systems in the human motor cortex. *Journal of Physiology 530*, 307–317.

Schuele SU, Kellinghaus C, Shook SJ, Boulis N, Bethoux FA, Loddenkemper T (2005). Incidence of seizures in patients with multiple sclerosis treated with intrathecal baclofen. *Neurology 64*, 1086–1087.

Schulman JA, Bloom FE (1981). Golgi cells of the cerebellum are inhibited by inferior olive activity. *Brain Research 210*, 350–355.

Schwarz C, Welsh JP (2001). Dynamic modulation of mossy fiber system throughput by inferior olive synchrony: a multielectrode study of cerebellar cortex activated by motor cortex. *Journal of Neurophysiology 86*, 2489–2504.

Siebner HR, Dressnandt J, Auer C, Conrad B (1998). Continuous intrathecal baclofen infusions induced a marked increase of the transcranially evoked silent period in a patient with generalized dystonia. *Muscle and Nerve 21*, 1209–1212.

Ugawa Y, Day BL, Rothwell JC, Thompson PD, Merton PA, Marsden CD (1991). Modulation of motor cortical excitability by electrical stimulation over the cerebellum in man. *Journal of Physiology 441*, 57–72.

Ugawa Y, Uesaka Y, Terao Y, Hanajima R, Kanazawa I (1995). Magnetic stimulation over the cerebellum in humans. *Annals of Neurology 37*, 703–713.

Ugawa Y, Terao Y, Hanajima R, *et al.* (1997). Magnetic stimulation over the cerebellum in patients with ataxia. *Electroencephalography and Clinical Neurophysiology 104*, 453–458.

Uno M, Yoshida M, Hirota I (1970). The mode of cerebello-thalamic relay transmission investigated with intracellular recording from cells of the ventrolateral nucleus of cat's thalamus. *Experimental Brain Research 10*, 121–139.

Valls-Sole J, Pascual-Leone A, Wassermann EM, Hallett M (1992). Human motor evoked responses to paired transcranial magnetic stimuli. *Electroencephalography and Clinical Neurophysiology 85*, 355–364.

Wassermann EM, Samii A, Mercuri B, *et al.* (1996). Responses to paired transcranial magnetic stimuli in resting, active, and recently activated muscles. *Experimental Brain Research 109*, 158–163.

Werhahn KJ, Kunesch E, Noachtar S, Benecke R, Classen J (1999). Differential effects on motorcortical inhibition induced by blockade of GABA uptake in humans. *Journal of Physiology (London) 517*, 591–597.

Zhang L, Jones EG (2004). Corticothalamic inhibition in the thalamic reticular nucleus. *Journal of Neurophysiology 91*, 759–766.

Ziemann U, Lonnecker S, Steinhoff BJ, Paulus W (1996). The effect of lorazepam on the motor cortical excitability in man. *Experimental Brain Research 109*, 127–135.

Ziemann U, Tergau F, Bruns D, Baudewig J, Paulus W (1997). Changes in human motor cortex excitability induced by dopaminergic and anti-dopaminergic drugs. *Electroencephalography and Clinical Neurophysiology 105*, 430–437.

Ziemann U, Hallett M, Cohen LG (1998). Mechanisms of deafferentation-induced plasticity in human motor cortex. *Journal of Neuroscience 18*, 7000–7007.

CHAPTER 13

Pharmacology of TMS measures

Ulf Ziemann

Summary – Application of a single dose of a CNS-active drug with a defined mode of action has been shown useful to explore pharmaco-physiological properties of TMS measures of motor cortical excitability. With this approach, it was possible to demonstrate that TMS measures reflect axonal, or excitatory or inhibitory synaptic excitability in distinct interneuron circuits. On the other hand, the array of pharmaco-physiologically well-defined TMS measures can now be employed to study the effects of a drug with unknown or multiple modes of action, and hence to determine its main mode of action at the systems level of motor cortex. Acute drug effects may be rather different from chronic drug effects, and these differences can also be studied in TMS experiments. Finally, TMS or repetitive TMS may induce changes in endogenous neurotransmitter or neuromodulator systems. This offers the opportunity to study neurotransmission along defined neuronal projections in health and disease. All these aspects of the pharmacology of TMS measures will be reviewed in this chapter.

Effects of CNS-active drugs with a defined mode of action on TMS measures of motor cortical excitability

All studies reviewed in the following sections were designed to compare TMS measures obtained at one or several time points after application of a single dose of the study drug with measures at baseline, i.e. before drug intake. Some studies employed a randomized placebo-controlled blinded parallel or crossover design to minimize experimenter bias. This chapter will review drug effects separately for various TMS measures of motor cortical and corticospinal excitability. Almost all reviewed studies were obtained in healthy subjects.

Motor threshold

Motor threshold is most often defined as the minimum intensity that is necessary to elicit a small (usually >50 μV) motor-evoked potential (MEP) in the target muscle in at least half of the trials (Rossini et al. 1999). Motor threshold is lower in the voluntarily contracting muscle (active motor threshold, AMT) compared to the resting muscle (resting motor threshold, RMT), usually by ~10% of maximum stimulator output (Devanne et al. 1997). It may be expected that motor threshold depends on the excitability of those elements which are activated by TMS. These are cortico-cortical axons (Amassian et al. 1987; Shimazu et al. 2004) and their excitatory synaptic contacts with the corticospinal neurons. Voltage-gated sodium channels are crucial in regulating axon excitability (Hodgkin and Huxley 1952), while ionotropic non-N-methyl-D-aspartate (NMDA) glutamate (GLU) receptors are responsible for fast excitatory synaptic neurotransmission in neocortex (Douglas and Martin 1998).

Accordingly, drugs which block voltage-gated sodium channels, in particular anticonvulsants such as carbamazepine (Ziemann *et al*. 1996c), oxcarbazepine (Kimiskidis *et al*. 2005), phenytoin (Mavroudakis *et al*. 1994; Chen *et al*. 1997), lamotrigine (Ziemann *et al*. 1996c; Boroojerdi *et al*. 2001; Tergau *et al*. 2003; Li *et al*. 2004), and losigamone (Ziemann *et al*. 1996c) elevate motor threshold (Table 13.1). Preliminary data support this notion by showing that the AMPA receptor antagonist talampanel increases motor threshold (Danielsson *et al*. 2004). The increase in motor threshold correlates with the serum level of the study drug (Chen *et al*. 1997; Tergau *et al*. 2003). In contrast, the NMDA receptor (NMDAR) antagonist ketamine increases neurotransmission indirectly through the non-NMDA AMPA receptor and, possibly through this action, decreases motor threshold (Di Lazzaro *et al*. 2003) (Table 13.1). Acute pharmacological modulation of other neurotransmitter (gamma-aminobutyric acid, GABA) and neuromodulator systems (dopamine, DA; norepinephrine, NE; serotonin, 5-HT; acetylcholine, ACh) does not produce consistent effects on motor threshold (Table 13.1).

Motor evoked potential (MEP) amplitude

MEP amplitude increases with stimulus intensity in a sigmoid fashion (Hess *et al*. 1987; Devanne *et al*. 1997) (cf. Chapter 9, this volume). At low stimulus intensity, the corticospinal volley resulting in the MEP often consists of only one single wave (I1-wave if the current induced by TMS in the brain runs in posterior–anterior direction), while the corticospinal volley becomes more complex and consists of multiple I-waves (I2–I4 in addition to I1) at higher stimulus intensity (Di Lazzaro *et al*. 2004) (cf. Chapter 14, this volume). In contrast to the I1-wave, the later I-waves are modifiable by many processes. Most likely, they are produced by activation of a chain of cortical excitatory interneurons (Amassian *et al*. 1987; Ziemann and Rothwell 2000). Accordingly, it may be expected that neurotransmitters (GLU, GABA) and neuromodulators (DA, NE, 5-HT, ACh) influence MEP amplitude, particularly in the high-amplitude MEP range. In order

to test a drug effect on MEP amplitude, it is important to record an MEP intensity curve to obtain information on the full range of MEP amplitudes.

Positive allosteric modulators of the GABA$_A$ receptor (GABA$_A$R), i.e. benzodiazepines (Schönle *et al*. 1989; Boroojerdi *et al*. 2001; Ilić *et al*. 2002b; Kimiskidis *et al*. 2006; Mohammadi *et al*. 2006) (Figure 13.1) and barbiturates (Inghilleri *et al*. 1996), the DA agonist cabergoline (Korchounov *et al*. 2007), and the NE antagonist guanfacine (Korchounov *et al*. 2003), depress high-amplitude but not low-amplitude MEPs (Table 13.1). The lack of effect of benzodiazepines on low-amplitude MEPs is consistent with findings of epidural recordings of the descending corticospinal volley that show a depression of late I-waves (I2–I4-waves) but not the I1-wave after a single dose of lorazepam (Di Lazzaro *et al*. 2000a). In contrast, the DA antagonist haloperidol (our unpublished observation), various NE agonists (methylphenidate, D-amphetamine, reboxetine, yohimbine) (Boroojerdi *et al*. 2001; Plewnia *et al*. 2001, 2002, 2004; Ilić *et al*. 2003), serotonin reuptake inhibitors (Ilić *et al*. 2002a; Gerdelat-Mas *et al*. 2005), and the muscarinic receptor (M1) antagonist scopolamine (Di Lazzaro *et al*. 2000b) increase MEP amplitude (Table 13.1). These effects are most likely explained by complex and hitherto only incompletely understood modulating effects of DA, NE, 5-HT, and ACh on inhibitory and excitatory synaptic transmission in neocortical neuronal networks (Hasselmo 1995). In several instances, changes in MEP amplitude occur without significant changes in motor threshold (Table 13.1), supporting the notion of an important difference in the mechanisms underlying motor threshold and high-amplitude MEPs (see above).

Cortical silent period (cSP)

The cSP (cf. Chapter 10, this volume) refers to a TMS-induced interruption of voluntary activity in the electromyogram of the target muscle. cSP duration increases with stimulus intensity and can reach a duration of 200–300 ms in hand muscles (Cantello *et al*. 1992; Kimiskidis *et al*. 2005, 2006). Whereas spinal inhibition contributes to the early part of the cSP (its first 50–75 ms), the late part originates in

Table 13.1 Acute drug effects on TMS measures of motor cortical excitability

Mode of action	Drugs	MT	MEP	cSP	SICI	ICF	SICF	LICI	SAI
Blockade of voltage-gated Na⁺ channels	Carbamazepine	▲○	○	○	○○	○▼	○		
	Oxcarbazepine	▲	○						◀
	Phenytoin	▲▲	○○	○○					
	Lamotrigine	▲▲▲▲	○	○	○○	○○	○		
	Losigamone	▲	○	○	○				
GABA_AR agonists	Diazepam	○○○○	▲▼▼	▼○	▲▲▲○	▼▼	▼	○	
	Lorazepam	○○○○○	▼▼	◀*	▲▲▲▲○	▼▼○	▼	▶▶	
	Zolpidem	○○	▼	◀	○◀	▶	◀		
	Midazolam	▼	▼						
	Thiopental	○		○					
	Phenobarbital	▼							
	Ethanol	○	○	◀	◀	▶	▶		
	Progesterone	○	◀	○					
	Vigabatrin	○○	○○	○○○◀	○○	▶○		◀	
	Tiagabine	○	○	◀	▶	◀	◀		
GABA_BR agonist	Baclofen	○○○	○○	○○○◀	◀▼	▶	○	◀	
GABA_AR antagonist	Flumazenil	○	○	○	○	○			
NMDA antagonists	Dextromethorphan	○	○	○	◀	▶▼			
	Memantine	○○	○○	○	◀	▶	○		
	Ketamine	▼	◀	○	○	○			
Anti-glutamatergic	Riluzole	○○	○○	○○	◀○	▶▶			
DA precursor	L-dopa	○	○	◀	○	○			

Table 13.1 (Cont.) Acute drug effects on TMS measures of motor cortical excitability

Mode of action	Drugs	MT	MEP	cSP	SICI	ICF	SICF	LICI	SAI
DA agonists	Bromocriptine	○	▲	○	◄				
	Pergolide	○	○	◄	◄	○			
	Cabergoline	○	▶	○	◄	▶	▶		
MAO-B inhibitor	Selegeline	○	○	○					
DA antagonists	Haloperidol	○○	▲	○	▼○	◄			
	Sulpiride	○	○	○					
DA/5HT₂A antagonist	Olanzapine	○	○	○					
NE agonists	Methylphenidate	○○	▲	○	▼▼○	◄◄○	◄		
	D-Amphetamine	○	▲	○	◄				
	Reboxetine	○○▼	▲	◄	▼○	◄◄◄			
	Atomoxetine	▼	▲						
α2R antagonist	Yohimbine	○	▲	○	◄				
α1R antagonist	Prazosin	○	○	○	○				
α2R agonist	Guanfacine	○	▶	◄	▶	▶			
SSRI	Sertraline	○	▲	○	○	▶	○		
	Paroxetine	○	▲	○	▶				
	Citalopram	○◄	○	◄	◄◄	○○			
5-HT₁B/₁D agonist	Zolmitriptan	○	○	○	▶	○			
ACh esterase inhibitor	Tacrine	○	○	○	▶	◄	◄		
M1/M2 antagonist	Atropine	○	○	○	▶	◄			
M1 antagonist	Scopolamine	▼	▲	○	○	○	▶		

MT, motor threshold; MEP, motor-evoked potential; cSP cortical silent period; SICI, short-interval intracortical inhibition; ICF, intracortical facilitation; SICI, short-interval intracortical inhibition; LICI, long-interval intracortical inhibition; SAI, short-latency afferent inhibition.
○, no effect; ▼, decrease; ▲, increase.
*Effect varies with stimulus intensity. For references of drug effects on TMS measures, see main text.

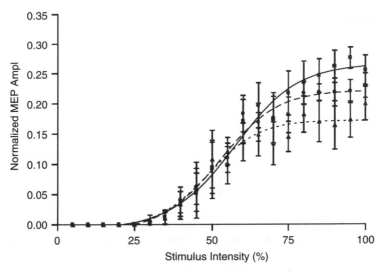

Fig. 13.1 Motor-evoked potential (MEP) intensity curves from the resting first dorsal interosseous muscle (means ± SD, 12 healthy subjects) at baseline (filled squares), under lorazepam (filled triangles, 0.045 mg/kg bolus + 2.6 µg/kg/h infusion), and followed by flumazenil administration (filled circles, 1 mg infusion of 3 min). MEP amplitude is normalized to the maximum M-wave obtained by supramaximal electrical stimulation of the ulnar nerve at the wrist. Boltzmann curves are fitted to the MEP intensity curves. Note significant depression of high- but not low-amplitude MEPs by lorazepam, and partial recovery by flumazenil. Reproduced with permission from Kimiskidis et al. (2006).

supraspinal structures, most likely in the motor cortex (Fuhr et al. 1991; Inghilleri et al. 1993; Ziemann et al. 1993; Chen et al. 1999). The duration of the cSP is consistent with the duration of the inhibitory postsynaptic potential (IPSP) elicited by $GABA_B R$ activation in pyramidal cells of animal preparations (Connors et al. 1988). Hence, it was proposed that the late part of the cSP is caused by a long-lasting cortical inhibition mediated by the $GABA_B R$. This notion is supported indirectly by the lengthening of the cSP by the GABA reuptake inhibitor tiagabine (Werhahn et al. 1999) and, in one study, by vigabatrin, an inhibitor of the GABA-degrading enzyme GABA transaminase (Pierantozzi et al. 2004). The resulting increase of GABA availability in the synaptic cleft is, however, not specific evidence in favor of a $GABA_B R$-mediated inhibition of the cSP because $GABA_A Rs$ are activated in addition. The effects of benzodiazepines, positive allosteric modulators at the $GABA_A R$, have been controversial. Both lengthening (Ziemann et al. 1996b) and shortening effects (Inghilleri et al. 1996) on cSP duration have been reported. One recent

study resolved this disparity by showing that the effect of the benzodiazepine lorazepam on cSP duration depends on stimulus intensity (Kimiskidis et al. 2006). Lorazepam lengthens cSP duration slightly but significantly if tested at low stimulus intensity, and shortens it if tested at high stimulation intensity (Figure 13.2). These effects were partially reversed by subsequent application of the benzodiazepine antagonist flumazenil (Figure 13.2). These data support the idea that cSP duration at the lower range of stimulus intensities reflects mainly the activation of $GABA_A Rs$, whereas it reflects mainly the activation of $GABA_B Rs$ at the higher range of stimulus intensities. Enhancement of neurotransmission through $GABA_A Rs$ by benzodiazepines can lead to depression of $GABA_B R$-mediated IPSP in slice experiments (Huguenard and Prince 1994). It is likely that this mechanism contributes to the observed shortening of cSP duration by benzodiazepines at the high range of stimulus intensities (Inghilleri et al. 1996; Kimiskidis et al. 2006). These findings clearly demonstrate that the neurophysiological mechanisms underlying the cSP

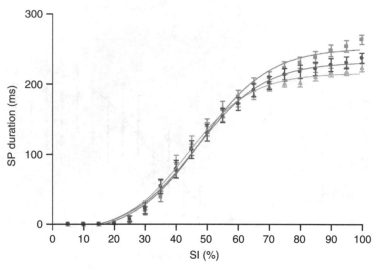

Fig. 13.2 Cortical silent period (cSP) intensity curves from the contracting first dorsal interosseous muscles (mean ± SD, 12 healthy subjects) at baseline (gray squares), under lorazepam (gray triangles, 0.045 mg/kg bolus + 2.6 μg/kg/h infusion), and followed by flumazenil administration (black circles, 1 mg infusion of 3 min). cSP duration is given in milliseconds, stimulus intensity in percentage of maximum stimulator output. Boltzmann curves are fitted to the cSP intensity curves. Note slight lengthening of cSP duration at low stimulus intensity (in the range of 30–50% maximum stimulator output) but shortening at high stimulus intensity (>70% of maximum stimulator output). Reproduced with permission from Kimiskidis *et al.* (2006).

vary with stimulus intensity. When effects of drugs or other experimental manipulations on cSP duration are studied, it is therefore important to obtain full cSP intensity curves. However, the direct evidence for a contribution of GABA$_B$R activation to the cSP at the high range of stimulus intensities is relatively weak (Table 13.1). In three studies, the GABA$_B$R agonist baclofen did not lead to a lengthening of the cSP (Inghilleri *et al.* 1996; McDonnell *et al.* 2006; Ziemann *et al.* 1996c), but the applied dosages may have been too low to result in an effective drug concentration in the brain to influence cSP duration. One patient with generalized dystonia who was treated with incremental doses of intrathecal baclofen showed a significant lengthening of the cSP, starting at 1.000 μg/day (Siebner *et al.* 1998). However, a contribution by an increase in GABA$_B$R-mediated spinal inhibition was not ruled out. Some studies reported a lengthening of the cSP by L-DOPA and DA agonists (Priori *et al.* 1994; Ziemann *et al.* 1996a). This is consistent with a DA-induced enhancement of postsynaptic

sensitivity to GABA in animal preparations (Beauregard and Ferron 1991), and with the shortened cSP duration in patients with a cortical dopaminergic deficit, such as in Parkinson's disease (Cantello *et al.* 2002).

One important potential confounding effect in cSP measurements is the influence of motor set and motor attention (Classen *et al.* 1997; Mathis *et al.* 1998, 1999). Disturbance of motor attention by sedative drugs such as ethanol or benzodiazepines may contribute to their lengthening effects on cSP duration (Ziemann *et al.* 1995, 1996b).

Short-interval intracortical inhibition (SICI)

SICI is measured in a paired-pulse TMS protocol that uses a subthreshold first (conditioning) pulse followed after a short interstimulus interval of 1–5 ms by a suprathreshold second (test) pulse (Kujirai *et al.* 1993; Ziemann *et al.* 1996d) (cf. Chapter 11, this volume). The first pulse probably produces an IPSP in corticospinal neurons through activation of a low-threshold cortical inhibitory circuit, which inhibits action potential generation in these neurons by the suprathreshold second pulse (Kujirai et al. 1993; Ilić et al. 2002b). In agreement with this hypothesis, benzodiazepines, positive modulators of the $GABA_AR$, enhance SICI (Ziemann et al. 1996b; Di Lazzaro et al. 2000a, 2005b,c; Ilić et al. 2002b). The $GABA_AR$ antagonist flumazenil does not change SICI, suggesting that there is no tonic activity at the benzodiazepine binding site of the $GABA_AR$ in normal human motor cortex (Jung et al. 2004). In contrast, the GABA reuptake inhibitor tiagabine (Werhahn et al. 1999) and the $GABA_BR$ agonist baclofen (McDonnell et al. 2006) decrease SICI. These findings support the idea that SICI is controlled by presynaptic $GABA_BR$-mediated auto-inhibition on inhibitory interneurons (Sanger et al. 2001), similar to presynaptic auto-inhibition revealed by paired intracellular recordings in slices of rat and human motor cortex (Deisz 1999a,b).

The SICI intensity curve is typically U-shaped (cf. Chapter 11, this volume). This implies that SICI is a net inhibition consisting of low-threshold inhibitory and high-threshold facilitatory effects of the conditioning pulse on the test MEP (Awiszus et al. 1999; Ilić et al. 2002b). Drugs that reduce facilitation, such as GLU receptor (GLUR) antagonist, may therefore lead to an apparent increase in SICI (Ziemann et al. 1998a; Schwenkreis et al. 1999, 2000). To clarify whether a drug truly affects SICI rather than short-interval intracortical facilitation (SICF), it is necessary to obtain the full SICI intensity curve (i.e. vary the intensity of the conditioning pulse). This has been done only rarely. The study by Ilić et al. (2002b) shows that the benzodiazepine diazepam enhances SICI most significantly with low intensities of the conditioning pulse, and only to a lesser extent at the range of high intensities, suggesting that the preponderant effect is a true enhancement of SICI.

Many animal experiments indicate that neuromodulators strongly influence GABA and GLU neurotransmission in cerebral cortex (Hasselmo 1995). In accordance, most TMS studies demonstrate significant effects of neuromodulators on SICI (Table 13.1). DA agonists

(Ziemann et al. 1996a, 1997; Korchounov et al. 2007) and the NE antagonist guanfacine (Korchounov et al. 2003) increase SICI, whereas the DA antagonist haloperidol (Ziemann et al. 1997), NE agonists (Gilbert et al. 2006; Herwig et al. 2002; Ilić et al. 2003), and the ACh esterase inhibitor tacrine (Korchounov et al. 2005) decrease SICI.

The effects of neuromodulators on TMS measures may depend on genetic polymorphisms of the receptors and transporters involved in the neuromodulator system under study, and this may explain some of the inconsistency of the reported data in Table 13.1. One first study in support of this notion demonstrates that the selective serotonin reuptake inhibitor citalopram increases SICI, but only in those subjects who are homozygotic for the long variant of the 5-HT transporter gene (Eichhammer et al. 2003). Homozygosity for the long variant of the 5-HT transporter gene is associated with a twofold more efficient serotonin reuptake compared to the short variant.

Finally, it should be noted that all findings reported here refer to SICI as tested at interstimulus intervals between conditioning and test pulse of 2–5 ms. Another type of inhibition at very short intervals of ~1 ms is most likely caused by different inhibitory circuits (Roshan et al. 2003) and by refractoriness of neural elements in the cortex activated by the conditioning pulse (Fisher et al. 2002; Hanajima et al. 2003; Vucic et al. 2006) (cf. Chapter 11, this volume). This process remained largely unaffected by neurotransmitter or neuromodulator systems in those studies that tested SICI separately at an interstimulus interval of 1 ms vs longer intervals of 2–5 ms.

Intracortical facilitation (ICF)

ICF is tested by the same protocol as SICI, but longer interstimulus intervals of 7–20 ms are used (Kujirai et al. 1993; Ziemann et al. 1996d) (cf. Chapter 11, this volume). Compared to SICI, the physiology of ICF is less clear. The leading hypothesis is that ICF tests excitability of an excitatory neuronal motor cortical circuit that is at least in part dissociable from the SICI network (Ziemann et al. 1996d). It is likely that ICF is a net facilitation consisting of prevailing facilitation and weaker inhibition. The inhibition probably comes from the tail of the GABA$_A$R-mediated IPSP, which has a duration of ~20 ms (Connors *et al.* 1988; Hanajima *et al.* 1998). EPSPs in neurons of rat sensorimotor cortex may consist of a fast component mediated by non-NMDA receptors, and a slower component mediated by NMDA receptors (Hwa and Avoli 1992). The latency to onset of the EPSP mediated by the NMDA receptor is in the order of 10 ms, which is consistent with the time course of ICF (Kujirai *et al.* 1993; Ziemann *et al.* 1996d). Therefore, it might be speculated that ICF reflects excitatory neurotransmission in human motor cortex largely mediated by NMDARs. This idea is supported by the majority of pharmacological studies, which demonstrated a decrease of ICF by NMDAR antagonists (Ziemann *et al.* 1998a; Schwenkreis *et al.* 1999), although this was not a unanimous finding (Di Lazzaro *et al.* 2003). Benzodiazepines, positive modulators at the GABA$_A$R, also decrease ICF (Inghilleri *et al.* 1996; Ziemann *et al.* 1996b; Mohammadi *et al.* 2006), supporting a contribution of inhibition through the GABA$_A$R to the magnitude of ICF. The synopsis of pharmacological modulation of SICI and ICF (Table 13.1) shows that the pharmacological profiles of ICF and SICI are similar but not identical, supporting the notion of partially dissociable contributing mechanisms (Ziemann *et al.* 1996d; Strafella and Paus 2001).

Short-interval intracortical facilitation (SICF)

SICF is also measured in a paired-pulse TMS protocol, but in contrast to SICI and ICF, the first pulse is suprathreshold and the second pulse is subthreshold (Ziemann *et al.* 1998b; Chen and Garg 2000; Hanajima *et al.* 2002), or both pulses are close to threshold intensity (Tokimura *et al.* 1996; Di Lazzaro *et al.* 1999) (cf. Chapter 11, this volume). SICF occurs at discrete interstimulus intervals of about 1.1–1.5, 2.3–2.9, and 4.1–4.4 ms. The interpeak latency between these facilitatory intervals is about 1.5 ms, which led to the hypothesis that SICF originates in those neural elements which are responsible for the generation of the I-waves (Tokimura *et al.* 1996; Ziemann *et al.* 1998b)

(cf. Chapter 14, this volume). It is currently thought that the second pulse excites directly the initial axon segments of those excitatory interneurons which have been depolarized by EPSPs from the first pulse but did not fire an action potential (Hanajima *et al.* 2002; Ilić *et al.* 2002b). Benzodiazepines and barbiturates, i.e. allosteric positive modulators at the GABA$_A$R, reduce SICF (Ziemann *et al.* 1998c; Ilić *et al.* 2002b). This supports the hypothesis that the first pulse elicits a GABA$_A$R-mediated short-latency IPSP in corticospinal and/or first-order excitatory interneurons that inhibits the facilitatory interaction with the second pulse. In contrast, the NMDAR antagonist memantine has no effect on SICF (Ziemann *et al.* 1998c). Other drug effects on SICF are summarized in Table 13.1.

Long-interval intracortical inhibition (LICI)

LICI is tested in a paired-pulse TMS protocol that employs two suprathreshold pulses at long interstimulus intervals of 50–200 ms (Claus *et al.* 1992; Valls-Sole *et al.* 1992; Nakamura *et al.* 1997; Di Lazzaro *et al.* 2002a) (cf. Chapters 11 and 12, this volume). The range of effective interstimulus intervals to produce LICI indicates a long-lasting inhibition that is distinct from SICI (Sanger *et al.* 2001) but similar to the cSP (Valls-Sole *et al.* 1992). Accordingly, it was proposed that LICI is mediated by slow IPSPs mediated by the GABA$_B$R (Werhahn *et al.* 1999; Sanger *et al.* 2001). This is directly supported by a study demonstrateding that baclofen, a specific GABA$_B$R agonist, increases the magnitude of LICI (McDonnell *et al.* 2006). A similar increase in LICI occurs after application of tiagabine (Werhahn *et al.* 1999) or vigabatrin (Pierantozzi *et al.* 2004), probably also through enhancement of GABA$_B$R-mediated inhibitory neurotransmission because both drugs increase the availability of GABA in the synaptic cleft. Other drug effects on LICI are summarized in Table 13.1.

Short-latency afferent inhibition (SAI)

SAI refers to an MEP inhibition produced by a conditioning afferent electrical stimulus applied to the median nerve at the wrist ~20 ms prior to TMS of the hand area of the contralateral motor cortex (Tokimura *et al.* 2000) (cf. Chapter 12, this volume).

Pharmacological experiments suggest that the neural circuit mediating SAI is distinct from the one mediating SICI because the ACh antagonist scopolamine reduces SAI but leaves SICI unaffected (Di Lazzaro *et al.* 2000b). In addition, the benzodiazepines diazepam and lorazepam both increase SICI but their effects on SAI are opposite: diazepam slightly increases SAI (Di Lazzaro *et al.* 2005c) whereas lorazepam decreases SAI (Di Lazzaro *et al.* 2005b,c). This dissociation suggests that TMS measures of cortical inhibition provide the opportunity to segregate differences of benzodiazepine action in human central nervous system circuits.

The suppression of SAI by scopolamine suggests that SAI may be a useful probe to test the integrity of central cholinergic neural circuits. This idea is supported by an abnormal reduction in SAI in patients with Alzheimer's disease (Di Lazzaro *et al.* 2002b; 2005a).

Summary

Pharmacological experiments have been extremely helpful in characterizing the physiology of TMS measures of motor cortical excitability. TMS allows measurements of axon-related excitability, and inhibitory and excitatory synaptic excitability. Synaptic excitability can be further dissected into, for instance, distinct forms of cortical inhibition, such as SICI (mediated by $GABA_A$Rs), cSP and LICI (mediated by $GABA_B$Rs), and SAI (modulated by ACh). These distinct forms of motor cortical inhibition may relate to distinct inhibitory neuronal circuits, and other recent experiments showed that these forms of inhibition interact in a complex manner (cf. Chapter 12, this volume). Therefore, pharmacological challenging of TMS measures has opened a broad window into human cortical physiology.

Effects of CNS-active drugs with incompletely known or multiple modes of action on TMS measures of motor cortical excitability

Many of the TMS measures are by now well defined in terms of their physiological and pharmacological properties (cf. 'Effects of CNS-active drugs with a defined mode of action on TMS measures of motor cortical excitability' above and Table 13.1, and Chapters 9–12 and 14, this volume). This knowledge can be utilized to identify, at the systems level of human motor cortex, the most prominent modes of action of CNS-active drugs with multiple or incompletely known mechanisms. The available data for this approach are summarized in Table 13.2.

One illustrative example along this avenue is the testing of the effects of the novel antiepileptic drug topiramate on a broad array of TMS measures (Reis *et al.* 2002). Topiramate demonstrated a wide spectrum of antiepileptic activities in preclinical animal experimental models of epilepsy and in clinical studies. Several different modes of action were identified at the cellular level (for review see Shank *et al.* 2000). These include: (1) blockade of voltage-gated sodium channels; (2) enhancement of neurotransmission through the $GABA_A$R; (3) decrease of neurotransmission through non-NMDARs of the kainate and AMPA subtypes; (4) blockade of voltage-gated calcium channels of the L-type. A single oral dose of 50 or 200 mg of topiramate dose dependently increased SICI and tended to decrease ICF, whereas RMT and CSP remained unaffected (Reis *et al.* 2002). From this pattern of effects it was concluded that a single dose of topiramate exerted significant effects at the level of human motor cortex through enhancement of $GABA_A$R and/or suppression of GLUR-dependent mechanisms, without detectable action on voltage-gated calcium channels or $GABA_B$R-mediated inhibition (Reis *et al.* 2002).

Issues of study design and pitfalls

The typical study design includes a baseline measurement before drug application and at least one measurement after drug application. The statistical analysis uses a repeated measures test, usually a paired Student's *t*-test for two time points, and a repeated measures analysis of variance with time as the within-subject effect for more than two time points, to evaluate the effect of the drug. This is a valid procedure because the within-subject test–retest reliability of TMS measures is high (Maeda *et al.* 2002; Wassermann 2002). Ideally, a

Table 13.2 Effects of CNS-active drugs with incompletely known or multiple modes of action on TMS measures of motor cortical excitability

Drug	Mode(s) of action	MT	MEP	cSP	SICI	ICF	SICF	Reference
Gabapentin	Blockade of calcium channels	○		▲	▲	▲		Ziemann et al. (1996c)
	Increase of GABA synthesis	○		▲	▲	▲		Rizzo et al. (2001)
	Increase of GLU release							
Levetiracetam	Reversal of negative allosteric	○	▼	○	○	○		Sohn et al. (2001)
	modulation at the $GABA_A R$	▲	(▼)	○	○	○		Reis et al. (2004)
	Blocker of N-type calcium channels							
	? Inhibition I_k							
Piracetam	?						▼	Wischer et al. (2001)
Acamprosate	NMDAR antagonist	▲			○	○		Wohlfarth et al. (2000)
	Blockade of calcium channels							
Topiramate	Blockade of sodium channels	○		○	▲	(▼)		Reis et al. (2002)
	Positive modulation at $GABA_A R$							
	Non-NMDAR antagonist							
	Blocker of L-type calcium channels							
Caffeine	Adenosine receptor antagonist	○	○	○	○	○		Orth et al. (2005)
			▲					Specterman et al. (2005)
Theophylline	Adenosine receptor antagonist	○		○	▼	○		Nardone et al. (2004)
Modafinil	? Modulation of NE, DA, 5-HT hypocretin,	○	○	○	○	○		Liepert et al. (2004)
	histamine, GLU, GABA							
Mirtazapine	α2R antagonist Indirect 5-HT agonist	○	(▲)	▲	○	○	○	Münchau et al. (2005)

MT, motor threshold; MEP, motor-evoked potential; cSP, cortical silent period; SICI, short-interval intracortical inhibition; ICF, intracortical facilitation; SICF, short-interval intracortical facilitation; LICI, long-interval intracortical inhibition.
GABA, gamma-aminobutyric acid; R, receptor; GLU, glutamine; NMDA, N-methyl-D-aspartate; NE, norepinephrine; DA, dopamine; 5-HT, serotonin.
○, no effect; ▼, decrease; ▲, increase.
?, mode of action unclear; I_k, delayed rectifier potassium current.

randomized and placebo-controlled study design should be used to exclude experimenter bias.

One important issue is to select an appropriate drug dose. The effect of the study drug on a TMS measure may increase with drug dose (Ziemann et al. 1995; Chen et al. 1997; Werhahn et al. 1999; Schwenkreis et al. 2000; Liepert et al. 2001; Plewnia et al. 2001, 2002; Di Lazzaro et al. 2003; Kimiskidis et al. 2006). One potential limitation of many of these studies is that the correlation between drug serum level and drug effect on a TMS measure was calculated not within individuals who received different doses of the study drug in different experimental sessions, but at the group level, i.e. all individuals received the same dose and only a single data point per individual entered the analysis. However, proper dose–response curves should be better obtained within individuals because the minimal dose to reach a threshold drug effect and the relation between drug dose and drug effect may vary considerably between subjects (Tergau et al. 2003), probably due to complex drug pharmacodynamics. Without dose response–curves, there is always a risk of missing a drug effect because the drug was dosed inappropriately low. Another critical issue concerning drug dose is to produce 'unwanted' drug effects when a drug has more than one mode of action ('dirty drug'). Usually, these different modes of action come into play at different drug doses due to differences in receptor density and/or drug-receptor affinity. This implies a risk of contaminating expected drug effects with unwanted drug effects because the drug was dosed inappropriately high. The only reasonable way to account for these potential problems in drug dosing is the inclusion of different or incremental doses into the study design (Tergau et al. 2003).

Another critical issue is the timing of TMS measurements. While most studies chose timing according to the pharmacokinetics, i.e. the course of the plasma level of the drug under study, this may be inappropriate for drugs which produce their effect through complex pharmacodynamic action. A good example is the antiepileptic drug vigabatrin. Its effect is exerted via irreversible inhibition of the GABA-degrading enzyme GABA transaminase, which has a much slower and longer time course than the vigabatrin plasma concentration

(Ben-Menachem 1995). GABA concentration in the brain peaks 24 h after intake of a single dose of vigabatrin (Petroff et al. 1996), while vigabatrin concentration in the plasma peaks after 1 h and the plasma half-life is 6–8 h (Ben-Menachem 1995). Accordingly, a significant vigabatrin-induced decrease in ICF was observed only 24 h after intake, but not after 6 h (Ziemann et al. 1996c). As a consequence, good knowledge about the pharmacokinetic/pharmacodynamic (PK/PD) actions of the drug under study is required in order not to miss drug effects by inappropriate timing of TMS testing.

Finally, the selection of subjects may play a pivotal role on the study results. It is possible that the level of a TMS measure at baseline influences its responsiveness to experimental manipulation, although this has not yet been directly explored in pharmacological TMS studies. The level of a TMS measure at baseline can be affected by many sources, such as age (Moll et al. 1999; Peinemann et al. 2001; Pitcher et al. 2003; Oliviero et al. 2006; Silbert et al. 2006), sex (Wassermann 2002; Pitcher et al. 2003), genetic endophenotype (Eichhammer et al. 2003), anxiety-related personality trait (Wassermann et al. 2001), or phase of the menstrual cycle (Smith et al. 1999). Patients with neurological or psychiatric disease may show drug effects on TMS measures that are different from healthy subjects. For instance, in patients with attention-deficit hyperactivity disorder, where there is pathologically reduced SICI, methylphenidate results in an increase of SICI towards normalization (Moll et al. 2000), whereas a significant decrease in SICI occurs in healthy subjects (Ilić et al. 2003; Gilbert et al. 2006). As a consequence, it is rather important to define carefully the characteristics of the study population.

Chronic versus acute effects of CNS-active drugs on TMS measures of motor cortical excitability

Chronic drug effects may be fundamentally different from acute ones. Several mechanisms exist that can alter the response of the human brain to a drug, if chronically administered.

1. Pharmacokinetic tolerance. This refers to changes in the distribution or metabolism of

a drug induced by repeated application. The most common mechanism is an increase in the rate of metabolism.

2. Pharmacodynamic tolerance. This refers to adaptive changes within the system affected by the drug. In the CNS, the most common mechanisms are drug-induced changes in receptor density, or efficiency of receptor coupling to signal transduction pathways.

3. Sensitization. This refers to an increase in drug effect with repeated application.

Only a few studies have investigated effects of chronically applied drugs on TMS measures in healthy subjects. Incremental doses of carbamazepine or lamotrigine given over a period of 5 weeks result in significant increases in RMT that correlate with drug serum levels (Lee *et al.* 2005), similar to the acute effects of these drugs observed after a single dose (see 'Motor threshold', above). In addition, chronic application of carbamazepine and lamotrigine results in a trend toward increased SICI, an effect that is not seen in the acute experiments, and which might be attributed to suppression of SICF by decrease of glutamate release in the chronic experiments (Lee *et al.* 2005). Another important aspect of this study is what happens to TMS measures of motor cortical excitability after acute drug withdrawal. Acute withdrawal of carbamazepine or lamotrigine is associated with a transient decrease in RMT (undershooting of RMT) in some subjects. It was speculated that this might reflect a physiological substrate predictive of antiepileptic drug withdrawal seizures (although these did not occur in this population of healthy subjects) (Lee *et al.* 2005).

The anti-GLU drug riluzole results in an increase in SICI (Schwenkreis *et al.* 2000) and a decrease in ICF after a single oral dose of 100–150 mg (Liepert *et al.* 1997; Schwenkreis *et al.* 2000). Both effects are maintained if riluzole is administered daily over a period of 1 week (Schwenkreis *et al.* 2000). In addition, a slight but significant increase in RMT occurs after 1 week that was not seen after the first day (Schwenkreis *et al.* 2000). These data suggest that pharmacodynamic tolerance does not develop during chronic riluzole treatment. This is a potentially important piece of information

for the clinical setting, where riluzole is used for treatment of various neurodegenerative disorders, such as amyotrophic lateral sclerosis.

More such studies are desirable in order to learn more about chronic versus acute drug effects at the systems level of the human motor cortex. This may also help to better understand apparently abnormal TMS measures of motor cortical excitability in neurological or psychiatric patients who are often on chronic treatment with CNS-active drugs.

TMS/rTMS-induced changes in endogenous neurotransmitters and neuromodulators

Endogenous neurotransmitters, such as GABA and GLU, and neuromodulators (DA, NE, 5-HT, ACh) play a fundamental role in the regulation of neuronal activity and plasticity in the cerebral cortex (McCormick *et al.* 1993; Hasselmo 1995; Gu 2002). The basis of many neurological and psychiatric disorders is thought to lie in abnormal neuronal network activity as a consequence of altered neurotransmitter or neuromodulator systems. For instance, DA is implicated in the control of fundamental processes such as movement, attention, and learning. Dysfunction of DA plays a pivotal role in disorders such as Parkinson's disease, schizophrenia, or drug addiction (Carlsson and Carlsson 1990).

TMS and rTMS may result in changes in endogenous neurotransmitters and neuromodulators, and these changes are of potential therapeutic interest. A wealth of studies have explored rTMS effects in the rat brain by using *ex vivo* analysis of brain homogenates or *in vivo* microdialysis techniques. For instance, several studies showed that rTMS leads to a DA increase in the striatum (Ben-Shachar *et al.* 1997; Belmaker and Grisaru 1998; Keck *et al.* 2002; Kanno *et al.* 2004). In healthy human subjects, by using [^{11}C]raclopride positron emission tomography (PET), a focal increase of DA in the striatum was demonstrated after subthreshold 10 Hz rTMS was applied to the ipsilateral primary motor cortex (Strafella *et al.* 2003) or dorsolateral prefrontal cortex (Strafella *et al.* 2001). In Parkinson's disease, abnormalities in cortico-striatal interactions are believed to play an important role in the pathophysiology of the disease (Carlsson and Carlsson 1990; Calabresi *et al.* 1993). This is supported by a recent raclopride PET study in seven patients with early and largely unilateral Parkinson's disease showing that 10 Hz rTMS of the primary motor cortex causes an increase in DA in the ipsilateral striatum of both hemispheres (Strafella *et al.* 2005). This effect is significantly smaller in the symptomatic hemisphere, but instead the area of DA increase is significantly larger than in the asymptomatic hemisphere. This finding of a spatially enlarged area of DA release, following motor cortical rTMS, may represent a possible *in vivo* expression of abnormal cortico-striatal neurotransmission in early Parkinson's disease (Strafella *et al.* 2005). To what extent this rTMS-induced cortico-striatal modulation of DA release has therapeutic implications in Parkinson's disease requires further investigation, given the hitherto moderate or inconsistent rTMS treatment effects in this disorder (for review see Cantello *et al.* 2002), and the considerable placebo effect in rTMS studies (Strafella *et al.* 2006).

Acknowledgement

Parts of this chapter are adapted from a previous review on the same topic (Ziemann 2004).

References

Amassian VE, Stewart M, Quirk GJ, Rosenthal JL (1987). Physiological basis of motor effects of a transient stimulus to cerebral cortex. *Neurosurgery 20*, 74–93.

Awiszus F, Feistner H, Urbach D, Bostock H (1999). Characterisation of paired-pulse transcranial magnetic stimulation conditions yielding inhibition or I-wave facilitation using a threshold-hunting paradigm. *Experimental Brain Research 129*, 317–324.

Beauregard M, Ferron A (1991). Dopamine modulates the inhibition induced by GABA in rat cerebral cortex: an iontophoretic study. *European Journal of Pharmacology 205*, 225–231.

Belmaker RH, Grisaru N (1998). Magnetic stimulation of the brain in animal depression models responsive to ECS. *Journal of ECT 14*, 194–205.

Ben-Menachem E (1995). Vigabatrin. Chemistry, absorption, distribution and elimination. In: RH Levy, RH Mattson and BS Meldrum (eds), *Antiepileptic drugs*, pp. 915–923. New York: Raven Press.

Ben-Shachar D, Belmaker RH, Grisaru N, Klein E (1997). Transcranial magnetic stimulation induces alterations in brain monoamines. *Journal of Neural Transmission 104*, 191–197.

Boroojerdi B, Battaglia F, Muellbacher W, Cohen LG (2001). Mechanisms influencing stimulus-response properties of the human corticospinal system. *Clinical Neurophysiology 112*, 931–937.

Calabresi P, Mercuri NB, Sancesario G, Bernardi G (1993). Electrophysiology of dopamine-denervated striatal neurons. Implications for Parkinson's disease. *Brain 116*(Pt 2), 433–452.

Cantello R, Gianelli M, Civardi C, Mutani R (1992). Magnetic brain stimulation: the silent period after the motor evoked potential. *Neurology 42*, 1951–1959.

Cantello R, Tarletti R, Civardi C (2002). Transcranial magnetic stimulation and Parkinson's disease. *Brain Research. Brain Research Reviews 38*, 309–327.

Carlsson M, Carlsson A (1990). Interactions between glutamatergic and monoaminergic systems within the basal ganglia–implications for schizophrenia and Parkinson's disease. *Trends in Neurosciences 13*, 272–276.

Chen R, Garg R (2000). Facilitatory I wave interaction in proximal arm and lower limb muscle representations of the human motor cortex. *Journal of Neurophysiology 83*, 1426–1434.

Chen R, Samii A, Canos M, Wassermann EM, Hallett M (1997). Effects of phenytoin on cortical excitability in humans. *Neurology 49*, 881–883.

Chen R, Lozano AM, Ashby P (1999). Mechanism of the silent period following transcranial magnetic stimulation. *Experimental Brain Research 128*, 539–542.

Classen J, Schnitzler A, Binkofski F, Werhahn KJ, Kim Y-S, Kessler K, *et al.* (1997). The motor syndrome associated with exaggerated inhibition within the primary motor cortex of patients with hemiparetic stroke. *Brain 120*, 605–619.

Claus D, Weis M, Jahnke U, Plewe A, Brunhölzl C (1992). Corticospinal conduction studied with magnetic double stimulation in the intact human. *Journal of the Neurological Sciences 111*, 180–188.

Connors BW, Malenka RC, Silva LR (1988). Two inhibitory postsynaptic potentials, and GABAA and GABAB receptor-mediated responses in neocortex of rat and cat. *Journal of Physiology (London) 406*, 443–468.

Danielsson I, SU K, Kauer L, Barnette L, Reeves-Tyer P, Kelley K, Theodore WH, Wassermann E, Rogawski MA (2004). Talampanel and human cortical excitability: EEG and TMS. *Epilepsia 45*(Suppl 7), 120–121.

Deisz RA (1999a). The GABA(B) receptor antagonist CGP 55845A reduces presynaptic GABA(B) actions in neocortical neurons of the rat in vitro. *Neuroscience 93*, 1241–1249.

Deisz RA (1999b). GABA(B) receptor-mediated effects in human and rat neocortical neurones in vitro. *Neuropharmacology 38*, 1755–1766.

Devanne H, Lavoie BA, Capaday C (1997). Input-output properties and gain changes in the human corticospinal pathway. *Experimental Brain Research 114*, 329–338.

Di Lazzaro V, Rothwell JC, Oliviero A, Profice P, Insola A, Mazzone P, *et al.* (1999). Intracortical origin of the short latency facilitation produced by pairs of threshold magnetic stimuli applied to human motor cortex. *Experimental Brain Research 129*, 494–499.

Di Lazzaro V, Oliviero A, Meglio M, Cioni B, Tamburrini G, Tonali P, *et al.* (2000a). Direct demonstration of the effect of lorazepam on the excitability of the human motor cortex. *Clinical Neurophysiology 111*, 794–799.

Di Lazzaro V, Oliviero A, Profice P, Pennisi MA, Di Giovanni S, Zito G, *et al.* (2000b). Muscarinic receptor blockade has differential effects on the excitability of intracortical circuits in human motor cortex. *Experimental Brain Research 135*, 455–461.

Di Lazzaro V, Oliviero A, Mazzone P, Pilato F, Saturno E, Insola A, *et al.* (2002a). Direct demonstration of long latency cortico-cortical inhibition in normal subjects and in a patient with vascular parkinsonism. *Clinical Neurophysiology 113*, 1673–1679.

Di Lazzaro V, Oliviero A, Tonali PA, Marra C, Daniele A, Profice P, *et al.* (2002b). Noninvasive in vivo assessment of cholinergic cortical circuits in AD using transcranial magnetic stimulation. *Neurology 59*, 392–397.

Di Lazzaro V, Oliviero A, Profice P, Pennisi MA, Pilato F, Zito G, *et al.* (2003). Ketamine increases motor cortex excitability to transcranial magnetic stimulation. *Journal of Physiology 547*, 485–496.

Di Lazzaro V, Oliviero A, Pilato F, Saturno E, Dileone M, Mazzone P, *et al.* (2004). The physiological basis of transcranial motor cortex stimulation in conscious humans. *Clinical Neurophysiology 115*, 255–266.

Di Lazzaro V, Oliviero A, Pilato F, Saturno E, Dileone M, Marra C, *et al.* (2005a). Neurophysiological predictors of long term response to AChE inhibitors in AD patients. *Journal of Neurology, Neurosurgery and Psychiatry 76*, 1064–1069.

Di Lazzaro V, Oliviero A, Saturno E, Dileone M, Pilato F, Nardone R, *et al.* (2005b). Effects of lorazepam on short latency afferent inhibition and short latency intracortical inhibition in humans. *Journal of Physiology 564*, 661–668.

Di Lazzaro V, Pilato F, Dileone M, Tonali PA, Ziemann U (2005c). Dissociated effects of diazepam and lorazepam on short latency afferent inhibition. *Journal of Physiology 569*, 315–323.

Douglas R, Martin K (1998). Neocortex. In: GM Sheperd (ed.), *The synaptic organization of the brain*, pp. 459–509. New York: Oxford University Press.

Eichhammer P, Langguth B, Wiegand R, Kharraz A, Frick U, Hajak G (2003). Allelic variation in the serotonin transporter promoter affects neuromodulatory effects of a selective serotonin transporter reuptake inhibitor (SSRI). *Psychopharmacology (Berlin) 166*, 294–297.

Fisher RJ, Nakamura Y, Bestmann S, Rothwell JC, Bostock H (2002). Two phases of intracortical inhibition revealed by transcranial magnetic threshold tracking. *Experimental Brain Research 143*, 240–248.

Fuhr P, Agostino R, Hallett M (1991). Spinal motor neuron excitability during the silent period after cortical stimulation. *Electroencephalography and Clinical Neurophysiology 81*, 257–262.

Gerdelat-Mas A, Loubinoux I, Tombari D, Rascol O, Chollet F, Simonetta-Moreau M (2005). Chronic administration of selective serotonin reuptake inhibitor (SSRI) paroxetine modulates human motor cortex excitability in healthy subjects. *NeuroImage 27*, 314–322.

Gilbert DL, Ridel KR, Sallee FR, Zhang J, Lipps TD, Wassermann EM (2006). Comparison of the inhibitory and excitatory effects of ADHD medications methylphenidate and atomoxetine on motor cortex. *Neuropsychopharmacology 31*, 442–449.

Gu Q (2002). Neuromodulatory transmitter systems in the cortex and their role in cortical plasticity. *Neuroscience 111*, 815–835.

Hanajima R, Ugawa Y, Terao Y, Sakai K, Furubayashi T, Machii K, *et al.* (1998). Paired-pulse magnetic stimulation of the human motor cortex: differences among I waves. *Journal of Physiology 509*, 607–618.

Hanajima R, Ugawa Y, Terao Y, Enomoto H, Shiio Y, Mochizuki H, *et al.* (2002). Mechanisms of intracortical I-wave facilitation elicited with paired-pulse magnetic stimulation in humans. *Journal of Physiology 538*, 253–261.

Hanajima R, Furubayashi T, Iwata NK, Shiio Y, Okabe S, Kanazawa I, *et al.* (2003). Further evidence to support different mechanisms underlying intracortical inhibition of the motor cortex. *Experimental Brain Research 151*, 427–434.

Hasselmo ME (1995). Neuromodulation and cortical function: modeling the physiological basis of behavior. *Behavioural Brain Research 67*, 1–27.

Herwig U, Bräuer K, Connemann B, Spitzer M, Schönfeldt-Lecuona C (2002). Intracortical excitability is modulated by a norepinephrine-reuptake inhibitor as measured with paired-pulse transcranial magnetic stimulation. *Psychopharmacology 164*, 228–232.

Hess CW, Mills KR, Murray NM (1987). Responses in small hand muscles from magnetic stimulation of the human brain. *Journal of Physiology (London) 388*, 397–419.

Hodgkin AL, Huxley AF (1952). A quantative description of membrane current and its application to conduction and excitation in nerve. *Journal of Physiology (London) 116*, 500–544.

Huguenard JR, Prince DA (1994). Clonazepam suppresses GABAB-mediated inhibition in thalamic relay neurons through effects in nucleus reticularis. *Journal of Neurophysiology 71*, 2576–2581.

Hwa GG, Avoli M (1992). Excitatory postsynaptic potentials recorded from regular-spiking cells in layers II/III of rat sensorimotor cortex. *Journal of Neurophysiology 67*, 728–37.

Ilić TV, Korchounov A, Ziemann U (2002a). Complex modulation of human motor cortex excitability by the specific serotonin re-uptake inhibitor sertraline. *Neuroscience Letters 319*, 116–120.

Ilić TV, Meintzschel F, Cleff U, Ruge D, Kessler KR, Ziemann U (2002b). Short-interval paired-pulse inhibition and facilitation of human motor cortex: the dimension of stimulus intensity. *Journal of Physiology 545.1*, 153–167.

Ilić TV, Korchounov A, Ziemann U (2003). Methylphenidate facilitates and disinhibits the motor cortex in intact humans. *Neuroreport 14*, 773–776.

Inghilleri M, Berardelli A, Cruccu G, Manfredi M (1993). Silent period evoked by transcranial stimulation of the human cortex and cervicomedullary junction. *Journal of Physiology (London) 466*, 521–534.

Inghilleri M, Berardelli A, Marchetti P, Manfredi M (1996). Effects of diazepam, baclofen and thiopental on the silent period evoked by transcranial magnetic stimulation in humans. *Experimental Brain Research 109*, 467–472.

Jung HY, Sohn YH, Mason A, Considine E, Hallett M (2004). Flumazenil does not affect intracortical motor excitability in humans: a transcranial magnetic stimulation study. *Clinical Neurophysiology 115*, 325–329.

Kanno M, Matsumoto M, Togashi H, Yoshioka M, Mano Y (2004). Effects of acute repetitive transcranial magnetic stimulation on dopamine release in the rat dorsolateral striatum. *Journal of the Neurological Sciences 217*, 73–81.

Keck ME, Welt T, Müller MB, Erhardt A, Ohl F, Toschi N, *et al.* (2002). Repetitive transcranial magnetic stimulation increases the release of dopamine in the mesolimbic and mesostriatal system. *Neuropharmacology 43*, 101–109.

Kimiskidis VK, Papagiannopoulos S, Sotirakoglou K, Kazis DA, Kazis A, Mills KR (2005). Silent period to transcranial magnetic stimulation: construction and properties of stimulus-response curves in healthy volunteers. *Experimental Brain Research 163*, 21–31.

Kimiskidis VK, Papagiannopoulos S, Kazis DA, Sotirakoglou K, Vasiliadis G, Zara F, *et al.* (2006). Lorazepam-induced effects on silent period and corticomotor excitability. *Experimental Brain Research 173*, 603–611.

Korchounov A, Ilić TV, Ziemann U (2003). The alpha2-adrenergic agonist guanfacine reduces excitability of human motor cortex through disfacilitation and increase of inhibition. *Clinical Neurophysiology 114*, 1834–1840.

Korchounov A, Ilić TV, Schwinge T, Ziemann U (2005). Modification of motor cortical excitability by an acetylcholine-esterase inhibitor. *Experimental Brain Research 164*, 399–405.

Korchounov A, Ilić TV, Ziemann U (2007). TMS-assisted neurophysiological profiling of the dopamine receptor agonist cabergoline in human motor cortex. *Journal of Neural Transmission 114*, 223–229.

Kujirai T, Caramia MD, Rothwell JC, Day BL, Thompson PD, Ferbert A, *et al.* (1993). Corticocortical inhibition in human motor cortex. *Journal of Physiology (London) 471*, 501–519.

Lee HW, Seo HJ, Cohen LG, Bagic A, Theodore WH (2005). Cortical excitability during prolonged antiepileptic drug treatment and drug withdrawal. *Clinical Neurophysiology 116*, 1105–1112.

Li X, Teneback CC, Nahas Z, Kozel FA, Large C, Cohn J, *et al.* (2004). Interleaved transcranial magnetic stimulation/functional MRI confirms that lamotrigine inhibits cortical excitability in healthy young men. *Neuropsychopharmacology 29*, 1395–1407.

Liepert J, Schwenkreis P, Tegenthoff M, Malin J-P (1997). The glutamate antagonist Riluzole suppresses intracortical facilitation. *Journal of Neural Transmission 104*, 1207–1214.

Liepert J, Schardt S, Weiller C (2001). Orally administered atropine enhances motor cortex excitability: a transcranial magnetic stimulation study in human subjects. *Neuroscience Letters 300*, 149–152.

Liepert J, Allstadt-Schmitz J, Weiller C (2004). Motor excitability and motor behaviour after modafinil ingestion – a double-blind placebo-controlled cross-over trial. *Journal of Neural Transmission 111*, 703–711.

McCormick DA, Wang Z, Huguenard J (1993). Neurotransmitter control of neocortical neuronal activity and excitability. *Cerebral Cortex 3*, 387–398.

McDonnell MN, Orekhov Y, Ziemann U (2006). The role of GABA(B) receptors in intracortical inhibition in the human motor cortex. *Experimental Brain Research 173*, 86–93.

Maeda F, Gangitano M, Thall M, Pascual-Leone A (2002). Inter- and intra-individual variability of paired-pulse curves with transcranial magnetic stimulation (TMS). *Clinical Neurophysiology 113*, 376–382.

Mathis J, de Quervain D, Hess CW (1998). Dependence of the transcranially induced silent period on the 'instruction set' and the individual reaction time. *Electroencephalography and Clinical Neurophysiology 109*, 426–435.

Mathis J, de Quervain D, Hess CW (1999). Task-dependent effects on motor-evoked potentials and on the following silent period. *Journal of Clinical Neurophysiology 16*, 556–565.

Mavroudakis N, Caroyer JM, Brunko F, Zegers de Beyl D (1994). Effects of diphenylhydantoin on motor potentials evoked with magnetic stimulation. *Electroencephalography and Clinical Neurophysiology 93*, 428–433.

Mohammadi B, Krampfl K, Petri S, Bogdanova D, Kossev A, Bufler J, *et al.* (2006). Selective and nonselective benzodiazepine agonists have different effects on motor cortex excitability. *Muscle and Nerve 33*, 778–784.

Moll GH, Heinrich H, Wischer S, Tergau F, Paulus W, Rothenberger A (1999). Motor system excitability in healthy children: developmental aspects from transcranial magnetic stimulation. *Electroencephalography and Clinical Neurophysiology 51*(Suppl), 243–249.

Moll GH, Heinrich H, Trott G, Wirth S, Rothenberger A (2000). Deficient intracortical inhibition in drug-naive children with attention-deficit hyperactivity disorder is enhanced by methylphenidate. *Neuroscience Letters 284*, 121–125.

Münchau A, Langosch JM, Gerschlager W, Rothwell JC, Orth M, Trimble MR (2005). Mirtazapine increases cortical excitability in healthy controls and epilepsy patients with major depression. *Journal of Neurology, Neurosurgery and Psychiatry 76*, 527–533.

Nakamura H, Kitagawa H, Kawaguchi Y, Tsuji H (1997). Intracortical facilitation and inhibition after transcranial magnetic stimulation in conscious humans. *Journal of Physiology (London) 498*, 817–823.

Nardone R, Buffone E, Covi M, Lochner PG, Tezzon F (2004). Changes in motor cortical excitability in humans following orally administered theophylline. *Neuroscience Letters* 355, 65–68.

Oliviero A, Profice P, Tonali PA, Pilato F, Saturno E, Dileone M, et al. (2006). Effects of aging on motor cortex excitability. *Neuroscience Research* 55, 74–77.

Orth M, Amann B, Ratnaraj N, Patsalos PN, Rothwell JC (2005). Caffeine has no effect on measures of cortical excitability. *Clinical Neurophysiology* 116, 308–314.

Peinemann A, Lehner C, Conrad B, Siebner HR (2001). Age-related decrease in paired-pulse intracortical inhibition in the human primary motor cortex. *Neuroscience Letters* 313, 33–36.

Petroff OA, Rothman DL, Behar KL, Collins TL, Mattson RH (1996). Human brain GABA levels rise rapidly after initiation of vigabatrin therapy. *Neurology* 47, 1567–1571.

Pierantozzi M, Marciani MG, Palmieri MG, Brusa L, Galati S, Caramia MD, et al. (2004). Effect of Vigabatrin on motor responses to transcranial magnetic stimulation: an effective tool to investigate in vivo GABAergic cortical inhibition in humans. *Brain Research* 1028, 1–8.

Pitcher JB, Ogston KM, Miles TS (2003). Age and sex differences in human motor cortex input-output characteristics. *Journal of Physiology* 546, 605–613.

Plewnia C, Bartels M, Cohen L, Gerloff C (2001). Noradrenergic modulation of human cortex excitability by the presynaptic alpha(2)-antagonist yohimbine. *Neuroscience Letters* 307, 41–44.

Plewnia C, Hoppe J, Hiemke C, Bartels M, Cohen L, Gerloff C (2002). Enhancement of human cortico-motoneuronal excitability by the selective norepinephrine reuptake inhibitor reboxetine. *Neuroscience Letters* 330, 231–234.

Plewnia C, Hoppe J, Cohen LG, Gerloff C (2004). Improved motor skill acquisition after selective stimulation of central norepinephrine. *Neurology* 62, 2124–2126.

Priori A, Berardelli A, Inghilleri M, Accornero N, Manfredi M (1994). Motor cortical inhibition and the dopaminergic system. Pharmacological changes in the silent period after transcranial brain stimulation in normal subjects, patients with Parkinson's disease and drug-induced parkinsonism. *Brain* 117, 317–323.

Reis J, Tergau F, Hamer HM, Muller HH, Knake S, Fritsch B, et al. (2002). Topiramate selectively decreases intracortical excitability in human motor cortex. *Epilepsia* 43, 1149–1156.

Reis J, Wentrup A, Hamer HM, Mueller HH, Knake S, Tergau F, et al. (2004). Levetiracetam influences human motor cortex excitability mainly by modulation of ion channel function–a TMS study. *Epilepsy Research* 62, 41–51.

Rizzo V, Quartarone A, Bagnato S, Battaglia F, Majorana G, Girlanda P (2001). Modification of cortical excitability induced by gabapentin: a study by transcranial magnetic stimulation. *Neurological Sciences* 22, 229–232.

Roshan L, Paradiso GO, Chen R (2003). Two phases of short-interval intracortical inhibition. *Experimental Brain Research* 151, 330–337.

Rossini PM, Berardelli A, Deuschl G, Hallett M, Maertens de Noordhout AM, Paulus W, et al. (1999). Applications of magnetic cortical stimulation. *Electroencephalography and Clinical Neurophysiology* 52(Suppl), 171–185.

Sanger TD, Garg RR, Chen R (2001). Interactions between two different inhibitory systems in the human motor cortex. *Journal of Physiology* 530.2, 307–317.

Schönle PW, Isenberg C, Crozier TA, Dressler D, Machetanz J, Conrad B (1989). Changes of transcranially evoked motor responses in man by midazolam, a short acting benzodiazepine. *Neuroscience Letters* 101, 321–324.

Schwenkreis P, Witscher K, Janssen F, Addo A, Dertwinkel R, Zenz M, et al. (1999). Influence of the N-methyl-D-aspartate antagonist memantine on human motor cortex excitability. *Neuroscience Letters* 270, 137–140.

Schwenkreis P, Liepert J, Witscher K, Fischer W, Weiller C, Malin J-P, et al. (2000). Riluzole suppresses motor cortex facilitation in correlation to its plasma level. *Experimental Brain Research* 135, 293–299.

Shank RP, Gardocki JF, Streeter AJ, Maryanoff BE (2000). An overview of the preclinical aspects of topiramate: pharmacology, pharmacokinetics, and mechanism of action. *Epilepsia* 41 Suppl 1, S3–9.

Shimazu H, Maier MA, Cerri G, Kirkwood PA, Lemon RN (2004). Macaque ventral premotor cortex exerts powerful facilitation of motor cortex outputs to upper limb motoneurons. *Journal of Neuroscience* 24, 1200–1211.

Siebner HR, Dressnandt J, Auer C, Conrad B (1998). Continuous intrathecal baclofen infusions induced a marked increase of the transcranially evoked silent period in a patient with generalized dystonia. *Muscle and Nerve* 21, 1209–1212.

Silbert LC, Nelson C, Holman S, Eaton R, Oken BS, Lou JS, et al. (2006). Cortical excitability and age-related volumetric MRI changes. *Clinical Neurophysiology* 117, 1029–1036.

Smith MJ, Keel JC, Greenberg BD, Adams LF, Schmidt PJ, Rubinow DA, et al. (1999). Menstrual cycle effects on cortical excitability. *Neurology* 53, 2069–2072.

Sohn YH, Kaelin-Lang A, Jung HY, Hallett M (2001). Effect of levetiracetam on human corticospinal excitability. *Neurology* 57, 858–863.

Specterman M, Bhuiya A, Kuppuswamy A, Strutton PH, Catley M, Davey NJ (2005). The effect of an energy drink containing glucose and caffeine on human corticospinal excitability. *Physiology and Behavior* 83, 723–728.

Strafella AP, Paus T (2001). Cerebral blood-flow changes induced by paired-pulse transcranial magnetic stimulation of the primary motor cortex. *Journal of Neurophysiology* 85, 2624–2629.

Strafella AP, Paus T, Barrett J, Dagher A (2001). Repetitive transcranial magnetic stimulation of the human prefrontal cortex induces dopamine release in the caudate nucleus. *Journal of Neuroscience* 21, RC157.

Strafella AP, Paus T, Fraraccio M, Dagher A (2003). Striatal dopamine release induced by repetitive transcranial magnetic stimulation of the human motor cortex. *Brain* 126, 2609–15.

Strafella AP, Ko JH, Grant J, Fraraccio M, Monchi O (2005). Corticostriatal functional interactions in Parkinson's disease: a rTMS/[11C]raclopride PET study. *European Journal of Neuroscience 22*, 2946–2952.

Strafella AP, Ko JH, Monchi O (2006). Therapeutic application of transcranial magnetic stimulation in Parkinson's disease: the contribution of expectation. *NeuroImage 31*, 1666–1672.

Tergau F, Wischer S, Somal HS, Nitsche MA, Joe Mercer A, Paulus W, *et al.* (2003). Relationship between lamotrigine oral dose, serum level and its inhibitory effect on CNS: insights from transcranial magnetic stimulation. *Epilepsy Research 56*, 67–77.

Tokimura H, Ridding MC, Tokimura Y, Amassian VE, Rothwell JC (1996). Short latency facilitation between pairs of threshold magnetic stimuli applied to human motor cortex. *Electroencephalography and Clinical Neurophysiology 101*, 263–72.

Tokimura H, Di Lazzaro V, Tokimura Y, Oliviero A, Profice P, Insola A, *et al.* (2000). Short latency inhibition of human hand motor cortex by somatosensory input from the hand. *Journal of Physiology 523*, 503–513.

Valls-Sole J, Pascual-Leone A, Wassermann EM, Hallett M (1992). Human motor evoked responses to paired transcranial magnetic stimuli. *Electroencephalography and Clinical Neurophysiology 85*, 355–64.

Vucic S, Howells J, Trevillion L, Kiernan MC (2006). Assessment of cortical excitability using threshold tracking techniques. *Muscle and Nerve 33*, 477–486.

Wassermann EM (2002). Variation in the response to transcranial magnetic brain stimulation in the general population. *Clinical Neurophysiology 113*, 1165–1171.

Wassermann EM, Greenberg BD, Nguyen MB, Murphy DL (2001). Motor cortex excitability correlates with an anxiety-related personality trait. *Biological Psychiatry 50*, 377–382.

Werhahn KJ, Kunesch E, Noachtar S, Benecke R, Classen J (1999). Differential effects on motorcortical inhibition induced by blockade of GABA uptake in humans. *Journal of Physiology (London) 517*, 591–597.

Wischer S, Paulus W, Sommer M, Tergau F (2001). Piracetam affects facilitatory I-wave interaction in the human motor cortex. *Clinical Neurophysiology 112*, 275–279.

Wohlfarth K, Schneider U, Haacker T, Schubert M, Schulze-Bonhage A, Zedler M, *et al.* (2000). Acamprosate reduces motor cortex excitability determined by transcranial magnetic stimulation. *Neuropsychobiology 42*, 183–186.

Ziemann U (2004). TMS and drugs. *Clinical Neurophysiology 115*, 1717–1729.

Ziemann U, Rothwell JC (2000). I-waves in motor cortex. *Journal of Clinical Neurophysiology 17*, 397–405.

Ziemann U, Netz J, Szelenyi A, Hömberg V (1993). Spinal and supraspinal mechanisms contribute to the silent period in the contracting soleus muscle after transcranial magnetic stimulation of human motor cortex. *Neuroscience Letters 156*, 167–171.

Ziemann U, Lönnecker S, Paulus W (1995). Inhibition of human motor cortex by ethanol. A transcranial magnetic stimulation study. *Brain 118*, 1437–1446.

Ziemann U, Bruns D, Paulus W (1996a). Enhancement of human motor cortex inhibition by the dopamine receptor agonist pergolide: evidence from transcranial magnetic stimulation. *Neuroscience Letters 208*, 187–190.

Ziemann U, Lönnecker S, Steinhoff BJ, Paulus W (1996b). The effect of lorazepam on the motor cortical excitability in man. *Experimental Brain Research 109*, 127–35.

Ziemann U, Lönnecker S, Steinhoff BJ, Paulus W (1996c). Effects of antiepileptic drugs on motor cortex excitability in humans: a transcranial magnetic stimulation study. *Annals of Neurology 40*, 367–78.

Ziemann U, Rothwell JC, Ridding MC (1996d). Interaction between intracortical inhibition and facilitation in human motor cortex. *Journal of Physiology (London) 496*, 873–81.

Ziemann U, Tergau F, Bruns D, Baudewig J, Paulus W (1997). Changes in human motor cortex excitability induced by dopaminergic and anti-dopaminergic drugs. *Electroencephalography and Clinical Neurophysiology 105*, 430–437.

Ziemann U, Chen R, Cohen LG, Hallett M (1998a). Dextromethorphan decreases the excitability of the human motor cortex. *Neurology 51*, 1320–1324.

Ziemann U, Tergau F, Wassermann EM, Wischer S, Hildebrandt J, Paulus W (1998b). Demonstration of facilitatory I-wave interaction in the human motor cortex by paired transcranial magnetic stimulation. *Journal of Physiology (London) 511*, 181–190.

Ziemann U, Tergau F, Wischer S, Hildebrandt J, Paulus W (1998c). Pharmacological control of facilitatory I-wave interaction in the human motor cortex. A paired transcranial magnetic stimulation study. *Electroencephalography and Clinical Neurophysiology 109*, 321–330.

CHAPTER 14

Transcranial stimulation measures explored by epidural spinal cord recordings

Vicenzo Di Lazzaro

Introduction

Initial studies in animals have shown that in response to a single electrical stimulus to the motor cortex, an electrode placed in the medullary pyramid or on the dorsolateral surface of the cervical spinal cord records a series of high-frequency waves (Adrian and Moruzzi 1939; Patton and Amassian 1954; Kernell and Wu 1967). The earliest wave that persisted after cortical depression and after cortical ablation was thought to originate from the direct activation of the axons of the fast pyramidal tract neurons (PTN) and was termed D-wave. The later waves that required intact gray matter were thought to originate from indirect, trans-synaptic activation of PTNs and were termed I-waves (Patton and Amassian 1954). Recordings from individual PTN axons showed that a given axon may produce both a D- and a subsequent I-wave discharge (Patton and Amassian 1954).

Direct evidence for the action of TMS and transcranial electrical stimulation (TES) on the human motor cortex was initially provided by recording from the surface of the spinal cord during spinal cord surgery (Boyd *et al.* 1986;

Berardelli *et al.* 1990; Thompson *et al.* 1991; Burke *et al.* 1993). Although the data were very useful, interpretation was limited because the patients were anesthetized. More recently, descending volleys have been recorded in conscious human subjects with no CNS abnormality who had electrodes implanted in the spinal cord for the relief of otherwise intractable pain (Kaneko *et al.* 1996; Nakamura *et al.* 1996; Di Lazzaro *et al.* 2004a for a review). These recordings provide detailed insights about the physiological basis of the excitatory and inhibitory phenomena produced by single, paired, and repetitive transcranial stimulation of the human brain.

Direct recording of the output of the motor cortex showed that transcranial stimulation can evoke several different kinds of descending activities depending on the type of stimulation (magnetic or electrical) and, in the case of magnetic stimulation, on the direction of the induced current in the brain, the intensity of the stimulus, the phases of the stimulating current (monophasic or biphasic), and the shape of the coil (Di Lazzaro *et al.* 2004a). The output also depends upon the representation of the motor cortex being stimulated (upper or lower limb area) (Di Lazzaro *et al.* 2004a).

Transcranial stimulation of the hand area of the motor cortex

Electrical stimulation

Anodal stimulation

Electrical (anodal) stimulation of the motor cortex at active motor threshold intensity (the minimum stimulus intensity that produces a response of ~100 µV in 50% of 10 trials) evokes in epidural recordings of the high cervical cord a single negative wave with a latency of 2.0–2.6 ms (Figure 14.1) (Di Lazzaro *et al.* 1998a). The short latency of this wave suggests that it originates from direct activation of corticospinal axons just below the gray matter, and that it is the equivalent of the D-wave described by Patton and Amassian (1954) after direct stimulation of the exposed motor cortex in monkeys. Voluntary contraction has no effect on the amplitude of the descending wave evoked by anodal stimulation (Di Lazzaro *et al.* 1999a) (Figure 14.2). The fact that this wave is not modified by changes in cortical excitability produced by voluntary contraction supports the hypothesis that it is due to activation of corticospinal axons in the subcortical white matter at some distance from the cell body. It presumably results from excitation of corticospinal axons at a particular low-threshold point (Amassian *et al.* 1992).

Cathodal stimulation

In one subject, the responses to cathodal stimulation were examined (Di Lazzaro *et al.* 1999a). The initial wave recruited by cathodal stimulation in this subject had a slightly longer latency than the one evoked by anodal stimulation, suggesting that it was recruited slightly closer to the cell body than the wave evoked by anodal stimulation.

Magnetic stimulation

Single-pulse TMS

Posterior–anterior induced current in the brain

With a monophasic waveform, the lowest threshold for evoking a motor-evoked potential (MEP) from the hand area of the cortex is

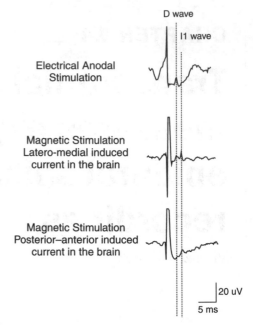

Fig. 14.1 Descending volleys evoked by different techniques of transcranial stimulation at around threshold intensity. The traces show averaged epidural recordings from the high cervical cord. Electrical anodal stimulation evokes the earliest volley (D-wave) indicated by the left vertical dotted line. The wave recruited at threshold intensity by magnetic stimulation with posterior–anterior (PA) induced current in the brain (I1-wave, indicated by the second dotted line) appears 1.4 ms later than the anodal D-wave. A clear D-wave and a small I1-wave are recruited by magnetic stimulation with latero-medial (LM) induced current in the brain.

obtained with a stimulus that induces a posterior–anterior (PA) current across the central sulcus in the brain. This lowest-threshold volley occurs at a latency 1.0–1.4 ms longer than the volley recruited by electrical anodal stimulation (Di Lazzaro *et al.* 1998a) (Figure 14.1). This volley increases in size, and is followed by later volleys as the intensity of stimulation increases (Figure 14.3). Since the earliest volley elicited by electrical stimulation is probably a D-wave, the later volleys recruited by magnetic stimulation represent I-waves. The interpeak interval between I-waves is ~1.4 ms, which indicates a discharge frequency of ~700 Hz. At a magnetic stimulus intensity of ~180–200% active motor

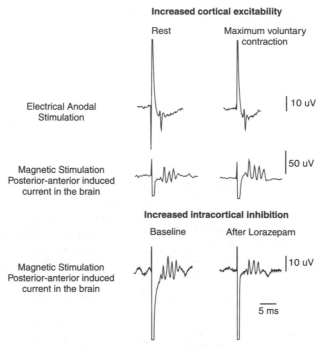

Fig. 14.2 Effects of changes in cortical excitability on the output of the motor cortex elicited by transcranial electrical and magnetic stimulation. *Upper traces*: descending volleys evoked by electrical anodal stimulation, at rest and during maximum voluntary contraction in one subject. Voluntary contraction at maximum strength does not modify D-wave amplitude. Adapted from Figure 1 in Di Lazzaro *et al.* (1999a). *Middle traces*: descending volleys evoked by TMS with a figure-8 coil and with posterior–anterior (PA) induced current in the brain at rest and during maximum voluntary contraction in one subject. At rest the magnetic stimulus evokes a small D-wave and three I-waves; during maximal voluntary contraction the size of the waves increases and a fourth I-wave becomes visible. Adapted from Figure 5 in Di Lazzaro *et al.* (1998b). *Lower traces*: descending volleys evoked by TMS with a figure-8 coil and with PA-induced current in the brain, at baseline and 2 h after a single oral dose of 2.5 mg of lorazepam. The earliest low-amplitude wave has the same latency as the lowest-threshold volley evoked by anodal stimulation (not illustrated) and therefore it is a D-wave; later waves represent I-waves. The D- and I1-waves are unaffected by lorazepam in contrast to later I-waves which are smaller after lorazepam. Adapted from Figure 1 in Di Lazzaro *et al.* (2000).

threshold, an earlier wave of small amplitude appears. This wave has the same latency as the D-wave evoked by electrical stimulation at threshold.

Changes in cortical excitability produce changes in the output of the motor cortex to TMS (cf. Chapters 13 and 15, this volume). An increase in cortical excitability produced by voluntary contraction of the target muscle results in an increase in the output of the motor cortex to TMS (Di Lazzaro *et al.* 1998b). Voluntary contraction increases the size and number of epidural volleys evoked by a given intensity of magnetic stimulation and the amplitude of the descending waves is higher during activity (particularly during maximum contractions) than at rest (Figure 14.2). This effect can be substantial: maximum contraction can increase the total amplitude of the volleys by 50%. This large effect on the size of the descending corticospinal waves is not paralleled by a comparable effect on the threshold for evoking recognizable activity after TMS. Voluntary activation of motor cortex has only a small effect on the threshold of

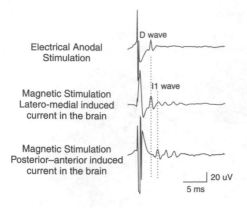

Electrical Anodal
Stimulation

D wave

Magnetic Stimulation
Latero-medial induced
current in the brain

I1 wave

Magnetic Stimulation
Posterior–anterior induced
current in the brain

20 uV

5 ms

Fig. 14.3 Averaged descending volleys evoked by electrical anodal, latero-medial (LM), and posterior–anterior (PA) magnetic stimulation at suprathreshold intensity in one conscious patient with no CNS abnormality. Electrical anodal stimulation evokes a single short-latency descending volley (D-wave). PA TMS recruits three descending volleys (I-waves), the earliest of which appears 1.3 ms later than the D-wave evoked by anodal stimulation. LM TMS recruits both D- and I-waves.

descending activity, and only in the minority of subjects. This suggests that the elements activated by PA magnetic stimulation of the upper limb area of motor cortex have a relatively constant threshold. The likely explanation is that magnetic stimulation activates axons at some distance from the cell body so that threshold is unaffected by synaptic activity. These axons cannot belong to corticospinal neurons since no D-wave is elicited at threshold intensity. Presumably they are cortico-cortical axons projecting onto corticospinal neurons.

The opposite effect on the output of the motor cortex is observed when the excitability of the motor cortex is decreased. A decrease in the level of cortical excitability can be produced by a pharmacological enhancement of inhibitory gamma-aminobutyric acid (GABA)ergic activity, for instance by benzodiazepine administration. After lorazepam, the output of the motor cortex to TMS is reduced (Di Lazzaro et al. 2000) (cf. Chapter 13, this volume). Neither the threshold nor the amplitude of the lowest-threshold volley evoked by TMS is affected by lorazepam. In contrast, lorazepam results

in a pronounced suppression of later waves (Figure 14.2).

This sensitivity of these later I-waves evoked by TMS to experimental manipulation of cortical excitability suggests that they are generated within cortical networks and that they are presynaptic in origin.

Latero-medial induced current in the brain

When the orientation of the figure-8 coil is changed, so that monophasic currents in the brain are induced in a lateral–medial (LM) direction, the MEP tends to have the same latency as the one evoked by electrical anodal stimulation (Werhahn et al. 1994). At liminal stimulus intensities, TMS with LM-induced current in the brain usually recruits two volleys (Di Lazzaro et al. 1998a) (Figure 14.1). The earliest volley has the same latency of the D-wave evoked by electrical anodal stimulation and the second one has the same latency of the I1-wave evoked by TMS with PA-induced current in the brain. This suggests that TMS with LM-induced current in the brain excites PTNs directly at their axons and, in addition, trans-synaptically.

With LM stimulation at suprathreshold intensity, the I1-wave may be relatively small, whereas the I3 may be much larger. Similar behavior can be seen in the experimental studies of Kernell and Wu (1967) on anodal stimulation of the exposed monkey motor cortex. This suggests that after direct activation of the corticospinal axons the stimulus spreads both orthodromically and antidromically, making the axon itself partially refractory to the I1-wave. When, with larger stimuli, most of the axons are activated directly there are no axons that can respond to indirect activation of PTNs in the latency range of the I1-wave.

Anterior–posterior-induced current in the brain

When reversing the direction of the induced current in the brain from the usual PA direction to the anterior–posterior (AP) direction, MEPs in small hand muscles have an onset latency that is ~3 ms longer than the one seen with PA stimulation (Sakai et al. 1997). It has been suggested, therefore, that stimulation with an AP-induced current can, at least in some individuals, preferentially recruit I3-waves in the pyramidal tract (Sakai et al. 1997). The direct recording of the output of the motor cortex produced by TMS with AP-induced current shows that reversing the direction of stimulating current does not simply reverse the order of recruitment of descending activity (i.e. an I3-wave before an I1-wave) in all subjects, but leads to a more complex pattern of activation (Di Lazzaro et al. 2001a). In some subjects, AP stimulation recruits a later I-wave rather than an I1-wave (Figure 14.4), while in other subjects, I1- or D-waves occur, but the latency of these waves is slightly longer than the corresponding waves evoked by PA magnetic stimulation. Also at suprathreshold intensity, the descending volleys evoked by AP stimulation have slightly different peak latencies and/or longer duration than those seen after PA stimulation (Di Lazzaro et al. 2001a).

These findings suggest that it is unlikely that AP and PA stimulation activate different proportions of the same cortical elements at the same sites. It appears more likely that they activate different sites, and perhaps even different populations of cortical neurons. Indeed, the folding of the cerebral cortex means that there will be many neurons with axons bent in a particular orientation. Some of these will be best stimulated by one direction of current flow, whereas others will be best stimulated by another current direction.

Biphasic magnetic stimulation

When using biphasic magnetic stimulation, the output is less consistent compared to monophasic stimulation (Di Lazzaro et al. 2001b). At active threshold intensity, PA–AP stimulation evokes either an I1-wave that has the same latency as the I1-wave after monophasic PA, or a delayed I1-wave together with a delayed D-wave, or a delayed I3-wave, each of which has the same latency as the I3-wave evoked by AP stimulation. At intermediate stimulus intensities, the recruitment of additional waves is observed, some of the waves are similar to those observed with monophasic AP stimulation, and some have the same latency of the waves evoked by PA stimulation. At high stimulus intensities, the I-waves evoked by PA–AP stimulation are similar to those evoked by monophasic PA stimulation. At all intensities the latency of the D-wave is longer than the latency of the D-wave from anodal stimulation. Using AP-PA stimulation, active threshold is lower than the one to PA–AP stimulation, and the pattern of recruitment of D- and I-waves at increasing stimulus intensities resembles the one with monophasic PA stimulation more than with AP stimulation. Therefore, the direct recording of epidural volleys suggests that both phases of the biphasic pulse are capable of activating descending motor output, and that biphasic magnetic pulses produce a more complex pattern of corticospinal activation than monophasic pulses. Both phases of the stimulus pulse can activate descending pathways, but the precise combination of neural elements activated by AP and PA directions depends on their relative threshold and the relative amplitude of the AP and PA phases.

Effects of fatigue on descending volleys

A brief period of strong muscle contraction can decrease the amplitude of MEPs in the resting muscle for several minutes after the end of the contraction, and this is associated with a decrease in the amplitude of all I-waves recruited by a standard LM magnetic pulse (Di Lazzaro et al. 2003). This depressive effect can be large, with the total volley being reduced by 40–50% for 2–3 min after a strong contraction (Figure 14.5). Interestingly, the D-wave is also

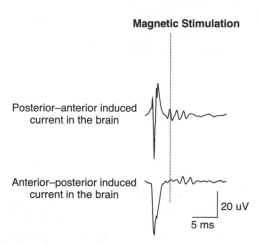

Magnetic Stimulation

Posterior–anterior induced current in the brain

Anterior–posterior induced current in the brain

20 uV

5 ms

Fig. 14.4 Averaged descending volleys evoked at suprathreshold intensity by posterior–anterior (PA) magnetic stimulation and anterior–posterior (AP) magnetic stimulation in one conscious patient with no abnormality of CNS. PA TMS recruits three descending volleys (I-waves). AP TMS recruits only later I-waves.

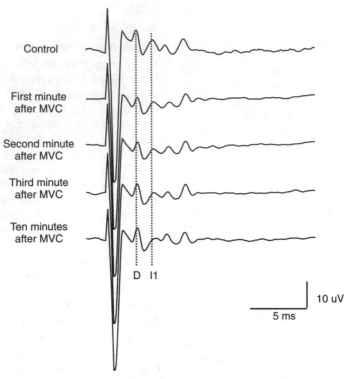

Fig. 14.5 Descending volleys recorded from the epidural surface after latero-medial TMS in baseline conditions and after 2 min of maximal voluntary contraction (MVC) in one conscious subject. Each trace is the mean of 12 trials. The peak latency of the volley originating from direct activation of corticospinal axons (D-wave) and the peak latency of the first volley evoked by indirect (trans-synaptic) activation of pyramidal cells (I1-wave) through the activation of cortico-cortical axons are indicated by the dotted lines. All the components of the descending volleys are reduced after the contraction. Ten minutes after the end of contraction the descending volleys have almost completely recovered to baseline values. Adapted from Figure 1 in Di Lazzaro *et al.* (2003).

reduced for 2 min in some of the patients. This would be consistent with a transient decrease in axon excitability of corticospinal neurons caused by fatiguing voluntary activation of these neurons.

Effects of brain lesions on the output of the motor cortex

The role of cortical and subcortical structures in the generation of the I-waves was further evaluated by recording directly with epidural electrodes from the spinal cord the output of the motor cortex in one patient with cerebral cortex atrophy (Di Lazzaro *et al.* 2004b) and in one patient with multiple vascular thalamic lesions (Di Lazzaro *et al.* 2002a). The patient with cerebral cortex atrophy had a history of >20 years of heavy alcohol abuse. In this patient, I-waves were lacking in the descending corticospinal discharge (Figure 14.6). Despite this, the D-wave was intact and of normal latency. The most likely explanation is that brain atrophy caused by the chronic alcohol abuse had compromised the function of the interneuronal circuitry responsible for I-wave generation. One possible candidate for the generation of I-waves after TMS is thalamocortical projections from the lateral and anterior ventral thalamic nuclei, known to activate large pyramidal tract neurons monosynaptically (Amassian and Weiner 1966; Ziemann and Rothwell 2000). However, experimental data in the cat suggest that projections from the thalamus are not essential for the

Magnetic Stimulation
Posterior–anterior induced current in the brain

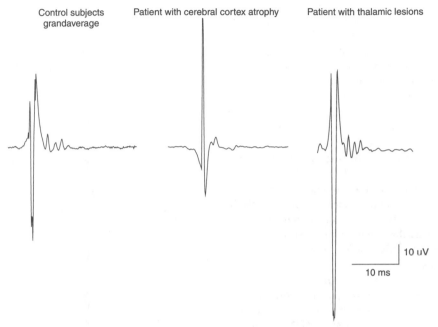

Control subjects   Patient with cerebral cortex atrophy   Patient with thalamic lesions
grandaverage

10 uV

10 ms

Fig. 14.6 Descending volleys evoked by posterior–anterior (PA) magnetic stimulation at suprathreshold intensity in five patients with no abnormality of the CNS (*left*), in one patient with cerebral cortex atrophy (*middle*). Adapted from Figure 1 in Di Lazzaro *et al*. 2004b) and in one patient with a vascular thalamic lesion (*right*). In the average of the control subjects, PA stimulation evokes only I-waves. In the patient with cerebral cortex atrophy, magnetic stimulation evokes only a large D-wave but no clear I-waves. Normal I-wave activity is recorded in the patient with the thalamic lesion.

production of I-waves (Amassian *et al.* 1987). In accord with this notion, in the patient with extensive vascular lesions in the thalamus, PA magnetic stimulation of the right motor cortex evoked a completely normal descending activity with three I-waves and a small D-wave (Figure 14.6). Cumulatively, the results from these two patients suggest that projections from the thalamus to the motor cortex are not involved in the generation of I-waves but that these much more likely originate through activation of cortico-cortical connections.

Paired-pulse TMS interactions

Paired-pulse experiments are designed to give insight into the nature of the cortical circuitry activated by TMS (cf. Chapter 11, this volume). A variety of different methods exist to examine the connections within the motor cortex itself,

or the connections to the motor cortex from other parts of the CNS. Direct recordings of the descending corticospinal volleys have not only confirmed the mechanisms of these paired-pulse TMS effects as they had been inferred from noninvasive MEP recordings, but in some cases revealed an unexpected selective involvement of different components (D- and/or I-waves) of the complex corticospinal volley.

Short-interval intracortical inhibition (SICI)

Kujirai *et al.* (1993) were the first to report SICI (Chapters 11–13, this volume). They found that a subthreshold conditioning stimulus could suppress an MEP to a later suprathreshold test stimulus if the interval between the stimuli was ≤5 ms. Since the conditioning stimulus was below active motor threshold, the authors suggested that the interaction occurred at the cortical level and that the conditioning stimulus suppressed the recruitment of descending volleys by the test stimulus. Direct recordings of the descending volleys have confirmed this (Figure 14.7) (Di Lazzaro *et al.* 1998c). A subthreshold conditioning stimulus that does not itself evoke descending corticospinal activity can produce very clear suppression of late I-waves if the interval to the suprathreshold test stimulus is between 1 and 5 ms (Figure 14.7). Interestingly, the I1-wave is virtually unaffected, while inhibition affects most conspicuously the I3-wave and later volleys. Hanajima *et al.* (1998), on the basis of studies of single motor units, obtained similar data (cf. Chapter 11, this volume). Since the I1-wave remains unaffected, this suggests that SICI does not modify directly the excitability of pyramidal neurons. It is more likely that the inhibition in this paradigm is due to activation of other intracortical elements.

Kujirai *et al.* (1993) originally suggested that SICI is GABAergic in origin (cf. Chapter 13, this volume). Administration of lorazepam, a positive allosteric modulator at the $GABA_A$ receptor ($GABA_A R$), increases the amount of SICI and also increases the inhibition of the late descending I-waves (Figure 14.8) (Di Lazzaro *et al.* 2000), thus supporting the notion that SICI is mediated through the $GABA_A R$.

Short-interval intracortical facilitation (SICF)

SICF was first described by Tokimura *et al.* (1996) and Ziemann *et al.* (1998) (cf. Chapters 11 and 13, this volume). If two stimuli are given at an intensity at or above active motor threshold, then MEP facilitation can be observed at discrete interstimulus intervals of around 1.3, 2.5 and 4.3 ms. This facilitatory interaction between the two pulses is thought to be due to the interaction of I-wave inputs in the periodic bombardment of pyramidal neurons. Epidural recordings of the descending corticospinal volleys from the spinal cord demonstrated the interaction on individual I-waves very clearly (Figure 14.9) (Di Lazzaro *et al.* 1999b). At an interval of 1.2 ms, later I-waves are clearly facilitated, whereas the I1-wave is unchanged. Ilić *et al.* (2002) reached a similar conclusion in SICF experiments by using electromyogram (EMG) surface and single motor unit recordings at an ISI of 1.5 ms.

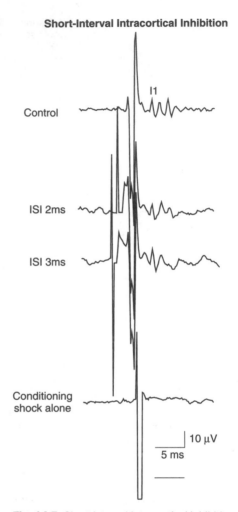

Short-Interval Intracortical Inhibition

Control

I1

ISI 2ms

ISI 3ms

Conditioning shock alone

10 µV

5 ms

Fig. 14.7 Short-interval intracortical inhibition. Epidural volleys evoked by the test stimulus alone (*upper trace*), conditioning stimulus alone (*lower trace*), and both stimuli at different interstimulus intervals (ISIs) (*middle traces*) in one subject. Recordings were performed at rest during test and paired magnetic stimulation, while the conditioning stimulus alone was delivered during voluntary contraction at ~20% of maximum. Each trace is the average of five sweeps. The test stimulus evokes multiple descending waves (four waves). The conditioning shock alone evokes no descending volleys. When both stimuli are delivered together, the later I-waves were markedly suppressed at ISIs of 2 and 3 ms. Adapted from Figure 1 in Di Lazzaro *et al.* (1998c).

Short-Interval Intracortical Inhibition

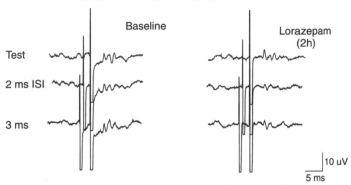

Fig. 14.8 Short-interval intracortical inhibition. *Left*: Epidural volleys evoked by the test stimulus alone and both conditioned and test stimuli at different interstimulus intervals (ISIs) in baseline conditions. Each trace is the average of 10 sweeps. The test stimulus evokes multiple descending waves (three waves). When both stimuli were delivered together, the epidural volleys were inhibited at ISIs of 2 and 3 ms. *Right*: After lorazepam the conditioning stimulus reduces the descending volleys more than at baseline. Adapted from Figure 2 in Di Lazzaro *et al.* (2000).

Long-interval intracortical inhibition (LICI)

Valls-Sole *et al.* (1992) were the first to describe long-interval intracortical inhibition (LICI) (cf. Chapters 11–13, this volume). They found that a conditioning stimulus strong enough to produce an MEP in the target muscle could suppress an MEP to a later stimulus of the same intensity if the interstimulus interval was 50–200 ms. Since the excitability of spinal H-reflexes had recovered at this interval, the effect was thought to originate at the cortical level. Nakamura *et al.* (1997) and Chen *et al.* (1999) recorded descending corticospinal volleys in this paired-pulse paradigm and showed consistently that the corticospinal volley to the second stimulus was considerably reduced. This was confirmed by another study in three subjects, in which it was demonstrated that later I-waves are reduced at interstimulus intervals of 100 and 150 ms, but that the D- and I1-waves remain unaffected (Di Lazzaro *et al.* 2002a) (Figure 14.10). In addition, facilitation of the late I-waves at an interstimulus interval of 50 ms occurs, suggesting that the suppression of the MEP at this interval is due to inhibition at spinal cord rather than cortex (Di Lazzaro *et al.* 2002a).

Interhemispheric inhibition (IHI)

Ferbert *et al.* (1992) showed that conditioning TMS of the motor cortex of one hemisphere can inhibit the MEP elicited in distal hand muscles by a test magnetic stimulus given 6–30 ms later over the motor cortex of the opposite hemisphere (cf. Chapters 11 and 12, this volume). On the basis of indirect arguments, Ferbert *et al.* (1992) suggested that this IHI is produced at the cortical level via a transcallosal route. This was confirmed by direct epidural recordings of the descending corticospinal volley elicited by the test stimulus showing that the I2-wave and later I-waves were clearly suppressed by a conditioning stimulus to the opposite hemisphere given 6–10 ms earlier (Di Lazzaro *et al.* 1999c) (Figure 14.11). In contrast, the D-wave and the I1-wave remained unaffected (Figure 14.11).

Short latency afferent inhibition (SAI)

Afferent inputs can modify the excitability of the motor cortex with a complex time-course. A short-latency inhibition of MEPs in hand muscles is produced by electrical stimulation of peripheral nerves innervating the hand (Delwaide and Olivier 1990; Tokimura *et al.* 2000). This SAI (cf. Chapter 13, this volume) requires a minimum interstimulus interval that is ~1 ms longer than the latency of the N20 component of the somatosensory-evoked potential, and lasts for ~7–8 ms. MEP studies comparing responses to TMS and TES suggested that SAI has a cortical origin, and this was

Short-Interval Intracortical Facilitation

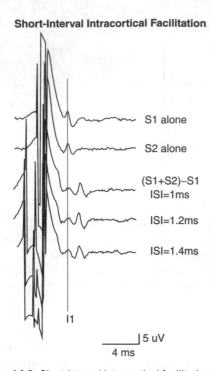

Long-Interval Intracortical Inhibition

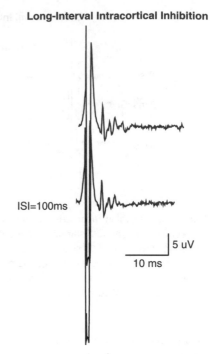

Fig. 14.9 Short-interval intracortical facilitation. Epidural volleys evoked by the conditioning magnetic stimulus alone (S1), test magnetic stimulus alone (S2), and both conditioning and test stimuli at different interstimulus intervals (ISIs, *lower traces*). Note that because the response to S1 alone has been subtracted from the combined response, the latter shows the portion of the response that is due to S2. By comparing them with the response to S2 alone it is possible to see how, using combined stimulation, the response to S2 is affected by a preceding S1. Recordings are performed during isometric voluntary contraction at ~20% of maximum. Each trace is the average of 10 sweeps. Conditioning magnetic stimulus alone evokes an I1-wave and a very small I2-wave. Test magnetic stimulus alone evokes one wave (I1). At ISIs of 1–1.4 ms, the I2-wave is clearly facilitated in the epidural recordings. Adapted from Figure 2 in Di Lazzaro *et al.* (1999b).

Fig. 14.10 Long-interval intracortical inhibition. Epidural volleys by single and paired cortical stimulation at long interstimulus interval (ISI) using a suprathreshold conditioning stimulus and a test stimulus of the same intensity given 100 ms later. The conditioning stimulus alone (control, *upper trace*) evokes multiple descending waves (four waves), while the output of the motor cortex produced by the test stimulus delivered 100 ms later is inhibited (*lower trace*). The latest I-wave (I4-wave) is suppressed and there is a slight inhibition of the I2- and I3-waves, whereas the I1-wave is not affected. Adapted from Figure 2 in Di Lazzaro *et al.* (2000).

confirmed in the epidural recordings of the descending corticospinal volley from the spinal cord (Tokimura *et al.* 2000). As with other forms of paired-pulse inhibition (see above), the I2-wave and later I-waves are strongly suppressed, whereas the D-wave and the I1-wave are unaffected (Figure 14.12).

Repetitive stimulation of the motor cortex

In recent years, several authors have used repetitive (r)TMS to produce changes in the excitability of the corticospinal system that outlast the period of stimulation (Pascual-Leone *et al.* 1994, 1998; Chen *et al.* 1997; Tergau *et al.* 1997; Berardelli *et al.* 1998; Maeda *et al.* 2000; Peinemann *et al.* 2000; Huang *et al.* 2005). For instance, 30 min of 1 Hz rTMS decreases the amplitude of the MEP elicited by single-pulse TMS for the next 30 min (Chen *et al.* 1997),

Interhemispheric Inhibition

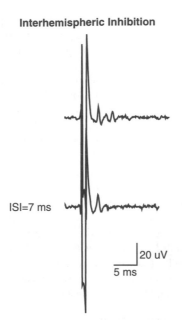

ISI=7 ms

|20 uV
5 ms

Short-Latency Afferent Inhibition

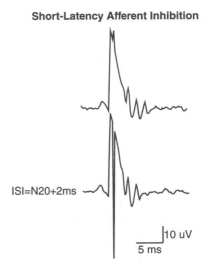

ISI=N20+2ms

|10 uV
5 ms

Fig. 14.12 Short-latency afferent inhibition. Average data from one patient with an implanted cervical epidural stimulator showing the effect of median nerve stimulation on descending volleys evoked by TMS. Each trace is the average of 10 trials recorded during muscle rest. The latency of the N20 somatosensory evoked potential has been subtracted from the interval between median nerve shock and cortical stimulus. The positive interval of 2 ms indicates that the cortical stimulus was applied after the presumed arrival of sensory input at the cortex. Note the inhibition of later waves of the corticospinal volley. The I1 component of the descending epidural volleys is not inhibited by median nerve stimulation. ISI, interstimulus interval. Adapted from Figure 4 in Tokimura et al. (2000).

Fig. 14.11 Interhemispheric Inhibition (IHI) is obtained when the magnetic test stimulus over one motor cortex is preceded by a conditioning stimulus applied to the opposite hemisphere 7 ms earlier. The test stimulus alone (control, *upper trace*) evokes multiple descending waves (three waves). When both stimuli are delivered, the later waves are suppressed, but the earliest (I1) wave is not modified (*lower trace*). Adapted from Figure 1 in Di Lazzaro et al. (1999c). ISI, interstimulus interval.

whereas higher rTMS frequencies may increase MEP amplitude (Berardelli et al. 1998; Maeda et al. 2000; Peinemann et al. 2000; Huang et al. 2005). Since spinal H-reflexes are unaffected, it is usually assumed that these rTMS after-effects are due to changes in neural circuits in the cortex, perhaps involving processes such as long-term depression (LTD) or potentiation (LTP) at cortical synapses (cf. Chapter 16, this volume). Two recent studies provided further evidence for the cortical origin of these rTMS effects by means of epidural recordings of the corticospinal volleys from the spinal cord (Di Lazzaro et al. 2002b,c). It was found that suprathreshold 5 Hz rTMS of motor cortex is accompanied by a gradual increase in the size and number of descending corticospinal volleys evoked by each

TMS pulse, and that this effect parallels the increase in MEP amplitude (Di Lazzaro et al. 2002b) (Figure 14.13). Subthreshold 5 Hz rTMS (total of 50 stimuli at an intensity of active motor threshold) has no effect on MEP amplitude but reduces SICI (Di Lazzaro et al. 2002c) (Figure 14.14). This suggests that low-intensity rTMS at 5 Hz can selectively modify the excitability of GABAergic networks in the human motor cortex.

Huang et al. (2005) have recently described a rapid method of reducing excitability in the motor cortex termed continuous theta burst stimulation (cTBS). The protocol uses a short burst of low-intensity (80% of active motor

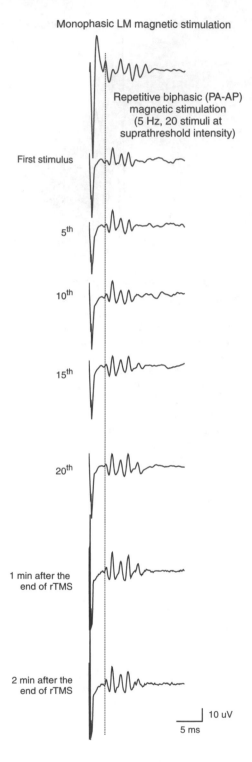

Monophasic LM magnetic stimulation

Repetitive biphasic (PA-AP)
magnetic stimulation
(5 Hz, 20 stimuli at
suprathreshold intensity)

First stimulus

5th

10th

15th

20th

1 min after the
end of rTMS

2 min after the
end of rTMS

10 uV

5 ms

threshold), high-frequency (50 Hz) pulses repeated at 5 Hz, the frequency of the theta rhythm in the electroencephalogram. Twenty seconds of cTBS reduce the amplitude of MEPs significantly for ~1 h. Recordings of motor cortical output in four conscious subjects who had cervical spinal electrodes implanted chronically for control of pain showed that cTBS leads to a pronounced decrease in the excitability of cortical circuits generating the I1-wave, whereas later I-waves are much less affected (Figure 14.15). This particular effect on the I1-wave contrasts with other TMS protocols (SICI, IHI, SAI, see above) that preferentially suppress the later I-waves but leave the I1-wave virtually unchanged (Di Lazzaro *et al.* 2005). This suggests that cTBS has its major effect on the synapse between the interneurons responsible for the I1-wave and the corticospinal neurons.

Nonfocal stimulation of the brain with a round coil

Studies in monkeys (Edgley *et al.* 1990) suggested that the lowest threshold volley evoked by a circular coil centered over the vertex originates from direct activation of the corticospinal axons at the initial segment. In anesthetized human subjects, Burke *et al.* (1993), by using a circular

Fig. 14.13 Epidural volleys evoked by monophasic latero-medial (LM) magnetic stimulation at baseline (*upper trace*) and by biphasic repetitive (r)TMS (20 stimuli at 5 Hz) and epidural volleys evoked 1 and 2 min after the end of rTMS. Each trace is the average of five sweeps. The first stimulus of the rTMS train evokes three I-waves and a small D-wave (slightly delayed of ~0.3 ms in respect to the wave evoked by monophasic LM magnetic stimulation indicated by the vertical dotted line). The amplitude of the D-waves increases with subsequent stimuli and remains substantially stable thereafter. At the 15th stimulus a later I-wave appears. The amplitude of this latest wave increases with subsequent stimuli. Both the D-wave and the latest I-wave are still facilitated 1 and 2 min after the end of rTMS. Adapted from Figure 1 in Di Lazzaro *et al.* (2002b).

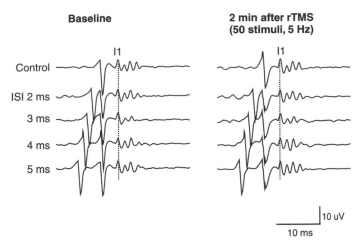

Fig. 14.14 Short-interval intracortical inhibition in a patient with high-cervical epidural electrode at baseline and after 5 Hz rTMS (50 stimuli at active motor threshold intensity). At baseline, the test stimulus evoked multiple descending waves (four waves); the first wave indicated by the dotted line is an I1-wave. When both stimuli are delivered the epidural volleys are suppressed at interstimulus intervals (ISIs) of 1–5 ms. After rTMS, the conditioning stimulus produces a lesser degree of inhibition of the descending volleys. rTMS seems to have the greatest influence on the last wave (I4) at an ISI of 5 ms. The I4-wave is suppressed at this ISI before rTMS, whereas it is no longer inhibited after rTMS. Adapted from Figure 2 in Di Lazzaro et al. (2002c).

coil centered over the vertex and recording from the spinal cord, confirmed that the D-wave is the component of lowest threshold.

The recording of epidural volleys evoked by stimulation with a circular coil centered over the vertex in conscious human subjects showed that this form of nonfocal TMS may produce different results in different subjects (Di Lazzaro et al. 2002d). The earliest volley evoked by a circular coil centered over the vertex is a descending wave that has an ~0.2 ms longer latency than the D-wave evoked by LM magnetic or anodal electrical stimulation. Either this wave or a wave corresponding to the I1-wave evoked by PA magnetic stimulation can be the lowest-threshold volley when using a circular coil. At suprathreshold intensities later waves are recruited; these waves, in some cases, may have a latency that is outside the periodicity of I-waves evoked by PA magnetic stimulation. At higher intensities the I-waves evoked by nonfocal magnetic stimulation are similar to those evoked by monophasic PA stimulation. Maximum voluntary

contraction increases the amplitude of all descending volleys including the earliest (D) wave (Figure 14.16). Similar to LM stimulation, the main I-wave recruited by nonfocal stimulation at suprathreshold intensity is in the latency range of the I3-wave, whereas the I1-wave is not prominent. The duration of the negative phase of the early (D-) wave is ~50% longer than that of the negative phase of a similar sized D-wave evoked by LM stimulation.

These data show that there is a major difference between nonfocal stimulation of the hand area with a circular coil and focal stimulation with a figure-8 coil, even when the induced current flows in the PA direction in both cases. Nonfocal stimulation is more likely to evoke a D-wave than PA focal stimulation. Moreover, the early (D-) volley recruited by nonfocal stimulation has a slightly longer latency than the LM or anodal D-wave. It is also facilitated by voluntary contraction. These features suggest that it is initiated closer to the cell body of the pyramidal neurons than the conventional D-wave evoked

Continuous theta burst stimulation (cTBS)

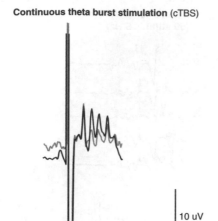

10 uV

5 ms

Fig. 14.15 Effects of continuous theta burst repetitive stimulation (cTBS) on epidural volleys evoked by posterior–anterior magnetic stimulation. When compared with baseline (black traces) the amplitude of the I1-wave is reduced after cTBS (gray traces). Adapted from Figure 1 in Di Lazzaro et al. (2005).

by LM magnetic or anodal stimulation, perhaps at the initial axon segment rather than at some distance down the axon.

Stimulation of the lower limb area of the motor cortex

Previous studies using surface EMG and single motor unit recordings suggest that magnetic stimulation over the leg area may recruit neurons in a different way to those of the hand area (Priori *et al.* 1993; Nielsen *et al.* 1995; Terao *et al.* 2000).

Electrical stimulation

Lateral anodal stimulation. At threshold intensity this kind of stimulation evokes the earliest volley and voluntary contraction has no effect on the amplitude of this earliest volley (Di Lazzaro *et al.* 2001c).

Vertex anodal stimulation. In some subjects, the earliest wave after vertex anodal stimulation has the same latency as the one after lateral anodal stimulation. In contrast, the earliest negative wave evoked in other subjects has a 1.1–1.4 ms longer

latency than the shortest-latency volley recruited by lateral anodal stimulation (Di Lazzaro *et al.* 2001c).

Magnetic stimulation

The lowest-threshold volley recruited by PA magnetic stimulation with a figure-8 coil has a latency ~1.4 ms longer than the earliest volley recruited by lateral anodal stimulation (Di Lazzaro *et al.* 2001c). The initial volley increases in size and is followed by later volleys as the intensity of stimulation is increased. The interpeak interval between the later waves is ~1.6 ms. At a stimulus intensity of ~150% active motor threshold an earlier small wave appears. This volley has the same latency as the earliest volley recorded following lateral anodal stimulation. The threshold of this volley is not modified by voluntary contraction.

Voluntary contraction has two effects on the volleys evoked by PA magnetic stimulation.

First, it appears that threshold for evoking recognizable activity is lower during voluntary contraction than at rest. During voluntary contraction a descending wave is seen at 5–10% of stimulator output below the threshold for recognizable activity at rest in almost all subjects. The second effect is that, at virtually all intensities, the amplitude of each volley, apart from the earliest volley evoked at high intensities of stimulation, is higher during activity than at rest. The shortest-latency volley evoked by lateral anodal stimulation at threshold and by magnetic stimulation at high intensity is not influenced by voluntary contraction of the target muscle, whereas the amplitude of the later volleys increases during tonic voluntary contraction. Data from stimulation of the hand area (see above) show that D-waves are unaffected by the level of voluntary contraction, whereas I-waves are usually facilitated. By analogy, the data obtained with stimulation of the lower limb area suggest that the earliest volley evoked by lateral anodal stimulation is a D-wave. Later volleys are I-waves. In contrast to the behavior seen after TES of the hand area, in some individuals, vertex anodal stimulation can preferentially recruit an I1-wave. Lateral anodal stimulation always recruits an initial D-wave that is insensitive to changes in cortical excitability.

Magnetic Stimulation

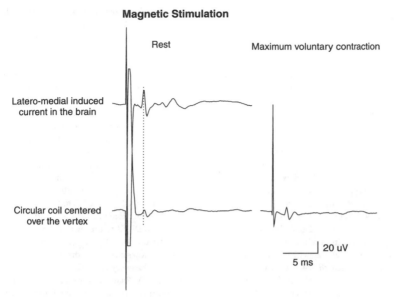

Fig. 14.16 Averaged descending volleys evoked by latero-medial magnetic stimulation (rest) at suprathreshold intensity and nonfocal magnetic stimulation (rest and active) at around active threshold intensity. The latency of the D-wave is indicated by the vertical dotted line. Nonfocal magnetic stimulation with a circular coil evokes a slightly delayed D-wave whose amplitude is increased during maximum voluntary contraction by ~100%. Adapted from Figure 1 in Di Lazzaro et al. (2002d).

References

Adrian ED, Moruzzi G (1939) Impulses in the pyramidal tract. *Journal of Physiology (London) 97*, 153–199.

Amassian VE, Weiner H (1966) Monsynaptic and polysynaptic activation of pyramidal tract neurons by thalamic stimulation. In: DP Purpura, MD Yahr (eds), *The thalamus*, pp. 255–282. New York: Columbia University Press.

Amassian VE, Stewart M, Quirk GJ, Rosenthal JL (1987) Physiological basis of motor effects of a transient stimulus to cerebral cortex. *Neurosurgery 20*, 74–93.

Amassian, VE, Eberle, LP, Maccabee, PJ, Cracco, RQ (1992) Modelling magnetic coil excitation of human cerebral cortex with a peripheral nerve immersed in a brain shaped volume conductor: the significance of fiber bending in excitation. *Electroencephalography and Clinical Neurophysiology 85*, 291–301.

Berardelli A, Inghilleri M, Cruccu G, Manfredi M (1990) Descending volley after electrical and magnetic transcranial stimulation in man. *Neuroscience Letters 112*, 54–58.

Berardelli A, Inghilleri M, Rothwell JC, Romeo S, Curra A, Gilio F, Modugno N, Manfredi M (1998) Facilitation of muscle evoked responses after repetitive cortical stimulation in man. *Experimental Brain Research 122*, 79–84.

Boyd SG, Rothwell JC, Cowan JM, Webb PJ, Morley T, Asselman P, Marsden CD (1986) A method of monitoring function in corticospinal pathways during scoliosis surgery with a note on motor conduction velocities. *Journal of Neurology, Neurosurgery and Psychiatry 49*, 251–257.

Burke D, Hicks R, Gandevia SC, Stephen J, Woodforth I, Crawford M (1993) Direct comparison of corticospinal volleys in human subjects to transcranial magnetic and electrical stimulation. *Journal of Physiology (London) 470*, 383–393.

Chen R, Classen J, Gerloff C, Celnik P, Wassermann EM, Hallett M, Cohen LG (1997) Depression of motor cortex excitability by low-frequency transcranial magnetic stimulation. *Neurology 48*, 1398–1403.

Chen R, Lozano AM, Ashby P (1999) Mechanism of the silent period following transcranial magnetic stimulation. Evidence from epidural recordings. *Experimental Brain Research 128*, 539–542.

Delwaide PJ, Olivier E (1990) Conditioning transcranial cortical stimulation (TCCS) by exteroceptive stimulation in parkinsonian patients. *Advances in Neurology 53*, 175–181.

Di Lazzaro V, Oliviero A, Profice P, Saturno E, Pilato F, Insola A, Mazzone P, Tonali P, and Rothwell JC (1998a) Comparison of descending volleys evoked by transcranial magnetic and electric stimulation in conscious humans. *Electroencephalography and Clinical Neurophysiology 109*, 397–401.

Di Lazzaro V, Restuccia D, Oliviero A, Profice P, Ferrara L, Insola A, Mazzone P, Tonali P, Rothwell JC (1998b) Effects of voluntary contraction on descending volleys evoked by transcranial stimulation in conscious humans. *Journal of Physiology (London) 508*, 625–633.

Di Lazzaro V, Restuccia D, Oliviero A, Profice P, Ferrara L, Insola A, Mazzone P, Tonali P, Rothwell JC (1998c) Magnetic transcranial stimulation at intensities below active motor threshold activates intracortical inhibitory circuits. *Experimental Brain Research* 119, 265–268.

Di Lazzaro V, Oliviero A, Profice P, Insola A, Mazzone P, Tonali P, Rothwell JC (1999a) Effects of voluntary contraction on descending volleys evoked by transcranial electrical stimulation over the motor cortex hand area in conscious humans. *Experimental Brain Research* 124, 525–528.

Di Lazzaro V, Rothwell JC, Oliviero A, Profice P, Insola A, Mazzone P, Tonali P (1999b) Intra-cortical origin of the short latency facilitation produced by pairs of threshold magnetic stimuli applied to human motor cortex. *Experimental Brain Research* 129, 494–499.

Di Lazzaro V, Oliviero A, Profice P, Insola A, Mazzone P, Tonali P, Rothwell JC (1999c) Direct demonstration of interhemispheric inhibition of the human motor cortex produced by transcranial magnetic stimulation. *Experimental Brain Research* 124, 520–524.

Di Lazzaro V, Oliviero A, Meglio M, Cioni B, Tamburrini G, Tonali P, Rothwell JC (2000) Direct demonstration of the effect of lorazepam on the excitability of the human motor cortex. *Clinical Neurophysiology* 111, 794–799.

Di Lazzaro V, Oliviero A, Saturno E, Pilato F, Insola A, Mazzone P, Profice P, Tonali P, Rothwell JC (2001a) The effect on corticospinal volleys of reversing the direction of current induced in the motor cortex by transcranial magnetic stimulation. *Experimental Brain Research* 138, 268–273.

Di Lazzaro V, Oliviero A, Mazzone P, Insola A, Pilato F, Saturno E, Accurso A, Tonali P, Rothwell JC (2001b) Comparison of descending volleys evoked by monophasic and biphasic magnetic stimulation of the motor cortex in conscious humans. *Experimental Brain Research* 141, 121–127.

Di Lazzaro V, Oliviero A, Profice P, Meglio M, Cioni B, Tonali P, Rothwell JC (2001c) Descending spinal cord volleys evoked by transcranial magnetic and electrical stimulation of the motor cortex leg area in conscious humans. *Journal of Physiology (London)* 537, 1047–1058.

Di Lazzaro V, Oliviero A, Mazzone P, Pilato F, Saturno E, Insola A, Visocchi M, Colosimo C, Tonali PA, Rothwell JC (2002a) Direct demonstration of long latency cortico-cortical inhibition in normal subjects and in a patient with vascular parkinsonism. *Clinical Neurophysiology* 113, 1673–1679.

Di Lazzaro V, Oliviero A, Berardelli A, Mazzone P, Insola A, Pilato F, Saturno E, Dileone M, Tonali PA, Rothwell JC (2002b) Direct demonstration of the effects of repetitive transcranial magnetic stimulation on the excitability of the human motor cortex. *Experimental Brain Research* 144, 549–553.

Di Lazzaro V, Oliviero A, Mazzone P, Pilato F, Saturno E, Dileone M, Insola A, Tonali PA, Rothwell JC (2002c) Short-term reduction of intracortical inhibition in the human motor cortex induced by repetitive transcranial magnetic stimulation. *Experimental Brain Research* 147, 108–113.

Di Lazzaro V, Oliviero A, Pilato F, Saturno E, Insola A, Mazzone P, Tonali PA, Rothwell JC (2002d) Descending volleys evoked by transcranial magnetic stimulation of the brain in conscious humans: effects of coil shape. *Clinical Neurophysiology* 113, 114–119.

Di Lazzaro V, Oliviero A, Tonali PA, Mazzone P, Insola A, Pilato F, Saturno E, Dileone M, Rothwell JC (2003) Direct demonstration of reduction of the output of the human motor cortex induced by a fatiguing muscle contraction. *Experimental Brain Research* 149, 535–538.

Di Lazzaro V, Oliviero A, Pilato F, Saturno E, Dileone M, Mazzone P, Insola A, Tonali PA, Rothwell JC (2004a) The physiological basis of transcranial motor cortex stimulation in conscious humans. *Clinical Neurophysiology* 115, 255–266.

Di Lazzaro V, Oliviero A, Pilato F, Saturno E, Dileone M, Meglio M, Cioni B, Colosimo C, Tonali PA, Rothwell JC (2004b) Direct recording of the output of the motor cortex produced by transcranial magnetic stimulation in a patient with cerebral cortex atrophy. *Clinical Neurophysiology* 115, 112–115.

Di Lazzaro V, Pilato F, Saturno E, Oliviero A, Dileone M, Mazzone P, Insola A, Tonali PA, Ranieri, Huang YZ, Rothwell JC (2005) Theta-burst repetitive transcranial magnetic stimulation suppresses specific excitatory circuits in the human motor cortex. *Journal of Physiology (London)* 565, 945–945.

Edgley SA, Eyre JA, Lemon RN, Miller S (1990) Excitation of the corticospinal tract by electromagnetic and electric stimulation of the scalp in the macaque monkey. *Journal of Physiology (London)* 425, 301–320.

Ferbert A, Priori A, Rothwell JC, Day BL, Colebatch JG, Marsden CD (1992) Interhemispheric inhibition of the human motor cortex. *Journal of Physiology (London)* 453, 525–546.

Hanajima R, Ugawa Y, Terao Y, Sakai K, Furubayashi T, Machii K, Kanazawa I (1998) Paired-pulse magnetic stimulation of the human motor cortex: differences among I waves. *Journal of Physiology (London)* 509, 607–618.

Huang YZ, Edwards MJ, Rounis E, Bhatia KP, Rothwell JC (2005) Theta burst stimulation of the human motor cortex. *Neuron* 45, 201–206.

Ilić TV, Meintzschel F, Cleff U, Ruge D, Kessler KR, Ziemann U (2002) Short-interval paired-pulse inhibition and facilitation of human motor cortex: the dimension of stimulus intensity. *Journal of Physiology (London)* 545, 153–167.

Kaneko K, Kawai S, Fuchigami Y, Shiraishi G, Ito T (1996) Effect of stimulus intensity and voluntary contraction on corticospinal potentials following transcranial magnetic stimulation. *Journal of the Neurological Sciences* 139, 131–136.

Kernell D, Chien-Ping Wu (1967) Responses of the pyramidal tract to stimulation of the baboon's motor cortex. *Journal of Physiology (London)* 191, 653–672.

Kujirai T, Caramia MD, Rothwell JC, Day BL, Thompson PD, Ferbert A, Wroe S, Asselman P, Marsden CD (1993) Corticocortical inhibition in human motor cortex. *Journal of Physiology (London)* 471, 501–519.

Maeda F, Keenan JP, Tormos JM, Topka H, Pascual-Leone A (2000) Interindividual variability of the modulatory effects of repetitive transcranial magnetic stimulation on cortical excitability. *Experimental Brain Research 133*, 425–430.

Nakamura H, Kitagawa H, Kawaguchi Y, Tsuji H (1996) Direct and indirect activation of human corticospinal neurons by transcranial magnetic and electrical stimulation. *Neuroscience Letters 210*, 45–48.

Nakamura H, Kitagawa H, Kawaguchi Y, Tsuji H (1997) Intracortical facilitation and inhibition after transcranial magnetic stimulation in conscious humans. *Journal of Physiology (London) 498*, 817–823.

Nielsen J, Petersen N, Ballegaard M (1995) Latency of effects evoked by electrical and magnetic brain stimulation in lower limb motoneurones in man. *Journal of Physiology (London) 484*, 791–802.

Pascual Leone A, Valls-Solè J, Wassermann EM, Hallett M (1994) Responses to rapid-rate transcranial magnetic stimulation of the human motor cortex. *Brain 117*, 847–858.

Pascual-Leone A, Tormos JM, Keenan J, Tarazona F, Canete C, Catala MD (1998) Study and modulation of human cortical excitability with transcranial magnetic stimulation. *Journal of Clinical Neurophysiology 15*, 333–343.

Patton HD, Amassian VE (1954) Single- and multiple-unit analysis of cortical stage of pyramidal tract activation. *Journal of Neurophysiology 17*, 345–363.

Peinemann A, Lehner C, Mentschel C, Munchau A, Conrad B, Siebner HR (2000) Subthreshold 5-Hz repetitive transcranial magnetic stimulation of the human primary motor cortex reduces intracortical paired-pulse inhibition. *Neuroscience Letters 296*, 21–24.

Priori A, Bertolasi L, Dressler D, Rothwell JC, Day BL, Thompson PD, Marsden CD (1993) Transcranial electric and magnetic stimulation of the leg area of the human motor cortex: single motor unit and surface EMG responses in the tibialis anterior muscle. *Electroencephalography and Clinical Neurophysiology 89*, 131–137.

Sakai K, Ugawa Y, Terao Y, Hanajima R, Furubayashi T, Kanazawa I (1997) Preferential activation of different I waves by transcranial magnetic stimulation with a figure-of-eight-shaped coil. *Experimental Brain Research 113*, 24–32.

Terao Y, Ugawa Y, Hanajima R, Machii K, Furubayashi T, Mochizuki H, Enomoto H, Shiio Y, Uesugi H, Iwata NK, Kanazawa I (2000) Predominant activation of I1-waves from the leg motor area by transcranial magnetic stimulation. *Brain Research 859*, 137–146.

Tergau F, Tormos JM, Paulus W, Pascual-Leone A, Ziemann U (1997) Effects of repetitive transcranial magnetic stimulation (rTMS) on cortico-spinal and cortico-cortical excitability. *Neurology 48*, A107.

Thompson PD, Day BL, Crockard HA, Calder I, Murray NM, Rothwell JC, Marsden CD (1991) Intra-operative recording of motor tract potentials at the cervico-medullary junction following scalp electrical and magnetic stimulation of the motor cortex. *Journal of Neurology, Neurosurgery and Psychiatry 54*, 618–623.

Tokimura H, Ridding MC, Tokimura Y, Amassian VE, and Rothwell JC (1996) Short latency facilitation between pairs of threshold magnetic stimuli applied to human motor cortex. *Electroencephalography and Clinical Neurophysiology 101*, 263–272.

Tokimura H, Di Lazzaro V, Tokimura Y, Oliviero A, Profice P, Insola A, Mazzone P, Tonali P, Rothwell JC (2000) Short latency inhibition of human hand motor cortex by somatosensory input from the hand. *Journal of Physiology (London) 523*, 503–513.

Valls-Sole J, Pascual-Leone A, Wassermann EM, Hallett M (1992) Human motor evoked responses to paired transcranial magnetic stimuli. *Electroencephalography and Clinical Neurophysiology 85*, 355–364.

Werhahn KJ, Fong JKY, Meyer BU, Priori A, Rothwell JC, Day BL, Thompson PD (1994) The effect of magnetic coil orientation on the latency of surface EMG and single motor unit responses in the first dorsal interosseous muscle. *Electroencephalography and Clinical Neurophysiology 93*, 138–146.

Ziemann U, Rothwell JC (2000) I-waves in motor cortex. *Journal of Clinical Neurophysiology 17*, 397–405.

Ziemann U, Tergau F, Wassermann EM, Wischer S, Hildebrandt J, Paulus W (1998) Demonstration of facilitatory I-wave interaction in the human motor cortex by paired transcranial magnetic stimulation. *Journal of Physiology (London) 511*, 181–190.

CHAPTER 15

TMS measures and voluntary motor function

John C. Rothwell

Summary – TMS can be viewed as interacting with voluntary movement in two ways: it can used to probe the excitability of CNS pathways before, during and after a movement; alternatively, it can be used to interfere with movement and give information about the role of different cortical areas in different aspects of a task. This chapter concentrates on the role of single- and double-pulse TMS methods that have been covered in detail in previous chapters. Long-lasting effects of repetitive (r)TMS are described in later chapters.

TMS studies of CNS pathways involved in voluntary movement

General considerations

Almost all of the TMS measures described so far (cf. Chapters 9–14, this volume) differ in subjects at rest and during tonic voluntary activity. Some of these effects occur because of changes in excitability of circuits in the cortex; others because of changes in spinal cord. This chapter reviews the work that has been done using TMS measures to probe excitability of central circuits before and after different types of real or imagined contraction in healthy subjects. In some cases, the results confirm previous work in animal preparations and allow us to use the same measures to investigate the pathophysiology of the same circuits in human neurological disease. Perhaps more unexpected is the fact that in many cases, the results reveal new information that had not previously been described in experiments on animals. For example, the idea that cortical inhibitory circuits can 'sculpt' patterns of motor cortex excitability is an idea that has been studied more in humans than in animal models, almost exclusively using the short-interval intracortical inhibition (SICI) method of TMS. Similarly, the role of interhemispheric inhibition (IHI) is rarely studied in animals, but in humans has led to the concept of an overactive 'normal' hemisphere inhibiting an already disabled lesioned hemisphere in patients after stroke. Such studies show that TMS measures can be more than tools to describe in humans what has already been described in animals; they can become drivers of new concepts as well.

There is one proviso that should be borne in mind when interpreting the results of these 'circuit-testing' experiments: the results only tell us about the *excitability* of the pathways at the time of testing, which is not necessarily the same as the *amount of activity* in a pathway. For example, if motor-evoked potentials (MEPs) increase in size prior to onset of a movement and if spinal cord excitability remains constant, we can conclude that a given TMS pulse results in more descending corticospinal activity. However, it does not necessarily mean that there is more ongoing descending activity when the TMS pulse is applied. A pathway can change excitability without changing ongoing activity.

As an example of this, take a neuron that has a resting potential below threshold. In this state it will not discharge impulses. Nevertheless, if the potential is close to threshold, we can say that the excitability is higher than if the potential is far from threshold. In both cases the ongoing output of the neuron is the same (i.e. zero), but the excitability is different. As another example, imagine a neuron at rest: the resting level of synaptic activity may be low, whereas during movement synaptic input might be much higher. However, the number of active synapses changes the total resistance of the membrane: the more synapses are active the lower the membrane resistance. Thus in the resting state, the resistance of the membrane is likely to be higher than during movement. The outcome is that any additional transmembrane currents generated by synaptic input at rest will have a larger effect on the membrane potential (and hence the overall neural discharge rate) than during movement.

It is difficult to take all these factors into account when interpreting the results of excitability measures. All that can be said is that conclusions about levels of activity are at best indirect, and at worst could be wrong.

Onset of a voluntary contraction: agonist muscle

Many studies have investigated how MEP amplitude changes prior to the onset of a movement. All of them agree that the MEP in a muscle increases prior to the onset of voluntary electromyographic (EMG) activity in the same muscle. However, there is a debate over exactly how long the motor cortex becomes excited in advance of changes in spinal cord. Studies in behaving primates have suggested that corticospinal neurons change their firing rates up to 400 ms prior to the onset of a self-initiated contraction whereas the interval is of the order of 100 ms in reaction-time tasks.

Early studies in humans suggested that excitability changes in the MEP also began about 100 ms before spinal motor neurons began to discharge impulses (Rossini *et al.* 1988;

Pascual-Leone *et al.* 1992a,b). The assumption was that this represented the sum of the time taken for rising depolarization to discharge corticospinal neurons, plus the time taken for sufficient corticospinal activity to reach the spinal cord and depolarize resting motor neurons to threshold. If this delay occurred in all contractions, it would represent a substantial fraction of the human subjects' minimum reaction times of ~130 ms (e.g. in a simple auditory reaction task).

All these studies had similar designs: TMS pulses were applied randomly in the reaction period of a task and the amplitude of the resulting MEP was measured. However, there is a problem in deciding how to measure the onset of the voluntary EMG in each trial. It is sometimes difficult to distinguish the onset of voluntary EMG if it starts just before the onset of the MEP; conversely if the voluntary EMG starts after the MEP, then its onset can be delayed by the presence of the silent period that follows the MEP. The latter tends to make it appear as if the voluntary reaction time is later than it should be and increases the measured interval between changes in MEP and EMG.

One way around this is to measure the amount of EMG activity immediately prior to the MEP without trying to distinguish whether this represents the start of the voluntary movement or not. As we change the timing of the TMS pulse, the level of EMG prior to the MEP follows the time course of the EMG bursts in the voluntary movement. This can then be directly compared with the time course of the changes in MEP. In experiments using this approach, MEPs increased in the target muscle only ~10 ms before onset of EMG changes (MacKinnon and Rothwell 2000; Schneider *et al.* 2004), indicating that the minimum delay between cortex and spinal cord is very short, at least in a rapid reaction time task (Figure 15.1).

Finally, several authors have examined how MEPs in nontarget muscles change during a focal reaction. The general pattern seems to be that they also tend to increase prior to contraction, even in the opposite limb, but at about the time of movement onset, they decrease (Gerloff *et al.* 1997; Leocani *et al.* 2000; Liepert *et al.* 2001; Sohn *et al.* 2003). The latter would be consistent with the existence of a mechanism that actively suppresses unintended contractions.

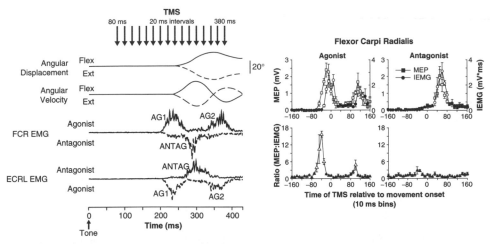

Fig. 15.1 Changes in corticospinal excitability as evidenced by changes in motor-evoked potential (MEP) amplitude before and during a simple ballistic wrist flexion movement. *Left panel*: Examples of the angular displacement, velocity, and rectified electromyogram (EMG) for flexor carpi radialis (FCR) and extensor carpi radialis longus (ECRL) muscles during wrist flexion (Flex) and extension (Ext) movements in a single subject. Each trace is the average of 15 trials. Solid lines are for flexion trials and dashed lines are for extension trials. Note the characteristic triphasic pattern of EMG activity (AG1, ANTAG and AG2) associated with movements in both directions. The arrows at the top of the figure show the times that TMS was applied across trials. The auditory tone (go signal) was presented at 0 ms. *Right panel*: Time-varying changes in the amplitude of the MEPs, background level of EMG (IEMG), and MEP:IEMG ratios in FCR during wrist flexion or extension movements in a single subject. Plots on the left show profiles when FCR functioned as the agonist (wrist flexion) and plots on the right show profiles when FCR functioned as the antagonist (wrist extension). Responses have been sorted into 10 ms time bins. In this subject, MEPs increased prior to the onset of the AG1 and AG2 bursts. The phase advance resulted in a marked increase in the MEP:IEMG ratio prior to both agonist bursts. In contrast, MEPs and IEMGs increased within the same time bins when the same muscle functioned as the antagonist resulting in no significant change in the MEP:IEMG ratio from baseline levels. Open symbols denote values that were significantly different from baseline ($P < 0.05$). Error bars are one standard error. Reproduced from MacKinnon and Rothwell (2000).

The question of whether there is active suppression of unwanted movement has been addressed more fully in studies using SICI. As noted in Chapter 11, this volume, SICI is less effective during voluntary contraction of a target muscle. Some of this change is due to changes in size of the test MEP, but some is due to removal of inhibition at the motor cortex. Interestingly, SICI may increase in nearby nontarget muscles as long as they remain silent during the contraction (Stinear and Byblow 2003). Thus SICI may be part of a mechanism of 'surround inhibition' that helps focus excitation onto the intended target of a movement whilst suppressing activity in nontarget muscles (Sohn and Hallett 2004). Like the MEP, SICI changes prior to the onset and offset of voluntary

contractions: it decreases in the agonist prior to onset of contraction (Reynolds and Ashby 1999), whereas it increases prior to offset of contraction (Buccolieri *et al.* 2004).

Suppression of unintended activity not only occurs within the motor cortex of the active hemisphere, but also involves interhemispheric connections with the opposite ('nonmoving') hemisphere. Interhemispheric inhibition using the paired-pulse paradigm (IHI) is usually performed at rest (cf. Chapter 11, this volume). However, early studies showed that it is modulated during tonic voluntary contraction (Ferbert *et al.* 1992): contraction of muscles in one hand increases the excitability of IHI to homologous muscles of the opposite (relaxed) hand. Recently it has been found that this change occurs during the reaction period prior to onset of a reaction-time movement (Murase et al. 2004).

In sum, these TMS studies show that excitability of circuits involved in MEP, SICI, and IHI all change in parallel before, during, and after voluntary activation of a target muscle. The data are compatible with a model in which the CNS focuses facilitation on the target muscle while at the same time actively suppressing unintended activity in nearby and contralateral muscles.

Onset of a voluntary contraction: antagonist muscle

Rapid isotonic voluntary reaction movements at the wrist joint are accompanied by a triphasic burst of activity that starts with a burst of EMG in the agonist (AG1). This is followed by a carefully timed burst in the antagonist (ANT) that slows down the initial movement, and then a second agonist burst (AG2) that prevents terminal oscillations of the joint (Berardelli et al. 1996). It is therefore possible to test how the excitability of the corticospinal pathway varies prior to the onset of ANT and compare it with the onset of AG1. An interesting result using MEP measures is that although there is a delay between increases in MEP evoked in the agonist and the onset of AG1 (see above), there is no delay between the increase in MEP in the antagonist muscle prior to ANT (MacKinnon and Rothwell 2000). Both MEP and ANT EMG increase at the same time (Figure 15.1). It may be that lack of a delay indicates that there is already substantial excitability in the corticospinal projections to the ANT, so that no subthreshold delays occur. In this experimental design, H-reflexes can be elicited in the wrist flexor muscles either when the muscles are acting as agonist (rapid wrist flexion) or antagonist (wrist extension). H-reflexes increase at the same time as the corresponding EMG bursts; during the AG1, H-reflex excitability in the ANT muscle is suppressed, consistent with presynaptic inhibition of the Ia afferents, perhaps reducing the effect of input from spindles in the stretched ANT muscle.

Preparation to move

These experiments test the situation when subjects are ready to make a movement in response to a 'go' signal that will occur at some time in the future. Early experiments on nonhuman primates performing such warned reaction tasks showed that during the interval between an instructional warning signal and a reaction signal, 61% of corticospinal neurons change discharge according to the nature of the instruction. For example, those that usually increased their firing rate during a push movement also increased their firing rate if the warning signal indicated that a push movement was to occur, whereas neurons that discharged in relation to a pull movement would decrease their firing (Evarts and Tanji 1976; Tanji and Evarts 1976). Such work suggests that MEPs in the target muscle of a forthcoming contraction should increase in the warning period of a reaction task. However, some studies have obtained the opposite results: MEPs may decrease in the target muscle, as if corticospinal excitability is actively reduced to prevent unintended escape of the intended reaction until appearance of the reaction signal (Touge et al. 1995, 1998; Hasbroucq et al. 1997). Spinal H-reflexes may also decrease in the target muscles prior to a warned reaction movement (Brunia and Vuister 1979; Brunia et al. 1982), but the effect is smaller and later than the reduction in MEP (Touge et al. 1998).

Offset/inhibition of voluntary contractions

There have been fewer studies of the changes in corticospinal excitability that accompany voluntary relaxation of a muscle, and none have employed the design used above to resolve the problem of identifying the time at which EMG activity begins to decrease. All that can be said with certainty is that MEPs in the contracting muscle decrease at approximately the same time as the EMG, slightly lagging an increase in SICI (Buccolieri et al. 2004). There is no indication that corticospinal excitability is reduced in advance of relaxation.

Cortical inhibitory mechanisms also seem to be involved in preventing the release of a prepared volitional movement in a Go–NoGo task. In the NoGo condition, when subjects were prepared to execute a movement but the cue instructed them to withhold it, MEPs in the target muscle, but not its near neighbors, decreased below baseline levels. Since F-waves were unchanged, this suggests that there was active and focal inhibition of the corticospinal output at the motor cortex. At about the same time, SICI in the target muscle increased, consistent with active inhibition at the cortex. Rather surprisingly, long-interval intracortical inhibition (LICI) (cf. Chapter 11, this volume) was reduced (Waldvogel et al. 2000; Sohn et al. 2002).

Fatiguing contractions

Fatigue is the gradual decline in muscle force that occurs during a sustained contraction. Much of the decline in force is caused by failure of contractile mechanisms in the muscle itself. However, this is compounded in many cases by a failure of voluntary motor commands to activate the muscle fully. The latter effect is termed 'central fatigue', and can be defined as an inability of central commands to recruit maximum evocable muscle force during voluntary contraction (Gandevia 2001). Central fatigue develops during sustained or intermittent isometric maximum voluntary contractions (MVCs), although the extent to which it develops depends on the muscle group, task, and subject. It can be demonstrated using the 'twitch interpolation technique' that was first developed by Merton (1954). Thus a single supramaximal stimulus to a muscle nerve does not evoke any additional force of contraction if a muscle is being maximally activated. During long contractions, however, a stimulus usually begins to produce a small increase in force, indicating that the muscle is no longer being activated maximally by the volitional effort. This represents 'central fatigue' (Herbert and Gandevia 1999; Gandevia 2001).

Many authors have used TMS to try to probe the contribution of motor cortical mechanisms to this effect by measuring the size of MEPs and their associated muscle contractions before, during, or after a fatigue contraction. Such measurements are usually corrected for changes in amplitude of the muscle M-wave that occur at the same time. During an MVC the MEP gradually increases in amplitude, whereas responses to transcranial electric stimulation (Brasil-Neto et al. 1993; Zanette et al. 1995) or

transmastoid electric stimulation (cervico-medullary motor-evoked potential; CEMP) (Ugawa et al. 1991; Taylor et al. 1996; Gandevia et al. 1999) are unaffected or even reduced. It is thought that intense voluntary input to motor cortex during the contraction increases its excitability to single TMS pulses.

A TMS pulse applied during an attempted maximum contraction is analogous to a 'central' version of the twitch interpolation technique. Effectively, if a TMS pulse can evoke more force from a muscle than maximal volitional effort, then the conclusion must be that voluntary drive is insufficient to activate maximally the available output from motor cortex. At the onset of a maximal contraction, a TMS pulse adds nothing, or very little, extra force. However, during the course of the contraction, as muscle force drops, TMS pulses begin to recruit force increments of increasing size (Gandevia et al. 1996). The conclusion is that some proportion of the fatigue is caused by failure of voluntary drive to activate motor cortex output, i.e. central fatigue must include processes upstream of the motor cortex.

It should be noted that although this statement is correct, the exact interpretation of the result is not easy. We do not know if the TMS pulse itself activates the motor cortex optimally, nor can we be certain that the TMS pulse has not in addition activated antagonist muscles that might cause us to underestimate the degree of force loss. All we can say is that if extra force is produced, then the voluntary motor system is unable to recruit it. It is an unexpected conclusion that could probably never have been addressed in animal experiments.

The results after the end of a fatiguing contraction depend on whether the measurements are made with the muscle active or at rest (Figure 15.2). If MEPs and CMEPs are tested during brief MVCs, both return to baseline levels within 15–30 s of the end of the fatiguing contraction. However, the results are quite different if measurements are taken with subjects relaxed. Following a short period of facilitation, MEPs are profoundly depressed for up to 30 min or more, whereas CMEPs from cervico-medullary stimulation follow the opposite time course, being briefly depressed and later facilitated (Gandevia 2001). Direct epidural recordings

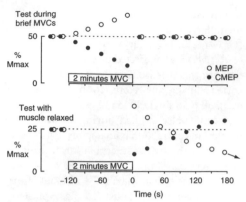

Fig. 15.2 Summary of typical changes in electromyogram responses to motor cortical stimulation (motor-evoked potential (MEP), open circles) and stimulation of descending motor paths including the corticospinal tracts (CMEP, solid circles) before, during, and after fatigue in a sustained maximum voluntary contraction (MVC). *Top panel*: responses observed with all stimuli delivered during MVCs; *bottom panel*: responses obtained with the muscle relaxed. The MEP increases during the sustained MVC, is not depressed when tested during brief MVCs in the recovery period (*top panel*), and is increased with a late prolonged depression when the muscle does not contract after the sustained MVC (arrow, *bottom panel*). In contrast, the response to a single corticospinal tract stimulus decreases during the sustained MVC and is depressed when tested during relaxation immediately after the contraction. The depression of the CMEP is removed by a brief 'recovery' MVC. Fatigue-induced changes are clearly documented at the motor cortical and motor neuronal level. Reproduced from Gandevia (2001).

of the corticospinal volley from the spinal cord have confirmed that corticospinal volleys evoked by TMS pulses in relaxed subjects are reduced after fatigue, supporting the conclusion that depression of the MEP is caused by reduced excitability at the motor cortex (Di Lazzaro *et al.* 2003) (cf. Chapter 14, this volume). It may then be that the ability of a brief MVC to restore the amplitude of depressed MEPs seen at rest is due to the fact that the excitability of motor cortical output neurons is reduced after fatigue and that it can be restored briefly by superimposing excitatory volitional drive. The reasons for the

changes in CMEPs are less clear. The initial reduction after fatigue seen at rest cannot have been due to axonal refractoriness of corticospinal axons active during fatigue (Vagg *et al.* 1998), since it disappears when the same axons are tested during brief MVCs. Other possibilities are that there is a change in efficacy of the corticomotor neuronal synapse or in the motor neuronal firing properties, but further work is needed before any definite conclusion can be drawn.

Imagined contractions

Motor imagery is conventionally defined as the covert rehearsal of movement. It is important to note, though, that most imagery, if not well controlled, will lead to small amounts of actual EMG activity in the target muscles (Gandevia *et al.* 1997). These can contaminate any TMS measures, and must be excluded by monitoring spinal excitability before any conclusions can be reached about the involvement of cortical circuits. Studies that have monitored EMG and H-reflex excitability during imagery have generally found changes in corticospinal and intracortical excitability that mirror those seen in actual movement. This supports the idea that imagined movements produce qualitatively similar, but quantitatively lesser, drive to muscles than overt movement. Thus, MEPs in the target muscle increase in amplitude compared to rest at the time of imagined contraction and decrease during imagined contraction of the antagonist (Yahagi *et al.* 1996; Fadiga *et al.* 1999; Hashimoto and Rothwell 1999; Facchini *et al.* 2002). SICI decreases during imagined contraction of the target muscle (Abbruzzese *et al.* 1999) and may even increase slightly in nontarget surrounding muscles (Stinear and Byblow 2004). Interestingly, visual imagery of the task may be rather less successful in modulating TMS measures than kinesthetic imagery (Fourkas *et al.* 2006; Stinear *et al.* 2006).

Observation of movement

There have been several studies on observation of movement that link with investigations of the 'mirror' neuron system that has been identified in primates. In general, MEPs seem to change in muscles homologous to those that are active in the observed task (Fadiga *et al.* 1995), and at approximately the same time (Gangitano *et al.* 2001). Unfortunately, few of these studies have controlled for possible changes in spinal excitability, but those that have (Baldissera *et al.* 2001; Montagna *et al.* 2005) suggest that even if there are changes in spinal excitability, there are additional cortical effects that confirm the basic observations derived from measurement of the MEP alone. Repeated observation of the same movement can even lead to a long-lasting change in the excitability of the corticospinal projections to that muscle. Stefan *et al.* (2005) asked subjects to observe another person making isolated movements of the thumb at the metacarpophalangeal joint. Importantly, the direction of the observed movement was the opposite to the usual direction of movements evoked in the thumb of the observer by a TMS pulse given at rest. At the end of observation, there was an increased probability that TMS-evoked thumb movements in the observer lay in the same direction as the observed movement. This is of course consistent with the idea that observing a person performing an action helps the CNS learn to perform the same action itself. The important point of the result, however, is that the learning appears to involve the primary motor cortex.

Single neuron recording experiments in monkeys have suggested that much of the facilitation of corticospinal output that occurs during observation is due to facilitatory inputs from the ventral premotor cortex (PMv) (Shimazu *et al.* 2004). Cattaneo *et al.* (2005) examined the modulation of short-interval intracortical facilitation (SICF, I-wave facilitation; cf. Chapter 3, this volume) as a possible surrogate marker of this input in humans. They found that facilitation between two TMS pulses given 2.5 ms apart was enhanced prior to reaching movements made by subjects to graspable objects. The facilitation was not observed if the subjects made the same movements without vision, or imagined the movement, and was specific to the muscles that would be needed to grasp the object in the usual grip used by subjects (Figure 15.3). The conclusion was that vision of the object produced an increase in excitability of the PMv projections to populations of corticospinal neurons in motor cortex that would be recruited in the forthcoming movement.

Contractions of different types

These experiments are examples in which the initial work simply served to confirm results already obtained from invasive recordings of neural activity in monkeys. However, later work then used the same approach to investigate related questions on control of gait (see below) that have not yet been examined in nonhuman primates. The latter give us novel information about the control of leg muscles.

There is good evidence from invasive recordings in monkeys that the activity of corticospinal projections to arm and hand muscles is affected by the type of contraction that the animal makes. For example, Muir and Lemon (1983) recorded task-related discharges of corticomotor neuronal cells which all facilitated muscle activity in at least one intrinsic hand muscle. Each of these cells discharged at a higher frequency during a precision grip of thumb and index than in a power grip involving all the fingers of the hand, even though the amount of muscle activation was usually much higher in the latter. They concluded that the corticomotor neuronal pathway had a particular role to play in fractionated finger movements.

Since the MEP is thought to reflect primarily activity in corticomotor neuronal connections, there have been several attempts to test whether the same differential involvement in power versus precision tasks can be seen in humans. The question has been whether the response to a given TMS pulse depends on the type of task being performed by a subject independent of any differences in ongoing EMG or spinal cord excitability. In general, most studies have found that there is some degree of task dependency of the responses to TMS that is likely to be due to changes in excitability of the corticospinal output during different tasks. Thus, Datta *et al.* (1989) found that the responses to TMS in the first dorsal interosseous muscle were larger during index finger abduction than during a power grip, whereas the responses to transcranial electrical stimulation did not differ. Others have gone on to investigate the time dependency of these changes during reach and grasping tasks (Lemon *et al.* 1995) and differences in proximal and distal muscle involvement in a task (Schieppati *et al.* 1996), again concluding that at

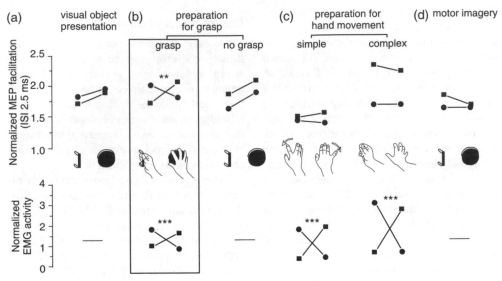

Fig. 15.3 Facilitation of motor-evoked potential (MEP) obtained with an interstimulus interval (ISI) of 2.5 ms (*upper*) and electromyogram (EMG) activation for the abductor digiti minimi (■) and first dorsal interosseous (●) (*lower*) with the key result highlighted with a box outline. (a) Object presentation alone. (b) Preparation for grasp (*left*) and no-grasp condition (*right*). (c) Preparation for simple movement (*left*) and complex movement (*right*). (d) Motor imagery. The values given in *upper* represent the ratio of the peak-to-peak amplitude of the MEPs obtained with paired pulses at 2.5 ms to that obtained with a single TMS pulse for the same condition. All values were obtained with the muscles at rest. *Lower* shows EMG activity during active grasp (b), a simple movement of the index or little finger, or a more complex hand movement (c). Values given represent the mean value of the normalized EMG recorded in the 300 ms preceding contact with the object (b) or onset of movement (c). Data were obtained from three different groups of subjects ($n = 10$ for (a) and (b); $n = 20$ for (c); $n = 9$ for (d); see text for methods). Significance of interactions: **$P < 0.001$, ***$P < 0.0001$. Reproduced from Cattaneo *et al.* (2005).

times when the greatest volitional control is required of a muscle, the level of corticospinal excitability is highest. The later studies did not use transcranial electrical stimulation to verify that the changes were due to cortical rather than spinal effects. However, they did point out that the clearest results were seen when TMS intensity was low, so as to maximize trans-synaptic activation of corticospinal neurons (I-wave inputs) (Lemon *et al.* 1995), and the H-reflexes in arm muscle were generally unchanged (Schieppati *et al.* 1996).

The most direct studies have recorded directly corticospinal volleys evoked by TMS pulses during different tasks. In both monkeys and humans, volleys recorded from the medullary pyramid or the epidural space of the upper cervical cord indicate that volitional contraction increases the amplitude and number of volleys

evoked by a given TMS pulse (Baker *et al.* 1995; Di Lazzaro *et al.* 1999). However, there have been no studies of the effects of different types of contraction for direct comparison with the MEP results above.

The role of the corticospinal system in control of leg movement, and particularly gait, is less well understood than its role in arm and hand movement. In cats, motor cortical activity is low during unobstructed walking, but becomes high when the cats are challenged to walk over difficult terrain, such as across the steps of a horizontal ladder, or during reactions to unexpected obstacles (Drew *et al.* 2002). Similarly, decorticate cats can walk over normal terrain but not across uneven or difficult surfaces, suggesting that the corticospinal system does not contribute greatly to walking unless the task becomes difficult. It is unknown to what extent such reasoning can be applied in humans. Spinal cats can walk if their body weight is supported, indicating that there must exist a powerful generator of walking rhythm in the spinal cord (the 'spinal locomotor generator'). Spinal humans cannot step even if fully supported. Even after extensive training only vestiges of the normal pattern of leg muscle activity are recovered (Dietz *et al.* 1998). Thus the question has been whether corticospinal inputs are important even in normal unobstructed human gait.

Capaday *et al.* (1999) found that MEPs were larger in the soleus muscle during voluntary plantar flexion of the ankle than they were at matched levels of EMG activity during walking on a treadmill. In contrast, MEPs in the tibialis anterior (TA) muscle had the same size during the swing phase of the gait cycle as during voluntary dorsiflexion. During the stance phase, the MEPs in the silent TA were larger than expected, sometimes being larger than those in the soleus muscle, which had much more background EMG activity. The conclusion was that corticospinal input to TA is at least as important during normal walking as it is in a focal volitional movement. In contrast, corticospinal input to the soleus is depressed (Figure 15.4).

However, these experiments did not control carefully for possible changes in spinal excitability. This was addressed later by Petersen *et al.* (2001) who made use of the fact that single pulses of low-intensity TMS can suppress EMG activity for a short period in actively contracting muscles. The effect may involve a mechanism similar to the inhibition more normally tested in SICI. Importantly, since the intensity of the TMS pulse is so low, it is unlikely to involve any direct spinal mechanism. Under these conditions the authors could record suppression of EMG activity in both TA and soleus muscles during unobstructed gait, indicating that cortical input must provide at least some excitatory input to ankle muscles in human walking.

TMS studies of interference with voluntary movement

Observations of the effects of stimulating the exposed cortex in conscious human patients undergoing neurosurgery show that stimulation of the motor cortex not only evokes contraction of contralateral muscles, but that it also interferes with voluntary movements of the same part of the body. TMS pulses can lead to similar interference, particularly if high-intensity stimulation is used. Such pulses synchronously activate a population of neurons and produce repetitive activity in cortical circuits that may last 10–15 ms. This is followed by a powerful GABAergic inhibitory postsynaptic potential that silences activity for the following 50–100 ms. In the motor cortex this produces the familiar succession of excitatory I-waves (cf. Chapter 14, this volume) followed by the cortical silent period (cf. Chapter 10, this volume). Together these events disrupt any ongoing processing that is occurring at the time, a phenomenon that has been termed a 'virtual lesion' (Cowey and Walsh 2001).

The first demonstration of this effect was in the visual cortex, where Maccabee *et al.* (1991) showed that a single TMS pulse over the occiput abolished perception of dim visual targets presented briefly ~80–120 ms earlier. A single pulse over the motor cortex, at ~120% resting motor threshold or above, has a slightly different effect. If given in the reaction period between a 'go' signal and a voluntary reaction, the pulse delays the onset of the movement (Day *et al.* 1989) (Figure 15.5). The delay is greater the higher the stimulus intensity, and the nearer the pulse is given to the expected time of onset of the movement. Unexpectedly, the movement is not abolished or even changed in form, as if the instructions for the task had been stored somewhere in the period of the delay. Note that electrical stimulation of a peripheral nerve to mimic the contraction evoked by a TMS pulse has no effect on the timing of movement even though the silent period that follows the stimulus interferes with the pattern of EMG activity. Delay is seen only when the stimulus involves motor cortex. This could not have been predicted from previous animal work and implies that the production of even a simple movement cannot proceed in serial fashion. Had it done so, then disruption of the motor cortex at the time of its involvement in the task would have abolished the movement, or at least severely disrupted it.

An interesting variation on this approach was used by Palmer *et al.* (1994, 1996) to investigate

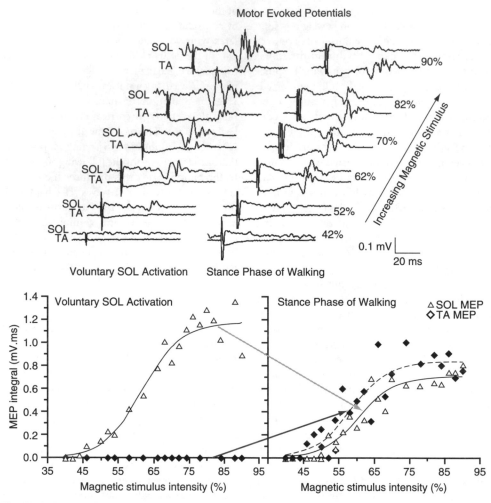

Fig. 15.4 *Top panel*: recordings of soleus (SOL) and tibialis anterior (TA) motor-evoked potentials (MEPs) during tonic voluntary SOL activation (*left panel*) and during the stance phase of walking (*right panel*) evoked by increasing magnetic stimulus intensity, from threshold to supramaximal levels. Mean ± SD rectified SOL electromyogram (EMG) is similar in both tasks [13.5 ± 2.3 and 8.6 ± 2.5 μV for voluntary SOL activation and stance, respectively). There was no activity in the TA in either task. Magnetic stimuli were delivered at 200 ms after heel contact, which corresponds to the early part of the stance phase of walking. *Bottom panel*: corresponding input–output curves of SOL and TA MEP integrals measured during tonic voluntary SOL activation and during the stance phase of walking. For each subject, the mean rectified SOL EMG was similar in both tasks. There was no activity in the TA in either task. Note how the curve for SOL is depressed during stance versus voluntary contraction whereas that for TA is enhanced. Thus corticospinal excitability to SOL is lower during stance than during matched levels of voluntary EMG whereas excitability to TA is higher. Reproduced from Capaday *et al.* (1999).

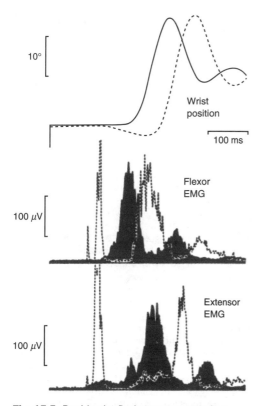

Fig. 15.5 Rapid wrist flexion movements in a single normal subject in response to an audio 'go' signal given at the start of the sweep, with (dotted lines) and without (solid traces) a TMS pulse delivered 100 ms after the start of the sweep. Shown are average wrist position (*upper traces*, flexion upwards), and wrist flexor (*middle traces*) and extensor (*lower traces*) rectified EMG activity. The control movement shows the usual triphasic pattern of EMG in agonist and antagonist muscles. The movement with the TMS pulse has the same form but is delayed by ~60 ms. Prior to onset of the delayed contraction the TMS artifact and the evoked MEP are visible. Reproduced from Day *et al.* (1989).

the contribution of the motor cortex to production of associated postural contractions. Abduction of one arm produces a triphasic ballistic pattern of EMG activation starting with a burst in the deltoid, followed by latissimus dorsi, and finally a second burst in the deltoid. In freely standing subjects this is accompanied by contralateral activity in latissimus dorsi,

pectoralis major, and abdominal muscles, which are thought to be involved in associated postural adjustments. TMS contralateral to the abducting arm delays the onset of the triphasic pattern as in the example above, but has little effect on the associated postural activity in the trunk muscles ipsilateral to TMS. Conversely, if the TMS pulse is applied contralateral to the associated postural activity, then it is delayed with respect to the focal abduction movement. The conclusion is that, in this task, the motor cortex is involved in control of postural contractions associated with activation of prime movers.

Interference with movement can be seen if TMS pulses are applied over other motor areas of cortex. For example, stimulation over the dorsal premotor cortex (PMd) delays the onset of movement in an arbitrarily cued choice-reaction task (Schluter *et al.* 1998). Such tasks involve instructions such as pressing one finger when a visual cue appears and a different finger when a different cue is given. PMd stimulation delays reaction time, but only if given early in the reaction period and has no effect late in the reaction when motor cortex disruption is most powerful. It appears that PMd processes data earlier in this task than the motor cortex.

This 'virtual lesion' approach has since been used in a large number of studies to investigate the involvement of different areas of frontal and parietal cortex in various movement tasks. They are not discussed further here since they have been reviewed elsewhere in some detail (Pascual-Leone *et al.* 2000).

References

Abbruzzese G, Assini A, Buccolieri A, Marchese R, Trompetto C (1999). Changes of intracortical inhibition during motor imagery in human subjects. *Neuroscience Letters 263*, 113–116.

Baker SN, Olivier E, Lemon RN (1995). Task-related modulation in the amplitude of the direct volley evoked by transcranial magnetic stimulation of the motor cortex and recorded from the medullary pyramid in the monkey. *Journal of Physiology (London)* 487P, 69–70.

Baldissera F, Cavallari P, Craighero L, Fadiga L (2001). Modulation of spinal excitability during observation of hand actions in humans. *European Journal of Neuroscience 13*, 190–194.

Berardelli A, Hallett M, Rothwell JC, Agostino R, Manfredi M, Thompson PD, Marsden CD (1996). Single-joint rapid arm movements in normal subjects and in patients with motor disorders. *Brain 119*, 661–674.

Brasil-Neto JP, Pascual-Leone A, Valls-Sole J, Cammarota A, Cohen LG, Hallett M (1993). Postexercise depression of motor evoked potentials: a measure of central nervous system fatigue. *Experimental Brain Research, 93*, 181–184.

Brunia CHM, Vuister FM (1979). Spinal reflexes as an indicator of motor preparation in man. *Physiological Psychology, 7*, 377–380.

Brunia CHM, Scheirs JGM, Haagh SAVM (1982). Changes of Achilles tendon reflex amplitudes during a fixed foreperiod of four seconds. *Psychophysiology 19*, 63–70.

Buccolieri A, Abbruzzese G, Rothwell JC (2004). Relaxation from a voluntary contraction is preceded by increased excitability of motor cortical inhibitory circuits. *Journal of Physiology (London), 558*, 685–695.

Capaday C, Lavoie BA, Barbeau H, Schneider C, Bonnard M (1999). Studies on the corticospinal control of human walking. I. Responses to focal transcranial magnetic stimulation of the motor cortex. *Journal of Neurophysiology 81*, 129–139.

Cattaneo L, Voss M, Brochier T, Prabhu G, Wolpert DM, Lemon RN (2005). A cortico-cortical mechanism mediating object-driven grasp in humans. *Proceedings of the National Academy Sciences of the USA 102*, 898–903.

Cowey A, Walsh V (2001). Tickling the brain: studying visual sensation, perception and cognition by transcranial magnetic stimulation. *Progress in Brain Research 134*, 411–425.

Datta AK, Harrison LM, Stephens JA (1989). Task-dependent changes in the size of response to magnetic brain stimulation in human first dorsal interosseous muscle. *Journal of Physiology (London) 418*, 13–23.

Day BL, Rothwell JC, Thompson PD, Maertens-de NA, Nakashima K, Shannon K, Marsden CD (1989). Delay in the execution of voluntary movement by electrical or magnetic brain stimulation in intact man. Evidence for the storage of motor programs in the brain. *Brain 112*, 649–663.

Dietz V, Wirz M, Curt A, Colombo G (1998). Locomotor pattern in paraplegic patients: training effects and recovery of spinal cord function. *Spinal Cord 36*, 380–390.

Di Lazzaro V, Oliviero A, Profice P, Insola A, Mazzone P, Tonali P, Rothwell JC (1999). Effects of voluntary contraction on descending volleys evoked by transcranial electrical stimulation over the motor cortex hand area in conscious humans. *Experimental Brain Research 124*, 525–528.

Di Lazzaro V, Oliviero A, Tonali PA, Mazzone P, Insola A, Pilato F, Saturno E, Dileone M, Rothwell JC (2003). Direct demonstration of reduction of the output of the human motor cortex induced by a fatiguing muscle contraction. *Experimental Brain Research, 149*, 535–538.

Drew T, Jiang W, Widajewicz W (2002). Contributions of the motor cortex to the control of the hindlimbs during locomotion in the cat. *Brain Research: Brain Research Reviews 40*, 178–191.

Evarts EV, Tanji J (1976). Reflex and intended responses in motor cortex pyramidal tract neurons of monkey. *Journal of Neurophysiology 39*, 1069–1080.

Facchini S, Muellbacher W, Battaglia F, Boroojerdi B, Hallett M (2002). Focal enhancement of motor cortex excitability during motor imagery: a transcranial magnetic stimulation study. *Acta Neurologica Scandinavica 105*, 146–151.

Fadiga L, Fogassi L, Pavesi G, Rizzolatti G (1995). Motor facilitation during action observation: a magnetic stimulation study. *Journal of Neurophysiology 73*, 2608–2611.

Fadiga L, Buccino G, Craighero L, Fogassi L, Gallese V, Pavesi G (1999). Corticospinal excitability is specifically modulated by motor imagery: a magnetic stimulation study. *Neuropsychologia 37*, 147–158.

Ferbert A, Priori A, Rothwell JC, Day BL, Colebatch JG, Marsden CD (1992). Interhemispheric inhibition of the human motor cortex. *Journal of Physiology (London) 453*, 525–546.

Fourkas AD, Ionta S, Aglioti SM (2006). Influence of imagined posture and imagery modality on corticospinal excitability. *Behavioral Brain Research 168*, 190–196.

Gandevia SC (2001). Spinal and supraspinal factors in human muscle fatigue. *Physiology Reviews 81*, 1725–1789.

Gandevia SC, Allen GM, Butler JE, Taylor JL (1996). Supraspinal factors in human muscle fatigue: evidence for suboptimal output from the motor cortex. *Journal of Physiology (London) 490*, 529–536.

Gandevia SC, Wilson LR, Inglis JT, Burke D (1997). Mental rehearsal of motor tasks recruits alpha-motoneurones but fails to recruit human fusimotor neurones selectively. *Journal of Physiology (London) 505*, 259–266.

Gandevia SC, Petersen N, Butler JE, Taylor JL (1999). Impaired response of human motoneurones to corticospinal stimulation after voluntary exercise. *Journal of Physiology (London) 521*, 749–759.

Gangitano M, Mottaghy FM, Pascual-Leone A (2001). Phase-specific modulation of cortical motor output during movement observation. *Neuroreport 12*, 1489–1492.

Gerloff C, Toro C, Uenishi N, Cohen LG, Leocani L, Hallett M (1997). Steady-state movement-related cortical potentials: a new approach to assessing cortical activity associated with fast repetitive finger movements. *Electroencephalography and Clinical Neurophysiology 102*, 106–113.

Hasbroucq T, Kaneko H, Akamatsu M, Possamai CA (1997). Preparatory inhibition of cortico-spinal excitability: a transcranial magnetic stimulation study in man. *Brain Research: Cognitive Brain Research 5*, 185–192.

Hashimoto R, Rothwell JC (1999). Dynamic changes in corticospinal excitability during motor imagery. *Experimental Brain Research 125*, 75–81.

Herbert RD, Gandevia SC (1999). Twitch interpolation in human muscles: mechanisms and implications for measurement of voluntary activation. *Journal of Neurophysiology 82*, 2271–2283.

Lemon RN, Johansson RS, Westling G (1995). Corticospinal control during reach, grasp, and precision lift in man. *Journal of Neuroscience 15*, 6145–6156.

Leocani L, Cohen LG, Wassermann EM, Ikoma K, Hallett M (2000). Human corticospinal excitability evaluated with transcranial magnetic stimulation during different reaction time paradigms. *Brain 123*, 1161–1173.

Liepert J, Dettmers C, Terborg C, Weiller C (2001). Inhibition of ipsilateral motor cortex during phasic generation of low force. *Clinical Neurophysiology 112*, 114–121.

MacKinnon CD, Rothwell JC (2000). Time-varying changes in corticospinal excitability accompanying the triphasic EMG pattern in humans. *Journal of Physiology (London) 528*, 633–645.

Maccabee PJ, Amassian VE, Cracco RQ, Cracco JB, Rudell AP, Eberle LP, Zemon V (1991). Magnetic coil stimulation of human visual cortex: studies of perception. *Electroencephalography and Clinical Neurophysiology, 43*(Suppl), 111–120.

Merton PA (1954). Voluntary strength and fatigue. *Journal of Physiology (London) 123*, 553–564.

Montagna M, Cerri G, Borroni P, Baldissera F (2005). Excitability changes in human corticospinal projections to muscles moving hand and fingers while viewing a reaching and grasping action. *European Journal of Neuroscience 22*, 1513–1520.

Muir RB, Lemon RN (1983). Corticospinal neurons with a special role in precision grip. *Brain Research 261*, 312–316.

Murase N, Duque J, Mazzocchio R, Cohen LG (2004). Influence of interhemispheric interactions on motor function in chronic stroke. *Annals of Neurology 55*, 400–409.

Palmer E, Cafarelli E, Ashby P (1994). The processing of human ballistic movements explored by stimulation over the cortex. *Journal of Physiology (London) 481*, 509–520.

Palmer E, Downes L, Ashby P (1996). Associated postural adjustments are impaired by a lesion of the cortex. *Neurology 46*, 471–475.

Pascual-Leone A, Brasil NJ, Valls SJ, Cohen LG, Hallett M (1992a). Simple reaction time to focal transcranial magnetic stimulation. Comparison with reaction time to acoustic, visual and somatosensory stimuli. *Brain, 115*, 109–122.

Pascual-Leone A, Valls SJ, Wassermann EM, Brasil NJ, Cohen LG, Hallett M (1992b). Effects of focal transcranial magnetic stimulation on simple reaction time to acoustic, visual and somatosensory stimuli. *Brain 115*, 1045–1059.

Pascual-Leone, A, Walsh, V, Rothwell, J. (2000). Transcranial magnetic stimulation in cognitive neuroscience – virtual lesion, chronometry, and functional connectivity. *Current Opinion in Neurobiology 10*, 232–237.

Petersen NT, Butler JE, Marchand-Pauvert V, Fisher R, Ledebt A, Pyndt HS, Hansen NL, Nielsen JB (2001). Suppression of EMG activity by transcranial magnetic stimulation in human subjects during walking. *Journal of Physiology (London) 537*, 651–656.

Reynolds C, Ashby P (1999). Inhibition in the human motor cortex is reduced just before a voluntary contraction. *Neurology 53*, 730–735.

Rossini PM, Zarola F, Stalberg E, Caramia MD (1988). Premovement facilitation of motor evoked potentials in man during transcranial stimulation of central motor pathways. *Brain Research 458*, 20–30.

Schieppati M, Trompetto C, Abbruzzese G (1996). Selective facilitation of responses to cortical stimulation of proximal and distal arm muscles by precision tasks in man. *Journal of Physiology (London) 491*, 551–562.

Schluter ND, Rushworth MF, Passingham RE, Mills KR (1998). Temporary interference in human lateral premotor cortex suggests dominance for the selection of movements. A study using transcranial magnetic stimulation. *Brain 121*, 785–799.

Schneider C, Lavoie BA, Barbeau H, Capaday C (2004). Timing of cortical excitability changes during the reaction time of movements superimposed on tonic motor activity. *Journal of Applied Physiology 97*, 2220–2227.

Shimazu H, Maier MA, Cerri G, Kirkwood PA, Lemon RN (2004). Macaque ventral premotor cortex exerts powerful facilitation of motor cortex outputs to upper limb motoneurons. *Journal of Neuroscience 24*, 1200–1211.

Sohn YH, Hallett M (2004). Surround inhibition in human motor system. *Experimental Brain Research 158*, 397–404.

Sohn YH, Wiltz K, Hallett M. (2002). Effect of volitional inhibition on cortical inhibitory mechanisms. *Journal of Neurophysiology 88*, 333–338.

Sohn YH, Jung HY, Kaelin-Lang A, Hallett M (2003). Excitability of the ipsilateral motor cortex during phasic voluntary hand movement. *Experimental Brain Research 148*, 176–185.

Stefan K, Cohen LG, Duque J, Mazzocchio R, Celnik P, Sawaki L, Ungerleider L, Classen J (2005). Formation of a motor memory by action observation. *Journal of Neuroscience 25*, 9339–9346.

Stinear CM, Byblow WD (2003). Role of intracortical inhibition in selective hand muscle activation. *Journal of Neurophysiology 89*, 2014–2020.

Stinear CM, Byblow WD (2004). Modulation of corticospinal excitability and intracortical inhibition during motor imagery is task-dependent. *Experimental Brain Research 157*, 351–358.

Stinear CM, Byblow WD, Steyvers M, Levin O, Swinnen SP (2006). Kinesthetic, but not visual, motor imagery modulates corticomotor excitability. *Experimental Brain Research 168*, 157–164.

Tanji J, Evarts EV (1976). Anticipatory activity of motor cortex neuron in relation to direction of an intended movement. *Journal of Neurophysiology 39*, 1062–1068.

Taylor JL, Butler JE, Allen GM, Gandevia SC (1996). Changes in motor cortical excitability during human muscle fatigue. *Journal of Physiology (London) 490*, 519–528.

Touge T, Werhahn KJ, Rothwell JC, Marsden CD (1995). Movement-related cortical potentials preceding repetitive and random-choice hand movements in Parkinson's disease. *Annals of Neurology 37*, 791–799.

Touge T, Taylor JL, Rothwell JC (1998). Reduced excitability of the cortico-spinal system during the warning period of a reaction time task. *Electroencephalography and Clinical Neurophysiology 109*, 489–495.

Ugawa Y, Rothwell JC, Day BL, Thompson PD, Marsden CD (1991). Percutaneous electrical stimulation of corticospinal pathways at the level of the pyramidal decussation in humans. *Annals of Neurology 29*, 418–427.

Vagg R, Mogyoros I, Kiernan MC, Burke D (1998). Activity-dependent hyperpolarization of human motor axons produced by natural activity. *Journal of Physiology (London) 507*, 919–925.

Waldvogel D, van Gelderen P, Muellbacher W, Ziemann U, Immisch I, Hallett M (2000). The relative metabolic demand of inhibition and excitation. *Nature 406*, 995–998.

Yahagi S, Shimura K, Kasai T (1996). An increase in cortical excitability with no change in spinal excitability during motor imagery. *Perceptual and Motor Skills 83*, 288–290.

Zanette G, Bonato C, Polo A, Tinazzi M, Manganotti P, Fiaschi A (1995). Long-lasting depression of motor-evoked potentials to transcranial magnetic stimulation following exercise. *Experimental Brain Research 107*, 80–86.

CHAPTER 16

Changes in TMS measures induced by repetitive TMS

Joseph Classen and Katja Stefan

Summary – Neuronal plasticity induced by repetitive (r)TMS protocols may offer important insights into the mechanisms of the CNS that support its remarkable flexibility. rTMS-induced plasticity has evolved into an important tool used to establish structure–function relationships. In addition, rTMS has raised considerable interest because of its therapeutic potential. This chapter reviews several protocols of rTMS-induced plasticity. One-hertz rTMS, the most widely used protocol, likely induces multiple effects of variable duration simultaneously at several sites within the nervous system although relatively little is known about the exact mechanisms involved. In two protocols the structure of rTMS trains is modified, informed by knowledge of physiological properties of the corticospinal system. Repetitive TMS, using pairs of TMS delivered at I-wave periodicity, and theta-burst stimulation display considerably enhanced intervention efficacy compared to uniform repetitive application of single TMS pulses. Changes of cortical excitability are induced if slow-rate TMS is combined with experimental deafferentation. Plasticity can also be induced by the combined action of TMS over the primary motor cortex and peripheral nerve stimulation ('paired associative stimulation'). In some of the latter rTMS protocols the direction of excitability changes depends on subtle details of the timing of the stimuli. The activation state of the motor cortex to which the rTMS protocol is directed has been identified as an important factor determining
the magnitude and even occasionally the polarity of induced effects. With increasing sophistication of the rTMS protocols, some of the mechanisms underlying rTMS-induced plasticity have been mapped to properties of intracortical inhibition and to long-term potentiation of synaptic efficacy of cortical connections.

Introduction

Repetitive (r)TMS, when applied to the motor cortex or other cortical regions of the brain, may induce effects that outlast the stimulation period. The neural plasticity, which emerges as a result of such interventions, has itself been studied to gain insight into plasticity mechanisms of the brain. Furthermore, rTMS-induced effects have been used as a tool to disrupt temporarily activity in local or remote cortical areas. This property has now widely been adopted to establish structure–function relationships. Finally, rTMS-induced effects have raised considerable interest because of their possible therapeutic potential in patients with neuropsychiatric disorders.

A large number of partially interdependent variables have to be taken into account when reviewing the physiological effects of rTMS. Among them are pulse configuration, stimulus frequency, stimulus intensity, duration of the application period, and the total number of stimuli. As a rule, short-lasting interventions trigger effects that outlast the intervention by only a short duration. However, there are important exceptions. Recent studies have identified the structure of the pulse train as a parameter of utmost importance determining the efficacy of the protocol. We will, therefore, term rTMS-protocols as 'simple' if individual stimuli are spaced apart by identical interstimulus intervals (ISIs). Conversely, protocols in which differing ISIs are employed will be termed 'patterned' rTMS. Only neurophysiological changes lasting at least several minutes will be considered in this overview.

Although corticospinal excitability is usually assessed by measuring the amplitude of motor-evoked potentials (MEPs) elicited by single-pulse TMS, a substantial number of other variables has been examined which may reflect changes in cortical or spinal neuronal elements and convey information about corticospinal excitability different from that contained in the motor response elicited by single-pulse TMS. As some of these variables are interdependent, a very complex pattern emerges whose understanding is currently rather limited.

Simple rTMS

One-hertz rTMS

One-hertz rTMS is by now the most frequently used protocol to establish structure–function relationships in brain areas outside the primary motor cortex. This is somewhat surprising because, as will be outlined below, the phenomena induced by 1 Hz rTMS in the motor system are already very complex and their physiological mechanisms are not well understood. It is likely that similarly complex physiological phenomena will have to be taken into account when applying 1 Hz rTMS to other brain areas. Chen et al. (1997) showed that low-frequency rTMS at 0.9 Hz with a stimulus intensity of 115% of the resting motor threshold (RMT) applied for 15 min led to a decrease in MEP amplitude of, on average, some 20% of the baseline amplitude. This decrease lasted for ≥15 min (Chen et al. 1997). Similar observations were made by several other groups (e.g. Siebner et al. 1999; Muellbacher et al. 2000; Gilio et al. 2003; Plewnia et al. 2003) although interindividual variability appears to be large (Maeda et al. 2000). Muellbacher et al. (2000) showed a

significant suppression of the MEP input–output curve after 1 Hz rTMS. Suppressive effects after 1 Hz may last for 30 min (Muellbacher et al. 2000). They were found to be specific for the hand motor representation, which was the target of the rTMS, whereas they were absent in adjacent muscle representations. This indicated a spatially relatively focused effect of rTMS. The suppressive effect of low-frequency rTMS was dependent on the intensity of the magnetic pulse (Fitzgerald et al. 2002). Fitzgerald et al. (2002) applied 15 min of 1 Hz rTMS at 85 and 115% RMT. rTMS at both intensities led to an increase in RMT but only the suprathreshold stimulation reduced the size of MEPs. In contrast, even subthreshold 1 Hz rTMS (10 min) at 90% of RMT led to a lasting suppression of MEP amplitudes which was present for ~10 min (Romero et al. 2002) suggesting that depression of MEP size by 1 Hz rTMS is not dependent on eliciting MEP responses during the conditioning period. Because corticospinal output neurons are active at 90% of RMT, this observation, however, does not exclude activity of cortical output neurons as a necessary precondition to inducing lasting depression of corticospinal excitability by 1 Hz rTMS. Intensities lower than those leading to activation of corticospinal output elements have not been reported to lead to lasting excitability changes. Longer trains of 1 Hz rTMS (applied at 90% (Maeda et al. 2000) or 95% RMT (Touge et al. 2001)) are more effective than shorter ones. Touge et al. (2001) observed that the 1 Hz rTMS-induced suppression of MEPs occurred only when the target muscle was at rest but not during voluntary activation. The authors concluded that rTMS was changing the level of excitability of the resting motor system rather than changing effectiveness of neuronal transmission in synaptic connections to pyramidal cells. However, voluntary activation may lead to recurrent activation of inhibitory intracortical circuits, which may stabilize the output from the primary motor cortex and may mask rTMS-induced changes in synaptic transmission.

As mentioned above, several recent studies have pointed out that, although 1 Hz rTMS leads to a suppression of MEP size across many subjects, results can be quite variable interindividually with some subjects even showing a facilitatory effect rather than suppression of corticospinal excitability (Maeda *et al.* 2000; Gangitano *et al.* 2002). The reasons for this interindividual response variability, which has also been noted in several other rTMS protocols, is not yet understood. It appears possible that some of the variability may be related to the activation history of the targeted motor cortex. This is suggested by findings demonstrating that a preconditioning of the motor cortex by high-frequency rTMS may enhance the inhibitory effect of the following 1 Hz rTMS (Iyer *et al.* 2003). Six-hertz rTMS at 90% RMT was applied to the motor cortex for 10 min which was then followed by 1 Hz rTMS at an intensity of 115% RMT for 10 min. With this protocol, the efficacy of 1 Hz rTMS could be improved as demonstrated by a stronger and longer-lasting depression of MEP amplitudes. That the state of the 1 Hz rTMS recipient cortex is an important determinant for the after-effects of 1 Hz rTMS is further suggested by results obtained by Siebner *et al.* (2004). These authors showed that preconditioning of the motor cortex by transcranial direct current stimulation modulated the effect of subsequently applied 1 Hz rTMS whose parameters were chosen to produce no effects when alone (i.e. without preconditioning stimulation). Preconditioning with anodal transcranial direct current stimulation, which by itself increases corticospinal excitability (cf. Chapter 19, this volume), led 1 Hz rTMS to induce a suppressive effect on subsequently recorded MEP. If, however, cathodal direct transcranial current stimulation was used as preconditioning stimulation, then 1 Hz rTMS even induced an increase in MEP size. Therefore, not only is the magnitude but also the direction of 1 Hz rTMS-induced effects determined by the activation state or the activation history of the targeted motor cortex.

Resting motor threshold may be regarded as a measure of neuronal membrane excitability (cf. Chapters 9 and 13, this volume). For reasons not well understood, its magnitude does not appear to be altered by changes in neurotransmission whereas it is subject to modification by pharmacological agents influencing ion channels (Ziemann *et al.* 1996). Muellbacher *et al.* (2000) reported an increase in RMT after 1 Hz rTMS. However, the RMT increase lasted for a shorter period than the MEP decrease (Muellbacher *et al.* 2000). Other 1 Hz rTMS studies reported no changes in RMT (Bagnato *et al.* 2005; Heide *et al.* 2006) or described an increase in RMT without MEP decrease (Fitzgerald *et al.* 2002). In view of these divergent observations it remains unclear whether changes in neuronal membrane excitability are induced by 1 Hz rTMS. In any case, if changes in RMT are present, they are likely to be independent from changes in mechanisms underlying depression of MEP size. Several studies addressed the question of whether the effects of rTMS on MEP size are due to effects originating at cortical, subcortical, or spinal level. H-reflex size remained unchanged (Gilio *et al.* 2003; Modunog *et al.* 2001; Touge *et al.* 2001); therefore effects are likely to be located supraspinally.

Short-interval intracortical inhibition (SICI) mediated by $GABA_A$ receptors may be tested by a paired-pulse protocol (Kujirai *et al.* 1993) (cf. Chapters 11–13, this volume) whereas $GABA_B$ receptor-mediated inhibition may be assessed by measuring the cortical silent period (cSP) (cf. Chapter 10, this volume). Studies employing a paired-pulse protocol reported no changes in SICI after 15 min of 1 Hz rTMS (Fitzgerald *et al.* 2002; Gilio *et al.* 2003). A reduction of intracortical facilitation (ICF) (cf. Chapters 11–13, this volume) was found in one study (Bagnato *et al.* 2005) but not in another (Heide *et al.* 2006). However, some of these studies were unable to reproduce the suppressive effect of 1 Hz rTMS on MEP size. One-hertz rTMS showed no effect on CSP duration (Fitzgerald *et al.* 2002; Gilio *et al.* 2003).

The contribution of GABAergic inhibition and *N*-methyl-D-aspartate (NMDA) receptors to 1 Hz rTMS-induced plasticity was probed by applying 1 Hz rTMS in subjects medicated with either lorazepam, an allosteric positive modulator at the $GABA_A$ receptor, or dextromethorphan, an NMDA receptor antagonist. Suprathreshold 1 Hz rTMS (15 min) resulted in a decrease in motor cortical excitability in subjects taking placebo but not in the medicated groups. These results suggested that cortical responses to 1 Hz rTMS are dependent on activity at both GABA and NMDA receptor systems (Fitzgerald *et al.* 2005).

The conditioning effects of rTMS are not limited to the cortical area targeted by rTMS but may also occur at distant interconnected sites in the brain. Wassermann *et al.* (1998) reported a depression of excitability in the homologous nonstimulated motor cortex with 15 min suprathreshold 1 Hz rTMS applied over the other primary motor cortex. However, other authors have not found any change in MEP amplitudes from the nonstimulated motor cortex (Plewnia *et al.* 2003), or even an increase in MEP size (Gilio *et al.* 2003; Schambra *et al.* 2003). The technique of transcallosal inhibition was used to examine the influence of 15 min 1 Hz suprathreshold rTMS on interhemispheric inhibition (IHI) (cf. Chapters 11 and 12, this volume). Furthermore, excitability of the stimulated and nonstimulated motor cortex was studied by measuring MEP amplitudes, the contra- and ipsilateral silent period (CSP, ISP), and SICI, and ICF (Gilio *et al.* 2003). After 1 Hz rTMS the authors found a reduction of IHI from the stimulated to the nonstimulated hemisphere and an increase of MEP amplitudes recorded from the nonstimulated motor cortex. CSP, ISP, SICI and ICF in the nonstimulated motor cortex were unaffected by rTMS. The authors concluded that 1 Hz rTMS might release the nonstimulated motor cortex from tonic inhibition by the opposite hemisphere which secondarily leads to an increased corticospinal excitability (Gilio *et al.* 2003). However, in the study by Gilio *et al.* there was no change in excitability of the stimulated hemisphere, since MEP amplitudes remained unaffected for this side. This suggested that the mechanisms underlying modulation of excitability in remote areas might be different from those leading to depression of MEP amplitudes. Pal *et al.* (2005) confirmed the finding of a decreased IHI from the stimulated to the nonstimulated hemisphere following 1 Hz rTMS. The inhibitory effect on IHI lasted for ≥15 min. In addition to the predominant reduction of IHI from the stimulated to the nonstimulated hemisphere they also found a reduction of IHI also from the nonstimulated to the stimulated motor cortex (Pal *et al.* 2005). Short-interval intracortical facilitation (SICF) measured in a paired-pulse TMS protocol probes the intrinsic neuronal excitability of neuronal elements mediating glutamatergic non-NMDA receptor-dependent excitation (Ziemann *et al.* 1998b) (cf. Chapters 11 and 13,

this volume). SICF was enhanced in the non-stimulated motor cortex after 1 Hz rTMS (Heide *et al.* 2006). The authors suggested that this was due to an increase in intrinsic excitability as a consequence of release from inhibition from the stimulated hemisphere.

Repetitive TMS of the premotor cortex, even at intensities below the RMT as assessed in the primary motor cortex, induced lasting changes of brain activity in remote brain areas including the ipsilateral primary motor cortex (Rollnik *et al.* 2000; Gerschlager *et al.* 2001; Paus *et al.* 2001; Münchau *et al.* 2002). Gerschlager *et al.* (2001) showed that low-frequency subthreshold rTMS of the lateral premotor cortex decreased excitability of the ipsilateral primary motor cortex which may have resulted from conduction of signals induced within the premotor cortex to the motor cortex by cortico-cortical connections. This finding was extended by Münchau *et al.* (2002) by demonstrating changes in SICI/ICF produced by 1 Hz conditioning of the premotor cortex. 1 Hz rTMS directed to the cerebellum increased the amplitudes of MEPs evoked from the motor cortex contralateral to the conditioning cerebellar stimulation (Oliveri *et al.* 2005). Moreover, left cerebellar rTMS increased the ICF of the right motor cortex as measured with paired pulses separated by an ISI of 15 ms. The effect lasted for up to 30 min afterward and was specific for the contralateral (right) motor cortex. The CSP duration was unaffected by cerebellar rTMS. The authors suggested that 1 Hz rTMS of the cerebellar cortex can reduce the flow of inhibition from Purkinje cells toward deep nuclei, thereby increasing the excitability of interconnected brain areas.

rTMS at frequencies higher than one hertz

Higher rTMS frequencies of 5–20 Hz tend to increase cortical excitability (Pascual-Leone *et al.* 1994; Maeda *et al.* 2000). In an initial rTMS study by Pascual-Leone *et al.* (1994) it was demonstrated that the excitability of the motor system following 10 pulses of TMS at 20 Hz and 150% threshold intensity facilitated the response to a smaller test stimulus (90% threshold intensity) for 3–4 min. Quartarone *et al.* (2005) found that long trains (≥900 pulses) of

5 Hz rTMS at 90% RMT facilitated MEP in the relaxed and tonically contracted hand muscle. Brain-stem electrical stimulation (BES) demonstrated that changes of excitability induced by 5 Hz rTMS were generated at cortical level (Quartarone et al. 2005). The authors concluded that prolonged 5 Hz rTMS may enhance synaptic transmission of connections onto pyramidal cells in the motor cortex (Quartarone et al. 2005). Furthermore, 5 Hz rTMS induced a decrease of SICI and a facilitation of the first SICF peak but had no influence on CSP duration. This might indicate a different responsiveness of distinct inhibitory motor cortical circuits to 5 Hz rTMS (Quartarone et al. 2005). The reduction of SICI after subthreshold 5 Hz rTMS is in good agreement with results from Di Lazzaro et al. (2002) who reported a decrease in SICI after a short train of subthreshold 5 Hz rTMS. In addition, epidural spinal recordings of the descending corticospinal volleys showed that the decrease in SICI was due to changes at a cortical level (Di Lazzaro et al. 2002) (cf. Chapter 12, this volume). Similar observations were made by Wu et al. (2000) using rTMS at 120% RMT. Following a train of 30 single TMS pulses at frequencies of 5 and 15 Hz, SICI was reduced for 3.2 min and ICF was enhanced for 1.5 min, whereas amplitudes of MEPs evoked by single TMS were increased in the first 30 s only (Wu et al. 2000). The different time course of the effects confirms the notion that the interneuronal apparatus generating SICI is differentially affected compared to the one generating ICF. Effects on either one of these circuits, in turn, may be at least partly distinct from the one generating the MEP elicited by single-pulse TMS. A train of 5 Hz rTMS (100 pulses) at active motor threshold did not change motor cortical excitability (Lang et al. 2004). However, when this short train of 5 Hz rTMS was delivered after pretreatment with transcranial direct current stimulation, lasting after-effects were induced whose polarity depended on the polarity of the conditioning current (Lang et al. 2004).

Studies that have investigated the interaction between frequency, intensity, and duration of rTMS showed that low-intensity stimuli tend to produce a post-train suppression (Todd et al. 2006), even if applied at frequencies >5 Hz (but see Quartarone et al. 2005), whereas suprathreshold stimuli tend to produce MEP facilitation (Modugno et al. 2001). Similar effects occurred with variation of the duration of the stimulus train. For short trains (<20) of any frequency the immediate after-effect on MEP amplitude is more likely to be inhibitory for low intensities, whereas facilitation becomes evident with longer trains. In a study by Modugno et al. (2001), it was shown that a short train at 5 Hz and 100% RMT produced post-train MEP inhibition, whereas longer trains with identical stimulus settings produced MEP facilitation. The authors concluded that inhibition and facilitation might build up gradually during the course of a stimulus train, with inhibition reaching its maximum after a lower dose of stimulation. Similar to low-frequency stimulation protocols, rTMS delivered over primary motor cortex at frequencies >3 Hz had no effect on the duration of the CSP (Peinemann et al. 2000; Quartarone et al. 2005).

Patterned rTMS

In contrast to 'simple rTMS', when TMS pulses are spaced apart by identical ISIs, several rTMS studies utilized trains of stimulus pairs consisting of stimuli with different ISIs and intensities.

Repetitive paired-pulse TMS at I-wave periodicity

Single-pulse TMS elicits a brief train of descending volleys at a periodicity of 1.5 ms (cf. Chapter 14, this volume). These volleys result from repeated indirect trans-synaptic activation of corticospinal neurons via interneurons. A suprathreshold stimulus followed by a subthreshold stimulus at intervals corresponding to the I-wave periodicity increases cortical excitability as measured by MEP amplitudes, which is thought to be due to facilitatory I-wave interactions (Tokimura 1996; Ziemann et al. 1998b). Thickbroom et al. (2006) demonstrated that facilitatory I-wave interactions set up by repetitive paired-pulse TMS corresponding to the I-wave periodicity may induce a lasting increase of excitability (Figure 16.1). Paired TMS pulses at equal strength and an ISI of 1.5 ms were delivered at a frequency of 0.2 Hz to

(a)

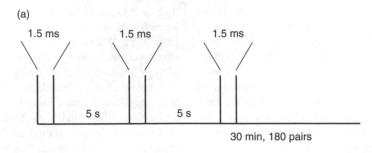

(b)

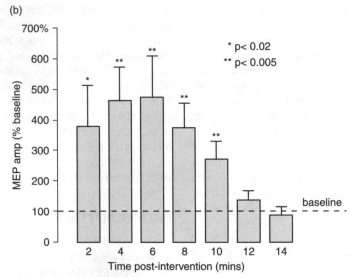

Fig. 16.1 Repetitive paired-pulse TMS at I-wave periodicity (iTMS). (a) Illustration of the general design principle. Paired stimuli of equal strength were delivered at an interstimulus interval of 1.5 ms, and at a repetition rate of 0.2 Hz. Stimulus intensity was set to the stimulus intensity that, when delivered as a pair, generated a motor-evoked potential (MEP) of between 0.5 and 1 mV. A period of 30 min of iTMS was administered. (b) MEP amplitude (percentage of pre-intervention baseline) across subjects at 2 min intervals after 30 min of iTMS, showing MEP amplitude significantly increased (~400%) for 10 min post-intervention. Modified from Thickbroom *et al.* (2006), with permission.

the primary motor cortex for 30 min. Stimulus intensity was set to the stimulus intensity that, when delivered as a pair, generated an MEP of between 0.5 and 1 mV. A total of 360 stimuli were delivered. Paired-pulse MEP amplitudes increased throughout the intervention up to fivefold of baseline amplitudes by the end of stimulation period, which was sustained for up to 10 min post-stimulation. Motor threshold and number of F-waves measured with electrical stimulation of the ulnar nerve remained unchanged by this intervention (Thickbroom *et al.* 2006). It is possible that this finding reflects an increase in synaptic efficacy in neuronal circuits underlying I-wave generation. However, to relate this phenomenon to long-term potentiation (LTP), rather than to post-tetanic potentiation, longer-lasting effects should be demonstrated.

Instead of applying paired TMS pulses of suprathreshold intensities, Khedr *et al.* (2004) used repeated pairs of stimuli with identical subthreshold (80% of active motor threshold) intensities to examine whether this would lead to long-term reduction of cortical excitability. Repetitive paired TMS was given to the motor cortex at a frequency of 0.6 Hz for 25 min (250 pairs). The ISI between paired TMS pulses was 3 ms.

Prolonged paired rTMS produced a decrease in cortical excitability as MEP amplitudes were reduced and RMT increased. The suppressive effect was present for ≥30 min. Furthermore, the authors found enhanced measures for cortical inhibition as intracortical inhibition tested with a paired-pulse paradigm and CSP increased. The authors concluded that pairs of low-intensity pulses can summate and activate cortical inhibitory elements more effectively than expected from one pulse alone (Khedr *et al.* 2004). The depression of single-pulse-TMS-evoked MEP size was thought to be due to a strengthening of synapses from inhibitory interneurons terminating at the cortical output neurons. The design of this protocol resembles that used by Thickbroom *et al.* (2006). The ISI of 3 ms may be viewed as twice the length of a single I-wave interval (~1.5 ms). Therefore, similar cortical targets may have been involved.

Theta-burst stimulation

Theta-burst stimulation (TBS) refers to an interventional protocol where brief trains of pulses are delivered at 5 Hz, i.e. at theta-frequency. Depending on the temporal ordering of the magnetic stimuli, TBS may induce long-lasting, reversible inhibitory, or facilitatory effects (Huang *et al.* 2005) (Figure 16.2). Continuous TBS (cTBS) refers to bursts containing three TMS pulses of 50 Hz (i.e. 20 ms between each stimulus) repeated at 200 ms intervals (i.e. 5 Hz) for a duration of 20 s (300 pulses), or for 40 s (600 pulses) at a stimulus intensity of 80% active motor threshold. In contrast, intermittent TBS consists of a 2 s train of TBS repeated every 10 s for a total of 190 s (600 pulses). TBS is particularly attractive because of the short duration of its intervention and because its stimulation design is similar to well-characterized protocols used in experimental animal physiology. This similarity once more suggests that TBS-induced excitability changes share properties with LTP and long-term depression (LTD) of synaptic efficacy, two principal candidate mechanisms of memory in the brain. The excitability changes induced by TBS outlast the effects seen in many other rTMS protocols. TBS is also remarkable because its stimulus intensities are clearly subthreshold for activation of descending pathways.

This low intensity also indicates that TBS-induced effects are likely generated exclusively cortically (Huang *et al.* 2005), a notion supported by the absence of changes in H-reflex amplitudes after TBS conditioning. cTBS is able to produce behavioral effects as demonstrated by changes in reaction time (Huang *et al.* 2005). To investigate the origin of depression of cortical excitability by cTBS further, corticospinal volleys evoked by single-pulse TMS over the primary motor cortex were recorded before and after a 20 s period of cTBS in patients undergoing cervical spinal surgery (Di Lazzaro *et al.* 2005). This study showed that cTBS preferentially decreases the amplitude of the corticospinal I1-wave, with approximately the same time course as the MEP size. This underlined a cortical origin of the effect on the MEP. The authors suggested that cTBS suppresses MEPs possibly through LTD of excitatory synaptic connections.

Other patterned rTMS protocols

Sommer *et al.* (2001) applied a train of 80 paired TMS pulses consisting of a subthreshold conditioning stimulus at 90% active motor threshold followed by a suprathreshold test stimulus. The ISIs between conditioning and test stimulus were chosen to be 2, 5, or 10 ms in order to induce either inhibitory or facilitatory effects, according to the effective intervals producing SICI and ICF (Kujirai *et al.* 1993). Pairs of stimuli were applied repetitively at frequencies ranging from 0.17 to 5 Hz. This was compared to simple rTMS. At 1 Hz, neither single nor paired pulse at a conditioning–test interval of 2 ms had any effect on MEP amplitude during or after the conditioning train. At 5 Hz, pairs of stimuli at 2 ms produced a small facilitatory effect during but not after conditioning. This was not substantially different from single-pulse rTMS, so the authors concluded that pairs of stimuli were not superior to single-pulse rTMS in producing inhibitory or facilitatory effects (Sommer *et al.* 2001). In a subsequent study, longer trains of unequal intensity paired-pulse rTMS with an ISI of 3 or 10 ms were applied at 1 Hz for 15 min (Sommer *et al.* 2002). The authors found a transient decrease in corticospinal excitability with the inhibiting conditioning–test interval of 3 ms but not after 10 ms. Simple 1 Hz rTMS did not produce any effects.

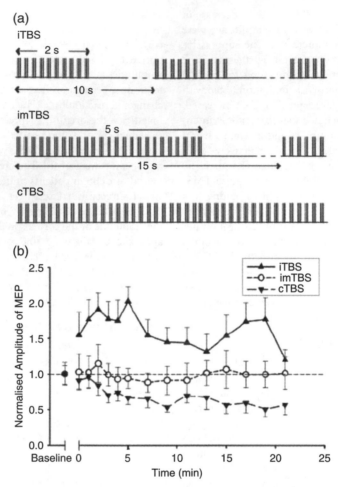

Fig. 16.2 Theta-burst stimulation (TBS). (a) Illustration of the three stimulation paradigms used. Each paradigm uses a TBS pattern in which three pulses of stimulation are given at 50 Hz, repeated every 200 ms. In the intermittent (i)TBS, a 2 s train of TBS is repeated every 10 s for a total of 190 s (600 pulses). In the intermediate (im)TBS paradigm, a 5 s train of TBS is repeated every 15 s for a total of 110 s (600 pulses). In the continuous (c)TBS paradigm, a 40 s train of uninterrupted TBS is given (600 pulses). (b) Time course of changes in MEP amplitude following conditioning with iTBS (closed up-triangle), cTBS (closed-down triangle), or imTBS (open circle). There is a significant facilitation of motor-evoked potential (MEP) size following iTBS lasting for ~15 min, and a significant reduction of MEP size following cTBS lasting for nearly 60 min. Intermediate TBS produces no significant changes in MEP size. Modified from Huang *et al.* (2005), with permission.

Repetitive TMS with ischemic nerve block

Ziemann *et al.* (1998a) combined 0.1 Hz rTMS to the contralateral motor cortex with temporary ischemic limb deafferentation (Figure 16.3). Repetitive TMS at 0.1 Hz, when given on its own, did not induce excitability changes.

Ischemic nerve block alone induced an increase in MEP size of muscles proximal to the level of deafferentation, such as the biceps brachii muscle, but the increase was only moderate. This increase was strongly enhanced when 0.1 Hz rTMS was applied to the motor cortex contralateral to the ischemic nerve block. After 0.1 Hz rTMS plus ischemic nerve block, SICI

(a)

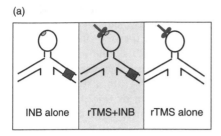

(b)

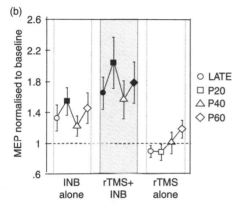

Fig. 16.3 Repetitive TMS at 0.1 Hz with ischemic nerve block (INB). (a) Principle of experimental design. For INB, a tourniquet was inflated above systolic blood pressure across the elbow distal to the target muscle (biceps brachii, BB). For rTMS plus INB, the motor cortex contralateral to the deafferented arm was stimulated at a rate of 0.1 Hz and an intensity of 120% of motor threshold for the BB from the onset of INB for a duration of ~30 min. As a control experiment, 0.1 Hz rTMS was applied for 30 min to the motor cortex contralateral to the target BB. (b) Changes in MEP size in the BB (means of six subjects) produced by the different interventions given on the x-axis. Data are expressed as values normalized to the pre-intervention baseline that has been assigned a value of 1. Different symbols refer to time points late into intervention (circles), and 20 min (squares), 40 min (triangles), and 60 min (diamonds) after the end of intervention. The gray box shows the test intervention (rTMS plus INB). Filled symbols indicate time points that were significantly different from baseline (P < 0.05). Note that rTMS alone that did not produce a change in BB MEP size when given in the absence of INB, but produced a significant increase in MEP size when given in the presence of INB. Panel (a) courtesy of Professor U. Ziemann; (b) modified from Ziemann et al. (1998a), with permission.

was reduced and ICF was enhanced, even when the TMS test-pulse intensity was adjusted to the altered MEP size after conditioning. RMT remained unaffected by rTMS plus ischemic nerve block. Because deafferentation induces GABAergic disinhibition in the motor cortex, as shown by magnetic resonance spectroscopy (Levy et al. 2002), the authors suggested that temporary ischemia renders the corticospinal system susceptible to rTMS by virtue of the deafferentation-induced release from inhibition (cooperativity). The effect of rTMS plus ischemic nerve block on excitability of the biceps muscle representation was elicited only when rTMS was directed to the upper arm representation, suggesting that the facilitatory effect of intervention was specific to the stimulated input (input specificity). Pretreatment of subjects with either lamotrigine, a sodium-channel blocker, or lorazepam, an allosteric positive modulator at the $GABA_A$ receptor, blocked this induction of an increase in MEP size by 0.1 Hz rTMS plus ischemic nerve block. In contrast, dextromethorphan, an NMDA receptor antagonist, had no effect on this MEP enhancement. However, all three pharmacological substances suppressed the changes of paired-pulse excitability induced by 0.1 Hz rTMS plus ischemic nerve block. Therefore, these observations suggest that multiple mechanisms, possibly including those involving excitatory synapse efficacy and excitability of GABAergic interneurons, had been operative. The moderate increase of cortical excitability induced by ischemic nerve block alone was blocked if 0.1 Hz rTMS was applied to the motor cortex ipsilateral to the nerve block. This observation raised the possibility that cortical excitability could be modulated bidirectionally by activating different pathways (Ziemann et al. 1998a).

Paired associative stimulation

Paired associative stimulation (PAS) refers to a paradigm consisting of low-frequency repetitive median nerve stimulation (typically 90–200) combined with timed TMS over the contralateral motor cortex. The design of this protocol resembles models of associative LTP as developed in animal studies. In many of these studies it has been found that if a weak excitatory synaptic input repeatedly arrives at a neuron shortly before the neuron has fired an action potential, then the strength of the synapse is increased. If the timing of the stimuli is reversed so that the input arrives just after neural discharge, then the strength of the synapse is reduced. This principle is known as spike-timing-dependent plasticity. Because TMS activates postsynaptic pyramidal output cells in the motor cortex mainly via intracortical fibers travelling 'horizontally' with respect to the cortical surface (Rothwell 1997) and because somatosensory information converges on pyramidal cells located within the motor cortex, it may be hypothesized that lasting excitability changes may be induced in the motor cortex by pairing afferent median nerve stimulation with magnetic stimulation over the motor cortex. In a study by Stefan et al. (2000), PAS was able to modulate the excitability of the motor system as probed by TMS elicited MEPs in the resting abductor pollicis brevis (APB) (Figure 16.4). The interval between peripheral nerve stimulation and TMS was set at 25 ms. Ninety pairs were applied at an interpair interval of 20 s. This protocol led to a lasting (≥60 min) increase in MEP amplitude recorded from the resting APB. The increase amounted to ~50% of the baseline value (Stefan et al. 2000). In subsequent studies (Wolters et al. 2003; Ziemann et al. 2004), the effect of PAS on MEP size was shown to be critically dependent on the timing of the TMS pulse with respect to the afferent median nerve stimulation. Effects were produced only when the events induced by the two stimulation modalities were nearly synchronous. Changing the interval between the two associative stimuli from 25 ms (PAS25) to 10 ms (PAS10) led to a depression of TMS-evoked motor-evoked potentials (MEPs) rather than the facilitation observed with PAS25 (Wolters et al. 2003) (Figure 16.4). This observation suggests that PAS-induced plasticity in human motor cortex is governed by strict temporal rules. It is likely that at an ISI of 25 ms the events induced by TMS in the primary motor cortex follow those induced in the primary motor cortex by median nerve stimulation, whereas the sequence of events induced by the two stimulation modalities is reversed with PAS10.

Further experiments employing brain-stem stimulation or assessments of the size of the F-waves suggested that the site of the plastic changes was within the motor cortex. Enhancement of cortical excitability induced by PAS evolved rapidly (within 30 min), was persistent (≥30–60 min duration), yet reversible, and was topographically specific. Involvement of NMDA receptors in both the facilitating and depressing PAS protocols was implicated by the fact that neither of the protocols led to a significant change of cortical excitability if subjects were premedicated with dextromethorphan, a blocker of NMDA receptors (Stefan et al. 2002; Wolters et al. 2003). In addition, PAS10-induced depression of cortical excitability was blocked by premediation with nimodipine, an L-type voltage-gated Ca-channel blocker. Excitability changes as probed by MEPs recorded from resting muscles were maximal in the muscle representation receiving homotopical input by afferent stimulation and TMS (Stefan et al. 2000; Ridding and Taylor 2001; Quartarone et al. 2003; Rosenkranz and Rothwell 2006). For example, MEP size increases following PAS were maximal in the APB muscle when median nerve stimulation was used in conjunction with TMS located at the optimal position over the motor cortex for exciting the APB muscle and was maximal in the ADM muscle when ulnar nerve stimulation was paired with TMS over the optimal position for ADM muscle activation (Weise et al. 2004). This property suggests that PAS-induced changes are spatially rather focused. The emerging physiological profile of PAS-induced plasticity resembles that observed for spike-timing-dependent associative LTP and LTD of cortical synapses in animal preparations. Because of this similarity, PAS-induced excitability changes have been suggested to represent signatures of LTP/LTD-like mechanisms in the human cortex (Stefan et al. 2000, 2002; Wolters et al. 2003). SICI remains unchanged after PAS25 (Stefan et al. 2002; Rosenkranz and Rothwell 2006; Kujirai et al. 2006) or a similar PAS protocol which combined trains of peripheral motor-point stimulation with TMS over the primary motor cortex (Ridding and Taylor 2001). This suggests that cortical elements underlying inhibition mediated by the $GABA_A$ receptor remain unchanged by PAS. Rosenkranz and

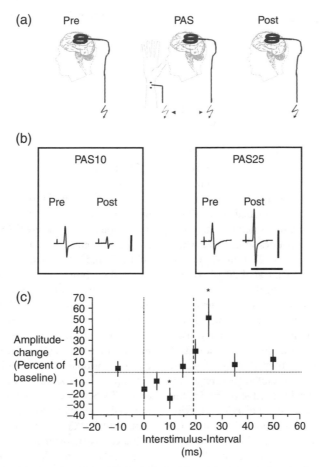

Fig. 16.4 Paired associative stimulation (PAS). (a) Principle of the stimulation protocol. Test motor-evoked potentials (MEPs) are evoked in the contralateral abductor pollicis brevis (APB) muscle by single pulses of TMS before and after the intervention. During the interventional stimulation, 90 pairs, consisting of an electric stimulus to the right median nerve followed by a magnetic pulse to the motor cortical representation of the right APB, were applied at a constant interstimulus interval (ISI) at a frequency of 0.05 Hz. The interval between the two associative stimulus modalities was varied in different sessions. (b) Effect of PAS utilizing ISIs of 10 ms (PAS10, left) or 25 ms (PAS25, right) on MEP amplitudes recorded from the right APB in one representative subject. Each tracing is the mean of 20 recorded potentials before and after PAS. Vertical bars 1 mV, horizontal bar 100 ms. (c) Group data (mean ± SEM). Asterisks indicate significant changes of MEP amplitudes ($P < 0.05$). Vertical broken line shows time of approximate arrival of afferent signal in primary somatosensory cortex. Reproduced from Classen *et al.* (2005), with permission.

Rothwell (2006) examined how sensorimotor organization might be modulated by PAS. Sensorimotor organization was tested by recording changes of MEP amplitudes recorded from intrinsic hand muscles induced by vibration applied to the same or adjacent muscle bellies. PAS led to an enhancement of MEP amplitudes, but did not influence sensorimotor organization. This finding suggested that PAS may not change the efficacy of the afferent pathway carrying proprioceptive information to the motor cortex, but may rather selectively change the efficacy of efferent motor cortical elements that are presynaptic to the cortical output neuron.

An alternative explanation of their findings would be that sensorimotor organization as probed with vibration-induced modulation of MEP amplitudes may be insensitive to changes in sensorimotor organization where mechanoreceptive signals are involved.

Apart from increasing the size of the MEP amplitude, PAS25 also led to an increase in the duration of the CSP recorded from the contracting APB muscle (Quartarone *et al.* 2003; Stefan *et al.* 2000, 2004). Therefore, PAS-induced plasticity, although induced at a resting state, may influence active neuronal circuits involved in GABA$_B$ receptor-mediated inhibition. This is an important conclusion because the size of MEP amplitudes after PAS, when recorded with the target muscle tonically active, remained unchanged (Stefan *et al.* 2000). Kujirai *et al.* (2006) used PAS with the ulnar nerve as the afferent stimulus to examine further which cortical elements might be responsible for the PAS-induced excitability changes. They used two modifications to maximize the chances of activating either preferentially I1- or I3-waves by TMS. At near-threshold intensities, reversing the direction of the induced current in the brain from the conventional posterior–anterior direction to an anterior–posterior direction has previously been shown to elicit a comparatively larger and relatively isolated I3-waves (cf. Chapters 11 and 14, this volume). I-waves are also facilitated when altering the state of the motor cortex from resting to active. Kujirai *et al.* (2006) showed that, when a comparatively small number (50) of pairs of stimuli (0.1 Hz) was applied using conventional PAS in resting subjects, there was no effect on MEPs in the first dorsal interosseous muscle. In contrast, MEPs were facilitated more readily if PAS was applied while the subjects made a small tonic contraction and at the same time the direction of the induced current was anterior–posterior. PAS was effective even with subthreshold TMS (95% of the active motor threshold) intensities. The authors also noted that with conventional PAS (employing suprathreshold TMS with the induced current directed posterior–anteriorly and the subject at rest) later components of the MEP were more likely to be facilitated than initial components. Thus, the evidence derived from these observations is that specific cortical

elements, namely those recruited at comparatively long latencies, are preferentially facilitated by PAS. These elements would likely correspond to those involved in generating late I-waves. In this way, PAS-induced plasticity may differ from TBS-induced plasticity which appears to rely on modulation of the first I-wave, at least as far as the depressing effect is concerned (Huang *et al.* 2005) (see above). However, in the study by Kujirai *et al.* (2006), there was no facilitation of MEP size if the compound nerve stimulus was replaced by digital nerve stimulation, whereas digital nerve stimulation plus conventionally oriented TMS had been found effective in a previous study where PAS was performed at rest (Stefan *et al.* 2000). In addition, after subthreshold PAS with anterior–posteriorly directed current, there was a fairly strong tendency for SICI to decrease. Therefore, mechanisms underlying plasticity induced by the PAS variant involving active muscle contraction and posterior–anteriorly directed induced current might differ slightly from those operative with the conventional PAS protocol.

Support for involvement of late, rather than early, TMS-induced cortical events in PAS-induced plasticity is also derived from two other arguments: PAS has been shown to lead to enhanced corticospinal excitability even when using ISIs shorter than the estimated arrival time of the afferent signal in the primary motor cortex (Wolters *et al.* 2003; Weise *et al.* 2004; Ziemann *et al.* 2004; Morgante *et al.* 2006). With these short intervals (ISI ~20–21.5 ms), a pre (afferent signal) → post (TMS-signal) sequence of events would be achieved only if the interaction between the afferent-stimulation-induced events were with late TMS-induced events. As noted above, such late events are likely to correspond to late I-waves and are believed to originate in upper cortical layers II/III. This conclusion gains further support from studies applying PAS with the magnetic coil centered over the primary somatosensory cortex (S1) (Litvak *et al.* 2005; Wolters *et al.* 2005). PAS-induced cortical excitability changes in S1, which were probed by median nerve somatosensory-evoked potentials (MN-SSEP), resembled those induced in the primary motor cortex in several respects. Similar to PAS targeting M1, bidirectional changes were induced and

PAS efficacy depended on near synchronicity between the median nerve pulse and the TMS pulse. PAS increased exclusively the amplitude of the P25 component of the MN-SSEP, which is probably generated in superficial cortical layers of area 3b. Indeed, changes of synaptic efficacy in superficially travelling horizontal cortical pathways are believed to underlie much of the experience-dependent plasticity in neocortex (Diamond et al. 1993; Rioult-Pedotti et al. 2000). In support of this conclusion, it was shown that training of ballistic (Ziemann et al. 2004) or dynamic (Stefan et al. 2006) thumb abductions led to a temporary blockade of PAS-induced plasticity similar to the blockade of LTP in the primary motor cortex of rats observed after training a new motor skill (Rioult-Pedotti et al. 2000; Monfils and Teskey 2004). Using a variety of different centrally acting substances, Ziemann et al. (2006) showed that the efficacy of PAS to induce plasticity may be modified by several modulating neurotransmitters, namely serotonin, dopamine, norepinephrine, and acetylcholine, compatible with the hypothesis that LTP/LTD-like processes are involved. The role of dopamine is implicated in studies on parkinsonian patients who show less than normal PAS-induced facilitation (Morgante et al. 2006; Ueki et al. 2006) which can be restored by substitution of the dopaminergic deficit. Modulation of cortical elements by acetylcholine may underlie the dramatic change of PAS-induced plasticity by attention (Stefan et al. 2004). Together, these observations suggest that PAS may generate cortical excitability changes with properties surprisingly similar to those of LTP as revealed by invasive animal studies. However, even more direct evidence for equivalence of mechanisms could be revealed only by studying PAS protocols in animals.

No difference in the efficacy of PAS was noted when comparing the dominant with the sub-dominant hemisphere (Ridding and Flavel 2006), despite subtle differences in SICI. Measures of intracortical excitability, therefore, do not appear to correlate strongly with the efficacy of PAS. PAS has also been shown to induce excitability changes in leg muscle representations (Stinear and Hornby 2005). PAS applied to the common peroneal nerve during the swing phase of walking induced bidirectional changes in corticospinal excitability. Tibialis anterior MEP amplitude increased when the magnetic stimulus was delivered over the primary motor cortex after the estimated arrival time of the afferent volley in sensorimotor cortex and decreased when the magnetic stimulus was delivered prior to the estimated afferent volley arrival.

Concluding remarks

This chapter has reviewed some of the changes in TMS measures induced by various rTMS protocols. Meanwhile, an impressive number of different rTMS protocols exist, and even with simple protocols a substantial number of variables span a large parameter space. Exploring this parameter space appears necessary to understand the mechanisms of rTMS-induced plasticity more fully and yet appears to be of only limited feasibility in humans. TMS is a relatively nonfocal technique. In most instances, rTMS will activate a mixture of systems that potentially could have interacting effects making the final outcome difficult to interpret. This is particularly evident with the most widely used rTMS protocol, 1 Hz rTMS, where multiple, apparently independent, effects are induced in cortical and possibly subcortical circuits. Plasticity induced by rTMS applied in the presence of ischemic nerve block may depend on unlocking GABAergic inhibition, one of the homeostatic mechanisms that counteract modification by simple rTMS. More specific changes appear to be induced when neuronal events are induced at near synchronicity, such as with protocols using very closely spaced TMS pulses, or with PAS. Although the design of some of the protocols has been guided by animal experiments, it would both increase confidence in the interpretation and facilitate the improvement of rTMS protocols if human studies were now complemented by invasive animal studies which use the parameters of actual rTMS protocols. If rTMS protocols are just one means to induce plasticity, it should not be neglected that TMS parameters are just one way of describing the consequences of plasticity-inducing protocols. For example, TMS parameters do not capture behavioral consequences of induced plasticity and are generally not suited to studying the effects of plasticity-inducing protocols outside the motor system. Therefore, much needs to be done before rTMS protocols can be developed into tailored therapeutic tools.

References

Bagnato S, Curra A, Modugno N, et al. (2005). One-hertz subthreshold rTMS increases the threshold for evoking inhibition in the human motor cortex. *Experimental Brain Research 160*, 368–374.

Chen R, Classen J, Gerloff C, et al. (1997). Depression of motor cortex excitability by low-frequency transcranial magnetic stimulation. *Neurology 48*, 1398–1403.

Classen J, Stefan K, Wolters A, et al. (2005). TMS-induzierte Plastizität: ein Fenster zum Verständnis des motorischen Lernens? *Klinische Neurophysiologie 36*, 178–185.

Diamond ME, Armstrong-James M, Ebner FF (1993). Experience-dependent plasticity in adult rat barrel cortex. *Proceedings of the National Academy of Sciences of the USA 90*, 2082–2086.

Di Lazzaro V, Oliviero A, Mazzone P, et al. (2002). Short-term reduction of intracortical inhibition in the human motor cortex induced by repetitive transcranial magnetic stimulation. *Experimental Brain Research 147*, 108–113.

Di Lazzaro V, Pilato F, Saturno E, et al. (2005). Theta-burst repetitive transcranial magnetic stimulation suppresses specific excitatory circuits in the human motor cortex. *Journal of Physiology 565*, 945–950.

Fitzgerald PB, Brown TL, Daskalakis ZJ, Chen R, Kulkarni J (2002). Intensity-dependent effects of 1 Hz rTMS on human corticospinal excitability. *Clinical Neurophysiology 113*, 1136–1141.

Fitzgerald PB, Benitez J, Oxley T, Daskalakis JZ, de Castella AR, Kulkarni J (2005). A study of the effects of lorazepam and dextromethorphan on the response to cortical 1 Hz repetitive transcranial magnetic stimulation. *Neuroreport 16*, 1525–1528.

Gangitano M, Valero-Cabre A, Tormos JM, Mottaghy FM, Romero JR, Pascual-Leone A (2002). Modulation of input-output curves by low and high frequency repetitive transcranial magnetic stimulation of the motor cortex. *Clinical Neurophysiology 113*, 1249–1257.

Gerschlager W, Siebner HR, Rothwell JC (2001). Decreased corticospinal excitability after subthreshold 1 Hz rTMS over lateral premotor cortex. *Neurology 57*, 449–455.

Gilio F, Rizzo V, Siebner HR, Rothwell JC (2003). Effects on the right motor hand-area excitability produced by low-frequency rTMS over human contralateral homologous cortex. *Journal of Physiology 551*, 563–573.

Heide G, Witte OW, Ziemann U (2006). Physiology of modulation of motor cortex excitability by low-frequency suprathreshold repetitive transcranial magnetic stimulation. *Experimental Brain Research 171*, 26–34.

Huang YZ, Edwards MJ, Rounis E, Bhatia KP, Rothwell JC (2005). Theta burst stimulation of the human motor cortex. *Neuron 45*, 201–206.

Iyer MB, Schleper N, Wassermann EM (2003). Priming stimulation enhances the depressant effect of low-frequency repetitive transcranial magnetic stimulation. *Journal of Neuroscience 23*, 10867–10872.

Khedr EM, Gilio F, Rothwell J (2004). Effects of low frequency and low intensity repetitive paired pulse stimulation of the primary motor cortex. *Clinical Neurophysiology 115*, 1259–1263.

Kujirai T, Caramia MD, Rothwell JC, et al. (1993). Corticocortical inhibition in human motor cortex. *Journal of Physiology (London) 471*, 501–519.

Kujirai K, Kujirai T, Sinkjaer T, Rothwell JC (2006). Associative plasticity in human motor cortex under voluntary muscle contraction. *Journal of Neurophysiology 96*, 1337–1346.

Lang N, Siebner HR, Ernst D, et al. (2004). Preconditioning with transcranial direct current stimulation sensitizes the motor cortex to rapid-rate transcranial magnetic stimulation and controls the direction of after-effects. *Biological Psychiatry 56*, 634–639.

Levy LM, Ziemann U, Chen R, Cohen LG (2002). Rapid modulation of GABA in sensorimotor cortex induced by acute deafferentation. *Annals of Neurology 52*, 755–761.

Litvak V, Zeller D, Oostenveld R, et al. (2005). Localization of cortical sources affected by associative plasticity in human primary somatosensory cortex. *Society for Neuroscience Abstracts*.

Maeda F, Keenan JP, Tormos JM, Topka H, Pascual-Leone A (2000). Interindividual variability of the modulatory effects of repetitive transcranial magnetic stimulation on cortical excitability. *Experimental Brain Research 133*, 425–430.

Modugno N, Nakamura Y, MacKinnon CD, et al. (2001). Motor cortex excitability following short trains of repetitive magnetic stimuli. *Experimental Brain Research 140*, 453–459.

Monfils MH, Teskey GC (2004). Skilled-learning-induced potentiation in rat sensorimotor cortex: a transient form of behavioural long-term potentiation. *Neuroscience 125*, 329–336.

Morgante F, Espay AJ, Gunraj C, Lang AE, Chen R (2006). Motor cortex plasticity in Parkinson's disease and levodopa-induced dyskinesias. *Brain 129*, 1059–1069.

Muellbacher W, Ziemann U, Boroojerdi B, Hallett M (2000). Effects of low-frequency transcranial magnetic stimulation on motor excitability and basic motor behavior. *Clinical Neurophysiology 111*, 1002–1007.

Münchau A, Bloem BR, Irlbacher K, Trimble MR, Rothwell JC (2002). Functional connectivity of human premotor and motor cortex explored with repetitive transcranial magnetic stimulation. *Journal of Neuroscience 22*, 554–561.

Oliveri M, Koch G, Torriero S, Caltagirone C (2005). Increased facilitation of the primary motor cortex following 1 Hz repetitive transcranial magnetic stimulation of the contralateral cerebellum in normal humans. *Neuroscience Letters 376*, 188–193.

Pal PK, Hanajima R, Gunraj CA, et al. (2005). Effect of low frequency repetitive transcranial magnetic stimulation on interhemispheric inhibition. *Journal of Neurophysiology 94*, 1668–1675.

Pascual-Leone A, Valls-Sole J, Wassermann EM, Hallett M (1994). Responses to rapid-rate transcranial magnetic stimulation of the human motor cortex. *Brain 117*, 847–858.

Paus T, Castro-Alamancos MA, Petrides M (2001). Cortico-cortical connectivity of the human mid-dorsolateral frontal cortex and its modulation by repetitive transcranial magnetic stimulation. *European Journal of Neuroscience 14*, 1405–1411.

Peinemann A, Lehner C, Mentschel C, Munchau A, Conrad B, Siebner HR (2000). Subthreshold 5-Hz repetitive transcranial magnetic stimulation of the human primary motor cortex reduces intracortical paired-pulse inhibition. *Neuroscience Letters 296*, 21–24.

Plewnia C, Lotze M, Gerloff C (2003). Disinhibition of the contralateral motor cortex by low-frequency rTMS. *Neuroreport 14*, 609–612.

Quartarone A, Bagnato S, Rizzo V, *et al.* (2003). Abnormal associative plasticity of the human motor cortex in writer's cramp. *Brain 126*, 2586–2596.

Quartarone A, Bagnato S, Rizzo V, *et al.* (2005). Distinct changes in cortical and spinal excitability following high-frequency repetitive TMS to the human motor cortex. *Experimental Brain Research 161*, 114–124.

Ridding MC, Flavel SC (2006). Induction of plasticity in the dominant and non-dominant motor cortices of humans. *Experimental Brain Research 171*, 551–557.

Ridding MC, Taylor JL (2001). Mechanisms of motor-evoked potential facilitation following prolonged dual peripheral and central stimulation in humans. *Journal of Physiology 537*, 623–631.

Rioult-Pedotti MS, Friedman D, Donoghue JP (2000). Learning-induced LTP in neocortex. *Science 290*, 533–536.

Rollnik JD, Schubert M, Dengler R (2000). Subthreshold prefrontal repetitive transcranial magnetic stimulation reduces motor cortex excitability. *Muscle and Nerve 23*, 112–114.

Romero JR, Anschel D, Sparing R, Gangitano M, Pascual-Leone A (2002). Subthreshold low frequency repetitive transcranial magnetic stimulation selectively decreases facilitation in the motor cortex. *Clinical Neurophysiology 113*, 101–107.

Rosenkranz K, Rothwell JC (2006). Differences between the effects of three plasticity inducing protocols on the organization of the human motor cortex. *European Journal of Neuroscience 23*, 822–829.

Rothwell JC (1997). Techniques and mechanisms of action of transcranial stimulation of the human motor cortex. *Journal of Neuroscience Methods 74*, 113–122.

Schambra HM, Sawaki L, Cohen LG (2003). Modulation of excitability of human motor cortex (M1) by 1 Hz transcranial magnetic stimulation of the contralateral M1. *Clinical Neurophysiology 114*, 130–133.

Siebner HR, Tormos JM, Ceballos-Baumann AO, *et al.* (1999). Low-frequency repetitive transcranial magnetic stimulation of the motor cortex in writer's cramp. *Neurology 52*, 529–537.

Siebner HR, Lang N, Rizzo V, *et al.* (2004). Preconditioning of low-frequency repetitive transcranial magnetic stimulation with transcranial direct current stimulation: evidence for homeostatic plasticity in the human motor cortex. *Journal of Neuroscience 24*, 3379–3385.

Sommer M, Tergau F, Wischer S, Paulus W (2001). Paired-pulse repetitive transcranial magnetic stimulation of the human motor cortex. *Experimental Brain Research 139*, 465–472.

Sommer M, Kamm T, Tergau F, Ulm G, Paulus W (2002). Repetitive paired-pulse transcranial magnetic stimulation affects corticospinal excitability and finger tapping in Parkinson's disease. *Clinical Neurophysiology 113*, 944–950.

Stefan K, Kunesch E, Cohen LG, Benecke R, Classen J (2000). Induction of plasticity in the human motor cortex by paired associative stimulation. *Brain 123*, 572–584.

Stefan K, Kunesch E, Benecke R, Cohen LG, Classen J (2002). Mechanisms of enhancement of human motor cortex excitability induced by interventional paired associative stimulation. *Journal of Physiology 543*, 699–708.

Stefan K, Wycislo M, Classen J (2004). Modulation of associative human motor cortical plasticity by attention. *Journal of Neurophysiology 92*, 66–72.

Stefan K, Wycislo M, Gentner R, *et al.* (2006). Temporary occlusion of associative motor cortical plasticity by prior dynamic motor training. *Cerebral Cortex 16*, 376–385.

Stinear JW, Hornby TG (2005). Stimulation-induced changes in lower limb corticomotor excitability during treadmill walking in humans. *Journal of Physiology 567*, 701–711.

Thickbroom GW, Byrnes ML, Edwards DJ, Mastaglia FL (2006). Repetitive paired-pulse TMS at I-wave periodicity markedly increases corticospinal excitability: A new technique for modulating synaptic plasticity. *Clinical Neurophysiology 117*, 61–66.

Todd G, Flavel SC, Ridding MC (2006). Low intensity repetitive transcranial magnetic stimulation decreases motor cortical excitability in humans. *Journal of Applied Physiology 101*, 500–505.

Tokimura H, Ridding, M.C., Tokimura, Y., Amassian, V.E., Rothwell, J.C. (1996). Short latency facilitation between pairs of threshold magnetic stimuli applied to human motor cortex. *Electroencephalography and Clinical Neurophysiology 101*, 263–272.

Touge T, Gerschlager W, Brown P, Rothwell JC (2001). Are the after-effects of low-frequency rTMS on motor cortex excitability due to changes in the efficacy of cortical synapses? *Clinical Neurophysiology 112*, 2138–2145.

Ueki Y, Mima T, Kotb MA, *et al.* (2006). Altered plasticity of the human motor cortex in Parkinson's disease. *Annals of Neurology 59*, 60–71.

Wassermann EM, Wedegaertner FR, Ziemann U, George MS, Chen R (1998). Crossed reduction of human motor cortex excitability by 1-Hz transcranial magnetic stimulation. *Neuroscience Letters 250*, 141–144.

Weise D, Schramm A, Stefan K, *et al.* (2004). Disturbance of associative motor cortical plasticity in focal hand dystonia. *Movement Disorders 19*, S94.

Wolters A, Sandbrink F, Schlottmann A, *et al.* (2003). A temporally asymmetric Hebbian rule governing plasticity in the human motor cortex. *Journal of Neurophysiology 89*, 2339–2345.

Wolters A, Schmidt A, Schramm A, *et al.* (2005). Timing-dependent plasticity in human primary somatosensory cortex. *Journal of Physiology 565*, 1039–1052.

Wu T, Sommer M, Tergau F, Paulus W (2000). Lasting influence of repetitive transcranial magnetic stimulation on intracortical excitability in human subjects. *Neuroscience Letters 287*, 37–40.

Ziemann U, Lönnecker S, Steinhoff BJ, Paulus W (1996). Effects of antiepileptic drugs on motor cortex excitability in humans: A transcranial magnetic stimulation study. *Annals of Neurology 40*, 367–378.

Ziemann U, Corwell B, Cohen LG (1998a). Modulation of plasticity in human motor cortex after forearm ischemic nerve block. *Journal of Neuroscience 18*, 1115–1123.

Ziemann U, Tergau F, Wassermann EM, Wischer S, Hildebrandt J, Paulus W (1998b). Demonstration of facilitatory I wave interaction in the human motor cortex by paired transcranial magnetic stimulation. *Journal of Physiology 511*, 181–190.

Ziemann U, Ilic TV, Pauli C, Meintzschel F, Ruge D (2004). Learning modifies subsequent induction of long-term potentiation-like and long-term depression-like plasticity in human motor cortex. *Journal of Neuroscience 24*, 1666–1672.

Ziemann U, Meintzschel F, Korchounov A, Ilic TV (2006). Pharmacological modulation of plasticity in the human motor cortex. *Neurorehabilitation and Neural Repair 20*, 243–251.

CHAPTER 17

Neuroplasticity induced by transcranial direct current stimulation

Michael A. Nitsche, Andrea Antal, David Liebetanz, Nicolas Lang, Frithjof Tergau, and Walter Paulus

Summary – Brain stimulation with weak direct currents has been known for about 40 years as a technique to generate prolonged modifications of cortical excitability and activity. Only recently has it been reintroduced as a method to elicit and modulate neuroplasticity of the human cerebral cortex. Primarily with the aid of TMS as a tool for monitoring cortical excitability, it was demonstrated that transcranial direct current stimulation (tDCS) generates modulations of excitability during, and up to an hour after the end of, stimulation depending on the duration of stimulation. Specific TMS protocols, partly combined with neuropharmacological intervention, revealed some of the basic mechanisms that underlie the effects of tDCS. tDCS has been shown to result in reversible modifications of perceptual, cognitive, and behavioral functions. Moreover, it possibly induces beneficial effects with regard to neurological diseases involving pathologically altered states of cortical activation. Thus brain stimulation with weak direct currents could evolve as a promising tool in neuroplasticity research. This overview describes how to elicit and monitor neuroplasticity by tDCS, the physiological foundations of the modulations achieved, and its functional consequences.

Introduction

One of the most challenging topics of brain research in recent decades has been the exploration of the properties and physiological as well as the molecular foundations of neuroplasticity (Bennett 2000; Malenka and Bear 2004). In recent years, newly developed techniques such as functional imaging, sophisticated electroencephalographic, and transcranial stimulation techniques have enabled researchers to study neuroplasticity, not only in animal and slice models, but also in humans. Brain stimulation with weak direct currents, although a relatively old method in strict terms, regained increasing interest as a potentially valuable tool for inducing and modulating neuroplasticity.

Some 40 years ago the application of weak direct currents was shown to result in neuroplastic modifications. In anesthetized rats, weak direct currents, delivered by intracerebral or epidural electrodes, induce activity and excitability diminutions or enhancements of the sensorimotor cortex, which can persist for hours after the end of stimulation (Bindman *et al.* 1964). Subsequent studies revealed that these effects were protein synthesis dependent (Gartside 1968) and accompanied by modifications of intracellular cAMP and calcium levels (Hattori et al. 1990; Islam et al. 1995). Thus they share some features with the nowadays more commonly known neuroplastic phenomena in animal experiments, namely long-term potentiation (LTP) and long-term depression (LTD). Furthermore, it was demonstrated that *transcranial* application of weak direct currents can also induce an intracerebral current flow sufficiently large to achieve the intended effects. In monkeys, ~50% of the transcranially applied currents enter the brain through the skull (Rush and Driscoll 1968) and these results have been replicated in humans (Dymond et al. 1975). Initial studies in humans have been dedicated to the treatment of psychiatric diseases, in particular depression. Although these early experiments included some cortical stimulation, most probably the positioning of the electrodes primarily resulted in brain-stem stimulation. Nevertheless, it was found (Pfurtscheller 1970) that this kind of stimulation changed electro-encephalogram (EEG) patterns and evoked potentials at the cortical level and was thus regarded as effective. Anodal stimulation was found to diminish depressive symptoms (Costain et al. 1964), and cathodal stimulation reduced manic symptoms (Carney 1969). Unfortunately, these results could not be replicated in all follow-up studies, possibly because of different patient subgroups, measures of changes, or other factors that were not controlled for systematically (for an overview see Lolas 1977). Apart from clinical studies, anodal stimulation of the motor cortex was reported to optimize performance in a choice reaction-time task in healthy subjects (Elbert et al. 1981; Jaeger et al. 1987). In the following years, noninvasive stimulation of the human brain via transcranial application of weak direct currents as a tool to influence brain function was nearly forgotten. This might have been due to the lack of methods to probe its effects beyond the phenomenological level. In the last few decades, however, transcranial electric and magnetic stimulation (TES, TMS), as well as functional imaging methods such as functional magnetic resonance imaging (fMRI) and positron emission tomography (PET), have evolved as suitable tools to monitor changes of brain activity and excitability. Consequently, transcranial direct current stimulation (tDCS)

has been re-evaluated and developed into a method that reliably induces and modulates neuroplasticity in the human cerebral cortex noninvasively and painlessly (Nitsche and Paulus, 2000, 2001; Nitsche et al. 2003a). In doing so, the polarity of stimulation determines whether tDCS will enhance or reduce excitability. This chapter offers an overview of the basic and functional effects of weak tDCS, focusing mainly on TMS studies.

Monitoring tDCS-induced excitability modulations of the human motor cortex by TMS

Physical parameters and practical application of tDCS

The combination of strength of current, size of stimulated area, and stimulation duration are thought to be the relevant parameters that determine the efficacy of electrical brain stimulation (Agnew and McCreery 1987). A formula representing these parameters is total charge (current strength (A)/area (cm²)× stimulation duration (s)) (Yuen et al. 1991). This formula was originally developed for suprathreshold electrical stimulation, but it also seems to be appropriate for weak subthreshold DC stimulation, because different current intensities per area will result in different amounts of neuronal de- or hyperpolarization, and it has been shown that different stimulation durations result in different time courses of the induced excitability shifts (Bindman et al. 1964; Nitsche and Paulus 2000). Preliminary limits for a safe total charge of stimulation have been given (Yuen et al. 1991). However, it should be borne in mind that different charges combined with different stimulation durations may have an identical total charge but still produce qualitatively quite different effects: a short strong stimulation may induce suprathreshold depolarization, whereas a weak prolonged stimulation may fail to elicit action potentials of a given neuron – though both have identical total charges. Thus, the comparability of different studies with regard to total charge is limited and should always be qualified by a separate equation of current density and stimulation duration.

Another parameter which seems to be important to achieve the intended electrical stimulation effects – most probably by determining the neuronal population stimulated – is the direction of current flow, which is defined generally by the electrode positions and polarity. As shown for the human motor cortex, only two of six different electrode-position combinations tested so far effectively influenced cortical excitability, and the effective combinations may have modulated different neuronal populations (Priori *et al.* 1998; Nitsche and Paulus, 2000; see below). Since in animal experiments differently oriented neuronal populations are influenced differently by a constant current flow direction (Creutzfeldt *et al.* 1962; Purpura and McMurtry 1965), the relation of current flow direction and neuronal orientation is crucial for the efficacy of stimulation and the direction of the current-induced changes of cortical excitability and activity.

Direct currents are routinely delivered via a pair of surface conductive rubber electrodes covered with water-soaked sponges (size between 25 and 35 cm^2 in different studies (e.g. Priori *et al.* 1998; Nitsche and Paulus 2000; Hummel *et al.* 2005; Iyer *et al.* 2005)). Alternatively the rubber electrodes can be spread with electrode cream and mounted directly on the head. The correct position of both electrodes is crucial for achieving the intended effects. Electrode positions tested as effective are available currently for the primary motor, dorsolateral prefrontal, frontopolar, and occipital cortices (see below). The electrodes are connected to a stimulator (e.g. available from Schneider Electronics, Gleichen, Germany; Eldith, Ilmenau, Germany). Since strength of current, not voltage, determines the effects of electrical stimulation, a stimulator delivering constant current is needed. The current strength delivered varies between 1 and 2 mA in most studies. The resulting current densities are sufficient to achieve the desired excitability, perceptual, and behavioral changes, and are regarded as safe if stimulation duration does not exceed ~15–20 min, as shown by behavioral measures, EEG, serum neuron-specific enolase concentration, and diffusion-weighted and contrast-enhanced MRI measures (Nitsche and Paulus 2001; Nitsche *et al.* 2003a, 2004a; Iyer *et al.* 2005). However, electrode positions above cranial foramina and fissures should be avoided because these could increase effective current density, and thus safety of stimulation may no longer be guaranteed. Most subjects stimulated will perceive a slight itching sensation at the beginning of stimulation, which then fades. With certain electrode positions (especially frontopolar), retinal phosphenes can be perceived at the start and end of stimulation. These are eliminated by starting and terminating the stimulation gradually (ramp switch), which also helps to avoid stimulation break effects (Roth 1994). Increased itching sensations during the experiment are in most cases caused by poor electrode contacts. If more than one tDCS application per person is planned within a given study, we separate the stimulations by at least 1 h for short-lasting after-effect experiments and by about a week for long-lasting effects, in order to avoid interference effects.

Time course of tDCS-induced modulations of motor cortex excitability

For the primary motor cortex, the effect of tDCS on cortical excitability is routinely tested by single-pulse TMS. The peak-to-peak amplitude of the motor-evoked potential (MEP) serves as an index of global corticospinal excitability (Rothwell 1993) (cf. Chapter 9, this volume). The reference muscles most often used are the abductor digiti minimi (ADM) and the first dorsal interosseous (FDI). Their motor cortical representations are situated on the lateral convexity and are easily stimulated by TMS and tDCS. Baseline TMS intensity is determined to elicit MEPs of moderate amplitude (~1 mV). This allows the monitoring of tDCS-driven MEP enhancements and reductions and avoids bottom or ceiling effects.

Intra-tDCS effects

Short-lasting anodal or cathodal tDCS results in a motor cortical excitability increase or decrease, respectively, which does not outlast the stimulation itself (Nitsche and Paulus 2000). These short-lived effects that only occur during tDCS will be referred to as intra-tDCS effects. The elicited excitability changes by ~30% as compared to values without tDCS, and thus is somewhat smaller compared to the long-lasting after-effects (see below). The standard protocol encompasses 4 s of anodal or cathodal tDCS (electrode size 35 cm^2, current strength 1 mA) repeated 12 times with an inter-tDCS interval of 10 s (to avoid interference effects) and the TMS stimulus to probe changes in corticospinal excitability applied immediately before the termination of each polarization. These stimuli are intermingled with single TMS stimuli (baseline stimuli) without preceding tDCS to enable the evaluation of tDCS-induced excitability changes by comparing tDCS with non-tDCS values. This protocol was consistently effective only for one electrode-position combination out of six tested (motor cortex vs contralateral over the orbit), underlining the importance of current flow direction relative to neuronal orientation.

Using a motor cortex–chin electrode montage, but in other aspects a similar protocol (25 cm^2 electrodes, current strength of up to 0.5 mA, pulse duration 7 s), a different pattern of motor cortical excitability change during stimulation was reported: here anodal stimulation *diminished* corticospinal excitability, as revealed by single-pulse TMS delivered immediately before the end of tDCS, but only if it was preceded by cathodal stimulation (Priori *et al.* 1998). Anodal or cathodal tDCS alone did not change MEP amplitudes. It was concluded that this effect might be caused by an anodal tDCS-elicited hyperpolarization of superficial cortical interneurons, presumably accentuated by prior cathodal stimulation via adaptive membrane alterations. The most likely reason for these divergent results is the difference in current flow direction. Additionally, some kind of short-term homeostatic mechanism might have contributed (see below).

After-effects of tDCS

In animals, it has been shown that DC stimulation of the brain, if applied for some minutes continuously, results in cortical excitability and activity modifications that are stable for hours (Bindman *et al.* 1964). Comparable – although somewhat weaker – results are achieved by tDCS of the human motor cortex. Here, tDCS induces after-effects for ~3 min or more, using the above-mentioned protocol with regard to electrode size and applied current strength.

These persist for a few minutes after 3–7 min tDCS, but are stable for up to ~1 h if prolonged protocols with a tDCS stimulation duration between 9 and 13 min (Nitsche and Paulus 2000, 2001; Nitsche *et al.* 2003a) (Figure 17.1) are applied. The respective excitability shifts are in the range of 30–50% compared to baseline. Anodal stimulation enhances, while cathodal tDCS diminishes, excitability. The strength and duration of the after-effects are determined not only by tDCS duration, but also by tDCS strength (Nitsche and Paulus 2000). The standard protocol to monitor these excitability shifts consists of a baseline single-pulse TMS measure before tDCS and repeated measures after tDCS. For a reliable and valid measure of the time course of motor cortex excitability modulations, it is essential that the state of muscle relaxation and alertness remains as constant as possible throughout the experiment. Muscle relaxation should therefore be monitored constantly by high-gain electromyogram. The series of TMS stimuli recorded at a certain time point should result in relatively constant MEP amplitudes without a clear trend in time, which in most cases would be caused by increasing tiredness or changes in alertness during the recording. In our laboratory, alertness is routinely maintained by low-volume music. Additionally, measures at one time point are restricted to ~20 MEP recordings and measurement sessions are separated by a break of ~3 min or more.

Apart from inducing neuroplasticity, tDCS of the primary motor cortex is suited to modulate it (metaplasticity). Afferent somatosensory input can induce prolonged excitability enhancements of the human primary motor cortex (Ridding *et al.* 2000). In a study demonstrating the modulating effects of tDCS on neuroplasticity, slightly suprathreshold ulnar nerve stimulation was applied in two blocks, each lasting for 10 min (10 Hz, 500 ms train duration, repeated every 10 s). This protocol does not evoke any overt excitability shifts of the motor cortex, as monitored by single-pulse TMS, if administered alone. However, when combined with 5 min anodal tDCS (1 mA, electrode size 25 cm^2) centered above the motor cortex representation of the FDI muscle, it resulted in motor cortex excitability enhancements that persisted for ≥30 min after stimulation (Uy and Ridding 2003).

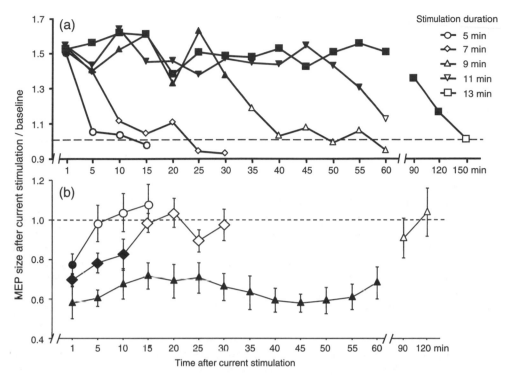

Fig. 17.1 Transcranial direct current stimulation (tDCS) of the human motor cortex modulates TMS-elicited motor-evoked potential (MEP) amplitudes after stimulation for up to an hour, depending on stimulation duration. Anodal stimulation (a) enhances, whereas cathodal (b) diminishes, cortical excitability. Note that stimulation for 5–7 min results in short-lasting after-effects, while prolonged tDCS increases the duration of the after-effects over-proportionally. Reproduced with permission from Nitsche and Paulus (2001) and Nitsche *et al.* (2003a).

Thus, in that study tDCS strengthened the effect of afferent sensory stimulation to induce neuroplastic shifts of motor cortex excitability.

The neurophysiological foundations of tDCS-elicited neuroplasticity, as revealed by TMS

MEP amplitudes elicited by single-pulse TMS of the motor cortex serve as a global measure of corticospinal excitability, but cannot localize the exact origin of the effects of tDCS. However, special TMS protocols are available to explore the involvement of cortical subsystems in changes of motor cortical excitability elicited by tDCS.

By comparing the size of MEPs elicited by TMS with those evoked by transcranial electrical stimulation (TES), intracortical effects can be separated from direct effects on pyramidal tract neurons (cf. Chapter 14, this volume). Responses to TES are dominated by direct stimulation of corticospinal axons whereas those evoked by TMS are dominated by trans-synaptic activation of corticospinal neurons (Edgley *et al.* 1997). Thus, comparing TMS- and TES-elicited MEPs before and after tDCS might give insight into the neuronal populations involved in the tDCS-induced excitability changes. For the after-effects of both cathodal and anodal tDCS, it has been demonstrated that TES, when applied at moderate intensity, does not reveal tDCS-induced changes in MEP amplitude – in contrast to MEPs elicited by TMS (Nitsche and Paulus 2000, 2001; Nitsche *et al.* 2003a). Thus, a predominantly intracortical origin of the effects of tDCS seems plausible. However, in another study, an impact on the TES-elicited MEP was reported for the after-effects of cathodal tDCS (Ardolino *et al.* 2005). Since low-intensity TES, as applied in that experiment, is thought to influence the proximal aspect of pyramidal tract axons, this might be an indication for an additional membrane effect of tDCS on corticospinal neurons. However, it cannot be ruled out that, in the latter experiment, deeper cortical layers were excited by tDCS compared to the former because current density was ~50% higher in that study.

Active and resting motor thresholds (MT) and MEP input/output curves (I/O curves) are global measures of corticospinal excitability (Chen 2000; Abbruzzese and Trompetto 2002) (cf. Chapter 9, this volume). MTs are defined as the minimum TMS intensity resulting in small MEPs, resting MT during muscle relaxation and active MT during moderate voluntary contraction. They likely reflect neuronal membrane rather than synaptic excitability, since they increase under voltage-gated sodium channel blocking drugs (Ziemann *et al.* 1996) (cf. Chapter 13, this volume). At stimulus intensities close to MT, TMS explores the excitability of lowest threshold corticospinal neurons in a central core region of the motor cortical body representation. The entire MEP I/O curve serves as an index of the excitability of the whole corticospinal population projecting to the target muscle (cf. Chapter 9, this volume). The slope of the I/O curve resulting from increasing MEP amplitude with increasing TMS intensity reflects the recruitment of corticospinal neurons. Similar to MT, the I/O curve depends on neuronal membrane excitability because its slope is decreased by sodium- and calcium-channel blockers (Ziemann *et al.* 1996) (cf. Chapter 13, this volume). However, synaptic mechan-isms are additionally involved, since the I/O curve is modulated by drugs influencing the GABAergic and the noradrenergic system (Boroojerdi 2002), and most likely by the glutamatergic system (Di Lazzaro *et al.* 2003) (cf. Chapter 13, this volume). Active and passive MTs were not modified during anodal and cathodal tDCS or during the after-effects in one study (Nitsche *et al.* 2005). However, they were found to increase during the after-effects of cathodal tDCS in another study (Ardolino *et al.* 2005),

probably caused by a higher current density applied in the second study. For the MEP I/O curve, anodal and cathodal tDCS influence the recruitment of neurons by TMS applied during DC stimulation as well as during the after-effects: anodal tDCS enhances, whereas cathodal tDCS diminishes, recruitment (Nitsche *et al.* 2005).

In order to further explore the intracortical mechanisms of tDCS, short-interval intracortical inhibition (SICI) and intracortical facilitation (ICF) were studied (Nitsche *et al.* 2005) by a paired-pulse TMS protocol (Kujirai *et al.* 1993) (cf. Chapters 11–13, this volume). Here, a subthreshold conditioning TMS pulse is followed by a suprathreshold test stimulus. The resulting inhibition or facilitation of the test MEP is determined by the interstimulus interval, is of intracortical origin, and reflects excitability of inhibitory and excitatory interneurons. Since SICI is enhanced and ICF diminished by GABAergic and anti-glutamatergic substances, but not influenced by ion-channel blockers (Ziemann *et al.* 1996, 1998a; Chen *et al.* 1997; Liepert *et al.* 1997), they likely reflect activity and/or excitability of GABAergic and glutamatergic motor cortical interneurons (cf. Chapter 13, this volume). ICF decreases during cathodal tDCS. For the after-effects, cathodal tDCS in addition enhances SICI while anodal stimulation results in opposite effects (Nitsche *et al.* 2005). Another study failed to show long-lasting after-effects of tDCS on SICI/ICF (Siebner *et al.* 2004). It is, however, difficult to compare the results of the two studies due to differences in the applied tDCS protocols.

In another study, the impact of tDCS on transcallosal inhibition was tested (Lang *et al.* 2004a). It was determined by the suppression of electromyographic activity of the voluntarily contracting FDI after single-pulse TMS of the ipsilateral motor cortex (TMS intensity 150% of resting MT). This measure is referred to as the ipsilateral silent period (iSP) (cf. Chapter 11, this volume). The iSP duration elicited by TMS of the right motor cortex increases by anodal tDCS of left motor cortex, but decreases by cathodal tDCS. This effect is likely explained by an enhancement of the activity of inhibitory interneurons in the left motor cortex by anodal tDCS, but reduction of their activity by cathodal tDCS.

Indirect waves (I-waves) refer to corticospinal discharge generated by motor cortex stimulation, which evolve after the first or direct corticospinal volley (D-wave) and are likely under control of intracortical neuronal circuits (Ziemann and Rothwell 2000) (cf. Chapter 14, this volume). I-waves can be tested noninvasively by a paired-pulse TMS protocol, where a suprathreshold TMS test stimulus is followed by a subthreshold stimulus. The resulting increase of the MEP amplitude occurs at discrete interstimulus intervals and likely reflects cortical interactions between the circuits responsible for the generation of I-waves (Ziemann *et al.* 1998b) (cf. Chapter 11, this volume). This short-interval intracortical facilitation (SICF) is controlled by the activity of the GABAergic system. Additionally the glutamatergic system may contribute (Ghaly *et al.* 2001) (cf. Chapter 13, this volume). SICF does not change during short-lasting tDCS. For the after-effects, the first I-wave peak increases by anodal and cathodal tDCS, while the fourth I-wave peak is facilitated by anodal but not cathodal tDCS (Nitsche *et al.* 2005). The lack of an effect on SICF by tDCS protocols which do not elicit after-effects (intra-tDCS condition) fits well with the notion that intra-tDCS modulation of excitability depends on membrane polarization, whereas SICF depends on largely synaptic mechanisms. The magnitude of the tDCS after-effects on SICF are relatively minor compared to those on SICI and ICF. This could be due to the more prominent dependence of SICF on GABAergic mechanisms, which seem not to be influenced by tDCS to a great extent (Nitsche *et al.* 2004b). At first sight, the facilitatory effect of both anodal *and* cathodal tDCS on the first I-wave peak during the long-lasting after-effects is surprising. Anodal tDCS may have caused an enhanced I-wave peak amplitude by increasing cortical facilitation. Cathodal tDCS may have caused a deactivation of inhibitory interneurons that control the first I-wave peak.

Taken together, these results suggest that modulation of corticospinal excitability induced during tDCS (intra-tDCS modulation) depends critically on membrane polarization. In contrast, the after-effects of tDCS on corticospinal excitability involve more prominently intracortical synaptic mechanisms. These latter effects are most likely explained by tDCS-generated modifications of N-methyl-D-aspartate (NMDA) receptor efficacy. The MEP I/O curve, SICI, ICF, and SICF are thought to be at least partly controlled by these receptors. Accordingly, NMDA receptor antagonists abolish the after-effects of tDCS (see below).

Mechanisms of tDCS explored by pharmacological intervention

Pharmacological intervention provides us with the opportunity to increase our knowledge about the mechanisms of action of tDCS – beyond usage of neurophysiological techniques alone (cf. Chapter 13, this volume). It can help to identify the involvement of ion channels or receptors in tDCS-induced changes of cortical excitability by blocking or enhancing their activity. At least two potential strategies exist to explore the impact of pharmacological intervention on the excitability changes elicited by tDCS: protocols in which a pharmacological agent is administered after tDCS, which would interfere with already established effects of stimulation, or alternatively, the administration of a pharmacological agent before tDCS to interfere with the development of the tDCS effects. Although attractive in principle, the first type of protocol is difficult to apply in noninvasive experiments in humans, since oral intake might result in delayed peak concentrations of many drugs, which outlast the effects of tDCS. Moreover, the dynamics of the increase of intracerebral drug concentration alone could result in adaptive modifications of the excitability and activity of the cerebral cortex. Intravenous application, which would be an alternative, is not applicable for many of the drugs tested. Thus the second approach was implemented in experimental tDCS protocols. Typically, tDCS is applied during the expected peak of the drug concentration in the CNS. Since direct information about the time at which a maximum cerebral drug concentration is achieved is usually not available, surrogate measures, such as peak serum level or behavioral/electrophysiological effects, are used in most cases. To control for any effect of the study drug on cortical excitability *per se*, the impact and/or time course of the study drug on cortical excitability (in the absence of tDCS) was also tested.

As suggested by animal experiments, the effects during short tDCS (duration of tDCS, 4 s), which elicits no after-effects, should depend primarily on polarity-specific shifts of the resting membrane potential (Purpura and McMurtry 1965): anodal stimulation should depolarize cortical neurons, whereas cathodal tDCS should hyperpolarize them. This hypothesis was tested by probing the effects of short tDCS under 600 mg of the voltage-dependent sodium-channel blocker carbamazepine or 10 mg of the voltage-dependent calcium-channel blocker flunarizine (Nitsche et al. 2003b). Both drugs diminish or completely prevent the excitability-enhancing effects of anodal tDCS. In contrast, cathodal tDCS-elicited excitability diminutions were not influenced. It was concluded that the primary effect of intra-tDCS is a neuronal depolarization which is prevented by blocking voltage-dependent sodium or calcium channels. The absence of an effect by the ion-channel blocker in the cathodal tDCS condition is most likely explained by the fact that hyperpolarization of neuronal membranes is not influenced by voltage-dependent channel-blocking drugs. Synaptic mechanisms seem not to contribute to the intra-tDCS effects, since application of the NMDA receptor-antagonist dextromethorphan, the positive allosteric modulator at the GABA$_A$ receptor lorazepam, and the monoamine reuptake inhibitor amfetaminil do not influence the tDCS-induced MEP amplitude changes (Nitsche et al. 2003b, 2004b,c).

For the after-effects of tDCS, the results of animal experiments imply an *inductive* effect of membrane polarization shifts, while the sustained neuroplastic excitability shifts themselves might depend on synaptic mechanisms (Frégnac et al. 1990; Froc et al. 2000). Pharmacological interventions suggest that, indeed, neuronal depolarization during tDCS is crucial for the induction of neuroplastic excitability enhancements by anodal tDCS, since its block by carbamazepine or flunarizine (same dosages as mentioned above; 5, 11, and 13 min tDCS) abolishes these after-effects. The lack of effect by these drugs on the cathodal tDCS-elicited prolonged excitability diminutions (5 and 9 min

tDCS) is in accordance with the crucial importance of a neuronal hyperpolarization during tDCS for the induction of these after-effects (Liebetanz et al. 2002; Nitsche et al. 2003b). At the synaptic level, activation of NMDA receptors seems to be important for inducing prolonged cortical excitability enhancements and reductions by tDCS, since blocking these receptors by dextromethorphan (5 and 9 min cathodal, 5 and 11 min anodal tDCS) prevents all after-effects (Liebetanz et al. 2002; Nitsche et al. 2003b). This is in line with other experiments in animals and humans, and highlights the important role of NMDA receptor activation in systems-level neuroplasticity (Malenka and Bear 2004; Ziemann 2004).

In contrast, the involvement of GABAergic receptors in the formation of tDCS-induced after-effects is not so clear. Although cathodal tDCS-induced excitability diminutions are not influenced by lorazepam, a positive modulator of the GABA$_A$ receptor, this drug alters the time course of the excitability enhancement elicited by anodal tDCS when compared to placebo; under lorazepam, MEP amplitudes remain close to the baseline level immediately after stimulation, but increase with a delay of a few minutes. This delayed excitability enhancement is – unlike the placebo condition – not accompanied by a decrease in SICI or an increase in SICF, and may thus not originate at the level of motor cortex (Nitsche et al. 2004b).

Two studies have been performed so far to evaluate the involvement of receptors in the stabilization of tDCS-generated neuroplasticity. D-Cycloserine, a partial NMDA receptor agonist, prolongs the long-lasting excitability enhancement caused by anodal tDCS until the following morning stimulation, as compared to a duration of ~1 h in the placebo condition, but leaves the duration of the decrease in corticospinal excitability elicited by cathodal tDCS unaffected (Nitsche et al. 2004d). Similarly, amfetaminil prolongs the long-lasting after-effects of anodal tDCS selectively and relevantly, while it has only a minor influence on short-lasting after-effects of tDCS (Nitsche et al. 2004c) (Figure 17.2). This result is consistent with animal experiments demonstrating a stabilizing effect of dopaminergic and β-adrenergic activity on neuroplasticity (Ikegaya et al. 1997;

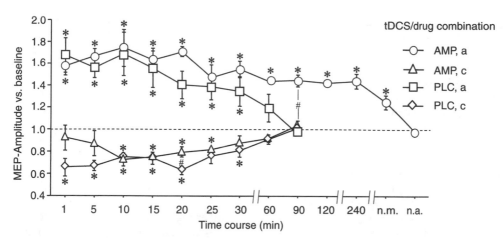

Fig. 17.2 Amphetamine selectively prolongs the anodal transcranial direct current stimulation (tDCS) after-effects (long-lasting enhancement of motor-evoked potential (MEP) amplitude). The graph shows the MEP amplitudes (mean ± SEM) normalized to baseline for different time points following anodal or cathodal tDCS, and for amphetamine (AMP) and placebo (PLC) conditions. On the morning following anodal tDCS, MEP amplitudes are still enhanced under amphetamine, whereas under placebo they returned to baseline values 60 min after tDCS. Asterisks indicate significant deviations of the post-tDCS MEP amplitudes from baseline; crosses mark significant deviations of amphetamine vs placebo conditions at that time point and tDCS polarity (Student's paired t-test, two-tailed, $P < 0.05$). n.m., next morning; n.a., next afternoon; a, anodal tDCS; c, cathodal tDCS. Reproduced with permission from Nitsche et al. (2004c).

Otani et al. 1998; Bailey et al. 2000). Co-administration of amfetaminil and dextromethorphan suggests that amphetamine indeed stabilizes, but does not induce, neuroplasticity, since under this drug combination anodal tDCS does not induce any after-effects. Moreover, the neuroplasticity-consolidating effect of amphetamine seems to be caused at least in part by β-adrenergic activity, since application of the β-adrenergic antagonist propanolol diminishes the duration of the tDCS-induced excitability shifts (Nitsche et al. 2004c).

Taken together, the findings of these pharmacological studies favor (i) a hyperpolarizing intra-tDCS effect of cathodal and a depolarizing intra-tDCS effect of anodal stimulation, whereas (ii) changes in synaptic strength do not play a prominent role in intra-tDCS effects. (iii) The after-effects of tDCS depend on the initial hyper- or depolarization, but (iv) the after-effects themselves are NMDA receptor dependent. (v) Consolidation of tDCS-elicited neuroplasticity can be accomplished by enhancement of neurotransmission through the NMDA receptor or

monoaminergic receptors. The latter effect is at least in part due to β-adrenergic activation.

Metaplasticity induced by combined tDCS and repetitive (r)TMS of the motor cortex

Beyond the induction of neuroplasticity itself, the ability of cortical neuronal networks to regulate neuronal activity within a useful dynamic range, called metaplasticity or homeostatic plasticity, has received increasing attention in recent years (Abraham and Tate 1997; Abbott and Nelson 2000). The mechanisms of metaplasticity were originally postulated from network modelling. A unidirectional and nonlimited increase in network activity and/or excitability induced, for instance by learning, would destabilize the system. Thus, algorithms were developed to limit these processes. One example is the Bienenstock–Cooper–Munro (BCM) rule, in which stabilization of neuronal activity is ensured by a dynamic adaptation of the modification threshold of long-term potentiation (LTP) and long-term depression (LTD) by the recent history of postsynaptic activity in the neuronal network (Bienenstock *et al.* 1982). A high level of previous activity enhances, while a low-level activity reduces, the modification threshold of LTP. Thus, a given stimulation protocol may result in LTP or LTD, depending on the history of postsynaptic activity in the stimulated network. Beyond network modelling, the existence of homeostatic mechanisms has already been demonstrated in animal models (Huang *et al.* 1992; Kirkwood *et al.* 1996; Wang and Wagner 1999).

To explore the existence of homeostatic plasticity in the human brain, the combination of tDCS and rTMS is a potentially well-suited experimental protocol. Both techniques, if applied alone, are capable of inducing enduring, yet reversible, increase or decrease of cortical excitability, depending on stimulation polarity (tDCS) or frequency (rTMS). tDCS might be especially well suited to modify the 'history of activation' and thus to be used as a preconditioning protocol (see above). If the tDCS-generated change in the history of activation influenced the modification threshold of synaptic plasticity, the effects of subsequently applied rTMS should critically depend on the polarity of the preconditioning tDCS. This hypothesis was tested in two studies. Anodal or cathodal tDCS of motor cortex was followed by rTMS protocols known to induce a long-lasting excitability enhancement (5 Hz rTMS) or reduction (1 Hz rTMS). The intensity of rTMS was subthreshold in both studies, and rTMS did not change corticospinal excitability when given alone. In accordance with the BCM rule, 1 Hz rTMS diminishes excitability, if conditioned by a high level of activity in the motor cortical network induced by prior anodal tDCS. In contrast, 1 Hz rTMS increases excitability when conditioned by a low level of activity in the motor cortical network induced by cathodal tDCS (Siebner *et al.* 2004). Similar results, consistent with the BCM theory, were obtained when 5 Hz rTMS is conditioned anodal or cathodal tDCS (Lang *et al.* 2004b) (Figure 17.3).

The results of both experiments show that, similar to animal experiments, homeostatic plasticity occurs in the human motor cortex and

influences magnitude and direction of stimulation-induced prolonged excitability changes. So far, the mechanisms of metaplasticity in the human brain are largely unclear. Changes in MT, SICI, and ICF seem not to be involved (Lang *et al.* 2004b; Siebner *et al.* 2004), and other candidate mechanisms as well as adaptive modifications of neuronal receptors and channels have not yet been tested. Possible implications of these experiments on metaplasticity, however, are clear: they offer a suitable tool to further explore metaplasticity in the human brain in healthy subjects and its possible pathological changes in neurological and psychiatric diseases. Moreover, they imply that the effects of neuroplasticity-inducing neurophysiological protocols are context dependent, i.e. are modified by the recent history of activity in the stimulated neural network. This should be taken into account for future therapeutic applications.

Induction of neuroplasticity by tDCS of the visual cortex and other tDCS applications

Although most of the basic studies concerning the effects of tDCS are performed in the motor cortex, the efficacy of this technique to induce neuroplasticity is not restricted to this area. In the occipital lobe, tDCS is capable of inducing bidirectional changes of excitability in the primary visual cortex (V1). Excitability of V1 can be monitored by the phosphene threshold. Phosphenes are subjective light sensations – static or dynamic, depending on the area stimulated – which are elicited by TMS of the visual cortex (Meyer *et al.* 1991). The phosphene threshold is the lowest TMS intensity which reliably induces phosphenes. In V1, 10 min of anodal tDCS (anode placed over Oz, cathodal electrode over Cz according to the international 10/20 EEG system) decreases the threshold for static phosphenes. This effect lasts for 10 min after the end of tDCS. In contrast, cathodal tDCS (electrode positions reversed) results in an increase in phosphene threshold (Antal *et al.* 2003a). Thus, as in the motor cortex, anodal tDCS enhances V1 excitability, whereas cathodal tDCS diminishes it. The duration of the after-effects, however, is shorter in the visual than in

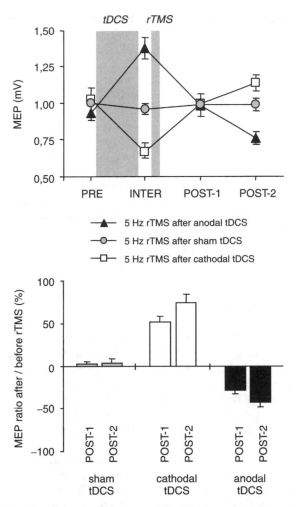

Fig. 17.3 Preconditioning with transcranial direct current stimulation (tDCS) results in homeostatic effects of subsequent repetitive (r)TMS-induced motor cortex excitability shifts. (a) Mean amplitudes of motor-evoked potentials (MEPs) elicited by single-pulse TMS before tDCS (PRE), between tDCS and repetitive (r)TMS (INTER), and after rTMS (POST-1 and POST-2). (b) Relative changes of MEP amplitudes within the first and second block of post-rTMS measurements (POST-1 and POST-2) compared with mean amplitudes immediately before rTMS (INTER). The type of preconditioning tDCS had a strong effect on the magnitude and direction of after-effects produced by subsequent 5 Hz rTMS. No effects on corticospinal excitability occurred after sham preconditioning. Depending on the polarity, effective tDCS resulted in a bidirectional modulation of the after-effects induced by subsequent 5 Hz rTMS. The 5 Hz rTMS given after 'inhibitory' preconditioning (cathodal tDCS) resulted in a significant increase of corticospinal excitability. Conversely, 5 Hz rTMS after 'facilitatory' preconditioning (anodal tDCS) caused a decrease in corticospinal excitability. The conditioning effect of rTMS gradually built up during the 20 min after the end of rTMS. Error bars represent SEM. Modified with permission from Lang *et al.* (2004b).

the motor cortex. Reasons for this might be regional differences of neuronal architecture, orientation, and receptor densities, as well as the fact that relevant parts of the visual cortex are more distant from the scalp than the motor cortex. tDCS of V1 shifts the phosphene threshold not only for stationary but also for moving phosphenes that are elicited by TMS of area V5 (Antal *et al.* 2003b). This suggests a modulating role of V1 in the perception of motion.

In addition, anodal tDCS of the left V5 (anode placed 4 cm above the mastoid–inion line and 7 cm lateral to midline, cathode placed at Cz) improves learning in a visuo-motor coordination task, while cathodal tDCS of the same region improved performance of the same task in an over-learned state (Antal *et al.* 2004a,b). It was speculated that different effects of tDCS on different learning phases might be due to phase-specificity of neuroplastic modifications. During learning an excitability enhancement should increase the strengthening of task-relevant synaptic connections; however, the benefit of suppressing task-irrelevant or distractive neuronal connections may be important during performance of an over-learned task to sharpen the task-relevant activation pattern.

tDCS can also influence the excitability of somatosensory cortex. Anodal tDCS applied over the hand motor cortex for 10 min increases the amplitudes of somatosensory potentials (P25/N33, N33/P40) evoked by contralateral median nerve stimulation for ≥60 min after the end of stimulation (Matsunaga *et al.* 2004). Cathodal tDCS was without effect in this protocol. However, cathodal tDCS delivered over C4 – and thus perhaps more directly over the somatosensory cortex – impairs the tactile discrimination threshold (Rogalewski *et al.* 2004), which can be taken as evidence for the efficacy of this tDCS polarity in the somatosensory cortex.

tDCS of prefrontal cortex has also proven to be effective: during anodal tDCS of the left dorsolateral prefrontal cortex with one electrode centered at F3 and the reference electrode over the contralateral orbit, performance in a three-back sequential-letter working memory task is improved (Fregni *et al.* 2005a). Using the same electrode position, a significant increase in word fluency occurs after 20 min of anodal tDCS, while cathodal tDCS decreases it mildly (Iyer *et al.* 2005). This effect was apparent only with a current strength of 2 mA, but not with 1 mA, indicating a stimulation-intensity-dependent effect of tDCS on the dorsolateral prefrontal cortex, similar to the one previously demonstrated for the primary motor cortex (Nitsche and Paulus 2000) (see above). Bilateral repetitive anodal and cathodal tDCS of the dorsolateral prefrontal cortex (30 min total stimulation duration, alternating phases of 15 s tDCS with 15 s without stimulation), applied via small electrodes (electrode diameter 8 mm, stimulation electrode positioned at F3 and F4, reference electrode at mastoids) with a relatively high current density (current strength of 0.26 mA), results in deficits of response selection and preparation in the Sternberg working memory task (Marshall *et al.* 2005). Conversely, anodal tDCS applied with the same protocol selectively improved declarative memory performance when applied during slow-wave sleep (Marshall *et al.* 2004). However, because of the small electrode size, the high current density, and the reference electrodes situated at the mastoids in that protocol, and because control experiments with different electrode positions were not performed, it cannot be concluded unequivocally that the dorsolateral prefrontal is crucial for performance in these experiments. For tDCS of the frontopolar cortex, the effects of left frontopolar tDCS (electrode position Fp3, reference electrode Cz, 1 mA, surface 35 cm^2) on implicit semantic learning were examined. Anodal stimulation improves performance in a probabilistic classification learning task (Knowlton *et al.* 1996), whereas cathodal tDCS tends to worsen it (Kincses *et al.* 2004). These findings indicate that stimulation of the frontopolar cortex is also effective in altering behavioral function.

The effect of tDCS on performance or excitability of other cortices has not been tested so far, but it is reasonable to assume that tDCS is also well suited to modulate other brain areas. Since the strength of the applied electrical field diminishes exponentially with distance from the electrodes (Rush and Driscoll 1968), it is likely that surface-near cortices are better reached by tDCS. However, it cannot be ruled out that beyond the direct influence of tDCS on the stimulated cortex, which seems to be fairly restricted to the area under the electrode

(Nitsche *et al.* 2003c; Lang *et al.* 2004a), indirect effects in distant, e.g. subcortical areas, occur, which might be connectivity driven (Lang *et al.* 2005).

Clinical applications

For the clinical application of tDCS, at least two main fields of interest can be identified: (i) exploration of pathological alterations of neuroplasticity in neurological and psychiatric diseases, and (ii) evaluation of a possible clinical benefit of tDCS in these diseases. Both lines of research are still in their early days. Pathological alterations of neuroplasticity can be studied by testing the inducibility of prolonged changes in excitability by anodal or cathodal tDCS. Moreover, combining tDCS and rTMS (Lang *et al.* 2004b; Siebner *et al.* 2004) (see above), might provide additional information about abnormalities of metaplasticity in certain neurological diseases. Disordered metaplasticity was explored in patients with focal hand dystonia. In this disease the excitability of inhibitory circuits is reduced at multiple levels of the sensorimotor system, including the hand area of the primary motor cortex. This might be caused or accompanied by deficient function of homeostatic mechanisms, which normally keep cortical excitability within a normal physiological range. This hypothesis was tested by application of the homeostatic plasticity protocol introduced by Siebner *et al.* (2004): anodal or cathodal tDCS of the primary motor cortex for 10 min is followed by 1 Hz rTMS (intensity 85% of resting motor threshold) of the same area. In healthy subjects, anodal tDCS enhances, while cathodal tDCS reduces, excitability, and this effect is reversed by subsequent rTMS. In patients with focal hand dystonia, however, 1 Hz rTMS failed to decrease corticospinal excitability when conditioned by an enhancement of excitability by prior anodal tDCS. Moreover, cathodal tDCS did not result in a significant inhibition, nor did it modify the effect of subsequent 1 Hz rTMS on excitability (Quartarone *et al.* 2005). Thus, homeostatic mechanisms, in particular those that prevent further increase in excitability after a recent history of high-level cortical activity, might be deficient in focal hand dystonia. The failure of cathodal tDCS alone to induce clear inhibition

in this patient group might be taken as evidence for an additional, generally reduced, ability to establish inhibitory neuroplasticity.

For the evaluation of clinical benefits from tDCS-induced neuroplasticity for patients with neurological and psychiatric diseases, a limited number of pilot studies have been performed. In chronic stroke patients with paresis, motor rehabilitation can be hampered by a kind of maladaptive plasticity called 'learned nonuse' (Liepert *et al.* 1998). Prolonged inactivity of the paretic extremity reduces its motor cortical representation and excitability, and thus compromises motor function. This may be partly caused by a hyperactivity of the contralateral nonlesioned motor cortex, which increases transcallosal inhibition of the lesioned hemisphere. Thus, enhancing excitability of the lesioned motor cortex by anodal tDCS, as well as reducing excitability of the nonlesioned contralateral one by cathodal tDCS, might improve motor function of the paretic extremity after stroke. Indeed, in stroke patients with residual paresis of the upper limb, 20 min of anodal tDCS of the hand area in the primary motor cortex improves hand motor performance (Hummel *et al.* 2005). In another study, it was demonstrated that 20 min of cathodal tDCS of the hand area situated within the nonlesioned cortex is equally effective (Fregni *et al.* 2005b).

With regard to focal epilepsy, a first pilot study has been performed to show the beneficial impact of excitability-diminishing cathodal tDCS on epileptic discharges and seizure frequency (Fregni *et al.* 2006c). Cathodal or sham stimulation was delivered once for 20 min over the EEG-determined epileptogenic focus. The anodal reference electrode was placed at a distant area without epileptic activity. Cathodal tDCS reduced the frequency of epileptic discharges significantly and resulted in a trend to reduce seizure frequency. This effect was most clear cut for a patient subgroup with a single epileptogenic focus. The impact of short-lasting tDCS on tinnitus was tested by applying 3 min anodal, cathodal, or sham stimulation to the left temporo-parietal area (Fregni *et al.* 2006b). Similar to high-frequency rTMS, anodal tDCS decreased tinnitus immediately after stimulation.

It is known that in depressed patients the left dorsolateral prefrontal cortex displays reduced

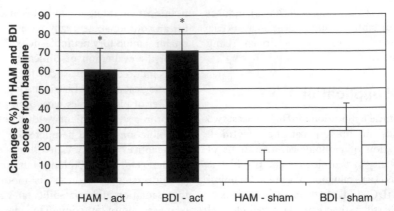

Fig. 17.4 Mean changes (%) in Hamilton (HAM) and Beck Depression Inventory (BDI) of depressive patients treated with active transcranial direct current stimulation (tDCS) (black columns) and sham tDCS (white columns). There was a significant improvement in depression scores measured by HAM and BDI after treatment only in the active tDCS group but not in the sham group. Error bars are SEM. Significance level (*$P < 0.05$) assessed by paired Student's two-tailed t-tests (comparison between HAM and BDI scores of baseline and post-treatment for both groups). Modified with permission from Fregni et al. (2006a).

activity, while the activity of the right prefrontal cortex might be increased (Schutter and van Honk 2005). Consequently, activity-enhancing rTMS of the left prefrontal cortex has been demonstrated to diminish depressive symptoms (Loo and Mitchell 2005). The ability of tDCS to produce similar effects was tested in a double-blind sham-controlled pilot study (Fregni et al. 2006a). Excitability-enhancing anodal tDCS of the left dorsolateral prefrontal cortex, combined with cathodal tDCS of the right frontopolar cortex (current strength 1 mA, electrode size 35 cm^2, stimulation duration 20 min), was applied for five consecutive days in a small group of patients suffering from major depression. Compared to baseline values and the sham-tDCS control group, the depression scores, as recorded by the Beck Depression Inventory and the HAMD, were significantly reduced in the treatment group (Figure 17.4).

Although research activities with regard to clinical applications of tDCS are still in their infancy, the results obtained so far show: (i) tDCS, e.g. in combination with TMS or rTMS, could evolve as a valuable technique to study abnormalities of neuroplasticity as well as metaplasticity in neurological and psychiatric diseases; (ii) a possible potential of tDCS would be to treat diseases of the CNS, which are

accompanied by alterations of cerebral excitability. However, these studies have been performed mainly to demonstrate in principle that tDCS can reduce symptoms. Thus the stimulation duration in most cases was relatively short, stimulation intensity was weak, and the effects were – perhaps with the exception of the depression study – mostly minor and short-lasting. Further studies are needed to explore if prolonged, repetitive, or stronger stimulation protocols, for which safety has to be assured, could evolve into clinically more relevant improvement.

References

Abbott LF, Nelson SB (2000). Synaptic plasticity: taming the beast. *Nature Neuroscience* 3(Suppl.), 1178–1183.

Abbruzzese G, Trompetto C (2002). Clinical and research methods for evaluating cortical excitability. *Journal of Clinical Neurophysiology* 19, 307–321.

Abraham WC, Tate WPC (1997). Metaplasticity: a new vista across the field of synaptic plasticity. *Progress in Neurobiology, 52*, 303–323.

Agnew WF, McCreery DB (1987). Considerations for safety in the use of extracranial stimulation for motor evoked potentials. *Neurosurgery 20*, 143–147.

Antal A, Kincses TZ, Nitsche MA, Paulus W (2003a). Manipulation of phosphene thresholds by transcranial direct current stimulation in man. *Experimental Brain Research 150*, 375–378.

Antal A, Kincses TZ, Nitsche MA, Paulus W (2003b). Modulation of moving phosphene thresholds by transcranial direct current stimulation of V1 in the human. *Neuropsychologia 41*, 1802–1807.

Antal A, Nitsche MA, Kinsces TZ, Kruse W, Hoffmann K-P, Paulus W (2004a). Facilitation of visuo-motor learning by transcranial direct current stimulation of the motor and extrastriate visual areas in humans. *European Journal of Neuroscience 19*, 2888–2892.

Antal A, Nitsche MA, Kruse W, Hoffmann K-P, Paulus W (2004b). Visuomotor coordination is improved by transcranial direct current stimulation of the human visual cortex. *Journal of Cognitive Neuroscience 16*, 521–527.

Ardolino G, Bossi B, Barbieri S, Priori A (2005). Non-synaptic mechanisms underlie the after-effects of cathodal transcutaneous direct current stimulation of the human brain. *Journal of Physiology 568*, 653–663.

Bailey CH, Giustetto M, Huang YY, Hawkins RD, Kandel ER (2000). Is heterosynaptic modulation essential for stabilizing Hebbian plasticity and memory? *Nature Reviews Neuroscience 1*, 11–20.

Bennett MR (2000). The concept of long term potentiation of transmission at synapses. *Progress in Neurobiology 60*, 109–137.

Bienenstock EL, Cooper LN, Munro PW (1982). Theory for the development of neuron selectivity: orientation specifity and binocular interaction in visual cortex. *Journal of Neuroscience 2*, 32–48.

Bindman LJ, Lippold OCJ, Redfearn JWT (1964). The action of brief polarizing currents on the cerebral cortex of the rat (1) during current flow and (2) in the production of long-lasting after-effects. *Journal of Physiology 172*, 369–382.

Boroojerdi B (2002). Pharmacologic influences on TMS effects. *Journal of Clinical Neurophysiology 19*, 255–271.

Carney MW (1969). Negative polarisation of the brain in the treatment of manic states. *Irish Journal of Medical Science 8*, 133–135.

Chen R (2000). Studies of human motor physiology with transcranial magnetic stimulation. *Muscle and Nerve 9*(Suppl.), S26–S32.

Chen R, Samii A, Canos M, Wassermann EM, Hallett M (1997). Effects of phenytoin on cortical excitability in humans. *Neurology 49*, 881–883.

Costain R, Redfearn JW, Lippold OC (1964). A controlled trial of the therapeutic effect of polarizazion of the brain in depressive illness. *British Journal of Psychiatry 110*, 786–799.

Creutzfeldt OD, Fromm GH, Kapp H (1962). Influence of transcortical d-c currents on cortical neuronal activity. *Experimental Neurology 5*, 436–452.

Di Lazzaro V, Oliviero A, Profice P, et al. (2003). Ketamine increases human motor cortex excitability to transcranial magnetic stimulation. *Journal of Physiology 547*, 485–496.

Dymond AM, Coger RW, Serafetinides EA (1975). Intracerebral current levels in man during electrosleep therapy. *Biological Psychiatry 10*, 101–104.

Edgley SA, Eyre JA, Lemon RN, Miller S (1997). Comparison of activation of corticospinal neurons and spinal motor neurons by magnetic and electrical transcranial stimulation in the lumbosacral cord of the anaesthetized monkey. *Brain 120*, 839–853.

Elbert T, Lutzenberger W, Rockstroh B, Birbaumer N (1981). The influence of low-level transcortical DC-currents on response speed in humans. *International Journal of Neuroscience 14*, 101–114.

Frégnac Y, Smith D, Friedlander MJ (1990). Postsynaptic membrane potential regulates synaptic potentiation and depression in visual cortical neurons. *Society for Neuroscience Abstracts 16*, 798.

Fregni F, Boggio PS, Nitsche MA, et al. (2005a). Anodal transcranial direct current stimulation of prefrontal cortex enhances working memory. *Experimental Brain Research 166*, 23–30.

Fregni F, Boggio PS, Mansur CG, et al. (2005b). Transcranial direct current stimulation of the unaffected hemisphere in stroke patients. *Neuroreport 16*, 1551–1555.

Fregni F, Boggio PS, Nitsche MA, Marcolin MA, Rigonatti SP, Pascual-Leone A (2006a). Treatment of major depression with transcranial direct current stimulation. *Bipolar Disorders 8*, 203–204.

Fregni F, Marcondes R, Boggio P, et al. (2006b). Transient tinnitus suppression induced by repetitive transcranial magnetic stimulation and transcranial direct current stimulation. *Journal of Neurology, 13*, 996–1001.

Fregni F, Thome-Souza S, Nitsche MA, Freedman S Valente KD, Pascual-Leone A (2006c). A sham-randomized clinical trial of transcranial direct current stimulation in patients with cortical malformations and refractory epilepsy. *Epilepsia 47*, 335–342.

Froc DJ, Chapman CA, Trepel C, Racine RJ (2000). Long-term depression and depotentiation in the sensorimotor cortex of the freely moving rat. *Journal of Neuroscience 20*, 438–434.

Gartside IB (1968). Mechanisms of sustained increases of firing rate of neurones in the rat cerebral cortex after polarization: Role of protein synthesis *Nature 220*, 383–384.

Ghaly RF, Ham JH, Lee JJ (2001). High-dose ketamine hydrochloride maintains somatosensory and magnetic motor evoked potentials in primates. *Neurology Research 23*, 881–886.

Hattori Y, Moriwaki A, Hori Y (1990). Biphasic effects of polarizing current on adenosine-sensitive generation of cyclic AMP in rat cerebral cortex. *Neuroscience Letters 116*, 320–324.

Huang YY, Colino A, Selig DK, Malenka RC (1992). The influence of prior synaptic activity on the induction of long-term potentiation. *Science 255*, 730–733.

Hummel F, Celnik P, Giraux P, et al. (2005). Effects of non-invasive cortical stimulation on skilled motor function in chronic stroke. *Brain 128*, 490–499.

Ikegaya Y, Nakanishi K, Saito H, Abe K (1997). Amygdala beta-noradrenergic influence on hippocampal long-term potentiation in vivo. *Neuroreport 8*, 3143–3146.

Islam N, Aftabuddin M, Moriwaki A, Hattori Y, Hori Y (1995). Increase in the calcium level following anodal polarization in the rat brain. *Brain Research 684*, 206–208.

Iyer MB, Mattu U, Grafman J, Lomarev M, Sato S, Wassermann EM (2005). Safety and cognitive effect of frontal DC brain polarization in healthy individuals. *Neurology 64*, 872–875.

Jaeger D, Elbert T, Lutzenberger W, Birbaumer N (1987). The effects of externally applied transcephalic weak direct currents on lateralization in choice reaction tasks. *Journal of Psychophysiology 1*, 127–133.

Kincses TZ, Antal A, Nitsche MA, Bártfai O, Paulus W (2004). Facilitation of probabilistic classification learning by transcranial direct current stimulation of the prefrontal cortex in the human. *Neuropsychologia 42*, 113–117.

Kirkwood A, Rioult MC, Bear MF (1996). Experience-dependent modification of synaptic plasticity in visual cortex. *Nature 381*, 526–528.

Knowlton BJ, Mangels JA, Squire LR (1996). A neostriatal habit learning system in humans. *Science 273*, 1399–1340.

Kujirai T, Caramia MD, Rothwell JC, *et al.* (1993). Corticocortical inhibition in human motor cortex. *Journal of Physiology 471*, 501–519.

Lang N, Nitsche MA, Paulus W, Rothwell JC, Lemon R (2004a). Effects of transcranial DC stimulation over the human motor cortex on corticospinal and transcallosal excitability. *Experimental Brain Research 156*, 439–443.

Lang N, Siebner HR, Ernst D, *et al.* (2004b). Preconditioning with transcranial direct current stimulation sensitizes the motor cortex to rapid-rate transcranial magnetic stimulation and controls the direction of after-effects. *Biological Psychiatry 56*, 634–639.

Lang N, Siebner HR, Ward NS, *et al.* (2005). How does transcranial DC stimulation of the primary motor cortex alter regional neuronal activity in the human brain? *European Journal of Neuroscience 22*, 495–504.

Liebetanz D, Nitsche MA, Tergau F, Paulus W (2002). Pharmacological approach to synaptic and membrane mechanisms of DC-induced neuroplasticity in man. *Brain 125*, 2238–2247.

Liepert J, Schwenkreis P, Tegenthoff M, Malin JP (1997). The glutamate antagonist riluzole suppresses intracortical facilitation. *Journal of Neural Transmission 104*, 1207–1214.

Liepert J, Miltner WH, Bauder H, *et al.* (1998). Motor cortex plasticity during constraint-induced movement therapy in stroke patients. *Neuroscience Letters 250*, 5–8.

Lolas F (1977). Brain polarization: behavioral and therapeutic effects. *Biological Psychiatry 12*, 37–47.

Loo CK, Mitchell PB (2005). A review of the efficacy of transcranial magnetic stimulation (TMS) treatment for depression, and current and future strategies to optimize efficiacy. *Journal of Affective Disorders 88*, 255–267.

Malenka RC, Bear MF (2004). LTP and LTD: an embarrassment of riches. *Neuron 44*, 5–21.

Marshall L, Molle M, Hallschmid M, Born J (2004). Transcranial direct current stimulation during sleep improves declarative memory. *Journal of Neuroscience 24*, 9985–9992.

Marshall L, Molle M, Siebner HR, Born J (2005). Bifrontal transcranial direct current stimulation slows reaction time in a working memory task. *BMC Neuroscience 6*, 23.

Matsunaga K, Nitsche MA, Tsuji S, Rothwell J (2004). Effect of trenscranial DC sensorimotor cortex stimulation on somatosensory evoked potentials in humans. *Clinical Neurophysiology 115*, 456–460.

Meyer BU, Diehl R, Steinmetz H, Britton TC, Benecke R (1991). Magnetic stimuli applied over motor and visual cortex: influence of coil position and field polarity on motor responses, phosphenes, and eye movements. *Electroencephalography and Clinical Neurophysiology 43*(Suppl), 121–123.

Nitsche MA, Paulus W (2000). Excitability changes induced in the human motor cortex by weak transcranial direct current stimulation. *Journal of Physiology 527*, 633–639.

Nitsche, MA, Paulus W (2001). Sustained excitability elevations induced by transcranial DC motor cortex stimulation in humans. *Neurology 57*, 1899–1901.

Nitsche, MA, Nitsche, MS, Klein, CC, Tergau, F, Rothwell, JC, Paulus W (2003a). Level of action of cathodal DC polarisation induced inhibition of the human motor cortex. *Clinical Neurophysiology 114*, 600–604.

Nitsche MA, Fricke K, Henschke U, *et al.* (2003b). Pharmacological modulation of cortical excitability shifts induced by transcranial DC stimulation. *Journal of Physiology 553*, 293–301.

Nitsche MA, Schauenburg A, Lang N, *et al.* (2003c). Facilitation of implicit motor learning by weak transcranial direct current stimulation of the primary motor cortex in the human. *Journal of Cognitive Neuroscience 15*, 619–626.

Nitsche MA, Niehaus L, Hoffmann KT *et al.* (2004a). MRI study of human brain exposed to weak direct current stimulation of the frontal cortex. *Clinical Neurophysiology 115*, 2419–2423.

Nitsche MA, Liebetanz D, Schlitterlau A, *et al.* (2004b). GABAergic modulation of DC-stimulation-induced motor cortex excitability shifts in the human. *European Journal of Neuroscience 19*, 2720–2726.

Nitsche MA, Grundey J, Liebetanz D, Lang N, Tergau F, Paulus W (2004c). Catecholaminergic consolidation of motor cortex plasticity in humans. *Cerebral Cortex, 14*, 1240–1245.

Nitsche MA, Jaussi W, Liebetanz D, Lang N, Tergau F, Paulus W (2004d). Consolidation of externally induced human motor cortical neuroplasticity by d-cycloserine. *Neuropsychopharmacology 29*, 1573–1578.

Nitsche MA, Seeber A, Frommann K, *et al.* (2005). Modulating parameters of excitability during and after transcranial direct current stimulation of the human motor cortex. *Journal of Physiology 568*, 291–303.

Otani S, Blond O, Desce JM, Crepel F (1998). Dopamine facilitates long-term depression of glutamatergic transmission in rat prefrontal cortex. *Neuroscience 85*, 669–676.

Pfurtscheller G (1970). [Changes in the evoked and spontaneous brain activity of man during extracranial polarization]. *Zeitschrift für die gesamte experimentelle Medizin einschliesslich experimentelle Chirurgie 152*, 284–293.

Priori A, Berardelli A, Rona S, Accornero N, Manfredi M (1998). Polarization of the human motor cortex through the scalp. *Neuroreport 9*, 2257–2260.

Purpura DP, McMurtry JG (1965). Intracellular activities and evoked potential changes during polarization of motor cortex. *Journal of Neurophysiology 28*, 166–185.

Quartarone A, Rizzo V, Bagnato S, *et al.* (2005). Homeostatic-like plasticity of the primary motor hand area is impaired in focal hand dystonia. *Brain 128*, 1943–1950.

Ridding MC, Brouwer B, Miles TS, Pitcher JB, Thompson PD (2000). Changes in muscle responses to stimulation of the motor cortex induced by peripheral nerve stimulation in human subjects. *Experimental Brain Research 131*, 135–143.

Rogalewski A, Breitenstein C, Nitsche MA, Paulus W, Knecht S (2004). Transcranial direct current stimulation disrupts tactile perception. *European Journal of Neuroscience 20*, 313–316.

Roth BJ (1994). Mechanisms for electrical stimulation of excitable tissue. *Critical Reviews in Biomedical Engineering 22*, 253–305.

Rothwell JC (1993). Evoked potentials, magnetic stimulation studies, and event-related potentials. *Current Opinion in Neurology 6*, 715–723.

Rush S, Driscoll DA (1968). Current distribution in the brain from surface electrodes. *Anaesthesia and Analgesia Current Research 47*, 717–7123.

Schutter DJ, van Honk J (2005). A framework for targeting alternative brain regions with repetitive transcranial magnetic stimulation in the treatment of depression. *Journal of Psychiatry and Neuroscience 30*, 91–97.

Siebner HR, Lang N, Rizzo V, Nitsche MD, Paulus W, Lemon RN, Rothwell JC (2004). Preconditioning of low-frequency repetitive transcranial magnetic stimulation: evidence for homeostatic plasticity in the human motor cortex. *Journal of Neuroscience 24*, 3379–3385.

Uy J, Ridding MC (2003). Increased cortical excitability induced by transcranial DC and peripheral nerve stimulation. *Journal of Neuroscience Methods 127*, 193–197.

Wang H, Wagner JJ (1999). Priming-induced shift in synaptic plasticity in the rat hippocampus. *Journal of Neurophysiology 82*, 2024–2028.

Yuen TGH, Agnew WF, Bullara LA, Jacques S, McCreery DB (1991). Histological evaluation of neural damage from electrical stimulation: considerations for the selection of parameters for clinical application. *Neurosurgery 9*, 292–298.

Ziemann U (2004). TMS induced plasticity in human cortex. *Reviews in Neuroscience 15*, 253–66.

Ziemann U, Rothwell JC (2000). I-waves in motor cortex. *Journal of Clinical Neurophysiology 17*, 397–405.

Ziemann U, Lonnecker S, Steinhoff BJ, Paulus W (1996). Effects of antiepileptic drugs on motor cortex excitability in humans: a transcranial magnetic stimulation study. *Annals of Neurology 40*, 367–378.

Ziemann U, Chen R, Cohen LG, Hallett M (1998a). Dextromethorphan decreases the excitability of the human motor cortex. *Neurology 51*, 1320–1324.

Ziemann U, Tergau F, Wassermann EM, Wischer S, Hildebrandt J, Paulus W (1998b). Demonstration of facilitatory I wave interaction in the human motor cortex by paired transcranial magnetic stimulation. *Journal of Physiology 511*, 181–190.

CHAPTER 18

Use-dependent changes in TMS measures

Cathrin M. Buetefisch and Leonardo G. Cohen

Introduction

Studies in animals and humans demonstrated that adult brains maintain the ability to reorganize throughout life. Cortical reorganization or plasticity, as defined by Donoghue *et al.* (1996), is 'any enduring changes in the cortical properties either morphological or functional.' The expression of these changes includes modification of synaptic efficacy as well as neuronal networks that carry behavioral implications.

In the primary motor cortex (M1), movement representations can reorganize rapidly in response to different stimuli or environmental modifications. Mapping experiments in rodents showed that trans-section of the facial nerve, which supplies the facial whisker musculature, leads to a loss of cortical representation of that body part in M1. The former cortical whisker representation is 'invaded' by the adjacent forelimb or eyelid areas (Sanes *et al.* 1988; Donoghue *et al.* 1990). Similar changes occur in response to changes in the position of a limb (Sanes *et al.* 1992), M1 lesions (Nudo and Milliken 1996), electrical stimulation of M1 (Nudo *et al.* 1990), or motor training (Nudo *et al.* 1996; Kleim *et al.* 1998) in animals.

Transcranial magnetic stimulation allows us to study M1 reorganization in humans. Single-pulse TMS can be used to measure changes in the human corticospinal motor output system (cf. Chapters 9 and 10) in different circumstances such as in response to practice or brain lesions. Paired-pulse TMS has provided means to study the differential regulation of intracortical inhibitory and excitatory networks (cf. Chapters 11 and 12) involved in reorganization of M1 in response to injury (Buetefisch *et al.* 2003), deafferentation (Ziemann *et al.* 1998a,b) and practice (Buetefisch *et al.* 2005). Further, combining TMS with CNS-active drugs (cf. Chapter 13) allows the investigation of some of the mechanisms underlying these plastic processes (Ziemann *et al.* 1998b; Buetefisch *et al.* 2000). In the first part of this chapter we will describe TMS measures used in different paradigms to characterize intracortical and corticospinal excitability changes in response to practice. In the second part we will discuss the underlying mechanisms and different approaches to modulate use-dependent reorganization of M1.

Effect of practice on TMS measures

Effect of practice on motor-evoked potential amplitudes

Motor-evoked potential (MEP) amplitudes change in response to practice. In general, the MEP amplitude recorded from muscles engaged in the training movement increases. The effect appears to be focal since it is not observed in muscles antagonistic to (Buetefisch *et al.* 2000) or uninvolved with (Muellbacher *et al.* 2001) the trained movements. For example, practicing thumb movements in an extension abduction direction results in an increase in the MEP amplitude of the extensor pollicis brevis (EPB) muscle, which supports the training movement relative to baseline. In contrast, the MEP amplitude of the muscle that is antagonistic to the training movement, the flexor pollicis brevis (FPB) muscle, decreases relative to baseline (Figure 18.1).

Because changes in MEP amplitudes depend on the excitability of the corticospinal neurons in M1 as well as the spinal motor neurons, any changes in MEP amplitudes could be caused by alterations in excitability at either cortical or spinal level or a combination of both. In order to localize the site of training-induced changes of excitability along the corticospinal system, additional measurements are needed. Direct recording of the TMS-evoked descending volleys with epidural electrodes is an invasive approach that can be used only in limited situations (Di Lazzaro *et al.* 1999) (cf. Chapter 14). Another approach involves the comparison of MEP amplitudes elicited by TMS and by transcranial electrical stimulation (TES). TMS activates pyramidal

tract neurons predominantly trans-synaptically (Day *et al.* 1987) whereas TES stimulates pyramidal tract neurons predominantly at their axon hillock (Amassian *et al.* 1990; Rothwell *et al.* 1991) (cf. Chapter 14). If training-induced changes in MEP amplitudes occur at a cortical site, one would expect to observe them with TMS but not with TES. On the other hand, changes at a spinal level would elicit comparable modifications in MEPs amplitudes to TMS and TES. TMS of the brain stem is another method to differentiate between changes at the cortical and spinal level. MEPs evoked by TMS of the brain stem should be relatively unaffected by changes in cortical excitability. Changes in MEPs after TMS in association with unaffected spinal excitability as tested by F-waves (Mercuri *et al.* 1996) or H-reflexes (Fuhr *et al.* 1992) would also support a cortical site for use-dependent plasticity. Training-induced increases in MEP amplitudes could reflect increases of excitability of the same number of pyramidal tract neurons in the motor cortex or an increase in the number of pyramidal tract neurons of similar excitability activated by the TMS pulse or a combination of both.

Increases in MEP amplitudes are often associated with improved performance or changes in the kinematics of movements elicited by TMS of

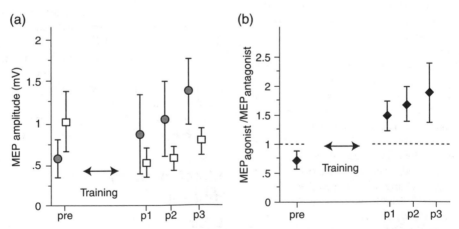

Fig. 18.1 Effect of training on motor-evoked potential (MEP) amplitude. (a) Practicing thumb movements for 30 min in an extension abduction direction results in an increase of MEP amplitude in a muscle supporting the training movement (extensor pollicis brevis (EPB), filled circle) but not in a muscle antagonistic to the training movement (flexor pollicis brevis (FPB), open square). (b) The MEP amplitude of the agonist muscle after training exceeds the amplitude of the antagonist muscle as indicated by the MEP agonist:MEP antagonist ratio of >1. Pre, testing prior to the training; p1–3, testing immediately (p1), 10–20 min (p2), and 20–30 min (p3) after the training. Modified after Buetefisch *et al.* (2000).

M1 after training protocols (Buetefisch *et al.* 2000; Muellbacher *et al.* 2001) and may reflect changes in the motor output zone related to motor learning (Muellbacher *et al.* 2001). These findings are consistent with concepts derived from animal experiments identifying multiple overlapping motor representations within M1 (Donoghue *et al.* 1992; Schieber and Hibbard 1993; Sanes *et al.* 1995), functionally connected through an extensive horizontal network (Huntley and Jones 1991). Whereas connections are abundant within somatic representations, they are sparse between them (Huntley and Jones 1991). By changing the strength of horizontal cortico-cortical connections between motor cortical neurons, functionally different neuronal assemblies can form, thereby providing a substrate to construct dynamic motor output zones.

The human primary motor cortex is involved in early aspects of motor skill acquisition, specifically early stages of motor consolidation (Muellbacher *et al.* 2002b). In these experiments increases in MEP amplitude are associated with skill acquisition, and MEP amplitudes return to their baseline amplitude after subjects had acquired the new skill. Training does not induce increases in MEP amplitude once the task is over-learned (Pascual-Leone *et al.* 1994; Muellbacher *et al.* 2002b).

Effect of practice on motor maps

TMS provides measures of cortical motor maps. In one study, it was shown that such cortical maps of muscles that support the training task increase (Pascual-Leone *et al.* 1994). During a serial reaction-time test, subjects performed a motor sequence repeatedly and subsequently developed implicit knowledge about this particular sequence. Motor maps of the muscle involved in the performance of the sequence increase until explicit knowledge is achieved, after which the maps return to their baseline (Pascual-Leone *et al.* 1994). Similar to the mechanisms discussed for training induced increases in MEP amplitudes (see above), this result illustrates the rapid functional reorganization of motor output zones associated with learning. Analogous to the return of MEP amplitudes to baseline once a task is over-learned, motor maps decrease with the transfer of knowledge from an implicit to an explicit state.

Because the absolute size of the motor map depends on the excitability and number of neurons that project to the target muscle but also on the stimulus intensity of the TMS pulse, the absolute size of the map is of limited value. Whereas using low-intensity TMS may result in an underestimation of the extent of the motor map because less excitable neurons will not be activated, using high-intensity TMS may result in an overestimation of motor maps because the stimulus current will spread to other areas (Thickbroom *et al.* 1998). Therefore, the changes in the topography of motor maps induced by experimental manipulation may be a more meaningful measurement.

In addition to measuring MEP amplitudes, which provide information on changes in M1 excitability or number of activated pyramidal tract neurons, motor maps also provide information about the topography of these changes. For example, in proficient blind Braille readers, the representation of the muscle involved in reading Braille was significantly larger in the reading hand than in the nonreading hand, or in either hand of the blind controls who used Braille reading nonproficiently. Conversely, the representation of a muscle not involved in the reading task was significantly smaller in the reading hand than in the nonreading hand, or in either hand of the controls. These results suggest that the cortical representation of the reading finger in proficient Braille readers is enlarged at the expense of the representation of other fingers (Pascual-Leone *et al.* 1993). In patients with phantom limb pain, motor mapping provided information on changes in the center of gravity of face and arm representations in M1 which correlated with the magnitude of painful phenomena (Karl *et al.* 2001).

Patients with upper limb paresis due to brain infarction who underwent a neurorehabilitative training program showed an increase in the number of cortical sites from where MEPs of the paretic hand could be elicited (Traversa *et al.* 1997; Liepert *et al.* 1998b; Wittenberg *et al.* 2003), indicating that the cortical representation of the target muscle (motor map) enlarged in size or increased in excitability compared to the pretraining mapping. A shift of the motor map (Liepert *et al.* 1998b; Wittenberg *et al.* 2003) or the increased number of abnormal locations of cortical stimulation sites (Traversa *et al.* 1997) indicates that cortical motor output zones may have expanded into the adjacent spared cortex, involving cortical areas previously not dedicated to this muscle, similar to the findings in M1 of nonhuman primates after lesion (Nudo and Milliken 1996). Directionally consistent changes in localization of motor maps could reveal true cortical reorganization (Karl *et al.* 2001) while inconsistent localization changes across different subjects are more difficult to interpret (Wolf *et al.* 2004).

Effect of practice on paired-pulse measures (SICI)

Derived from animal data, one of the main mechanisms that has been suggested for mediating reorganization in M1 involves the unmasking of existing, but latent, horizontal connections (for review see Sanes and Donoghue 2000). It relies on the previously discussed concept that M1 contains multiple overlapping motor representations (Donoghue *et al.* 1992; Schieber and Hibbard 1993; Sanes *et al.* 1995) functionally connected through an extensive horizontal network (Huntley and Jones 1991). One mechanism of rapid plastic changes in the motor cortex is the removal of local GABAergic inhibition (Jacobs and Donoghue 1991; Hess *et al.* 1996).

In humans, short-interval intracortical inhibition (SICI) using a paired-pulse TMS technique provides an electrophysiological means to evaluate a form of intracortical inhibition that is influenced by GABAergic mechanisms (cf. Chapters 11, 13, and 14).

Corresponding to the data from animal studies, SICI decreases in muscles acting as agonist and synergist in a motor task (Liepert *et al.* 1998a; Buetefisch *et al.* 2005). Specifically, nonselective training of thumb movements in an abduction direction (Figure 18.2a) results in decreases in SICI in muscles supporting the training movement either as an agonist (abductor pollicis brevis muscle; APB) or synergist (fourth dorsal interosseus muscle; 4DIO, Figure 18.2b). In contrast, when the subject was asked to abduct the thumb but consciously inhibit the synergistic muscle (4DIO, Figure 18.2a), SICI decreases in the agonist muscle but increases in

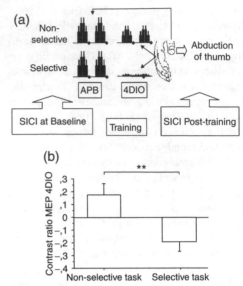

Fig. 18.2 (a) Experimental set-up for evaluating the effect of different training tasks on short-interval intracortical inhibition (SICI) in the fourth dorsal interosseus (4DIO) muscle. In the nonselective task, the abductor pollicis brevis (APB) muscle that mediates thumb movements in an abduction direction (open arrow) acts as agonist while the 4DIO, a muscle usually co-activated during the abduction of the thumb acts as synergist (indicated by the schematic drawing on electromyogram (EMG) bursts with each movement in APB and 4DIO). In the selective task, subjects abducted the thumb (open arrow) but were asked to consciously inhibit the 4DIO by means of EMG feedback, indicated by the lack of EMG burst in 4DIO. (b) Effect of training task on SICI of 4DIO. The effect of the selective and nonselective tasks on SICI of 4DIO was expressed by calculating the contrast ratio between the conditioned MEP amplitudes at baseline and at the end of the training (post-training) for 4DIO. The following formula was used: (conditioned MEP_{4DIO} post-training – conditioned MEP_{4DIO} baseline)/(conditioned MEP_{4DIO} post-training + conditioned MEP_{4DIO} baseline). Training of the selective task, but not the non-selective task resulted in an increase in SICI as indicated by the negative MEP_{4DIO} contrast ratio. Data are means and SE. $**P = 0.01$. Modified after Buetefisch *et al.* (2005).

the consciously inhibited muscle, which no longer acts as a synergist (Buetefisch *et al*. 2005).

Effect of practice on behavioral measures

The effects of motor practice and action observation on various measures of cortical excitability have been evaluated in many different protocols. In one of them, developed to study a form of use-dependent plasticity in humans, TMS of M1 was used to evoke isolated and directionally consistent thumb movements (baseline direction) (Classen *et al*. 1998b; Buetefisch *et al*. 2000; Stefan *et al*. 2005). Subsequently, voluntary thumb movements were practiced or observed in a direction opposite to the baseline direction for 30 min, after which TMS-evoked movements were in or near the recently practiced movement direction (Figure 18.3). This form of use-dependent

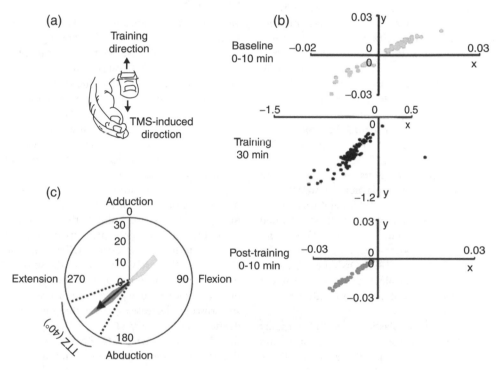

Fig. 18.3 (a) The direction of TMS-evoked or voluntary movement was derived from the first-peak acceleration in the two major axes of the movement (extension/flexion and abduction/adduction) measured by a two-dimensional accelerometer mounted on the dorsum of the proximal phalanx of the thumb. (b) Peak acceleration in the two main movement axes y (extension/flexion) and x (abduction/adduction) of a single representative subject shown to illustrate the main effect of training. Before training (baseline), TMS evokes predominantly flexion and adduction thumb movements (*upper*: 60 trials, 10 min). Training consisted of repetitive stereotyped brisk thumb movements in an extension and abduction direction (*middle*: 180 representative movements, 30 min). Post-training, the direction of TMS-evoked thumb movements changed from the baseline direction to the trained direction (*lower*: 60 trials, 10 min). (c) Circular frequency histogram of this subject. Baseline TMS-induced movement directions are a combination of flexion and adduction (band in upper right quadrant). Training movements were performed in a direction approximately opposite to baseline (arrow). The mean training direction is at the centre of the training target zone (TTZ, dotted lines, mean angle ± 20°). The scale indicates the number of TMS-evoked movements that fall in each 10° bin. TMS-induced movement directions after training largely fall within the TTZ, close to a 180° change from the baseline direction. Modified after Buetefisch *et al*. (2000).

plasticity occurs predominantly in the motor cortex as demonstrated by experiments using TES and TMS (Classen *et al.* 1998b).

Mechanisms of underlying practice-induced plasticity (use-dependent plasticity)

The mechanisms underlying use-dependent plasticity as tested in the previously described protocol in humans were studied with TMS in combination with different CNS-active drugs (Buetefisch *et al.* 2000) that interfere with synaptic plasticity (Riches and Brown 1986; Wong *et al.* 1988): (i) lorazepam, a drug that enhances GABA$_A$ receptor function (Macdonald and Kelly 1995) and blocks the induction of long-term potentiation (LTP) (Riches and Brown 1986); (ii) dextromethorphan, a drug that blocks *N*-methyl-D-aspartate (NMDA) receptors (Wong *et al.* 1988) required for LTP in the motor cortex (Wong *et al.* 1988; Hess *et al.* 1996) and experience-dependent plasticity in somatosensory cortex (Rema *et al.* 1998); and (iii) lamotrigine, a drug that blocks voltage-activated sodium and calcium channels (Lees and Leach 1993) without affecting LTP induction (Xiong and Stringer 1997; Otsuki *et al.* 1998). Under this protocol, blockade of use-dependent plasticity by a specific drug identifies a contributing mechanism.

Use-dependent plasticity is substantially reduced by dextromethorphan and lorazepam but not by lamotrigine (Figure 18.4). These results are consistent with animal studies as they identify NMDA receptor activation and GABAergic inhibition as mechanisms operating in use-dependent plasticity in the intact human motor cortex. Because LTP in the motor cortex requires activation of NMDA receptors (Hess *et al.* 1996) and because down-regulation of the inhibitory neurotransmitter GABA facilitates LTP in motor cortex slices (Hess and Donoghue 1994; Hess *et al.* 1996), these pharmacological experiments suggest the involvement of an activity-dependent LTP-like mechanism in use-dependent plasticity of human motor cortex similar to LTP and use-dependent plasticity of the motor system in animal preparations.

Modulation of use-dependent changes in TMS measures

Because use-dependent plasticity may play a beneficial role in the functional recovery following injury to the CNS (Buetefisch *et al.* 1995; Nudo and Milliken 1996), and in motor learning (Donoghue *et al.* 1996), enhancing this process would be of significant clinical interest.

Anesthesia

Similar to the expansion of representation of adjacent forelimb or eyelid following transsection of the facial nerve in rats (Sanes *et al.* 1988; Donoghue *et al.* 1990), transient deafferentation of the forearm and hand induced by a blood pressure cuff that is inflated above systolic blood pressure across the elbow results in a rapid increase of the motor cortical output to muscles proximal to the ischemic block (Brasil-Neto *et al.* 1993). This indicates that distal arm deafferentation could result in a relative expansion of the representation of the nearby proximal arm. Consistent with this view, SICI as measured with paired-pulse TMS is decreased in motor cortical representations of muscles proximal to the ischemic block (Ziemann *et al.* 1998b) and the levels of GABA in the sensorimotor cortex contralateral to ischemic nerve block, as measured with magnetic resonance spectroscopy, decrease rapidly within minutes of deafferentation (Levy *et al.* 2002), suggesting that the rapid removal of GABA-related inhibition is a mechanism also involved in this type of plasticity.

These findings led to the proposal that disinhibition within the upper arm representation could result in behavioral gains, for example after motor training of brisk elbow flexion movements. It was shown that motor training of the proximal arm during deafferentation of the distal forearm and hand results in improved motor performance with execution of brisk elbow flexions and a remarkable increase in motor output of the cortical representation of the muscle involved in motor training (Ziemann *et al.* 2001). Similarly, preliminary evidence showed that brachial plexus anesthesia of the nerves innervating the proximal arm in chronic stroke patients enhances the beneficial

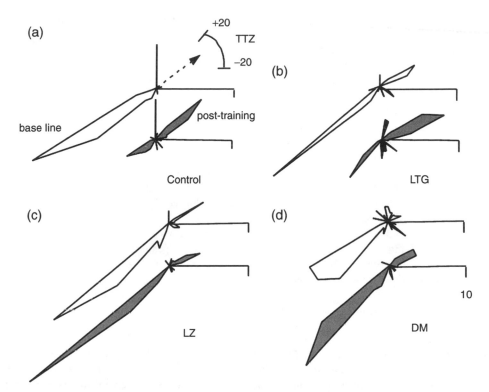

Fig. 18.4 Mechanisms of use-dependent plasticity. Circular histograms show the effects of drugs (lamotrigine, LTG; lorazepam, LZ; dextromethorphan, DM) on directional distribution of TMS-evoked thumb movements in a single subject at baseline (*upper*, white) and post-training (*lower*, gray). Frequencies are plotted on the same scale. Directions are grouped in bins of 10°. Mean training angle (arrow) and training target zone (TTZ) for all conditions are shown in the upper left diagram (Control). In the Control and LTG conditions, TMS-evoked movements at baseline were mainly in the extension/abduction direction. Post-training, the majority of TMS-evoked thumb movements occurred in TTZ, in the flexion/adduction direction. LZ and DM blocked this training effect. TMS-evoked thumb movements remained in the extension/abduction direction after training. Modified after Buetefisch *et al.* (2000).

effects of motor training on hand function and increases the excitability of the motor cortical representation of hand muscles involved in the motor training (Muellbacher *et al.* 2002a).

In healthy volunteers, ischemic deafferentation of one hand also results in increased corticomotor excitability targeting muscles in the other hand (Werhahn *et al.* 2002a). This phenomenon was documented in a similar way in the somatosensory system leading to improvements in tactile spatial acuity in one hand with anesthesia of the other hand (Werhahn *et al.* 2002b). Therefore, in both motor and somatosensory domains, unilateral hand anesthesia results

in a clear increase in cortical excitability in the motor and somatosensory cortices ipsilateral to ischemic hand deafferentation. The most likely explanation for these findings is a modulation of interhemispheric interactions (Werhahn *et al.* 2002a). This particular anesthetic procedure was applied to the intact hand of patients with chronic stroke in an attempt to ameliorate functions of the paretic hand. Initial experiments reported encouraging results in the motor (Floel *et al.* 2004) and somatosensory (Voller *et al.* 2006) domains, characterized by performance improvements in the paretic hand with anesthesia of the intact hand.

Somatosensory stimulation

Somatosensory input is required for skill acquisition (Pavlides *et al.* 1993) and accurate motor control (Pearson 2000). In healthy subjects, somatosensory stimulation, which activates group Ia large muscle afferents, group Ib afferents from Golgi organs, group II afferents from slow and rapidly adapting skin afferents, and cutaneous afferent fibers results in an increase in motor cortical excitability of body part representations that control the stimulated body part (Hamdy *et al.* 1998a,b; Ridding *et al.* 2000; Kaelin-Lang *et al.* 2002; Luft *et al.* 2002; Kobayashi *et al.* 2003) and in reorganization of the motor and somatosensory cortices (Golaszewski *et al.* 2004; Wu *et al.* 2005). Somatosensory stimulation elicits task-related increases in functional magnetic resonance imaging activity that outlast the stimulation period in M1 and in the primary somatosensory cortex (Golaszewski *et al.* 2004; Wu *et al.* 2005). Direct connections between S1 and M1 could provide the anatomical substrate for the influence of electrical somatosensory stimulation on motor cortical organization and on TMS measures (Wu and Kaas 2002). In the human motor cortex, peripheral nerve stimulation results in increased motor cortical excitability of the stimulated body part that outlasts the stimulation period (Hamdy *et al.* 1998a,b; Stefan *et al.* 2000; Ridding and Taylor 2001; Kaelin-Lang *et al.* 2002; Luft *et al.* 2002). For example, Stefan *et al.* (2000) showed that peripheral nerve stimulation time-locked with TMS to the contralateral motor cortex induces an increment in motor cortical excitability, as demonstrated by larger MEP amplitudes (cf. Chapter 16). In a different protocol, Kaelin-Lang *et al.* (2002) showed that a period of 2 h of somatosensory stimulation results in characteristic changes in corticomotor excitability that develop rapidly, persist for ~60 min, and have topographic specificity. Similarly, prolonged ulnar and radial nerve stimulation lead to enlarged MEPs of the first dorsal interosseous muscle, and to a lesser extent also of other hand muscles (Ridding and Taylor 2001). This effect takes place at least in part at the level of motor cortex, and it is blocked by lorazepam but not by dextromethorphan or placebo (Kaelin-Lang *et al.* 2002).

Altogether, this evidence supports the hypothesis of GABAergic involvement as an operating mechanism.

Somatosensory stimulation, in addition to influencing cortical excitability, modulates motor behavior, for example in stroke patients (Johansson *et al.* 1993; Powell *et al.* 1999; Wong *et al.* 1999; Conforto *et al.* 2002; Dobkin 2004) in whom application to the paretic hand improves aspects of motor performance for variable periods of time (Conforto *et al.* 2002; Struppler *et al.* 2003a,b; Wu 2006). A recent study showed that somatosensory stimulation administered immediately preceding motor training enhances the beneficial effects of training on use-dependent plasticity tested with TMS (Sawaki *et al.* 2006) and can enhance learning of activities of daily living in patients with chronic stroke (Celnik *et al.* in press).

Neuromodulators

Use-dependent plasticity may contribute to the recovery of motor function after injury to the brain (Nudo 1999). This led to increased interest in developing strategies to enhance these processes. Combining specific training with pharmacological agents represents another way to accomplish this goal. Dextro-amphetamine (AMPH), a drug that exerts its effect through the presynaptic release of the monoamines norepinephrine, dopamine, and serotonin, and inhibition of their reuptake from the synaptic cleft (Creese and Iversen 1975; Boyeson and Feeney 1990; Goldstein 1993; Boyeson *et al.* 1994), and scopolamine, a drug that exerts its effect through the blockade of muscarinic receptors, enhance use-dependent plasticity in M1 (Buetefisch *et al.* 2002; Sawaki *et al.* 2002a,b). TMS of M1 was used to evoke isolated and directionally consistent thumb movements (baseline direction, see 'Effect of practice on behavioral measures' above). Voluntary thumb movements were then practiced in a direction opposite to the baseline direction for 30 min in blocks of 5 min duration. Following each training block of 5 min, TMS of the motor cortex was used to evoke 10 isolated thumb movements. At the end of the 30 min training TMS-evoked movements were recorded for 60 min.

The endpoint measure of the study was the magnitude of training-induced directional changes in TMS-evoked thumb movements.

Under AMPH, the training time required to elicit directional changes in TMS-evoked movement directions is consistently reduced to 5–10 min as compared to 25 min in the placebo condition, indicating that AMPH accelerates the induction of use-dependent plasticity (Figure 18.5). This conclusion is underscored by

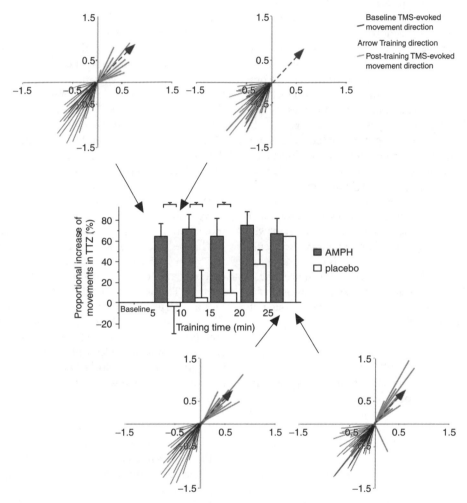

Fig. 18.5 Modulation of use-dependent plasticity. Effect of D-amphetamine (AMPH) on training time required eliciting directional changes in TMS-evoked thumb movements. Bar graph shows the group data of six subjects during the 30 min of training. After completion of each 5 min block of training, direction of 10 evoked thumb movements were recorded. Training for 5–10 min sufficed to induce a proportional increase of >60% in the TMS evoked movements falling in the training target zone (TTZ) in the AMPH condition (dark bars) while training effects develop more slowly in the placebo condition (open bars) (mean ± SE). The other diagrams show the effects of AMPH on directional distribution of TMS-evoked thumb movements in a single representative subject after 5 min (*top*) and 25 min (*bottom*) of training. Note that 5 min of training resulted in substantial directional changes in the AMPH condition but not in the placebo condition (*top right*). By 25 min of training similar directional changes were seen in both conditions (*bottom, left and right*). Directions are grouped in bins of 10°. Mean training angle is indicated by the dotted arrow. Modified after Buetefisch *et al.* (2002).

the findings of two additional experiments using a similar experimental design (Buetefisch *et al.* 2002; Sawaki *et al.* 2002b). First, regardless of total training time (10 or 30 min), when pre-medicated with AMPH, training-induced directional changes are long-lasting (>30 min). Second, this strategy enhances the training effects in a subgroup of subjects unresponsive to training alone. Following completion of motor training, these changes last for 10–15 min in the placebo condition but for >30 min in the AMPH condition. Similar effects are accomplished using dopaminergic agents (Floel *et al.* 2005; Meintzschel and Ziemann in press).

Directional changes are accompanied by increases in MEP amplitudes of the muscles supporting the training movements. Concurrent with the duration of the directional changes, MEP training agonist amplitudes remain high for >30 min in the AMPH condition. Therefore, AMPH substantially prolongs the duration of use-dependent plasticity, an effect consistent with the action of this drug on memory (Reus *et al.* 1979; Soetens *et al.* 1995). It is conceivable that this drug enhances recovery of function after cortical lesions (Feeney *et al.* 1982; Crisostomo *et al.* 1988; Walker-Batson *et al.* 1995) through its effects on use-dependent plasticity.

Cortical stimulation

Several reports demonstrated that cortical stimulation can enhance the beneficial effects of motor training on performance, cortical plasticity and motor cortical excitability (Muellbacher *et al.* 2000; Buetefisch *et al.* 2004; Hummel and Cohen 2005; Hummel *et al.* 2005; Khedr *et al.* 2005). Low-frequency repetitive TMS at a rate of 0.1 Hz applied to the motor representation of muscles proximal to a deafferented forearm enhances deafferentation-induced cortical reorganization (Ziemann *et al.* 1998a). In the protocol described above (Figure 18.3; 'Effect of practice on behavioral measures'), use-dependent plasticity can be enhanced by concomitant application of TMS in synchrony with the thumb-training movements (Buetefisch *et al.* 2004) in a way similar to the *in vitro* experiments in which stimulation of cortical afferents was paired with depolarization of the synaptic target neuron in a specific temporal relationship (Baranyi and Szente 1987; Baranyi *et al.* 1991). This effect appears to be more prominent if TMS is applied synchronously to the training of the motor cortex involved in the training movements (Buetefisch *et al.* 2004) (Figure 18.6).

In healthy volunteers, training coupled with TMS at 0.1 Hz time-locked to the training movements and applied to the contralateral motor cortex increased the longevity of the plastic changes when compared to training alone or to TMS at 0.1 Hz that was applied to the motor cortex at random (Buetefisch *et al.* 2004). In contrast, TMS at 0.1 Hz applied to the ipsilateral motor cortex abolished plastic changes. Therefore, use-dependent encoding of a motor memory can be enhanced by synchronous Hebbian stimulation (Hebb 1949) of the 'training' motor cortex and blocked by stimulation of the homologous opposite motor cortex, a result relevant to studies of cognitive and physical rehabilitation.

Anodal transcranial direct current stimulation (tDCS) appears to exert effects comparable to those of excitatory TMS when applied to cortical regions engaged in a practice or learning

---

Fig. 18.6 (a) Schematic diagram of the experimental set-up during training. TMS application to motor cortex was triggered by the electromyographic (EMG) activity of the muscle supporting the training movement of the thumb (training agonist). In the subject illustrated in this figure, the training direction was flexion; therefore the flexor pollicis brevis (FPB) operated as training agonist during training movements. EMG activity was recorded from the FPB. The potentiometer was adjusted in such a way that TMS was triggered when the EMG amplitude of the training agonist (FPB) was 10–20% of the maximal EMG amplitude during a ballistic movement. Accordingly, the TMS to the contralateral motor cortex occurred within the first half of the EMG burst that was generated by the training agonist. EPB, extensor pollicis brevis. (b) Modulation of use-dependent plasticity by TMS. Diagrams show the TMS effects on directional distribution of TMS-evoked thumb movements in a single subject. Directions of TMS-evoked thumb movements are shown in pairs of circular histograms, baseline (*upper*, white) and

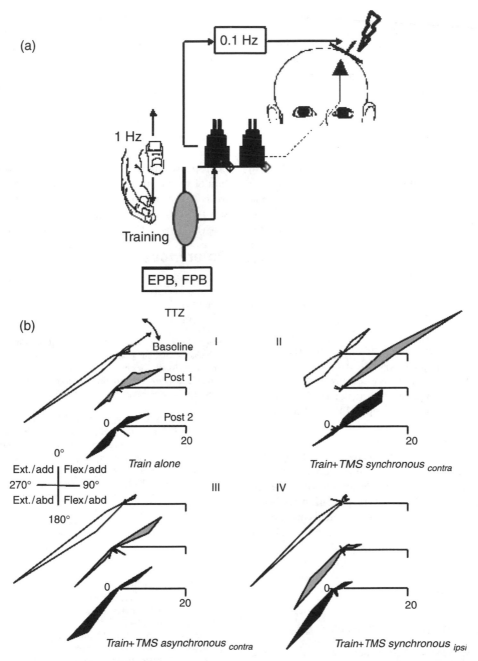

(a)

0.1 Hz

1 Hz

Training

EPB, FPB

(b)

TTZ

I

II

Baseline

Post 1

Post 2

0

20

0°

Ext./add | Flex/add
270° ——|—— 90°
Ext./abd | Flex/abd

180°

Train alone

0

20

Train+ TMS synchronous contra

III

IV

0

20

Train+ TMS asynchronous contra

0

20

Train+ TMS synchronous ipsi

post-training (*lower*, gray). Frequencies are plotted on the same scale. Directions are grouped in bins of 10°. *Upper left*: mean training angle (arrow) and training target zone (TTZ) for all conditions (Train alone (I)). In Train alone, Train + TMS synchronous$_{contra}$ (II), and Train + TMS asynchronous$_{contra}$ (III), TMS-evoked movements at baseline were mainly in the extension/abduction direction. Post-training, the majority of TMS-evoked thumb movements occurred in the TTZ, in the flexion/adduction direction, with Train + TMS synchronous$_{contra}$ having the greatest effect. Train + TMS synchronous$_{ipsi}$ blocked the training effect: TMS-evoked thumb movements remained in the extension/abduction direction after training (IV). Modified after Buetefisch et al. (2004).

task (cf. Chapter 17). Several reports have recently documented performance improvements in visuo-motor coordination (Antal *et al.* 2004a,b), implicit motor learning (Nitsche *et al.* 2003), and probabilistic classification learning (Kincses *et al.* 2004). Altogether, it appears that tDCS and TMS may represent useful tools to modulate motor cortical excitability in regions engaged in practice or learning tasks. This approach has also been used in patients with chronic stroke. tDCS can be applied continuously and safely for up to 30 min (Ardolino *et al.* 2005; Iyer *et al.* 2005; Hummel *et al.* 2005; Nitsche *et al.* 2005). Anodal tDCS applied to the affected hemisphere of chronic stroke patients, preferably in association with motor training, appears to benefit aspects of motor performance (Hummel and Cohen 2005; Hummel *et al.* 2005). Improvement of motor performance in the paretic hand is associated with increments in MEP recruitment curves to TMS and decreased SICI, suggesting the involvement of GABAergic mechanisms (Hummel and Cohen 2005; Hummel *et al.* 2005).

Finally, down-regulation of activity in one motor cortex influences corticomotor excitability in the opposite motor cortex. As described above, 0.1 Hz TMS applied to M1 time-locked to the training movement generated by the opposite M1 blocked training-induced increases in MEP amplitudes and changes in the direction of TMS evoked movements (Buetefisch *et al.* 2004; Figure 18.6B, IV). Several reports have now demonstrated that 1 Hz stimulation applied to one hemisphere results in increased corticomotor excitability in the opposite hemisphere (Plewnia *et al.* 2003; Schambra *et al.* 2003), an effect with beneficial behavioral implications in healthy subjects (Kobayashi *et al.* 2004), and in stroke patients (Mansur *et al.* 2005).

Concluding remarks

In summary, there is now extensive evidence that practice tends to enhance motor excitability in the cortical representations involved in the practice task. These changes may be accompanied by down-regulation of activity in nearby body part representations within the same hemisphere and in homonymous regions of the opposite hemisphere, likely mediated by interhemispheric interactions (Werhahn *et al.* 2002a).

The specific mechanisms that are up- and down-regulated by practice are incompletely understood and remain under investigation but appear to involve alpha- and beta-adrenergic, cholinergic, GABAergic, and dopaminergic neurotransmission (Ziemann *et al.* 2006). Overall, this emerging evidence points to a more extensive influence of practice (as well as disuse) on a distributed network of cortical representations within multiple regions of both cerebral hemispheres. This view has led to the formulation of various interventional strategies geared to enhance training effects by cortical or soma-tosensory stimulation in health and disease (Buetefisch 2004; Ward and Cohen 2004).

References

Amassian VE, Quirk GJ, Stewart M (1990). A comparison of corticospinal activation by magnetic coil and electrical stimulation of monkey motor cortex. *Electroencephalography and Clinical Neurophysiology 77*, 390–401.

Antal A, Kincses TZ, Nitsche MA, Bartfai O, Paulus W (2004a). Excitability changes induced in the human primary visual cortex by transcranial direct current stimulation: direct electrophysiological evidence. *Investigative Ophthalmology and Visual Science 45*, 702–707.

Antal A, Nitsche MA, Kincses TZ, Kruse W, Hoffmann KP, Paulus W (2004b). Facilitation of visuo-motor learning by transcranial direct current stimulation of the motor and extrastriate visual areas in humans. *European Journal of Neuroscience 19*, 2888–2892.

Ardolino G, Bossi B, Barbieri S, Priori A (2005). Non-synaptic mechanisms underlie the after-effects of cathodal transcutaneous direct current stimulation of the human brain. *Journal of Physiology 568*, 653–663.

Baranyi A, Szente MB (1987). Long-lasting potentiation of synaptic transmission requires postsynaptic modifications in the neocortex. *Brain Research 423*, 378–384.

Baranyi A, Szente MB, Woody CD (1991). Properties of associative long-lasting potentiation induced by cellular conditioning in the motor cortex of conscious cats. *Neuroscience 42*, 321–334.

Boyeson MG, Feeney DM (1990). Intraventricular norepinephrine facilitates motor recovery following sensorimotor cortex injury. *Pharmacology, Biochemistry and Behavior 35*, 497–501.

Boyeson MG, Jones JL, Harmon RL (1994). Sparing of motor function after cortical injury A new perspective on underlying mechanisms. *Archives of Neurology 51*, 405–414.

Brasil-Neto JP, Valls-Sole J, Pascual-Leone A, *et al.* (1993). Rapid modulation of human cortical motor outputs following ischaemic nerve block. *Brain 116*, 511–525.

Buetefisch CM (2004). Plasticity in the human cerebral cortex: lessons from the normal brain and from stroke. *Neuroscientist 10*, 163–173.

Buetefisch C, Hummelsheim H, Denzler P, Mauritz KH (1995). Repetitive training of isolated movements improves the outcome of motor rehabilitation of the centrally paretic hand. *Journal of the Neurological Sciences 130*, 59–68.

Buetefisch CM, Davis BC, Wise SP, *et al.* (2000). Mechanisms of use-dependent plasticity in the human motor cortex. *Proceedings of the National Academy of Sciences of the USA 97*, 3661–3665.

Buetefisch CM, Davis BC, Sawaki L, *et al.* (2002). Modulation of use-dependent plasticity by d-amphetamine. *Annals of Neurology 51*, 59–68.

Buetefisch CM, Netz J, Wessling M, Seitz RJ, Homberg V (2003). Remote changes in cortical excitability after stroke. *Brain 126*, 470–481.

Buetefisch CM, Khurana V, Kopylev L, Cohen LG (2004). Enhancing encoding of a motor memory in the primary motor cortex by cortical stimulation. *Journal of Neurophysiology 91*, 2110–2116.

Buetefisch CM, Boroojerdi B, Chen R, Battaglia F, Hallett M (2005). Task-dependent intracortical inhibition is impaired in focal hand dystonia. *Movement Disorders 20*, 545–551.

Celnik P, Hummel F, Harris-Love M, Wolk R, Cohen LG. Somatosensory stimulation enhances the effects of training functional hand tasks in patients with chronic storke. *Archives of Physical Medicine and Rehabilitation* in press.

Classen J, Knorr U, Werhahn KJ, *et al.* (1998a). Multimodal output mapping of human central motor representation on different spatial scales. *Journal of Physiology 512*, 163–179.

Classen J, Liepert J, Wise SP, Hallett M, Cohen LG (1998b). Rapid plasticity of human cortical movement representation induced by practice. *Journal of Neurophysiology 79*, 1117–1123.

Conforto AB, Kaelin-Lang A, Cohen LG (2002). Increase in hand muscle strength of stroke patients after somatosensory stimulation. *Annals of Neurology 51*, 122–125.

Creese I, Iversen SD (1975). The pharmacological and anatomical substrates of the amphetamine response in the rat. *Brain Research 83*, 419–436.

Crisostomo EA, Duncan PW, Propst M, Dawson DV, Davis JN (1988). Evidence that amphetamine with physical therapy promotes recovery of motor function in stroke patients. *Annals of Neurology 23*, 94–97.

Day BL, Thompson PD, Dick JP, Nakashima K, Marsden CD (1987). Different sites of action of electrical and magnetic stimulation of the human brain. *Neuroscience Letters 75*, 101–106.

Di Lazzaro V, Oliviero A, Profice P, *et al.* (1999). Direct demonstration of interhemispheric inhibition of the human motor cortex produced by transcranial magnetic stimulation. *Experimental Brain Research 124*, 520–524.

Dobkin BH (2004). Strategies for stroke rehabilitation. *Lancet Neurology 3*, 528–536.

Donoghue JP, Suner S, Sanes JN (1990). Dynamic organization of primary motor cortex output to target muscles in adult rats. II. Rapid reorganization following motor nerve lesions. *Experimental Brain Research 79*, 492–503.

Donoghue JP, Leibovic S, Sanes JN (1992). Organization of the forelimb area in squirrel monkey motor cortex: representation of digit, wrist, and elbow muscles. *Experimental Brain Research 89*, 1–19.

Donoghue JP, Hess G, Sanes JN (1996). Substrates and mechanisms for learning in motor cortex. In: JR Bloedel, TJ Ebner, SP Wise (eds), *Acquisition and mechanisms for learning in motor cortex.* Cambridge, MA: MIT Press, 363–386.

Feeney DM, Gonzalez A, Law WA (1982). Amphetamine, haloperidol, and experience interact to affect rate of recovery after motor cortex injury. *Science 217*, 855–857.

Floel A, Nagorsen U, Werhahn KJ, *et al.* (2004). Influence of somatosensory input on motor function in patients with chronic stroke. *Annals of Neurology 56*, 206–212.

Floel A, Breitenstein C, Hummel F, *et al.* (2005). Dopaminergic influences on formation of a motor memory. *Annals of Neurology 58*, 121–130.

Fuhr P, Cohen LG, Dang N, *et al.* (1992). Physiological analysis of motor reorganization following lower limb amputation. *Electroencephalography and Clinical Neurophysiology 85*, 53–60.

Golaszewski SM, Siedentopf CM, Koppelstaetter F, *et al.* (2004). Modulatory effects on human sensorimotor cortex by whole-hand afferent electrical stimulation. *Neurology 62*, 2262–2269.

Goldstein LB (1993). Basic and clinical studies of pharmacologic effects on recovery from brain injury. *Journal of Neural Transplantation and Plasticity 4*, 175–192.

Hamdy S, Aziz Q, Rothwell JC, *et al.* (1998a). Recovery of swallowing after dysphagic stroke relates to functional reorganization in the intact motor cortex. *Gastroenterology 115*, 1104–1112.

Hamdy S, Rothwell JC, Aziz Q, Singh KD, Thompson DG (1998b). Long-term reorganization of human motor cortex driven by short-term sensory stimulation. *Nature Neuroscience 1*, 64–68.

Hebb DO (1949). *The organization of behavior. A neuropsychological theory.* New York: Wiley.

Hess G, Donoghue JP (1994). Long-term potentiation of horizontal connections provides a mechanism to reorganize cortical motor maps. *Journal of Neurophysiology 71*, 2543–2547.

Hess G, Aizenman CD, Donoghue JP (1996). Conditions for the induction of long-term potentiation in layer II/III horizontal connections of the rat motor cortex. *Journal of Neurophysiology 75*, 1765–1778.

Hummel F, Cohen LG (2005). Improvement of motor function with noninvasive cortical stimulation in a patient with chronic stroke. *Neurorehabilitation and Neural Repair 19*, 14–19.

Hummel F, Celnik P, Giraux P, *et al.* (2005). Effects of non-invasive cortical stimulation on skilled motor function in chronic stroke. *Brain 128*, 490–499.

Huntley GW, Jones EG (1991). Relationship of intrinsic connections to forelimb movement representations in monkey motor cortex: a correlative anatomic and physiological study. *Journal of Neurophysiology 66*, 390–413.

Iyer MB, Mattu U, Grafman J, Lomarev M, Sato S, Wassermann EM (2005). Safety and cognitive effect of frontal DC brain polarization in healthy individuals. *Neurology 64*, 872–875.

Jacobs KM, Donoghue JP (1991). Reshaping the cortical motor map by unmasking latent intracortical connections. *Science 251*, 944–947.

Johansson K, Lindgren I, Widner H, Wiklund I, Johansson BB (1993). Can sensory stimulation improve the functional outcome in stroke patients? *Neurology 43*, 2189–2192.

Kaelin-Lang A, Luft AR, Sawaki L, Burstein AH, Sohn YH, Cohen LG (2002). Modulation of human corticomotor excitability by somatosensory input. *Journal of Physiology 540*, 623–633.

Karl A, Birbaumer N, Lutzenberger W, Cohen LG, Flor H (2001). Reorganization of motor and somatosensory cortex in upper extremity amputees with phantom limb pain. *Journal of Neuroscience 21*, 3609–3618.

Khedr EM, Ahmed MA, Fathy N, Rothwell JC (2005). Therapeutic trial of repetitive transcranial magnetic stimulation after acute ischemic stroke. *Neurology 65*, 466–468.

Kincses TZ, Antal A, Nitsche MA, Bartfai O, Paulus W (2004). Facilitation of probabilistic classification learning by transcranial direct current stimulation of the prefrontal cortex in the human. *Neuropsychologia 42*, 113–117.

Kleim JA, Barbay S, Nudo RJ (1998). Functional reorganization of the rat motor cortex following motor skill learning. *Journal of Neurophysiology 80*, 3321–3325.

Kobayashi M, Ng J, Theoret H, Pascual-Leone A (2003). Modulation of intracortical neuronal circuits in human hand motor area by digit stimulation. *Experimental Brain Research 149*, 1–8.

Kobayashi M, Hutchinson S, Theoret H, Schlaug G, Pascual-Leone A (2004). Repetitive TMS of the motor cortex improves ipsilateral sequential simple finger movements. *Neurology 62*, 91–98.

Lees G, Leach MJ (1993). Studies on the mechanism of action of the novel anticonvulsant lamotrigine (Lamictal). using primary neurological cultures from rat cortex. *Brain Research 612*, 190–199.

Levy LM, Ziemann U, Chen R, Cohen LG (2002). Rapid modulation of GABA in sensorimotor cortex induced by acute deafferentation. *Annals of Neurology 52*, 755–761.

Liepert J, Classen J, Cohen LG, Hallett M (1998a). Task-dependent changes of intracortical inhibition. *Experimental Brain Research 118*, 421–426.

Liepert J, Miltner WH, Bauder H, et al. (1998b). Motor cortex plasticity during constraint-induced movement therapy in stroke patients. *Neuroscience Letters 250*, 5–8.

Luft AR, Kaelin-Lang A, Hauser TK, et al. (2002). Modulation of rodent cortical motor excitability by somatosensory input. *Experimental Brain Research 142*, 562–569.

Macdonald RL, Kelly KM (1995). Antiepileptic drug mechanisms of action. *Epilepsia 36*(Suppl. 2), S2–12.

Mansur CG, et al. (2005) A sham stimulation-controlled trial of rTMS of the unaffected hemisphere in stroke patients. *Neurology 64*, 1802–1804

Mercuri B, Wassermann EM, Manganotti P, Ikoma K, Samii A, Hallett M (1996). Cortical modulation of spinal excitability: an F-wave study. *Electroencephalography and Clinical Neurophysiology 101*, 16–24.

Muellbacher W, Ziemann U, Boroojerdi B, Hallett M (2000). Effects of low-frequency transcranial magnetic stimulation on motor excitability and basic motor behavior. *Clinical Neurophysiology 111*, 1002–1007.

Muellbacher W, Ziemann U, Boroojerdi B, Cohen L, Hallett M (2001). Role of the human motor cortex in rapid motor learning. *Experimental Brain Research 136*, 431–438.

Muellbacher W, Richards C, Ziemann U, et al. (2002a). Improving hand function in chronic stroke. *Archives of Neurology 59*, 1278–1282.

Muellbacher W, Ziemann U, Wissel J, et al. (2002b). Early consolidation in human primary motor cortex. *Nature 415*, 640–644.

Nitsche MA, Schauenburg A, Lang N, et al. (2003). Facilitation of implicit motor learning by weak transcranial direct current stimulation of the primary motor cortex in the human. *Journal of Cognitive Neuroscience 15*, 619–626.

Nitsche MA, Seeber A, Frommann K, et al. (2005). Modulating parameters of excitability during and after transcranial direct current stimulation of the human motor cortex. *Journal of Physiology 568*, 291–303.

Nudo RJ (1999). Recovery after damage to motor cortical areas. *Current Opinion in Neurobiology 9*, 740–747.

Nudo RJ, Milliken GW (1996). Reorganization of movement representations in primary motor cortex following focal ischemic infarcts in adult squirrel monkeys. *Journal of Neurophysiology 75*, 2144–2149.

Nudo RJ, Jenkins WM, Merzenich MM (1990). Repetitive microstimulation alters the cortical representation of movements in adult rats. *Somatosensory and Motor Research 7*, 463–483.

Nudo RJ, Milliken GW, Jenkins WM, Merzenich MM (1996). Use-dependent alterations of movement representations in primary motor cortex of adult squirrel monkeys. *Journal of Neuroscience 16*, 785–807.

Otsuki K, Morimoto K, Sato K, Yamada N, Kuroda S (1998). Effects of lamotrigine and conventional antiepileptic drugs on amygdala- and hippocampal-kindled seizures in rats. *Epilepsy Research 31*, 101–112.

Pascual-Leone A, Cammarota A, Wassermann EM, Brasil-Neto JP, Cohen LG, Hallett M (1993). Modulation of motor cortical outputs to the reading hand of braille readers. *Annals of Neurology 34*, 33–37.

Pascual-Leone A, Grafman J, Hallett M (1994). Modulation of cortical motor output maps during development of implicit and explicit knowledge. *Science 263*, 1287–1289.

Pavlides C, Miyashita E, Asanuma H (1993). Projection from the sensory to the motor cortex is important in learning motor skills in the monkey. *Journal of Neurophysiology 70*, 733–741.

Pearson KG (2000). Plasticity of neuronal networks in the spinal cord: modifications in response to altered sensory input. *Progress in Brain Research 128*, 61–70.

Plewnia C, Lotze M, Gerloff C (2003). Disinhibition of the contralateral motor cortex by low-frequency rTMS. *Neuroreport 14*, 609–612.

Powell J, Pandyan AD, Granat M, Cameron M, Stott DJ (1999). Electrical stimulation of wrist extensors in poststroke hemiplegia. *Stroke 30*, 1384–1389.

Rema V, Armstrong-James M, Ebner FF (1998). Experience-dependent plasticity of adult rat S1 cortex requires local NMDA receptor activation. *Journal of Neuroscience 18*, 10196–10206.

Reus VI, Silberman E, Post RM, Weingartner H (1979). d-Amphetamine: effects on memory in a depressed population. *Biological Psychiatry 14*, 345–356.

Riches IP, Brown MW (1986). The effect of lorazepam upon hippocampal long-term potentiation. *Neuroscience Letters 24*(Suppl), S42.

Ridding MC, Taylor JL (2001). Mechanisms of motor-evoked potential facilitation following prolonged dual peripheral and central stimulation in humans. *Journal of Physiology 537*, 623–631.

Ridding MC, Brouwer B, Miles TS, Pitcher JB, Thompson PD (2000). Changes in muscle responses to stimulation of the motor cortex induced by peripheral nerve stimulation in human subjects. *Experimental Brain Research 131*, 135–143.

Rothwell JC, Thompson PD, Day BL, Boyd S, Marsden CD (1991). Stimulation of the human motor cortex through the scalp. *Experimental Physiology 76*, 159–200.

Sanes JN, Donoghue JP (2000). Plasticity and primary motor cortex. *Annual Review of Neuroscience 23*, 393–415.

Sanes JN, Suner S, Lando JF, Donoghue JP (1988). Rapid reorganization of adult rat motor cortex somatic representation patterns after motor nerve injury. *Proceedings of the National Academy of Sciences of the USA 85*, 2003–2007.

Sanes JN, Wang J, Donoghue JP (1992). Immediate and delayed changes of rat motor cortical output representation with new forelimb configurations. *Cerebral Cortex 2*, 141–152.

Sanes JN, Donoghue JP, Thangaraj V, Edelman RR, Warach S (1995). Shared neural substrates controlling hand movements in human motor cortex. *Science 268*, 1775–1777.

Sawaki L, Boroojerdi B, Kaelin-Lang A, et al. (2002a). Cholinergic influences on use-dependent plasticity. *Journal of Neurophysiology 87*, 166–171.

Sawaki L, Cohen LG, Classen J, Davis BC, Butefisch CM (2002b). Enhancement of use-dependent plasticity by D-amphetamine. *Neurology 59*, 1262–1264.

Sawaki L, Wu CW, Kaelin-Lang A, Cohen LG (2006). Effects of somatosensory stimulation on use-dependent plasticity in chronic stroke. *Stroke 37*, 246–247.

Schambra HM, Sawaki L, Cohen LG (2003). Modulation of excitability of human motor cortex (M1). by 1 Hz transcranial magnetic stimulation of the contralateral M1. *Clinical Neurophysiology 114*, 130–133.

Schieber MH, Hibbard LS (1993). How somatotopic is the motor cortex hand area? *Science 261*, 489–492.

Soetens E, Casaer S, D'Hooge R, Hueting JE (1995). Effect of amphetamine on long-term retention of verbal material. *Psychopharmacology (Berlin) 119*, 155–162.

Stefan K, Kunesch E, Cohen LG, Benecke R, Classen J (2000). Induction of plasticity in the human motor cortex by paired associative stimulation. *Brain 123*, 572–584.

Stefan K, Cohen LG, Duque J, et al. (2005). Formation of a motor memory by action observation. *Journal of Neuroscience 25*, 9339–9346.

Struppler A, Angerer B, Havel P (2003a). Modulation of sensorimotor performances and cognition abilities induced by RPMS: clinical and experimental investigations. *Clinical Neurophysiology 56*(Suppl), 358–367.

Struppler A, Havel P, Muller-Barna P (2003b). Facilitation of skilled finger movements by repetitive peripheral magnetic stimulation (RPMS) – a new approach in central paresis. *Neurorehabilitation 18*, 69–82.

Thickbroom GW, Sammut F, Mastaglia R (1998). Magnetic stimulation mapping of motor cortex: factors contributing to map area. *Electroencephalography and Clinical Neurophysiology 109*, 79–84.

Traversa R, Cicinelli P, Bassi A, Rossini PM, Bernardi G (1997). Mapping of motor cortical reorganization after stroke. A brain stimulation study with focal magnetic pulses. *Stroke 28*, 110–117.

Voller B, Floel A, Werhahn KJ, Ravindran S, Wu CW, Cohen LG (2006). Contralateral hand anesthesia transiently improves poststroke sensory deficits. *Annals of Neurology 59*, 385–388.

Walker-Batson D, Smith P, Curtis S, Unwin H, Greenlee R (1995). Amphetamine paired with physical therapy accelerates motor recovery after stroke. Further evidence. *Stroke 26*, 2254–9.

Ward NS, Cohen LG (2004). Mechanisms underlying recovery of motor function after stroke. *Archives of Neurology 61*, 1844–1848.

Werhahn KJ, Mortensen J, Kaelin-Lang A, Boroojerdi B, Cohen LG (2002a). Cortical excitability changes induced by deafferentation of the contralateral hemisphere. *Brain 125*, 1402–1413.

Werhahn KJ, Mortensen J, Van Boven RW, Zeuner KE, Cohen LG (2002b). Enhanced tactile spatial acuity and cortical processing during acute hand deafferentation. *Nature Neuroscience 5*, 936–938.

Wittenberg GF, Chen R, Ishii K, et al. (2003). Constraint-induced therapy in stroke: magnetic-stimulation motor maps and cerebral activation. *Neurorehabilitation and Neural Repair 17*, 48–57.

Wolf SL, Butler AJ, Campana GI, et al. (2004). Intra-subject reliability of parameters contributing to maps generated by transcranial magnetic stimulation in able-bodied adults. *Clinical Neurophysiology 115*, 1740–1747.

Wong AM, Su TY, Tang FT, Cheng PT, Liaw MY (1999). Clinical trial of electrical acupuncture on hemiplegic stroke patients. *American Journal of Physical Medicine and Rehabilitation 78*, 117–122.

Wong BY, Coulter DA, Choi DW, Prince DA (1988). Dextrorphan and dextromethorphan, common antitussives, are antiepileptic and antagonize N-methyl-D-aspartate in brain slices. *Neuroscience Letters 85*, 261–266.

Wu C (2006). Influence of somatosensory stimulation on paretic hand motor function in chronic stroke. *Archives of Physical Medicine and Rehabilitation 87*, 351–357.

Wu CW, Kaas JH (2002). The effects of long-standing limb loss on anatomical reorganization of the somatosensory afferents in the brainstem and spinal cord. *Somatosensory and Motor Research 19*, 153–163.

Wu CW, Van Gelderen P, Hanakawa T, Yaseen Z, Cohen LG (2005). Enduring representational plasticity after somatosensory stimulation. *NeuroImage 27*, 872–884.

Xiong ZQ, Stringer JL (1997). Effects of felbamate, gabapentin and lamotrigine on seizure parameters and excitability in the rat hippocampus. *Epilepsy Research 27*, 187–194.

Ziemann U, Corwell B, Cohen LG (1998a). Modulation of plasticity in human motor cortex after forearm ischemic nerve block. *Journal of Neuroscience 18*, 1115–1123.

Ziemann U, Hallett M, Cohen LG (1998b). Mechanisms of deafferentation-induced plasticity in human motor cortex. *Journal of Neuroscience 18*, 7000–7007.

Ziemann U, Muellbacher W, Hallett M, Cohen LG (2001). Modulation of practice-dependent plasticity in human motor cortex. *Brain 124*, 1171–1181.

Ziemann U, Meintzschel F, Korchounov A, Ilic TV (2006). Pharmacological modulation of plasticity in the human motor cortex. *Neurorehabilitation and Neural Repair 20*, 243–251.

The Motor-evoked Potential in Health and Disease

Eric M. Wassermann

The muscle twitch from motor cortex stimulation was the first phenomenon observed with TMS. Not only was this unambiguous evidence that the brain could be stimulated safely and painlessly through the head, but subsequent work demonstrated that it could be a quantitative measure of the state of a set of neurons in the CNS. It was soon clear that voluntary muscle activation or even the intention to move could enhance the motor-evoked potential (MEP). Later, the description of the cortical silent period and advent of paired-pulse techniques enabled the study of inhibitory processes and the effects of drugs, hormones, and diseases on cortical circuits. Changes observed in the motor cortex may reflect global alterations in cortical function and we can sometimes infer important facts about the mechanisms of disorders, for example migraine, and hormones with behavioral impacts that express themselves primarily through other regions. Additionally, the MEP promises to reveal highly specific information on the expression of the some of the genotypes associated with neurological and neurobehavioral disorders,

Motor-evoked potential conduction studies have enhanced our understanding of neurodevelopment in healthy children and those with developmental disorders. They have proven useful in the diagnosis of lesions in the central and proximal nerves and, while occupying a narrow diagnostic niche, provide a valuable tool for clinical neurophysiologists and anesthesiologists monitoring the integrity of central pathways. They have also been used to locate the motor cortex for neurosurgical planning.

Nearly all we know about the mechanisms of transcranial brain stimulation in its many forms has been inferred from this deceptively simple phenomenon, so distant from its source, and an understanding of its origin and the many factors that can affect it is important for any investigator who plans to use it.

CHAPTER 19

The MEP in clinical neurodiagnosis

Friedhelm Sandbrink

MEP parameters: methodology and physiology

Introduction

Transcranial stimulation of the cerebral cortex to elicit motor-evoked potentials (MEPs) is a noninvasive method for assessing the integrity of the central motor pathway function. An MEP may be defined as the electrical muscular response elicited by artificially stimulating the motor cortex or motor pathway above the spinal motor neuron. TMS was introduced by Barker *et al.* (1985) and since then has largely replaced transcranial electrical stimulation (TES) (Merton and Morton 1980) as a diagnostic clinical tool, mainly due to its much better tolerability for the awake subject. In TMS, a pulse of large electrical current flows through a coated coil placed on the scalp, inducing a transient magnetic field at right angles that penetrates readily into the superficial brain layers. This very brief magnetic field induces a secondary electric field within the cortex in a direction opposite to the originating current within the coil. The scalp and skull have only minimal impedance to the passage of magnetic pulses; therefore, much less electrical current is being induced in the scalp by TMS in comparison to TES, and contraction of scalp muscles and activation of pain receptors is minimal.

TMS is used in diagnosing and monitoring neurological disorders with upper motor neuron involvement and has a high sensitivity in detecting abnormalities of corticospinal pathways in various diseases (Di Lazzaro *et al.* 1999a). This chapter outlines the standard application of TMS as a neurodiagnostic tool, providing first an overview of the routine technique and MEP parameters, and subsequently describing the experience in neurological disorders.

Safety and side-effects of TMS

TMS using single pulses is safe if used with basic precautions. It has become a standard diagnostic tool for neurologists in many countries, with the exception of the USA, where its approval by the Food and Drug Administration (FDA) has been limited to peripheral nerve stimulation. The FDA considers stimulation of the CNS (including motor cortex) with single-pulse TMS (≤ 1 Hz) an investigational diagnostic procedure with 'nonsignificant risk' that requires an informed consent and tracking of adverse events.

In this chapter, only the safety aspects of single-pulse TMS as related to routine clinical use will be outlined, with more extensive review including repetitive (t) TMS elsewhere in this book, or as reviewed by Wassermann (1998). Contraindications to TMS are generally related to exposure to the magnetic field and therefore similar to those for magnetic resonance imaging (MRI) and apply to both the patient and the examiner. The large magnetic pulse may damage electronic devices, and metal objects will be subject to mechanical forces and may become hot. The subject needs to be asked specifically about the following exclusion criteria before proceeding with magnetic stimulation.

1. Implanted metal devices such as cardiac pacemakers or defibrillators, intrathecal drug delivery pumps, or spinal cord, vagus nerve, or similar stimulators. The risk of damage to the internal electronics of a programmable device is related to the distance between the stimulating coil and the device. Several TMS studies have been published in patients with deep brain stimulation of globus pallidus, thalamus, and subthalamic nucleus (Kumar *et al.* 1999; Chen *et al.* 2001; Cunic *et al.* 2002; Dauper *et al.* 2002; Kuhn *et al.* 2002, 2003; Pierantozzi *et al.* 2002; Molnar *et al.* 2005). In 11 patients with continuous intrathecal baclofen perfusion pumps, TMS of the cortex was performed up to maximal stimulator output with no evidence of interference on the drug delivery system (Auer *et al.* 1999), and also in another patient with baclofen pump (Siebner *et al.* 1998). While these studies indicate relative safety of TMS if the coil is not placed in the vicinity of such a device, its presence is considered a relative contraindication for routine TMS applications.

2. Acoustic devices such as cochlear implants.

3. Presence of intracranial metal such as aneurysm clips that might be dislodged by high-intensity TMS. It is prudent to ask about any prior neurosurgical procedure.

4. History of epileptic seizures. There are a few reports of seizures occurring at or shortly after single-pulse magnetic stimulation, mostly in patients with epilepsy (Homberg and Netz 1989; Hufnagel *et al.* 1990a; Classen *et al.* 1995), but also in a patient with multiple sclerosis (MS) (Haupts *et al.* 2004) and bipolar disease (Tharayil *et al.* 2005). One patient with a large middle cerebral artery infarction developed the first seizure during TMS and subsequently required anticonvulsant treatment for epilepsy (Homberg and Netz 1989). There is no report of single-pulse TMS inducing a seizure in a normal subject. The history of seizures is not an absolute contraindication for single-pulse TMS, and single and repetitive TMS techniques have been used successfully in the study and treatment of epilepsy (Hufnagel and Elger 1991; Theodore *et al.* 2002) and without worsening of epilepsy. A recent meta-analysis of epilepsy

patients estimated that the risk of a TMS-associated seizure ranges from 0.0 to 2.8% for single-pulse TMS, with higher risk for medically intractable epilepsy patients and after recent lowering of antiepileptic drugs (Schrader *et al.* 2004). The concern of triggering epileptiform activity is much greater in rTMS that may, depending on stimulation parameters, evoke seizures even in normal subjects, with published safety guidelines (Wassermann 1998).

Other safety concerns relate to the high-intensity impulse noise artifact associated with the discharge of the magnetic coil that is often placed in close proximity to the ear. No acoustic damage has been documented in humans from this acoustic click even after exposure for years (Pascual-Leone *et al.* 1992). Nevertheless, many investigators use hearing protectors including earplugs for the subjects (Rossini *et al.* 1994). The acoustic output depends on the coil design and manufacturer. Large round coils as used for routine TMS are generally less noisy than small-diameter coils, and the limited exposure with standard single-pulse TMS applications also makes this a lesser concern for routine studies. Side-effects from magnetic stimulation may include transient local discomfort due to contracting scalp muscles near the stimulation site, especially if high-intensity TMS close to the maximal stimulator output is applied. Occasionally subjects report mild headache during or after prolonged stimulation studies that is usually self-limited and can be treated with over-the-counter analgesics. If stimulating near the masseter muscle, forceful jaw closing may occur, and use of a mouth guard for teeth protection may be considered, in particular when using a large double (figure-8) coil. In sensitive subjects, single-pulse TMS of the motor cortex may also elicit tingling paresthesias in the contralateral hand near the target muscle that appear to originate in the precentral rather than parietal cortex (Amassian *et al.* 1991; Rossini and Rossi 1998). There is no evidence of any significant adverse effect of single TMS on cognition, hormone release, or the cardiovascular system.

Metal objects in the stimulation area (such as electroencephalogram (EEG) recording electrodes or jewelry) should be removed due to danger of heating and burning the patient

(Pascual-Leone *et al.* 1990), with safety limits published for standard silver chloride EEG electrodes: stimulation rate <0.4 Hz or total number of consecutive stimuli <20 (Roth *et al.* 1992). Before beginning the examination, both the patient and the operator(s) should also move watches and magnetic-sensitive devices (credit cards etc.) to a safe place at least 50 cm away from the magnetic coil (Rossini *et al.* 1994).

Neurophysiological background

A single high-intensity transcranial stimulation pulse to the motor cortex triggers repetitive discharges along the corticospinal neurons. TMS and TES act on different structures within the cerebral cortex, at least in low-intensity stimulation of the cortical hand region. In TMS, only the superficial cortical layers are stimulated and the induced current flows parallel to the surface of the brain, thus preferentially exciting horizontally oriented neurons, whereas in TES, the current flow is in all directions both parallel and radial to the surface. TES preferentially activates axons of large corticospinal tract cells *directly*, at the proximal portion close to the axon hillock. Low-intensity TES results in a single descending volley that is correspondingly termed the D-wave or direct wave, thus bypassing the neural network within the cortex. With higher intensity, *indirect* trans-synaptic activation of pyramidal cells via cortico-cortical connections also occurs, leading to a series of synchronized descending volleys that are termed I-waves or indirect waves and which follow the initial D-wave. These I-waves are separated by intervals of ~1.5 ms and are named I1, I2, and so on. The summation of these multiple volleys at the anterior horn cells in the spinal cord results in firing of the alpha motor neurons.

In contrast to electrical stimulation, TMS over the cortical hand region preferentially activates pyramidal neurons via trans-synaptic inputs from excitatory interneurons, likely through activation of cortico-cortical axons (Ziemann and Rothwell 2000). With a monophasic magnetic stimulator, the lowest threshold occurs when the direction of the induced electrical current within the cortex flows from posterior to the anterior direction, perpendicular to the line of the central sulcus.

TMS applied in such a way near threshold produces a single I-wave (I1) without a preceding D-wave (Kaneko *et al.* 1996; Nakamura *et al.* 1996; Di Lazzaro *et al.* 1999b). Thus, depolarization of the spinal motor neuron in TMS occurs with a slight delay compared to TES, and correspondingly onset latency of the magnetically induced MEP is longer by ~1.5–2 ms compared to TES.

Orientation of the magnetic coil is of particular importance in magnetic stimulators with monophasic discharge characteristics, and as mentioned above the lowest MEP threshold for TMS occurs with the cortically induced current in postero-anterior orientation and perpendicular to the central sulcus, thus producing the first I-wave (I1). Even a slight positional change or rotation of the coil on the scalp can alter the size and latency of the MEP significantly, in particular when using a focal figure-8-shaped coil. If the coil is rotated so that the induced current direction is from lateral to medial, along the line of the central sulcus, both D- and I-waves may be elicited near threshold. With further coil rotation by another 90° so that the induced current within the cortex is directed from anterior to posterior, again perpendicular to the central sulcus, the third I-wave is favored at the lowest threshold (Rothwell *et al.* 1999). With higher stimulation intensities, both TES and TMS will result in formation of both D- and I-waves. High-intensity stimulation is needed to activate the leg region, and in this case TMS similar to TES tends to elicit D-waves in the corticospinal tract; MEP latencies are comparable (Rothwell *et al.* 1987a).

At the level of the spinal alpha motor neuron, the descending D- and/or I-wave volleys result in progressive depolarization until there are action potentials. Whether the initial part of the descending bursts (D- or I1-wave) or later waves (I2 or higher) will activate the motor neuron depends not only on the strength and number of the descending volleys, but also on the excitability status of the spinal alpha motor neuron. During voluntary muscle contraction of the target muscle, the resting potential of inactive spinal motor neurons is closer to threshold, and a single descending volley may provide sufficient synaptic input to discharge the spinal motor neurons. When relaxed, several descending volleys may be needed to provide the necessary temporal summation for motor neuron firing. Voluntary muscle contraction therefore provides *facilitation* for MEP recordings: MEP latency is reduced by ~2–3 ms compared to the relaxed state; MEP amplitude is larger; and MEP threshold – the intensity of cortical stimulation needed to evoke a motor response – is reduced (Thompson *et al.* 1991).

TMS and TES activate not only excitatory but also inhibitory circuits in the cortex, producing sequential excitation and inhibition of pyramidal cell firing. The excitatory effects are reflected in the MEP, whereas the inhibitory effects are noted as the cortical silent period (cSP, see below).

Technical requirements for MEP recordings

For routine MEP studies, the magnetic stimulator is connected to a standard electromyography (EMG) machine to synchronize the recording with the TMS pulse. Measuring MEPs from the upper limbs requires post-stimulus analysis time of ~50 ms, and 100 ms for the lower limbs. If the cSP following the MEP is also analyzed, the recording time is typically extended to 300–500 ms (Rossini *et al.* 1994). A brief pre-stimulus time may be included in the recording to monitor EMG activity prior to TMS stimulation to assure complete muscle inactivity in studies done at rest or to monitor baseline EMG activity as for cSP studies.

MEPs are usually recorded with surface electrodes in a belly-tendon configuration taped to the skin overlying the target muscle, such as when recording compound muscle action potentials (CMAPs) for motor nerve conduction studies. Filter setting should be relatively open (~1–2000 Hz), and a low-pass filter of ≤1 Hz is recommended to minimize the duration of the stimulus artifact during magnetic stimulation (Rossini *et al.* 1994).

The subject should be seated comfortably, with easy access to the subject's head and spine for stimulation of these areas. During MEP recordings, the subject should be relaxed with eyes open. It has been recommended to ask the subject to perform simple calculations (such as subtracting serially 7 from 100) in order to minimize

threshold and MEP amplitude variability (Rossini *et al.* 1994; Rossini and Rossi 1998).

After localizing the optimal stimulation site, this coil position is usually marked with a pen on the scalp and used for the remainder of the testing for this muscle. The magnetic coil may be fixed with a coil holder or other stabilization device to ensure stable recordings without excessive coil movements.

Magnetic stimulators that are commercially available mainly induce two types of pulses. In the (near-) *monophasic* stimulator (such as the Magstim 200), the initial current flow is rapid, then decays slowly and is followed by a long-duration low-amplitude current of opposite duration. Stimulation occurs primarily during the initial phase. In the monophasic magnetic stimulator, the current direction depends on the coil's orientation. A large round stimulator centered on the vertex with a clockwise current within the coil (Magstim coil side B uppermost) has an anterior–posterior current direction in the part of the coil overlying the right hemisphere (convention that current flows from positive to negative). The induced electrical current direction in the motor cortex is always opposite to the direction in the coil, and therefore in this condition counterclockwise, and in the right motor cortex postero-anteriorly. This is optimal for eliciting MEPs with lowest threshold over the cortical hand area. Turning the coil over to a counterclockwise coil current orientation (Magstim coil side A uppermost) reverses the direction of the induced current to clockwise and better stimulates the left hemisphere (Meyer *et al.* 1991c; Brasil-Neto *et al.* 1992; Trompetto *et al.* 1999). In the *biphasic* or *polyphasic* stimulator (Cadwell, also the newer magnetic stimulators capable of rTMS) the time course of the initial and the reverse current flow are similar, and there are two or more oscillatory current phases. Stimulation is less dependent on coil orientation. Because direction and phases of current flow determine which neural elements are activated within the cortex, different populations of cells may be activated by monophasic and biphasic stimulation. In clinical routine, however, there is no significant difference of MEP latency between the different stimulator types, but other MEP parameters may vary (Claus 1990; Rossini *et al.* 1994; Rothwell *et al.* 1999).

For routine diagnostic TMS studies, using a large circular coil (diameter 8–12 cm) is preferable due to its strong magnetic field. These round coils induce a circular current loop under the coil wiring that is maximal under the mean diameter of the coil. In the center of the coil, the magnetic field strength is zero and no stimulation occurs. A large volume of brain tissue can be activated with a large-diameter round coil, resulting in nonfocal stimulation. It is suitable for MEP recordings from upper and lower extremities.

For more focal stimulation, 'butterfly' or 'figure-8' coils are used that consist of two adjacent round coils with opposite current direction. These are also called 'twin coils' or 'double coils' and have maximal magnetic field strength (twice that at the wings) at a point slightly anterior to the junction region of the '8'. A small figure-8-shaped coil is used primarily for focal stimulation of the motor cortex hand area (such as for mapping studies). Stimulating the lower extremity motor cortex is facilitated by using a figure-8-shaped coil with large wings that are slightly angulated, providing a very strong magnetic pulse penetrating deeply into the cortex.

Coil positioning for transcranial stimulation

A large circular coil is appropriate for standard MEP recordings. It stimulates a large brain volume and thus allows less precise coil placement than with a focal figure-8 coil. With a circular coil (diameter 8–12 cm) a standard position may therefore be used instead of defining the optimal stimulation site in each subject.

Upper-extremity recordings

It is usually recommended to place the large round coil with its center at vertex (Cz, the intersection of the nasion–inion and tragus–tragus lines) flat on top of the head. The coil edge with maximum magnetic field strength under the middle coil windings thus overlies the motor cortex region for the hand and arm. This position may be used for both distal and proximal upper-extremity recordings. Recording during slight muscle contraction and stimulating near vertex with a large circular coil, MEP onset

latency was stable within an area of 6 × 6 cm (Hess et al. 1987a) making more precise placement unnecessary for routine measurement of this parameter. Some authors recommend placement of the coil center at a point 2 cm posterior and 2 cm lateral to Cz (overlying the target hemisphere) for optimal stimulation of the first dorsal interosseus muscle (large round coil with 11.6 cm outer diameter) (Meyer et al. 1991c).

Using the standard position at vertex is sufficient for routine studies with a nonfocal circular coil, but is likely not exactly over the *hotspot* for all individuals. This is defined as the coil position that results in minimal MEP threshold and maximal amplitude (Wassermann 2002). Mapping studies with a focal coil indicate that the distal upper-limb region on average is best stimulated ~5 cm lateral and 1–1.5 cm anterior to Cz and the proximal upper limb at ~3.5–4 cm lateral and 0–0.5 cm anterior to Cz (Wassermann 2002). The exact location of the optimal stimulation site shows interindividual variability by ~2 cm (Meyer et al. 1991c), with slightly greater variability on the left than right hemisphere (Wassermann 2002) and may even be different for each intrinsic hand muscle (Brasil-Neto et al. 1992). The optimal stimulation site for each individual may be determined by a slight variation of coil positioning until the site with lowest MEP threshold is identified (Meyer et al. 1991c). A practical approach is to stimulate at vertex and then 1 cm away in the four quadrants. MEP threshold when stimulating at the hotspot is likely lower than when using the standard circular coil position, but MEP amplitudes at higher stimulation intensities and MEP latency will not differ significantly (Conforto et al. 2004). This makes such optimization of coil positioning optional for a circular coil (Conforto et al. 2004) whereas it is imperative if a focal figure-8-shaped coil is used (Fuhr et al. 1991b).

The coil current direction is important for monophasic stimulators. For upper-extremity TMS of the right hemisphere, the current direction within the circular coil needs to be clockwise so that the induced cortical current (with opposite direction) is perpendicular to the precentral gyrus in posterior–anterior direction, and vice versa for the left hemisphere. With biphasic stimulation, the current direction is not important and a large circular coil at vertex activates both hemispheres simultaneously. When a figure-8-shaped coil is used for upper-extremity recordings, the center of the coil should be directly over the target region in the motor cortex. The handle is often pointed backwards (Rosler *et al.* 1989; Wassermann *et al.* 1992). Largest responses are obtained when the coil axis is ~45–50° to the parasagittal plane with a backward-flowing current in the coil so that the induced current in the brain is perpendicular to the precentral gyrus flowing posterior–anteriorly (Brasil-Neto *et al.* 1992; Rossini *et al.* 1994; Rothwell *et al.* 1999; Weber and Eisen 2002).

Lower-extremity recordings

The motor cortex region for the leg is located rather midline, close to Cz. For optimal stimulation, the induced cortical current direction should be latero-lateral perpendicular to the midsagittal line. This is achieved by placing a large circular coil ~2–4 cm anteriorly to Cz, so that the posterior segment lies midline over the precentral leg region. Some authors recommend the center of a large round coil to be positioned 4 cm lateral and 4 cm anterior to Cz over the predominantly activated hemisphere for optimal tibialis anterior MEPs (Meyer *et al.* 1991c; Rossini *et al.* 1994). A large figure-8-shaped coil with the center near Cz or slightly posteriorly is particularly well suited for leg recordings due to its more focal and deeper penetration than a circular coil, especially if the wings are slightly angulated. A lateral position of the handle in order to assure latero-lateral current direction is recommended (Rosler *et al.* 1989; Hess 2005). Large angulated double coils often have the handle pointing upwards toward the ceiling.

Facial muscle recordings

The cortical stimulation area is a few centimeters laterally to the hand region. Coil orientation is the same for the upper extremity stimulation (Dubach *et al.* 2004), with the exception of the masseter muscle that is best stimulated on the cortex when the current direction is parallel rather than perpendicular to the precentral gyrus (Guggisberg *et al.* 2001).

Coil positioning for spinal nerve root and facial nerve stimulation

The large circular coil is also appropriate for spinal nerve root stimulations that are performed to measure peripheral delay from the spinal cord to the muscle. Because electrical or magnetic stimulations stimulate the nerve roots *directly* at the same site, they usually result in the same peripheral latencies (Schmid *et al.* 1990; Macdonell *et al.* 1992). Most authors agree that excitation of the spinal motor root occurs in the region of the intervertebral foramen, based on anatomical and magnetic stimulation studies. Cervical nerve roots have a short segment of upward direction within the neural foramen that likely is the site of excitation (Mills *et al.* 1993). The electrical field induced by the magnetic stimulator has its highest current density in the region of the neural foramina, whereas the field is only small within the spinal canal and the spinal cord is not stimulated (Maccabee *et al.* 1996). Nerve roots of adjacent levels are also stimulated by magnetic root stimulation, and target muscles at several levels may be recorded simultaneously.

The large circular coil is usually placed in the midline or slightly lateral to this (up to 2 cm) toward the site under investigation. Many studies vary with regard to recommended spinal level for nerve root stimulation, depending on coil characteristics, in particular coil diameter. There is agreement that if the nerve root is stimulated magnetically with a reproducible response, this typically occurs in the region of the intervertebral foramen with rather stable peripheral latency. Largest response in a particular limb muscle occurs usually when the coil windings overlie the appropriate nerve roots tangentially and with parallel orientation of the induced electrical current.

When using a monophasic stimulator, the current direction is less important for nerve root stimulation compared to cortical stimulation, but a direction of the induced current from medial to lateral has been suggested for both upper and lower extremities, i.e. clockwise orientation of the coil current (looking from behind) for the right side and vice versa for the left (Britton *et al.* 1990; Schmid *et al.* 1990;

Samii et al. 1998; Attarian et al. 2005). Other authors did not report any significant differences for either latency or amplitude related to direction of current flow (Macdonell et al. 1992).

Cervical nerve root stimulation

Many investigators place the center of the coil over the C7 spinous process for the commonly studied hand muscles (Garassus et al. 1993; Samii et al. 1998; Attarian et al. 2005); this is often also recommended for more proximal arm muscles that may thus be recorded simultaneously (Schmid et al. 1990). A slightly more caudal positioning was also recommended, with the upper inner edge over the C7 spinous process (Chokroverty et al. 1991; Rossini et al. 1994). Others report a good response in first dorsal interosseus muscle with the coil centered over C7, but an even larger response with the center over C3 (Britton et al. 1990) or over C5 (Schmid et al. 1990). The coil may also be placed lower at the T3 level at ~2 cm laterally, thus placing the C8/T1 nerve roots under the upper quadrant of the coil for optimal abductor digiti minimi muscle recordings (Furby et al. 1992).

For recordings from proximal arm muscles (biceps and deltoid, C5 and C6 innervated), the optimal coil position is 2–3 cm above C7, midline or 2 cm lateral to this position (Chokroverty et al. 1991). In other studies, the largest response in the biceps brachii muscle was with the coil centered slightly higher over C5 (Britton et al. 1990) or slightly lower over C7 (Schmid et al. 1990).

An alternative suggestion for spinal nerve root stimulation is to place the coil center 5–7 cm lateral to the midline toward the recording side and then determine the best stimulation point by rostro-caudal displacements (Rossini et al. 1994), but this is not generally recommended because nerve stimulation may then occur more distally.

When using a figure-8-shaped coil for cervical root stimulation, it is best oriented when the coil current flows slightly downward toward the midline at a mean angle of 15–28° horizontally (Mills et al. 1993).

Lumbar nerve root stimulation

The large round coil typically activates the lumbosacral nerve roots also at the level of the neural foramina, in the distal cauda equina. For S1-innervated muscles (abductor hallucis, soleus), the large round coil is placed with its center slightly lateral of the L4–5 spinous process and adjusted slightly rostrally and caudally (Di Lazzaro et al. 2004). Others place the center of the round coil directly over the S1 spinous process (Han et al. 2004) or midline 10 cm below the L4 lumbar level marked by the horizontal line connecting the iliac crests (Banerjee et al. 1993), and others recommend the proximal inner edge at the level of the S1 vertebra (Chokroverty et al. 1993).

For L5 muscles (tibialis anterior, extensor digitorum brevis) the large round coil is placed midline slightly higher, with the inner edge at the level of L5 vertebra (Chokroverty et al. 1993; Weber and Eisen 2002). Larger responses with the coil centered lower, maximally over S2–S3, have also been reported (Britton et al. 1990; Furby et al. 1992).

For quadriceps muscle, the same authors noted optimal placement with the coil center over L1–L2 spinous process (Britton et al. 1990). For lower-extremity recordings using a figure-8-shaped coil, Maccabee et al. (1996) noted that the induced monophasic current was usually most effective when directed toward the spinal fluid-filled thecal sac and away from the most distal root segments covered by epidural fat, inward for lumbar roots and usually rostral for sacral roots.

Cranial nerve stimulation: facial nerve

The facial nerve is the only cranial nerve that can be magnetically stimulated at a constant site within the skull. The facial nerve is stimulated within the facial canal proximally in the petrous bone by positioning the round coil posteriorly to the ear, ipsilateral to the recording site (Dubach et al. 2004).

MEP measurements in clinical neurodiagnosis: overview

MEP recordings after cortical magnetic stimulation have an inherent variability, even with constant stimulation and recording techniques. Of all the different MEP parameters that can be measured by TMS, the latency of the MEP is generally regarded as the most reliable and useful. If combined with a measure of the peripheral motor conduction time (PMCT), a calculation of the central motor conduction time (CMCT) is possible. In routine clinical practice, this is the most important MEP parameter for evaluation of pyramidal tract function.

Additional information may come from MEP amplitude as a marker for the degree of cortical and pyramidal tract activation, especially if correlated to the size of the CMAP obtained with supramaximal peripheral nerve stimulation. In addition, the MEP threshold obtained during MEP studies and the cSP may be useful as parameters of cortical excitability. These parameters, however, not only depend greatly on the technique, but also show great interindividual variability, making them more suitable for statistically comparing subject groups (as in research projects) than for assessing individual pathology. Since bilateral MEP responses are usually strongly correlated, a significant side-to-side difference in one of these parameters may help detect a unilateral or asymmetrical pyramidal tract lesion in an individual subject. There are many additional MEP parameters that are used for research applications but not for routine diagnostic applications and therefore are not discussed in this chapter.

The stimulation details and recording techniques vary greatly between laboratories. MEP may be obtained with recording from a variety of muscles in face, upper and lower extremities, as well as trunk including diaphragm (Lissens 1994; Zifko *et al.* 1996), abdominal wall muscles (Lissens *et al.* 1995), and pelvic muscles (Opsomer *et al.* 1989). For routine applications, recording from a hand muscle and from a leg muscle (often tibialis anterior) allows determination of the central motor conduction to the cervical and lumbosacral cord. Additional muscles may be recorded depending on the clinical situation, such as several upper-extremity recordings for cervical radiculopathy or myelopathy including muscles innervated from above and below the suspected lesion. The addition of a proximal upper-extremity muscle may sometimes help in the detection of abnormality in motor neuron disorder (Di Lazzaro *et al.* 1999a). Recording from the lower extremity improves sensitivity in many spinal cord disorders (such as hereditary spastic paraparesis).

Standard experimental protocol (for each target muscle):

1. MEP threshold determination, usually at rest. Optional is threshold determination also with facilitation.

2. MEP recording during facilitation with mild voluntary muscle contraction, for cortical MEP latency and amplitude measurement. Optional is MEP recording at rest.

3. Optional: MEP with cSP recording during moderate-to-high level of voluntary contraction.

4. Magnetic root stimulation over the cervical and lumbar spine for peripheral latency.

Alternatively, or in addition:

5. Peripheral nerve conduction studies with recording of compound muscle action potential (CMAP) and F-wave for calculation of peripheral latency. This step may also be performed as an initial step prior to magnetic stimulation.

6. Calculation of CMCT by subtracting the peripheral latency from the cortical MEP latency. The peripheral latency may be obtained from the root stimulation (step 4) and/or F-wave latency (step 5).

MEP threshold

MEP or motor threshold is the lowest stimulus intensity of TMS that gives a recordable MEP in a target muscle. The motor threshold is usually determined at the beginning of the MEP recordings, as it provides a reference for setting the stimulation intensity for recording other parameters. There are different ways to measure this excitability threshold, which is usually determined during complete relaxation (resting motor threshold). A common definition of the MEP threshold at rest is the stimulus intensity required to elicit reproducible MEPs of ~100 µV in ~50% of 10–20 consecutive trials (Rossini *et al.* 1994). The resting motor threshold is not significantly different if determined as at least three of six stimulations resulting in an MEP (Conforto *et al.* 2004). Sometimes rectified and averaged recordings are used to determine the threshold especially when determining the 'active' threshold (Wassermann 2002). It is practical to start the stimulation below the expected threshold intensity and increase stimulator output progressively by 5% steps (absolute percentage of stimulator output) (Rossini et al. 1994), and then near threshold successively decrease the stimulation intensity in 1% or 2% steps until <50% of 10 stimulations produce a measurable response (Reid 2003).

Mills and Nithi (1997) proposed a measure of corticomotor threshold with better reproducibility that consists of lower and upper threshold. The lower threshold is obtained by decreasing stimulus intensity in 1% increments until 10 stimuli fail to give a response. The upper threshold is then determined by increasing in 1% steps until all 10 stimuli induce an MEP response of >20 μV amplitude. They reported normal threshold values of 38.0 ± 8.6% (lower) and 46.6 ± 9.4% (upper) for first dorsal interosseus muscle using a figure-8 coil (Mills and Nithi 1997).

Using large round coils for stimulation and recording from abductor digiti minimi (ADM), MEP threshold was 47.7 ± 7.5% with a 1.5 T coil, and 41.2 ± 7.3% with a 2 T coil (Triggs et al. 1999a). In longitudinal studies, the test–retest MEP threshold difference was 2.5 ± 2.6% of the stimulator output, and a threshold change of ≥10% was considered pathological (Triggs et al. 1999a). Side-to-side variation should be <5% (Eisen 2001).

MEP threshold is generally lower for distal than proximal muscles; lowest threshold values are reported for intrinsic hand muscles and finger extensors, in keeping with their large cortical representation (Rossini et al. 1994; Mills et al. 1997). Lower-extremity muscles and pelvic muscles have higher thresholds. MEP threshold varies widely in the healthy population, with high correlation between siblings (Wassermann 2002). A slight increase with aging has been reported for hand and foot muscles (Rossini et al. 1992), but not found by others (Mills et al. 1997; Wassermann 2002). There is no significant difference by gender (Furby et al. 1992; Mills et al. 1997; Wassermann 2002). A lower threshold has been reported for the dominant hemisphere by some authors and handedness should be documented (Macdonell et al. 1991; Triggs et al. 1994, 1999b), but any difference between the two hemispheres, if at all present, is physiologically minimal. Side-to-side threshold comparison is therefore useful in the diagnosis of monohemispheric lesions of various etiologies (Traversa et al. 1998).

MEP threshold is a measure of global motor system excitability at both cortex and spinal level. It reflects in particular neuronal and interneuronal membrane excitability, because it is increased by drugs such as sodium-channel blockers (Ziemann et al. 1996). It is influenced by other factors including posture (lower when sitting or standing vs lying supine (Ackermann et al. 1991) and mental activity. If subjects are allowed to close their eyes and be mentally inactive, MEP amplitude decreases and the threshold is slightly higher, whereas MEP threshold is lower if the eyes are open and subjects are asked to perform simple calculations, as has been recommended for routine studies (Rossini and Rossi 1998). Similarly, mental activity that includes 'thinking about the movement' of the target muscle increases rather selectively the MEP amplitude of the 'prime movers' for the selected movement (Izumi et al. 1995; Abbruzzese et al. 1996). MEP threshold also shows spontaneous physiological fluctuations during serial MEP recordings (Rossini et al. 1994).

Any slight contraction of the target muscle decreases MEP threshold, and it is therefore important to assure complete muscle relaxation when determining the resting threshold. EMG feedback over the loudspeaker and incorporation of prestimulus time interval into the recording are helpful in this regard.

An interstimulus interval of >3 s has been recommended for determination of MEP threshold to prevent any facilitatory or inhibitory influence on the subsequent stimulation (Rothwell et al. 1999).

As mentioned, the resting motor threshold is used as a reference to set the stimulation intensity during subsequent MEP recordings and may serve as a physiological measure for upper and lower motor neuron excitability. Less commonly, as an alternative or additional value, the threshold for obtaining MEPs during facilitation (active motor threshold) is measured. This may be defined as the stimulus intensity required to elicit reproducible MEPs of ~200–300 μV (minimum amount to distinguish the MEP from background muscle activity) in ~50% of consecutive trials, while the subject maintains a slight voluntary contraction at a level of 5–15% of the maximum voluntary force (Rossini *et al.* 1994), or by averaging and setting a criterion of amplitude relative to baseline.

MEP latency

The latency of the MEP is the time between the cortical stimulation and the onset of an evoked potential in the target muscle. Latency depends on whether the recording is at rest or with muscle activation. In routine studies, MEP recordings for latency and amplitude measurements are done with facilitation for several practical reasons. Patients often have difficulty completely relaxing the target muscles, and recording during slight voluntary muscle contraction avoids the need to achieve complete relaxation. More importantly, during facilitation MEP threshold is decreased and MEP amplitudes are larger, thus allowing lower stimulation intensities. This is particularly helpful in the lower extremities where MEPs are sometimes difficult to obtain even at maximal stimulator output. MEP latency during facilitation is typically 2–3 ms shorter than during complete relaxation, sometimes up to 6 ms (Rossini *et al.* 1994), and recording during muscle contraction results in the shortest reproducible MEP latency. Normal values are typically reported for this condition, and different normal values must be used for recordings during complete muscle relaxation.

In intrinsic hand muscles, facilitation of amplitude and latency is usually complete with a background force of 10% (Hess *et al.* 1987a; Kischka *et al.* 1993). Facilitation effects from increasing muscle contraction rise more gradually in proximal arm muscles and even more so in leg muscles. These effects are more important with regard to MEP amplitudes than for onset latencies. In a study of abductor digiti minimi, biceps brachii, tibialis anterior, and soleus muscles at rest and during muscle contraction at 10% of maximum force, the MEP onset latency shortened in all muscles by ~3 ms, with only minimal additional decrease in latency when the background contraction was increased to 60% of maximal force (Kischka *et al.* 1993).

In practice, the subject is asked to moderately contract the target muscle for a few seconds and the TMS is applied when the subject maintains a tonic contraction. In between each stimulus, the subject rests in order to avoid fatigue. Ideally, the background level of muscle contraction is monitored by a force transducer or handheld dynamometer. Most investigators use values between 10% background muscle contraction (especially for distal muscles in hand) and 20% of maximal voluntary force (especially for proximal arm and leg muscles). Alternatively, feedback of the background EMG may be provided and a voluntary contraction with 2–6% of the maximum surface EMG is recommended for MEP recordings in order to maximize facilitation, or, if the 'root mean square value' of the background EMG is used, then ~15% is recommended (Philipson and Larsson 1988; Ravnborg *et al.* 1991). If latency and central motor conduction time are the main considerations (and not MEP amplitude or cSP), the force level does not need to be accurately controlled. Appropriate muscle contraction may be achieved by pinching thumb and fingers against each other for intrinsic hand muscles, or by making an isotonic muscle movement against gravity and without resistance (proximal arm muscles and leg muscles). Higher degrees of muscle contraction should be avoided, since excessive EMG background makes the identifying MEP onset rather difficult (Rossini *et al.* 1994).

There are likely peripheral and central mechanisms involved in producing facilitation. Reafferent facilitation of spinal motor neurons from muscle receptors (Rossini *et al.* 1994) contribute to increased motor neuron excitability. The increased number and greater synchronization of the descending volleys in conjunction with the greater excitability of the alpha motor neuron result in faster depolarization of the motor neuron by the D- or I1-wave instead of later I-waves, thus explaining the latency decrease by ~2–3 ms compared to relaxation.

If the subject cannot maintain a steady low level muscle contraction in the target muscle, e.g. in a patient with hemiplegia from cerebral infarction, then other techniques may be used, including reflex activation, or contraction of contralateral muscles, proximal ipsilateral muscles, or facial muscles. The subject may also be asked to imagine the target muscle movement. Other facilitatory maneuvers include vibration of the examined muscle (Rossini *et al.* 1994) or prestimulation of the mixed nerve innervating the target muscle (Mariorenzi *et al.* 1991). Paired-pulse stimulation may also augment MEPs in the leg and can be used in addition to other methods of facilitation (Sandbrink *et al.* 1998). These techniques may be helpful in increasing MEP amplitude and result in some shortening of MEP latency compared to full relaxation. For contralateral muscle contraction, MEP latency shortening was reported to be similar to ipsilateral background muscle contraction (Hess *et al.* 1987a). Other authors report an overall lower degree of facilitation in the contralateral condition, and in one study MEP latency and CMCT to the abductor pollicis brevis (APB) muscle was 1.7 ms longer in the contralateral compared to ipsilateral target muscle contraction (Zwarts 1992). The stimulation intensity of the TMS applied for MEP latency measurements should be well above the resting MEP threshold. When using low stimulation intensities close to threshold, MEP latencies of repeat trials are more variable, and higher stimulation intensities should therefore be used for clinical testing (van der Kamp *et al.* 1996). Conventionally, the stimulation intensity is set in reference to MEP threshold value, and intensities of 120–150% of the resting threshold are used (Rothwell *et al.* 1999). With recording during slight voluntary background contraction, MEP latency from APB was little affected over a wide range of stimulus intensities (Hess *et al.* 1987a). Occasionally, stimulation intensity is reported in relation to the active motor threshold (Claus 1990). A fixed absolute stimulator output level is sometimes used for hand muscle recordings (Fuhr *et al.* 2001), and some investigators reported using 100% of maximum stimulator output for MEP recordings and reference values (Furby *et al.* 1992). This appears particularly convenient for lower-extremity MEPs where higher stimulation intensities than for the upper extremities are generally needed, often close to maximal stimulator output (Di Lazzaro *et al.* 2004).

Most investigators use four or five well-reproducible large MEPs that may be superimposed to measure the shortest-onset latency of the MEP. In practice, the shortest of four MEPs is similar to the shortest of 15 recordings (Hess *et al.* 1988). Even fewer recordings may be necessary in an individual when the MEP latency of the first trials is noted to be well

within normal limits. Other investigators record rectified averages of five MEPs, and take the average latency of two separate sets (M.K. Floeter, personal communication). Only rarely is the MEP so small or the background EMG so large and irregular that averaging of multiple recordings becomes a necessity.

Central motor conduction time

MEP latency described above is the time from the motor cortex stimulation to MEP response in the target muscle and therefore includes both a central component (time from cortex to activation of the spinal motor neurons) and a peripheral component (time from activation of the spinal motor neurons to the muscle response). A more sensitive measure of pyramidal tract abnormalities is the CMCT that can be obtained by subtracting from MEP latency the time for the peripheral segment, the peripheral motor conduction time (PMCT), following the formula

$$CMCT = MEP \; latency - PMCT.$$

PMCT is usually measured either by F-wave method (PMCT-F) or by magnetic stimulation of the spinal nerve roots (PMCT-M), with similar degree of discomfort for the subject and equivalent between-trial variability (Samii *et al.* 1998). Alternatives are electrical stimulation of the nerve roots by surface or needle stimulation – significantly more painful for the subject (Samii *et al.* 1998) – or measurement of the tendon reflex latency.

The *F-wave method* consists of conventional electrical stimulation of the peripheral nerve innervating the target muscle and measuring the minimal F-wave latency of (usually) 20 recordings. The peripheral motor conduction time PMCT-F is then calculated as

$$PMCT\text{-}F = \frac{(F + M - 1)}{2}$$

where F is the minimal F-wave latency, M the latency of the compound muscle action potential or M-wave, and 1 (in ms) is the estimated turnaround time for antidromic activation of the spinal motor neuron (Rossini *et al.* 1994). This method is only applicable for muscles and nerves with recordable F-waves, i.e. most appropriate for distal muscles in hand and foot.

For proximal muscles such as the biceps brachii, a *tendon reflex method* has been described (Ofuji *et al.* 1998). The PMCT-T is derived from the tendon reflex latency T in analogy to the F-wave method as

$$\text{PMCT-T} = \frac{(T-1)}{2}$$

where the synaptic delay at the alpha motor neuron is estimated as 1 ms.

Many laboratories use the method of *magnetic stimulation of the spinal nerve root* and less commonly *electrical nerve root stimulation* to measure the peripheral motor conduction time, PMCT-M. The technique and coil placements for cervical and lumbar magnetic stimulation are described above. Whether magnetic or electrical stimulation is used, the nerve roots are typically stimulated in the region of the intervertebral foramen. The PMCT-M therefore does not include conduction time from anterior horn cells to the intervertebral foramen. As a result, the calculated central motor conduction time is estimated as slightly too long, as it includes this proximal segment of the spinal nerve root. Cervical root stimulation results in overestimation of the CMCT by 0.5–1.4 ms, corresponding to a calculated distance of ~3–4 cm (Claus 1990; Chokroverty *et al.* 1991; Furby *et al.* 1992; Samii *et al.* 1998) and up to 7 cm (Garassus *et al.* 1993) from the anterior horn cells to the intervertebral foramen. For the lumbar region, the difference between the CMCT by F-wave method (CMCT-F) and by lumbosacral root stimulation (CMCT-M) is much greater, ~3.0–3.1 ms (Britton *et al.* 1990; Garassus *et al.* 1993; Di Lazzaro *et al.* 2004) to 3.9–4.1 ms (Chokroverty *et al.* 1991, 1993; Furby *et al.* 1992; Samii *et al.* 1998) which represents a distance of ~17–20 cm (assuming a conduction velocity of 50 m/s along the cauda equina). Whenever there is significant slowing across this proximal motor root segment, the technique of spinal nerve root stimulation may result in false-positive slowing of central motor conduction due to inclusion of this proximal root segment in the CMCT-M measurement. This was shown for peripheral nerve disorders and lumbar radiculopathies due to disk herniation and lumbar spinal stenosis and does not occur with the F-wave method; this difference between the two techniques may be used diagnostically (Banerjee *et al.* 1993; Di Lazzaro *et al.* 2004).

Magnetic root stimulations are performed with the target muscles relaxed. Higher stimulation intensities are recommended that elicit several superimposable responses from which a reproducible peripheral onset latency can be determined. Stimulation intensities range from 30% above the excitability threshold, ~60–80% for first dorsal interosseus (FDI) and EDB recordings (Britton *et al.* 1990), to maximal stimulator output (Banerjee *et al.* 1993; Samii *et al.* 1998). In a systematic analysis, latency values reached a plateau at intensities above 60% maximal stimulator output (Schmid *et al.* 1990). Stimulation with a fixed high-stimulation intensity makes determination of threshold unnecessary and readily results in reproducible potentials. A practical approach is using a stimulation intensity of 60% of maximum magnetic stimulator output for upper-limb muscles and 80% of maximum stimulator output for lower-limb muscles (Di Lazzaro *et al.* 1999a) with further increase if necessary. There is considerably more patient discomfort at stimulation levels close to maximal stimulator output (Britton *et al.* 1990), and some authors have suggested that with maximal stimulation intensities the stimulation point may migrate distally beyond the neural foramina (Schmid *et al.* 1991).

Even at maximal stimulation intensities, it is usually not possible to stimulate the nerve root supramaximally and magnetic stimulation is therefore not suitable to assess proximal conduction block. Advantages of magnetic root stimulation vs the F-wave method include the ability to record from proximal and distal muscles, often simultaneously. In addition, F-wave latencies are inherently more variable and thus a larger number of electrical stimulations are needed for this method to obtain a reliable minimal F-wave latency. It may also become difficult in peripheral nerve disorders to obtain F-waves at all.

The CMCT is the most important MEP parameter in clinical practice and normative values for a large variety of muscles have been published, with values for more common muscles provided in Table 19.1. It is important to use

Text continued on page 253

Table 19.1 Central motor conduction time (CMCT) to upper-extremity (A) and lower-extremity (B) muscles

Muscle Height (cm)	Facilitation	MEP lat. (ms)	R/L diff. (ms)	CMCT-F (ms)	R/L diff. (ms)	CMCT-M (ms)	R/L diff. (ms)	CMCT-ES (ms)	CMCT-T (ms)	Reference
(A) Upper extremity										
Biceps	+					5.1 ± 1.0				Dvorak et al. 1990
	+	12.5 ± 1.2	0.65 ± 0.6			7.1 ± 1.1	0.60 ± 0.51			Furby et al. 1992
	+	9.4 ± 1.7				6.0 ± 1.2				Eisen et al. 1990
	+					4.6 ± 0.9	0.5 ± 0.4			Bischoff et al. 1993
	+	11.5 ± 0.8							3.8 ± 0.5	Ofuji et al. 1998
	+					7.6 (3 SD)				Di Lazzaro et al. 1999a
	+	10.2 ± 0.5				5.1 ± 0.3				Abbruzzese et al. 1993
	+	11.4 ± 1.2	0.6 ± 0.63							Osei-Lah and Mills 2004
EDC	+	14.2 ± 1.7				6.7 ± 1.0				Eisen et al. 1990
	+					5.6 ± 0.9	0.5 ± 0.4			Bischoff et al. 1993
APB	–	22.6 ± 1.2				9.5 ± 1.1				Barker et al. 1987
	+	21.1 ± 1.5				8.0 ± 1.2				Barker et al. 1987
	+			4.3 ± 0.8		5.2 ± 0.6				Dvorak et al. 1990
	+	20.2 ± 1.6		6.5 ± 2.0		7.9 ± 2.1				Eisen et al. 1990
	+	20.0 ± 1.7				6.9 ± 0.6				Abbruzzese et al. 1993
	+	20.3 ± 1.2		5.8 ± 1.0		6.7 ± 1.0				Garassus et al. 1993
ADM	–	22.5 ± 1.5				9.4 ± 1.0				Barker et al. 1987
	+	20.5 ± 1.7				7.4 ± 1.2				Barker et al. 1987
	+			5.8 ± 0.8	1.8 (99th)	6.0 ± 0.9	2.4 (99th)	6.2 ± 0.7		Claus 1990 (Magstim 200)

Table 19.1 (*contd.*) Central motor conduction time (CMCT) to upper-extremity (A) and lower-extremity (B) muscles

Muscle Height (cm)	Facilitation	MEP lat. (ms)	R/L diff. (ms)	CMCT-F (ms)	R/L diff. (ms)	CMCT-M (ms)	R/L diff. (ms)	CMCT-ES (ms)	CMCT-T (ms)	Reference
	+			5.8 ± 0.8	1.8 (99th)	6.0 ± 0.9	2.4 (99th)	6.4 ± 0.9		Claus 1990 (Digitimer D190)
	+			5.8 ± 0.8	1.8 (99th)	6.0 ± 0.9	2.4 (99th)	6.3 ± 0.8		Claus 1990 (Cadwell MES-10)
	+			4.0 ± 0.8		5.2 ± 0.9				Dvorak et al. 1990
	+	20.5 ± 1.2	0.69 ± 0.52	6.1 ± 1.0	0.72 ± 0.48	7.0 ± 0.9	0.66 ± 0.43			Furby et al. 1992
	+	20.2 ± 1.1		5.8 ± 1.0		6.7 ± 1.3				Garassus et al. 1993
	+	19.5 ± 1.5							4.0 ± 1.0	Ofuji et al. 1998
	+			6.3 (3 SD)		7.7 (3 SD)				Di Lazzaro et al. 1999a
FDI	–	23.0 ± 1.7								Mills and Nithi 1997
	+	21.1 ± 1.6				5.7 ± 1.1				Mills and Nithi 1997
	+	19.7 ± 1.2	0.8 ± 0.75	5.5 ± 0.7	1.0 ± 0.95					Osei-Lah and Mills 2004
	+					6.0 ± 1.0	0.6 ± 0.4			Bischoff et al. 1993
	+			5.5 ± 1.1	0.8 ± 0.9					Lo et al. 2004

(B) Lower extremity

Quadriceps

150–174	+	20.9 ± 1.2			13.0 ± 1.4				Dvorak et al. 1991
175–191	+	23.4 ± 1.9							Dvorak et al. 1991
R. femoris	+				16.6 (3 SD)	1.8			Di Lazzaro et al. 1999a
	+	21.5 ± 1.7	0.88 ± 0.85		14.2 ± 1.5		0.93 ± 0.90		Furby et al. 1992
V. medialis	+	21 ± 1.9	1.2 ± 1.08						Osei-Lah and Mills 2004
Tib. ant.	+				* = 0.076 × ht + 3.4 (2.5 SD)			12.5 ± 1.7	Claus 1990
150–174	+	28.5 ± 2.5			12.8 ± 1.4				Dvorak et al. 1991
175–191	+	30.7 ± 1.8			14.0 ± 1.3				Dvorak et al. 1991
	+	26.1 ± 1.9	0.82 ± 0.98	9.9 ± 1.8	0.99 ± 0.87	13.8 ± 1.5	0.93 ± 0.87	* = 0.083 × ht + 3.28 (2.5 SD)	Furby et al. 1992
168 ± 8.8	+	29.1 ± 1.4			14.4 = 0.9				Abbruzzese et al. 1993
155–187	+	27.4 ± 2.6		10.7 ± 1.77	14.2 ± 1.7				Garassus et al. 1993
	+				14.3 ± 1.7	0.7 ± 0.6			Bischoff et al. 1993
	+			14.7 (3 SD)	2.1	17.1 (3 SD)	2.0		Di Lazzaro et al. 1999a

EDB

150–174	+	36.9 ± 2.4	11.3 ± 1.7	13.4 ± 1.7		Dvorak et al. 1991
175–191	+	39.6 ± 2.0				Dvorak et al. 1991

Table 19.1 *(contd.)* Central motor conduction time (CMCT) to upper-extremity (A) and lower-extremity (B) muscles

Muscle Height (cm)	Facilitation	MEP lat. (ms)	R/L diff. (ms)	CMCT-F (ms)	R/L diff. (ms)	CMCT-M (ms)	R/L diff. (ms)	CMCT-ES (ms)	CMCT-T (ms)	Reference
155–187	+	36.3 ± 2.6		11.1 ± 1.4		14.5 ± 1.5				Garassus et al. 1993
Gastroc.	+					14.2 ± 1.5	0.8 ± 0.5			Bischoff et al. 1993
Abd. hall.	-	43.3 ± 3.0				18.8 ± 2.0				Barker et al. 1987
	+	41.2 ± 3.4				16.7 ± 2.4				Barker et al. 1987
	+					15.9 ± 2.0	0.9 ± 0.6			Bischoff et al. 1993
	+			15.9 (3 SD)	2.2	18.2 (3 SD)	1.9			Di Lazzaro et al. 1999a
	+	39.1 ± 2.5	1.5 ± 1.13	12.5 ± 2.2	1.6 ± 1.02					Osei-Lah and Mills 2004
	+	39.4 ± 2.7	0.8 ± 0.6	12.4 ± 1.2	0.9 ± 0.4	16.9 ± 0.9	0.5 ± 0.4			Di Lazarro et al. 2004
	+			12.3 ± 1.9	1.3 ± 1.1					Lo et al. 2004
Flexor hall.	+	38.5 ± 2.4				15.2 ± 1.7				Miscio et al. 1999

MEP lat., motor-evoked potential latency; R/L diff., side-to-side difference; CMCT-F, CMCT recorded by F-wave method, CMCT-M by magnetic nerve root stimulation, CMCT-ES by electrical nerve root stimulation and CMCT-T by tendon reflex are reported as either mean ± SD or as normal limit (with indication of SD or percentile).
*Equation reported as the upper normal limit for the CMCT-M to the tibialis anterior muscle; height (ht) in centimeters.
EDC, extensor digitorum communis; APB, abductor pollicis brevis; ADM, abductor digiti minimi; FDI, first dorsal interosseus; R. femoris, rectus femoris; V. medialis, vastus medialis; Tib. ant., tibialis anterior; EDB, extensor digitorum brevis; Gastroc., gastrocnemius; Abd. hall., abductor hallucis; Flexor hall., flexor hallucis.

reference values obtained with the same or a comparable technique. The stimulator and coil type are of lesser importance for MEP latency and CMCT values (Claus 1990), but the absence or presence of facilitation by voluntary muscle contraction and the method of measuring the peripheral motor conduction time (CMCT-F vs CMCT-M) are critical.

The central motor conduction is composed of the time to activate the corticospinal system at the cortex, conduction time along the pyramidal tract, plus the spinal delay time. The physiological conduction velocity along the corticospinal tract, has been reported as 65 m/s (Eisen and Shtybel 1990), confirmed in another study as 68 ± 5 m/s (Ugawa et al. 1995). The spinal delay time consists of at least one synaptic delay to the anterior horn cells and the time to reach the firing threshold by temporal summation, and was estimated as 0.5 ± 0.3 ms with the target muscle activated (Ugawa et al. 1995).

The CMCT is dependent on height, but mainly only for the lower extremity and pelvic muscles. When recording from leg muscles, adjustment for height is advisable by using separate subgroups of progressively greater height or normograms (Furby et al. 1992; Rossini et al. 1994). In the tibialis anterior muscle, the upper normal limit of CMCT for two different subject groups depending on height differed by 1 ms (Dvorak et al. 1991). The upper limit of CMCT to the tibialis anterior (2.5 SD) may be calculated as CMCT = 0.076 × height (cm) + 3.4, based on recordings during 10% isometric muscle contraction and with the PMCT determined by electrical nerve root stimulation (Claus 1990). An alternate formula for the upper limit of CMCT to the tibialis anterior (2.5 SD) was given by Furby et al. (1992) as CMCT = 0.083 × height (cm) + 3.28, based on recordings during 10% muscle contraction and with PMCT measured by magnetic root stimulation. CMCT is slightly faster in women then men, but only as a result of lower body height (Furby et al. 1992). There is a slight increase of CMCT with age, but the correlation is weak and may be neglected for routine studies (Eisen et al. 1990; Weber and Eisen 2002). Kloten et al. (1992) found that the latency difference between age groups 19–29 and ≥60 years was 0.1 ms for the biceps brachii, 0.7 ms for the first dorsal interosseus (5.8 ± 1.0 ms

vs 6.5 ± 1.1 ms), and 2.1 ms for the tibialis anterior muscle (14.0 ± 1.3 ms vs 16.1 ± 1.9 ms). The authors used magnetic root stimulation and matched the groups for height.

MEP amplitude

MEP amplitude provides an estimate of the extent of corticospinal tract and motor neuron activation by magnetic stimulation, with limitations (described below). The absolute MEP amplitude reflects both upper and lower motor neuron activity and is affected by peripheral nerve disorders. Correlating MEP amplitude with the amplitude of the compound muscle action potential (CMAP) that is obtained by conventional electrical nerve stimulation of the peripheral nerve is a more useful measure of upper motor neuron function. It estimates the portion of the pool of spinal motor neurons that is activated by TMS (Rossini et al. 1994).

The ratio of MEP/CMAP amplitudes may also be expressed as a percentage:

$$MEP\% = \frac{MEP \text{ amplitude}}{CMAP \text{ amplitude}} \times 100.$$

MEP amplitude is measured peak-to-peak (Rossini et al. 1994) or baseline-to-peak (Triggs et al. 1999a); CMAP amplitude should be measured the same way. Alternatively, the area under the curve is used for both measures. MEP amplitude (and therefore the MEP:CMAP ratio or MEP%) increases with higher cortical stimulation intensity in a sigmoid input–output relation. Intrinsic hand muscles have the steepest input–output relation and the highest MEP% values, ~50–60% (Hess et al. 1987a; Rothwell et al. 1987b). For proximal arm muscles and trunk and leg muscles, MEP amplitude increases more gradually with increasing stimulation intensity, and MEP% with maximal stimulation intensity is lower (Hess et al. 1987a; Rothwell et al. 1987b).

Facilitation maneuvers such as voluntary contraction of the target muscle also increase MEP amplitude, likely based on similar mechanisms as for shortening MEP latency. Even slight muscle contraction may significantly increase MEP amplitude. For intrinsic hand muscles, MEP amplitude reaches the maximal value by ~10% of maximal voluntary contraction force, whereas this is more gradual for proximal arm and leg muscles. In the tibialis anterior muscle, MEP amplitude increases progressively up to 60% of maximal force (Kischka *et al.* 1993). Therefore, MEP amplitudes of proximal muscles or the leg muscles may only be used for clinical evaluation if the degree of voluntary contraction is monitored and the control values were obtained under the same condition. For intrinsic hand muscles, however, maximal MEP amplitude is rather constant as long as the contraction level is maintained at >10% maximal force. The clinical use of MEP amplitudes is limited due to its variability. In addition to interindividual differences there is great intertrial variability even in the same subject. Many technical factors influence MEP amplitude, such as coil positioning and amount of muscle contraction. Even with all conditions stable, however, there remains a considerable between-trial variability that is essentially random (Kiers *et al.* 1993). It may be reduced to some degree by stimulating with higher stimulation intensity, using a larger or nonfocal coil and recording during background muscle contraction instead of relaxation (Kiers *et al.* 1993; van der Kamp *et al.* 1996). Due to this rather random variability of MEP amplitude, the mean value of several individual trials may be preferred over a single MEP result or the largest MEP amplitude of several trials, but overall the differences between the various methods of assessing MEP amplitudes are minor as long as patients and controls are studied the same way (McDonnell *et al.* 2004).

In general, an MEP amplitude side-to-side difference of ≥50% can be regarded as abnormal in patients without lower motor neuron or peripheral nerve disease (Weber and Eisen 2002). In small hand muscles, an MEP% of ≤15% of the CMAP amplitude is always abnormal, and ≥20% must be considered normal (Hess *et al.* 1987a; Rothwell *et al.* 1987a). Other authors considered an MEP% abnormal in intrinsic hand muscle if <10% (Eisen *et al.* 1990) or <11% (Triggs *et al.* 1999a).

As mentioned, mean MEP% that can be obtained from an intrinsic hand muscle with maximal stimulator output and during facilitation is usually ~50–60% of the CMAP amplitude, with a broad range of interindividual

variability from 10% to 100% (Hess *et al.* 1987a; Rothwell *et al.* 1987a). Recent studies have shown that conventional MEP recordings underestimate the portion of the corticospinal system activated by TMS primarily due to phase cancellation in the peripheral motor pathway (Magistris *et al.* 1999). Using the technique of triple stimulation that involves a cortical magnetic stimulation coupled with two electrical collision stimuli over the peripheral nerve (described elsewhere in this book in detail), they documented that TMS achieves activation of nearly all spinal motor neurons (99 ± 2%) supplying the abductor digiti minimi when the conventionally recorded MEP% was 65 ± 16% (Magistris *et al.* 1999). Similarly, TMS activates virtually all neurons innervating the abductor hallucis muscle in the lower extremities (Buhler *et al.* 2001). The complexity of the descending volleys from the motor cortex (D- and I-waves) along the long motor pathway to the muscle and intercalated spinal anterior horn synapse results in desynchronized firing of the spinal motor neuron and repetitive discharging of some motor neurons, especially at higher stimulation intensities. This causes phase cancellation phenomena and temporal dispersion of the biphasic CMAP. The triple-stimulation technique eliminates these sources of variability in the lower motor neuron pathway and thus provides a more accurate estimate of central motor system activation. Recording from hand muscle, it was shown to be 2.75 times more sensitive than conventional MEPs in detecting conduction failure of the corticospinal tract in a study of various central motor disorders (Magistris *et al.* 1999). In the lower extremity, it was reported as 2.54 times more sensitive compared to conventional recordings (Buhler *et al.* 2001). The TST technique, however, is technically more demanding, suitable only for distal muscles, and less well tolerated by patients, limiting its usefulness in routine clinical applications.

Silent period

With TMS during a sustained muscle contraction, the MEP is immediately followed by a period of electromyographic silence that is termed the cortical silent period (cSP). Assessment of the cSP as a measure of cortical inhibition has provided insights into the patho-physiology of many disorders. Its value for routine clinical application is limited due to the very large interindividual variability and its dependence on a variety of technical factors. The cSP is described in detail elsewhere in this book and discussed only briefly here.

The cSP represents an inhibitory phenomenon produced by cortical magnetic stimulation that occurs with a threshold at or slightly lower than MEP threshold. It may be observed without a preceding MEP when stimulating just below MEP threshold (Triggs et al. 1992). The early part (the first 50–60 ms) of the cSP is likely caused by spinal mechanisms including Renshaw inhibition; the later part is of cortical origin and likely mediated by cortical inhibitory interneurons that are activated by TMS (Fuhr et al. 1991a; Inghilleri et al. 1993; Ziemann et al. 1993; Manconi et al. 1998; Chen et al. 1999). The cSP duration thus provides a measure of cortical inhibitory mechanisms including gamma-aminobutyric acid $(GABA)_B$ function, and is influenced by GABAergic medication, dopamine, and ethanol (Ziemann et al. 1996).

Cortical silent period duration increases progressively with stimulation intensity; stimulation with high intensities reduces variability and is usually recommended. For quantitative assessment, the stimulation intensity needs to be standardized to the individual MEP threshold (or SP threshold). cSP durations are longest in intrinsic hand muscles (200–300 ms), and shorter and less prominent in proximal arm and leg muscles. The SP is best measured with strong voluntary background muscle contraction, higher than commonly used for MEP latency measurement. cSP duration becomes rather independent of the degree of muscle activation, as long as maintained at >60% of maximum force (Prout and Eisen 1994). At lower muscle contraction levels, cSP duration becomes more variable and more dependent on the level of contraction.

When recording the cSP, the exact measurement criteria for determining the duration need to be specified. Most commonly, the cSP duration is measured from the beginning of the preceding MEP to the return of voluntary EMG activity. Sometimes the cSP onset is measured from the beginning of the EMG silence.

Recording during a higher level of muscle contraction may also help determine the end of the cSP that is marked by a somewhat gradual return of EMG activity.

Despite the limited value of cSP duration as a diagnostic tool in the individual patient due to its high interindividual variability, it is usually symmetric between hemispheres, allowing potential use in the diagnosis of unilateral or asymmetric cerebral lesions (Fritz et al. 1997). High test–retest reliability intra-individually has been reported and, in longitudinal studies, change of cSP duration of >30 ms is considered pathological (Triggs et al. 1999a), thus making this parameter suitable for monitoring changes over time including treatment effects.

The MEP in clinical disorders

Indications for MEP testing

Since its introduction, TMS studies have been described in many neurological disorders, with an overview provided in Table 19.2. While TMS frequently demonstrates corticospinal tract involvement in patients with clinically evident upper motor neuron lesions, the value of TMS in improving the diagnostic process depends on its ability to detect subclinical lesions of the pyramidal tract in patients who lack clear upper motor neuron signs. In this way, it may establish motor system involvement in suspected MS or document central motor abnormalities in patients with suspected amyotrophic lateral sclerosis, when upper motor neuron signs may be equivocal or lacking at the time of initial presentation, possibly masked by lower motor neuron findings. The sensitivity of TMS in detecting subclinical upper motor neuron lesion varies considerably in different disorders, and depends to a large degree on the extent of testing, including the number of muscles and the different parameters studied. In many disorders (including MS), the inclusion of lower extremity muscles in the study may improve the sensitivity greatly, in part due to the longer spinal pathway to the lower-extremity motor neurons.

TMS may be used to quantify damage to the pyramidal tract system and may thus contribute to prognosis, as in patients with cerebral infarctions where it may also monitor recovery.

Table 19.2 Motor-evoked potential (MEP) abnormalities in various diseases

Disease	Sensitivity (+:~25%)	Subclinical (+:~10%)	CMCT prolongation	Amplitude reduction	Threshold	cSP duration
Multiple sclerosis	+++	+	+++	++	↑↑	↑ (↓ relapse)
Motor neuron disease						
Sporadic ALS	+++	++	+	+++/absent	early ↓/ later ↑↑	↓↓
Familial (SOD1 mutation)			+++	++		
Stroke		–	+	++/absent	↑	↑
Epilepsy			nl		↓ or nl/↑	→
Movement disorders						
Parkinson disease	–	–	nl	nl/↑ at rest (?)	nl	→
Multiple system atrophy	+ (legs)	–	+	(+)/nl/large	nl	↑ or ↓
Progressive supranuclear palsy	–	–	nl	nl/large	nl	↑
Corticobasal degneration	–	–	nl/+	nl/large	↑ or ↓	→
Dystonia	–	–	nl/(+)	nl/large	nl	→
Huntington's disease	(+)	–	nl/(+)	nl/(+)	nl/↑	↑/Westphal ↓
Progressive ataxias	++	(+)	+	large?	↑	→

Disorder					
Unilateral cerebellar lesion				↑	
Wilson's disease	++	+	++	↑	↑ / −
Stiff-person syndrome		nl	nl	→	→
Restless legs syndrome		nl	nl	nl	(↓)
Spinal disorders:					
Spond. cervical myelopathy	++	+++	++	↑	↓/↑/absent
Spinal cord injury	+++	++	++/absent	↑↑	onset lat ↑
Hereditary spastic paraparesis	+++ (legs)	(+)	+++/absent		
Syringomyelia	++	++	++/absent		
Myopathy:					
Mitochondrial	+	+	+		
Myotonic dystrophy	+	+	nl	nl	

The table presents an overview of the typical MEP findings for common parameters as reported in the literature.

'Sensitivity' refers to the frequency of abnormal central motor conduction time (CMCT) and/or MEP amplitude findings in clinically affected patients, and each + indicates about 25% on a semi-quantitative scale, (+) about half of this.

'Subclinical' indicates the frequency of abnormal CMCT and/or MEP amplitude in patients without upper motor neuron findings on examination, and each + indicates about 10% of patients.

cSP, cortical silent period; nl, normal; lat, latency.

Other indications for TMS testing include monitoring disease progression or treatment effect. The routine parameters of TMS, CMCT, and MEP amplitude are informative in many disorders and are described in detail below. The addition of MEP threshold and cSP duration, or technically more difficult and time-consuming nonstandard parameters such as paired-pulse TMS, may improve sensitivity in the early stages of various disease states (including motor neuron disease). These nonstandard measures, however, have uncertain value for the diagnosis of the individual patient and thus are only mentioned here briefly.

Pathophysiology

MEP abnormalities with routine TMS may be categorized broadly into changes affecting the MEP amplitude and/or slowing of the central motor conduction. In addition, alterations of MEP threshold or cSP duration as markers of excitability may be noted. These MEP abnormalities are not disease specific, and the correlation between clinical deficit and degree of MEP abnormalities is rather poor. In general, demyelinating disorders are associated with slowing of motor conduction, whereas neuronal disorders may show more prominent amplitude reduction or absence of a recordable potential. These general rules, however, apply to a lesser degree to the MEP in the diagnosis of central disorders than to conventional nerve conduction studies in the diagnosis of peripheral nerve disorders.

Prolongation of CMCT occurs with slowing of the rapidly conducting thickly myelinated corticospinal fibers that contribute to MEP onset, as is typically seen in demyelinative disorders. CMCT prolongation in MS is greater in progressive than in relapsing remitting forms. Significant delay may occasionally occur in disorders of axonal degeneration including motor neuron disease, especially the familial form of amyotrophic lateral sclerosis (ALS) that carries the D90A superoxide dismutase 1 mutation (Andersen *et al.* 1996; Osei-Lah *et al.* 2004). In sporadic ALS, the CMCT is usually normal to only modestly prolonged. In cerebrovascular disorders, the CMCT is usually within normal limits or only mildly prolonged.

There are several mechanisms contributing to prolongation of CMCT in axonal loss. If conduction along the large myelinated corticospinal fibers fails due to axon loss or demyelination, transmission occurs along the smaller, slowly conducting fibers and possibly also along alternative oligosynaptic pathways, such as the corticorubro-spinal tracts (Weber *et al.* 2000). Loss of corticospinal fibers reduces the size and synchrony of the descending volleys reaching the spinal motor neuron and increases the time to activation by temporal summation. Activation of the spinal motor neuron not by the D- or I1-wave, but by later I-waves, adds ~1.5 ms or so for each additional wave, possibly accounting for a total delay of up to 5 ms. This process is also operative in patients with cortical or spinal hypoexcitability.

Reduction in MEP amplitude: the MEP amplitude (or area) is physiologically lower than the peripheral CMAP as a result of phase cancellation, as explained above, with a typical MEP% in intrinsic hand muscles of 50–60%. Neuronal disorders typically result in MEP amplitude reduction or absent MEP response. In demyelinative disorders, reduced MEP amplitude may also occur as a result of additional temporal dispersion and phase cancellation. Severe depression of cortical or spinal excitability may result in reduced or absent MEP amplitude. In general, absence of MEP is rare in MS and cervical spondylosis with myelopathy, whereas it is a common finding in motor neuron disorders and cerebrovascular disease.

Multiple sclerosis

Since the first clinical studies of TMS, prolongation of CMCT has been reported as a frequent finding in MS (Barker *et al.* 1986; Hess *et al.* 1986) confirming prior observations with electrical motor cortex stimulation. These and multiple subsequent studies documented the typical findings in MS patients as variable CMCT prolongation (ranging from normal to markedly prolonged latency) that may be accompanied by reduced or dispersed MEP amplitude. In addition, increase in threshold and changes of cSP duration, most commonly prolongation, are noted.

The first study comparing different evoked potential modalities including MEPs in MS reported prolonged CMCT from intrinsic hand muscle in 72% of 83 patients, and 10% had CMCT more than three times the upper normal limit (Hess *et al.* 1987a). In comparison, visual-evoked potentials (VEPs) were abnormal in 67%, somatosensory-evoked potential (SEPs) in 59%, and brain-stem auditory-evoked potentials (BAER) in 39%. MEP amplitude reduction was found in about one-half of the patients, in some patients without CMCT prolongation.

Abnormal CMCTs are more likely found if multiple target muscles are studied, especially if this includes one or more leg muscles (Ingram *et al.* 1988; Ravnborg *et al.* 1992; Mathis and Hess 1996). In a prospective study of 68 patients (Ravnborg *et al.* 1992) with suspected demyelinating disease, MEPs recorded from three upper-extremity and two lower-extremity muscles were positive in 83% of patients diagnosed subsequently with MS, compared to 67% with abnormal VEPs and 63% SEPs. Brain MRI was the most sensitive test with 88%. There was a good correlation between MRI and MEP, with 85% concordance.

In multiple additional studies, the sensitivity of MEP testing for detecting abnormalities in MS patients ranged between 57% and 89% (Rossini *et al.* 1989; Britton *et al.* 1991; Caramia *et al.* 1991; Jones *et al.* 1991; Mayr *et al.* 1991; Andersson *et al.* 1995), and generally higher than afferent evoked potentials when compared (Beer *et al.* 1995). The sensitivity of CMCT, MEP amplitude and threshold measurements increased further when interside asymmetries in MEP latency and/or in CMCT were considered (Cruz-Martinez *et al.* 2000; Sahota *et al.* 2005).

MEP abnormalities are usually present in muscles that show clinical weakness and also common when upper motor neuron signs are present, but only occasionally in extremities with normal clinical examination. MEP recordings therefore are more helpful in confirming abnormalities in patients with clinically equivocal motor findings than establishing motor lesions in patients with normal examination. The sensitivity in detecting subclinical motor involvement in demyelinating disorders was 13.5% in 162 patients when studying CMCT to five muscles including two in the lower extremity (Di Lazzaro *et al.* 1999a) and similar in other studies, clearly lower than for VEPs (Hess *et al.* 1987b; Kandler *et al.* 1991).

Both CMCT delay and amplitude reduction correlate to some degree with clinical deficit and the presence of upper motor neuron signs. This was shown for presence of brisk finger flexor reflexes in regard to hand muscle (Hess *et al.* 1987b), and in the leg for hyper-reflexia and presence of the Babinski sign (Ingram *et al.* 1988; Jones *et al.* 1991). Correlation between CMCT prolongation and clinical deficit in MS patients has also been shown in multiple other studies (Kandler *et al.* 1991; Facchetti *et al.* 1997; Fuhr *et al.* 2001), but has not been consistent. In a study of progressive MS patients, CMCT to upper extremities and clinical measurements did not correlate. Correlation between CMCT to the anterior tibialis muscle and disability was weak, whereas a correlation to the lesion load in the cervical cord (on MRI) was noted for the upper extremities only (Kidd *et al.* 1998). In a large study separating patients with relapsing–remitting MS (RR-MS) from progressive forms, CMCT did not relate to clinical motor deficit in either group (Humm *et al.* 2003). No significant change in CMCT was found in a large group of RR-MS patients studied during relapse and then again in remission (Caramia *et al.* 2004). Differences between types of MS, in particular longer CMCT in secondary progressive MS compared to RR-MS, were already noted previously and felt to be secondary to increased spinal conduction time (Facchetti *et al.* 1997). The triple-stimulation TMS (TST) study by Humm *et al.* (2003) demonstrated that CMCT was markedly more prolonged in patients with progressive MS who were either secondary progressive (SP-MS) or primary progressive (PP-MS) compared to RR-MS patients, even in patients with similar clinical motor deficit and disease duration. The difference was not attributed to differential involvement of the spinal cord, but to a pathophysiological difference of corticospinal conduction between these two disease states. The marked prolongation of CMCT in progressive MS was felt to be due to persistent demyelination of fibers when the capacity for remyelination has been exhausted. The TST technique documented conduction failure in RR-MS patients believed to be the cause of clinical deficit in this group, whereas amplitude reduction in progressive MS patients was felt to be secondary to axonal loss primarily. In a study of MS patients and temperature-associated changes (Uhthoff phenomena), patients with prolonged CMCT were more likely to have subjective vulnerability to temperature and corresponding TST amplitude changes (reflecting conduction block), whereas CMCT was not significantly affected by temperature (Humm *et al.* 2004).

Evoked potentials have a limited role in predicting future disease course. In 30 patients with RR-MS or SP-MS, VEP and MEP abnormalities correlated with Expanded Disability Status Scale (EDSS) scores within the 2 year observation period, and VEP and MEP at baseline correlated with the EDSS 2 years later (Fuhr *et al.* 2001). The correlation with future course is limited to group data; individual prediction with reliability is certainly not possible. In another study, a combination of pathological SEP and MEP findings (but not VEP) in patients studied early after diagnosis (within 2 years of disease onset) predicted clinical disability with EDSS ≥3.5 after 5 years (Kallmann *et al.* 2006). In this study, however, evoked potential (EP) data and EDSS at first presentation were not significantly correlated, and this was interpreted by the authors to mean that EP abnormalities in part represented clinically silent lesions not mirrored by EDSS.

Evaluating other parameters in addition to CMCT and MEP amplitude may increase sensitivity and provide further pathophysiological insight. MEP threshold is often increased in advanced MS (Caramia *et al.* 1991). cSP duration may be prolonged (Haug and Kukowski 1994), and this parameter was more sensitive than CMCT prolongation in subclinical disease (Tataroglu *et al.* 2003). Extremely prolonged cSP duration was noted in MS patients with cerebellar dysfunction (Tataroglu *et al.* 2003).

A recent study compared excitability parameters in RR-MS patients during stable (remitting) phase vs clinical exacerbation (relapse) (Caramia *et al.* 2004). Patients in relapse showed increased threshold and reduced cSP duration and lack of intracortical inhibition on paired-pulse testing, consistent with cortical hyperexcitability. By contrast, patients in remission showed a significant cSP prolongation with normal motor thresholds.

A particularly sensitive nonstandard MEP measure in MS patients is the transcallosal inhibition (TI), a marker for corpus callosum and periventricular white matter involvement in MS. Stimulating with a focal coil over the cerebral hemisphere ipsilateral to the voluntarily contracted target muscle induces a brief interruption in EMG activity, the 'ipsilateral cSP', whose onset latency and duration is measured (Wassermann *et al.* 1991). In MS patients, abnormal TI is more frequently observed than prolongation of CMCT or MEP reduction, especially in early stages of MS (Hoppner *et al.* 1999; Schmierer *et al.* 2000). The transcallosal inhibition had better correlation than CMCT with clinical disability (Schmierer *et al.* 2002) and corpus callosum lesion load on MRI (Hoppner *et al.* 1999).

Similar to other evoked potentials, MEP abnormalities vary considerably in individuals with MS when measured longitudinally and may improve with treatment. TMS may be an objective physiological tool to monitor treatment responses in MS. MEP threshold, CMCT and motor strength improved in patients with RR-MS treated with high-dose intravenous methylprednisolone for 5 days (more with 2 g daily vs 1 g daily) (Fierro *et al.* 2002). In a study of 0.5 g methylprednisolone daily for 5 days CMCT did not change, but the amplitude ratio (measured with TST) improved in RR-MS and SP-MS together with clinical improvement, with no change noted in PP-MS (Humm *et al.* 2006). In a study of eight women on interferon beta-1a treatment over the course of 1 year, MEP latency did not change, whereas MEP amplitude at 6 months was less reduced and recovery from post-exercise depression of MEP amplitudes was faster on treatment compared to baseline (White and Petajan 2004). Treatment with 3,4-diaminopyridine, a potassium-channel blocker that improves fatigue and motor function in MS, did not change CMCT, whereas intracortical excitability was increased (paired-pulse testing) (Mainero *et al.* 2004).

Motor neuron disease

As expected for an axonal disorder, the diagnostic value of routine CMCT measurements in motor neuron disease (MND) is limited.

The first TMS study in this disorder reported normal CMCT to hand muscle in five patients with definite upper motor neuron signs (Barker *et al.* 1986, 1987). Subsequent studies documented similar findings as with electrical stimulation (Ingram and Swash 1987), mainly mild to moderate CMCT prolongation in some patients, and more frequently reduction or dispersion of MEP amplitude, or complete absence of MEP response (Schriefer *et al.* 1989; Eisen *et al.* 1990; Berardelli *et al.* 1991; Uozumi *et al.* 1991; Claus *et al.* 1995; Mills and Nithi 1998). By measuring MEPs from three upper-extremity muscles (thenar, forearm extensors, biceps brachii), Eisen *et al.* (1990) reported a sensitivity of abnormal MEPs approaching 100%. Had they limited the recording site to the thenar muscle, the yield would have been reduced by 25%. Whereas CMCT was not significantly different in their 40 patients with MND compared to controls, the total MEP delay was frequently abnormal and MEP amplitude often reduced or absent, including in several patients with MEP% of <15% (Eisen *et al.* 1990). Absent MEP was particularly noted in patients with upper motor neuron and pseudobulbar signs, but not lower motor neuron patients. In another study, reduced MEP% was more closely correlated with pyramidal tract involvement than prolonged MEP latency (Uozumi *et al.* 1991).

In a large group of 129 patients, the sensitivity of MEPs in documenting corticospinal tract involvement was 74% in patients with MND, defined as prolonged CMCT or absent MEP in any of five muscles recorded (Di Lazzaro *et al.* 1999a). In this large study of various disorders, TMS had the greatest sensitivity for detection of subclinical central motor pathway abnormalities in MND, with a sensitivity of 26%. An interesting finding in this study was frequent isolated abnormality of CMCT to the biceps brachii that was never found in other disorders. In follow-up, all patients with suspected MND and subclinical abnormalities were later diagnosed with ALS (Di Lazzaro *et al.* 1999a).

Several groups have evaluated MND patients with particular emphasis on different stages or forms including pure lower motor neuron presentation. Recording from the ADM and flexor hallucis muscle (FH), prolonged CMCT or absent MEP was noted in 95% of patients with definite ALS, 72% of suspected ALS with probable upper motor neuron signs, 50% of pure lower motor neuron syndrome, and 20% of pseudobulbar presentation (Miscio *et al.* 1999). The authors concluded that TMS allowed early prediction of ALS in patients with clinically equivocal upper motor neuron impairment, because all patients that developed definite ALS within the subsequent year had abnormalities on the initial TMS study. In another study, CMCT prolongation or MEP absence in either ADM or tibialis anterior (TA) muscle was noted in 71% definite, 71% probable, 81% possible, and 25% suspected ALS patients, with 63% overall and somewhat more sensitive than proton resonance spectroscopy (Pohl *et al.* 2001). A recent study documented decreased MEP:CMAP ratio in ALS patients with rather intact muscle strength, whereas the amplitude ratio was increased in limbs with no upper motor neuron signs (de Carvalho *et al.* 2003).

Detecting upper motor neuron dysfunction in ALS patients may be enhanced by studying a minimum of three muscles, by selecting weak muscles or distal muscles, and by measuring the interside differences (Osei-Lah and Mills 2004). In this study, 44% of patients or 25 of 200 muscles were abnormal for MEP latency, CMCT or MEP absence, with 30% of patients demonstrating the sole abnormality of increased interside difference of CMCT to the abductor hallucis (AH) muscle.

Measurement of threshold and cSP duration in addition to CMCT and MEP amplitude may increase the sensitivity of MEP testing in motor neuron disorders. A normal or lower than normal MEP threshold was reported for early stages of disease, well compatible with the theory of cortical hyperexcitability (Caramia *et al.* 1991; Eisen *et al.* 1993). Mills and Nithi (1997) confirmed a decreased corticomotor threshold when recording from FDI muscle in ALS patients with no abnormal physical signs other than fasciculations in intrinsic hand muscles, whereas increased threshold was noted in hands with lower motor neuron signs only or mixed upper and lower motor neuron signs. Evidence for cortical hyperexcitability in ALS was also derived from the shortened cSP duration, in particular when testing at higher stimulation intensity (Prout and Eisen 1994; Salerno *et al.* 1996) and this was felt to be a neurophysiological marker of ALS (Desiato and Caramia 1997).

Several studies combined testing of CMCT, MEP amplitude, MEP threshold, and cSP duration in MND. In a large group of 121 patients, 69% had neurophysiological evidence for upper motor neuron dysfunction by at least one abnormal parameter, with 83% for definite ALS, 75% in patients with probable upper motor neuron signs, and 27% of patients with purely lower motor neuron syndrome clinically (Triggs et al. 1999a). In this study, increased MEP threshold was the most sensitive parameter, and longitudinal assessment including progressive inexcitability of central motor pathways and loss of the normal cSP provided additional diagnostic value. They also noted that when administering TMS during voluntary contraction, MEP sometimes failed to facilitate and became obscured in the background contraction, leaving only the cSP. Another study also looked at these four MEP parameters and found at least one abnormal value in 84% of 54 ALS patients, with shortened cSP duration as the most sensitive parameter, abnormal in 70% (Pouget et al. 2000).

The same group recently provided a prospective study of 40 patients with lower MND, over a time frame of 1–4 years with visits every 3 months (Attarian et al. 2005). Six of the seven patients who later developed ALS had abnormality of cSP duration on initial testing associated or not with abnormal MEP threshold and MEP/CMAP ratio. When combining cSP duration and amplitude ratio, the overall sensitivity for diagnosis of ALS among the lower motor neuron patients was 86% and its specificity was 94%. Patients with pure motor neuropathy had normal cSP duration and MEP amplitude ratio, with 5/14 mildly increased CMCT (likely due to slowing in the proximal nerve root segment, with the peripheral latency determined by root stimulation).

As was noted already in the initial TMS studies in MND, the correlation between the degree of MEP abnormalities including CMCT prolongation and MEP amplitude reduction with the extent of clinical upper motor neuron signs is rather poor (Schriefer et al. 1989; Claus et al. 1995). Mills et al. (2003) described a large longitudinal study in 76 patients with idiopathic ALS,

testing the FDI muscles bilaterally with TMS, in many patients from presentation to death. Patients were classified according to region of onset and the physical signs in the hands. MEP threshold and CMCT showed no change as the disease evolved except for patients with mixed signs, who had a terminal increase in threshold and CMCT prolongation. MEP amplitude declined over the course of the disease. cSP duration was shorter than in controls early in the disease and then showed progressive lengthening throughout the illness, but generally values remained within the normal range. They concluded that none of the measures of central motor function are likely to be useful for monitoring ALS patients in a clinical trial setting. In agreement with this assessment, recent recommendations for neurophysiological measures in ALS in clinical trials included MEP testing only as a secondary parameter due to its limited value (de Carvalho et al. 2005).

Using nonstandard MEP techniques has provided further evidence for cortical hyperexcitability in ALS, in particular by demonstrating decreased inhibition on paired pulse TMS (Yokota et al. 1996; Ziemann et al. 1997; Hanajima and Ugawa 1998; Salerno and Georgesco 1998). The triple stimulation technique (TST) was found to be more sensitive in detecting subclinical upper motor neuron involvement in suspected or possible ALS cases, with evidence for conduction failure in many patients who had normal MEPs on conventional testing (Rosler et al. 2000; Komissarow et al. 2004). Recording MEPs from cranial muscles is another approach that may increase the detection of central motor abnormalities early in the disease. Absent or delayed MEPs from the masseter muscle were found in 63% of ALS patients including four (22%) without clinical bulbar signs (Trompetto et al. 1998). Urban et al. (1998) reported MEP abnormalities in the tongue in 50%, and in the orofacial muscles in 57% of ALS patients, with 70% being abnormal when combining both sites, whereas only 7% of their patients had clinical evidence of upper motor neuron involvement in the cranial nerves (Urban et al. 1998, 2001). Another study reported that recording MEPs from the trapezius muscle was abnormal in all of 10 patients with ALS tested, making this particularly useful when compared with cervical myelopathy (Truffert *et al.* 2000).

Unlike those with sporadic ALS, patients with slowly progressive familial ALS with the autosomal recessive D90A superoxide dismutase mutation have rather markedly prolonged central motor conduction (Andersen *et al.* 1996; Weber *et al.* 2000; Osei-Lah *et al.* 2004).

Patients with *primary lateral sclerosis* (PLS) often have absent MEPs, especially if symptoms begin in the legs and then progress in an ascending fashion (Zhai *et al.* 2003). Patients with multifocal onset had rather heterogeneous MEP findings in this study. Others have reported increased thresholds and prolonged CMCT (Weber and Eisen 2002).

Myelopathy

MEP studies are sensitive in detecting *compressive cervical myelopathy* due to *spondylosis or disk herniation*. The typical findings include prolongation of CMCT (that may be rather marked), with MEP amplitude reduction and dispersion (Dvorak *et al.* 1990; De Noordhout *et al.* 1991, 1998; Di Lazzaro *et al.* 1992a; Travlos *et al.* 1992; De Mattei *et al.* 1993; Brunholzl and Claus 1994; Tavy *et al.* 1994; Lo *et al.* 2004; Chan and Mills 2005). The mechanisms of conduction slowing in cervical spondylotic myelopathy is not a focal conduction delay across the compression site, but rather corticospinal conduction block and impaired temporal summation of the descending volleys at the spinal motor neuron resulting in delayed motor neuron firing (Kaneko *et al.* 2001; Nakanishi *et al.* 2006).

If lower-extremity recordings are included in MEP studies, sensitivity as high as 100% has been reported (Lo *et al.* 2004). In this prospective study, 141 patients with a clinical diagnosis of cervical spondylotic myelopathy were classified into four groups based on severity of cervical cord changes on MRI. All patients with cervical spondylosis but no cord deformity (group 1) had normal CMCT to both FDI and AH muscles. In patients with cord lesions on MRI, CMCT increased with severity of MRI cord compression. All patients with higher-grade cord compression, defined as less than two-thirds of its original antero-posterior diameter and without

T2 cord signal change (group 3) or with cord signal change (group 4), had abnormal CMCT or absent MEPs in at least one limb, more frequently in the lower than upper extremities. Mean CMCT to the FDI muscle was 12.1 ± 3.7 ms in group 3 and 12.9 ± 3.1 ms in group 4 (vs 5.5 ± 1.1 ms in the control group), and to the AH muscle 21.3 ± 4.6 ms in group 3 and 21.5 ± 4.6 ms in group 4 (vs 12.3 ± 1.9 ms in controls). The sensitivity and specificity for TMS for differentiating the presence from absence of MRI cord abnormality were 100% and 85%, respectively; TMS was considered an effective screening technique for cervical cord abnormalities.

In a study of patients with clinically suggestive and myelographically documented spondylotic cord compression, MEP was more sensitive than somatosensory-evoked potentials (SEPs), possibly because bony spurs projecting from the vertebral bodies predominantly affect the antero-lateral portion of the spinal cord (De Noordhout *et al.* 1998). In this study of 55 patients, MEPs were abnormal in the biceps brachii muscle in 38%, FDI 89%, and anterior tibialis muscle 85%, with at least one muscle abnormal in 93%. SEPs recorded from ulnar, median, and tibial nerves were abnormal in at least one nerve in 73%. Other studies also indicate greater sensitivity of MEP recordings than clinical upper motor neuron signs in diagnosing myelopathy, and slowing of CMCT or reduction of MEP/CMAP ratio may precede clinical evidence for myelopathy (De Noordhout *et al.* 1991, 1998; Travlos *et al.* 1992; Brunholzl and Claus 1994). In 28 patients with cervical spondylotic myelopathy, MEPs were abnormal in 27 patients, the responses in the leg muscles being affected most often. Clinically asymptomatic motor lesions were detected in 25%. CMCT for the ADM correlated significantly with the clinical disability, whereas the radiological findings did not (Tavy *et al.* 1994).

In combination with SEP testing, neurophysiological abnormalities were reported in up to 50% of patients with 'silent' compression, and one-third of patients with abnormal EPs at baseline deteriorated over the next 2 years, whereas no patient with normal electrophysiological studies at baseline worsened during the observation period (Bednarik *et al.* 1998). On the other hand, in a large study of asymptomatic spondylotic cord compression, MEPs were normal in 92% (Tavy *et al.* 1999).

MEP testing may be particularly helpful in deciding upon surgery in clinically equivocal patients, and serial recording might be useful in ascertaining progressive myelopathy (De Noordhout *et al.* 1998; Di Lazzaro *et al.* 1999a). After surgical decompression, MEPs normalize only in a few patients, independently of clinical outcome (De Mattei *et al.* 1993; de Noordhout *et al.* 1998). Improved MEPs after surgery were noted in patients with severe, but not mild, myelopathy (Bednarik *et al.* 1999).

Recording MEPs from several muscles with different spinal segmental innervation may help localize the level of myelopathy by demonstrating normal CMCT above and prolonged CMCT below the relevant segment (Di Lazzaro *et al.* 1992b; Tavy *et al.* 1994; Mathis and Hess 1996; Chan *et al.* 1998). Typically, a normal recording from the biceps brachii muscle together with a prolonged CMCT to the intrinsic hand muscles may document a midcervical lesion level in the appropriate setting (Mathis *et al.* 1996). This may be particularly helpful when the spinal MRI shows lesions at multiple levels with unclear clinical significance (Chan and Mills 2005).

MEP study is also sensitive for *other extramedullary cervical cord compression* including changes from rheumatoid arthritis, atlantoaxial subluxation, and meningeoma and other tumors (Dvorak *et al.* 1990; De Mattei *et al.* 1993; Brunholzl *et al.* 1994).

In *intramedullary cervical cord tumors* and *syringomyelia*, MEP findings are similar and include CMCT prolongation and reduced/absent MEP amplitude, again more frequently found in the recordings from the lower extremities (Nogues *et al.* 1992; Brunholzl *et al.* 1994). The sensitivity is somewhat lower compared to extramedullary cord compression. In 22 patients with syringomyelia, both MEP and SEP showed a close relation between clinical symptoms and electrophysiological data and detected subclinical deficits (Masur *et al.* 1992). Improvement in CMCT after syringomyelia shunting surgery was noted (Robinson and Little 1990).

In patients with traumatic *spinal cord injury* (SCI) in chronic phase, MEPs correlate well with the motor function on clinical examination. Patients with *chronic motor complete SCI* have generally absent MEPs below the lesion level. When recording MEPs from paravertebral musculature in thoracic myelopathy, MEPs may be elicited below the level of a complete SCI, likely due to multisegmental innervation of these muscles and the long muscle fiber conduction (Cariga *et al.* 2002; Taniguchi *et al.* 2002). Rarely, the combined electrical cutaneous stimulation of peripheral afferents and TMS timed to provide convergent input to the target spinal motor neuron pool resulted in modified segmental reflexes and thus provided electrophysiological evidence of conduction across the lesion (Hayes *et al.* 1991). Above the lesion of complete SCI in chronic phase, the MEP threshold is decreased, suggesting corticospinal hyperexcitability as a result of central plasticity (Cariga *et al.* 2002). In *chronic motor incomplete SCI*, the presence of MEPs is correlated with voluntary motor control (McKay *et al.* 2005). In patients with American Spinal Injury Association (ASIA) Impairment Scale C and D, MEPs if present have higher thresholds, reduced amplitudes, and mild to moderately prolonged MEP latency and CMCT. Patients with chronic motor incomplete SCI of the cervical or thoracic cord are more likely to show preserved volitional contractions and MEPs in distal than proximal lower-limb muscles (Calancie *et al.* 1999). The response latencies are delayed equally for persons with injury to the cervical or thoracic cord, suggesting normal central conduction velocity in motor axons caudal to the lesion (Calancie *et al.* 1999). MEPs can be absent in some incomplete SCI patients who may be able to volitionally activate muscles below the lesion level (McKay *et al.* 1997).

Conditioning the MEP with paired-pulse stimulation at intervals of 15–50 ms increased the likelihood of obtaining an MEP in incomplete SCI patients classified as ASIA C and D (McKay *et al.* 2005).

In *acute SCI*, MEPs are often absent initially, and as predictors of outcome, results have been mixed. In cervical SCI patients, MEP recordings from the ADM muscle correlated to outcome hand function, and patients who lacked both ADM and biceps brachii MEPs in the initial study developed no or only a passive hand function (Curt *et al.* 1998). MEP values of the lower limbs are about as predictive for recovery of ambulatory capacity as the clinical motor scores in the legs (Curt and Dietz 1999), and only a few (<20%) acute SCI patients with initial loss of lower-limb MEPs finally achieved a functional ambulatory capacity (Curt et al. 1998). A study with very early TMS showed that MEPs are no better than clinical motor examination in documenting residual motor function. In this study of 25 quadriplegic and paraplegic patients studied within 6 h of injury by recording from the ADM, biceps brachii, flexor hallucis brevis, and tibialis anterior muscles on each side, MEPs were not obtained in patients without preceding clinical evidence of voluntary activation (Macdonell and Donnan 1995). This was the case even for muscles that later had motor recovery after an initial paralysis. MEP thresholds at rest in ADM and biceps were elevated acutely even when above the level of injury (Macdonell and Donnan 1995). In a longitudinal study of 21 SCI patients over several years, clinical improvement was noted up to ~300 days post-injury, whereas the prolongation of MEP latency did not change over time and there was no correlation between clinical assessment and electrophysiological data (Smith et al. 2000). This study also showed that within a few days of spinal trauma the latency to the onset of the cSP is prolonged to a greater degree than the MEP latency, indicative of reduced corticospinal inhibition and in agreement with earlier findings in chronic SCI patients (Davey et al. 1998). MEPs in acute SCI may be most useful for prognosis in comatose patients or subjects otherwise unable to cooperate with volitional motor activation.

In hereditary spastic paraparesis (HSP), a group of genetically heterogeneous disorders leading to axonal degeneration of central pathways, MEPs of the legs are often abnormal with either delayed and reduced or absent responses, whereas responses in the upper limbs are often normal or show only mildly increased CMCT (Claus et al. 1990; Pelosi et al. 1991; Schady et al. 1991; Polo et al. 1993; Cruz Martinez and Tejada 1999; Nielsen et al. 2001). Findings vary according to genetic subtype, and considerable delay of central motor conduction including in upper-extremity recordings has been described (Schady et al. 1991). Studies in autosomal dominant HSP linked to chromosome 2p document central motor conduction velocities ranging from normal to greatly reduced and increased cortical excitability (Nielsen et al. 1998, 2001; Nardone and Tezzon 2003) that was not present in patients with linkage to chromosome 16q (Nardone and Tezzon 2003).

Radiation therapy of the spine may cause radiation injury to the cord with prolonged CMCT even in asymptomatic patients (de Scisciolo et al. 1991).

In 39 patients with acute transverse myelitis, CMCT to the abductor digiti minimi muscle was abnormal in 30% and to the anterior tibialis muscle in 90%, and CMCT prolongation correlated with muscle power, tone, reflex, and MRI changes (Kalita and Misra 2000).

Patients with human T-cell lymphotrophic virus type I-associated tropical spastic paraparesis (HAM/TSP) have markedly prolonged CMCT and reduced MEP amplitudes to the legs, and less slowing or normal responses to the arms. CMCT increased as the disease progressed from mild to moderate, thereafter remaining largely unchanged (Young et al. 1998).

Cerebral infarction

In the acute stage of cerebral infarcts, reduced or absent MEP amplitude and mildly prolonged CMCT are common findings over the affected hemisphere. In patients tested within 24 h of stroke onset, ~20% have absent MEPs, usually in subjects with severe motor paralysis (Catano et al. 1995; D'Olhaberriague et al. 1997). In less severely affected patients, MEPs often have reduced amplitude, prolonged latency, and increased threshold (Homberg et al. 1991; Ferbert et al. 1992; Heald et al. 1993a). If only the CMCT is assessed, the false-negative rate is one-third of patients with middle cerebral artery infarctions (Di Lazzaro et al. 1999a).

Many studies have indicated prognostic value of early TMS studies in predicting outcome in ischemic stroke patients (Arac et al. 1994; Catano et al. 1995; Traversa et al. 1997; Escudero et al. 1998; Trompetto et al. 2000; Steube et al. 2001; Hendricks et al. 2003). In general, the presence of MEPs on early testing correlates with a good functional recovery in most reports, whereas absent MEPs predict poor function. In a large study of 118 patients with first-ever stroke of mixed etiology and location (excluding subarachnoid hemorrhage) who underwent TMS within 72 h and serially up to a year, patients with MEPs present on initial testing had consistently higher functional scores throughout the study and better recovery at 12 months, whereas those with absent MEPs had high probability of death and poor functional outcome (Heald *et al.* 1993a). Of the patients with initially present MEPs, patients with prolonged CMCT recovered more slowly than those with normal CMCT, but were similar after 1 year. In another study, patients with normal CMCT recovered better by 6 months than those with delayed CMCT, and the presence of MEPs provided information on motor recovery regardless of initial strength (Escudero *et al.* 1998).

The recovery of MEP latency on serial testing is highly correlated with return of muscle strength and hand function scores (Turton *et al.* 1996). For patients tested within 24 h with recordings from FDI muscle, relatively preserved MEP amplitude (>5% of maximum motor response with electrical ulnar nerve stimulation) was a better prognostic factor than normal CMCT (Rapisarda *et al.* 1996). MEP testing was more sensitive than clinical examination to detect residual corticospinal function, and nearly all muscles with present MEPs on initial testing ultimately showed motor recovery (Hendricks *et al.* 2003). They reported correlation between abductor digiti minimi MEP amplitude and the hand motor score 6 months later, and between biceps brachii amplitude and the subsequent arm motor score, but no clear correlation between MEP from leg muscles and lower-extremity motor scores. The presence of MEPs in a patient with locked-in syndrome also indicates better prognosis (Bassetti *et al.* 1994).

MEPs are best recorded during facilitation (by voluntary contraction of target muscle, or if not possible contralateral muscle), as MEPs are often absent at rest in affected limbs of acute stroke patients who may have good outcome (Heald *et al.* 1993b). Significant motor recovery occurs in some patients with absent MEPs on initial evaluation, even when assessed during facilitation (Hendricks *et al.* 2003).

MEPs recorded from patients with superficial (cortical) infarcts have smaller amplitudes and longer CMCTs than those in deep infarcts, in correlation with worse prognosis in the cortical infarct group (D'Olhaberriague *et al.* 1997). Serial MEP recordings over 3 months documented gradual recovery of MEP amplitudes, to a greater degree for the deeper infarcts, whereas CMCT changes over time were generally variable. On initial testing (day 1) MEPs were evenly absent in upper- and lower-extremity muscles, with subsequently earlier recovery in proximal arm muscle (biceps) and in lower-limb muscles than in hand muscle (D'Olhaberriague *et al.* 1997). Normalization of MEP is greatest in the first 80 days, but continues for many months (Heald *et al.* 1993a; Cicinelli *et al.* 1997; Traversa *et al.* 2000).

The cSP duration is often prolonged in patients with cerebral infarcts (Haug and Kukowski 1994; Braune and Fritz 1995; Ahonen *et al.* 1998), but may also be shorter than normal in patients with cortical infarcts (Werhahn *et al.* 1995). The length of the cSP duration on day 7 had predictive value in one study: patients with good recovery on follow-up 3 months later and normal controls typically had longer SPs with increasing stimulation intensity, whereas in acute patients with subsequently poor outcome and in chronic patients with spasticity there was shortening of the cSP with increasing volitional isometric muscle contraction (Catano *et al.* 1997).

Ipsilateral MEPs are sometimes present in stroke patients when stimulating the unaffected cerebral hemisphere and recording from the paretic muscle. These ipsilateral MEPs are more easily elicited in proximal than in distal muscles, and may have poor prognostic implications according to some authors (Turton *et al.* 1996), or be important for recovery (Caramia *et al.* 1996).

Movement disorders

MEP studies in movement disorders are discussed elsewhere in this volume in greater detail. The study of excitability parameters has provided valuable pathophysiological insights into various movement disorders, whereas the diagnostic value of routine MEP parameters (CMCT, MEP amplitude) is rather limited. The TMS literature in regard to cortical excitability in movement disorders is extensive and not reviewed here; for an excellent review see Cantello (2002).

CMCT is normal in most movement disorders including idiopathic *Parkinson's disease* (Valls-Sole *et al.* 1994; Abbruzzese *et al.* 1997). A few studies have indicated a shortening of CMCT (Kandler *et al.* 1990; Dioszeghy *et al.* 1999; Hu *et al.* 1999) explained subsequently as likely secondary to increased background EMG in Parkinson patients due to difficulty with relaxation (Cantello 2002). Several studies reported increased MEP amplitudes 'at rest' in Parkinson patients (Kandler *et al.* 1990; Cantello *et al.* 1991) and voluntary muscle contraction elicited smaller MEP increase in patients than in controls (Valls-Sole *et al.* 1994). Abnormal excitability parameters include a shortened cSP (Cantello *et al.* 1991).

Some patients with atypical parkinsonian syndrome due to *multisystem atrophy* (MSA) may have prolonged CMCT to the lower limbs (Abbruzzese *et al.* 1997), but this was not confirmed in another study (Abele *et al.* 2000), and recordings in upper limbs have been normal (Cruz Martinez *et al.* 1995; Abbruzzese *et al.* 1997). Studies in *progressive supranuclear palsy* (PSP) (Molinuevo *et al.* 2000) and *corticobasal degeneration* (CBD) (Valls-Sole *et al.* 2001) reported normal CMCT. Motor cortex disinhibition in MSA and PSP was noted, and hypoexcitability in CBD (Kuhn *et al.* 2004). In a study of transcallosally mediated cortical inhibition, the ipsilateral cSP was abnormal in five CBD and five PSP patients, but normal in five patients with MSA and 10 patients with idiopathic parkinsonian syndrome (Wolters *et al.* 2004).

In *dystonia* including focal dystonia (writer's cramp) and spasmodic torticollis the CMCT is normal. Nearly all studies reported normal MEP amplitude at rest, with several authors indicating enhanced MEP facilitation during contraction and shortening of the cSP (Mavroudakis *et al.* 1995; Ikoma *et al.* 1996).

According to several studies of *Huntington's chorea*, CMCT and MEP amplitude are normal. Discrete alterations of special MEP measurements (latency variability, latency:amplitude ratio) and increased MEP thresholds in some patients were reported by Meyer *et al.* (1992b), and in one study also decreased MEP amplitude (Priori *et al.* 1994). Huntington patients have motor cortex hypoexcitability as documented by prolonged cSP duration (Priori *et al.* 1994; Modugno *et al.* 2001) except in patients exhibiting

muscle rigidity who had abnormally short cSP (Tegenthoff *et al.* 1996).

Among patients with cerebellar disorders, patients with *Friedreich's ataxia* have moderate to marked prolongation of CMCT and decreased MEP amplitudes, to a slightly lesser degree also noted in patients with early-onset cerebellar ataxia with retained reflexes (Claus *et al.* 1988; Cruz Martinez and Anciones 1992; Lanzillo *et al.* 1994; Mondelli *et al.* 1995; Schwenkreis *et al.* 2002). MEP threshold is increased.

In late-onset hereditary cerebellar degenerations, a heterogeneous group with variable pyramidal tract involvement depending on subtype, mild to moderate CMCT prolongation was noted in some patients (Claus *et al.* 1988; Mondelli *et al.* 1995). In particular, patients with *spinocerebellar ataxia* type 1 (SCA1) have prolonged CMCT, sometimes rather markedly, and increased MEP threshold (Schols *et al.* 1995, 1997; Perretti *et al.* 1996; Abele *et al.* 1997; Yokota *et al.* 1998; Schwenkreis *et al.* 2002). In SCA2 and SCA3 (Machado–Joseph disease) central motor conduction was mostly normal, but prolongation of CMCT to the lower limbs was reported in one study (Restivo *et al.* 2000).

In subjects with *unilateral cerebellar lesions* the MEP threshold is increased in the motor cortex contralateral to the impaired hemicerebellum (Di Lazzaro *et al.* 1994), similar to the increased MEP threshold described for Friedreich's ataxia and SCA1.

TMS studies in *Wilson's disease* indicate prolonged CMCT in ~30–70% of patients, often in combination with reduced or absent MEP amplitude (Chu 1990; Meyer *et al.* 1991b; Hefter *et al.* 1994). In a few patients with Wilson's disease who had abnormal MEPs by TMS, the TES resulted in normal responses, interpreted as evidence of intracortical presynaptic motor dysfunction (Perretti *et al.* 2001). MEP changes are potentially reversible, at least in the early stages of the disorder (Meyer *et al.* 1991a; Hefter *et al.* 1994).

In *stiff-person syndrome*, CMCT and MEP amplitudes are normal, but excitability parameters including shortened cSP indicate cortical hyperexcitability (Sandbrink *et al.* 2000).

Epilepsy

TMS is not used in the routine diagnosis of patients with epilepsy, and a history of seizures is a relative contraindication for this procedure. Nevertheless, there are multiple studies of TMS in epilepsy that indicate a rather low risk, in particular for single-pulse studies (see 'Safety and side-effects of TMS' above).

Many TMS studies in seizure patients have focused on cortical MEP thresholds with mixed results in part explained by concurrent antiepileptic drugs (AEDs). In untreated patients with idiopathic generalized epilepsy, MEP threshold was reduced, but the difference from controls was small (Reutens and Berkovic 1992; Reutens et al. 1993), and treatment with AEDs reversed cortical hyperexcitability and resulted in higher threshold than controls (Reutens et al. 1993). In contrast, 18 untreated patients with first grand mal seizure and normal brain MRI studied within 48 h had increased resting and active MEP threshold (Delvaux et al. 2001). A study in patients with absence epilepsy and typical 3 Hz spike-and-wave EEG complexes also indicated higher thresholds than in controls, regardless of whether patients were treated with AEDs (Gianelli et al. 1994). MEP thresholds were normal in studies of untreated patients with partial epilepsy (Werhahn et al. 2000) and benign rolandic epilepsy (Nezu et al. 1997). AEDs, especially sodium-channel-blocking medication, increase the motor threshold in patients with epilepsy (Hufnagel et al. 1990b; Tassinari et al. 1990; Reutens et al. 1993) and in normal volunteers (Ziemann et al. 1996).

In patients with focal epilepsy, an abnormal interhemispheric asymmetry of cSP duration was found with shorter duration at higher stimulation intensities (Cicinelli et al. 2000). TMS as a technique to localize epileptic foci in partial epilepsy provided complementary information in some patients, but this usually requires serial (repetitive) stimulations and its practical use is limited mainly by its low ability to induce seizures (Hufnagel and Elger 1991). Repetitive TMS is also being studied as a potential treatment in epilepsy. A separate chapter in this volume is devoted to these applications.

Lumbar spinal stenosis and radiculopathies

Lumbar spinal stenosis may cause focal demyelination and conduction slowing of motor nerve roots within the cauda equina and thus result in prolonged cortical MEP latency. The sensitivity of cortical MEPs to detect conduction slowing within the cauda equina is limited due to the short length of the involved segment compared to the long length over which conduction is measured. Recording cortical MEPs after the onset of spinal claudication may document functional slowing or blocking of conduction including in patients who have normal baseline cortical MEP latency (Saadeh et al. 1994; Baramki et al. 1999; Lang et al. 2002). In patients with symptomatic neurogenic claudication on treadmill testing, reversible slowing of latency was noted in the patients with objective exercise-induced neurological deficit and not in patients without neurological deficit (Baramki et al. 1999; Lang et al. 2002). In another study, however, no diagnostic benefit was noted from MEP recordings before and after exercise treadmill testing (Adamova et al. 2005).

Several methods have been reported to analyze conduction time across the cauda equina in order to improve the diagnostic sensitivity of MEP testing in lumbosacral root disorders. One method is based on the difference between the peripheral motor conduction time calculated by F-wave method (PMCT-F) and by magnetic root stimulation (PMCT-M) to sacrally innervated muscle (see 'Central motor conduction time'). With the F-wave method, PMCT-F includes the time for conduction along the cauda equina, and cauda equina lesions including lumbar spinal stenosis may prolong the PMCT-F, whereas the CMCT-F calculated as the difference between cortical MEP latency and PMCT-F remains normal (Baramki et al. 1999). Magnetic spinal nerve root stimulation activates the nerve roots in the region of the intervertebral foramen at the distal end of the cauda equina, and the conduction time along the cauda equina is not included in the PMCT-M, but in the CMCT-M calculated as the difference between cortical MEP latency and PMCT-M. The conduction time along the lumbosacral nerve roots or the cauda equina to the abductor hallucis muscle may therefore be measured as the difference between PMCT-F and PMCT-M to this muscle. This has been termed 'motor root conduction time' (MRCT) (Banerjee et al. 1993) and corresponds to the difference between the

CMCT-M and the CMCT-F to this muscle. This method depends on the presence of F-waves and is therefore limited to distal muscles. In addition, cauda equina lesions and peripheral neuropathies may reduce F-wave persistence.

Lumbar spinal stenosis causes conduction slowing within the cauda equina and may prolong the CMCT-M but not CMCT-F (Dvorak et al. 1991; Di Lazzaro et al. 2004). In a large study of patients with spinal disorders, 28 of the 43 patients with lumbar spinal stenosis showed increased CMCT-M in at least one muscle (quadriceps, anterior tibial and extensor digitorum brevis muscles) of at least one leg (Dvorak et al. 1991). Patients with motor or sensory deficit on examination had prolonged CMCT-M in 82% of limbs studied, vs 45% in patients without any significant clinical deficit. F-waves were recorded in only some patients, but when obtainable allowed localization of the lesion to the cauda equina fibers in 75% of patients.

Determination of the CMCT from the abductor hallucis muscles bilaterally with both techniques allows discrimination between spinal cord involvement and cauda equina lesions in patients with equivocal clinical presentation (Di Lazzaro et al. 2004). All 10 patients with a lesion of the spinal cord (confirmed by MRI) had prolongation of both CMCT-F and CMCT-M indicative of corticospinal tract lesion. In contrast, 16 of 27 patients with a cauda equina lesion on MRI had prolonged CMCT-M combined with normal CMCT-F in at least one lower extremity. The normal result in the remaining patients was explained by involvement of a single nerve root (that usually does not result in delayed conduction because muscles are supplied by more than one root) or by recording only from sacrally innervated abductor hallucis muscle. The authors noted reduced F-wave persistence in the patients with cauda equina lesions and recorded 50 trials for each F-wave analysis. They concluded that such an MEP study may be particularly useful in the diagnosis of patients with acute cauda equina syndrome due to central disk herniation who require immediate surgery (Di Lazzaro et al. 2004).

Another method for measuring conduction time across the lumbosacral nerve roots in order to improve the diagnostic sensitivity in lumbar spinal stenosis was reported by Han et al. (2004).

They recorded MEPs simultaneously from the rectus abdominis (RA) muscle (one channel, midline recording) and AH muscles bilaterally, with magnetic stimulation of the cerebral cortex and paraspinal lower thoracic (T12) and sacral (S1) nerve roots. The difference between the CMCTs to the AH muscles bilaterally and to the rectus abdominis muscle is a measure of the conduction time across the cauda equina that the authors termed 'caudal motor conduction time' (caudal MCT). They found abnormal caudal MCT in 11 of 16 patients with MRI-proven and clinically symptomatic lumbar spinal stenosis (18 of 32 limbs studied, sensitivity 56%), compared with abnormal CMCT to the AH in 28% and prolonged cortical MEP latency in 22% of limbs studied. In four limbs, delayed peripheral motor latency to the AH muscle upon sacral root stimulation was noted in addition to prolonged CMCT and prolonged caudal MCT, explained by the authors as likely chronic nerve root compression resulting in a long-segment demyelination in the cauda equina with extension beyond the site of magnetic stimulation.

Maccabee et al. (1996) reported a method of direct magnetic stimulation of the proximal cauda equina, which, in combination with distal cauda equina stimulation, allowed direct calculation of the cauda equina conduction time to several lower-limb muscles. They used large figure-8-shaped coils placed over the proximal cauda equina with a vertically oriented coil junction and cranially directed induced current. These proximal responses were similar in appearance to those with distal cauda equina stimulation, but with longer-onset latencies, typically 1.9 ms for vastus medialis, 2.3 ms for tibialis anterior, and 3.5 ms for abductor hallucis. The authors postulate that the proximal cauda equina responses arise near or at the rootlet exit zone of the conus medullaris. Preliminary results using this technique in patients with demyelinating neuropathy revealed markedly slowed conduction times restricted to, or predominantly within, the cauda equina (Maccabee et al. 1995, 1996). A study using this technique in lumbar spinal stenosis has not been published.

The diagnostic value of MEPs in *cervical and lumbosacral radiculopathies* is controversial (Wilbourn and Aminoff 1998). Some of the initial studies in patients with lumbosacral radiculo-pathies indicated prolonged peripheral motor conduction time (PMCT-M) and decreased amplitudes after magnetic spinal nerve root stimulation in small case series (Chokroverty et al. 1989, 1993). Larger and more recent stud-ies, however, revealed heterogeneous results, and magnetic nerve root stimulation alone is not appropriate for the diagnosis of radiculopathies. As outlined above for lumbar spinal stenosis, magnetic spinal stimulation activates the nerve roots in the region of the intervertebral fora-men. Disk herniations affecting the nerve roots proximal to the neural foramen would be expected to prolong CMCT-M rather than PMCT-M, whereas more distal root lesions may cause prolongation of PMCT-M. In a study of monoradicular lumbosacral nerve root com-pression syndromes, 15 out of 30 patients showed increased CMCT-M to at least one mus-cle of the affected leg (Dvorak et al. 1991). In patients with cervical radiculopathy studied by MEP recordings from biceps brachii muscle (C6 radiculopathy patients) and abductor polli-cis brevis muscle (C7 radiculopathy patients), CMCT-M was delayed in the patient group with clinical paresis, and total MEP latency was prolonged in eight of 11 patients with paresis and five of 14 patients without clinical paresis (Wehling et al. 1995). Bischoff et al. (1993) studied 42 patients with monoradiculopathies (12 cervical and 30 lumbar) and recorded MEP responses from either deltoid, biceps brachii, extensor digitorum communis, and first dorsal interosseus muscles or the vastus medialis, tibialis anterior, gastrocnemius, and abductor hallucis muscles simultaneously, on both sides of every patient. Of the 26 patients with muscle weakness on examination, 24 had prolonged CMCT-M or PMCT-M by absolute value or side-to-side difference, compared with only five of the 16 patients without weakness. MEP find-ings were highly correlated with the presence of spontaneous activity on needle EMG and of overall similar sensitivity. In the 20 patients with postero-medial and postero-lateral disk hernia-tions, the most frequent MEP finding was prolonged CMCT-M (16 patients) in three of these patients combined with prolonged PMCT-M, and one patient had prolonged PMCT-M only. In the 12 patients with lateral

disk herniation or osteophytes encroached into the intervertebral root canal, seven patients had prolonged PMCT-M with prolonged CMCT-M also in one patient, and one patient had only abnormal CMCT-M. Linden and Berlit (1995) reported MEP abnormalities (latency prolonga-tion or amplitude reduction as side-to-side dif-ference > 50%) in 10 of 19 patients (53%) with L5 or S1 radiculopathy, a diagnostic yield lower than needle EMG (79%). Prolonged motor con-duction times were infrequent, but if present correctly identified the site of lesion with increased CMCT-M ($n = 3$), indicating medial root involvement, and increased PMCT-M ($n = 3$), indicating lateral root involvement. Several other MEP studies reported rather low diagnostic value of MEP studies in radicu-lopathies (Herdmann et al. 1992), lower than conventional electrodiagnostic testing including needle EMG (Ertekin et al. 1994; Weber and Albert 2000). The combination of MEP testing with F-wave recording may improve the diag-nostic yield (Chistyakov et al. 1995). The cal-culation of a 'motor root conduction time' based on MEP testing and F-wave analysis, with emphasis on side-to-side comparison, identified all patients with spondylotic lum-bosacral radiculopathy and clinical weakness and 35% of root compression patients without weakness (Banerjee et al. 1993). For lumbar nerve roots, a side-to-side difference of 1–1.5 ms is considered borderline abnormal, and ≥1.5 ms as definitely abnormal (Chokroverty et al. 1993).

In conclusion, the diagnostic value of MEP studies in the routine clinical evaluation of radiculopathies is uncertain, with likely lower sensitivity than conventional needle EMG even if MEPs are recorded from multiple key muscles in side-to-side comparison. It is necessary to understand the influence of nerve root lesions on MEP parameters, depending on the localiza-tion relative to the neural foramen, with proxi-mal motor root lesions (i.e. more medial disk herniations) potentially mimicking a central lesion due to prolongation of CMCT-M.

Peripheral nerve disorders

The technique of peripheral nerve stimulation with a magnetic stimulator is beyond the scope of this chapter. In the present context, the impact of peripheral nerve disorders with significant nerve root involvement on routine MEP parameters should be noted. The total MEP latency may be prolonged and the MEP may be of reduced amplitude and polyphasic configuration, as shown for patients with *Guillain–Barré syndrome* (GBS) (Benecke 1996). If the PMCT is measured by magnetic nerve root stimulation, the PMCT-M may be prolonged as expected, but also the calculated CMCT-M due to proximal nerve root slowing (see 'Central motor conduction time' and 'Lumbar spinal stenosis and radiculopathies' above). In a mixed group of peripheral nerve disorders, CMCT-M was prolonged in 18% of patients, but only 8% truly had central conduction delay with prolonged CMCT-F also (Di Lazzaro *et al.* 1999a). Pronounced proximal slowing in some patients with demyelinating neuropathy restricted to, or predominantly within, the cauda equina, up to 10–12 ms, has been reported preliminarily (Maccabee *et al.* 1995). If there is no evidence for any lesion of the corticospinal tract, then prolongation of CMCT-M may be interpreted as indication of nerve root involvement in patients with peripheral neuropathies (Benecke 1996). Presence of a proximal conduction block may be assumed in GBS when, in contrast to peripheral nerve or plexus stimulation, cortex or root stimulation does not induce a measurable response (Benecke 1996), but it needs to be kept in mind that magnetic root stimulation is usually not supramaximal. One patient with severe GBS had on initial testing normal spinal MEPs and absent cortical MEPs indicating conduction block proximal to the intervertebral foramen (Dillmann *et al.* 1998).

In a study of 21 patients with *acute inflammatory demyelinating polyneuropathy* (AIDP) or *chronic inflammatory demyelinating polyneuropathy* (CIDP), the frequency of pathological results obtained with cortical and spinal magnetic stimulation was 60% for CMCT-M and 70% for PMCT-M (Benecke 1996), and similar sensitivity was reported by others (Wohrle *et al.* 1995). In 30 patients with GBS, mild prolongation of CMCT was frequently noted and correlated with muscle strength, disability scored, and outcome (Kalita *et al.* 2001).

In *hereditary motor and sensory neuropathy* (HMSN), CMCT was reported as normal in patients classified as HMSN I or II who did not have pyramidal tract features, with few exceptions (Claus *et al.* 1990; Cruz Martinez and Tejada 1999). Other authors reported rather markedly prolonged CMCT in two of 13 HMSN I patients without pyramidal signs, which, in the authors' interpretation, may have been masked by prominent lower motor neuron weakness and atrophy (Mano *et al.* 1993). In another group of HMSN I patients without upper motor neuron signs, Sartucci *et al.* (1997) noted prolonged CMCT measured not only by magnetic nerve root stimulation, but also by the F-wave method (with this technique abnormal in 22% of upper extremities, 28% of lower extremities). Patients reported as HMSN I with pyramidal signs had rather greatly prolonged central motor conduction time (Claus *et al.* 1990; Cruz Martinez and Tejada 1999), whereas patients with the axonal variant HMSN II and accompanying pyramidal features had only slightly prolonged CMCT (Claus *et al.* 1990). In this study, electrical nerve root stimulation was used to determine the peripheral conduction time, and the authors introduced a correction factor to the CMCT for patients with decreased peripheral conduction velocity in order to compensate for any prolonged conduction in proximal motor roots. Studies of patients reported with hereditary motor and sensory neuropathy with pyramidal tract findings (HMSN V) (Schnider *et al.* 1991) and patients with hereditary motor neuronopathy type V also revealed discretely abnormal central motor conduction (Auer-Grumbach *et al.* 2000), as well as X-linked Charcot–Marie–Tooth disease (CMT1X) with connexin 32 mutation (Bahr *et al.* 1999).

Miscellaneous disorders

In patients with *diabetes mellitus*, CMCT may be mildly prolonged (Tchen *et al.* 1992; Abbruzzese *et al.* 1993; Comi 1997; Dolu *et al.* 2003; Kucera *et al.* 2005). CMCT-M was slightly longer for a group of 70 patients compared to controls, and CMCT-M above the normal limit was noted in 21 of the patients (Abbruzzese *et al.* 1993). Even if a possible contribution of proximal nerve conduction abnormalities to CMCT evaluation could not be ruled out, the authors concluded that subclinical impairment of central motor conduction was present in 30% of diabetes mellitus patients, as CMCT abnormalities also occurred in diabetic patients without evidence for peripheral neuropathy and there was no relationship between central and peripheral nervous system impairment. Others report more frequent central nervous system abnormalities in patients with peripheral diabetic neuropathy (Comi 1997). Correlation of CMCT prolongation with disease duration, not metabolic control, has been noted (Dolu et al. 2003).

Prolongation of CMCT occurs in *subacute combined degeneration from vitamin B$_{12}$ deficiency* and documents early involvement of higher cervical segments (Di Lazzaro et al. 1992b). MEPs often normalize after replacement therapy (Hemmer et al. 1998).

MEP abnormalities with prolonged CMCT were also described in patients with *human immunodeficiency virus (HIV) infection* (Zandrini et al. 1990; Moglia et al. 1991), to a greater degree in patients with immunodeficiency than in asymptomatic subjects. Other studies indicated normal CMCT (Connolly et al. 1995) and rather slowed peripheral conduction at the root level (Arendt et al. 1992).

In *hypothyroidism* (~20%) and *hyperthyroidism* (10%), a few patients were reported to have mildly abnormal CMCT (Ozata et al. 1996).

While patients with most muscle disorders have normal MEPs, subclinical upper motor neuron dysfunction in the form of prolonged CMCT and decreased MEP amplitude may be noted in patients with *mitochondrial myopathy* (Schubert et al. 1994; Di Lazzaro et al. 1999a), and prolonged CMCT in some patients with *myotonic dystrophy* (Oliveri et al. 1997).

Psychogenic weakness

Patients with psychogenic limb weakness have normal MEPs (Meyer et al. 1992a). In particular, a normal MEP in a patient with complete paresis (plegia) confirms the nonorganic etiology, at least as a major contributing factor to the paralysis. An underlying mild weakness from organic pathology cannot be excluded by a normal MEP. The magnetic stimulation procedure can be used therapeutically (Hess 2005). The presence of the MEP and the visible muscle jerking induced by the procedure should be used to reassure the subject that all the important motor pathways are functional and that recovery is to be expected with appropriate retraining.

Acknowledgment

The author gratefully acknowledges Mary Kay Floeter for helpful discussions and comments on the manuscript.

References

Abbruzzese G, Schenone A, Scramuzza G, et al. (1993). Impairment of central motor conduction in diabetic patients. *Electroencephalography and Clinical Neurophysiology 89*, 335–340.

Abbruzzese G, Trompetto C and Schieppati M (1996). The excitability of the human motor cortex increases during execution and mental imagination of sequential but not repetitive finger movements. *Experimental Brain Research 111*, 465–472.

Abbruzzese G, Marchese R, Trompetto C (1997). Sensory and motor evoked potentials in multiple system atrophy: a comparative study with Parkinson's disease. *Movement Disorders 12*, 315–321.

Abele M, Burk K, Andres F et al. (1997). Autosomal dominant cerebellar ataxia type I. Nerve conduction and evoked potential studies in families with SCA1, SCA2 and SCA3. *Brain 120*, 2141–2148.

Abele M, Schulz JB, Burk K, Topka H, Dichgans J, Klockgether T (2000). Evoked potentials in multiple system atrophy (MSA). *Acta Neurologica Scandinavica 101*, 111–115.

Ackermann H, Scholz E, Koehler W, Dichgans J (1991). Influence of posture and voluntary background contraction upon compound muscle action potentials from anterior tibial and soleus muscle following transcranial magnetic stimulation. *Electroencephalography and Clinical Neurophysiology 81*, 71–80.

Adamova B, Vohanka S, Dusek L (2005). Dynamic electrophysiological examination in patients with lumbar spinal stenosis: is it useful in clinical practice? *European Spine Journal 14*, 269–276.

Ahonen JP, Jehkonen M, Dastidar P, Molnar G, Hakkinen V (1998). Cortical silent period evoked by transcranial magnetic stimulation in ischemic stroke. *Electroencephalography and Clinical Neurophysiology 109*, 224–229.

Amassian VE, Somasundaram M, Rothwell JC, et al. (1991). Paraesthesias are elicited by single pulse, magnetic coil stimulation of motor cortex in susceptible humans. *Brain 114*, 2505–2520.

Andersen PM, Forsgren L, Binzer M, et al. (1996). Autosomal recessive adult-onset amyotrophic lateral sclerosis associated with homozygosity for Asp90Ala CuZn-superoxide dismutase mutation. A clinical and genealogical study of 36 patients. *Brain 119*, 1153–1172.

Andersson T, Siden A, Persson A (1995). A comparison of motor evoked potentials and somatosensory evoked potentials in patients with multiple sclerosis and potentially related conditions. *Electromyography and Clinical Neurophysiology 35*, 17–24.

Arac N, Sagduyu A, Binai S, Ertekin C (1994). Prognostic value of transcranial magnetic stimulation in acute stroke. *Stroke 25*, 2183–2186.

Arendt G, Maecker HP, Jablonowski H, Homberg V (1992). Magnetic stimulation of motor cortex in relation to fastest voluntary motor activity in neurologically asymptomatic HIV-positive patients. *Journal of Neurological Sciences 112*, 76–80.

Attarian S, Azulay JP, Lardillier D, Verschueren A, Pouget J (2005). Transcranial magnetic stimulation in lower motor neuron diseases. *Clinical Neurophysiology 116*, 35–42.

Auer C, Siebner HR, Dressnandt J, Conrad B (1999). Intrathecal baclofen increases corticospinal output to hand muscles in multiple sclerosis. *Neurology, 52*, 1298–1299.

Auer-Grumbach M, Loscher WN, Wagner K, *et al.* (2000). Phenotypic and genotypic heterogeneity in hereditary motor neuronopathy type V: a clinical, electrophysiological and genetic study. *Brain, 123*, 1612–1623.

Bahr M, Andres F, Timmerman V, Nelis ME, Van Broeckhoven C, Dichgans J (1999). Central visual, acoustic, and motor pathway involvement in a Charcot-Marie-Tooth family with an Asn205Ser mutation in the connexin 32 gene. *Journal of Neurology, Neurosurgery and Psychiatry 66*, 202–206.

Banerjee TK, Mostofi MS, Us O, Weerasinghe V, Sedgwick EM (1993). Magnetic stimulation in the determination of lumbosacral motor radiculopathy. *Electroencephalography and Clinical Neurophysiology 89*, 221–226.

Baramki HG, Steffen T, Schondorf R, Aebi M (1999). Motor conduction alterations in patients with lumbar spinal stenosis following the onset of neurogenic claudication. *European Spine Journal 8*, 411–416.

Barker AT, Jalinous R, Freeston IL (1985). Non-invasive magnetic stimulation of human motor cortex. *Lancet i*, 1106–1107.

Barker AT, Freeston IL, Jabinous R, Jarratt JA (1986). Clinical evaluation of conduction time measurements in central motor pathways using magnetic stimulation of human brain. *Lancet i*, 1325–1326.

Barker AT, Freeston IL, Jalinous R, Jarratt JA (1987). Magnetic stimulation of the human brain and peripheral nervous system: an introduction and the results of an initial clinical evaluation. *Neurosurgery 20*, 100–109.

Bassetti C, Mathis J, Hess CW (1994). Multimodal electrophysiological studies including motor evoked potentials in patients with locked-in syndrome: report of six patients. *Journal of Neurology, Neurosurgery and Psychiatry 57*, 1403–1406.

Bednarik J, Kadanka Z, Vohanka S, *et al.* (1998). The value of somatosensory and motor evoked evoked potentials in pre-clinical spondylotic cervical cord compression. *European Spine Journal 7*, 493–500.

Bednarik J, Kadanka Z, Vohanka S, Stejskal L, Vlach O, Schroder R (1999). The value of somatosensory- and motor-evoked potentials in predicting and monitoring the effect of therapy in spondylotic cervical myelopathy. Prospective randomized study. *Spine 24*, 1593–1598.

Beer S, Rosler KM and Hess CW (1995). Diagnostic value of paraclinical tests in multiple sclerosis: relative sensitivities and specificities for reclassification according to the Poser committee criteria. *Journal of Neurology, Neurosurgery and Psychiatry 59*, 152–159.

Benecke R (1996). Magnetic stimulation in the assessment of peripheral nerve disorders. *Baillière's Clinical Neurology 5*, 115–128.

Berardelli A, Inghilleri M, Cruccu G, Mercuri B, Manfredi M (1991). Electrical and magnetic transcranial stimulation in patients with corticospinal damage due to stroke or motor neurone disease. *Electroencephalography and Clinical Neurophysiology 81*, 389–396.

Bischoff C, Meyer BU, Machetanz J, Conrad B (1993). The value of magnetic stimulation in the diagnosis of radiculopathies. *Muscle and Nerve 16*, 154–161.

Brasil-Neto JP, Cohen LG, Panizza M, Nilsson J, Roth BJ, Hallett M (1992). Optimal focal transcranial magnetic activation of the human motor cortex: effects of coil orientation, shape of the induced current pulse, and stimulus intensity. *Journal of Clinical Neurophysiology 9*, 132–136.

Braune HJ and Fritz C (1995). Transcranial magnetic stimulation-evoked inhibition of voluntary muscle activity (silent period) is impaired in patients with ischemic hemispheric lesion. *Stroke 26*, 550–553.

Britton TC, Meyer BU, Herdmann J, Benecke R (1990). Clinical use of the magnetic stimulator in the investigation of peripheral conduction time. *Muscle and Nerve 13*, 396–406.

Britton TC, Meyer BU, Benecke R (1991). Variability of cortically evoked motor responses in multiple sclerosis. *Electroencephalography and Clinical Neurophysiology 81*, 186–194.

Brunholzl C, Claus D (1994). Central motor conduction time to upper and lower limbs in cervical cord lesions. *Archives of Neurology, 51*, 245–249.

Buhler R, Magistris MR, Truffert A, Hess CW, Rosler KM (2001). The triple stimulation technique to study central motor conduction to the lower limbs. *Clinical Neurophysiology 112*, 938–949.

Calancie B, Alexeeva N, Broton JG, Suys S, Hall A, Klose KJ (1999). Distribution and latency of muscle responses to transcranial magnetic stimulation of motor cortex after spinal cord injury in humans. *Journal of Neurotrauma 16*, 49–67.

Cantello R (2002). Applications of transcranial magnetic stimulation in movement disorders. *Journal of Clinical Neurophysiology 19*, 272–293.

Cantello R, Gianelli M, Bettucci D, Civardi C, De Angelis MS, Mutani R (1991). Parkinson's disease rigidity: magnetic motor evoked potentials in a small hand muscle. *Neurology 41*, 1449–1456.

Caramia MD, Cicinelli P, Paradiso C, *et al.* (1991). Excitability changes of muscular responses to magnetic brain stimulation in patients with central motor disorders. *Electroencephalography and Clinical Neurophysiology 81*, 243–50.

Caramia MD, Iani C, Bernardi G (1996). Cerebral plasticity after stroke as revealed by ipsilateral responses to magnetic stimulation. *Neuroreport 7*, 1756–1760.

Caramia MD, Palmieri MG, Desiato MT, *et al.* (2004). Brain excitability changes in the relapsing and remitting phases of multiple sclerosis: a study with transcranial magnetic stimulation. *Clinical Neurophysiology 115*, 956–965.

Cariga P, Catley M, Nowicky AV, Savic G, Ellaway PH, Davey NJ (2002). Segmental recording of cortical motor evoked potentials from thoracic paravertebral myotomes in complete spinal cord injury. *Spine 27*, 1438–1443.

Catano A, Houa M, Caroyer JM, Ducarne H, Noel P (1995). Magnetic transcranial stimulation in non-haemorrhagic sylvian strokes: interest of facilitation for early functional prognosis. *Electroencephalography and Clinical Neurophysiology 97*, 349–354.

Catano A, Houa M, Noel P (1997). Magnetic transcranial stimulation: dissociation of excitatory and inhibitory mechanisms in acute strokes. *Electroencephalography and Clinical Neurophysiology 105*, 29–36.

Chan KM, Nasathurai S, Chavin JM, Brown WF (1998). The usefulness of central motor conduction studies in the localization of cord involvement in cervical spondylytic myelopathy. *Muscle and Nerve 21*, 1220–1223.

Chan YC, Mills KR (2005). The use of transcranial magnetic stimulation in the clinical evaluation of suspected myelopathy. *Journal of Clinical Neuroscience 12*, 878–881.

Chen R, Lozano AM, Ashby P (1999). Mechanism of the silent period following transcranial magnetic stimulation. Evidence from epidural recordings. *Experimental Brain Research 128*, 539–542.

Chen R, Garg RR, Lozano AM, Lang AE (2001). Effects of internal globus pallidus stimulation on motor cortex excitability. *Neurology 56*, 716–723.

Chistyakov AV, Soustiel JF, Hafner H, Feinsod M (1995). Motor and somatosensory conduction in cervical myelopathy and radiculopathy. *Spine 20*, 2135–2140.

Chokroverty S, Sachdeo R, Dilullo J, Duvoisin RC (1989). Magnetic stimulation in the diagnosis of lumbosacral radiculopathy. *Journal of Neurology, Neurosurgery and Psychiatry 52*, 767–772.

Chokroverty S, Picone MA, Chokroverty M (1991). Percutaneous magnetic coil stimulation of human cervical vertebral column: site of stimulation and clinical application. *Electroencephalography and Clinical Neurophysiology, 81*, 359–365.

Chokroverty S, Flynn D, Picone MA, Chokroverty M, Belsh J (1993). Magnetic coil stimulation of the human lumbosacral vertebral column: site of stimulation and clinical application. *Electroencephalography and Clinical Neurophysiology 89*, 54–60.

Chu NS (1990). Motor evoked potentials in Wilson's disease: early and late motor responses. *Journal of Neurological Sciences 99*, 259–269.

Cicinelli P, Traversa R, Rossini PM (1997). Post-stroke reorganization of brain motor output to the hand: a 2–4 month follow-up with focal magnetic transcranial stimulation. *Electroencephalography and Clinical Neurophysiology 105*, 438–450.

Cicinelli P, Mattia D, Spanedda F, *et al.* (2000). Transcranial magnetic stimulation reveals an interhemispheric asymmetry of cortical inhibition in focal epilepsy. *Neuroreport 11*, 701–707.

Classen J, Witte OW, Schlaug G, Seitz RJ, Holthausen H, Benecke R (1995). Epileptic seizures triggered directly by focal transcranial magnetic stimulation. *Electroencephalography and Clinical Neurophysiology 94*, 19–25.

Claus D (1990). Central motor conduction: method and normal results. *Muscle and Nerve 13*, 1125–1132.

Claus D, Harding AE, Hess CW, Mills KR, Murray NM, Thomas PK (1988). Central motor conduction in degenerative ataxic disorders: a magnetic stimulation study. *Journal of Neurology, Neurosurgery and Psychiatry 51*, 790–795.

Claus D, Waddy HM, Harding AE, Murray NM, Thomas PK (1990). Hereditary motor and sensory neuropathies and hereditary spastic paraplegia: a magnetic stimulation study. *Annals of Neurology 28*, 43–49.

Claus D, Brunholzl C, Kerling FP, Henschel S (1995). Transcranial magnetic stimulation as a diagnostic and prognostic test in amyotrophic lateral sclerosis. *Journal of Neurological Sciences 129*(Suppl.), 30–34.

Comi G (1997). Evoked potentials in diabetes mellitus. *Clinical Neuroscience 4*, 374–379.

Conforto AB, Z'Graggen WJ, Kohl AS, Rosler KM, Kaelin-Lang A (2004). Impact of coil position and electrophysiological monitoring on determination of motor thresholds to transcranial magnetic stimulation. *Clinical Neurophysiology 115*, 812–819.

Connolly S, Manji H, McAllister RH, *et al.* (1995). Neurophysiological assessment of peripheral nerve and spinal cord function in asymptomatic HIV-1 infection: results from the UCMSM/Medical Research Council neurology cohort. *Journal of Neurology 242*, 406–414.

Cruz Martinez A, Anciones B (1992). Central motor conduction to upper and lower limbs after magnetic stimulation of the brain and peripheral nerve abnormalities in 20 patients with Friedreich's ataxia. *Acta Neurologica Scandinavica 85*, 323–326.

Cruz Martinez A, Tejada J (1999). Central motor conduction in hereditary motor and sensory neuropathy and hereditary spastic paraplegia. *Electromyography and Clinical Neurophysiology 39*, 331–335.

Cruz Martinez A, Arpa J, Alonso M, Palomo F, Villoslada C (1995). Transcranial magnetic stimulation in multiple system and late onset cerebellar atrophies. *Acta Neurologica Scandinavica 92*, 218–224.

Cruz-Martinez A, Gonzalez-Orodea JI, Lopez Pajares R, Arpa J (2000). Disability in multiple sclerosis. The role of transcranial magnetic stimulation. *Electromyography and Clinical Neurophysiology 40*, 441–447.

Cunic D, Roshan L, Khan FI, Lozano AM, Lang AE, Chen R (2002). Effects of subthalamic nucleus stimulation on motor cortex excitability in Parkinson's disease. *Neurology 58*, 1665–1672.

Curt A, Dietz V (1999). Neurologic recovery in SCI. *Archives of Physical Medicine and Rehabilitation 80*, 607–608.

Curt A, Keck ME, Dietz V (1998). Functional outcome following spinal cord injury: significance of motor-evoked potentials and ASIA scores. *Archives of Physical Medicine and Rehabilitation 79*, 81–86.

Dauper J, Peschel T, Schrader C, *et al.* (2002). Effects of subthalamic nucleus (STN) stimulation on motor cortex excitability. *Neurology 59*, 700–706.

Davey NJ, Smith HC, Wells E, *et al.* (1998). Responses of thenar muscles to transcranial magnetic stimulation of the motor cortex in patients with incomplete spinal cord injury. *Journal of Neurology, Neurosurgery and Psychiatry 65*, 80–87.

De Carvalho M, Turkman A, Swash M (2003). Motor responses evoked by transcranial magnetic stimulation and peripheral nerve stimulation in the ulnar innervation in amyotrophic lateral sclerosis: the effect of upper and lower motor neuron lesion. *Journal of Neurological Sciences 210*, 83–90.

De Carvalho M, Chio A, Dengler R, Hecht M, Weber M, Swash M (2005). Neurophysiological measures in amyotrophic lateral sclerosis: markers of progression in clinical trials. *Amyotrophic Lateral Sclerosis and Other Motor Neuron Disorders 6*, 17–28.

De Mattei M, Paschero B, Sciarretta A, Davini O, Cocito D (1993). Usefulness of motor evoked potentials in compressive myelopathy. *Electromyography and Clinical Neurophysiology 33*, 205–216.

De Noordhout AM, Remacle JM, Pepin JL, Born JD, Delwaide PJ (1991). Magnetic stimulation of the motor cortex in cervical spondylosis. *Neurology 41*, 75–80.

De Noordhout AM, Myressiotis S, Delvaux V, Born JD, Delwaide PJ (1998). Motor and somatosensory evoked potentials in cervical spondylotic myelopathy. *Electroencephalography and Clinical Neurophysiology 108*, 24–31.

De Scisciolo G, Bartelli M, Magrini S, Biti GP, Guidi L, Pinto F (1991). Long-term nervous system damage from radiation of the spinal cord: an electrophysiological study. *Journal of Neurology 238*, 9–15.

Delvaux V, Alagona G, Gerard P, De Pasqua V, Delwaide PJ, Maertens de Noordhout A (2001). Reduced excitability of the motor cortex in untreated patients with de novo idiopathic 'grand mal' seizures. *Journal of Neurology, Neurosurgery and Psychiatry 71*, 772–776.

Desiato MT, Caramia MD (1997). Towards a neurophysiological marker of amyotrophic lateral sclerosis as revealed by changes in cortical excitability. *Electroencephalography and Clinical Neurophysiology 105*, 1–7.

Di Lazzaro V, Restuccia D, Colosimo C, Tonali P (1992a). The contribution of magnetic stimulation of the motor cortex to the diagnosis of cervical spondylotic myelopathy. Correlation of central motor conduction to distal and proximal upper limb muscles with clinical and MRI findings. *Electroencephalography and Clinical Neurophysiology 85*, 311–320.

Di Lazzaro V, Restuccia D, Fogli D, Nardone R, Mazza S, Tonali P (1992b). Central sensory and motor conduction in vitamin B12 deficiency. *Electroencephalography and Clinical Neurophysiology 84*, 433–439.

Di Lazzaro V, Restuccia D, Molinari M, *et al.* (1994). Excitability of the motor cortex to magnetic stimulation in patients with cerebellar lesions. *Journal of Neurology, Neurosurgery and Psychiatry 57*, 108–110.

Di Lazzaro V, Oliviero A, Profice P, *et al.* (1999a). The diagnostic value of motor evoked potentials. *Clinical Neurophysiology 110*, 1297–1307.

Di Lazzaro V, Oliviero A, Profice P, *et al.* (1999b). Direct recordings of descending volleys after transcranial magnetic and electric motor cortex stimulation in conscious humans. *Electroencephalography and Clinical Neurophysiology 51*(Suppl), 120–126.

Di Lazzaro V, Pilato F, Oliviero A, Saturno E, Dileone M, Tonali PA (2004). Role of motor evoked potentials in diagnosis of cauda equina and lumbosacral cord lesions. *Neurology 63*, 2266–2271.

Dillmann U, Ohlmann D, Hamann GF, Schimrigk K (1998). [Value of magnetic stimulation and F-wave determination in diagnosis of proximal demyelinating lesions. Follow-up of acute Guillain-Barre polyradiculitis]. *Nervenarzt 69*, 338–341.

Dioszeghy P, Hidasi E, Mechler F (1999). Study of central motor functions using magnetic stimulation in Parkinson's disease. *Electromyography and Clinical Neurophysiology 39*, 101–105.

D'Olhaberriague L, Espadaler Gamissans JM, Marrugat J, Valls A, Oliveras Ley C, Seoane JL (1997). Transcranial magnetic stimulation as a prognostic tool in stroke. *Journal of Neurological Sciences 147*, 73–80.

Dolu H, Ulas UH, Bolu E *et al.* (2003). Evaluation of central neuropathy in type II diabetes mellitus by multimodal evoked potentials. *Acta Neurologica Belgica 103*, 206–211.

Dubach P, Guggisberg AG, Rosler KM, Hess CW, Mathis J (2004). Significance of coil orientation for motor evoked potentials from nasalis muscle elicited by transcranial magnetic stimulation. *Clinical Neurophysiology 115*, 862–870.

Dvorak J, Janssen B, Grob D (1990). The neurologic workup in patients with cervical spine disorders. *Spine 15*, 1017–1022.

Dvorak J, Herdmann J, Theiler R, Grob D (1991). Magnetic stimulation of motor cortex and motor roots for painless evaluation of central and proximal peripheral motor pathways. Normal values and clinical application in disorders of the lumbar spine. *Spine 16*, 955–961.

Eisen A (2001). Clinical electrophysiology of the upper and lower motor neuron in amyotrophic lateral sclerosis. *Seminars in Neurology 21*, 141–154.

Eisen AA, Shtybel W (1990). AAEM minimonograph #35, Clinical experience with transcranial magnetic stimulation. *Muscle and Nerve 13*, 995–1011.

Eisen A, Shytbel W, Murphy K and Hoirch M (1990). Cortical magnetic stimulation in amyotrophic lateral sclerosis. *Muscle and Nerve 13*, 146–151.

Eisen A, Pant B, Stewart H (1993). Cortical excitability in amyotrophic lateral sclerosis: a clue to pathogenesis. *Canadian Journal of Neurological Sciences 20*, 11–16.

Ertekin C, Nejat RS, Sirin H, *et al.* (1994). Comparison of magnetic coil stimulation and needle electrical stimulation in the diagnosis of lumbosacral radiculopathy. *Clinical Neurology and Neurosurgery 96*, 124–129.

Escudero JV, Sancho J, Bautista D, Escudero M, Lopez-Trigo J (1998). Prognostic value of motor evoked potential obtained by transcranial magnetic brain stimulation in motor function recovery in patients with acute ischemic stroke. *Stroke 29*, 1854–1859.

Facchetti D, Mai R, Micheli A, *et al.* (1997). Motor evoked potentials and disability in secondary progressive multiple sclerosis. *Canadian Journal of Neurological Sciences 24*, 332–337.

Ferbert A, Vielhaber S, Meincke U, Buchner H (1992). Transcranial magnetic stimulation in pontine infarction: correlation to degree of paresis. *Journal of Neurology, Neurosurgery and Psychiatry 55*, 294–299.

Fierro B, Salemi G, Brighina F, *et al.* (2002). A transcranial magnetic stimulation study evaluating methylprednisolone treatment in multiple sclerosis. *Acta Neurologica Scandinavica 105*, 152–157.

Fritz C, Braune HJ, Pylatiuk C, Pohl M (1997). Silent period following transcranial magnetic stimulation: a study of intra- and inter-examiner reliability. *Electroencephalography and Clinical Neurophysiology 105*, 235–240.

Fuhr P, Agostino R, Hallett M (1991a). Spinal motor neuron excitability during the silent period after cortical stimulation. *Electroencephalography and Clinical Neurophysiology 81*, 257–262.

Fuhr P, Cohen LG, Roth BJ, Hallett M (1991b). Latency of motor evoked potentials to focal transcranial stimulation varies as a function of scalp positions stimulated. *Electroencephalography and Clinical Neurophysiology 81*, 81–89.

Fuhr P, Borggrefe-Chappuis A, Schindler C, Kappos L (2001). Visual and motor evoked potentials in the course of multiple sclerosis. *Brain 124*, 2162–2168.

Furby A, Bourriez JL, Jacquesson JM, Mounier-Vehier F, Guieu JD (1992). Motor evoked potentials to magnetic stimulation: technical considerations and normative data from 50 subjects. *Journal of Neurology 239*, 152–156.

Garassus P, Charles N, Mauguere F (1993). Assessment of motor conduction times using magnetic stimulation of brain, spinal cord and peripheral nerves. *Electromyography and Clinical Neurophysiology 33*, 3–10.

Gianelli M, Cantello R, Civardi C, *et al.* (1994). Idiopathic generalized epilepsy: magnetic stimulation of motor cortex time-locked and unlocked to 3-Hz spike-and-wave discharges. *Epilepsia 35*, 53–60.

Guggisberg AG, Dubach P, Hess CW, Wuthrich C, Mathis J (2001). Motor evoked potentials from masseter muscle induced by transcranial magnetic stimulation of the pyramidal tract: the importance of coil orientation. *Clinical Neurophysiology 112*, 2312–2319.

Han TR, Paik NJ, Lee SJ, Kwon BS (2004). A new method to measure caudal motor conduction time using magnetic stimulation. *Muscle and Nerve 30*, 727–731.

Hanajima R, Ugawa Y (1998). Impaired motor cortex inhibition in patients with ALS: evidence from paired transcranial magnetic stimulation. *Neurology 51*, 1771–1772.

Haug BA, Kukowski B (1994). Latency and duration of the muscle silent period following transcranial magnetic stimulation in multiple sclerosis, cerebral ischemia, and other upper motor neuron lesions. *Neurology 44*, 936–940.

Haupts MR, Daum S, Ahle G, Holinka B, Gehlen W (2004). Transcranial magnetic stimulation as a provocation for epileptic seizures in multiple sclerosis. *Multiple Sclerosis 10*, 475–476.

Hayes KC, Allatt RD, Wolfe DL, Kasai T, Hsieh J (1991). Reinforcement of motor evoked potentials in patients with spinal cord injury. *Electroencephalography and Clinical Neurophysiology 43*(Suppl), 312–329.

Heald A, Bates D, Cartlidge NE, French JM, Miller S (1993a). Longitudinal study of central motor conduction time following stroke. 1. Natural history of central motor conduction. *Brain 116*, 1355–1370.

Heald A, Bates D, Cartlidge NE, French JM, Miller S (1993b). Longitudinal study of central motor conduction time following stroke. 2. Central motor conduction measured within 72 h after stroke as a predictor of functional outcome at 12 months. *Brain 116*, 1371–1385.

Hefter H, Roick H, von Giesen HJ, *et al.* (1994). Motor impairment in Wilson's disease. 3, The clinical impact of pyramidal tract involvement. *Acta Neurologica Scandinavica 89*, 421–428.

Hemmer B, Glocker FX, Schumacher M, Deuschl G, Lucking CH (1998). Subacute combined degeneration: clinical, electrophysiological, and magnetic resonance imaging findings. *Journal of Neurology, Neurosurgery and Psychiatry 65*, 822–827.

Hendricks HT, Pasman JW, Merx JL, van Limbeek J, Zwarts MJ (2003). Analysis of recovery processes after stroke by means of transcranial magnetic stimulation. *Journal of Clinical Neurophysiology 20*, 188–195.

Herdmann J, Dvorak J, Bock WJ (1992). Motor evoked potentials in patients with spinal disorders: upper and lower motor neurone affection. *Electromyography and Clinical Neurophysiology 32*, 323–330.

Hess CW (2005). Central motor conduction and its clinical application. In: M Hallett and S Chokroverty (eds), *Magnetic stimulation in clinical neurophysiology*, pp. 83–103. Boston: Butterworth Heinemann.

Hess CW, Mills KR, Murray NM (1986). Measurement of central motor conduction in multiple sclerosis by magnetic brain stimulation. *Lancet ii*, 355–358.

Hess CW, Mills KR and Murray NM (1987a). Responses in small hand muscles from magnetic stimulation of the human brain. *Journal of Physiology 388*, 397–419.

Hess CW, Mills KR, Murray NM, Schriefer TN (1987b). Magnetic brain stimulation: central motor conduction studies in multiple sclerosis. *Annals of Neurology 22*, 744–752.

Hess CW, Mills KR, Murray NM (1988). Methodological considerations for magnetic brain stimulation. In: C Barber and T Blum (eds), *Evoked potentials, III*, pp. 456–461. London: Butterworth.

Homberg V, Netz J (1989). Generalised seizures induced by transcranial magnetic stimulation of motor cortex. *Lancet ii*, 1223.

Homberg V, Stephan KM, Netz J (1991). Transcranial stimulation of motor cortex in upper motor neurone syndrome: its relation to the motor deficit. *Electroencephalography and Clinical Neurophysiology 81*, 377–388.

Hoppner J, Kunesch E, Buchmann J, Hess A, Grossmann A, Benecke R (1999). Demyelination and axonal degeneration in corpus callosum assessed by analysis of transcallosally mediated inhibition in multiple sclerosis. *Clinical Neurophysiology 110*, 748–756.

Hu MT, Bland J, Clough C, Ellis CM, Chaudhuri KR (1999). Limb contractures in levodopa-responsive parkinsonism: a clinical and investigational study of seven new cases. *Journal of Neurology 246*, 671–66.

Hufnagel A, Elger CE (1991). Responses of the epileptic focus to transcranial magnetic stimulation. *Electroencephalography and Clinical Neurophysiology 43*(Suppl), 86–99.

Hufnagel A, Elger CE, Durwen HF, Boker DK, Entzian W (1990a). Activation of the epileptic focus by transcranial magnetic stimulation of the human brain. *Annals of Neurology 27*, 49–60.

Hufnagel A, Elger CE, Marx W, Ising A (1990b). Magnetic motor-evoked potentials in epilepsy: effects of the disease and of anticonvulsant medication. *Annals of Neurology 28*, 680–686.

Humm AM, Magistris MR, Truffert A, Hess CW, Rosler KM (2003). Central motor conduction differs between acute relapsing-remitting and chronic progressive multiple sclerosis. *Clinical Neurophysiology 114*, 2196–2203.

Humm AM, Beer S, Kool J, Magistris MR, Kesselring J, Rosler KM (2004). Quantification of Uhthoff's phenomenon in multiple sclerosis: a magnetic stimulation study. *Clinical Neurophysiology 115*, 2493–2501.

Humm AM, Z'Graggen WJ, Buhler R, Magistris MR, Rosler KM (2006). Quantification of central motor conduction deficits in multiple sclerosis patients before and after treatment of acute exacerbation by methylprednisolone. *Journal of Neurology, Neurosurgery and Psychiatry 77*, 345–350

Ikoma K, Samii A, Mercuri B, Wassermann EM, Hallett M (1996). Abnormal cortical motor excitability in dystonia. *Neurology 46*, 1371–1376.

Inghilleri M, Berardelli A, Cruccu G, Manfredi M (1993). Silent period evoked by transcranial stimulation of the human cortex and cervicomedullary junction. *Journal of Physiology 466*, 521–534.

Ingram DA, Swash M (1987). Central motor conduction is abnormal in motor neuron disease. *Journal of Neurology, Neurosurgery and Psychiatry 50*, 159–166.

Ingram DA, Thompson AJ, Swash M (1988). Central motor conduction in multiple sclerosis: evaluation of abnormalities revealed by transcutaneous magnetic stimulation of the brain. *Journal of Neurology, Neurosurgery and Psychiatry 51*, 487–494.

Izumi S, Findley TW, Ikai T, Andrews J, Daum M, Chino N (1995). Facilitatory effect of thinking about movement on motor-evoked potentials to transcranial magnetic stimulation of the brain. *American Journal of Physical Medicine and Rehabilitation 74*, 207–213.

Jones SM, Streletz LJ, Raab VE, Knobler RL, Lublin FD (1991). Lower extremity motor evoked potentials in multiple sclerosis. *Archives in Neurology 48*, 944–948.

Kalita J, Misra UK (2000). Neurophysiological studies in acute transverse myelitis. *Journal of Neurology 247*, 943–948.

Kalita J, Misra UK, Bansal R (2001). Central motor conduction studies in patients with Guillain Barre syndrome. *Electromyography and Clinical Neurophysiology 41*, 243–246.

Kallmann BA, Fackelmann S, Toyka KV, Rieckmann P, Reiners K (2006). Early abnormalities of evoked potentials and future disability in patients with multiple sclerosis. *Multiple Sclerosis 12*, 58–65.

Kandler RH, Jarratt JA, Sagar HJ, *et al.* (1990). Abnormalities of central motor conduction in Parkinson's disease. *Journal of Neurological Sciences 100*, 94–97.

Kandler RH, Jarratt JA, Gumpert EJ, Davies-Jones GA, Venables GS, Sagar HJ (1991). The role of magnetic stimulation in the diagnosis of multiple sclerosis. *Journal of Neurological Sciences 106*, 25–30.

Kaneko K, Kawai S, Fuchigami Y, Morita H, Ofuji A (1996). The effect of current direction induced by transcranial magnetic stimulation on the corticospinal excitability in human brain. *Electroencephalography and Clinical Neurophysiology 101*, 478–482.

Kaneko K, Taguchi T, Morita H, Yonemura H, Fujimoto H, Kawai S (2001). Mechanism of prolonged central motor conduction time in compressive cervical myelopathy. *Clinical Neurophysiology 112*, 1035–1040.

Kidd D, Thompson PD, Day BL, *et al.* (1998). Central motor conduction time in progressive multiple sclerosis. Correlations with MRI and disease activity. *Brain 121*, 1109–1116.

Kiers L, Cros D, Chiappa KH, Fang J (1993). Variability of motor potentials evoked by transcranial magnetic stimulation. *Electroencephalography and Clinical Neurophysiology 89*, 415–423.

Kischka U, Fajfr R, Fellenberg T, Hess CW (1993). Facilitation of motor evoked potentials from magnetic brain stimulation in man: a comparative study of different target muscles. *Journal of Clinical Neurophysiology 10*, 505–512.

Kloten H, Meyer BU, Britton TC, Benecke R (1992). [Normal values and age-related changes in magneto-electric evoked compound muscle potentials]. *EEG-EMG Zeitschrift für Elektroenzephalographie, Elektromyographie und verwandte Gebiete 23*, 29–36.

Komissarow L, Rollnik JD, Bogdanova D, *et al.* (2004). Triple stimulation technique (TST) in amyotrophic lateral sclerosis. *Clinical Neurophysiology* 115, 356–360.

Kucera P, Goldenberg Z, Varsik P, Buranova D, Traubner P (2005). Spinal cord lesions in diabetes mellitus. Somatosensory and motor evoked potentials and spinal conduction time in diabetes mellitus. *Neuroendocrinology Letters* 26, 143–147.

Kuhn AA, Trottenberg T, Kupsch A, Meyer BU (2002). Pseudo-bilateral hand motor responses evoked by transcranial magnetic stimulation in patients with deep brain stimulators. *Clinical Neurophysiology* 113, 341–345.

Kuhn AA, Meyer BU, Trottenberg T, Brandt SA, Schneider GH, Kupsch A (2003). Modulation of motor cortex excitability by pallidal stimulation in patients with severe dystonia. *Neurology* 60, 768–774.

Kuhn AA, Grosse P, Holtz K, Brown P, Meyer BU, Kupsch A (2004). Patterns of abnormal motor cortex excitability in atypical parkinsonian syndromes. *Clinical Neurophysiology* 115, 1786–1795.

Kumar R, Chen R, Ashby P (1999). Safety of transcranial magnetic stimulation in patients with implanted deep brain stimulators. *Movement Disorders* 14, 157–158.

Lang E, Hilz MJ, Erxleben H, Ernst M, Neundorfer B, Liebig K (2002). Reversible prolongation of motor conduction time after transcranial magnetic brain stimulation after neurogenic claudication in spinal stenosis. *Spine* 27, 2284–2290.

Lanzillo B, Perretti A, Santoro L, *et al.* (1994). Evoked potentials in inherited ataxias: a multimodal electrophysiological study. *Italian Journal of Neurological Sciences* 15, 25–37.

Linden D, Berlit P (1995). Comparison of late responses, EMG studies, and motor evoked potentials (MEPs) in acute lumbosacral radiculopathy. *Muscle and Nerve* 18, 1205–1207.

Lissens MA (1994). Motor evoked potentials of the human diaphragm elicited through magnetic transcranial brain stimulation. *Journal of Neurological Sciences* 124, 204–207.

Lissens MA, De Muynck MC, Decleir AM, Vanderstraeten GG (1995). Motor evoked potentials of the abdominal muscles elicited through magnetic transcranial brain stimulation. *Muscle and Nerve* 18, 1353–1354.

Lo YL, Chan LL, Lim W, *et al.* (2004). Systematic correlation of transcranial magnetic stimulation and magnetic resonance imaging in cervical spondylotic myelopathy. *Spine* 29, 1137–1145.

Macdonell RA, Donnan GA (1995). Magnetic cortical stimulation in acute spinal cord injury. *Neurology* 45, 303–306.

Macdonell RA, Shapiro BE, Chiappa KH, *et al.* (1991). Hemispheric threshold differences for motor evoked potentials produced by magnetic coil stimulation. *Neurology* 41, 1441–1444.

Macdonell RA, Cros D, Shahani BT (1992). Lumbosacral nerve root stimulation comparing electrical with surface magnetic coil techniques. *Muscle and Nerve* 15, 885–890.

Maccabee PJ, Lipitz ME, Desudchit T, *et al.* (1995). Detection of proximal demyelinating neuropathy in cauda equina by neuromagnetic stimulation. *Neurology* 45(Suppl 4), A170 (abstract).

Maccabee PJ, Lipitz ME, Desudchit T, *et al.* (1996). A new method using neuromagnetic stimulation to measure conduction time within the cauda equina. *Electroencephalography and Clinical Neurophysiology* 101, 153–166.

McDonnell MN, Ridding MC, Miles TS (2004). Do alternate methods of analysing motor evoked potentials give comparable results? *Journal of Neuroscience Methods* 136, 63–67.

McKay WB, Stokic DS, Dimitrijevic MR (1997). Assessment of corticospinal function in spinal cord injury using transcranial motor cortex stimulation: a review. *Journal of Neurotrauma* 14, 539–548.

McKay WB, Lee DC, Lim HK, Holmes SA, Sherwood AM (2005). Neurophysiological examination of the corticospinal system and voluntary motor control in motor-incomplete human spinal cord injury. *Experimental Brain Research* 163, 379–387.

Magistris MR, Rosler KM, Truffert A, Landis T, Hess CW (1999). A clinical study of motor evoked potentials using a triple stimulation technique. *Brain* 122, 265–279.

Mainero C, Inghilleri M, Pantano P, *et al.* (2004). Enhanced brain motor activity in patients with MS after a single dose of 3,4-diaminopyridine. *Neurology* 62, 2044–2050.

Manconi FM, Syed NA, Floeter MK (1998). Mechanisms underlying spinal motor neuron excitability during the cutaneous silent period in humans. *Muscle and Nerve* 21, 1256–1264.

Mano Y, Nakamuro T, Ikoma K, Takayanagi T, Mayer RF (1993). A clinicophysiologic study of central and peripheral motor conduction in hereditary demyelinating motor and sensory neuropathy. *Electromyography and Clinical Neurophysiology* 33, 101–107.

Mariorenzi R, Zarola F, Caramia MD, Paradiso C, Rossini PM (1991). Non-invasive evaluation of central motor tract excitability changes following peripheral nerve stimulation in healthy humans. *Electroencephalography and Clinical Neurophysiology* 81, 90–101.

Masur H, Oberwittler C, Fahrendorf G, *et al.* (1992). The relation between functional deficits, motor and sensory conduction times and MRI findings in syringomyelia. *Electroencephalography and Clinical Neurophysiology* 85, 321–330.

Mathis J, Hess C (1996). Motor-evoked potentials from multiple target muscles in multiple sclerosis and cervical myelopathy. *European Journal of Neurology* 3, 567–573.

Mavroudakis N, Caroyer JM, Brunko E, Zegers de Beyl D (1995). Abnormal motor evoked responses to transcranial magnetic stimulation in focal dystonia. *Neurology* 45, 1671–1677.

Mayr N, Baumgartner C, Zeitlhofer J, Deecke L (1991). The sensitivity of transcranial cortical magnetic stimulation in detecting pyramidal tract lesions in clinically definite multiple sclerosis. *Neurology* 41, 566–569.

Merton PA, Morton HB (1980). Stimulation of the cerebral cortex in the intact human subject. *Nature* 285, 227.

Meyer BU, Britton TC, Benecke R (1991a). Wilson's disease: normalisation of cortically evoked motor responses with treatment. *Journal of Neurology 238*, 327–330.

Meyer BU, Britton TC, Bischoff C, Machetanz J, Benecke R, Conrad B (1991b). Abnormal conduction in corticospinal pathways in Wilson's disease: investigation of nine cases with magnetic brain stimulation. *Movement Disorders 6*, 320–323.

Meyer BU, Britton TC, Kloten H, Steinmetz H, Benecke R (1991c). Coil placement in magnetic brain stimulation related to skull and brain anatomy. *Electroencephalography and Clinical Neurophysiology 81*, 38–46.

Meyer BU, Britton TC, Benecke R, Bischoff C, Machetanz J, Conrad B (1992a). Motor responses evoked by magnetic brain stimulation in psychogenic limb weakness: diagnostic value and limitations. *Journal of Neurology 239*, 251–255.

Meyer BU, Noth J, Lange HW, *et al.* (1992b). Motor responses evoked by magnetic brain stimulation in Huntington's disease. *Electroencephalography and Clinical Neurophysiology 85*, 197–208.

Mills KR, Nithi KA (1997). Corticomotor threshold to magnetic stimulation: normal values and repeatability. *Muscle and Nerve 20*, 570–576.

Mills KR, Nithi KA (1998). Peripheral and central motor conduction in amyotrophic lateral sclerosis. *Journal of Neurological Sciences 159*, 82–87.

Mills KR, McLeod C, Sheffy J, Loh L (1993). The optimal current direction for excitation of human cervical motor roots with a double coil magnetic stimulator. *Electroencephalography and Clinical Neurophysiology 89*, 138–144.

Miscio G, Pisano F, Mora G, Mazzini L (1999). Motor neuron disease: usefulness of transcranial magnetic stimulation in improving the diagnosis. *Clinical Neurophysiology 110*, 975–981.

Moglia A, Zandrini C, Alfonsi E, Rondanelli EG, Bono G, Nappi G (1991). Neurophysiological markers of central and peripheral involvement of the nervous system in HIV-infection. *Clinical Electroencephalography 22*, 193–198.

Molinuevo JL, Valls-Sole J, Valldeoriola F (2000). The effect of transcranial magnetic stimulation on reaction time in progressive supranuclear palsy. *Clinical Neurophysiology 111*, 2008–2013.

Molnar GF, Sailer A, Gunraj CA, *et al.* (2005). Changes in cortical excitability with thalamic deep brain stimulation. *Neurology 64*, 1913–1919.

Modugno N, Curra A, Giovannelli M, *et al.* (2001). The prolonged cortical silent period in patients with Huntington's disease. *Clinical Neurophysiology 112*, 1470–1474.

Mondelli M, Rossi A, Scarpini C, Guazzi GC (1995). Motor evoked potentials by magnetic stimulation in hereditary and sporadic ataxia. *Electromyography and Clinical Neurophysiology 35*, 415–424.

Nakamura H, Kitagawa H, Kawaguchi Y, Tsuji H (1996). Direct and indirect activation of human corticospinal neurons by transcranial magnetic and electrical stimulation. *Neuroscience Letters 210*, 45–48.

Nakanishi K, Tanaka N, Fujiwara Y, Kamei N, Ochi M (2006). Corticospinal tract conduction block results in the prolongation of central motor conduction time in compressive cervical myelopathy. *Clinical Neurophysiology 24*, 24.

Nardone R and Tezzon F (2003). Transcranial magnetic stimulation study in hereditary spastic paraparesis. *European Neurology 49*, 234–237.

Nezu A, Kimura S, Ohtsuki N, Tanaka M (1997). Transcranial magnetic stimulation in benign childhood epilepsy with centro-temporal spikes. *Brain Development 19*, 134–137.

Nielsen JE, Krabbe K, Jennum P *et al.* (1998). Autosomal dominant pure spastic paraplegia: a clinical, paraclinical, and genetic study. *Journal of Neurology, Neurosurgery and Psychiatry 64*, 61–66.

Nielsen JE, Jennum P, Fenger K, Sorensen SA, Fuglsang-Frederiksen A (2001). Increased intracortical facilitation in patients with autosomal dominant pure spastic paraplegia linked to chromosome 2p. *European Journal of Neurology 8*, 335–339.

Nogues MA, Pardal AM, Merello M, Miguel MA (1992). SEPs and CNS magnetic stimulation in syringomyelia. *Muscle and Nerve 15*, 993–1001.

Ofuji A, Kaneko K, Taguchi T, Fuchigami Y, Morita H, Kawai S (1998). New method to measure central motor conduction time using transcranial magnetic stimulation and T-response. *Journal of Neurological Sciences 160*, 26–32.

Oliveri M, Brighina F, La Bua V, Aloisio A, Buffa D, Fierro B (1997). Magnetic stimulation study in patients with myotonic dystrophy. *Electroencephalography and Clinical Neurophysiology 105*, 297–301.

Opsomer RJ, Caramia MD, Zarola F, Pesce F, Rossini PM (1989). Neurophysiological evaluation of central-peripheral sensory and motor pudendal fibres. *Electroencephalography and Clinical Neurophysiology 74*, 260–270.

Osei-Lah AD, Mills KR (2004). Optimising the detection of upper motor neuron function dysfunction in amyotrophic lateral sclerosis – a transcranial magnetic stimulation study. *Journal of Neurology 251*, 1364–1369.

Osei-Lah AD, Turner MR, Andersen PM, Leigh PN, Mills KR (2004). A novel central motor conduction abnormality in D90A-homozygous patients with amyotrophic lateral sclerosis. *Muscle and Nerve 29*, 790–794.

Ozata M, Ozkardes A, Dolu H, Corakci A, Yardim M, Gundogan MA (1996). Evaluation of central motor conduction in hypothyroid and hyperthyroid patients. *Journal of Endocrinological Investigation 19*, 670–677.

Pascual-Leone A, Dhuna A, Roth BJ, Cohen L, Hallett M (1990). Risk of burns during rapid-rate magnetic stimulation in presence of electrodes. *Lancet 336*, 1195–1196.

Pascual-Leone A, Cohen LG, Shotland LI, *et al.* (1992). No evidence of hearing loss in humans due to transcranial magnetic stimulation. *Neurology 42*, 647–651.

Pelosi L, Lanzillo B, Perretti A, Santoro L, Blumhardt L, Caruso G (1991). Motor and somatosensory evoked potentials in hereditary spastic paraplegia. *Journal of Neurology, Neurosurgery and Psychiatry 54*, 1099–1102.

Perretti A, Santoro L, Lanzillo B *et al.* (1996). Autosomal dominant cerebellar ataxia type I: multimodal electrophysiological study and comparison between SCA1 and SCA2 patients. *Journal of Neurological Sciences 142*, 45–53.

Perretti A, Pellecchia MT, Lanzillo B, Campanella G, Santoro L (2001). Excitatory and inhibitory mechanisms in Wilson's disease: investigation with magnetic motor cortex stimulation. *Journal of Neurological Sciences 192*, 35–40.

Philipson L, Larsson PG (1988). The electromyographic signal as a measure of muscular force: a comparison of detection and quantification techniques. *Electromyography and Clinical Neurophysiology 28*, 141–150.

Pierantozzi M, Palmieri MG, Mazzone P *et al.* (2002). Deep brain stimulation of both subthalamic nucleus and internal globus pallidus restores intracortical inhibition in Parkinson's disease paralleling apomorphine effects: a paired magnetic stimulation study. *Clinical Neurophysiology 113*, 108–113.

Pohl C, Block W, Traber F, *et al.* (2001). Proton magnetic resonance spectroscopy and transcranial magnetic stimulation for the detection of upper motor neuron degeneration in ALS patients. *Journal of Neurological Sciences 190*, 21–27.

Polo JM, Calleja J, Combarros O, Berciano J (1993). Hereditary 'pure' spastic paraplegia: a study of nine families. *Journal of Neurology, Neurosurgery and Psychiatry 56*, 175–181.

Pouget J, Trefouret S, Attarian S (2000). Transcranial magnetic stimulation (TMS): compared sensitivity of different motor response parameters in ALS. *Amyotrophic Lateral Sclerosis and Other Motor Neuron Disorders 1*(Suppl 2), S45–49.

Priori A, Berardelli A, Inghilleri M, Polidori L, Manfredi M (1994). Electromyographic silent period after transcranial brain stimulation in Huntington's disease. *Movement Disorders 9*, 178–182.

Prout AJ, Eisen AA (1994). The cortical silent period and amyotrophic lateral sclerosis. *Muscle and Nerve 17*, 217–223.

Rapisarda G, Bastings E, de Noordhout AM, Pennisi G, Delwaide PJ (1996). Can motor recovery in stroke patients be predicted by early transcranial magnetic stimulation? *Stroke 27*, 2191–2196.

Ravnborg M, Blinkenberg M, Dahl K (1991). Standardization of facilitation of compound muscle action potentials evoked by magnetic stimulation of the cortex. Results in healthy volunteers and in patients with multiple sclerosis. *Electroencephalography and Clinical Neurophysiology 81*, 195–201.

Ravnborg M, Liguori R, Christiansen P, Larsson H, Sorensen PS (1992). The diagnostic reliability of magnetically evoked motor potentials in multiple sclerosis. *Neurology 42*, 1296–1301.

Reid V (2003). Transcranial magnetic stimulation. *Physical Medicine and Rehabilitation Clinics of North America 14*, 307–325, ix.

Restivo DA, Giuffrida S, Rapisarda G *et al.* (2000). Central motor conduction to lower limb after transcranial magnetic stimulation in spinocerebellar ataxia type 2 (SCA2). *Clinical Neurophysiology 111*, 630–635.

Reutens DC, Berkovic SF (1992). Increased cortical excitability in generalised epilepsy demonstrated with transcranial magnetic stimulation. *Lancet 339*, 362–363.

Reutens DC, Berkovic SF, Macdonell RA, Bladin PF (1993). Magnetic stimulation of the brain in generalized epilepsy: reversal of cortical hyperexcitability by anticonvulsants. *Annals of Neurology 34*, 351–355.

Robinson LR, Little JW (1990). Motor-evoked potentials reflect spinal cord function in post-traumatic syringomyelia. *American Journal of Physical Medicine and Rehabilitation 69*, 307–310.

Rosler KM, Hess CW, Heckmann R, Ludin HP (1989). Significance of shape and size of the stimulating coil in magnetic stimulation of the human motor cortex. *Neuroscience Letters 100*, 347–352.

Rosler KM, Truffert A, Hess CW, Magistris MR (2000). Quantification of upper motor neuron loss in amyotrophic lateral sclerosis. *Clinical Neurophysiology 111*, 2208–2218.

Rossini PM, Rossi S (1998). Clinical applications of motor evoked potentials. *Electroencephalography and Clinical Neurophysiology 106*, 180–194.

Rossini PM, Zarola F, Floris R, *et al.* (1989). Sensory (VEP, BAEP, SEP) and motor-evoked potentials, liquoral and magnetic resonance findings in multiple sclerosis. *European Neurology 29*, 41–47.

Rossini PM, Desiato MT, Caramia MD (1992). Age-related changes of motor evoked potentials in healthy humans: non-invasive evaluation of central and peripheral motor tracts excitability and conductivity. *Brain Research 593*, 14–19.

Rossini PM, Barker AT, Berardelli A, *et al.* (1994). Non-invasive electrical and magnetic stimulation of the brain, spinal cord and roots: basic principles and procedures for routine clinical application. Report of an IFCN committee. *Electroencephalography and Clinical Neurophysiology 91*, 79–92.

Roth BJ, Pascual-Leone A, Cohen LG, Hallett M (1992). The heating of metal electrodes during rapid-rate magnetic stimulation: a possible safety hazard. *Electroencephalography and Clinical Neurophysiology 85*, 116–123.

Rothwell JC, Day BL, Thompson PD, Dick JP, Marsden CD (1987a). Some experiences of techniques for stimulation of the human cerebral motor cortex through the scalp. *Neurosurgery 20*, 156–163.

Rothwell JC, Thompson PD, Day BL, *et al.* (1987b). Motor cortex stimulation in intact man. 1. General characteristics of EMG responses in different muscles. *Brain 110*, 1173–1190.

Rothwell JC, Hallett M, Berardelli A, Eisen A, Rossini P, Paulus W (1999). Magnetic stimulation: motor evoked potentials. The International Federation of Clinical Neurophysiology. *Electroencephalography and Clinical Neurophysiology 52*(Suppl), 97–103.

Saadeh IK, Illis LS, Jamshidi Fard AR, Hughes PJ, Sedgwick EM (1994). Reversible motor and sensory neurophysiological abnormalities in cauda equina claudication. *Journal of Neurology, Neurosurgery and Psychiatry 57*, 1252–1254.

Sahota P, Prabhakar S, Lal V, Khurana D, Das CP, Singh P (2005). Transcranial magnetic stimulation: role in the evaluation of disability in multiple sclerosis. *Neurology India 53*, 197–201.

Salerno A, Georgesco M (1998). Double magnetic stimulation of the motor cortex in amyotrophic lateral sclerosis. *Electroencephalography and Clinical Neurophysiology 107*, 133–139.

Salerno A, Carlander B, Camu W, Georgesco M (1996). Motor evoked potentials (MEPs): evaluation of the different types of responses in amyotrophic lateral sclerosis and primary lateral sclerosis. *Electromyography and Clinical Neurophysiology 36*, 361–368.

Samii A, Luciano CA, Dambrosia JM, Hallett M (1998). Central motor conduction time: reproducibility and discomfort of different methods. *Muscle and Nerve 21*, 1445–1450.

Sandbrink F, Syed N, Floeter MK (1998). Paired pulse magnetic stimulation to augment motor evoked potentials in the leg. *Muscle and Nerve 21*, 1578 (abstract).

Sandbrink F, Syed NA, Fujii MD, Dalakas MC, Floeter MK (2000). Motor cortex excitability in stiff-person syndrome. *Brain 123*, 2231–2239.

Sartucci F, Sagliocco L, Murri L (1997). Central motor pathway evaluation using magnetic coil stimulation in hereditary motor and sensory neuropathy type I (HMSN type I, Charcot–Marie–Tooth disease). *International Journal of Neuroscience 92*, 145–159.

Schady W, Dick JP, Sheard A, Crampton S (1991). Central motor conduction studies in hereditary spastic paraplegia. *Journal of Neurology, Neurosurgery and Psychiatry 54*, 775–779.

Schmid UD, Walker G, Hess CW, Schmid J (1990). Magnetic and electrical stimulation of cervical motor roots: technique, site and mechanisms of excitation. *Journal of Neurology, Neurosurgery and Psychiatry 53*, 770–777.

Schmid UD, Walker G, Schmid-Sigron J, Hess CW (1991). Transcutaneous magnetic and electrical stimulation over the cervical spine: excitation of plexus roots – rather than spinal roots. *Electroencephalography and Clinical Neurophysiology 43*(Suppl), 369–384.

Schmierer K, Niehaus L, Roricht S, Meyer BU (2000). Conduction deficits of callosal fibres in early multiple sclerosis. *Journal of Neurology, Neurosurgery and Psychiatry 68*, 633–638.

Schmierer K, Irlbacher K, Grosse P, Roricht S, Meyer BU (2002). Correlates of disability in multiple sclerosis detected by transcranial magnetic stimulation. *Neurology 59*, 1218–1224.

Schnider A, Hess CW, Koppi S (1991). Central motor conduction in a family with hereditary motor and sensory neuropathy with pyramidal signs (HMSN V). *Journal of Neurology, Neurosurgery and Psychiatry 54*, 511–515.

Schols L, Riess O, Schols S, *et al.* (1995). Spinocerebellar ataxia type 1. Clinical and neurophysiological characteristics in German kindreds. *Acta Neurologica Scandinavica 92*, 478–485.

Schols L, Amoiridis G, Langkafel M, Schols S, Przuntek H (1997). Motor evoked potentials in the spinocerebellar ataxias type 1 and type 3. *Muscle and Nerve 20*, 226–228.

Schrader LM, Stern JM, Koski L, Nuwer MR, Engel J, Jr (2004). Seizure incidence during single- and paired-pulse transcranial magnetic stimulation (TMS) in individuals with epilepsy. *Clinical Neurophysiology 115*, 2728–2737.

Schriefer TN, Hess CW, Mills KR, Murray NM (1989). Central motor conduction studies in motor neurone disease using magnetic brain stimulation. *Electroencephalography and Clinical Neurophysiology 74*, 431–437.

Schubert M, Zierz S, Dengler R (1994). Central and peripheral nervous system conduction in mitochondrial myopathy with chronic progressive external ophthalmoplegia. *Electroencephalography and Clinical Neurophysiology 90*, 304–312.

Schwenkreis P, Tegenthoff M, Witscher K, *et al.* (2002). Motor cortex activation by transcranial magnetic stimulation in ataxia patients depends on the genetic defect. *Brain 125*, 301–309.

Siebner HR, Dressnandt J, Auer C, Conrad B (1998). Continuous intrathecal baclofen infusions induced a marked increase of the transcranially evoked silent period in a patient with generalized dystonia. *Muscle and Nerve 21*, 1209–1212.

Smith HC, Savic G, Frankel HL, *et al.* (2000). Corticospinal function studied over time following incomplete spinal cord injury. *Spinal Cord 38*, 292–300.

Steube D, Wietholter S, Correll C (2001). Prognostic value of lower limb motor evoked potentials for motor impairment and disability after 8 weeks of stroke rehabilitation–a prospective investigation of 100 patients. *Electromyography and Clinical Neurophysiology 41*, 463–469.

Taniguchi S, Tani T, Ushida T, Yamamoto H (2002). Motor evoked potentials elicited from erector spinae muscles in patients with thoracic myelopathy. *Spinal Cord 40*, 567–573.

Tassinari CA, Michelucci R, Forti A, *et al.* (1990). Transcranial magnetic stimulation in epileptic patients: usefulness and safety. *Neurology 40*, 1132–1133.

Tataroglu C, Genc A, Idiman E, Cakmur R, Idiman F (2003). Cortical silent period and motor evoked potentials in patients with multiple sclerosis. *Clinical Neurology and Neurosurgery 105*, 105–110.

Tavy DL, Wagner GL, Keunen RW, Wattendorff AR, Hekster RE, Franssen H (1994). Transcranial magnetic stimulation in patients with cervical spondylotic myelopathy: clinical and radiological correlations. *Muscle and Nerve 17*, 235–241.

Tavy DL, Franssen H, Keunen RW, Wattendorff AR, Hekster RE, Van Huffelen AC (1999). Motor and somatosensory evoked potentials in asymptomatic spondylotic cord compression. *Muscle and Nerve 22*, 628–634.

Tchen PH, Fu CC, Chiu HC (1992). Motor-evoked potentials in diabetes mellitus. *Journal of the Formosan Medical Association 91*, 20–23.

Tegenthoff M, Vorgerd M, Juskowiak F, Roos V, Malin JP (1996). Postexcitatory inhibition after transcranial magnetic single and double brain stimulation in Huntington's disease. *Electroencephalography and Clinical Neurophysiology 101*, 298–303.

Tharayil BS, Gangadhar BN, Thirthalli J, Anand L (2005). Seizure with single-pulse transcranial magnetic stimulation in a 35-year-old otherwise-healthy patient with bipolar disorder. *Journal of ECT 21*, 188–189.

Theodore WH, Hunter K, Chen R, *et al.* (2002). Transcranial magnetic stimulation for the treatment of seizures: a controlled study. *Neurology 59*, 560–562.

Thompson PD, Day BL, Rothwell JC, Dressler D, Maertens de Noordhout A, Marsden CD (1991). Further observations on the facilitation of muscle responses to cortical stimulation by voluntary contraction. *Electroencephalography and Clinical Neurophysiology 81*, 397–402.

Traversa R, Cicinelli P, Bassi A, Rossini PM and Bernardi G (1997). Mapping of motor cortical reorganization after stroke. A brain stimulation study with focal magnetic pulses. *Stroke 28*, 110–117.

Traversa R, Cicinelli P, Pasqualetti P, Filippi M, Rossini PM (1998). Follow-up of interhemispheric differences of motor evoked potentials from the 'affected' and 'unaffected' hemispheres in human stroke. *Brain Research 803*, 1–8.

Traversa R, Cicinelli P, Oliveri M, *et al.* (2000). Neurophysiological follow-up of motor cortical output in stroke patients. *Clinical Neurophysiology 111*, 1695–1703.

Travlos A, Pant B, Eisen A (1992). Transcranial magnetic stimulation for detection of preclinical cervical spondylotic myelopathy. *Archives of Physical Medicine and Rehabilitation 73*, 442–446.

Triggs WJ, Macdonell RA, Cros D, Chiappa KH, Shahani BT, Day BJ (1992). Motor inhibition and excitation are independent effects of magnetic cortical stimulation. *Annals of Neurology 32*, 345–351.

Triggs WJ, Calvanio R, Macdonell RA, Cros D, Chiappa KH (1994). Physiological motor asymmetry in human handedness: evidence from transcranial magnetic stimulation. *Brain Research 636*, 270–276.

Triggs WJ, Menkes D, Onorato J *et al.* (1999a). Transcranial magnetic stimulation identifies upper motor neuron involvement in motor neuron disease. *Neurology 53*, 605–611.

Triggs WJ, Subramanium B, Rossi F (1999b). Hand preference and transcranial magnetic stimulation asymmetry of cortical motor representation. *Brain Research 835*, 324–329.

Trompetto C, Caponnetto C, Buccolieri A, Marchese R, Abbruzzese G (1998). Responses of masseter muscles to transcranial magnetic stimulation in patients with amyotrophic lateral sclerosis. *Electroencephalography and Clinical Neurophysiology 109*, 309–314.

Trompetto C, Assini A, Buccolieri A, Marchese R, Abbruzzese G (1999). Intracortical inhibition after paired transcranial magnetic stimulation depends on the current flow direction. *Clinical Neurophysiology 110*, 1106–1110.

Trompetto C, Assini A, Buccolieri A, Marchese R, Abbruzzese G (2000). Motor recovery following stroke: a transcranial magnetic stimulation study. *Clinical Neurophysiology 111*, 1860–1867.

Truffert A, Rosler KM, Magistris MR (2000). Amyotrophic lateral sclerosis versus cervical spondylotic myelopathy: a study using transcranial magnetic stimulation with recordings from the trapezius and limb muscles. *Clinical Neurophysiology 111*, 1031–1038.

Turton A, Wroe S, Trepte N, Fraser C, Lemon RN (1996). Contralateral and ipsilateral EMG responses to transcranial magnetic stimulation during recovery of arm and hand function after stroke. *Electroencephalography and Clinical Neurophysiology 101*, 316–328.

Ugawa Y, Genba-Shimizu K, Kanazawa I (1995). Electrical stimulation of the human descending motor tracts at several levels. *Canadian Journal of Neurological Sciences 22*, 36–42.

Uozumi T, Tsuji S, Murai Y (1991). Motor potentials evoked by magnetic stimulation of the motor cortex in normal subjects and patients with motor disorders. *Electroencephalography and Clinical Neurophysiology 81*, 251–256.

Urban PP, Vogt T, Hopf HC (1998). Corticobulbar tract involvement in amyotrophic lateral sclerosis. A transcranial magnetic stimulation study. *Brain 121*, 1099–1108.

Urban PP, Wicht S, Hopf HC (2001). Sensitivity of transcranial magnetic stimulation of cortico-bulbar vs. cortico-spinal tract involvement in amyotrophic lateral sclerosis (ALS). *Journal of Neurology 248*, 850–855.

Valls-Sole J, Pascual-Leone A, Brasil-Neto JP, Cammarota A, McShane L, Hallett M (1994). Abnormal facilitation of the response to transcranial magnetic stimulation in patients with Parkinson's disease. *Neurology 44*, 735–741.

Valls-Sole J, Tolosa E, Marti MJ, *et al.* (2001). Examination of motor output pathways in patients with corticobasal ganglionic degeneration using transcranial magnetic stimulation. *Brain 124*, 1131–1137.

Van der Kamp W, Zwinderman AH, Ferrari MD, van Dijk JG (1996). Cortical excitability and response variability of transcranial magnetic stimulation. *Journal of Clinical Neurophysiology 13*, 164–171.

Wassermann EM (1998). Risk and safety of repetitive transcranial magnetic stimulation: report and suggested guidelines from the International Workshop on the Safety of Repetitive Transcranial Magnetic Stimulation, June 5–7, 1996. *Electroencephalography and Clinical Neurophysiology 108*, 1–16.

Wassermann EM (2002). Variation in the response to transcranial magnetic brain stimulation in the general population. *Clinical Neurophysiology 113*, 1165–1171.

Wassermann EM, Fuhr P, Cohen LG, Hallett M (1991). Effects of transcranial magnetic stimulation on ipsilateral muscles. *Neurology 41*, 1795–1799.

Wassermann EM, McShane LM, Hallett M, Cohen LG (1992). Noninvasive mapping of muscle representations in human motor cortex. *Electroencephalography and Clinical Neurophysiology 85*, 1–8.

Weber F, Albert U (2000). Electrodiagnostic examination in lumbosacral radiculopathies. *Electromyography and Clinical Neurophysiology 40*, 231–236.

Weber M, Eisen AA (2002). Magnetic stimulation of the central and peripheral nervous systems. *Muscle and Nerve 25*, 160–175.

Weber M, Eisen A, Stewart HG, Andersen PM (2000). Preserved slow conducting corticomotor neuronal projections in amyotrophic lateral sclerosis with autosomal recessive D90A CuZn-superoxide dismutase mutation. *Brain 123*, 1505–1515.

Wehling P, Cleveland S, Reinecke J, Schulitz KP (1995). Magnetic stimulation as a diagnostic tool in cervical nerve root compression and compression-induced neuropathy. *Journal of Spinal Disorders 8*, 304–307.

Werhahn KJ, Classen J, Benecke R (1995). The silent period induced by transcranial magnetic stimulation in muscles supplied by cranial nerves: normal data and changes in patients. *Journal of Neurology, Neurosurgery and Psychiatry 59*, 586–596.

Werhahn KJ, Lieber J, Classen J, Noachtar S (2000). Motor cortex excitability in patients with focal epilepsy. *Epilepsy Research 41*, 179–89.

White AT, Petajan JH (2004). Physiological measures of therapeutic response to interferon beta-1a treatment in remitting-relapsing MS. *Clinical Neurophysiology 115*, 2364–2371.

Wilbourn AJ, Aminoff MJ (1998). AAEM minimonograph 32, the electrodiagnostic examination in patients with radiculopathies. American Association of Electrodiagnostic Medicine. *Muscle and Nerve 21*, 1612–1631.

Wöhrle JC, Kammer T, Steinke W, Hennerici M (1995). Motor evoked potentials to magnetic stimulation in chronic and acute inflammatory demyelinating polyneuropathy. *Muscle and Nerve 18*, 904–906.

Wolters A, Classen J, Kunesch E, Grossmann A, Benecke R (2004). Measurements of transcallosally mediated cortical inhibition for differentiating parkinsonian syndromes. *Movement Disorders 19*, 518–528.

Yokota T, Yoshino A, Inaba A, Saito Y (1996). Double cortical stimulation in amyotrophic lateral sclerosis. *Journal of Neurology, Neurosurgery and Psychiatry 61*, 596–600.

Yokota T, Sasaki H, Iwabuchi K, *et al.* (1998). Electrophysiological features of central motor conduction in spinocerebellar atrophy type 1, type 2, and Machado-Joseph disease. *Journal of Neurology, Neurosurgery and Psychiatry 65*, 530–534.

Young RE, Morgan OS, Forster A (1998). Motor pathway analysis in HAM/TSP using magnetic stimulation and F-waves. *Canadian Journal of Neurological Sciences 25*, 48–54.

Zandrini C, Ciano C, Alfonsi E, Sandrini G, Minoli L, Moglia A (1990). Abnormalities of central motor conduction in asymptomatic HIV-positive patients. Significance and prognostic value. *Acta Neurologica 12*, 296–300.

Zhai P, Pagan F, Statland J, Butman JA, Floeter MK (2003). Primary lateral sclerosis: a heterogeneous disorder composed of different subtypes? *Neurology 60*, 1258–1265.

Ziemann U and Rothwell JC (2000). I-waves in motor cortex. *Journal of Clinical Neurophysiology 17*, 397–405.

Ziemann U, Netz J, Szelenyi A, Homberg V (1993). Spinal and supraspinal mechanisms contribute to the silent period in the contracting soleus muscle after transcranial magnetic stimulation of human motor cortex. *Neuroscience Letters 156*, 167–171.

Ziemann U, Lonnecker S, Steinhoff BJ, Paulus W (1996). Effects of antiepileptic drugs on motor cortex excitability in humans: a transcranial magnetic stimulation study. *Annals of Neurology 40*, 367–378.

Ziemann U, Winter M, Reimers CD, Reimers K, Tergau F, Paulus W (1997). Impaired motor cortex inhibition in patients with amyotrophic lateral sclerosis. Evidence from paired transcranial magnetic stimulation. *Neurology 49*, 1292–1298.

Zifko U, Remtulla H, Power K, Harker L, Bolton CF (1996). Transcortical and cervical magnetic stimulation with recording of the diaphragm. *Muscle and Nerve 19*, 614–620.

Zwarts MJ (1992). Central motor conduction in relation to contra- and ipsilateral activation. *Electroencephalography and Clinical Neurophysiology 85*, 425–428.

CHAPTER 20

TMS in the perioperative period

Laverne D. Gugino, Rafael Romero, Marcella Rameriz, Marc E. Richardson, and Linda S. Aglio

Historical perspectives

Since Barker *et al.* (1985) introduced a practical noninvasive stimulation technique for harnessing pulsed magnetic fields in 1985, research has focused on both understanding the effects of this form of stimulation on the central nervous system, as well as developing new clinical applications for this technique. This chapter will review the use of TMS for monitoring the functional integrity of the descending motor systems during surgery, as well as a brief discussion of a potential role in the preoperative period for conscious patients planning to undergo neurosurgical procedures involving the cerebral cortex.

Intraoperative use of somatosensory-evoked potentials (SSEPs) for monitoring central nervous system function has been a popular technique for more than 25 years (Gugino and Chabot 1990). Several studies have shown that monitoring SSEPs results in acceptable levels of postoperative outcome prediction, as well as possibly improving surgical outcomes (Brown *et al.* 1984; John *et al.* 1988; Nuwer *et al.* 1995; McCaffrey 1997; York 1997). In a multicenter survey of 173 orthopedic spinal surgeons, 8% used SSEP monitoring for a total of 51 263 procedures (Nuwer 1995). Based on surgical outcomes, a neurological morbidity rate of 0.6% was found (i.e. 343 postoperative deficits). False-negative predictions occurred in 0.63% of cases. False-positive outcomes were found in

1.51% of the monitored cases. The negative prediction rate was 99.93%. The higher false-positive rate implied a positive prediction value of 42%, suggesting a tendency for false alarms (McCaffrey 1997; York 1997).

In spite of the low false-negative rate, the appearance of several case reports describing new postoperative motor deficits with unchanged intraoperative SSEP responses served as the stimulus for finding a direct motor system monitor (Ginsburg *et al.* 1985; Chatrian *et al.* 1988; Crawford *et al.* 1988; Zornow *et al.* 1990). The false negatives presumably occur because SSEP monitoring, particularly during spinal cord procedures, assesses the dorsal column pathway, which is anatomically restricted to the posterior columns (Barker *et al.* 1985). Descending motor paths, however, are located in the lateral and anterior spinal cord white columns (McCaffrey 1997). Therefore, for SSEPs to adequately predict motor deficits, the spinal cord injury must affect both lateral, anterior as well as posterior, spinal regions. The false-negative rate can then be expected to correlate with the frequency with which surgically induced injury is restricted to the anterior and/or lateral spinal cord regions (Barker *et al.* 1985).

Two stimulation approaches have been developed for selectively exciting descending motor pathways. They are transcranial electrical (TES) and transcranial magnetic (TMS) stimulation. Both excite corticospinal tract neurons of origin using a transcranial technique. Levy and York (1983), Levy et al. (1984), and Levy (1987) were the first to show that TES could be used for eliciting averaged spinal cord and peripheral nerve responses. Their stimulation protocol involved a 'comb-like' scalp electrode consisting of numerous gold-plated electrodes as the anode with a single cathode fixed to the roof of the mouth (Levy and York 1983; Levy et al. 1984). Another group (Zentner et al. 1988, 1989; Zentner and Ebner 1989; Zentner 1989a,b; Zentner and Rieder 1990) used a single pulse (i.e. nonaveraging technique) with the anode placed at Cz and the cathode 6 cm anterior along the scalp midline. These two groups of investigators can be credited with generating interest in the use of transcranial intraoperative stimulation for monitoring descending motor systems during spinal cord surgery.

The use of magnetic stimulation for exciting CNS structures is a more recent advance (Barker et al. 1985). Although Michael Faraday described the phenomenon of electromagnetic induction in the early 1800s, it remained for Barker et al. in 1985 to rekindle interest in magnetic stimulation by demonstrating the ability of pulsed magnetic fields to produce transcranial motor cortex stimulation with acquisition of myogenic responses in a relatively painless fashion (Cushing 1909; Gugino et al. 1998). Since Barker's work, numerous refinements in equipment have led to studies exploring the many potential uses for TMS in humans. Edmonds and colleagues are credited with the first demonstration of spinal cord motor function monitoring using TMS in anesthetized patients (Edmonds et al. 1989; Glassman et al. 1996). TMS has been used for monitoring the functional integrity of descending motor paths during spinal cord surgery; the fact that it painlessly produces cortical excitation has led to the development of other perioperative applications in which TMS is used for studying the functional integrity of the motor in conscious patients. The section 'Contraindications for TMS in conscious and anesthetized patients' below is devoted to a discussion of a potential role for producing functional maps of the cerebral cortex relative to cortical lesions. This information may be useful as an aid for presurgical planning (Gugino et al. 1998).

Anatomical and physiological considerations for motor system excitation

A predominant descending motor path originating in the motor cortex which is excited by TMS is the corticospinal tract (CST) (Boyd et al. 1986; Day et al. 1986, 1988; Burke et al. 1990; Edgley et al. 1990; Rothwell et al. 1991; Fujiki et al. 1996). Amassian and Cracco (1987) and Amassian et al. (1987a, 1991) studied the activation of the cells of origin of the CST using cortical surface anodal electrical stimulation. Recording from the CST at the medullary pyramids and spinal cord demonstrated a train of descending waves (Boyd et al. 1986; Fujiki et al., 1996; Amassian and Cracco 1987; Amassian and Cracco 1987; Amassian et al. 1987a–c; Burke et al. 1990; Edgley et al. 1990). The first wave in the train, called the D-wave, was shown to be elicited by direct activation of either the axon hillock region or the first node of Ranvier of CST cells (Amassian et al. 1987a,b). The waves following the D-wave were referred to as I-waves, as they were believed to be generated by recurrent cortical interneuronal excitation of CST neurons (Boyd et al. 1986; Amassian and Cracco 1987; Beradelli et al. 1991; Deletis 1993; Fujiki et al. 1996). Similar descending CST activity has been recorded from the human spinal cord epidural space with TES and TMS (Boyd et al. 1986; Amassian and Cracco 1987; Beradelli et al. 1991; Deletis 1993; Fujiki et al. 1996).

The neuronal cell bodies of the CST axons are located in widespread areas of the cortex including premotor, motor, and somatosensory cortices (Lawrence et al. 1968, 1972). TMS or TES can be expected to excite many of these CST neurons. Lawrence, however, has demonstrated using selective lesions of these cortical areas that the spinal cord terminals of CST cells originating in premotor and motor cortex tend to pass anteriorly toward spinal cord anterior horn cell columns within the spinal cord gray area (Gilman and Marco 1971). Brouwer and Asby (1990) have emphasized that projection of the cortical neurons to spinal anterior horn cells in the spinal cord gray matter is characteristic of the CST descending cortical motor path.

In primates the CST axons have a 'strong' monosynaptic input to contralateral anterior horn cells supplying the distal small muscles of the upper and lower limbs. Brouwer *et al.* (1990) experimentally showed that TMS activates the same descending pathway (i.e. CST path) in humans based on the relative amplitudes of several upper and lower limb muscle responses. In addition, they showed that TMS activates the larger diameter CST neurons and these neurons, in turn, project in a monosynaptic fashion to the same muscles in the baboon and monkey using cortical surface anodal electrical stimulation. Phillips (1969), using intracellular recordings, demonstrated that a train of CST pulses (as occurs during cortical anodal surface stimulation) leads to a significant post-tetanic potentiation of anterior horn cell postsynaptic excitatory potentials (PSEPs). This property of the CST synaptic input onto anterior horn cells led to more efficacious excitation of these cells when compared to Group 1a (i.e. stretch reflex sensory input) monosynaptic train input onto the same cells.

Initial studies concerning the mechanism of TMS-induced excitation of CST neurons suggested that it differed from the mechanism of TES-induced excitation. Using single-fiber recordings it was independently demonstrated by two research groups that CST excitation by TMS was probably mediated through cortical interneuronal excitation of CST neurons (see Figure 20.1) (Day *et al.* 1986, 1987a, 1989; Roth *et al.* 1991; Rothwell 1991; Rothwell *et al.* 1991). TMS at all stimulus intensities produced single-muscle-fiber excitation at a longer latency than seen using TES. Investigators reasoned that because TMS-induced current loops were parallel to the inner cranial surface, they would likewise parallel the surface of the cortical gyri. The current loops would, therefore, be oriented along the axis of cortical interneurons, an orientation appropriate for cortical interneuron excitation. The induced current loops, however, would be orthogonal to CST neurons which are radially oriented within the cortical gyral caps. This current loop orientation is much less efficient for direct stimulation of CST cortical neurons. Day *et al.* (1987b, 1989) centered the round magnetic stimulating coil on the vertex. Amassian and Cracco (1987) and Amassian

et al. (1987a–c, 1991, 1992) later demonstrated that direct TMS activation of CST neurons was possible with appropriate tilting of the stimulating coil on the scalp. The change in coil orientation on the scalp presumably led to induced current loops which paralleled the orientation of CST neurons within the cortex. Jalinous (1991) has shown that large-diameter stimulating coils produce electric fields of greater intensity as a function of cortical depth than smaller coils. Theoretically, a scalp vertex orientation of a large round coil should directly excite CST neurons located within the anterior bank of the central sulcus. At this location, these cells are oriented in a radial direction to both the cortical surface and induced current loops.

Amassian *et al.* (1987a) studied the site of action potential initiation in monkey cortical CST neurons using surface anodal electrical stimulation. They recorded the CST epidural responses at the pyramids and lateral columns of the spinal cord. By studying the occurrence and latencies of both D- and I-waves, they believed that cortical surface anodal electrical stimulation initiated CST action potentials at the first or second node of Ranvier. Cathodal stimulation was believed to initiate CST I-waves through synaptic activation by excitatory cortical interneurons. Because appropriate tilting of a circular coil on the scalp could lead to direct excitation of motor cortex CST neurons, Amassian *et al.* (1990, 1992) reasoned that TMS caused CST activation by exciting axons within the subcortical white matter. In this regard, Amassian and Cracco (1987) and Amassian *et al.* (1987a–c) used a second approach for studying the location along axons where action potentials were initiated. In these studies, a long peripheral nerve was immersed in a saline-filled plastic model of the human skull. Figure-8 and circulator coils were placed along the outside of the plastic model (Figure 20.2). The peripheral nerve was arranged within the skull with a bend between the site of magnetic stimulation and the recording electrodes placed distal to the site of stimulation. Bends in the nerve were placed in order to mimic the upper limb and lower limb CST axonal trajectories after they emerged from the cortex *en route* to the internal capsule. The results of this study suggested that action potential initiation occurred at bends in the

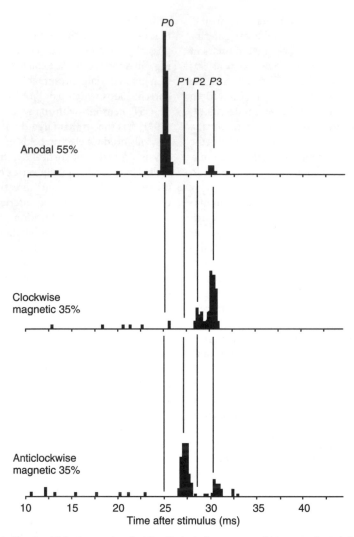

Fig. 20.1 Post-stimulus histograms showing the timing of responses from recordings taken from a first dorsal interosseous motor unit driven either by an anodal TES (at 55% of full strength) at the vertex (top plot) or by TMS applied with a vertex-centered round magnetic coil (middle and lower plots). The middle plot was obtained using a clockwise coil current at 35% of full power, whereas the lowest plot represents the response latencies acquired with a counterclockwise coil current. The horizontal axis of each histogram represents time in milliseconds after 100 of the three types of stimuli were applied to the scalp. The vertical axis represents the number of responses that responded with a latency as shown along the horizontal time axis. It should be noted that anodal TES produced the shortest latency responses when compared with both TMS stimuli. Data of this type were used to hypothesize that TMS activates corticospinal tract neurons indirectly (i.e. through interneuronal synaptic excitation). Reprinted with permission from Day *et al.* (1989).

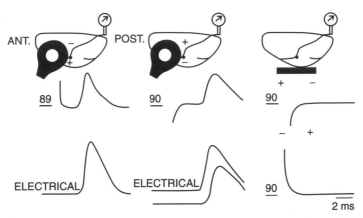

Fig. 20.2 Experimental protocol for determining the TMS site of induced excitation of a phrenic nerve immersed in an inverted saline-filled plastic model of the skull. The round magnetic coil is used for eliciting an action potential at a site along the phrenic nerve. This site is determined by comparing the latencies of TMS-acquired action potentials (recorded in air after the nerve emerges from the saline-conducting media) with the latencies of direct electrical excitation in which the cathode is moved along the nerve until the latencies of both stimulation modalities match. Electrical stimulation initiates action potentials beneath the cathode. It should be noted that the skull is inverted so that the mediolateral axis is oriented upwards, the opposite direction to that seen in erect volunteers. The site of the coil's minimal field gradient should represent approximately the 'virtual cathode' (i.e. site for action potential initiation). Thus, a 'virtual anode to cathode' orientation for the round magnetic coil can be predicted as a function of the polarity of the monophasic current pulse fed through the coil. The upper row represents three trials in which the TMS-induced 'virtual cathode' is closest to the recording electrode (left), the 'virtual anode' is closest to the recording electrode (middle), and the coil is placed symmetrically, parallel to the vertex (right). The middle row shows the responses to TMS stimulation (at the percentage of full power indicated to the left of each trace) for the trial indicated in the diagrams above. The bottom row shows the responses obtained with the recording electrode using electrical stimulation with a cathode at the 'bend' in the nerve within the skull (left lower trace) and at two locations further distal to the recording electrode and bend (middle lower trace). The rightmost TMS-induced trials in the second and third rows show the absence of TMS-initiated action potentials using either coil current pulse polarity when the round coil is symmetrically tangent to the vertex. There is a latency match between direct electrical and magnetic stimulation when the electrical stimulating cathode is placed at the bend in the nerve (left response column). This suggests that TMS excites intracranial axons at the sites where the axon changes direction as opposed to the site along an axon where the maximum negative spatial derivative of the induced electrical field exists. This figure represents a model for TMS excitation of corticospinal tract neurons projecting to the lower limb anterior horn cells spatial cord gray matter. Reprinted with permission from Amassian et al. (1992).

peripheral nerve within the skull model as opposed to the maximum negative spatial derivative of the magnetically induced electrical field. Amassian et al. (1992) reasoned that magnetically induced currents flowing within the axons would tend to exit at the axonal bend, leading to a situation favorable for axonal excitation.

Epstein et al. (1990) used two different-sized figure-8 coils to study the location of TMS-induced direct excitation of CST neurons in humans.

A small magnetic coil produces a more intense magnetic field close to the coil surface than is the case with a larger coil. However, the magnetic field attenuates faster with distance from the smaller coil. Because CST neurons have the same threshold for activation, the locus for CST action potential initiation is the same for both types of coil. The magnetic coil current intensities, however, are greater for threshold impulse initiation when using the smaller coil.

Epstein *et al.* determined the depth within the cortex where CST impulse initiation occurred by knowing the degree of the induced electric field attenuation with distance and the threshold for producing a thumb twitch. Their results were consistent with CST activation deep within the cortex located at the junction of the cortex and underlying white matter. They hypothesized that impulse initiation occurred near the axon hillock region or the CST basal dentrites.

Recent studies comparing the latencies of the CST epidural responses along the spinal cord in humans have shown that both TMS and TES cause direct activation of CST neurons (Boyd *et al.* 1986; Edgley *et al.* 1990; Fujuki *et al.* 1990;

Hicks *et al.* 1993). As the intensity of TMS was increased, the latency of the CST epidural D-wave response decreased but was still consistent with activation within the deep layer of the cortex (Fujiki *et al.* 1996). In addition, I-waves appeared when the TMS intensity was increased (Figure 20.3) (Fujiki *et al.* 1996). Increasing TES intensity produced an abrupt decrease in the spinal cord epidural D-wave latency, which was shown to be consistent with activation of CST axons at a midbrain level (Hicks *et al.* 1993). A shift in CST axon location of impulse initiation with increasing TES intensities could explain the earlier results discussed above, where TES caused an earlier activation of peripheral single

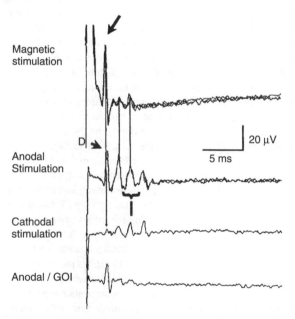

Fig. 20.3 Examples of corticospinal tract (CST) responses acquired from the human epidural space at an upper cervical location. The uppermost response was TMS-induced at 100% of full power. The arrow points to the CST D-wavelet. The second and third responses were induced by scalp surface anodal and cathodal TES, respectively. It should be noted that the anodal TES trial produced a CST D-wavelet with the same latency as TMS. Cathodal TES produces a smaller D-wavelet. The wavelets following the D-wavelets in the upper three trials are I-wavelets (see text). The lowest response in this experiment was obtained from anodal TES during general anesthesia. It should be noted that the D-wavelet (due to direct CST neuronal excitation) is still present, despite the fact that the patient is anesthetized, whereas the I-waves, produced by trans-synaptic excitation of CST neurons, succumb to the depressant effects of general anesthesia. Results shown in this figure do not support the earlier hypothesis that TMS excites CST neurons only through a trans-synaptic mechanism (see text). The general oral tracheal inhalational anesthesia was 0.5% isoflurane, 50.0% nitrous oxide by volume in oxygen. Negative voltage excursion values are raised in these recordings. Reprinted with permission from Fujiki (1996).

muscle fibers than that seen with TMS (Burke et al. 1990; Edgley et al. 1990). Thus, direct activation of CST neurons by TMS was compatible with the earlier single muscle fiber studies if TES caused earlier activation because of a shift in CST impulse initiation closer to the spinal cord.

Anesthetic considerations for monitoring descending motor systems

Selective monitoring of spinal cord motor function involves acquisition of TMS-induced epidural and/or myogenic responses (i.e. muscle compound action potentials). Because spinal cord surgery is usually performed in anesthetized patients, it is important to discuss the effect of general anesthetic agents on these responses. Hicks et al. (1993) reported that general anesthesia using isoflurane obliterated single pulse TMS-induced myogenic responses. Since that report, several research groups have sought for anesthetic agents that permit TMS-induced CST responses (Karama et al. 1972; Zentner 1989a,b; Zenter and Ebner 1989; Ghaly et al. 1990a–c, 1991a–c; Haghighi et al. 1990; Sloan et al. 1990; Sloan and Levin 1991; Kalkman et al. 1992). General anesthesia is expected to fulfill the following requirements for spinal cord surgery: loss of consciousness, amnesia, analgesia, absence of surgically induced movements, homeostasis of vital organ function, and expeditious postoperative arousal, allowing early neurological assessment (Stanski 1994).

Early studies concerning the depth of anesthesia using inhalational agents suggested that the loss of function seen with increasing concentration of inhalation agents followed an orderly sequence. Increasing anesthetic concentrations produce analgesia and amnesia, followed by loss of consciousness (Thornton and Jones 1993; Stanski 1994). Much larger concentrations were required to depress surgically induced movements. This led to the concept of a minimum alveolar concentration in which half of the population of patients would move in response to a surgical stimulus. An analogous concept was developed for intravenous agents (Schwinn et al. 1994). By achieving a minimum alveolar concentration, the anesthesiologist was usually assured of analgesia, amnesia, and hypnosis in the patient. The introduction of muscle relaxants allowed using lower concentrations of general anesthetic agents, leading to an increased incidence of patient awareness under anesthesia (Sloan 1990; Heneghan, 1993; Stanski 1994; Stinson et al. 1994). Recent studies suggest that general anesthetics inhibit surgically induced movements by depressing spinal cord interneural circuits (Rampil and Laster 1992; Rampil 1993, 1994; Rampil et al. 1993). The depression of the spinal gray matter in part explains the depressive effects of most general anesthesia agents on single-pulse TMS-induced myogenic responses (Zhou et al. 1997; Pereon et al. 1999). Studies concerning anesthetic effects on CST responses have also shown minimal effects on the D-wave, which is thought to be elicited by direct activation of CST cells of origin in the cortex (Sloan and Levin 1991; Yamada et al. 1994). Occasional I-waves are reported when lower concentrations of inhalational agents are used (Burke 1990; Deletis 1993). Thus, a second site of anesthetic agent depression of TMS-induced myogenic responses is at the cortical interneuronal level (Woodforth et al. 1999). Rampil and Laster (1992) have shown the necessity of trains of CST volleys (i.e. both D- and I-waves) for producing myogenic responses. It is reasonable to assume that, under anesthesia, the lack of I-waves leads to loss of temporal summation at the spinal cord interneuron and anterior horn cell level, precluding the consistent acquisition of myogenic responses (Woodforth et al. 1999; Zhou and Zhu 2000; Sloan and Heyer 2002; Kawaguchi and Juruya 2004). Therefore, some clinicians rely on monitoring the epidural TMS-induced D-wave because it is minimally affected by general anesthetic agents (Burke et al. 1990; Deletis 1993).

Myogenic responses should be monitored during surgery on the descending aorta. Using epidural CST responses in this type of surgery leads to a high false-negative rate for predicting new postoperative motor deficits (Elmore et al. 1991; deHann et al. 1996). Recent clinical experience has suggested a few anesthetic techniques that permit acquisition of single-pulse TMS-induced myogenic responses. A nitrous oxide narcotic technique has been successfully used: nitrous oxide is used at a 50% by volume concentration ratio in oxygen (Aglio *et al.* 2002b). Narcotics in high concentrations minimally depress TMS-induced myogenic responses and have been used successfully as the main anesthetic when monitoring myogenic responses during descending aortic surgery (Gugino *et al.* 1992). However, this technique does not permit early arousal after surgery. Lower concentrations of narcotics (i.e. for analgesia) can be successfully combined with ketamine, etomidate, methohexital, and propofol in oxygen for spinal cord surgery requiring early postoperative arousal (Ghaly *et al.* 1990a–c, 1991a–c 2000; Kalkman, *et al.* 1992; Ubags *et al.* 1997; Aglio *et al.* 2002a). Etomidate, in particular, has been a useful hypnotic agent for successfully using single-pulse TMS techniques. Etomidate is a unique anesthetic agent in that it causes a disinhibition of cortical neurons, an effect which is not seen in the thalamic-specific somatosensory nucleus (Ghaly *et al.* 1990a–c, 1991a–c; Losasso *et al.* 1991; Sloan and Heyer 2002).

In this regard, Gugino *et al.* (2001a) studied the changes in the median nerve SSEP scalp distribution in patients before and during an etomidate anesthetic consisting of etomidate, a narcotic, and oxygen (Figure 20.4). Scalp multi-recording electrode techniques using 20 active recording sites referenced to average linked ears were acquired. In most conscious adult patients, median nerve somatosensory input is thought to reach the cortex in 19–20 ms (Kelly *et al.* 1965; Golding *et al.* 1970; Arezzo *et al.* 1981; Broughton *et al.* 1981; Desmedt and Bourquet 1985; Deiber *et al.* 1986). During the initial arrival of SSEP input to the cortex, investigators have defined an initial negative deflection at 19–20 ms (n19) which localizes on the scalp overlying the central and parietal areas of the contralateral cortex. Positive voltage excursions with latencies of 20 (p20), 23 (p22), and 27 (p27) ms localize topographically on the scalp overlying (1) the bilateral frontal, (2) the contralateral central, and (3) the parietal cortical areas, respectively (Desmedt *et al.* 1987; Kelly *et al.* 1965; Deiber *et al.* 1986; Yamada 1988) (Figure 20.5). Several investigators have shown that the component SSEP wavelets described above are also present in cortical surface recordings (Broughton *et al.* 1981). The fact that these waveforms are volume-recorded potentials implies

that the cortex immediately below each recording electrode may or may not contain the cortical generators that produce these SSEP components. This has led to a controversy involving two schools of thought (Pakakostopoulos and Crow 1984; Yamada *et al.* 1984, 1985, 1988; Slimp *et al.* 1986; Yamada 1988). The first places a cortical generator within area 3b, located on the posterior bank of the central sulcus. This location can be shown to become active with timing appropriate to the surface-recorded p20–n20 components. The generator is thought to produce a tangential voltage dipole that leads to a positive component recorded frontally and a negative component over the posterior sensory cortex (Pakakostopoulos and Crow 1984). However, lesion studies have shown that in some cases, when area 3b is ablated, a p20 and p22 can still be recorded frontally, suggesting separate generators located in the frontal cortex (Pakakostopoulos and Crow 1984). The alternate theory, that a separate generator produces the p20 (located in the frontal areas) and an n20 located in area 3b, was for the most part based on lesion studies. Yamada *et al.* (1985) showed that the various levels of normal physiological sleep appear to affect the amplitude and latencies of the p20, p22 and n20 differently, supporting separate cortical generators for these components. The p22 generator site has, likewise, led to controversy, with one school placing it in area 3b of the postcentral gyrus and the other in area 3a or motor cortex. With regard to somatosensory input reaching both motor and somatosensory cortex, investigators have demonstrated a direct thalamic cortical SSEP input to both areas (Yamada *et al.* 1985).

Initial results from studies focused on anesthetic effects on cortical scalp components have suggested that etomidate does have a differential effect on the early cortical SSEP wavelets (Figures 20.4 and 20.5) (Gugino *et al.* 2001a). This effect is seen as a selective augmentation of the p20 and p22 amplitudes with a negligible effect on the contralateral parietal cortex n19 component. These results suggest that the cortical generator sites for the p20 and n19 differ. The augmentation of the p20 implies an initial disinhibition of perhaps the motor cortex which may account for the permissive effect of etomidate for acquiring myogenic responses using the single TMS technique (Ghaly *et al.* 1990a–c,

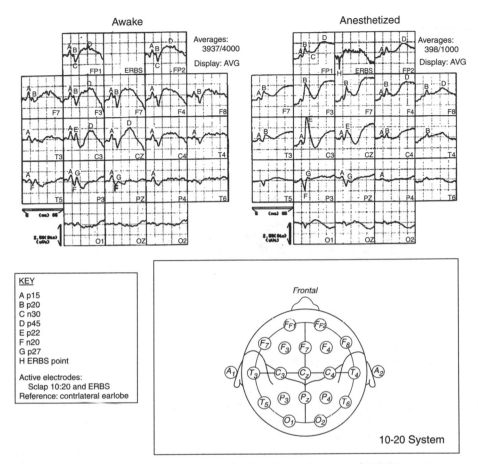

Fig. 20.4 Morphological distribution of scalp somatosensory-evoked potential (SSEP) waveforms generated by averaging right median-induced responses from 20 electrodes placed on the scalp according to the 10–20 recording electrode convention. The responses obtained from an awake patient are displayed with those obtained from the anterior scalp at the top; those from posterior scalp regions are at the bottom of the display on the left. The voltage and time scales are shown at the lower left corner of the display. Positive voltage excursion values are raised in both the awake and anesthetized recordings. The typical wavelets noted for scalp SSEP responses are labeled A–H (Erbs response, H, is not shown at the top of the awake display). In the anesthetized recording, the scalp distribution was obtained from the same patient during the anesthetic maintenance phase using etomidate as the primary hypnotic in oxygen with fentanyl (for analgesia) and vecuronium for muscle relaxation. Fentanyl has minimal amplitude effects on the scalp SSEP distribution. Note the lack of significant change for the contralateral n20 (labelled F) with a concomitant major increase in the amplitude of p20 and p22 (labelled B and E, respectively). The selective augmentation of p20 and p22 argues against a common cortical generator producing the n20, p20, and p22 SSEP wavelets. Modified with permission from Gugino (2001a).

1991a–c; Losasso *et al.* 1991; Ubags *et al.* 1997; Lotto *et al.* 2004). A shortcoming of the etomidate technique is that many of the patients emerged from general anesthesia in a confused state (Gugino *et al.* 2004). To determine if this confused state was due to single-pulse TMS or the etomidate anesthetic, seven volunteers were anesthetized with etomidate, oxygen, and a narcotic. On emergence, the volunteers were obviously awake (sitting up on a stretcher with

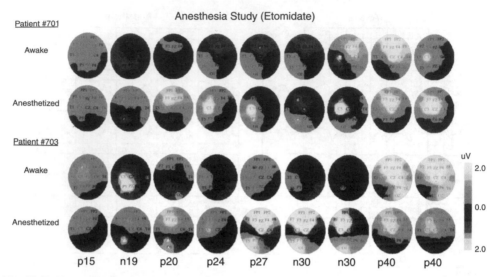

Fig. 20.5 Changes in the right median nerve-derived somatosensory-evoked potential (SSEP) wavelet amplitude distribution resulting from use of an etomidate anesthetic. The anesthetic consisted of etomidate in hypnotic doses, fentanyl, and oxygen. Fentanyl has minor effects on the amplitude of p15, n19, p20, p24, p27, and n30. Each cartoon-like head depicts the amplitude distribution of the scalp SSEP waveforms as a function of latency. The amplitude distributions across the scalp are viewed from above looking down on the scalp, with the front of the scalp above and the posterior scalp at the bottom of each cartoon. Each map (or cartoon) is constructed by noting the amplitude of the SSEP wavelets from all 20 recording electrodes on the scalp (using the 10–20 electrode system) at a single time point after the stimulus and color-coding the values according to the color chart at the lower right of the figure. Scalp amplitude values between depicted recording electrodes are derived from an interpolation technique. The amplitude distribution of p15 (the positive scalp wave occurring 15 ms after the electrical stimulus) shows little amplitude change from the awake to anesthetized states for the two patients shown. In contrast, the amplitude and scalp area giving rise to the negative component at 19 ms after the stimulus (n19) decreases. The positive components at 20 ms (p20), 24 ms (p22), and 27 ms (p27) increased in amplitude, as well as retrieved scalp area, in the anesthetized compared with the awake state. Evidence of this type argues against a single cortical generator site producing the n19, p20, and p22 SSEP wavelets because etomidate affects each wavelet differently. Unpublished observations. (Plate 4)

eyes open). However, they were unable to complete simple neurocognitive tasks, which was not a problem when they emerged from an inhalation or propofol total intravenous anesthetic (TIVA) anesthetic technique. The electroencephalogram acquired from the volunteers during the first 30 min after emergence from etomidate showed a diffuse gamma frequency activity, as well as superimposed frontal slow waves (Gugino et al. 2004). The diffuse gamma activity has been recently correlated with the conscious state, whereas the frontal slowing suggested frontal lobe dysfunction caused by residual etomidate effects (John and Prichep 2005). The frontal slowing typically wore off within 30–45 min into the postanesthetic recovery period. Nevertheless, the facilitating effect of etomidate for acquiring myogenic responses using a single-pulse TMS technique was felt to compensate for the transient postoperative confusional state.

Each of these anesthetic techniques requires using muscle relaxants (Ghaly et al. 2000; Sloan et al. 1990). TMS-induced myogenic responses can be acquired with a controlled muscle-relaxant-induced myoneural junction blockade of 50–70% (Sloan and Heyer 1990).

An automated train-of-four monitor can be used to determine the degree of muscle relaxation. The potent inhalational agents, barbiturates (except for methohexital), and benzodiazepines should be avoided when monitoring TMS-induced myogenic responses using single-pulse TMS (Tung et al. 1988; Haghighi et al. 1990; Thornton and Jones 1993; McPherson 1994; Yamada et al. 1994).

As will be discussed below, the use of the train TMS increases the list of permissive anesthetic techniques (Kalkman et al. 1993; Jennum et al. 1995). The probable reason for this is that train TMS produces, in general, repetitive D waves, which are minimally depressed by general anesthetic agents. This attribute in effect removes one anatomical region of anesthetic depression – the cortical interneurons responsible for producing I-waves so necessary for adequate excitation of anterior horn cells using the single-pulse TMS technique. Thus, using TMS trains has eliminated the need for anesthetic agents like etomidate which, due to their pharmacokinetics, cause an initial period of confusion in the anesthetic recovery period presumably due to a lingering effect of prefrontal lobe dysfunction. Intravenous propofol and/or methohexital narcotic and nitrous oxide in oxygen at >50% by volume and potent inhalational agents using <0.5 MAC concentrations have been used successfully for acquiring responses with train magnetic and/or electrical stimulation (Lotto et al. 2004). These anesthetic techniques, suitable for train TMS, presumably produce more depression of the cortical interneurons than the anesthetic techniques which permit motor response acquisition using single-pulse TMS (Kalkman et al. 1993; Taniguichi et al. 1993; Jennum et al. 1995; Jones et al. 1996; Watt et al. 1996; Lotto et al. 2004).

Intraoperative monitoring with single-pulse TMS techniques

The early experience of monitoring spinal cord motor function with TMS used single-pulse magnetic stimulation. As discussed in the previous section, successful monitoring of myogenic responses requires repetitive activation of spinal cord anterior horn cells as afforded by the D- and repetitive I-wave volleys generated by single transcranial stimulation. Because potent general anesthetics tend to suppress the cortical synapses producing the CST I-waves, two modifications were necessary to successfully use the single-pulse technique. The first modification was the design of an improved stimulation coil for producing TMS excitation of the motor cortex.

The improved coil was called the 'cap coil' (Cadwell Laboratories Inc., Kennewick, WA, USA), which is shown in Figure 20.6. One characteristic of this coil that makes it efficacious for TMS is its increased size due to an increase in the number of wire turns (Kraus et al. 1992). The larger diameter (i.e. 17 cm), as discussed previously, induces more intensive magnetic fields (Jalinous 1991). This leads to induction of greater electric fields deeper in the motor cortex. This is useful because it leads to an electric field that parallels CST neurons located in the anterior bank of the central sulcus. These CST neurons excite anterior horn cells activating distal upper extremity muscles, as well as CST neurons located within the medial aspects of motor cortex. The latter neurons excite distal lower extremity muscles. Intracortical currents, which flow parallel to the long axis of CST neurons, are the most efficacious for producing excitation (Amassian et al. 1992; Gugino et al. 1998). The anterior horn cells that supply distal extremity muscles typically receive monosynaptic CST synaptic excitation. The decreased requirement for spinal cord interneuron excitation of these anterior horn cells may lead to a decreased general anesthetic depression, since synapses are involved in the CST anterior horn call pathway.

The second useful characteristic is the saddle shape of the cap coil (Figure 20.6). This shape brings the wire turns of the coil closer to the scalp as compared to the rigid planar round coil. Figure 20.7a shows the induced electric field strength as a function of location along a reference plane 1 cm below the cap coil. The cap coil can be seen to have a 'volcano appearance' with a fore-and-aft peak above the remainder of the peak on top of the activation area. This induced voltage boost is a manifestation of the improved magnetic field coupling to the motor cortex afforded by the saddle shape. Figure 20.7b shows the electric field as a function of increasing depths from the surface of a

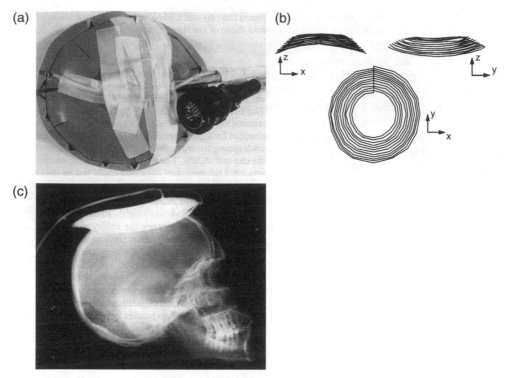

Fig. 20.6 (a) Prototype of the cap coil (17 cm in diameter). (b) Line segment representation of the cap coil showing the saddle-like design of this coil. (c) Radiograph of the cap coil placed on a volunteer's head showing the potential for ease of secure placement which is important during monitoring sessions. Reprinted with permission from Kraus *et al.* (1992b).

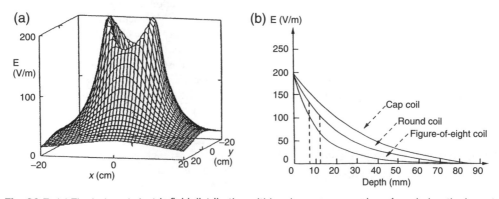

Fig. 20.7 (a) The induced electric field distribution within a homogeneous plane 1 cm below the lowest edges of the cap coil. Note the peaks of accentuated field intensity caused by the curved edges of the cap coil, which approach the reference plane at a reduced distance compared with the remainder of the coil. (b) A comparison of the attenuation of the induced electric fields beneath the cap coil (17 cm in diameter), the planar round coil (9 cm in diameter), and the figure-8 (5 cm per wing) coil. Note the faster rate of field attenuation with axial distance for the smaller coils. See text for further information. Reprinted with permission from Kraus *et al.* (1992a).

three-concentric-spherical model of the head and brain, for the cap coil, a 7 cm planar round coil, and the use of figure-8 coil. The cap coil induces the greatest electric field intensity at all depths within the model (Kalkman *et al.* 1993). A second benefit of the cap coil's saddle shape is the ease of attaching the coil to the patient's head during monitoring sessions. It is important to know how to place the cap coil on a patient's head in order to maximize the amplitude of acquired motor-evoked potentials (MEPs). To determine the optimum placement of the cap coil with respect to the underlying motor cortex, the cortices of five volunteers were mapped using a figure-8 stimulating coil (Aglio *et al.* 2002a). Figure 20.8 shows the typical amplitude

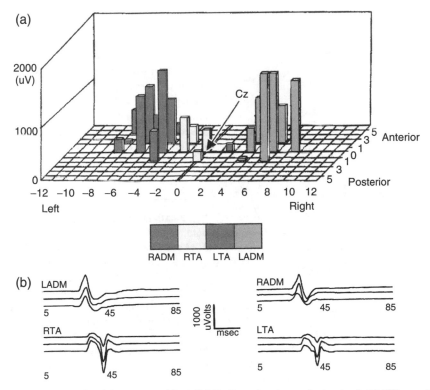

Fig. 20.8 (a) An example of the topographical distribution of motor-evoked potential (MEP) amplitudes acquired using focal TMS. The distribution is displayed as a pseudo-three-dimensional (3D) plot of four bar graphs, one for each of the muscles used for locating the motor cortex optima. The location on the scalp where focal TMS elicited MEP responses is indicated by bars plotted on the 1 cm × 1 cm grid that is referenced to the vertex and centered in the graph. Each group of bars representing MEP amplitudes was derived as a function of the scalp location used for stimulating the left and right abductor digiti minimi (LADM and RADM) and left and right tibialis anterior muscles (LTA and RTA). Color scale coding of the bar graphs is used to identify each of the four muscles studied according to the color scale below. (b) Below the pseudo-3D plot are examples of MEP responses acquired from the four muscles when local TMS was applied at the scalp locations resulting in the largest response amplitude for each muscle. These scalp locations are marked by asterisks within each scalp optimum indicated on the plot in (a). Each set of three consecutively acquired responses is shown to indicate the characteristic stability of the replicated responses when focal stimulation was applied to the scalp locations indicated. The voltage calibration for responses was (1) left and right ADM 3000 µV, (2) for LTA, RTA, 1000 µV. Time sweep calibration is 10 ms. All traces begin at 8 ms for artifact elimination. Negativity is up for all responses. Reprinted with permission from Aglio (2002a). (Plate 5)

distribution for the left and right adductor digiti minimi and tibialis anterior muscles. The distributions are plotted as a pseudo-three-dimensional (3D) plot where the x–y plane represents stimulated locations along the scalp with the z-axis showing the amplitude of the compound muscle action potentials for the muscles studied. Figure 20.9 shows the same mapping data redrawn to scale on a diagram of

the scalp illustrating the locations where a compound muscle action potential of >50 μV was acquired for each muscle. Figure 20.10 shows the cap coil superimposed on the 'motor map' at four different positions. As can be seen in Figure 20.10, responses from all four limbs occur when the cap's coil overlies all four optimum scalp loci. This generally occurs when the anterior edge of the cap coil is located 2 cm above the nasion, allowing the cap's posterior edge to overlie all four scalp loci.

Figure 20.11 shows the change in MEP amplitude and response area (the product of voltage and time duration of the MEP) as a function of stimulation strength. In general, as the stimulus intensity increases, a hyperbolic relationship is seen in which MEP amplitude and response areas approach a plateau in the awake volunteer. Takeoff latency, on the other hand, shows a modest decrease (Figure 20.12).

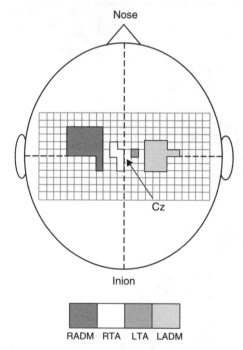

RADM RTA LTA LADM

Fig. 20.9 Map of the scalp showing the location of the scalp optima of the four muscles is displayed. The data for producing this map were taken from the pseudo-3D plot shown in Figure 20.8. This alternate format was constructed to show the scalp motor optimal locations from the viewer's perspective of looking down on a 2D model of the scalp surface. The anterior aspect of the scalp is at the top of the map. The 1 cm × 1 cm grid drawn on the subject's scalp during the mapping study is drawn on the scalp map centered at the vertex (Cz). Identification of each muscle optimum location uses the same color scale used for the construction of the pseudo-3D plot in Figure 20.8. Asterisks mark the location within each map optimum where TMS resulted in MEPs of maximum amplitude. Reprinted with permission from Aglio (2002a). (Plate 6)

Fig. 20.10 Demonstration of motor-evoked potentials (MEPs) acquired from left and right ADM and TA as a function of cap locations for the same volunteer whose data are shown in Figures 20.8 and 20.9. In the center are depicted four cap positions on the scalp: (a) anterior with cap rim 2 cm above the nasion; (b) cap coil centered on vertex; (c) anterior rim at the vertex; (d) cap anterior rim 3 cm posterior to the vertex. The cap coil locations are depicted as a coil of wire superimposed on replicas of the scalp model shown in Figure 20.9, which identify the scalp areas where focal TMS produced MEPs for the left and right abductor digiti minimi (LADM, RADM) and tibialis anterior (LTA, RTA) muscles. The grid has been removed for purposes of clarity. *Lower left*: lateral view of the head that shows the relative position on the head of the cap coil placements (a)–(d) from this perspective. *Bottom right*: color scale used for identifying the scalp muscle focus defined by the figure-8 coil. To the left and right of each scalp model are examples of MEPs obtained from each of the four muscles when simulated at 80% full power of the MES-10. Negativity is raised in all the traces. In each set of responses, the sweep begins at 8 ms for artifact elimination. Reprinted with permission from Aglio *et al.* (2002a).

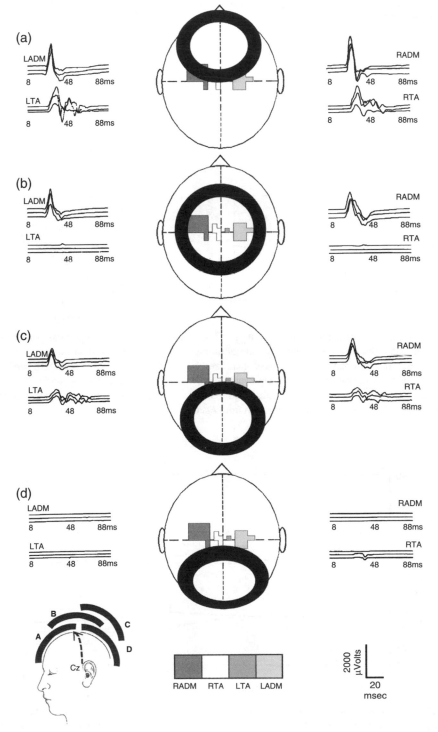

See also Plate 7.

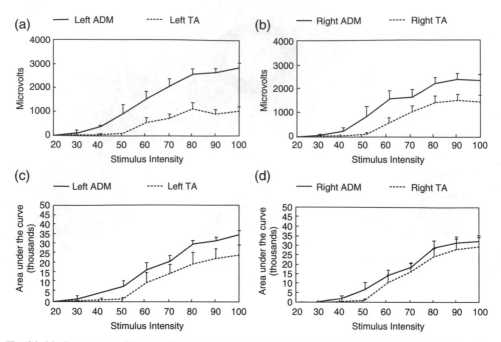

Fig. 20.11 Responses of the amplitude and area of the motor-evoked potentials (MEPs) to increasing stimulus intensity. The cap coil was placed anterior on the scalp 2 cm above the nasion in all five awake volunteers. Stimulus intensity along the x-axis increased from 20% to 100% of the MES-10 power. (a, b) Amplitude response for the left abductor digiti minimi (ADM) and tibialis anterior (TA), and right ADM and MEPs, respectively. (c, d) Area under the curve response for the same MEPs, respectively. Error bars represent SEM. Reprinted with permission from Aglio *et al.* (2002a).

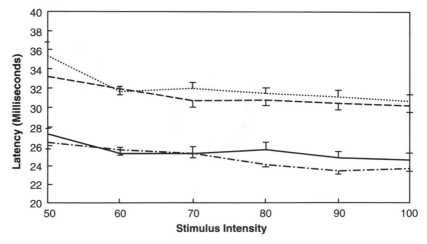

Fig. 20.12 Relationship of mean latency across volunteers of the left and right abductor digiti minimi (ADM) and tibialis anterior (TA) motor-evoked potentials as a function of stimulus intensity. Stimulus intensity axis is represented as percentage of maximum MES-10 power. Error bars represent SEM. Reprinted with permission from Aglio *et al.* (2002a).

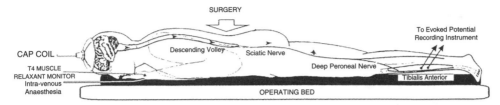

Fig. 20.13 Diagrammatic representation of the monitoring protocol used for acquiring motor evoked potentials (MEPs) during spinal cord procedures. A cap-shaped magnetic stimulating coil is secured to the patient's head for providing a single transcranial magnetic stimulus to the underlying cortex. MEPs were acquired from the tibialis anterior muscle for monitoring spinal cord motor function during procedures involving the thoracic and lumbar spine ($n = 27$). MEPs were acquired from the first dorsal interossei muscle for a procedure on the cervical cord (not shown). The degree of muscle relaxation was assessed with an automated train of four monitors placed on a muscle of the upper limb. The protocol for SSEP acquisition is omitted for ease of visualization. Reprinted with permission from Aglio *et al.* (2002b).

Figure 20.13 illustrates the monitoring protocol used for acquiring intraoperative MEPs. All monitored procedures ($n = 27$) were carried out on posterior spinal cord procedures (Aglio *et al.* 2002b). Scalp recording electrodes were placed at the internationally 10–20 defined C3′ and C4′ for recording SSEP electrical stimulation. MEPs were acquired from the adductor digiti minimi and tibialis anterior muscles using silver–silver chloride stick-on skin electrodes placed over the muscles of interest in a muscle 'belly-tendon' fashion. As discussed in the two preceding sections, the use of a single-pulse TMS requires efficacious excitation of both the CST cells of origin and the cortical interneurons responsible for re-excitation of the same population of cortical neurons. In this way, a volley of descending CST activity is obtained for producing temporal summation of CST synaptic activation of anterior horn cells. The second structure of concern for anesthetic depression is the CST–anterior horn cell synapse. In our experience, the CST–anterior horn cell pathway was least suppressed using one of three anesthetic techniques (see Table 20.1). Etomidate is a hypnotic agent thought to cause disinhibition of the cerebral cortex. Thus, the excitability of cortical interneurons required for repetitive reactivation of the CST neurons is better maintained using this agent. It is given as a constant infusion together with a narcotic infusion and nitrous oxide in oxygen ≤50% by volume. Muscle relaxation is maintained relatively constant at a 50% block with a relaxant infusion controlled

Table 20.1 Single-pulse technique: distribution of anesthetic techniques and response failures for monitoring MEPs[a]

Type of procedure	Number of cases	Type of anesthetic	% Non-monitorable	% Preoperative deficits/nonmonitorable
Vascular	2	High dose narcotic, etomidate	0%	0%
Spinal column, spinal cord	4	Ketamine, propofol, narcotics	50% (2)[b]	100% (2)
	21	Etomidate, narcotics, nitrous oxide	19% (4)	25% (1)
Total	27		22% (6)	50% (3)

[a]One of the 27 patients had the upper limb monitored, the other 26 patients had lower limb monitoring.
[b]Numbers in parentheses Indicate number of patients.

by a train of four monitors. A second anesthetic technique was a propofol TIVA with propofol infused at 75–100 mg/kg/min supplemented with ketamine (40 mg bolus once an hour). No nitrous oxide was used for this technique. These two anesthetic techniques were developed in order to assure patient loss of consciousness and a rapid patient emergence at the end of the procedure (Aglio *et al.* 2002a).

In procedures where immediate postprocedural emergence was not required (i.e. aortic surgery), a high-dose narcotic and an etomidate infusion technique were administered with oxygen (Gugino *et al.* 1992). Table 20.1 also shows the percentage of patients in whom responses suitable for monitoring were not obtained and the percentage of monitoring failures who had perioperative motor deficits. Although the number of patients within each anesthetic category is small, the anesthetic most permissive for acquiring MEPs is the high-dose narcotic-in-oxygen technique.

Figure 20.14 shows the distribution, in bar graph form, of the tibialis anterior MEP latency, baseline-to-peak amplitude, maximum peak-to-trough amplitude, and area under the response for 17 of the patients anesthetized with the etomidate, nitrous oxide, narcotic technique (Aglio *et al.* 2002a). The data were collected 20 min after anesthetic induction, but before surgical manipulation of the spinal column began. Note that latency appears to have a Gaussian distribution, whereas amplitude or response area

do not. Logarithmic transformation produces a near-Gaussian distribution for these MEP parameters. Figure 20.15 shows the distribution for the lower limb cortical SSEPs for comparison. Tables 20.2 and 20.3 contain descriptive statistics for the MEPs and SSEPs acquired prior to surgical manipulation of the spinal column. Based on these values, it was shown that a change of 40.6% decrease in amplitude and an 8.6% increase in latency would define a significant change from baselines for the SSEPs (Aglio *et al.* 2002a). The variability of MEP parameters is two to three times that for averaged SSEPs. Similar calculations for determining a statistically significant change in MEPs would be a complete loss of MEPs over two sequential updates during a surgical procedure (Aglio *et al.* 2002a).

Figure 20.16 is an example of MEP monitoring for a spinal cord decompression procedure. The patient had a history of a previous motor vehicle accident in which he sustained multiple thoracic and lumbar spine fractures. He was left with a T10 sensory pinprick level and absence of voluntary movement below the knees. Within the month prior to his recent admission, he complained of increasing weakness in the proximal lower limb muscles. A new bony callus had formed at one of his previous fracture sites, which caused compression of the L1 spinal cord. The patient was brought to the operating room for a decompression of his lumbar spine. Initially, the left lower limb MEPs were absent. During the decompression, the left rectus femoris

Table 20.2 Means and coefficients of variation for MEP response characteristics ($n = 17$)

	Takeoff latency (ms)	Amplitude (mV)	Area (mV/ms)
Mean ± SD[a]	29.6 ± 3.1	264.0 ± 152.8	1789.1 ± 1025.2
Coefficient of variation[b] (%)	11.2	53.6	58.8
Range of coefficient of variation[c] (%)	1.2 – 53.6	17.0 – 99.2	26.3 – 105.7
Required change[d] (%)	20.2	113.4	112.3
Required change after log transformation[d] (%)	44.1	54.4	41.5

[a]Mean and SD are average mean and SD values of 17 patients.
[b]The coefficient of variation is the average coefficient of variations of 17 patients.
[c]Range of coefficient of variation is the minimum and maximum value of CV seen in 17 patients.
[d]Required change is the percentage change necessary before an individual value would be significiently different from the preoperative control ($p<0.05$). It is calculated as follows: required change = (preoperative SD*100*1.96)/(preoperative

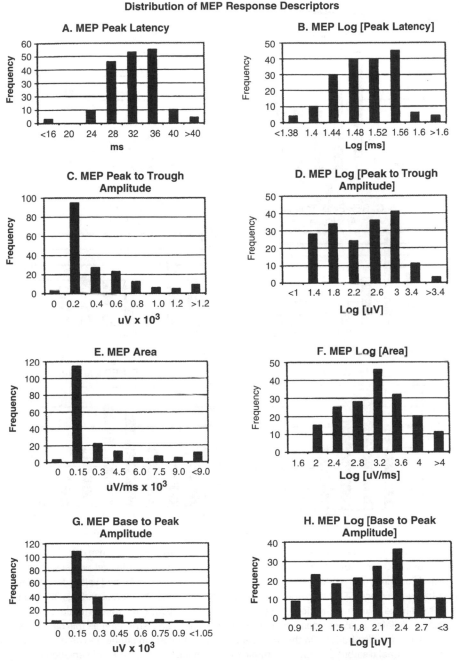

Fig. 20.14 Distribution of TMS-induced motor-evoked potential (MEP) response descriptors acquired from 17 patients, 20 min after induction of an etomidate, narcotic, nitrous oxide general anesthetic. Myoneural junction block was maintained at 50% during acquisition of the MEPs. The left column of each histogram shows the distribution of nontransformed values for latency, first peak-to-trough MEP amplitude, and baseline to first peak amplitude for the MEP responses. The corresponding histograms along the right-hand column show the same distribution after logarithmic transformation. Note that the transformation of latency values adds little to producing a Gaussian distribution as this parameter already has a normal distribution. On the other hand, peak-to-trough amplitude incompletely approaches a normal distribution with logarithmic transformation. Reprinted with permission from Aglio et al. (2002b).

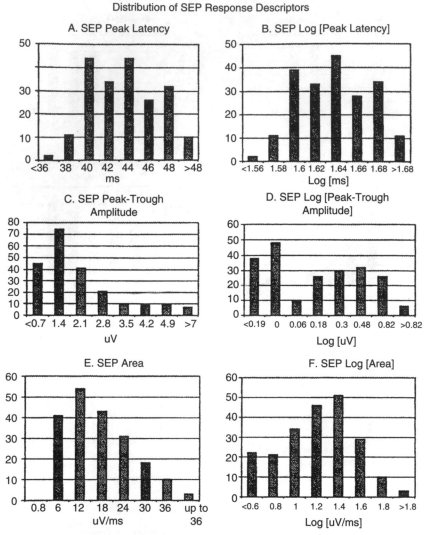

Fig. 20.15 Distribution of cortical somatosensory-evoked potential (SSEP) response descriptors acquired from 17 patients, 20 min after induction of an etomidate, narcotic, nitrous oxide general anesthetic. The left-hand column of each histogram shows the distribution of nontransformed values for peak latency, first peak-to-trough cortical SSEP amplitude, and SSEP response area. The corresponding histograms along the right-hand column show the same distributions after logarithmic transformation. Note that transformation of latency values adds little to producing a Gaussian distribution as this parameter already has a normal distribution. On the other hand, peak-to-trough amplitude and area in completely approaches a normal distribution with logarithmic transformation. Unpublished observations.

motor responses appeared and remained to the end of the procedure. The patient's motor examination improved after the procedure.

Table 20.4 shows outcome data for the patients monitored with single-pulse TMS (Aglio *et al.* 2002b). The state of the MEPs at the end of the procedure were compared to the motor function examination after the procedure. Note the incidence of one false positive and no false negatives using the criterion discussed above for predicting postoperative motor function with single-pulse TMS monitoring (Aglio *et al.* 2002b).

Table 20.3 Means and coefficients of variation for SEP response characteristics ($n = 17$)

	Peak latency (ms)	Amplitude (mV)	Area (mV/ms)
Mean ± SD[a]	42.5 ± 1.9	1.5 ± 0.3	15.0 ± 3.6
Coefficient of variation[b] (%)	4.4	25.6	29.9
Range of coefficient of variation[c] (%)	0.5 – 10.5	6.6 – 57.2	8.3 – 67.1
Required change[d] (%)	8.6	40.6	46.5

[a]Mean and SD are average mean and SD values of 17 patients.
[b]The coefficient of variation is the average coefficient of variations of 17 patients.
[c]Range of coefficient of variation is the minimum and maximum value of CV seen in 17 patients.
[d]Required change is the percent change necessary before an individual value would be significiently different from the preoperative control [$p<0.05$]. It is calculated as follows: required change = (preoperative SD*100*1.96)/(preoperative

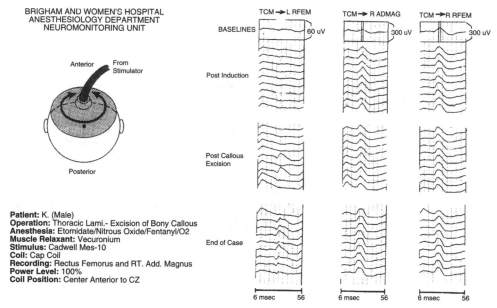

Fig. 20.16 Example of a case demonstrating single-pulse TMS-induced myogenic responses obtained from the left rectus femoris muscle (RFEM), right adductor magnus muscle (ADMAG), and right RFEM. The patient had a preoperative T10 sensory pinprick level and an absence of voluntary movement beneath the knees due to a previous spinal cord injury. The patient presented with an anterior spinal column outgrowth compressing the L1 lumbar spinal cord. Initially, myogenic responses from the left RFEM muscles were absent. After excision of the callus, the left RFEM myogenic responses began to appear. The patient demonstrated an improved postoperative motor examination. The uppermost response represents a postanesthetic induction baseline from the three muscles monitored. The two vertical lines superimposed on the right, second, and third baseline responses represent the 90% confidence limits for latency of the myogenic peaks monitored. Voltage and time calibration are shown at top and bottom of each response column. Negativity is raised. Reprinted with permission from Gugino *et al.* (1997).

Table 20.4 MEP outcome prediction[a]

N	21
Postoperative changes	3[b]
True positives	3
True negatives	17
False positives	1
False negatives	0
Sensitivity	100%
Specificity	94%
Positive predictive value	75%
Negative predictive value	100%

[a]Sensitivity = (true positive)/(true positive + false negative).
Specificity = (true negative)/(true negative + false positive).
Positive predictive value = (true positive)/(true positive + false positive).
Negative predictive value = (true negative)/(true negative + false negative).
[b]The 4th patient did not survive the surgical procedure (patient #3, see Table 20.1).

Train TMS

The introduction of train stimuli was a major improvement for intraoperative noninvasive excitation of the cortex (Kalkman *et al.* 1993; Taniguichi *et al.* 1993; Jennum *et al.* 1995; Jones *et al.* 1996; Watt *et al.* 1996). The use of stimulating trains in which the train pulses are set to an appropriate interpulse interval (i.e. usually from 2.0 to 5.0 ms) can produce repetitive direct excitation of cortical neurons. This eliminates the reliance on intracortical interneurons for producing the repetitive synaptic re-excitation of upper motor neurons. As discussed in a previous section, volleys of descending CST activity are required for successful excitation of lower motor neurons, particularly during general anesthesia. We have had experience with a magnetic train stimulator able to produce four monophasic pulses at a frequency up to 1000 Hz (Quadropulse™, Magstim Ltd, Wales, UK) which has the capability to reverse the current polarity within the stimulating coil (Gugino *et al.* 1997). This feature enables reversing the induced intracortical current's

polarity for more efficient unilateral motor cortex excitation, particularly of the upper limb area (Gugino *et al.* 1997).

Although several stimulating coil geometries have been tried, the currently used coil for acquiring lower limb responses is an oversized figure-8 design, which has a bend along the body of the coil. The coil is referred to as the cone coil (Magstim Ltd, Wales, UK). As was the case for the cap coil, this geometry serves to bring the induced magnetic flux from the coil wings as well as the center of the coil closer to the scalp (see Figure 20.17). This improves the magnetic flux coupling with the underlying cortex. A curved round coil has also been used for procedures in which only the upper limb MEPs were monitored (Gugino *et al.* 1997).

Both coils are fixed to the scalp with the posterior edge overlying the vertex, such that the lateral and longitudinal midpoint of the coil straddles the vertex. Fine coil adjustment involved an anterior or posterior displacement from the initial position for optimizing the recorded responses. Bite blocks are placed in order to prevent potential tongue lacerations from possible masseter contractions.

MEPs are recorded from lower extremity muscles with the same techniques used for single-pulse TMS. Figure 20.18 shows the appearance of bilateral tibialis anterior myogenic responses during a propofol-based anesthetic in which the stimulating trains increase from two to four pulses. The responses appear at a train of three pulses with a minor increase in response amplitude at four pulses (Gugino *et al.* 1997). Figure 20.19 shows the effect of stimulating the motor cortex with the curved round coil using trains differing in coil current polarity. When the coil current induced an anteriorly directed cortical current through the right hemisphere, a left adductor digiti minimi response was obtained. Reversing the polarity of the coil current produced a contralateral response. Figure 20.20 shows the use of the cone coil for acquiring bilateral tibialis anterior responses. A posterior–anterior orientation of induced cortical current leads to bilateral lower extremity response (Gugino *et al.* 1997).

Figure 20.21 shows the posterior tibial cortical SSEPs and first dorsal interossei (FDI) and tibialis

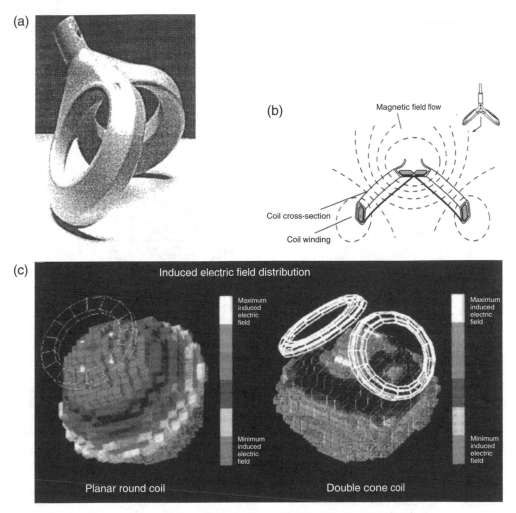

(a)

(b)

Magnetic field flow

Coil cross-section

Coil winding

(c) Induced electric field distribution

Maximum induced electric field

Minimum induced electric field

Planar round coil

Maximum induced electric field

Minimum induced electric field

Double cone coil

Fig. 20.17 (a) The double-cone coil. (b) Cross-section of the coil at a segment of each component circular coil and the region where the circular coil converge. (c) Induced magnetic field around these regions of the coil. Reprinted with permission from Magstim Ltd, Wales (1996).

anterior (TA) MEPs acquired during a cervical intramedullary tumor excision. All responses were present after induction of anesthesia. During the midline longitudinal myelotomy, the lower limb SSEPs and right FDI motor responses were significantly attenuated (>80% amplitude attenuation). The tibialis anterior responses were also attenuated but still had amplitudes >20% of baseline values. The patient emerged from anesthesia with normal muscle strength but abnormal proprioception. This case represents a false positive with respect to the right upper limb postoperative motor examination (Gugino *et al.* 1997).

Table 20.5 demonstrates the success rate for acquiring responses as a function of anesthetic technique. Note that the majority of procedures were monitored during a propofol-based anesthetic. There was a lower proportion of failed monitoring due to preoperative deficits than was the case for the single-shock technique (unpublished data).

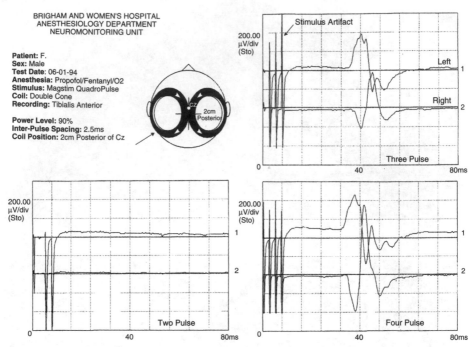

BRIGHAM AND WOMEN'S HOSPITAL
ANESTHESIOLOGY DEPARTMENT
NEUROMONITORING UNIT

Patient: F.
Sex: Male
Test Date: 06-01-94
Anesthesia: Propofol/Fentanyl/O2
Stimulus: Magstim QuadroPulse
Coil: Double Cone
Recording: Tibialis Anterior

Power Level: 90%
Inter-Pulse Spacing: 2.5ms
Coil Position: 2cm Posterior of Cz

Fig. 20.18 TMS-induced myogenic responses obtained using a two (*lower left*), three (*upper right*), and four magnetic pulse train. There is a stimulus artifact to the left of each trace. The left and right anterior tibialis motor-evoked potentials appeared with only the three and four magnetic pulse trains. The TMS train consisted of monophasic pulses with an interpulse interval of 2.5 ms. These responses were obtained from a patient under a propofol-based total intravenous anesthetic. The horizontal axis is measured in milliseconds. The vertical axis is measured in microvolts per division. Negativity is raised. Reprinted with permission from Gugino *et al.* (1997).

Table 20.5 Four-pulse TCMS technique

Operation	No.	Anesthetic	Failed responses	Preoperative deficit/failed responses
Vascular	3	High dose Fentanyl Etomidate	0	0
Spinal column/cord	6	Ketamine Propofol Fentanyl	33 (2)[a]	100 (2)
	49	Propofol Fentanyl	14 (7)	57 (4)
Total	58		16 (9)	67 (6)

[a]Number in parenthesis equals number of patients

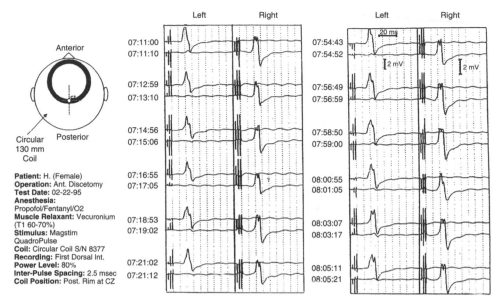

Fig. 20.19 Example of upper limb myogenic responses obtained with a four-pulse TMS train during anesthesia. A curved round coil centered left to right on the scalp with the posterior overlying the vertex was used for obtaining these responses. The interpulse interval was 2.5 ms with a four-monophasic-pulse train. Time and voltage calibration bars are shown at upper right of figure. Note the two different display gains for left and right responses. Each set of responses represents and alternating left and right first dorsal interossei motor-evoked potentials obtained with reversal of the stimulating coil current direction. Response negativity is raised. Reprinted with permission from Gugino et al. (1997).

Table 20.6 shows statistical descriptors of the responses acquired with the train technique for both upper and lower limb responses (unpublished data). Note that the smaller coefficient of variation (compared to the single shock technique (see Table 20.2)) decreased the change necessary for defining a significant statistical change during monitoring. The data for this table was collected in the same fashion as those for Table 20.2.

Finally, Table 20.7 shows the relationship for the status of the last MEP and outcomes for 54 cases in which motor function was monitored with train TMS. Based on the results in

Table 20.6 Four-pulse TCMS technique MEP response characteristics

	Upper N = 25			Lower N = 17		
	Mean	CV %	% Change needed	Mean	CV (%)	% Change needed
Takeoff Latency (ms)	23.48	4.86	9.52	34.64	3.52	6.89
Peak-to-peak amplitude (µV)	1388.2	31.98	62.68	555.74	32.61	63.91
Area (µV x ms)	6661.1	33.98	66.59	3641.0	36.69	71.91

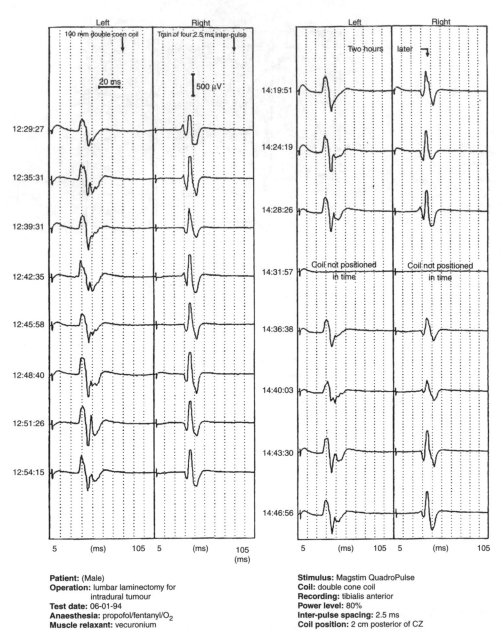

Patient: (Male)
Operation: lumbar laminectomy for
intradural tumour
Test date: 06-01-94
Anaesthesia: propofol/fentanyl/O$_2$
Muscle relaxant: vecuronium

Stimulus: Magstim QuadroPulse
Coil: double cone coil
Recording: tibialis anterior
Power level: 80%
Inter-pulse spacing: 2.5 ms
Coil position: 2 cm posterior of CZ

Fig. 20.20 Stability of myogenic response sequence from left and right tibialis anterior muscles is illustrated using the double-cone coil and a TMS train stimuli during a monitored case. The sequence on the right was obtained 45 min after the sequence illustrated on the left and demonstrates improved response variability using TMS trains compared with single-pulse techniques. Voltage and amplitude calibrations shown at the top of the left response columns. Negativity is raised. Reprinted with permission from Gugino *et al.* (1997). (Plate 6)

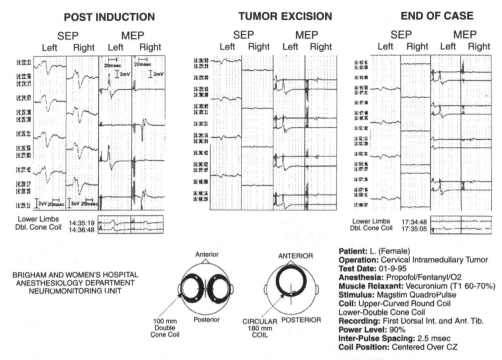

Fig. 20.21 Illustration of a monitored case that represented a false-positive result with respect to new postoperative motor deficits. Control median nerve cortical SSEPs and first dorsal interossei (FDI) myogenic responses using the curved round coil are shown at left. Beneath the control FDI responses are tibialis anterior motor-evoked potentials (MEPs) acquired using the double-cone coil. The center response sequence shows changes in SSEPs bilaterally and for the right FDI response after spinal cord (midline) longitudinal myelotomy and during excision of an intramedullary tumor. The right sequence shows the responses at the end of the case before the anesthetic was discontinued. Voltage calibrations are shown within the left response sequence. After recovery from anesthesia, the patient showed new proprioceptive sensory deficits but had a normal motor 'examination'. Negativity is raised in all records. Reprinted with permission from Gugino et al. (1997).

Table 20.6, a significant MEP amplitude change was defined as an 80% attenuation from baseline values. As was the case for single-pulse TMS monitoring, there was a conspicuous absence of false-negative outcomes (Gugino et al. 1997).

Table 20.7 Four-pulse TCMS technique

Prediction	N = 54 cases monitored
True positives	2
True negatives	51
False positives	1
False negatives	0

Comparisons between electrical and magnetic transcranial stimulation

Magnetic stimulation is a less popular technique compared to electrical techniques for monitoring descending motor function during surgery. There are several reasons for this. First, the coils are large and bulky and, in general, preclude monitoring supraspinal intracranial procedures. Train magnetic stimulators are three to six times more expensive than electrical train stimulation, in part due to the requirement of expensive capacitors for energy storage. Cadwell (1992) has discussed the theoretical reasons why magnetic

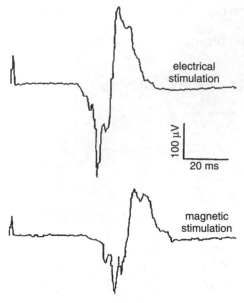

electrical
stimulation

100 μV

20 ms

magnetic
stimulation

Fig. 20.22 Comparison of lower limb motor-evoked potentials using TES and TMS. Note that the TES-derived response is larger than the response obtained with TMS. Negativity is raised. Reprinted with permission from Ubags *et al.* (1999).

stimulators are less efficient than electrical transcranial stimulators. For these latter two reasons the cost for producing equivalent stimulation strength is greater for magnetic than electrical stimulators. Also, magnetic stimulators produce

a characteristic noise, as well as a tendency to overheat, particularly if the stimulation rate is >1 stimulus every 30 s.

Figure 20.22 shows a comparison of myogenic pulse response amplitude for lower extremity motor responses using single-pulse electrical or magnetic pulse stimulation techniques (Ubags *et al.* 1999). Note that, in general, TES-induced myogenic responses tend to be larger (Figure 20.23). Our group compared magnetic to electrical train stimulation techniques in patients randomized to an intravenous methohexital or propofol anesthetic technique (unpublished observations). The experimental protocol is shown in Figure 20.24. A Quadropulse™ magnetic train stimulator was used to produce a four-pulse train with the technique described in 'Train TMS' above. A curved round coil was used for eliciting upper extremity (first dorsal interossei) responses. The cone coil was used to acquire anterior tibialis responses. The Quadropulse™ coil current output was then led into a magnetic to electrical stimulus adapter. The adapter converted a magnetic train surrounding the adapter's primary coil into an electrical train by taking the voltage drop from a secondary coil within the adapter. The dimensions of the secondary coil were designed to produce, at full Quadropulse™ output, an electrical train amplitude of 500 V (unpublished data). Figures 20.25 and 20.26 show examples of the typical results seen when

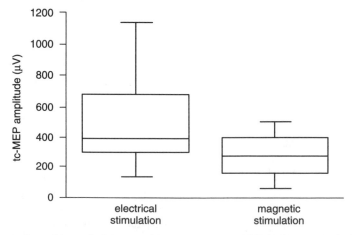

Fig. 20.23 Comparison of lower limb motor-evoked potentials (tc-MEP) acquired with TES (left) and TMS (right) using box plots. Redrawn with permission from Ubags *et al.* (1999).

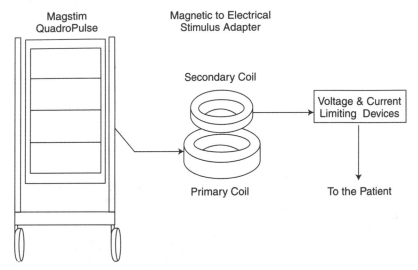

Fig. 20.24 Diagram of the modification to the Quadropulse™ used for producing both train TMS and TES. Train TMS is produced without the magnetic-to-electrical stimulus adapter in the center of the figure. In order to produce train TES, the voltage drop is taken across the secondary coil of the adapter and applied between two scalp needle electrodes. Reprinted with permission from Magstim Ltd, Wales (1997).

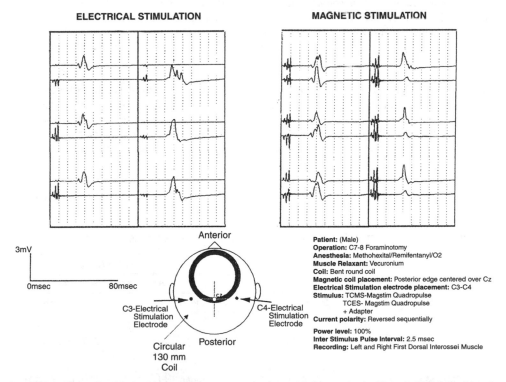

Fig. 20.25 Example of typical upper limb motor-evoked potentials acquired with train TES (left side of figure) and Train TMS (right side of figure). The curved round coil was used for eliciting magnetic motor responses whereas the adapter (see Figure 20.24) was used for stimulating across the C3′ and C4′ scalp needle electrodes. Negativity is raised. Calibration bar to the left for both response types. Unpublished observations.

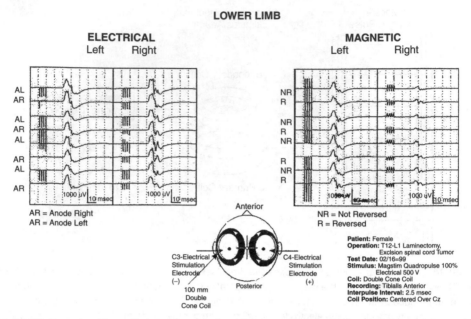

Fig. 20.26 Example of lower limb motor-evoked potentials produced by TES and TMS; same format as used for Figure 20.25. The cone coil was used for acquiring train TMS-evoked tibialis anterior myogenic responses. The adapter (see Figure 20.24) was used for applying train TES across C3 and C4 scalp needle electrodes. Negativity is raised. Unpublished observations.

comparing the upper and lower extremity myogenic responses acquired using electrical and magnetic stimulation. Note that while the response amplitudes are similar, the electrical technique did a better job of isolating excitation to one hemisphere for the upper extremity responses (Figure 20.25). Both techniques produced bilateral lower extremity response despite stimulation

current polarity reversal. For the lower limb, however, the magnetic technique produced an obvious asymmetry when comparing the left- and right-sided responses (Figure 20.26). Tables 20.8 and 20.9 show the results obtained for seven trials involving the upper extremity, as well as two electrical and one magnetic trial for the lower extremities. Five responses were collected for

Table 20.8 Comparison of TCES and TCMS induced characteristics for the upper limb

	TCES *N* = 7		TCMS *N* = 7	
	Anodal		**Anterior directed current**	
TOL	Mean ± SD	22.8 ± 0.68	Mean ± SD	22.5 ± 0.68
	CV %	3	CV %	3
PPA	Mean ± SD	3370.4 ± 404	Mean ± SD	2973 ± 743
	CV %	12	CV %	25
AREA	Mean ± SD	15881 ± 2382	Mean ± SD	13276.7 ± 4115
	CV %	15	CV %	31

Anesthesia: brevital + propofol TIVA TCMS 100%, ISI 2.5 ms.
Muscle relaxation: 40 – 60% TCES 500V, ISI 2.5 ms.

Table 20.9 Comparison of TCES and TCMS induced characteristics in the lower limb

	TCES N = 2			TCMS N = 1	
	Anodal			Anterior directed current	
TOL	Mean ± SD	30.3 ± 0.90		Mean ± SD	30.3 ± 0.45
	CV %	3		CV %	1.5
PPA	Mean ± SD	1210.8 ± 133		Mean ± SD	1050 ± 346
	CV %	11		CV %	33
AREA	Mean ± SD	8840 ± 1502		Mean ± SD	6147 ± 1844
	CV %	17		CV %	30

Anesthesia: brevital + propofol TIVA TCMS 100%, ISI 2.5 ms.
Muscle relaxation: 40 – 60% TCES 500V, ISI 2.5 ms.

each stimulus polarity and stimulation technique for computing the results shown in the two tables. Although the sample numbers are small, it is apparent that while the takeoff latencies are similar, the electrical technique produced larger-amplitude responses with smaller variability (i.e. smaller coefficient of variation) compared to magnetic-induced responses. Thus, preliminary data support those of others suggesting that TES is superior to TMS for monitoring the descending motor systems intraoperatively in anesthetized patients.

Presurgical planning: motor cortex mapping using TMS

TMS is safe, noninvasive, and allows for assessment of various brain functions. It has been increasingly used over the past decade, especially to obtain functional maps of the motor cortex. When TMS is applied to restricted areas of the scalp, it elicits MEPs in contralateral muscles, which may be used to map the location and extent of motor cortical outputs for several muscles. Several studies have proven the value of TMS to assess normal motor cortical function as well as brain plasticity as a result of skill acquisition or pathological lesions (Levy *et al.* 1991; Wassermann *et al.* 1992; Liepert *et al.* 1995; Gugino *et al.* 2001a). There are several other instances when the use of TMS for motor cortical mapping may provide substantial contributions to the medical field. These include measuring brain responses to injury (as in stroke recovery), providing a tool to evaluate interventions for enhancing recovery, evaluating cortical excitability in different diseases affecting the nervous system, and assessing patients with lesions near eloquent cortex (e.g. speech and primary motor cortices) prior to surgical interventions (i.e. presurgical planning).

The importance of mapping motor cortex, as well as other functional cortical areas, with regard to presurgical planning includes the well-known variability of functional patterns of the cortical surface (Penfield and Welch 1949; Black and Ronner 1987). A cortical mapping technique, which relates cortical surface anatomy, pathology, and function, is invaluable for aiding the surgeon's decision regarding the feasibility of a surgical procedure. It also adds important insight as to the optimum approach for surgical therapy. Finally, the technique to be described has added appeal because of the fact that it best mimics the intraoperative cortical surface electrical stimulation technique used in neurosurgical centers for locating eloquent cortical areas (Penfield and Welch 1949; Black and Ronner 1987).

The utility of a mapping technique for presurgical planning will depend on two essential considerations: (i) the precision of a mapping technique for targeting specific functional brain regions; (ii) given a stable pathological lesion, the mapping results should remain accurate over the short term because the mapping technique will typically be carried out several days prior to a surgical procedure. These two considerations have recently been addressed experimentally. The following sections review this work (Penfield and Welch 1949; Gugino *et al.* 2001c).

Precision of TMS for brain mapping

Traditionally TMS has been delivered using skull or scalp landmarks in a 'blind fashion', holding the TMS coil in place while stimulation is delivered. Some have used head frames to immobilize the person's head (Nielsen 1996); others mark the coil contour on a swimming cap worn by the individual being stimulated. However, displacement of the coil may occur, resulting in stimulation of different brain regions from those intended, and potentially affecting the results of the intervention.

Gugino and colleagues have recently described a TMS mapping technique using stereotactic optic guidance (Amassian *et al.* 1991; Gugino *et al.* 1998). The purpose of this technique is to facilitate visualization of the cortical surface, and to guide placement of the TMS coil relative to the cortical surface of an individual. Thus, specific brain regions can be stimulated, while controlling for coil movements relative to the individual's brain. The system is described below.

First, a 3D model of the brain and head of the person undergoing TMS is obtained (Figure 20.27). An anatomical magnetic resonance imaging scan of the brain is acquired using a 1.5 T scanner (TR 35 ms, TE 5 ms, slice thickness of 1.5 mm, FOV 24 cm, acquisition matrix of 256×256 voxels) resulting in 124 contiguous double-echo slices. Using multistep algorithms developed at our institution (Kapur *et al.* 1995; Shenton *et al.* 1995; Ettinger *et al.* 1996; Gugino *et al.* 2001b) the 3D reconstructions of head and brain are obtained. The 3D model displays the skin surface of the scalp and face as well as the underlying cortical surface of the brain. The center of the 3D reconstructed model is used as the origin of a Cartesian system that serves to locate any coordinate in the scalp or cortical surface.

Second, a co-registration procedure is performed in order to align the subject's head with the virtual 3D model. This step involves using a stereotactic optical tracking system. Using three cameras mounted on a mobile arm, the system detects and tracks infrared light-emitting diodes (LEDs) when placed under their view. The LEDs are used to track a subject's head and TMS coil under the cameras' view, which is done by securely taping LEDs to both the subject and the coil. Then, with a probe that has two LEDs and is a component of the tracking system, we record the position of the LEDs in the subject, and several landmarks easily identified in the subject's head. Next, several points are traced over the skin surface and a matching procedure is carried out to evaluate the accuracy of the co-registration procedure. This is followed by calibration of the coil, so that the focal point of a

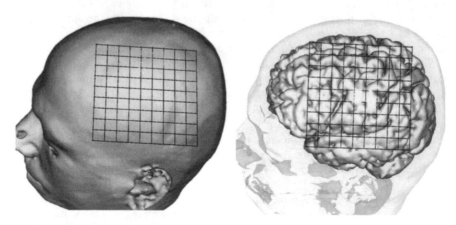

Fig. 20.27 Example of three-dimensional model of the head and brain of a volunteer used for brain mapping. Unpublished observations.

figure-8 coil (i.e. the center point of the bottom surface) can be tracked. The focal point of a figure-8 coil is selected for tracking purposes because this is where the peak magnetic field and induced electric field occurs (Cohen *et al.* 1990). Using the above-described probe, the location of the LEDs attached to the coil is recorded as well as several predetermined points on the coil's surface, including the focal point. The result is that the optical tracking system can track the subject's head and the TMS coil under the cameras' view, and relative to each other. The system has an accuracy of ~1 mm when the coil and subject are within 1 m distance from the cameras (Ettinger *et al.* 1996). Detailed description of the co-registration algorithm can be found elsewhere (Grimson *et al.* 1994; Ettinger *et al.* 1996; Gugino *et al.* 2001b). The system detects head and coil movements in three dimensions, relative to each other, including detection of coil tilt and rotation (referred to as 3D angle). Details of the geometry premise

used to calculate the 3D angle are available elsewhere (Horn 1987). Coil placement is guided in real time: the image is available instantaneously and is updated at a rate of five times per second.

In a previous study, the precision attained using this optical tracking system was evaluated (Gugino *et al.* 2001c). In this study, single-pulse TMS was delivered over the optimal location to obtain first dorsal interosseus muscle responses in normal volunteers. A blind technique, using markings of the figure-8 contour on a swimming cap worn by the volunteer to guide coil placement over the scalp, was compared to a guided technique, using the optical tracking system described. The authors also compared the effects of three levels of stimulation intensity. They attempted to reproduce the coil position at the optimal location using a blind vs guided approach. It was found that using the guided technique resulted in a smaller number of cortical regions stimulated (Figure 20.28). In addition, the distances between the cortical optimal

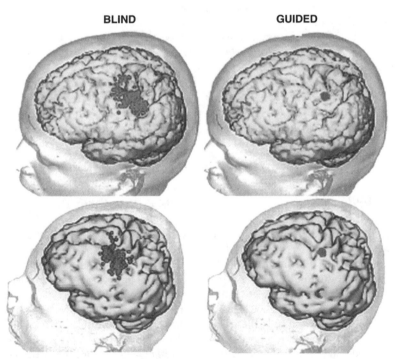

BLIND　　　　　　　　　**GUIDED**

Fig. 20.28 Cortical surface area stimulated using blind vs guided stimulation when attempting to stimulate the optimal location for the contralateral first dorsal interosseus muscle during several trials. Note the larger area stimulated when using blind stimulation compared to guided. Reprinted with permission from Gugino *et al.* (2001d). (Plate 8)

locus and loci stimulated during subsequent stimulation trials were significantly shorter using the guided technique (<2 mm compared to 10–14 mm using a blinded technique). The differences of 3D angle achieved during the initial stimulation at the optimal location and subsequent trials attempting to reproduce it were also significantly smaller using the guided technique. These findings were more evident at lower stimulation intensities, as expected, since higher stimulation intensity is associated with current spread (Pascual-Leone et al. 1994a) and responses may be elicited by stimulating a location further from the intended target cortical locus. This study also showed that guided stimulation resulted in higher probability of eliciting MEPs, and that the MEPs obtained were of higher amplitudes and areas. The coefficients of variation did not decrease significantly. In addition, a significant improvement in the precision to deliver TMS targeted cortical regions was noted and illustrated some of the physiological consequences.

Other authors have also found improvement in precision for scalp placement of a coil by using other methods such as a 3D digitizer (Miranda et al. 1997) or a chin rest and clamping the coil (Nielsen 1996). However, the system developed by our group allows quantification of error in coil placement and real time optical guidance for stimulation, and is frameless, allowing for greater subject comfort. The conclusion reached was that optical guidance to assist TMS studies improves the precision for coil placement.

Short-term replication of TMS functional maps

A second important consideration when obtaining motor cortical maps is the reproducibility of the maps over time. This is relevant for TMS in presurgical planning and studying brain plasticity, for example. Investigors have looked into this aspect (Levy et al. 1991; Mortifee et al. 1994; Krings et al. 1996, 1998; Uy et al. 2002); some limitations of these studies include the use of blind techniques (Levy et al. 1991; Mortifee et al. 1994; Uy et al. 2002), limiting the mapping procedure to the pericentral cortex in a single session (Krings et al. 1996, 1998), and limited control of coil placement (Miranda et al. 1997).

Romero et al. (2006) evaluated short-term variability of TMS functional maps in a recent study, using the above-described optical tracking system. In two mapping sessions performed 1 month apart, eight healthy volunteers were stimulated using single-pulse TMS to obtain the functional representation for the FDI muscle. After performing a co-registration procedure and calibration of the coil, a grid with intersections every 1 cm was displayed and projected onto the cortical surface (Figure 20.27). Although the grid was flat, the resulting distance between stimulated cortical loci was <1 cm. The coil was placed at each grid intersection, visually guided by the tracking system. TMS was delivered and MEPs recorded for offline analysis until no additional responses were obtained. We used a stimulation intensity set at 85% of stimulator output for mapping, based on prior experience showing that this intensity is greater than the resting motor threshold for reliably eliciting responses from distal upper limb muscles. At the same time, less intracortical current spread was expected compared with the use of 100% of stimulator output. Three TMS stimuli were applied at each grid intersection point and the coil position was recorded. The MEPs were averaged at each intersection and a four-point interpolated color cortical surface and scalp topographical maps were constructed using the MEP amplitudes as a function of cortical loci stimulated (Figure 20.29).

The second mapping session followed the same process, i.e. co-registration of the subject's head and calibration of the coil, followed by retrieval of the initial map. The display in the monitor showed the coil orientation achieved during the initial map every time TMS stimuli were delivered. The coil position was shown represented by vectors (Figure 20.30). Next, the coil was placed on the scalp, and as soon as it was detected by the optical tracking cameras, vectors representing the coil location were displayed. An attempt was then made to match the coil orientation achieved during the first mapping session (by superimposing the vectors in the display). Once matching of coil placement occurred, TMS was delivered and the MEPs obtained recorded as well as coil position information. This procedure was repeated until all the loci stimulated in the first session were stimulated.

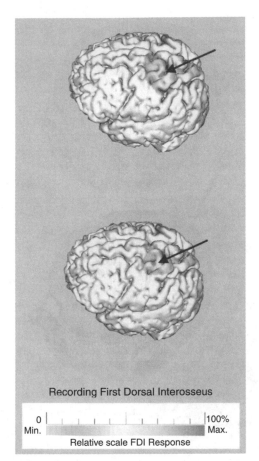

Recording First Dorsal Interosseus

0 |___|___|___|___|___|___|___|___| 100%
Min.   Max.
Relative scale FDI Response

Fig. 20.29 Example of an interpolated color cortical surface map of the motor representation for the contralateral first dorsal interosseus muscle. The arrow points to the optimal location. Unpublished observations. (Plate 9)

Using this system, it was found that accurate coil repositioning could be achieved. Comparison of the coordinates of cortical loci stimulated in the two mapping sessions demonstrated excellent prediction of loci stimulated during a second mapping session based on the coordinates achieved in the first session. The system achieved an error of <3 mm (distance between actual cortical loci stimulated in the two sessions) in ~95% of the stimulation trials in the study, with a 3D angle difference of <3° in ~85% of the trials. Evaluation of the physiological responses, by comparing the MEPs obtained in the two mapping sessions, revealed no significant differences between MEP latency, amplitude, and area between the two sessions. The system again increased the probability of obtaining MEPs. Map response concordance rate was 0.86, reflecting the probability of either eliciting or failing to elicit MEPs at cortical loci stimulated in the two mapping sessions. In addition, comparing the motor map parameters revealed no significant differences between map area, volume, and amplitude at the optimal location in the two mapping sessions (Figure 20.31). The resting motor threshold at the optimal location was not significantly different between mapping sessions. The distance between the amplitude weighted center of gravity between the two maps was 2.6 ± 1.3 mm (mean ± SD) across subjects. Table 20.10 summarizes the map parameters obtained in the two mapping sessions.

There were a number of interesting observations in this study. First, some of the subjects showed several cortical loci where large MEPs were obtained, with amplitudes close to that obtained at the optimal location. This is likely a reflection of the microscopic organization of the motor cortex in which CST neurons projecting to the same anterior horn cell columns are arranged as several closely spaced neuronal clusters (Asanuma and Rosen 1961; Kwan et al. 1978; Sato and Tanji 1989). This is a relevant consideration because using TMS in a blind fashion may target different brain areas which may elicit similar responses (at least in the motor cortex). Second, there were prominent interindividual differences in the area of motor maps across subjects. This is consistent with prior studies (Levy et al. 1991; Wassermann et al. 1992; Mortifee et al. 1994) and is an important consideration supporting the use of a guided mapping technique.

Guided TMS using an optical tracking system reliably reproduced coil placement and motor maps that were stable over the short term in a group of normal volunteers (Mortifee et al. 1994). Other studies support these findings. Mortifee et al. (1994) reported that motor maps were replicable over a range of 21–132 days on six normal volunteers. Thus, maps with greater variability might represent abnormalities or plastic changes in the brain as a result of adaptation to underlying pathology or physiological adaptation,

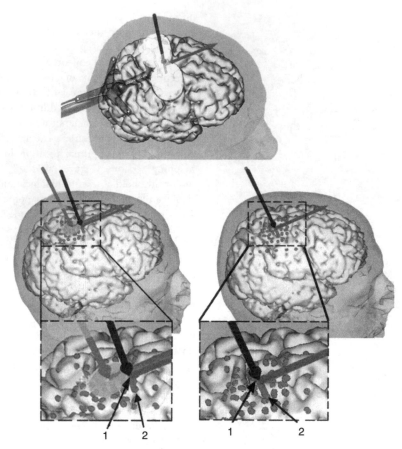

Fig. 20.30 Coil placement and repositioning. *Top*: the coil superimposed on the scalp surface and vectors represent coil placement in the monitor. The blue vector projects through the coil focal point onto the scalp and cortical surface. The green vector aids to control for coil rotation. *Lower left and insert*: the vectors representing coil placement during a stimulation trial in the first (blue vector) and second (red vector) mapping sessions. Note the error circles surrounding the vertical vector during the second mapping session (red vector), which guide coil repositioning by decreasing in size and eventually disappearing when complete matching of vectors (coil repositioning) occurs (*lower right and insert*). Note the coil focal point projection onto the scalp (1) and cortical surface (2). The red spheres represent cortical loci stimulated. Unpublished observations. (Plate 10)

for instance during skill learning (Pascual-Leone *et al*. 1994b; Chokroverty *et al*. 1995).

Future applications of the technique described for mapping the motor cortex include TMS applied to brain regions other than the motor cortex. In addition, correlation with other mapping methods will provide further clarification for the best method of obtaining functional cortical maps, useful for presurgical planning.

Contraindications for TMS in conscious and anesthetized patients

Both single-pulse and train TMS have been shown to minimally affect cognitive function or stress hormone levels in conscious patients. Single-pulse TMS has been reported to cause

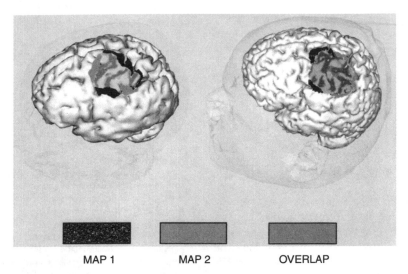

MAP 1   MAP 2   OVERLAP

Fig. 20.31 Motor cortical surface map area in two subjects. The motor map areas in the first mapping session (black) and second session (orange) are shown overlapped (gray). Note the overlap of map areas in the two mapping sessions. Unpublished observations. (Plate 11)

Table 20.10 Motor map characteristics

	Map 1	Map 2
Area (cm²)*ᵃ	14.63 ± 8.52	15.75 ± 8.88
Volume*ᵇ	384.98 ± 391.06	381.23 ± 330.71
Maximal amplitude (mV)*ᶜ	2.23 ± 1.25	2.82 ± 1.59
Resting MT*ᵈ	70.2 ± 3.7	69 ± 3.38
Distance between amplitude-weighted center of gravity (mm)ᵉ	2.6 ± 1.3	
Distance between optimal loci (mm)ᶠ	10 ± 5.5	

Values are mean of all subjects ± SD.
*No statistically significant differences were found across sessions.
ᵃArea is calculated from the number of excitable grid loci stimulated, expressed in cm².
ᵇVolume of the map refers to the sum of the relative motor-evoked potential (MEP) amplitudes across all excitable locations.
ᶜMaximal amplitude refers to the highest peak-to-peak amplitude MEP elicited.
ᵈThe resting motor threshold (MT) is expressed as percentage of magnetic stimulator output.
ᵉThe x, y, and z coordinates of the center of gravity were calculated by multiplying the coordinate at each position by its amplitude weight and summing over all positions.
ᶠThe reported distance between optimal loci is the calculated distance between the loci where maximal amplitude MEP is elicited in the two mapping sessions.
Unpublished observations.

seizures in conscious patients with a seizure history. Train TMS, on the other hand, has been reported to elicit seizures in conscious patients with or without a seizure history (Hufnagel *et al.* 1990; Dhuna *et al.* 1991).

For anesthetized patients, the authors know of only one case of TMS seizure induction. The procedure involved a 46-year-old female undergoing a laminectomy for a thoracic spine mass excision. The patient was anesthetized with a methohexital-based total intravenous technique.

The patient had a history of seizures with the last occurring 10 years prior to her hospital admission. She had not taken antiseizure medication for >8 years prior to admission. The patient emerged from anesthesia with a generalized seizure, which was treated successfully with intravenous benzodiazepines. It is instructive to note that even with the use of methohexital, an anesthetic known to support seizure activity during anesthesia, this patient only demonstrated seizures after emergence from anesthesia. This suggests that anesthetic depression of the CNS may be responsible for the rare incidence of intraoperative seizure activity during motor system monitoring. Intracranial metallic devices (i.e. aneurysm clips, deep brain stimulators) as well as electronic devices within a patient's body (i.e. pacemaker) represent an absolute contraindication to using TMS (Pascual-Leone *et al.* 1993; Chokroverty *et al.* 1995).

It should be pointed out that magnetic stimulators are still not approved by the US Food and Drug Administration (FDA). The use of these stimulators therefore requires an FDA Investigational Device Exemption (IDE).

Conclusion

Although experience has demonstrated that transcranial magnetic single-pulse and train stimuli can produce myogenic responses suitable for intraoperative monitoring, there is little doubt that this application is better handled by transcranial electrical techniques. However, because magnetic stimulation is relatively painless and can be used for producing focal excitation of the CNS, it is the stimulating technique of choice for studying motor system function and functional cortex mapping in the perioperative conscious patient.

References

Aglio LS, Kraus KH, Desai S, *et al.* (2002a). Efficacious use of a cap shaped coil for transcranial magnetic stimulation of descending motor paths. *Electroencephalography and Clinical Neurophysiology* 33, 21–29.

Aglio LS, Romero R, Desai S, *et al.* (2002b). The use of transcranial magnetic stimulation for monitoring descending spinal cord motor function. *Electroencephalography and Clinical Neurophysiology* 33, 30–41.

Allison T, McCarthy G, Wood CC, *et al.* (1989a). Human cortical potentials evoked by stimulation of the median nerve. I. Cytoarchitectonic areas generating short-latency activity. *Journal of Neurophysiology 62*, 694–710.

Allison T, McCarthy G, Wood CC (1989b). Human cortical potentials evoked by stimulation of the median nerve. II. Cytoarchitectonic areas generating short-latency activity. *Journal of Neurophysiology 62*, 712.

Amassian VE, Cracco RQ (1987). Human cerebral cortical responses in contralateral transcranial stim

Amassian VE, Quirk GJ, Sewart M (1987a). Magnetic coil versus electrical stimulation of monkey motor cortex. *Journal of Physiology 394*, 119P.

Amassian VE, Stewart M, Quirk GJ, *et al.* (1987b). Pathological basis of motor effect on a transient sti

Amassian VE, Stewart M, Quirk GJ, *et al.* (1987c). Physiological basis of motor effect on a transient stimulation to cerebral cortex. *Neurosurgery 20*, 74–93.

Amassian VE, Quirk GJ, Stewart M (1990). A comparison of corticospinal activation by magnetic coil and electrical stimulation of monkey motor cortex. *Electroencephalography and Clinical Neurophysiology 77*, 390–401.

Amassian VE, Cracco RO, Maccabee PJ, *et al.* (1991). Matching focal and non-focal magnetic coil stimulation to properties of human nervous system: Mapping motor unit fields in motor cortex contrasted with alternating sequential digit movements by pre-motor SMA stimulation. *Electroencephalography and Clinical Neurophysiology, Supplement 43*, 3–28.

Amassian VE, Eberle L, Maccabee PJ, *et al.* (1992). Modeling magnetic coil excitation of human cerebral cortex with a peripheral nerve immersed in a brain-shaped volume conductor. The significance of fiber bending in excitation. *Electroencephalography and Clinical Neurophysiology 85*, 291–301.

Arezzo JC, Vaughan HG, Legatt AD (1981). Topography and intracranial sources of somatosensory evoked potentials in the monkey. II. Critical components. *Electroencephalography and Clinical Neurophysiology 51*, 11–18.

Asanuma H, Rosen I (1961). Topographical organization of cortical efferent zones projecting to distal forelimb muscles in the monkey. *Experimental Brain Research 14*, 243–256.

Barker AT, Jalinous R, Freeston IL (1985). Noninvasive magnetic stimulation of the human motor cortex. *Lancet i*, 1106–1107.

Barker AT, Jalinous R, Freeson IL, *et al.* (1986). Clinical evaluation of conduction time measurements in central motor pathways using magnetic stimulation of human brain. *Lancet i*, 1325–1326.

Beradelli A, Inghiller M, Cracco G, *et al.* (1991). Corticospinal potentials after electrical and magnetic stimulation in man. *Electroencephalography and Clinical Neurophysiology, Supplement 43*, 147–154.

Black PM, Ronner SF (1987). Cortical mapping for defining the limits of tumor resection. Cortical mapping for defining the limits of tumor resection. *Neurosurgery 20*, 914–919.

Boyd S, Rothwell JC, Cowan JMA, *et al.* (1986). A method of monitoring function in cortical pathways during scoliosis surgery with a note on motor conduction velocities. *Journal of Neurology, Neurosurgery and Psychiatry 69*, 251–257.

Brasil-Neto JP, Pascual-Leone A, *et al.* (1993). Plasticity of cortical motor output organization following deafferentation, cerebral lesions, and skill acquisition. *Advances in Neurology 63*, 187–182.

Broughton R, Rasmussen T, Branch C, (1981). Scalp and direct cortical recordings of somatosensory evoked potentials in man. *Canadian Journal of Psychological Reviews 35*, 136–158.

Brouwer B, Asby P (1990). Corticospinal projections to upper and lower limb spinal motoneurons in man. *Electroencephalography and Clinical Neurophysiology 76*, 509–519.

Brown RH, Nash CL, Jr, Berilla JA (1984). Cortical evoked potential monitoring: a system for intraoperative monitoring of spinal cord function. *Spine 9*, 256–261.

Burke D, Hicks RG, Stephen JPH (1990). Corticospinal volleys evoked by anodal and cathodal stimulation of the human motor cortex. *Journal of Physiology 425*, 283–299.

Cadwell J (1992). Optimizing magnetic stimulator design. *Electroencephalography and Clinical Neurophysiology 85*, 291–301.

Chatrian GE Berger MS, Wirch AL (1988). Discrepancy between intraoperative SSEPs and postoperative function. Case Report. *Journal of Neurosurgery 69*, 450–454.

Chokroverty S, Hening W, Wright D, *et al.* (1995). Magnetic brain stimulation: safety studies. *Electroencephalography and Clinical Neurophysiology 97*, 36–42.

Classen J, Witte OW, Schlaug G, *et al.* (1995). Epileptic seizures triggered directly by focal transcranial magnetic stimulation. *Electroencephalography and Clinical Neurophysiology 94*, 19–25.

Cohen LG, Hallett M (1987). Cortical stimulation does not cause short-term changes in the electroencephalogram. *Annals of Neurology 21*, 512–513.

Cohen LG, Roth BJ, Nilsson J, *et al.* (1990). Effects of coil design on delivery of focal magnetic stimulation. Technical considerations. *Electroencephalography and Clinical Neurophysiology 75*, 350–357.

Crawford ES, Mizrahi EM, Hess KR, *et al.* (1988). The impact of distal aortic perfusion and somatosensory evoked potential monitoring on prevention of paraplegia after aortic aneurysm operation. *Thoracic and Cardiovascular Surgery 95*, 357–367.

Cushing H (1909). A note upon the faradic stimulation of the post-cerebral gyrus in conscious patients. *Brain 32*, 44–53.

Day BL, Dick JPR, Marsden CD, *et al.* (1986). Differences between electrical and magnetic stimulation of the human brain. *Journal of Physiology (London) 378*, 36.

Day BL, Dressler D, de Noordhout A, *et al.* (1987a). Different sites of action of electrical and magnetic simulation of the human brain. *Neuroscience Letters 75*, 101–106.

Day BL, Maertsens de Noordhut A, Marsden CD, *et al.* (1987b). A comparison of the effects of cathodal and anodal stimulation of the human cortex through the intact scalp. *Journal of Physiology (London) 118*.

Day BL, Dressler D, Maertens de Noordhout A, *et al.* (1988). Differential effect of cutaneous stimuli on responses to electrical or magnetic stimulation of the human brain. *Journal of Physiology (London) 399*, 68.

Day BL, Dressler D, de Noordhout A, *et al.* (1989). Electric and magnetic stimulation of the human motor cortex: surface EMG and single motor unit responses. *Journal of Physiology (London) 412*, 449–473.

deHann P, Kalkman CJ, Ubags LH, *et al.* (1996). A comparison of the sensitivity of epidural and myogenic transcranial motor evoked responses in the detection of acute spinal cord ischemia in the rabbit. *Anesthesia and Analgesia 83*, 1022–1027.

Deiber MP, Giard MH, Mauguiere F (1986). Separate generators with distinct orientation for N_{20} and P_{22} somatosensory evoked potentials to finger stimulation. *Electroencephalography and Clinical Neurophysiology 65*, 321–324.

Deletis V (1993). Intraoperative monitoring of the functional integrity of the motor pathways. *Advances in Neurology 6*, 201–214.

Desmedt JE, Bourquet M (1985). Color imaging of parietal and frontal somatosensory potential fields evoked by stimulation of median or posterior tibial nerve in man. *Electroencephalography and Clinical Neurophysiology 62*, 1–17.

Desmedt JE, Nguyen T, Bourquet M (1987). Bit-mapped color imaging of human evoked potentials with reference to the N_{20}, P_{22}, P_{27} and N_{30} somatosensory responses. *Electroencephalography and Clinical Neurophysiology 68*, 1–9.

Dhuna A, Pascual-Leone A (1992). Lack of pathological changes in human temporal lobe after transcranial magnetic stimulation. *Epilepsia 33*, 504–508.

Dhuna AK, Gates JR, Pascual-Leone A (1991). Transcranial magnetic stimulation in patients with epilepsy. *Neurology 41*, 1067–1072.

Ebner A, Deuschl G (1988). Frontal and parietal components of enhanced somatosensory evoked potentials: a comparison between pathological and pharmacologically induced conditions. *Electroencephalography and Clinical Neurophysiology 71*, 170–179.

Edgley SA, Eyre JA, Lemon RN, *et al.* (1990). Excitation of the corticospinal tract by electromagnetic and electrical stimulation of the scalp in the macque monkey. *Journal of Physiology (London) 425*, 301–320.

Edmonds HL, Poloheimo MP, Backman MH, *et al.* (1989). Transcranial magnetic motor evoked potentials (tc MMEP) for functional monitoring of motor pathways during scoliosis. *Spine 14*, 683–686.

Elmore JR, Gloviczki P, Harper CM, *et al.* (1991). Failure of motor evoked potentials to predict neurologic outcome in experimental thoracic aortic occlusion. *Journal of Vascular Surgery 14*, 131–139.

Epstein CM, Schwartzberg DA, Dandy KE, *et al.* (1990). Localizing the site of magnetic brain stimulation in humans. *Neurology 40*, 666–670.

Ettinger GJ, Grimson WEL, Leventon ME, *et al.* (1996). Non-invasive functional brain mapping using registered transcranial magnetic stimulation. IEEE Workshop on Mathematical Methods in Biomedical Image Analysis, June 21–22, San Francisco.

Fujiki M, Isono M, Hori S, *et al.* (1996). Corticospinal direct response to transcranial magnetic stimulation in humans. *Electroencephalography and Clinical Neurophysiology 101*, 48–57.

Ghaly RF, Stone JL, Levy WJ, *et al.* (1990a). The effects of nitrous oxide on transcranial magnetic induced electromyographic responses in the monkey. *Journal of Neurosurgical Anesthesiology 2*, 175–181.

Ghaly RF, Stone JL, Levy WJ, *et al.* (1990b). The effect of etomidate on transcranial magnetic-induced motor evoked potentials in the monkey. *Neurosurgery 20*, 184.

Ghaly RF, Stone JL, Levy WJ, *et al.* (1990b). The effect of etomidate on motor evoked potentials induced by transcranial magnetic stimulation in the monkey. *Neurosurgery 27*, 936–942.

Ghaly RF, Stone JL, Levy WJ, *et al.* (1991a). The effect of an anesthetic induction dose of midazolam on motor evoked potentials evoked by transcranial magnetic stimulation in the monkey. *Journal of Neurosurgical Anesthesiology 3*, 20–27.

Ghaly RF, Stone JL, Levy WJ, *et al.* (1991b). The effect of etomidate or midazolam hypnotic doses on motor evoked potentials in the monkey. *Journal of Neurosurgical Anesthesiology 3*, 20–27.

Ghaly RF, Stone JL, Levy WJ, *et al.* (1991c). The effect of neuroleptanalgesia (droperidol-fentanyl) on motor potentials evoked by transcranial magnetic stimulation in the monkey. *Journal of Neurosurgical Anesthesiology 3*, 117–123.

Ghaly RF, Stone JL, Aldrete A, *et al.* (2000). Motor evoked potentials in primates, anesthetic considerations. *Journal of Neuroanesthesia 182*, 199.

Gilman S, Marco TA (1971). Effects of medullary pyramidotomy in the monkey. Clinical and electromyography abnormalities. *Brain 94*, 495–514.

Ginsburg HH, Shetter AG, Raudzens PA (1985). Postoperative paraplegia with preserved intraoperative somatosensory evoked potentials: a case report. *Journal of Neurosurgery 63*, 296–300.

Glassman SD, Zhang YP, Shields CB, *et al.* (1996). Transcranial magnetic motor evoked potentials in scoliosis surgery. *Orthopedics 18*, 1017–1023.

Golding S, Aras E, Weber PC (1970). Comparative study of sensory input to motor cortex in animals and man. *Electroencephalography and Clinical Neurophysiology 29*, 537–350.

Grimson WEL, Lozano-Perez T, Weiss III WM, *et al.* (1994). Autonomic registration method for frameless stereotaxy, image guided surgery, and enhanced reality visualization. Proceedings of IEEE Computer Vision and Pattern Recognition Conference, Seattle, WA, June, pp. 430–436.

Gugino V, Chabot RJ (1990). Somatosensory evoked potentials. *International Anesthesiology Clinics 28*, 154–164.

Gugino LD, Kraus K, Heino R, *et al.* (1992). Peripheral ischemia as a complicating factor during somatosensory and motor evoked potential monitoring of aortic surgery. *Journal of Cardiovascular Anesthesia 6*, 715–719.

Gugino LD, Aglio LS, Segal ME, *et al.* (1997). Transcranial magnetic stimulation for monitoring spinal cord motor paths. *Seminars in Spine Surgery 9*, 315–336.

Gugino LD, Potts G, Aglio LS, *et al.* (1998). Localization of eloquent cortex using transcranial magnetic stimulation. In: E Alexander, RJ Maciunas (eds), *Advanced Neurosurgical Navigation*. New York: Thieme.

Gugino LD, Aglio LS, Raymond SA, *et al.* (2001a). Intraoperative cortical function localization techniques. *Techniques in Neurosurgery 7*, 19–32.

Gugino LD, Aglio LS, Potts G, *et al.* (2001b). Perioperative use of transcranial magnetic stimulation. *Techniques in Neurosurgery 7*, 33–51.

Gugino LD, Romero JR, Aglio LS, *et al.* (2001c). Transcranial magnetic stimulation coregistered with MRI: A comparison of a guided vs. blind stimulation technique and its effect on evoked compound muscle action potentials. *Clinical Neurophysiology 112*, 1781–1792.

Gugino LD *et al.* (2001d). *Electroencephalography and Clinical Neurophysiology 112*, 1781–1792.

Gugino LD, Aglio LS, Yli-Hankala A, *et al.* (2004). Monitoring the electroencephalogram during bypass surgery. *Seminars in Cardiothoracic and Vascular Anesthesia 8*, 61–83.

Haghighi SS, Madsen R, Green D, *et al.* (1990). Suppression of motor evoked potentials by inhalational anesthetics. *Journal of Neurosurgical Anesthesiology 2*, 75–78.

Heneghan C (1993). Clinical and medicolegal aspects of conscious awareness during anesthesia. *International Anesthesiology Clinics 31*, 1–12.

Hicks DBR, Gandevia SC, Stephen J, *et al.* (1993). Direct comparison of corticospinal volleys in human subjects to transcranial magnetic and electrical stimulation. *Journal of Physiology (London) 470*, 383–392.

Homberg V, Netz J (1989). Generalized seizures induced by transcranial magnetic stimulation of motor cortex (letter). *Lancet ii*, 1223.

Horn BKP (1987). Closed-form solution of absolute orientation using unit quaternions. *Journal of the Optical Society of America A4*, 629–642.

Hufnagel A, Elger CE, Klingmüller D, *et al.* (1960). Activation of epileptic foci by transcranial magnetic stimulation: effects on secretion of prolactin and luteinizing hormone. *Journal of Neurology 237,* 242–246.

Hufnagel A, Elger CE, Durwen HF *et al.* (1990). Activation of the epileptic focus by transcranial magnetic stimulation of the human brain. *Annals of Neurology 27,* 49–60.

Jalinous R (1991). Technical and practical aspects of magnetic nerve stimulation. *Journal of Clinical Neurophysiology 8,* 10–25.

Jennum P, Winkel H, Fuglsang-Frederiksen A-F (1995). Repetitive magnetic stimulation and motor evoked potentials. *Electroencephalography and Clinical Neurophysiology.*

John ER, Prichep L (2005). The anesthetic cascade. *Anesthesiology 102,* 447–471.

John ER, Chabot RJ, Prichep LS, *et al.* (1988). Intraoperative monitoring during neurosurgical and neuroradiological procedures. *Journal of Clinical Neurophysiology 6,* 125–158.

Jones SJ, Harrison R, Koh KF, *et al.* (1996). Motor evoked potential monitoring during spinal surgery: Responses of distal limb muscles to transcranial cortical stimulation with pulse trains. *Electroencephalography and Clinical Neurophysiology 100,* 375–383.

Kalkman CJ, Drummond JC, Patel PM, *et al.* (1992). Effects of droperidol, pentobarbital and ketamine on myogenic transcranial motor evoked responses in humans. *Anesthesiology 77(3A),* A163.

Kalkman CJ, Been HD, Ubags JH *et al.* (1993). Improved amplitude of intraoperative myogenic evoked responses after paired and transcranial electrical stimulation. *Anesthesiology 79 (3A),* A176.

Kandler R (1990). Safety of transcranial magnetic stimulation (letter). *Lancet 335,* 469–470.

Kapur T, Grimson WEL, Kikinis R (1995). Segmentation of brain tissue from MR images. *Proceedings of First International Conference on Computer Vision, Virtual Reality and Robotics in Medicine,* Nice, France, April, pp. 429–433.

Karama Y, Swama K (1972). The EEG evoked potentials and single unit activity during ketamine anesthesia in cats. *Anesthesiology 36,* 316–328.

Kawaguchi M, Juruya H (2004). Intraoperative spinal cord monitoring of motor function with myogenic motor evoked potentials: a consideration in anesthesia. *Journal of Anesthesiology 18,* 18–28.

Kelly DL, Goldring S, O'Leary JL (1965). Averaged evoked somatosensory responses from exposed cortex of man. *Archives of Neurology 13,* 1–9.

Krain L, Kimura J, Yamada T *et al.* (1990). Consequences or cortical magnetoelectric stimulation. In: S Chokroverty (ed.), *Magnetic Stimulation in Clinical Neurophysiology,* pp. 157–163. Boston: Butterworth.

Kraus KH, Gugino LD, Levy WJ, *et al.* (1992). The use of a cap shaped coil for transcranial magnetic stimulation of the motor cortex. *Journal of Clinical Neurophysiology 10,* 353–362.

Krings T, Buchbinder BR, Butler WE, *et al.* (1996). Stereotactic transcranial magnetic stimulation: correlation with direct electrica cortica stimulation. *Neurosurgery 41,* 1319–1326.

Krings T, Naujokat C, von Keyerslingk DG, *et al.* (1998). Representation of cortical motor function as revealed by stereotactic transcranial magnetic stimulation. *Electroencephalography and Clinical Neurophysiology 109,* 85–93.

Kwan HC, MacKay WA, Murphy JT, *et al.* (1978) An intracortical microstimulation study of output organization in precentral cortex of awake primates. *Journal of Physiology (Paris) 74,* 231–233.

Lawrence DG, Kuypers HG (1968). The functional organization of the motor systems in the monkey. I. The effects of bilateral pyramidal lesions. *Brain 91,* 1–14.

Lawrence DG, Hopkins DH (1972). Development aspects of pyramidal control in the rhesus monkey. *Brain Research 40,* 117–118.

Levy WJ (1987). Transcranial stimulation of the motor cortex to produce motor evoked potentials. *Medical Instrumentation 21,* 248–254.

Levy WJ, York DH (1983). Evoked potentials from the motor tracts in humans. *Neurosurgery 12,* 422–429.

Levy WJ, York DH, McCaffrey M, *et al.* (1984). Motor evoked potentials from transcranial electrical stimulation of the motor cortex in humans. *Neurosurgery 15,* 287–302.

Levy WJ, Amassian VE, Schmid UD, *et al.* (1991). Mapping of motor cortex gyral sites non-invasively by transcranial magnetic stimulation in normal subjects and patients. *Electroencephalography and Clinical Neurophysiology, Supplement 43,* 51–75.

Liepert J, Tegenthoff M, Malin JP (1995). Changes of cortical motor area size during immobilization. *Electroencephalography and Clinical Neurophysiology 97,* 3820–386.

Losasso TJ, Boudreaux JB, Muzzi DA, *et al.* (1991). The effect of anesthetic agents on transcranial magnetic motor evoked potentials (TMEP) in neurosurgical patients. *Journal of Neurosurgical Anesthesiology 3,* 200 (abstract).

Lotto M, Banoub N, Schubert A (2004). Effects and anesthetic agents on physiological changes on intraoperative motor evoked potentials. *Journal of Neurosurgical Anesthesiology 16,* 32–42.

McCaffrey M (1997). Somatosensory evoked potential monitoring during spinal surgery. *Seminars in Spine Surgery 4,* 309–314.

McPherson RW (1994). General anesthetic considerations in intraoperative monitoring: Effects of anesthetic agents and neuromuscular blockade on evoked potentials, EEG and cerebral blood.flow. In: CM Loftus (ed.), *Intraoperative monitoring Techniques in neurosurgery,* pp 97–106. New York: McGraw-Hill.

Miranda PC, deCarvalho M, Conseicao I, *et al.* (1997). A new method for reproducible coil-positioning in transcranial magnetic stimulation mapping. *Electroencephalography and Clinical Neurophysiology 105,* 116–123.

Mortifee P, Stewart H, Schulzer M, et al. (1994). Reliability of transcranial magnetic stimulation for mapping the human motor cortex. Electroencephalography and Clinical Neurophysiology 93, 131–137.

Nadstawek J, Rechstein U, Taniguchi M, et al. (1993). Repetitive transcranial electrical stimulation for myogenic motor evoked potentials under balanced anesthesia (BA), with isoflurane, nitrous oxide, and alfentanil versus total intravenous anesthesia (TIVA), with propofol and alfentanil. Anesthesiology 79(3A), A461.

Nielsen JF (1996). Improvement of amplitude variability of motor evoked potentials in multiple patients and in healthy subjects. Electroencephalography and Clinical Neurophysiology 101, 404–411.

Nuwer MR, Dawson EG, Carlsen, et al. (1995). Somatosensory evoked potential spinal cord monitoring reduces neurologic deficits are scoliosis surgery: results of a large multicenter survey. Electroencephalography and Clinical Neurophysiology 96, 6–11.

Pakakostopoulos D, Crow HJ (1984). The precentral somatosensory evoked potential. Annals of the New York Academy of Sciences 425, 265–261.

Pascual-Leone A, Houser CM, Reese K, et al. (1993). Safety of rapid-rate transcranial magnetic stimulation in normal volunteers. In: WJ Levy, WJ Cracco (eds), Magnetic motor stimulation: basic principles and clinical neurophysiology, pp. 120–130. New York: Elsevier.

Pascual-Leone A, Valls-Solé, Wassermann EM, et al. (1994a). Responses to rapid-rate transcranial magnetic stimulation of the human motor cortex. Brain 117, 847–858.

Pascual-Leone A, Grafman J, Hallett M, et al. (1994b). Modulation of cortical motor output maps during development of implicit and explicit knowledge. Science 263, 1287–1289.

Penfield W, Welch K (1949). Instability of response to stimulation of the sensorimotor cortex of man. Journal of Physiology 109, 358–365.

Pereon Y, Bernard JM, Nguyen S, et al. (1999). The effects of desflurane on the nervous system: From spinal cord to muscles. Anesthesia and Analgesia 89, 490–495.

Phillips CG (1969). Motor apparatus of the baboon's hand. Proceedings of the Royal Sociey of London B 173, 141–174.

Rampil IJ (1993). Is MAC testing a spinal reflex? Anesthesiology 79, A422.

Rampil IJ (1994). 'F waves' – a nonsynaptic but sensitive indicator of anesthetic effect in rats. Anesthesia and Analgesia 78, S1–S503.

Rampil IJ, Laster MJ (1992). No correlation between quantitative electroencephalographic measurements and movement response to noxious stimuli during isoflurane anesthesia in rats. Anesthesiology 77, 920–925.

Rampil IJ, Mason P, Singh H (1993). Anesthetic potency (MAC) is independent of forebrain structures in the rat. Anesthesiology 78, 707–712

Romero JR, Ramirez M, Aglio L, et al. (2006). Benefit of stereotactic optic guidance for reproducibility of motor cortex functional maps using transcranial magnetic stimulation (TCMS). 58th Annual Meeting of the American Academy of Neurology.

Roth BJ, Cohen LG, Hallett M (1991). The electric field induced during magnetic stimulation. Electroencephalography and Clinical Neurophysiology, Supplement 43, 268–278.

Rothwell JC (1991). Physiological studies of electric and magnetic stimulation of the human brain. Electroencephalography and Clinical Neurophysiology, Supplement 43, 29–35.

Rothwell JC, Thompson PD, Day BL, et al. (1991). Stimulation of the human motor cortex through the scalp. Experimental Physiology 76, 159–200.

Sato KC, Tanji J (1989). Digit-muscle responses evoked from multiple intracortical foci in monkey precentral motor cortex. Journal of Neurophysiology 62, 959–970.

Schwinn DA, Watkins W, Leslie JB (1994). Basic principles of pharmacology related to anesthesia. In: RD Miller (ed.), Anesthesia, 4th edn, p. 43. New York: Churchill Livingstone.

Shenton ME, Kikinis R, McCarley RW, et al. (1995). Harvard brain atlas: a teaching and visualization tool. Proceedings of the '95 Biomedical Visualization, Atlanta GA, October 30, pp. 10–17.

Slimp JC, Tamas LB, Stolov WC, et al. (1986). Somatosensory evoked potentials after removal of somatosensory cortex in man. Electroencephalography and Clinical Neurophysiology 65, 111–111.

Sloan TB (1990). The vecuronium alters cortical magnetic motor evoked potentials. Journal of Neurosurgical Anesthesiology 2, 251.

Sloan T, Heyer E (2002). Anesthesia for intraoperative neurophysiologic monitoring of the spinal cord. Clinical Neurophysiology 19, 430–443.

Sloan TB, Levin D (1991). Effect of enflurane, halothane, isoflurane and nitrous oxide on cortical magnetic motor evoked potentials. Journal of Neurosurgical Anesthesiology 3, 201 (Abstract).

Stanski D (1994). Monitoring depth of anesthesia. In: RD Miller (ed.), Anesthesia, 4th edn, p. 1127. New York: Churchill Livingstone.

Stinson W, Murray MJ, Jones KA (1994). A computer-controlled, closed-loop infusion system for infusing muscle relaxants: its use during motor evoked potential monitoring. Journal of Cardiothoracic and Vascular Anesthesia 8, 40–44.

Stohr PE, Goldring S (1969). Origin of somatosensory evoked scalp responses in man. Journal of Neurosurgery 31, 117–127.

Taniguichi M, Cedzich C, Schramm J (1993). Modification of cortical stimulation for motor evoked potentials under general anesthesia: technical description. Neurosurgery 32, 219–226.

Thornton C, Jones JG (1993). Evaluating depth of anesthesia: review of methods. International Anesthesiology Clinics 13, pp. 67–68.

Tung HL, Drummond JC, Bickford RG (1988). The effects of anesthetic and sedative agents on magnetic motor evoked responses. Anesthesiology 69, 313A (Abstract)

Ubags LH, Kalkman CJ, Been HD, *et al.* (1997). The use of ketamine on etomidate to supplement sufentanil/N20 anesthesia does not disrupt monitoring of myogenic transcranial motor evoked responses. *Journal of Neurosurgical Anesthesiology 9*, 228–233.

Ubags LH, Kalkman CJ, Been HD, *et al.* (1999). A comparison of myogenic motor evoked responses to electrical and magnetic transcranial stimulation during nitrous oxide/ opioid anesthesia. *Anesthesia and Analgesia 88*, 568–572.

Uy J, Ridding MC, Miles TS, *et al.* (2002). Stability of maps of human motor cortex made with transcranial magnetic stimulation. *Brain Topography 14*, 293–297.

Wassermann EM, McShane LM, Hallet M (1992). Non-invasive mapping of muscle representations in human motor cortex. *Electroencephalography and Clinical Neurophysiology 85*, 1–8.

Watt JWH, Fraser MH, Soni BM, *et al.* (1996). Total IV anaesthesia for transcranial magnetic evoked potential spinal cord monitoring. *British Journal of Anaesthesia 76*, 876–871.

Woodforth IJ, Hicks RG, Crawford MR, Stephen JPH, Burke D (1999). Depression of I waves in corticospinal volleys by sevoflurane, thiopental and propofol. *Anesthesia and Analgesia 89*, 1182–1187.

Yamada H, Torres F, Tarnaki T, *et al.* (1994). Effects of halothane, isoflurane, and enflurane on motor potentials evoked by transcranial magnetic stimulation in cats. *Anesthesia and Analgesia 78*, S1–S492.

Yamada T (1988). The autonomic and physiologic bases of median nerve somatosensory evoked potentials. *Neurologic Clinics 6*, 705–730.

Yamada T, Kayamori R, Kimura J, *et al.* (1984). Topography of somatosensory evoked potentials after stimulation of the median nerve. *Electroencephalography and Clinical Neurophysiology 59*, 29–43.

Yamada T, Graff-Radford NR, Kimura J, *et al.* (1985). Topographic analysis of somatosensory evoked potentials in patients with well-localized thalamic infarctions. *Journal of Neurological Sciences 68*, 31–46.

Yamada T, Karmeyama S, Fuchigami Y, *et al.* (1988). Changes in short-latency somatosensory evoked potential in sleep. *Electroencephalography and Clinical Neurophysiology 70*, 1267–136.

York D (1997). A review of neurophysiology monitoring of spinal cord function in relation to changes in the marketplace. *Seminars in Spine Surgery 9*, 295–301.

Zentner J (1989a). Noninvasive motor evoked potential monitoring during neurosurgical operations on the spinal cord. *Neurosurgery 23*, 709–712.

Zentner J (1989b). Influence of anesthetics on the electromyographic response evoked by transcranial electrical cortex stimulation (letter). *Functional Neurology 4*, 299–300.

Zentner J, Ebner A (1989). Nitrous oxide suppresses the electromyographic response evoked by electrical stimulation of the motor cortex. *Neurosurgery 24*, 60–62.

Zentner J, Rieder G (1990). Diagnostic significance of motor evoked potentials in space-occupying lesions of the brainstem and spinal cord. *European Archives of Psychiatry and Clinical Neurological Science 239*, 285–289.

Zentner J, Schumacher M, Bien S (1988). Motor evoked potentials during interventional neuroradiology. *Neuroradiology 30*, 252–255.

Zentner J, Kiss I, Ebner A (1989). Influence of anesthetics – nitrous oxide in particular – on electromyographic response evoked by transcranial electrical stimulation of the cortex. *Neurosurgery 24*, 253–256.

Zhou H, Zhu C (2000). Comparison of isoflurane effects on motor evoked potential and F wave. *Anesthesiology 93*, 32–38.

Zhou HH, Mehta M, Leis A (1997). Spinal cord motoneuron excitability during isoflurane and nitrous oxide anesthesia. *Anesthesiology 86*, 302–307.

Zornow MH, Grafe MR, Tybor C, *et al.* (1990). Preservation of evoked potentials in a case of anterior spinal aneurysm syndrome. *Electroencephalography and Clinical Neurophysiology 77*, 137–139.

CHAPTER 21

TMS in movement disorders

Alfredo Berardelli and Mark Hallett

Introduction

Since TMS was first applied to the motor cortex, its use in studying patients with movement disorders was an obvious and valuable application. In this chapter, we will review findings obtained in patients with Parkinson's disease, dystonia, Huntington's disease, Tourette's syndrome, and essential tremor (Table 21.1). TMS is also being applied as therapy for movement disorders and that topic is covered elsewhere.

Parkinson's disease

The first study describing the effects of TMS in patients with Parkinson's disease (PD) 'on' and 'off' medication showed that the latency and threshold of the motor-evoked potentials (MEPs) elicited by electrical stimulation of motor cortex were normal (Dick et al. 1984). These early findings have been confirmed by a number of studies using TMS (see Currà et al. 2002, 2005 for references).

Cortical motor excitability has been investigated with various TMS techniques in PD patients. Motor threshold which is considered the lowest stimulation intensity able to evoke an MEP of minimal size at rest and during contraction is generally normal.

Experiments investigating changes in MEP size with stimulus intensity and different degrees of voluntary muscle contraction (input–output curves) showed that in PD, cortical stimuli of equal intensity given at rest induced abnormally large amplitude MEPs; stimuli of similar intensity delivered during a muscle contraction elicit abnormally small amplitude MEPs (Valls-Sole et al. 1994). These studies suggest changes in cortical motor excitability.

Cortical motor excitability can be tested by recording the cortical silent period (cSP) in response to TMS. The cSP is a pause lasting ~100–200 ms in the ongoing voluntary electromyogram (EMG) activity elicited by a single magnetic stimulus. This electrical silence originates mainly from cortical mechanisms (Inghilleri et al. 1993) and is mediated by gamma-aminobutyric acid $(GABA)_B$ receptors (Werhahn et al. 1999). A prolonged cSP indicates hyperactivity whereas shortened cSP suggests hypoactivity of inhibitory interneurons in the motor cortex. A prolonged cSP can also be due to loss of excitatory interneurons. The cSP is shorter in patients with PD than in normal subjects (Cantello et al. 1991; Priori et al. 1994a; Berardelli et al. 1996) and its duration is increased by dopaminergic treatment and by surgical interventions involving the basal ganglia (Priori et al. 1994a; Strafella et al. 1997; Young et al. 1997; Chen et al. 2001).

Another method to test cortical motor excitability is by delivering paired-pulse TMS. An MEP evoked by a suprathreshold test stimulus is preceded at different interstimulus intervals (ISIs) by a conditioning subthreshold stimulus unable to evoke any motor potential.

Table 21.1 Central motor conduction time, short-interval intracortical inhibition, and cortical silent period in movement disorders

	Central motor conduction time	Short-interval intracortical inhibition	Cortical silent period
Primary dystonia	Normal	Decreased	Shortened
Parkinson's disease	Normal	Decreased	Shortened
Huntington's disease	Normal	Normal	Lengthened
Tourette's syndrome	Normal	Decreased	Shortened
Essential tremor	Normal	Normal	Normal

With this method, an initial conditioning sub-threshold stimulus activates cortical neurons and modulates the size of a second suprathreshold test stimulus (Kujirai et al. 1993; Ziemann et al. 1996). At short interstimulus intervals (short-interval intracortical inhibition; SICI) (<5 ms), the inhibition of the test response is largely a GABAergic effect, specifically GABA$_A$ (Di Lazzaro et al. 2000). Patients with PD tested at rest and with short ISIs have reduced inhibition of the test response (Ridding et al. 1995a). Ridding et al. (1995a) suggested that the reduced SICI could be ascribed to decreased cortical inhibition or to excess cortical excitation.

To explore this question, MacKinnon et al. (2005) investigated intracortical inhibition in PD over a wide range of conditioning stimulus intensities and found that a reduction in SICI and a switch to facilitation occurred when the conditioning stimulus intensities were 90% and 100% of resting motor threshold. The authors concluded that the threshold of intracortical facilitatory pathways is decreased in PD. Studies on the connectivity between premotor and motor cortex showed that, in PD patients, low-frequency trains of repetitive (r)TMS delivered at 1 Hz over the premotor cortex normalized motor cortex excitability tested with paired-pulse techniques (Buhmann et al. 2004). Hence, the increased excitability of the interneurons mediating intracortical inhibition, tested at short ISI, presumably reflects an increased facilitatory input from the premotor to the motor cortex. Additionally, in a study of facilitatory premotor–motor interactions, premotor stimulation with trains of rTMS failed to increase

motor cortex excitability. This change was restored to normal by dopaminergic stimulation (Mir et al. 2005a).

Changes in cortical motor excitability have been also demonstrated by applying suprathreshold stimuli at long conditioning–test intervals (100–200 ms). With these stimulus variables the test MEP is inhibited at ISIs of 100–200 ms. The MEP inhibition is considered to be mediated by cortical mechanisms. With this technique, PD patients have an increased inhibition of the test response (Berardelli et al. 1996), suggesting depressed excitability of the motor cortex. Decreased cortical motor excitability in PD is also supported by the reduced MEP facilitation seen in response to rTMS applied to the motor cortex (Gilio et al. 2002).

Changes in cortical motor plasticity have been reported in PD patients. Studies investigating cortical topography have shown that in PD the corticomotor area is larger and extends more toward the anterior regions in parkinsonian patients than in normal subjects (Kargerer et al. 2003). Specific topographic changes can also be studied with paired associative stimulation (PAS) (Stefan et al. 2000). This technique entails delivering low-frequency electrical stimulation to the right median nerve paired with single-pulse TMS of the motor cortex. Previous studies suggested that in normal subjects PAS induces a stimulation of the motor cortex similar to that induced in protocols of experimental models inducing long-term potentiation (Stefan et al. 2000). In normal subjects, PAS facilitated MEP amplitude only in the APB muscle and not in the abductor digiti minimi (ADM) muscle, whereas in patients off therapy it strongly facilitated MEP amplitudes in the abductor pollicis brevis (APB) muscle and also increased MEP size in the ADM muscle. In normal subjects after PAS, the duration of the cSP increased but in patients off therapy, it remained unchanged.

The abnormal responsiveness of sensorimotor cortex to PAS in PD patients could reflect disordered plasticity within the motor cortex and an abnormal long-term potentiation (LTP) like mechanism (Bagnato et al. 2006). Changes in cortical plasticity have been also demonstrated by Ueki et al. (2006) in patients with PD with a protocol of associative plasticity similar to that proposed by Stefan et al. (2000). In this study, however, the PAS effect was less in patients when off compared with normal subjects. The explanation for the difference is still unclear. The studies of Ueki et al. (2006) and Bagnato et al. (2006) both demonstrate that dopamine deficiency can modify plasticity of motor cortex and suggest that the abnormal plasticity in the motor cortex of PD patients might be associated with higher motor dysfunction, including motor learning.

Cortical excitability has also been studied during the preparation and execution of a voluntary movement. The increased cortical excitability before EMG onset begins to increase earlier and rises at a slower rate (Pascual-Leone et al. 1994; Chen et al. 2001) in patients with PD than in normal subjects. This suggests an explanation for slowness of reaction time since it takes longer than normal to increase excitability to the level to produce a motor command.

In summary the most consistent findings of PD are characterized by shortening of the cSP, reduction of short intracortical inhibition, increase in the long-lasting intracortical inhibition, and reduction of the normal MEP facilitation after single and repetitive TMS stimuli. Studies of connectivity between premotor and motor cortex demonstrated that some of the changes in cortical excitability described in PD are due to abnormal inputs from the premotor cortex.

Putting all these facts together into a coherent story is not easy. There is a mixture of increases and decreases in excitability, but most of the deviations from normal are ameliorated by deep brain stimulation or dopaminergic therapy (Cunic et al. 2002; Lefaucheur 2005). Perhaps the basal ganglia influences on the cortex are mixed excitatory and inhibitory, depending on the neural circuit. This produces a situation where movements are produced sluggishly and plastic processes are also impaired.

Dystonia

In primary dystonia, central conduction time is normal whereas in cases of secondary dystonias it is often abnormal (Thompson et al. 1986; Mavroudakis et al. 1995). In primary dystonia, cortical motor excitability has been investigated with various TMS techniques (Berardelli et al. 1998). Studies exploring the input–output relationship in patients with dystonia showed that the size of MEPs increased more steeply with increased levels of background contraction or stimulus intensities (Mavroudakis et al. 1995; Ikoma et al. 1996). A study mapping the cortical sites from which specific MEPs can be elicited by brain stimulation showed an increase in the size of the cortical motor areas in patients with hand dystonia (Byrnes 1998). These changes provide evidence of an altered corticomotor representation in dystonia.

Studies using TMS of cortical hand motor areas show that the duration of the cSP is shortened in hand dystonia (Ikoma et al. 1996; Filipovic et al. 1997; Rona et al. 1998). TMS also elicits a short-lasting cSP in the facial muscles of patients with cranial dystonia (Currà et al. 2000). The shortened duration of the cSP suggests that dystonia is characterized by an abnormality of cortical inhibitory interneurons, probably through an abnormal control exerted by the frontal non-primary motor cortex to the primary motor cortex (Murase et al. 2005). In patients with hand dystonia, a study investigating the effects of subthreshold low-frequency rTMS demonstrated that stimulation of premotor areas prolonged cSP duration and also improved the quality of handwriting. This finding, together with the shortened cSP, suggests that in dystonia the lack of inhibition in primary motor cortex is secondary to hyperactivity of the premotor cortex.

Abnormalities of cortical excitability have been also demonstrated with paired-pulse TMS. With short ISI, the conditioning stimulus inhibits the test stimulus less in patients with dystonia than in normal subjects (Ridding et al. 1995b; Gilio et al. 2000). The reduced inhibition of the test response suggests increased excitability of the cortical hand motor area. Abnormalities of intracortical inhibition studied on test responses recorded in the hand muscle are also present in patients with blepharospasm but without hand dystonia (Sommer et al. 2002). Additionally, in generalized dystonia abnormalities are present in clinically unaffected carriers of the DYT1 gene mutation (Edwards et al. 2003).

The excitability in a cortical area surrounding an activated neural network is increased in patients with hand dystonia compared to normal subjects (Sohn and Hallett 2004a,b; Buetefisch et al. 2005). When subjects have to move one finger alone there is widespread cortical inhibition of muscles in the contralateral limb and the ipsilateral limb that are not involved in the motor act (Sohn et al. 2003; Sohn and Hallett 2004a,b). An abnormal surround inhibition is pathophysiologically important in dystonia (Sohn et al. 2003; Sohn and Hallett 2004a,b) because insufficient surround inhibition could predispose patients with dystonia to excessive movements (Berardelli et al. 1998).

In experiments investigating the effects of peripheral stimulation on cortical excitability, it was found that the normal inhibitory effects induced by conditioning nerve stimulation are lost in patients with arm dystonia (Abbruzzese et al. 2001). Stimulation of the skin contiguous or noncontiguous to the recorded muscle produces an abnormal somato-topical arrangement of sensorimotor integration (Tamburin et al. 2002). A study using the PAS technique detected abnormal plasticity in cortical motor areas in hand dystonia (Quartarone et al. 2003). PAS produces a larger and more diffuse increase in MEP size in patients with hand dystonia than in normal controls. In our opinion, an altered pattern of sensorimotor plasticity may favor maladaptive plasticity during repetitive skilled hand movements in focal task-specific hand dystonia.

Abnormalities in cortical plasticity are present not only when dystonic patients perform a movement but also if they imagine the movement without actually performing it (Quartarone et al. 2005a). When the same investigators conducted experiments combining low-frequency rTMS with transcranial direct current stimulation (tDCS) they found that the normal homeostatic effects produced by TDCS are lost in patients

with hand dystonia (Quartarone et al. 2005b). In normal subjects, a facilitatory conditioning of the M1 with anodal TDCS enhanced the inhibitory effects of a subsequent 1 Hz rTMS train on corticospinal excitability, whereas inhibitory preconditioning with cathodal tDCS reversed the after-effects of 1 Hz rTMS. In patients with writer's cramp, tDCS stimulation induced no consistent changes in corticospinal excitability. Collectively, these findings suggest that in dystonia there is abnormal cortical plasticity and that the normal homeostatic mechanisms that stabilize excitability levels within a useful dynamic range are lost. Investigating the effects of focal muscle vibration on the corticospinal excitability in writer's cramp and musician's dystonia, it was found that sensory information plays a smaller role in provoking pathological changes in writer's cramp than in musician's dystonia (Rosenkranz et al. 2005).

Cortical motor excitability has been tested before the execution of voluntary wrist movements in patients with dystonia. In normal subjects before movement onset, MEP size increased progressively and intracortical inhibition decreased, whereas in patients with hand dystonia, MEP size remained unchanged from resting values and intracortical inhibition decreased less than it did in patients (Gilio et al. 2003). These results suggest abnormal premovement motor cortex excitability in dystonia. Using a model for studying excitability changes in cortical motor areas, task-specific impairment of motor cortical excitation and inhibition is present in patients with writer's cramp (Tinazzi et al. 2005).

The most consistent findings in dystonia are: reduction of the short intracortical inhibition, shortening of the cortical silent period, and abnormalities in experiments designed to investigate sensorimotor integration. Abnormal inputs from the premotor cortex are relevant for the changes in cortical excitability of dystonic patients. Changes in cortical excitability can also be demonstrated in clinically unaffected carriers of the DYT1 gene mutation (Edwards et al. 2003).

The findings in dystonia are more consistent than in PD, in that the abnormalities of excitability are largely losses of inhibitory mechanisms. This general tendency presumably leads to excessive movement and the specific loss of surround inhibition leads to overflow of the motor command. The more recent findings of increased and abnormal homeostatic plasticity are likely also important in understanding the genesis of dystonic manifestations.

Paroxysmal kinesigenic dyskinesia is characterized by brief episodes of involuntary movements (very often dystonic) precipitated by sudden movement. In these patients, short intracortical inhibition and the early phase of transcallosal inhibition are reduced. Because intracortical inhibition and transcallosal inhibition are both mediated by $GABA_A$ergic mechanisms, the authors suggested that in paroxysmal kinesigenic dystonia there is abnormal $GABA_A$-ergic inhibition (Mir et al. 2005b).

Huntington's disease

Early studies showed that patients with Huntington's disease (HD) have normal corticospinal conduction (Thompson et al. 1986) but may have abnormal variability of MEP latency, size, and threshold (Meyer et al. 1992). This variability may be due to the underlying chorea. Cortical motor excitability has been investigated in HD with the cSP and with paired-pulse techniques (Berardelli et al. 1999). An early study found that some HD patients have a prolonged cSP (Priori et al. 1994b), whereas a later study determined that the cSP abnormalities can be better disclosed by measuring the SP from unselected traces (Modugno et al. 2001). In addition to the changes in duration, the cSP is more variable in HD patients than in normal subjects, again likely due to the chorea. In the Westphal variant, the cSP can be shortened (Tegenthoff et al. 1996). Studies with paired-pulse TMS have provided controversial information on cortical motor excitability in HD. One study reported reduced cortico-cortical inhibition at short interstimulus intervals, and an increased facilitation at longer intervals (Abbruzzese et al. 1997), whereas other studies found normal cortico-cortical inhibition at short ISIs (Hanajima et al. 1996; Priori et al. 2000).

In patients with HD, controversial and variable TMS results are probably due to phenotypic heterogeneity, various stages of the disease, and different methods used by the different investigators.

In recent years, cortical motor excitability has been studied with rTMS. Whereas in normal subjects, rTMS elicits MEPs that increase in amplitude over the course of the trains, in HD patients, rTMS leaves the MEP size almost unchanged (Lorenzano et al. 2006). The loss of the normal rTMS-induced MEP facilitation suggests that in HD the excitability of the motor cortex is globally reduced. In HD, changes in motor cortex excitability may be due to an abnormal basal ganglia projection to the motor cortex but may also reflect cortical degeneration.

Tourette's syndrome

The threshold for evoking an MEP is normal, but the cSP is short and intracortical inhibition is decreased in patients with Tourette's syndrome (Ziemann et al. 1997). These abnormalities are mainly found in patients whose tics involve the EMG target muscle. Similar changes have also been found in patients with obsessive–compulsive disorders and comorbid tics (Greenberg et al. 2000). Changes in motor cortical excitability in Tourette's syndrome are likely due to an abnormal input from basal ganglia to motor cortex (Berardelli et al. 2003), and may be indicative of a loss of inhibition that may be etiological in allowing tics to be released.

Essential tremor

Patients with essential tremor have normal corticospinal conduction, normal duration of the cSP, and normal intracortical inhibition (Romeo et al. 1998). Results similar to those reported in patients with essential tremor have been described in patients with primary writing tremor (Modugno et al. 2002). Finally, single TMS pulses reset essential tremor, supporting the role of the primary motor cortex in the pathophysiology of essential tremor (Britton et al. 1993).

Conclusions

As a diagnostic tool in movement disorders, TMS can supply some information by demonstrating central motor conduction abnormalities in patients with secondary forms of movement disorders when pyramidal signs are clinically equivocal. Otherwise, there is little demonstrated routine clinical use for TMS in this area. On the other hand, TMS has provided an enormous amount of new important data in our understanding of movement disorders and promises to continue to do so.

References

Abbruzzese G, Buccolieri A, Marchese R, Trompetto C, Mandich P, Schieppati M (1997). Intracortical inhibition and facilitation are abnormal in Huntington's disease: a paired magnetic stimulation study. *Neuroscience Letters 228*, 87–90.

Abbruzzese G, Marchese R, Buccolieri A, Gasparetto B, Trompetto C (2001). Abnormalities of sensorimotor integration in focal dystonia. A transcranial magnetic stimulation study. *Brain 124*, 537–545.

Bagnato S, Agostino R, Modugno N, Quartarone A, Berardelli A (2006). Plasticity of the motor cortex in Parkinson's disease patients On and Off therapy. *Movement Disorders 21*, 639–645.

Berardelli A, Rona S, Inghilleri M, Manfredi M (1996). Cortical inhibition in Parkinson's disease. A study with paired magnetic stimuli. *Brain 119*, 71–77.

Berardelli A, Rothwell JC, Hallett M, Thompson PD, Manfredi M, Marsden CD (1998). The pathophysiology of primary dystonia. *Brain 121*, 1195–1212.

Berardelli A, Noth J, Thompson PD et al. (1999). Pathophysiology of chorea and bradykinesia in Huntington's disease. *Movement Disorders 14*, 398–403.

Berardelli A, Currà A, Fabbrini G, Gilio F, Manfredi M (2003). Pathophysiology of tics and Tourette syndrome. *Journal of Neurology 250*, 781–787.

Britton TC, Thompson PD, Day BL et al. (1993). Modulation of postural wrist tremors by magnetic stimulation of the motor cortex in patients with Parkinson's disease or essential tremor and in normal subjects mimicking tremor. *Annals of Neurology 33*, 473–479.

Buetefisch CA, Boroojerdi B, Chen R, Battaglia F, Hallett M (2005). Task-dependent intracortical inhibition is impaired in focal hand dystonia. *Movement Disorders 20*, 545–551.

Buhmann C, Gorsler A, Baumer T, et al. (2004). Abnormal excitability of premotor–motor connections in de novo Parkinson's disease. *Brain 127*, 2732–2746.

Byrnes ML, Thickbroom GW, Wilson SA, et al. (1998). The corticomotor representation of upper limb muscles in writer's cramp and changes following botulinum toxin injection. *Brain 121*, 977–988.

Cantello R, Gianelli M, Bettucci D, Civardi D, De Angelis MS, Mutani R (1991). Parkinson's disease rigidity: magnetic motor evoked potentials in a small hand muscle. *Neurology 41*, 1449–1456.

Chen R, Garg RR, Lozano AM, Lang AE (2001). Effects of internal globus pallidus stimulation on motor cortex excitability. *Neurology 56*, 716–723.

Cunic D, Roshan L, Khan FI, Lozano AM, Lang AE, Chen R (2002). Effects of subthalamic nucleus stimulation on motor cortex excitability in Parkinson's disease. *Neurology 58*, 1665–1672.

Currà A, Romaniello A, Berardelli A, Cruccu G, Manfredi M (2000). Shortened cortical silent period in facial muscles of patients with cranial dystonia. *Neurology 54*, 130–135.

Currà A, Modugno N, Inghillieri M, Manfredi M, Hallett M, Berardelli A (2002). Transcranial magnetic stimulation techniques in clinical investigation. *Neurology 59*, 1851–1859.

Currà A, Agostino R, Berardelli A (2005). Neurophysiology of Parkinson's disease, levodopa-induced dyskinesias, dystonia, Huntington's disease and myoclonus. *Neurodegenerative Diseases 18*, 227–250.

Dick JPR, Cowan JMA, Day BL et al. (1984). Corticomotoneurone connection is normal in Parkinson's disease. *Nature 310*, 407–409.

Di Lazzaro V, Oliviero M, Meglio M et al. (2000). Direct demonstration of the effect of lorazepam on the excitability of the human motor cortex. *Clinical Neurophysiology 111*, 794–799.

Edwards M, Huang YZ, Wood NW, Rothwell JC, Bathia K (2003). Different patterns of electrophysiological deficits in manifesting and non-manifesting carriers of the DTY1 gene mutation. *Brain 126*, 2074–2080.

Filipovic SR, Ljubisavljevic M, Svetel M, Milanovic S, Kacar A, Kostic VS (1997). Impairment of cortical inhibition in writer's cramp as revealed by changes in electromyographic silent period after transcranial magnetic stimulation. *Neuroscience Letters 222*, 167–170.

Gilio F, Currà A, Lorenzano C, Modugno N, Manfredi M, Berardelli A (2000). Effects of botulinum toxin type A on intracortical inhibition in patients with dystonia. *Annals of Neurology 48*, 20–26.

Gilio F, Currà A, Inghilleri M, Lorenzano C, Manfredi M, Berardelli A (2002). Repetitive magnetic stimulation of cortical motor areas in Parkinson's disease: implications for the pathophysiology of cortical function. *Movement Disorders 17*, 467–473.

Gilio F, Currà A, Inghilleri M et al. (2003). Abnormalities of motor cortex excitability preceding movement in patients with dystonia. *Brain 126*, 1745–1754.

Greenberg BD, Ziemann U, Cora-Locatelli G et al. (2000). Altered cortical excitability in obsessive compulsive disorder. *Neurology 54*, 142–147.

Hanajima R, Ugawa Y, Terao Y, Ugata K, Kanazawa I (1996). Ipsilateral cortico-cortical inhibition of the motor cortex in various neurological disorders. *Journal of Neurological Science 140*, 109–116.

Ikoma K, Samii A, Mercuri B, Wassermann EM, Hallett M (1996). Abnormal cortical motor excitability in dystonia. *Neurology 46*, 1371–1376.

Inghilleri M, Berardelli A, Cruccu G, et al. (1993). Silent period evoked by transcranial stimulation of the human cortex and cervicomedullary junction. *Journal of Physiology 466*, 521–534.

Kargerer FA, Summers JJ, Byblow WD, Taylor B (2003). Altered corticomotor representation in patients with Parkinson's disease. *Movement Disorders 18*, 919–927.

Kujirai T, Caramia MD, Rothwell JC, et al. (1993). Corticocortical inhibition in human motor cortex. *Journal of Physiology 471*, 501–519.

Lefaucheur JP (2005). Motor cortex dysfunction revealed by cortical excitability studies in Parkinson's disease: influence of antiparkinsonian treatment and cortical stimulation. *Clinical Neurophysiology 116*, 244–253.

Lorenzano C, Dinapoli L, Gilio F, et al. (2006). Motor cortical excitability studied with repetitive transcranial magnetic stimulation in patients with Huntington's disease. *Clinical Neurophysiology 117(8)*,1677–81.

MacKinnon C, Gilley E, Weis-McNulty A, Simuni T (2005). Pathways mediating abnormal intracortical inhibition in Parkinson's disease. *Annals of Neurology 58*, 516–524.

Mavroudakis N, Caroyer JM, Brunko E, Zegers de Beyl D (1995). Abnormal motor evoked responses to transcranial magnetic stimulation in focal dystonia. *Neurology 45*, 1671–1677.

Meyer BU, Noth J, Lange HW, et al. (1992). Motor responses evoked by magnetic brain stimulation in Huntington's disease. *Electroencephalography and Clinical Neurophysiology 85*, 197–208.

Mir P, Huang YZ, Gilio F, et al. (2005a). Abnormal cortical and spinal inhibition in paroxysmal kinesigenic dyskinesia. *Brain 128*, 291–299.

Mir P, Matsunaga K, Gilio F, Quinn NP, Siebner HH, Rothwell JC (2005b). Dopaminergic drugs restore facilitatory premotor–motor interactions in Parkinson's disease. *Neurology 64*, 1906–1912.

Modugno N, Currà A, Giovannelli M, et al. (2001). The prolonged cortical silent period in patients with Huntington's disease. *Clinical Neurophysiology 112*, 1470–1474.

Modugno N, Nakamura Y, Bestmann S, et al. (2002). Neurophysiological investigations in patients with primary writing tremor. *Movement Disorders 17*, 1336–1340.

Murase N, Rothwell J C, Kaji R, et al. (2005). Subthreshold low-frequency repetitive transcranial magnetic stimulation over the premotor cortex modulates writer's cramp. *Brain 128*, 104–115.

Pascual-Leone A, Valls-Sole J, Brasil-Neto JP, Cohen LG, Hallett M (1994). Akinesia in Parkinson's disease. I. Shortening of simple reaction time with focal, single-pulse transcranial magnetic stimulation. *Neurology 44*, 884–891.

Priori A, Berardelli A, Inghilleri M, Accornero N, Manfredi M (1994a). Motor cortical inhibition and the dopaminergic system. *Brain 117*, 317–323.

Priori A, Berardelli A, Inghilleri M, Polidori L, Manfredi M (1994b). Electromyographic silent period after transcranial brain stimulation in Huntington's disease. *Movement Disorders 9*, 178–182.

Priori A, Polidori L, Rona S, Manfredi M, Berardelli A (2000). Spinal and cortical inhibition in Huntington's disease. *Movement Disorders 15*, 938–946.

Quartarone A, Bagnato S, Rizzo V, et al. (2003). Abnormal associative plasticity of the human motor cortex in writer's cramp. *Brain 126*, 2586–2596.

Quartarone A, Bagnato S, Rizzo V, et al. (2005a). Corticospinal excitability during motor imagery of a simple tonic finger movement in patients with writer's cramp. *Movement Disorders 20*, 1488–1495.

Quartarone A, Rizzo V, Bagnato S, et al. (2005b). Homeostatic-like plasticity of the primary motor hand area is impaired in focal hand dystonia. *Brain 128*, 1943–1950.

Ridding MC, Inzelberg R, Rothwell JC (1995a). Changes in excitability of motor circuitry in patients with Parkinson's disease. *Annals of Neurology 37*, 181–188.

Ridding MC, Sheean G, Rothwell JC, Inzelberg R, Kujirai T (1995b). Changes in the balance between motor cortical excitation and inhibition in focal, task specific dystonia. *Journal of Neurology, Neurosurgery and Psychiatry 53*, 493–498.

Romeo S, Berardelli A, Pedace F, et al. (1998). Cortical excitability in patients with essential tremor. *Muscle and Nerve 21*, 1304–1308.

Rona S, Berardelli A, Vacca L, Inghilleri M, Manfredi M (1998). Alterations of motor cortical inhibition in patients with dystonia. *Movement Disorders 13*, 118–124.

Rosenkranz K, Williamon A, Butler K, Cordivari C, Lees AJ, Rothwell JC (2005). Pathophysiological differences between musician's dystonia and writer's cramp. *Brain 128*, 918–931.

Sohn Y, Hallett M (2004a). Disturbed surround inhibition in focal hand dystonia. *Annals of Neurology 56*, 595–599.

Sohn Y, Hallett M (2004b). Surround inhibition in human motor system. *Experimental Brain Research 158*, 397–404.

Sohn Y, Jung HY, Kaelin-Lang A, Hallett M (2003). Excitability of the ipsilateral motor cortex during phasic voluntary hand movement. *Experimental Brain Research 148*, 176–185.

Sommer M, Ruge D, Tergau F, Beuche W, Altenmuller E, Paulus W (2002). Intracortical excitability in the hand motor representation in hand dystonia and blepharospasm. *Movement Disorders 17*, 1017–1025.

Stefan K, Kunesch E, Cohen LG, Benecke L, Classen J (2000). Induction of plasticity in the human motor cortex by paired associative stimulation. *Brain 123*, 572–584.

Strafella A, Ashby P, Lozano A, Lang AE (1997). Pallidotomy increases cortical inhibition in Parkinson's disease. *Canadian Journal of Neurological Science 24*, 133–136.

Tamburin S, Manganotti P, Marzi CA, Fiaschi A, Zanette G (2002). Abnormal somatotopic arrangement of sensorimotor interactions in dystonic patients. *Brain 125*, 2719–2730.

Tegenthoff M, Vorgerd M, Juskowiak F, Roos V, Malin JP (1996). Postexcitatory inhibition after transcranial magnetic single and double brain stimulation in Huntington's disease. *Electroencephalography and Clinical Neurophysiology 101*, 298–303.

Thompson PD, Dick JP, Day BL, et al. (1986). Electrophysiology of the corticomotoneurone pathways in patients with movement disorders. *Movement Disorders 1*, 113–117.

Tinazzi M, Farina S, Edwards M, *et al.* (2005). Task-specific impairment of motor cortical excitation and inhibition in patients with writer's cramp. *Neuroscience Letters 378*, 55–58.

Ueki Y, Mima T, Ali Kotb M, *et al.* (2006). Altered plasticity of the human motor cortex in Parkinson's disease. *Annals of Neurology 59*, 60–71.

Valls-Sole J, Pascual-Leone A, Brasil-Neto JP, Cammarota A, McShane L, Hallett M (1994). Abnormal facilitation of the response to transcranial magnetic stimulation in patients with Parkinson's disease. *Neurology 44*, 735–741.

Werhahn KJ, Kunesch E, Noachtar S, Benecke R, Classen J (1999). Differential effects on motorcortical inhibition induced by blockade of GABA uptake in humans. *Journal of Physiology 517*, 591–597.

Young MS, Triggs WJ, Bowers D, Greer M, Friedman WA (1997). Stereotactic pallidotomy lengthens the transcranial magnetic cortical stimulation silent period in Parkinson's disease. *Neurology 49*, 1278–1283.

Ziemann U, Rothwell JC, Ridding MC (1996). Interaction between intracortical inhibition and facilitation in human motor cortex. *Journal of Physiology 496*, 873–881.

Ziemann U, Paulus W, Rothenberger A (1997). Decreased motor inhibition in Tourette disorder: evidence from transcranial magnetic stimulation. *American Journal of Psychiatry 154*, 142–147.

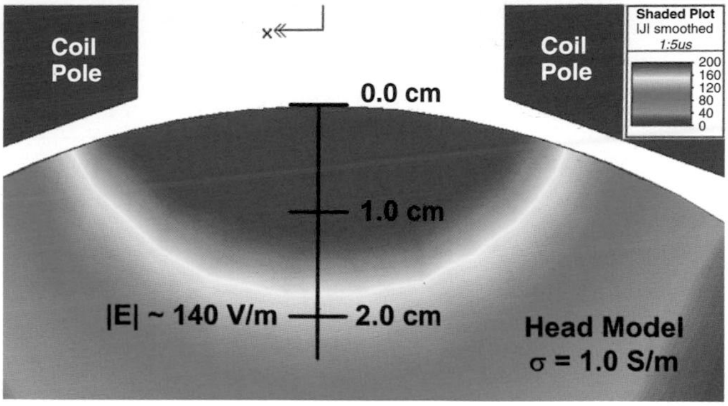

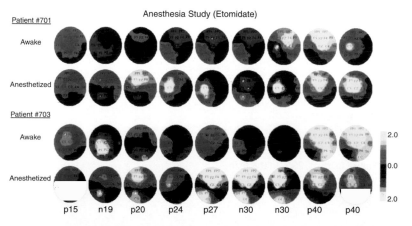

Plate 1 See also Figure 3.7, p. 20, in the text.

Coil Pole

Coil Pole

0.0 cm

1.0 cm

|E| ~ 140 V/m — 2.0 cm

Head Model
σ = 1.0 S/m

Shaded Plot
|J| smoothed
1:5us
200
160
120
80
40
0

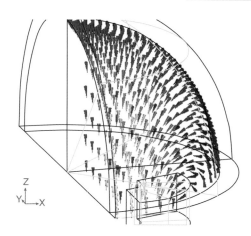

Plate 2 See also Figure 5.2, p. 35, in the text.

Z
Y
X

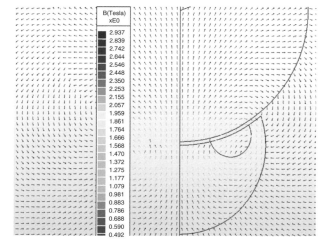

Plate 3 See also Figure 5.10, p. 44, in the text.

B(Tesla)
xE0

2.937
2.839
2.742
2.644
2.546
2.448
2.350
2.253
2.155
2.057
1.959
1.861
1.764
1.666
1.568
1.470
1.372
1.275
1.177
1.079
0.981
0.883
0.786
0.688
0.590
0.492

Anesthesia Study (Etomidate)

Patient #701

Awake

Anesthetized

Patient #703

Awake

Anesthetized

Plate 4 See also Figure 20.5, p. 294, in the text.

2.0

0.0

2.0

p15   n19   p20   p24   p27   n30   n30   p40   p40

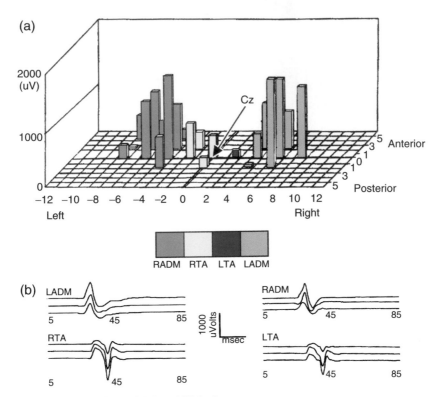

Plate 5 See also Figure 20.8, p. 297, in the text.

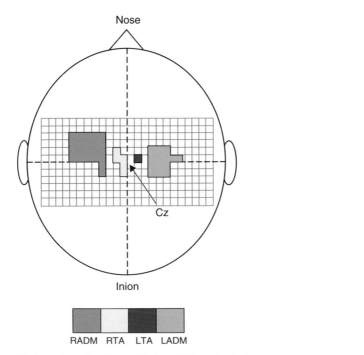

Plate 6 See also Figure 20.9, p. 298, in the text.

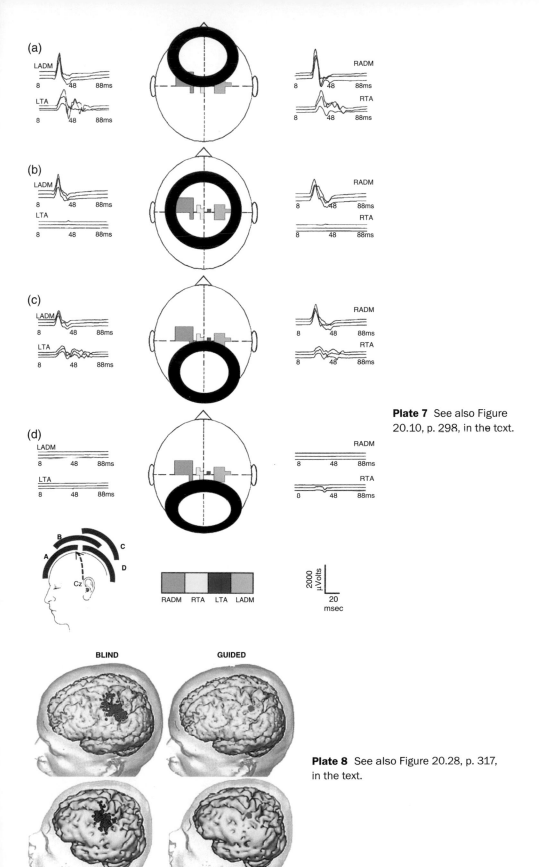

Plate 7 See also Figure 20.10, p. 298, in the text.

Plate 8 See also Figure 20.28, p. 317, in the text.

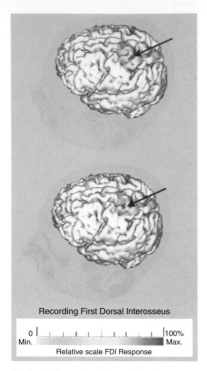

Recording First Dorsal Interosseus

0 |———————————| 100%
Min.   Max.
Relative scale FDI Response

Plate 9 See also Figure 20.29, p. 319, in the text.

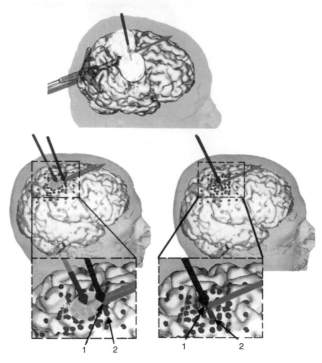

1   2   1   2

Plate 10 See also Figure 20.30, p. 320, in the text.

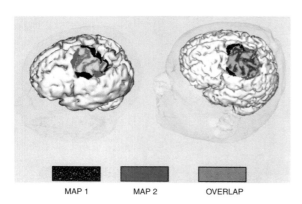

MAP 1   MAP 2   OVERLAP

Plate 11 See also Figure 20.31, p. 321, in the text.

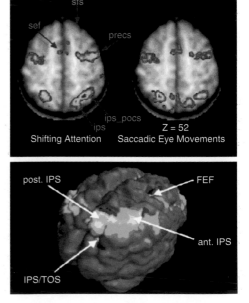

sfs

sef   precs

ips_pocs
ips   Z = 52
Shifting Attention   Saccadic Eye Movements

post. IPS   FEF

ant. IPS

IPS/TOS

Plate 12 See also Figure 28.1, p. 432, in the text.

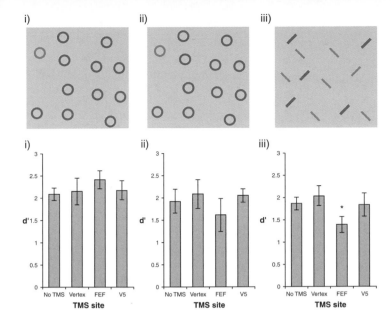

i)

ii)

iii)

Plate 13 See also Figure 28.5, p. 438, in the text.

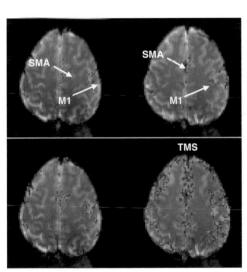

TMS

Plate 14 See also Figure 36.3, p. 574, in the text.

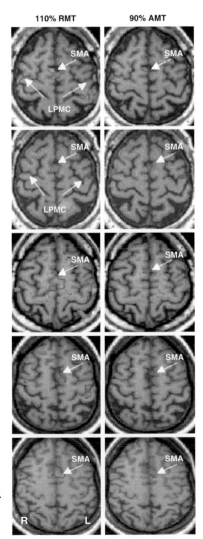

Plate 15 See also Figure 36.6, p. 579, in the text.

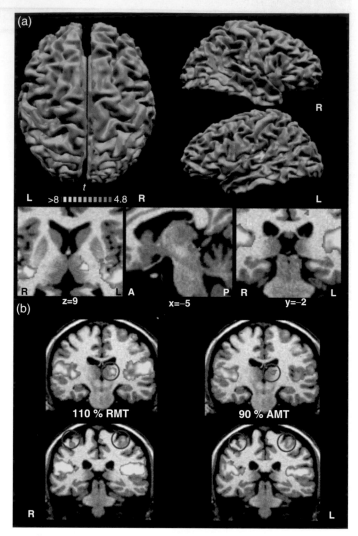

Plate 16 See also Figure 36.7, p. 582, in the text.

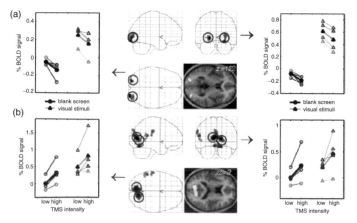

Plate 17 Please see Figure 36.8, p. 583, in the text also.

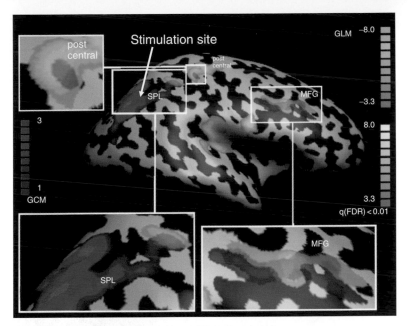

Plate 18 Please see Figure 36.9, p. 585, in the text also.

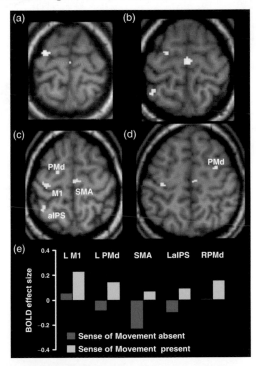

Plate 19 See also Figure 36.10, p. 587, in the text.

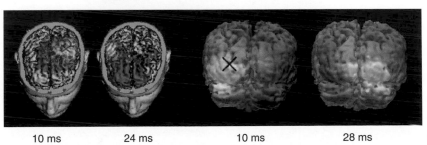

10 ms   24 ms   10 ms   28 ms

Plate 20 See also Figure 37.3, p. 598, in the text.

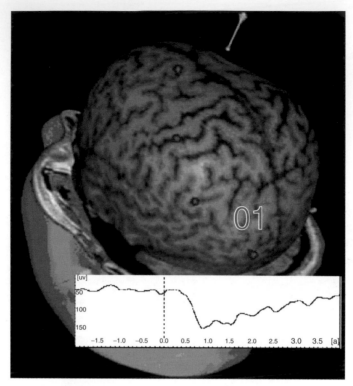

Plate 21 See also Figure 37.4, p. 599, in the text.

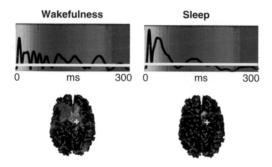

Plate 22 See also Figure 37.5, p. 604, in the text.

Plate 23 See also Figure 29.2, p. 451, in the text.

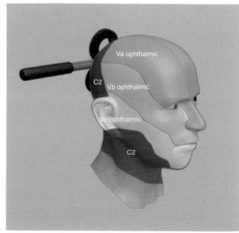

CHAPTER 22

TMS: neurodevelopment and perinatal insults

Marjorie A. Garvey

Introduction

The truism 'children are not little adults' denotes many important aspects about children. One of these is that their motor skills are not static but undergo continual change until the late adolescent and early adult years. It is reasonable to assume that the neural substrate for changes in neuromotor skills of typically developing children is the complex and organized maturation of underlying brain structures. In like manner, abnormalities of the brain maturation may be responsible for the anomalous maturation of motor skills in children with developmental disabilities. TMS has shown potential as a useful tool to test these hypotheses.

This chapter will start with a brief overview of the changes that occur in motor function as children get older and those aspects of central nervous development which may form the neural substrates of motor function development. The chapter will then describe those TMS-evoked parameters, related to the motor system, that have been studied in both typically developing children and in those who have suffered perinatal insults to the CNS.

Understanding neurodevelopment

Clinical measures of neurodevelopment

Motor control has been the most thoroughly examined area of child development because of its ease of observation and the creation of tools that can objectively measure its change (Lipkin 2005). Spontaneous movements in infants are the first good measure of motor skills (Einspieler and Prechtl 2005). The majority of gross motor milestones (sitting, standing, walking) are achieved through the preschool years. Fine motor skills, which first appear in a rudimentary fashion (e.g. grasping, pointing, pincer grip) during the first year of postnatal life (Illingsworth 1983), make noticeable gains (e.g. drawing and writing skills, manipulating objects such as scissors) through the early school years (Yule *et al.* 1967; Frankenburg *et al.* 1992; Croce *et al.* 2001; Plubrukarn and Theeramanoparp 2003; Majnemer and Snider, 2005) After this time, they continue to improve in a more subtle way (Denckla 1973, 1974; Wolff *et al.* 1998; Largo *et al.* 2001a,b) (increase in rhythmicity, speed, and coordination of movements) and reach a plateau during mid- to late-adolescence depending on the complexity of the motor skill (more complex motor skills reach maturity later). Associated movements (synkinesis, mirror movements, etc.) are present in most children under 10 years of age, and their intensity and frequency vary according to the motor task's complexity. By the time puberty is reached, associated movements have diminished in intensity although they continue to accompany complex or forceful movements throughout adult life (Lazarus 1992; Largo *et al.* 2003).

Structural changes in the brain during neurodevelopment

During the first decade of life, when fine motor skills show the highest rates of change, the structure of motor cortex is also rapidly changing. Although a comprehensive review of cortical maturation is beyond the scope of this chapter, a brief review is in order since knowledge of these processes is essential when interpreting TMS-evoked parameters in children. The reader who wishes in-depth descriptions will find an excellent overview of structural brain maturation in Unit 1 of the *Neurology of the newborn* (Volpe 2001).

Postnatal CNS maturation is a complex process which can be traced using both histopathology and *in vivo* neuroimaging. In the first 2 years of postnatal life there is overproduction of synapses. This is followed by a longer process, lasting through mid-adolescence, consisting of pruning of excessive synapses and activity-dependent refinement of synaptic connections (Johnston 2003, 2004). The visual cortex completes this process of pruning earlier than the motor cortex (Huttenlocher and Dabholkar 1997).

Myelination is virtually complete in the corticospinal tracts by the end of the second year of postnatal life. In contrast, intracortical and callosal white matter myelination mature more gradually (Yakovlev and Lecours 1967; Giedd *et al.* 1999).

Using magnetic resonance imaging (MRI), it is possible to map the structural changes that occur in the brain during neurodevelopment. Using standard MRI techniques, it is possible to see maturational changes in central white matter up until the preschool years. From 4 years of age onwards, quantitative evaluation of MR images is necessary to detect the temporal sequence of central myelination which occurs over the first two to three decades of postnatal life. For an excellent review of the literature regarding the uses of MRI in understanding neurodevelopment, see Paus *et al.* (2001).

Diffusion tensor imaging (DTI) is becoming an important tool in the evaluation of brain development. Water apparent diffusion coefficient and fractional diffusion anisotropy appear to be the DTI parameter most sensitive to change during development, reflecting underlying changes in tissue water content and cytoarchitecture (Neil *et al.* 1998; Barnea-Goraly *et al.* 2005; Ben Bashat *et al.* 2005). This imaging technique may provide a clearer picture of intracortical white matter maturation than MRI and has the potential to reveal subtle white matter abnormalities in developmental disorders which show few abnormalities on structural MRI. However, much work is required to refine the technique and establish norms.

Insights into maturation of cortical function using functional imaging

Functional imaging of the motor areas of the cortex during development is still in its infancy. Despite this, it has the potential to provide key insights to neuromotor maturation. Positron emission tomography (PET) has examined motor control in the contralesional hemisphere in children with unilateral cerebral lesions (Muller *et al.* 1998). An important finding in this study was that, while cortical activation in the primary motor areas is identical in adults and children when they perform the same motor task, children show greater activation in nonmotor areas of the brain compared with adults. These results suggest that the increased coordination and decreased unintentional movements seen during maturation of neuromotor skills may be associated with increasing lateralization and intrahemispheric focalization of motor control during motor cortex maturation. However, motor function in the 'unaffected' hand in children with hemiplegia is abnormal, which makes it difficult to generalize these results to typically developing children (Brown *et al.* 1989; Gordon *et al.* 1999).

In contrast, a functional (f)MRI study examining maturation of brain activation during motor movements in healthy children (Mall *et al.* 2005) showed that, compared with adults, children had *decreased* activation of motor areas in both hemispheres. Although this study was performed in healthy children, it is difficult to reconcile the results with the many behavioral studies showing an increase in associated movements in younger children (see for example Denckla 1973, 1974). Neurophysiological studies (Mayston *et al.* 1999; Mall *et al.* 2004) (reviewed below) suggest that both intracortical and transcallosal inhibition are decreased during voluntary movement in young children when compared with adults. In addition, developmental mirror movements in children arise as a result of cortical activity in both hemispheres (Mayston *et al.* 1999). Together, these studies appear to predict increased amount of fMRI activation during movement in children. Unfortunately, the authors of the fMRI study do not comment on why their results might differ from the behavioral and neurophysiological studies.

Cross-modality studies using TMS and functional imaging may be most useful in the study of brain function during normal development and in developmental disorders (Staudt *et al.* 2002). Functional MRI is likely to be of more use than PET in elucidating *in vivo* measures of cortical function, but, as with DTI, numerous methodological issues must be resolved before it can become a reliable neurophysiological tool (Hodics and Cohen 2005).

TMS-evoked parameters in children

TMS has allowed access to the functional changes of the brain that occur during neuromaturation. Though it is restricted in its scope, TMS is a powerful tool when its limitations are kept in mind. This section will review what is known about maturation of motor cortex from studies using TMS. In particular, the focus will be on how TMS parameters in children differ from those in adults.

Motor-evoked potential threshold

Motor-evoked potential (MEP) thresholds are higher in children than in adults and they gradually decrease to adult levels by mid-adolescence (Nezu *et al.* 1997; Moll *et al.* 1999; Garvey *et al.* 2003). In children as in adults, the MEP threshold is higher when the target muscle is at rest (resting motor threshold: RMT) than when there is background muscle activation (active motor threshold: AMT) (Garvey *et al.* 2003). It may not be possible to elicit, even using maximal stimulator output, reliable MEP responses when muscles are at rest in children before 6 years of age (Koh and Eyre, 1988). When the target

muscle is active, MEP responses can be elicited even in neonates when the stimulus intensity is set to 100% of stimulator output.

MEP latency

An estimate of central motor conduction time (CMCT) can be assessed by measuring the latency of a TMS-evoked MEP. There are two distinct developmental patterns of CMCT depending on whether it is measured while muscles are at rest ('resting' CMCT) or with background muscle activation ('active' CMCT). Active CMCT reaches maturity in children by 2 years of age; in contrast, resting CMCT does not reach maturity until early adolescence (Koh and Eyre 1988; Eyre *et al.* 1991; Müller *et al.* 1994; Fietzek *et al.* 2000) (Figure 22.1). In adults, the 'latency jump' (i.e. the difference between MEP latencies evoked at rest and those evoked with background muscle activation) is thought to reflect trans-synaptic activation of cortical motor neurons via interneurons and recruitment of faster pyramidal neurons at higher levels of muscle activation (Rossini *et al.* 1994; Abbruzzese and Trompetto 2002). The 'latency jump' is four times greater in preschool children than in adults and it gradually decreases in magnitude until mid-adolescence, at which time it reaches maturity. Mechanisms responsible for the gradual decrease of this latency jump in children are still unclear, but may include neuronal and synaptic maturation within the motor cortex, maturation of central myelination, and developmental aspects of motor neuronal recruitment which are at present unknown (Caramia *et al.* 1993) (Figure 22.2).

MEP amplitude and stimulus response curves

The suprathreshold MEP amplitude can be studied using the stimulus response curve. Stimulus–response, or input–output, curves are thought to assess those neurons that are intrinsically less excitable or spatially further from the center of activation. In adults, the shape of the curve is usually sigmoidal and its features are represented by the threshold, the steepness of the slope, and the plateau level (Chen 2000).

A recent study from our laboratory examined maturation of the stimulus–response curves

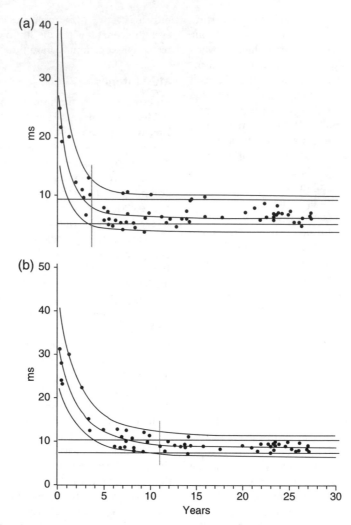

Fig. 22.1 Maturation profile for TMS-evoked motor-evoked potential (MEP) latency measured with (a) and without (b) background muscle activation of the first dorsal interosseus (FDI). Profiles were calculated according to an exponential equation, and determine dynamic and stable phases in development of each parameter. The vertical line in each graph indicates the age at which TMS-evoked MEP latency reaches maturity. Note that TMS-evoked MEP latency measured with background activation reaches maturity earlier than that measured in resting muscles. Modified from Fietzek *et al.* (2000). © Mac Keith Press, London. Reproduced with the permission of Cambridge University Press.

(SRCs) in healthy right-handed adults and children (Figure 22.3). The target muscles were the right and left first dorsal interosseus (FDI). We used a focal coil to deliver five stimuli (from threshold to 30% above threshold) to the optimal sites for eliciting an MEP in the right and left FDI muscles at rest. Stimuli were delivered using the posterior–anterior (PA) orientation for the entire stimulus–response curve and also using the latero-medial (LM) orientation at 30% above threshold to examine maturation of indirect (PA) vs direct (LM) stimulation of the corticospinal neurons.

Peak-to-peak MEP amplitudes were measured in those single trials which showed no evidence of background electromyographic (EMG) activity.

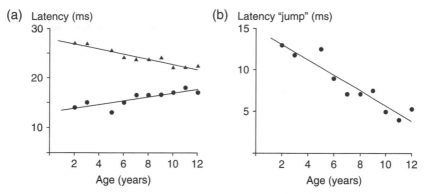

Fig. 22.2 Latency 'jump' between TMS-evoked motor-evoked potential (MEP) latencies measured with and without background muscle activation. (a) Age-related changes in TMS-evoked MEP latencies measured with (closed circles) and without (closed triangle) background muscle activation in the thenar eminence. Measurements were made in 14 subjects between 2 and 12 years of age. Each symbol represents the mean value of the two limbs or more, whenever different subjects of the same age were present. Correlation coefficients for MEP latency measured with muscle activation: $r = 0.83$; $P < 0.001$; without muscle activation: $r = 0.92$; $P < 0.001$. (b) Age-related changes in the latency 'jump' (i.e. the difference between the TMS-evoked MEP latency measured with and without background muscle activation). Correlation coefficient for this parameter: $r = 0.95$; $P < 0.001$. Note that the latency 'jump' decreases with age. Reprinted from Caramia *et al.* with permission from the International Federation of Clinical Neurophysiology.

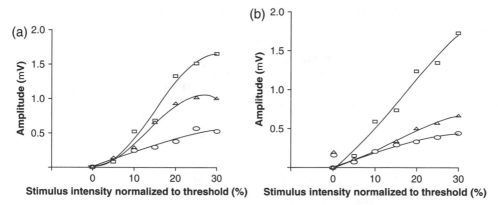

Fig. 22.3 Maturation of the TMS-evoked stimulus–response curve (SRC) when stimulating the dominant hemisphere (a) and non-dominant hemisphere (b). Percentage of maximum stimulator output (%) normalized to threshold (= 0) is represented on the y axes; mean motor-evoked potential (MEP) amplitude (mV) is represented on the x axes. The SRCs of three different age groups are shown: adults, open squares; 12–14-year-old children, open triangles; 6–11-year-old children, open circles. The slope and plateau of the dominant hemisphere SRC differ between the age groups. The SRC of the children <12 years of age does not reach a plateau; the slope of the SRC of the adolescent subjects is steeper than that seen in younger children but is less steep than that seen in the SCR of adults. There is no plateau in the non-dominant SRC for any age-group. No difference was found in the slopes of the SRC between the child and adolescent age-groups; however, both of these were different from the slope of the SRC in the adult groups.

When plotting the SRC, stimulus intensities for each subject were normalized relative to their MEP threshold in order to control for the expected age-related decrease in RMT. To avoid a direct comparison between the absolute MEP amplitudes in different individuals we examined the *rates of change* of MEP amplitudes in individual SRCs and compared these rates across age groups. Our first step was to normalize MEP amplitudes for each individual by subtracting the amplitude identified at threshold from the amplitude obtained at every stimulus intensity (including threshold). In this way, the data for each individual started from a zero point. We then performed a repeated measures analysis of variance (ANOVA) with MEP amplitude as the outcome variable, age group (<12 years vs 12–14 years vs adults) as the between-subjects variable, and relative stimulus intensity [–10 vs –5 vs 0 (threshold) vs 5 vs 10 vs 15 vs 20 vs 25 vs 30] and side (left vs right) as the within-subject variables.

In the dominant motor cortex, rate of change of the MEP amplitude was greatest in the adult subjects and smallest in the children under 12 years of age. Although RMT did not differ significantly between the adults and the 12–14-year-old children, there was a significant difference between rates of change of MEP amplitude in the two groups.

Findings for the nondominant hemisphere were significantly different from the dominant hemisphere only in the 12–14-year-old children: the nondominant SRCs of these children had a slower rate of change, similar to that seen in younger children.

These findings concur with earlier studies of the suprathreshold MEP amplitude and indicate that, unlike motor threshold, maturation of the representation of the FDI in the dominant motor cortex is not complete at puberty (Garvey *et al.* 2003). In addition, maturation of the suprathreshold MEP amplitude in the nondominant motor cortex lags behind that in the dominant motor cortex. The absence of side-to-side asymmetry in the adult group suggests that maturation of the nondominant cortex is complete by early adulthood.

Effect of coil orientation

As has been previously reported in adults (Werhahn *et al.* 1994), the latency of MEP,

evoked using the LM (direct) coil orientation was shorter than the MEP latency for the PA (indirect) coil orientation for all age groups. This provided an opportunity to examine maturation of the cortical interneurons. Indirect stimulation, using PA coil orientation, is the most efficient method for stimulating the motor cortex and evokes MEPs with the greatest amplitudes compared with direct stimulation using the LM coil orientation. This is thought to be due to the summation of impulses on the pyramidal tract neuron via the fastest-conducting interneurons.

Since myelination of the central white matter and intracortical interneurons is not complete until early adulthood, it is possible that indirect stimulation in children would be less efficient than in adults and that this would manifest as a decrease in the relative amplitude of the MEP evoked using the PA coil orientation (PA-MEP) when compared with the MEP evoked using the LM coil orientation (LM-MEP). To study maturation of central white matter we used repeated measures ANOVA to compare the amplitudes of the PA-MEP and the LM-MEP. The outcome variable was the difference between the MEP amplitudes of two coil orientations (LM-MEP minus PA-MEP (LM-PA)), the between-subjects variable was age group (<12 vs 12–14 years vs adults) and the within-subject variable was side of stimulation (right cortex vs left cortex).

LM-MEP amplitudes were larger than PA-MEP amplitudes in younger children (<12 years of age), but were smaller than PA-MEP amplitudes in older children (12–14 years of age) and adults (Figure 22.4). There were no side-to-side differences in any of the age groups. This age-related change in the relative amplitude of the LM-MEP and PA-MEP, which reflects maturation of intracortical neurons, may explain certain aspects of the maturational trajectory of the SRCs. Since no asymmetry was found in the LM-PA amplitude differences in 12–14-year-old children, other factors must also play a role in the maturation process.

Cortical maps

Cortical maps give an estimate of the somatotopic representations of muscles within the motor cortex. By stimulating at a number of

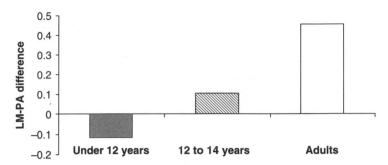

Fig. 22.4 Difference in latero-medial (LM)–posterior-anterior (PA) motor-evoked potential (MEP) amplitudes when stimulating the dominant cortex with the coil in the LM or PA orientations. Age groups: children (<12 years of age), adolescents (12–14 years), and adults. Note that the LM-PA difference is negative (TMS-evoked MEP amplitude is larger using the LM coil orientation than the PA coil orientation) in children, but is positive (TMS-evoked MEP using LA coil orientation smaller than PA coil orientation) in adolescents and adults. There was a significant difference between the groups.

different scalp positions and measuring the amplitude of the MEP at each site, it is possible to assess the location of the optimal position for stimulation and the center of gravity, which defines the mean position of the map (Chen 2000; Abbruzzese and Trompetto, 2002). Only one study has examined cortical maps in typically developing children as a comparison for children with cerebral palsy (Maegaki *et al.* 1999) (Figure 22.5). Cortical representation sites for the tibialis anterior, biceps brachialis, and abductor pollicis brevis muscle were identified between 1 and 4 cm, 4 and 6 cm, and 5 and 8 cm lateral to the cranial vertex, respectively. Although the authors did not report the optimal stimulation site or the center of gravity, these data suggest that it is feasible to study the somatotopic representations of muscles within the motor cortex in children. It is not possible to determine from this report whether motor maps change during development.

Ipsilateral MEP

Stimulation of the motor cortex with TMS may evoke an MEP in the homologous target muscle ipsilateral to the stimulation (iMEP). This iMEP indicates the presence of an ipsilateral corticofugal motor projection which may be the means by which the motor cortex controls ipsilateral movements in healthy subjects (Ziemann *et al.* 1999). Investigators also speculate that these ipsilateral projections may mediate the ability of the less-affected hemisphere to control movement in the affected limb after injury to the developing brain (see below) (Carr *et al.* 1993; Staudt *et al.* 2002). Different types of TMS-evoked ipsilateral projections can be identified by comparing the characteristics of the iMEP with those of the contralateral MEP. Further information can also be acquired when TMS is combined with cross-correlation analysis of EMG activity or, in distal muscles, the long-latency reflex (LLR) (Mayston *et al.* 1997; Eyre *et al.* 2001; Staudt *et al.* 2002).

Cross-correlation analysis examines functional coupling between different muscles by demonstrating synchronization of motor neurons with a short-duration central peak in the cross-correlogram (Harrison *et al.* 1991). The LLR is a cortically mediated muscle response that can be elicited either by electrical stimulation or by muscle stretch. Although the LLR is found only in the muscle that is stimulated in neurologically intact individuals (Deuschl and Lucking 1990; Fellows *et al.* 1996), it is present in both the muscle stimulated and in the homologous muscle of the opposite hand in some subjects with mirror movements. These individuals also show a central peak in the cross-correlogram of the two homologous muscles (Mayston *et al.* 1997).

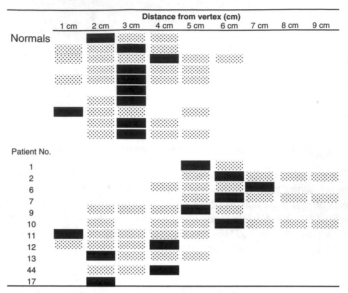

Fig. 22.5 Mapping of the TMS-evoked motor-evoked potential (MEP) 'hot spot' for the tibialis anterior in typically developing children (upper) and children with cerebral palsy (lower). Cortical maps for each subject are indicated by bars arranged in columns which represent 1 cm intervals between the vertex and the auricle along the inter-auricular line. Each bar represents four MEPs evoked by TMS in the resting tibialis anterior muscle at a specific site between the vertex and the auricle. Solid bars indicate the 'hot spot', i.e. those scalp sites at which TMS stimulation evoked MEPs with the highest amplitude; shaded bars indicate those scalp sites at which TMS stimulation evoked MEPs with an amplitude >50% of maximum. Patients 1, 2, 6, 7, and 9 were all born prematurely and had spastic diplegia or quadriplegia; the remaining patients had a dyskinetic form of cerebral palsy. Note that the 'hot spots' in the prematurely born children show a marked lateral shift compared with the typically developing children. This is not seen in those children with dyskinetic types of cerebral palsy. Reproduced with permission from Maegaki *et al.* (1997).

In a subgroup of persons who demonstrate 'fast-conducting' TMS-evoked iMEP (identical threshold, onset latency, and amplitude as the contralateral MEP) there is also a central peak in the cross-correlograms or an anomalous LLR in the homologous muscle of the opposite hand. This type of ipsilateral projection is found in individuals with congenital mirror movements and in certain children with cerebral palsy (Carr *et al.* 1993; Mayston *et al.* 1997) The central peak on the cross-correlogram and the anomalous LLR provide supporting evidence that these 'fast-conducting' iMEPs originate from shared presynaptic fibers in the ipsilateral primary motor cortex which branch and innervate the motor neuron pools of both ipsilateral and contralateral homologous muscles. Based on experimental models in animals, investigators speculate that these ipsilateral projections cross the midline with the corticospinal tract, and then re-cross to the ipsilateral side. The point at which they re-cross in human subjects is unclear, but the most likely place is within the spinal cord (Staudt *et al.* 2002).

A second subgroup of individuals with a 'fast-conducting' iMEP demonstrates no central peak on the cross-correlogram. These fast-conducting ipsilateral projections are present in healthy neonates soon after birth (Eyre *et al.* 2001) (Figure 22.6). Animal studies have shown that the cells of origin of these fast-conducting projections are distinct from, and are more widely distributed than, the cells of origin of the contralateral corticospinal tract projections. In human subjects, the absence of a central peak using EMG cross-correlation supports the separate origin of cell bodies.

Older, typically developing, children demonstrate TMS-evoked iMEPs that have a higher threshold, longer latency, and smaller amplitude compared with the contralateral MEPs. These 'slow-conducting' iMEPs are present from before 18 months of age and continue to be accessible in ~60% of healthy children up to 10 years of age. They are also present in a smaller proportion of healthy adults, but only under certain conditions (Müller *et al.* 1997; Ziemann *et al.* 1999). 'Slow-conducting' iMEPs have also been identified in certain children with cerebral palsy and adults following a stroke. Clinical studies using TMS have identified cortical map locations for these projections

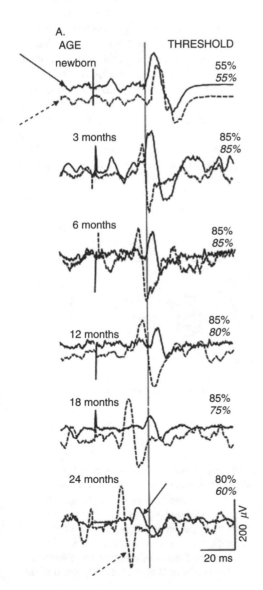

Fig. 22.6 Serial ipsilateral and contralateral TMS-evoked motor-evoked potential (MEP) responses following stimulation of the left hemisphere recorded in the biceps in the same healthy subject from birth to 24 months of age. The continuous line traces (continuous arrows) are recorded from ipsilateral (left) biceps, and dashed line traces (dashed arrows) are from contralateral (right) biceps. The stimulus artifact marks the point at which the TMS stimulus occurred. The vertical line indicates the onset of the ipsilateral response when the subject was newborn. Thresholds for the responses are recorded on the right above the traces: those in italics are for contralateral responses. Note that the ipsilateral TMS-evoked MEP response at 24 months of age shows a delayed latency and smaller amplitude compared with those found in the newborn period. Modified from Eyre *et al.* (2001). Reprinted with permission from Lippincott–Williams & Wilkins.

different from those of the contralateral corticospinal tract projections in the same hemisphere (Chen *et al.* 2002), thus ruling out the possibility of shared presynaptic input. This is reflected in the absence of a central peak on the cross-correlogram (Carr *et al.* 1993).

Cortical inhibition

Inhibitory functions of the motor cortex have been examined using different TMS paradigms. Two methods have been used in children: the paired-pulse paradigm and the silent period. The paired-pulse paradigm examines the effect of a subthreshold conditioning pulse on the amplitude of the MEP evoked by a suprathreshold test stimulus in the target muscle at rest. A TMS-evoked silent period may be present in the active target muscle and homologous ipsilateral muscle when the TMS stimulus is delivered to the motor cortex contralateral to the target muscle.

Paired-pulse studies

Intracortical excitability and inhibition can be assessed delivering two stimuli in a condition-test paradigm (Kujirai *et al.* 1993). Here the investigator delivers two pulses with varying time intervals between the pulses, ranging from 1 to 70 ms. Several paradigms using different inter-stimulus intervals (ISIs) have been described in adult subjects, but only one has been used in children. This is the short-interstimulus intra-cortical inhibition (SICI). The physiology of this technique has been discussed in detail elsewhere in this volume.

A recent study examined maturation of ICI in a large number of subjects ranging in age from 6 to 34 years of age (Mall *et al.* 2004) (Figure 22.7). The paradigm was limited to testing inhibition using a 2 ms ISI since this ISI reliably approximates gamma-aminobutyric acid-mediated intracortical inhibition (Ziemann *et al.* 2001). The authors of this study point out a possible confound when using this paradigm in children. In the original SICI paired-paradigm the test stimulus is determined by the level of the motor threshold. However, motor threshold is higher in children than in adults and it changes with age until maturation is complete.

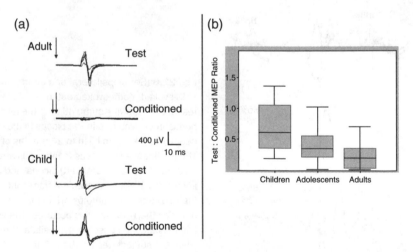

Fig. 22.7 Maturation of intracortical inhibition (ICI). (a) Five superimposed TMS-evoked motor-evoked potentials (MEPs) of one adult subject (top two tracings) and one child subject (bottom two tracings). For each subject, the upper tracings show the unconditioned MEP evoked following a single test stimulus (one down-arrow); the lower tracings show the conditioned MEP evoked following paired-pulse stimulus (two down-arrows). The interstimulus interval was set to 2 ms. The attenuated response in the conditioned MEP seen in the adult subject is not present in the child. (b) Comparison of the ICI ratio (unconditioned: conditioned MEP amplitudes) in children, adolescents, and adults. There was a significant difference between the ICI ratios of children and adults ($P < 0.001$). Modified from Mall *et al.* (2004); reproduced by permission.

Therefore, differences in the ICI between adults and children could be influenced by the developmental changes in motor threshold. The authors address this confound by first setting the test stimulus to 20% above resting motor threshold and then, in a repeat experiment, varying the level of the test stimulus so that the amplitude of the MEP evoked by the test stimulus was at the same level for every child (unfortunately, they do not mention the range of stimulus intensities that were used in this second experiment).

This study demonstrated that ICI is nearly four times greater in adults compared with children aged <10 years. This difference is the same whether the intensity of test stimulus is fixed according to the level of the motor threshold or is varied according to the amplitude of the MEP evoked by test stimulus. Since previous studies have shown that decreased levels of SICI are associated with increased practice-dependent plasticity (Ziemann et al. 1999), these results are particularly interesting in that they provide neurophysiological evidence for the greater capacity for plastic changes following CNS injury known to be present in children (Kennard 1936).

The two-coil conditioned-test paradigm has also been used to study intercortical inhibition (Ferbert et al. 1992). Here, a conditioning stimulus is delivered to the motor cortex of one hemisphere and the test stimulus is delivered to the homologous motor cortex of the other hemisphere. Although there is a clear pattern of inhibition and facilitation in neurologically intact adult subjects when ISIs are set between 5 and 15 ms (with maximal inhibition at 7 ms), no clear developmental pattern is present in typically developing children (Mayston et al. 1999).

Silent periods

The ontogeny of the ipsilateral silent period (SP) is thought to reflect maturation of cortical inhibitory neurons and myelination of the midbody of the corpus callosum. The ipsilateral SP is absent in preschool children and can be first consistently evoked in 6–7-year-old children (Heinen et al. 1998b). At this point, latency is delayed and duration shortened compared with the mature ipsilateral SP. Over the ensuing years latency decreases and duration increases; both

are close to maturity by early adolescence (Garvey et al. 2003) (Figure 22.8).

Unlike the ipsilateral SP, the contralateral SP is present in preschool children (Heinen et al. 1998b) This TMS-evoked parameter shows no age-related changes in children 6–14 years of age suggesting that maturation occurs in very young children (Garvey et al. 2003). Future studies may be helpful in determining the maturational trajectory of this TMS-evoked inhibitory parameter.

Age-related changes in asymmetry

Mechanisms underlying changes in cortical asymmetry during development are unclear. Hand preference becomes apparent as early as 2 years of age and is firmly established by 4 years of age (Illingsworth 1983). Abnormalities of cortical function are associated with anomalous patterns of hand preference as seen in many developmental disorders, including those that do not appear to have clear structural abnormalities of either hemisphere (Hauck and Dewey 2001; Niederhofer 2005). These data suggest that genetic and early environmental factors (e.g. in utero hormonal levels) may establish dominance (Geschwind and Galaburda 1985a–c; Hammond 2002) while structural and functional asymmetries of the primary motor cortex may act to maintain the stability of this dominance through mechanisms which, at present, remain obscure (Muellbacher et al. 2000; Facchini et al. 2002).

The most extensively studied phenomenon of cortical asymmetry is the MEP threshold. In young adults, a mean difference of 5% (ranging between −10% and +15%) has been consistently found between the dominant (lower) and nondominant MEP threshold (Triggs et al. 1994, 1997, 1999). Asymmetry has also been examined in other TMS-evoked parameters: cortical mapping (larger maps on the dominant hemisphere) (Cicinelli et al. 1997); the cSP (shorter duration when the dominant motor cortex is stimulated) (Priori et al. 1999); the SICI/ICF (reports show conflicting data) (Cicinelli et al. 2000; Civardi et al. 2000; Maeda et al. 2002); and the two-coil paired-pulse paradigm testing interhemispheric inhibition (greater inhibition when the conditioning stimulus is delivered to the dominant hemisphere) (Netz et al. 1995).

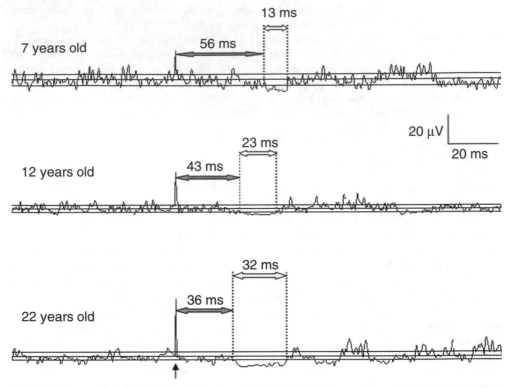

Fig. 22.8 Maturation of the ipsilateral silent period (SP). Three electromyogram (EMG) tracings of the ipsilateral SP from a 7-year-old boy (*upper*), a 12-year-old girl (*middle*), and a 22-year-old man (*lower*) are shown. Each tracing is an average of 10 single-trial rectified sweeps. The vertical arrow represents the stimulus which occurred 100 ms after the start of each sweep. There are three horizontal solid lines in each tracing: the middle line represents the mean EMG amplitude (µV) during the prestimulus period; the upper and lower lines represent the upper and lower 95% variation limits of the mean prestimulus EMG amplitude. These limits were calculated using statistical process control. The vertical dotted lines represent the onset latency and end of the ipsilateral SP. For each tracing, the shaded arrow heads and accompanying number indicates the onset latency, and the open arrow heads and accompanying number indicate the ipsilateral SP duration. Note that latency is longest and duration shortest in the 7-year-old child, latency and duration are intermediate in the 12-year-old child, and latency is shortest and duration is longest in the adult subject.

Few TMS studies have examined cortical asymmetry in children. Side-to-side differences in MEP threshold (lower in the dominant motor cortex) and duration of contralateral SP (longer when evoked in the nondominant hemisphere) are similar to those seen in adults. However, the side-to-side difference in MEP threshold is five times larger in 6–8-year-old children than in adults, gradually decreasing as children get older (Garvey *et al.* 2003) (Figure 22.9). This decrease in side-to-side asymmetry continues throughout the adult years; no significant asymmetry of MEP threshold is found in elderly subjects (Matsunaga *et al.* 1998).

Maturation of the ipsilateral SP also shows side-to-side differences. In 6-year-old children, stimulation of the dominant hemisphere evokes an ipsilateral SP more consistently than the stimulation of the nondominant hemisphere. If both are present, duration of the dominant ipsilateral SP is shorter than the nondominant ipsilateral SP which is coherent with asymmetry of the

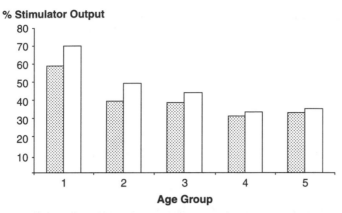

Fig. 22.9 Asymmetry of motor-evoked potential (MEP) thresholds. The active MEP thresholds (AMTs) in a group of right-handed subjects obtained with background muscle activation of the first dorsal interosseus. Shaded bars represent the AMT when the left (dominant) cortex is stimulated; white bars represent the AMT when the right cortex is stimulated. Group 1, 6–7-year-old children; Group 2, 8–9-year-old children; Group 3, 10–11-year-old children; Group 4, 12–13-year-old children; Group 5, adults. Note that AMT requires higher stimulation intensities when the right (nondominant) cortex is stimulated. Asymmetry in the AMT is greater in the youngest children (Group 1) and smallest in the adults (Group 5).

contralateral SP. These asymmetries are not present in older children and in adults (Garvey et al. 2003).

Correlation between TMS-evoked parameters and neurodevelopment

Areas of investigation that have recently received more attention are the association between the TMS-evoked parameters and motor function in neurologically intact individuals, and TMS studies of the motor system in children with developmental disabilities (such as attention-deficit hyperactivity disorder). These latter studies are undertaken with the underlying hypothesis that the status of motor function acts as a 'biomarker' for neighboring systems and circuits which are responsible for the behavioral anomalies in these children (Denckla 2005).

To date, those studies examining maturation of normal motor function have concentrated on showing similarities between the maturational profiles of behavioral measures (e.g. finger tapping) and TMS-evoked parameters (Müller and Hömberg 1992; Heinen et al. 1998a; Fietzek et al. 2000). Maturational profiles of acoustic reaction times, finger-tapping movements, and ballistic movements are similar to the developmental

trajectory of resting CMCT but not of active CMCT which matures at an earlier age. This implies that neuromotor development progresses at the same rate as intracortical maturation (synaptic proliferation and pruning, central and callosal myelination, etc.) rather than at the rate of maturation of the corticospinal tracts.

However, since age is the strongest predictor of change in behavioral measures of motor function and in TMS-evoked parameters, the finding of similarities in the developmental trajectories of the two latter parameters may not indicate a real relationship between them (Dawson-Saunders and Trapp 1994). One way to avoid the confounding effect of age is to include it in a multiple regression model analyzing relationships between motor behavior and TMS-evoked parameters (Kleinbaum et al. 1998; Garvey et al. 2003). Such analyses indicate that finger-tapping speed is only related to ipsilateral SP latency, but not to motor threshold, contralateral SP, or ipsilateral SP duration. The correlation between finger speed and ipsilateral SP latency suggests that development of certain fine motor skills may be related to maturation of callosal myelination. These data concur with a prior study which has shown that mirror movements in typically developing children <10 years of age result from physiological immaturity of intra-cortical inhibition (Mayston *et al.*).

TMS-evoked parameters in perinatal insults

It is well known that functional recovery from a perinatal insult may be quite good. Unfortunately, this is by no means universal and, at times, the outcome can be devastating (Johnston 2003). There are many factors which may determine this variability, including the type of process that causes the original insult, and the extent and timing of the lesion (Staudt *et al.* 2004). Cerebral palsy is a group of disorders arising from a static injury to the developing brain perinatally. There are many different etiologies, including cerebral dysplasia, middle cerebral artery infarct, infection, and trauma. Although children with cerebral palsy may manifest different functional deficits, it is possible to study carefully selected groups of children whose cerebral palsy arises from the same or similar etiologies. These studies may be especially helpful in determining the association between functional outcome in these children and the underlying pattern of brain reorganization.

Enhanced plasticity of the developing brain

To date, investigators have studied children with hemiplegic cerebral palsy more frequently than any other type. The hypothesis underlying these studies is that better functional outcome may occur when the contralesional hemisphere contributes greater control of the paretic limb. Although prominent mirror movements consistently indicate the presence of this enhanced participation of the contralesional hemisphere, neurophysiological tools have since shown that it may also exist in the absence of mirror movements (Carr *et al.* 1993; Eyre *et al.* 2001; Staudt *et al.* 2002).

The TMS-evoked ipsilateral MEP points to the presence of an ipsilateral corticofugal projection arising from areas of the motor cortex in the contralesional hemisphere (see section above regarding these iMEP). Three different types of ipsilateral projections have been identified in children with cerebral palsy (Carr *et al.* 1993; Eyre *et al.* 2001). Fast-conducting ipsilateral projections (the ipsilateral MEP is identical

to the contralateral MEP) are present in two groups of patients with cerebral palsy distinguished from each other by the presence or absence of mirror movements. Those with mirror movements show a central peak on the cross-correlogram (or an anomalous LLR), whereas those without mirror movements show no central peak (Carr *et al.* 1993; Eyre *et al.* 2001). Slow-conducting ipsilateral projections (where the ipsilateral MEP has delayed-onset latency, smaller amplitude, and higher threshold compared with the contralateral MEP) may be found in some individuals with hemiplegic cerebral palsy who do not have prominent mirror movements (Carr *et al.* 1993).

Staudt *et al.* (2002) published an excellent study of enhanced participation of the contralesional hemisphere using fMRI and TMS in a group of subjects with congenital hemiplegic cerebral palsy. The strength of this study was that the authors carefully controlled for both the site and timing of the lesion, so were able to study the effect of lesion size. Those subjects who had small lesions had no mirror movements of the paretic hand during unimanual movements of the unaffected hand. TMS demonstrated residual activity in the affected motor cortex and no evidence of ipsilateral projections from the contralesional cortex.

Some subjects with large lesions had intense mirror movements of the paretic hand during unimanual movements of the unaffected hand; others only showed EMG activity with no visible movements of the paretic hand. Despite this, all showed evidence of fast-conducting ipsilateral projections from the contralesional motor cortex. Those subjects who had intense mirror movements showed no residual activity of the affected hemisphere suggesting that motor control in the paretic hand was mediated via ipsilateral projections from the contralesional hemisphere. Those who had minimal or no mirror movements showed residual activity in the affected motor cortex suggesting that, although these subjects had ipsilateral projections arising from the contralesional motor cortex, motor control of the paretic hand arose from the affected cortex.

It is possible to surmise that EMG cross-correlation (not performed in this study) would have shown a central peak in the subjects with mirror movements, and that no central peak would have been found in subjects without mirror movements (Farmer *et al.* 1991; Carr *et al.* 1993; Maegaki *et al.* 1997).

Altered cortical maps

TMS has been used to examine the effect of perinatal insults on the motor cortex representation of lower extremity muscles in children with periventricular leukomalacia. This perinatal insult results from ischemia of the central white matter (especially the motor and sensory fibers of the lower extremities; Volpe 1998; Hoon *et al.* 2002) is accompanied by abnormalities of the corpus callosum, and is frequently manifest clinically as the spastic diplegia type of cerebral palsy (a bilateral form of cerebral palsy where the lower extremities are more severely affected than the upper). This study demonstrated lateral displacement of the cortical maps of lower extremity muscles in some children with periventricular leukomalacia (see cortical maps from patients 1, 2, 6, 7, and 9 in the lower half of Figure 22.5). The degree of lateral displacement in these subjects did not correlate with the severity of the white matter lesion; no correlation was performed between these findings and the severity of the lower limb involvement. All the children with the dyskinetic form of cerebral palsy with normal central white matter had normally placed cortical maps for the tibialis anterior.

Transcallosal inhibition

The severe callosal abnormalities in children with spastic diplegia prompted an investigation of transcallosal inhibition using the ipsilateral SP. The study compared the iSP in typically developing children, in children with paraplegia (due to spinal cord injury), and in children with spastic diplegia (Heinen *et al.* 1999). An ipsilateral SP was readily seen in both typically developing children and those with paraplegia, but was not apparent in children with spastic diplegia (Figure 22.10). This suggests that the midbody of the corpus callosum in spastic diplegia

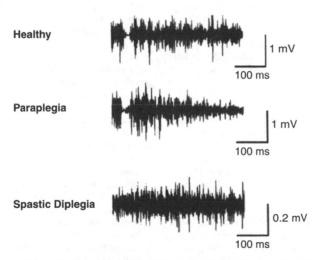

Fig. 22.10 Absence of the ipsilateral silent period (SP) in children with spastic diplegia type of cerebral palsy. Each tracing represents four superimposed electromyogram (EMG) tracings from the first dorsal interosseus muscle ipsilateral to site of TMS during background activation. The TMS stimulus occurred at the start of the EMG recording. *Upper* and *middle* tracings: ipsilateral SP is present in a healthy adolescent and a patient with hereditary spastic paraplegia, respectively. *Lower* tracing: ipsilateral SP is not present in a patient with spastic diplegia type of cerebral palsy. The contralateral SP was present in all three subjects. Modified from Heinen *et al.* (1999). Reproduced with permission of John Wiley & Sons Inc.

is abnormal. It would be interesting to examine the clinical correlates of an abnormal ipsilateral SP in spastic diplegia.

Summary

TMS is an important tool to examine cortical function in typically developing children and in those with perinatal insults. It has its limitations and is especially useful when used in combination with other neurophysiological modalities (e.g. EMG, fMRI). Future directions should focus on correlating TMS-evoked parameters with behavioral measures, including motor performance, visual–motor integration, etc., in typically developing children, and further elucidation of the neural substrates of the motor abnormalities in children with perinatal insults and developmental disabilities. An important goal from these latter studies is the development of biomarkers of treatment-related improvement which can serve as quantitative measures of the efficacy of novel interventions.

References

Abbruzzese G, Trompetto C (2002). Clinical and research methods for evaluating cortical excitability. *Journal of Clinical Neurophysiology 19*, 307–321.

Barnea-Goraly N, Menon V, Eckert M, et al. (2005). White matter development during childhood and adolescence: a cross-sectional diffusion tensor imaging study. *Cerebral Cortex 15*, 1848–1854.

Ben Bashat D, Ben Sira L, Graif M, et al. (2005). Normal white matter development from infancy to adulthood: comparing diffusion tensor and high b value diffusion weighted MR images. *Journal of Magnetic Resonance Imaging 21*, 503–511.

Brown JV, Schumacher U, Rohlmann A, Ettlinger G, Schmidt RC, Skreczek W (1989). Aimed movements to visual targets in hemiplegic and normal children: is the "good" hand of children with infantile hemiplegia also normal? *Neuropsychologia 27*, 283–302.

Caramia MD, Desiato MT, Cicinelli P, Iani C, Rossini PM (1993). Latency jump of 'relaxed' versus 'contracted' motor evoked potentials as a marker of cortico-spinal maturation. *Electroencephalography and Clinical Neurophysiology 89*, 61–66.

Carr LJ, Harrison LM, Evans AL, Stephens JA (1993). Patterns of central motor reorganization in hemiplegic cerebral palsy. *Brain 116*(Pt 5), 1223–1247.

Chen R (2000). Studies of human motor physiology with transcranial magnetic stimulation. *Muscle and Nerve 23* (Suppl 9), S26–S32.

Chen R, Cohen LG, Hallett M (2002). Nervous system reorganization following injury. *Neuroscience 111*, 761–773.

Cicinelli P, Traversa R, Bassi A, Scivoletto G, Rossini PM (1997). Interhemispheric differences of hand muscle representation in human motor cortex. *Muscle and Nerve 20*, 535–542.

Cicinelli P, Traversa R, Oliveri M, et al. (2000). Intracortical excitatory and inhibitory phenomena to paired transcranial magnetic stimulation in healthy human subjects: differences between the right and left hemisphere. *Neuroscience Letters 288*, 171–174.

Civardi C, Cavalli A, Naldi P, Varrasi C, Cantello R (2000). Hemispheric asymmetries of cortico-cortical connections in human hand motor areas. *Clinical Neurophysiology 111*, 624–629.

Croce RV, Horvat M, McCarthy E (2001). Reliability and concurrent validity of the movement assessment battery for children. *Perception and Motor Skills 93*, 275–80.

Dawson-Saunders B, Trapp RG (1994). *Basic and clinical biostatistics*. Norwalk, CT: Appleton & Lange.

Denckla MB (1973). Development of speed in repetitive and successive finger-movements in normal children. *Developmental Medicine and Child Neurology 15*, 635–645.

Denckla MB (1974). Development of motor co-ordination in normal children. *Developmental Medicine and Child Neurology 16*, 729–741.

Denckla MB (2005). Why assess motor functions 'early and often?' *Mental Retardation and Developmental Disability Research Reviews 11*, 3.

Deuschl G, Lucking CH (1990). Physiology and clinical applications of hand muscle reflexes. *Electroencephalography and Clinical Neurophysiology, Supplement 41*, 84–101.

Einspieler C, Prechtl HF (2005). Prechtl's assessment of general movements: a diagnostic tool for the functional assessment of the young nervous system. *Mental Retardation and Developmental Disability Research Reviews 11*, 61–67.

Eyre JA, Miller S, Ramesh V (1991). Constancy of central conduction delays during development in man: investigation of motor and somatosensory pathways. *Journal of Physiology 434*, 441–452.

Eyre JA, Taylor JP, Villagra F, Smith M, Miller S (2001). Evidence of activity-dependent withdrawal of corticospinal projections during human development. *Neurology 57*, 1543–1554.

Facchini S, Muellbacher W, Battaglia F, Boroojerdi B, Hallett M (2002). Focal enhancement of motor cortex excitability during motor imagery: a transcranial magnetic stimulation study. *Acta Neurologica Scandinavica 105*, 146–151.

Farmer SF, Harrison LM, Ingram DA, Stephens JA (1991). Plasticity of central motor pathways in children with hemiplegic cerebral palsy. *Neurology 41*, 1505–1510.

Fellows SJ, Topper R, Schwarz M, Thilmann AF, Noth J (1996). Stretch reflexes of the proximal arm in a patient with mirror movements: absence of bilateral long-latency components. *Electroencephalography and Clinical Neurophysiology 101*, 79–83.

Ferbert A, Priori A, Rothwell JC, Day BL, Colebatch JG, Marsden CD (1992) Interhemispheric inhibition of the human motor cortex. *Journal of Physiology 453*, 525–546.

Fietzek UM, Heinen F, Berweck S, *et al.* (2000). Development of the corticospinal system and hand motor function: central conduction times and motor performance tests. *Developmental Medicine and Child Neurology 42*, 220–227.

Frankenburg WK, Dodds J, Archer P, Shapiro H, Bresnick B (1992). The Denver II: a major revision and restandardization of the Denver Developmental Screening Test. *Pediatrics 89*, 91–97.

Garvey MA, Ziemann U, JJ B, Denckla MB, Barker CA, Wassermann EM (2003). Cortical correlates of neuromotor development in healthy children. *Clinical Neurophysiology 114*,1662–1670.

Geschwind N, Galaburda AM (1985a). Cerebral lateralization. Biological mechanisms, associations, and pathology: I. A hypothesis and a program for research. *Archives of Neurology 42*, 428–459.

Geschwind N, Galaburda AM (1985b). Cerebral lateralization. Biological mechanisms, associations, and pathology: II. A hypothesis and a program for research. *Archives of Neurology 42*, 521–552.

Geschwind N, Galaburda AM (1985c). Cerebral lateralization. Biological mechanisms, associations, and pathology: III. A hypothesis and a program for research. *Archives of Neurology 42*, 634–654.

Giedd JN, Blumenthal J, Jeffries NO, *et al.* (1999) Brain development during childhood and adolescence: a longitudinal MRI study. *Nature Neuroscience 2*, 861–863.

Gordon AM, Charles J, Duff SV (1999). Fingertip forces during object manipulation in children with hemiplegic cerebral palsy. II: bilateral coordination. *Developmental Medicine and Child Neurology 41*, 176–185.

Hammond G (2002). Correlates of human handedness in primary motor cortex: a review and hypothesis. *Neuroscience and Biobehavioral Reviews 26*, 285–292.

Harrison LM, Ironton R, Stephens JA (1991). Cross-correlation analysis of multi-unit EMG recordings in man. *Journal of Neuroscience Methods 40*, 171–179.

Hauck JA, Dewey D (2001). Hand preference and motor functioning in children with autism. *Journal of Autism and Developmental Disorders 31*, 265–77.

Heinen F, Fietzek UM, Berweck S, Hufschmidt A, Deuschl G, Korinthenberg R (1998a). Fast corticospinal system and motor performance in children: conduction proceeds skill. *Pediatric Neurology 19*, 217–221.

Heinen F, Glocker FX, Fietzek U, Meyer B-U, Lücking CH, Korinthenberg R (1998b). Absence of transcallosal inhibition following focal magnetic stimulation in preschool children. *Annals of Neurology 43*, 608–612.

Heinen F, Kirschner J, Fietzek U, Glocker FX, Mall V, Korinthenberg R (1999). Absence of transcallosal inhibition in adolescents with diplegic cerebral palsy. *Muscle and Nerve 22*, 255–257.

Hodics T, Cohen LG (2005). Functional neuroimaging in motor recovery after stroke. *Topics in Stroke Rehabilitation 12*, 15–21.

Hoon AH Jr, Lawrie WT Jr, Melhem ER, Reinhardt EM, Van Zijl PC, Solaiyappan M, *et al.* (2002) Diffusion tensor imaging of periventricular leukomalacia shows affected sensory cortex white matter pathways. *Neurology 59*, 752–756.

Huttenlocher PR, Dabholkar AS (1997). Regional differences in synaptogenesis in human cerebral cortex. *Journal of Comparative Neurology 387*, 167–178.

Illingsworth RS (1983). *The development of the infant and young child: normal and abnormal.* Edinburgh: Churchill Livingstone.

Johnston MV (2003). Injury and plasticity in the developing brain. *Experimental Neurology 184*(Suppl 1), S37–41.

Johnston MV (2004). Clinical disorders of brain plasticity. *Brain Development 26*, 73–80.

Kennard MA (1936). Age and other factors in motor recovery from precentral lesions in monkeys. *American Journal of Physiology 115*, 137–146.

Kleinbaum DG, Kupper LL, Muller KE, Nizam A (1998). *Applied regression analysis and other multivariate methods.* Pacific Grove: Brooks Cole.

Koh TH, Eyre JA (1988). Maturation of corticospinal tracts assessed by electromagnetic stimulation of the motor cortex. *Archives of Disease in Childhood 63*, 1347–1352.

Kujirai T, Caramia MD, Rothwell JC, Day BL, Thompson PD, Ferbert A, *et al.* (1993) Corticocortical inhibition in human motor cortex. *Journal of Physiology 471*, 501–519.

Largo RH, Caflisch JA, Hug F, *et al.* (2001a) Neuromotor development from 5 to 18 years. Part 1: timed performance. *Developmental Medicine and Child Neurology 43*, 436–443.

Largo RH, Caflisch JA, Hug F, Muggli K, Molnar AA, Molinari L (2001b). Neuromotor development from 5 to 18 years. Part 2: Associated movements. *Developmental Medicine and Child Neurology 43*, 444–453.

Largo RH, Fischer JE, Rousson V (2003). Neuromotor development from kindergarten age to adolescence: developmental course and variability. *Swiss Medicine Weekly 133*, 193–199.

Lazarus JC (1992). Associated movement in hemiplegia: the effects of force exerted, limb usage and inhibitory training. *Archives of Physical Medicine and Rehabilitation 73*, 1044–1049.

Lipkin PH (2005). Towards creation of a unified view of the neurodevelopment of the infant. *Mental Retardation and Developmental Disability Research Reviews 11*, 103–106.

Maeda F, Gangitano M, Thall M, Pascual-Leone A (2002). Inter- and intra-individual variability of paired-pulse curves with transcranial magnetic stimulation (TMS). *Clinical Neurophysiology 113*, 376–382.

Maegaki Y, Maeoka Y, Ishii S, *et al.* (1997) Mechanisms of central motor reorganization in pediatric hemiplegic patients. *Neuropediatrics 28*, 168–174.

Maegaki Y, Maeoka Y, Ishii S, *et al.* (1999) Central motor reorganization in cerebral palsy patients with bilateral cerebral lesions. *Pediatric Research 45*, 559–567.

Majnemer A, Snider L (2005). A comparison of developmental assessments of the newborn and young infant. *Mental Retardation and Developmental Disability Research Reviews 11*, 68–73.

Mall V, Berweck S, Fietzek UM, *et al.* (2004) Low level of intracortical inhibition in children shown by transcranial magnetic stimulation. *Neuropediatrics 35*, 120–125.

Mall V, Linder M, Herpers M, *et al.* (2005) Recruitment of the sensorimotor cortex – a developmental FMRI study. *Neuropediatrics 36*, 373–379.

Matsunaga K, Uozumi T, Tsuji S, Murai Y (1998). Age-dependent changes in physiological threshold asymmetries for the motor evoked potential and silent period following transcranial magnetic stimulation. *Electroencephalography and Clinical Neurophysiology 109*, 502–507.

Mayston MJ, Harrison LM, Quinton R, Stephens JA, Krams M, Bouloux PM (1997). Mirror movements in X-linked Kallmann's syndrome. I. A neurophysiological study. *Brain 120*(Pt 7), 1199–1216.

Mayston MJ, Harrison LM, Stephens JA (1999). A neurophysiological study of mirror movements in adults and children. *Annals of Neurology 45*, 583–594.

Moll GH, Heinrich H, Wischer S, Tergau F, Paulus W, Rothenberger A (1999). Motor system excitability in healthy children: developmental aspects from transcranial magnetic stimulation. *Electroencephalography and Clinical Neurophysiology Supplement 51*, 243–249.

Muellbacher W, Facchini S, Boroojerdi B, Hallett M (2000). Changes in motor cortex excitability during ipsilateral hand muscle activation in humans. *Clinical Neurophysiology 111*, 344–349.

Müller K, Hömberg V (1992). Development of speed of repetitive movements in children is determined by structural changes in corticospinal efferents. *Neuroscience Letters 144*, 57–60.

Müller K, Ebner B, Hömberg V (1994). Maturation of fastest afferent and efferent central and peripheral pathways: no evidence for a constancy of central conduction delays. *Neuroscience Letters 166*, 9–12.

Müller K, Kass-Iliyya F, Reitz M (1997). Ontogeny of ipsilateral corticospinal projections: a developmental study with transcranial magnetic stimulation. *Annals of Neurology 42*, 705–711.

Muller RA, Rothermel RD, Behen ME, Muzik O, Mangner TJ, Chugani HT (1998). Developmental changes of cortical and cerebellar motor control: a clinical positron emission tomography study with children and adults. *Journal of Child Neurology 13*, 550–556.

Neil JJ, Shiran SI, McKinstry RC, *et al.* (1998). Normal brain in human newborns: apparent diffusion coefficient and diffusion anisotropy measured by using diffusion tensor MR imaging. *Radiology 209*, 57–66.

Netz J, Ziemann U, Hömberg V (1995). Hemispheric asymmetry of transcallosal inhibition in man. *Experimental Brain Research 104*, 527–533.

Nezu A, Kimura S, Uehara S, Kobayashi T, Tanaka M, Saito K (1997). Magnetic stimulation of motor cortex in children: maturity of corticospinal pathway and problem of clinical application. *Brain Development 19*, 176–180.

Niederhofer H (2005). Hand preference in attention deficit hyperactivity disorder. *Perceptual and Motor Skills 101*, 808–810.

Paus T, Collins DL, Evans AC, Leonard G, Pike B, Zijdenbos A (2001). Maturation of white matter in the human brain: a review of magnetic resonance studies. *Brain Research Bulletin 54*, 255–266.

Plubrukarn R, Theeramanoparp S (2003). Human figure drawing test: validity in assessing intelligence in children aged 3–10 years. *Journal of the Medical Association of Thailand 86*(Suppl 3), S610–617.

Priori A, Oliviero A, Donati E, Callea L, Bertolasi L, Rothwell JC (1999). Human handedness and asymmetry of the motor cortical silent period. *Experimental Brain Research 128*, 390–396.

Rossini PM, Barker AT, Berardelli A, *et al.* (1994). Non-invasive electrical and magnetic stimulation of the brain, spinal cord and roots: basic principles and procedures for routine clinical application. Report of an IFCN committee. *Electroencephalography and Clinical Neurophysiology 91*, 79–92.

Staudt M, Grodd W, Gerloff C, Erb M, Stitz J, Krageloh-Mann I (2002). Two types of ipsilateral reorganization in congenital hemiparesis: a TMS and fMRI study. *Brain 125*, 2222–2237.

Staudt M, Gerloff C, Grodd W, Holthausen H, Niemann G, Krageloh-Mann I (2004). Reorganization in congenital hemiparesis acquired at different gestational ages. *Annals of Neurology 56*, 854–63.

Triggs WJ, Calvanio R, Macdonell RA, Cros D, Chiappa KH (1994). Physiological motor asymmetry in human handedness: evidence from transcranial magnetic stimulation. *Brain Research 636*, 270–276.

Triggs WJ, Calvanio R, Levine M (1997). Transcranial magnetic stimulation reveals a hemispheric asymmetry correlate of intermanual differences in motor performance. *Neuropsychologia 35*, 1355–1363.

Triggs WJ, Subramanium B, Rossi F (1999). Hand preference and transcranial magnetic stimulation asymmetry of cortical motor representation. *Brain Research 835*, 324–329.

Volpe JJ (1998). Brain injury in the premature infant: overview of clinical aspects, neuropathology, and pathogenesis. *Seminars in Pediatric Neurology 5*, 135–151.

Volpe JJ (2001). Human brain development. In: *Neurology of the newborn*, pp. 1–102. Philadelphia: WB Saunders.

Werhahn KJ, Fong JK, Meyer BU, *et al.* (1994). The effect of magnetic coil orientation on the latency of surface EMG and single motor unit responses in the first dorsal interosseous muscle. *Electroencephalography and Clinical Neurophysiology 93*, 138–146.

Wolff PH, Kotwica K, Obregon M (1998). The development of interlimb coordination during bimanual finger tapping. *International Journal of Neuroscience 93*, 7–27.

Yakovlev PI, Lecours A-R (1967). The myelogenetic cycles of regional maturation of the brain. In: Minkowski A (ed.), *Regional development of the brain in early life*, pp. 3–70. Oxford: Blackwell.

Yule W, Lockyer L, Noone A (1967). The reliability and validity of the Goodenough–Harris drawing test. *British Journal of Educational Psychology 37*, 110–111.

Ziemann U, Ishii K, Borgheresi A, *et al.* (1999) Dissociation of the pathways mediating ipsilateral and contralateral motor-evoked potentials in human hand and arm muscles. *Journal of Physiology 518*(Pt 3), 895–906.

Ziemann U, Muellbacher W, Hallett M, Cohen LG (2001). Modulation of practice-dependent plasticity in human motor cortex. *Brain 124*, 1171–1181.

CHAPTER 23

Using the TMS-induced motor-evoked potential to evaluate the neurophysiology of psychiatric disorders

Bertram Möller, Andrea J. Levinson, and Zafiris J. Daskalakis

Introduction

TMS represents an important neurophysiological tool to assess a variety of cortical neurophysiological processes including excitability, inhibition, and plasticity. Assessing such phenomena to investigate neuropsychiatric disorders, therefore, may provide valuable insights into the neurophysiological substrates that may be perturbed in these disorders. In the following sections, we review these studies and discuss how TMS has helped to enhance our understanding of the neurobiology and treatment of a variety of psychiatric disorders including schizophrenia (SCZ), major depressive disorder (MDD), bipolar disorder (BD), obsessive–compulsive disorder (OCD), and Tourette's disorder (TD). The following sections will review such studies, which are also summarized in Tables 23.1 and 23.2.

Schizophrenia

Schizophrenia is one of the major unsolved problems of modern medicine. Nearly 51 million people worldwide suffer from this illness, which is characterized by delusions, hallucinations, disorganized thinking, and life-long disability. Implications for healthcare systems are enormous: patients with SCZ occupy 10% of all hospital beds and, despite treatment efforts, 10% lose their lives to the disorder. Fortunately, tremendous inroads have been made in understanding the neurobiology of this debilitating condition. Weinberger *et al.* (1988) reported that patients with SCZ demonstrate hypofunctioning of their frontal lobes. Others have reported that patients with SCZ have widespread gray matter deficits compared with healthy controls (Zipursky *et al.* 1992). Moreover, a meta-analysis reviewing cognition and SCZ revealed that these patients consistently demonstrate impaired memory, attention, and motor skills (Heinrichs and Zakzanis 1998). As such, SCZ is now widely considered as a neurophysiological disorder. However, these lines of evidence only suggest neurophysiological impairment in SCZ. Using TMS to further our understanding of the disturbed pathophysiological mechanisms in this disorder may be the key to better understand this illness and may help develop novel treatment approaches in the future.

Table 23.1 Review of studies examining cortical excitability in psychotic disorders

Study	Objectives	Medication status	No. of subjects	Findings
Abarbanel *et al.* 1996	MT, MEP amplitude, total and central conduction time in MDD/SCZ vs healthy subjects	Medicated	10 MDD, 10 SCZ 10 healthy subjects	↓ MT in SCZ vs MDD and healthy subjects ↔ MT in MDD vs healthy subjects ↑ MEP amplitude in SCZ vs MDD and healthy subjects ↔ total and central conduction time in SCZ/MDD vs healthy subjects
Puri *et al.* 1996	MT, latency in MEP production and latency of EMG suppression in SCZ vs healthy subjects	Unmedicated	9 SCZ 9 controls	↓ MEP latency in SCZ, ↔ MT ↔ latency of EMG suppression
Davey *et al.* 1997	MT, MEP latency; SP latency; EMG suppression for early and late part of SP in medicated and unmedicated SCZ	Medicated Unmedicated	9 unmedicated SCZ 9 medicated SCZ	↓ suppression in the early part of SP in medicated patients ↔ MT differences ↔ MEP latency differences
Boroojerdi *et al.* 1999	MT, MEP latency iSP latency, iSP and TCT in SCZ vs healthy subjects	Medicated	10 SCZ 10 healthy subjects	↑ TCT and iSP in SCZ ↔ in MT or MEP latency
Hoppner *et al.* 2001	MT, iSP latency, iSP in SCZ vs healthy subjects	Medicated	12 SCZ 12 healthy subjects	↑ iSP in SCZ ↔ iSP latency
Daskalakis *et al.* 2002	RMT, SP, SICI/ICF, IHI in SCZ vs healthy subjects	Medicated Unmedicated	15 unmedicated SCZ 15 medicated SCZ 15 healthy subjects	↓ RMT in unmedicated SCZ vs medicated SCZ and healthy subjects ↓ SICI in unmedicated SCZ vs healthy subjects ↓ SP in unmedicated SCZ vs medicated SCZ and healthy subjects ↓ in unmedicated SCZ vs healthy subjects
Fitzgerald *et al.* 2002	MT, IHI and iSP in SCZ vs healthy subjects	Medicated	25 SCZ, 20 healthy subjects	↑ iSP in SCZ vs healthy subjects ↓ IHI in SCZ vs healthy subjects

Table 23.1 (*cont.*) Review of studies examining cortical excitability in psychotic disorders

Study	Objectives	Medication status	No. of subjects	Findings
				↔ MT differences between SCZ vs healthy subjects no differences in IHI latency and TCT in SCZ vs controls, no difference in EMG amplitude reduction in SCZ vs controls
Fitzgerald *et al.* 2002	MT, MEP size, MEP latency, SP, SICI, ICF and IHI in SCZ vs healthy subjects	Medicated	22 SCZ 21 healthy subjects	↓ SICI in SCZ vs healthy subjects ↓ SP in SCZ vs healthy subjects ↔ in MEP size and MEP latency in SCZ vs healthy subjects ↔ in MT in SCZ vs healthy subjects
Fitzgerald *et al.* 2002	MT, MEP size, SP, SICI/ICF, IHI and iSP in SCZ vs healthy subjects	Medicated	40 SCZ (20 treated with olanzapine and 20 with risperidone) 20 healthy subjects	↑ MT in risperidone vs olanzapine ↓ SP in risperidone and olanzapine treated groups vs healthy controls ↔ SICI/ICF between patient groups vs healthy subjects ↑ iSP in olanzapine-treated patients than in risperidone-treated patients or healthy subjects ↔ iSP latency between risperidone and olanzapine treated patients vs healthy subjects ↓ IHI in olanzapine-treated and risperidone-treated patients than in healthy subjects but no differences between the onlazapine and risperidone group

Table 23.1 (*cont.*) Review of studies examining cortical excitability in psychotic disorders

Study	Objectives	Medication status	No. of subjects	Findings
Pascual-Leone *et al.* 2002	RMT, SICI/ICF in SCZ vs healthy subjects	Medicated Unmedicated	7 unmedicated SCZ 7 medicated SCZ 7 healthy subjects	↑ RMT in medicated patients vs unmedicated patients and healthy subjects for both hemispheres ↑ RMT in right hemisphere in medicated patients vs unmedicated patients and healthy subjects ↑ right RMT in healthy subjects ↑ left RMT in medicated and unmedicated patients ↓ SICI in medicated patients vs unmedicated patients and healthy subjects ↑ ICF in medicated patients vs unmedicated patients and healthy subjects
Fitzgerald *et al.* 2003	RMT, LICI, I-wave facilitation in SCZ vs healthy subjects	Medicated Unmedicated	9 unmedicated SCZ 9 medicated SCZ 8 healthy subjects	↔ RMT between groups ↔ LICI between groups ↑ I-wave facilitation in both patients groups vs healthy subjects
Bajbouj *et al.* 2004b	MT, SP and iSP in SCZ vs healthy subjects	Medicated	16 SCZ 16 healthy subjects	↔ MT between groups ↑ SP in SCZ vs healthy subjects ↑ iSP in SCZ vs healthy subjects
Eichhammer *et al.* 2004	MT, SICI/ICF in SCZ vs healthy subjects	Unmedicated	21 unmedicated SCZ 21 healthy subjects	↓ MT in patients vs healthy subjects ↔ in SICI and ICF between groups

MT, motor threshold; MEP, motor-evoked potential; MDD, major depressive disorder; SCZ, schizophrenia; EMG, electromyogram; SP, silent period; iSP, ipsilateral silent period; **TCT**,; RMT, resting motor threshold; SICI, short-interval intracortical inhibition; ICF, intracortical facilitation; IHI, interhemispheric inhibition; RMT, resting motor threshold; LICI, long-interval intracortical inhibition.

Table 23.2 Summary of studies examining cortical excitability in disorders and associated nonpsychotic disorders

Study	Objectives	Medication status	No. of subjects	Findings
Samii et al. 1996	Post-exercise MEP facilitation in MDD and CFS vs healthy subjects	Medicated	12 CFS 10 MDD 18 controls	↓ MEP facilitation in MDD and CFS vs healthy subjects ↔ between MDD and CFS
Shajahan et al. 1999a	Post-exercise MEP facilitation in MDD vs healthy subjects	Medicated	10 MDD 10 controls	↓ MEP facilitation in MDD vs healthy subjects
Shajahan et al. 1999b	Post-exercise MEP facilitation in MDD, recovered MDD patients vs healthy subjects	Medicated	10 MDD 10 recovered MDD 10 controls	↓ MEP facilitation in MDD vs healthy subjects but no difference between recovered MDD and healthy subjects
Reid et al. 2002	Post-exercise MEP facilitation and depression in MDD and SCZ	Medicated	10 MDD 11 SCZ 13 healthy subjects	↓ MEP amplitude in MDD and SCZ vs healthy subjects ↔ between MDD and SCZ
Maeda et al. 2000	Interhemispheric MT; SICI/ICF in MDD	Unmedicated	8 treatment-resistant MDD 8 healthy subjects	↓ left MT vs right MT in MDD not demonstrated in healthy controls ↓ left SICI/ICF and ↑ right SICI/ICF in MDD at 6 ms ISI
Steele et al. 2000	Cortical SP in MDD	Medicated	16 MDD (unipolar or bipolar) 19 controls	↑ SP in MDD
Grunhaus et al. 2003	MT, MEP amplitude in MDD	Unmedicated	19 MDD 13 controls	↔ between MDD vs healthy subjects
Fitzgerald et al. 2004	RMT, AMT, MEP amplitude, SP, SICI/ICF in right and left motor cortex in MDD	Medicated	60 MDD	↓ Left RMT vs right RMT (trend level) ↓ SICI in right hemisphere in MDD
Manganotti et al. 2001	MT; MEP amplitude; SP; SICI/ICF post-IV clomipramine in MDD	Clomipramine	6 MDD	↑ MT, ↑ SICI and ↓ ICF in MDD post clomipramine ↔ MEP amplitude and SP

Table 23.2 (*cont.*) Summary of studies examining cortical excitability in disorders and associated nonpsychotic disorders

Study	Objectives	Medication status	No. of subjects	Findings
Munchau et al. 2005	MT, MEP amplitude, SP and SICI/ICF in epilepsy + MDD before and after mirtazapine treatment vs healthy subjects	Mirtazapine	7 patients with epilepsy and MDD 6 healthy subjects	↑ RMT, AMT and SP/MEP ratio in MDD + epilepsy compared to healthy subjects ↓ AMT in MDD following mirtazapine not demonstrated in healthy subjects ↔ MEP amplitude, SP/MEP ratio, SICI/ICF in healthy subjects following mirtazapine
Gerdelat-Mas et al. 2005	MT, MEP recruitment curve, SP, SICI/ICF in healthy subjects on paroxetine or placebo for 30 days separated by 3 month washout	Paroxetine	21 healthy subjects	↑ ICF and ↓ MEP recruitment curve after paroxetine ↔ in MT, SICI, SP
Ziemann et al. 1997a	MT, SP, SICI/ICF in TD vs healthy subjects	Medicated Unmedicated	20 TD 21 controls	↔ in MT ↓ SP and ↓ SICI in TD vs healthy subjects
Gilbert et al. 2005	SICI/ICF in TD with comorbid ADHD	Medicated	28 TD	↓ SICI significantly correlated with ↑ADHD symptoms in TD
Gilbert et al. 2006	SICI/ICF in healthy subjects receiving single dose methylphenidate or atomoxitine	Methylphenidate Atomoxitine	21 healthy subjects	↑ ICF ↓ SICI
Buchmann et al. 2003	RMT, MEP amplitude, SP, iSP vs healthy subjects	Unmedicated	13 ADHD 13 healthy subjects	↔ in MT, MEP amplitude, SP ↑ iSP latency and ↓ iSP in ADHD vs healthy subjects
Garvey et al. 2005	RMT, AMT, MEP amplitude and iSP in ADHD	Unmedicated	12 ADHD 12 healthy subjects	↔ in MT, iSP iSP latency did not decrease with age in ADHD as in healthy subjects
Greenberg et al. 2000	MT, SP, SICI/ICF in OCD vs healthy controls	Medicated Unmedicated	16 OCD 11 healthy subjects	↓ SICI (ISI 2–5 ms) in OCD vs healthy subjects ↓ MT in OCD (greatest in subjects with tics)

Table 23.2 (*cont.*) Summary of studies examining cortical excitability in disorders and associated nonpsychotic disorders

Study	Objectives	Medication status	No. of subjects	Findings
Wassermann et al. 2001	MT, MEP amplitude, SICI/ICF vs NEO-PI-R scores in healthy subjects	Unmedicated	46 healthy subjects	Pooled SICI/ICF correlated with neuroticism across entire sample of men and women but was not significant for women when sex was analyzed separately

MEP, motor-evoked potential; MDD, major depressive disorder; CFS, chronic fatigue syndrome; SCZ, schizophrenia; MT, motor threshold; SICI, short-interval intracortical inhibition; ISI, interstimulus interval; ICF, intracortical facilitation; SP, silent period; iSP, ipsilateral silent period; RMT, resting motor threshold; AMT, active motor threshold; IV, intravenous; TD, Tourette's disorder; ADHD, attention deficit hyperactivity disorder; OCD, obsessive–compulsive disorder; NEO-PI-R, Revised NEO Personality Inventory.

One of the first neurophysiological studies using TMS to evaluate the neurophysiology of psychiatric disorders was by Abarbanel *et al.* (1996) who studied cortical excitability with TMS in 10 patients with MDD and 10 patients with SCZ. Both groups were medicated. Patients with SCZ were treated with antipsychotics, antidepressants, and benzodiazepines. Patients with MDD were treated with mood stabilizers, antidepressants, and benzodiazepines. Ten non-age- or gender-matched healthy controls were used as a comparison group. Dependent variables included:

- total conduction time (calculated from TMS onset to motor-evoked potential (MEP) production)
- central conduction time (calculated by subtracting peripheral conduction time from total conduction time)
- MEP ratio (obtained from TMS and peripheral nerve root stimulation)
- motor threshold (MT).

The latter two represent measures of excitability in the cortex.

MT is conventionally defined as the minimum intensity of the stimulator output needed to elicit a minimal MEP response in the target muscle. It is generally accepted that the MT is largely mediated by ion channel conductivity of neural membranes but may also depend on

inhibitory tone in the cortex (Ziemann *et al.* 1996; Chen *et al.* 1997b; Mavroudakis *et al.* 1997; Boroojerdi *et al.* 2001). By contrast, MEP size is obtained by stimulating the cortex with intensities above the MT and its size is typically measured by recording peak-to-peak amplitude (Figure 23.1). In this study, it was demonstrated that the MT was lower and MEPs larger in the SCZ patients compared with MDD patients and healthy controls, the latter two not differing significantly. There were no significant differences in conduction time or left–right differences in MT or MEP amplitude between groups.

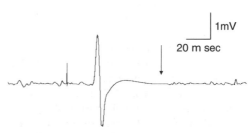

Fig. 23.1 Surface electromyogram recordings from the tonically active first dorsal interosseus muscle following 40% suprathreshold TMS. Each waveform represents the average of 15 trials. The motor-evoked potential (MEP) occurs ~20 ms after the TMS stimulus. The silent period starts at the onset of the MEP and ends with return of motor activity marked by the arrow.

These findings provided initial evidence for increased cortical excitability and perhaps inhibitory impairments in SCZ.

In another early study, Puri et al. (1996) demonstrated a shorter MEP latency in patients with SCZ compared with healthy age- and sex-matched controls. In contrast to the findings of Abarbanel et al. (1996), they found no differences in either MT or MEP amplitude. Furthermore, Davey et al. (1997) explored the influence of antipsychotic medication on some of the aforementioned TMS measures in SCZ. Medicated patients showed a weaker period of electromyogram (EMG) suppression after the MEP and a longer latency to maximum EMG suppression than drug-naïve patients. They found no difference in the MT or MEP latency.

However, the comparability of these early studies was limited in several important ways. For example, unlike the majority of TMS studies examining motor physiology by stimulating the motor cortex with a figure-8-shaped coil, Puri et al. (1996) and Davey et al. (1997) applied TMS over the vertex using a round coil. Furthermore, while Abarbanel et al. (1996) examined medicated patients, Puri et al. (1996) included only unmedicated patients and Davey et al. (1997) studied unmedicated patients but with no control group. In addition, the sample sizes of the studies by Abarbanel et al. (1996), Puri et al. (1996), and Davey et al. (1997) were relatively small ($n < 20$). Finally, Abarbanel et al. (1996) included patients with chronic SCZ, making illness duration an important differentiating variable from the latter two studies.

Consistent with Abrabanel et al. (1996), several studies have provided further evidence for increased MT (i.e. increased cortical excitability) in SCZ. In particular, two studies reported reduced left hemisphere rest MT (RMT) in unmedicated patients. RMT is defined as the minimum intensity sufficient to elicit an MEP of 50 µV in five of 10 trials during complete muscle relaxation. Daskalakis et al. (2002b) reported a significantly lower RMT in unmedicated patients compared with medicated patients and healthy controls. There was no RMT difference between medicated patients and controls. In line with these results, Eichhammer et al. (2004) reported a lower RMT in drug-naïve first-onset patients with SCZ.

By contrast, Pascual-Leone et al. (2002) observed higher RMT in medicated patients compared with unmedicated patients and healthy controls without an RMT difference between the unmedicated patients and controls. Furthermore, Pascual-Leone et al. (2002) described a higher left- than right-hemisphere MT in both patient groups, whereas healthy controls exhibited the opposite pattern. Several additional studies have reported no MT difference between medicated SCZ patients and healthy controls (Boroojerdi et al. 1999; Kubota et al. 1999; Fitzgerald et al. 2002b,c; Bajbouj et al. 2004a). However, small sample sizes, differing coil types (i.e. round vs figure-8), illness duration, and treatment with antipsychotic medications may account for the discrepant MT findings in SCZ patients.

Several studies have employed TMS to evaluate TMS-induced silent period (SP) in SCZ patients. Studies using single-pulse TMS have demonstrated a shorter SP (i.e. TMS-induced cessation of the EMG activity during sustained voluntary activation of the targeted muscle) compared with healthy subjects (Daskalakis et al. 2002a; Fitzgerald et al. 2002b–d, 2004). For example, Daskalakis et al. (2002a) reported that unmedicated SCZ patients had a significantly shorter SP than medicated patients and healthy controls, but observed no SP differences between medicated patients and controls. Fitzgerald et al. (2004) demonstrated shorter SP in unmedicated and medicated patients than in controls, with no differences between the patient groups. By contrast, Bajbouj et al. (2004a) found a prolonged SP in a group of mostly medicated SCZ patients compared with age- and sex-matched controls. Different methodology of SP evaluation may account for these discrepant findings. Bajbouj et al. (2004a) used a stimulation intensity of 80% of the maximal stimulator output during maximal voluntary contraction of the hand muscles, whereas all other cited studies employed stimulation intensities of 10%, 20%, 30%, and 40% over the MT in moderately activated hand muscles. Although the results regarding SP are inconsistent, it can be concluded that most studies have found reduced SP duration in SCZ patients and that antipsychotic medications may normalize this deficit. This latter evidence has been substantiated in a recent study (Daskalakis *et al.* 2005b) which found that treatment with the atypical antipsychotic clozapine further prolongs the SP, suggesting that clozapine increases cortical inhibition in SCZ patients. Since the SP is thought to be linked to gamma-aminobutyric acid $(GABA)_B$ receptor-mediated neurotransmission (Siebner *et al.* 1999), this increase may be related to potentiation of this inhibitory neurotransmitter system.

Paired-pulse TMS has also been used to evaluate short-interval intracortical inhibition (SICI) and intracortical facilitation (ICF) in SCZ. Through paired-pulse TMS, pairing a subthreshold conditioning stimulus (CS) (e.g. typically 80% of the MT) with a suprathreshold test stimulus (TS) (e.g. typically 120% of the MT) with an interstimulus interval of 1–5 ms and 10–20 ms leads to MEP inhibition/enhancement, commonly referred to as SICI/ICF, respectively (Kujirai *et al.* 1993) (Figure 23.2). Several pharmacological studies have emphasized the

pivot role of $GABA_A$ and *N*-methyl-D-aspartate (NMDA) receptor-mediated neurotransmission in SICI/ICF (Ziemann *et al.* 1996; Nakamura *et al.*, 1997; Di Lazzaro *et al.* 2000; Ilic *et al.* 2002). Fitzgerald *et al.* (2004) reported reduced SICI in medicated patients compared with healthy controls. Daskalakis *et al.* (2002a) found less SICI in unmedicated patients with SCZ compared with healthy controls, but they observed no differences between unmedicated and medicated patients or between medicated patients and healthy controls.

Moreover, Daskalakis *et al.* (2002a) found that the severity of psychotic symptoms related to the extent of inhibitory deficits. In contrast, Pascual-Leone *et al.* (2002) reported reduced SICI in medicated patients compared with unmediated patients and healthy controls. Furthermore, one study in medicated patients (Fitzgerald *et al.* 2002d) and one study in drug-naïve patients (Eichhammer *et al.* 2004) found no differences in SICI. These findings might

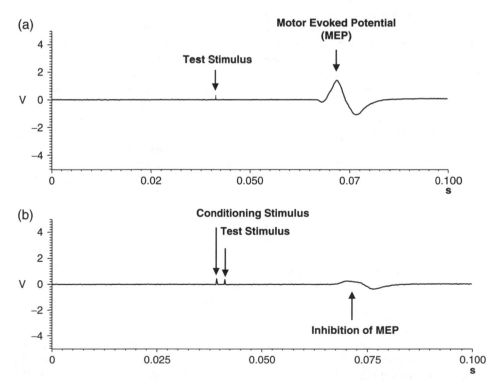

Fig. 23.2 (a) TMS test stimulus applied to the motor cortex producing a motor-evoked potential (MEP) following a 25 ms latency. (b) Conditioning stimulus preceding the test stimulus by 2 ms produces inhibition of the MEP.

provide evidence of reduced SICI in SCZ patients at a slightly later stage or in the more severely affected, and the increased SICI in medicated patients could be seen as evidence for normalization of SICI by medication. With respect to ICF, Pascual-Leone *et al.* (2002) reported enhancement of this measure in medicated SCZ patients relative to unmedicated patients and healthy controls. However, other studies did not demonstrate any difference in ICF between patients and healthy controls or between different medication groups (Daskalakis *et al.* 2002a,c,d; Eichhammer *et al.* 2004).

As it is often unclear whether group differences (i.e. patients and healthy subjects) are a consequence of illness or medications, several studies have evaluated the effects of psychotropic medications on some of the aforementioned TMS paradigms. For example, a few studies have directly investigated the effects of dopaminergic agents on cortical inhibitory neurotransmission. For example, Ziemann *et al.* (1997b), in a group of healthy control subjects, demonstrated that 5 mg of bromocriptine, a dopamine agonist, increased SICI whereas 2.5 mg of haloperidol, a dopamine antagonist, decreased SICI. Moreover, 200 mg of sulpiride, an atypical dopaminergic antagonist, had no effect on SICI. In contrast, Daskalakis *et al.* (2003) demonstrated that neither olanzapine (an atypical dopamine antagonist) nor haloperidol (a typical dopamine antagonist) resulted in any significant changes to SICI, ICF, or SP. Generalizing these findings to SCZ, however, is limited in several important ways. First, the effects of single-dose administration of antipsychotic medications on cortical inhibitory neurotransmission may not adequately capture what occurs with repeated administration. Second, the effects of psychotropic medications on these measures may differ between patients and healthy subjects. Therefore, to best address the medications as confounding variables when interpreting illness effects, an optimal study design should involve evaluation of the aforementioned measures prior to, and after, a course of pharmacological treatment in patients.

Long-interval cortical inhibition (LICI) refers to the reduction of a suprathreshold test stimulus MEP when preceded by a suprathreshold conditioning stimulus with an interstimulus interval between 50 and 200 ms (Claus *et al.* 1992; Valls-Sole *et al.* 1992). It is likely that LICI is mediated by slow IPSP via activation of $GABA_B$ receptors similar to the SP (Werhahn *et al.* 1999). In this regard, Fitzgerald *et al.* (2003) evaluated LICI in nine medicated and nine unmedicated patients with SCZ compared with eight healthy control subjects. They also evaluated RMT and I-wave facilitation because I-wave facilitation has also been shown to be closely associated with GABAergic inhibition in the motor cortex (Ziemann *et al.* 1998b). As previously described, TMS of the human motor cortex results in multiple discharges in the corticospinal tract (Ziemann *et al.*, 1998b). The initial descending volley is produced by direct neuronal stimulation (D-wave) which is followed by waves of activation from stimulation of cortical interneurons (I-waves) Ziemann *et al.*, 1998a). Through a modified paired-pulse protocol, I-waves can be recorded in peripheral hand muscles. This modification involves paired stimulation with a suprathreshold conditioning stimulus followed by stimulation with a subthreshold test stimulus (Ziemann *et al.* 1998b). It has been shown that I-wave production is dependent largely on GABA inhibitory neurotransmission (Ziemann *et al.* 1998b; Wischer *et al.*, 2001). Medicated patients had significantly more I-wave facilitation than healthy subjects, while unmedicated patients demonstrated a trend toward greater I-wave facilitation than healthy subjects. By contrast, there were no differences between these three groups on measures of LICI or RMT. It was concluded that these findings were consistent with others suggesting that SCZ is associated with deficient inhibitory neurotransmission in the cortex. More specifically, since I-wave facilitation can be suppressed with $GABA_A$-potentiating drugs, such as benzodiazepines (Ziemann *et al.* 1998b), these findings suggest a relative deficit in $GABA_A$ inhibitory function in SCZ. Nevertheless, the possibility remains that these inhibitory deficits may also be related to antipsychotic medications.

Interhemispheric inhibition (IHI) represents another TMS inhibitory paradigm. IHI can be evaluated through either single-pulse TMS or paired-pulse TMS. In single-pulse TMS, the ipsilateral SP (iSP), or the SP generated by stimulating the motor cortex ipsilateral to the target hand muscle, is the principal dependent variable (Wassermann *et al.* 1991). To generate the iSP, stimulation is typically delivered at very high intensities (e.g. typically 100% of stimulator output) (Wassermann *et al.* 1991). There is compelling evidence to suggest that this process is, in part, processed through callosal fibers (Meyer *et al.* 1998). IHI can also be assessed through paired-pulse TMS. In this paradigm, the CS is applied to the motor cortex ipsilateral to the hand muscle being recorded several milliseconds prior to a test stimulus delivered to the contralateral motor cortex (Figure 23.3). This CS inhibits the size of the MEPs produced by contralateral motor cortex stimulation by ~50% (Ferbert *et al.* 1992). Employing these techniques can be useful in deriving information regarding both interhemispheric connectivity and the extent to which one hemisphere inhibits the other. Moreover, there is evidence to suggest that this form of cortical inhibition is coordinated by neurocircuitry similar to that of LICI (i.e. GABA$_B$ receptor-mediated neurotransmission) (Daskalakis *et al.* 2002b). Several paired-pulse TMS studies have reported interhemispheric inhibition (IHI) deficits in both medicated (Fitzgerald *et al.* 2002b,c) and unmedicated (Daskalakis *et al.* 2002a) SCZ patients. Daskalakis *et al.* (2002a) demonstrated that unmedicated patients with SCZ had significantly less IHI than healthy controls. However, they found no differences between unmedicated and medicated patients. Four studies using single-pulse TMS to evaluate the IHI-mediated iSP in patients with SCZ found that it was significantly longer in SCZ patients compared with healthy controls (Boroojerdi *et al.* 1999; Hoppner *et al.* 2001; Fitzgerald *et al.* 2002b,d). This effect may be due to medications, since Fitzgerald *et al.* (2002b) demonstrated a significant positive relationship between the dose of antipsychotic medication and IHI duration. Taken together, the majority of these studies have demonstrated impairment of several aspects of inhibitory interhemispheric connectivity in patients with SCZ. Moreover, some limited evidence suggests that medications may normalize this impairment.

In SCZ, abnormalities in cortical inhibitory neurotransmission have been identified as central to understanding its pathophysiology. Benes (1998) has demonstrated that the density of GABAergic interneurons, which mediate cortical inhibitory neurotransmission, is reduced in the anterior cingulate cortex in SCZ patients. Moreover, these authors postulate that excessive activation of dopaminergic afferents leads to damage of GABAergic interneurons and, consequently, disrupted GABAergic neurotransmission. In this model, increased dopamine activity or loss of GABAergic cells would result in decreased inhibitory modulation of pyramidal neurons (Benes 1998). Decreased inhibitory modulation of cortical pyramidal neurons may result in unmodulated stimulatory activity which would flood corticolimbic brain regions – a process that could produce both psychotic and neuromotor abnormalities. This model also predicts that dopamine antagonism (i.e. antipsychotics) may reverse these effects, thereby restoring GABAergic function (Benes 1998).

In the cerebellum, inhibitory Purkinje cell output results in a reduction of excitatory output from the deep cerebellar nuclei (DCN) through the thalamus to the cortex leading to modification of cortical control (Allen and Tsukahara 1974). Therefore, TMS has been used to explore the inhibitory influence of the cerebellum on the cortex. Magnetic stimulation of the cerebellum prior to TMS of the motor cortex results in suppression of the MEP induced by TMS applied to the motor cortex ipsilateral to the side of cerebellar stimulation

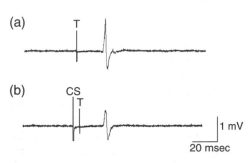

(a) T

(b) CS
T

1 mV

20 msec

Fig. 23.3 (a) Test stimulus (TS) applied to the left motor cortex produces a motor-evoked potential (MEP) in the right hand. (b) Conditioning stimulus (CS) applied to the right motor cortex ~6 ms prior to a test stimulus applied to the left motor cortex inhibits the size of the MEP produced by left motor cortex stimulation.

(Ugawa *et al.* 1995). This paradigm is commonly referred to as cerebellar inhibition (CBI). Recent studies employing CBI have demonstrated that the cerebellum forms inhibitory connections with both pyramidal and interneurons in the human motor cortex (Daskalakis *et al.* 2004b). Therefore, CBI may be used to test the inhibitory influence of the cerebellum on the motor cortex but is also valuable in providing information regarding cerebello-thalamo-cortical connectivity (Daskalakis *et al.* 2004c). Daskalakis *et al.* (2005a) recently demonstrated that SCZ patients have a deficit in CBI compared with healthy controls. It was concluded that this deficit could be due either to abnormal cerebellar Purkinje cell inhibitory output or to disrupted cerebellar–cortical connectivity. This finding is consistent with the neuro-anatomic findings of Reyes and Gordon (1981) who demonstrated that patients with SCZ had a reduced number of Purkinje cells per millimeter. On the other hand, it has also been reported that during recall of novel and practiced word lists in neuroleptic-free SCZ patients, there was reduced regional cerebral blood flow in anterior cingulate, thalamus, and cerebellum compared with healthy controls (Crespo-Facorro *et al.* 1999), suggesting that SCZ patients have altered cerebellar-thalamic-prefrontal connectivity. Therefore, the finding of deficient CBI in patients with SCZ is consistent with either possibility.

Plasticity in the human cortex involves functional reorganization of synaptic connections in an effort to change or to adapt throughout life. Evidence suggests that neural plasticity may also be a corollary of cortical inhibition. That is, the mechanisms mediating plasticity include unmasking of existent cortico-cortical connections (Schieber and Hibbard 1993) by removing cortical inhibitory neurotransmission (Jacobs and Donoghue 1991). For example, in humans administration of a GABAergic agonist disrupts plasticity (Bütefisch *et al.* 2000) while physiological plasticity following lower-limb amputation may occur through reduced cortical GABAergic inhibition (Chen *et al.* 1998).

Several TMS paradigms have been used to evaluate neural plasticity. Use-dependent plasticity is one TMS paradigm that can directly measure neural plasticity in the cortex (Classen

et al. 1998). Use-dependent plasticity is accomplished in several steps.

1. The spontaneous direction of TMS-induced thumb movements is measured in two axes (x and y).

2. Individuals are then trained to perform a simple motor task opposite to the direction of TMS-induced thumb movement.

3. TMS is then reapplied to the cortex while evaluating the direction of induced thumb movement.

4. Directional changes in thumb movement are evaluated over time (Classen *et al.* 1998) (Figure 23.4).

Classen *et al.* (1998) demonstrated that, immediately after training, the direction of TMS-induced movements follows the direction of training. It has been suggested that both GABA and NMDA receptor-mediated neurotransmission play an important role in use-dependent plasticity (Bütefisch *et al.* 2000).

Neural plasticity can also be evaluated by inducing a change in local cortical excitability, as indexed through a change in MEP amplitudes and/or MT, after repetitive (r)TMS at stimulus frequencies of 1 Hz or higher (Chen *et al.* 1997a; Siebner *et al.* 1999; Romeo *et al.* 2000; Fitzgerald *et al.* 2002a). Several neurotransmitter systems have also been shown to be involved in this form of plasticity. For example, Fitzgerald *et al.* (2005) demonstrated that following 1 Hz rTMS to M1 there was decreased motor excitability following placebo administration but not following the administration of lorazepam or dextromethorphan. These results suggest that this form of plasticity is coordinated, in part, through GABA and NMDA receptor-mediated neurotransmission.

In SCZ, Fitzgerald *et al.* (2004) applied 900 rTMS stimuli with a stimulus frequency of 1 Hz and an intensity of 110% of the RMT over the primary motor cortex. It was demonstrated that healthy subjects displayed an anticipated increase in the RMT after rTMS, whereas medicated and unmedicated patients demonstrated an RMT decrease. Furthermore, there was a significant increase in active motor threshold (AMT) in healthy participants whereas unmedicated patients demonstrated an AMT decrease. The AMT is described as the minimum intensity

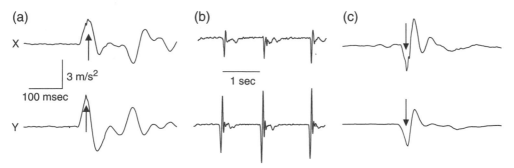

Fig. 23.4 TMS measures of use-dependent plasticity. Waveforms represent recordings from thumb movement in two dimensions measured with accelerometers positioned over the dorsum of the thumb in the *x*- and *y*-directions. (a) Spontaneous direction of TMS-induced movement is measured; (b) individuals are then trained to perform a simple motor task opposite to the direction of TMS-induced thumb movement; (c) TMS is reapplied to the cortex while evaluating the direction of induced thumb movement. The shift toward the direction of training (i.e. A–C) represents the reorganization in the cortex or plasticity.

to evoke an MEP of 200 µV in at least five of 10 trials during isometric contraction of the tested muscle at ~20% maximum (Rossini *et al.* 1994). Finally, it was demonstrated that the SP was decreased in healthy subjects after rTMS, while neither patient group displayed such changes. In a similar study, Oxley *et al.* (2004) examined changes in TMS measures after rTMS of the supplementary motor cortex in 12 SCZ patients and 12 healthy participants. While the control group demonstrated increased RMT and decreased MEP sizes, the opposite pattern was found in the SCZ patients. As has been previously indicated, these findings of abnormal excitability responses to rTMS in SCZ may reflect altered GABA or NMDA receptor-mediated neurotransmission. Recent studies have also used TMS to evaluate use-dependent plasticity in 14 medicated and six unmedicated patients with SCZ compared with 12 healthy subjects (Daskalakis *et al.* 2004a). A significant reduction of use-dependent plasticity was demonstrated in SCZ compared with healthy control subjects. That is, SCZ patients demonstrated significantly smaller angular deviations in the 5–10 min post-training period versus pretraining compared with healthy controls (e.g. patients 40.99° ± 12.13, controls 145.83° ± 17.87; $F = 24.72, P < 0.001$).

Taken together, these studies provide preliminary evidence for a diminution of the neurophysiological processes that mediate neural plasticity (i.e. a cortical adaptation in response to external cortical influences) in SCZ. Furthermore, this evidence suggests that the decreased neural plasticity is even more pronounced in patients with impaired cortical inhibition. Dysfunctional neural plasticity has been proposed as a pathophysiological mechanism in SCZ. For example, neuropathological studies have demonstrated reduced molecular markers of neural plasticity in various brain regions (Weinberger 1999). Moreover, Ruppin *et al.* (1996) suggested that deficient inhibitory neurotransmission results in the erroneous and excessive growth of synapses. These altered synaptic connections, in turn, distort retrieval of information and are spontaneously activated (Ruppin *et al.* 1996). It has been postulated that these erroneously activated circuits could co-opt cortical speech production and result in inner speech or thoughts, which may be experienced by the patient as alien or inserted by a nonself force (Ruppin, 2000).

Major depressive disorder

Major depressive disorder is characterized by one or more major depressive episodes and, left untreated, can be associated with prolonged and recurrent symptoms. MDD is identified as the fourth-ranked cause of disability and premature death worldwide (Murray and Lopez 1996). It is frequently associated with a tremendous impact on the mental health of family members and caregivers, often with an increased presence of depression and anxiety symptoms (Howarth and Weissman 1995). Despite this, little information is known about the neurophysiology of this illness.

As previously discussed, cortical excitability can be loosely defined as the responsivity of the cortex to either endogenous or exogenous stimuli. It is often the product of a net balance between inhibitory circuits which suppress cortical activity and facilitatory circuits that enhance cortical activity (Nakamura *et al.* 1997). Indicators of increased cortical excitability include, but are not limited to, increased MEP amplitude, decreased MT, short SP, and decreased SICI. The latter two examples represent heightened excitability secondary to decreased cortical inhibition (Triggs *et al.* 1993). The following sections will review the evidence for abnormal excitability in MDD, as well as in other psychiatric disorders where such research has been conducted.

It was previously reported in imaging studies that MDD may involve dysregulation of cortical activity with lower activity in the left dorsolateral prefrontal cortex and higher activity in the right dorsolateral prefrontal cortex (Baxter *et al.* 1989; Abou-Saleh *et al.* 1999; Davidson and Meltzer-Brody 1999). However, these imaging studies provide little additional information regarding actual neurophysiological processes that may underlie such abnormalities. Here, TMS can be used to evaluate excitability in the cortex that may inform us about the pathophysiological and treatment mechanisms involved in these disorders. Moreover, despite the fact that MDD is most commonly associated with abnormalities in nonmotor brain regions (i.e. prefrontal cortex) (Campbell and MacQueen, 2006), the clinical presentation of these disorders suggests that motor abnormalities are indeed present, thus providing further rationale for evaluating the motor system as a window to evaluate their pathophysiology. For example, neurovegetative signs of MDD often include psychomotor changes (either agitation or retardation) as well as changes in energy levels

associated with marked fatigue. Furthermore, with antidepressant treatment, there is often marked improvement in motor function. Also, patients with mood disorders can present with catatonia, a complex motor disorder, providing further evidence for motor system involvement in mood disorders.

Several TMS studies have explored corticospinal excitability as a potential biological correlate of illness and recovery in MDD. For example, Samii *et al.* (1996) evaluated the phenomenon of post-exercise facilitation and post-exercise MEP depression in 18 healthy subjects, 12 patients with chronic fatigue syndrome (CFS), and 10 patients with MDD. Studies in healthy subjects have shown that moderate voluntary exercise results in an enhanced post-exercise TMS-induced MEP facilitation (Samii *et al.* 1996). When exercise continues beyond fatigue, MEPs decrease in amplitude, referred to as a post-exercise MEP depression. In these studies, exercise involves maintaining isometric contraction of the extensor carpi radialis muscle until 50% of maximum force is unsustainable. Patients with MDD and CFS demonstrated significantly less post-exercise MEP facilitation than healthy subjects, although there were no significant differences between clinical groups. However, MEP facilitation that was demonstrated early in the first post-exercise period decayed faster in CFS patients compared with MDD patients and healthy subjects. The authors postulated that the cortical mechanisms responsible for reduced post-exercise MEP facilitation are similar to those in CFS and related to deficient cortical excitability.

In a second study, Shajahan *et al.* (1999a) tested the hypothesis that medicated MDD patients would show less post-exercise MEP facilitation than healthy subjects. Ten patients with MDD and 10 healthy subjects were recruited. MEPs were elicited from the contralateral abductor pollicis brevis muscle at rest and after exercise. It was demonstrated that the mean MEP amplitude after exercise was significantly greater after exercise in healthy subjects (210%) than in MDD patients (130%). Similar to Samii *et al.* (1996), MDD patients showed initial facilitation, followed by a rapid return to baseline, suggesting earlier and increased post-exercise depression. However, in both studies the possibility that medications were partly responsible for the lack of post-exercise facilitation could not be excluded. Therefore, as a follow-up, Shajahan et al. (1999b) evaluated post-exercise facilitation in 10 patients with MDD, 10 recovered MDD patients, and 10 healthy subjects. Both patient groups were medicated. The authors reported that recovered MDD patients showed similar post-exercise facilitation compared with healthy subjects, while depressed patients showed no post-exercise facilitation. Collectively, these studies show that MDD patients demonstrate little or no post-exercise facilitation of MEPs, possibly suggesting that this deficit is related to depressive symptom severity. In addition, this finding of a reduced post-exercise facilitation in MDD suggests that there are motor cortical excitability deficits in this disorder.

Reid et al. (2002) conducted similar studies to determine whether such findings were specific to MDD or could also be found in patients with other psychiatric illnesses. Ten MDD patients, 11 SCZ patients, and 13 healthy subjects were evaluated. TMS-induced MEPs were elicited pre- and post-exercise from the contralateral abductor pollicis brevis. It was demonstrated that post-exercise facilitation, expressed as a percentage of baseline, was 510% in healthy subjects compared with 110% in MDD and 190% in SCZ patients. Of note, however, is the fact that the dependent variable (i.e. MEP change) was derived in two ways. In the first, MEP change was found by subtracting post-exercise MEP from pre-exercise MEP. In the second, the MEP change was expressed as the ratio of the post-exercise MEP to the pre-exercise MEP. Using the first method, both patient groups had reduced post-exercise facilitation compared with healthy subjects, while MDD patients also demonstrated significantly reduced post-exercise facilitation compared with SCZ patients. However, using the second method, the difference between the depressed and SCZ groups did not reach statistical significance. They concluded that post-exercise facilitation was reduced in MDD and SCZ, suggesting impaired cortical excitability in both disorders. Moreover, the authors suggested that reduced post-exercise facilitation was not specific to MDD. The authors highlighted two important limitations in this study. The first was that groups were not matched for age or sex and the second was the relatively small sample size, which may have limited the power to find significant differences between patient groups.

Several studies have used alternative methods to evaluate cortical excitability in MDD patients. For example, Grunhaus et al. (2003) assessed cortical excitability by comparing MT and MEP amplitude in 19 MDD patients, who were participating in an rTMS treatment study, and 13 age- and sex-matched controls. MT and the MEP amplitude were similar between patients and healthy subjects. Maeda et al. (2000) compared interhemispheric motor cortical excitability (i.e. MT and SICI/ICF) differences between a small group (n = 8) of unmedicated treatment-resistant MDD patients and healthy subjects. Subjects with MDD showed lower MT in the left compared with right hemisphere, while control subjects failed to demonstrate this hemispheric difference. There were also differences in SICI/ICF with smaller left-sided and greater right-sided SICI/ICF in the patients compared with the healthy subjects at an interstimulus interval of 6 ms.

Steele et al. (2000) hypothesized that motor retardation in MDD may be caused by neurophysiological mechanisms similar to those that mediate bradykinesia in Parkinson's disease. Since Parkinson's disease has been associated with an abnormally short SP (Cantello et al. 1991), it was anticipated that MDD patients would also demonstrate similar short SPs. Sixteen MDD patients and 19 age-, sex-, and IQ-matched healthy subjects were evaluated. Their results demonstrated that, contrary to their original hypothesis, the SP was significantly increased in MDD patients. Moreover, there was no correlation between the SP and the severity of depressive symptoms. The authors concluded that MDD is associated with increased cortical inhibition.

Fitzgerald et al. (2004) assessed whether cortical excitability could predict the response to rTMS treatment in MDD. The study included 60 treatment-resistant patients (54 MDD, six BD). Of these, 46 were medicated during the trial (antidepressants, mood stabilizers, and antipsychotics). Measures of excitability included RMT, AMT, MEP amplitude, SP, and SICI/ICF.

The authors reported: (i) a trend toward lower RMT in the left hemisphere in subjects with MDD; (ii) less SICI at an ISI of 1 ms over the right motor cortex; (iii) a significant positive correlation between ICF in the right hemisphere and the severity of symptoms (as measured by the Beck Depression Inventory, Brief Psychotic Rating Scale, and the CORE Rating Scale for Psychomotor Retardation); (iv) increased inhibition (i.e. increased cortical SP) in the left motor cortex predicted a poorer response to rTMS treatment, especially in subjects who had a preponderance of melancholic symptoms. The authors suggest that greater inhibition may be associated with greater brain blood flow and, thus, increased cortical activation. This latter finding is consistent with previous positron emission tomography activation studies (Kimbrell et al. 1999). An important limitation noted by the authors was the heterogeneous nature of the sample with respect to the psychotropic medications used.

Several studies have examined the effect of drugs on motor cortex excitability in mood disorder patients. Manganotti et al. (2001) measured MT, MEP amplitude, SP, SICI, and ICF after a single intravenous dose of clomipramine, a tricyclic antidepressant. They reported a temporary, but significant, increase in MT and SICI and a decrease in ICF 4 h after medication administration. Their findings indicate that clomipramine can exert a significant transient suppression of cortical excitability in MDD, possibly through enhanced inhibition, in the cortex. These data provide preliminary evidence that clomipramine may have effects on GABA$_A$ receptor-mediated inhibition, and glutamatergic neurotransmission either directly or indirectly, through its effects on serotonergic neurotransmission (Bhagwagar et al. 2004).

Munchau et al. (2005) examined the effects of mirtazapine, a potentiator of both noradrenergic and serotonergic systems, on motor cortex excitability in patients with epilepsy and comorbid MDD compared with healthy controls. Seven epilepsy patients with MDD on anticonvulsants were started on mirtazapine treatment and six healthy volunteers received a single oral dose. The AMT, RMT, MEP amplitude, SP, and SICI/ICF were measured before treatment and at 24 h in both groups and at 3 weeks in the patients.

They found that baseline AMT, RMT, and the ratio of SP duration:MEP area were significantly higher in the patients. In patients, but not in healthy subjects, AMT was significantly lower at 24 h. Mirtazapine had no significant effect on MEP size, SP duration, or SP duration:MEP area ratio in patients. In the healthy subjects, the duration of the SP was significantly longer and there was a trend toward increased MEP size. However, the SP duration:MEP area ratio did not change after mirtazapine. There were no significant changes in SICI/ICF in either group or differences between groups at baseline or 24 h. The authors concluded that in medicated epilepsy patients, mirtazapine increases cortical excitability. Combined with the finding in healthy subjects that SP duration and MEP size were enhanced, this suggests that in both patients and healthy subjects mirtazapine likely increases excitability when the cortex is in an activated state (i.e. during voluntary muscle contraction), which the authors suggest may occur as a result of enhanced noradrenergic and/or serotonergic neurotransmission by mirtazapine.

Gerdelat-Mas et al. (2005) investigated the effect of chronic administration of paroxetine on TMS measures of excitability/inhibition in healthy subjects. In a crossover study, 21 subjects received 20 mg daily of paroxetine or placebo for 30 days separated by a period of 3 months. They tested MT, MEP recruitment curve (average MEP sizes and resultant area under the curve of the MEP), SP, and SICI/ICF. Paroxetine resulted in increased ICF and decreased MEP amplitude, but did not alter SICI, MT, or SP. This finding was unexpected, because an increase in the ICF suggests increased cortical excitability whereas a decrease in the recruitment curve suggests decreased cortical excitability. However, as ICF may be more closely associated with NMDA and GABA receptor mediated neurotransmission (Ziemann et al. 1998a; Di Lazzaro et al. 2003), it is likely that the effect of paroxetine on ICF and MEP recruitment curve may be associated with potentiation of both these neurotransmitter systems rather than with a direct effect on voltage-gated ion channel neurotransmission. However, these findings also underscore the fact that the short-term changes in physiological measures induced by psychotropic medications may be quite different from long-term changes which correspond better to the timing of the therapeutic response.

Bipolar disorder

Bipolar disorder is a serious neuropsychiatric disorder which affects 1.5% of the population. It is characterized by episodes of mania and depression (Goodwin 1990), with an onset usually in early adulthood and a chronic relapsing course in ≥80% of patients (Winokur et al. 1993). It is associated with a high mortality (15–20% suicide rate) and morbidity (at least two-thirds have markedly impaired quality of life; Cooke et al. 1996). Despite these sobering statistics, relatively little work has been done to understand the neurophysiological changes occurring in this disorder. Limited neuro-anatomic evidence suggests that BD patients have impaired cortical inhibitory neurotransmission (Benes et al. 1998). However, there is little in vivo neurophysiological evidence supporting such impairments in this disorder. Levinson et al. (2006) used TMS to compare SICI and SP in a group of 15 BD patients (13 medicated with a single mood stabilizer, two unmedicated) with 15 healthy subjects. Based on neuroanatomic findings, it was hypothesized that BD patients would show a deficit in SICI and SP. Their results demonstrated that BD patients demonstrated significant deficits in both SICI and SP compared with healthy volunteers with a large effect size (SICI: Cohen's $d = 0.86$; SP: Cohen's $d = 0.84$). It was concluded that inhibitory neurotransmission in the cortex was deficient in BD. Moreover, while the majority of BD patients were medicated, the evidence suggested that these mood-stabilizing medications enhanced cortical GABAergic inhibitory neurotransmission. Therefore, it would be anticipated that any inhibitory deficits would be magnified without these medications.

Obsessive–compulsive disorder

Obsessive–compulsive disorder is associated with intrusive thoughts and behaviors that have been speculated to be related to dysfunctional inhibitory neurotransmission possibly originating from subcortical circuits (Rapoport and Wise, 1988). Tic disorders, closely linked clinically and genetically with OCD, are associated with decreased ICI/increased ICF. Therefore, Greenberg et al. (2000) compared the RMT and AMT, SP and SICI/ICF in 16 OCD patients (seven unmedicated) and 11 age-matched healthy volunteers. The patients had significantly less SICI than the healthy subjects at 2–5 ms interstimulus interval. They also reported that the RMT and AMT were decreased in OCD patients. Medications did not appear to account for deficits in SICI because there were no differences in SICI between medicated and unmedicated patients. The SICI and MT differences were greatest in the patients with comorbid tics, but remained significant in patients without tics. No differences in SP duration were found. They concluded that OCD was associated with deficient cortical inhibition and that their findings provided support for the theory that tic disorders and OCD share a dysfunction in corticobasal circuits that may mediate the inhibitory deficit.

Personality

Neuroticism is a personality dimension characterized by inability to control emotions and behaviors. Deficient brain inhibition has been suggested as a possible mechanism leading to impulsive behavior, often a behavioral subtype of neuroticism. For example, deficits in inhibition, assessed via the Go—NoGo task, are associated with impulsivity (Horn et al. 2003). Barratt and Patton (1983) demonstrated that impulsivity is inversely related to cognitive inhibition, as measured using the Stroop task. In addition, Visser et al. (1996) reported that negative priming, yet another measure of brain inhibition, is related to social impulsivity in children. Wassermann et al. (2001) used the TMS ICI/ICF paradigm to evaluate the relationship between cortical excitability (i.e. inhibition and facilitation) and impulsivity in a group of 46 healthy volunteers who were given the Revised NEO Personality Inventory. They demonstrated that when data for SICI (ISI: 3, 4 ms) and ICF (ISI: 10, 15 ms) were pooled, there was a significant relationship between these pooled values and neuroticism scores across the entire sample and for men. Interestingly, these TMS measures did not correlate with any of the other dimensions of personality.

In conclusion, the findings from this review demonstrate that TMS is a useful tool to evaluate several neurophysiological processes that may be altered in psychiatric illness. Collectively, evidence suggests that disorders including SCZ, MDD, BD, and OCD may, in part, be associated with deficient inhibition, altered cortical excitability, and disrupted neural plasticity, although studies often report conflicting findings. Moreover, evidence also suggests that psychotropic medications alter the aforementioned mechanisms, often in a direction opposite to that of illness, which could explain some of their therapeutic effects. A significant confound in most of these aforementioned studies, however, includes heterogeneity in the treatment with psychotropic medications, TMS parameters used, sample size, and illnesses being studied. Future studies designed to evaluate these aforementioned paradigms before and after treatment with psychotropic medications in a sufficiently large and homogeneous sample of patients may help rectify these abnormalities and clarify the neurophysiological mechanisms that either mediate these disorders or are involved in their treatment.

Acknowledgments

This work was funded, in part, by awards from the Alfried Krupp von Bohlen und Halbach Foundation (B.M.), the Ontario Mental Health Foundation (A.J.L.), and the Canadian Institutes of Health Research (CIHR) Clinician Scientist Award (Z.J.D.), and by Constance and Stephen Lieber through a National Alliance for Research on Schizophrenia and Depression (NARSAD) Young Investigator award (Z.J.D.).

References

Abarbanel JM, Lemberg T, Yaroslavski U, Grisaru N, Belmaker RH (1996). *Biological Psychiatry 40*, 148–150.

Abou-Saleh MT, Al Suhaili AR, Karim L, Prais V, Hamdi E (1999). *Journal of Affective Disorders 55*, 115–23.

Allen GI, Tsukahara N (1974). *Physiology Reviews 54*, 957–1006.

Bajbouj M, Gallinat J, Lang UE, Roricht S, Meyer BU (2004a). *Pharmacopsychiatry 37*, 74–80.

Bajbouj M, Gallinat J, Niehaus L, Lang UE, Roricht S, Meyer BU (2004b). *Pharmacopsychiatry 37*, 74–80.

Barratt ES, Patton J (1983). In: M Zukerman (ed.), *The biological bases of sensation seeking, impulsivity and anxiety*, pp. 77–122. London: Erlbaum.

Baxter LR, Jr, Schwartz JM, Phelps ME, et al. (1989). *Archives of General Psychiatry 46*, 243–250.

Benes FM (1998). *Schizophrenia Bulletin, 24*, 219–230.

Benes FM, Kwok EW, Vincent SL, Todtenkopf MS (1998). *Biological Psychiatry 44*, 88–97.

Bhagwagar Z, Wylezinska M, Taylor M, et al. (2004). *Am Journal of Psychiatry 161*, 368–370.

Boroojerdi B, Topper R, Foltys H, Meincke U (1999). *British Journal of Psychiatry, 175*, 375–379.

Boroojerdi B, Phipps M, Kopylev L, Wharton CM, Cohen LG, Grafman J (2001). *Neurology 56*, 526–528.

Buchmann J, Wolters A, Haessler F, Bohne S, Nordbeck R, Kunesch E (2003). *Clinical Neurophysiology 114*, 2036–2042.

Bütefisch CM, Davis BC, Wise SP, et al. (2000). *Proceedings of the National Academy of Sciences of the USA 97*, 3661–3665.

Campbell S, MacQueen G (2006). *Current Opinion in Psychiatry 19*, 25–33.

Cantello R, Gianelli M, Bettucci D, Civardi C, De Angelis MS, Mutani R (1991). *Neurology 41*, 1449–1456.

Chen R, Classen J, Gerloff C, Celnik, P, et al. (1997a). *Neurology 48*, 1398–1403.

Chen R, Samii A, Canos M, Wassermann EM, Hallett M (1997b). *Neurology 49*, 881–883.

Chen R, Corwell B, Yaseen Z, Hallett M, Cohen LG (1998). *Journal of Neuroscience 18*, 3443–3450.

Classen J, Liepert J, Wise SP, Hallett M, Cohen LG (1998). *Journal of Neurophysiology 79*, 1117–1123.

Claus D, Weis M, Jahnke U, Plewe A, Brunholzl C (1992). *Journal of the Neurological Sciences 111*, 180–188.

Cooke RG, Robb JC, Young LT, Joffe RT (1996). *Journal of Affective Disorders 39*, 93–97.

Crespo-Facorro B, Paradiso S, Andreasen NC, et al. (1999). *American Journal of Psychiatry 156*, 386–392.

Daskalakis ZJ, Christensen BK, Chen R, Fitzgerald PB, Zipursky RB, Kapur S (2002a). *Archives of General Psychiatry, 59*, 347–54.

Daskalakis ZJ, Christensen BK, Fitzgerald PB, Roshan L, Chen R (2002b). *Journal of Physiology 543*, 317–26.

Daskalakis ZJ, Christensen BK, Chen R, Fitzgerald PB, Zipursky RB, Kapur S (2003). *Psychopharmacology 170*, 255–262.

Daskalakis ZJ, Christensen BK, Fitzgerald PB, Chen R (2004a). In: *Society for Biological Psychiatry*. New York, NY.

Daskalakis ZJ, Paradiso GO, Christensen BK, Fitzgerald PB, Gunraj C, Chen R (2004b). *Journal of Physiology (London) 557*, 689–700.

Daskalakis ZJ, Paradiso GO, Christensen BK, Fitzgerald PB, Gunraj C, Chen R (2004c). *Journal of Physiology 557*, 689–700.

Daskalakis ZJ, Christensen BK, Fitzgerald PB, Fountain SI, Chen R (2005a). *American Journal of Psychiatry 162*, 1203–1205.

Daskalakis ZJ, Fountain SI, Chen R (2005b). In: CB Nemeroff (ed.), *American College of Neuropsychopharmacology, Annual Meeting, Waikoloa, HI*.

Davey NJ, Puri BK, Lewis HS, Lewis SW, Ellaway PH (1997). *Journal of Neurology, Neurosurgery and Psychiatry 63*, 468–473.

Davidson JR, Meltzer-Brody SE (1999). *Journal of Clinical Psychiatry 60*(Suppl. 7), 4–9; discussion 10–11.

Di Lazzaro V, Oliviero A, Meglio M, *et al.* (2000). *Clinical Neurophysiology 111*, 794–799.

Di Lazzaro V, Oliviero A, Profice P, *et al.* (2003). *Journal of Physiology 547*, 485–496.

Eichhammer P, Wiegand R, Kharraz A, Langguth B, Binder H, Hajak G (2004). *Schizophrenia Research 67*, 253–259.

Ferbert A, Priori A, Rothwell JC, Day BL, Colebatch JG, Marsden CD (1992). *Journal of Physiology 453*, 525–546.

Fitzgerald PB, Brown TL, Daskalakis ZJ, Chen R, Kulkarni J (2002a). *Clinical Neurophysiology 113*, 1136–1141.

Fitzgerald PB, Brown TL, Daskalakis ZJ, deCastella A, Kulkarni J (2002b). *Schizophrenia Research 56*, 199–209.

Fitzgerald PB, Brown TL, Daskalakis ZJ, Kulkarni J (2002c). *Psychiatry Research 114*, 11–22.

Fitzgerald PB, Brown TL, Daskalakis ZJ, Kulkarni J (2002d). *Psychopharmacology 162*, 74–81.

Fitzgerald, PB, Brown, TL, Marston, NA, *et al.* (2003). *Psychiatry Research 118*, 197–207.

Fitzgerald PB, Brown TL, Marston NA, *et al.* (2004). *Journal of Affective Disorders 82*, 71–76.

Fitzgerald PB, Benitez J, Oxley T, Daskalakis JZ, de Castella AR, Kulkarni J (2005). *Neuroreport 16*, 1525–1528.

Garvey MA, Barker CA, Bartko JJ, *et al.* (2005). *Clinical Neurophysiology 116*, 1889–1896.

Gerdelat-Mas A, Loubinoux I, Tombari D, Rascol O, Chollet F, Simonetta-Moreau M (2005). *NeuroImage 27*, 314–322.

Gilbert, DL, Sallee, FR, Zhang, J, Lipps, TD, Wassermann, EM (2005). *Biological Psychiatry 57*, 1597–1600.

Gilbert, DL, Ridel, KR, Sallee, FR, Zhang, J, Lipps, TD, Wassermann, EM (2006). *Neuropsychopharmacology 31*, 442–449.

Goodwin KJ (ed.) (1990). *Manic-depressive illness.* New York: Oxford University Press.

Greenberg BD, Ziemann U, Cora-Locatelli, *et al.* (2000). *Neurology 54*, 142–147.

Grunhaus L, Polak D, Amiaz R, Dannon PN (2003). *International Journal of Neuropsychopharmacology 6*, 371–378.

Heinrichs RW, Zakzanis KK (1998). *Neuropsychology 12*, 426–445.

Hoppner J, Kunesch E, Grossmann A, *et al.* (2001). *Acta Psychiatrica Scandinavica 104*, 227–235.

Horn NR, Dolan M, Elliott R, Deakin JF, Woodruff PW (2003). *Neuropsychologia, 41*, 1959–1966.

Howarth E, Weissman MM (1995). In: MT Tsuang, Tohen M, GEP Zahner (eds), *Textbook in psychiatric epidemiology*, pp. 317–344. New York: Wiley–Liss.

Ilic TV, Meintzschel F, Cleff U, Ruge D, Kessler KR, Ziemann U (2002). *Journal of Physiology 545*, 153–167.

Jacobs KM, Donoghue JP (1991). *Science 251*, 944–947.

Kimbrell TA, Little JT, Dunn RT, *et al.* (1999). *Biological Psychiatry 46*, 1603–1613.

Kubota F, Miyata H, Shibata N, Yarita H (1999). *Biological Psychiatry 45*, 412–416.

Kujirai T, Caramia MD, Rothwell JC, *et al.* (1993). *Journal of Physiology (London) 471*, 501–519.

Levinson AJ, Young LT, Daskalakis ZJ (2006). In *Collegium International Neuro-Psychopharmacologicum.* Chicago.

Maeda F, Keenan JP, Pascual-Leone A (2000). *British Journal of Psychiatry 177*, 169–173.

Manganotti P, Bortolomasi M, Zanette G, Pawelzik T, Giacopuzzi M, Fiaschi A (2001). *Journal of the Neurological Sciences 184*, 27–32.

Mavroudakis N, Caroyer JM, Brunko E, Zegers de Beyl D (1997). *Electroencephalography and Clinical Neurophysiology 105*, 124–127.

Meyer BU, Roricht S, Woiciechowsky C (1998). *Annals of Neurology 43*, 360–369.

Munchau A, Langosch JM, Gerschlager W, Rothwell JC, Orth M, Trimble MR (2005). *Journal of Neurology, Neurosurgery and Psychiatry 76*, 527–533.

Murray CJL, Lopez AD (1996). *The global burden of disease: a comprehensive assessment of mortality and disability from diseases, injuries, and risk factors in 1990 and projected to 2020.* Cambridge, MA: Harvard University Press.

Nakamura H, Kitagawa H, Kawaguchi Y, Tsuji H (1997). *Journal of Physiology 498*, 817–823.

Oxley T, Fitzgerald PB, Brown TL, de Castella A, Daskalakis ZJ, Kulkarni J (2004). *Biological Psychiatry 56*, 628–633.

Pascual-Leone A, Manoach DS, Birnbaum R, Goff DC (2002). *Biological Psychiatry 52*, 24–31.

Puri BK, Davey NJ, Ellaway PH, Lewis SW (1996). *British Journal of Psychiatry 169*, 690–695.

Rapoport JL, Wise SP (1988). *Psychopharmacology Bulletin 24*, 380–384.

Reid PD, Daniels B, Rybak M, Turnier-Shea Y, Pridmore S (2002). *Australia and New Zealand Journal of Psychiatry 36*, 669–673.

Reyes MG, Gordon A (1981). *Lancet ii*, 700–701.

Romeo S, Gilio F, Pedace F, *et al.* (2000). *Experimental Brain Research 135*, 504–510.

Rossini PM, Barker AT, Berardelli A, *et al.* (1994). *Electroencephalography and Clinical Neurophysiology 91*, 79–92.

Ruppin E (2000). *Medical Hypotheses 54*, 693–697.

Ruppin E, Reggia JA, Horn D (1996). *Schizophrenia Bulletin 22*, 105–123.

Samii, A, Wassermann, EM, Ikoma, K, *et al.* (1996). *Neurology 47*, 1410–1414.

Schieber MH, Hibbard LS (1993). *Science 261*, 489–492.

Shajahan PM, Glabus MF, Gooding PA, Shah PJ, Ebmeier KP (1999a). *British Journal of Psychiatry 174*, 449–454.

Shajahan PM, Glabus MF, Jenkins JA, Ebmeier KP (1999b). *Neurology, 53*, 644–646.

Siebner HR, Auer C, Conrad B (1999). *Neuroscience Letters 262*, 133–136.

Steele JD, Glabus MF, Shajahan PM, Ebmeier KP (2000). *Psychological Medicine, 30*, 565–570.

Triggs WJ, Cros D, Macdonell RA, Chiappa KH, Fang J, Day BJ (1993). *Brain Research 628*, 39–48.

Ugawa Y, Uesaka Y, Terao Y, Hanajima R, Kanazawa I (1995). *Annals of Neurology 37*, 703–713.

Valls-Sole J, Pascual-Leone A, Wassermann EM, Hallett M (1992). *Electroencephalography and Clinical Neurophysiology 85*, 355–364.

Visser M, Das-Smaal E, Kwakman H (1996). *British Journal of Psychology 87*(Pt 1), 131–140.

Wassermann EM, Fuhr P, Cohen LG, Hallett M (1991). *Neurology 41*, 1795–1799.

Wassermann EM, Greenberg BD, Nguyen MB, Murphy DL (2001). *Biological Psychiatry 50*, 377–382.

Weinberger DR (1999). *Biological Psychiatry 45*, 395–402.

Weinberger DR, Berman KF, Illowsky BP (1988). *Archives of General Psychiatry 45*, 609–615.

Werhahn KJ, Kunesch E, Noachtar S, Benecke R, Classen J (1999). *Journal of Physiology (London) 517*, 591–597.

Winokur G, Coryell W, Keller M, Endicott J, Akiskal H (1993). *Archives of General Psychiatry 50*, 457–465.

Wischer S, Paulus W, Sommer M, Tergau F (2001). *Clinical Neurophysiology 112*, 275–279.

Ziemann U, Lonnecker S, Steinhoff BJ, Paulus W (1996). *Experimental Brain Research 109*, 127–135.

Ziemann U, Paulus W, Rothenberger A (1997a). *American Journal of Psychiatry 154*, 1277–1284.

Ziemann U, Tergau F, Bruns D, Baudewig J, Paulus W (1997b). *Electroencephalography and Clinical Neurophysiology 105*, 430–437.

Ziemann U, Tergau F, Wassermann EM, Wischer S, Hildebrandt J, Paulus W (1998a). *Journal of Physiology (London) 511*, 181–190.

Ziemann U, Tergau F, Wischer S, Hildebrandt J, Paulus W (1998b). *Electroencephalography and Clinical Neurophysiology 109*, 321–330.

Zipursky RB, Lim KO, Sullivan EV, Brown BW, Pfefferbaum A (1992). *Archives of General Psychiatry 49*, 195–205.

CHAPTER 24

TMS in migraine

Jean Schoenen, Valentin Bohotin, and
Alain Maertens de Noordhout

Introduction

Migraine pathogenesis is complex and involves
the trigeminovascular pain system, the brain
stem pain control centers, and the cerebral cortex.
It is generally accepted that migraine patients
present an abnormal cortical activation pattern,
but it is still controversial whether this is hyper-,
hypo, or another type of dysexcitability. TMS
has the advantage of being able to assess cortical
excitability in an atraumatic and repeatable
manner and repetitive (r)TMS can modulate it.
TMS has thus been used to search for cortical
dysfunction in migraine. Both the motor and
the visual cortices have been explored. The former has the advantage of offering an objective
measure, the motor-evoked potential (MEP),
but the motor system is little involved in
migraine except in familial hemiplegic migraine.
The latter has the advantage that the visual cortex plays a major role in migraine, especially in
migraine with aura, but unfortunately it is usually explored with a subjective measure of activation, the (magneto)phosphene. Moreover, new
data show that the response patterns observed
with TMS may differ between motor and visual
cortex. It is well known, for instance, that in
healthy controls, stimulation of the motor cortex with paired pulses with short interstimulus
intervals (ISIs), between 2 and 5 ms, produces
intracortical inhibition, whereas pairs with
longer ISIs (6–12 ms) produce facilitation. In a
recent study, Sparing et al. (2005) used the
paired-pulsed TMS paradigm to explore whether

this principle can be extended to the visual
system. Unexpectedly, they found a global facilitation of magnetophosphenes for ISIs ranging
from 2 to 12 ms (Sparing et al. 2005). This
suggests that the mechanisms underlying
phosphene induction in the visual cortex are
different from those underlying the MEP in the
motor cortex. Another interesting study (Fierro
et al. 2005) showed that during light deprivation, which decreases the magnetophosphene
threshold, 10 Hz rTMS partly reverses and
shortens the effect, while 1 Hz rTMS has no significant effect, suggesting that the modulatory
effects of different rTMS frequencies on the
visual cortex critically depend on the pre-existing
excitability state of inhibitory and facilitatory
circuits. We will see that this could be of importance for the interpretation of the rTMS effects
on the visual cortex observed in migraineurs.

In this chapter we will review and discuss the
results of the various studies performed in
migraine patients with TMS of motor or occipital
cortices.

Motor cortex

Data from literature in this area are summarized
in Table 24.1. The first study of motor cortex
TMS in migraine was published in 1992
(Maertens de Noordhout et al. 1992). In this
pilot study conducted between attacks in unilateral migraine with aura (MA) patients, it was
noticed that motor-evoked potential threshold

Table 24.1 Transcranial magnetic stimulation – motor cortex

Authors	Diagnosis (N° subjects)	Mean age	Attack control		Methods		Results					
			Before	After	coil shape	Max. output	Motor threshold %		MEP amplitude	Central motor conduction time (ms)	Cortical silent period (ms)	Intracortical inhibition
							Rest	Active				
Maertens de Noordhout et al. (1992)	MA (10)	36	1 week	1 week	circular 130 mm	1.5 Tesla	55* (↑)	41±8* (↑)	↓	6.6		
	MO (10)	39					45	33±5	normal	6.5		
	HV (20)	40					48	33±3	normal	6.5		
Bettucci et al. (1992)	MO (10)	33	no°	no°		1.9 Tesla	58* (↑)			5.7		
	HV (10)	31					48			5.7		
Van der Kamp et al. (1996)	MA (10)	35	3 days	no	circular 130 mm	?	37		↑	5.7		
	MO (10)	50					38		↑	6.6		
	HV (10)	30					36			5.7		
Van der Kamp et al. (1997)	MA (10)	35	3 days	no	circular 130 mm	?	37		↑	5.7		
	FHM (10)	30					44* (↑)		normal	6.8		
	HV (6)	30					36		normal	5.7		
Afra et al. (1998)	MA (25)	36	3 days	3 days	circular 130 mm	2.5 Tesla	54	43* (↑)	normal		101	normal
	MO (33)	36					52	41			100	normal
	HV (27)	33					47	36			101	normal

Aurora et al. (1999)	MA (10)	36	1 week	none	circular 95 mm	2 Tesla	63	63*
	HV (10)	38					58	1070
Werhahn et al. (2000)	MA (12)	38	2 days	none	8-coil 90 mm	2.2 Tesla	61	183 normal
	FHM (9)	37					60	178 normal
	HV (17)	29					55	179 normal
Brighina et al. (2002)	MA (13)	39	2 days	2 days	8-coil 45 mm	?	58	
	HV (15)	32					55	
Brighina et al. (2005)	MA (9)	35.1	2 days	2 days	8-coil	?	50	normal
	HV (8)	30.4					50	Baseline ↓ After 1 Hz rTMS ↑ ICF(MA) ↓ ICF(HV)

(MT) was significantly increased on the affected hemisphere compared with normal subjects or with the unaffected side in the patients. No MT differences were observed between normal subjects and patients with unilateral migraine without aura (MO) or between the painful and nonpainful hemispheres of MO patients. Moreover, the maximal amplitude of MEPs expressed as a ratio to the maximal response to peripheral nerve stimulation (MEP_{max}/M_{max}) was significantly reduced on the body side of the auras in MA patients. Abnormally high MT was reported later in menstrual MO patients (Bettucci et al. 1992), not only in the interictal phase, but also during attacks.

These results were not confirmed in a study by van der Kamp et al. (1996) who found increased MEP amplitudes and reduced MT between attacks in MA and MO patients. They also reported a positive correlation between MEP amplitude and attack frequency, but they did not control for the occurrence of an attack in the days following the recordings. In a subsequent paper (van der Kamp et al. 1997), the same authors reported increased interictal MT and reduced MEP amplitude on the side of motor deficits in patients with familial hemiplegic migraine (FHM). These results were very similar to those obtained in our first study of patients with unilateral MA (Maertens de Noordhout et al. 1992).

In a subsequent paper, Afra et al. (1998b) studied a larger group of MA ($n = 12$) and MO ($n = 19$) patients with attacks occurring on either side, ensuring that TMS was performed at least 3 days after the previous, or before the next, attack. Here, significantly higher mean MTs were observed during voluntary muscle contraction in MA patients than in controls. Maximal MEP/M_{max} values were normal in MA as well as in MO patients, whose attacks were not always located on the same side. Other measures were also recorded: electromyogram silent period (SP) elicited by motor cortex stimulation and intracortical inhibition (ICI) and facilitation (ICF) tested with paired-pulse TMS (Kujirai et al. 1993; Ziemann et al. 1996). All were normal in MA and MO patients. We replicated these results in a recent study using a more focal stimulation with a figure-8 coil (Bohotin et al. 2003), but this time the trend for an MT

increase in migraineurs did not reach the level of statistical significance.

By contrast, Aurora et al. (1999b) found that the cortical SP was significantly shorter in MA patients than in controls. There was, however, no control in this study for the possible occurrence of a migraine attack within 24 h after the recordings.

Finally, others found no significant changes in MT to paired stimulation (Brighina et al. 2002) or SP (Werhahn et al. 2000) in patients with MA or FHM.

In a more recent paired-pulse study, Brighina et al. (2005) found that in basal conditions (between attacks) migraineurs have significantly reduced ICI compared with controls. In addition, after 1 Hz rTMS, ICF decreased significantly in controls, while it increased in migraineurs. ICI was not significantly affected by low-frequency stimulation in either group. They interpreted their results as showing that the motor cortex in migraine patients presents an abnormal modulation of cortical excitability, where an inefficiency of inhibitory circuits likely plays a relevant role.

Visual cortex

The conflicting results obtained in studies of *phosphenes* induced by occipital TMS are summarized in Table 24.2. Aurora et al. (1998), using TMS over the occipital lobe, reported an increased occurrence of magnetophosphenes in MA patients between attacks (11/11 MA compared with 3/11 control subjects). A similar prevalence difference (100% in MA, 47% in controls) was reported by Brighina et al. (2002). Moreover, the phosphene threshold (PT), was lower in MA patients than in controls (Aurora et al. 1998). The authors confirmed these results in a subsequent study of 15 migraineurs (14 MA) and eight controls (Aurora et al. 1999a). Aguggia et al. (1999) also found a significant PT decrease in MA (58.3%) compared with controls (83.7%) or tension-type headache (81.9%). Another study (Mulleners et al. 2001a) reported a reduced phosphene in MA and in MO, but the proportion experiencing TMS-induced phosphenes, although lower in migraineurs, was not significantly different from controls. The same group found that prophylactic treatment

Table 24.2 Transcranial magnetic stimulation – visual cortex

Authors	Diagnosis (N° subjects)	Mean age	Attack control		Methods		Results		
			Before	After	Coil shape	Max. output	Phosphene prevalence	Phosphene threshold	Others
Aurora et al. (1998)	MA (11)	37	1 week	No	circular coil 95 mm	2 Tesla	100	* (↑)	44* (↓)
	HV (11)	36					27	68.7	
Afra et al. (1998)	MA (25)	36	3 days	3 days	circular 130 mm	2.5 Tesla	56* (↑)	46	
	MO (33)	36					82	50	
	HV (27)	33					89	48	
Aurora et al. (1999)	MA (14) +MO (1)	40	1 week	No	circular coil 95 mm	2 Tesla	86.7* (↑)	45* (↓)	
	HV (8)	37					25	81	
Mulleners et al. (2001a)	MA (16)	43	1 day	No	circular coil 130 mm	2 Tesla	75	47* (↓)	
	MO (12)	46					83	46* (↓)	
	HV (16)	43					94	66	
Mulleners et al. (2001b)	MA (7)	34	No	No	circular coil 130 mm	2 Tesla	No	No	↓ visual extinction
	HV (7)	36							
Battelli et al. (2002)	MA (16)	42	2 weeks	No	Figure-of-eight 70 mm	2 Tesla	65 (left)	?80 (↓)	
	MO (9)	35					67 (left)	?83 (↓)	
	HV (16)	40					6 (left)	?110	

Table 24.2 (*cont.*) Transcranial magnetic stimulation – visual cortex

Authors	Diagnosis (N° subjects)	Mean age	Attack control		Methods		Results		
			Before	After	Coil shape	Max. output	Phosphene prevalence	Phosphene threshold	Others
Brighina et al. (2002)	MA (13) HV (15)	39 32	2 days	2 days	Figure-of-eight 45 mm	?	100 47	56 57	↓ by 1 Hz rTMS ↑ by 1 Hz rTMS
Young et al. (2004)	MA (11) MO (10) MM (9) HV (15)	?	-	-	Figure-of-eight 90 mm	?	?	37* 35* 39* 51	
Aurora et al. (2005)	CM (5) EM (5) HV (5)		3 days	3 days	circular coil 95 mm	2 Tesla	?	?	↓ magnetic suppr. of perceptual accuracy
Gerwig et al. (2005)	MA (19) MO (19) HV (22)	32 39 30	3 days	3 days	Figure-of-eight 100 mm	?	100 (?)	53*/40.3* 58/40* 64/45 (single/paired)	

with valproate increased PT in MA, but not in MO (Mulleners *et al.* 2002), and the ability of a TMS pulse over the occipital cortex to suppress visual perception was reduced in MA patients (Mulleners *et al.* 2001b), which was interpreted as reflecting reduced activity of inhibitory circuits in the occipital cortex of migraineurs. Significantly lower PT for TMS over visual area V5 was found in MA and MO compared with controls (Battelli *et al.* 2002). A sequential study by Young *et al.* (2004) also concluded that PTs for occipital TMS were lower in migraine with or without aura, as well as in menstrual migraine compared with healthy subjects. Taken together, these studies were thought to favor the hypothesis of visual cortex hyperexcitability in migraine.

By contrast, Afra *et al.* (1998a) have obtained opposite results using a similar methodology and a circular coil: the prevalence of phosphenes was significantly lower in MA patients than in controls (10/18 versus 17/19), while no differences were found between controls and MO patients (18/22). Among subjects reporting phosphenes, mean PT was similar in all groups. We replicated these findings using more focal visual cortex stimulation with a figure-8 coil (Bohotin *et al.* 2003). Others failed to find significant differences between migraineurs and healthy subjects in PT for the primary visual cortex (Brighina *et al.* 2002; Valli *et al.* 2002). Interestingly, in the latter study, PT also tended to be higher in MA (71.04%) and MO (74.21% of maximum stimulator output) than in controls (62.51%).

Paired-pulse TMS of the visual cortex with an ISI of 50 ms may enhance the detection of PT differences between migraineurs and controls, and decrease variability (Gerwig *et al.* 2005). This technique allowed demonstration of significantly decreased PT in MO patients compared with controls, while the difference with single pulses was only significant between MA patients and controls. One must, however, take into account the global facilitatory effect of the paired-pulse paradigm in the visual system (Sparing *et al.* 2005).

A paradoxical effect of *rTMS* in migraine was reported by Brighina *et al.* (2002): 1 Hz rTMS, expected to inhibit the underlying cortex, actually decreased the PT. This may be due to the fact that the effects of rTMS on the cortex depend on its ongoing activation level (Kimbrell *et al.* 1999; Fierro *et al.* 2005), which may be decreased in migraineurs (Schoenen *et al.* 2003).

Although several studies confirm the reliability of PT measurements (Fumal *et al.* 2002) there is an important subjective component in the reporting of phosphenes, as well as a learning effect (Young *et al.* 2004). Moreover, the data from subjects reporting no phosphenes are not handled uniformly across studies. More objective measures are therefore needed.

Magnetic suppression of perceptual accuracy (MSPA) is a new technique combining TMS and a visual perception task in order to assess cortical excitability. Briefly, a TMS pulse is delivered over the occipital cortex 40–190 ms after the visual presentation of a letter trigram. Performance is decreased when the magnetic pulse is delivered after 100 ms and the relation between the perceptual performance and the TMS delay is U-shaped (Amassian *et al.* 1989; Corthout *et al.* 1999; Thut *et al.* 2003). In a reproducibility study, MSPA showed similar results at a 2 week interval in healthy volunteers (Custers *et al.* 2005). In a small group of migraine patients the MSPA showed no suppression of visual perception at any time interval (Aurora *et al.* 2005). This was attributed to increased cortical excitability due to decreased efficacy of inhibitory neuronal networks, although alternative mechanisms, such as of lack of habituation, could, in theory, produce similar results.

Visual-evoked potentials (VEP) represent a standard method for the investigation of visual pathway. When the amplitude change of the VEP N1-P1 and P1-N2 components are measured sequentially during sustained stimulation, migraineurs between attacks are characterized by a deficit in physiological habituation, i.e. amplitude reduction of the cortical response. This is replaced in most migraineurs by potentiation, i.e. an amplitude increase (Schoenen *et al.* 1995). This habituation deficit has been demonstrated for several other sensory modalities, but its underlying mechanisms are not known. Although increased cortical excitability and/or decreased intracortical inhibition may be responsible, the finding of decreased amplitude in the first blocks of averaged evoked responses favors another interpretation: decreased cortical preactivation level (Schoenen *et al.* 2003). Repetitive TMS offers a unique possibility to test this hypothesis. We have thus used the pattern reversal VEP (PR-VEP) amplitude and habituation as a more objective measure than phosphenes to assess excitability changes induced in the visual cortex by rTMS at high- and low-stimulation frequencies (Bohotin *et al.* 2002). In normal subjects, 1 Hz rTMS, which is known to inhibit the underlying cortex (Chen *et al.* 1997), reduced amplitude in the first block of 100 averaged responses and induced lack of habituation over successive blocks. By contrast, in MA and MO patients between attacks, 10 Hz rTMS, which most often activates the underlying cortex (Pascual-Leone *et al.* 1994), increased first block VEP amplitude in migraineurs and transformed their lack of habituation into a normal habituation pattern. This study therefore strongly supports the concept that the preactivation level in the visual cortex is low, rather than high, in migraineurs between attacks. In a more recent study (Fumal *et al.* 2006), we showed that the habituation-enhancing effect of 10 Hz rTMS increases in duration when the stimulation is repeated on 5 consecutive days.

Discussion

The majority of evoked and event-related potential studies in migraine have shown two abnormalities: increased amplitude of grand averaged responses and lack of habituation in successive blocks of averaged responses with decreased amplitude in the first block. These abnormalities suggest that the excitability state of the cerebral cortex, particularly of the visual cortex, is abnormal in migraineurs between attacks. TMS studies in migraineurs have not allowed definitive identification of the cortical dysfunction, but they have produced interesting data and stimulated research.

The use of TMS to assess motor and visual cortex excitability has yielded conflicting results. Some of the discrepancies could be due to methodological differences which may be device and patient group dependent. Most studies were performed with a circular coil, and some with a figure-8 coil (see Tables 24.1 and 24.2). The two types of coils differ substantially, since a figure-8 coil produces a focal stimulation under the center of the coil while a circular coil causes diffuse stimulation of the underlying cortical area (Cohen *et al.* 1990; Hallett 2000). It is thus likely that a larger cortical area was stimulated with the circular coils. In addition, the human cortex is sensitive to the direction of current flow in the coil, and with the circular coil this effect is more pronounced (Niehaus *et al.* 2000; Tings *et al.* 2005). Other technical differences, including biphasic versus monophasic magnetic pulses (Antal *et al.* 2002) and maximum stimulator output, for example, may contribute to the varying results, but these are sometimes difficult to evaluate based on available information.

With regard to patient selection, one must keep in mind that dramatic changes in evoked cortical responses, and thus of cortical excitability, occur 24 h before and during a migraine attack (Kropp and Gerber 1995; Judit *et al.* 2000). While the occurrence of the last attack before the recording can be checked by history, occurrence of an attack within 24 h after the recording has to be controlled by other means, such as telephone follow-up. This was done in a restricted number of studies. In addition, cortical excitability fluctuates with hormonal variations during the menstrual cycle (Smith *et al.* 2002). In our studies we avoided such hormonal influences by recording all females at midcycle.

Nevertheless, some general lines can be drawn, particularly for *motor cortex* excitability. With the notable exception of one study (van der Kamp *et al.* 1996) the general finding was reduced interictal motor cortex excitability in several forms of migraine: unilateral or bilateral MA (Maertens de Noordhout *et al.* 1992; Afra *et al.* 1998b), menstrual MO (Bettucci *et al.* 1992), and familial hemiplegic migraine (Van der Kamp *et al.* 1997). These findings do not favor the hypothesis (Welch *et al.* 1990) of a permanent cortical hyperexcitability in migraine. Excitability changes may also result from dysfunction of cortical inhibitory interneurons (Chronicle and Mulleners 1994). ICI was found to be normal in MA and MO patients (Afra *et al.* 1998b; Werhahn *et al.* 2000), although decreased ICI was found by Brighina *et al.* (2005). Moreover, in all studies except one, SPs to motor cortex stimulation were also normal. Although the origin of the SP remains debatable (Hallett 1996), this finding also argues against abnormalities of inhibitory output pathways of the motor cortex. Thus, it seems that the increase in MT in several subtypes of migraine arises from decreased excitability of the large pyramidal neurons that contribute to the MEP, as decreased spinal motoneuron excitability is unlikely in migraine.

Published studies on *visual cortex* excitability give contradictory results. Although their generator remains a matter of debate, phosphenes elicited by brief intense magnetic pulses directed to the occipital area of the brain are probably due to activation of the primary visual cortex and/or of subcortical sites such as the optic radiations adjacent to the posterior tip of the lateral ventricle (Mulleners et al. 2001a). One puzzling result in several studies (Aurora et al. 1998, 1999a; Battelli et al. 2002; Brighina et al. 2002) is the very low prevalence of phosphenes in the control groups (3/11, 2/8, 1/16, and 7/15 respectively), while all previous studies conducted in normal subjects report a much higher prevalence, usually between 60 and 80% (Maccabee et al. 1991; Meyer et al. 1991; Kastner et al. 1998; Stewart et al. 2001). Although one cannot exclude the presence of migraineurs among the 'healthy' volunteers in the latter studies, the phosphene prevalence rates found in our studies (Afra et al., 1998b; Bohotin et al. 2003), where subjects with a personal or family history of migraine were excluded, were similar (89% and 64%).

Contrasting with the findings from our group (Afra et al. 1998b; Bohotin et al. 2003) and others (Brighina et al. 2002) of a significantly higher or normal PT in MO and MA patients, Aurora et al. (1998, 1999), Mulleners et al. (2001a), and Young et al. (2004) reported a significantly lower PT. In another recent study, where TMS was applied laterally over visual area V5 (Battelli et al. 2002), PT was also lower in migraine patients. Results of the latter study are difficult to compare with previous studies in which midline occipital TMS, which likely stimulates V1 or subjacent white matter, was used. It must be pointed out that in studies (Aurora et al. 1998, 1999a; Mulleners et al. 2001a; Young et al. 2004) reporting reduced PT, subjects who had no phosphenes were not included numerically in the average threshold calculations. As a matter of fact, if one assumes that subjects without phosphenes have at least a 100% threshold, a recalculation of the figures in Mulleners et al.'s (2001a) paper, for instance, increases the mean PT in the MO group from 46% to 55%, in the MA group from 47% to 60.2%, while in the healthy volunteer group it would increase only from 66% to 68%. Needless to say, the differences between groups would lose statistical significance after such recalculations. On the other hand, some subjects may be resistant to induction of phosphenes by TMS and arbitrarily assigning them a PT of 100% may increase the risk of type 2 error. This was done, nevertheless, in Battelli et al.'s (2002) study, in which a threshold value of 110% was arbitrarily assigned to the 94% of control subjects who reported no phosphenes. Considering that only one-third of migraineurs had no phosphenes in that study, there is little doubt that the results of PT were biased towards higher values in the controls. Moreover, interpretation of Battelli et al.'s (2002) results may be confounded by the modulatory effect of attention and expectation on extrastriate visual areas (Walsh et al. 1998).

It may be premature to consider decreased PT in migraine as evidence for increased excitability of the visual cortex. The fact that opposite results were obtained in different laboratories with similar methods suggests that phosphenes are too subjective and variable to be used to measure excitability of the visual cortex. Alternative, more objective and reliable methods should be used. We have shown, for instance, that PR-VEP can be used to assess excitability changes of the visual cortex induced by rTMS (Bohotin et al. 2002). Using this method and analyzing PR-VEP habituation, we have obtained evidence in favor of hypoexcitability, not hyperexcitabilty. Others (Aurora et al. 2005; Custers et al. 2005) have assessed visual cortex excitability by measuring TMS-induced visual extinction and have concluded that inhibitory mechanisms are deficient in migraineurs. Taken together, these studies indicate that the changes in cortical reactivity are more complex in migraineurs than initially thought and suggest that both larger multidisciplinary studies and focused analyses of subgroups of patients with more refined clinical phenotypes are necessary to disentangle the role of the cerebral cortex in migraine pathophysiology.

References

Afra J, Cecchini AP, De Pasqua V, Albert A, Schoenen J (1998a). Visual evoked potentials during long periods of pattern-reversal stimulation in migraine. *Brain 121*, 233–241.

Afra J, Mascia A, Gérard P, Maertens de Noordhout A and Schoenen J (1998b). Interictal cortical excitability in migraine: a study using transcranial magnetic stimulation of motor and visual cortices. *Annals of Neurology 44*, 209–215.

Amassian VE, Cracco RQ, Maccabee PJ, Cracco JB, Erberle L, Rudell A (1989). Suppression of visual perception by magnetic cortical stimulation of human occipital cortex. *Electroencephalography and Clinical Neurophysiology 74*, 458–462.

Antal A, Kincses TZ, Nitsche MA, *et al.* (2002). Pulse configuration-dependent effects of repetitive transcranial magnetic stimulation on visual perception. *Neuroreport 13*, 2229–2233.

Aurora SK, Ahmad BK, Welch KMA, *et al.* (1998). Transcranial magnetic stimulation confirms hyperexcitability of visual cortex in migraine. *Neurology 50*, 1105–1110.

Aurora SK, Cao Y, Bowyer SM, Welch KM (1999a). The occipital cortex is hyperexcitable in migraine: experimental evidence. *Headache 39*, 469–476.

Aurora SK, al-Sayeed F, Welch KM (1999b). The cortical silent period is shortened in migraine with aura. *Cephalalgia 19*, 708–712.

Aurora SK, Barrodale P, Chronicle EP, Mulleners WM (2005). Cortical inhibition is reduced in chronic and episodic migraine and demonstrates a spectrum of illness. *Headache 45*, 546–552)

Battelli L, Black KR, Wray SH (2002). Transcranial magnetic stimulation of visual area V5 in migraine. *Neurology 58*, 1066–1069.

Bettucci D, Cantello M, Gianelli M, *et al.* (1992) Menstrual migraine without aura: cortical excitability to magnetic stimulation. *Headache 32*, 345–347.

Bohotin V, Fumal A, Vandenheede M, *et al.* (2002). Effects of repetitive transcranial magnetic stimulation on visual evoked potentials in migraine. *Brain 125*, 1–11.

Bohotin V, Fumal A, Vandenheede M, Bohotin C, Schoenen J (2003). Excitability of visual V1-V2 and motor cortices to single transcranial magnetic stimuli in migraine: a reappraisal using a figure-of-eight coil. *Cephalalgia 23*, 264–270.

Brighina F, Piazza A, Daniele O, Fierro B (2002). Modulation of visual cortical excitability in migraine with aura: effects of 1 Hz repetitive transcranial magnetic stimulation. *Experimental Brain Research 145*, 177–181.

Brighina F, Giglia G, Scalia S, Francolini M, Palermo A, Fierro B (2005). Facilitatory effects of 1 Hz rTMS in motor cortex of patients affected by migraine with aura. *Experimental Brain Research 161*, 34–38.

Chen R, Classen J, Gerloff C, *et al.* (1997). Depression of motor cortex excitability by low-frequency transcranial magnetic stimulation. *Neurology 48*, 1398–1403.

Chronicle EP, Mulleners W (1994). Might migraine damage the brain? *Cephalalgia 14*, 415–418.

Cohen LG, Roth BJ, Nilsson J, *et al.* (1990). Effects of coil design on delivery of focal magnetic stimulation. Technical considerations. *Electroencephalography and Clinical Neurophysiology 75*, 350–357.

Corthout E, Uttl B, Ziemann U, Cowey A, Hallett M (1999). Two periods of processing in the (circum)striate visual cortex as revealed by transcranial magnetic stimulation. *Neuropsychologia 37*, 137–145.

Custers A, Mulleners WM, Chronicle EP (2005). Assessing cortical excitability in migraine: reliability of magnetic suppression of perceptual accuracy technique over time. *Headache 45*, 1202–1207.

Fierro B, Brighina F, Vitello G, *et al.* (2005). Modulatory effects of low- and high-frequency repetitive transcranial magnetic stimulation on visual cortex of healthy subjects undergoing light deprivation. *Journal of Physiology 565*, 659–665.

Fumal A, Bohotin V, Vandenheede M, Seidel L, Maertens de Noordhout A, Schoenen J (2002). Motor and phosphene thresholds to transcranial magnetic stimuli: a reproducibility study. *Acta Neurologica Belgica 102*, 171–175.

Fumal A, Coppola G, Bohotin V, *et al.* (2006). Induction of long-lasting changes of visual cortex excitability by five daily sessions of repetitive transcranial magnetic stimulation (rTMS) in healthy volunteers and migraine patients. *Cephalalgia 26*, 143–149.

Gerwig M, Niehaus L, Kastrup O, Stude P, Diener HC (2005). Visual cortex excitability in migraine evaluated by single and paired magnetic stimuli. *Headache 45*, 1394–1399.

Hallett M (1996). Transcranial magnetic stimulation: a useful tool for clinical neurophysiology. *Annals of Neurology 40*, 344–345.

Hallett M (2000). Transcranial magnetic stimulation and the human brain. *Nature 406*, 147–150.

Hegerl U, Juckel G (1993). Intensity dependence of auditory evoked potentials as an indicator of central serotonergic neurotransmission: a new hypothesis. *Biological Psychiatry 33*, 173–187.

Judit A, Sandor P, Schoenen J (2000). Habituation of visual and intensity dependence of auditory evoked cortical potentials tends to normalize just before and during the migraine attack. *Cephalalgia 20*, 714–719.

Kastner S, Demmer I, Zieman U (1998). Transient visual field defects induced by transcranial magnetic stimulation over human occipital lobe. *Experimental Brain Research 118*, 19–26.

Kimbrell TA, Little JT, Dunn RT, *et al.* (1999). Frequency dependence of antidepressant response to left prefrontal repetitive transcranial magnetic stimulation (rTMS) as a function of baseline cerebral glucose metabolism. *Biological Psychiatry 46*, 1603–1613.

Kropp P, Gerber WD (1995). Contingent negative variation during migraine attack and interval: evidence for normalization of slow cortical potentials during the attack. *Cephalalgia 15*, 123–128.

Kujirai T, Caramia MD, Rothwell JC *et al.* (1993). Corticocortical inhibition in human motor cortex. *Journal of Physiology (London) 471*, 501–519.

Maccabee PJ, Amassian VE, Cracco JB *et al.* (1991). Magnetic coil stimulation of human visual cortex: studies of perception. *Electroencephalography and Clinical Neurophysiology 43*, 111–120.

Maertens de Noordhout A, Pepin JL, Schoenen J, Delwaide PJ (1992). Percutaneous magnetic stimulation of the motor cortex in migraine. *Electroencephalography and Clinical Neurophysiology 85*, 110–115.

Meyer BU, Diehl R, Steinmetz H, *et al.* (1991). Magnetic stimuli applied over motor and visual cortex: influence of coil position and diled polarity on motor responses, phosphenes and eye movements. *Electroencephalography and Clinical Neurophysiology 43*, 121–134.

Mulleners WM, Chronicle EP, Palmer JE, Koehler PJ, Vredeveld JW (2001a). Visual cortex excitability in migraine with and without aura. *Headache 41*, 565–572.

Mulleners WM, Chronicle EP, Palmer JE, Koehler PJ, Vredeveld JW (2001b). Suppression of perception in migraine: evidence for reduced inhibition in the visual cortex. *Neurology 56*, 178–183.

Mulleners WM, Chronicle EP, Vredeveld JW, Koehler PJ (2002). Visual cortex excitability in migraine before and after valproate prophylaxis: a pilot study using TMS. *European Journal of Neurology 9*, 35–40.

Niehaus L, Meyer BU, Weyh T (2000). Influence of pulse configuration and direction of coil current on excitatory effects of magnetic motor cortex and nerve stimulation. *Clinical Neurophysiology 111*, 75–80.

Pascual-Leone A, Valls Sole J, Wassermann EM, Hallett M (1994). Responses to rapid-rate transcranial magnetic stimulation of the human motor cortex. *Brain 117*, 847–858.

Schoenen J (1994). Pathogenesis of migraine: the biobehavioural and hypoxia theories reconciled. *Acta Neurologica Belgica 94*, 79–86.

Schoenen J (1996). Abnormal cortical information processing between migraine attacks. In: M Sandler, M Ferrari, S Harnett (eds), *Migraine: pharmacology and genetics*, pp. 233–253. London: Altman.

Schoenen J (1997). Clinical neurophysiology of headache. *Neurology Clinical 15*, 85–105.

Schoenen J (1998). Cortical electrophysiology in migraine and possible pathogenetic implications. *Clinical Neuroscience 5*, 10–17.

Schoenen J, Wang W, Albert A, Delwaide PJ (1995). Potentiation instead of habituation characterizes visual evoked potentials in migraine patients between attacks. *European Journal of Neurology 2*, 115–122.

Schoenen J, Ambrosini A, Sandor PS, Maertens de Noordhout A (2003). Evoked potentials and transcranial magnetic stimulation in migraine: published data and viewpoint on their pathophysiologic significance. *Clinical Neurophysiology 114*, 955–972.

Smith MJ, Adams LF, Schmidt PJ, Rubinow DR, Wassermann EM (2002). Effects of ovarian hormones on human cortical excitability. *Annals of Neurology 51*, 599–603.

Sparing R, Dambeck N, Stock K, Meister IG, Huetter D, Boroojerdi B (2005). Investigation of the primary visual cortex using short-interval paired-pulse transcranial magnetic stimulation (TMS). *Neuroscience Letters 382*, 312–316.

Stewart LM, Walsh V, Rothwell JC (2001). Motor and phosphene thresholds: a transcranial magnetic stimulation correlation study. *Neuropsychologia 39*, 415–419.

Thut G, Northoff G, Ives JR, *et al.* (2003). Effects of single-pulse transcranial magnetic stimulation (TMS) on functional brain activity: a combined event-related TMS and evoked potential study. *Clinical Neurophysiology 114*, 2071–2080.

Tings T, Lang N, Tergau F, Paulus W, Sommer M (2005). Orientation-specific fast rTMS maximizes corticospinal inhibition and facilitation. *Experimental Brain Research 164*, 323–333.

Valli G, Cappellari A, Zago S, Ciammola A, De Benedittis G (2002). Migraine is not associated with hyperexcitability of the occipital cortex. A transcranial magnetic stimulation controlled study. *Proceedings of the 10th World Congress of Pain, 2002*, 1580-P128, p. 525. Seattle, WA IASP Press.

van der Kamp W, Maassen van den Brink, Ferrari M, van Dijck JG (1996). Interictal cortical hyperexcitability in migraine patients demonstrated with transcranial magnetic stimulation. *Journal of the Neurological Sciences 139*, 106–110.

van der Kamp W, Maassen van den Brink A, Ferrari M, van Dijck JG (1997). Interictal cortical excitability to magnetic stimulation in familial hemiplegic migraine. *Neurology 48*, 1462–1464.

Walsh V, Ellison A, Battelli L, Cowey A (1998). Task-specific impairments and enhancements induced by magnetic stimulation of human visual area V5. *Proceedings of the Royal Society of London B 265*, 537–543.

Welch KMA, D'Andrea G, Tepley N, Barkley G, Ramadan NM (1990). The concept of migraine as a state of central neuronal hyperexcitability. *Neurology Clinics 8*, 817–828.

Werhahn KJ, Wiseman K, Herzog J, Forderreuther S, Dichgans M, Straube A (2000). Motor cortex excitability in patients with migraine with aura and hemiplegic migraine. *Cephalalgia 20*, 45–50.

Young WB, Oshinsky ML, Shechter AL, Gebeline-Myers C, Bradley KC, Wassermann EM (2004). Consecutive transcranial magnetic stimulation: phosphene thresholds in migraineurs and controls. *Headache 44*, 131–135.

Ziemann U, Lönnecker S, Steinhof B, Paulus W (1996). Effects of antiepileptic drugs on motor cortex excitability in humans: a transcranial magnetic stimulation study. *Annals of Neurology 40*, 367–378.

CHAPTER 25

Design and analysis of motor-evoked potential data in pediatric neurobehavioral disorder investigations

Donald L. Gilbert

Introduction

Transcranial magnetic stimulation has been used in children and adolescents to study and quantify neurophysiological properties in motor cortex. There are special challenges in pediatric TMS research, including limited co-operation of subjects, human subject protection regulations in special populations, and parental concerns about research safety. Here, the term 'neurobehavioral disorders' refers to conditions where patients have difficulty regulating cognition, behavior, or mood, or have subnormal function in these areas. Pathophysiological investigations in this area require quantitative biological markers. However, considerable barriers exist to identifying markers for these clinically diagnosed disorders.

With these issues in mind, this chapter aims to instruct readers, including those with little knowledge of neurophysiological research, about how TMS can be used to study the pathophysiological substrata of pediatric neurological and neurobehavioral disorders and to provide practical guidance for future research. It is organized into four sections. First, for context and to aid new researchers in this area, I will sketch briefly the substantial challenges inherent in studying *in vivo* the neurobiology of pediatric neurobehavioral disorders. These challenges are broadly applicable to the use of quantitative CNS physiology in neurobehavioral disorders in children or adults. Second, I will discuss ways in which TMS generates quantitative measures that may function as endophenotypes for neurobehavioral disorders. Third, I will review the small amount of published research using single- and paired-pulse TMS in motor cortex in pediatric neurobehavioral disorders and show, using examples from our work, methods by which some of the research challenges may be addressed statistically. Fourth, I will conclude by discussing recent studies that may point to new and more productive directions for TMS research in children.

Challenges in using quantitative neurophysiological techniques to study neurobehavioral disorders

Safety

The first challenge is that TMS imparts energy to the human body. Therefore, the first issue that must be confronted when putting a proposal before an ethical review board is that of safety. A full discussion of this issue lies outside the scope of this chapter, but can be found elsewhere (Gilbert et al. 2004b). To summarize, after reviewing (i) the United States Code of Federal Regulations involving protection of pediatric populations in research, (ii) theoretical and biological considerations based on energy imparted by the TMS coil, (iii) animal studies using magnetic stimulation at high intensities and frequencies, (iv) published research using single- and paired-pulse TMS in children (28 studies, >850 children prior to 2001), and (v) our own safety data in children, we concluded that many well-designed studies using single- and paired-pulse TMS in children could be classified by US Institutional Review Boards as having minimal risk (Gilbert et al. 2004b).

Quantitative versus categorical measures

When a TMS coil is discharged over motor cortex and an evoked response produced in muscle, a series of quantitative continuous measures can be derived including stimulus intensity or threshold, latency, silent period, and motor-evoked potential (MEP) amplitude. These quantitative measures can be recorded digitally and analyzed automatically without recourse to subjective visual interpretation. Thus, TMS does not suffer from the wide interobserver variation found in, for example, interictal electroencephalograms (van Donselaar et al. 1992; Gilbert et al. 2003).

However, it is important to note that for a quantitative measure to be 'abnormal' a threshold level of abnormality has to be defined. Abnormality can be defined for a specific disorder, based on data from a large number of normal and affected volunteers, using receiver operating characteristic analysis (Connell and Koepsell 1985; Sackett et al. 1991). This can be problematic for several reasons. One is that, if the variation within normals is extremely wide, there may be significant overlap between affected and unaffected groups. Then, for any threshold abnormality there will be a large tradeoff between sensitivity and specificity (highly sensitive values will have low specificity

and vice versa). Another is that, in children, age or maturation may affect these measures, and therefore 'normal' would have to be based on a large body of normative pediatric data. No such data exist, and acquiring it would be time consuming and difficult.

With this in mind, there are two common and relatively simple study designs to consider for the use of TMS in studying behavioral disorders. In the first approach, neurobehavioral phenotypes, e.g. disease cases vs normal controls, are treated as class variables and the TMS measures are suitable for group-mean comparisons. In the second approach, scalar measures or ratings of behavioral symptoms may be compared to TMS measures using correlational statistics or regression. Investigators should consider carefully the strengths and weaknesses of these approaches and make explicit the implications of the study designs in interpreting their data. We will consider each of these approaches in turn.

Case–control studies

In this type of design, the disorder is a class variable, indicated as *present* or *absent*. The study hypothesis is that a particular neurophysiological measure is different when the disorder is present versus absent and can be tested with a parametric Student's *t*-test or a more conservative nonparametric test, if necessary. If two or more disease groups are compared with normals, analysis of variance would be used. Since there is not a large body of normative data, case–control studies should ideally match age, gender, and possibly other demographic factors as closely as possible.

The first challenge of this study design occurs because most neurobehavioral diagnoses are based on the presence of signs or symptoms, without independent biological validation, and because, prior to initiating the study, it is unknown if a neurophysiological measure will correlate with (i) an underlying disease process, or (ii) a symptom. This problem is best understood by considering a specific hypothetical example.

Suppose investigators wish to know whether the sign of chorea corresponds to a particular measure of disinhibition in motor cortex. A case–control study is proposed to compare chorea patients with normal controls. Suppose, moreover, that the chorea group includes individuals with benign hereditary chorea and individuals with Huntington's disease. Here are three possible outcomes.

1. The group with chorea has significantly different cortical inhibition than the normal controls. This supports an interpretation that the measure of inhibition reflects the clinical presence of chorea.

2. The TMS measure of inhibition does not distinguish between Huntington's chorea and normal controls, but inhibition is significantly different in the group with benign hereditary chorea. This supports the interpretation that the abnormal inhibition is specific for one form of chorea, but the presence of chorea alone is not linked to the abnormal cortical inhibition.

3. There is no difference between either form of chorea and controls.

In this example, there are clinical features and genetic tests which can be used to distinguish the two chorea diagnoses, so both outcome 1 and outcome 2 would be interesting and informative. Suppose, however, that the two forms of chorea were not in any way distinguishable. Then, the results of a comparison of 'chorea' versus nonchorea patients would be difficult to interpret. If the groups have significantly different cortical inhibition, it could be because (i) the chorea symptom correlates with the inhibitory physiology, irrespective of the cause, or (ii) one of the two forms of chorea had extremely abnormal results which pulled the group mean away from the control mean, leading to a statistically significant difference. In contrast, if the study failed to identify a significant difference, it could be because (i) the chorea symptom did not correlate with the inhibitory physiology, or (ii) despite the fact that one of the two forms of chorea did differ significantly from the normal controls, the difference was too small to pull the entire chorea group's mean away from the control mean.

This situation may apply to several neurobehavioral disorders. For any condition that is not a high-penetrance single-gene disorder, or that lacks some other means of biological confirmation, in a case–control design we cannot be sure whether the 'cases' represent one disease or not. Even behavioral disorders that appear highly heritable are likely the product of multiple genes, each of which makes a small contribution to the risk of developing the clinical condition. If multiple genes and environmental factors contribute, or if there are multiple disorder 'phenocopies', a neurophysiological measure linked tightly to a neurobiological process may correlate with some disease subgroups, but not others. However, if the TMS measure relates to a biological process responsible for the core symptoms of the disorder, irrespective of cause, then there may be a statistically significant difference in a case-controlled study. In either circumstance, case mix may exert a large effect on study results and the interpretation of positive and negative results may be problematic.

The second challenge relates to the method of diagnosis of pediatric neurobehavioral disorders, which relies on a classification scheme such as the *Diagnostic and statistical manual* (American Psychiatric Association 2000) and observer judgment. These disorder phenotypes are constellations of overlapping symptoms placed into consensus-driven categories, which have evolved over time and will change again with the next edition of the manual. Diagnosis requires the use of a symptom rating scale that reduces patterns of behavior to descriptive sentences ('often has difficulty organizing tasks and activities') to be rated on a Likert Scale (e.g. 'not at all', 'just a little', 'pretty much', 'very much'). There may be substantial inter-rater variation between parents, teachers, clinicians, and researchers using these scales. As a result, even if there are biological entities captured by these behavioral syndromic categories, there may be diagnostic misclassification and, therefore, ambiguous class variables. In addition, for many behavioral disorders, symptoms at the mild end of the spectrum blend into age-appropriate normal behavior, leading to failure to find significant differences in group mean comparisons.

The third challenge relates to deciding how to analyze common occurrence of multiple diagnoses in the same child. For example, children with attention deficit hyperactivity disorder (ADHD) may have comorbid obsessive–compulsive disorder (OCD), children with Tourette's syndrome (TS) may have comorbid

ADHD and OCD, and children with Asperger's disorder may have comorbid TS, ADHD, and OCD. This may be conceptualized as multiple separate disorders occurring in the same individual or as a single disorder with a spectrum of presentations. Those with a large number of symptoms may not have more diagnoses, but rather a more severe phenotype. Investigators must decide how to classify individuals who meet criteria for multiple disorders. A decision to exclude such children from study may enhance validity, but a filtered group of cases may poorly represent the clinical condition being studied.

A fourth challenge involves medications, which may affect many TMS measures. If only unmedicated children are included, a potential confounder may be removed. However, the cases may then represent only the mild end of the disorder's severity spectrum. Not only does this reduce generalizability, but it may also reduce the chance of a positive finding, since these patients may be physiologically more like normal controls than those more severely affected. On the other hand, if medicated patients are included, medications may exert known or unanticipated effects on TMS measures, biasing results. If medications are to be discontinued for a short time in order to allow for study participation, the child may suffer unnecessarily, a particular difficulty during the academic year. Most likely, it will only be ethical to discontinue short-acting medications, such as stimulants. Other patients may have to be studied at the time of diagnosis or during scheduled medication holidays or changes.

Finally, there is the issue of blinding. Ideally, the TMS operators would perform measures without knowing the diagnostic status of the subjects. In some cases, it may be obvious which children are affected, introducing the possibility of bias in TMS performance or data analysis.

Challenges in correlational studies

In this type of study, patients with the disorder under investigation may be enrolled without any healthy control subjects. The study hypothesis is that severity of a disorder's symptoms correlates with a physiological measure. This hypothesis can be tested with a parametric or nonparametric correlation statistic, or with a regression if age, sex, current medications, or other characteristics may be important.

For this approach to work well, the disorder of interest must have a clear phenotype, but the distinction between mild cases and controls may not be as important, since a control group is not necessary in this design. The most significant challenge still involves the use of rating scales. Inter-rater variation and nonspecificity of symptom ratings make it unlikely that tight correlations between neurophysiological measures and symptom severity will be found. More importantly, the correlational design presupposes that the scalar data produced by a symptom rating scale fit the proposed model, for instance a linear one, where increases from 1 to 2 and from 2 to 3 represent the same quantity in biological terms. In reality, however, data from clinical rating scales only approximate true measures and there is no reason to assume that an increase on a particular domain on a Likert scale from 1 to 2 is biologically equivalent to an increase from 3 to 4. Worse still, numerically equivalent increases in separate domains may be treated equally in this kind of design. Yet it is not likely that each change would correlate with the same quantitative change in a biological measurement.

To see this problem more clearly, take the example of ratings within the inattentive and hyperactive domains of ADHD. Individuals who are 'moderately inattentive *and* minimally hyperactive' and individuals who are 'minimally inattentive *and* moderately hyperactive' may have the same ADHD rating scale score. If distinct neurobiological processes underlie the domains of inattention and hyperkinetic behavior, then the correlation between overall ADHD severity and a neurophysiological measure may be weak. One possible solution to this problem is to model the neurophysiological measure as a dependent variable and consider each symptom domain scale as an independent continuous variable, and then perform a stepwise regression to determine which domain accounts for most variance in the neurophysiological measure. If one domain accounts for significantly greater variance, then the neurophysiological measure may be an important endophenotype for the symptoms in that domain.

To reduce the likelihood of bias in this study design, the rater performing the diagnostic rating scales should be blinded to the TMS results, and the investigator and technician performing the TMS should blinded to the clinical rating scale results.

Use of TMS to identify endophenotypes

The challenges inherent in the study of neurobehavioral disorders, some of which have just been described, have resulted in a quest for *endophenotypes*, which are objectively measurable and quantifiable traits intermediate between the clinical expression of a disorder its underlying biology (Castellanos and Tannock 2002). Endophenotypes may bridge clinical symptoms and the genetic factors that confer risk for the disorder. Ideally, endophenotypes should be measures that correlate with a disorder's severity or that distinguish between clinical subtypes. TMS studies yield a variety of quantitative measures, which may potentially serve as endophenotypes in pediatric and adult neurobehavioral disorders.

The MEP as endophenotype

The MEP amplitude that results from TMS of motor cortex is the most commonly reported measure in neurobehavioral studies. Many pediatric neurobehavioral disorders putatively involve abnormal signalling within frontal cortex or striatal circuits (Penney and Young 1983; Alexander *et al.* 1986; Singer 2000; Mostofsky *et al.* 2002). These disorders may include ADHD, OCD, and TS (Singer 1997; Saxena *et al.* 1999; Mink 2001; Durston *et al.* 2003). The neurophysiology of subcortical structures cannot currently be measured noninvasively, but TMS can measure properties of the motor cortex node of these circuits.

Use of TMS to study properties of the cortical node of cortical–subcortical circuits

TMS measures of developmental changes in the stimulation thresholds, latencies, and amplitudes of MEPs during development (Muller *et al.* 1991;

Eyre *et al.* 2001; Garvey *et al.* 2003, 2005) are relevant to behavioral disorders but are reviewed elsewhere in this volume (Chapter 22). The remainder of this chapter will focus instead on protocols using paired stimuli from the same coil, over motor cortex. This procedure has been used to assess excitatory and inhibitory function in motor cortex. For example, motor cortex inhibitory function has been studied using TMS in ADHD, OCD, and TS. The most studied measure has been short-interval intracortical inhibition (SICI) (Kujirai *et al.* 1993).

A change in SICI, expressed as a paired-pulse to single-pulse MEP amplitude ratio, could be an endophenotype for behavioral disorders that result from either cortical or subcortical pathological processes. Used alone, motor cortex TMS cannot distinguish between these possibilities. For subcortical processes, for example, if a hyperkinetic disorder results from striatal dysfunction, there may be increased excitatory output from the disinhibited thalamus to motor cortex pyramidal cells, to cortical interneurons, or to both. Over-driven motor cortex pyramidal cells might be less susceptible to inhibition by cortical interneurons, resulting in larger conditioned:unconditioned MEP amplitude ratios (less SICI or greater intracortical facilitation; ICF). For cortical processes, for example, if a hyperkinetic disorder results from dysfunctional cortical inhibitory interneurons, there may be deficient inhibitory capacity within cortex. The disinhibited motor cortex might be less susceptible to inhibition or more susceptible to excitation, resulting in larger conditioned: unconditioned MEP amplitude ratios. Since the paired and single pulse outputs reflect both cortical and subcortical function, behavioral-disorder-related differences in SICI cannot be accurately localized anatomically in simple case–control or correlational studies.

MEP studies in behavioral disorders

Case-control studies: motor cortex physiology in ADHD and TS

Several studies have assessed MEP amplitudes in children with ADHD and TS using case-comparison designs. Such studies are thus prone to some of the design issues discussed in 'Challenges in using quantitative neurophysiological techniques to study neurobehavioral disorders' above. Some studies show that the prevalence of undiagnosed tic disorders is quite high (Snider *et al*. 2002). In addition, myoclonus or stereotypies may be misdiagnosed as tics. Thus, there exists the potential for diagnostic misclassification. However, since tics are directly observable behaviors, as compared with many of the clinical phenomena of ADHD, correct diagnosis and inter-rater variation are probably not a major challenge for tic disorder studies.

In contrast, the other caveats discussed probably apply. For example, if a case–control study shows abnormal SICI in tic disorders, does this indicate that the function assessed with paired-pulse TMS correlates with tic behavior? Or, if there are multiple clinically indistinguishable etiologies of tic disorders, perhaps SICI correlates with the presence of one these entities, but not with the tic symptom *per se*. Defining the phenotypic spectrum is also important. Most individuals with tics also have ADHD or OCD/anxiety symptoms. Are these distinct diagnoses or part of a spectrum?

In studies comparing healthy children with children aged 10–16 years with tic disorders, with ADHD excluded or analyzed as a separate categorical variable, there was no reduced SICI in children with tics (Moll *et al*. 1999, 2001). In contrast, two studies in adults have reported reduced SICI in TS (Ziemann *et al*. 1997; Orth *et al*. 2005). In one study, the presence of ADHD appears not to have been assessed (Orth *et al*. 2005). Only the earlier adult study assessed patients for ADHD and overall it is unclear what role, if any, ADHD symptoms played in the findings of reduced cortical inhibition. The differences in these studies of SICI in tic disorders may be a function of patient age or case mix and decisions about including individuals with comorbid disorders.

In another case–control study in children, SICI was compared between children with ADHD ($n = 18$) and normal volunteers ($n = 18$) (Moll *et al*. 2000). SICI was significantly reduced in the ADHD children. Retesting the ADHD children after a single 10 mg dose of methylphenidate showed increased SICI, although still significantly less than the normal group. A subsequent study comparing children with ADHD only ($n = 16$), ADHD plus tics ($n = 16$), tics only ($n = 16$), and children with no tics or ADHD ($n = 16$) (Moll *et al*. 2001) showed that SICI was significantly reduced only in the ADHD groups. Another measure, the cortical silent period (cSP; see Chapter 10) was significantly shorter only in the tic disorder groups.

Attention deficit hyperactivity disorder has also been studied by measuring the latency between the TMS pulse and the MEP. Under various conditions, the central conduction time is longer, on average, than in healthy controls. The results suggest that ADHD may involve delayed maturation of white matter, particularly transcallosal pathways (Buchmann *et al*. 2003; Garvey *et al*. 2005). One additional study purported to show right–left differences in central motor conduction latencies (Ucles *et al*. 1996), but inappropriate (paired, parametric) statistical tests were used and the results have not been verified by other investigators.

Correlational studies: regression analysis of tics, ADHD and OCD symptoms, and cortical inhibition in a TS cohort and methods for interpreting data in complex patient cohorts

In this section we discuss our results but also give examples of how challenges discussed above may be addressed.

Because case–control TMS studies in children and adults with ADHD, TS, and OCD led to contradictory results, we chose to clarify the SICI–behavior relationship by studying 36 patients, aged 8–47 years (28 children, eight adults) with TS and a broad spectrum of tic, ADHD, and OCD symptoms (Gilbert *et al*. 2004a). In order to obtain a representative cohort that included severely affected patients, we chose to include patients taking symptom-suppressing medication and had them take their medications on the day of the study as they normally would. Thus our neurophysiological measures represented that patient's typical baseline state on the day of the study visit.

To reduce bias, a nurse clinician experienced in the use of validated clinical rating scales

(Leckman *et al.* 1989; Scahill *et al.* 1997; DuPaul *et al.* 1998) graded tic, ADHD, and OCD symptoms independently, without knowledge of the TMS data.

We used correlation and regression analyses to estimate the strength of the relationship between current symptom severity and SICI.

In preparation for this analysis, we performed univariate analyses for all medications to determine if medication use, by class (e.g. stimulants) accounted for significant variance in SICI. No medication exerted a statistically significant effect.

Next, we looked at simple correlations between the severity of the three symptoms of interest and SICI. In a univariate analysis, motor tic severity correlated inversely with SICI ($r = -0.43$, $P = 0.02$; Figure 25.1a): patients with more severe motor tics had less SICI. *Post hoc* analysis suggested that the motor tic frequency subscale accounted for the most tic-related variance in SICI. However, we found a more robust univariate, inverse correlation between SICI and higher ADHD scores ($r = -0.53$, $P = 0.003$), particularly among subjects not taking D_2 receptor-blocking agents ($r = -0.72$, $P = 0.003$). In contrast, there was no significant univariate correlation between OCD severity and SICI. Stratifying this cohort, ADHD + TS children and adults had significantly less SICI than TS-only subjects, consistent with results reported by Moll *et al.* (2001).

Because all patients had a TS diagnosis, we could not compare TS vs non-TS. However, not all patients met criteria for ADHD or OCD as categorical diagnoses. Therefore we could estimate the SICI by categorical diagnoses.

1. For TS only, SICI = 0.27 (0.17–0.41).
2. For TS + OCD, SICI = 0.28 (0.08–1.06).
3. For TS + OCD + ADHD, SICI = 0.48 (0.28–0.83).
4. For TS+ ADHD, SICI = 0.70 (0.47–1.03).

Consistent with results reported by Moll *et al.* (2001) children and adults with ADHD had significantly less SICI (larger MEP amplitude ratios).

Since our cohort was complex and many factors could be important, we then assessed the influence of multiple factors on SICI simultaneously. The model was constructed using all demographic and clinical rating scale variables and medications by class (e.g. stimulants), with selection of the best-fit variables using the maximal r^2 improvement method. This showed that the only significant factors were ADHD and motor tic severity. Age, medications, and OCD were not significant, although there was a trend toward an interaction with the use of D_2 receptor-blocking medications and ADHD. The severities of ADHD symptoms and motor tics were significantly and independently associated with reduced SICI ($r^2 = 0.50$, $F(2,27) = 13.7$, $P < 0.001$).

Post hoc analyses of medication and symptom domain effects

Univariate analyses had identified no significant medication effects on SICI; however, the multivariate analysis suggested there was an interaction between D_2 receptor-blocking agents and ADHD. Therefore, we assessed this relationship and found that, particularly in subjects not taking D_2 receptor blockers, symptom severity accounted for an even higher proportion of the variation in SICI ($r^2 = 0.68$, $F(2,17) = 17.8$, $P < 0.0001$). We assessed symptom domains in both the tic (number, frequency, intensity, complexity, interference) and ADHD (attention, hyperactivity/impulsivity) scales. Since ADHD symptoms are rated in two major domains (inattention and hyperactivity), we wondered which might be more important. *Post hoc* analysis suggested that tic frequency, not the other scale domains, and hyperactivity, rather than inattention, accounted for most of the tic- and ADHD-related variance in SICI.

Within this cohort, the range of cSP duration values, measured after a pulse at 30% above the active MEP threshold, was consistent with the range reported in the prior case–control studies of patients with tic disorders. However, in contrast to other studies (Ziemann *et al.* 1997; Moll *et al.* 2001), we did not identify any clinical or demographic factors that accounted for variance in cSP. In particular, we found no correlation between tic severity and cSP duration.

Internal validation: reproducibility and reliability of the relationship between SICI and ADHD

Theoretically, since TMS generates quantitative measures that do not rely on visual interpretation,

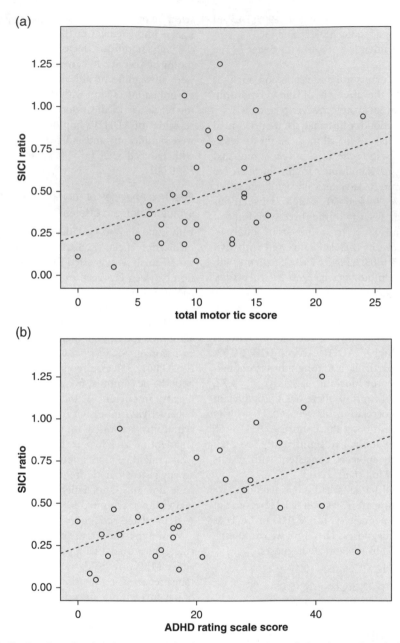

Fig. 25.1 Scatterplots showing the relationship of symptom severity and short-interval cortical inhibition (SICI). Each dot represents data from a single individual. SICI is represented as the *ratio* of MEP amplitude after 3 ms paired-pulse TMS over the motor-evoked potential (MEP) amplitude after single-pulse TMS (see text). Larger MEP ratios indicate less inhibition. A ratio of 1.0 indicates no inhibition. (a) As the motor tic score of the Yale Global Tic Severity Scale (YGTSS) increases, the MEP amplitude ratio increases (SICI decreases). YGTSS motor tic score rates number, frequency, intensity, complexity, and interference from 0 to 5 each. (b) As the severity of the attention deficit hyperactivity disorder (ADHD) symptoms increases on the ADHD rating scale, the MEP amplitude ratio also increases (SICI decreases). The ADHD rating scale rates inattention and hyperactive/impulsive symptoms based on parental ratings of the frequency of nine symptoms of inattention and nine symptoms of hyperactive/impulsive behavior. Reproduced from Gilbert *et al.* (2004a) with permission from Wiley–Liss.

findings should be technically replicable across laboratories. Unfortunately, the intensity of the conditioning pulse and the interstimulus intervals have been different in different SICI studies. In addition, behavioral diagnostic ratings, demographics including age, phenotypic spectrum, and other unmeasured differences may also confound interpretation of the consistency of findings across laboratories. Some of the reported inconsistencies between laboratories comparisons of SICI and ADHD, tic, or OCD symptoms may be due to such factors. However, at least two groups have found consistent reductions in SICI in ADHD children (Moll et al. 2001; Gilbert et al. 2004a).

A technique, such as measurement of SICI, that is valuable for clinical care or research should be replicable. Laboratories rarely publish data replicating their findings, however. We assessed the stability of the correlation between ADHD severity and reduced SICI by performing TMS on TS subjects at two separate visits, separated by at least one month (Gilbert et al. 2005). We found that SICI, in patients not taking dopamine receptor blockers, correlated consistently with ADHD severity ($r = -0.60$, $P = 0.002$ at visit 1; and $r = -0.58$, $P = 0.005$ at visit 2; Figure 2a and b). As in our prior study, hyperactivity, not inattention, scores accounted for ADHD-related variance in SICI. The intraclass coefficient for SICI from visit 1 to visit 2 was 0.76 ($P < 0.0001$), again showing this measure was reproducible and stable.

External validation of SICI–ADHD associations: pharmacological and genetic studies

If reduced SICI is an endophenotype for ADHD, then medications that treat ADHD should increase, or normalize, SICI. The standard medical treatment for ADHD is stimulants. TMS has been used to probe the effects of stimulants, as well as many other classes of medications (Chapter 13, this volume). Most TMS studies of the effects of stimulants have been performed in healthy adults. Pooled analysis of these studies and work done in our laboratory show that, in healthy adults, both stimulants and selective norepinephrine reuptake inhibitors decrease, not increase, SICI (Figure 25.3) (Gilbert et al. 2006a).

The relevance of these findings to children with ADHD is unclear. In the only published TMS study of stimulants in children with ADHD, methylphenidate increased SICI toward more normal levels (Moll et al. 2000). In addition to age and diagnosis, genetic variation in dopamine transporters may affect the stimulant response (Gilbert et al. 2006b).

More work is needed to determine whether TMS-generated measures, such as SICI, are valid endophenotypes for pediatric behavioral disorders. In the final section, some useful directions for this research to pursue are outlined.

New directions

There are substantial but surmountable obstacles to using TMS, or any neurophysiological technique, to understand neurobehavioral disorders. The following criteria are suggested as reasonable for future TMS studies in children to identify behavioral endophenotypes.

1. Group mean comparisons of cases vs controls should show a statistically significant difference in the TMS measure. If the measure is not distributed normally or the sample size in the study is small, a nonparametric test should be used to compare group means.

2. The case–control, group mean differences should be validated by studies in two different laboratories using similar techniques.

3. When possible, TMS investigators and technicians should be blinded to the diagnosis. This will not always be achievable. Behavioral ratings and TMS measures should be performed blinded by independent members of the research team.

4. The TMS measure should be stable over time, demonstrated by assessing the intraclass correlation coefficient across visits.

5. The TMS measure should have at least a modest ($r \geq 0.25$), but consistent and statistically significant, correlation with symptom severity rating scale scores. Very strong statistical correlations between behavioral scale scores and TMS measures are unlikely because of the inter-rater variability in behavioral ratings, because behavioral ratings encompass multiple possibly biologically distinct domains, and because numeric

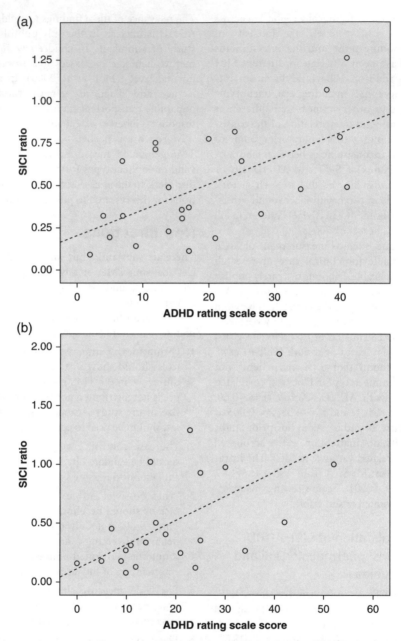

Fig. 25.2 The relationship between attention deficit hyperactivity disorder (ADHD) severity and short-interval cortical inhibition (SICI) as in Figure 25.1b. In this follow-up study, patients not taking D_2 receptor-blocking agents were studied at visits separated by 1 month. (a) Data from visit 1. (b) Data from visit 2. Reproduced from Gilbert *et al.* (2005) with permission from Elsevier.

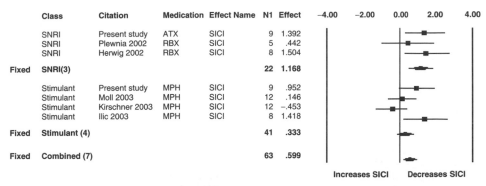

Fig. 25.3 Meta-analysis of attention deficit hyperactivity disorder (ADHD) drug effects on short-interval cortical inhibition (SICI): the stimulant methylphenidate (MPH) (Ilic *et al.* 2003; Kirschner *et al.* 2003; Moll *et al.* 2003; Gilbert *et al.* 2006a) and the selective norepinephrine reuptake inhibitors (SNRIs) atomoxetine (ATX) (Gilbert *et al.* 2006a) and reboxetine (RBX) (Herwig *et al.* 2002; Plewnia *et al.* 2002). Pooled analysis shows that both medications, in healthy adult volunteers, decrease SICI. Reproduced from (Gilbert *et al.* 2006a) with permission from Nature Publishing Group.

intervals in scalar data differ from intervals among measures.

6. The measure should be sensitive to one or more pharmacological agents or other therapeutic modalities shown to be effective in treating the disorder.

7. Clinical or genetic factors that account for variation in disease severity or treatment response should also account for variation in the neurophysiological measure.

Research using TMS has emphasized that the balance of inhibition and excitation in motor cortex is altered in some pediatric neurobehavioral disorders. The strongest and most consistent correlations have been between ADHD symptoms and SICI. SICI abnormalities are not specific to ADHD. SICI is also reduced in other neurobehavioral disorders, including schizophrenia (Daskalakis *et al.* 2002) and Parkinson's disease (Ridding *et al.* 1995). A number of other measures, including surround inhibition (Sohn and Hallett 2004), paired associative stimulation (Stefan *et al.* 2000, 2004), and short-interval afferent inhibition (Stefan *et al.* 2002; Orth *et al.* 2005) await application to pediatric and adult neurobehavioral disorders.

Combining TMS with other modalities may also be informative. For example, cortical networks studied by functional magnetic resonance imaging (fMRI) could be probed with TMS (Desmond *et al.* 2005; Thiel *et al.* 2005). Repetitive TMS can be used to activate or suppress neuronal activity transiently in a region where there is task-related change in the fMRI bold signal, supporting the inference that a certain cortical or cerebellar region subserves the task. Little information is available on the use of repetitive TMS in children (Morales *et al.* 2005), but protocols that inhibit, rather than activate, cortex should fall within TMS safety guidelines (Wassermann 1998) and may ultimately be acceptable to ethics boards.

In summary, single- and paired-pulse TMS is safe and well tolerated in children (Garvey et al. 2001; Gilbert *et al.* 2004b). The most consistently reliable data involve the diagnosis of ADHD and SICI. The application of rigorous experimental designs and combination of TMS with other research methods may increase our knowledge of pathophysiology and treatment of pediatric neurobehavioral disorders.

Acknowledgments

This research was supported by a generous grant from the Tourette Syndrome Association, and by NINDS K23 NS41920. I thank Jie Zhang for expert technical assistance.

References

Alexander GE, DeLong MR, Strick PL (1986). *Annual Review of Neuroscience 9*, 357–81, 357–381.

American Psychiatric Association (2000). *DSM-IV-TR*. Washington, DC: American Psychiatric Association.

Buchmann J, Wolters A, Haessler F, Bohne S, Nordbeck R, Kunesch E (2003). *Clinical Neurophysiology 114*, 2036–2042.

Castellanos FX, Tannock R (2002). *Nature Reviews Neuroscience 3*, 617–28.

Connell FA, Koepsell TD (1985). *Americal Journal of Epidemiology 121*, 744–753.

Daskalakis ZJ, Christensen BK, Chen R, Fitzgerald PB, Zipursky RB, Kapur S (2002). *Archives of General Psychiatry 59*, 347–354.

Desmond JE, Chen SH, Shieh PB (2005). *Annals of Neurology 58*, 553–560.

DuPaul GJ, Power TJ, Anastopoulos AD, Reid R (1998). *ADHD Rating Scale-IV: Checklist, norms, and clinical interpretations*. New York: Guilford Press.

Durston S, Tottenham NT, Thomas KM, et al. (2003). *Biological Psychiatry 53*, 871–878.

Eyre JA, Taylor JP, Villagra F, Smith M, Miller S (2001). *Neurology 57*, 1543–1554.

Garvey MA, Kaczynski KJ, Becker DA, Bartko JJ (2001). *Journal of Child Neurology 16*, 891–894.

Garvey MA, Ziemann U, Bartko JJ, Denckla MB, Barker CA, Wassermann EM (2003). *Clinical Neurophysiology 114*, 1662–1670.

Garvey MA, Barker CA, Bartko JJ, et al. (2005). *Clinical Neurophysiology 116*, 1889–1896.

Gilbert DL, Sethuraman G, Kotagal U, Buncher CR (2003). *Neurology 60*, 564–570.

Gilbert DL, Bansal AS, Sethuraman G, et al. (2004a). *Movement Disorders 19*, 416–425.

Gilbert DL, Garvey MA, Bansal AS, Lipps T, Zhang J, Wassermann EM (2004b). *Clinical Neurophysiology 115*, 1730–1739.

Gilbert DL, Sallee FR, Zhang J, Lipps TD, Wassermann EM (2005). *Biological Psychiatry 57*, 1597–1600.

Gilbert DL, Ridel KR, Sallee FR, Zhang J, Lipps T, Wassermann E (2006a). *Neuropsychopharmacology 31*, 442–449.

Gilbert DL, Wang Z, Ridel KR, Merhar SL, Sallee FR, Zhang J, et al. (2006b). Dopamine transporter genotype influences the physiological response to medication in ADHD. *Brain 129*, 2038–2046.

Herwig U, Brauer K, Connemann B, Spitzer M, Schonfeldt-Lecuona, C (2002). *Psychopharmacology 164*, 228–232.

Ilic TV, Korchounov A, Ziemann U (2003). *Neuroreport 14*, 773–776.

Kirschner J, Moll GH, Fietzek UM, et al. (2003). *Pharmacopsychiatry 36*, 79–82.

Kujirai T, Caramia MD, Rothwell JC, et al. (1993). *Journal of Physiology 471*, 501–19, 501–519.

Leckman JF, Riddle MA, Hardin MT, et al. (1989). *Journal of the American Academy of Child and Adolescent Psychiatry 28*, 566–573.

Mink JW (2001). *Pediatric Neurology 25*, 190–198.

Moll GH, Wischer S, Heinrich H, Tergau F, Paulus W, Rothenberger A (1999). *Neuroscience Letters 272*, 37–40.

Moll GH, Heinrich H, Trott G, Wirth S, Rothenberger A (2000). *Neuroscience Letters 284*, 121–125.

Moll GH, Heinrich H, Trott G, Wirth S, Bock N, Rothenberger A (2001). *Annals of Neurology 49*, 393–396.

Moll GH, Heinrich H, Rothenberger A (2003). *Acta Psychiatrica Scandinavica 107*, 69–72.

Morales OG, Henry ME, Nobler MS, Wassermann EM, Lisanby SH (2005). *Child and Adolescent Psychiatric Clinics of North America 14*, 193–210.

Mostofsky SH, Cooper KL, Kates WR, Denckla MB, Kaufmann WE (2002). *Biological Psychiatry 52*, 785–794.

Muller K, Homberg V, Lenard HG (1991). *Electroencephalography and Clinical Neurophysiology 81*, 63–70.

Orth M, Amann B, Robertson MM, Rothwell JC (2005). *Brain 128*, 1292–1300.

Penney JB, Jr, Young AB (1983). *Annual Review of Neuroscience 6*, 73–94, 73–94.

Plewnia C, Hoppe J, Hiemke C, Bartels M, Cohen LG, Gerloff C (2002). *Neuroscience Letters 330*, 231–4.

Ridding MC, Inzelberg R, Rothwell JC (1995). *Annals of Neurology 37*, 181–188.

Sackett DL, Haynes RB, Guyatt G, Tugwell P (1991). In: *Clinical epidemiology: a basic science for clinical medicine*, pp. 69–152. Boston: Little, Brown.

Saxena S, Brody AL, Maidment KM, et al. (1999). *Neuropsychopharmacology 21*, 683–693.

Scahill L, Riddle MA, McSwiggin-Hardin M, et al. (1997). *Journal of the American Academy of Child and Adolescent Psychiatry 36*, 844–852.

Singer HS (1997). *Neurologic Clinics 15*, 357–379.

Singer HS (2000). *Movement Disorders 15*, 1051–1063.

Snider LA, Seligman LD, Ketchen BR, et al. (2002). *Pediatrics 110*, 331–336.

Sohn YH, Hallett M (2004). *Annals of Neurology 56*, 595–599.

Stefan K, Kunesch E, Cohen LG, Benecke R, Classen J (2000). *Brain 123*(Pt 3), 572–584.

Stefan K, Kunesch E, Benecke R, Cohen LG, Classen J (2002). *Journal of Physiology 543*, 699–708.

Stefan K, Wycislo M, Classen J (2004). *Journal of Neurophysiology 92*, 66–72.

Thiel A, Haupt WF, Habedank B, et al. (2005). *NeuroImage 25*, 815–823.

Ucles P, Lorente S, Rosa F (1996). *Child's Nervous System 12*, 215–217.

van Donselaar CA, Schimsheimer RJ, Geerts AT, Declerck AC (1992). *Archives of Neurology 49*, 231–237.

Wassermann EM (1998). *Electroencephalography and Clinical Neurophysiology 108*, 1–16.

Ziemann U, Paulus W, Rothenberger A (1997). *American Journal of Psychiatry 154*, 1277–1284.

CHAPTER 26

Inter- and intra-individual variation in the response to TMS

Eric M. Wassermann

Introduction

A striking aspect of the muscle response to TMS of the motor cortex is its variable amplitude. It is easy to produce large motor-evoked potentials (MEPs) in some healthy subjects, while others' cortico-muscular pathways seem barely excitable, even by the strongest available stimuli. MEP amplitude and other measures also vary widely within individuals over time. Hence, there are no useful population norms for these measures. We can only assume that susceptibility varies similarly in other brain areas.

Much of the TMS literature is devoted to statistical comparisons between groups of patients and healthy subjects or to measuring the effects of manipulations and treatments on the MEP, with the tacit assumption that differences between individuals within groups (other than, perhaps, age and sex) are negligible and/or randomly distributed with respect to the independent variables under study. However, this assumption has rarely been challenged or validated.

When encountered, individual differences are often treated as 'noise'. However, differences among healthy subjects and patients are important sources of data in their own right. The physiological correlates of dimensional variables, such as behavioral scales, disease severity, or performance scores can be explored with correlational and regression-type statistical designs.

Other differences, genotypes for example, can be treated as categorical variables that differentiate groups of subjects. TMS has been used successfully in studies of both types to identify physiological traits or 'endophenotypes' associated with behavioral and other characteristics.

Differences between individuals

Age

Two small studies, each comparing pairs of groups with different mean ages, have produced disparate results. In one (Peinemann *et al.* 2001), an elderly group showed significantly less paired-pulse intracortical inhibition (ICI) than a group of young adults. In the other (Kossev *et al.* 2002), a 'middle-aged' group showed more ICI than young adults. Neither study controlled for any differences between the groups other than age. In our own larger studies, we found no age-related differences in paired-pulse responses ($n = 53$) or MEP threshold ($n = 138$) (Wassermann 2002).

Gross anatomy

Another factor proposed as a source of individual variation in MEP amplitude and threshold is the distance of the coil from the motor cortex, which increases with cerebral atrophy, skull thickness, etc. One might expect MEP amplitude to be sensitive to this measure, since the intensity of the induced magnetic field falls exponentially with distance from the source. McConnell et al. (2001) found, in 17 healthy individuals aged 19–75 years, that the MEP threshold increased with the distance from scalp to cortex as determined from magnetic resonance imaging. Earlier, the same group had found that age and scalp–motor cortex distance were highly correlated in a sample of depressed patients (Kozel et al. 2000). Most recently, Stokes et al. (2005) reproduced this finding of increased threshold with scalp–cortex distance, and showed experimentally that removing the coil incrementally from the scalp surface resulted in a linear increase in twitch threshold over a 1 cm range. The basis of this useful set of findings is not entirely clear. None of the studies controlled for potential confounds, such as age, which might raise the motor threshold on their own, and none used functional means of locating the motor cortex on the anatomical images. This is important, since there is a suggestion from some studies combining motor cortex TMS and functional and anatomical imaging (Wassermann et al. 1996; Speer et al. 2003) that the site of neuronal activation for the muscle twitch may actually lie anterior to the primary motor area and that the distribution of atrophy may not be uniform.

Genetic factors

The corticospinal response to TMS is subject to genetic influences and, in some instances, can be considered an endophenotype or physiological intermediate between a genetic variation and a behavioral or clinical manifestation. For example, increased corticospinal excitability (Ikoma et al. 1996) and decreased ICI (Ridding et al. 1995; Berardelli, 1999) are well described in symptomatic dystonia. However, clinically normal individuals heterozygous for the DYT1 mutation, a dominant gene with only 30–40% phenotypic penetrance, also show physiological abnormalities, including reduced ICI and shorter cortical silent periods compared to subjects without the mutation (Edwards et al. 2003).

Genes regulating the plasticity of the nervous system would be expected to influence the response of the brain to TMS, particularly after training. Kleim et al. (2006) recently showed that individuals with the va166met polymorphism in the brain-derived neurotrophic factor (BDNF) gene show less increase in the MEP after motor training. This common polymorphism is associated with decreased memory, increased anxiety, and susceptibility to psychiatric disorders (Bath and Lee 2006).

Polymorphisms in genes that produce small effects on behavior can have physiological endophenotypes that are easier to detect, particularly as they interact with exogenous factors, such as drugs. The serotonin transporter promoter gene has two allelic forms, short (s) and long (l) that result in variations in the efficiency of synaptic uptake of serotonin in the general population. A large population-based study ($n = 505$) (Lesch et al. 1996) showed that this polymorphism makes a small, but measurable, contribution to the expression of anxiety-related personality traits. Eichhammer et al. (2003) showed that individuals homozygous for the l allele, which is associated with more efficient uptake of serotonin, had a significantly greater increase in ICI after a single dose of the serotonin reuptake inhibitor citalopram, compared with those heterozygous or homozygous for the s allele.

Finally, unknown genetic influences may exert strong effects on the response to TMS. In a set of 20 healthy sibling pairs (Wassermann 2002), we found a significant correlation between the MEP thresholds in the right (dominant) hand among sib pairs ($r = 0.55$; $P < 0.01$; Figure 26.1). There was no association between age or any other measure and MEP threshold in this sample. Although the similarity in threshold between siblings could have been due to an anatomical factor, such as scalp–cortex distance (see above), it was considerably stronger for the right hand, suggesting that the inherited factor might be related to the organization of the hand representation in the dominant hemisphere.

Physiological differences associated with behavioral and other traits

Personality

Differences in the response to TMS are associated with some surprising traits. In 46 healthy subjects, whom we studied for differences in ICI and intracortical facilitation (ICF), we also

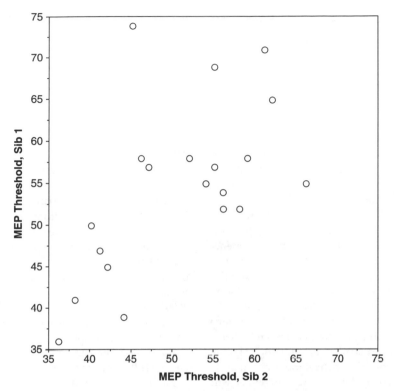

Fig. 26.1 Plot showing the correlation of resting motor-evoked potential (MEP) thresholds between 20 pairs of siblings. Axes are in percentage of maximum stimulator output.

looked for associated cognitive and emotional traits with tests of verbal and visual–spatial intelligence, psychomotor speed, and personality (Wassermann *et al.* 2001a). We found a correlation ($r = 0.48$; $P = 0.0006$) between paired-pulse excitability (lower ICI/higher ICF) and the tendency to experience anxiety and other negative emotions, i.e. 'Neuroticism', a dimension in the five-factor model of personality as tested with the Revised NEO Personality Inventory (Costa and McCrae 1992). Various other behavioral measures failed to show associations either with personality or with physiological measures. However, in a subset ($n = 34$) of the participants, who were given the Revised Wechsler Adult Inteligence Scale Block Design test as a measure of visual–spatial function, performance was inversely correlated with resting ($r = -0.51$, $P = 0.002$) and active ($r = -0.5$, $P = 0.003$) MEP threshold. The bases of these associations are open to speculation (Wassermann *et al.* 2001a), but the findings

indicated that differences in the response to TMS may have functional implications.

Attention deficit hyperactivity disorder (ADHD)

Patients with tic disorder are known to show decreased ICI on average (Ziemann *et al.* 1997). However, in two cohorts of children and adults diagnosed with Tourette's syndrome (Gilbert *et al.* 2004, 2005), ADHD symptomatology was the strongest predictor of ICI and showed a strong negative correlation with ADHD severity, even in individuals who did not reach the threshold for diagnosis of ADHD. Interestingly, ICI is increased by the ADHD medications methylphenidate and atomoxetine, which operate through different neurotransmitter systems (Gilbert *et al.* 2006a), and the TMS response is strongly influenced by dopamine transporter (DAT1) phenotype (Gilbert *et al.* 2006b). ADHD, with a lifetime diagnosis rate of nearly 8% in US children (Centers for Disease Control

2005), is common in the general population. At the milder end of the spectrum, ADHD may blend indistinguishably with normal human behavior and the subclinical trait is common.

Migraine

Migraine is another condition where the milder end of the severity spectrum merges with the population norm. Here, TMS studies in the visual and motor systems (Aurora and Welch 1998; Maertens de Noordhout and Schoenen 1999; Ambrosini *et al.* 2003; Fumal *et al.* 2003; Schoenen *et al.* 2003) have generated some controversy and have underlined the importance of disease subclassification, but strongly suggest differences from screened 'healthy' groups (see Chapter 24). Migraine is associated, clinically and genetically, with other disorders and traits (Scher *et al.* 2005), including depression and anxiety (Merikangas *et al.* 1993; Merikangas and Stevens 1997), which may have their own neurophysiological endophenotypes (Wassermann *et al.* 2001b). Migraine and its companion behavioral traits may be clinical markers for genetic differences with subtle, but measurable, effects on the brain response to TMS.

Systematic intra-individual variation

The MEP varies over time within individuals at rest under laboratory conditions. While much of this variability is related to random factors (Magistris *et al.* 1998) and experimental error, not all of it is noise.

Short-term variation

MEP amplitude often appears to vary cyclically, with a period of seconds or minutes (Figure 26.2). The basis of this type of variation is not known. However, it seems to arise from the brain and is not clearly related to the cardiac (Ellaway *et al.* 1998; Filippi *et al.* 2000) or respiratory (Filippi *et al.* 2000) cycles. There are regular 0.1–0.4 Hz oscillations in cerebral blood flow, particularly at the capillary level (Mayhew *et al.* 1996; Obrig *et al.* 2000; Schroeter *et al.* 2004). These may have neural or vascular origin, but might affect cortical excitability in either case.

Some aperiodic factors are known to influence MEP amplitude in resting subjects. Removal of visual input by eye closure or blindfolding with eyes open causes an immediate increase in MEP amplitude and a decrease in ICI (Leon-Sarmiento *et al.* 2005). Reading single

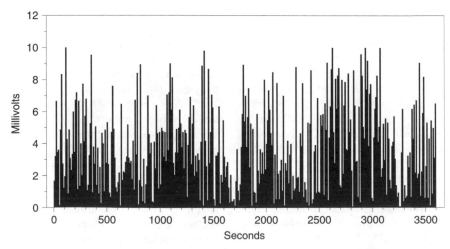

Fig. 26.2 Amplitude of motor-evoked potentials (MEPs) evoked every 10 s over a period of 1 h in a healthy subject.

words, or merely anticipating the task, also increases the amplitude of MEPs evoked from the left hemisphere (Seyal *et al.* 1999).

Long-term variation

The menstrual cycle

Steroid hormones are potent modulators of neuronal excitability. Estradiol is a facilitator of glutamatergic transmission (Wong *et al.* 1996; Woolley, 1999), while progesterone and cortisol are metabolized to neurosteroids that bind to a site on the alpha-subunit of the gamma-aminobutyric acid (GABA)$_A$ receptor, increasing its activity in a manner analogous to the action of the benzodiazepines. Effects of androgens on this system are less well defined, but also likely to exist. Testosterone can decrease neuronal excitability, but it is metabolized *in vivo* to estradiol and dihydrotestosterone, which may have the opposite effect (Edwards *et al.* 1999; Beyenburg *et al.* 2001). Drugs affecting the balance of GABA and glutamate activity in the cortex increase ICI and decrease ICF at moderate doses (Ziemann *et al.* 1998a,b; Schwenkreis *et al.* 1999; see Chapter 13, this volume). Therefore, one might expect that the paired TMS response would also be sensitive to the actions of neurosteroids in healthy individuals. In women immediately after menstruation, the circulating levels of estradiol and progesterone are low. Estradiol rises gradually throughout the follicular phase and progesterone secretion begins in the luteal phase, while estrogen remains high.

In two studies of carefully screened and monitored regularly ovulating women (Smith *et al.* 1999, 2002), we performed paired-pulse TMS experiments across the menstrual cycle and found that ICI decreased and ICF increased late in the follicular phase, when high estradiol levels were unopposed by progesterone. Then, there was a drop in ICF and increase in ICI in the luteal phase, when progesterone was present. The magnitudes of these changes were comparable to effects described for behaviorally significant doses of drugs, such as lorazepam (Ziemann *et al.* 1998b). Data from animal experiments also indicate that progesterone withdrawal can decrease GABA-mediated paired-pulse inhibition (Hsu and Smith 2003).

Not all neurologically normal women show the expected decrease in excitability in the luteal phase. In a sample of women with normal menstrual cycles and hormone levels, but meeting *Diagnostic and Statistical Manual – IV* criteria for premenstrual syndrome and premenstrual dysphoric disorder, ICI actually *decreased* and ICF *increased* in the luteal phase (Smith *et al.* 2003). This aberrant neuronal response to a normal circulating progesterone level is consistent with current theories on the pathogenesis of this disorder (Schmidt *et al.* 1998) and could be caused by an alteration in the GABA$_A$ receptor complex or in the cerebral conversion of progesterone to neurosteroids. Three to eight per cent of women may meet strict criteria for this condition (Halbreich *et al.* 2003) and could express other physiological variations.

In an experiment in healthy men (Bonifazi *et al.* 2004), a single dose of human chorionic gonadotrophin, which stimulates both testosterone and estrogen production in the testis, lowered the MEP threshold and shifted the MEP recruitment curve upward (Bonifazi *et al.* 2004). This suggests that some differences among men in the response to TMS may be related to hormone levels.

It is notable that our studies in healthy women (Smith *et al.* 1999, 2003) and women with premenstrual syndrome (Smith *et al.* 2002) failed to find any significant effects of the menstrual cycle on MEP amplitude or threshold (Figure 26.3). Bonifazi *et al.* (2004) cite experiments suggesting sex differences in the neuronal response to estradiol in animals that might provide an explanation.

A study using repetitive (r)TMS (Inghilleri *et al.* 2004) found that MEPs enlarged during a 5 Hz train in men and women on day 14 of the menstrual cycle, but not on day 1. However, the basis of the incrementing response to 5 Hz rTMS has not been extensively investigated.

While the effects of cortisol-derived neurosteroids on human cortical excitability have not been measured with TMS, there is reason to believe that they act very similarly to the neurosteroid metabolites of progesterone (Majewska 1992). Cortisol levels vary in a circadian pattern, which is lost in clinical depression, and cortisol levels rise with exercise, illness, smoking, and other stressors. These factors should be taken

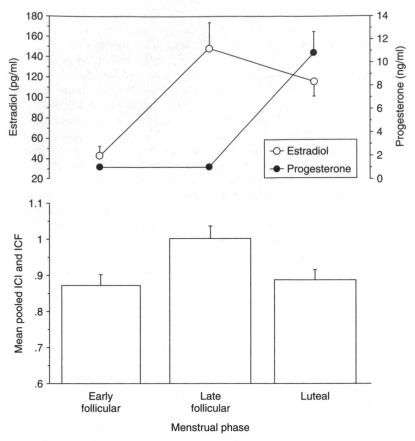

Fig. 26.3 Intracortical inhibition (ICI) and intracortical facilitation (ICF) as they change with hormone levels across the menstrual cycle in a sample of 14 healthy women (all interstimulus intervals pooled and expressed as the ratio of the conditioned to the unconditioned motor-evoked potential). High estrogen in the late follicular phase appears to increase paired-pulse excitability, whereas the addition of progesterone in the luteal phase appears to reduce it. Adapted from Smith *et al.* (2002).

into account in the composition of experimental groups and the timing of experiments and in explaining differences in the responses to paired-pulse TMS.

Conclusions

TMS of the motor cortex is sensitive to underlying time-varying modulations of neuronal excitability. Some of these may be random and some related to changes in external factors, such as visual input (Leon-Sarmiento *et al.* 2005). Others, however, may reflect underlying cycles that modulate behavior and could be altered in neurobehavioral disorders (Castellanos *et al.* 2005).

Systematic, but often unobvious, differences among neurologically normal individuals have important implications for research using TMS. First, they should alert investigators to the importance of screening experimental subjects for individual factors known to influence cortical excitability, e.g. psychopathology, and balancing experimental groups for sex and demographic factors, such as age and education, that might confound results. At the same time, this sensitivity to individual differences also opens new fields of study to neurophysiologists. For example, individual variations in the response to substances and the environment are increasingly acknowledged as important factors in the treatment and etiology of brain disease.

Many of these are genetically determined. Physiological studies with TMS could prove particularly useful in identifying the physiological phenotypes associated with genetic variations that affect behavior, the susceptibility to disease, and the response to chemical agents.

References

Ambrosini A, De Noordhout AM, Sandor PS, Schoenen J (2003). Electrophysiological studies in migraine: a comprehensive review of their interest and limitations. *Cephalalgia 23*(Suppl. 1), 13–31.

Aurora SK, Welch KM (1998). Brain excitability in migraine: evidence from transcranial magnetic stimulation studies. *Current Opinion in Neurology 11*, 205–209.

Bath KG, Lee FS (2006). Variant BDNF (Va166 met) impact on brain structure and function. *Cognitive, Affective and Behavioral Neuroscience 6*, 79–85.

Berardelli A (1999). Transcranial magnetic stimulation in movement disorders. *Electroencephalography and Clinical Neurophysiology Suppl 51*, 276–280.

Beyenburg S, Stoffel-Wagner B, Bauer J, et al. (2001). Neuroactive steroids and seizure susceptibility. *Epilepsy Research 44*, 141–153.

Bonifazi M, Ginanneschi F, Della Volpe R, Rossi A (2004). Effects of gonadal steroids on the input–output relationship of the corticospinal pathway in humans. *Brain Research 1011*, 187–194.

Castellanos FX, Sonuga-Barke EJ, Scheres A, Di Martino A, Hyde C, Walters JR (2005). Varieties of attention-deficit/hyperactivity disorder-related intra-individual variability. *Biological Psychiatry 57*, 1416–1423.

Centers for Disease Control (2005). Mental Health in the United States: Prevalence of Diagnosis and Medication Treatment for Attention-Deficit/Hyperactivity Disorder — United States, 2003. *Morbidity and Mortality Weekly Report 54*, 842–847.

Costa PT, McCrae RR (1992). *Revised NEO Personality Inventory (NEO-PI-R) and NEO Five-Factor Inventory (NEO-FFI) Professional Manual.* Odessa, FL: Psychological Assessment Resources.

Edwards HE, Burnham WM, MacLusky NJ (1999). Testosterone and its metabolites affect afterdischarge thresholds and the development of amygdala kindled seizures. *Brain Research 838*, 151–157.

Edwards MJ, Huang YZ, Wood NW, Rothwell JC, Bhatia KP (2003). Different patterns of electrophysiological deficits in manifesting and non-manifesting carriers of the DYT1 gene mutation. *Brain 126*, 2074–2080.

Eichhammer P, Langguth B, Wiegand R, Kharraz A, Frick U, Hajak G (2003). Allelic variation in the serotonin transporter promoter affects neuromodulatory effects of a selective serotonin transporter reuptake inhibitor (SSRI). *Psychopharmacology 166*, 294–297.

Ellaway PH, Davey NJ, Maskill DW, Rawlinson SR, Lewis HS, Anissimova NP (1998). Variability in the amplitude of skeletal muscle responses to magnetic stimulation of the motor cortex in man. *Electroencephalography and Clinical Neurophysiology 109*, 104–113.

Filippi MM, Oliveri M, Vernieri F, Pasqualetti P, Rossini PM (2000). Are autonomic signals influencing cortico-spinal motor excitability? A study with transcranial magnetic stimulation. *Brain Research 881*, 159–164.

Fumal A, Bohotin V, Vandenheede M, Schoenen J (2003). Transcranial magnetic stimulation in migraine: a review of facts and controversies. *Acta Neurologica Belgica 103*, 144–154.

Gilbert DL, Bansal AS, Sethuraman G, et al. (2004). Association of cortical disinhibition with tic, ADHD, and OCD severity in Tourette syndrome. *Movement Disorders 19*, 416–425.

Gilbert DL, Sallee FR, Zhang J, Lipps TD, Wassermann EM (2005). Transcranial magnetic stimulation-evoked cortical inhibition: a consistent marker of attention-deficit/hyperactivity disorder scores in Tourette syndrome. *Biological Psychiatry 57*, 1597–1600.

Gilbert DL, Ridel KR, Sallee FR, Zhang J, Lipps TD, Wassermann EM (2006a). Comparison of the inhibitory and excitatory effects of ADHD medications methylphenidate and atomoxetine on motor cortex. *Neuropsychopharmacology 31*, 442–449.

Gilbert DL, Wang Z, Sallee FR, et al. (2006b). Dopamine transporter genotype influences the physiological response to medication in ADHD. *Brain 129*, 2038–2046.

Halbreich U, Borenstein J, Pearlstein T, Kahn LS (2003). The prevalence, impairment, impact, and burden of premenstrual dysphoric disorder (PMS/PMDD). *Psychoneuroendocrinology 28*(Suppl. 3), 1–23.

Hsu FC, Smith SS (2003). Progesterone withdrawal reduces paired-pulse inhibition in rat hippocampus: dependence on GABA(A) receptor alpha4 subunit upregulation. *Journal of Neurophysiology 89*, 186–198.

Ikoma K, Samii A, Mercuri B, Wassermann EM, Hallett M (1996). Abnormal cortical motor excitability in dystonia. *Neurology 46*, 1371–1376.

Inghilleri M, Conte A, Curra A, Frasca V, Lorenzano C, Berardelli A (2004). Ovarian hormones and cortical excitability. An rTMS study in humans. *Clinical Neurophysiology 115*, 1063–1068.

Kleim JA, Chan S, Pringle E, et al. (2006). BDNF va166met polymorphism is associated with modified experience-dependent plasticity in human motor cortex. *Nature Neuroscience 9*, 735–737.

Kossev AR, Schrader C, Dauper J, Dengler R, Rollnik JD (2002). Increased intracortical inhibition in middle-aged humans; a study using paired-pulse transcranial magnetic stimulation. *Neuroscience Letters 333*, 83–86.

Kozel FA, Nahas Z, Debrux C et al. (2000). How coil–cortex distance relates to age, motor threshold, and antidepressant response to repetitive transcranial magnetic stimulation. *Journal of Neuropsychiatry and Clinical Neuroscience 12*, 376–384.

Leon-Sarmiento FE, Bara-Jimenez W, Wassermann EM (2005). Visual deprivation effects on human motor cortex excitability. *Neuroscience Letters 389*, 17–20.

Lesch KP, Bengel D, Heils A, et al. (1996). Association of anxiety-related traits with a polymorphism in the serotonin transporter gene regulatory region. *Science 274*, 1527–1531.

McConnell KA, Nahas Z, Shastri A, *et al.* (2001). The transcranial magnetic stimulation motor threshold depends on the distance from coil to underlying cortex: a replication in healthy adults comparing two methods of assessing the distance to cortex. *Biological Psychiatry* 49, 454–459.

Maertens De Noordhout A, Schoenen J (1999). Transcranial magnetic stimulation in migraine. *Electroencephalography and Clinical Neurophysiology Supplement 51*, 260–264.

Magistris MR, Rosler KM, Truffert A, Myers JP (1998). Transcranial stimulation excites virtually all motor neurons supplying the target muscle. A demonstration and a method improving the study of motor evoked potentials [see comments]. *Brain 121*, 437–450.

Mayhew JE, Askew S, Zheng Y, *et al.* (1996). Cerebral vasomotion: a 0.1-Hz oscillation in reflected light imaging of neural activity. *NeuroImage 4*, 183–193.

Majewska MD (1992). Neurosteroids: endogenous bimodal modulators of the GABAA receptor. Mechanism of action and physiological significance. *Progress in Neurobiology 38*, 379–395.

Merikangas KR, Stevens DE (1997). Comorbidity of migraine and psychiatric disorders. *Neurology Clinics 15*, 115–123.

Merikangas KR, Merikangas JR, Angst J (1993). Headache syndromes and psychiatric disorders: association and familial transmission. *Journal of Psychiatric Research 27*, 197–210.

Obrig H, Neufang M, Wenzel R, *et al.* (2000). Spontaneous low frequency oscillations of cerebral hemodynamics and metabolism in human adults. *NeuroImage 12*, 623–639.

Peinemann A, Lehner C, Conrad B, Siebner HR (2001). Age-related decrease in paired-pulse intracortical inhibition in the human primary motor cortex. *Neuroscience Letters 313*, 33–36.

Ridding MC, Sheean G, Rothwell JC, Inzelberg R, Kujirai T (1995). Changes in the balance between motor cortical excitation and inhibition in focal, task specific dystonia. *Journal of Neurology, Neurosurgery and Psychiatry 59*, 493–498.

Scher AI, Bigal ME, Lipton RB (2005). Comorbidity of migraine. *Current Opinion in Neurology 18*, 305–310.

Schmidt PJ, Nieman LK, Danaceau MA, Adams LF, Rubinow DR (1998). Differential behavioral effects of gonadal steroids in women with and in those without premenstrual syndrome. *New England Journal of Medicine 338*, 209–216.

Schoenen J, Ambrosini A, Sandor PS, Maertens De Noordhout A (2003). Evoked potentials and transcranial magnetic stimulation in migraine: published data and viewpoint on their pathophysiologic significance. *Clinical Neurophysiology 114*, 955–972.

Schroeter ML, Schmiedel O, Von Cramon DY (2004). Spontaneous low-frequency oscillations decline in the aging brain. *Journal of Cerebral Blood Flow and Metabolism 24*, 1183–1191.

Schwenkreis P, Witscher K, Janssen F, *et al.* (1999). Influence of the *N*-methyl-D-aspartate antagonist memantine on human motor cortex excitability. *Neuroscience Letters 270*, 137–140.

Seyal M, Mull B, Bhullar N, Ahmad T, Gage B (1999). Anticipation and execution of a simple reading task enhance corticospinal excitability. *Clinical Neurophysiology 110*, 424–429.

Smith MJ, Keel JC, Greenberg BD, *et al.* (1999). Menstrual cycle effects on cortical excitability. *Neurology 53*, 2069–2072.

Smith MJ, Adams LF, Schmidt PJ, Rubinow DR, Wassermann EM (2002). Ovarian hormone effects on human cortical excitability. *Annals of Neurology 51*, 599–603.

Smith MJ, Adams LF, Schmidt PJ, Rubinow DA, Wassermann EM (2003). Abnormal luteal phase excitability of the motor cortex in women with PMS. *Biological Psychiatry 54*, 757–762.

Speer AM, Willis MW, Herscovitch P, *et al.* (2003). Intensity-dependent regional cerebral blood flow during 1-Hz repetitive transcranial magnetic stimulation (rTMS) in healthy volunteers studied with H215O positron emission tomography: I. Effects of primary motor cortex rTMS. *Biological Psychiatry 54*, 818–25.

Stokes MG, Chambers CD, Gould IC, *et al.* (2005). Simple metric for scaling motor threshold based on scalp–cortex distance: application to studies using transcranial magnetic stimulation. *Journal of Neurophysiology 94*, 4520–4527.

Wassermann EM (2002). Variation in the response to transcranial magnetic brain stimulation in the general population. *Clinical Neurophysiology 113*, 1165–1171.

Wassermann EM, Wang B, Zeffiro TA, *et al.* (1996). Locating the motor cortex on the MRI with transcranial magnetic stimulation. *NeuroImage 3*, 1–9.

Wassermann EM, Greenberg BD, Nguyen MB, Murphy DL (2001a). Motor cortex excitability correlates with an anxiety-related personality trait. *Biological Psychiatry 50*, 377–382.

Wassermann EM, Greenberg BD, Nguyen MB, Murphy DL (2001b). Motor cortex excitability correlates with an anxiety-related personality trait. *Biological Psychiatry 50*, 377–382.

Wong M, Thompson TL, Moss RL (1996). Nongenomic actions of estrogen in the brain: physiological significance and cellular mechanisms. *Critical Reviews in Neurobiology 10*, 189–203.

Woolley CS (1999). Electrophysiological and cellular effects of estrogen on neuronal function. *Critical Reviews in Neurobiology 13*, 1–20.

Ziemann U, Paulus W, Rothenberger A (1997). Decreased motor inhibition in Tourette's disorder: evidence from transcranial magnetic stimulation. *American Journal of Psychiatry 154*, 1277–1284.

Ziemann U, Chen R, Cohen LG, Hallett M (1998a). Dextromethorphan decreases the excitability of the human motor cortex. *Neurology 51*, 1320–1324.

Ziemann U, Steinhoff BJ, Tergau F, Paulus W (1998b). Transcranial magnetic stimulation: its current role in epilepsy research. *Epilepsy Research 30*, 11–30.

SECTION IV

TMS in Perception and Cognition

Vincent Walsh

The ability to interfere with the activity of cortical regions by using TMS in cognitive neuroscience experiments has opened a new world of lesion methodology. In describing the functions of brain areas in cognition, lesions have the last word: whatever correlations can be made between cognitive functions and brain activity measured by electrode recordings or brain imaging, the test of the contribution of a brain area to a task lies in either testing patients who have damage to specific areas, removing or deactivating the area in animals, or using TMS to disrupt the normal functioning of the area.

Since Amassian's experiments on vision (e.g. Amassian *et al.* 1989), TMS has been seized upon as a means of making inferences about brain function and testing neuropsychological theories. The advantages of TMS over testing patients and lesions in animals are many. The ability to use subjects as their own controls allows one to study the effects of learning and plasticity while avoiding the problems caused by reorganization in patients. An important consequence of this difference between TMS interference and real lesions is that it leads to ways in which TMS can give usefully different results from the findings with neuropsychological patients. For example, despite years of assumption that the right posterior parietal cortex is important for visual search, it is clear that this is only the case when the search is novel rather

than familiar to the subjects (Walsh *et al.* 1998). To discover this requires being able to train subjects – almost impossible for experimental purposes with neuropsycholgical patients – and test them at different levels of expertise. Equally important to the value of TMS is the temporal precision with which interference can be induced. To test psychological or physiologically inspired theories of cognitive functions requires one to be able to dissect the time course of neural processes. This is all but impossible with patients, but a range of paradigms have been developed for TMS timing studies (see Walsh and Pascual-Leone 2003).

The chapters in this section address most of the core subject areas of cognitive neuroscience – language, vision, awareness, numerical processing, action, plasticity, and memory – and they provide a snapshot of where the field stands at this point in time. For anyone embarking on TMS studies in cognition this survey is a good place to start.

References

Amassian VE, Cracco RQ, Maccabee PJ, Cracco JB, Rudell A, Eberle L (1989) *Electroencephalography and Clinical Neurophysiology 74*, 458–462.

Walsh V, Pascual-Leone A (2003) *Transcranial magnetic stimulation: a neurochronometrics of mind*. Cambridge, MA: MIT Press.

Walsh V, Ashbridge E, Cowey A (1998) *Neuropsychologia 36*, 363–367.

CHAPTER 27

TMS and visual awareness

Alan Cowey

Introduction

'The highest activities of consciousness have their origins in physical occurrences of the brain just as the loveliest melodies are not too sublime to be expressed by notes' (Somerset Maugham, *A writer's notebook*, 1949, p. 78). If Maugham is correct, TMS should be a means of studying consciousness by interfering with these physical occurrences. This chapter describes ways in which this can be done, while drawing attention to limitations of the technique. As we know from personal experience, one can be conscious yet unaware. Indeed it is important that at any moment we are not aware of most of the events happening inside and outside of us, not only because most of these events are perceptually unimportant (breathing, heart rate, the ticking of the clock) but also because the brain seems to be severely limited in the information it can consciously register. It is this aspect of consciousness, namely being aware or unaware and its cerebral basis, that is the focus of this chapter.

Background

The first clear and, as it proved, seminal demonstration that TMS could be used to study awareness was provided by Amassian *et al.* (1989). Four of the authors tested themselves on a task in which they had to identify briefly presented (17 ms) trigrams (three alphabetical letters). Without any TMS, letter recognition

was almost perfect. When a single pulse from a circular coil was delivered 2 cm above the inion, where the central retina is represented in striate cortex (V1), at various times after the display, performance could be disrupted. When the pulse was delivered 80–100 ms after the display appeared the subjects reported 'either a blur of dots or nothing' and performance was actually 0% correct, presumably because they did not even try to guess (Figure 27.1, *upper*). Amassian *et al.*'s paper remains important because, among other things, they proposed that the pulse had inhibited activity in visual cortex by eliciting inhibitory postsynaptic potentials (IPSPs), that the effect was cortical rather than subcortical, that the phenomenon shares some of the characteristics of visual masking, and that the effect of TMS might be understood by considering it in relation to physiological properties of the visual pathways as studied in animals. Their final and masterly understated sentence could hardly be bettered: 'We believe that the MC [magnetic coil] will be of considerable value in analysing this [masking] and other visual phenomena' (Amassian *et al.* 1989, p. 462). Much of the work since then with respect to TMS and awareness has followed their lead.

TMS above V1: when and why is it effective?

The investigation by Amassian *et al.* (1989) revealed a single period after the presentation of

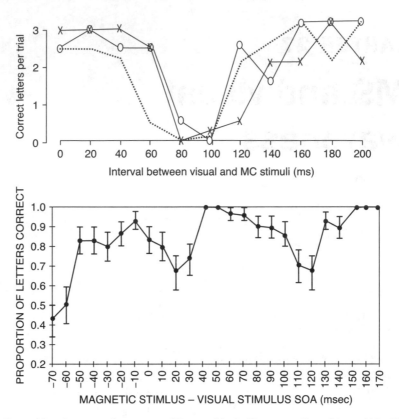

Fig. 27.1 *Upper:* Visual suppression curves of three subjects. The proportion of correct identifications of three dark letters briefly flashed on a bright background is shown as a function of the delay between presentation of the visual stimulus and the subsequent application of a TMS pulse delivered over the primary visual cortex. Reproduced from Amassian *et al.* (1989) with permission from Elsevier Ltd. *Lower:* Identification of a single letter chosen at random from a possible five letters. The letter was present at the fixation point for roughly 4 ms. Graph shows the proportion of letters correctly identified as a function of the delay for a subject and shows three dips in performance, as well as the impairment caused by blinking when the TMS pulse was delivered 70 ms before the letter. Each point is the mean of 30 trials and error bars denote the SEM. Reproduced from Cowey (2005).

a brief visual display in which TMS diminished or abolished awareness of the stimulus. Might there be other periods in which TMS impairs performance with even briefer or slightly different displays and many more trials at each stimulus-onset asynchrony (SOA)? Corthout *et al.* (1999a,b, 2000, 2002, 2003) carried out a finer-grained analysis of the temporal properties of single TMS pulse delivered over the same part of the head. Using both negative and positive SOAs from −70 to +200 ms they found that correct identification of just one letter out of five possible letters, presented for as little as 4 ms, was impeded at no fewer than four SOAs

(Figure 27.1, *lower*). How could these multiple disruptions of awareness arise? The most severe impairment occurred when the pulse was delivered −70 ms before the presentation of the visual stimulus, and the reason for that apparently paradoxical result is now clear. By measuring the temporal dynamics of the eye blink, Corthout (2002) showed that both eyes had usually closed sufficiently far to cover the pupil ~70 ms after the pulse, almost certainly as a result of stimulating the efferent pathway to the eyelids. Although this might seem trivial, it is a lesson about evaluating the effects of TMS on visual perception: check the eyes!

The other dips in recognition with respect to the timing of the TMS pulse are more interesting but much more difficult to explain. At an SOA of −20 ms there is a shallow dip, which initially was overlooked but proved to be repeatable. A plausible explanation is that TMS delivered to primary visual cortex has two effects on the thalamus. The first is orthodromic activation of the cortico-fugal projection from V1 to the thalamic dorsal lateral geniculate nucleus (dLGN) which out-numbers tenfold the thalamo-cortical projection to the cortex and could disrupt the signals from the retina that reach the dLGN from the subsequent brief visual stimulus. The second, as noted by Amassian et al. (1989), is that TMS will antidromically affect the dLGN by eliciting impulses that collide with those from the thalamus to the cortex. These likely effects of cortical TMS are still mostly ignored. The third dip is deepest at an SOA of about +20–30 ms, coinciding with the arrival in V1 of retino-cortical impulses. It is the fourth dip, first shown by Amassian et al., that remains difficult to understand. At an SOA of about +90 ms the visual information from the dLGN has long since reached primary (and probably secondary) visual cortex (Corthout et al. 2007), contrary to the view incautiously expressed by Chambers and Mattingley (2005). Either the TMS has back-wardly masked the processing of the stimulus or it coincides with information projecting back to primary visual cortex that is concerned with

visual identification rather than visual registration. The following sections provide evidence for the latter. But the general message with respect to the use of single-pulse TMS is that, at least when delivered over V1, its effects are more complex than first envisaged. But why was the dip in performance demonstrated by Corthout et al. (2002) at an SOA of +20–30 ms absent in the original experiment by Amassian et al. (1989)? Possible explanations are that in the latter the visual display was longer, stimulus contrast was different, coil orientation was different, and very few trials were given at brief delays. However, as both experiments revealed a dip at ~80–90 ms it seems that two processes necessary for visual awareness have different properties. The precise nature of these processes remains unclear.

One lively field of investigation concerns the possibility that TMS delivered over the primary visual cortex has effects on awareness that are no different from those of a visual mask delivered before or after the target stimulus. If TMS diminishes or even abolishes awareness because its effects on the visual cortex are similar to those of a visual mask, it should be possible to compare the two disruptive processes by calculating the time taken for information from a mask to reach the cortex and appropriately adjusting the curve of performance at different SOAs so that TMS and a visual mask effectively have the same SOA. A recent example is shown in Figure 27.2 (Breitmeyer et al. 2003).

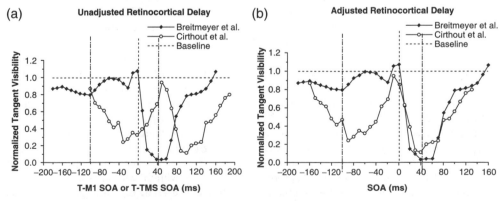

Fig. 27.2 Comparison of a typical visual masking function and a typical masking function using single-pulse TMS. The ordinate shows the negative or positive stimulus-onset asynchrony (SOA). On the left the results are not corrected for a 60 ms delay from retina to V1 whereas on the right they are so adjusted. Reprinted from Breitmeyer et al. (2003) with permission from Elsevier Ltd.

Target visibility is plotted on the left as a function of SOA without any adjustment for retino-cortical delay and on the right with an adjustment of 60 ms. When this adjustment is made, the late TMS dip at ~80–90 ms corresponds well to the trough in target visibility produced by a retinal mask 20–60 ms after the visual target. Whether the less prominent paracontrast masking effect (forward masking), greatest at –100 ms, corresponds to the TMS dip at –40 ms has not yet been explored. Nonetheless, the authors plausibly argue that the TMS-induced dips at –10 and + 90 ms correspond to those produced by forward and backward masking respectively, which in turn reflect feed-forward and feed-backward (re-entrant) processes in occipital visual cortex. TMS exerts its effects by interfering with these cortical processes, which are important for awareness and selective attention, as discussed below.

Despite these complementary findings with TMS and visual masks there are troublesome differences. First, the strength of the TMS pulse used by Corthout *et al.* was deliberately chosen *not* to produce a visual phosphene, i.e. an invisible event 'masked' a visual one whereas in metacontrast masking the mask is perceived. This problem, if indeed it is a problem, is lessened by the finding that a visual metacontrast mask retains its effectiveness even when its own visibility is greatly reduced by a second mask that follows it (Breitmeyer *et al.* 1981). Second, Kammer *et al.* (2005) discovered that TMS and visual masks have different effects in an orientation discrimination. Subjects had to report target orientation in a four-alternative forced-choice task (4AFC) with small targets near the fovea and positioned where the figure-8 coil produced a phosphene at high magnetic intensities. Luminance increment thresholds for orientation discrimination were determined at different SOAs from –125 to +205 ms. TMS significantly reduced the slope of the psychometric function in all four subjects whereas a real visual mask (light flash) did not. The authors conclude that occipital TMS, unlike a visual mask, probably reduces visual awareness by inhibitory processes and that excitation and inhibition, both of which accompany TMS, at least in motor cortex (Hallett 2000), must have different temporal properties. There is a deeply

regrettable scarcity of evidence with respect to the last point but an experiment by Moliadze *et al.* (2003) is relevant. The authors studied the effects of a biphasic pulse on the visual responses of single cells in striate cortex of anesthetized cats. One pulse at the high intensity typically used in studying the effects of TMS on visual performance suppressed excitability for ~100–200 ms, followed by a period of increased responsiveness. The authors suggest that these two periods might reflect the different effects of TMS on excitatory and inhibitory cortical neurons. The issue will only be decided when physiological recordings are made from unanesthetized animals and, as Breitmeyer *et al.* (2003) point out, systematic studies of the effects of visual and TMS masking on psychophysical thresholds for awareness are carried out on the same subjects.

P and M visual pathways

The existence and properties of magnocellular (M) and parvocellular (P) retinal ganglion cells and neurons of the dorsal lateral geniculate nucleus (dLGN) have been described innumerable times (for recent review see Kaplan 2004). As both channels project to V1 but have different temporal and spectral properties it should be possible to use TMS to interfere with them and confirm their contrasting functional roles in perceptual awareness. In one of the first systematic attempts to do so, Paulus *et al.* (1999) asked subjects to discriminate randomly arranged Gaussian blobs that varied in chromatic or luminance contrast and were present for a maximum of 22 ms. A single TMS pulse was delivered above the central representation of the retina in V1 (and perhaps V2 as well) with a round coil placed just above the inion. The results are shown in Figure 27.3. Color discrimination was maximally impaired at an SOA of 90 ms whereas there were two peaks of disruption for luminance discrimination, at ~30 and ~90 ms. This statistically highly significant result is consistent with the early arrival in V1 of information conveyed by the M luminance-channel and the later arrival of chromatic information in the P channel. What is still uncertain is the reason for a single peak with colored targets and twin peaks with luminance targets. The twin

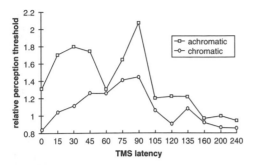

Fig. 27.3 Performance, expressed in relation to the normalized scores of subjects without TMS, when TMS was delivered above V1 at various stimulus-onset asynchronies. The value on the ordinate is the threshold contrast with TMS divided by the threshold contrast without TMS. Positive values therefore indicate an impairment. Note the two peaks for luminance contrast and a single peak for chromatic contrast. Reprinted from Paulus *et al.* (1999) with permission from Elsevier Ltd.

peaks have already been discussed above in relation to the experiments by Corthout *et al.* (2002, 2003), which used only luminance targets, and it now seems likely that the peak at 90 ms does not represent the disruption of thalamo-cortical signals arriving in V1. Rather, it unveils subsequent activity associated with awareness associated with feedback from extrastriate cortex. Why this is not apparent for colored targets is mysterious, as is the contribution of the K pathway, about which much less is known but which is believed to be blue–yellow opponent.

Given the initial success in distinguishing between P and M contributions to visual awareness in V1, it might be reckoned that by disrupting different extrastriate target areas of P and M projections from V1, a similar dissociation could be revealed. This has not transpired, probably because the P and M distinction is swiftly blurred beyond V1 and even within the supragranular layers of V1, and that the P and M latency differences are no longer sharp, presumably as a result of connections between different cortical laminae and different extrastriate visual areas and extensive intra-area processing of visual information, whether or not it is predominantly P or M. However, even though the

temporal segregation is debased it should still be possible to demonstrate the functional specificity of different extrastriate visual with respect to awareness of different visual qualities by using more focal TMS. This is discussed in subsequent sections.

Is V1 necessary for phenomenal visual awareness?

When part of the primary visual cortex V1 is destroyed the patient is clinically blind in the corresponding visual field defect in the sense that visual stimuli entirely confined to the defect are not reported. However, when forced-choice guessing is used, some patients perform well and can detect, localize, and discriminate among visual stimuli. This is termed 'blindsight' (Weiskrantz *et al.* 1974; for review see Cowey 2004). There is debate about whether the patients are really unaware that anything has occurred or whether they might be nonvisually aware in the sense of being able to detect that an event without sensory quale has occurred. The latter was called blindsight type II by Weiskrantz (1998). In some patients a weak visual percept is reported or can be inferred despite the almost total destruction of V1 (e.g. Barbur *et al.* 1993: Stoerig and Barth 2001). Whether or not the residual visual processing can be consciously accessed it should be possible to study it by applying TMS to different parts of extrastriate cortex that are candidates for the surviving processing. Figure 27.4 shows results from one of the first experiments. When rTMS is delivered at various positions above the medial occipital cortex (Figure 27.4, *left*) a reproducible and often faintly colored phosphene is elicited and can be drawn by the subject (Cowey and Walsh 2000). When the TMS is applied above the motion area MT/V5, larger, colorless, and moving phosphenes are reported. With hemianopic subject GY, the phosphenes were restricted to the hemifield contralateral to the normal hemisphere. Prima facie this is good evidence that V1 is indispensable for phenomenal visual awareness. However, more recently we have studied the effect of stimulating MT/V5 in both hemispheres at slightly different times. When the rTMS on the normal side slightly preceded and

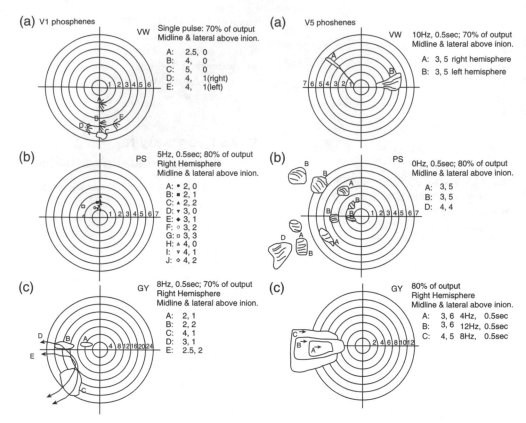

Fig. 27.4 *Left*: Phosphenes elicited by rTMS above V1 in (a) a normally sighted observer, (b) a retinally blind subject PS, and (c) hemianopic patient GY. The coordinates give the site of stimulation, for example, '2, 1' indicates that the figure-8 coil was centered 2 cm rostral to the inion and 1 cm lateral to the midline. Note that as the coil is moved rostrally the phosphenes migrate inferiorly and that as it is moved away from the midline the phosphenes migrate further into the contralateral visual field.
In subject PS, the phosphenes are confined to the central few degrees of the visual field despite stimulation being delivered between 2 and 5 cm above the inion and up to 2 cm lateral. *Right*: Moving phosphenes elicited in the same subjects by rTMS delivered above area MT/V5, roughly 3–4 cm above inion and 3.5–5.0 cm lateral. Reproduced from Cowey and Walsh (2000) with permission.

overlapped that on the damaged side, bilateral phosphenes were reported and drawn (Silvanto *et al.* 2007, in press). As areas MT/V5 are known to be connected transcallosally, the TMS to the normal side is presumably exciting the damaged side and rendering the latter more susceptible to TMS. Alternatively, the bilateral stimulation may be more effective at exciting the back-projections from MT/V5 on the normal side to V1 on that side. In other words perhaps it is V1 that is essential for a conscious visual percept. Evidence for the latter is provided by Silvanto *et al.* (2005). They first

determined, for each subject separately, the threshold magnetic intensity for producing a phosphene by stimulating V1 or MT/V5. They then showed that if MT/V5 were stimulated with a single pulse below phosphene threshold and then followed (10–40 ms) by a suprathreshold pulse to V1, the subjects experienced a phosphene whose characteristics (size, motion) took on those of phosphenes characteristic of MT/V5. The authors interpreted this as evidence that exciting V1 at the appropriate moment enables the information in the back-projection from MT/V5 to enter awareness.

Whether one or both explanations for the results of stimulating MT/V5 are correct, the results suggest that conscious visual awareness in blindsight might indeed be present if the remaining cortex is physiologically in an appropriate condition and that prolonged training with visual stimuli in the blind field might enable such conditions to develop (Sabel and Trauzettel-Klosinski 2005).

There are claims that even in the absence of V1 a hemianopic subject can discriminate the direction of visual motion in the blind field (e.g. Zeki and ffytche 1998). In fact the direction of motion of a single object can be inferred as long as its start and end points are capable of being registered and temporally separated, and this is really a positional discrimination, which is known from other evidence to be spared in blindsight. When global motion is used instead, for example the motion of a group of random dots within a stationary window (a kinematogram), the direction of motion of the array of dots – so perceptually conspicuous in normal vision – is absent in blindsight (Azzopardi and Cowey, 2001). However, even with global motion the discrimination between motion and nonmotion is excellent in blindsight and this might be mediated by area MT/V5. We tested this in the much-studied hemianopic subject GY by applying rTMS during both intervals in one of which the global display moved. The result is shown in Figure 27.5. When repetitive (r)TMS (10 Hz, nine pulses) was applied above area MT/V5 of the damaged hemisphere while GY reported whether the display moved in the first or second of two consecutive intervals, his performance was ignificantly impaired (A. Cowey and I. Alexander, unpublished data). We therefore conclude that, irrespective of whether GY has phenomenal consciousness of the moving stimuli, he cannot discriminate between them when area MT/V5 is disrupted by TMS. His verbal reports confirmed that all awareness of events in his blind hemifield was abolished by the TMS.

Figure 27.4 also illustrates the phosphenes elicited in subject PS by occipital TMS. With respect to visual awareness the results are important because PS was totally ocularly blinded by an accident that destroyed the vision in both eyes ~10 years before he was studied with TMS.

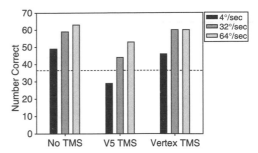

Fig. 27.5 The effect of stimulating the motion area MT/V5 (TMS 10 Hz, 0.6 s, 0.5 s visual display) in the 'blind' hemisphere of subject GY on his discrimination of globally moving vs stationary dots. At all three speeds (4, 32, and 64°/s) he was impaired by the TMS above V5/MT but not above the vertex. The dashed line indicates chance performance out of 70 trials and at the two lower speeds he did not differ significantly from chance in the TMS condition. Reproduced from Cowey (2005).

Despite his total blindness he experienced vivid and frequently colored phosphenes when rTMS was delivered above his medial occipital lobes, and moving but grey phosphenes when it was delivered above area MT/V5 (Cowey and Walsh 2000). These results indicate that as long as primary visual cortex is intact the conscious experience of visual qualia is possible, whereas in cortically blind subjects like GY, who lack area V1, it is either absent or weak. PS is equally interesting in a further respect. A few years after his accident he became aware of phosphenes provoked by sudden, unexpected, loud, and meaningful noises, like the squeal of car brakes or thunder. Is this because his visual cortex is now excited by sounds, as in subjects with auditory/visual synesthesia? This was examined by presenting a variety of meaningful sounds and recording his auditory-evoked potentials. It was found (Rao *et al.* 2007) that when he experienced a phosphene there was an early evoked potential above OZ, PO, and P1, i.e. above V1. But there was no such potential or a much smaller potential when he did not experience a phosphene and there was never a potential in control sighted subjects (Figure 27.6). Moreover, the phosphene was always perceived directly in front of his eyes (even if the sound

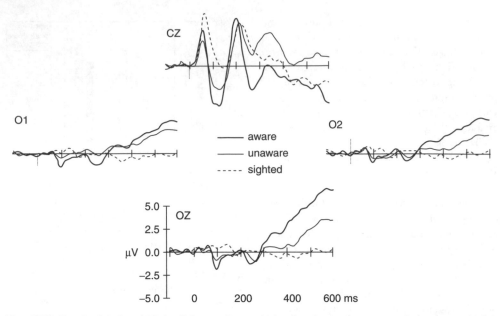

Fig. 27.6 Event-related scalp potentials over the posterior visual areas in response to loud meaningful stimuli. There was no early response in normal subjects, including subject AC whose results are shown by the dashed line. However, there was an early and statistically significant evoked response in the same occipital region in the retinally blind subject, corresponding to his synesthetic experience of visual phosphenes in his central visual field, in response to sudden and meaningful sounds (thick black line). The thin black line shows a smaller occipital response to sounds that failed to produce a phosphene. At the top are shown the evoked responses in auditory cortex recorded at position CZ in both subjects. Reproduced from Cowey (2005).

was to one side or behind him) in the position elicited by TMS and shown in Figure 27.4. The simplest explanation for these induced visual experiences is that his occipital visual areas have become excitable to auditory stimuli, as in sighted synesthetes, albeit it in a nonretinotopic or spatiotopic manner.

The importance of V1 to phenomenal visual consciousness has also been investigated by the effects of TMS on reaction times. Ro *et al.* (2004) measured the saccadic reaction time to a target presented to the left or right of the fixation point. When a brief distracting target was presented at the fixation point, reaction time was lengthened. However, when a single pulse of TMS was delivered above V1 from 83 to 116 ms after the distractor, none of the six subjects saw the distractor but all of them continued to show its effect on saccadic reaction time. The effect of

unseen targets in a blind field on manual reaction time to seen targets in the normal field is well known but the novelty of the experiment by Ro *et al.* is that TMS centered on V1 rather than a lesion of V1 briefly instigates the lack of phenomenal consciousness. Unfortunately they did not use a control condition in which the subjects' ability to process a distractor was also assessed by forced-choice guessing. However, the omission was subsequently dealt with by Boyer *et al.* (2005). Subjects performed an orientation or color discrimination for targets close to the fovea. When a single TMS pulse was delivered close to the inion and therefore above V1, at a positive SOA of 100–128 ms (orientation) or 86–114 ms (color), the subjects no longer perceived the stimulus. But they still performed well above chance levels when forced to guess the orientation (horizontal or vertical) or the color (red or green). Why they found no suppression of awareness at even shorter SOAs in pilot experiments is unclear.

Awareness and discrimination of visual motion

Several studies show that a single pulse of TMS delivered above the cortical motion area MT/V5 impairs the discrimination of the direction of visual motion (Beckers and Homberg 1992; Hotson *et al.* 1994; Beckers and Zeki 1995; Anand *et al.* 1998). The conclusions of such studies about the timing of the involvement of area MT/V5 in motion perception are vigorously contested (Amassian *et al.* 2002) but the importance of MT/V5 to awareness of motion is not disputed. However, there are several types of visual motion and two of them are particularly different. We can perceive the direction of motion of a multi-element display even when there is no overall change in the position of the elements, for example rain seen through a window. This is known as global motion. But there are two kinds of moving elements in global motion. Usually the elements – raindrops – can be defined with respect to their mean luminance contrast with that of the background (perceptually brighter or dimmer) and in most circumstances this suffices. But the average luminance of an element can also be the same as that of its background – like mottled leaves falling against a background of similar but still attached leaves in the autumn – and if we were able to compute only luminance differences we would not see their global motion. Yet when displays are constructed in which the moving elements have the same mean luminance as the background, their direction of motion remains obvious. These two types of global motion are termed first-order and second-order, respectively, and this distinction is one of the most important in the study of the neural basis of visual motion processing, but the many proposed computational explanations for the distinction will not be discussed here. There are even more psychophysical studies of the properties of these two motion systems and their possible interaction but they do not directly address

the problem of how and where the distinction arises and whether first-order precedes and leads to second-order. There are four ways of investigating the two types of motion processing: single-cell recording; neuroimaging; the study of patients with cortical damage; and TMS. First, there is physiological evidence for neurons in areas MT and in the superior polysensory (STP) areas in macaque monkeys that are especially sensitive to second-order motion (Albright 1992; O'Keefe *et al.* 1993; O'Keefe and Movshon 1996). Second, several functional neuroimaging studies of normal subjects show that the relative activity in different extrastriate visual areas depends on whether the display consists of first- or second-order motion (e.g. Dumoulin *et al.* 2003), although no study has shown any area to be solely activated by one or the other type of motion. Third, psychophysical studies of neurological patients with unilateral brain lesions have shown that second-order motion can be selectively impaired in the contralateral visual field (Plant and Nakayama 1993; Vaina and Cowey 1996) and that the obverse also occurs, i.e. awareness of second-order motion is normal, but that of first-order motion is impaired (Vaina *et al.* 1998, 1999). For example, patient RA, impaired only on first-order motion, has a focal lesion in the medial part of his occipital lobe within dorsal aspects of areas V2/V3, whereas patient FD, selectively impaired on second-order motion, has a lesion adjacent to and dorsal to area MT/V5. Such functional imaging and neuropsychological studies are evidence for gross, although not complete, regional segregation of the two mechanisms.

If the latter conclusion is correct, TMS might be able to uncover the regional functional segregation by disruptively stimulating the two regions in normal observers. We tested six normal observers with the two types of motion (Cowey *et al.* 2006). After determining the threshold coherence for performance at 75–80% correct when the display was presented in the right lower quadrant (exactly where visual phosphenes occurred with high current rTMS over MT/V5 or V2/V3) they were then given blocks of trials at this threshold coherence while TMS *below* phosphene threshold was

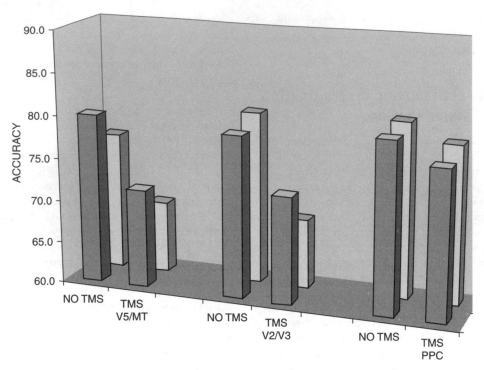

Fig. 27.7 Results of testing six normal subjects on first- (foreground) and second-order (background) global motion tasks. TMS delivered above area V5/MT or V2/V3 impaired performance on both tasks whereas it had no effect when delivered over the vertex. PPC, posterior parietal cortex. Modified from Cowey (2005).

delivered to the left hemisphere over areas MT/V5 or V2/V3 or the posterior parietal cortex (PPC) centered on the angular gyrus. The display was presented for 500 ms and coincided with TMS at 10 Hz for the same duration. The result is shown in Figure 27.7. TMS over PPC had no significant effect on performance, whereas performance was significantly impaired for *both* first- and second-order motion by rTMS over V5 or V2/V3.

This result appears to show that both regions are equally involved in processing first-order and second-order motion, as suggested by some functional neuroimaging experiments where the two kinds of motion have similar effects with respect to regional cerebral blood flow (e.g. Dupont *et al.* 2003). However, an alternative explanation is that the two regions are indeed carrying out different neural computations but that additionally they are in functional communication with respect to the motion the observer

is attempting to discriminate and that stimulating either of them by TMS will activate the other. This possibility will be unwelcome to those who use TMS as if its effects on the brain were confined to the tissue directly excited by the TMS. A lesion will remove this communication and might be better able to reveal their contrasting contributions to motion perception. This is why the increasingly common description of TMS as producing a 'virtual lesion' and 'virtual patients' might be seriously misleading.

Although all subjects, including two of the experimenters, reported that during rTMS they remained aware of the display in the contralateral visual field and that its elements moved, they no longer perceived, or were not confident of, its overall direction of global motion. Even earlier studies of the disruptive effects of single-pulse TMS on the perception of direction of motion have been published (Hotson *et al.* 1992, 1994; Beckers and Zeki 1995; Anand *et al.* 1998) but they did not study areas V2/V3 or describe the subjects' conscious experiences. The alleged timing of the effect in area MT/V5, which suggested that this cortical area receives a retinal input even sooner than that to V1 and is, with respect to conscious awareness, independent of V1, has also been queried on methodological and analytical grounds (Amassian *et al.* 2002).

TMS and visual perceptual priming

In visual perceptual priming, presentation of a visual stimulus reduces the reaction time to respond to the *same* stimulus over a subsequent period of seconds or even minutes, regardless of whether the subject was aware of the repetition. The locus and mechanism of this perceptual priming are assumed to be cortical, although it could involve a wide distribution (Tulving and Schacter 1990) of areas beyond V1 but prior to regions involved in object recognition (Magnussen and Greenlee 1999; Magnussen 2000), or predominantly the parietal lobes (Farah *et al.* 1993; Marangolo *et al.* 1998). TMS proffers a means of exposing the locus of the priming by briefly disrupting cortical activity *after* the stimulus has first been seen but before its next presentation. Several candidates have been suggested for the site of the priming effect, notably V1 or one or more of the extrastriate visual areas whose neurons are selectively responsive to the relevant visual features of the stimuli, e.g. areas MT/V5 and PPC for stimulus motion, ventrolateral area LOC for shape, and the caudal ventromedial areas V4/V8 for color. Some of these regions are inaccessible to focal TMS, but not the dorsolateral areas. Campana *et al.* (2002) therefore presented subjects with short trains of rTMS over regions V1, PPC, or MT/V5 while they performed a visual motion direction discrimination task. If a cortical area is the site of the neural record that underlies priming, the latter should be impaired by the TMS.

Subjects viewed a display consisting of four small virtual squares (each 2° by 2°) symmetrically arranged around the fixation point on a VDU. Each notional square contained 100 spatially randomized bright dots, moving horizontally or vertically at 3°/s in the same direction

within a square, providing a powerful percept of global motion. In three of the squares the direction of motion was identical and in the fourth square it was orthogonal to the other three. This odd-one-out direction was the target. The sequence of events is shown in Figure 27.8. The subject had to press one of four response buttons to indicate which square contained the target. A train of TMS pulses was applied *after* the subject had responded, i.e. before the next trial. Unknown to the subject the target never appeared in the same square on any two consecutive trials, thereby excluding spatial priming. Otherwise, the direction of the moving dots and the location of the target were randomized. Before the main experiment the subjects practised without TMS and all showed a significant priming effect, i.e. when the same odd-one-out direction of motion appeared as in the preceding trial, irrespective of its spatial location, reaction time to identify it was faster.

The results are shown in Figure 27.8. Although rTMS had no significant effect on percentage correct performance (a common finding with rTMS), the priming effect on same-target and different-target trials was eliminated by TMS over area V5/MT but was unperturbed by TMS at other sites. In a control condition the dots moved in the same direction in each square but had the same color in three squares and a different color in the target square. TMS over V5/MT failed to abolish the color-priming effect. More recently the authors found that the priming effect of position of the target is also resistant to TMS over any of these dorsolateral extrastriate visual areas. The results indicate that whatever the neurophysiological and pharmacological nature of the 'record' of the defining feature of the target stimulus on the previous trial, that record resides in the extrastriate visual area that is specialized for processing that feature, in this case area MT/V5 and the direction of first-order global motion. Similar priming experiments can and presumably will be performed with other areas and appropriate stimuli, e.g. shape and position. Finally, the results complement demonstrations, using lesions in monkeys, of the importance of areas V4 and TEO for priming in the color and form domains (Walsh *et al.* 2000).

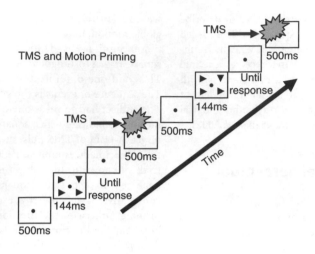

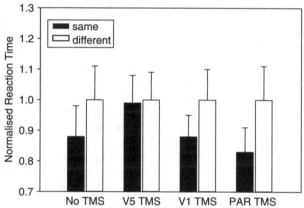

Fig. 27.8 The sequence of events in a priming task where the subjects had to judge the direction of global motion. Subjects fixated a small spot, which was followed for 500 ms by four virtual squares, each containing prominent random dots moving in the same direction in three squares and a different direction in the fourth. The stimuli were presented for 144 ms followed by a blank screen until the subject responded. Response was followed by a further 500 ms of blank screen after which rTMS was applied for 500 ms before the fixation point for the next trial. The histogram shows that the priming effect is abolished by TMS delivered over the motion area V5/MT but not over V1 or parietal cortex. Modified from Campana *et al.* (2002) and Cowey (2005).

Visual neglect

Visual left-sided hemineglect in neurological patients follows extensive damage to the right parietal cortex, especially the angular and supramarginal gyri. It is one of the best known and most intensively studied disorders of visual awareness. Yet, after a century of investigations the crucial cortical areas are still contested (e.g. Karnath *et al.* 2001). Although there are many ways of demonstrating the neglect (e.g. item cancellation tasks, and extinction of the detection of targets on the left by a simultaneous target on the right) a swift clinical method is to ask subjects to bisect a horizontal line or to indicate whether an objectively bisected line looks longer to the right or to the left of the bisection. When normal subjects are given the latter task there is a misjudgment in that the left half of a bisected line is erroneously perceived to be slightly longer than the right, termed pseudo-neglect.

However, patients with large right dorsolateral parietal damage judge a bisected line to be much longer on the right. Moreover, the distortion can be much greater in near-space (within reach) than beyond it, or vice versa (Halligan and Marshall 1991; Cowey *et al.* 1999). When transected or bisected horizontal lines were briefly presented within reaching distance to normal subjects and rTMS (10 Hz, 0.5 s) was delivered over the intraparietal sulcus the subjects misperceived the bisected lines as being longer on the right, i.e. the opposite of the normal pseudo-neglect but in accordance with the spatial distortion or spatial unawareness of patients with parietal neglect (Bjoertomt *et al.* 2002). The effect was not present with displays in far space, well beyond reach (Figure 27.9), nor was it produced by TMS at other sites. It was effective only when delivered over the right parietal region that is activated in functional neuroimaging of subjects while making visual–spatial judgments

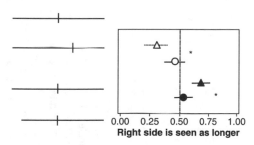

Fig. 27.9 Effects of viewing distance and parietal TMS on neglect, using displays like those shown on the left. The results on the right, top, show the proportion of 'right side is longer' responses for bisected lines in near space (unfilled triangle) and far space (unfilled circle) with normal subjects. Significantly fewer of the responses were 'right is longer', which demonstrates the well-known right pseudo-neglect. The difference in the subjects' responses at the two viewing distances was significant (P < 0.05). Below, right, are shown the effects of stimulating the right posterior parietal cortex on the same subjects' responses. They are again normalized and show significant shifts in the perceived midpoint during TMS; in near space the right side was now perceived as longer, filled symbols (P < 0.05). Error bars show ±1 SD. Reproduced from Cowey (2005).

of the kind that are severely impaired after right parietal lesions. This is an example of a defect that cannot be attributed to the local effects of TMS being propagated to distant and more important sites. It also indicates that the damage that leads to parietal neglect is the cortex ventral to the intraparietal sulcus and not the cortex at the parieto-temporal junction as suggested by Karnath *et al.* (2001).

TMS and visual search: parietal cortex and the frontal eye-fields

Parietal cortex

One of the commonest things we do in everyday life is look for something in a cluttered environment. The 'something' might be an article on a desk, a kitchen utensil in a crowded kitchen drawer, a face in a crowd. We are good at doing this and there is much discussion of how we do it. This kind of perceptual task is called visual search. Figure 27.10 shows two kinds of visual display often used to study it. On the left the target is instantly perceived because its orientation pops out from that of all the other items on the screen, termed distractors. On the right the target does not pop out and must be found by searching the display for the sole item with a unique conjunction of two features, one of which occurs in every distractor. In real life most visual search is of the latter kind. In laboratory experiments designed to simulate the real-life experience the subject is usually asked whether or not a display actually contains the target and the dependent variable is the time taken to decide. The target is customarily present on half the trials. When Ashbridge *et al.* (1997) delivered a single magnetic pulse over the right posterior parietal cortex of normal subjects they discovered the effect shown in Figure 27.10. TMS had no significant effect on pop-out tasks but on a conjunction task it significantly slowed the reaction time on target-present trials when delivered ~90 ms after the display appeared and at ~160 ms on target-absent trials. The former is believed to be the effect of TMS on this extrastriate region when information about the display is first processed

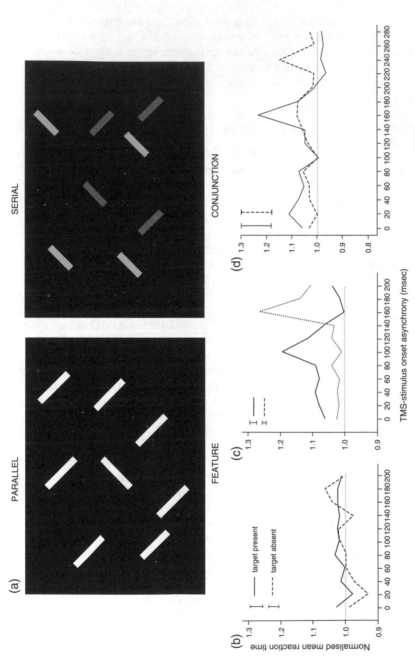

Fig. 27.10 (a) Example of the difference between parallel (pop-out) and serial (conjunction) search. In practice many more items are displayed. (b–d) Mean normalized response time (± SE) of five subjects when a single pulse of TMS was applied over the right parietal cortex at various delays after the stimulus was presented. For each subject, the average response for each delay was normalized to the no-TMS condition. (b) There was no effect of TMS on reaction times in the pop-out task. (c) The solid line shows that stimulation at ~100 ms after presenting the display during a conjunction task resulted in significantly longer reaction times when the target was present (compared to the no-stimulation condition, dotted line). Other TMS delays had no significant effect. The dashed line shows that TMS on the conjunction task, when the target was absent, significantly increased reaction times when applied 160 ms after the stimulus onset. (d) After over-training on the task, which rendered TMS ineffective, the effect was restored when new conjunction stimuli were introduced. Reproduced from Cowey (2005).

there, whereas TMS applied at the later disruptive period is more likely to reflect the time period during which the top-down cognitive decision to end the search is relayed back, from prefrontal cortex, to the parietal cortex.

The finding shown in Figure 27.10 has been confirmed many times. It highlights the role of the right parietal cortex in visually searching the environment and is consistent with functional activations found in parietal and prefrontal cortex during visual search tasks and with the well-known effects of parietal lesions on visuo-spatial perception. But TMS demonstrated a hitherto unsuspected property of visual search: its disruption of performance is only conspicuous while the conjunction search task is being learned. With extensive practice subjects became progressively faster at recognizing the presence or absence of the target and increasingly immune to the TMS. However, a new conjunction of features reinstates the effect of TMS (Ashbridge et al. 1997). It is as if the right posterior parietal cortex is necessary for learning to identify visual conjunctions but not when performance becomes automatized. This is echoed in later demonstrations, using functional magnetic resonance imaging (fMRI), of changes in regional blood flow during the *learning* of tasks involving the discrimination of visual motion but not during subsequent stable performance (e.g. Vaina et al. 2001). TMS could therefore be used to explore the relative importance of different activated areas before, during, and after perceptual learning. Further discussion and examples of the importance of parietal cortex to visual awareness are given by O'Shea and Rushworth (Chapter 28, this volume).

Prefrontal cortex: frontal eye-fields

The frontal eye-fields (FEFs), which coincide with Brodmann's area 8, were, a trifle dismissively, for many years thought to be the motor cortex for voluntary eye movements. But experiments on monkeys showed that unilateral removal of FEF produced a strange contralateral neglect of visual stimuli that was so severe as to be operationally indistinguishable from a hemianopia produced by unilateral removal of primary visual cortex (Kennard 1939, Latto and Cowey 1971a,b, 1972). The article by Latto and

Cowey (1972) is especially relevant to awareness because it was the first demonstration that when monkeys have to find a target in a complex display (serial visual search) they took much longer after one or both FEFs were removed, an effect first thought to result from their impoverished eye movements. However, the latter are short-lived (reviewed by Dias and Segraves 1996), unlike the field defects or the disorganized pattern of searching a display for hidden food pellets (Collin et al. 1982). Might the FEF be important and even necessary for top-down influences in exploring a visual scene for a target?

The effects of parietal TMS on visual search have already been described. Does TMS delivered above the FEF have similar effects? This was investigated by Muggleton et al. (2003), who showed that rTMS (10 Hz for 0.5 s) over right, but not left, FEF impairs conjunction visual search on target-present trials even though searching eye movements were neither necessary nor made. There was no effect on simple visual search. The reduction in d' was a consequence of making more false-positive responses, with no change in target hit rate, a result that resembles the consequences of right parietal lesions in patients who experience illusory conjunctions (Friedman-Hill et al. 1995). As the experiment by Muggleton et al. (2003) involved rTMS for 500 ms it could not reveal when the TMS was maximally effective. O'Shea et al. (2004) accordingly repeated the experiment but used only two pulses, separated by 40 ms, at a variety of SOAs. Only when the two pulses occurred 40 and 80 ms after the display appeared was performance impaired (a reduction in d'). As pulses at an SOA of 0 and 40 or 80 and 120 ms did not impair performance, the crucial timing must be centered at ~60 ms, which is even sooner than the greatest effect of a pulse above V1 or the parietal cortex! Why?

The first niggling problem raised by the latter discovery is that all the well-known hierarchical schemes of visual cortical processing placed the FEF at or near the hierarchical pinnacle. A resolution appeared with the work of Bullier (2004) and others. Contrary to what had been widely assumed, the median response latency of neurons in FEF with visual receptive fields is *shorter* than that of neurons in any other visual area except V1 and MST. Furthermore, anatomical studies have shown that the FEF have a feed-forward projection to various extrastriate visual areas rather than a feed-back (re-entrant) projection, i.e. the projection is generated in supra-granular layers and terminates in layer 4 of the recipient area. The effects of TMS above FEF on complex visual search may therefore arise either by disrupting this feed-forward projection or by impairing planned search (whether covert or overtly implemented by eye movements) or both. One unforeseen outcome of these recent results is that they support the suggestion that when awareness of visual targets is impaired by a single TMS pulse above primary and second-ary visual cortex at an SOA of ~80–90 ms it is because the pulse is disrupting the arrival of top-down information from the frontal eye-fields rather than the arrival of information from the eye itself.

The second problem is that when a single pulse of TMS is applied to FEF 40 ms *before* presentation of a target in a simple detection task, visual sensitivity, i.e. awareness of the tar-get, is improved (Grosbras and Paus 2003). The effect did not occur at control sites and cannot therefore be attributed to the TMS acting as a temporal cueing signal. Apart from providing yet another example of how TMS can actually improve performance on certain perceptual tasks, their paper indicates that visual awareness can be improved or impaired depending on its timing with respect to the visual display. Sadly, the explanation is unknown.

Synchronized cortical discharge

Whether synchronized cortical activity at par-ticular frequency bands is important for con-scious awareness is a lively topic of debate (Singer 1993, 2004). As TMS can be delivered with millisecond precision and, briefly, at high temporal frequencies, it should be possible to impose a particular frequency in the underlying cortex and thereby investigate any effect on sen-sory awareness of the frequency and phase-locking of the electroencephalogram (EEG). An example is provided by Klimesch *et al.* (2003), who studied the performance of subjects on a mental rotation task in which on each trial they viewed a two-dimensional representation of a cube with figures on its sides and then had to select the only cube that matched it from an array of six cubes presented at different rota-tions. The EEG was recorded and the alpha rhythm (~10 Hz) for each subject was first determined. The visual display on each trial was then presented after a train of 24 TMS pulses at a rate of alpha +1, alpha −3, or 20 Hz. The train was delivered over medial frontal cortex, posterior parietal cortex, or with the coil tilted away from the cortex as a sham condition. At alpha −3 or 20 Hz there were no differences in percentage correct performance on the mental rotation task with respect to the baseline scores with sham stimulation. But at an imposed frequency of alpha +1 there was a statistically significant improvement of up to 15% after stimulation of the frontal and parietal sites. The improvement mirrors previous demonstra-tions on a variety of similar tasks that perform-ance correlates with endogenous modulations of the alpha rate (Klimesch 1999) and with results of functional neuroimaging studies showing that both of these cortical areas are activated during mental rotation (Richter *et al.* 2000). It does not explain why a particular alpha frequency just before carrying out the mental rotation is so important but it does suggest that it is causative.

Even a single pulse can briefly synchronize cortical activity beneath it. An example is pro-vided by Paus *et al.* (2001) where the pulse induced synchronized activity in the beta range (15–30 Hz) for ~300 ms. There was no measure of performance on any behavioral task but the results are relevant to theories of perceptual awareness that attribute awareness to phase-locked oscillations in the 40–70 Hz band (Singer 1993). If, as the authors propose, TMS resets the neural oscillations at a lower frequency and pos-sibly abolishes phase-locking between neuronal discharge, the disruptive effect of single-pulse TMS on letter recognition in the experiments of Amassian *et al.* (1989) and Corthout *et al.* (1999a, 2002, 2003) is not surprising. A more convincing interpretation requires studies of the effects of TMS on the temporal properties of individual neurons in behaving animals, which have scarcely begun.

The future

Only a fool, it is often said, predicts the future. Despite not having a crystal ball, let me risk being foolish.

1. After at least 15 years of using TMS to investigate cognitive abilities we still do not understand why TMS causes many of its effects. It is uncertain why monophasic and biphasic pulses have rather different effects, why TMS can improve or impair performance depending on its timing, or why it impairs identification at several times before and after a stimulus is presented. Further work must address these problems or the technique will become disreputable.

2. TMS can briefly impose (or disrupt) rhythmic discharge in the underlying cortex, e.g. alpha or theta, and some of these rhythms are thought to be important for selective attention and awareness. These TMS-imposed rhythms will surely be further investigated.

3. Performance on psychophysical tasks depends on the criterion used by a subject in deciding whether a stimulus is present or not. The criterion, known as β, is bread and butter to generations of psychophysicists steeped in signal detection theory, but is almost ignored by cognitive psychologists. Future experiments on the effects of TMS on awareness must take account of the possibility the TMS might alter β rather than altering awareness *per se*. This applies to a vast range of investigations.

4. TMS can disrupt activity in underlying brain tissue with millisecond precision but thus far it is usually used in isolation. When combined with event-related potentials and fMRI its usefulness will expand.

5. There is great interest in regional cortical functional specialization and TMS has been used to demonstrate it. However, many of the demonstrations amount to little more than a confirmation of previous accounts of the effects of localized brain lesions, a charge initially levelled against functional brain imaging. Yet TMS could do much more than this, for example by comparing in the same subject its effects on left and right hemispheres, or the time course of interactions between different visual areas.

6. TMS can reveal functional connections between brain areas by ortho- or antidromically exciting distant regions, the latter being revealed in turn by the elicited changes in the EEG. Diffusion tensor imaging has better spatial resolution but tells us nothing about temporal properties. Together they might provide the longed-for 'dream team'.

7. Most TMS experiments study its effects on one cortical area at a time. But the interactions between cortical areas are indubitably complex. A few investigators have successfully and productively delivered TMS to two areas in close temporal conjunction in attempts to study the interaction. This is still a largely untapped but potentially rich vein of research and, even though present stimulating coils are physically large, the use of three or even four figure-8 coils is possible.

However, as David Hubel (1963) noted concerning his discoveries with Torsten Wiesel of the receptive field properties of neurons in different visual areas and their attempts to explain them, which subsequently led to their Nobel Prize, 'Our ignorance of CNS processes is such that the best predictions stand a good chance of being wrong'!

References

Albright TO (1992). Form-cue invariant motion processing in primate visual cortex. *Science 255*, 1141–1143.

Amassian VE, Cracco RQ, Maccabee PJ, Cracco JB, Rudell AP, Eberle L (1989). Suppression of visual perception by magnetic coil stimulation of human occipital cortex. *Electroencephalography and Clinical Neurophysiology 74*, 458–462.

Amassian VE, Cracco RQ, Maccabee PJ, Cracco JB (2002). Visual system. In: A Pascuale-Leone *et al.* (eds), *Handbook of transcranial magnetic stimulation*, pp. 323–334. London: Arnold.

Anand S, Olson JD, Hotson DD (1998). Tracing the timing of human analysis of motion and chromatic signals from occipital to temporo-parietal-occipital cortex. *Vision Research 38*, 2617–2627.

Ashbridge E, Walsh V, Cowey A (1997). Temporal aspects of visual search studied by transcranial magnetic stimulation. *Neuropsychologia 35*, 1121–1131.

Azzopardi P, Cowey A (2001). Motion discrimination in cortically blind patients. *Brain 124*, 30–46.

Barbur JL, Watson JD, Frackowiak RSJ, Zeki S (1993). Conscious visual perception without V1. *Brain 116*, 1293–1302.

Beckers G, Homberg V (1992). Cerebral visual motion blindness: transitory akinatopsia induced by transcranial magnetic stimulation of human area V5. *Proceedings of the Royal Society of London B 249*, 173–178.

Beckers G, Zeki S (1995). The consequences of inactivating V1 and V5 on visual motion perception. *Brain 118*, 49–60.

Bjoertomt O, Cowey A, Walsh V (2002). Spatial neglect in near and far space investigated by repetitive transcranial magnetic stimulation. *Brain 125*, 2012–2022.

Boyer JL, Harrison, Ro T (2005). Unconscious processing of orientation and color without primary visual cortex. *Proceedings of the National Academy of Sciences of the USA 102*, 16875–16879.

Breitmeyer BG, Rudd M, Dunn K (1981). Metacontrast investigations of sustained-transient channel inhibitory interactions. *Journal Experimental Psychology: Human Perception and Performance 7*, 770–779.

Breitmeyer BG, Ro T, Ogmen H (2003). A comparison of masking by visual and transcranial magnetic stimulation: implications for the study of conscious and unconscious visual processing. *Consciousness and Cognition 13*, 829–843.

Bullier J (2004). Communications between cortical areas of the visual system. In LM Chalupa, JS Werner (eds), *The visual neurosciences*, pp. 522–540. Cambridge, MA: MIT Press.

Campana G, Cowey A, Walsh V (2002). Priming of motion direction and area V5/MT: a test of perceptual memory. *Cerebral Cortex 12*, 663–669.

Chambers CD, Mattingley JB (2005). Neurodisruption of selective attention: insights and implications. *Trends in Cognitive Science 10*, 1364–1366.

Collin N, Cowey A, Latto R, Marzi, C(1982) The role of frontal eye-fields and superioe colliculi in visual search and non-visual search in rhesus monkeys. *Behavioural Brain Research 4*, 177–193.

Corthout E (2002). Vision investigated with transcranial magnetic stimulation. Unpublished D.Phil. thesis, University of Oxford.

Corthout E, Uttl B, Ziemann U, Cowey A, Hallett M (1999a). Two periods of processing in the (circum)striate visual cortex as revealed by transcranial magnetic stimulation. *Neuropsychologia 37*, 137–145.

Corthout E, Uttl B, Walsh V, Hallett M, Cowey A (1999b). Timing of activity in early visual cortex as revealed by transcranial magnetic stimulation. *Neuroreport 10*, 2631–2634.

Corthout E, Uttl B, Chi-Hung J, Hallett M, Cowey A (2000). Suppression of vision by transcranial magnetic stimulation: a third mechanism. *Neuroreport 11*, 2345–2349.

Corthout E, Hallett M, Cowey A (2002). Early visual cortical processing suggested by transcranial magnetic stimulation. *Neuroreport 13*, 1163–1166.

Corthout E, Hallett M, Cowey A (2003). Interference with vision by TMS over the occipital pole: a fourth period. *Neuroreport 14*, 651–655.

Corthout E, Hallett M, Cowey A. (2007) TMS-induced suppression of visual perception: another kind of scotoma. *Clinical Neurophysiology 118*, 1895–1898.

Cowey A (2004). Fact, artefact and myth about blindsight. *Quarterly Journal of Experimental Psychology 57A*, 577–609.

Cowey A, Walsh V (2000). Magnetically induced phosphenes in sighted, blind and blindsighted observers. *Neuroreport 11*, 3269–3273.

Cowey A, Campana G, Walsh V, Vaina LM (2006) The role of human extra-striate visual areas V5/MT and V2/V3 in the perception of the direction of global motion: a transcranial magnetic stimulation study. *Experimental Brain Research 171*, 558–62.

Cowey A, Small M, Ellis S (1999). No abrupt change in visual hemineglect from near to far space. *Neuropsychologia 37*, 1–6.

Dias EC, Segraves MA (1996). The primate frontal eye field and the generation of saccadic eye movements: comparison of lesion and acute inactivation studies. *Revista Brasileira de Biologia 56*, 239–255.

Dumoulin SO, Baker CL, Hess RF, Evans AC (2003). Cortical specialization for processing first- and second-order motion. *Cerebral Cortex 13*, 1375–1385.

Dupont P, Sary G, Peuskens H, Orban GA (2003). Cerebral regions processsing first- and higher-order motion in an opposed-direction discrimination task. *European Journal of Neuroscience 17*, 1509–1517.

Farah M, Wallace M, Vecera SP (1993). 'What' and 'where' in visual attention: evidence from the neglect syndrome. In IH Robertson, JC Marshall (eds), *Unilateral neglect: clinical and experimental studies*, pp. 123–137. Hillsdale, NJ: Erlbaum.

ffytche DH, Guy CN, Zeki S (1995). The parallel visual motion inputs into areas V1 and V5 of human cerebral cortex. *Brain 118*, 1375–1394.

Friedman-Hill SR, Robertson LC, Treisman A (1995) Parietal contributions to visual feature binding: evidence from a patient with bilateral lesions. *Science 269*, 853–855.

Grosbras M, Paus T (2003) Transcranial magnetic stimulation of the humanfrontal eye field facilitates visual awareness. *European Journal of Neuroscience 18*, 3121–3125

Hallett M(200) Transcranial magnetic stimulation and the human brain. *Nature 406*, 147–150.

Halligan PW, Marshall JC (1991). Left neglect for near but not far space in man. *Nature 350*, 498–500.

Hotson J, Braun D, Herzberg W, Boman D. (1994). Transcranial magnetic stimulation of extrastriate cortex degrades human motion perception. *Vision Research 14*, 2115–2123.

Hubel DH (1963). Integrative processes in central visual pathways of the cat. *Journal of the Optical Society of America 53*, 58–66.

Kammer T (1999). Phosphenes and transient scotomas induced by magnetic stimulation of the occipital lobe: their topographic relationship. *Neuropsychologia 37*, 191–199.

Kammer T, Puls K, Strasburger H, Hill NJ, Wichmann FA (2005). Transcranial magnetic stimulation in the visual system. 1. The psychophysics of visual suppression. *Experimental Brain Research 160*, 118–128.

Kaplan E (2004). The M, P, and K pathways of the primate visual system. In: LM Chalupa and JS Werner (eds), *The visual neurosciences*, pp. 481–493. MIT Press, Cambridge.

Karnath H-O, Ferber S, Himmelbach M (2001). Spatial awareness is a function of the temporal not the posterior parietal lobe. *Nature 411*, 950–953.

Kennard MA (1939) Alterations in response to visual stimuli following lesions of frontal lobe in monkeys. *Archives of Neurology and Psychiatry 41*,1153–1165.

Klimesch W (1999). EEG alpha and theta oscillations reflect cognitive and memory performance: a review and analysis. *Brain Reserch Reviews 29*, 169–195.

Klimesch W, Sauseng P, Gerloff G (2003). Enhancing cognitive pereformance with repetitive transcranial magnetic stimulation and human individual alpha frequency. *European Journal of Neuroscience 17*, 1129–1133.

Latto RM, Cowey A (1971a). Visual field defects after frontal eye-field lesions in monkeys. *Brain Research 30*, 1–24.

Latto RM, Cowey A (1971b). Fixation changes after frontal eye-field lesions in monkeys. *Brain Research 30*, 25–36.

Latto R, Cowey A (1972). Frontal eye-field lesions in monkeys. *Bibliotheca Ophthalmologica 82*, 159–168.

Magnussen S (2000). Low level memory processes in vision. *Trends in Neuroscience 23*, 247–251.

Magnussen S, Greenlee MW (1999). The psychophysics of perceptual memory. *Psychological Research 62*, 81–92.

Marangolo P, Di Pace E, Rafal R, Scabini D (1998). Effects of parietal lesions in humans on color and location priming. *Journal of Cognitive Neuroscience 10*, 704–716.

Moliadze V, Zhao Y, Eysel U, Funke K (2003). Effect of transcranial magnetic stimulation on single unit activity in cat primary visual cortex. *Journal of Physiology 553*, 665–679.

Muggleton N, Chi-Hung J, Cowey A, Walsh V (2003). Human frontal eye fields and visual search. *Journal of Neurophysiology 89*, 3340–3343.

O'Keefe LP, Carandini M, Beusmans IMH, Movshon IA (1993). MT neuronal responses to 1st- and 2nd-order motion. *Society for Neuroscience Abstracts 19*, 1283.

O'Keefe LP, Movshon IA (1996). Processing of first- and second- order motion signals by neurons in area MT of the macaque monkey. *Visual Neuroscience 15*, 305–317.

O'Shea J, Muggleton NG, Cowey A, Walsh V (2004). Timing of target discrimination in human frontal eye-fields. *Journal of Cognitive Neuroscience 16*, 1060–1067.

Pascual-Leone A, Walsh V (2001). Fast back-projections from the motion area to the primary visual area necessary for visual awareness. *Science 292*, 510–512.

Paulus W, Korinth S, Wischer S, Tergau F (1999). Differentiation of parvo- and magno-cellular pathways by TMS at the occipital cortex. *Electroencephalography and Clinical Neurophysiology Supplement 51*, 351–360.

Paus T, Sipila PK, Strafella AP (2001). Synchronization of neuronal activity in the human primary motor cortex by transcranial magnetic stimulation: an EEG study. *Journal of Neurophysiology 86*, 1983–1990.

Plant GT, Laxer KD, Barbaro NM, Schiffman IS, Nakayama K (1993). Impaired visual motion perception in the contralateral hemifield following unilateral posterior cerebral lesions. *Brain 116*, 1337–1353.

Rao A, Nobre AC, Alexander I, Cowey A (2007) Auditory evoked visual awareness in a late blind human subject: an electrophysiological investigation. *Experimental Brain Research 176*, 288–298

Richter W, Somorjai R, Summers R *et al.* (2000). Motor area activity during mental rotation studied by time-resolved single-trial fMRI. *Journal of Cognitive Neuroscence 12*, 310–320.

Ro T, Shelton D, Lee OL, Chang E (2004). Extrageniculate mediation of unconscious vision in transcranial magnetic stimulation-induced blindsight. *Proceedings of the National Academy of Sciences of the USA 101*, 9933–9935.

Sabel BA, Trauzettel-Klosinski S (2005). Improving vision in a patient with homonymous hemianopia. *Journal of Neuroophthalmology 25*, 143–149.

Silvanto J, Cowey A, Lavie N, Walsh V (2005). Striate cortex activity gates awareness of motion. *Nature Neuroscience 8*, 143–144.

Silvanto J, Cowey A, Lavie N, Walsh V (2007) Making the blindsighted see. *Neuropsychologia*, in press.

Singer W (1993). Synchronization of cortical activity and its putative role in information processing and learning. *Annual Review of Physiology 55*, 349–374.

Singer W (2004). Synchrony, oscillations, and relational clocks. In: LM Chalupa, JS Werner (eds), *The visual neurosciences*, pp. 1665–1681. Cambridge, MA: MIT Press.

Stoerig P, Barth E (2001). Low-level phenomenal vision despite unilateral destruction of primary visual cortex. *Consciousness and Cognition 10*, 574–587.

Tulving E, Schacter DL (1990). Priming and human memory systems. *Science 247*, 301–306.

Vaina LM, Cowey A (1996). Impairment of the perception of second order motion but not first order motion in a patient with unilateral focal brain damage. *Proceedings of the Royal Society of London B 263*, 1225–1232.

Vaina LM, Makris N, Kennedy D, Cowey A (1998). The selective impairment of the perception of first-order motion by unilateral cortical brain damage. *Visual Neuroscience 15*, 333–348.

Vaina LM, Cowey A, Kennedy D (1999). Perception of first- and second-order motion: separable neurological mechanisms? *Human Brain Mapping 7*, 67–77.

Vaina LM, Solomon J, Chowdhury S, Sinha P, Belliveau JW (2001). Functional neuroanatomy of biological motion pereption in humans. *Proceedings of the National Academy of Sciences of the USA 98*, 11656–11661.

Walsh V, Pascual-Leone A (2003). *Transcranial magnetic stimulation: a neurochronometrics of mind*. Cambridge, MA: MIT Press.

Walsh V, Ellison A, Battelli L, Cowey A (1998). Task-specific impairments and enhancements induced by magnetic stimulation of human visual area V5. *Proceedings of the Royal Society of London B 265*, 537–543.

Walsh V, Le Mare C, Blaimire A, Cowey A (2000). Normal discrimination performance accompanied by priming deficits in monkeys with V4 or TEO lesions. *Neuroreport 11*, 1459–1462.

Weiskrantz L (1998). Consciousness and commentaries. In: SR Hameroff, AW Kaszniak, AC Scott (eds), *Towards a science of consciousness*, Vol. II, pp. 371–377. Cambridge, MA: MIT Press.

Weiskrantz L, Warrington EK, Sanders MD, Marshall J (1974). Visual capacity in the hemianopic field following a restricted occipital ablation. *Brain 97*, 709–728.

Zeki S, ffytche DH (1998). The Riddoch syndrome: insights into the neurobiology of conscious vision. *Brain 121*, 25–45.

CHAPTER 28

Higher visual cognition: search, neglect, attention, and eye movements

Jacinta O'Shea and Matthew F. S. Rushworth

Introduction

Amassian *et al.*'s (1989) pioneering paper, on the role of striate cortex in letter identification, paved the way for the current generation of TMS studies of visual cognition. This chapter reviews the contribution of TMS research to our understanding of attention, eye movements, visual search, and neglect. It considers how, indeed whether, TMS studies have confirmed, refined, or challenged prevailing ideas about the neural basis of higher visual cognition. In particular, it asks how TMS has enhanced our understanding of the location, timing, and functional roles of visual cognitive processes in the human brain. The main focus is on studies of posterior parietal cortex (PPC), with reference to some more recent work on the frontal eye fields (FEFs).

It has long been argued that the parietal cortex plays a critical role in the determination of what is attended, and what is ignored, in the representation of space, and in aspects of sensorimotor transformation (Ungerleider and Mishkin 1982; Milner and Goodale 1995; Driver and Mattingley 1998). More recently, it has been argued that the FEFs, which share dense reciprocal connections with PPC (area 7a), play a critical role in visuo-spatial processing in addition to their oculomotor functions (Latto 1986; Murthy *et al.* 2001). Our understanding of how this fronto-parietal cortical network functions (Figure 28.1) has largely been derived from

animal lesion and neuropsychological patient studies, together with evidence about the intact brain derived from single-unit recordings and functional imaging. TMS offers a number of advantages to complement these techniques.

Trauma-induced lesions typically compromise an extended region of the brain. For instance, strokes of the middle cerebral artery damage not only parietal cortex but a number of other regions served by that blood supply. It is unusual to find cases where damage is restricted to circumscribed anatomical or functional regions (Miklossy 1993; Wolpert *et al.* 1998). The excellent spatial resolution of TMS offers the experimenter a degree of control over the location and extent of neural interference. In principle, this confers the potential to make more refined claims about functional localization within subregions of the fronto-parietal network.

Unlike lesions, the interference caused by TMS is temporary, and this has been exploited to make claims about when fronto-parietal cortex contributes to a sensorimotor process (Ashbridge *et al.* 1997; Fierro *et al.* 2001; Leff *et al.* 2001). The rapid nature of many sensorimotor processes makes them particularly amenable to investigation using TMS. A number of studies concerned with covert visuo-spatial orienting and overt oculomotor function have illustrated an important functional role of the parietal cortex in redirecting and updating

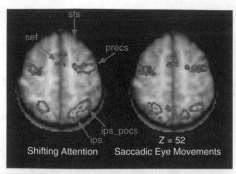

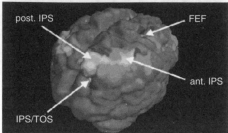

Fig. 28.1 The fronto-parietal attention network. Functional magnetic resonance activation maps showing overlapping networks in fronto-parietal cortex that are typically activated during tasks involving eye movements, visual search, or covert visuo-spatial attention. *Upper panel*: Data reproduced with permission from Corbetta *et al.* (1998) showing activations elicited by saccadic eye movements (on the right), and activations associated with covert visuo-spatial attention (on the left). sfs=superior frontal sulcus; precs=precentral sulcus; sef=supplementary eye fields; ips=intra-parietal sulcus; ips-pocs=intra-parietal and post-central sulci. *Lower panel*: Results reproduced with permission from Donner *et al.* (2002) showing (color-coded) areas activated by visual search tasks of varying difficulty. FEF=frontal eye fields; ant./post. IPS=anterior/posterior intra-parietal sulcus; IPS/TOS=intra-parietal transverse occipital sulcus. (Plate 12)

information in response to changes in the environment, or changes caused by earlier movements (Rushworth *et al.* 2001a; van Donkelaar and Müri 2002; Tobler and Müri 2002; Chambers *et al.* 2004). In the FEFs, the temporal resolution of TMS has been exploited to try to tease apart earlier visuo-spatial from later oculomotor processes (O'Shea *et al.* 2004). Such studies have been feasible because it has been possible to

temporarily disrupt cortical functioning at specific times with respect to sensorimotor events.

Localizing fronto-parietal cortical sites for TMS

Prior to the widespread availability of frameless stereotactic devices for TMS coil positioning, areas were targeted for TMS using functional criteria (Thickbroom *et al.* 1996; Ashbridge *et al.* 1997). Unlike studies of motor and primary visual cortex, which can take advantage of hand twitches, motor-evoked potentials or visual phosphenes to confirm effective stimulation, TMS studies of 'association' cortex have attempted to replicate the sensorimotor deficits of patients with brain damage.

As would be predicted from lesion data (Schiller and Chou 1998), TMS over the FEFs slows saccadic reaction times to the contralateral hemifield. Early studies of FEF function used TMS-elicited hand twitches to first identify the motor cortex, followed by an exploratory mapping procedure to locate the probabilistic FEFs (Thickbroom *et al.* 1996). TMS pulses were applied successively over scalp coordinates in a 1 × 1 cm grid anterior to the motor hand area, until a site was found where TMS delayed contralateral saccades (Ro *et al.* 2002). The anatomical location was subsequently confirmed using magnetic resonance imaging (MRI) (Ro *et al.* 1997) and shown to correspond to the caudal part of the middle frontal gyrus, at the junction of the precentral and superior frontal sulci. Imaging and intracranial recordings concur in suggesting that this anatomical site is the probabilistic human FEFs (Paus 1996; Lobel *et al.* 2001; Grosbras *et al.* 2005) (Figure 28.2).

A similar mapping procedure was used in early work on parietal cortex (Ashbridge *et al.* 1997). The TMS coil was positioned over a grid on the scalp surrounding parietal electrode sites, and single pulses were successively administered until a location was found which delayed manual reaction times in a visual search task. Studies that have examined the relationship between scalp landmarks and underlying brain anatomy (Herwig *et al.* 2003; Okamoto *et al.* 2004) have confirmed that a given scalp landmark lies above an approximately similar brain region

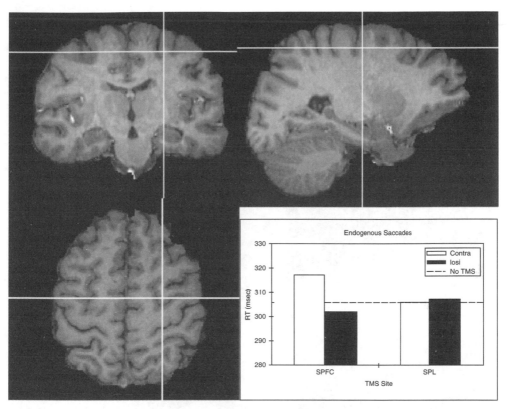

Fig. 28.2 Location of the human frontal eye fields (FEFs). Slice views of a single subject with cross-hairs illustrating the probabilistic location of the human FEFs. The FEFs are situated in the caudal middle frontal gyrus at the junction of the superior frontal sulcus and the precentral sulcus. The graph (*lower right*) shows data reproduced with permission from Ro *et al*. (1997). TMS applied over the FEFs delayed saccades to targets located in the contralateral but not the ipsilateral hemifield. The dashed line shows median saccadic reaction times (RT) without TMS.

across subjects. However, it is clear that there is individual variability. For example, the parietal P4 electrode position is usually situated close to the posterior intra-parietal sulcus (IPS) and the adjacent inferior parietal lobule (IPL), but in some subjects it lies closer to the posterior superior parietal lobule (SPL). In these early studies of parietal cortex, subsequent anatomical localization confirmed the effective site to be located in the dorsal posterior part of the angular gyrus in the IPL (electrode position P4) (AEEGS 1991; Okamoto *et al*. 2004) (Figure 28.3). However, a posterior part of the descending IPS is also situated beneath the coil at this location, only slightly further away from the center-point of stimulation (P6). It is likely that the interference

effects resulted from stimulation of both of these regions.

Functional localization using TMS remains an important method for targeting stimulation sites and for validating the effectiveness of a stimulation protocol. It can, however, lead to the charge of circularity. If a potential site for TMS is first identified by disrupted performance on a visuo-spatial task, then it is clear that follow-up manipulations must address more specific hypotheses (Rushworth *et al*. 2001a).

It is probably a fair criticism to say that TMS studies have yet to deliver on the promise of more precise functional localization within subregions of parietal cortex. Functional MRI studies have shown that regional differences in

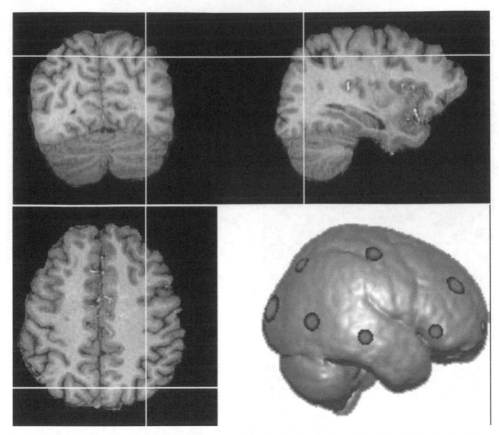

Fig. 28.3 Location of the human angular gyrus. Slice views of a single subject illustrating the location in the dorsal part of the right hemisphere angular gyrus over which TMS was applied in a number of studies cited in the text. The inset figure (*lower right*) shows the mean locations on the brain that correspond to electrode positions on the scalp using the 10–20 electrode placement system. The area under the P4 electrode is shown at the top left, just below the posterior intra-parietal sulcus in the posterior inferior parietal lobule. Reproduced with permission from Okamoto *et al.* (2004).

activation within the parietal cortex can be reliably observed and interpreted (Culham and Kanwisher 2001; Rushworth *et al.* 2001b; Astafiev *et al.* 2003). The relative dearth of TMS-based hypotheses compares poorly, even with patient studies. Systematic MRI-based investigations of lesion differences and overlap in patients with parietal damage have led to very specific hypotheses about localization within the parietal cortex (e.g. Mort *et al.* 2003a; Mannan *et al.* 2005). Several studies have replicated the basic finding that TMS over the PPC disrupts several aspects of visuo-spatial attention and they have been consistent in reporting right parietal dominance on these visuo-spatial tasks (Walsh *et al.* 1999; Rushworth *et al.* 2001a; Müri *et al.* 2002). They have also shown that reliable disruption issues from posterior rather than more anterior parietal regions (Chambers *et al.* 2004). This is consistent with evidence that damage to the angular gyrus and/or the IPS is the critical locus underlying the syndrome of neglect (Mort *et al.* 2003a). However, single-unit recording and human fMRI studies indicate that although the functional roles of the IPS and the IPL are complementary, they are distinct (Constantinidis and Steinmetz 2001a,b; Shulman *et al.* 2003). It seems feasible for carefully targeted future TMS work to dissociate these processes and to clarify their functional segregation in the PPC.

Finally, an inherent risk in functional localization approaches to cognition (using TMS, fMRI or single-unit recording) is reasoning about cortical sites as though they were isolated modules. Combined imaging and TMS studies (Section V, this volume) have shown that TMS changes brain activity not only in the area underneath the coil, but also in anatomically interconnected regions (e.g. Paus et al. 1997). The presence of large oriented fiber bundles located underneath the coil may be an important factor in several of the TMS effects reported here. As data on human anatomy emerge from diffusion imaging (Rushworth et al. 2005), future TMS investigators should be in a position to test more anatomically constrained hypotheses, to extend our understanding of functional localization beyond isolated cortical regions to the circuits in which they operate.

Neglect

Patients with right parietal damage typically present with a range of visuo-motor (Pierrot-Deseilligny et al. 1986; Mattingley et al. 1998) and visuo-spatial deficits including neglect, extinction, and oculomotor and spatial memory impairments (Critchley 1953; Mesulam 1981). The neglect syndrome itself is characterized by several deficits and typologies. Here we will restrict our focus to visuo-spatial hemineglect, its salient feature, which is a failure of patients to attend to the hemispace opposite the lesion. A related phenomenon, extinction, is characterized by a failure to detect a stimulus in the contralesional hemifield when two stimuli are presented simultaneously, one in each hemifield. The predominant view is that neglect and extinction are a function of abnormal competitive interactions between the hemispheres. Normally, the two hemispheres compete to direct spatial attention to the contralateral hemispace. When this balance is disrupted by a lesion, the intact hemisphere generates an unopposed orienting response to the ipsilesional side.

Although the neuroanatomy of neglect remains a matter of debate (Karnath et al. 2001; Mort et al. 2003a), deficits most commonly result when damage compromises the right angular gyrus and/or the IPS (Vallar 1993; Mannan et al. 2005). Patients and monkeys with damage that includes the FEFs often present with the same impairments as right parietal patients (Eglin et al. 1991; Schiller and Chou 1998; Maguire and Ogden 2002; Peers et al. 2005). Unilateral FEF damage induces neglect and extinction with both visuo-spatial and oculomotor aspects (Kennard and Ectors 1938; Kennard 1939; Silberpfennig 1941; Rizzolatti et al. 1983). While the oculomotor deficits tend to be mild or transient, visual target detection, selection, and spatial memory processes show more severe long-lasting impairments (Gaymard and Pierrot-Deseilligny 1999; Schiller and Chou 2000a,b). A study by Latto and Cowey (1971) showed that monkeys with unilateral FEF lesions exhibited raised detection thresholds for flashed luminance stimuli presented in the contralesional field. This field defect persisted, despite recovery of exploratory eye movements. Although the oculomotor deficits of FEF patients are well characterized, a tendency to group their damage with that of other frontal patients has led to the role of the FEFs in neglect going largely unexplored.

Modelling neglect with TMS of the parietal cortex

The visuo-spatial neglect- and extinction-like deficits incurred by parietal damage have been modelled successfully using TMS (Pascual-Leone et al. 1994; Hilgetag et al. 2001; Bjoertomt et al. 2002). Parietal TMS not only diminishes the probability of detecting contralateral stimuli but it also enhances the probability of detecting ipsilateral stimuli (Hilgetag et al. 2001). This can be explained if the parietal cortices in the two hemispheres are competing to direct attention contralaterally; interfering with one parietal cortex leaves the remaining parietal cortex undisturbed and better able to direct attention ipsilaterally. In the tactile domain, TMS pulses applied to the intact hemisphere have been shown to diminish the degree of neglect associated with lesions in the other hemisphere (Oliveri et al. 1999). This can be further augmented by a prior short-interval conditioning pulse which facilitates the effect of the test pulse (Oliveri et al. 2000).

In the first TMS study to model visuo-spatial extinction (Pascual-Leone *et al.* 1994), 25 Hz TMS applied over the right PPC extinguished left visual field stimuli when two targets were presented, but left PPC stimulation also produced the same phenomenon in the right hemifield. This is at odds with patient data indicating a dominant role for the right hemisphere in directing visuo-spatial attention. The discrepancy may in part reflect longer-term functional brain reorganization in response to injury.

TMS modelling of visual neglect, however, does support the view that the right hemisphere has a special role in visuo-spatial orienting. Fierro *et al.* (2000) presented pre-bisected lines, and subjects were required to judge whether the left, right, or neither side was longer. In the baseline condition, subjects exhibited 'pseudo-neglect', a tendency to report the left side as longer. TMS over the right PPC corrected this pseudo-neglect, whereas left PPC stimulation had no effect. These disruptive effects appear to be restricted to particular spatial reference frames. Bjoertomt *et al.* (2002) presented the line bisection task at a distance of either 50 or 150 cm, in order to dissociate near space from far space. They found a double dissociation where TMS over right PPC only disrupted performance in the near-space condition, and TMS over the right ventral occipital lobe only disrupted a far-space task. The PPC deficit was similar to that demonstrated by patients with right parietal hemineglect, who often bisect the line incorrectly but only when it is in near space (Halligan and Marshall 1991). The finding is also consistent with neuroimaging demonstrations that the IPS is activated when making judgments in near space (Weiss *et al.* 2000).

Recently, Ellison *et al.* (2004) reported distinct patterns of disruption on line bisection and visual search after TMS over the posterior angular gyrus/IPS under P4 or to a region in the superior temporal gyrus. As previously observed (Fierro *et al.* 2000; Ellison *et al.* 2003), while the parietal region was important for line bisection and conjunction search regardless of task difficulty (Figure 28.4), the superior temporal gyrus was implicated in difficult visual search tasks regardless of the need for feature conjunction. As discussed below ('Visual search'), the involvement of PPC in feature conjunction regardless of task difficulty is at odds with fMRI modulations of activity in that region related to task difficulty (Nobre *et al.* 2003). The involvement of the superior temporal gyrus is surprising because the region is not normally activated during visual search. In a controversial paper, however, it has been argued that superior temporal gyrus lesions are associated with neglect (Karnath *et al.* 2001), and there are reports of activation in adjacent areas when invalid warning information means that subjects have to reorient to detect targets (Corbetta *et al.* 2000). Discrepancies between imaging results on the one hand and TMS lesion results on the other may be attributable to the impact of TMS, like the impact of a lesion, needing to be understood in terms of white matter fiber bundles and not just the overlying cortex (Thiebaut de Schotten *et al.* 2005).

Visual search

In a visual search task, a subject is presented with a brief display and must press a button to indicate the presence or absence of a predefined target among distractors. When the target differs from the distractors along a single feature dimension, detection is usually rapid. But when the difference between target and distractor features is more subtle, or when the target is defined by a conjunction of two feature dimensions (e.g. color and orientation), target discrimination becomes less efficient, as subjects must search the array in a way that taxes attention. Feature integration theory claims that attention is required in conjunction search to bind the different features of a single target together (Treisman and Gelade 1980; Treisman and Sato 1990). Patients with right posterior parietal lesions perform poorly on conjunction search, frequently reporting illusory conjunctions – a kind of error in which features of distinct stimuli are reported as features of a single stimulus (Arguin *et al.* 1994; Friedman-Hill *et al.* 1995). This has led to the view that one functional role of the parietal cortex is feature binding, which enables target selection during conjunction search.

One of the most well-replicated findings in the visual cognition literature is that TMS applied over the right angular gyrus disrupts

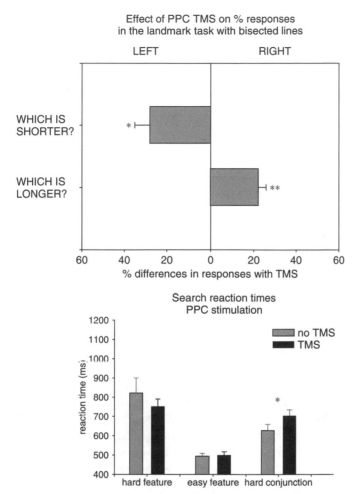

Fig. 28.4 Effect of angular gyrus TMS on visual search and line bisection performance. *Upper panel:* Subjects were presented with centrally bisected lines and asked to indicate which side was longer or shorter. When TMS was applied over the right angular gyrus subjects reported that the left side was shorter on 20% more trials and that the right side was longer on 22% more trials compared to baseline performance without TMS. Thus TMS of the right angular gyrus induced a tendency to underestimate the left segment of a line, as seen in patients with contalateral neglect. *Lower panel:* TMS over the right angular gyrus significantly increased subjects' manual reaction times on a difficult conjunction search task. There was no effect on feature search performance. PPC=posterior parietal cortex. Reproduced with permission from Ellison *et al.* (2004).

performance on conjunction search. By contrast, TMS does not affect search for targets defined by single visual features that 'pop out' (Ellison *et al.* 2003, 2004). A similar pattern of results has been demonstrated in the FEFs. TMS applied over the right FEFs impairs conjunction search accuracy, but has no effect on popout search performance (Muggleton *et al.* 2003) (Figure 28.5).

These TMS studies both replicate basic findings from the patient literature and confirm their relevance for understanding the intact brain. Together with related findings in the fMRI literature, they suggest that the nature of a search task and its experimental implementation are important factors to bear in mind when attempting to understand the functional roles of distinct fronto-parietal areas. Whilst confirming

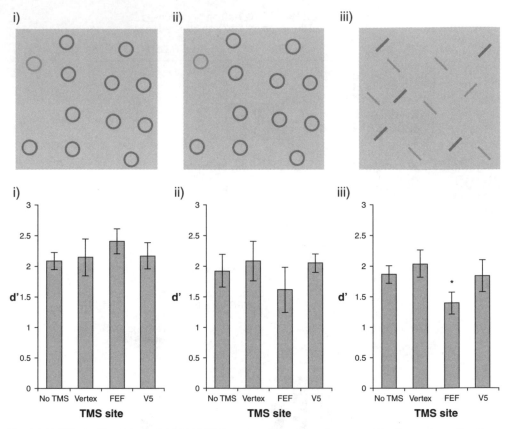

Fig. 28.5 Effect of frontal eye field (FEF) TMS on visual search performance. The figure shows the three types of search task used by Muggleton *et al.* (2003): (i) simple feature search; (ii) the same feature search task as in (i) except that on 50% of trials the color pairings switched, with the target in red and the distractors in blue; (iii) conjunction search. Data from TMS applied over the right FEFs and data from control conditions (no TMS, TMS over right V5, TMS over the vertex) are shown in the lower panel. (i) TMS over the right FEFs degraded performance (*d'*) on the conjunction search task. (ii) A trend was observed in the feature search task, in which target and distractor color pairings switched on a trial-by-trial basis. (iii) There was no effect on simple feature search performance. Reproduced with permission from Muggleton *et al.* (2003). Asterisk denotes significant effect of TMS (P<0.05). (Plate 13)

the importance of a right-lateralized fronto-parietal network underlying search, TMS studies have questioned earlier hypotheses about the functional roles of these areas.

The effect of learning

After extensive experience of searching for a given target among distractors, subjects' search performance becomes faster and is less influenced by the number of distractors that are present. Walsh *et al.* (1998) trained subjects on a

conjunction visual search task until their search slopes became flat (an index of search expertise). After training, parietal TMS no longer disrupted performance, but a deficit re-emerged upon transfer to a novel search array. The interference effect could also be reinstated if the subject was tested on the trained array but the learned target/distractor features or the stimulus–response mapping rules were inverted. As TMS ceased to be disruptive after only a small amount of training (~250 trials) (Walsh *et al.* 1999), the authors concluded that the critical role of the right angular gyrus in search may be that of learning the association between a novel visual target and the requisite manual response (Ellison *et al.* 2003). The authors rejected an interpretation based on visual feature binding because TMS also disrupted performance on trials when there was no target with features to bind (Ashbridge *et al.* 1997), and because when subjects had foreknowledge of the location of a conjunction target, TMS had no effect (Ellison *et al.* 2003). The claim that one role for the parietal cortex is to link the target stimulus with the detection response has received additional support from a patient study. The automaticity of the detection response was shown to be a determinant of search performance in a neglect patient (Humphreys and Riddoch 2001). The patient was quick to find targets but only when they automatically afforded actions.

Why is the parietal cortex less important after learning? It may be that training strengthens feature representations in infero-temporal cortex, causing conjunction targets that were previously difficult to detect to pop out (Chelazzi *et al.* 1993; Jagadeesh *et al.* 2001). Thus, the formation of a target representation may no longer need to be sustained by parietal activity in order for search to proceed and a detection response to be made. It might be argued that if a biasing signal related to the target's identity is at first used to guide attentional selection, then it might no longer be necessary after extensive experience of searching for the same target. The absence of any parietal deficit after learning is consistent with the notion that the parietal cortex may contribute to such a bias signal (Kastner *et al.* 1999; Kastner and Ungerleider 2001). Alternatively, if it is the case that the parietal cortex is normally needed in order to establish a single focus of behavioral salience before an action, whether a saccade or limb movement, is selected and initiated, then the absence of a parietal TMS deficit after learning suggests that learning establishes an alternative route by which stimulus representations in extrastriate cortex are able to trigger actions.

The role of the frontal eye fields

A robust finding of neuroimaging studies is that search activates a fronto-parietal network with nodes in the PPC and the FEFs (Corbetta *et al.* 1998). While parietal activations are usually ascribed to visuo-spatial processes (coordinate transformation, binding, 'top-down control', distractor filtering, etc.) (Corbetta *et al.* 1995; Shulman *et al.* 2002), FEF activations are typically identified with saccade programs (Buchel *et al.* 1998; Gitelman *et al.* 1999; but see Hung *et al.* 2005). According to a commonly invoked explanatory scheme (Mesulam 1981, 1990), parietal cortex integrates information from feature-selective extrastriate areas to select targets that are behaviorally relevant. This information is then relayed to the FEFs, where saccades are programmed to these targets in an obligatory manner. Inhibition prevents these programs being executed during covert search tasks. This interpretation concurs with the classical view of the FEFs as an oculomotor executor. Against this view, together with single-unit data (Thompson and Schall 1999) and a small number of lesion studies (e.g. Latto and Cowey 1971), recent TMS studies of visual search have argued that the FEFs also have visuo-spatial functions, and that these are dissociable from their oculomotor functions.

By using small foveal search arrays ($2° \times 2°$), brief display durations that were subsequently masked, and by monitoring fixation, Muggleton *et al.* (2003) argued that the search deficits incurred by right FEF TMS could not be explained in terms of latent saccade programs. This argument was strengthened by a follow-up study in which it was shown that dual pulses of TMS applied between 40 and 80 ms after search array onset degraded search performance (O'Shea *et al.* 2004). This early and discrete effect, when the mean array-viewing duration was 180 ms, does not readily permit an interpretation in terms of saccade execution. Instead, relying on single-unit evidence (Sato *et al.* 2001), the authors argued that TMS had interfered with the generation of a visuo-spatial signal in the FEFs that is required for accurate target discrimination (Figure 28.6).

The similar functional profile of areas FEF and PPC in visuo-spatial processing raises the question of the distinct causal roles of these nodes within 'the fronto-parietal attention network' (O'Shea and Walsh 2006). TMS timing data suggest a primary role for the FEFs in target

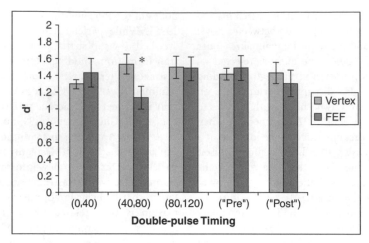

Fig. 28.6 Timing of frontal eye field (FEF) TMS interference during visual search performance. Dual pulses of TMS applied over the right FEFs at 40 and 80 ms after search array onset degraded conjunction search performance (d′). There was no effect of TMS in any of the later time bins. Reproduced with permission from O'Shea *et al.* (2004). Asterisk denotes significant effect of TMS (P<0.05).

detection that is not dependent in a simple way on prior activity in parietal and other regions. Whereas TMS pulses over the FEFs just 40–80 ms after search array onset degrade search performance, parietal disruption occurs significantly later. One research group applied single-pulse TMS to the parietal cortex at 20 different stimulus-onset asynchronies between 0 and 200 ms (Ashbridge *et al.* 1997; Walsh *et al.* 1998, 1999; Rosenthal *et al.* 2006). Trials on which conjunction targets were present were affected by TMS at 100 ms and target-absent trials were affected by TMS delivered 160 ms after array onset. The critical period for the parietal cortex overlaps with the relatively late critical time period of primary visual cortical involvement in conjunction visual search (Juan and Walsh 2003). Taken together with single-unit data, these results are consistent with models that envisage rapid activation of the FEFs leading to first-pass global 'gist' processing of the visual scene, which then alerts extrastriate areas to potential target loci via fast feedback signals onto retinotopically organized cortex (Bullier 2001; Hochstein and Ahissar 2002). The excellent temporal resolution of TMS offers scope for follow-up studies to pursue the comparative project of distinguishing the relative contributions of the PPC and the FEFs to visual target selection (Hung *et al.* 2005).

Priming

One visual search phenomenon that has only recently begun to be explored by TMS is priming. Priming refers to the enhanced speed and accuracy of search when an observer is presented with a fixed combination of target and distractor features, compared with more inefficient performance when target/distractor features change across trials (Maljkovic and Nakayama 1994, 1996, 2000). That is, if the current trial target has the same features or is in the same location as the target on the previous trial, then the observer's detection response is facilitated. There have been very few studies of the neural basis of priming in visual search. One study has shown that patients with right parietal damage, compromising the inferior parietal lobe (including the angular gyrus) and the superior temporal gyrus, show intact feature and spatial priming (Kristjansson *et al.* 2005). Whilst there have not been any lesion studies of priming in the FEFs, single-unit recordings have shown that target discrimination processes in FEF visual neurons are influenced by long-term visual experience (Bichot *et al.* 1996), short-term visual priming (Bichot and Schall 2002), and visual feature similarity between targets and distractors (Bichot and Schall 1999).

The feasibility of TMS studies of priming has been demonstrated in area V5/MT, where 10 Hz TMS applied in the inter-trial interval abolished priming for motion direction (Campana *et al.* 2002). The authors argued that TMS had disrupted a memory trace for motion direction that was stored in V5/MT during the inter-trial interval, which facilitated performance across trials. A suggestion that priming may be one mechanism contributing to FEF TMS interference effects comes from the findings of Muggleton *et al.* (2003). Whereas TMS disrupted conjunction search and had no effect on popout search, a trend was observed in a third popout condition, where target/distractor color pairings randomly switched across trials (Figure 28.5). The null effects of parietal and FEF TMS on fixed popout arrays may require further validation. More recent human fMRI and macaque temporary inactivation studies suggest that various regions in the inferior parietal lobule, the angular gyrus, and the IPS are important when subjects are searching for single features that are difficult to discern (Nobre *et al.* 2003; Wardak *et al.* 2004). In the few fMRI studies where search difficulty has been manipulated parametrically, PPC and FEF activations have been modulated accordingly (Donner *et al.* 2002; Marois *et al.* 2004).

Visuo-spatial attention and eye movements

Visuo-spatial attention: the PPC

A number of studies have been concerned with the movement of the focus of covert attention using variations of a paradigm popularized by Posner and colleagues. In one version of the task, subjects indicated detection of a target stimulus in one of two possible locations by pressing a button. On some trials a warning cue instructed subjects where the stimulus was likely to appear but on a few trials the warning was invalid. Subjects responded more slowly on such invalid trials but the effect was particularly pronounced in patients with parietal lesions (Posner *et al.* 1984). The results were interpreted as indicating an important role for the parietal cortex in the reorienting of attention. By using short trains of TMS it has been possible to

confirm that the parietal cortex is particularly vulnerable to disruption when subjects are redirecting attention from one location to another. TMS application over the posterior IPS/angular gyrus of the IPL disrupts performance if it is delivered shortly after target presentation, but only if the target has been invalidly precued so that the focus of attention must be moved and redirected (Rushworth *et al.* 2001a; Chambers *et al.* 2004). Activation in the posterior IPS region has been reported in imaging studies on trials on which subjects have to redirect attention, but activity in the posterior angular gyrus and adjacent superior occipital gyrus is especially closely related to reorienting (Mort *et al.* 2003b; Thiel *et al.* 2004; Kincade *et al.* 2005).

Intriguingly it has been argued that there may be two separate brief periods when the parietal cortex is active in reorienting attention. Chambers *et al.* (2004) have shown that single TMS pulses are most disruptive when they are applied either 90–120 or 210–240 ms after the invalidly cued target. They argue that the two critical time periods can be explained if the visual information that guides attentional reorienting arrives at two different times via two anatomical pathways. The interesting, but unproven, suggestion is that a fast route may run from the retina via the superior colliculus and pulvinar nucleus of the thalamus while a slower route may course through striate and extrastriate cortex. This hypothesis about a fast collicular pathway could be tested using S-cone stimuli. S-cones do not project to the colliculus, so if the early TMS effect disappeared, this would substantiate the hypothesis (Sumner *et al.* 2002).

The parietal contribution to orienting does not seem to be restricted to the case of redirecting attention after an invalid cue. Thut *et al.* (2004) administered trains of 1 Hz TMS for 25 min to the same right parietal region and then tested performance on a similar paradigm. Such 1 Hz trains are thought to affect activity for many minutes after application. It is possible that the effects of long 1 Hz trains are not as spatially circumscribed as the effects of single pulses or short trains, and they are limited in what they can reveal about the timing of a cognitive process. Nevertheless, after TMS, when subjects were cued to the right field, the worst performance was on trials where subjects were miscued and were then required to redirect attention to the left field contralateral to the TMS site. Performance, however, was also noticeably poor on all trials when subjects oriented to the left field contralateral to the site of TMS application.

While TMS investigations of the temporal course of events in attentional paradigms have tended to concentrate on events that occur over a very brief time scale, it is clear that other cognitive processes within the parietal cortex take place over slightly longer time scales. Patient studies have been important in elucidating these processes. For example, it is becoming clear that the poor search performance of some parietal patients in cancellation tasks is due to an impairment of spatial memory so that the same area in the stimulus array is searched more than once (Husain *et al.* 2001; Malhotra *et al.* 2005; Mannan *et al.* 2005). High-resolution MRI mapping of the lesions in these patients suggests that the critical lesion was to the IPS or adjacent parietal white matter. The application of TMS to posterior parietal sites during memory delays has been reported to affect the maintenance of information in memory because it affects the speed of subsequent responses (Oliveri *et al.* 2001; Koch *et al.* 2005). Similar patterns of impairments have been observed regardless of whether the final response is oculomotor or manual (Müri *et al.* 1996, 2000; Smyrnis *et al.* 2003). Müri and colleagues have used paradigms with short delays and have concluded that the performance is most vulnerable to parietal TMS at very short time periods, just 150 ms, after stimulus presentation. It is clear from the patient studies, however, that the parietal cortex, perhaps together with interconnected areas, is essential for mediating memory over longer time scales; the chances of returning to a search area increase over time following damage to the IPS, but not after damage to the frontal regions associated with visual search impairments (Mannan *et al.* 2005).

Visuo-spatial attention: the FEFs

A small number of recent studies have begun to assess visuo-spatial function in the human FEFs using TMS. In a Posner paradigm, single-pulse TMS applied during the cue-target period speeded subjects' manual reaction times. When TMS was applied over left FEF, this effect occurred for validly, invalidly, and neutrally cued trials, but only when the target was in the contralateral hemifield. TMS over right FEF facilitated reaction time to targets in either hemifield on all but invalidly cued trials (Grosbras and Paus 2002). The authors concluded that TMS had enhanced target detection processes in the FEFs on the trials specified above, except on the invalidly cued trials when TMS over right FEF disrupted spatial attentional reorienting.

In a follow-up study, the authors tested the detection interpretation using a backward masking paradigm (Grosbras and Paus 2003). Single-pulse TMS applied 40–100 ms prior to the onset of a single target (that was subsequently masked) lowered subjects' luminance detection thresholds. The hemifield asymmetry was replicated: TMS over right FEF improved detection in both hemifields; TMS over left FEF facilitated detection of only contralateral targets. This asymmetry concurs with patient, imaging and TMS studies claiming a right hemisphere bias in the fronto-parietal network subserving visuo-spatial attention (Mesulam 1981; Kim *et al.* 1999; Walsh *et al.* 1999) (Figure 28.7).

Grosbras and Paus interpreted these results by analogy with a single-unit detection and backward masking study (Thompson and Schall 1999) and with microstimulation studies of the macaque FEFs (Moore and Armstrong 2003; Moore and Fallah 2004). Moore and Armstrong have shown that stimulation of FEF neurons is associated with an enhanced response to visual stimuli in extrastriate neurons with spatially overlapping receptive fields. Grosbras and Paus concluded that detection sensitivity had improved in their study because TMS had 'preactivated' FEF neurons, raising their baseline activation level such that incoming visual stimulation could reach the detection threshold more easily. However, TMS should affect both the signal and noise distributions in the FEFs, making it difficult to see how a selective enhancement could occur. The key factor may be the presence of a single stimulus – if TMS increases neural activity in an area carrying a single visual representation, then it is conceivable that this signal could be enhanced.

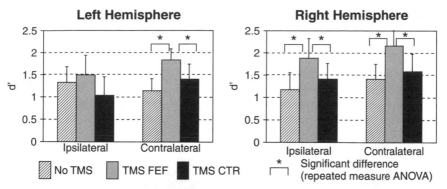

Fig. 28.7 Effect of frontal eye field (FEF) TMS on visual target detection. TMS over the right hemisphere FEFs enhanced detection (*d'*) of masked targets in the contralateral and ipsilateral visual fields. TMS over the left FEFs enhanced detection only when targets were located in the contralateral visual field. These effects were obtained relative to baseline performance without TMS and detection performance with TMS applied over a control site in the temporal lobe (TMS CTR). Reproduced with permission from Grosbras and Paus (2003).

The addition of a distractor, making the task one of discrimination rather than detection, would predict a disruptive effect of TMS (O'Shea and Walsh 2004).

Parietal cortex and eye movements

The same regions of parietal cortex that are involved in covert attentional processes are also active when overt eye movements are made (Corbetta 1998; Gitelman *et al.* 1999; Nobre *et al.* 2000), so it is not surprising that parietal TMS also affects eye movements (Zangemeister *et al.* 1995; Terao *et al.* 1998; Kapoula *et al.* 2001). TMS does not evoke saccades (Müri *et al.* 1991), nor does it disrupt express saccades, which are not cortically mediated (Müri *et al.* 1999). However, TMS applied over the PPC has been shown to transiently disrupt performance on visually guided (Yang and Kapoula 2004), memory-guided (Müri *et al.* 2000), and double-step saccade tasks (van Donkelaar and Müri 2002). The timing of application appears to be crucial. Several studies converge on the need to stimulate during a time window yoked to the expected saccadic reaction time (Priori *et al.* 1993; Grosbras and Paus 2003). Single pulses applied within 70–50 ms prior to saccade onset (but not earlier or later than this) have been shown to disrupt saccade latency, while effects on accuracy have been less commonly reported (Zangemeister *et al.* 1995).

Just as TMS over the posterior IPS/IPL region disrupts updating of the direction of attention, it also seems to disrupt the updating of visual space, often described as spatial remapping, that occurs when a saccade is made (van Donkelaar and Müri 2002). In the double-step saccade task, subjects make two consecutive saccades to two targets that are flashed so briefly that they disappear even before the subject completes the first eye movement. It is necessary to compensate for the change in eye position entailed by the first saccade if the second target is to be reached successfully. Normal compensation for the first saccade does not occur if TMS is applied after it is completed, just before the second saccade. The spatial remapping of receptive fields just before saccades are made has been recorded in lateral intra-parietal neurons in the macaque (Duhamel *et al.* 1992) and activation in the human IPS has been linked to saccadic updating (Heide *et al.* 2001).

The FEFs and eye movements

Transient disruption of antisaccades (Müri *et al.* 1991), saccade triggering (Wipfli *et al.* 2001; Nyffeler *et al.* 2004), double-step saccades (Tobler and Müri 2002), and inhibition of return (Ro *et al.* 2003) can be induced by applying TMS over the FEFs. Various lines of evidence indicate that the FEFs make a critical contribution to the generation of intentional but not reflexive saccades, which appears to rely more on parietal cortex and the superior colliculus (Gaymard *et al.* 2003). Patients with frontal lobe damage often exhibit deficits in inhibiting reflexive saccades to salient targets (Guitton *et al.* 1985). Consistent with this, TMS studies have reported disruptive effects on inhibiting reflexive attentional shifts (Ro *et al.* 2003; Smith *et al.* 2005), but whether the FEFs are critical for reflexive saccade inhibition is disputed (Gaymard *et al.* 1999).

A systematic investigation of the relative roles of the FEFs and the PPC in oculomotor control using TMS remains to be carried out. A small number of studies have addressed this issue (e.g. Müri *et al.* 1996). In an antisaccade task, Terao *et al.* (1998) applied single-pulse TMS over two scalp positions: one 2–4 cm anterior and 2–4 cm lateral of the motor hand area (likely to be the FEFs), and a second site 6–8 cm posterior and 0–4 cm lateral to the hand area (a site they called PPC). Single TMS pulses applied 80 ms (PPC) or 100 ms (FEFs) before cue onset delayed antisaccade latencies. This occurred irrespective of the direction of the saccade. There was a contralateral increase in the number of erroneous prosaccades (subjects look towards the target instead of away): left hemisphere stimulation increased errors to the right and vice versa. Erroneous prosaccades were induced by TMS of the occipital cortex, the motor cortex, and the anterior or posterior parietal cortex as well as the FEFs. They were most commonly induced by early (80 ms prior to the cue) rather than later (100 or 120 ms) applications of TMS. The nonspecificity of the effects suggests that TMS likely disrupted a variety of processes related to inhibiting an orienting response towards the visual cue.

Conclusion

TMS studies of search, neglect, attention, and eye movements have largely replicated the kinds of deficits that result from damage to the PPC or the FEFs. By exploiting the temporal resolution of TMS, these studies have also offered chronometric insights not possible with patients, which have in turn suggested new hypotheses about the underlying cognitive operations. Future work might be directed at teasing apart the distinct functional roles of nodes within this fronto-parietal network in different sensorimotor contexts.

References

AEEGS (1991). *Journal of Clinical Neurophysiology 8*, 200–2002.

Amassian VE, Cracco RQ, Maccabee PJ, Cracco JB, Rudell, Eberle L (1989). *Electroencephalography and Clinical Neurophysiology 74*, 458–462.

Arguin M, Cavanagh P, Joanette Y (1994). *Brain and Cognition 24*, 44–56.

Ashbridge E, Walsh V, Cowey A (1997). *Neuropsychologia 35*, 1121–1131.

Astafiev SV, Shulman GL, Stanley CM, Snyder AZ, Van Essen DC, Corbetta M (2003). *Journal of Neuroscience 23*, 4689–4699.

Bichot NP, Schall JD (1999). *Nature Neuroscience 2*, 549–554.

Bichot NP, Schall JD (2002). *Journal of Neuroscience 22*, 4675–4685.

Bichot NP, Schall JD, Thompson KG (1996). *Nature 381*, 697–699.

Bjoertomt O, Cowey A, Walsh V (2002). *Brain 125*, 2012–2022.

Buchel C, Josephs O, Rees G, Turner R, Frith CD, Friston KJ (1998). *Brain 121*(Pt 7), 1281–1294.

Bullier J (2001). *Brain Research Reviews 36*, 96–107.

Campana G, Cowey A, Walsh V (2002). *Cerebral Cortex 12*, 663–669.

Chambers CD, Payne JM, Stokes MG, Mattingley JB (2004). *Nature Neuroscience 7*, 217–218.

Chelazzi L, Miller EK, Duncan J, Desimone R (1993). *Nature 363*, 345–347.

Constantinidis C, Steinmetz MA (2001a). *Cerebral Cortex 11*, 592–597.

Constantinidis C, Steinmetz MA (2001b). *Cerebral Cortex 11*, 581–591.

Corbetta M (1998). *Proceedings of the National Academy of Sciences of the USA 95*, 831–838.

Corbetta M, Shulman GL, Miezin FM, Petersen SE (1995). *Science 270*, 802–805.

Corbetta M, Akbudak E, Conturo TE, *et al.* (1998). *Neuron 21*, 761–773.

Corbetta M, Kincade JM, Ollinger JM, McAvoy MP, Shulman GL (2000). *Nature Neuroscience 3*, 292–297.

Critchley M (1953). *The parietal lobes.* London: Arnold.

Culham JC, Kanwisher NG (2001). *Current Opinion in Neurobiology 11*, 157–163.

Donner TH, Kettermann A, Diesch E, Ostendorf F, Villringer A, Brandt SA (2002). *NeuroImage 15*, 16–25.

Driver J, Mattingley JB (1998). *Nature Neuroscience 1*, 17–22.

Duhamel JR, Colby CL, Goldberg ME (1992). *Science 255*, 90–92.

Eglin M, Robertson LC, Knight RT (1991). *Cerebral Cortex 1*, 262–272.

Ellison A, Rushworth M, Walsh V (2003). *Clinical Neurophysiology Supplement 56*, 321–330.

Ellison A, Schindler I, Pattison LL, Milner AD (2004). *Brain 127*, 2307–2315.

Fierro B, Brighina F, Oliveri M, *et al.* (2000). *Neuroreport 11*, 1519–1521.

Fierro B, Brighina F, Piazza A, Oliveri M, Bisiach E (2001). *Neuroreport 12*, 2605–2607.

Friedman-Hill SR, Robertson LC, Treisman A (1995). *Science 269*, 853–855.

Gaymard B, Pierrot-Deseilligny C (1999). *Current Opinion in Neurology 12*, 13–19.

Gaymard B, Ploner CJ, Rivaud-Pechoux S, Pierrot-Deseilligny C (1999). *Experimental Brain Research 129*, 288–301.

Gaymard B, Lynch J, Ploner CJ, Condy C, Rivaud-Pechoux S (2003). *European Journal of Neuroscience 17*, 1518–1526.

Gitelman DR, Nobre AC, Parrish TB *et al.* (1999). *Brain 122*(Pt 6), 1093–1106.

Grosbras MH, Paus T (2002). *Journal of Cognitive Neuroscience 14*, 1109–1120.

Grosbras MH, Paus T (2003). *European Journal of Neuroscience 18*, 3121–3126.

Grosbras MH, Laird AR, Paus T (2005). *Human Brain Mapping 25*, 140–154.

Guitton D, Buchtel HA, Douglas RM (1985). *Experimental Brain Research 58*, 455–472.

Halligan PW, Marshall JC (1991). *Nature 350*, 498–500.

Heide W, Binkofski F, Seitz RJ, *et al.* (2001). *European Journal of Neuroscience 13*, 1177–1189.

Herwig U, Satrapi P, Schoenfeldt-Leucuona C (2003). *Brain Topography 16*, 95–99.

Hilgetag CC, Theoret H, Pascual-Leone A (2001). *Nature Neuroscience 4*, 953–957.

Hochstein S, Ahissar M (2002). *Neuron 36*, 791–804.

Humphreys GW, Riddoch MJ (2001). *Nature Neuroscience, 4*, 84–88.

Hung J, Driver J, Walsh V (2005). *Journal of Neuroscience 25*, 9602–9612.

Husain M, Mannan S, Hodgson T, Wojciulik E, Driver J, Kennard C (2001). *Brain 124*, 941–952.

Jagadeesh B, Chelazzi L, Mishkin M, Desimone R (2001). *Journal of Neurophysiology 86*, 290–303.

Juan CH, Walsh V (2003). *Experimental Brain Research 150*, 259–263.

Kapoula Z, Isotalo E, Müri RM, Bucci MP, Rivaud-Pechoux S (2001). *Neuroreport 12*, 4041–4046.

Karnath HO, Ferber S, Himmelbach M (2001). *Nature 411*, 950–953.

Kastner S, Ungerleider LG (2001). *Neuropsychologia 39*, 1263–1276.

Kastner S, Pinsk MA, De Weerd P, Desimone R, Ungerleider LG (1999). *Neuron 22*, 751–761.

Kennard MA (1939). *Archives of Neurology and Psychiatry (Chicago) 41*, 1153–1165.

Kennard MA, Ectors L (1938). *Journal of Neurophysiology 1*, 45–54.

Kim Y-H, Gitelman DR, Nobre AC, Parrish TB, LaBar KS, Mesulam M-M (1999). *NeuroImage 9*, 269–277.

Kincade JM, Abrams RA, Astafiev SV, Shulman GL, Corbetta M (2005). *Journal of Neuroscience 25*, 4593–4604.

Koch G, Oliveri M, Torriero S, Carlesimo GA, Turriziani P, Caltagirone C (2005). *NeuroImage 24*, 34–39.

Kristjansson A, Vuilleumier P, Malhotra P, Husain M, Driver J (2005). *Journal of Cognitive Neuroscience 17*, 859–873.

Latto R (1986). Behavioral Brain Research 22, 41–52.

Latto R, Cowey A (1971). *Brain Research 30*, 1–24.

Leff AP, Scott SK, Rothwell JC, Wise RJ (2001). *Cerebral Cortex 11*, 918–923.

Lobel E, Kahane P, Leonards U, *et al.* (2001). *Journal of Neurosurgery 95*, 804–815.

Maguire AM, Ogden JA (2002). *Neuropsychologia 40*, 879–887.

Malhotra P, Jager HR, Parton A, *et al.* (2005). *Brain 128*, 424–435.

Maljkovic V, Nakayama K (1994). *Memory and Cognition 22*, 657–672.

Maljkovic V, Nakayama K (1996). *Perception and Psychophysics 58*, 977–991.

Maljkovic V, Nakayama K (2000). *Visual Cognition 7*, 571–595.

Mannan S, Mort D, Hodgson TL, Driver J, Kennard C, Husain M (2005). *Journal of Cognitive Neuroscience 17*, 340–354.

Marois R, Chun MM, Gore JC (2004). *Journal of Neurophysiology 92*, 2985–2992.

Mattingley JB, Husain M, Rorden C, Kennard C, Driver J (1998). *Nature 392*, 179–182.

Mesulam MM (1981). *Annals of Neurology 10*, 309–325.

Mesulam MM (1990). *Annals of Neurology 28*, 597–613.

Miklossy J (1993). In: B Gulyas, D Ottoson, PE Roland (eds), *Functional organization of the human visual cortex*, Vol. 61, pp. 123–136. Oxford: Pergamon Press.

Milner AD, Goodale MA (1995). *The visual brain in action*. Oxford: Oxford University Press.

Moore T, Armstrong KM (2003). *Nature 421*, 370–373.

Moore T, Fallah M (2004). *Journal of Neurophysiology 91*, 152–162.

Mort DJ, Malhotra P, Mannan SK, *et al.* (2003a). *Brain 126*, 1986–1997.

Mort DJ, Perry RJ, Mannan, SK, *et al.* (2003b). *NeuroImage 18*, 231–246.

Muggleton NG, Juan C-H, Cowey A, Walsh V (2003). *Journal of Neurophysiology 89*, 3340–3343.

Müri RM, Hess CW, Meienberg O (1991). *Experimental Brain Research 86*, 219–223.

Müri RM, Vermersch AI, Rivaud S, Gaymard B, Pierrot-Deseilligny C (1996). *Journal of Neurophysiology 76*, 2102–2106.

Müri RM, Rivaud S, Gaymard B, *et al.* (1999). *Neuropsychologia 37*, 199–206.

Müri RM, Gaymard B, Rivaud S, Vermersch A, Hess CW, Pierrot-Deseilligny C (2000). *Neuropsychologia 38*, 1105–1111.

Müri RM, Buhler R, Heinemann D, *et al.* (2002). *Experimental Brain Research 143*, 426–430.

Murthy A, Thompson KG, Schall JD (2001). *Journal of Neurophysiology 86*, 2634–2637.

Nobre AC, Gitelman DR, Dias EC, Mesulam MM (2000). *NeuroImage 11*, 210–216.

Nobre AC, Coull JT, Walsh V, Frith CD (2003). *NeuroImage 18*, 91–103.

Nyffeler T, Bucher O, Pflugshaupt T, *et al.* (2004). *European Journal of Neuroscience 20*, 2240–2244.

Okamoto M, Dan H, Sakamoto K, *et al.* (2004). *NeuroImage 21*, 99–111.

Oliveri M, Rossini PM, Traversa R, *et al.* (1999). *Brain 122*(Pt 9), 1731–1739.

Oliveri M, Rossini PM, Filippi MM, *et al.* (2000). *Brain 123*(Pt 9), 1939–19347.

Oliveri M, Turriziani P, Carlesimo GA, *et al.* (2001). *Cerebral Cortex 11*, 606–618.

O'Shea J, Walsh V (2004). *Current Biology 14*, R279–R281.

O'Shea J, Muggleton NG, Cowey A, Walsh V (2004). *Journal of Cognitive Neuroscience 16*, 1060–1067.

O'Shea J, Muggleton N, Cowey A, Walsh V (2006). *Visual Cognition 14*, 934–957.

Pascual-Leone A, Gomez-Tortosa E, Grafman J, Alway D, Nichelli P, Hallett M (1994). *Neurology 44*, 494–498.

Paus T (1996). *Neuropsychologia 34*, 475–483.

Paus T, Jech R, Thompson CJ, Comeau R, Peters T, Evans AC (1997). *Journal of Neuroscience 17*, 3178–3184.

Peers PV, Ludwig CJ, Rorden C, *et al.* (2005). *Cerebral Cortex 15*, 1469–1484.

Pierrot-Deseilligny C, Gray F, Brunet P (1986). *Brain 109*(Pt 1), 81–97.

Posner MI, Walker JA, Friedrich FJ, Rafal RD (1984). *Journal of Neuroscience 4*, 1863–1874.

Priori A, Bertolasi L, Rothwell JC, Day BL, Marsden CD (1993). *Brain 116*(Pt 2), 355–367.

Rizzolatti G, Matelli M, Pavesi G (1983). *Brain 106*(Pt 3), 655–673.

Ro T, Henik A, Machado L, Rafal R (1997). *Journal of Cognitive Neuroscience 9*, 433–440.

Ro T, Farne A, Chang E (2002). *Journal of Clinical and Experimental Neuropsychology 24*, 930–940.

Ro T, Farne A, Chang E (2003). *Experimental Brain Research 150*, 290–296.

Rosenthal CR, Walsh V, Mannan SK, Anderson EJ, Hawken MB, Kennard C (2006). *Neuropsychologia 44*, 731–743.

Rushworth MF, Ellison A, Walsh V (2001a). *Nature Neuroscience 4*, 656–661.

Rushworth MF, Paus T, Sipila PK (2001b). *Journal of Neuroscience 21*, 5262–5271.

Rushworth MFS, Behrens T, Johansen-Berg H (2005). *Cerebral Cortex 16*, 1418–1430.

Sato T, Murthy A, Thompson KG, Schall JD (2001). *Neuron 30*, 583–591.

Schiller PH, Chou IH (1998). *Nature Neuroscience 1*(3), 248–253.

Schiller PH, Chou I (2000a). *Vision Research 40*, 1609–1626.

Schiller PH, Chou I (2000b). *Vision Research 40*, 1627–1638.

Shulman GL, d'Avossa G, Tansy AP, Corbetta M (2002). *Cerebral Cortex 12*, 1124–1131.

Shulman GL, McAvoy MP, Cowan MC, *et al.* (2003). *Journal of Neurophysiology 90*, 3384–3397.

Silberpfennig J (1941). *Confinia Neurologica (Basel) 4*, 1–13.

Smith DT, Jackson SR, Rorden C (2005). *Neuropsychologia 43*, 1288–1296.

Smyrnis N, Theleritis C, Evdokimidis I, Müri RM, Karandreas N (2003). *Journal of Neurophysiology 89*, 3344–3350.

Sumner P, Adamjee T, Mollon JD (2002). *Current Biology 12*, 1312–1316.

Terao Y, Fukuda H, Ugawa Y, *et al.* (1998). *Journal of Neurophysiology 80*, 936–946.

Thickbroom GW, Stell R, Mastaglia FL (1996). *Journal of Neurological Sciences 144*, 114–118.

Thiebaut de Schotten MT, Urbanski M, Duffau H, *et al.* (2005). *Science 309*, 2226–2228.

Thiel CM, Zilles K, Fink GR (2004). *NeuroImage 21*, 318–328.

Thompson KG, Schall JD (1999). *Nature Neuroscience 2*, 283–288.

Thut G, Nietzel A, Pascual-Leone A (2004). *Cerebral Cortex 15*, 628–638.

Tobler PN, Müri RM (2002). *Neuroreport 13*, 253–255.

Treisman AM, Gelade G (1980). *Cognitive Psychology 12*, 97–136.

Treisman A, Sato S (1990). *Journal of Experimental Psychology: Human Perception and Performance 16*, 459–478.

Ungerleider LG, Mishkin M (1982). In: DJ Ingle, MA Goodale, RJW Mansfield (eds), *Analysis of visual behavior*, pp. 549–586. Cambridge, MA: MIT Press.

Vallar G (1993). In: IHM Robertson (ed.), *Unilateral neglect: clinical and experimental findings*. Hove, UK: Erlbaum.

van Donkelaar P, Müri R (2002). *Proceedings of the Royal Society of London B 269*, 735–739.

Walsh V, Ashbridge E, Cowey A (1998). *Neuropsychologia 36*, 363–367.

Walsh V, Ellison A, Ashbridge E, Cowey A (1999). *Neuropsychologia 37*, 245–251.

Wardak C, Olivier E, Duhamel JR (2004). *Neuron 42*, 501–508.

Weiss PH, Marshall JC, Wunderlich G, *et al.* (2000). *Brain 123*, 2531–2541.

Wipfli M, Felblinger J, Mosimann UP, Hess CW, Schlaepfer TE, Müri RM (2001). *European Journal of Neuroscience 14*, 571–575.

Wolpert DM, Goodbody SJ, Husain M (1998). *Nature Neuroscience 1*, 529–533.

Yang Q, Kapoula Z (2004). *Investigative Ophthalmology and Visual Science 45*, 2231–2239.

Zangemeister WH, Canavan AG, Hoemberg V (1995). *Journal of Neurological Sciences 133*, 42–52.

CHAPTER 29

Studies of crossmodal functions with TMS

Lotfi Merabet and Alvaro Pascual-Leone

Our experience of the world is multisensory

We are endowed with receptors responding to different types of sensory information coming from our surrounding environment. Some detect pressure, others vibrations of waves of different frequency bands, chemicals, or light. Thanks to the diversity of our sensory receptors, we are able to capture different aspects of our environment and convey it, once translated into a unified electrical signal, to be processed by our brains. In the brain, information from all the senses interacts and, ultimately, is integrated in order to create a unified sensory percept. Some percepts appear uniquely unimodal – consider chromatic information that can only be comprehended by vision, pitch information that is unique to hearing, or a tickle which is purely tactile. Nevertheless, our appreciation of reality is itself intrinsically multisensory (Stein *et al.* 1993; Ghazanfar and Schroeder 2006), and even apparently unimodal percepts can be modified by crossmodal interactions given that our brains process multiple streams of sensory information in parallel and promote extensive interactions. In fact, the ability to integrate crossmodal sensory information confers clear behavioral advantages, such as perceptual enhancement and reduction in the ambiguity of a perceived sensory event (Welch *et al.* 1986; Calvert *et al.* 2000). Furthermore, in situations where a sensory modality never develops or is no longer present (such as in congenital or late blindness, respectively), crossmodal sensory interactions may underlie a neuroplastic framework allowing for brain structures, normally associated with the deprived sense, to become recruited by the remaining sensory modalities in an adaptive and compensatory manner (Bavelier and Neville 2002; Pascual-Leone *et al.* 2005).

The neural substrate of sensory interactions

The pervasive view regarding the organization of sensory interactions purports that sensory systems are arranged in a hierarchal fashion of increasing functional complexity (Mesulam 1998). For each sensory modality, we assume a hierarchically organized system beginning with specialized receptors that 'feed' to unimodal primary sensory cortical areas. A series of secondary areas then integrate different aspects of the processed information within the same modality (for example, the color and motion of a viewed object). Eventually, multimodal association areas integrate the processed signals with information derived from other sensory modalities (for example, the texture or sound made by the same object). In this scheme, sensory information is believed to remain isolated by modality within primary sensory areas and the merger of sensory experiences is the result of processing occurring within higher-order multimodal (or 'heteromodal') associative areas of the brain (Stein *et al.* 1993; Mesulam 1998; Calvert *et al.* 2000).

Recent experimental evidence is now forcing us to re-evaluate this view. In particular, the question now arises as to whether or not crossmodal processing is a unique property of higher-order multimodal sensory areas. Results from both anatomical (Falchier *et al.* 2002; Rockland and Ojima 2003; Cappe and Barone 2005) and physiological (Sathian *et al.* 1997; Giard and Peronnet 1999; Zangaladze *et al.* 1999; Calvert *et al.* 2000; Foxe *et al.* 2000; Laurienti *et al.* 2002; Merabet *et al.* 2004; Schroeder and Foxe 2005) studies suggest that crossmodal interactions can occur not only within regions deemed multimodal, but also within unimodal cortical areas; that is, regions traditionally considered to be located very early within the cortical processing stream and responsive to only one particular sensory modality. Further strengthening this conclusion are findings emerging from studies of sensory impaired subjects such as the blind. Results from these investigations demonstrate that crossmodal processing occurs not only within early levels of the cortical processing stream but also within cortical regions typically associated with the processing of the sensory modality that has been lost. Together, these findings have led us to revise our assumptions regarding the strict organization of the brain as a hierarchal yet parallel system that only integrates senses within higher-order brain centers (Pascual-Leone and Hamilton 2001).

The underlying neurophysiological substrate for crossmodal interactions is still a subject of intense debate. It has recently been suggested that different sensory systems interact with each other through feed-back and feed-forward interactions communicating through multisensory integration areas (Driver and Spence 1998; Macaluso *et al.* 2000; Cappe and Barone 2005). Seminal work carried out by Macaluso *et al.* (2000) has provided insight as to the spatiotemporal constraints underlying these crossmodal interactions. Using functional magnetic resonance imaging (fMRI), these authors

demonstrated that a tactile stimulus can modulate activity within cortical areas deemed strictly visual in response to a simultaneously presented visual target. More importantly, these interactive effects were both spatially and temporally specific; that is, the observed increase in modulatory effect was maximal when the tactile and visual stimuli were in close spatial register and presented within a short temporal delay. The authors further proposed that back projections (likely originating from multisensory parietal areas) may represent the neurophysiological substrate underlying these crossmodal sensory effects, further implying that multimodal perception is itself the result of sensory interactions taking place along a complex functional network that includes unimodal sensory areas (Macaluso *et al.* 2000).

Fundamentally, there are two, though not mutually exclusive, possible mechanisms to implement interactions obtained by distinct sensory modalities. The first suggests that the processing of a given modality can be merged in dedicated sensory integration areas containing neurons that receive multiple sensory inputs. It is possible that all stages of brain processing participate and interact in information processing presumably via recurrent interactive connections linking distinct sensory systems. An alternative and provocative conceptualization is that of an amodal, or 'metamodal', system of operators, all of which receive inputs from different sensory modalities but each operator preferentially selects the one that is best suited for the given operation to be done (Pascual-Leone and Hamilton 2001). It is important to realize that according to this model, the predilection of a given sensory processing area may be based on its relative suitability for certain kinds of computations (Pascual-Leone and Hamilton 2001). It is this predilection (along with the preponderance of its sensory afferents) that determines its function, its overall contribution to the percepts generated by crossmodal processing, and its developmental fate following sensory loss and deprivation. The computational processing, ability of a given area may itself lead to an 'operator-specific' selectivity that is in turn reinforced and eventually generates the impression of a parallel and highly specialized processing brain with segregated systems processing different sensory signals (Pascual-Leone and amilton 2001). According to this view, a sense-specific brain region, such as the visual cortex, may be visual only because we have sight and because the kinds of computations performed by the striate cortex are best suited for retinal captured or internally generated visual information. It would seem reasonable that the visual cortex with its combined relatively high spatial and temporal processing capability is involved with the discriminating and precise spatiotemporal relations of local detailed features within an object or scene being viewed. However, in the face of visual deprivation (or under carefully controlled experimental conditions), the visual cortex may become recruited selectively for the processing of other forms of sensory information, thus revealing or 'unmasking' its inputs from other modalities (such as touch and hearing) implementing nonvisual sensory processing. Obviously, these different theoretical conceptualizations of brain organization and multisensory processing have fundamental implications towards our understanding of brain organization, as well as a practical relevance to neurorehabilitation. A number of neurological and medical conditions are associated with altered sensory inputs and the profound neuroplastic changes that follow sensory deprivation and/or focal cerebral damage must be taken into consideration in order to develop effective rehabilitative treatment strategies (Merabet *et al.* 2005; Pascual-Leone *et al.* 2005).

The tools of study

As with any phenomenon or scientific question under investigation, the use of appropriate instruments and methodologies is critical. Arguably, a key goal in cognitive neuroscience is to establish a functional or causal link between areas of brain activity and behavioral manifestations (Robertson *et al.* 2003). Most of our knowledge regarding cerebral function has been from the analysis of patients suffering focal brain lesions and, more recently, from noninvasive neuroimaging methodologies. Functional MRI and positron emission tomography (PET) have provided invaluable insight as to *where* certain

visual tasks are carried out within the brain. On the other hand, event-related potential measurements (such as electroencephalography (EEG) and visual-evoked potentials (VEPs)) have provided information regarding the *timing* of information processing. The respective superior spatial and temporal resolution capabilities of fMRI/PET and EEG/VEP have rendered these techniques into methods of choice in the study of crossmodal sensory processing. This represents a distinct advantage over the inferences made from brain-damaged patients whose lesions are often highly variable and not always well circumscribed. Furthermore, the extent of neural reorganization (compensatory plasticity) after the insult is largely unknown in these patients, making conclusions even more difficult to ascertain (Bavelier and Neville 2002; Pascual-Leone *et al.* 2005).

However, it is crucial to underline the fact that conclusions drawn from lesion and neuroimaging methodologies do not always provide definitive causal evidence that a specific function is supported by that brain area or given network of activity. Many alternative or confounding explanations for the findings may exist or the activity may in itself be epiphenomenal. TMS provides a novel approach to overcome these limitations and expand functional imaging results, adding information about causality. Applied as single pulses appropriately delivered in time and space, or as trains of repetitive stimuli at appropriate frequency and intensity, TMS can be used to transiently disrupt the function of the targeted brain area, functionally map cortical areas, and assess, or even transiently modulate, cortical excitability (Pascual-Leone *et al.* 1999, 2000; Walsh and Rushworth 1999; Walsh and Cowey 2000; Pascual-Leone 2002). By establishing a causal link between brain activity and behavior, TMS can be used to identify which regions of a neural network are critical in behavioral task performance. Furthermore, since TMS can also modulate cortical excitability, areas that are anatomically connected to the site of stimulation should manifest changes in cortical excitability as well, thus revealing potential differences in connectivity. More recently, technical advances have permitted the use of TMS in combination with established neuroimaging methodologies such as PET, fMRI, and EEG (Bohning *et al.* 1997; Paus *et al.* 1997; Siebner *et al.* 2003; Thut *et al.* 2005). The combination of TMS with concurrent neuroimaging technologies adds the exciting possibility of supplementing and refining conclusions drawn from these investigative techniques, further elucidating the functional networks implicated with behavior.

In this chapter, we discuss the role of TMS in investigating underlying mechanisms associated with crossmodal sensory processing. We will first outline the fact that TMS is itself inherently multimodal and discuss the potential confounds this creates for studies of crossmodal interactions. Possible experimental designs and strategies aimed at overcoming these limitations will also be suggested. Finally, we will highlight studies that have exploited the strengths of combining TMS with other neuroimaging methodologies to address questions regarding crossmodal sensory processing.

TMS is itself multimodal

TMS is commonly described as a 'relatively painless' method of stimulating the brain noninvasively (Pascual-Leone 2002; Robertson *et al.* 2003). However, anyone who has participated in a TMS study knows that the technique itself elicits significant sensory sensations including auditory and somatosensory stimulation. Relevant to this discussion is the fact that the same TMS device used to establish causality by interfering with task performance can itself nonspecifically interfere with task performance. This is because TMS can essentially be considered as a method of multimodal sensory stimulation. In addition to the actual brain site-specific effects that it can induce, stimulation from TMS further gives rise to (among others) auditory sensations, somatosensory and tactile stimulation, potential startle effects, and even visual percepts in the form of phosphenes. All of these effects are dependent largely upon the site and intensity of the stimulation being delivered. Thus, the multisensory nature of TMS needs to be considered and controlled for. Furthermore, experiments and subsequent analysis must be carefully designed and executed in order to circumnavigate these potential confounds.

TMS-induced auditory sensations

As electrical current passes through the coil generating a time-varying magnetic field, the copper wire windings within the coil tense, giving rise to a brief, ringing 'click' sound that may exceed 120 dB. The characteristics of this click depend on the type of stimulator used, the stimulation parameters delivered, and the coil design itself (Pascual-Leone *et al.* 1992). It is important to realize that, even though the click may not always be perceived as loud, it is deceptive and its very short duration renders it potentially dangerous and liable to generate potential confounds (Pascual-Leone *et al.* 1992). In some cases, the generated noise is loud enough to cause transient auditory threshold shifts (Pascual-Leone *et al.* 1993) and can potentially damage cochlear hair cells (Counter *et al.* 1991; Counter 1993) unless certain safety precautions are adhered to. Specifically, the use of earplugs for TMS studies (both for the subjects and the investigators applying the stimulation) is strongly recommended (Wassermann 1998) and does effectively prevent the risk of hearing damage. Despite these safety precautions, the click sound remains only attenuated and thus substantial auditory stimulation with every TMS stimulus still takes place. This can be objectively demonstrated by the robust activation of bilateral auditory cortices in studies combining TMS with functional imaging (Siebner *et al.* 2003; Bestmann *et al.* 2004, 2005) (see Figure 29.1).Therefore, in studies employing TMS to study crossmodal interactions, there remains the possibility of auditory contamination resulting in activation of the auditory cortex and this needs to be taken into consideration and appropriately controlled for. For example, consider an experiment aimed at exploring interactions between visual and auditory inputs. Visual and auditory stimuli might be presented with variable intervals such that crossmodal interactions are induced. In a typical TMS study, the delivered pulse could be applied at various time intervals targeting various cortical areas in order to assess brain regions causally engaged in a crossmodal task, or even to investigate the chronometry of the effects. However, with each TMS pulse (and regardless of the cortical target), robust auditory stimulation and activation

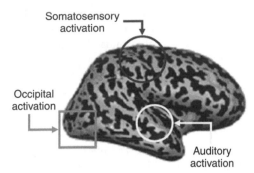

Fig. 29.1 Areas of the face and scalp typically stimulated during TMS experiments. Dermatomes of the head and neck are derived from the trigeminal nerve (V) and the cervical spinal nerves (C2). Dermatomes corresponding to the three branching divisions of the trigeminal nerves are also shown. Inspired by Gosling *et al.* (1995).

introduce differences across TMS conditions that would be independent of the brain effects.

TMS-induced somatosensory sensations

TMS is generally applied with the coil rested flat and tangential to the scalp surface (see coil positioning over the occipital pole in Figure 29.2). As current passes through the coil, there is an electrical tension that generates a click noise (discussed above) as well as a mechanical displacement that impacts onto the hard plastic casing, causing a tapping sensation upon the surface of the scalp. Somatosensory innervation of the scalp is differentially transmitted via high cervical roots (largely through the two major occipital nerves), the facial nerve and in particular the trigeminal cranial nerve (Figure 29.2). Thus, the tapping sensation eventually generates a somatosensory input that can be demonstrated,

within auditory cortex would result, which itself must be expected to affect crossmodal interactions. It might be argued that the auditory stimulus would be the same regardless of the cortical target, and thus could be considered as a non-specific confound built into the study. The simple solution might be just to run more trials and with more subjects to properly control for this. Unfortunately, things are not that simple. Depending on the critical timing, an auditory stimulus may induce differential modulation of visuo-auditory interactions, e.g. creating the illusion of a visual percept induced by appropriately timed auditory stimuli (Shimojo and Shams 2001; Shams *et al.* 2002). Therefore, it is clear that chronometric single-pulse TMS studies of crossmodal interactions represent a technical challenge and require carefully designed psychophysical controls to ensure that the percepts a subject perceives (or the stimuli that a subject responds to) are in fact real and not illusory confounds.

Finally, the relative proximity of the TMS coil to the ear may effectively change the magnitude and the lateralization of the auditory stimulus. Bone conduction may differ depending on where and how the coil is rested on the scalp. These two later effects could in turn influence the degree of auditory activation confounds and potentially

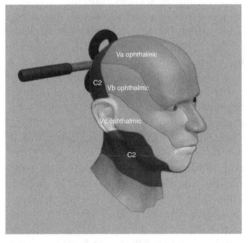

Fig. 29.2 Associated activation of cortical areas revealed using combined fMRI–TMS. Changes in measured BOLD responses are overlaid upon an inflated projection of the right hemisphere of a representative subject. TMS applied to occipital cortex shows areas of direct activation (increase in measured BOLD signal; red box) accompanied by areas of crossmodal activation within somatosensory cortex (blue circle) and auditory cortex (yellow circle). Analysis contrast: TMS stimulation > no TMS stimulation: threshold of $P < 0.05$. Modified from Amedi *et al.* (2005). (Plate 23)

for example, by activation within somatosensory cortex in studies combining TMS with functional imaging (see Figure 29.1) despite the fact that TMS is delivered to sites distal to somatosensory cortex itself.

In addition, on its way through the scalp, the TMS-induced magnetic field induces a current within the muscles and nerve endings that lead to their contraction. The distribution of muscles over the scalp is not uniform and innervation patterns are variable across subjects. Therefore, depending on the positioning of the TMS coil, variable muscle contraction and activation of one or a number of nerve distribution patterns might ensue. The induced muscle contraction itself will contribute to the activation of the somatosensory input and the activation of somatosensory cortical areas. Furthermore, activation of trigeminal, facial, or high spinal afferents may differentially affect activity in neural structures, which may modify the TMS-induced or task-related brain activity. In some cases, a study subject might find the stimulation of facial musculature somewhat distracting, thus hampering task performance unrelated to the disruptive effects of TMS delivered to the cortical locus in question. Therefore, as with accompanying auditory stimulation, these potential confounds are also site and intensity dependent and thus need to be controlled for in studies exploring crossmodal brain interactions employing TMS. Finally, it is important to note that these TMS-induced somatosensory sensations are different from TMS-induced paresthesias or dysesthesias, which can be induced (though rarely) by stimulation of the somatosensory cortex (Amassian *et al.* 1991). Of course, even when TMS is directly targeting somatosensory cortex and is aimed at inducing paresthesias, nonspecific somatosensory stimulation of scalp receptors and muscles can still take place and interact with the brain activation induced by TMS.

Startle effects

A startle can be considered as a rapid response to a sudden and intense stimulus. Presumably, its evolutionary significance is protective in nature, leading to avoidance behaviors from potential causes of injury or a predator. In general,

the startle reflex is elicited by intense tactile, acoustic, or vestibular stimuli. The acoustic startle response (ASR) of mammals is mediated by a relatively simple neuronal circuit located in the lower brain stem. Neurons of the caudal pontine reticular nucleus (PnC) are key elements of this primary ASR pathway. The ASR in humans and animals has a nonzero baseline, i.e. the response magnitude can be increased or decreased by a variety of pathological conditions and experimental manipulations. For example, sensitization and fear can enhance the ASR, while habituation, prepulse inhibition, and positive affect can reduce the ASR. Medications can also increase or suppress the ASR depending on their effects. The ASR has been used as a behavioral tool to assess the neuronal basis of behavioral plasticity and to model neuropathological dysfunctions of sensorimotor information processing (Koch 1999). With respect to TMS, the loud, transient, and brief TMS-associated auditory stimulus is a strong and reliable trigger for the ASR, as carefully examined by Valls-Sole *et al.* (1994, 1999). In addition to the auditory startle response, it is important to realize that mechanoreceptors in each modality (auditory, somatosensory, or vestibular) can also respond to skin or head displacement. In each modality, stimulation of cranial nerves or primary sensory nuclei evokes startle-like responses. Furthermore, crossmodal summation is stronger than intramodal temporal summation, eliciting enhanced startle responses. This summation is maximally sharp if the crossmodal stimuli are synchronous. Head impact stimuli activate trigeminal, acoustic, and vestibular systems together, suggesting that the startle response may serve to protect the body from impact stimuli. However, in the context of TMS experiments, this adaptive phenomenon becomes an important potential confounder to control for. As mentioned earlier, auditory, somatosensory, and vestibular activation occurs practically synchronously following the delivery of a TMS pulse and each stimulus is quite strong, so that robust and reliable startle reactions can be generated. For example, consider that the startle reaction is one of the fastest human movements in response to a sensory stimulus. In fact, it is likely that the central nervous system uses the circuits of the startle reactions for the execution of a ballistic movement. In situations when subjects are highly prepared for movement execution (for example, in the setting of reaction-time experiments), a startle reaction will accelerate the response even though the movement pattern is not modified. This acceleration is limited to the nature of the movement *per se*, and does not involve the subjective perception of the movement. Since TMS can induce a robust startle reaction, reaction-time measures (particularly if ballistic movements are required) in the context of a TMS experiment may be significantly affected. In general, reaction times appear to be accelerated and this issue necessitates that careful controls are built into the study design to avoid erroneous conclusions (Pascual-Leone *et al.* 1992a,b). Furthermore, impairment of the central control of the startle reaction may lead to an excess or reduction of the response. Therefore, TMS experiments with reaction-time measures comparing normal subjects to patients with disorders with altered startle responses (e.g. Parkinson's disease or other movement disorders; Kofler *et al.* 2006) need to consider the TMS-induced startle as a possible contributor to the findings (Valls-Sole *et al.* 1994).

Finally, habituation is also a strong modulator of the startle response. As such, one could anticipate that over the course of a given experiment, the magnitude of the startle will tend to decrease, particularly in non-TMS naïve subjects. Furthermore, the rate of habituation may not be easily characterized or predictable. It is also for this reason that the use of non-TMS naïve subjects is desirable for certain cognitive neuroscience experiments utilizing TMS (Robertson *et al.* 2003).

TMS-induced visual sensations

Earliest experiences with TMS in the mid-to-late 1800s and early 1900s reported the possibility of inducing visual sensations, referred to as 'phosphenes', by noninvasive brain stimulation. It is indeed possible to induce phosphenes in certain subjects with TMS appropriately delivered over the occipital pole of the brain. For example, simple patterned phosphenes can be induced from stimulation of striate cortical areas, and moving phosphenes can also be induced by stimulation of visual area V5/MT

(Kammer 1999; Fernandez *et al.* 2002; for review see Merabet *et al.* 2003). Furthermore, the specific topography of the elicited phosphenes follows the known retinotopic organization of visual cortex, and their characteristics (such as brightness or intensity) depend on the orientation and localization of the TMS coil. Interestingly, phosphene perception can also be induced by targeting circumscribed regions of the occipital cortex in recently blind, though not in congenitally blind, subjects (Gothe *et al.* 2002). However, in addition to these TMS-induced phosphenes from stimulation of specific brain regions under specific conditions, visual sensations can be induced by TMS affecting the retina. Indeed, as discussed by Marg and Rudiak (1994), the phenomena described by the early pioneers of TMS were the consequence of electrical stimulation of the retina and not the brain (Pascual-Leone 2002; Merabet *et al.* 2003). This means that TMS applied close to the orbit or frontal areas might give rise to visual percepts which would interact with other sensory stimuli and possibly have an impact on the results of crossmodal interaction experiments. A detailed questioning of study participants regarding their sensory experiences after TMS can help to identify any confounding visual sensations caused by stimulating the retina directly. For example, visual sensations resulting from retinal stimulation should diminish with increasing distance from the orbit and also follow an ocular centered retinotopic pattern rather than cortical retinotopic coordinates.

Controlling for the multisensory stimuli

Completely eliminating the multimodal nature of stimulation associated with TMS is not possible. Therefore, strategies need to be developed to control for them in order to minimize the risk of misinterpretation of findings and drawing erroneous conclusions (Robertson *et al.* 2003). One approach is to compare the effects of stimulation at several sites versus the effects of multiple tasks at the same site of stimulation. If the effects of stimulation are observed exclusively at one site, this gives some reassurance that the differences across sites are due to the specific effects of neuronal disruption. This later strategy assumes that the nonspecific effects of TMS are equivalent not only across sites but also across subjects, an assumption that is not easy to validate. In fact, even relatively small changes in position can cause substantial changes in the sensory effects of stimulation and thus many studies have also taken the approach of observing behavior performance across several distinct tasks with respect to the same site of stimulation (e.g. Beckers and Zeki 1995). 'Sham' stimulation, in various forms (e.g. using specially designed sham stimulation coils, or tilting the coil away from the scalp) have also been used with the aim of diminishing (or even effectively negating) the amount of overall current delivered to the brain while still giving the study subject the impression of receiving real stimulation. While potentially effective in non-TMS, naïve subjects, the subtleties of the click generated, the lack of induced facial muscular contractions and, in some cases, the different orientation and positioning of the coil, can be appreciated by the study subject, thus negating the overall 'sham' effect.

Finally, single-pulse TMS, short trains of 'on-line' repetitive (r)TMS, or varying the delay between a particular event (e.g. the presentation of a stimulus) and stimulation can be also be incorporated in the study design. The assumption is that nonspecific distracting effects of TMS will be independent, whereas the behavioral effects will be highly dependent on the precise interval between the event and the stimulation. Unfortunately, even such approaches remain potentially confounded by crossmodal effects, which might, for example, be chronometrically specific (such as induced illusory effects discussed above). These issues notwithstanding, it has been possible using TMS to provide novel and important insights into the neural substrates of crossmodal interactions and we shall now summarize briefly some exemplary experiments.

TMS in the investigation of crossmodal sensory integration

As mentioned in the Introduction, current evidence suggests that crossmodal sensory processing implicates even the earliest stages of cortical processing. Furthermore, loss of sensory input (such as sight or hearing) leads to profound neuroplastic changes in sensory cortical areas involved with not only the processing of the remaining senses, but also those implicated with the deprived sensory modality. It thus appears that sensory processing can be reorganized and redistributed along functional networks based on task demands, in a compensatory manner, which exploits the strengths of the neuronal processing machinery available.

Marked anatomical and functional changes are particularly evident in subjects who suffer sensory deprivation early in life, that is to say, during the critical period of development when cerebral plasticity is at its greatest (for reviews, see Bavelier and Neville 2002; Theoret et al. 2004). However, growing evidence also suggests that crossmodal processing also occurs within the intact brain. It is possible that an unmasking of existing and/or latent connections between sensory processing areas underlie the transfer of crossmodal information. These very same networks may also be in part responsible for the overall functional recruitment of particular cortical areas in response to sensory deprivation (Pascual-Leone and Hamilton 2001; Pascual-Leone et al. 2005).

Neuroplastic changes following sensory deprivation

Are these shifts in activation and modifications in functional connectivity behaviorally relevant? This remains a fundamental question in the study of crossmodal sensory interactions. One of the earliest examples of crossmodal recruitment of a sensory cortical region has been from studies following visual deprivation. In blind subjects, Braille reading is associated with a variety of neuroplastic changes such as larger sensorimotor representation of the reading finger (Pascual-Leone and Torres 1993) and recruitment of the occipital cortex (including areas corresponding to primary visual cortex) during tactile discrimination (Sadato et al. 1996, 1998; Buchel et al. 1998; Burton et al. 2002). Using TMS, crossmodal somatosensory-to-visual plasticity has been shown to be behaviorally relevant given that reversible disruption of occipital cortex disrupts Braille reading in blind participants (Cohen *et al.* 1997). In line with evidence that can be drawn from lesion data, these findings are also in agreement with observations from a patient who became alexic for Braille reading following a bilateral occipital stroke (Hamilton *et al.* 2000).

It is important to underline that a similar account also appears to emerge regarding the recruitment of occipital visual cortical areas for the processing of auditory information. Activation localized to occipital cortical areas has been demonstrated in congenitally blind subjects during auditory localization tasks (ERP: Kujala *et al.* 1992; PET: Weeks *et al.* 2000). Other groups have further addressed this issue by mapping language-related brain activity implicated in speech processing (Roder *et al.* 2000, 2002) and auditory verb generation (Burton *et al.* 2002; Amedi *et al.* 2003). TMS has also been used to address the functional causality related to the recruitment of occipital cortex in processing auditory and linguistic information. In a recent study, rTMS was used to investigate the functional role of occipital cortex in the use of a prosthesis substituting vision with audition (PSVA) device for the visually impaired designed to assist in the recognition of bidimensional shapes (Collignon *et al.* 2007). Targeting dorsal extrastriate cortex, rTMS interfered with both the PSVA use and an auditory spatial location task in early blind subjects but not in sighted controls (Collignon *et al.* 2007). Amedi *et al.* (2004) also used TMS to investigate high-level verbal processing in the blind. Using rTMS delivered to the occipital cortex, the authors noted a reduction in accuracy on a verb-generation task in blind subjects, but not in sighted controls. Further analysis of error types revealed that the most common error produced by rTMS was semantic in nature rather than phonological (Amedi *et al.* 2004).

All these studies contribute to our current notions that early visual deprivation leads to functional crossmodal reorganization of cerebral areas and, in particular, the functional recruitment of occipital cortex in the processing of tactile, auditory and linguistic information. However, can this same crossmodal reorganization and changes in connectivity be revealed without the necessity of carrying out a behavioral task?

Combining rTMS with neuroimaging to probe functional connectivity

A novel approach by Wittenberg *et al.* (2004) was used to address this issue of functional connectivity in response to sensory deprivation. Given accumulating evidence of occipital recruitment for the processing of tactile information following visual deprivation, the group hypothesized that a more robust physiological linkage between primary somatosensory cortex (S1) and primary visual cortex should manifest itself in early blind subjects compared to sighted and even late blind controls. This hypothesis has been previously alluded to from studies using fMRI demonstrating greater magnitudes of occipital activation (Burton 2003) and selective task disruption using TMS (Cohen *et al.* 1997) in early blind compared to late bind subjects. However, in light of the potential limitations and behavioral confounds associated with these techniques, Wittenberg *et al.* sought a more direct demonstration of differences in connectivity across sighted, late blind, and early blind individuals. Using rTMS combined with PET, the group was able to measure the effect of rTMS applied to somatosensory cortex on the cerebral blood flow (rCBF) in the visual cortex. Rather than seeking a behavioral association between task performance and activation of occipital cortical areas, this group exploited the effects of rTMS in order to probe the connection between primary somatosensory cortex (S1) and early visual cortex using PET imaging to further reveal these connections. The investigators found that baseline rCBF in occipital cortex was highest in early blind and lowest in late blind individuals; furthermore, only the early blind group showed significant activation of early visual areas when rTMS was delivered over S1 (Figure 29.3). The fact that activation was significantly higher in early compared to late blind subjects is consistent with the hypothesis that tactile information may reach visual areas through cortico-cortical pathways. The overall magnitude of the activation (presumably reflecting the overall strength of the connections) may itself be a reflection of the degree of functional recruitment.

Taken together, long-term sensory deprivation leads to anatomical, behavioral, and functional

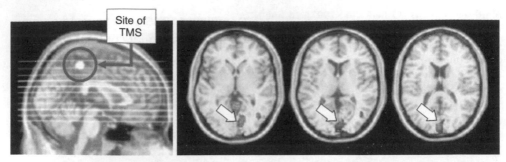

Site of
TMS

Fig. 29.3 Functional connectivity between somatosensory and occipital cortex in early blind humans revealed with combined TMS and positron emission tomography. Circle denotes site of repetitive TMS stimulation. Axial slices reveal areas of significant changes in regional cerebral blood flow ($P < 0.05$) within occipital cortical areas (arrows) following TMS stimulation of somatosensory cortex. Modified from Wittenberg *et al.* (2004).

changes that implicate the remaining unimodal sensory processing areas as well as the processing machinery of the sensory modalities that have been lost. However, this poses the question as to what extent these anatomical and functional changes are implicated in crossmodal processing in the intact brain?

Revealing crossmodal interactions in normal subjects

As stated earlier, activation of the occipital cortex during processing of nonvisual information may not necessarily represent the establishment of new connections, but rather the unmasking of latent pathways that participate in the multisensory perception. Viewed in this way, unimodal sensory areas must be part of a network of areas subserving multisensory integration, rather than mere processors of a single modality. In sighted subjects, Sathian *et al.* (1997) used a tactile version of a gratings orientation task (GOT) to demonstrate that tactile discrimination leads to increased metabolic activity (as measured with PET) in particular cortical locus located within the parieto-occipital (PO) area. The investigators then performed a follow-up study demonstrating that occipital-delivered TMS could actually interfere with GOT performance (Zangaladze *et al.* 1999). The authors interpreted the observed activations to visual imagery and purported that this process was an obligatory component of spatial discrimination skills (Zangaladze *et al.* 1999). Could visual imagery alone account for the functional

recruitment of visual areas in crossmodal sensory processing? Visual imagery can be viewed as a 'top-down' phenomenon (Kosslyn *et al.* 2001) and thus the recruitment of higher-order associative visual areas would be expected. What remains less clear is the involvement of cortical areas early along the visual processing pathway in crossmodal sensory processing and whether visual imagery could also account for their involvement.

Given the fine spatial processing ability and topographic organization of early visual areas, it would seem reasonable that these areas would be implicated in tactile tasks requiring fine spatial discriminations. In order to test this further, Merabet *et al.* (2004) conducted a study using tactile arrays of raised dot patterns of varying inter-dot distances in normally sighted subjects. Stimuli consisted of eight tactile patterns of raised embossed dots arranged in tetragonal arrays with constant inter-dot spacing varying from 1 to 7 mm. Subjects performed two different tactile tasks using identical stimuli. This task design and stimuli have been used in various neurophysiology and behavioral studies and have been described in detail previously (Johnson and Hsiao 1992; Merabet *et al.* 2004). Briefly, the task was to judge either the perceived roughness or the perceived inter-dot distance spacing of the tactile patterns. Normally sighted subjects were blindfolded during the task, and, using their index finger, judged either the roughness or the spacing distance between the dots for a given array. When judging roughness, closely spaced dots lead to an impression of little roughness, and very widely spaced dots also give the impression of reduced roughness. Conversely, patterns of intermediate spacing were perceived as most rough. Plotting perceived roughness against inter-dot spacing yields an inverted U-shaped psychophysical curve (see 'roughness', Figure 29.4b). However, distance-spacing judgments (using the identical texture patterns) generated a linear relation when subjective reports were plotted against actual dot spacing, since the greater the gap between dots, the greater the distance perceived (see 'distance' Figure 29.4b).

In order to identify the neural correlates associated with the tactile task, the group employed fMRI while blindfolded sighted subjects explored and rated the tactile stimuli. Areas of activity were then identified relative to individually

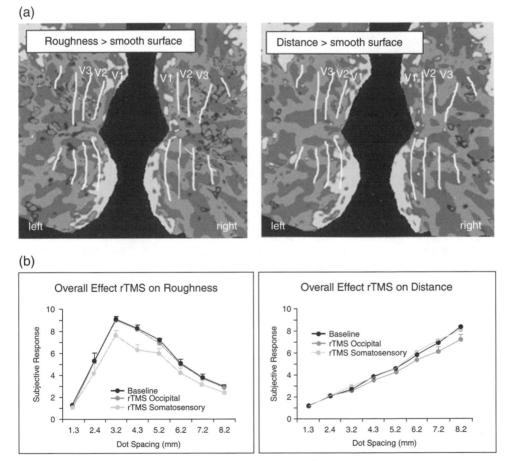

Fig. 29.4 (a) Occipital activation in normally sighted (blindfolded) subjects performing two different tactile discrimination tasks (see text for explanation of task). Activity associated with tactile roughness determinations and spatial distance determinations (compared to a smooth surface) are shown on an occipital patch projection for a representative subject overlaid upon their identified retinotopic borders. Similar patterns of activation within early visual cortex (bilaterally and upper and lower visual fields) are evident regardless of the tactile task being performed. (b) Functional contribution of occipital cortex on tactile task performance revealed using repetitive (r)TMS. Note an overall dampening of the roughness judgment curve following somatosensory cortex disruption (dotted line) but not occipital cortex (dashed line) compared to baseline (solid line). Conversely, for distance determination, rTMS applied to occipital cortex impairs distance perception at increasing dot spacing (dark gray symbols) but has no effect following somatosensory cortex disruption (light gray symbols). Modified from Merabet *et al.* (2004).

defined retinotopic maps. Comparison of tactile task performance and an active control condition (exploring a smooth surface) revealed significant activation in primary visual cortex (V1) and deactivation in extrastriate cortical regions V2 and V3. Strong modulation of parietal and somatosensory areas was also observed (Figure 29.4a). When patterns of activation associated with roughness judgments were contrasted with distance-spacing judgments, no differences in activation patterns were observed within early retinotopic visual areas.

The equal activation of early visual cortex for both tactile tasks suggests that within the resolution limits of fMRI, the inputs to local circuits within primary visual cortex may not themselves be sensitive to tactile task requirements. The group then employed rTMS delivered to either occipital or somatosensory cortex in order to inquire whether these same activations observed during tactile performance might play a similar functionally relevant role in the roughness and/or distance-spacing judgments (for experimental details see Merabet *et al.* 2004). Following rTMS of somatosensory cortex (contralateral to the exploring finger), there was an overall disruption of roughness perception (Figure 29.4b). Subjects still perceived intermediate spacings as the maximal impression of roughness for the array, but the amount of roughness experienced was suppressed. Remarkably, distance-spacing judgment was not affected at all. On the other hand, rTMS to the occipital cortex did not affect roughness judgments, rather disrupting distance-spacing perception. Specifically, subjects tended to scale increasing inter-dot spacing with less perceived distance than prior to the suppression of the striate and peristriate cortex (Figure 29.4b).

The results from the TMS study suggest that tasks requiring precise reconstruction of spatial patterns might engage topographically organized cortical areas including striate cortex. The results from this investigation also highlight the advantage of exploiting the strengths of multiple methodologies to help disentangle questions relating to crossmodal sensory processing. Here, fMRI was used to identify the neural correlates associated with tactile processing. Retinotopic mapping techniques delineated areas within occipital visual cortex implicated in task

performance and, in turn, identified potential sites to be targeted using TMS in order to probe their causal role in behavioral performance. Furthermore, the identified pattern of activation from fMRI was not entirely consistent with a conclusion that visual imagery alone could explain the recruitment of early visual areas. Imagery- and attention-modulated activation of V1 typically occurs only in concert with even stronger activation of extrastriate cortex (Kosslyn and Thompson 2003), a pattern very different from what was reported here. In this study, the activation of area V1 was associated with a *suppression* of extrastriate visual areas and perhaps more likely reflects an underlying network possibly mediated by direct long-range cortico-cortical connections from somatosensory cortex and higher-order multimodal associative areas.

The potential confounds associated with the multimodal nature of TMS were avoided in this study by demonstrating a functional resolution between two distinct tactile tasks and two different cortical sites. Specifically, the paradigm took advantage of the fact that tactile roughness and distance-spacing determinations were dissociable by psychophysical performance. Furthermore, by using an 'off-line' stimulation paradigm, the nonspecific effects due to direct TMS are minimized, thereby contributing to the conclusion that occipital cortex and somatosensory cortex contribute differently to behavioral performance.

Finally, it is worth noting that neither technique alone could provide a full and definitive explanation regarding the role of early retinotopic areas in tactile processing in normally sighted subjects. Specifically, fMRI with its high spatial resolution and localization ability, was able to identify the areas of retinotopic cortex implicated in the behavioral task and overall patterns of cortical network activity. Based on the patterns of activation alone, one would presume that early retinotopic cortex contributes equally, or at least in a similar fashion, to both tactile tasks being performed. Conversely, the use of TMS allowed for the causal establishment of function, demonstrating that activation in occipital cortex (specifically, within primary visual cortex) is preferentially implicated with tactile tasks requiring fine spatial discriminations (as suggested by the selective interference of the spatial distance task rather than roughness determinations following occipital cortex disruption). The demonstration of crossmodal connectivity in the early visual cortex of sighted individuals supports the concept that the recruitment of visual cortex observed in blind subjects may represent the expansion of a pre-existing multimodal capacity.

Summary

The different senses capture different aspects of reality and our brain merges these different sources of information into a rich and unitary multisensory percept of reality. Exactly how our brains accomplish this remains the subject of intense investigation and TMS can provide valuable insights on the neural substrates associated with multisensory processing in humans. However, it is important to realize that TMS itself is strongly multisensory, and this needs to be considered in the interpretation of results of TMS experiments, and appropriate controls need to be implemented. With regard to the crossmodal sensory changes that follow sensory deprivation, these changes can be revealed using a variety of methods including the combination of TMS with neuroimaging. The conclusions drawn from these studies can be confirmed or refuted by exploiting the strengths and controlling for the weaknesses inherent in the investigative tools being used.

References

Amassian VE, Somasundaram M, et al. (1991). Paraesthesias are elicited by single pulse, magnetic coil stimulation of motor cortex in susceptible humans. Brain 114, 2505–2520.

Amedi A, Raz N, et al. (2003). Early 'visual' cortex activation correlates with superior verbal memory performance in the blind. Nature Neuroscience 6, 758–766.

Amedi A, Floel A, et al. (2004). Transcranial magnetic stimulation of the occipital pole interferes with verbal processing in blind subjects. Nature Neuroscience 7, 1266–1270.

Bavelier D, Neville HJ (2002). Cross-modal plasticity: where and how? Nature Reviews Neuroscience 3, 443–452.

Beckers G, Zeki S (1995). The consequences of inactivating areas V1 and V5 on visual motion perception. Brain 118(Pt 1), 49–60.

Bestmann S, Baudewig J, et al. (2004). Functional MRI of the immediate impact of transcranial magnetic stimulation on cortical and subcortical motor circuits. European Journal of Neuroscience 19, 1950–1962.

Bestmann S, Baudewig J, et al. (2005). BOLD MRI responses to repetitive TMS over human dorsal premotor cortex. NeuroImage 28, 22–29.

Bohning DE, Pecheny AP, et al. (1997). Mapping transcranial magnetic stimulation (TMS) fields in vivo with MRI. Neuroreport 8, 2535–2538.

Buchel C, Price C, et al. (1998). Different activation patterns in the visual cortex of late and congenitally blind subjects. Brain 121(Pt 3), 409–419.

Burton H (2003). Visual cortex activity in early and late blind people. Journal of Neuroscience 23, 4005–4011.

Burton H, Snyder AZ, et al. (2002). Adaptive changes in early and late blind: a FMRI study of verb generation to heard nouns. Journal of Neurophysiology 88, 3359–3371.

Calvert GA, Campbell R, et al. (2000). Evidence from functional magnetic resonance imaging of crossmodal binding in the human heteromodal cortex. Current Biology 10, 649–657.

Cappe CP, Barone P (2005). Heteromodal connections supporting multisensory integration at low levels of cortical processing in the monkey. European Journal of Neuroscience 22, 2886–2902.

Cohen LG, Celnik P, et al. (1997). Functional relevance of cross-modal plasticity in blind humans. Nature 389, 180–183.

Collignon O, Lassonde M, et al. (2007). Functional cerebral reorganization for auditory spatial processing and auditory substitution of vision in early blind subjects. Cerebral Cortex 17, 457–465.

Counter SA (1993). Electromagnetic stimulation of the auditory system: effects and side-effects. Scandinavian Audiology Supplementum 37, 1–32.

Counter SA, Borg E, et al. (1991). Acoustic trauma in extracranial magnetic brain stimulation. Electroencephalography and Clinical Neurophysiology 78, 173–184.

Driver J, Spence C (1998). Crossmodal attention. Current Opinion in Neurobiology 8, 245–253.

Falchier A, Clavagnier S, et al. (2002). Anatomical evidence of multimodal integration in primate striate cortex. Journal of Neuroscience 22, 5749–5759.

Fernandez E, Alfaro A, et al. (2002). Mapping of the human visual cortex using image-guided transcranial magnetic stimulation. Brain Research: Brain Research Protocols 10, 115–124.

Foxe JJ, Morocz IA, et al. (2000). Multisensory auditory–somatosensory interactions in early cortical processing revealed by high-density electrical mapping. Brain Research: Cognitive Brain Research 10(1–2), 77–83.

Ghazanfar AA, Schroeder CE (2006). Is neocortex essentially multisensory? Trends in Cognitive Sciences.

Giard MH, Peronnet F (1999). Auditory–visual integration during multimodal object recognition in humans: a behavioral and electrophysiological study. Journal of Cognitive Neuroscience 11, 473–90.

Gothe J, Brandt SA, et al. (2002). Changes in visual cortex excitability in blind subjects as demonstrated by transcranial magnetic stimulation. Brain 125(Pt 3), 479–490.

Hamilton R, Keenan JP, et al. (2000). Alexia for Braille following bilateral occipital stroke in an early blind woman. Neuroreport 11, 237–240.

Johnson KO, Hsiao SS (1992). Neural mechanisms of tactual form and texture perception. Annual Review of Neuroscience 15, 227–250.

Kammer T (1999). Phosphenes and transient scotomas induced by magnetic stimulation of the occipital lobe: their topographic relationship. Neuropsychologia 37, 191–198.

Koch M (1999). The neurobiology of startle. Progress in Neurobiology 59, 107–128.

Kofler M, Muller J, et al. (2006). Auditory startle responses as a probe of brainstem function in healthy subjects and patients with movement disorders. Supplement to Clinical Neurophysiology 58, 232–248.

Kosslyn SM, Thompson WL (2003). When is early visual cortex activated during visual mental imagery? Psychological Bulletin 129, 723–746.

Kosslyn SM, Ganis G, et al. (2001). Neural foundations of imagery. Nature Review Neuroscience 2, 635–642.

Kujala T, Alho K, et al. (1992). Neural plasticity in processing of sound location by the early blind: an event-related potential study. Electroencephalography and Clinical Neurophysiology 84, 469–472.

Laurienti PJ, Burdette JH, et al. (2002). Deactivation of sensory-specific cortex by cross-modal stimuli. Journal of Cognitive Neuroscience 14, 420–429.

Macaluso E, Frith CD, et al. (2000). Modulation of human visual cortex by crossmodal spatial attention. Science 289(5482), 1206–1208.

Marg E, Rudiak D (1994). Phosphenes induced by magnetic stimulation over the occipital brain: description and probable site of stimulation. Optometry and Vision Science 71, 301–311.

Merabet LB, Theoret H, et al. (2003). Transcranial magnetic stimulation as an investigative tool in the study of visual function. Optometry and Visual Science 80, 356–368.

Merabet L, Thut G, et al. (2004). Feeling by sight or seeing by touch? Neuron 42, 173–179.

Merabet LB, Rizzo JF, et al. (2005). What blindness can tell us about seeing again: merging neuroplasticity and neuroprostheses. Nature Reviews Neuroscience 6, 71–77.

Mesulam MM (1998). From sensation to cognition. Brain 121(Pt 6), 1013–1052.

Pascual-Leone A (2002). Handbook of transcranial magnetic stimulation. London/New York: Arnold/Oxford University Press.

Pascual-Leone A, Hamilton R (2001). The metamodal organization of the brain. Progress in Brain Research 134, 427–445.

Pascual-Leone A, Torres F (1993). Plasticity of the sensorimotor cortex representation of the reading finger in Braille readers. Brain 116(Pt 1), 39–52.

Pascual-Leone A, Cohen LG, et al. (1992a). No evidence of hearing loss in humans due to transcranial magnetic stimulation. Neurology 42(3 Pt 1), 647–651.

Pascual-Leone A, Houser CM, et al. (1992b). Reaction time and transcranial magnetic stimulation. Lancet 339, 1420.

Pascual-Leone A, Valls-Sole J, et al. (1992c). Effects of focal transcranial magnetic stimulation on simple reaction time to acoustic, visual and somatosensory stimuli. Brain 115(Pt 4), 1045–1059.

Pascual-Leone A, Houser CM, et al. (1993). Safety of rapid-rate transcranial magnetic stimulation in normal volunteers. Electroencephalography and Clinical Neurophysiology 89, 120–130.

Pascual-Leone A, Bartres-Faz D, et al. (1999). Transcranial magnetic stimulation: studying the brain-behaviour relationship by induction of 'virtual lesions'. Philosophical Transactions of the Royal Society of London B 354(1387), 1229–1238.

Pascual-Leone A, Walsh V, et al. (2000). Transcranial magnetic stimulation in cognitive neuroscience–virtual lesion, chronometry, and functional connectivity. Current Opinion in Neurobiology 10, 232–237.

Pascual-Leone A, Amedi A, et al. (2005). The plastic human brain cortex. Annual Review of Neuroscience 28, 377–401.

Paus T, Jech R, et al. (1997). Transcranial magnetic stimulation during positron emission tomography: a new method for studying connectivity of the human cerebral cortex. Journal of Neuroscience 17, 3178–3184.

Robertson EM, Theoret H, et al. (2003). Studies in cognition: the problems solved and created by transcranial magnetic stimulation. Journal of Cognitive Neuroscience 15, 948–960.

Rockland KS, Ojima H (2003). Multisensory convergence in calcarine visual areas in macaque monkey. International Journal of Psychophysiology 50(1–2), 19–26.

Roder B, Rosler F, et al. (2000). Event-related potentials during auditory language processing in congenitally blind and sighted people. Neuropsychologia 38, 1482–1502.

Roder B, Stock O, et al. (2002). Speech processing activates visual cortex in congenitally blind humans. European Journal of Neuroscience 16, 930–936.

Sadato N, Pascual-Leone A, et al. (1996). Activation of the primary visual cortex by Braille reading in blind subjects. Nature 380(6574), 526–528.

Sadato N, Pascual-Leone A, et al. (1998). Neural networks for Braille reading by the blind. Brain 121(Pt 7), 1213–1229.

Sathian K, Zangaladze A, et al. (1997). Feeling with the mind's eye. Neuroreport 8, 3877–3881.

Schroeder CE, Foxe J (2005). Multisensory contributions to low-level, 'unisensory' processing. Current Opinion in Neurobiology 15, 454–458.

Shams L, Kamitani Y, et al. (2002). Visual illusion induced by sound. Brain Research: Cognitive Brain Research 14, 147–152.

Shimojo S, Shams L (2001). Sensory modalities are not separate modalities: plasticity and interactions. *Current Opinion in Neurobiology 11*, 505–509.

Siebner HR, Peller M, *et al.* (2003). Applications of combined TMS-PET studies in clinical and basic research. *Supplement to Clinical Neurophysiology 56*, 63–72.

Stein BE, Meredith MA, *et al.* (1993). The visually responsive neuron and beyond: multisensory integration in cat and monkey. *Progress in Brain Research 95*, 79–90.

Theoret H, Merabet L, *et al.* (2004). Behavioral and neuroplastic changes in the blind: evidence for functionally relevant cross-modal interactions. *Journal of Physiology, Paris 98*, 221–233.

Thut G, Ives JR, *et al.* (2005). A new device and protocol for combining TMS and online recordings of EEG and evoked potentials. *Journal of Neuroscience Methods 141*, 207–217.

Valls-Sole J, Pascual-Leone A, *et al.* (1994). Abnormal facilitation of the response to transcranial magnetic stimulation in patients with Parkinson's disease. *Neurology 44*, 735–741.

Valls-Sole J, Valldeoriola F, *et al.* (1999). Prepulse modulation of the startle reaction and the blink reflex in normal human subjects. *Experimental Brain Research 129*, 49–56.

Walsh V, Cowey A (2000). Transcranial magnetic stimulation and cognitive neuroscience. *Nature Review Neuroscience 1*, 73–79.

Walsh V, Rushworth M (1999). A primer of magnetic stimulation as a tool for neuropsychology. *Neuropsychologia 37*, 125–135.

Wassermann EM (1998). Risk and safety of repetitive transcranial magnetic stimulation: report and suggested guidelines from the International Workshop on the Safety of Repetitive Transcranial Magnetic Stimulation, June 5–7, 1996. *Electroencephalography and Clinical Neurophysiology 108*, 1–16.

Weeks R, Horwitz B, *et al.* (2000). A positron emission tomographic study of auditory localization in the congenitally blind. *Journal of Neuroscience 20*, 2664–2672.

Welch RB, DuttonHurt LD, *et al.* (1986). Contributions of audition and vision to temporal rate perception. *Perception and Psychophysics 39*, 294–300.

Wittenberg GF, Werhahn KJ, *et al.* (2004). Functional connectivity between somatosensory and visual cortex in early blind humans. *European Journal of Neuroscience 20*, 1923–1927.

Zangaladze A, Epstein CM, *et al.* (1999). Involvement of visual cortex in tactile discrimination of orientation. *Nature 401*, 587–590.

CHAPTER 30

Motor cognition: TMS studies of action generation

Simone Schütz-Bosbach, Patrick Haggard, Luciano Fadiga, and Laila Craighero

Introduction

The discovery and earliest applications of TMS both involved the motor system (Barker *et al.* 1985). Since then, TMS has been used in three quite different ways to study motor cognition. First, TMS can be used to provide a controllable and physiologically specified input to the skeletomotor system. Several sensory studies, for example, have used TMS to generate muscle contractions in the absence of volition and movement preparation. This allows controlled psychophysical studies of the perception of bodily movement (Haggard *et al.* 2002; Ellaway *et al.* 2004; Haggard and Whitford 2004). In other studies, TMS-evoked movements are used as perturbations of the motor apparatus. Here the focus is on preparatory and reactive adjustment for the perturbation (Bonnard *et al.* 2003, 2004). In this method, TMS is generally delivered over the primary motor cortex, but effects on the brain are less important than the effects on the body. Although this use of TMS has great value as a peripheral stimulus for studying kinesthesis, it is logically quite different from the use of TMS to study specific brain areas and processes, and so is not considered further here.

A second, very important, use of TMS has been as an online probe of cortical motor excitability. This is reviewed in detail elsewhere

(e.g. Chapter 9, this volume). A TMS test pulse can provide a known, if artificial, input to the motor cortex. This will cause a twitch in target muscles (motor-evoked potential, MEP) whose amplitude can be precisely measured. It may also cause an inhibition of ongoing electromyogram (EMG) (silent period, SP). In cognitive-motor studies, the size of these excitatory or inhibitory effects is measured as a function of cognitive factors like task, expectancy, and so forth. Changes in the motor output for a constant TMS input are interpreted in terms of differences between conditions, or across time, in motor system excitability. Importantly, this method can provide a completely implicit and on-line measure of the state of the cortical action system. Often a test pulse is preceded by a conditioning stimulus such as a sensory input or a conditioning TMS pulse to the same or another brain area.

Third, TMS can be used to interfere with cognitive-motor processes involved in action control, and widely described throughout this volume. Because the brain processes involved in generating a simple action are essentially serial, a single TMS pulse delivered at an appropriate time over an actively involved brain area may disrupt action control. Such single-pulse effects tend to be highly informative, because of their temporal and spatial specificity. On the other hand, their interpretation rests on a serial model of action control, which may not be sufficient for all situations. Other studies have used off-line TMS effects, as a short-term virtual lesion. This approach may be more powerful than single-pulse approaches, since it does not depend on precisely timing a single pulse with respect to the underlying brain processes. However, by the same token, it cannot clarify *at what stage* of the action control process a particular brain area makes its contribution.

TMS allows the experimenter to selectively interfere with a specific brain process. It is therefore particularly adapted to testing serial models of cognitive processing (Donders 1868; Sternberg 1969). In these models, processing is assumed to occur in a serial sequence of independent modules, which implement distinct and independent operations. The successful completion of each operation allows the next module to begin its operation. The value of these models is widely debated. Recent studies view the visual system as a parallel rather than serial architecture, involving multiple interconnected processing streams (Milner and Goodale 1993).

In contrast, the brain's action system can be viewed in two distinct ways. Voluntary actions involve a clearly serial process (cf. Figure 30.1). Volition or intention can be seen as the input to the process. These are followed by action selection or specification. At this stage, a specific set

of motor commands generating an appropriate movement pattern must be retrieved from the many alternatives, thus achieving the desired goal. This stage corresponds to the inverse model or planner of computational models (Ghahramani *et al.* 1996). Preparation for action then follows. This may involve further elaboration of the motor command itself, but also more general anticipatory modulation of reflex pathways and sensory areas likely to receive afferent feedback as a result of the impending action (Voss *et al.* 2006). A key moment in the serial control of action is the release of the motor command from the motor cortex, down the corticospinal tract (CT). The corticospinal volley drives the actual contraction of the muscles, and is the proximate cause of the movement itself. This point therefore marks the transition between action preparation and action execution. For some very simple 'ballistic' actions, the model may be considered to stop here. In most cases, however, afferent feedback from the moving effectors, and also internal feedback from predictions based on efference copy, are used to monitor the progress of the movement. Monitoring allows the motor command to be adjusted if it is incorrect, thus reiterating the model. It also allows the successful completion of one movement to serve as the trigger for the next movement in a sequence. Finally, action monitoring may be used for

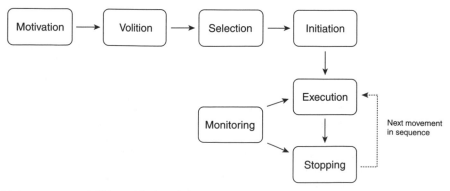

Fig. 30.1 A simple serial hierarchical model of action control suitable for interpreting TMS studies. Note the increasing quantity and specificity of information as the action is elaborated.

perceptual processes beyond the immediate motor control system, such as self-recognition and agency (Haggard 2005).

Not all action research fits well with this serial model. Several neurophysiological, neuroimaging, and behavioral studies have focused on the sensory guidance and internal representation of action by a network of parietal and premotor regions (for a review see Freund *et al.* 2005). These studies focus on the transformation of sensory representations into motor codes, and the commonality between visual and motor representations of action. However, the results do not always support a simple serial flow from sensation to action. Neurophysiological and neuroimaging results often reveal visual responses in 'motor' areas, while responses in early sensory areas can show dramatic top-down modulation according to current motor task (Ruff *et al.* 2006). A recurrent feedback model may therefore be more appropriate than a strictly serial model for those sensorimotor actions that involve relatively direct responses to environmental stimuli. In the following we first review studies which can be situated within a simple serial model of action generation. These studies have in common that they largely focus on the control of voluntary or internally generated actions. Here TMS has predominantly been used as a technique to temporally disrupt specific cognitive processes at particular times. Second, we will discuss the use of TMS in research focusing on perception–action linkage, such as reaction and interaction with the environment, including the social environment. In that tradition, parallel and interactive models dominate over serial models, but TMS has still proved to be an important research tool, notably in measuring cortical excitability.

Intentional actions and the serial model of action generation

Context and motivation for action

Human action is generally goal directed. Our actions therefore occur in the context of our internal environment (e.g. needs and desires) and the opportunities afforded by the current external environment. The agent's emotional and motivational states therefore constitute a reason for action. Neurophysiological evidence supports the existence of anatomical–functional links from the limbic system to premotor areas, mediated by connections to the cingulated and prefrontal cortical regions. These earliest contextual antecedents of action have proved difficult to study with TMS for two reasons. First, many motivational and limbic structures lie deep within the brain and cannot be stimulated externally. Second, antecedent states such as motivations and drives provide a tonic background to action rather than a single neural event. They are not therefore amenable to investigation using phasic interventions such as TMS.

In one of the few TMS studies to investigate action antecedents, Oliveri *et al.* (2003) used TMS to investigate the role of the supplementary motor area (SMA) as a mediator between emotion and action. They accordingly measured cortical excitability of primary motor cortex (M1) during processing of emotional versus nonemotional visual stimuli. Subjects were required to perform arbitrary movements in response to unpleasant or neutral pictures of people, animals, or landscapes. The subjects received a single TMS pulse over the left M1, which was randomly preceded by paired TMS over the ipsilateral left SMA, left premotor cortex (PM), or right M1. The amplitudes of motor-evoked potentials (MEPs) recorded from subject's right first dorsal interosseus (FDI) after conditioning TMS were compared against those obtained after single TMS over the left M1. The authors showed that conditioning TMS to SMA selectively enhanced MEP amplitudes when subjects responded to emotionally unpleasant pictures, and not when neutral visual cues were presented. However, conditioning TMS of PM or of the contralateral primary motor area did not show this effect. This finding confirms a specific functional link between SMA and primary motor areas in the control of movements that are triggered by emotional processing of certain visual cues. SMA seems to serve as a key area for transforming motivations, such as emotional states, into motor responses.

Intention and planning of action

Once a contextual reason for action exists, and a particular action goal is identified, the brain faces a series of computational problems before the action itself can be initiated. In particular, most goals can be achieved by a number of different means. The brain must identify just one of the possible movements consistent with an action goal before a detailed motor command can be assembled. In computational motor control this is known as the inverse or planning problem (Wolpert 1997). The premotor areas immediately in front of the primary motor strip play a particular role in selecting the specific movement that will be made given a general action goal.

The dorsal premotor cortex (PMd) seems to be particularly concerned with the selection of movements according to learned associations. Schluter *et al.* (1998) showed that stimulation of the left PMd disrupts the selection of movements that will be made with either hand. In their study subjects were asked to perform a choice reaction with either their index or middle finger of one hand in response to a shape presented on-screen. Reaction times were measured while subjects received a TMS pulse over the contralateral cortex. TMS over premotor areas, when applied at intervals close to movement onset, significantly delayed response time. This effect was present both in a group of subjects who used their left hand to respond while being stimulated over the right hemisphere, and in a second group who used the right hand while being stimulated over the left hemisphere. In contrast, TMS over the primary motor cortex delayed responses only at longer cue–stimulus intervals (>300 ms). Moreover, a second experiment stimulating either left or right PMd suggested that the neural system for action selection was lateralized to the left hemisphere. Stimulation of the left hemisphere replicated the delay found when subjects used two digits of the right hand to respond. It also delayed responses in a separate block using the left hand. Right hemisphere stimulation affected only contralateral and not ipsilateral responses. These findings show that premotor cortex is functionally relevant in an early stage of movement selection whereas the motor cortex comes into play at a later stage. In particular, they fit well with the serial model of action: disruption of action selection was found at an earlier time and an anatomically upstream location compared

disruption of final motor output. This study indicates the high scientific potential of combining single-pulse TMS and precisely timed task in order to investigate classic serial models of cognitive psychology. Once an action is selected, it may be initiated immediately, or maintained in a state of preparation to be finally triggered at a later time. The phrase 'motor attention' has been used for this state of preparedness occurring between selection and execution of action. In the primate, cells active in the delay period between a selection cue and a go-stimulus are widely found in both premotor (Boussaoud and Wise 1993) and parietal (Goldberg *et al.* 1990; Li *et al.* 1999) regions. Human studies suggest that premotor and parietal regions may act in concert to prepare a selected action. For instance, Rushworth *et al.* (2001) have shown that redirecting of covert orienting is impaired when the parietal cortex is temporarily disrupted by TMS. In their experiment a visual precue preceded the presentation of an imperative stimulus indicating which of two manual responses to execute. On infrequent trials, the precue provided invalid information so that the subject had to shift from one intended movement to making a quite different movement. A brief train of rTMS was applied over the left anterior parietal region after target presentation but prior to response execution. Reaction times were impaired but only in invalid trials. This result suggests that the parietal cortex contributes either to the processes of reselection of a novel motor response or to preparation to perform this response. This motor attentional effect was distinct from a spatial orienting effect tested by the same authors using a conventional orienting paradigm, and found to be localized more posteriorly within the parietal cortex.

Selection and specification of intentional action

TMS can also be used to influence movement selection externally and even independently from a person's conscious movement intention. For instance, Brasil-Neto *et al.* (1992) studied the influence of TMS on forced-choice response times. Subjects were required to extend their index finger in response to the click of a

TMS pulse. Moreover, they were instructed to freely choose either their right or left finger for making the response, but this decision was only to be made after hearing the TMS click. Subthreshold TMS was delivered over the prefrontal or motor area. In a control situation subjects were stimulated peripherally. Hand preference was only then affected when TMS was delivered over the motor cortex; when being stimulated over this area, subjects more often chose the hand contralateral to the stimulated site. The effect of hand preference was most pronounced in responses with very short latency (<200 ms). This bias replicated Ammon and Gandevia (1990) but suggested that the effect was focal and restricted to motor, not frontal, areas of the brain. Another important observation is that in both studies subjects were unaware of the preference in their responses. They felt that their decisions were entirely made in a neutral way. This finding could also be interpreted as evidence suggesting that motor selection can precede the conscious intention to select a given response. Selection could even generate conscious intentions (Haggard and Eimer 1999).

Movement execution

Once a movement is selected, the motor command must be sent down the CT for its execution. Day et al. (1989) showed that a single magnetic stimulus can interfere with processes controlling the initiation of simple reaction movements. Subjects were trained to flex or extend their wrist following an auditory signal. In one-third of the trials, subjects received a single magnetic stimulus to the contralateral motor cortex of varying intensity but sufficiently strong to produce a flexor muscle response. The pulse was delivered at a predefined onset time after the tone and just before the expected onset of the wrist movement. When TMS was delivered, the execution of the movement was delayed up to 150 ms. Furthermore, the amount of delay turned out to be a function of both stimulus intensity and the onset time prior to the movement: the delay increased with increasing stimulus intensity and the closer the stimulus was to the expected onset of the voluntary action. However, the brain stimulus did not

show any effect on the organization of the pattern of the agonist (flexor)/antagonist (extensor) muscle activities. Thus the form of the response still remained intact. In contrast, stimulation of peripheral nerves did not lead to similar effects of delaying movement onset. Interestingly and in line with a serial model of action control, TMS led only to a delay, not to an abolition, of a voluntary action. Thus, it seems that only the selective part of the central motor program, probably the release of the motor command, was transiently disrupted or rather temporarily inhibited.

In a similar vein, Pascual-Leone et al. (1992a) compared simple reaction times (RTs) to go-signals of different modalities and investigated the effects of TMS on RTs. In their study, subjects were asked to flex their right elbow as rapidly as possible in order to touch the shoulder with their right hand in response to a go-signal. Shortest RTs occurred for auditory go-signals, followed by somatosensory, and then visual. In all cases RTs were shorter with increasing intensity of the signal. This effect is probably due to different recognition time for the different go-signal modalities. However, in line with Day et al. (1989), longest RTs (i.e. longer than the RTs to any other go-signal) were found to TMS over the contralateral motor cortex at above threshold intensity, i.e. at an intensity to induce an MEP in the responding arm. In contrast, shortest RTs, even shorter than reactions to auditory signals, occurred to either TMS at subthreshold intensity over the contralateral hemisphere or to TMS over the ipsilateral motor cortex. TMS over parietal and frontal areas did not have an effect on RTs. The effect of shortening of RTs by TMS over the contralateral motor cortex at subthreshold intensity was also replicated in a further study by Pascual-Leone et al. (1992b).

Goal-directed movements require frequent updating of the movement trajectory via feedback loops throughout its execution. A key brain area mediating these processes seems to be the posterior parietal cortex (PPC). Desmurget et al. (1999) tested the hypothesis that the PPC supports on-line motor adjustment by computing the instantaneous differences between hand and target locations. Subjects pointed to visual targets in the peripheral visual field which either remained stationary or changed position during saccadic eye movements. Subjects could not visually monitor their pointing movement. Just after movement onset, TMS was applied over the left PPC. This intervention abolished on-line trajectory adjustments. Moreover, this effect occurred only when the visual target jumped to different positions but not when it remained stationary. This finding supports PPC involvement in on-line movement corrections. PPC might serve as a 'neural comparator' which computes a current motor error. Johnson and Haggard (2005) were unable to replicate these effects, though their TMS intensities were lower than those that appear to have been used by Desmurget *et al.* (1999).

Motor awareness

Neuroscientists have recently shown a developing interest in the conscious experience of action. Several studies have used TMS to investigate what processes within the motor system are associated with consciousness and which are not. These studies have typically used TMS paradigms developed for investigating motor execution, and assessed how they influence motor awareness.

For example, Haggard and Magno (1999) used Day *et al.*'s (1989) method (see above) to delay simple RT movements to auditory stimuli. They also asked subjects to judge the time at which they felt they reacted by indicating the position that a clock hand had occupied at the time of their response. Single-pulse TMS was delivered over contralateral motor cortex 75 ms before the expected reaction. This intervention delayed voluntary reactions by >200 ms. However, subjects' reports of *when* they reacted suggested that less than half of this delay entered into awareness. Stimulation over a more anterior location (electrode site FCz) produced shorter delays in actual RT, of which a relatively *larger* proportion entered into awareness. The authors concluded that intervening on the voluntary motor system at the M1 level had only minor effects on awareness, because an important component of motor awareness is generated upstream of M1, in the premotor areas.

Voss *et al.* (2006) used the same method of TMS-induced delay, but focused on awareness

of sensory events during movement. They measured the well-known sensory suppression effect: sensitivity to electrocutaneous stimuli on a moving body part is reduced relative to sensitivity when the same body part is at rest (Angel and Malenka 1982). Voss *et al.* found that this sensory suppression was also present during an RT task in the time window when a voluntary action was expected, but had been artificially delayed by TMS over contralateral M1 (Day *et al.* 1989). Controls showed that the suppression during the TMS-induced delay period could not be attributed to direct masking of the electrocutaneous stimulus by TMS effects on SI. Instead, the finding of sensory suppression during TMS-induced delays was used to localize the signals involved in sensory suppression. The signals that produce sensory suppression must originate upstream of the primary motor cortex.

A more precise localization was proposed by Haggard and Whitford (2004). They asked subjects to judge whether the first or second of two involuntary movements (MEPs produced by M1 TMS) was larger. When the first, test MEP occurred during a self-generated voluntary movement, it was less likely to be judged larger than the second, reference, MEP compared with test MEPs delivered at rest. This effectively replicates previous sensory suppression results. However, a conditioning TMS pulse delivered over the SMA 10 ms before the test stimulus abolished the sensory suppression effect. The authors concluded that the SMA is actively involved in generating the efferent signals that modulate afferent input through sensorimotor gating.

Motor sequencing

In order to perform a goal-directed behaviour, we have to organize actions in a specific spatiotemporal order. A number of studies confirmed that the medial frontal cortex and, in particular, the human pre-supplementary motor area (pre-SMA), plays an important role in the sequencing of actions. More precisely, SMA seems to be particularly involved in both the encoding of movement sequences and in the planning of forthcoming movements in a motor sequence retrieved from memory.

Müri *et al.* (1995) investigated the role of SMA in the cortical control of sequences of memory-guided saccades. Subjects were asked to fixate a central point while four different targets appeared laterally on either one or both sides of a screen. The task was to remember the order of target appearance without looking directly at them. Then the fixation point disappeared and subjects were required to make saccades successively to the targets in the same order in which they appeared. While subjects performed the task, TMS was delivered over SMA or as a control over the occipital cortex at random time intervals during three different phases: the target presentation phase, the memorization phase, or the phase in which the saccades were executed. Stimulation over SMA and not over the occipital cortex induced an increase in error rates but only when TMS was delivered during the phase of target presentation. This indicates that the learning phase was selectively disturbed and that SMA appears to be functionally relevant in memory encoding. The finding that performance was not affected when SMA was stimulated during the execution phase indicates – in line with a serial model of motor control – that once the motor program is initiated it is no longer under control of the SMA region.

Gerloff *et al.* (1997) asked subjects to learn playing three finger-sequences of different complexities for ~8 s periods with their right hand following a metronome beat of 2 Hz. Task complexity was varied as follows. In a 'simple' sequence they repetitively (16 times) pressed one key using their index finger. In a 'scale' sequence they used four fingers and pressed consecutively four different notes but always in the same order (i.e. 5–4–3–2–5–4–3–2 etc.). Finally, in a 'complex' sequence subjects played a nonrepetitive and nonconsecutive order by using four fingers. Subjects practised the sequence until they could play it from memory 10 times consecutively without making any errors. During the actual experiment subjects were asked to play a certain sequence (complexity varied randomly). Two seconds after the first key press, high-frequency (15–20 Hz) rTMS was delivered over the fronto-central midline including SMA. When subjects performed complex movement sequences, TMS led to interference with the organization of the future components

in this sequence. In contrast, stimulation over M1 induced accuracy errors in both the complex and scale sequences, whereas stimulation over other control regions (F3, F4, FCz, P3, P4) did not cause interference at all. Moreover, rTMS over SMA and M1, respectively, led to different timing patterns of error induction: error induction following stimulation over SMA occurred ~1 s later than with stimulation over M1. The result of this study suggests that SMA is of critical importance for the time-dependent organization of *future* elements in complex sequential actions retrieved from memory. Thus, before sending movement commands to primary motor areas for execution, SMA seems to be a key area for organizing upcoming movements in a complex motor sequence.

Kennerly *et al.* (2004) investigated the role of pre-SMA in the internal organization of motor elements within a sequence organization and initiation (cf. Sternberg 1969). In their first experiment the authors asked subjects to learn a bimanual sequence of 12 alternating movements so that they could perform the sequence from memory. In line with several behavioral studies on sequence learning, subjects showed a spontaneous organization of the long sequences of finger key-press movements into smaller component units or 'chunks'. With practice, subjects typically executed short phrases within the overall motor sequence as a single 'chunk' characterized by a low interval between successive movements. In contrast, the interval between some successive elements consistently showed a higher RT, suggesting a chunk boundary (cf. Sternberg 1969). The authors used 0.5 s trains of 10 Hz repetitive TMS to transiently disrupt pre-SMA activity at three different stimulation times: just prior to the first movement, at the chunk point, i.e. the movement with the highest RT within the sequence, and finally at nonchunk points, i.e. a low RT movement in the middle of a pre-organized chunk. Repetitive TMS over pre-SMA disrupted performance, i.e. caused significantly longer RTs, when it was applied at the initiation of a new sequence chunk but not during the course of an ongoing chunk. This effect was specific to pre-SMA since no disruptive effect of TMS was seen when it was applied over PMd. One elegant feature of this study was the clear separation between the cognitive and motor components of the task. The motor execution for a key press that marked a chunk point was similar to that for one that did not. However, the movements clearly differed with regard to the cognitive organization of the sequence as a whole. Within the serial model of action (cf. Figure 30.1), the chunk point was more strongly associated than nonchunk points with a number of cognitive processes. These included stopping the previous chunk, retrieval of the next chunk from motor memory, and preparation of the motor programs required for the next chunk.

In a study by Müri *et al.* (1996) subjects were asked to fixate a central point on-screen. Two seconds later, a target appeared laterally for 50 ms with unpredictable position and randomized amplitude. A go-signal indicated to perform a saccade to the remembered position of the flash. After 2 s the target reappeared and subjects made a corrective saccade, if necessary. A single TMS pulse was delivered over the right PPC or the dorsolateral prefrontal cortex (DLPFC) randomly at different time intervals in relation to the target's appearance: between 160 and 360 ms after target presentation, during the encoding phase, i.e. between 700 and 1500 ms, and finally at 2100 ms, i.e. 100 ms after the fixation point disappeared. TMS showed both temporal and topographic specific effects. Stimulation over PPC and not over DLPFC significantly affected contralateral saccade accuracy and bilateral saccade latency. This effect was present when TMS was delivered 260 ms after target presentation indicating that PPC is functionally relevant especially during early phases of encoding and sensorimotor integration processes (cf. Goldberg *et al.* 1990). Additionally, the latency of saccades increased when TMS over PPC was delivered 2100 ms after target presentation. This later effect was attributed to a second function of PPC in triggering saccade execution. In contrast, stimulation over DLPFC selectively affected contralateral saccade accuracy, but only when the pulse was applied during later periods of encoding, i.e. between 700 and 1500 ms after target presentation. This study therefore is evidence that the prefrontal cortex plays a crucial role in the preparation of memory-guided movements. Moreover, it does so later than the PPC. Whereas the PPC seems to be more relevant in early sensorimotor integration processes, the DLPFC seems to control memory processes relevant to the subsequent action. Taken together these studies indicate that the DLPFC seems to control memory retrieval, whereas composing the retrieved items in memory into an appropriate composite action sequence seems to be a main function of the SMA.

TMS studies of intentional action: concluding remarks and future prospects

TMS can be used to clarify the relationship between cognition and action in the human brain. TMS works well for testing serial models of cognitive processing because it can selectively and temporarily disrupt identified brain functions. In this way one can prove whether a certain brain area carries out the cognitive operation that is essential for a certain task at a given time point. Here, we have approached voluntary, goal-directed action as a computational problem involving a sequence of several separate modular processing stages or components. Voluntary action starts with an abstract description of the goal. This abstract task description has then to be translated into a detailed movement pattern. Only when an appropriate movement has been selected from many alternatives can the motor plan be sent to the output areas of the motor cortex for final execution. Feedback from execution allows monitoring and correction of ongoing actions, and may also contribute to chaining successive movements into an overall action sequence. Neuropsychological studies of the cognitive-motor functions of the frontal lobe (for an overview see Stuss and Knight 2002) confirm that inhibition of action is at least as important as generation of action in these brain regions. The nature of inhibitory components within the action control system as a whole is not yet well understood, and remains an important area for future research. TMS has the potential to measure effects of inhibition directly within the cortex using paired-pulse (Chapter 11, this volume) and double-pulse (Haggard and Whitford 2004) techniques. This avoids the key problem of psychological studies of inhibition, namely that inhibition cannot be easily quantified because it does not produce overt behavior.

TMS studies of perception–action linkage

Not all research on motor cognition is in line with this serial model. A large body of evidence suggests a strong linkage or even communality between sensory, notably visual, and motor representations of action. Direct reactions to environmental stimuli, and reciprocal interactions with the environment, may be better explained by parallel models of cognitive processing. The traditional view of perception and action in terms of two independent processing systems has been challenged by research showing that the properties of a visual stimulus constrain the motor process of generating a response to that stimulus, and vice versa. An area of special interest has been the brain's 'mirror systems' that respond to both self-generated actions and also to observing actions of others. TMS has proved a valuable tool for testing parallel models of perception–action linkage, because it can be used to measure cortical excitability and thus the involvement of the motor system in a temporally precise way during action observation. Here we review a series of TMS studies that provide convincing evidence for the tight coupling between perception and action. Taken as a group, these studies suggest that the actions of others are covertly resonated or re-enacted online, with high temporal fidelity. The effects on the observer's motor system revealed by TMS are even somatotopically specific.

Action representation

The view on the motor system that dominated during the last century has been challenged in the last 20 years. The classical view was based on the existence of two complete representations of body movements located in the posterior part of the frontal lobe (Woolsey et al. 1952). The first representation was located on the lateral cortical convexity, and included Brodmann's area 4 and part of area 6. This representation was called 'primary motor cortex' or M1. The second representation, smaller and less precise than M1,

was located on the mesial cortical surface, and was named supplementary motor area (SMA; Woolsey et al. 1952). A series of anatomical and functional studies have shown, first in non-human primates and more recently in humans, that this picture of the motor cortex is too simplistic. First, area 4 is functionally distinct from area 6. Second, area 6 is not homogeneous but is formed by a multiplicity of distinct anatomical areas. Third, these various motor areas are characterized by peculiar afferent and efferent connections and seem to play different functional roles in motor control (see above). The organization of the motor system in the frontal cortex is mirrored in the posterior parietal lobe. Again, several independent areas are involved in different aspects of the sensorimotor transformation processes. Frontal and parietal lobes are reciprocally connected according to the following rule. Each frontal motor area receives its main afferents from one single parietal area, which is also the main target for its efferent projections. In this way, the reciprocally connected motor and parietal areas constitute series of specialized circuits working in parallel. These circuits transform sensory information into a specific action and form the basic elements of the motor system. It is important to note that neural activity associated with action execution has also been observed in many posterior parietal areas and that somatosensory, visual and acoustic stimulations evoke responses in many frontal regions.

Linkage between visual and motor representations of actions: mirror neurons

This neural organization of the motor system could hardly have been represented by a model describing a simple serial flow from sensation to action. Indeed, one of most fascinating discoveries of recent decades is that some premotor neurons, in addition to their motor discharge, respond also to the presentation of purely visual stimuli. This functional property led to substantial change in views of motor system organization. Neurons with this property belong to different parieto-frontal circuits, such as the LIPFEF circuit (Bruce and Golberg 1985), which is essentially involved in the control of eye movements, the VIP–F4 circuit that plays a role in encoding the peripersonal space and in transforming object locations into appropriate reaching movements, and also the AIPF5 circuit, in which hand and mouth goal-directed actions are represented. The discovery of the AIP–F5 circuit and the functional properties of its neurons stimulated the idea that the motor system is also involved in high-level cognitive functions such as the understanding of others' actions and social communication.

From a motor point of view, neurons in F5 seem to code especially the *goal* of the actions. This evidence comes from electrophysiological studies indicating that neurons fire during object-directed actions such as grasping, holding, and manipulating, whereas they do not fire during actions that involve a similar muscular pattern but do not aim at manipulating a certain object (e.g. scratching or grooming). Moreover, some F5 neurons discharge independently from the acting effector: they fire when the monkey grasps an object with its right or left hand or with its mouth.

From a sensory point of view, area F5 contains two different categories of visuo-motor neurons. Neurons of the first category discharge when the monkey observes graspable objects, and they have been called 'canonical neurons' (Rizzolatti and Fadiga 1998). These neurons discharge at the mere presentation of objects whose shape and size are congruent with the type of grip motorically coded by the same neurons: neurons that are active during observation of small objects are also active during precision grip (Murata *et al.* 1997). These functional properties indicate a close link between graspable objects and the respective actions that they afford: whenever a graspable object is perceived, the most suitable grasping action is automatically evoked. Neurons of the second category discharge when the monkey observes hand actions performed by other individuals and have been called 'mirror neurons'. These neurons discharge when the monkey manipulates objects, as well as when it observes another individual making similar goal-directed actions (di Pellegrino *et al.* 1992). In contrast to canonical neurons, mirror neurons do not discharge by the mere visual presentation of objects. An interaction between a biological effector and an object is a necessary condition for mirror

neuron activity. The mirror neuron response is not affected according to whether the actions are executed by a human or by another monkey, nor whether the action occurs near or far from the observing monkey (implying that the size of the observed hand is unimportant). Typically, mirror neurons show congruence between the observed and executed action. That is, the neuron's visual response occurs selectively when viewing the same kind of action which selectively evokes motor responses in the neuron when the monkey performs it. That is, the effective motor action coincides with the action that, when seen, triggers the neurons. The most likely interpretation for visual discharge in mirror neurons is that the observed action automatically evokes an internal motoric representation of the same action. In other words, the properties of mirror neurons seem to suggest that an observed action is covertly re-enacted by the observer's motor system.

The human mirror system as investigated by TMS

In recent years, a series of brain-imaging studies has investigated whether a mirror neuron system is also present in the human brain. Indeed, it has been demonstrated that when an individual observes an action, a network of cortical areas is activated, including the ventral premotor cortex, the inferior frontal gyrus, the inferior parietal lobule, and the superior temporal cortex (see for review Rizzolatti and Craighero 2004). This network is also involved when an individual executes the action. However, given the limited temporal resolution of brain-imaging studies, it is still unclear whether the internal replication of an observed action reflects an *online* or *off-line* process. TMS can provide an alternative technique to tackle this question. Single- or paired-pulse TMS allows measurement of cortical excitability during different phases of an observed action. Moreover, this technique is able to verify a specific involvement of the motor system by discriminating those muscles that are involved in the motor replica. The first attempt to study corticospinal (CS) excitability during action observation was made by Fadiga *et al.* (1995). Single-pulse TMS was applied over the hand motor representation in

M1 and MEPs were recorded from four intrinsic hand muscles. Participants were tested under four experimental conditions: observation of an experimenter grasping different objects; observation of an experimenter drawing geometric shapes in the air; observation of different objects; and a dimming detection task. The study showed three main results. First, CS excitability is modulated by action observation, indicating that the human motor system is concretely involved during the perception of others' action. Second, modulation of CS excitability is also present during observation of intransitive actions (arm movements). This finding may reflect a main difference between the human mirror neuron system and that of monkeys. In the latter, mirror neurons only respond during the observation of transitive actions (see above). Third, motor excitability is limited only to those muscles that are specifically involved in the observed action. In fact, MEPs recorded from the opponens pollicis (OP) muscle were modulated only during observation of grasping movements and not during arm movements, whereas MEPs recorded from the other three muscles (extensor digitorum communis (EDG), flexor digitorum superficialis (FDS), and FDI) were modulated during both action observation conditions. The latter finding might be due to the fact that during the actual execution of arm movements, the OP muscle is not contracted.

Recently, Montagna et al. (2005) elegantly showed that, during the observation of action, those specific muscles are activated in the observer which he/she would also recruit for overtly executing the observed movement. MEPs were recorded from three forearm muscles (FDI, flexor carpis radialis (FCR), and FDS) while subjects were watching a human hand performing a reaching and grasping movement on a screen. The excitability time-course during the observed action was explored at four different phases of the movement: at mid-hand opening; at the end of hand opening; at the mid-hand closing on the object; and when fingers contacted the object. In a separate block of experimental trials, subjects overtly imitated the reaching and grasping movement in synchrony with the observed action in order to show the temporal pattern of activation of the same muscles that had been investigated during

action viewing. In this condition, EMG signals were selectively recorded from the same forearm muscles recorded during the observation condition. The results showed a remarkable correlation between the temporal pattern of EMG recruitment in the imitation condition and the time course of MEP modulation in the observation condition. In other words, each subject's MEP facilitation resembled the idiosyncratic EMG patterns that they produced when asked to make overt imitative movements. This indicates that during the observation of a specific action, the same muscles are activated as the observer would use in their own execution of that action. Most importantly, the modulation followed the same temporal order as when they would have been recruited for overtly executing the observed movement. This suggests that 'motor resonance' really means that an observed action is re-enacted in terms of the observer's own motor control strategy adapted to the same task.

Several TMS studies have been carried out which aimed at investigating the nature of the 'human mirror system'. One major aim was to explore whether muscle facilitation has a cortical origin or not. A series of experiments (Strafella and Paus 2000; Baldissera et al. 2001; Patuzzo et al. 2003) have demonstrated that the facilitation of MEPs induced by action observation is due to the enhancement of M1 excitability produced through excitatory cortico-cortical connections. The double-stimulus TMS technique has mainly been used to determine the origin of CS facilitation. This technique consists of a subthreshold conditioning TMS pulse followed by a suprathreshold TMS test pulse at various delays. By considering different delays between the two pulses it is possible to investigate changes in the excitability of excitatory or inhibitory interneurons within M1 itself. In fact, intracortical inhibition is usually observed for short (1–5 ms) or long (50–200 ms) intervals between conditioning and test TMS pulses, whereas intracortical facilitation is usually observed for 8–20 ms intervals. Strafella and Paus (2000) used this technique and stimulated left M1 during action observation. Results showed a decreased intracortical inhibition at the 3 ms interstimulus interval, indicating that CS facilitation is attributable to cortico-cortical facilitating connections.

Another field of investigation was devoted to understand whether the specific activation of the observer's muscles is temporally coupled to the dynamics of the observed action. Gangitano et al. (2001) used TMS to stimulate the left hemisphere and evoke MEPs in the contralateral FDI muscle, while subjects were watching a video clip of a hand approaching and grasping a ball. TMS pulse was delivered at five different times, covering all different movement phases. The results showed that response facilitation was differently tuned depending on the different phases of the grasping action. MEP amplitude became larger with increasing finger aperture and became smaller during the closure phase, indicating that the mirror system compares the observed action with the internal correspondent also in terms of a temporal coding. In a more recent study Gangitano et al. (2004) investigated whether this pattern of modulation was the consequence of a 'resonant plan' evoked at the beginning of the observation phase or whether the plan was fractioned in different phases sequentially recruited during the course of the ongoing action. The authors therefore used the same procedure as in Gangitano et al. (2001) with the following exception: subjects were shown video clips representing an unnatural movement, in which the temporal coupling between reaching and grasping components was disrupted, either by changing the time of appearance of maximal finger aperture, or by substituting it with an unpredictable closure. In the first case, the observation of the uncommon movements did not exert any modulation in motor excitability. In the second case, the modulation was limited to the first time-point. Modulation of motor excitability was clearly suppressed by the appearance of the sudden finger closure and was not substituted by any other pattern of modulation. This finding suggests that a motor plan, which includes the temporal features of the natural movement, is activated immediately after the observed movement onset and is discarded when these features cease to match the visual properties of the observed movement. Thus the human mirror system seems to be able to infer the goal and the probability of an action during the development of its ongoing features.

In a very accurate and precise experiment Borroni et al. (2005) aimed at verifying the degree of correspondence, especially with respect to a fine temporal resolution, between the observation of prolonged movements and its modulatory effects in the observer. For this purpose the authors asked subjects to watch a cyclic flexion–extension movement of the wrist. The same sinusoidal function was used to fit both observed wrist oscillation and motor resonance effects on the observer's wrist motor circuits. In this way the authors could describe a continuous time course of the two events and precisely determine their phase relation. MEPs were elicited in the right forearm muscle (FCR) of subjects who were observing a 1 Hz cyclic oscillation of the right prone hand executed by another person. The results indicated that movement observation elicited a parallel cyclic excitability modulation of the observer's MEP responses following the same period as the observed movement. Interestingly, the MEP modulation preceded the muscle activation of the observed movement, indicating that the mirror system anticipates the movement trajectory, rather than simply reacting to visual events in the movement. The same results were obtained when the observed hand oscillation was executed with different frequency (1.6 Hz) and when the hands of the actor and observer were supine. In a control condition subjects were confronted with an oscillatory movement of the metal platform itself, without the actor's hand resting on it. The platform was oscillated by a human actor hidden behind a screen, so that the movement profile was identical to the flexion–extension movement of the visible actor's hand. However, this condition did not evoke any resonant response in the observer. These findings suggest that during observation, motor pathways are modulated so that the motor command which is needed to execute the observed movement is reproduced with high temporal fidelity. Romani et al. (2005) demonstrated that motor excitability can also be modulated by the observation of biomechanically impossible movements. Participants observed sequences of abduction/adduction movements of a right index finger and a right little finger, and of extension/flexion movements of a right index finger. Based on the angular displacements of the fingers, movements were defined as biomechanically possible or impossible. The results showed a selective motor facilitation of the muscle that would be involved in actual execution of the observed movement for possible movements and, most interestingly, also for movements well beyond the normal range of joint mobility. This finding seems to suggest that the human mirror system does not differentiate biologically possible and impossible movements. Rather, it seems that even impossible movements are coded in the frontal mirror system, suggesting that observation-related motor facilitation is not due to coding muscles *per se* but to coding the role a specific muscle plays in a given overall action.

TMS study of frames of reference for mirror systems

A different field of research on this topic addresses the question of whether postural congruency between the observer and the model modulates CS facilitation. For instance, Maeda *et al.* (2002) investigated the role of visual perspective on movement observation-induced motor excitability. Subjects viewed a model's right hand abducting either the thumb or index finger, or vertically moving the index finger. Critically, subjects saw the model's hand either from a first-person or a third-person perspective. The results first of all confirmed that action observation enhances motor output to the muscles involved in the observed movement regardless of its orientation. However, the degree of modulation depended on the hand orientation. Greater modulation of motor excitability was observed for movements in first-person than in third-person perspective. In contrast to this study, Urgesi *et al.* (2006) recently obtained different results by slightly modifying the same paradigm. They recorded MEPs from the FDI and abductor digiti minimi muscles during observation of right index and little finger abduction/adduction movements of models keeping their hands in a palm-down or palm-up position. In different conditions observers were also asked to keep their right-hand palm down or up so that the observer's posture was congruent or not congruent to the observer's posture. The authors found that mirror motor activation is more influenced by the topographic matching of the model's movement on the observer's motor system than by the spatial and postural congruency between the model and observer's hand. The authors attributed the discrepancy to the fact that in Maeda *et al.*'s (2002) paper the inversion of the hand orientation not only changed the side of space where the finger movements were directed, but also the perspective from which the hand stimuli were viewed. In a similar vein, Patuzzo *et al.* (2003) investigated whether the observation of one's own or another's action influences CS excitability differently. The authors confronted subjects with videos of their own or another's hand performing the same movements. No significant differences between the self and other conditions were found (but see Schütz-Bosbach *et al.* 2006). It is interesting to note that high-functioning individuals with autism spectrum disorder, when tested with a paradigm very similar to that used by Maeda *et al.* (2002), lack muscle-specific facilitation only during observation of moving hands presented from a first-person perspective (Theoret *et al.* 2005).

Action representation beyond the visual modality

Finally, several studies investigated motor excitability to TMS during acoustic, rather than visual action perception. In fact, action-generated sounds and noises are also very common in our daily environment. Monkey studies show that a proportion of mirror neurons not only respond to visual stimuli, but also become active when the monkey is listening to an action-related sound (Kohler 2002). Aziz-Zadeh *et al.* (2004) used TMS to explore whether an equivalent effect is also present in humans. The authors stimulated the left and right hemispheres and recorded MEPs from the contralateral FDI muscle while subjects were listening to bimanual hand action sounds (e.g. typing or tearing a paper) or to control sounds (e.g. walking, thunder). The results showed that sounds associated with hand actions produced greater CS excitability than the control sounds. Moreover, this facilitation was exclusively lateralized to the left hemisphere. Fadiga *et al.* (2002) investigated whether speech listening is also able to increase MEPs recorded from the listeners' tongue muscles. Subjects were instructed to listen carefully to a sequence of acoustically presented verbal and nonverbal stimuli, while their left motor cortex was magnetically stimulated in correspondence with tongue movement representations. The embedded consonants in the middle of the verbal stimuli determined whether the pronunciation required either slight tongue tip movement (e.g. double 'f') or strong tip movement (e.g. double 'r'). The results showed that listening to words containing, for instance, a double 'r' consonant led to an increase of tongue MEPs relative to all the other experimental stimuli. This finding seems to suggest that listening to speech leads to specific activation of speech-related motor areas in the listener.

Summary and conclusion

To conclude, TMS has been a key methodological tool for studying motor cognition. In studies of the serial processes of action generation, TMS has been used to identify and describe the individual processes that extend along the motor processing chain from motivation and volition to muscle contraction. Here, TMS has been used as both an excitability measure and a transient inactivation. In studies of the parallel loops linking perception to action, TMS has been used primarily, though not exclusively, as a probe to measure excitability. In the future, double-pulse approaches may offer the interesting possibility of disrupting one arm of such loops in order to modulate effects of TMS, including excitability effects, elsewhere in the loop. In both cases, the high temporal resolution of TMS has been important in giving precise information about the time course of neural information underlying action. Finally, TMS offers a conceptual as well as a methodological advance. Scientific knowledge of action systems has lagged behind knowledge of perceptual systems because it is easy to deliver a controlled input to perceptual systems, but harder to deliver a controlled input to the action system. TMS has allowed neuroscientists to activate or inactivate the brain's action systems artificially. This has provided key insights into normal motor function.

Acknowledgments

This work has been supported by EC grants ROBOT-CUB, NEUROBOTICS, and CONTACT to L.F. and L.C. and by Italian Ministry of Education grants to L.F. S.S.-B. was supported by a fellowship from the Max Planck Society.

References

Ammon K, Gandevia SC (1990) Transcranial magnetic stimulation can influence the selection of motor programmes. *Journal of Neurology, Neurosurgery and Psychiatry 53*, 705–707.

Angel RW, Malenka RC (1982) Velocity-dependent suppression of cutaneous sensitivity during movement. *Experimental Neurology 77*, 266–274.

Aziz-Zadeh L, Iacoboni M, Zaidel E, Wilson S, Mazziotta J (2004) Left hemisphere motor facilitation in response to manual action sounds. *European Journal of Neuroscience 19*, 2609–2612.

Baldissera F, Cavallari P, Craighero L, Fadiga L (2001) Modulation of spinal excitability during observation of hand actions in humans. *European Journal of Neuroscience 13*, 190–194.

Barker AT, Jalinous R, Freeston IL (1985) Non-invasive magnetic stimulation of human motor cortex. *Lancet ii*, 1106–1107.

Bonnard M, Camus M, de Graaf M, Pailhous J (2003) Direct evidence for a binding between cognitive and motor functions in humans: a TMS study. *Journal of Cognitive Neuroscience 15*, 1207–1216.

Bonnard M, de Graaf J, Pailhous J (2004) Interactions between cognitive and sensorimotor functions in the motor cortex: evidence from the preparatory motor sets anticipating a perturbation. *Reviews in the Neurosciences 15*, 371–382.

Borroni P, Montagna M, Cerri G, Baldissera F (2005) Cyclic time course of motor excitability modulation during the observation of a cyclic hand movement. *Brain Research 1065*, 115–124.

Boussaoud D, Wise SP (1993) Primate frontal cortex: neuronal activity following attentional versus intentional cues. *Experimental Brain Research 95*, 15–27.

Brasil-Neto JP, Pascual-Leone A, Valls-Sole J, Cohen LG, Hallett M (1992) Focal transcranial magnetic stimulation and response bias in a forced-choice task. *Journal of Neurology, Neurosurgery and Psychiatry 55*, 964–966.

Bruce CJ, Golberg ME (1985) Primate frontal eye field I Single neurons discharging before saccades. *Journal of Neurophysiology 53*, 603–635.

Day BL, Rothwell JC, Thompson PD *et al.* (1989) Delay in the execution of voluntary movement by electrical of magnetic brain stimulation in intact man. *Brain 112*, 649–663.

Desmurget M, Epstein CM, Turner RS, Prablanc C, Alexander GE, Grafton ST (1999) Role of the posterior parietal cortex in updating reaching movements to a visual target. *Nature Neuroscience 2*, 563–567.

di Pellegrino G, Fadiga L, Fogassi L, Gallese V, Rizzolatti G (1992) Understanding motor events: a neurophysiological study. *Experimental Brain Research* 91, 176–180.

Donders FC (1868) On the speed of mental processes. [Translated by WG Koster, 1969.] *Acta Psychologica 30*, 412–431.

Ellaway PH, Prochazka A, Chan M, Gauthier MJ (2004) The sense of movement elicited by transcranial magnetic stimulation in humans is due to sensory feedback. *Journal of Physiology 556*, 651–660.

Fadiga L, Fogassi L, Pavesi G, Rizzolatti G (1995) Motor facilitation during action observation: a magnetic stimulation study. *Journal of Neurophysiology 73*, 2608–2611.

Fadiga L, Craighero L, Buccino G, Rizzolatti G (2002) Speech listening specifically modulates the excitability of tongue muscles: a TMS study. *European Journal of Neuroscience 15*, 399–402.

Freund H-J, Jeannerod M, Hallett M, Leiguarda R (2005) *Higher-order motor disorders*. Oxford: Oxford University Press.

Gangitano M, Mottaghy FM, Pascual-Leone A (2001) Phase-specific modulation of cortical motor output during movement observation. *Neuroreport 12*, 1489–1492.

Gangitano M, Mottaghy FM, Pascual-Leone A (2004) Modulation of premotor mirror neuron activity during observation of unpredictable grasping movements. *European Journal of Neuroscience 20*, 2193–2202.

Gerloff C, Corwell B, Chen R, Hallett M, Cohen LG (1997) Stimulation over the human supplementary motor area interferes with the organization of future elements in complex motor sequences. *Brain 120*, 1587–1602.

Ghahramani Z, Wolpert DM, Jordan MI (1996) Generalization to local remappings of the visuomotor coordinate transformation. *Journal of Neuroscience 16*, 7085–7096.

Goldberg ME, Colby CL, Duhamel J-R (1990) The presentation of visuomotor space in the parietal lobe of the monkey. *Cold Spring Harbor Symposia on Quantitative Biology 55*, 729–239.

Haggard P (2005) Conscious intention and motor cognition. *Trends in Cognitive Sciences 9*, 290–295.

Haggard P, Eimer M (1999) On the relation between brain potentials and the awareness of voluntary movements. *Experimental Brain Research 126*, 128–133.

Haggard P, Magno E (1999) Localising awareness of action with transcranial magnetic stimulation. *Experimental Brain Research 127*, 102–107.

Haggard P, Whitford B (2004) Supplementary motor area provides an efferent signal for sensory suppression. *Cognitive Brain Research 19*, 52–58.

Haggard P, Clark S, Kalogeras J (2002) Voluntary action and conscious awareness. *Nature Neuroscience 5*, 382–385.

Johnson H, Haggard P (2005) Motor awareness without perceptual awareness. *Neuropsychologia 43*, 227–237.

Kennerly SW, Sakai K, Rushworth MFS (2004) Organization of action sequences and the role of the pre-SMA. *Journal of Neurophysiology 91*, 978–993.

Kohler E, Keysers C, Umilta MA, Fogassi L, Gallese V, Rizzolatti G (2002) Hearing sounds, understanding actions: action representation in mirror neurons. *Science 297*, 846–848.

Li CS, Mazzoni P, Andersen RA (1999) Effect of reversible inactivation of macaque lateral intraparietal area on visual and memory saccades. *Journal of Neurophysiology 81*, 1827–1837.

Maeda F, Kleiner-Fishman G, Pascual-Leone A (2002) Motor facilitation while observing hand actions: specificity of the effect and role of observer's orientation. *Journal of Neurophysiology 87*, 1329–1335.

Milner AD, Goodale MA (1993) Visual pathways to perception and action. *Progress in Brain Research 95*, 317–337.

Montagna M, Cerri G, Borroni P, Baldissera F (2005) Excitability changes in human corticospinal projections to muscles moving hand and fingers while viewing a reaching and grasping action. *European Journal of Neuroscience 22*, 1513–1520.

Murata A, Fadiga L, Fogassi L, Gallese V, Raos V, Rizzolatti G (1997) Object representation in the ventral premotor cortex (area F5) of the monkey. *Journal of Neurophysiology 78*, 2226–2230.

Müri RM, Rivaud S, Vermersch AI, Leger JM, Pierrot-Deseilligny C (1995) Effects of transcranial magnetic stimulation over the region of the supplementary motor area during sequences of memory-guided saccades. *Experimental Brain Research 104*, 163–166.

Müri RM, Vermersch A-I, Rivaud S, Gaymard B, Pierrot-Deseilligny C (1996) Effects of single-pulse transcranial magnetic stimulation over the prefrontal and posterior parietal cortices during memory-guided saccades in humans. *Journal of Neurophysiology 76*, 2102–2106.

Oliveri M, Bablioni C, Filippi MM, *et al.* (2003) Influence of the supplementary motor area on primary motor cortex excitability during movements triggered by neutral or emotionally unpleasant visual cues. *Experimental Brain Research 149*, 214–221.

Pascual-Leone A, Brasil-Neto JP, Valls-Sole J, Cohen LG, Hallett M (1992a) Simple reaction time to focal transcranial magnetic stimulation. *Brain 115*, 109–122.

Pascual-Leone A, Valls-Sole J, Wassermann EM, Brasil-Neto JP, Cohen LG, Hallett M (1992b) Effects of focal transcranial magnetic stimulation on simple reaction time to acoustic, visual and somatosensory stimuli. *Brain 115*, 1045–1059.

Patuzzo S, Fiaschi A, Manganotti P (2003) Modulation of motor cortex excitability in the left hemisphere during action observation: a single- and pairedpulse transcranial magnetic stimulation study of self- and non-self action observation. *Neuropsychologia 41*, 1272–1278.

Rizzolatti G, Craighero L (2004) The mirror-neuron system. *Annual Review of Neuroscience 27*, 169–192.

Rizzolatti G, Fadiga L (1998) Grasping objects and grasping action meanings: the dual role of monkey rostroventral premotor cortex (area F5). *Novartis Foundation Symposium 218*, 81–95.

Romani M, Cesari P, Urgesi C, Facchini S, Aglioti SM (2005) Motor facilitation of the human cortico-spinal system during observation of biomechanically impossible movements. *NeuroImage 26*, 755–763.

Ruff CC Blankenburg F, Bjoertomt O *et al.* (2006) Frontal influences on human retinotopic visual cortex revealed by concurrent TMS-fMRI and psychophysics. *Current Biology 16*, 1479–1488.

Rushworth MFS, Ellison A, Walsh V (2001) Complementary localization and lateralization of orienting and motor attention. *Nature Neuroscience 4*, 656–660.

Schluter ND, Rushworth MFS, Passingham RE, Mills KR (1998) Temporary interference in human lateral premotor cortex suggests dominance for the selection of movements. *Brain 121*, 785–799.

Schütz-Bosbach S, Mancini B, Aglioti SM, Haggard P (2006). Self and other in the human motor system. *Current Biology 16*, 1830–1834.

Sternberg S (1969) Memory scanning: mental processes revealed by reaction time experiments. *American Scientist 57*, 421–457.

Strafella AP, Paus T (2000) Modulation of cortical excitability during action observation: a transcranial magnetic stimulation study. *Neuroreport 11*, 2289–2292.

Stuss DT, Knight RT (2002) *Principles of frontal lobe function*. New York: Oxford University Press.

Theoret H, Halligan E, Kobayashi M, Fregni F, Tager-Flusberg H, Pascual-Leone A (2005) Impaired motor facilitation during action observation in individuals with autism spectrum disorder. *Current Biology 15*, R84–85.

Urgesi C, Candidi M, Fabbro F, Romani M, Aglioti SM (2006) Motor facilitation during action observation: topographic mapping of the target muscle and influence of the onlooker's posture. *European Journal of Neuroscience 23*, 2522–2530.

Voss M, Ingram JN, Haggard P, Wolpert DM (2006) Sensorimotor attenuation by central motor command signals in the absence of movement. *Nature Neuroscience 9*, 26–27.

Wolpert DM (1997) Computational approaches to motor control. *Trends in Cognitive Sciences 1*, 209–216.

Woolsey CN, Settlage PH, Meyer DR, Sencer W, Pinto Hamuy T, Travis AM (1952) Patterns of localization in precentral and 'supplementary' motor areas and their relation to the concept of a premotor area. *Research Publications – Association for Nervous and Mental Disease 30*, 238–264.

CHAPTER 31

Investigating language organization with TMS

Joseph T. Devlin and Kate E. Watkins

Summary – Since its invention in 1985, TMS has become an increasingly important tool for investigating the neurological basis of language. Here we review the brief history of language studies that span a range of TMS methodologies with the aim of highlighting both the novel insights as well as areas of controversy. When used in its 'virtual lesion' mode, TMS offers the spatial precision to explore causal relations between focal brain regions and specific language functions such as the role of the different subregions of the left inferior frontal gyrus, the relationship between premorbid language organization and susceptibility to unilateral lesions, and the contribution of both left and right hemisphere language areas to recovery following aphasic brain injury. When TMS is used to investigate functional connectivity, it demonstrates a close link between action words and motor programs, it suggests a potential evolutionary link between hand gestures and language, and it demonstrates a clear linking between speech perception and those parts of the motor system used to produce speech. Finally, TMS even offers the potential for enhancing recovery processes and aiding rehabilitation. We conclude by highlighting future directions in language research that are likely to benefit from further developments in TMS.

Introduction

The earliest insights into the neurological basis of language come from the seminal work of Pierre Paul Broca and Carl Wernicke, who first recognized the relation between left hemisphere brain regions and language functions. Broca examined the patient Leborgne who had a long-term deficit in producing speech despite intact language comprehension, unimpaired motor function of the mouth and tongue, and preserved general intelligence (Broca 1861). At autopsy, the lesion focus was located in the posterior portion of the third frontal convolution. Subsequent patients with similar lesions confirmed the original report and led Broca to postulate that the organ for the articulation of speech was located in the left inferior frontal gyrus, an area which now bears his name (Broca 1865). Wernicke (1874) studied patients with lesions affecting the posterior portion of the left superior temporal gyrus, who presented with speech comprehension deficits and fluent, if relatively empty, speech. These studies came to define the lesion-deficit model in which behavioral deficits are causally linked to lesions of specific brain areas, or in other words, damage to a region demonstrates that the region is necessary for the impaired functions. Modern functional neuroimaging techniques such as event-related potentials (ERPs), magnetoencephalograhy (MEG), positron emission tomography (PET), and functional magnetic resonance imaging (fMRI) complement neuropsychological studies by providing whole-brain information about regions engaged by a particular task, but unlike patient studies, they cannot demonstrate that a particular region is necessary for a specific function (Price and Friston 2002b).

Transcranial magnetic stimulation, in contrast, can be used to draw causal inferences, as the cortical disruption induced by stimulation can act like a 'virtual lesion' lasting from tens of milliseconds up to ~1 h, depending on the specific type of stimulation (Pascual-Leone *et al.* 2000). Moreover, TMS avoids some of the well-known difficulties of patient studies which limit their interpretation, including potential differences in premorbid ability, compensatory plasticity following the lesion, the large and varied extents of naturally occurring lesions, and damage to subjacent fibers-of-passage. By comparing stimulated with unstimulated trials, participants in TMS experiments act as their own controls, avoiding the potential confound of premorbid differences. In addition, there is insufficient time for functional reorganization to occur during single TMS events. Consequently, the results should not be substantially confounded by any recovery processes (Walsh and Cowey 1998). Finally, the induced disruption is generally more focal than naturally occurring lesions and does not affect deep white matter pathways. There are, of course, disadvantages to TMS, such as the need to choose appropriate stimulation parameters (single vs repetitive stimulation, intensity, timing, location, etc.), restricted access limited to surface structures only, and the fact that TMS produces both sound and somatosensory stimulation which can also influence behavior (Walsh and Rushworth 1999). Nonetheless, TMS offers a powerful methodology for investigating the anatomy of cortical language processing. It is not, however, a replacement for patient studies nor is it limited to lesion-deficit testing. Unlike patient studies, TMS can provide information regarding the time course of information processing within a region, it can demonstrate functional relations between cognitive systems, it can be used with neurological patients to test both normal and compensatory organization of brain functions, and it may even be useful for enhancing recovery following brain damage.

In this chapter we review these functions of TMS and their role in developing a more complete understanding of the neurological basis of language. Although the aim is to provide a broad overview of the current state of the art, we have not attempted to include every study. Instead we emphasize the novel contributions of TMS and use specific studies to highlight important methodological issues. The first section focuses on the 'virtual lesion' approach, and the second introduces TMS as a measure of functional connectivity. The final section discusses TMS studies with neurological patients, both as a method for investigating the mechanisms of recovery and as a tool for enhancing recovery. We conclude by highlighting future directions in language research that are likely to benefit from further developments in TMS.

Virtual lesions

In language studies, by far the most commonly used form of TMS is to generate 'virtual patients' by temporarily disrupting cortical processing. TMS uses a rapidly changing current within a conducting coil to induce a strong, but relatively focal, magnetic field. When placed on the scalp, the magnetic field induces a physiological response (i.e. depolarization and/or spiking) in the underlying neural tissue (Jahanshahi and Rothwell 2000; Pascual-Leone *et al.* 2000). This introduces transient noise into the neural computation being performed and, when the tissue is required to perform a task, this can lead to longer reaction times (RTs) or even errors (Pascual-Leone *et al.* 1999; Walsh and Rushworth 1999), suggesting a causal link between the stimulated region and the behavioral task being performed.

To successfully induce a 'virtual lesion' three criteria must be met: (i) the coil must be positioned correctly to stimulate the desired site, (ii) the stimulation must be delivered at the right time, and (iii) it must be sufficiently intense to introduce noise into the regional information processing. The first hurdle is to successfully target a cortical region for stimulation. Most studies locate their stimulation site based on functional, anatomical, or heuristic criteria. One approach is to empirically identify the stimulation site using a 'functional localizer' task over various sites within a predefined region. When an effect of TMS is observed, the site is marked and used for subsequent testing (e.g. Gough *et al.* 2005). Alternatively, functional imaging can be used to identify activated brain regions in individuals which can then be targeted with TMS (e.g. Kohler *et al.* 2004). Another method uses high-resolution structural images (e.g. T1-weighted MRI scans) to position the coil based on the subject's underlying cortical anatomy (e.g. Sakai *et al.* 2002; Nixon *et al.* 2004). Finally, a heuristic approach can also be taken in which population-based estimates of the underlying anatomy are used. These include positioning the coil based on the International 10–20 system commonly used for electrode placement in ERP studies (e.g. Drager *et al.* 2004) or using scalp-based measurements from clear external landmarks such as the vertex, inion, or canthus–tragus line (e.g. Stewart *et al.* 2001a). The second hurdle is to correctly target the stimulation in time. Most studies avoid this problem by using trains of repetitive (r)TMS which last for hundreds of milliseconds – long enough to guarantee that stimulation disrupts processing even if the precise timing of the processing is unknown. It is possible, however, to use single pulses of TMS and investigate both the spatial and temporal aspects of cortical information processing (Walsh and Pascual-Leone 2003). Finally, even if the stimulation is targeted correctly in both space and time, it stills needs to be strong enough to induce a physiological response and disrupt processing. This may be the most arbitrary decision because there is no systematic method for determining an appropriate intensity. Some studies base the intensity on an individual's motor threshold (MT), which is the level of stimulator output necessary to elicit a motor-evoked potential (MEP) with a single pulse over motor cortex 50% of the time in a hand muscle as recorded with electromyography (e.g. Wassermann *et al.* 1999). There is, however, no systematic relationship between the threshold needed to evoke an MEP and the threshold needed to evoke flashes of light (phosphenes) with visual cortex stimulation, suggesting that outside of the motor system,

MTs may not be particularly appropriate (Stewart *et al.* 2001c). In the end, most studies simply choose a stimulation intensity that appears to work, presumably based on pilot testing. Because these choices vary considerably from study to study, we include these details (parenthetically) when discussing each study to highlight the range of options available.

In contrast to patient studies, which typically measure behavioral deficits in terms of reduced accuracies, the virtual lesions induced by TMS generally manifest as changes in reaction times (RTs), rather than accuracy. TMS does not inactivate a region in the same way that a lesion does – instead it introduces random transient neural firing (i.e. 'noise') into the computation being performed. In most cases, this leads to an increase in RTs rather than actual errors. Presumably this is because the information that remains intact in the neural circuit is sufficient to overcome the noise, but this process requires additional time and manifests as longer RTs. Consequently, a TMS-induced increase in RT can indicate that the stimulated region is 'necessary' for performing a task, even in the absence of errors. For example, Stewart *et al.* (2001a) used short (600 ms) trains of rTMS (10 Hz at 75% maximum stimulator output) to stimulate either left or right posterior infero-temporal cortex in a set of healthy volunteers while they named pictures, read words, or named patches of color aloud. TMS selectively increased RTs in the picture-naming task relative to trials without stimulation but did not affect either word reading or color naming, suggesting that the stimulated region was necessary for picture naming.

Behavioral consequences of TMS do not always manifest as RT increases, however; sometimes TMS can produce facilitation instead. In another picture-naming study, Mottaghy *et al.* (1999) observed the opposite behavioral pattern to Stewart *et al.* (2001a). They found that stimulation (20 Hz rTMS for 2 s at 55% maximum output) of Wernicke's area *decreased* naming latencies without affecting accuracies. The fact that neither RTs nor accuracies were affected by stimulation of Broca's area or visual cortex led the authors to claim that stimulation at a low intensity 'pre-activated' Wernicke's area, demonstrating its involvement in picture naming.

In other words, stimulation may potentiate a region, effectively priming it to respond. Although speculative, this explanation for a TMS-induced facilitation effect may explain why several other studies have also reported facilitation of picture naming for Wernicke area stimulation (Topper *et al.* 1998; Wassermann *et al.* 1999; Andoh *et al.* 2006).

Sometimes the pattern of results is even more subtle, and the interpretation depends on having multiple baselines for comparison. For instance, Dräger *et al.* (2004) were able to identify specific and nonspecific effects of TMS by including multiple control conditions in their design. In their experiment, participants performed a picture–word verification task either before or after a 10 min session of 1 Hz rTMS delivered to one of five locations (Broca's and Wernicke's areas, their right hemisphere homologs, or midline occipital cortex). One-hertz rTMS reduces the excitability of the stimulated region with the effects lasting roughly as long as the stimulation period. Sham TMS, i.e. stimulation with the coil turned 90° away from the scalp, was included as a further control condition, though it should be noted that this does not typically involve the same somatosensation as real TMS. After 10 min of rTMS, RTs were consistently faster than those before stimulation, regardless of stimulation site, suggesting a nonspecific arousal effect, possibly due to intersensory facilitation (Figure 31.1a). This type of nonspecific TMS effect is fairly common (Flitman *et al.* 1998; Wassermann *et al.* 1999; Shapiro *et al.* 2001; Kohler *et al.* 2004; Nixon *et al.* 2004) and can be problematic if it hides region-specific effects. For instance, when the regional TMS effects were recalculated relative to the mean RT across regions following stimulation (rather than relative to no stimulation), then an inhibitory effect of stimulation in Wernicke's area and a further facilitation effect in Broca's area were revealed (Figure 31.1b). It is interesting to note that, had there been only a single control condition (e.g. midline occipital or sham stimulation), it would have been difficult to separate these regional effects from the overall facilitation effect.

In summary, a region-specific inhibitory effect of TMS demonstrates that the area makes a necessary contribution to the task, even if stimulation only affects RTs and not accuracies. Behavioral facilitation, on the other hand, is more difficult to interpret because there is no convincing physiological interpretation of the effect. Because stimulation can sometimes increase arousal via intersensory facilitation, detecting region-specific effects may require additional control conditions beyond a 'no-stimulation' condition. In practice, inhibitory effects may be particularly small relative to no stimulation but much clearer relative to stimulation of a control site or relative to sham stimulation. Consequently, the choice of baseline conditions is critical for detecting effects of interest.

Speech arrest

The earliest use of TMS to investigate language was by Pascual-Leone *et al.* (1991) who induced speech arrest in pre-surgical epilepsy patients in order to determine whether TMS could be used as a noninvasive alternative to intracarotid amobarbital testing (IAT) (Wada and Rasmussen 1960). Trains (10 s) of rTMS were delivered at rates of 8, 16, or 25 Hz over 15 different scalp positions surrounding peri-sylvian cortex in each hemisphere. The stimulation sites were defined by the International 10–20 electrode system and patients were asked to count aloud from 'one'. After 4–6 s of stimulation over left inferior frontal cortex, a reproducible speech arrest was observed in each of the six patients. One patient noted, 'I could move my mouth and I knew what I wanted to say, but I could not get the numbers to my mouth' (Pascual-Leone *et al.* 1991, p. 699). In contrast, no speech arrest was seen during any right hemisphere stimulation. IAT with these same patients revealed left hemisphere language dominance in all six, suggesting that the TMS-induced speech arrest offered a noninvasive alternative for determining language dominance.

Subsequent studies, however, have called into question the usefulness of rTMS in pre-surgical planning. Jennum *et al.* (1994) were only able to induce complete speech arrest in 14/21 patients, although when slowed speech was included there was still a 95% concordance between the rTMS and IAT findings. In contrast, Michelucci *et al.* (1994) produced speech arrest in only 7/14 subjects which, they argued, called into

(a) Non-specific facilitation effect of TMS

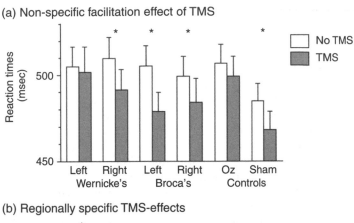

(b) Regionally specific TMS-effects

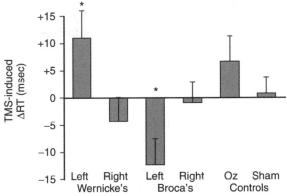

Fig. 31.1 Generic and specific effects of stimulation on reaction times (RTs) in a picture–word verification task. (a) Across regions TMS had a facilitory effect on RTs relative to no stimulation which was even present for sham stimulation indicating a nonspecific arousal effect. (b) Relative to mean stimulation, however, two regionally specific effects of TMS become apparent. Stimulation of Wernicke's area significantly slowed performance, whereas stimulation of Broca's area decreased RTs even further. Adapted from Dräger *et al.* (2004), Specific and non-specific effects of transcranial magnetic stimulation on picture-word verification. *European Journal of Neuroscience.* Reproduced with permission from Blackwell Publishing.

question the reliability of rTMS for determining language dominance. Epstein *et al.* (1996) hypothesized that some of this inter-study variability may have been due to the specific stimulation parameters chosen, and systematically investigated the effects of both intensity and rate of stimulation on speech arrest in normals. As expected, higher intensities led to stronger speech arrest effects, but surprisingly it was the lower rates of stimulation (4–8 Hz) that were more reliable at inducing speech arrest than those used in previous studies (16–32 Hz; Jennum *et al.* 1994; Michelucci *et al.* 1994;

Pascual-Leone *et al.* 1991). Higher frequencies led to prominent facial and laryngeal muscle contractions and significantly increased the discomfort or pain associated with stimulation, making speech arrest more difficult to determine. Stimulation at 4 Hz, on the other hand, not only consistently disrupted highly overlearned speech such as counting, but also interfered with reading aloud and spontaneous speech (Epstein *et al.* 1999). Consequently, this paradigm was used to test the reliability of rTMS relative to IAT in 16 presurgical epilepsy patients (Epstein *et al.* 2000). rTMS indicated left hemisphere language dominance for 12 patients and right dominance for the remaining four, whereas IAT indicated that all 16 patients were left dominant. Despite the significant correlation between IAT and rTMS ($r = 0.57$, $P < 0.05$), rTMS overestimated right hemisphere involvement in a considerable proportion of patients. Moreover, the IAT findings were a better predictor of postoperative language difficulties than the rTMS. As the purpose of determining language dominance is to minimize the impact of the surgical resection on language abilities, rTMS appears to be less reliable than IAT.

Nonetheless, there is still reason to be optimistic regarding the possibility of using TMS as a reliable alternative to IAT. IAT affects the functioning over a large region of one hemisphere for several minutes, whereas TMS disruption is far more focal and transient, particularly with rTMS at rates of >1 Hz. Studies by Stewart et al. (2001b) and Aziz-Zadeh et al. (2005) showed that speech arrest can be induced from two different sites within the inferior frontal cortex. At the more posterior site, stimulation of either the left or right hemisphere induced speech arrest, although the effect was typically stronger on the left. In addition, stimulation at these sites evoked a clear facial muscle response as measured with electromyography (EMG). In contrast, only left hemisphere stimulation of the more anterior site led to speech arrest and it did not evoke an EMG response. The authors hypothesized that the posterior site may correspond to the ventral limb of the precentral gyrus where both motor and premotor regions enervate the mouth and jaw. Consequently, stimulation of either hemisphere can induce speech arrest by interfering with the motor output of speech. The more anterior site may correspond to prefrontal cortex (i.e. Broca's area in the left hemisphere) where stimulation would be expected to interfere with language production per se. These findings may help to explain the relatively high concordance between rTMS and IAT in some studies (Pascual-Leone et al. 1991; Jennum et al. 1994) but not others (Michelucci et al. 1994; Epstein et al. 2000). Further work is clearly warranted to determine whether anterior inferior frontal stimulation corresponds more closely with IAT and can accurately predict postoperative language

deficits following surgical intervention in intractable epilepsy.

Semantic and phonological processing

Broca's area is not limited to speech production – in fact, it is part of a larger region that plays an important role in processing meaning, sounds and syntax as well as many nonlinguistic functions (Zurif et al. 1972; Levy and Anderson 2002; Hagoort et al. 2004). Functional imaging studies have suggested that within the left inferior frontal gyrus (LIFG), the site of Broca's area, there is a rostro-caudal division of labor for semantic and phonological processing that was not apparent from previous neuropsychological studies (Buckner et al. 1995; Fiez 1997). This claim has received considerable support from recent TMS studies that not only confirm this division of labor, but also clarify the specific regional contributions to semantic and phonological processing. Devlin et al. (2003) investigated whether stimulation (10 Hz for 300 ms at 110% MT) of rostral LIFG interfered with simple semantic decisions such as deciding whether a visually presented word referred to a man-made (e.g. 'kennel') or natural object (e.g. 'dog'). Relative to no stimulation, TMS significantly increased RTs in the semantic task, but not when participants focused on visual properties of the presented words. Similarly, Kohler et al. (2004) used fMRI-guided rTMS (7 Hz for 600 ms at 100% MT) to stimulate rostral LIFG and showed that decisions based on their meaning were significantly slowed by TMS, consistent with the claim that rostral LIFG is necessary for semantic processing. The other half of this division of labor was investigated by Nixon et al. (2004) who examined whether stimulation of caudal LIFG interfered with a phonological working memory task. Participants saw a word on a computer screen (e.g. 'knees') and then held it in memory during a 1–2 s delay before deciding whether it sounded the same as a subsequently presented nonword (e.g. 'neaze'). rTMS (10 Hz for 500 ms at 120% MT) during the delay period selectively increased the error rate during the phonological task, but not of a comparable visual working memory task. Aziz-Zadeh et al. (2005) also included a phonological task in their study of speech arrest by investigating 'covert speech arrest', which was measured by participants silently reading a visually presented word and counting its syllables. Once again, rTMS (5 Hz for 2.4 s at 115% MT) over caudal LIFG increased RTs relative to unstimulated trials, consistent with a role in phonological processing. Taken together these studies significantly extend the previous neuroimaging results by demonstrating that rostral LIFG is necessary for semantic processing while caudal LIFG is necessary for phonological processing.

A number of functional imaging studies, however, have shown that phonological and semantic tasks commonly engage both rostral and caudal LIFG (Barde and Thompson-Schill 2002; Gold and Buckner 2002; Devlin et al. 2003), raising the possibility that both regions are necessary for both types of processing. In other words, LIFG may act as a single functional region that is required for semantic and phonological processing with regional shifts in the peak activation, or there may be subregions within LIFG specialized for semantic and

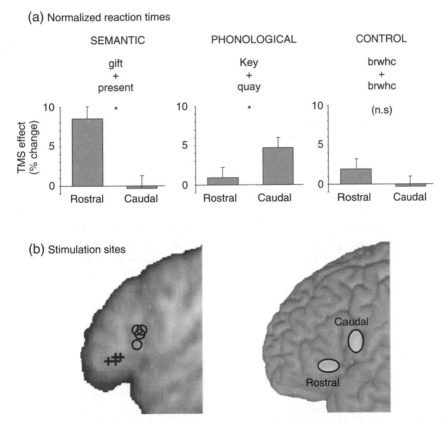

(a) Normalized reaction times

SEMANTIC   PHONOLOGICAL   CONTROL

Fig. 31.2 Effects of stimulation on rostral and caudal left inferior frontal gyrus. (a) The bar plots show the mean normalized TMS effects as percentage change from the non-TMS baseline during synonym judgments (*left*), homophone judgments (*middle*), and visual matching (*right*). Error bars indicate the standard error of the mean. (b) Location of stimulation sites for four participants on their mean structural image with rostral locations marked with crosses and caudal locations marked with circles. Adjacent is a three-dimensional rendering with the stimulation sites shown as ovals representing the spatial 85% confidence interval. Stimulation sites were on average 2.5 cm apart on the cortical surface. Adapted from Gough et al. (2005), by permission of the Society for Neuroscience.

phonological processing but co-activated due to incidental processing (MacLeod 1991; Price *et al.* 1996). The TMS results do not distinguish between these possibilities as each of the single dissociations would be predicted by both accounts. Consequently, Gough *et al.* (2005) designed a TMS experiment to test for a double dissociation between semantic and phonological processing in LIFG. Participants saw two letter strings presented simultaneously on a computer screen and had to decide whether they meant the same (e.g. 'idea–notion'), sounded the same (e.g. 'nose–knows'), or looked the same (e.g. 'fwtsp–fwtsp'). Relative to no stimulation, TMS of rostral LIFG selectively increased response latencies when participants focused on the meaning of simultaneously presented words (i.e. semantics) but not when they focused on the sound pattern of the words (i.e. phonology). In contrast, the opposite dissociation was observed with stimulation of caudal LIFG, where stimulation selectively interfered with the phonological, but not the semantic, task (Figure 31.2). Neither site of stimulation affected the RTs in the visual control task. In other words, the authors demonstrated a functional double dissociation for semantic and phonological processing within LIFG in sites separated by <3 cm. Although this double dissociation was first suggested by functional imaging, it required the spatial precision of TMS to independently disrupt the regions and clarify their distinct contributions to word processing.

Reading

Despite considerable interest in the neurobiology of reading and reading disorders in the functional imaging literature, there has been relatively little TMS work in this area. Unsurprisingly, visual cortex stimulation interferes with reading (Amassian 1989; Beckers and Homberg 1991; Lavidor and Walsh 2003), but Lavidor and Walsh (2003) further demonstrated that stimulating (8 Hz for 500 ms at 65% maximum output) right, but not left, visual cortex abolished the normal orthographic neighborhood effect. Typically, words with large neighborhood size (defined as the number of same-length words which differ by only a single letter; Coltheart *et al.* 1977) such as 'grape' have shorter RTs than

those with a low neighborhood size such as 'group'. This effect is most robust when the words are presented to the left visual hemifield (Lavidor and Walsh 2003). The fact that right visual cortex stimulation disrupted the orthographic neighborhood effect was seen as evidence that the ability to respond to the initial letters of a word relies on right visual cortex, supporting a split-fovea model of reading.

Other studies have investigated the contribution to reading of nonlinguistic faculties such as eye movements, spatial attention, or visual motion processing. For example, Leff *et al.* (2001) demonstrated that when stimulation (10 Hz for 1.5 s at 110% MT) of the left frontal eye field (FEF) preceded presentation of a list of visual words, the initial saccade was slowed relative to no stimulation. Stimulation of left posterior parietal cortex (PPC), however, only slowed eye movements between words. The authors suggested that FEF is crucial for planning reading saccades whereas PPC is necessary for guiding them. Braet and Humphreys (2006) also investigated PPC involvement in reading, although they focused on the right hemisphere as patients with damage to this area sometimes present with reading impairments subsequent to spatial attention deficits (e.g. Riddoch *et al.* 1990). They found that words presented in mIxEd cAsE showed the most robust disruption with right PPC stimulation (20 Hz for 150 ms at 90% MT), suggesting that the region mediates attentionally demanding visual word recognition. Finally, Liederman *et al.* (2003) examined the role of the extrastriate motion area V5/MT+ in reading in order to determine whether it may contribute to rapid grapheme-to-phoneme conversions. One-hertz stimulation (7.5 min at 70% maximum) successfully disrupted moving after-images, consistent with V5/MT+ stimulation, but had very little effect on the reading tasks. Following stimulation, the performance on the two 'phonological' tasks was unimpaired. The only task that showed an effect of stimulation was nonword reading, but the errors were not primarily phonological. Indeed, the authors suggest that V5/MT+'s involvement in reading may be to bias spatial attention via anatomical links to PPC or in stabilizing the visual image.

Finally, some studies have begun to explore the time course of visual word recognition using single-pulse TMS. Stimulation of visual cortex occurring between 40 and 120 ms after the onset of the visual stimulus interferes with letter recognition (Amassian 1989; Beckers and Homberg 1991) whereas pulses over LIFG at 250 ms post-stimulus onset slowed responses in a reading-for-meaning task (Devlin *et al.* 2003). Such 'neurochronometric TMS' studies are still relatively rare in language research, but offer considerable potential for addressing central questions regarding the nature of neural information processing. For instance, one theory of reading posits that the left posterior fusiform cortex is the first stage in a serially ordered reading system (Dehaene *et al.* 2004); others suggest that it is one part of a highly interactive system (Price and Devlin 2003; Devlin *et al.* 2006). By identifying the critical time windows for processing in specific cortical regions, the flow of information (both forwards and backwards, e.g. Pascual-Leone and Walsh 2001) can be clarified and used to address these fundamental questions. TMS offers two advantages relative to ERP or MEG in this regard. It is not only anatomically more precise, but the temporal information is also potentially more accurate. This is because the synchronized neuronal activity necessary to produce an ERP or MEG signal may be delayed relative to the physiological source (Walsh and Cowey 2000). If so, systematic comparisons between the various neurophsyiological measures (electrical and magnetic) of the time course of information processing will be necessary to determine the temporal signatures underlying language processes.

Syntax

Patient studies demonstrate a clear link between grammatical impairments and left prefrontal lesions (Zurif *et al.* 1972; Shapiro and Caramazza 2003) – a relationship which TMS studies have also begun to explore. Sakai *et al.* (2002) investigated whether, and when, Broca's area was involved in syntactic processing using a sentence validation task. Participants viewed sentences and had to identify each as either correct, grammatically incorrect, or semantically incorrect. All sentences used a simple noun phrase–verb phrase (NP–VP) construction, with the VP

appearing 200 ms after the NP. TMS (two pulses at 42% maximum output, 2 ms apart) to Broca's area was delivered either 0, 150, or 350 ms after the VP onset. Relative to sham stimulation, TMS selectively facilitated RTs for syntactic, but not semantic, decisions and the effect was specific to the 150 ms time window, which the authors interpreted as strong evidence that Broca's area is causally involved in syntactic processing.

What is particularly intriguing about this study is the timing of the TMS effect: 150 ms after the onset of the VP. Electrophysiological studies often report a waveform called the early left anterior negativity (ELAN) which is sensitive to syntactic processes and occurs in roughly that same time window (150–250 ms) (Friederici 2002). The ELAN effect, however, is generally associated with morpho-syntactic violations such as disagreement between a noun and verb form (e.g. 'he lie' rather than 'he lies'), whereas in Sakai *et al.*'s (2002) study the syntactic violation was a disturbance of verb argument structure (e.g. 'someone lies snow'). Here, 'lies' does not take a direct object because one cannot 'lie snow' and this type of syntactic violation is more typically associated with an N400–P600 complex, occurring much later (Friederici and Kotz 2003). In other words, this study may be an example of a timing mismatch between TMS and ERP with TMS revealing effects that occur considerably before they are seen in ERP. On the other hand, it is possible that TMS primed the region *before* it was required for the syntactic judgments. Clearly further studies will be necessary to determine whether there really is a systematic timing difference between TMS and ERP or whether apparent discrepancies are in fact artifacts of particular types of stimulation.

The other aspect of grammar that has been investigated with TMS is that of grammatical class. Both neuropsychological (Caramazza and Hillis 1991) and electrophysiological studies (Federmeier *et al.* 2000) suggest that there may be different neural substrates for processing nouns and verbs. Cappa *et al.* (2002) postulated that verb specificity may be due to the close relation between verbs and actions, and used TMS to investigate the role of left dorso-lateral prefrontal cortex (DLPFC) in action naming. A set of Italian-speaking participants were shown pictures of common objects and asked to name either the object (e.g. 'telefono' (a telephone)) or the associated action ('telefonare' (to telephone)). Repetitive TMS (20 Hz for 500 ms at 90% MT) of left DLPFC decreased naming latencies for verbs relative to right DLPFC and sham stimulation. In contrast, the latencies for object naming were unaffected. The authors suggested that verbs may be preferentially impaired by left frontal lesions because damage to DLPFC affects action observation and representation which are more tightly linked with verbs than with nouns.

The work of Shapiro *et al.* (2001), however, calls this interpretation into question. In their study, participants were asked to inflect nouns and verbs (e.g. 'song' → 'songs' or 'sing' → 'sings') either before or after 10 min of 1 Hz stimulation (100% MT) over left DLPFC. RTs were significantly slowed for verbs, but not nouns. In order to determine whether this effect was due to the action-related meaning of the verbs, a second experiment used pseudo-words (e.g. 'flonk') treated as either nouns or verbs. Because pseudo-words do not have any associated meaning, the authors reasoned that TMS would only affect RTs if the region was important for processing the grammatical class of verbs rather than words with action-related meanings. Once again, DLPFC stimulation selectively slowed RTs only in the verb condition – a finding interpreted as evidence for a neuroanatomical basis for grammatical categories *per se* rather than a byproduct of the differences in meaning between nouns and verbs.

These findings provide strong evidence that left DLPFC is important for processing verbs, but its precise role remains unclear. It is theoretically possible that a particular patch of left DLPFC is dedicated to processing the grammatical class of verbs but this seems unlikely given the wide variety of nonlinguistic tasks that also engage the region (Duncan and Owen 2000; Petrides 2000). In contrast, the relation between verbs and actions is appealing, if for no other reason than verbs imply acts via their thematic roles, even when the action is either unspecified ('flonks') or not particularly active ('sleeps'). That is, 'he flonks' suggests someone (an agent) who is flonking (an action) whereas 'the flonks' suggests multiple somethings (no action).

Additional evidence for this relation between verbs and actions comes from studies highlighting functional connections between language and the motor system (Aziz-Zadeh *et al.* 2004; Buccino *et al.* 2005; Pulvermüller *et al.* 2005), reviewed below. Nonetheless, there remain at least two distinct explanations for left DLPFC involvement in verb processing: (i) there may be functional specialization within the region for processing the grammatical class of verbs or (ii) verbs may be preferentially linked to actions in a manner not found for other parts of speech. If the former, it may be possible to use TMS to selectively disrupt verb processing without affecting any other DLPFC-mediated functions. For the latter, stimulation might be expected to have a parametric effect related to the 'activeness' of the action with pseudo-words showing relatively smaller effects than verbs such as 'kick' or 'throw'. Either way, TMS offers a potential method for systematically examining these possibilities.

Motor excitability, speech, and language

TMS has its origins and earliest uses in clinical neurophysiology, providing measurements that test the integrity of the motor system, in particular corticospinal processing. Single pulses of TMS are applied over motor cortex to elicit changes in the electromyographic (EMG) recording from the muscle of interest. In the studies reviewed here, the same methods are used to examine the effects of perceptual and cognitive tasks on the motor system. Consequently, it is worth briefly explaining the method.

The most commonly used dependent measure in motor cortex studies is the MEP measured using EMG by electrodes placed peripherally over the muscle of interest. When used to probe muscles of interest to speech and language, such as face and tongue muscles, several specific issues arise. In the hand muscle, the latency of the MEP response is ~20 ms post stimulation whereas in face and tongue muscles these latencies are considerably shorter (~10 ms) and can be obscured by an EMG artifact caused by the discharge of the TMS coil (see similar artifacts and their solutions in EEG/TMS studies, Chapter 37, this volume). To circumvent this problem one can use special circuitry to clamp and hold the EMG signal during stimulation (e.g. Watkins *et al.* 2003). Also problematic for EMG recordings from face muscles is that stimulation over the scalp can induce direct peripheral stimulation of head muscles. These responses are ipsilateral to the side of stimulation but if recording from a central muscle, like the tongue, or a circular muscle, like the lips, they can appear in the EMG recording. In this case they are distinguished from centrally mediated responses by their even shorter latency (i.e. <10 ms). Finally, when recording MEPs, one typically wishes to record from a completely relaxed muscle, since MEP size increases with muscle contraction. The muscles of the face, however, are rarely at rest. Also, the stimulator-output intensity required to elicit an MEP in face muscles is often higher than that required to elicit an MEP in hand muscles. This may reflect the anatomical location of the face area of motor cortex, which is more lateral and ventral to the hand area, and lies beneath thicker skull or deeper inside the anterior lip of the central sulcus. For these two reasons, therefore, subjects may be trained to produce a constant level of contraction of the muscle of interest using visual biofeedback, allowing normalization for levels of contraction across conditions and the use of lower intensities of stimulation. Under such conditions, it is necessary also to ensure that there is no effect across conditions on the baseline EMG activity recorded prior to the TMS pulse. One potentially useful side-effect of a voluntary contraction of the muscles is that it provides another dependent measure, namely the cortical silent period (Lo and Fook-Chong 2004), to which we will return shortly.

In most of the studies reviewed below, MTs are initially determined for the area stimulated. Variability in MTs might reflect relative levels of excitability of the motor systems, but across individuals, it most likely reflects variability in skull thickness. Some consensus has been reached in MT measurement: the level of stimulator output required to elicit at least 5/10 MEPs of >50 μV amplitude. In order to guarantee an MEP response on every stimulation trial,

researchers then choose an arbitrary suprathreshold level of stimulation (e.g. 120% MT). Others determine the stimulator output required to elicit an MEP of a certain amplitude (e.g. 0.3–1 mV). Both serve to normalize stimulator output levels across subjects. For analytical purposes, the dependent measure is MEP size, measured either in peak-to-peak amplitude or by the area under the curve in a rectified EMG trace. These values are then normalized usually with reference to some baseline condition expressed as a proportion or as Z-scores.

Finally, it is important to consider what is being measured by the MEP variable. Researchers consider increased MEP size to reflect increased excitability of the cortex via cortical inputs, but these changes could also arise via facilitation of neurons in the brain stem or spinal cord. A study by Strafella and Paus (2000) suggests that the facilitation of the MEP is due to changes in cortico-cortical inputs resulting in increased primary motor cortex excitability rather than subcortical excitability changes. They first demonstrated with single-pulse stimulation (set to elicit MEP amplitude of 0.5 mV at rest) over motor cortex that visual action observation increased MEP sizes in the muscle observed making the action. Then they used paired-pulse TMS where a subthreshold conditioning pulse (95% active MT) was followed by a suprathreshold test pulse (set to elicit MEP amplitude of 0.5 mV at rest). With a 3 ms interval between the conditioning and test pulses, they found reduced MEP sizes, attributed to intracortical inhibition. With a 12 ms interval, on the other hand, they observed enhanced MEPs due to intracortical facilitation. Critically, during action observation, both inhibition and facilitation effects in the paired-pulse stimulations were reduced, suggesting that the change in excitability due to action observation had a cortical, rather than subcortical, origin. In another visual observation of actions study, the excitability of spinal-cord motor neurons was measured directly and found to show the opposite pattern of facilitation seen at the cortex, suggesting that the spinal response may serve to prevent overt action production during perception (Baldissera *et al.* 2001). In view of these findings, and corroboration from other imaging modalities, most researchers assume that the changes in MEP reflect changes in cortical excitability.

Speech production and the motor system

As in the 'virtual lesion' studies of language, the earliest work examining cortical excitability during speech processing was aimed at assessing lateralization of function (Tokimura et al. 1996). MEPs in a hand muscle were measured in response to single pulses of TMS over the hand area of the contralateral motor cortex while subjects read aloud, read silently, spoke spontaneously or made nonspeech vocal sounds. The MEP size was facilitated equally for left and right hemisphere stimulations during spontaneous speech, whereas this effect was lateralized to the dominant hemisphere during reading aloud. In contrast, the silent reading condition and the nonspeech sound production did not result in changes in excitability. In other words, the authors demonstrated a functional connection between speech output and the hand area of the left motor cortex.

Recent studies confirm and refine this finding. The basic laterality effect has been replicated in several studies (Seyal et al. 1999; Lo et al. 2003; Meister et al. 2003; but see Floel et al. 2003) demonstrating that the functional link between speech production and the hand motor area is left lateralized. In addition, Meister et al. (2003) found evidence that the increased motor excitability during reading aloud was restricted to the hand area as MEPs from the leg area were unchanged by the task. In contrast, Lo et al. (2003) reported that reading aloud also facilitated MEPs recorded from leg muscles. Thus, the anatomical specificity of this function link remains unclear; it may be specific to the hand area or it may extend to the motor cortex more generally.

As mentioned above, MEPs elicited in a contracted muscle provide another useful dependent measure, namely the cortical silent period. Lo and Fook-Chong (2004) used single-pulse TMS during maximal voluntary contraction of the hand muscles to assess ipsilateral and contralateral silent periods during singing and reading aloud. Increased interhemispheric inhibition (longer ipsilateral silent period) was seen following right hemisphere stimulation during singing. Decreased intrahemispheric inhibition (shorter contralateral silent period) was seen following stimulation of either the right or the left motor cortex during singing and following right hemisphere stimulation during reading. This suggests that during singing, the hand area of right motor cortex increases its inhibitory effect on the hand area of left motor cortex and both hemispheres show decreased local inhibition. During reading, local inhibition only is decreased in the right hemisphere.

It seems, then, that speech production increases motor excitability not only in the face area of the left hemisphere, but also in the hand area. This may be evidence for a functional link between the hand area and language, either due to the irrepressible use of hand gestures during speech or reflecting an evolutionary link in the development of speech and language through hand gestures (Rizzolatti and Arbib 1998; Corballis 2003). Then again, it is less clear why there should be a functional connection between language and the leg motor area (Lo et al. 2003). One possibility is that simply producing speech increases the excitability of the entire motor cortex, possibly extending even as far as the leg area, the most distal motor representations from those of the articulators. Clearly, this is an area where further studies will be necessary to determine more precisely the anatomical boundaries of enhanced motor excitability as well as their functional significance.

Speech perception and the motor system

It may seem obvious that speech *production* affects motor system excitability, but it is potentially surprising that perception should also do this. Initial studies focused on visual perception of actions, aiming to providing evidence for an action observation–execution matching mechanism in the human brain (e.g. Strafella and Paus 2000; Gangitano et al. 2004) akin to the 'mirror neuron' system in the macaque brain (di Pellegrino et al. 1992; Gallese et al. 1996). As in the case of mirror neurons recorded from macaque premotor cortex during auditory perception of actions (Kohler *et al.* 2002), Aziz-Zadeh *et al.* (2004) used TMS to show increased excitability of the motor system underlying hand actions while subjects listened to sounds associated with actions performed by the hands. Interestingly, the facilitation of MEP size seen for these bimanual sounds (tearing paper, typing) was exclusively lateralized to the left hemisphere. In other words, in both humans and monkeys, hearing a sound known to be related to a particular hand action activates that corresponding motor system.

In the speech domain, the suggestion that perception involves the motor system was not new; Liberman *et al.* (1967) had proposed that auditory speech tokens were perceived by mapping them onto the articulatory gestures used in speech production (Liberman and Mattingly 1985). Three studies have examined the effects of speech perception on the motor system underlying speech production. The first of these found increased MEP size in the lip muscles (orbicularis oris) during visual observation of speech that required lip movements (e.g. /ba/ vs /ta/; Sundara *et al.* 2001). Auditory perception of the same syllables, however, did not facilitate motor excitability. In contrast, Fadiga *et al.* (2002) found that auditory presentation of specific phonemes activated the corresponding speech motor centre. They measured MEPs from the tongue while subjects passively listened to stimuli with or without consonant sounds that required tongue movements in their production such as the labiodental 'rr' in the Italian word 'terra' or the lingua-palatal fricative 'ff' in the Italian word 'zaffo'. MEP size was facilitated when subjects listened to words and nonwords containing the 'rr' phoneme (i.e. requiring tongue movements in their production) but not the 'ff' phoneme (i.e. which did not require tongue movement in production). Finally, Watkins *et al.* (2003) found that both visual and auditory perception of speech facilitated MEP responses (Figure 31.3a). Subjects listened to continuous prose passages and viewed noise, or viewed speech-related lip movements and listened to white noise, while EMG was recorded from the lips (orbicularis oris). MEP size was significantly facilitated during both auditory and visual speech perception, but only for stimulation over the left hemisphere, not for the right. In short, the passive perception of speech (either visually or auditorily) induces activation in motor centres.

TMS studies of this type reveal a functional connection between speech comprehension and specific components of the motor system but they do not provide any anatomical information to suggest how this link is mediated. By combining TMS and PET, Watkins and Paus (2004) investigated the brain regions that mediate the change in motor excitability during speech perception. They obtained measures of motor excitability by TMS over the face area of left motor cortex eliciting MEPs from orbicularis oris muscle during auditory speech perception. These measures were then correlated with regional cerebral blood flow measures across the whole brain obtained simultaneously. Increased motor excitability during speech perception correlated with blood flow increases in the posterior part of the left inferior frontal gyrus (Broca's area; see Figure 31.3b), the homolog of the region in the macaque containing mirror neurons (Kohler *et al.* 2002). In other words, Broca's area plays a central role in linking speech perception with speech production, consistent with theories that emphasize the importance of integration of sensory and motor representations in understanding speech (Liberman and Mattingly 1985; Hickok and Poeppel 2000; Scott and Wise 2004). The increased motor excitability of the speech production system could reflect covert imitative mechanisms or internal speech, which might, in turn, improve comprehension of the percept. From this limited set of initial studies in speech perception, two conclusions arise: increased motor excitability during auditory perception of actions is specific to the effector used to produce the sound (Fadiga *et al.* 2002), and understanding actions through sound is lateralized to the left hemisphere (Watkins *et al.* 2003; Aziz-Zadeh *et al.* 2004).

Semantics and the motor system

The above studies demonstrate a link between the motor system and speech but a few studies have examined whether more abstract representations of actions, provided by the semantics of language, also activate motor systems. For instance, Hauk *et al.* (2004) used fMRI to

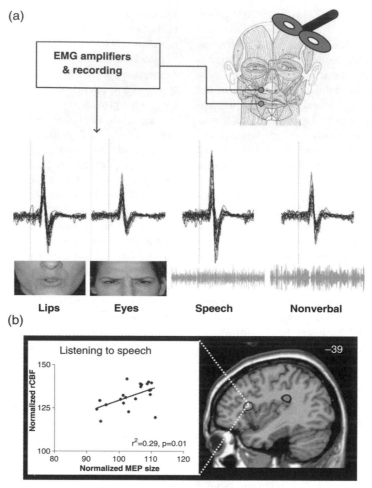

Fig. 31.3 Experimental set-up and typical motor-evoked potential (MEP) data. (a) *Upper*: position of the stimulation coil over the mouth region of motor cortex and the placements of recording electrodes on the contralateral orbicularis oris muscle. *Lower*: data from stimulation of the left primary motor face area in a single subject when listening to speech, listening to nonverbal sounds, viewing speech, and viewing eye movements. Electromyographics recordings from individual trials are superimposed and the dotted line indicates the time of stimulation. Modified and reprinted from *Neuropsychologia*, 41, Watkins, K.E., Strafella, A.P. and Paus, T. 'Seeing and hearing speech excites the motor system involved in speech production', pp. 990–2, © 2003 Elsevier. (b) *Left*: relation between regional cerebral blood flow (rCBF) in Broca's area and the size of the MEP evoked by single-pulse TMS over the mouth region of primary motor cortex during speech perception. *Right*: activation map showing the anatomical location of the significant positive relationship illustrated in the graph. Reprinted from the *Journal of Cognitive Neuroscience 16(6)*, Watkins, K.E. and Paus, T. 'Modulation of motor excitability during speech perception: the role of Broca's area', pp. 990–2, © 2004 MIT Press.

demonstrate somatotopic brain activation when subjects passively read words typically executed by different effectors such as the feet ('kick'), hands ('pick'), and mouth ('lick'). The findings suggest a functional link between the meaning of the words and specific motor centres that would be used to execute corresponding actions. Buccino *et al.* (2005) used TMS to measure this even more directly. In their experiment, MEPs were measured from the hand and foot muscles while subjects listened to sentences related to hand actions (e.g. 'he sewed the skirt'), foot actions (e.g. 'he jumped the rope'), or more abstract actions (e.g. 'he forgot the date'). MEPs recorded from hand muscles were significantly modulated by sentences referring to hand actions but not foot or abstract actions. Similarly, sentences with foot, but not hand or abstract, actions modulated MEP responses in the foot muscle. In other words, the size of the MEP in each effector muscle was only affected when listening to sentences containing actions related to the that effector. A similar study by Pulvermüller et al. (2005) showed that single pulse TMS over the arm or leg motor cortex in the left hemisphere decreased reaction times to lexical decisions for actions related to arms (e.g. 'folding') and legs (e.g. 'stepping'), respectively. Subthreshold stimulation (90% MT) may have 'primed' subjects, perhaps by partially activating the representation of actions related to the specific effector being stimulated.

These studies provide additional evidence for the close relation between verbs and actions by demonstrating a functional link between the meaning of individual verbs and regionally specific enhanced excitability of the motor cortex. As mentioned previously, stimulation of left DLPFC preferentially affected verbs relative to nouns (Shapiro et al. 2001; Cappa et al. 2002). This may indicate that the functional link with hand or leg regions of motor cortex is mediated via DLPFC in much the same way as the ventro-lateral prefrontal cortex mediates the link between perceiving speech and the mouth region of motor cortex (Watkins and Paus 2004).

TMS with neurological patients

In combination with neurological patients, TMS offers a very powerful tool for investigating the effects of brain damage. Questions of central importance to both cognitive and clinical neuroscience include: 'What are the mechanisms that support recovery following damage?' and 'Can these be enhanced to improve outcomes?' In both cases, TMS is providing new insights.

It is often assumed that following left hemisphere damage, homolog areas in the right hemisphere are recruited to (at least partially) take over lost functions. For instance, Coltheart

(1980) suggested that following extensive left hemisphere lesions, the right hemisphere is capable of supporting partial reading ability primarily limited to high-frequency concrete nouns (e.g. 'apple' but not 'cognition'). Functional imaging studies confirm that such patients activate their right hemisphere when reading, but the activation is also present in neurologically normal control subjects (Price et al. 1998). To explore whether right hemisphere involvement in reading was qualitatively different between 'right hemisphere readers' and controls, Coslett and Monsul (1994) delivered a single TMS pulse (60% max output) to the right temporo-parietal junction at 145 ms after the onset of a visual word. In the patient, but not controls, this significantly reduced the number of correctly read words from 17/24 without TMS to 5/24 with TMS. Such a dramatic effect on accuracy using single-pulse stimulation is rare, if not unique, and suggests that the reading processes in this patient were particularly fragile. Moreover, the pattern of errors induced by TMS was an exaggeration of normal difficulty effects (i.e. more errors for low- than for high-frequency items) – exactly opposite to what one would expect for 'right hemisphere reading'. If the right hemisphere selectively supports highly frequent concrete items, then stimulation of this hemisphere should preferentially impair those items. In fact, right hemisphere stimulation affected low-frequency items to a greater extent. So while the findings do not support the right hemisphere reading hypothesis, they do provide clear evidence that, in this patient, reading relied on the right temporo-parietal junction to a greater extent than in normals.

Other studies have investigated the role of the right inferior frontal gyrus in patients recovering from either left hemisphere strokes (Winhuisen et al. 2005) or brain tumours (Thiel et al. 2005) using a combination of PET and TMS. In both cases, PET was used to identify activation in the left and right inferior frontal gyri for each subject as they performed either a verb–picture matching or a word generation task. The activations were then used to individually target TMS to the site of activated tissue in each hemisphere. In one study, all 11 stroke patients showed LIFG activation and stimulating (4 Hz at 20% max output for 10 s) this region increased latencies or errors in 10/11 patients (Winhuisen *et al.* 2005). Five out of 11 patients showed stronger activation in RIFG than LIFG and four of these had longer response latencies with RIFG stimulation. In the other study, all 14 of the tumor patients showed LIFG activation with 7/14 also showing RIFG activation (Thiel *et al.* 2005). Once again, a majority of patients (11/14) showed latency increases with LIFG stimulation (4 Hz at 100% MT) while the five patients with the most rightward activation asymmetry also showed a latency increase with RIFG stimulation. Taken together, these findings demonstrate that, in most patients, LIFG remains essential for word generation tasks even after left hemisphere damage. Moreover, in a subset of patients RIFG is also essential, but only in those showing the strongest rightward asymmetries.

Why do only some patients show evidence of compensatory right hemisphere involvement? One possibility is that this reflects differences in premorbid language organization. To investigate this, Knecht *et al.* (2002) first identified a set of neurologically normal participants who varied in their degree of language lateralization. Functional transcranial Doppler sonography (fTCD) was used to measure hemispheric perfusion increases during a word generation task across a large sample of the population (Knecht *et al.* 2000) and then 20 subjects were selected who covered the full range of hemispheric dominance from strongly left to strongly right lateralized. Each performed a picture–word verification task before and after 10 min of 1 Hz stimulation (110% MT) over either left or right Wernicke's area. Participants with left, but not right, language dominance were significantly slowed by left hemisphere stimulation whereas the opposite pattern was observed for right hemisphere stimulation. In addition, the amount of interference correlated with the degree of language lateralization – in other words, strongly lateralized subjects were more severely affected by unilateral TMS than those with more bilateral language organization. Premorbid differences, therefore, render the right hemisphere more or less receptive for language before any reorganization takes place and may play an important role in determining the likelihood of right hemisphere compensation following

left-sided damage. In addition, these findings provide strong evidence for the hypothesis that crossed aphasia – that is, aphasia resulting from purely right hemisphere lesions – is a result of atypical premorbid organization.

Finally, among the most intriguing patient studies are those in which TMS is used to enhance recovery. Hoffman *et al.* (1999), for example, hypothesized that auditory hallucinations are due to overactivation in auditory cortex and used a 4 day schedule of 1 Hz stimulation (80% MT) over left auditory association cortex to suppress activity in the region. After treatment, all three patients reported reduced auditory hallucinations, and in two of the patients the effect lasted for ≥2 weeks. A similar approach was adopted in a series of studies investigating whether rTMS could be used to improve recovery in a set of chronic nonfluent aphasics (Martin *et al.* 2004; Naeser *et al.* 2005a,b). In such patients, strong right hemisphere activation is often observed, even in the absence of behavioral improvements (Naeser *et al.* 2004). In order to reduce this potentially maladaptive right hemisphere activation, 10 min of 1 Hz rTMS (90% MT) was delivered to four different right hemisphere peri-sylvian sites including rostral RIFG, caudal RIFG, posterior superior temporal gyrus, and the mouth area of primary motor cortex. Following rostral RIFG stimulation, the patients were able to correctly name more pictures than before TMS (Martin *et al.* 2004). Consequently, this region was targeted with a 10 day regime of 1 Hz rTMS (20 min at 90% MT) in four of the patients to determine whether a lasting facilitation could be achieved. Immediately following the final rTMS session, picture-naming performance was significantly enhanced in each patient but the critical finding was that these effects were still present 2 months later without any additional TMS sessions or intervening speech therapy (Naeser *et al.* 2005b). One patient was seen again 8 months after TMS 'treatment' and her performance remained stable at a level significantly better than before treatment (Naeser *et al.* 2005a). Admittedly these are preliminary findings, but they are nonetheless remarkable in that they demonstrate a very long-lasting effect of TMS – far beyond transient first-order effects of stimulation. Presumably TMS modulated activity throughout the language system via cortico-cortical spreading (Ilmoniemi *et al.* 1997; Paus *et al.* 1997), and this activity was sufficient to promote plastic changes (i.e. reorganization) that improved performance. Although speculative, the underlying mechanism is similar to Ramachandran *et al's.* (1995) explanation of why visual perception of limb movements can help patients recover from phantom-limb syndrome. In that case, the induced sensorimotor plasticity was driven by an endogenous top-down signal due to the illusion of moving the phantom arm, whereas in the aphasic patients, TMS was an external source of input driving the plasticity. Regardless of the mechanism, the results certainly suggest that, in some aphasic patients, rTMS might be successfully employed as part of rehabilitative therapy to improve recovery.

In summary, TMS studies of patients challenge the notion that homologous regions assume lost language functions following left hemisphere lesions – at the very least, the story is considerably more complicated. For one thing, TMS confirms the importance of residual left hemisphere function (Thiel *et al.* 2005; Winhuisen *et al.* 2005), as suggested by previous functional imaging studies (e.g. Karbe *et al.* 1998; Musso *et al.* 1999; Warburton *et al.* 1999). In fact, left hemisphere stimulation interfered with performance more consistently than right hemisphere stimulation, which only affected the subset of patients with the strongest rightward asymmetries. The reason for these asymmetries remains unclear, but one factor likely to play a role is premorbid language organization (Knecht *et al.* 2002). Another complicating finding is that in some cases right hemisphere stimulation interfered with performance (Coslett and Monsul 1994; Thiel *et al.* 2005; Winhuisen *et al.* 2005), whereas in others it improved performance (Martin *et al.* 2004; Naeser *et al.* 2005a,b). There are, of course, significant differences between the studies including the types of patients and the type of TMS; nonetheless, the findings demonstrate that one cannot draw a simple conclusion regarding right hemisphere involvement in recovery. Understanding these differences poses a major challenge for cognitive neuroscience and may require adopting more sophisticated models of recovery that move beyond the simple notions of 'homologous transfer of function' and 'necessary and sufficient' brain regions (Price and Friston 2002a; Friston and Price 2003).

Summary and future directions

After 15 years of language research using TMS, the field is still in its infancy; nevertheless several important themes have begun to emerge. When used in its virtual lesion mode, TMS offers the spatial precision to explore causal relations between focal brain regions and specific language functions which may not be feasible with patient studies. Such studies have helped to clarify the role of the different regions of LIFG in semantic and phonological processing (Gough *et al.* 2005), to illustrate the critical relationship between premorbid language organization and susceptibility to unilateral lesions (Knecht *et al.* 2002), and to demonstrate that left hemisphere activation in aphasic patients is more consistently critical for performance than right hemisphere activation (Thiel *et al.* 2005; Winhuisen *et al.* 2005). When TMS is used as a measure of functional connectivity, it demonstrates a close link between action words and motor programs (Pulvermüller *et al.* 2005), it suggests a potential evolutionary link between hand gestures and language (Corballis 2003), and it demonstrates that speech comprehension potentiates the specific parts of the motor system engaged to produce equivalent movements (Fadiga *et al.* 2002; Watkins *et al.* 2003; Watkins and Paus 2004). Finally, TMS even offers the potential for enhancing recovery processes and aiding rehabilitation (Matsumoto *et al.* 2004; Naeser *et al.* 2005a,b). A number of existing TMS techniques have not been applied to language, despite their obvious potential. These include combining TMS and PET to investigate the anatomical connectivity of language processing (e.g. Fox *et al.* 1997; Paus *et al.* 1997) and using multiple stimulation sites to explore both functional connectivity (e.g. Munchau *et al.* 2002) and the specific mechanisms of recovery (Price and Friston 2002a). Over the next 15 years, the field seems poised to expand enormously in virtually all areas of language research, building on the early successes and developing novel methods capable of answering an even wider range of questions.

References

Amassian VE (1989). Suppression of visual perception by magnetic coil stimulation of human occipital cortex. *Electroencephalography and Clinical Neurophysiology 74*, 458–462.

Andoh J, Artiges E, Pallier C, *et al.* (2006). Modulation of language areas with functional MR image-guided magnetic stimulation. *NeuroImage 29*, 619–627.

Aziz-Zadeh L, Iacoboni M, Zaidel E, Wilson S, Mazziotta J (2004). Left hemisphere motor facilitation in response to manual action sounds. *European Journal of Neuroscience 19*, 2609–2612.

Aziz-Zadeh L. Cattaneo L, Rochat M, Rizzolatti G (2005). Covert speech arrest induced by rTMS over both motor and nonmotor left hemisphere frontal sites. *Journal of Cognitive Neuroscience 17*, 928–938.

Baldissera F, Cavallari P, Craighero L, Fadiga L (2001). Modulation of spinal excitability during observation of hand actions in humans. *European Journal of Neuroscience 13*, 190–194.

Barde LH, Thompson-Schill SL (2002). Models of functional organization of the lateral prefrontal cortex in verbal working memory: evidence in favor of the process model. *Journal of Cognitive Neuroscience 14*, 1054–1063.

Beckers G, Homberg V (1991). Impairment of visual perception and visual short term memory scanning by transcranial magnetic stimulation of occipital cortex. *Experimental Brain Research 87*, 421–432.

Braet W, Humphreys GW (2006). Case mixing and the right parietal cortex: evidence from rTMS. *Experimental Brain Research 168*, 265–271.

Broca P (1861). Remarques sur le siège de la faculté du langage articulé suivies d'une observation d'aphemie. *Bulletin de la Société Anatomique de Paris 6*, 330–357.

Broca P (1865). Sur le siège de la faculté du langage articulé. *Bulletin of the Society of Anthropology 6*, 377–396.

Buccino G, Riggio L, Melli G, Binkofski F, Gallese V, Rizzolatti G (2005). Listening to action-related sentences modulates the activity of the motor system: a combined TMS and behavioral study. *Brain Research: Cognitive Brain Research 24*, 355–363.

Buckner RL, Raichle ME, Petersen SE (1995). Dissociation of human prefrontal cortical areas across different speech production tasks and gender groups. *Journal of Neurophysiology 74*, 2163–2173.

Cappa SF, Sandrini M, Rossini PM, Sosta K, Miniussi C (2002). The role of the left frontal lobe in action naming: rTMS evidence. *Neurology 59*, 720–723.

Caramazza A, Hillis A (1991). Lexical organization of nouns and verbs in the brain. *Nature 349*, 788–790.

Coltheart M (1980). Deep dyslexia: a review of the syndrome. In: M Coltheart, K Patterson,

J Marshall (eds), *Deep dyslexia*. London: Routledge & Kegan Paul.

Coltheart M, Davelaar E, Jonasson JT, Besner D (1977). Access to the internal lexicon. In: S Dornic (ed.), *Attention and performance VI*, pp. 535–555.

Corballis MC (2003). From mouth to hand: gesture, speech, and the evolution of right-handedness. *Behavioral Brain Science 26*, 199–208; discussion 208–160.

Coslett HB, Monsul N (1994). Reading with the right hemisphere: evidence from transcranial magnetic stimulation. *Brain and Language 46*, 198–211.

Dehaene S, Jobert A, Naccache L, *et al.* (2004). Letter binding and invariant recognition of masked words: behavioral and neuroimaging evidence. *Psychological Science 15*, 307–313.

Devlin JT, Matthews PM, Rushworth MF (2003). Semantic processing in the left inferior prefrontal cortex: a combined functional magnetic resonance imaging and transcranial magnetic stimulation study. *Journal of Cognitive Neuroscience 15*, 71–84.

Devlin JT, Jamison HL, Gonnerman LM, Matthews PM (2006). The role of the posterior fusiform in reading. *Journal of Cognitive Neuroscience 18*, 911–922.

di Pellegrino G, Fadiga L, Fogassi L, Gallese V, Rizzolatti G (1992). Understanding motor events: a neurophysiological study. *Experimental Brain Research 91*, 176–180.

Dräger B, Breitenstein C, Helmke U, Kamping S, Knecht S (2004). Specific and nonspecific effects of transcranial magnetic stimulation on picture–word verification. *European Journal of Neuroscience 20*, 1681–1687.

Duncan J, Owen AM (2000). Common regions of the human frontal lobe recruited by diverse cognitive demands. *Trends in Neurosciences 23*, 475–483.

Epstein CM, Lah JK, Meador K, Weissman JD, Gaitan LE, Dihenia B (1996). Optimum stimulus parameters for lateralized suppression of speech with magnetic brain stimulation. *Neurology 47*, 1590–1593.

Epstein CM, Meador KJ, Loring DW *et al.* (1999). Localization and characterization of speech arrest during transcranial magnetic stimulation. *Clinical Neurophysiology 110*, 1073–1079.

Epstein CM, Woodard JL, Stringer AY *et al.* (2000). Repetitive transcranial magnetic stimulation does not replicate the Wada test. *Neurology 55*, 1025–1027.

Fadiga L, Craighero L, Buccino G, Rizzolatti G (2002). Speech listening specifically modulates the excitability of tongue muscles: a TMS study. *European Journal of Neuroscience 15*, 399–402.

Federmeier KD, Segal JB, Lombrozo T, Kutas M (2000). Brain responses to nouns, verbs and class-ambiguous words in context. *Brain 123*(Pt 12), 2552–2566.

Fiez JA (1997). Phonology, semantics, and the role of the left inferior prefrontal cortex. *Human Brain Mapping 5*, 79–83.

Flitman SS, Grafman J, Wassermann EM, *et al.* (1998). Linguistic processing during repetitive transcranial magnetic stimulation. *Neurology 50*, 175–181.

Floel A, Ellger T, Breitenstein C, Knecht S (2003). Language perception activates the hand motor cortex: implications for motor theories of speech perception. *European Journal of Neuroscience 18*, 704–708.

Fox P, Ingham R, George MS, *et al.* (1997). Imaging human intra-cerebral connectivity by PET during TMS. *Neuroreport 8*(12), 2787–2791.

Friederici AD (2002). Towards a neural basis of auditory sentence processing. *Trends in Cognitive Sciences 6*, 78–84.

Friederici AD, Kotz SA (2003). The brain basis of syntactic processes: functional imaging and lesion studies. *NeuroImage 20*(Suppl. 1), S8–17.

Friston KJ, Price CJ (2003). Degeneracy and redundancy in cognitive anatomy. *Trends in Cognitive Sciences 7*, 151–152.

Gallese V, Fadiga L, Fogassi L, Rizzolatti G (1996). Action recognition in the premotor cortex. *Brain 119*, 593–609.

Gangitano M, Mottaghy FM, Pascual-Leone A (2004). Modulation of premotor mirror neuron activity during observation of unpredictable grasping movements. *European Journal of Neuroscience 20*, 2193–2202.

Gold BT, Buckner RL (2002). Common prefrontal regions coactivate with dissociable posterior regions during controlled semantic and phonological tasks. *Neuron 35*, 803–812.

Gough PM, Nobre AC, Devlin JT (2005). Dissociating linguistic processes in the left inferior frontal cortex with transcranial magnetic stimulation. *Journal of Neuroscience 25*, 8010–8016.

Hagoort P, Hald L, Bastiaansen M, Petersson KM (2004). Integration of word meaning and world knowledge in language comprehension. *Science 304*, 438–441.

Hauk O, Johnsrude I, Pulvermüller F. (2004). Somatotopic representation of action words in human motor and premotor cortex. *Neuron 41*, 301–307.

Hickok G, Poeppel D (2000). Towards a functional neuroanatomy of speech perception. *Trends in Cognitive Sciences 4*, 131–138.

Hoffman RE, Boutros NN, Berman RM *et al.* (1999). Transcranial magnetic stimulation of left temporoparietal cortex in three patients reporting hallucinated 'voices'. *Biological Psychiatry 46*, 130–132.

Ilmoniemi RJ, Virtanen J, Ruohonen J, *et al.* (1997). Neuronal responses to magnetic stimulation reveal cortical reactivity and connectivity. *Neuroreport 8*, 3537–3540.

Jahanshahi M, Rothwell J (2000). Transcranial magnetic stimulation studies of cognition: an emerging field. *Experimental Brain Research 131*, 1–9.

Jennum P, Friberg L, Fuglsang-Frederiksen A, Dam M (1994). Speech localization using repetitive transcranial magnetic stimulation. *Neurology 44*, 269–273.

Karbe H, Thiel A, Weber-Luxenburger G, Herholz K, Kessler J, Heiss WD (1998). Brain plasticity in poststroke aphasia: what is the contribution of the right hemisphere? *Brain and Language 64*, 215–230.

Knecht S, Drager B, Deppe M, *et al.* (2000). Handedness and hemispheric language dominance in healthy humans. *Brain 123*, 2512–2518.

Knecht S, Floel A, Drager B, *et al.* (2002). Degree of language lateralization determines susceptibility to unilateral brain lesions. *Nature Neuroscience 5*, 695–699.

Kohler E, Keysers C, Umilta MA, Fogassi L, Gallese V, Rizzolatti G (2002). Hearing sounds, understanding actions: action representation in mirror neurons. *Science 297*(5582), 846–848.

Kohler S, Paus T, Buckner RL, Milner B (2004). Effects of left inferior prefrontal stimulation on episodic memory formation: a two-stage fMRI-rTMS study. *Journal of Cognitive Neuroscience 16*, 178–188.

Lavidor M, Walsh V (2003). A magnetic stimulation examination of orthographic neighborhood effects in visual word recognition. *Journal of Cognitive Neuroscience 15*, 354–363.

Leff AP, Scott SK, Rothwell JC, Wise RJ (2001). The planning and guiding of reading saccades: a repetitive transcranial magnetic stimulation study. *Cerebral Cortex 11*, 918–923.

Levy BJ, Anderson MC (2002). Inhibitory processes and the control of memory retrieval. *Trends in Cognitive Sciences 6*, 299–305.

Liberman AM, Mattingly IG (1985). The motor theory of speech perception – revised. *Cognition 21*, 1–36.

Liberman AM, Cooper FS, Shankweiler D, Studdert-Kennedy M (1967). Perception of the speech code. *Psychological Review 74*, 431–461.

Liederman J, McGraw Fisher J, Schulz M, Maxwell C, Theoret H, Pascual-Leone A (2003). The role of motion direction selective extrastriate regions in reading: a transcranial magnetic stimulation study. *Brain and Language 85*, 140–155.

Lo YL, Fook-Chong S (2004). Ipsilateral and contralateral motor inhibitory control in musical and vocalization tasks. *Experimental Brain Research 159*, 258–262.

Lo YL, Fook-Chong S, Lau DP, Tan EK (2003). Cortical excitability changes associated with musical tasks: a transcranial magnetic stimulation study in humans. *Neuroscience Letters 352*, 85–88.

MacLeod CM (1991). Half a century of research on the Stroop effect: an integrative review. *Psychological Bulletin 109*, 163–203.

Martin PI, Naeser MA, Theoret H, *et al.* (2004). Transcranial magnetic stimulation as a complementary treatment for aphasia. *Seminars in Speech and Language 25*, 181–191.

Matsumoto R, Nair DR, LaPresto E, *et al.* (2004). Functional connectivity in the human language system: a cortico-cortical evoked potential study. *Brain 127*, 2316–2330.

Meister IG, Boroojerdi B, Foltys H, Sparing R, Huber W, Topper R (2003). Motor cortex hand area and speech: implications for the development of language. *Neuropsychologia 41*, 401–406.

Michelucci R, Valzania F, Passarelli D, *et al.* (1994). Rapid-rate transcranial magnetic stimulation and hemispheric language dominance: usefulness and safety in epilepsy. *Neurology 44*, 1697–1700.

Mottaghy FM, Hungs M, Brugmann M, *et al.* (1999). Facilitation of picture naming after repetitive transcranial magnetic stimulation. *Neurology 53*, 1806–1812.

Munchau A, Bloem BR, Irlbacher K, Trimble MR, Rothwell JC (2002). Functional connectivity of human premotor and motor cortex explored with repetitive transcranial magnetic stimulation. *Journal of Neuroscience 22*, 554–561.

Musso M, Weiller C, Kiebel S, Muller SP, Bulau P, Rijntjes M (1999). Training-induced brain plasticity in aphasia. *Brain 122*, 1781–1790.

Naeser MA, Martin PI, Baker EH, *et al.* (2004). Overt propositional speech in chronic nonfluent aphasia studied with the dynamic susceptibility contrast fMRI method. *NeuroImage 22*, 29–41.

Naeser MA, Martin PI, Nicholas M, *et al.* (2005a). Improved naming after TMS treatments in a chronic, global aphasia patient – case report. *Neurocase 11*, 182–193.

Naeser MA, Martin PI, Nicholas M, *et al.* (2005b). Improved picture naming in chronic aphasia after TMS to part of right Broca's area: an open-protocol study. *Brain and Language 93*, 95–105.

Nixon P, Lazarova J, Hodinott-Hill I, Gough P, Passingham R (2004). The inferior frontal gyrus and phonological processing: an investigation using rTMS. *Journal of Cognitive Neuroscience 16*, 289–300.

Pascual-Leone A, Walsh V (2001). Fast backprojections from the motion to the primary visual area necessary for visual awareness. *Science 292*, 510–512.

Pascual-Leone A, Gates JR, Dhuna A (1991). Induction of speech arrest and counting errors with rapid-rate transcranial magnetic stimulation. *Neurology 41*, 697–702.

Pascual-Leone A, Bartres-Faz D, Keenan JP (1999). Transcranial magnetic stimulation: studying the brain–behaviour relationship by induction of 'virtual lesions'. *Philosophical Transactions of the Royal Society of London B 354*, 1229–1238.

Pascual-Leone A, Walsh V, Rothwell J (2000). Transcranial magnetic stimulation in cognitive neuroscience – virtual lesion, chronometry, and functional connectivity. *Current Opinion in Neurobiology 10*, 232–237.

Paus T, Jech R, Thompson CJ, Comeau R, Peters T, Evans AC (1997). Transcranial magnetic stimulation during positron emission tomography: a new method for studying connectivity of the human cerebral cortex. *Journal of Neuroscience 17*, 3178–3184.

Petrides M (2000). The role of the mid-dorsolateral prefrontal cortex in working memory. *Experimental Brain Research 133*, 44–54.

Price CJ, Devlin JT (2003). The myth of the visual word form area. *Neuroimage 19*, 473–481.

Price CJ, Friston KJ (2002a). Degeneracy and cognitive anatomy. *Trends in Cognitive Sciences 6*, 416–421.

Price CJ, Friston KJ (2002b). Functional imaging studies of neuropsychological patients: applications and limitations. *Neurocase 8*, 345–354.

Price CJ, Wise RJS, Frackowiak RSJ (1996). Demonstrating the implicit processing of visually presented words and pseudowords. *Cerebral Cortex 6*, 62–70.

Price CJ, Howard D, Patterson K, Warburton EA, Friston KJ, Frackowiak RSJ (1998). A functional neuroimaging description of two deep dyslexic patients. *Journal of Cognitive Neuroscience 10*, 303–315.

Pulvermüller F, Hauk O, Nikulin VV, Ilmoniemi RJ (2005). Functional links between motor and language systems. *European Journal of Neuroscience 21*, 793–797.

Ramachandran VS, Rogers-Ramachandran D, Cobb S (1995). Touching the phantom limb. *Nature 377*, 489–490.

Riddoch MJ, Humphreys G, Cleton P, Fery P (1990). Interaction of attention and lexical processes in neglect dyslexia. *Cognitive Neuropsychology 7*, 479–517.

Rizzolatti G, Arbib MA (1998). Language within our grasp. *Trends in Neurosciences 21*, 188–194.

Sakai KL, Noguchi Y, Takeuchi T, Watanabe E (2002). Selective priming of syntactic processing by event-related transcranial magnetic stimulation of Broca's area. *Neuron 35*, 1177–1182.

Scott SK, Wise RJ (2004). The functional neuroanatomy of prelexical processing in speech perception. *Cognition 92*, 13–45.

Seyal M, Mull B, Bhullar N, Ahmad T, Gage B (1999). Anticipation and execution of a simple reading task enhance corticospinal excitability. *Clinical Neurophysiology 110*, 424–429.

Shapiro KA, Caramazza A (2003). Grammatical processing of nouns and verbs in left frontal cortex? *Neuropsychologia 41*, 1189–1198.

Shapiro KA, Pascual-Leone A, Mottaghy FM, Gangitano M, Caramazza A (2001). Grammatical distinctions in the left frontal cortex. *Journal of Cognitive Neuroscience 13*, 713–720.

Stewart L, Meyer BU, Frith U, Rothwell J (2001a). Left posterior BA37 is involved in object recognition: a TMS study. *Neuropsychologia 39*, 1–6.

Stewart L, Walsh V, Frith U, Rothwell JC (2001b). TMS produces two dissociable types of speech disruption. *NeuroImage 13*, 472–478.

Stewart L, Walsh V, Rothwell JC (2001c). Motor and phosphene thresholds: a transcranial magnetic stimulation correlation study. *Neuropsychologia 39*, 415–419.

Strafella AP, Paus T (2000). Modulation of cortical excitability during action observation: a transcranial magnetic stimulation study. *Neuroreport 11*, 2289–2292.

Sundara M, Namasivayam AK, Chen R (2001). Observation–execution matching system for speech: a magnetic stimulation study. *Neuroreport 12*, 1341–1344.

Thiel A, Habedank B, Winhuisen L *et al.* (2005). Essential language function of the right hemisphere in brain tumor patients. *Annals of Neurology 57*, 128–131.

Tokimura H, Tokimura Y, Oliviero A, Asakura T, Rothwell JC (1996). Speech-induced changes in corticospinal excitability. *Annals of Neurology. 40*, 628–634.

Topper R, Mottaghy FM, Brugmann M, Noth J, Huber W (1998). Facilitation of picture naming by focal transcranial magnetic stimulation of Wernicke's area. *Experimental Brain Research 121*, 371–378.

Wada J, Rasmussen T (1960). Intracarotid injection of sodium amytal for the lateralization of cerebral speech dominance: experimental and clinical observations. *Journal of Neurosurgery 17*, 266–282.

Walsh V, Cowey A (1998). Magnetic stimulation studies of visual cognition. *Trends in Cognitive Science 2*, 103–110.

Walsh V, Cowey A (2000). Transcranial magnetic stimulation and cognitive neuroscience. *Nature Reviews Neuroscience 1*, 73–79.

Walsh V, Pascual-Leone A (2003). *Transcranial magnetic stimulation: a neurochronometrics of mind*. London: MIT Press.

Walsh V, Rushworth MFS (1999). The use of transcranial magnetic stimulation in neuropsychological testing. *Neuropsychologia 37*, 125–135.

Warburton E, Price CJ, Swinburn K, Wise RJ (1999). Mechanisms of recovery from aphasia: evidence from positron emission tomography studies. *Journal of Neurology, Neurosurgery and Psychiatry 66*, 155–161.

Wassermann EM, Blaxton TA, Hoffman EA, *et al.* (1999). Repetitive transcranial magnetic stimulation of the dominant hemisphere can disrupt visual naming in temporal lobe epilepsy patients. *Neuropsychologia 37*, 537–544.

Watkins KE, Paus T (2004). Modulation of motor excitability during speech perception: the role of Broca's area. *Journal of Cognitive Neuroscience 16*, 978–987.

Watkins KE, Strafella AP, Paus T (2003). Seeing and hearing speech excites the motor system involved in speech production. *Neuropsychologia 41*, 989–994.

Wernicke C (1874). *Der aphasische Symptomenkomplex.* Breslau: Cohen & Weigert.

Winhuisen L, Thiel A, Schumacher B, *et al.* (2005). Role of the contralateral inferior frontal gyrus in recovery of language function in poststroke aphasia: a combined repetitive transcranial magnetic stimulation and positron emission tomography study. *Stroke 36*, 1759–1763.

Zurif EB, Caramazza A, Myerson R (1972). Grammatical judgements of agrammatic aphasics. *Neuropsychologia 10*, 405–417.

CHAPTER 32

Higher cognitive functions: memory and reasoning

Simone Rossi, Stefano F. Cappa, and Paolo Maria Rossini

Introduction

According to the Latin poet Horace (Orazio 2000), during legal disputes at the time of ancient Rome, the judge used to pull several times on the witness's earlobe in order to invite him to remember a particular event. It was thus believed (provided that this approach was successful) that the material to be remembered, stored within the earlobe (Plinius 2000), could be made more readily accessible for retrieval. Contemporary neuroscientists have recently developed a more sophisticated approach for interfering with human memory and reasoning than the one based on the symbolic anatomophysiology just described: this is TMS.

The present chapter reviews the TMS studies dealing with short-term retention and manipulation of information (working memory), as well as with the episodic component of declarative memory. Selected aspects of semantic memory and nonverbal reasoning are also considered here. Nondeclarative aspects of memory functions are dealt with in other sections of this volume.

Before discussing the relevant literature, it is useful to delineate some basic methodological considerations about the experimental designs which can be used for the investigation of human higher-cognitive functions with TMS.

Rationale for the use of TMS

TMS interference versus focal brain lesions

The relevance that TMS has gained in recent years for investigation of the neural mechanisms of memory and reasoning depends mainly on its unique ability to transiently interfere with the functions of the specialized cortical network, especially when applied as repetitive (r)TMS. rTMS can be applied either on-line (with high-frequency trains), coincident with the execution of a given cognitive task, or as an off-line (with low-frequency trains) inhibitory preconditioning of a given area before task completion (Pascual-Leone et al. 2000; Walsh and Cowey 2000; Rossi and Rossini 2004). As a general concept, the use of TMS or rTMS to study higher cognitive functions is therefore not different from the application of this technique for the investigation of motor, perceptual, or attentional domains, as discussed in detail in other chapters of this volume. The transient interference of rTMS with neural information processing led some researchers to introduce the concept of the so called 'virtual patient', i.e. of a virtual lesion induced by rTMS in an otherwise healthy subject (Walsh and Cowey 1998; Pascual-Leone et al. 2000). This concept of a 'virtual lesion' is intriguing, though it may also be somewhat misleading when used to interpret rTMS effects. Indeed, as discussed below, emerging evidence indicates that, if appropriately timed to the demands of the cognitive task, rTMS may induce facilitatory behavioral effects, rather than disruption of a given function. It remains true, however, that the TMS approach to the neural mechanisms of human memory and reasoning represents a valid, and sometimes superior, alternative to classical lesion studies in patients for a number of reasons. Indeed, even discrete lesions are chronic processes, or have chronic consequences after an acute presentation. The resulting behavioral effects thus reflect both the specific information provided by the lesion itself, and the plastic adaptive changes of the surviving brain. Moreover, TMS can safely be repeated in subjects on different occasions, eventually allowing an intra- or between-laboratory retest of a given experimental hypothesis. Finally, in the specific case of long-term memory, TMS allows one to disentangle the effects on encoding and retrieval more easily than in the case of lesion studies.

TMS is complementary to other neuroimaging procedures

Many complementary methods are used to investigate the working brain, such as positron emission tomography (PET), functional magnetic resonance imaging (fMRI), magnetoencephalography, and high-resolution electroencephalography (EEG). Basically, the neuroimaging techniques based on *in vivo* measurements of local changes of metabolism and blood flow, such as PET and fMRI, despite allowing the best available spatial resolution, offer a temporal resolution (from seconds to minutes) which is insufficient to covary with cognitive processes – such as memory or reasoning – occurring in a few tens or hundreds of milliseconds. Because of these limitations, hemodynamic methods cannot disentangle the functional hierarchy of a given area, when several cortical regions are active simultaneously, or the cross-talk between specific nodes of a network. They also do not usually allow one to know the net effect – either inhibitory or functional – of each node of the general network function.

More importantly, brain rhythms, event-related potentials, PET, or fMRI cannot offer definitive answers about the causal relationships between an active area and a particular function or behavior (Price and Friston 1999). Nevertheless, the correct anatomical targeting of a specific brain region with TMS must take into account what independent neuroimaging studies have previously indicated in relation to the task under investigation. The best approach in this sense is represented by neuronavigation of TMS towards active regions already detected by PET of fMRI scanning in the same subject and in the same experimental context (Herwig *et al.* 2003). Other approaches are based on the digitalization of the coil position in relation to the target region, known to be active in a given task from independent studies, and the anatomical matching in the Talairach space with the subject's MR or with an accurate template brain (Rossi *et al.* 2001, 2004, 2006). Another valid but less accurate method is the use of the International 10–20 EEG system as a reference for coil positioning (Herwig *et al.* 2002).

Possible sources of bias in the interpretation of TMS results

In general, behavioral changes induced by rTMS are transient and subtle, and clearly emerge only when appropriate group analyses are performed. In the field of higher cognitive functions, rTMS effects may induce overt changes in accuracy that can be scored according to measures derived from signal detection theory (i.e. discrimination d', criterion C) or with the more conventional measures such as error rate, hits, and hits minus false alarms, yes/no recognition scores etc. These changes can take place together, or independently of, modifications of the reaction times (RTs) associated with the task. However, modification of RTs may be the only effect of the application of TMS during a given task, and this is the case in most studies. Finally, TMS-induced behavioral changes can be accompanied by modifications of metabolic and neurophysiological indexes taking place (locally or trans-synaptically) in the stimulated regions.

rTMS-induced changes must therefore be clearly distinguished from behavioral modifications obtained with a sham stimulation of the same target area or with active stimulation of a brain region which, on the basis of independent neuroimaging studies, is known not to be involved in the task under investigation. Such control conditions are mandatory, since rTMS induces many multisensory and alerting effects, which may modify behavior *per se*. A recent paper (Abler *et al.* 2005) highlighted the role of rTMS-induced subjective discomfort in producing errors during a verbal delayed match to sample task. Moreover, high intensities of stimulation produce a disturbing discharging noise which may be sufficient to distract subjects during demanding cognitive tasks (Rossi *et al.* 2006). A baseline condition (i.e. without stimulation) is also necessary to verify that the task is not too easy or too difficult: in the former case rTMS may fail to induce any behavioral change, and the latter condition may introduce a floor effect in the performance that cannot be further affected by rTMS.

TMS can produce local, trans-synaptic, and system-level effects, and this should always be kept in mind when interpreting behavioral results after interference with a given target region. Here, we only reiterate that the target region is usually 'the most active' (or the centre of gravity) of a network subserving the function under investigation, or the most easily reachable (i.e. a cortical relay of a network including deep, subcortical, relays). Moreover, local rTMS-induced effects usually produce larger metabolic or neurophysiological changes than those detectable in distant but interconnected brain areas.

TMS studies of working memory

Definition of WM and neuroimaging background

Working memory (WM) is the ability to retain and manipulate verbal, visuo-spatial, or other information over a short time period (Baddeley 1992). The main function of WM is to keep the information on-line, readily available for a particular task or goal, often in the context of a relatively complex problem-solving task. While an extensive experimental literature indicates that the prefrontal cortex (PFC) plays a crucial role

in WM function, there is an ongoing debate about the organization of WM processes within this region (Petrides 2005). Neuroimaging and electrophysiological studies have consistently shown discrete activations of different subregions of the PFC in humans engaged in WM tasks. The review by Fletcher and Henson (2001) concludes that ventrolateral PFC is more often activated by tasks requiring maintenance, whereas dorsolateral PFC is engaged when manipulation is required. Moreover, the nature of the memoranda plays a key role in the lateralization of the PFC activations. A direct comparison of verbal and visuo-spatial tasks indicates, in agreement with classical neuropsychology, that the former are left lateralized, while the latter involve mainly the right hemisphere. The dorsolateral PFC, together with the anterior PFC, is associated with executive control of WM, especially in the case of task-switching (Fletcher and Henson 2001). These PFC systems are physiologically capacity constrained (Callicott *et al.* 1999), making the WM processes particularly suitable for rTMS interference. The first study in this field dealt with nondeclarative motor memory, and showed an increased error rate in a delayed-response motor task after rTMS to the left and right dorsolateral PFC, but not after stimulation of the motor cortex or in absence of stimulation (Pascual-Leone and Hallett 1994). The PFC and its subregions are densely and reciprocally interconnected with parietal cortex, and fronto-parietal networks are functionally associated in WM operations, both in the verbal and in the visuo-spatial domain (Fletcher and Henson 2001). EEG studies during WM paradigms indicate that activity in these distributed networks is likely to differ in terms of activation timing (Babiloni *et al.* 2004a,b); thus fronto-parietal chronometry is another topic that can appropriately be tested by rTMS.

Studies of verbal WM

Many studies have employed the *n*-back task, which requires the retention and continuous updating of incoming information. In the verbal domain, Mottaghy *et al.* (2000) assessed the effects of the continuous application of 4 Hz rTMS (30 s at 110% of individual motor threshold (MT)) during a 2-back verbal WM task in 14 healthy males. Repetitive TMS applied to the right and, to a lesser degree, to the left dorsolateral PFC (corresponding to the middle frontal gyrus, BA 9/46), but not to the midline frontal cortex, significantly worsened accuracy in the WM task, while inducing significant reductions in regional cerebral blood flow (rCBF), as revealed by PET scans, at the stimulation site and in interconnected brain regions. More specifically, left stimulation decreased rCBF in the left PFC, while right stimulation decreased rCBF in the right PFC and parietal areas (Mottaghy et al. 2000). These findings were later better detailed (Mottaghy et al. 2003a) by suggesting that adjacent subregions of the stimulated PFC could be able to react to rTMS interference, probably through short-term functionally relevant plastic changes in these interconnected areas. Both studies, however, failed to observe a clear-cut left-sided PFC for verbal WM operation, as would be expected on the basis of previous neuroimaging and neuropsychological studies (see Fletcher and Henson 2001; Rajah and D'Esposito 2005).

Mull and Seyal (2001) used a 3-back letter task in nine subjects. During letter presentation, single-pulse TMS (15% above MT) applied over the left PFC resulted in increased errors relative to the same task performed without TMS. On the other hand, TMS over the right dorsolateral PFC did not affect WM performance. There is also evidence that both preconditioning of the left dorsolateral PFC with 1 Hz rTMS and on-line high-frequency rTMS of the same area failed to induce behavioral changes during verbal WM, as assessed by digit span and letter–number sequencing tasks of the Wechsler Adult Intelligence Scale (Rami et al. 2003).

The role of phonologically based WM mechanisms has been addressed in a recent TMS study in which stimulation was applied over the frontal operculum while subjects performed a delayed phonological matching task (Nixon et al. 2004). Repetitive TMS (trains of 375–550 ms at 10–13.5 Hz, ~120% of the MT) was delivered either during the delay (memory) phase or at the response (decision) phase of the task. rTMS delivered at a time when subjects were required to remember the sound of a visually presented word impaired the accuracy with which they subsequently performed the task. However, rTMS did not impair accuracy when delivered later in the trial, as the subjects compared the remembered word with a given pseudo-word. Performance on a control task, requiring the processing of nonverbal visual stimuli, was unaffected by rTMS. Similarly, rTMS delivered over a more anterior site (pars triangularis) had no effect on the performance in both tasks. Therefore, in agreement with the results of neuroimaging studies (Paulesu et al. 1997; Poldrack et al. 1999) the opercular region of the inferior frontal gyrus (Broca's area) seems to play a causal role in the normal operation of phonologically based WM mechanisms, linked to subvocal articulatory rehearsal.

The causal role in verbal WM operations of other neocortical regions, such as the parietal cortex, has been addressed by TMS with conflicting results. Repetitive TMS at 50 Hz delivered on other scalp regions (left or right anterolateral parietal, but also superior and posterior temporal) did not impair verbal and visuo-spatial memory span performance (Hufnagel et al. 1993), as could be mechanistically expected on the basis of the lack of significant neuroimaging-derived activations for these regions in short-term memory. Conversely, single-pulse TMS to the posterior third of bilateral intraparietal suclus resulted in increased errors during a 2-back verbal WM task (Mottaghy et al. 1999a). Increased errors after left compared with right rTMS of BA 7/40 (15 Hz for 3 s, 110% of MT) were found during a verbal Sternberg task (Herwig et al. 2003).

A weak behavioral effect (i.e. significant increase in RTs on correct trials with no change in overall accuracy) was found when single-pulse TMS (120% of MT, double-cone-shaped coil, fMRI-guided scalp position for stimulation) was applied to the right superior cerebellum during a Sternberg-type verbal WM task immediately after letter presentation. This confirmed previous neuroimaging results (see Paulesu et al. 1993) suggesting that cerebellum may have a causal role in WM operations under these circumstances (Dresmond et al. 2005). However, these changes, obtained with single-pulse TMS, are difficult to reconcile with the finding that high-frequency rTMS of the cerebellum, likely to be more 'disrupting' than a single pulse, did not result in overt WM behavioral changes (Rami et al. 2003).

The functional coupling of parieto-frontal networks, and their chronometry of activation, during verbal WM were addressed by another TMS investigation. Mottaghy et al. (2002a) tested the effect of rTMS (4 Hz, 110% of individual MT) on the performance of a verbal WM both during and after the stimulation, which was delivered over nine different bilateral scalp locations including the middle frontal gyrus (BA 9/46), the supramarginal gyrus, the inferior parietal cortex, and three different midline control sites. A significant impairment of performance, lasting only for the time of stimulation, was observed during rTMS over the left and right middle frontal gyri and inferior parietal sites. These data provide evidence for a parallel, bilateral parieto-frontal network underlying verbal WM information processing (Mottaghy et al. 2002a).

In order to investigate the chronometry of activation of these parieto-frontal neural networks underpinning verbal WM tasks, Mottaghy et al. (2003b) additionally used an interferential approach based on single-pulse TMS (120% of individual MT), delivered over the middle frontal gyrus (BA 9/46) and the inferior parietal cortex in each hemisphere at 10 different time-points spanning 140–500 ms with respect to the onset of two different 2-back verbal tasks. The task assessed different aspects of information storage, manipulation, and updating. The accuracy of the responses was affected by TMS earlier in the parietal cortex than in PFC, and earlier in the case of right- rather than left-sided stimulation. Thus, the chronometry of the interference supported the notion that the information processing flow during verbal WM is faster in the right hemisphere, and directed from posterior to anterior sites, with the left PFC playing a key role in the final processing stages (Mottaghy et al. 2003b).

Studies of visual and spatial WM

The functional relevance of prefrontal and parietal regions, their hemispheric specialization, and the chronometry of their activation have been investigated by a number of TMS studies in spatial WM tasks.

There are converging TMS results, generally based on changes in RTs rather than in accuracy, showing that parietal cortex plays a key role in visuo-spatial WM operations, with a right-sided functional predominance (Hong et al. 2000; Kessels et al. 2000). A key role of the PFC was originally suggested by the study of memory-guided saccades of Müri et al. (1996), which found a specific effect of dorsolateral PFC single-pulse TMS applied during the delay interval, suggesting a functional role on the memorization process. Subsequent TMS studies aimed to test the hypothesis that WM and PFC function might be segregated according to informational domains (Mottaghy et al. 2002b). Here, a pre-conditioning train (600 pulses at 1 Hz, 90% of MT) was applied in separate sessions to the left dorsomedial, dorsolateral, and ventral PFC of eight subjects, with the aim of inhibiting the function of these regions during the subsequent delayed WM spatial or face-recognition task. The error rate significantly increased after stimulation of the dorsomedial PFC in the spatial task, and after stimulation of the ventral PFC in the face-recognition (visual) task. Both WM tasks were impaired after rTMS to the dorsolateral PFC. These results confirmed a domain-specific functional segregation within the left PFC for WM operations (Mottaghy et al. 2002b). They should, however, be considered with caution, due to the possibility of trans-synaptic effects between the stimulated region and the other two PFC subregions, which are known to be densely and reciprocally interconnected.

Oliveri et al. (2001) investigated 35 healthy subjects performing two visual n-back tasks, in which they were asked to memorize spatial locations or abstract patterns. In a first series of experiments, unilateral or bilateral TMS was delivered to the left and right posterior parietal and middle temporal regions after various delays during the WM task. The accuracy of the performance was not affected, but bilateral temporal TMS increased RTs in the visual–object task, whereas bilateral parietal TMS selectively increased RTs in the visual–spatial WM task. These effects were evident at a delay of 300 ms. In a second group of experiments, bilateral TMS was applied over the superior frontal gyrus or the dorsolateral PFC. Frontal TMS selectively increased RTs in the visual–spatial WM task, whereas TMS to the dorsolateral PFC interfered with both WM tasks, in terms of both accuracy and RTs. These effects were evident when TMS was applied after a delay of 600 ms, but not at 300 ms. Results provided direct evidence for a segregation of WM buffers for object and spatial information in the posterior cortical regions, and confirmed the causal role of the dorsolateral PFC for WM computations, regardless of the stimulus material (Oliveri et al. 2001).

As in the case of verbal WM processes, the temporal dynamics and the reciprocal interactions of the different areas of the widely distributed fronto-parietal network for visuo-spatial WM have recently been addressed by rTMS (Koch et al. 2005). Here, nine healthy subjects received rTMS trains (300 ms at 25 Hz, 110% of MT) over the right posterior parietal cortex (PPC), the premotor cortex, and the dorsolateral PFC during the delay phase or the decision process of a spatial WM task (encoding, short-term retention, and matching of a sequence of six black squares displayed in different positions on a white screen). rTMS did not impair accuracy, but RTs significantly increased when rTMS was applied during the delay phase to PPC and dorsolateral PFC. In the decision phase of the task, interference was observed only after dorsolateral PFC stimulation. Premotor cortex rTMS had no effect on RTs. These findings provide new evidence for the existence of parallel processing in the parieto-frontal networks of the right hemisphere during the delay phase of spatial WM. The leading role of the dorsolateral PFC for both delay and decision processes suggests the segregation of different neural populations at this level: a local one engaged in decisional processes, and a second, interconnected with PPC, for short-term maintenance.

TMS studies of episodic memory

Definition of episodic memory and neuroimaging background

A symbolic representation of the concept of episodic memory can be found in the *Odyssey*, when Homer describes the 'mnemones' (see Vernant 2000) as individuals taking the essential responsibility of remembering all the material loaded in a cargo-boat as forms of 'living archives'. Episodic memory is a complex set of human cognitive processes that allows the encoding, long-term storage, and intentional recollection (retrieval) of unique events associated with the context in which they occurred (Baddeley et al. 2001). Therefore, the correct functioning of encoding/retrieval mechanisms is vital to form the conscious story of our existence (Tulving 2002). Studies in neurological patients indicate that episodic memory is dramatically disrupted by lesions of the medial temporal lobe, and in particular of the hippocampal formation and enthorinal cortex (Squire et al. 2001), whereas lesions in other neocortical areas – such as the PFC – are associated only with more subtle dysfunctions of learning and memory (Alexander et al. 2003). Neuroimaging investigations of mechanisms underlying encoding and retrieval have confirmed the crucial role of medial temporal lobe structures in episodic memory (Buckner and Wheeler 2001). In addition, they have disclosed the unexpected activation of many regions of the PFC (Buckner and Wheeler 2001; Fletcher and Henson 2001), previously thought to be involved mainly in working memory and self-monitoring processes (see Petrides 1994). A neuroimaging-based model, conceptualizing PFC activations during episodic memory tasks, is the so-called HERA (Hemispherical Encoding Retrieval Asymmetry) theory of functional brain organization during long-term memorization (Tulving et al. 1994). According to its original formulation, the left PFC plays a crucial role in encoding, while the right PFC is essential for retrieval processes. In its most recent revision (Habib et al. 2003) the HERA pattern is not considered as an absolute feature of cortical activity during long-term memorization, since the asymmetry level is affected – and in some cases abolished – by the nature of the material to be memorized, and by the strategy of the memorization process. Besides the verbal/nonverbal distinction, other factors, such as task difficulty, familiarity with memoranda, and age can modify the pattern of functional asymmetry (Ranganath et al. 2000; Buckner and Wheeler 2001; Johnson et al. 2003; Miniussi et al. 2003). The interferential approach with rTMS is providing important clues to a better understanding of the causal relationships of such hemispheric functional asymmetries with long-term memory performance.

Studies of visuo-spatial episodic memory

The first controlled study systematically addressing the causal role of PFC asymmetries by the interferential rTMS approach during encoding and retrieval operations appeared a few years ago (Rossi *et al.* 2001). Here, 13 right-handed healthy volunteers received left or right active or sham rTMS of the dorsolateral PFC in a six-block encoding/retrieval experimental design (see Figure 32.1) based on a visuo-spatial indoor/outdoor paradigm representing real-life environments. The paradigm allowed us to investigate both encoding and retrieval processes, minimizing the WM load, since subjects were asked to identify the nature of memoranda (i.e. indoor and outdoor distinction in encoding and test/distractor indoors recognition during retrieval) only during the display of the pictures. This experimental detail is important, since it is known that PFC regions engaged in WM and episodic memory processes are largely overlapping, as suggested by activation studies (Fletcher and Henson 2001). Behavioral results of this study showed that rTMS delivered to the left dorsolateral PFC affected the encoding of memoranda, since subjects made more errors (vs right stimulation and vs reference blocks sham and baseline) during the corresponding retrieval blocks. Conversely, when encoding processes were completed in the absence of stimulation but rTMS was applied in the corresponding retrieval blocks, the opposite result emerged: right stimulation disrupted the retrieval more than left rTMS and reference blocks (sham and baseline). According to the experimental timing, the interferential effect of rTMS lasted ~1.5 s, that is the time elapsed from the picture presentation (corresponding also to the beginning of the rTMS train) to the instant in which the required motor response was completed after the go-signal (Figure 32.1). This pattern of functional PFC asymmetries (Figure 32.2) therefore directly demonstrated predictions of the HERA model for brain activity during episodic memorization (Habib *et al.* 2003).

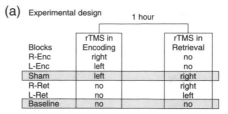

(a) Experimental design

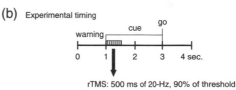

(b) Experimental timing

rTMS: 500 ms of 20-Hz, 90% of threshold

Fig. 32.1 (a) The six-block encoding/retrieval experimental design utilized for visuo-spatial episodic memory investigations with repetitive (r)TMS delivered to the dorsolateral prefrontal cortex (PFC) (Rossi *et al.* 2001, 2004) and to the parietal cortex in the region of the intraparietal sulcus (Rossi *et al.* 2006) and investigation of verbal episodic memory with rTMS to the dorsolateral PFC (Sandrini *et al.* 2003). R-Enc, right rTMS during encoding – no stimulation during retrieval; L-Enc, left rTMS during encoding – no stimulation during retrieval; Sham, sham rTMS (left during encoding and right during retrieval); R-Ret, right rTMS during retrieval – no stimulation during encoding; L-Ret, left rTMS during retrieval – no stimulation during encoding; Baseline, no stimulation. The gray rectangles indicate the 'reference' blocks (i.e. Sham and Baseline). The design made it possible to test interference with rTMS during both encoding and retrieval. (b) The experimental timing: note that the protocol minimized the WM load, since there was no interval between cue presentation and subjects' response.

A comparable rTMS sham-controlled approach (trains of 500 ms, 20 Hz, subthreshold rTMS) delivered to left or right PFC (BA 45/47) simultaneously with memoranda presentation in encoding did not confirm that left PFC was functionally prevalent over the right PFC in the case of abstract shapes (Floel *et al.* 2004). In this study, 15 subjects had to perform a recognition task in which they were presented with pairs of abstract shapes, in which a stimulus previously seen in encoding (during which active or sham rTMS was delivered) was coupled with a new

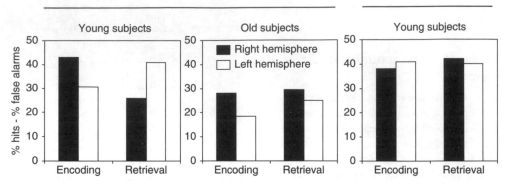

Fig. 32.2 Results of the interaction (right vs left hemisphere and TMS stimulation side during encoding vs retrieval as within-subject factors), showing a clear HERA (Hemispherical Encoding Retrieval Asymmetry) pattern in subjects aged <45 years (37 subjects; Rossi *et al.* 2001, 2004). The asymmetry is lost in the older group (29 subjects; Rossi *et al.* 2004) and in the 10 subjects who received the same rTMS stimulation to the intraparietal sulcus (IPS) (Rossi *et al.* 2006). PFC, prefrontal cortex.

shape that served as a distractor. Recognition errors, based on a subjective rating of each picture of the presented couple as 'well remembered', 'familiar', or 'new', were significantly higher when the right dorsolateral PFC had been stimulated (versus left and sham), suggesting a specific role of this region for encoding of visuo-spatial material (Floel *et al.* 2004). These results are substantially in agreement with another TMS study (Epstein *et al.* 2002) in which 10 healthy subjects were tested using an associative memory task involving six trials of three pairs of Kanji (Chinese) pictographs and unfamiliar abstract patterns (displayed for 500 ms), so that only the abstract pattern represented novel material. Here, TMS interference consisted of two suprathreshold TMS pulses at delays of 140 and 180 ms delivered during the blackout interval (lasting 500 ms) following the presentation of each Kanji character and its matching pattern. Subjects were instructed to remember all three pairs of word–pattern association. Immediately after the end of each presentation set, the subject was handed three cards labelled with the three Kanji characters and a sheet containing six abstract patterns, half of which had previously appeared during the encoding phase. Finally, the subject was asked to lay the Kanji cards directly on the patterns with which they had been paired, therefore

introducing a WM load concerning the maintenance of the associations. However, the percentage of correct responses in subsequent recall of new associations was significantly lower when TMS was applied to the right dorsolateral PFC (~40%), and active TMS delivered to left dorsolateral PFC or on the vertex (as well as a sham TMS in five additional subjects) led to correspondingly better performances (~60% of correct responses) (Epstein *et al.* 2002).

By using the same episodic memory paradigm and rTMS parameters successfully used to confirm functional asymmetries of PFC during memorization of scenes (Rossi *et al.* 2001, 2004), the same authors addressed the functional relevance of parietal cortices during episodic memorization (Rossi *et al.* 2006). The work was motivated by converging neuroimaging findings that demonstrated increased activations in the intraparietal sulcus (IPS) region during episodic encoding and retrieval (Wagner *et al.* 2005) and by recent EEG investigations carried out with the same set of memoranda utilized for rTMS studies on PFC function (Rossi *et al.* 2001, 2004); indeed, these high-resolution EEG studies showed specific changes in the gamma band of left parietal reactivity during encoding and in right parietal areas during retrieval (Babiloni *et al.* 2004c), mainly related to an increase of fronto-parietal connectivity in the left hemisphere during the encoding and in the right hemisphere during the retrieval (Babiloni *et al.* 2006), in line with the prefrontal HERA model predictions (Habib *et al.* 2003). The interference of rTMS (90% of MT) on encoding/retrieval performance was negligible (Figure 32.2), lacking specificity even when higher intensities of stimulation (120% of MT) were applied. However, rTMS at 90% of MT delivered on the right IPS lengthened RTs more than left stimulation and more than sham rTMS, in the context of a purely attentive visuo-spatial task, in line with other studies on the same topic (cf. Chapters 27, 28, and 33, this volume). These results suggested that the activity of the intraparietal sulci shown in several fMRI studies on episodic memory, unlike that of dorsolateral PFCs, is not causally engaged to a useful degree in memory encoding and retrieval of visual scenes representing real-life environments. Therefore, the parietal activations accompanying the memorization processes could reflect the engagement of a widespread brain attentional network, in which interference on a single 'node' is probably insufficient for an overt disruption of memory performance, which requires an overall scan of the picture rather than selective attention to details (Rossi *et al.* 2006).

Studies of verbal episodic memory

Two sham-uncontrolled studies have suggested the possibility that high-frequency rTMS to the PFC might have been able to interfere with free recall of a word list in normal subjects (Grafman *et al.* 1994; Flitman *et al.* 1998). More recently, verbal free recall was also impaired by low-frequency rTMS (0.9 Hz, 110% of individual MT) of the left dorsolateral PFC (Skrdlantova *et al.* 2005). A more systematic approach to encoding and episodic retrieval processes of verbal material was carried out by Sandrini *et al.* (2003). They investigated, using an event-related active and sham rTMS design, the role of PFCs in a verbal episodic memory task on 12 healthy young Italian-speaking subjects, who were tested according to the six-block (and same rTMS parameters) encoding/retrieval protocol previously utilized for visuo-spatial material (Rossi *et al.* 2001) (Figure 32.1). For each of the six encoding phase blocks, 16 word pairs (eight semantically related and eight unrelated) were randomly presented on the monitor for 2000 ms, with two inter-trial intervals. During the encoding phase the subjects were asked to classify word pairs as highly associated (e.g. bread–butter, garlic–onion) or nonassociated (cow–table), according to norms collected for this experiment. One hour later, they were presented with the first word of each pair and a choice between the second and a novel word, which served as distractor. At retrieval, subjects indicated by a mouse key-press the position of the word seen during encoding. A significant interference by active, but not sham, rTMS was produced by right and left dorsolateral PFC stimulation, resulting in a higher error rate during retrieval of the cue words. More importantly, the effect was specific for semantically unrelated word pairs, thereby suggesting that novelty for, but not familiarity with, memoranda specifically engaged dorsolateral PFC(s) (Sandrini *et al.* 2003), in line with a previous pilot study with visuo-spatial memoranda (Miniussi *et al.* 2003). The bilateral reliance on PFCs in encoding was somewhat unexpected, since most neuroimaging studies indicated a more strict left-lateralization for verbal materials (Fletcher and Henson 2001; Floel *et al.* 2004). A tentative speculation was that the task might have required a deep manipulation, in agreement with the 'dual-coding theory' regarding processing of abstract nouns (Paivio 1986), which would rely almost exclusively on left-sided verbal code representations, whereas concrete nouns (which require a relatively high imagery content) additionally would access a second image-based right-sided processing system. The results for retrieval were more predictable, and in agreement with the classic HERA model (Habib *et al.* 2003), with a definite right-sided prevalence suggested by the rTMS interference on the behavioral performance.

The causality of PFC regions in verbal encoding operations has been addressed by TMS in other studies. Floel *et al.* (2004) examined 15 subjects while performing a recognition task in which they were presented with pairs of words containing a set of words previously seen in encoding (during which active or sham rTMS was delivered simultaneously with word presentation) coupled with the same number of new words, which served as distractors. Recognition errors, based on a subjective rating of each word of the presented couple as 'well remembered', 'familiar', or 'new', were significantly higher when the left PFC (BA 45/47) had been stimulated (versus right and sham stimulations), suggesting a specific role of this region for encoding of verbal material (Floel *et al.* 2004). The key role of the left PFC (dorsolateral) for verbal encoding operations was confirmed in another rTMS study (Rami *et al.* 2003) examining the performance of 16 healthy subjects (along a double-blind crossover within-subject repeated measures design) in different memory subtypes.

Two recent studies have investigated the role of the ventrolateral PFC in verbal encoding with TMS. An fMRI-guided TMS study examined the role of the left inferior PFC regions during a recognition memory task (Kohler *et al.* 2004). Two additional sites (right inferior PFC and left parietal) were used as active controls. Trains of rTMS (600 ms of 7 Hz rTMS) were applied during the encoding of words, which were intermixed with words presented in the absence of stimulation. Subsequently, subjects performed a recognition memory task for the encoded words. In contrast with the other studies suggesting an interference of rTMS on verbal encoding (Rami *et al.* 2003; Sandrini *et al.* 2003; Floel *et al.* 2004), words encoded during left rTMS were subsequently recognized with higher accuracy than words encoded under stimulation of the two cortical control sites, while no performance difference emerged when the two control sites were compared with each other. These results led the authors to hypothesize that rTMS to the left inferior PFC might have induced a physiological facilitation of the encoding process, resulting in a behavioral gain due to increased item distinctiveness. Another fMRI-guided study used single-pulse TMS applied to the right and left posterior ventrolateral PFC (Kahn *et al.*, 2005). The results indicated that TMS stimulation 380 ms after stimulus onset applied to the left side impaired recognition performance; an opposite facilitation effect, reflected by increases in ratings of recognition confidence, was observed in the case of right-sided stimulation. These contrasting

results may be due to the different stimulation parameters, as well as to the heterogeneity of the memory tasks. However, they might also be due partly to the poor spatial selectivity of the large round coil utilized for stimulation (Kohler *et al.* 2004), despite the fMRI-neuronavigation approach for its correct positioning on the left ventrolateral PFC.

Episodic memory in aging

The ability to learn and remember new information declines with age, as suggested by both cross-sectional and longitudinal studies (Grady and Craick 2000). These behavioral changes are associated with functional changes: indeed, there is neuroimaging evidence suggesting that PFC activations in elders tend to be less asymmetric, as conceptualized by the so-called HAROLD model (Hemispheric Asymmetry Reduction in OLDer adults (Cabeza 2002)). Due to the lack of causality between neuroimaging findings and performance, it remains an open question whether asymmetry reductions reflect a compensation mechanism, helping to counteract the age-related cognitive decline, or a dedifferentiation process, reflecting the age-related failure to allocate functional resources in a selective manner. This issue has been recently addressed using TMS (Rossi *et al.* 2004) by comparing the effects of a rapid-rate train occurring simultaneously with the presentation of memoranda, applied to the left or right dorsolateral PFC, on visuo-spatial recognition memory in 66 healthy subjects divided into two age groups (<45 and >50 years). Baseline memory performance in the two age groups was comparable. Experimental timing and settings were the same as previously utilized for healthy young subjects (Rossi *et al.* 2001; see Figure 32.1). The study confirmed that in young subjects left and right dorsolateral PFC are functionally necessary for encoding and retrieval, respectively, in line with the HERA model. More importantly, results extended the predictions of the neuroimaging-based HAROLD model, by showing differential effects on encoding and retrieval. In particular, aging did not influence the role of the left dorsolateral PFC during encoding where left rTMS consistently decreased the probability of a successful retrieval in both young and elderly groups.

This pointed to a clear functional relevance of the left dorsolateral PFC for encoding throughout the life-span, at least in this experimental context. Conversely, the functional asymmetries of the dorsolateral PFC during retrieval progressively vanished with aging (see Figure 3 in Rossi et al. 2004), since recognition errors produced by left or right rTMS were similar. This strongly suggested that the predominant role of the right dorsolateral PFC during retrieval was age dependent. Hence, rTMS interference provided direct evidence that the age-related reduction of hemispheric asymmetry, with the loss of the HERA pattern (Figure 32.2), reflects a compensatory engagement of the left hemisphere to support a comparable level of performance (Cabeza et al. 2002), rather than a dedifferentiation process in the right hemisphere, as predicted by the neuroimaging-derived right hemi-aging model (Dolcos et al. 2002). Neither neuroimaging nor rTMS approaches, however, can definitely clarify whether the compensatory changes in the PFCs reflect an anatomo-functional reorganization or the age-related utilization of different cognitive strategies, including distinct developmental experiences acquired earlier in life by people born in the 1930s (i.e. more verbally oriented) compared with those born in the 1980s (i.e. more visually oriented).

The relevance of functional changes accompanying the aging process in other PFC subregions, besides dorsolateral PFC (see Rajah and D'Esposito 2005), still needs to be investigated by TMS, in the case of both working memory and episodic memory processes.

Lexical–semantic processing and analogical reasoning: TMS-induced facilitatory effects?

Lexical–semantic processing

Several studies have investigated the effects of TMS on the processes involved in word retrieval, which are considered to engage a specific aspect of semantic memory, i.e. conceptual knowledge and lexical representations. It is noteworthy that facilitation, rather than inhibition, has been reported after rTMS applied to several cerebral regions involved in language processing, both at high and low frequency of stimulation. Mottaghy et al. (1999b) asked healthy subjects to name pictures as quickly as possible, immediately after rTMS to Wernicke's area of the dominant hemisphere (20 Hz, 2 s, estimated prefixed subthreshold intensity). The stimuli were 10 black-and-white line drawings (representing common objects of everyday life), standardized for frequency and visual complexity, presented twice. Spatial and temporal control conditions included stimulation of Wernicke's and Broca's areas of the right hemisphere and the primary visual cortex, and a retest of picture naming after 2 min. A further control condition included a sham stimulation in a subgroup of subjects. Accuracy of naming was never affected by rTMS, while RTs for naming significantly decreased only after stimulation of the dominant Wernicke's area, compared to naming without rTMS and to the other conditions. Such facilitatory effects persisted no longer than 30 s after rTMS (Mottaghy et al. 1999b). A similar shortening of naming objects after stimulation of Wernicke's area, but not of the motor cortex of the nondominant temporal lobe, was found by Topper et al. (1998) in a large sample of healthy subjects who received single-pulse TMS. The facilitatory effect on naming latencies appeared only when TMS preceded the picture presentation by 500 or 1000 ms, and occurred only within a certain range of stimulation intensity (33–55% of the stimulator output), therefore minimizing the possibility that intersensory facilitatory mechanisms might interfere with the task. Such intensity-dependent facilitation of picture naming was also confirmed in the case of high- and low-frequency preconditioning of Wernicke's area (Sparing et al. 2001). Dräger et al. (2004) found contrasting effects of 1 Hz rTMS for 600 s at 110% of subjects' resting MTs to Wernicke's and Broca's area. Stimulation of the former resulted in inhibition during a picture–word verification task, while facilitation was present after stimulation of the latter. Low-frequency stimulation to Wernicke's area has also been recently reported to facilitate perceptual linguistic processing in an fMRI guided study (Andoh et al. 2006). In an investigation of object and action naming, Cappa et al. (2002) tested nine healthy subjects who were required to name as quickly as possible different sets, matched for lexical frequency, of objects belonging to different semantic categories or actions implying tool utilization. Naming latencies were determined as the initial wave envelope of the microphone-recorded verbal response. Trains of active rTMS (500 ms, 20 Hz, 90% of MT), coincident with picture presentation, were randomly delivered to the left and right dorsolateral PFC; the sham stimulation was applied with the coil on the vertex. Accuracy was not affected in any condition, whereas verbal RTs showed faster naming latency for actions only after stimulation of the left dorsolateral PFC, confirming the key role of this region in verb production.

Nonverbal reasoning

Nonverbal analogical reasoning is a complex high-level cognitive function which allows one to establish conceptual analogies between different stimuli, scenes, or events. There is substantial lesion and neuroimaging evidence supporting the role of the left PFC in reasoning processes (Wharton and Grafman 1998). This issue was recently addressed with an on-line rTMS approach (Boroojerdi et al. 2001). The authors applied focal active or sham rTMS (10% below MT, three trains of 10 s duration and 5 Hz frequency) to the left PFC in 16 healthy individuals while they were shown two sets of colored geometric shapes presented in a match-to-sample (or literal) and analogy condition. The latter was based on shapes with the same system of abstract visuo-spatial relations, but of different geometric form (see Figure 1 of Boroojerdi et al. 2001). The two conditions included four blocks of 16 randomly ordered trials (run in separate sessions with at least 1 day in between), each one containing a sequential or simultaneous presentation of two pictures, the former being used to control for eventual effects on WM processes. Sequential and simultaneous trials included a source and a target picture. Subjects were asked to judge, in an RT paradigm, whether the presented pictures were analogous (analogy condition) or identical (literal condition). Active rTMS to the left PFC, whose coordinates were determined by an independent PET investigation, significantly reduced RTs in the analogy

condition versus sham for both sequential and simultaneous trials. RTs were neither affected in the literal condition nor following rTMS of the right PFC, whereas stimulation of the left motor cortex nonspecifically shortened RTs, since this effect was present in all conditions. Error rate was not affected by rTMS. These results suggest that a form of analogical reasoning can be facilitated by low-intensity rTMS to the left PFC, thus implying that the functioning of this region is crucial for this cognitive process. The mechanisms explaining such facilitation remain unclear, also considering that lesions of the left PFC impair inductive reasoning (Reverberi et al. 2005); however, if replicated, these results may open new neuromodulatory strategies for research in the field of rehabilitation of conceptual impairments.

Rigorous explanations for the prevalence of facilitatory effects of rTMS on linguistic processing and nonverbal reasoning are still lacking. It can be hypothesized that, since the effect may persist shortly after the end of the rTMS train, it is due to an increase in the strength of synaptic connections among neurons induced to fire simultaneously by rTMS (i.e. Hebbian learning), as postulated for motor cortex circuitry (Lee et al. 2003). An emerging issue in this sense is the use of EEG individual alpha frequency (IAF) as a 'physiological template' to set the best tuning of rTMS for inducing transient improvements of cognitive performance. When rTMS was delivered on the right parietal cortex at IAF (usually 10–12 Hz) + 1 Hz, the combined increase of the alpha power (which is generally related to good cognitive performance) was paralleled by a shortening of the time required to successfully complete a psychophysical task of mental rotation (Klimesh et al. 2003). Further studies are required to adequately answer these questions, which may have extremely interesting practical implications.

Safety aspects

A question of safety arises when considering the short-lived impairments on different aspects of memory performance induced by rTMS, as well as the interference with the other cognitive functions: is there any detectable long-term neuropsychological consequence of these 'experimental intrusions' in our brains? To the best of our current knowledge, the answer is negative for a wide sample of healthy subjects who received, for experimental purposes, a single session of 1 Hz rTMS (Koren *et al.* 2001) or 25 min rTMS of the left dorsolateral PFC (i.e. 50 × 2 s train of 40 pulses, 20 Hz at 80% MT) (Roth *et al.* 2004). In the latter study, several declarative memory domains were not affected by rTMS, despite MT changes occurring in the homologous motor cortex.

The answer is again negative when considering the relatively large groups of patients with major depression who have received weekly sessions of rTMS for therapeutic purposes: in these patients, no neuropsychological deterioration was found for several neurocognitive variables, with the exception of a mild impairment in verbal memory in a single study (Fabre *et al.* 2004). Conversely, mild improvements of cognitive functions, including verbal and visuo-spatial memory, executive functions, and verbal fluency, were generally detected on long-term follow-up, which appear to be independent of the improved mood (Martis *et al.* 2003; Fabre *et al.* 2004; Hausmann *et al.* 2004; Schulze-Rauschenbach *et al.* 2005). It is also noteworthy that positive effects of rTMS on naming abilities have recently been reported in patients with chronic aphasia (Naeser *et al.*, 2005). If these effects are confirmed in double-blind placebo-controlled studies, then we may enter an exciting new chapter of the TMS story, aimed at understanding the potential of TMS for the rehabilitation of cognitive functions (Rossi and Rossini 2004).

Concluding remarks

Interference with rTMS provides relevant information, complementary to neuroimaging and EEG findings, about the neural mechanisms of short-term and declarative memory. The most consistent result, at the moment, is the direct demonstration of the key role of the PFC in episodic encoding and WM processes (here coupled with fronto-parietal networks), with a right prevalence in retrieval operations of young subjects, irrespective of the nature of the memorized material. However, some discrepancies also emerged between different TMS studies and between TMS and neuroimaging investigations,

particularly regarding the expected role of the left PFC in verbal WM processes.

Moreover, many aspects of declarative memory remain to be investigated by rTMS, the most relevant being compensation/dedifferentiation issues during WM tasks in the elderly, physiological mechanisms of source memory, recency and autobiographic memory, and functional correlates of memory recovery after discrete PFC lesions. Improvement of cognitive functions, such as lexical–semantic processing and nonverbal reasoning, by rTMS seems to represent an exciting challenge for the near future, taking into account encouraging findings from other sources on procedural and nondeclarative learning.

It is clear that higher cognitive functions provide a good example as to how underlying physiological mechanisms cannot be fully disclosed by investigations based on a single technique. Future studies will need to take advantage of a true multimodal approach to the working brain, with the aim of combining the relative advantages provided by neuroimaging techniques such as PET or fMRI (with their superior spatial resolution) and by human electrophysiological methods such as EEG and ERPs (which have comparable temporal resolution to TMS). In this context, behavioral interference studies will gain new power in combination with disruptive and correlational methodologies, establishing causality in a more sophisticated manner than has been possible hitherto.

References

Abler B, Walter H, Wunderlich A, *et al.* (2005). Side effects of transcranial magnetic stimulation biased task performance in a cognitive neuroscience study. *Brain Topography 17*, 193–196.

Alexander MP, Stuss DT, Fansabedian N (2003). California Verbal Learning Test: performance by patients with focal frontal and non-frontal lesions. *Brain 126*, 1493–1503.

Andoh J, Artiges E, Pallier C, *et al.* (2006). Modulation of language areas with functional MR image-guided magnetic stimulation. *NeuroImage 29*, 616–627.

Babiloni C, Babiloni F, Carducci F, *et al.* (2004a). Human cortical responses during one-bit short-term memory. A high-resolution EEG study on delayed choice reaction time tasks. *Clinical Neurophysiology 115*, 161–170.

Babiloni C, Babiloni F, Carducci F, *et al.* (2004b). Functional frontoparietal connectivity during short-term memory as revealed by high-resolution EEG coherence analysis. *Behavioral Neuroscience 118*, 687–697.

Babiloni C, Babiloni F, Carducci F, *et al.* (2004c). Human cortical EEG rhythms during long-term episodic memory. A high-resolution EEG study of the HERA model. *NeuroImage 21*, 1576–1584.

Babiloni C, Vecchio F, Cappa S, *et al.* (2006) Functional frontoparietal connectivity during encoding and retrieval processes follows the HERA model. A high-resolution EEG study. *Brain Research Bulletin 68*, 203–212.

Baddeley A (1992). Working memory. *Science 255*, 556–559.

Baddeley A, Conway M, Aggleton J (eds) (2001) Episodic memory. *Philosophical Transactions of the Royal Society of London B 356*, 1341–1515.

Boroojerdi B, Phipps M, Kopylev L, Wharton CM, Cohen LG, Grafman J (2001). Enhancing analogic reasoning with rTMS over the left prefrontal cortex. *Neurology 56*, 526–568.

Buckner RL, Wheeler ME (2001). The cognitive neuroscience of remembering. *Nature Neuroscience Reviews 2*, 624–634.

Cabeza R (2002). Hemispheric asymmetry reduction in older adults: the HAROLD model. *Psychology and Ageing 17*, 85–110.

Callicott JH, Mattay VS, Bertolino A, *et al.* (1999). Physiological characteristics of capacity constraints in working memory as revealed by functional MRI. *Cerebral Cortex 9*, 20–26.

Cappa SF, Sandrini M, Rossini PM, Sosta K, Miniussi C (2002). The role of the left frontal lobe in action naming: rTMS evidence. *Neurology 59*, 720–723.

Dolcos F, Rice HJ, Cabeza R (2002). Hemispheric asymmetry and aging: right hemisphere decline or asymmetry reduction. *Neuroscience and Biobehavioral Reviews 26*, 819–825.

Dräger B, Breitenstein C, Helmke U, Kamping S, Knecht S (2004). Specific and nonspecific effects of transcranial magnetic stimulation on picture–word verification. *European Journal of Neuroscience 20*, 1681–1687.

Dresmond JE, Chen SH, Shieh PB (2005). Cerebellar transcranial magnetic stimulation impairs verbal working memory. *Annals of Neurology 58*, 553–560.

Epstein CM, Sekino M, Yamaguchi K, Kamiya S, Ueno S (2002). Asymmetries of prefrontal cortex in human episodic memory: effects of transcranial magnetic stimulation on learning abstract patterns. *Neuroscience Letters 320*, 5–8.

Fabre I, Galinowski A, Oppenheim, C *et al.* (2004). Antidepressant efficacy and cognitive effects of repetitive transcranial magnetic stimulation in vascular depression: an open trial. *International Journal of Geriatric Psychiatry 19*, 833–842.

Fletcher PC, Henson RN (2001). Frontal lobes and human memory: insights from functional neuroimaging. *Brain 124*, 849–881.

Flitman SS, Grafman J, Wassermann EM, *et al.* (1998). Linguistic processing during repetitive transcranial magnetic stimulation. *Neurology 50*, 175–181.

Floel A, Poeppel D, Buffalo EA, *et al.* (2004). Prefrontal cortex asymmetry for memory encoding of words and abstract shapes. *Cerebral Cortex 14*, 404–409.

Grady CL, Craik IM (2000). Changes in memory processing with age. *Current Opinion in Neurobiology 10*, 224–231.

Grafman J, Pascual-Leone A, Alway D, Nichelli P, Gomez-Tortosa E, Hallett M (1994). Induction of a recall deficit by rapid-rate transcranial magnetic stimulation. *Neuroreport 9*, 1157–1160.

Habib R, Nyberg L, Tulving E (2003). Hemispheric asymmetries of memory: the HERA model revisited. *Trends in Cognitive Sciences 7*, 241–245.

Herwig U, Satrapi P, Schonfeldt-Lecuona C (2002). Using the international 10–20 EEG system for positioning of transcranial magnetic stimulation. *Brain Topography 16*, 95–99.

Herwig U, Abler B, Schonfeldt-Lecuona C, *et al.* (2003). Verbal storage in a premotor-parietal network: evidence from fMRI-guided magnetic stimulation. *NeuroImage 20*, 1032–1041.

Hausmann A, Pascual-Leone A, Kemmler G, *et al.* (2004). No deterioration of cognitive performance in an aggressive unilateral and bilateral antidepressant rTMS add-on trial. *Journal of Clinical Psychiatry 65*, 772–782.

Hong KS, Lee SK, Kim JY, Kim KK, Nam H (2000). Visual working memory revealed by repetitive transcranial magnetic stimulation. *Journal of Neurological Sciences 181*, 50–55.

Hufnagel A, Claus D, Brunhoetzl C, Sudhop T (1993). Short-term memory: no evidence of effect of rapid-repetitive transcranial magnetic stimulation in healthy individuals. *Journal of Neurology 240*, 373–376.

Johnson MK, Raye CL, Mitchell KJ, Greene EJ, Anderson AW (2003). fMRI evidence for an organization of prefrontal cortex by both type of process and type of information. *Cerebral Cortex 13*, 265–273.

Kahn I, Pascual-Leone A, Theoret H, Fregni F, Clark D, Wagner AD (2005). Transient disruption of ventrolateral prefrontal cortex during verbal encoding affects subsequent memory performance. *Journal of Neurophysiology 94*, 688–698.

Kessels RP, d'Alfonso AA, Postma A, de Haan EH (2000). Spatial working memory performance after high-frequency repetitive transcranial magnetic stimulation of the left and right posterior parietal cortex in humans. *Neuroscience Letters 287*, 68–70.

Klimesh W, Sauseng P, Gerloff C (2003). Enhancing cognitive performance with repetitive transcranial magnetic stimulation at human individual alpha frequency. *European Journal of Neuroscience 17*, 1129–1133.

Koch G, Oliveri M, Torriero S, Carlesimo GA, Turriziani P, Caltagirone C (2005). rTMS evidence of different delay and decision processes in a fronto-parietal neuronal network activated during spatial working memory. *NeuroImage 24*, 34–39.

Kohler S, Paus T, Buckner RL, Milner B (2004). Effects of left inferior prefrontal stimulation on episodic memory formation: a two-stage fMRI-rTMS study. *Journal of Cognitive Neuroscience 16*, 178–188.

Koren D, Shefer O, Chistyakov A, Kaplan B, Feinsod M, Klein E (2001). Neuropsychological effects of prefrontal slow rTMS in normal volunteers: a double-blind sham-controlled study. *Journal of Clinical and Experimental Neuropsychology 23*, 424–430.

Lee L, Siebner HR, Rowe JB, et al. (2003). Acute remapping within the motor system induced by low-frequency repetitive transcranial magnetic stimulation. *Journal of Neuroscience 23*, 5308–5318.

Martis B, Alam D, Dowd SM, et al. (2003). Neurocognitive effects of repetitive transcranial magnetic stimulation in severe major depression. *Clinical Neurophysiology 114*, 1125–32

Miniussi C, Cappa SF, Sandrini M, Rossini PM, Rossi S (2003). The causal role of the prefrontal cortex in episodic memory as demonstrated with rTMS. *Clinical Neurophysiology 56*(Suppl), 312–320.

Mottaghy FM, Gangitano M, Krause BJ, Pascual-Leone A (1999a). Chronometry of parietal and prefrontal activation in verbal working memory revealed by transcranial magnetic stimulation. *NeuroImage 18*, 565–575.

Mottaghy FM, Hungs M, Brugman M, et al. (1999b). Facilitation of picture naming after repetitive transcranial magnetic stimulation. *Neurology 53*, 1806–1812.

Mottaghy FM, Krause BJ, Kemna LJ et al. (2000). Modulation of the neuronal circuitry subserving working memory in healthy human subjects by repetitive transcranial magnetic stimulation. *Neuroscience Letters 280*, 167–170.

Mottaghy FM, Doring T, Muller-Gartner HW, Topper R, Krause BJ (2002a). Bilateral parieto-frontal network for verbal working memory: an interference approach using repetitive transcranial magnetic stimulation (rTMS). *European Journal of Neuroscience 16*, 1627–1632.

Mottaghy FM, Gangitano M, Sparing R, Krause BJ, Pascual-Leone A (2002b). Segregation of areas related to visual working memory in the prefrontal cortex revealed by rTMS. *Cerebral Cortex 12*, 369–375.

Mottaghy FM, Pascual-Leone A, Kemna LJ et al. (2003a). Modulation of a brain–behavior relationship in verbal working memory by rTMS. *Brain Research and Cognitive Brain Research 15*, 241–249.

Mottaghy FM, Gangitano M, Krause BJ, Pascual-Leone A (2003b). Chronometry of parietal and prefrontal activations in verbal working memory revealed by transcranial magnetic stimulation. *NeuroImage 18*, 565–575.

Mull BR, Seyal M (2001). Transcranial magnetic stimulation of left prefrontal cortex impairs working memory. *Clinical Neurophysiology 112*, 1672–1675.

Müri RM, Vermersch AI, Rivaud S, Gaymard B, Pierrot-Deseilligny C (1996). Effects of single-pulse transcranial magnetic stimulation over the prefrontal and posterior parietal cortices during memory-guided saccades in humans. *Journal of Neurophysiology 76*, 2102–2106.

Naeser MA, Martin PI, Nicholas M et al. (2005). Improved picture naming in chronic aphasia after TMS to part of right Broca's area: an open-protocol study. *Brain and Language 93*, 95–105.

Nixon P, Lazarova J, Hodinott-Hill I, Gough P, Passingham R (2004). The inferior frontal gyrus and phonological processing: an investigation using rTMS. *Journal of Cognitive Neuroscience 16*, 289–300.

Oliveri M, Turriziani P, Carlesimo GA et al. (2001). Parieto-frontal interactions in visual–object and visual–spatial working memory: evidence from transcranial magnetic stimulation. *Cerebral Cortex 11*, 606–618.

Orazio (2000). Satire, *1*, 9, 75 sg. In: M Bettini (ed.), *Le orecchie di Hermes: luoghi e simboli della comunicazione nella cultura antica.* Torino: Einaudi.

Paivio A (1986). *Mental representations: a dual coding theory.* Oxford: Oxford University Press.

Pascual-Leone A, Hallett M (1994). Induction of errors in a delayed response task by repetitive transcranial magnetic stimulation of the dorsolateral prefrontal cortex. *Neuroreport 5*, 2517–2520.

Pascual-Leone A, Walsh V, Rothwell J (2000). Transcranial magnetic stimulation in cognitive neuroscience – virtual lesion, chronometry, and functional connectivity. *Current Opinion in Neurobiology 10*, 232–237.

Paulesu E, Frith CD, Frackowiack RS (1993). The neural correlates of the verbal component of working memory. *Nature 362*, 342–345.

Paulesu E, Goldacre B, Scifo P et al. (1997). Differential activation of left frontal cortex during phonemic and semantic word fluency. An epi-fMRI activation study. *Neuroreport 8*, 2011–2016.

Petrides M (1994). Frontal lobes and working memory: evidence from investigation of the effects of cortical excisions in nonhuman primates. In: F Boller and J Grafmann (eds), *Handbook of neuropsychology.* Amsterdam: Elsevier.

Petrides M (2005). Lateral prefrontal cortex: architectonic and functional organization. *Philosophical Transactions of the Royal Society of London B 360*, 781–795.

Plinius (2000). Storia naturale, *11*, 251. In: M Bettini (ed.), *Le orecchie di Hermes: luoghi e simboli della comunicazione nella cultura antica.* Torino: Einaudi.

Poldrack RA, Wagner AD, Prull MW, Desmond JE, Glover GH, Gabrieli JDE (1999). Functional specialisation for semantic and phonological processing in the left inferior prefrontal cortex. *NeuroImage 10*, 15–35.

Price CJ and Friston KJ (1999). Scanning patients with tasks they can perform. *Human Brain Mapping 8*, 102–108.

Rajah MN and D'Esposito M (2005). Region-specific changes in prefrontal function with age: a review of PET and fMRI studies on working and episodic memory. *Brain 128*, 1964–1983.

Rami L, Gironell A, Kulisevsky J, Garcia-Sanchez C, Berthier M, Estevez-Gonzalez A (2003) Effects of repetitive transcranial magnetic stimulation on memory subtypes: a controlled study. *Neuropsychologia 41*, 1877–1883.

Ranganath C, Johnson MK, D'Esposito M (2000). Left anterior prefrontal activation increases with demands to recall specific perceptual information. *Journal of Neuroscience 20*, RC108.

Reverberi C, D'Agostini S, Skrap M, Shallice T (2005). Generation and recognition of abstract rules in different frontal lobe subgroups. *Neuropsychologia 43*, 1924–1937.

Rossi S and Rossini PM (2004) TMS in cognitive plasticity and the potential for rehabilitation. *Trends in Cognitive Sciences 8*, 273–279.

Rossi S, Cappa SF, Babiloni C et al. (2001). Prefrontal cortex in long-term memory: an 'interference' approach using magnetic stimulation. *Nature Neuroscience 4*, 948–952.

Rossi S, Miniussi C, Pasqualetti P, Babiloni C, Rossini PM, Cappa SF (2004). Age-related functional changes of prefrontal cortex in long-term memory. A repetitive transcranial magnetic stimulation (rTMS) study. *Journal of Neuroscience 24*, 7939–7944.

Rossi S, Pasqualetti P, Zito GC et al. (2006). Prefrontal and parietal cortex in episodic memory. An interference study with rTMS. *European Journal of Neuroscience 23*, 793–800.

Roth HL, Nadeau SE, Triggs WJ (2004). Effect of repetitive transcranial magnetic stimulation on rate of memory acquisition. *Neurology 63*, 1530–1531.

Sandrini M, Cappa SF, Rossi S, Rossini PM, Miniussi C (2003) The role of prefrontal cortex in verbal episodic memory: rTMS evidence. *Journal of Cognitive Neuroscience 15*, 855–861.

Schulze-Rauschenbach SC, Harms U, Schlaepfer TE, Maier W, Falkai P, Wagner M (2005). Distinctive neurocognitive effects of repetitive transcranial magnetic stimulation and electroconvulsive therapy in major depression. *British Journal of Psychiatry 186*, 410–416.

Skrdlantova L, Horacek J, Dockery C, et al. (2005). The influence of low-frequency left prefrontal repetitive transcranial magnetic stimulation on memory for words but not for faces. *Physiological Research 54*, 123–128.

Sparing R, Mottaghy FM, Hungs M, et al. (2001). Repetitive transcranial magnetic stimulation effects on language function depend on the stimulation parameters. *Journal of Clinical Neurophysiology 18*, 326–330.

Squire LR, Clark RE, Knowlton BJ (2001). Retrograde amnesia. *Hippocampus 11*, 50–55.

Topper R, Mottaghy FM, Brugmann M, Noth J, Huber W (1998) Facilitation of picture naming by focal transcranial magnetic stimulation. *Experimental Brain Research 121*, 371–378.

Tulving E (2002) Episodic memory: from mind to brain. *Annual Review of Psychology 53*, 1–25.

Tulving E, Kapur S, Craik FI, Moscovitch M, Houle S (1994). Hemispheric encoding/retrieval asymmetry in episodic memory: positron emission tomography findings. *Proceedings of the National Academy of Sciences of the USA 91*, 2016–2020.

Vernant JP (2000). Myth and memory in Greece. In: M Bettini (ed.), *Le orecchie di Hermes: luoghi e simboli della comunicazione nella cultura antica*. Torino: Einaudi, pp. 2–37.

Wagner AD, Shannon BJ, Kahn I, Buckner R (2005). Parietal lobe contributions to episodic memory retrieval. *Trends in Cognitive Sciences 9*, 445–453.

Walsh V, Cowey A (1998). Magnetic stimulation studies of visual cognition. *Trends in Cognitive Sciences 2*, 103–110.

Walsh V, Cowey A (2000). Transcranial magnetic stimulation in cognitive neuroscience. *Nature Reviews Neuroscience 1*, 73–79.

Wharton CM, Grafman J (1998). Deductive reasoning and the brain. *Trends in Cognitive Science 2*, 54–59.

CHAPTER 33

Mathematics and TMS

Elena Rusconi and Carlo Umiltà

The mental number line

Mathematical cognition is concerned with a number of more or less related issues: for example, various types of enumeration procedures, such as subitizing, counting, and estimating, the way numbers are represented in the human mind, with spatial, symbolic, or abstract codes, and how arithmetical operations are performed, either by memory retrieval of overlearnt facts or by performing actual calculations (for a recent overview see Campbell 2005). For most of the issues listed above there are relevant studies with functional magnetic resonance imaging (fMRI). In contrast, studies with TMS are much fewer and have a narrower scope. The present chapter therefore focuses only on the analog representation of numbers, i.e. the so-called mental number line in the parietal lobe.

The first to propose that number magnitude is represented through the location of numbers along a mental line oriented from left to right was Sir Francis Galton at the end of the nineteenth century (Galton 1880a,b). Galton's proposal was based on introspective reports obtained from healthy subjects. The notion that number magnitude is spatially represented and the notion of a mental number line were taken up again by Restle (1970), after a seminal paper by Moyer and Landauer (1967) who showed the so-called distance effect. Moyer and Landauer asked their participants to decide which of two visually presented numbers was larger by pressing one of two response buttons. They found that speed of response increased as a function of the numerical distance between the numbers (e.g. participants were faster at deciding that 8 was larger than 2 than they were at deciding it was larger than 6, and the speed was intermediate for 8 and 4). The explanation provided was that in performing the comparison task the two numbers were located in their correct positions on the number line and the speed of the comparison process was modulated by the distance that separated them on the mental number line. A related effect was the size effect, according to which the speed of the comparison process decreases as a function of the size of the two numbers to be compared, when distance between them is equated (Strauss and Curtis 1981; Antell and Keating 1983). That is to say, deciding that 15 is larger than 11 is faster than deciding that 25 is larger than 21. The size effect is different from the *problem size effect* (Groen and Parkman 1972; Zbrodoff and Logan 2005), which concerns arithmetical operations and is due to the fact that the time needed to solve an arithmetical problem increases as a function of the size of the correct result. For the size effect, two explanations have been proposed. One of them evokes scalar variability (e.g. Cordes *et al.* 2001) and states that, regardless of the existence of the mental number line, magnitude representation is approximate and the variability of the distribution increases proportionally with magnitude, so that there is more overlap between larger than smaller numbers. The alternative explanation instead invokes the mental number line by stating that it is compressed in such a way that, for equal numerical intervals, distance on the number line would be smaller as the size of the numbers increases (e.g. Dehaene *et al.* 1990).

The most compelling evidence in favor of the existence of a mental number line was provided by Dehaene *et al.* (1993). Their participants were asked to decide whether a visual number, centrally presented, was even or odd by pressing one of two lateralized keys. Thus, magnitude of the number was irrelevant to the task. Large numbers were responded to faster with the right than with the left key and small numbers were responded to faster with the left than with the right key. This effect was named Spatial–Numerical Association of Response Codes (SNARC) and was explained by assuming that, although magnitude was task irrelevant, the mere presentation of the number activated the mental number line on which the number was then located. Therefore, smaller numbers were on the left side of the mental representation whereas larger numbers were on the right side. It is worth stressing that the numbers were in fact presented at fixation, i.e. in the center of the display.

It emerged that what mattered was not the hand executing the response but rather the left–right location of the response device. This was demonstrated by asking the subjects to operate the response keys with their arms crossed so that the right hand operated in left space and pressed the left key, whereas the left hand operated in right space and pressed the right key. Despite the crossing of the hands the SNARC effect did not reverse (but see Wood *et al.* (2006) for the finding of a null SNARC effect when hands are crossed). How the SNARC effect is affected by the nature of the effector that executes a response is an important issue in its own right. This is because the fact that the SNARC effect can be found with different effectors and different response modalities (e.g. manual pointing and saccadic responses; Fischer 2003; Schwarz and Keus 2004) may be taken as evidence that the SNARC effect is not actually produced by the spatial code of the response but rather by an attentional orienting. What we have in mind (Rusconi *et al.* 2005a) is that the presentation of the number triggers an attentional orienting to the left or to the right

depending on its magnitude, which in turn produces the spatial code of the response; i.e. the spatial code that enters the response selection stage is produced at a stage that precedes the one in which the task-specific response is selected.

Another interesting finding was that the left–right location of a given number on the mental number line depended on the range of numbers presented in the experiment. For example, number 8 was on the right side when the range was 0 to 9, whereas it was on the left side when the range was 5 to 14.

In addition, it is known that the SNARC effect can be found regardless of whether numbers are presented as arabic digits or as written number words (e.g. Dehaene *et al.* 1993).

The SNARC effect can also be found when the task requires the participant to compare the magnitude of the number presented at fixation with a fixed standard, very often with number 5 or 65, as in the original paper by Dehaene *et al.* (1990). The fact that the SNARC effect manifests itself when the task explicitly requires the use of number magnitude is, however, less surprising than when it manifests itself in a task that ostensibly has nothing to do with number magnitude. In fact, the SNARC effect obtained with a parity judgment task is often taken as evidence not only of the existence of the mental number line, but also of the fact that it is automatically activated by the mere presentation of a numeral.

The neural basis of the mental number line

At the beginning of this chapter we stated that the mental number line is located in the parietal lobe. In the present section we will discuss in more detail the evidence that led to such a conclusion. As we will see, there are still aspects of the neural basis of the mental number line that are in need of clarification and provide a useful contribution.

However, the first question to ask when discussing the issue of how magnitude is represented in the brain is what neuronal mechanisms are involved. Sawamura *et al.* (2002) trained macaque monkeys to perform an action a given number of times and then to switch to

Table 33.1 TMS protocol details

Study	Stimulation sites	Localization method	Frequency of stimulation	Intensity of stimulation	Motor threshold (MT)	Duration of stimulation	Stimulator model
Göbel et al. 2001	AGs vs SMGs	Functional and Brainsight	10 Hz	105% of individual MT	Min. % of stimulator output evoking visually detectable twitch in a tense muscle of the hand	500 ms, 100 ms before stimulus onset	Magstim Super Rapid
Oliveri et al. 2004	P3 vs P4	10/20	1 Hz offline	90% of individual MT	See Rossini et al. 1994	5 min (30–60 min rest before the next session)	Magstim Rapid
Sandrini et al. 2004	1 cm lateral to CP3 vs 1 cm lateral to CP4	10/20 and Polhemus	15 Hz	110% of individual MT	Min. intensity inducing a visible contraction in the first dorsal interosseus	225 ms, from stimulus onset	Magstim Super Rapid
Andres et al. 2005	P3 vs P4 vs P3 + P4	10/20 and Polhemus	Single pulse	130% of individual MT	Min. intensity required to generate 5/10 visible finger movements	Pulses at 150, 200, 250 ms after stimulus onset	Magstim 200
Rusconi et al. 2005	AGs	Brainsight	10 Hz	60% of max. stimulator output	–	500 ms, from stimulus onset	Magstim Super Rapid
Göbel et al. 2006	AG vs OCC	Functional and Brainsight	5 Hz	110% of individual MT	See Göbel et al. 2001	1000 ms, from stimulus offset	Magstim Super Rapid
Rusconi et al. 2005a, 2007	AGs vs SMGs	Polhemus	10 Hz	62% of max stimulator output	–	500 ms, from stimulus onset	Magstim Rapid
Rusconi et al. 2006	Right FEF and IFg vs vertex	Brainsight	12 Hz	110% of individual MT	See Göbel et al. 2001	400 ms, from stimulus onset	Magstim Super Rapid

In all studies a 70 mm figure-8 coil was used, except for Rusconi et al. (2006) who used a 50 mm figure-8 coil.

SMG, supramarginal gyrus; AG, angular gyrus; OCC, occipital cortex,; FEF, frontal eye field; IFg, inferior frontal gyrus.

Table 33.2 Experimental paradigm details

Study	Set of stimuli	Task	Stimuli per trial	Critical area	TMS effect
Göbel et al. 2001	Range: 31 to 99	Comparison with 65	1 central	left AG	Slowing of RTs for numbers larger than and close to 65 with compatible mapping; generalized slowing of RTs with incompatible mapping
Oliveri et al. 2004	Range: 1–99	Comparison of two numerical intervals	3 in horizontal alignment	P4	Elimination of pseudo-neglect
Sandrini et al. 2004	1–9	Comparison of two digits	2 in horizontal alignment	1 cm lateral to CP3	Generalized slowing of RTs
Andres et al. in press	1–9, except 5	Comparison with 5	1 central	P3 and P3 + P4	Slowing of RTs to numbers close to 5 with TMS on P3 and P3 + P4; slowing of RTs to numbers far from 5 with TMS on P3 + P4
Rusconi et al. 2005b	1–9, except 5 + task-irrelevant double-digit numbers	Magnitude matching + arithmetic prime	2 horizontal and then 1 central	Left AG	Slowing of RTs in trials with unrelated primes (magnitude-matching task)
Göbel et al. in press	Three-digit spoken numbers	Mental number bisection	Utterance with 2 numbers separated by 500 ms	Right AG	Reduction of pseudo-neglect
Rusconi et al. 2005a, 2007	1–9, except 5	Parity judgment	1 to the right or left of fixation	Left and right AG/posterior IPS	Elimination of SNARC interference
Rusconi et al. 2006	1–9, except 5	Parity judgment magnitude comp.	1 central	Right FEF and right IFg	In magnitude comparison only: elimination of SNARC interference for small numbers on the FEF; for both small and large numbers on IFg

AG, angular gyrus; RT, reaction time; IPS, intraparietal sulcus; SNARC, Spatial–Numerical Association of Response Codes; FEF, frontal eye field; IFg, inferior frontal gyrus.

another action. In the intraparietal sulcus (IPS), in a location that likely corresponds to the human horizontal intraparietal sulcus (HIPS) (see also Nieder and Miller 2003), they found neurons that were tuned to a particular number of repetitions of the same action. Different populations of neurons were tuned to a different number of repetitions from 1 to 5. Neurons tuned to different numbers were reported also by Nieder et al. (2002) (see also Nieder and Miller 2003) but they were mainly located in a prefrontal area. Their macaque monkeys were trained to perform a match-to-sample task on two visual displays presented in succession which showed from one to five randomly arranged items. In the displays, number was confounded with physical variables such as total area or total length of the contours. However, the role of these confounds was excluded because after training the monkeys generalized to novel displays in which all non-numerical variables were disentangled. In accordance with Sawamura et al. (2002), Nieder et al. (2002) found that each neuron was tuned to a specific numerosity between 1 and 5. Interestingly, Nieder et al.'s neurons showed broad tuning curves, which suggests approximate numerosity coding. In addition the breadth of the tuning curves increased as a function of the neuron's preferred number and was skewed toward the right, in accordance with the notion of a compressed number line. Nieder (2004) suggests that the neurons in the prefrontal region might have been flexibly adapted, during the training phase, to perform the particular numerical task at hand. They may constitute adjustable neuronal ensembles that reorganize according to different requirements (a prefrontal neural circuitry guiding executive functions). 'Such neuronal numerical representations may not be established automatically (as seems to be the case in IPS), but they are nevertheless genuine and absolutely necessary for the monkey's behaviour' (Nieder 2004, p. 408).

Based on the evidence available, it would seem that the site of the mental number line in the human brain is the HIPS. Converging evidence comes mainly from brain-imaging studies that have used fMRI. Before proceeding with a brief review of the relevant studies, the Triple Code model of number processing must be introduced (Dehaene 1992; Dehaene and Cohen 1997). The crucial aspect of the model is that three distinct codes are involved in processing numbers: a magnitude code, a verbal code, and a visual code. The magnitude code would be a nonverbal semantic representation of the number size and distance among numbers in the form of a mental number line. In the verbal code, numerals are represented like other words, having lexical, phonological, and syntactical representations. In the visual code, numbers are represented as strings of arabic numerals. Each code is thought to underlie different operations involving numbers. The visual arabic code is used for accessing the other two through the visual modality, for performing arithmetical complex calculations which require procedures such as borrowing and carrying over in the written form, and for operations that do not depend on the magnitude code, such as parity judgments (Dehaene et al. 1993). The verbal code is used for rote learning, storage, and retrieval of arithmetical facts, the best known example being that of multiplication tables. The magnitude code is used for manipulating quantities, as happens in number comparison tasks or in performing arithmetical operations when the result cannot be retrieved from memory, the best known example being subtraction. In contrast, multiplication and small addition facts are stored in rote verbal memory and minimally require magnitude manipulation. However, the magnitude code is also used with multiplications and additions when participants are asked to provide the approximate result rather than computing the exact solution. Moreover, as we noted when we introduced the SNARC effect, the magnitude code can become automatically available even though its use is not required by the task at hand (e.g. in a parity judgment).

To date, all the studies that have employed TMS explored issues related to the mental number line; therefore in what follows we will confine ourselves to studies centered on the magnitude code. In agreement with the Triple Code model (e.g. Dehaene and Cohen 1997), several studies have shown that the HIPS is active in many varieties of the number comparison task (for review see Dehaene et al. 2003). This region shows greater activation for subtraction than for multiplication (e.g. Chochon *et al.* 1999; Simon *et al.* 2002), and with addition it is more active when participants estimate the approximate result than when they compute the exact solution (Dehaene *et al.* 1999; Stanescu-Cosson *et al.* 2000). Of course, it was important to demonstrate that the activation of the HIPS was confined to tasks in which numbers were compared or was greater when numbers were compared (Le Clec'h *et al.* 2000; Pesenti *et al.* 2000). To this end, participants were asked to perform not only a number comparison task but also comparisons regarding ferocity of animals, relative positions of body parts, and orientation of characters (Thioux *et al.* 1998; Le Clec'h *et al.* 2000; Pesenti *et al.* 2000). It was shown that the activation of HIPS was greater when the task involved numbers. Other studies showed that the activation of the HIPS is modulated by the absolute magnitude of the numbers or their relative magnitude with reference to a given standard (Kiefer and Dehaene 1997; Stanescu-Cosson *et al.* 2000). The HIPS is not only activated by arabic numerals in tasks that require magnitude manipulation, but is also specifically activated in magnitude comparison tasks that are performed with number words or sets of dots or tones (e.g. Piazza *et al.* 2002).

Even though the parietal activation is nearly always present in both hemispheres, it is often asymmetric, being greater in the right hemisphere when quantification of nonverbal and nonsymbolic material is required (e.g. Piazza *et al.* 2003). Studies employing arabic digits concluded that there was a major involvement of the left hemisphere (especially in the posterior IPS; e.g. Pesenti *et al.* 2000), an equal contribution from the left and right hemispheres (and in the posterior IPS, e.g. Pinel *et al.* 2001), or even a major role for the right hemisphere (e.g. Chochon *et al.* 1999).

Neuropsychological studies confirm the relation between the magnitude code and the parietal lobe, with specific reference to a region in the vicinity of the IPS (Gerstmann 1940; Benton 1992). Historically, discussions about the relation between deficits in numerical processing and the parietal lobe were often framed in the context of the Gerstmann's syndrome (Gerstmann 1940), which comprises calculation impairments (acalculia), writing deficits (agraphia), left–right disorientation,

and difficulties in recognizing fingers (finger agnosia). A debated question concerns where the lesions that cause acalculia in Gerstmann's syndrome are located and whether the syndrome has a real functional meaning or is just the by-product of a impairments to contiguous but dissociable neural circuits. The classical notion is that the lesions affect the left angular gyrus or its underlying white matter (e.g. Strub and Geschwind 1983; Martory *et al.* 2003). In contrast, some authors have maintained that the lesions affect the depth of the left IPS, because number magnitude representation is severely compromised (e.g. Takayama *et al.* 1994; Mayer *et al.* 1999). Depending on the site where the putative association of symptoms originates, it is possible to draw different predictions as to its functional locus (see the discussion of Rusconi *et al.* 2005b below).

An interesting double dissociation between grossly deteriorated semantic processing and spared number processing on the one hand, and grossly deteriorated number processing with spared semantic processing on the other was described. Cipolotti *et al.* (1991) reported a patient who showed a nearly complete deficit in all aspects of number processing after a cerebrovascular accident in the region of the left middle cerebral artery. In contrast, patients with semantic dementia still having good number comprehension skills were described by Diesfeldt (1993), Thioux *et al.* (1998), and Butterworth *et al.* (2001).

The deficits in number processing can affect tasks different from calculation and number comparison; for example, Dehaene and Cohen (1997) described a patient who failed in a number bisection task, that is he was unable to decide what number occupied the middle position in the interval defined by two other numbers. The deficit was confined to number intervals and the patient could perform well in bisection tasks concerning non-numerical domains, such as days of the week, months of the year, and letters of the alphabet.

Dehaene and Cohen (1997) considered the deficit in the number bisection task as evidence of a specific impairment of the semantic representation and manipulation of analog magnitude. However, deficits in the number bisection task can also index an impairment in spatial processing. Zorzi et al. (2002, 2006) asked neglect patients to perform the number bisection task and found that they systematically shifted the midpoint toward larger numbers (i.e. to the right on the mental number line), except for very short intervals for which there was a shift of the midpoint to the left. The pattern of performance in the number bisection task mirrored that shown by the same patients in a physical line bisection task (Zorzi et al. 2006). The deficit was specific to numbers because the same patients did not show systematic biases when asked, for example, to bisect letters of the alphabet. It is important to note that, at variance with the patient described by Dehaene and Cohen (1997), the neglect patients studied by Zorzi et al. showed no impairment in number processing. This means that the deficit of Zorzi et al.'s patients was spatial in nature and concerned with the way spatial attention was deployed to perform the bisection task. An alternative interpretation of the bias in the number bisection task shown by neglect patients was proposed by Doricchi et al. (2005), who maintain that the deficit is attributable to spatial working memory rather than to spatial attention. Of course the bias is easily explained on the basis of the attentional hypothesis, because larger numbers are located to the right side of the mental number line. On the other hand, an ipsilesional bias was also present after disruption of neuronal populations which, in the human vertrolateral prefrontal cortex, are involved in the short-term retention of contralateral spatial positions (Rizzuto et al. 2005).

Finding a relation between number processing and spatial attention is not surprising if one considers that both systems are centered in the parietal lobe (see also Walsh 2003). We have already pointed out that number processing is mainly located in the HIPS. Spatial attention depends on two parietal–frontal channels, one ventral and one dorsal, that connect the superior and inferior parietal lobules to corresponding areas in the prefrontal cortex (Corbetta and Shulman 2002). Of these channels, the ventral one, which is mainly in charge of automatic orienting, is largely lateralized to the right hemisphere, whereas the dorsal channel, which is in charge of intentional orienting, is present in both hemispheres. Note that whereas acalculia is

caused by parietal lesions located in the left hemisphere (Cipolotti and van Harshkamp, 2001), the parietal structure that specializes in semantic number processing, i.e. the HIPS, is thought to be bilateral (Dehaene et al. 2003). This represents a sort of contradiction, as the left but not the right HIPS seems to be strictly necessary for semantic number processing according to lesion studies, whereas both the left and the right HIPS seem to contribute to normal number semantic processing. The lateralization issue will be addressed by the majority of TMS studies discussed in the next section. TMS allows one to distinguish in normal participants whether an area plays either a necessary or a secondary role in the process which is being investigated, which is not possible simply on the basis of correlational techniques.

TMS and number representations

There is a common denominator in all the TMS studies addressing numerical processing; the pioneering work by Göbel et al. (2001) set the stage for subsequent research by focusing on the posterior parietal lobe (see Figure 33.1) and the processing of number magnitude. Göbel et al. (2001) predicted that the same parietal circuits that play a crucial role in visuo-spatial attention might contribute to the internal spatial representation of numbers. Therefore they first performed a functional localization of the parietal site over which rTMS interfered with a visual conjunction search task (Ashbridge et al. 1997). Unlike Ashbridge et al. (1997), who used a single-pulse protocol with intensity set at 80% of the maximum stimulator output, and who had reported interference only from TMS on the right superior parietal lobule (but see Walsh et al. 1999), Göbel et al. found interference by stimulating with 10 Hz rTMS at 105% of active motor threshold (MT) both the left and the right angular gyrus (AG) near the superior temporal sulcus. The same participants were then asked to perform a comparison task between target numbers ranging from 31 to 99 and the reference 65. Targets were visually presented at the center of a screen and responses were given by pressing a left button for 'smaller than 65'

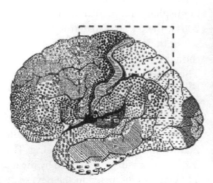

Fig. 33.1 Lateral views of the parietal lobe showing the maps of the posterior parietal cortex according to Brodmann (1909). Posterior parietal lobe (that is posterior to the post-central sulcus), is divided into upper and lower parts, namely the superior parietal lobule and the inferior parietal lobule. The latter includes two gyri that turn around corners: the supramarginal gyrus (SMG, BA40), around the end of the sylvian fissure, and the angular gyrus (AG, BA39), around the end of the superior temporal sulcus. Medially, the posterior intraparietal sulcus divides the inferior from the superior parietal lobule and, with its descending ramus (the sulcus primus of Jensen), the angular from the supramarginal gyrus. The anterior horizontal branch of the intraparietal sulcus marks the boundary between the supramarginal and post-central gyrus (Duvernoy 1999). Reproduced with permission from Rizzolatti and Matelli (2003).

and a right button for 'larger than 65'. When the target was larger and close to the standard, RTs were slower in trials where rTMS was delivered on the left AG relative to no rTMS trials in the same block. Right AG rTMS tended to slow down RTs ($P = 0.06$) but did not interact with either response or numerical distance (see Figure 33.2a). A further experiment was performed on a different group of participants, with the opposite mapping of stimuli to response keys (i.e. participants pressed a left key to respond 'larger than 65' and a right key to respond 'smaller than 65') to discard the hypothesis that left AG rTMS interfered with contralateral motor responses rather than with the processing of number magnitude, and on this occasion left AG rTMS slowed down RTs but did not interact with either response or distance. Right AG rTMS effect approached a trend level ($P = 0.10$; see Figure 33.2b). No effect was found; instead, when participants were stimulated on either left or right supramarginal gyrus (SMG), the authors concluded that the AGs played a fundamental role in the spatial representation of number magnitude. The left AG, in particular, would contain a spatiotopic

representation akin to a mental number line. The lack of interaction between left AG rTMS and number magnitude in the reverse-response version of the task was attributed to the remapping of magnitudes onto a spatially incongruent response array. In addition, it was claimed that these effects could not be attributed to interference with nonspecific spatial functions, since they would have been disrupted more by right than by left rTMS.

Sandrini *et al.* (2004) employed a number comparison task in which two single-digit numbers (range 1–9, 5 excluded) were displayed at the same time, one to the left and one to the right of fixation, and the participant had to select the larger number by pressing the corresponding response key (either left or right). They intended to interfere with number magnitude representations, and as a consequence with the number comparison task, by applying rTMS over the HIPS region. Two homologous locations were chosen, one in the left and one in the right hemisphere, corresponding to the HIPS and neighboring BA40 (SMG). Subjects' performance was significantly slower when rTMS was delivered over the left hemisphere but not

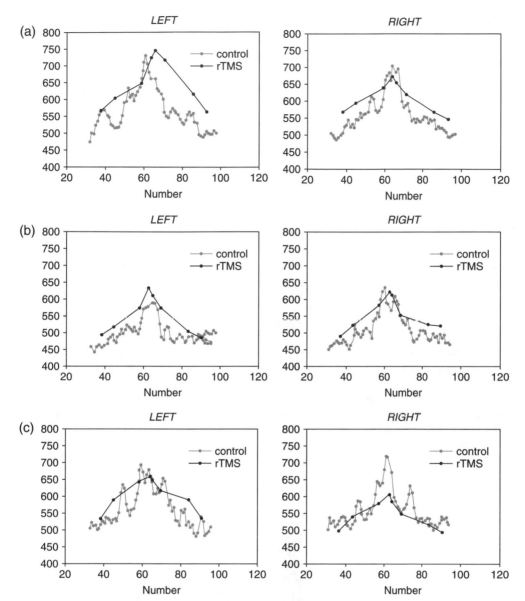

Fig. 33.2 Mean reaction time (RT), in ms, in the two-digit number comparison with a standard (65) in Göbel *et al.*'s (2001) study. The black line describes performance for trials with rTMS and the gray line describes performance for trials without rTMS. (a) Results for stimulation over left vs right AG when responses were given by pressing a left button for 'smaller than 65' and a right button for 'larger than 65'. (b) Performance for stimulation over left vs right AG when responses were given by pressing a left button for 'larger than 65'. (c) Performance for stimulation over left vs right SMG (control experiment) when responses were given by pressing a left button for 'smaller than 65' and a right button for 'larger than 65'.

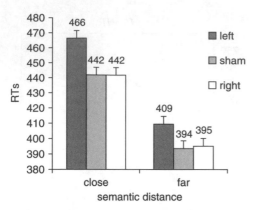

Fig. 33.3 Effects of TMS on mean reaction times (RTs) as a function of stimulation site and semantic distance between the two numbers to be compared in Sandrini *et al.*'s (2004) study. rTMS significantly delayed performance, for both close (i.e. differing by 1, 2, or 3 units) and far (i.e. differing by 5, 6, or 7 units) numbers, only when applied to left inferior parietal lobule. It did not impair performance when applied to right inferior parietal lobe or in the sham condition (coil perpendicular to the scalp, over CPZ).

when over the right hemisphere (see Figure 33.3). Since the interaction between site of stimulation and numerical distance is insignificant, it is also possible to interpret this result as interference on motor response selection, a process in which

the left SMG is thought to play a fundamental role (e.g. Rushworth *et al.* 2001). The same alternative account, however, does not hold for the study by Andres *et al.* (2005), in which the TMS condition did interact with hemisphere and numerical distance. Centrally presented single digits were compared with 5 and responses given by pressing one of two keys. Unlike previous studies, a single-pulse protocol was employed and a double stimulation condition (i.e. with pulse delivery on the left and on the right hemisphere at the same time) was included in the design. Trials in which the target was 3, 4, 6, or 7 were classified as 'close'; trials in which the target was 1, 2, 8, or 9 were classified as 'far'. Responses to close trials were significantly slower in the unilateral left and in the bilateral stimulation condition than in the unilateral right stimulation and sham condition. Responses to far trials, instead, were significantly slower in the bilateral stimulation only than in all the other conditions (see Figure 33.4). In other words, left and right parietal cortices seemed to have a different resolution power in number comparison, the left being strictly necessary for fine discrimination, and both the right and the left being able to support easier comparisons (far trials).

Direct access to symbolic tools such as language and finger counting might be at the

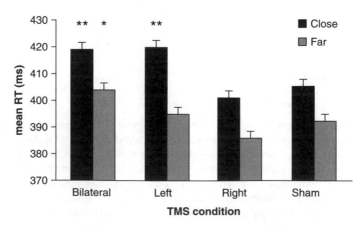

Fig. 33.4 Mean reaction time (RT) in each TMS condition is shown as a function of distance from the standard in Andres *et al.*'s (2005) study. Asterisks signal TMS conditions where RTs were significantly increased compared with the control condition (*$P < 0.01$, **$P < 0.001$). Trials in which the target was 3, 4, 6, or 7 were classified as 'close'; trials in which the target was 1, 2, 8, or 9 were classified as 'far' from the standard (5).

origin of the left hemisphere superiority for fine discrimination in the number domain (Butterworth 1999). Language, which would favor exact as opposed to approximate numerical processing (e.g. Pica *et al.* 2004), and praxis (e.g. Sirigu *et al.* 1996), which may favor the mapping of analog magnitude representations onto discrete sets of body parts like fingers, reside in circuits adjacent to or intermingled with numerical representations in the left hemisphere. At the ontogenetic level, for example, there seems to be a very tight connection between the development of numerical skills and the use of fingers (e.g. Fuson 1988). Some dyscalculic children suffer from Gerstmann's tetrad of symptoms (Rourke 1993), and performance in tests of finger knowledge revealed the best predictor of arithmetical achievement in 5–6-year-old children (Kinsbourne and Warrington 1963; Fayol *et al.* 1998).

Rusconi *et al.* (2005b) sampled two main regions in the inferior parietal lobule (AG and adjacent posterior IPS, IPS vs SMG and adjacent HIPS) bilaterally to test for an area which contributed both to finger gnosis and to number processing. Electrical stimulation data (e.g. Roux *et al.* 2003) and neuropsychological evidence (e.g. Tucha *et al.* 1997) were taken to guide anatomical and functional predictions for the finger agnosia site. The first experiment, as a sort of functional localization, allowed the authors to restrict the regions of interest to the AGs, and the second experiment aimed to test the putative functional link between finger gnosis and numerical processing in the left AG. Participants were asked to perform a parity or a magnitude-matching task on pairs of single digits (e.g. 3, 9) in the context of arithmetically related (e.g. 27) or unrelated (e.g. 25) numerical primes. A 10 Hz, 500 ms train of rTMS at 60% of the maximum stimulator output was delivered (see Figure 33.5a), either on the left or on the right AG, at the onset of the target digits pair. rTMS exerted a disruptive effect when delivered on the left AG and during the magnitude-matching task only. Interestingly, there was also an interaction with arithmetical relatedness, as RTs were slower on left AG rTMS trials having an unrelated prime only (see Figure 33.5b). Non-specific factors such as the encoding of arabic numerals or response selection processes

can thus be discarded, as they are common to the parity and the magnitude-matching tasks. If the left AG was crucial for either stimulus encoding or response selection, and the most difficult task (in this case, the magnitude-matching task) was particularly sensitive to its effects, no interaction would be expected between rTMS condition and trial type (unrelated vs related prime). In other terms, the left AG appears to contribute specifically to the processing of number magnitude, when it is relevant to the task at hand. rTMS over the left AG, instead, does not seem to prevent the implicit retrieval of arithmetic facts, as a significant difference was found between trials with related and trials with unrelated primes. It was thus proposed that the left AG could be involved in a spatial representation of number magnitude that develops in relation to the body, a sort of *mental digit line*, which might represent the adult substitute of early finger counting and be supported by an area which also supports finger gnosis (Rusconi *et al.* 2005b).

On the whole, the studies reviewed so far point to the left inferior parietal lobule as a necessary substrate for the explicit processing of exact number magnitude, when manual keypress responses to arabic numbers are required by the task. The functional role of the posterior parietal cortex seems to reside at the level of either core or auxiliary number magnitude representations: Göbel *et al.* (2001) and Andres *et al.* (2005) found disruptive effects of rTMS especially on close numerical distances, and Sandrini *et al.*'s (2004) results, although the interaction between rTMS and distance was not significant, show the same pattern as the results of Andres *et al.* (2005). Rusconi *et al.* (2005b) found disruptive rTMS effects when the task required explicit judgments on the magnitude but not on the parity of digit pairs; within the magnitude task, in addition, only trials with unrelated primes were affected. Some weaker evidence relating the right posterior parietal lobule to the comparison of arabic numerals was found by Göbel *et al.* (2001). Andres *et al.* (2005) showed that the comparison task at far numerical distances was not affected by stimulation on the left HIPS only, but was affected by bilateral stimulation. It was therefore inferred that the right HIPS can carry out the comparison

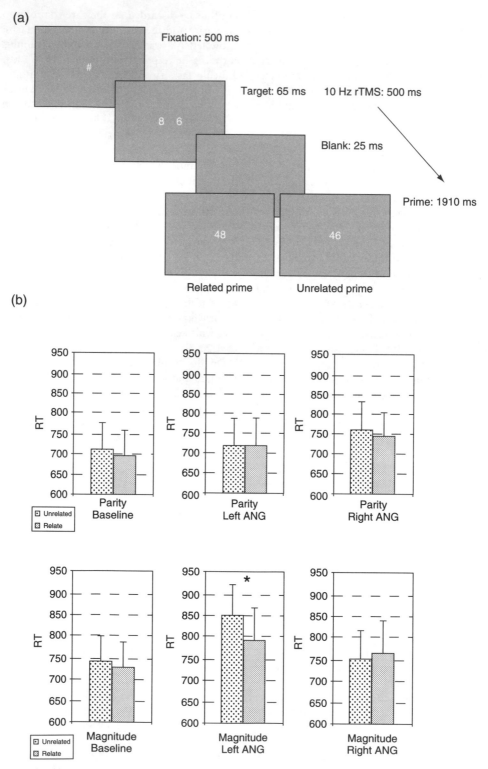

Fig. 33.5 See legend opposite page.

Fig. 33.5 (a) Sequence of events on each trial in the parity- and magnitude-matching tasks employed by Rusconi et al. (2005b). Participants were presented with pairs of single-digit numbers between 1 and 9 (except 5) and were asked to indicate with one of two possible key presses whether the numbers had the same or different parity in the parity-matching task and whether they had the same or different magnitude compared to 5 in the magnitude-matching task. Related ('48' in the present example) or their paired unrelated primes ('46' in the present example) followed the target digits in critical trials. Of the total trials, 75% were fillers, in which primes were unrelated to the target digits, and were not included in the analysis. (b) Means of median correct reaction times (RT) for trials with unrelated (light bar) and trials with related (dark bar) primes in each block. *Upper row*: plots for the parity-matching task. *Lower row*: plots for the magnitude-matching task. ANG, angular gyrus.

task with arabic digits, provided that the target–standard distance is large enough.

Deficits in spatial processing can interfere with numerical processing depending on the specific task one is asked to perform, as reported by Zorzi et al. (2002, 2006). Four patients with right fronto-(temporo)-parietal lesion and left neglect were asked to perform a number bisection task (e.g. to state what number is halfway between 2 and 6, with oral presentation of stimuli and verbal response production) on the assumption that number bisection is a representational homolog of the classical physical bisection task, where neglect patients show a typical rightward bias, which becomes larger as line length increases. Their patients were not acalculic (e.g. they could perform arithmetic tasks and number magnitude comparison normally) but they showed a systematic bias towards larger magnitudes (i.e. the right part of the mental number line) in the bisection task, that increased with increasing numerical distance. After this finding, Oliveri et al. (2004) tested whether right posterior parietal cortex (PPC) can exert an influence on numerical processing in normal healthy adults. They employed a visual variant of the number bisection task and asked their participants whether the numerical distance between a middle number and two outer numbers was larger on the right or on the left side. In the baseline, participants tended to overestimate the difference between the middle and the left-side number (pseudo-neglect); the bias disappeared after 5 min of 1 Hz rTMS over P4 (right PPC) but not after rTMS over P3 (left PPC). The authors therefore claimed that right PPC is responsible for

pseudo-neglect in the number domain (see Oliveri et al. 2001, for a homolog task and identical conclusions with physical line segments). Recently, Göbel et al. (2006) tested the same hypothesis by using auditory stimuli in the form of two successive English verbal numerals (three-digit numbers in ascending order, e.g. '117 166', or '959 984'). Soon after stimuli presentation, their participants were stimulated for 1000 ms with a 5 Hz rTMS train at 110% of the motor threshold and had to name the number lying in the middle of the interval. A first group of participants received stimulation on the right and on the left AG in different blocks (the same functional localization procedure as in Göbel et al. (2001) was employed). A second group of participants received stimulation centrally, on the occipital cortex. Parietal rTMS did not affect latencies but shifted the perceived midpoint by ~0.3 units to the right relative to trials without rTMS, in which participants showed an average deviation of ~0.5 units toward the left of the actual midpoint (Figure 33.6). The interaction between stimulation site (left AG vs right AG) and rTMS condition (present vs absent) was not significant. However, when separate analyses were performed for the left and the right AG, it turned out that the difference between trials with rTMS and trials without rTMS approached significance for the right (0.39 units, $P = 0.05$) but not for the left AG (0.23 units, $P = 0.16$). The difference had the same sign (i.e. the average deviation from baseline performance was toward the right) for the left and the right AG blocks, which might explain the lack of interaction in the two-way analysis of variance. In the group with stimulation on the occipital site,

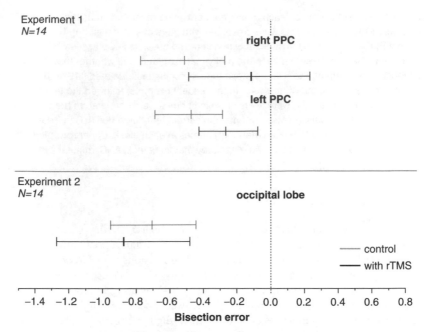

Fig. 33.6 rTMS effects on number bisection errors in Göbel *et al.*'s (2006) study. On control trials (gray lines), participants underestimated the midpoint of the numerical interval. On trials with rTMS (black lines), participants showed a significant shift in the midpoint estimation towards the right for parietal rTMS, whereas occipital rTMS exerted no effect. PPC, posterior parietal cortex.

rTMS did not exert any effect on accuracy, whereas it speeded up performance. The authors therefore concluded that TMS over the right parietal cortex was altering a representation of numbers that is likely to follow a spatial format, in line with evidence from number bisection in neglect patients and from physical line bisection in normal participants after TMS (e.g. Oliveri *et al.* 2001; Zorzi *et al.* 2002). A possible alternative view of Oliveri *et al.*'s (2004) and Göbel *et al.*'s (2006) results would maintain that rTMS over AGs actually improved accuracy in number bisection by counteracting the original bias. Given that the same functional localization as in Göbel *et al.* (2001) was employed, the absence of a disruptive effect from stimulation of the left hemisphere is quite surprising. Considering the fact that both number bisection and number comparison are thought to rely on the same magnitude representation, the possibility remains that the region including the left AG and posterior IPS is crucial when number magnitudes are to be mapped onto manual rather than verbal motor responses, which would also

predict a functional role of the left AG in the relation between fingers and numbers (see Rusconi *et al.* 2005b).

Recently, Rusconi *et al.* (2005a, 2007) found bilateral effects (left and right posterior IPS) of 10 Hz rTMS on a parity judgment task with lateralized target digits. A single-digit number was presented either on the left or on the right of a central fixation point and the participant had to categorize the digit as odd or even by pressing one of two lateralized response keys. It is worth noting that the parity judgment task was just a pretext and that the rTMS manipulation was not intended to interfere with number parity processing. The rTMS condition, instead, was expected to suppress the SNARC effect (difference between noncorresponding and corresponding trials) if it was delivered to an essential substrate. The results showed a suppression of the SNARC effect when rTMS was delivered either on the left or on the right posterior IPS (see Figure 33.7). The suppression was due to elimination by rTMS of interference in the non-corresponding trials, while the

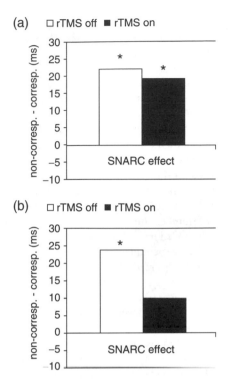

Fig. 33.7 Mean SNARC (Spatial–Numerical Association of Response Codes) effects (reaction time difference between the corresponding and the noncorresponding condition): (a) for the anterior and (b) for the posterior stimulation sites of Rusconi *et al.* (2005a).

corresponding trials remained unvaried. Fischer *et al.* (2003) suggested that numbers can cause attentional orienting even in a simple detection task. According to the premotor hypothesis of attention (Rizzolatti *et al.* 1987; Umiltà *et al.* 1991) an attention shift occurs because a motor program for the corresponding saccade is prepared, regardless of whether the saccade is subsequently executed (overt orienting) or not (covert orienting). The motor program specifies the direction of the saccade; hence the stimulus spatial code is formed and might generate the SNARC effect by priming one of the response alternatives. Rizzolatti and Matelli (2003) noticed that in the macaque different regions of the IPS show preparatory activity that is selective for different effectors and that neurons of the lateral intraparietal area (area LIP, just lateral to the monkey homolog of the AG) code for

impending saccadic eye movements. Corbetta and Shulman (2002) proposed the existence, in humans, of a bilateral parieto-frontal network for the top-down control of visuo-spatial attention, which also carries neuronal signals that are related to the preparation of eye and arm movements. The results of Rusconi *et al.* (2005a, 2007) seem consistent with the hypothesis of a direct involvement of this bilateral attentional network in the origin of SNARC interference.

In summary, TMS studies suggest that the right hemisphere is involved in number processing when the input is arabic or verbal. However, its contribution seems much less necessary than the contribution of the left hemisphere in numerical comparison and bisection tasks. A crucial distinction might be between the HIPS and the area including AG and posterior IPS, the former being able to perform 'approximate' number comparison (Andres *et al.* 2005) and the latter originating a spatial bias in number bisection tasks or producing the SNARC interference, most likely via the mechanisms of attentional orienting, in a numerical task with lateralized responses (Oliveri *et al.* 2004; Rusconi *et al.* 2005a; Göbel *et al.* 2006).

Very recently, we have started to investigate the possible role of anterior attentional circuits in exploring mental number space and generating the SNARC effect (Rusconi *et al.* 2006). It is indeed generally claimed that the spatial representation of number is an exclusive function of the posterior parietal cortex; however, the representation of space and covert orienting to it can involve frontal brain areas. Five-hertz rTMS was delivered in each trial at stimulus onset for 400 ms (110% MT) with 50 mm coils, while participants performed magnitude comparison or parity judgment on centrally presented digits. Responses were given by pressing a left or a right key with the right index and middle finger, respectively. SNARC effect was used to probe the spatial representation of numbers during rTMS over right frontal eye fields (FEFs) and over right inferior frontal gyrus (IFg; Figure 33.8). To control for nonspecific effects, the baseline was collected with rTMS over the vertex. As shown in Table 33.3, for the number representations elicited by magnitude comparison, spatial coding (as indexed by SNARC effect) was completely eliminated by stimulating right IFg,

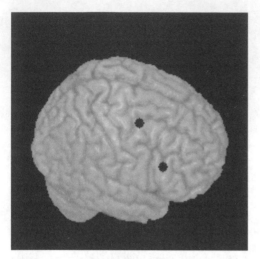

Fig. 33.8 Stimulation sites were localized in Brainsight on individual scans which were co-registered with the participants' head by frameless stereotaxy. Coil center was held tangential to the scalp in constant position over vertex, right frontal eye field (FEF), or right inferior frontal gyrus (IFg). MNI coordinates were adapted to individual scans through SPM.

while stimulating right FEF eliminated spatial coding for small numbers, i.e. those in left representational space. In parity judgment, the SNARC effect survived stimulation at both sites. TMS suppressed the link between numbers and space rather than core magnitude or parity

Table 33.3 Mean SNARC effect (ms) for small and large numbers in magnitude comparison

Magnitude comparison	Stimulated site		
	Vertex	FEF	VLPFC
Small numbers	29*	5	2
Large numbers	34*	22*	8

Right FEF rTMS eliminated the SNARC effect for small numbers and right inferior frontal gyrus eliminated the SNARC effect for both small and large numbers.

*$P < 0.05$.

SNARC, Spatial–Numerical Association of Response Codes; FEF, frontal eye field; VLPFC, ventrolateral prefrontal cortex.

representations, since overall performance was not affected by TMS in either task. These results add two important qualifications to standard models of number space. First, the spatial coding for numbers is not a purely parietal matter. Second, in contrast with the claim that a number stimulus always elicits the same representations, our results show that representations are task dependent, such that the spatial coding elicited when magnitude is relevant to behavior depends on the contribution of frontal cortex, while that elicited when it is irrelevant does not. This brings the analysis of mental representation of numbers into agreement with the other spatially coded mental representations.

Finally, we would like to mention a recent study by Knops *et al.* (2006), which draws attention to possible gender differences in the effects of rTMS over the left HIPS by pointing to differences in the level of hemispheric specialization and interhemispheric transfer. Clear predictions about the incidence of primary acalculia in males and females suffering from left parietal lesion could easily be drawn and tested on the basis of Knops *et al.*'s claim, thus reversing the usual direction of the link between neuropsychological data and TMS research.

Conclusion

The potential contribution of TMS to the study of mathematical cognition is extensive and many opportunities remain to be exploited. In the short term there is much to be resolved that requires temporary lesion effects. The relative involvement or relative timing of involvement of the left and right parietal cortices in number tasks is one important area of future study. The extent to which number-related processes are number specific, and the extent to which they overlap with other aspects of spatial or magnitude representation, is currently a burgeoning area of research. The combination of TMS with measures of brain activity is likely to be an important development, and current work is aimed to disrupt numerical processes and observe concomitant changes in brain activation. Finally, given the emphasis of some other chapters in this volume on possible facilitatory effects of TMS, there is also the possibility that disrupting competitive processes in numerical processing may lead to ways of studying temporary improvements in number-related tasks.

References

Andres M, Seron X, Olivier E (2005). Hemispheric lateralization of number comparison. *Cognitive Brain Research 25*, 283–290.

Antell SE, Keating DP (1983). Perception of numerical invariance in neonates. *Child Development 54*, 695–701.

Ashbridge E, Walsh V, Cowey A (1997). Temporal aspects of visual search studied by transcranial magnetic stimulation. *Neuropsychologia 35*, 1121–1131.

Benton AL (1992). Gerstmann's syndrome. *Archives of Neurology 49*, 445–447.

Brodmann K (1909). *Vergleichende Lokalisationslehre der Grosshirnrinde in ihren prinzipien Dargerstellt auf Grund des Zellenbaues.* Leipzig: Barth.

Butterworth B (1999). *The mathematical brain.* London: Macmillan.

Butterworth B, Cappelletti M, Kopelman M (2001). Category specificity in reading and writing: the case of number words. *Nature Neuroscience 4*, 784–786.

Campbell JID (ed.) (2005). *Handbook of mathematical cognition.* New York: Psychology Press.

Chochon, F, Cohen L, van de Moortele PF, Dehaene S (1999). Differential contributions of the left and right inferior parietal lobules to number processing. *Journal of Cognitive Neuroscience 11*, 617–630.

Cipolotti L, van Harskamp N (2001). Disturbances of number processing and calculation. In: RS Berndt (ed.), *Handbook of neuropsychology,* Vol. 3, pp. 305–331. Amsterdam: Elsevier Science.

Cipolotti L, Butterworth B, Denes G (1991). A specific deficit for numbers in a case of dense acalculia. *Brain 114*, 2619–2637.

Corbetta M, Shulman (2002). Control of goal-directed and stimulus-driven attention in the brain. *Nature Reviews Neuroscience 3*, 201–215.

Cordes S, Gelman R, Gallistel CR, Whalen J (2001). Variability signatures distinguish verbal from nonverbal counting for both large and small numbers. *Psychonomic Bulletin and Review 8*, 698–707.

Dehaene S (1992). Varieties of numerical abilities. *Cognition 44*, 1–42.

Dehaene S, Cohen L (1997). Cerebral pathways for calculation: Double dissociation between rote verbal and quantitative knowledge of arithmetic. *Cortex 33*, 219–250.

Dehaene S, Dupoux E, Mehler J (1990). Is numerical comparison digital? Analogical and symbolic effects in two-digit number comparison. *Journal of Experimental Psychology: Human Perception and Performance 16*, 626–641.

Dehaene S, Bossini S, Giraux P (1993). The mental representation of parity and number magnitude. *Journal of Experimental Psychology: General 122*, 371–396.

Dehaene S, Spelke E, Stanescu R, Pinel P, Tsivikin S (1999). Sources of mathematical thinking: Behavioural and brain-imaging evidence. *Science 284*, 970–974.

Dehaene S, Piazza M, Pinel P, Cohen L (2003). Three parietal circuits for number processing. *Cognitive Neuropsychology 20*, 487–506.

Diesfeldt HFA (1993). Progressive decline of semantic memory with preservation of number processing and calculation. *Behavioral Neurology 6*, 239–242.

Doricchi F, Guariglia P, Gasparini M, Tomaiuolo F (2005). Dissociation between physical and mental line bisection in right hemisphere brain damage. *Nature Neuroscience 8*, 1663–1665.

Duvernoy HM (1999). *The human brain.* Wien: Springer-Verlag.

Fayol M, Barrouillet P, Marinthe C (1998). Predicting arithmetical achievement from neuropsychological performance: a longitudinal study. *Cognition 68*, 63–70.

Fischer M (2003). Spatial representations in number processing – evidence from a pointing task. *Visual Cognition 10*, 493–508.

Fischer MH, Castel AD, Dodd MD, Pratt J (2003). Perceiving numbers causes spatial shifts of attention. *Nature Neuroscience 6*, 555–556.

Fuson K (1988). *Children's counting and concepts of number.* New York: Springer.

Galton F (1880a). Visualised numerals. *Nature 21*, 252–256.

Galton F (1880b). Visualised numerals. *Nature 21*, 494–495.

Gerstmann J (1940). Syndrome of finger agnosia: Disorientation for right and left, agraphia and acalculia. *Archives of Neurology and Psychiatry 44*, 398–408.

Göbel SM, Walsh V, Rushworth MFS (2001). The mental number line and the human angular gyrus. *NeuroImage 14*, 1278–1289.

Göbel SM, Calabria M, Farnè A, Rossetti Y (2006). Parietal rTMS distorts the mental number line: Simulating 'spatial' neglect in healthy subjects. *Neuropsychologia 44*, 860–868.

Groen GJ, Parkman JM (1972). A chronometric analysis of simple addition. *Psychological Review 79*, 329–343.

Kiefer M, Dehaene S (1997). The time course of parietal activation in single-digit multiplication: Evidence from event-related potentials. *Mathematical Cognition 3*, 1–30.

Kinsbourne M, Warrington E (1963). The developmental Gerstmann syndrome. *Annals of Neurology 8*, 490–501.

Knops A, Nuerk H-C, Sparing R, Foltys H, Willmes K (2006). On the functional role of human parietal cortex in number processing: how gender mediates the impact of a 'virtual lesion' induced by rTMS. *Neuropsychologia 44*, 2270–2283.

Le Clec'h G, Dehaene S, Cohen L *et al.* (2000). Distinct cortical areas for name of numbers and body parts independent of language and input modality. *NeuroImage 12*, 381–391.

Martory MD, Mayer E, Pegna AJ, Annoni JM, Landis T, Khateb A (2003). Pure global acalculia following a left subangular lesion. *Neurocase 9*, 319–328.

Mayer E, Martory MD, Pegna AJ, Landis T, Delavelle J, Annoni JM (1999). A pure case with Gerstmann syndrome with a subangular lesion. *Brain 122*, 1107–1120.

Moyer RS, Landauer TK (1967). Time required for judgments of numerical inequality. *Nature 215*, 1519–1520.

Nieder A (2004). The number domain – can we count on parietal cortex? *Neuron 44*, 407–409.

Nieder A, Miller EK (2003). Coding of cognitive magnitude. Compressed scaling of numerical information in the primate prefrontal cortex. *Neuron 37*, 149–157.

Nieder A, Miller EK (2004). A parieto-frontal network for visual numerical information in the monkey. *Proceedings of the National Academy of Sciences of the USA 101*, 7457–7462.

Nieder A, Friedman DJ, Miller EK (2002). Representation of the quantity of visual items in the primate prefrontal cortex. *Science 297*, 1708–1711.

Oliveri M, Bisiach E, Brighina F *et al.* (2001). rTMS of the unaffected hemisphere transiently reduces contralesional visuospatial hemineglect. *Neurology 57*, 1338–1340.

Oliveri M, Rausei V, Koch G, Torriero S, Turriziani P, Caltagirone C (2004). Overestimation of numerical distances in the left side of space. *Neurology 63*, 2139–2141.

Pesenti M, Thioux M, Seron X, De Volder A (2000). Neuroanatomical substrates of Arabic number processing, numerical comparison, and simple addition: a PET study. *Journal of Cognitive Neuroscience 12*, 461–479.

Piazza M, Mechelli A, Butterworth B, Price C (2002). Are subitizing and counting implemented as separate or functional overlapping processes? *NeuroImage 15*, 435–446.

Piazza M, Giacomini E, Le Bihan D, Dehaene S (2003). Single-trial classification of parallel pre-attentive and serial attentive processes using functional magnetic resonance imaging. *Proceedings of the Royal Society of London B 270*, 1237–1245.

Pica P, Lemerc C, Izard V, Dehaene S (2004). Exact and approximate arithmetic in an Amazonian indigene group. *Science 306*, 499–503.

Pinel P, Dehaene S, Riviere D, LeBihan D (2001). Modulation of parietal activation by semantic distance in a number comparison task. *NeuroImage 14*, 1013–1026.

Restle F (1970). Speed of adding and comparing numbers. *Journal of Experimental Psychology 83*, 274–278.

Rizzolatti G, Matelli M (2003). Two different streams from the dorsal visual system: anatomy and functions. *Experimental Brain Research 153*, 146–157.

Rizzolatti G, Riggio L, Dascola I, Umiltà C (1987). Reorienting attention across the horizontal and vertical meridians: evidence in favor of a premotor theory of attention. *Neuropsychologia 25*, 31–40.

Rizzuto DS, Mamelak AN, Sutherling WW, Fineman I, Andersen RA (2005). Spatial selectivity in human ventrolateral prefrontal cortex. *Nature Neuroscience 8*, 415–417.

Rossini PM, Barker AT, Berardelli A, *et al.* (1994). Non-invasive electrical and magnetic stimulation of the brain, spinal cord and roots: basic principles and procedures for routine clinical application. Report of an IFCN committee. *Electroencephalography and Clinical Electrophysiology 91*, 79–92.

Rourke BP (1993). Arithmetic disabilities, specific and otherwise: a neuropsychological perspective. *Journal of Learning Disabilities 26*, 226–241.

Roux F-E, Boetto S, Sacko O, Chollet F, Trémoulet M (2003). Writing, calculating and finger recognition in the region of the angular gyrus: a cortical stimulation study of Gerstmann syndrome. *Journal of Neurosurgery 99*, 716–727.

Rusconi E, Turatto M, Umiltà C (2005a). The Simon and the SNARC effect in posterior parietal lobule: an rTMS study. *Journal of Cognitive Neuroscience Supplement*, 153.

Rusconi E, Walsh V, Butterworth B (2005b). Dexterity with numbers: rTMS over left angular gyrus disrupts finger gnosis and number processing. *Neuropsychologia 43*, 1609–1624.

Rusconi E, Bueti D, Walsh V, Butterworth B (2006). Frontal lobe shapes number space. *XVII Meeting of the International Society for Brain Electromagnetic Topography*, Chieti, Italy.

Rusconi E, Turatto M, Umiltà C (2007). Two orienting mechanisms in posterior parietal lobule: an rTMS study of the Simon and SNARC effect. *Cognitive Neuropsychology 24*, 373–392.

Rushworth MF, Ellison A, Walsh V (2001). Complementary localization and lateralization of orienting and motor attention. *Nature Neuroscience 4*, 656–661.

Sandrini M, Rossini PM, Miniussi C (2004). The differential involvement of inferior parietal lobule in number comparison: a rTMS study. *Neuropsychologia 42*, 1902–1909.

Sawamura H, Shima K, Tanji J (2002). Numerical representation for action in the parietal cortex of the monkey. *Nature 415*, 918–922.

Schwarz W, Keus I (2004) Moving the eyes along the mental number line: comparing SNARC effects with manual and saccadic responses. *Perception and Psychophysics 66*, 651–664.

Simon O, Cohen L, Mangin JF, Bihan DL, Dehaene S (2002). Topographical layout of hand, eye, calculation and language-related areas in the left parietal lobe. *Neuron 33*, 475–487.

Sirigu A, Duhamel JR, Cohen L, Pinon B, Dubois B, Agid Y (1996). The mental representation of hand movements after pariental cortex damage. *Science 273*, 1564–1568.

Stanescu-Cosson R, Pinel P, van de Moortele P-F, Le Bihan D, Cohen L, Dehaene S (2000). Cerebral bases of calculation processes: impact of number size on the cerebral circuits for exact and approximate calculation. *Brain 123*, 2240–2255.

Strauss MS, Curtis LE (1981). Infant perception of numerosity. *Child Development 52*, 1146–1152.

Strub R, Geschwind N (1983). Localization in Gerstmann syndrome. In: A Kertesz (ed.), *Localization in neuropsychology*, pp. 295–322. New York: Academic Press.

Takayama Y, Sugishita M, Akiguchi I, Kimura J (1994). Isolated acalculia due to left parietal lesion. *Archives of Neurology 51*, 286–291.

Thioux M, Pillon A, Samson D *et al.* (1998). The isolation of numerals at the semantic level. *Neurocase 4*, 371–389.

Tucha O, Steup A, Smely C, Lange KW (1997). Toe agnosia in Gerstmann's syndrome. *Journal of Neurology, Neurosurgery and Psychiatry 63*, 399–403.

Umiltà C, Riggio L, Dascola I, Rizzolatti G (1991). Differential effects of central and peripheral cues on the orienting of spatial attention. *European Journal of Cognitive Psychology 3*, 247–267.

Walsh V (2003). A theory of magnitude: common cortical metrics of time, space and quantity. *Trends in Cognitive Science 7*, 483–488.

Walsh V, Ellison A, Ashbridge E, Cowey A (1999). The role of parietal cortex in visual attention, hemispheric asymmetries and the effects of learning: a magnetic stimulation study. *Neuropsychologia 37*, 245–251.

Wood G, Nuerk HC, Willmes K (2006). Crossed hands and the SNARC effect: a failure to replicate. *Cortex 42*, 1069–1079.

Zbrodoff NJ, Logan GD (2005). What everyone finds: the problem-size effect. In JID Campbell (ed.), *Handbook of mathematical cognition*. New York: Psychology Press.

Zorzi M, Priftis K, Umiltà C (2002). Brain damage: neglect disrupts the mental number line. *Nature 417*, 138–139.

Zorzi M, Priftis K, Meneghello F, Marenzir R, Umiltà C (2006). The spatial representation of numerical and non-numerical sequences: evidence from neglect. *Neuropsychologia 44*, 1061–1067.

SECTION V

TMS and Brain Mapping

Tomáš Paus

Why should we combine TMS with brain mapping? The answer is threefold: (1) to localize the target of stimulation; (2) to measure local and distal response of the brain to the stimulation; and (3) to assess long-term (hours, days, weeks) effects of repetitive TMS. The following four chapters provide a detailed overview of the approaches in this domain, focusing on the combination of TMS with three imaging modalities, namely positron emission tomography (PET), functional magnetic resonance imaging (fMRI), and electroencephalography (EEG).

Over 20 years ago, two very different approaches to brain mapping emerged. In 1985, Barker and colleagues described their system for noninvasive stimulation of the human cerebral cortex. At the same time, the first series of PET studies appeared in which various perceptual, motor, and cognitive processes were mapped into distinct regions using regional cerebral blood flow as an index of brain activity.

In the next 10 years, these two brain-mapping approaches lived rather independent lives. For a variety of reasons and motivations, however, they began to converge in the mid-1990s. The functional neuroimaging community perhaps realized the need for injecting causality into their otherwise correlational approach afforded by measuring changes in brain activity during a task performance, while the TMS community was interested in other-than-motor measures of TMS effects on the brain.

Introduction of repetitive TMS made it possible to test behavioral effects of brain stimulation in healthy volunteers. But those interested in targeting nonmotor cortical regions in their cognitive studies needed a tool allowing them to localize precisely the region of interest. This need was answered by the use of frameless stereotaxy, which uses structural MRI for this purpose.

The interest of the TMS community in measuring more than motor-evoked potentials coincided with the growing interest of the functional neuroimaging community in developing new techniques for studying functional (or effective) connectivity. This common goal was achieved by combining TMS with PET and EEG and, a few years later, with fMRI. An overview of these initial efforts, as well as the description of the technical know-how and the latest developments in this area, is the main focus of this section.

CHAPTER 34

Combining brain imaging with brain stimulation: causality and connectivity

Tomáš Paus

Introduction

The chief objective of cognitive neuroscience is identification, in time and space, of neural circuits underlying particular sensory, motor, cognitive, and affective processes. As the term 'circuit' implies, one of the assumptions we make is that a given process is supported by a set of spatially segregated and functionally specialized modules sharing information via interconnecting pathways. In the human brain, this framework reflects an extensive body of data accumulated over the past 20 years with various neuroimaging techniques, most commonly with positron emission tomography (PET) and functional magnetic resonance imaging (fMRI). Using changes in regional cerebral blood flow (PET) and blood oxygenation level dependent signal (fMRI) as a proxy of neural activity, nearly all experimental paradigms reveal task-related changes of neural activity in multiple brain regions. Two questions arise: (1) Are all such regions necessary for the performance of a given task? (2) Are all such regions indeed interconnected? To answer these questions, many investigators have turned to TMS, a tool that allows them to *perturb* neural activity, in time and space, in a noninvasive manner.

Causality

Perturbation as an approach employed in studies of brain–behavior relationships has a long history. Irreversible perturbations, i.e. brain lesions, told us, for example, that the inferior frontal cortex is essential for language production (Broca 1861) and that the hippocampal system is necessary for declarative memory (Scoville and Milner 1957). Reversible perturbation of neural activity with direct electrical stimulation revealed the somatotopic organization of the motor cortex (Fritsch and Hitzig 1870; Leyton and Sherington 1917; Penfield and Rasmussen 1950). Electrical stimulation is still used by neurosurgeons to identify 'eloquent' cortex to be avoided during surgery carried out, for example, to treat epilepsy. Today, TMS is employed as a noninvasive perturbation tool in healthy adult volunteers; it allows investigators to manipulate brain activity in spatially distinct cortical regions for a brief (milliseconds) or extended (minutes) period of time.

To answer the first question, namely 'Are all regions [revealed with functional neuroimaging] necessary for the performance of a given task?', an increasing number of investigators use the following 'two-stage' approach. In Stage 1,

PET or fMRI is used to identify a cortical region with statistically significant task-related changes in local hemodynamics. In Stage 2, TMS is utilized to manipulate neural activity in this 'target' region during performance of the same task. In this way, TMS is employed to test whether or not a region in which change in neural activity is *associated* with a given task is also *necessary* for the performance of this task (see Section IV, this volume).

Identification of the target region can be based on a single-subject imaging study or on group averages of PET/fMRI data. In the former case, the same individual participates in both the imaging and TMS stage. For example, in a recent study (Gagnon *et al.* 2006), we used fMRI to identify, in each individual, a subregion of the frontal eye field (FEF) in which the fMRI signal increased during smooth-pursuit eye movements. Subsequently, each subject participated in a TMS experiment where the coil was positioned, with frameless stereotaxy (e.g. Paus 1999), over the 'activated' region and a single pulse of TMS was applied during smooth-pursuit eye movements. TMS was applied to the left and right FEF, as well as to SMA and a control region in the somatosensory cortex. The results showed an increase and decrease in gain when the eyes were, respectively, speeding up and slowing down during the pursuit of a visual target.

Considering a relatively high consistency in the location of task-related 'activations' across individuals, single-subject fMRI studies may not always be required to identify the target region. An alternative approach takes advantage of standardized stereotaxic space and uses X, Y, and Z coordinates of 'activation' peaks identified in previous group-based PET or fMRI studies (Paus *et al.* 1997). For example, we have used a probabilistic location of the FEF in a TMS study of visual awareness (Grosbras and Paus 2003). The FEF was identified through a meta-analysis of previous oculomotor blood-flow activation studies (Paus 1996; for a recent meta-analysis see Grosbras *et al.* 2005), its X, Y, and Z coordinates were transformed from the standardized stereotaxic space to the subject's brain coordinate ('native') space, and the coil was positioned over this location with frameless stereotaxy. To verify accuracy of the positioning, we used a 'functional probe' based on the known effect of TMS on the latency of eye movements directed to the contralateral hemifield; only subjects who demonstrated such an effect were included in the actual study. The main finding of this experiment was an *increased* visual awareness in trials with single-pulse TMS applied 40 ms before the onset of the visual stimulus.

The above examples of the two-stage approach of combining neuroimaging with TMS suggest that individual- and group-based coordinates of target areas can be used with equal success, and that single-pulse TMS applied at a particular time may have a facilitatory rather than interfering effect on behavior. The latter observation is particularly intriguing. What might the neural mechanisms underlying such facilitation be? One way to address this issue is to turn to the motor system and its physiology, as studied with various TMS paradigms (see Section II, this volume). Another approach would be to carry out such an experiment in the scanner and measure changes in brain activity, for example with event-related fMRI, while the subject performs the task and the experimenter applies single pulses of TMS at the appropriate times. Chapter 36 of this volume describes this approach in detail and illustrates it with an example of a recent study on the role of the parietal cortex in visuo-spatial judgments (Sack *et al.* 2005). Such an on-line combination of TMS and functional neuroimaging may provide insight regarding the causality in inter-regional interactions during the performance of a task. Let us turn now to a more general question, namely neural connectivity and its assessment in the living human brain.

Connectivity

Typically, region A is said to be connected with region B only if neurons A possess synaptic connections with neurons B. But current techniques for studying neural connectivity in the human brain do not provide such a level of spatial neuron-to-neuron specificity. With the exception of post-mortem studies of short-range cortical connectivity with the carbocyanine tracer DiI (e.g. Tardif and Clarke 2001), most current research focuses on *in vivo* studies of structural and functional connectivity at the macroscopic level.

This work is carried out with a variety of brain-mapping techniques, including structural and functional MRI, PET, TMS, electroencephalography (EEG), and magnetoencephalography. In this context, the term 'connectivity' refers either to the structural properties of white matter and major fiber tracts, namely *structural connectivity*, or to the statistical relationship in neural activity recorded simultaneously in a number of spatially distinct regions, namely *functional connectivity*. Under certain circumstances, we can also evaluate how one region influences another, that is *effective connectivity*.

Studies of *structural connectivity* in the healthy human brain focus on assessing the volume and structural properties of major white matter pathways. The former can be measured with a computational analysis of regular (e.g. T1- and/or T2-weighted) anatomical images whereas the latter is most often captured with diffusion tensor imaging (DTI) or magnetization transfer imaging (MTI). The main advantage of the 'anatomical' approach is the ease of data acquisition: 15–30 min of scanning time provides a wealth of data covering the entire brain and, hence, all major pathways. The main drawback is the lack of information about the point-to-point neural connectivity (but see recent successes of certain versions of fiber tractography (e.g. Johansen-Berg *et al.* 2005)) and, by definition, absence of information about the functional state of a given pathway.

Functional connectivity can be defined operationally as the extent of correlation in brain activity measured across a number of spatially distinct brain regions (e.g. Friston 1994; Horwitz 2003; Sporns *et al.* 2004). When discussing various approaches to the study of functional connectivity in the human brain, Horwitz (2003) pointed out that conclusions reached by different investigators regarding the presence or absence of functional connectivity between a set of brain regions depend on the type of measurement (e.g. functional MRI, EEG, MEG), type of analysis (e.g. correlation, structural equation modelling), and, most importantly, the state of the subject during the recording of brain activity (rest, type of stimulation/task). The main advantage of this statistical approach to the assessment of connectivity is the fact that it can be readily applied to almost any dataset acquired with the current brain-mapping tools. The main disadvantage is the complexity of its interpretation. Perhaps, functional connectivity should be viewed as yet another way to represent functional neuroimaging data rather than to indicate neural connectivity *per se*.

Effective connectivity attempts to describe causal effects exerted by one brain region onto another (Friston 1994). As pointed out by Sporns *et al.* (2004), effective connectivity can be inferred through a perturbation or through the observation of temporal order of neural events (Granger causality) (Granger 1969). At least theoretically, the latter approach is possible in the case of electrophysiological signals. It is still unclear, however, whether relatively short (a few milliseconds) delays in monosynaptic pathways can be discerned using EEG or MEG measures, which are based on a spatially integrated response of a large population of neurons. For this reason, we would argue that the perturbation approach seems to be the only technique available today to assess truly effective connectivity. This can be achieved by combining brain imaging with brain stimulation.

The 'perturb-and-record' approach has been used before. Dusser de Barenne and McCulloch (1936) developed a technique called neuronography to map functional connections in the monkey sensorimotor cortex by tracing the spread of seizures induced by a local application of strychnine. TMS obviously provides a non-invasive tool to perturb brain activity in time and space. The main challenge is measuring TMS-induced effects. In its simplest form, we can assess effective connectivity by measuring motor-evoked potentials (MEPs) elicited by a single TMS pulse applied over the motor cortex; in this way, we can make inferences about the state of effective connectivity in the corticospinal system, as well as that between the left and right primary motor cortex or between the premotor and primary motor cortex (Münchau *et al.* 2002; Chen *et al.* 2003; Chouinard and Paus 2006). On the other hand, being able to measure brain activity directly (with PET, fMRI, EEG) gives the investigator the necessary flexibility to study neural circuits other than those giving rise to overt behavior. Clearly, the main advantages of the combined stimulation/

imaging approach are the proximity of the cause and effect and the degree of experimental control not available when subject's behavior is the main independent variable. What we need to overcome, however, is the technical complexity of the combined studies. Future studies may benefit from focusing on neurochemical transmission in specific neural circuits and on temporal dynamics of cortico-cortical interactions.

If we were to take a 'bird's-eye' view of the past 10 years of work that combined on-line brain imaging with brain stimulation, what would we see? Table 34.1 provides a few statistics. First of all, the total number of published studies is relatively low (~60). Most likely, this reflects high technical demands associated with the concurrent use of two very different methods (i.e. TMS and PET/fMRI/EEG), each

Table 34.1 Overview of studies that combined TMS with functional neuroimaging

Imaging method	Subjects	Cortical target	TMS frequency (Hz)	TMS intensity (% MT)	Reference
PET–CBF	Healthy	PFC–L	1, 10	100	Barrett et al. 2004
PET–CBF	Stroke	M1	10	95	Chouinard et al. 2006
PET–CBF	Healthy	M1–L	1	90	Chouinard et al. 2003
PET–CBF	Healthy	PMCd–L	1	90	Chouinard et al. 2003
PET–CBF	Healthy	M1–L	1	120	Ferrarelli et al. 2004
PET–CBF	Healthy	V1–R	1	120	Ferrarelli et al. 2004
PET–CBF	Healthy	PFC–L	1	120	Ferrarelli et al. 2004
PET–CBF	Healthy	PFC–R	1	120	Ferrarelli et al. 2004
PET–CBF	Healthy	M1–L	1	120	Fox et al. 1997
PET–CBF	Healthy	M1–L	3	125	Fox et al. 2004
PET–CBF	Healthy	M1–L	3	75–125	Fox et al. 2006
PET–CBF	Healthy	M1–L	1	90	Lee et al. 2003
PET–CBF	Healthy	PFC–L	4	110	Mottaghy et al. 2003b
PET–CBF	Healthy	PFC–R	4	110	Mottaghy et al. 2003b
PET–CBF	Healthy	PFC–L	10, PP	100	Paus et al. 2001a
PET–CBF	Healthy	M1–L	10	70 (MSO)	Paus et al. 1998
PET–CBF	Healthy	FEF–L	10	70 (MSO)	Paus et al. 1997
PET–CBF	Healthy	M1–L	1, 5	90	Rounis et al. 2005
PET–CBF	Focal dystonia	PMCd	1	90	Siebner et al. 2003
PET–CBF	Healthy	M1–L	1–5	90 (active)	Siebner et al. 2001b
PET–CBF[a]	Depression	PFC–L	1, 20	100	Speer et al. 2000
PET–CBF	Healthy	M1–L	1	80–120	Speer et al. 2003a
PET–CBF	Healthy	PFC–L	1	80–120	Speer et al. 2003b
PET–CBF	Healthy	M1–L	SP, PP	52, 39 (MSO)	Strafella and Paus 2001
PET–CBF	Healthy	M1–L	5	90 (active)	Takano et al. 2004
PET–CBF	Blind	S1	10	90	Wittenberg et al. 2004

Table 34.1 (*contd.*) Overview of studies that combined TMS with functional neuroimaging

Imaging method	Subjects	Cortical target	TMS frequency (Hz)	TMS intensity (% MT)	Reference
fMRI	Healthy	M1–L	10	110	Baudewig et al. 2001
fMRI	Healthy	PMCd–L	10	90, 110	Baudewig et al. 2001
fMRI	Healthy	M1–L	3.1	110; 90 (active)	Bestmann et al. 2004
fMRI	Healthy	M1–L	4	110; 90, 110 (active)	Bestmann et al. 2003
fMRI	Healthy	PMCd–L	3	110; 90 (active)	Bestmann et al. 2005
fMRI	Amputee	M1			Bestmann et al. 2006
fMRI	Healthy	M1–L	1	80, 110	Bohning et al. 1999
fMRI	Healthy	M1–L	0.83	110	Bohning et al. 1998
fMRI	Healthy	M1–L	1	110	Bohning et al. 2000a
fMRI	Healthy	M1–L	SP	120	Bohning et al. 2000b
fMRI	Healthy	M1–L	1	110	Bohning et al. 2003
fMRI	Healthy	M1–L	1	110	Denslow et al. 2005
fMRI	Healthy	M1–L	1	110	Denslow et al. 2005
fMRI	Healthy	M1–L	1	110	Denslow et al. 2005
fMRI	Healthy	M1–L	1	110	Denslow et al. 2004
fMRI	Healthy	M1–L	4	150	Kemna and Gembris 2003
fMRI	Healthy	M1–L	1	110, 120	Li et al. 2004b
fMRI	Healthy	PFC–L	1	100, 120	Li et al. 2004b
fMRI	Depression	PFC–L	1	100	Li et al. 2004a
fMRI	Healthy old	M1–L	1	110	McConnell et al. 2003
fMRI	Healthy	PFC–L	1	80, 100, 120	Nahas 2001
fMRI	Healthy	S1–L	5	90	Pleger et al. 2006; off-line
EEG	Healthy	PPC–R	SP	85 (MSO)	Fuggetta et al. 2006
EEG	Healthy	M1–L	0.5	90	Nikouline et al. 1999
EEG	Healthy	M1–L	SP, PP	45–65 (MSO)	Paus et al. 2001b
EEG[a]	Healthy	M1–L	1	90 (active)	Strens et al. 2002
EEG	Healthy	FEF–R	10	110	Taylor et al. 2007
EEG	Healthy	M1–L	SP 0.6	90, 115	van der Werf and Paus 2006
EEG	Healthy	PMCd–L	0.6	90	van der Werf and Paus 2006
EEG	Parkinson/ thalamotomy	M1	SP	120	van der Werf et al. 2006

Table 34.1 (*contd.*) Overview of studies that combined TMS with functional neuroimaging

Imaging method	Subjects	Cortical target	TMS frequency (Hz)	TMS intensity (% MT)	Reference
SPECT–CBF	Depression	PFC–L	20	90	Catafau *et al.* 2001
SPECT–CBF[a]	Depression	PFC–L	20	80	Nadeau *et al.* 2002
SPECT–CBF	Healthy	M1–L	1	110 (active)	Okabe *et al.* 2003
PET– dopamine	Healthy	M1–L	10	90	Strafella *et al.* 2003
PET– dopamine	Healthy	PFC–L	10	100	Strafella *et al.* 2001
PET–FDG	Healthy	M1–L	5	90	Siebner 2000
PET–FDG	Healthy	M1–L	2	140	Siebner 2001a
MRI– diffusion	Healthy	M1–L	1	90	Mottaghy 2003a

M1, primary motor cortex; PMCd, dorsal premotor cortex; PFC, prefrontal cortex; V1, primary visual cortex; S1, primary somatosensory cortex; FEF, frontal eye field; PPC, posterior parietal cortex; SP, single pulse; PP, paired pulse; MSO, maximum stimulator output; active, active motor threshold; MT, resting motor threshold.
[a]Studies that employed off-line TMS.
PET, positron emission tomography; CBF, cerebral blood flow; fMRI, functional magnetic resonance imaging; EEG, electroencephalography; SPECT, single-photon emission computed tomography; FDG, [^{18}F]deoxyglucose.

requiring a particular set of skills. Second, the overwhelming majority of previous studies targeted the primary motor cortex and the prefrontal cortex (PFC). The former choice may reflect the usefulness of having an additional dependent variable, namely MEPs, to be used in parallel with the imaging data. But it also likely relates to the relative ease with which the primary motor cortex can be localized and accessed in the scanner. On the other hand, the choice of the PFC as a target is mostly motivated by the desire to explain beneficial effects of rTMS applied over the left PFC in patients with major depression (Paus and Barrett 2004; see Section VI, this volume). Other targets included dorsal premotor cortex, FEF, primary visual and somatosensory cortex, and the posterior parietal cortex. Only a handful of studies were carried out in patients with psychiatric (depression) or neurological (stroke, Parkinson's disease) disorders. Finally, the great majority of published studies used hemodynamics as the proxy of neural activity (CBF–PET: $n = 25$; fMRI: $n = 21$), followed by EEG ($n = 8$) and glucose metabolism measured with PET ($n = 4$).

Conclusion

The last 10 years have seen the birth and growth of a new methodological approach in cognitive neuroscience, namely the combination of brain imaging with brain stimulation. The two main contributions of this new paradigm have been the incorporation of causality into the interpretation of correlation-based functional imaging studies, and the description of effective connectivity of the motor and prefrontal cortex. The following three chapters provide a detailed overview of the technical and methodological issues involved in carrying out such studies, as well as a comprehensive review of the findings obtained to date.

Acknowledgments

I thank my colleagues at the Montreal Neurological Institute for their contributions in carrying out our TMS/PET and TMS/EEG studies of cortical excitability and connectivity. The author's research is supported by the Canadian Institutes of Health Research, Canadian Foundation for Innovation, the National Science and Engineering Research Council of Canada, and the Royal Society (UK).

References

Barrett J, Della-Maggiore V, Chouinard P, Paus T (2004). Mechanisms of action underlying the effect of repetitive transcranial magnetic stimulation on mood: behavioral and imaging studies. *Neuropsychopharmacology 29*, 1172–1189.

Baudewig J, Siebner HR, Bestmann S, *et al.* (2001). Functional MRI of cortical activations induced by transcranial magnetic stimulation (TMS). *Neuroreport 12*, 3543–3548.

Bestmann S, Baudewig J, Siebner HR, Rothwell JC, Frahm J (2003). Subthreshold high-frequency TMS of human primary motor cortex modulates interconnected frontal motor areas as detected by interleaved fMRI-TMS. *NeuroImage 20*, 1685–1696.

Bestmann S, Baudewig J, Siebner HR, Rothwell JC, Frahm J (2004). Functional MRI of the immediate impact of transcranial magnetic stimulation on cortical and subcortical motor circuits. *European Journal of Neuroscience 19*, 1950–1962.

Bestmann S, Baudewig J, Siebner HR, Rothwell JC, Frahm J (2005). BOLD MRI responses to repetitive TMS over human dorsal premotor cortex. *NeuroImage 28*, 22–29.

Bestmann S, Oliviero A, Voss M, *et al.* Concurrent TMS and fMRI in an amputee reveals cortical correlates of TMS-induced phantom hand movements. *Neuropsychologia,* in press.

Bohning DE, Shastri A, Nahas Z, *et al.* (1998). Echoplanar BOLD fMRI of brain activation induced by concurrent transcranial magnetic stimulation. *Investigations in Radiology 33*, 336–340.

Bohning DE, Shastri A, McConnell KA, *et al.* (1999). A combined TMS/fMRI study of intensity-dependent TMS over motor cortex. *Biological Psychiatry 45*, 385–394.

Bohning DE, Shastri A, McGavin L, *et al.* (2000a). Motor cortex brain activity induced by 1-Hz transcranial magnetic stimulation is similar in location and level to that for volitional movement. *Investigations in Radiology 35*, 676–683.

Bohning DE, Shastri A, Wassermann EM, *et al.* (2000b). BOLD-f MRI response to single-pulse transcranial magnetic stimulation (TMS). *Journal of Magnetic Resonance Imaging 11*, 569–574.

Bohning DE, Shastri A, Lomarev MP, Lorberbaum JP, Nahas Z, George MS (2003). BOLD-fMRI response vs. transcranial magnetic stimulation (TMS) pulse-train length: testing for linearity. *Journal of Magnetic Resonance Imaging 17*, 279–290.

Broca P (1861). Nouvelle observation d'aphémie produite par une lésion de la moitié postérieure des deuxième et troisième circonvolution frontales gauches. *Bulletin de la Société Anatomique 36*, 398–407.

Catafau AM, Perez V, Gironell A, *et al.* (2001). SPECT mapping of cerebral activity changes induced by repetitive transcranial magnetic stimulation in depressed patients. A pilot study. *Psychiatry Research 106*, 151–160.

Chen R, Yung D, Li JY (2003). Organization of ipsilateral excitatory and inhibitory pathways in the human motor cortex. *Journal of Neurophysiology 89*, 1256–1264.

Chouinard PA, Paus T (2006). The primary motor and premotor areas of the human cerebral cortex. *Neuroscientist 2*, 143–152.

Chouinard PA, Van Der Werf YD, Leonard G, Paus T (2003). Modulating neural networks with transcranial magnetic stimulation applied over the dorsal premotor and primary motor cortices. *Journal of Neurophysiology 90*, 1071–1083.

Chouinard PA, Leonard G, Paus T (2006). Changes in effective connectivity of the primary motor cortex in stroke patients after rehabilitation. *Experimental Neurology 201*, 375–387.

Denslow S, Bohning DE, Lomarev MP, George MS (2004). A high resolution assessment of the repeatability of relative location and intensity of TMS-induced and volitionally-induced BOLD response in the motor cortex. *Cognitive and Behavioral Neurology 17*, 163–173.

Denslow S, Lomarev M, George MS, Bohning DE (2005). Cortical and subcortical brain effects of transcranial magnetic stimulation (TMS)-induced movement: an interleaved TMS/functional magnetic resonance imaging study. *Biological Psychiatry 57*, 752–760.

Dusser de Barenne JG, McCulloch WS (1936). Functional boundaries in the sensori-motor cortex of the monkey. *Proceedings of the Society of Experimental Biology and Medicine 35*, 329–331.

Ferrarelli F, Haraldsson HM, Barnhart TE, *et al.* (2004). A [17F]-fluoromethane PET/TMS study of effective connectivity. *Brain Research Bulletin 64*, 103–113.

Fox P, Ingham R, George MS, *et al.* (1997). Imaging human intra-cerebral connectivity by PET during TMS. *Neuroreport 8*, 2787–2791.

Fox PT, Narayana S, Tandon N, *et al.* (2004). Column-based model of electric field excitation of cerebral cortex. *Human Brain Mapping 22*, 1–16.

Fox PT, Narayana S, Tandon N, *et al.* (2006). Intensity modulation of TMS-induced cortical excitation: primary motor cortex. *Human Brain Mapping 27*, 478–487.

Friston KJ (1994). Functional and effective connectivity in neuroimaging: a synthesis. *Human Brain Mapping 2*, 56–78.

Fritsch G, Hitzig E (1870). Über die elektrische Erragbarkeit des Grosshirns. *Archiv für Anatomie und Physiologie (Leipzig) 37*, 300–332.

Fuggetta G, Pavone EF, Walsh V, Kiss M, Eimer M (2006). Cortico-cortical interactions in spatial attention: A combined ERP/TMS study. *Journal of Neurophysiology 95*, 3277–3280.

Gagnon D, Paus T, Grosbras MH, Pike GB, O'Driscoll GA (2006). Transcranial magnetic stimulation of frontal oculomotor regions during smooth pursuit. *Journal of Neuroscience 26*, 458–466.

Granger CWJ (1969). Investigating causal relations by econometric models and cross-spectral methods. *Econometrica 37*, 424–438.

Grosbras MH, Paus T (2003). Transcranial magnetic stimulation of the human frontal eye field facilitates visual awareness. *European Journal of Neuroscience 18*, 3121–3126.

Grosbras MH, Laird AR, Paus T (2005). Cortical regions involved in eye movements, shifts of attention, and gaze perception. *Human Brain Mapping 25*, 140–154.

Horwitz B (2003). The elusive concept of brain connectivity. *NeuroImage 19*, 466–470.

Johansen-Berg H, Behrens TE, Sillery E, et al. (2005). Functional–anatomical validation and individual variation of diffusion tractography-based segmentation of the human thalamus. *Cerebral Cortex 15*, 31–39.

Kemna LJ, Gembris D (2003). Repetitive transcranial magnetic stimulation induces different responses in different cortical areas: a functional magnetic resonance study in humans. *Neuroscience Letters 336*, 85–88.

Lee, L, Siebner, HR, Rowe, JB, et al. (2003) Acute remapping within the motor system induced by low-frequency repetitive transcranial magnetic stimulation. *Journal of Neuroscience 23*, 5308–5318.

Leyton ASF, Sherington CS (1917). Observations on the excitable cortex of the chimpanzee, orang-utan and gorilla. *Quarterly Journal of Experimental Physiology 11*, 135–222.

Li X, Nahas Z, Kozel FA, Anderson B, Bohning DE, George MS (2004a). Acute left prefrontal transcranial magnetic stimulation in depressed patients is associated with immediately increased activity in prefrontal cortical as well as subcortical regions. *Biological Psychiatry 55*, 882–890.

Li X, Teneback CC, Nahas Z, et al. (2004b). Interleaved transcranial magnetic stimulation/functional MRI confirms that lamotrigine inhibits cortical excitability in healthy young men. *Neuropsychopharmacology 29*, 1395–1407.

McConnell KA, Bohning DE, Nahas Z, et al. (2003). BOLD fMRI response to direct stimulation (transcranial magnetic stimulation) of the motor cortex shows no decline with age. *Journal of Neural Transmission 110*, 495–507.

Mottaghy FM, Gangitano M, Horkan C, Chen Y, Pascual-Leone A, Schlaug G (2003a). Repetitive TMS temporarily alters brain diffusion. *Neurology 60*, 1539–1541.

Mottaghy FM, Pascual-Leone A, Kemna LJ, et al. (2003b). Modulation of a brain–behavior relationship in verbal working memory by rTMS. *Brain Research: Cognitive Brain Research 15*, 241–249.

Münchau A, Bloem BR, Irlbacher K, Trimble MR, Rothwell JC (2002). Functional connectivity of human premotor and motor cortex explored with repetitive transcranial magnetic stimulation. *Journal of Neuroscience 22*, 554–561.

Nadeau SE, McCoy KJ, Crucian GP, et al. (2002) Cerebral blood flow changes in depressed patients after treatment with repetitive transcranial magnetic stimulation: evidence of individual variability. *Neuropsychiatry, Neuropsychology, and Behavioral Neurology 15*, 159–175.

Nahas Z, Lomarev M, Roberts DR, et al. (2001). Unilateral left prefrontal transcranial magnetic stimulation (TMS) produces intensity-dependent bilateral effects as measured by interleaved BOLD fMRI. *Biological Psychiatry 50*, 712–720.

Nikouline V, Ruohonen J, Ilmoniemi RJ (1999). The role of the coil click in TMS assessed with simultaneous EEG. *Clinical Neurophysiology 110*, 1325–1328.

Okabe S, Hanajima R, Ohnishi T, et al. (2003). Functional connectivity revealed by single-photon emission computed tomography (SPECT) during repetitive transcranial magnetic stimulation (rTMS) of the motor cortex. *Clinical Neurophysiology 114*, 450–457.

Paus T (1996). Location and function of the human frontal eye-field: a selective review. *Neuropsychologia 34*, 475–483.

Paus, T (1999). Imaging the brain before, during, and after transcranial magnetic stimulation. *Neuropsychologia 37*, 219–224.

Paus T, Barrett J (2004). Transcranial magnetic stimulation of the human frontal cortex: implications for rTMS treatment of depression. *Journal of Psychiatry and Neuroscience 29*, 268–277.

Paus T, Jech R, Thompson CJ, Comeau R, Peters T, Evans AC (1997). Transcranial magnetic stimulation during positron emission tomography: a new method for studying connectivity of the human cerebral cortex. *Journal of Neuroscience 17*, 3178–3184.

Paus T, Jech R, Thompson CJ, Comeau R, Peters T, Evans AC (1998). Dose-dependent reduction of cerebral blood flow during rapid-rate transcranial magnetic stimulation of the human sensorimotor cortex. *Journal of Neurophysiology 79*, 1102–1107.

Paus T, Castro-Alamancos M, Petrides M (2001a). Cortico-cortical connectivity of the human mid-dorsolateral frontal cortex and its modulation by repetitive transcranial magnetic stimulation. *European Journal of Neuroscience 14*, 1405–1411.

Paus T, Sipila PK, Strafella AP (2001b). Synchronization of neuronal activity in the human sensori-motor cortex by transcranial magnetic stimulation: a combined TMS/EEG study. *Journal of Neurophysiology 86*, 1983–1990.

Penfield W, Rasmussen AT (1950). *The cerebral cortex of man*. New York: Macmillan.

Pleger B, Blankenburg F, Bestmann S, et al. (2006). Repetitive transcranial magnetic stimulation-induced changes in sensori-motor coupling parallel improvements of somatosensation in humans. *Journal of Neuroscience 26*, 1945–1952.

Rounis E, Lee L, Siebner HR, et al. (2005). Frequency specific changes in regional cerebral blood flow and motor system connectivity following rTMS to the primary motor cortex. *NeuroImage 26*, 164–176.

Sack AT, Kohler A, Bestmann S, et al. (2005a). Visualizing virtual brain lesions. *Society of Neuroscience 934.10*.

Scoville WB, Milner B (1957). Loss of recent memory after bilateral hippocampal lesions. *Journal of Neurology, Neurosurgery and Psychiatry 20*, 11–21.

Siebner HR, Peller M, Willoch F, et al. (2000). Lasting cortical activation after repetitive TMS of the motor cortex: a glucose metabolic study. *Neurology 54*, 956–963.

Siebner H, Peller M, Bartenstein P, et al. (2001a). Activation of frontal premotor areas during suprathreshold transcranial magnetic stimulation of the left primary sensorimotor cortex: a glucose metabolic PET study. *Human Brain Mapping 12*, 157–167.

Siebner HR, Takano B, Peinemann A, Schwaiger M, Conrad B, Drzezga A (2001b). Continuous transcranial magnetic stimulation during positron emission tomography: a suitable tool for imaging regional excitability of the human cortex. *NeuroImage 14*, 883–890.

Siebner HR, Filipovic SR, Rowe JB, et al. (2003). Patients with focal arm dystonia have increased sensitivity to slow-frequency repetitive TMS of the dorsal premotor cortex. *Brain 126*, 2710–2715.

Speer AM, Kimbrell TA, Wassermann EM, et al. (2000). Opposite effects of high and low frequency rTMS on regional brain activity in depressed patients. *Biological Psychiatry 48*, 1133–1141.

Speer AM, Willis MW, Herscovitch P, et al. (2003a). Intensity-dependent regional cerebral blood flow during 1-Hz repetitive transcranial magnetic stimulation (rTMS) in healthy volunteers studied with H215O positron emission tomography: I. Effects of primary motor cortex rTMS. *Biological Psychiatry 54*, 818–825.

Speer AM, Willis MW, Herscovitch P, et al. (2003b). Intensity-dependent regional cerebral blood flow during 1-Hz repetitive transcranial magnetic stimulation (rTMS) in healthy volunteers studied with H215O positron emission tomography: II. Effects of prefrontal cortex rTMS. *Biological Psychiatry 54*, 826–832.

Sporns O, Chialvo DR, Kaiser M, Hilgetag CC (2004). Organization, development and function of complex brain networks. *Trends in Cognitive Sciences 8*, 418–425.

Strafella A and Paus T (2001). Cerebral blood-flow changes induced by paired-pulse transcranial magnetic stimulation of the primary motor cortex. *Journal of Neurophysiology 85*, 2624–2629.

Strafella AP, Paus T, Barrett J, Dagher A (2001). Repetitive transcranial magnetic stimulation of the human prefrontal cortex induces dopamine release in the caudate nucleus. *Journal of Neuroscience 21*, RC157.

Strafella AP, Paus T, Fraraccio M, Dagher A (2003). Striatal dopamine release induced by repetitive transcranial magnetic stimulation of the human motor cortex. *Brain 126*, 2609–2615.

Strens LH, Oliviero A, Bloem BR, Gerschlager W, Rothwell JC, Brown P (2002). The effects of subthreshold 1 Hz repetitive TMS on cortico-cortical and interhemispheric coherence. *Clinical Neurophysiology 113*, 1279–1285.

Takano B, Drzezga A, Peller M, et al. (2004). Short-term modulation of regional excitability and blood flow in human motor cortex following rapid-rate transcranial magnetic stimulation. *NeuroImage 23*, 849–859.

Tardif E, Clarke S (2001). Intrinsic connectivity of human auditory areas: a tracing study with DiI. *European Journal of Neuroscience 13*, 1045–1050.

Taylor PC, Nobre AC, Rushworth MF (2007). FEF TMS affects visual cortical activity. *Cerebral Cortex 17*, 391–399.

van der Werf Y, Paus T (2006). The neural response to transcranial magnetic stimulation of the human motor cortex. I. Intracortical and cortico-cortical contributions. *Experimental Brain Research 175*, 231–245.

van der Werf Y, Strafella A, Sadikot A, Paus T (2006). The neural response to transcranial magnetic stimulation of the human motor cortex. II. Thalamocortical contributions. *Experimental Brain Research 175*, 246–255.

Wittenberg GF, Werhahn KJ, Wassermann EM, Herscovitch P, Cohen LG (2004). Functional connectivity between somatosensory and visual cortex in early blind humans. *European Journal of Neuroscience 20*, 1923–1927.

CHAPTER 35

TMS and positron emission tomography: methods and current advances

Hartwig R. Siebner, Martin Peller, and Lucy Lee

Introduction

The combined use of positron emission tomography (PET) and focal TMS was established about 10 years ago. In 1997, two independent studies showed that PET measurements of TMS-induced changes in regional cerebral blood flow (rCBF) could be used to map the connectivity of the stimulated cortex *in vivo* (Fox *et al.* 1997; Paus *et al.* 1997). These studies prompted a series of combined TMS–PET studies which have significantly advanced our understanding of how TMS interacts with the brain. This chapter provides an overview of how TMS and PET can be combined to investigate human brain function. First, we address methodological issues concerning the combination of TMS with PET. Second, we discuss possible applications of the combined TMS–PET approach for studying human brain function. We outline how this approach can be used (i) to examine the regional responsiveness of the cortex to TMS as well as inter-regional coupling in the intact human brain; (ii) to explore inter-regional patterns of acute reorganization at the systems level; (iii) to further our understanding of potential therapeutic effects of TMS conditioning; and (iv) to trace dynamic aspects of the

pathophysiology in neuropsychiatric disorders. Finally, we address future perspectives of the combined TMS–PET approach.

It is worth remembering that TMS represents a nonphysiological means of producing or modulating neuronal activity in the human brain. Therefore, it is important for the future use of TMS to improve our understanding of how this nonphysiological mode of brain stimulation interacts with 'normal' activity in the human brain. There are currently three methods available with which to examine the effects of TMS on human brain function. First, it is possible to measure the effects of TMS on distinct aspects of cortical excitability. This approach has been used extensively in the primary motor cortex (M1) where it is possible to assess the excitatory and inhibitory effects of TMS further downstream by recording the motor-evoked responses. Second, the behavioral sequelae of TMS can be determined by examining how TMS interferes with behavior during a given task. Third, the effects of TMS on measures of regional activity can be examined with functional imaging techniques such as PET. These approaches offer complementary information and have different methodological strengths and weaknesses. Combined TMS–PET studies should incorporate concurrent measurements of behavior and cortical excitability whenever possible. By using different sources of information, the TMS–PET approach will provide deeper insight into the neurophysiological effects of TMS on human brain function.

There are two ways to combine TMS and PET (Figure 35.1). TMS and PET can be given simultaneously (referred to as on-line PET imaging) or separated in time (referred to as off-line PET imaging). On-line PET imaging can be used to investigate the immediate neurophysiological effects of TMS. Off-line PET

(a) "on-line" PET imaging
"map-during-stimulation" approach:
TMS during PET mapping

(b) "off-line" PET imaging
1. "map-and-perturb" approach:
PET mapping followed by TMS lesioning

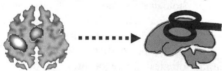

2. "condition-and-map" approach:
TMS conditioning followed by PET mapping

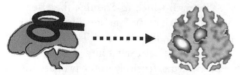

Fig. 35.1 The relative timing of TMS and positron emission tomography (PET) determines the scientific or clinical application of the combined TMS–PET approach. (a) PET can be performed during TMS (i.e. on-line approach) to investigate the immediate effects of TMS on brain function. (b) Alternatively, TMS can precede or follow PET imaging (i.e. off-line approach). The off-line approach can be used to identify appropriate sites for focal TMS (i.e. PET preceding TMS) or to probe the lasting effects of TMS conditioning on brain function (i.e. TMS preceding PET).

imaging can be used to study the enduring effects of repetitive (r)TMS on brain function (TMS preceding PET) or to define appropriate cortical sites to be targeted by TMS (PET preceding TMS). In the latter case, PET is used to delineate the cortical activation pattern during the performance of a given task. This topographic information can be used to target a distinct cortical area in a subsequent experiment in which focal TMS is given to interfere with task performance. The advantage of previous neuroimaging is that the *a priori* knowledge about the task-related activation patterns enables more precise positioning of the TMS coil.

A short outline of positron emission tomography

PET imaging is based on the intravenous administration of radioisotopes. PET maps the *in vivo* binding and metabolism of compounds that have been tagged with short-lived positron-emitting isotopes such as ^{15}O, ^{11}C or ^{18}F. These radio-labelled biological compounds emit positrons, which produce pairs of gamma rays that are detected by the PET scanner. Based on the amount and origin of radiation emitted, a quantitative three-dimensional image of tracer distribution can be generated. Importantly, the degree of binding or metabolism of the biologically active compound depends on the functional state of the underlying tissue. This enables assessment of regional patterns of neuronal activity at rest or during performance of specific tasks.

PET offers a range of well-established techniques to study human brain function (Piccini and Whone 2004). Various radioactive tracers (radioligands) can be used to quantify changes in rCBF or regional cerebral metabolic rate of glucose (rCMRglc) (Herholz and Heiss 2004). Since rCBF and rCMRglc are tightly coupled with regional synaptic activity, metabolic PET imaging enables assessment of the level of regional brain activation at rest and during specific tasks. Because changes in local field potential show a strong positive correlation with changes in rCBF, it has been suggested that local processing of inputs is the main source of hemodynamic changes in the cortex (Logothetis 2002; Logothetis and Wandell 2004). Other important applications of PET are the study of specific neurotransmitter and receptor systems, amino acid uptake, or microglial activation (Cagnin et al. 2002; Herholz and Heiss 2004). The radioisotope-based imaging technique with the highest resolution and greatest sensitivity to differentiate between normal and abnormal functional states is three-dimensional PET. The most readily available technique, however, is single-photon emission computed tomography (SPECT), because it uses radioisotopes with a long half-life and does not require an on-site cyclotron. With regard to the combined use of SPECT and TMS, the general issues are identical to those encountered when combining TMS with PET. Therefore, TMS–SPECT studies are discussed together with TMS–PET studies.

PET imaging versus functional magnetic resonance imaging (fMRI)

Before considering how PET imaging may be used to probe TMS-induced changes in synaptic activity, it is useful to consider briefly the specific limitations and advantages of the combined TMS–PET approach as opposed to the combined TMS–fMRI approach. PET imaging of rCBF or rCMRglc provides indirect measures of regional synaptic activity over several tens of seconds ($H_2^{15}O$ PET) or minutes ([^{18}F] deoxyglucose PET), the latter being well below the temporal resolution of fMRI. Because of the limited temporal resolution, combined TMS–PET experiments require an epoch-related rather than an event-related design. This has important implications regarding on-line PET imaging because TMS needs to be given repeatedly during a PET scan, using a continuous train or intermittent bursts of rTMS. Therefore, a single PET scan will represent the summation of the effects of individual stimuli on regional synaptic activity. The lower temporal resolution of PET has both advantages and disadvantages for on-line TMS–PET. PET is not a suitable modality with which to examine the effects of a single pulse or a short train of TMS on regional neuronal activity. However, the repeated application of TMS during PET scanning allows for a temporal summation of the neuronal effects, thereby increasing the sensitivity of the combined TMS–PET approach to detect cumulative changes in regional neuronal activity in the stimulated cortex during TMS.

Although PET has a poorer spatial resolution than fMRI, this may not be critical because currently available TMS devices do not produce highly focal cortical stimulation requiring high spatial resolution. The main drawback of PET imaging is of course the exposure to radiation, which limits the number of measurements per subject.

There are several features which distinguish PET from fMRI and make PET imaging attractive to researchers interested in TMS. In contrast to the interleaved TMS–fMRI approach, there are no constraints on the relative timing of TMS during PET imaging. The PET scanner also imposes fewer spatial constraints for reaching the target site than the MRI head coils. Therefore, all currently available TMS protocols can be given during PET imaging and all cortical areas that have been studied with TMS are accessible to on-line TMS–PET imaging. In addition, PET allows for direct comparisons of synaptic activity between different scanning sessions. This makes metabolic PET imaging of rCBF or rCMRglc an ideal tool with which to study how changes in regional neuronal activity depend on the TMS protocol (e.g. the intensity or frequency of TMS) or the site of stimulation. Serial PET scanning can also be used to track the time-course of rTMS conditioning effects, at rest or during a task, for at least 1 h after rTMS. Finally, there are several radiotracers available that can be used to investigate the effects of TMS on specific neurotransmitter systems (e.g. dopaminergic system) or cell populations (e.g. microglia).

Methodological considerations

Although the prerequisites for on-line PET imaging are relatively easy to establish, several methodological issues need to be considered when TMS is applied during PET imaging. For off-line PET imaging, no specific methodological precautions are required because TMS can be given outside the scanner before or after the PET measurements.

Technical considerations

An initial concern was that the strong phasic magnetic field could compromise the function of the PET detectors. Thompson et al. (1998)

showed that TMS can seriously perturb the operation of the photomultiplier of the PET detectors not only during the TMS pulse, but for a period of 100 ms after the pulse. They also demonstrated that four layers of well-grounded mu-metal were sufficient to protect the photomultipliers in the PET detectors from the effects of the TMS-induced magnetic field (Thompson et al. 1998). Therefore, a well-grounded cylindrical shield consisting of four layers of mu-metal was inserted into the PET scanner in initial studies to ensure normal function of the detectors (Paus et al. 1997, 1998, 2001; Siebner et al. 2001b; Takano et al. 2004). The use of a cylindrical mu-metal shield resulted in a ~15–20% decrease in coincidence counts due to increased attenuation. Other groups have opted not to use shielding because TMS did not interfere with the quality of PET scanning (Fox et al. 1997; Speer et al. 2000, 2001; Ferrarelli et al. 2004).

When TMS is applied in the PET scanner, the presence of the coil causes significant attenuation of radiation. Pre-processing of the PET images needs to correct for the coil-induced attenuation. For attenuation correction, it is necessary to acquire a transmission scan with the TMS coil *in situ*. If two cortical areas are sequentially targeted during the same PET experiment (Mottaghy et al. 2000, 2003; Ferrarelli et al. 2004; Wittenberg et al. 2004), separate transmission scans need to be obtained for each coil position (Paus et al. 1997, 1998).

Localizing and verifying the cortical target area

In combined TMS–PET studies, the exact location of the target site is critical. For the M1, the TMS-induced motor response provides a physiological measure with which to guide placement of the TMS coil. The location of the M1 can also be used as an 'anchor' point for localizing cortical areas that are close to the M1 such as the lateral premotor cortex. The majority of cortical areas require frameless stereotaxy to localize the target site (Paus et al. 1997; Paus and Wolforth 1998). Spatial accuracy of frameless stereotaxy is in the order of several millimeters which is sufficient for current TMS applications. An additional advantage of frameless stereotaxy is that the

position of the coil can be monitored throughout the PET session. While frameless stereotaxy can be easily applied outside the scanner in studies using off-line PET imaging, the use of a frameless stereotaxy system is difficult during on-line PET imaging given the limited space within the PET scanner. One approach is to determine the correct coil position outside the PET scanner with frameless stereotaxy prior to the experiment and mark the correct coil position on the subject's head. The correct coil position can then be verified using the transmission scan, which is acquired before the first emission scan. The coil is clearly visible on the transmission scan and the anatomical location of the coil can be determined when the transmission scan is co-registered on the individual structural MRI scan (Paus and Wolforth, 1998). Alternatively, a vitamin E capsule can be taped on the scalp under the center of the coil immediately after PET; correct placement of the coil can be confirmed with standard T1-weighted structural MRI.

The position of the TMS coil needs to be kept constant across consecutive PET scans. An image-guided robotically positioned TMS system has been developed for this purpose (Lancaster et al. 2004). For most applications, a custom-made mechanical fixation unit is sufficient to position and fix the coil over the cortical target area and to monitor coil position throughout the experiment.

Nonspecific rTMS effects

'Focal' TMS always produces significant auditory and somatosensory stimulation in addition to direct transcranial cortical stimulation. The loud click produced by the discharging magnetic coil activates the auditory system (Siebner et al. 1999b; Takano et al. 2004). The rapidly changing magnetic field also activates the somatosensory system by stimulating afferent trigeminal nerve fibers (Siebner et al. 1999a). TMS may also induce an emotional response (e.g. unpleasantness or discomfort) that may influence regional activity in limbic brain areas. The sensory stimulation caused by TMS may also act as a lateralized attentional cue that may modify neuronal activity related to spatial attention. These nonspecific effects need to be taken into account when interpreting

TMS-induced changes in regional neuronal activity. When TMS is given to the M1 at intensities that elicit muscle twitches, reafferent feedback activation will make a substantial contribution to the activity pattern of the sensorimotor system observed during PET (Siebner et al. 1998; Speer et al. 2003a).

How is it possible to distinguish specific from nonspecific TMS effects on rCBF or rCMRglc? Some researchers have attempted to mask TMS-related acoustic input with white noise. Since the coil is in contact with the head during rTMS, the click of the discharging coil is transmitted via bone conduction; therefore, white noise may not be sufficient to mask effectively the TMS-related acoustic input (Siebner et al. 1999b). In addition, auditory masking does not control for concurrent somatosensory stimulation. An alternative approach to control for TMS-induced auditory stimulation is to use a control condition in which a second coil is discharged at the same rate some centimeters away from the scalp. In this case, it is difficult to match exactly the intensity of acoustic stimulation because there is no bone conduction when the coil has no contact with the scalp. This control condition also produces no somatosensory stimulation and therefore cannot control for TMS-induced somatosensory stimulation.

The minimum requirement is a control condition that mimics both the auditory and somatosensory input produced by TMS. Electrical stimulation of the skin at the site of TMS has been used to mimic the peripheral somatosensory input produced by TMS (Okabe et al. 2003). Although this is a reasonable approach, the sensation induced by electrical scalp stimulation is qualitatively different from the sensation induced by TMS. Alternatively, TMS can be given to two different cortical areas using the same stimulation protocol (Strafella et al. 2001). In this scenario, one area is the target area; the other area is only stimulated to control for the nonspecific effects of TMS. Since the non-specific effects are comparable between both TMS conditions, the differences in rCBF, rCMRglc, or receptor binding can be attributed to cortical stimulation. However, this experimental approach may complicate a straightforward interpretation of relative differences in regional neuronal activity between the two TMS conditions.

If subtraction analysis reveals increased activity in area C during TMS of area A than during TMS of area B, it is impossible to distinguish between increased activity in area C caused by TMS to area A or decreased activity in area B caused by TMS to area C. A priori knowledge of the anatomical connectivity of areas A and B with area C can facilitate the interpretation of such results (Strafella et al. 2001). To avoid this potential problem, one can introduce an additional control condition into the experimental design during which no rTMS is given.

A parametric modulation of a distinct parameter of TMS such as the frequency or intensity of rTMS may help to separate specific and non-specific effects of TMS because a parametric study design enables inferences about the dose dependency of the neurophysiological effects induced by rTMS (Paus et al. 1997, 1998; Siebner et al. 2001b; Speer et al. 2003a,b). However, a note of caution is warranted. Nonspecific rCBF changes caused by auditory and somatosensory stimulation or discomfort during rTMS may show a dose-dependent pattern similar to that of rCBF changes induced by cortical stimulation (Siebner et al. 2001b).

Interference with neurovascular coupling

It has been suggested that TMS may have a direct effect on vascular tone, leading to problems with interpreting rCBF changes because any change in rCBF might result from increased or decreased vascular tone, and therefore be unrelated to TMS-induced modulations of neuronal activity. This concern mainly applies to rCBF changes occurring at the site of stimulation. Although this issue remains to be clarified, there are several lines of evidence suggesting that the rCBF changes in the stimulated cortex can be attributed to changes in neuronal activity. First, TMS produces well-defined changes in regional glucose metabolism (as indexed by a change in rCMRglc) in the stimulated cortex that cannot be explained by a change in vascular tone. Second, identical rTMS protocols can produce an increase or decrease in rCBF depending on the cortical target area. If TMS-induced rCBF changes were caused by a change in vascular tone, a given TMS protocol should produce a stereotypic effect on rCBF in all stimulated areas. This is not the case: identical rTMS protocols cause different patterns of rCBF change depending on the site of stimulation (Paus *et al*. 1997, 1998; Speer *et al*. 2003a,b). Third, there are several studies demonstrating a close correlation between rTMS-induced changes in excitability measured with muscle-evoked potentials and changes in rCBF, suggesting that the changes in rCBF are closely related to changes in synaptic activity (Chouinard *et al*. 2003; Takano *et al*. 2004).

Relationship between measures of regional activity and cortical excitability

Interpreting data from combined TMS–PET experiments is complicated by the absence of a direct relationship between changes in regional synaptic activity (as indexed by rCBF or rCMRglc) and regional cortical excitability (as indexed by TMS measures of cortical excitability). In the M1, TMS measures of cortical excitability depend on synaptic efficacy and membrane excitability of excitatory and inhibitory neuronal populations in the stimulated area and reflect distinct aspects of cortical excitability. Because changes in rCBF or rCMRglc reflect overall changes in regional neuronal activity, PET cannot differentiate between changes in regional activity of excitatory or inhibitory neurons. Additional TMS measurements of cortical excitability enable a more complete interpretation of TMS-induced rCBF changes. This applies to combined TMS–PET studies of the M1, where corticospinal and intracortical excitability can be readily assessed by measuring transcranially evoked motor responses (Rothwell *et al*. 1999).

On-line PET imaging during TMS

TMS can be combined with concurrent measurement of regional neural activity to visualize the acute effects of TMS on regional synaptic activity at the site of stimulation (Siebner *et al*. 2001b) and in connected brain regions (Paus *et al*. 1997). As such, on-line PET imaging of immediate changes in rCBF or rCMRglc provides an indication of changes in regional

activity and connectivity, which can be independent of behavior. PET imaging can identify distributed changes in synaptic activity across the whole brain, including cortical and subcortical areas. In particular, PET can detect regional changes in neuronal activity in brain regions that cannot be stimulated directly with TMS such as the mesial temporal cortex or basal ganglia (Strafella *et al*. 2001; Siebner *et al*. 2003a).

When using the standard figure-8 coil, TMS only produces direct neuronal excitation in a limited area close to the centre of the coil. This is because the magnetic field produced during a TMS pulse is maximal at the centre of the coil and the strength of the magnetic field decreases rapidly with increasing distance from the coil. If PET reveals a change in rCBF or rCMRglc in the cortical target area, one can infer that the TMS protocol has produced a change in the overall level of regional synaptic activity during the PET measurement.

It remains to be clarified how TMS causes an increased regional synaptic activity in the stimulated cortex. There are several possible mechanisms. First, the tissue current induced by TMS may directly induce action potentials in intracortical neurons which results in synaptic neurotransmission in the stimulated cortex. Second, the TMS-induced tissue current may induce action potentials in cortico-cortical axons in white matter underlying the cortex or in the axon terminals within the stimulated cortex. Action potentials in cortico-cortical projections can influence synaptic activity in the stimulated cortex if these projections synapse onto intracortical neurons in the target area. Third, TMS may also induce action potentials in corticofugal pyramidal axons and produce local synaptic activity in the stimulated cortex through recurrent axon collaterals. The excitation of corticofugal axons may also activate cortico-subcortico-cortical re-entry loops and may cause neuronal activity through excitation of subcortico-cortical neurons that project back to the stimulated cortex. All these mechanisms may contribute, to some extent, to changes in synaptic activity at the site of stimulation. The relative contribution of a specific mechanism may differ across cortical areas. It also depends on extrinsic variables (i.e. the TMS protocol) and intrinsic variables such as the functional state of the cortex at the time of stimulation or neuroanatomical factors. The different mechanisms outlined above will have different effects on cortical excitability at the site of stimulation, but those which induce less activation of local synapses will result in smaller changes in rCBF or rCMRglc at the site of stimulation. Therefore, it is not surprising that there is a dissociation between changes in local excitability and changes in rCBF/rCMRglc.

If a focal figure-8 shaped coil is used for TMS, remote effects in connected areas can be attributed to a spread of excitation via cortico-cortical and cortico-subcortical connections. Depending on the site and intensity of stimulation, TMS may cause both indirect (trans-synaptic) and direct (antidromic or orthodromic) activation of cortico-cortical and cortico-subcortical pathways. If TMS induces a consistent modulation of activity in cortico-cortical and cortico-subcortical connections, this may influence intracortical processing in connected areas and lead to changes in regional synaptic activity. Polysynaptic spread of excitation through multiple modules of the network may cause changes in activity in remote brain areas that have no point-to-point connection with the cortical target area, but are connected indirectly through multisynaptic loops. Although the prevailing interpretation assigns remote changes in regional neuronal activity to 'active' trans-synaptic spread of activity from the stimulated region to distant brain regions, TMS may suppress or increase the physiological output from the stimulated region to remote areas without directly exciting the corticofugal pathway. For instance, if the stimulated cortex exerts a tonic inhibitory effect on a remote brain region, a suppression of intracortical activity in the stimulated cortex may release the remote area from the inhibitory drive (Gilio et al. 2003). In summary, PET is a sensitive probe of the TMS-induced changes in regional synaptic activity, but the physiological mechanisms causing local and remote changes cannot be dissociated using PET.

Synaptic activity in the stimulated cortex

The effects of TMS on neuronal excitability depend on the functional state of the cortex at the time of stimulation (Siebner et al. 2004). Most on-line PET imaging studies have given TMS while participants lay in the scanner with eyes closed (i.e. at rest). The state dependency of the neurophysiological effects of TMS is likely to account for some of the inter-regional differences in the functional response to identical rTMS protocols, for instance between motor and prefrontal areas (Speer et al. 2003a,b), because different brain regions will be more or less 'active' during rest, i.e. motor areas will be less active during rest than movement, whereas the state of prefrontal areas is harder to control.

On-line PET imaging can be used very effectively to examine interactions between focal TMS and task-related activity (Figure 35.2). Mottaghy et al. (2000, 2003) studied the effects of a continuous train of 4 Hz rTMS to dorsolateral prefrontal cortex (PFC) at 110% of resting motor threshold while participants performed an 'n-2 back' working memory task. Though focal rTMS to right and left dorsolateral PFC impaired task performance, rTMS-induced effects on the task-related activity pattern were different depending on the site of rTMS. Four-hertz rTMS to left dorsolateral PFC reduced task-related increases in rCBF in the left dorsolateral PFC alone, whereas rTMS to the right dorsolateral PFC reduced task-related rCBF increases in right dorsolateral PFC and the parietal cortex bilaterally (Mottaghy et al. 2000). The same rTMS protocol also changed the relationship between task performance and rCBF changes in the PFC (Mottaghy et al. 2003).

Another way to assess the influence of TMS on task-related activation is to give single pulses or a short train of TMS at a specific time during the performance of a certain task. This transient perturbation induced by TMS may change the activation pattern relative to a condition in which no TMS is given. TMS may also produce a change in behavior and disrupt task performance depending on the timing and intensity of TMS. If the experimental design includes a parametric modulation of these variables, 'on-line' PET imaging may help to determine how strongly the activation pattern needs to be altered by TMS in order to cause a functional lesion. This approach may yield relevant clues about how the cognitive anatomy of the brain acutely compensates for the TMS-induced lesion effect. Although no such

(a) TMS during "steady state"
(e.g. "resting state", tonic pain)
Mapping TMS-induced changes
in "steady state" activity

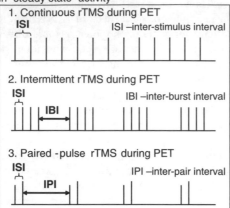

1. Continuous rTMS during PET

ISI   ISI –inter-stimulus interval

2. Intermittent rTMS during PET

ISI   IBI –inter-burst interval

IBI

3. Paired -pulse rTMS during PET

ISI   IPI –inter-pair interval

IPI

(b) Task-triggered TMS
   TMS-induced changes in task-related activity

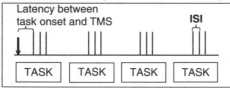

Latency between
task onset and TMS   ISI

| TASK | TASK | TASK | TASK |

Fig. 35.2 There are several complementary paradigms that can be used to study the acute effects of TMS on brain function during on-line positron emission tomography (PET) imaging. (a) If the functional state of the brain is kept constant during PET, it is possible to investigate how a given TMS protocol changes steady-state activity in the brain. (b) TMS can also be given while the subject performs a cognitive task. This approach enables investigators to explore interactions between TMS and task-related activation in the stimulated areas.

TMS–PET study has been conducted so far, we anticipate that 'on-line' PET imaging of precisely timed TMS lesions during a task will help to understand the mechanisms that create a 'virtual lesion' (Pascual-Leone *et al.* 1999; Jahanshahi and Rothwell 2000).

On-line PET imaging during intermittent rTMS

Paus *et al.* (1997) were the first to map the acute effects of rTMS on rCBF in healthy volunteers.

They gave intermittent trains of 10 Hz rTMS to the left frontal eye field (FEF). Using a parametric study design, the number of stimulus trains given during $H_2{}^{15}O$ PET was gradually varied, ranging from five to 30 trains per PET scan (Figure 35.2). This 'burst-like' pattern of intermittent rTMS induced rate-dependent increases in rCBF in the stimulated FEF and in anatomically connected visual areas in the superior parietal and medial parieto-occipital cortex (Paus *et al.* 1997). The pattern of remote changes agreed with the known anatomical connections of the stimulated FEF, showing that 'on-line' PET imaging can be used to map functional cortico-cortical connectivity (Paus *et al.* 1997). A second experiment was performed on the same subjects where an identical rTMS protocol was used to stimulate the left primary motor hand area ($M1_{HAND}$) (Paus *et al.* 1998). In contrast to stimulation of the FEF, TMS of the $M1_{HAND}$ provoked rCBF changes in the stimulated $M1_{HAND}$ and connected areas that were negatively correlated with the number of TMS trains per PET scan. Thus, an identical rTMS protocol delivered to the FEF or $M1_{HAND}$ induced either relative increases or decreases in rCBF at both the site of stimulation and in connected areas.

These regional differences in the stimulus–response relationship have important implications for using TMS in humans. First, the response to TMS differs between areas, and thus the functional response evoked in the $M1_{HAND}$ should be used with caution when predicting the responsiveness of other nonmotor areas. Second, intrinsic properties of the stimulated cortex can influence local and remote effects on regional synaptic activity. Regionally specific properties that define the functional response of the cortex to TMS may change over time, e.g. the level of excitability, or remain constant, e.g. anatomical differences in cortical microarchitecture and preferential fiber orientation with respect to the stimulating coil.

On-line PET imaging during continuous rTMS

PET has also been used to examine the immediate effects of continuous trains of TMS (Fox *et al.* 1997; Siebner *et al.* 1998, 2001a,b;

Speer et al. 2003a,b) (Figure 35.2). An early study using [^{18}F]deoxyglucose PET showed that irregular 2 Hz rTMS given to the left $M1_{HAND}$ at suprathreshold intensity led to an acute increase in rCMRglc in the stimulated cortex, caudal supplementary motor area (SMA), and contralateral right dorsal premotor cortex (PMd) (Siebner et al. 1998, 2001a). The increases in rCMRglc in the left $M1_{HAND}$ were smaller in magnitude than rCMRglc increases during voluntary imitation of rTMS-induced arm movements, indicating that rTMS-induced changes in regional activity were within the normal physiological range (Siebner et al. 1998). Focal rTMS of the $M1_{HAND}$ increased activity in executive frontal motor areas, whereas voluntary movements were associated with additional increases in activity in more rostral motor areas involved in cognitive aspects of motor control (Siebner et al. 1998). This finding suggests that focal rTMS can alter synaptic activity in subsets of areas that form a functional network.

At suprathreshold intensities, rTMS consistently evokes muscle twitches; therefore reafferent somatosensory activation contributes to the increase in activity within the $M1_{HAND}$ during suprathreshold rTMS (Siebner et al. 2001a; Speer et al. 2003a). Consistent increases in synaptic activity in the stimulated $M1_{HAND}$ have also been demonstrated during continuous rTMS at subthreshold intensities (Siebner et al. 2001b, 2003b; Speer et al. 2003a). Siebner et al. (2001b) investigated changes of regional neuronal activity in the stimulated left $M1_{HAND}$ during subthreshold rTMS (90% of active motor threshold). Continuous rTMS at frequencies ranging from 1 to 5 Hz led to a rate-dependent increase in rCBF in the stimulated $M1_{HAND}$ (Figure 35.3). Because the intensity of stimulation was very low, rTMS caused no spread of activation to connected areas. Another parametric TMS–PET study gave continuous 1 Hz rTMS to left $M1_{HAND}$ and varied the intensity of stimulation, ranging from 80% to 120% of motor threshold (Speer et al. 2003a). One-hertz rTMS caused an intensity-dependent increase in rCBF. In contrast to the study by Siebner et al. (2001), linear increases in synaptic activity were also observed in remote anatomically connected areas. Taken together, these results suggest that the immediate changes in synaptic activity

remain limited to the site of stimulation when very low stimulation intensities are used for rTMS. As stimulus intensity increases, the impact on synaptic activity in distant connected areas gradually increases and may be even stronger in magnitude than the local activity changes in the stimulated cortex (Okabe et al. 2003).

Study designs that parametrically modulate the frequency or intensity of rTMS have been successfully used to obtain a 'response profile' of the stimulated cortex. Using different intensities of rTMS, Speer et al. gave focal 1 Hz rTMS to the $M1_{HAND}$ (Speer et al. 2003a) and dorsolateral PFC (Speer et al. 2003b) and found an inverse intensity–response pattern. The $M1_{HAND}$ and connected areas showed an intensity-dependent increase in rCBF during a continuous train of 1 Hz rTMS (Speer et al. 2003a), whereas the left dorsolateral PFC and connected areas showed a progressive decrease in rCBF with increasing stimulus intensity (Speer et al. 2003b). A recent $H_2^{15}O$ PET study showed that 0.75 Hz rTMS to rostral PMd produced a maximal decrease in rCBF in the stimulated cortex at 80% of resting motor threshold, with no further increase in effect size at higher stimulus intensities (Peller et al. 2004).

One of the practical problems of selecting stimulation intensities for TMS relates to the fact that the threshold for activating cortico-cortical or cortico-subcortical connections is unknown for nonmotor areas. Parametric PET-rTMS studies that map the functional brain response across a range of intensities can help to determine the intensity at which rTMS starts to induce a significant modulation of neuronal activity in distant areas. This may provide a means of individually adjusting the intensity of rTMS in 'silent' cortical areas.

On-line PET imaging during paired-pulse rTMS

Over the last 10 years, paired-pulse TMS has been used extensively to quantify the excitability of the primary motor cortex in health and disease (Rothwell 1999; Ziemann 1999). Using a conditioning–test stimulus paradigm, paired-pulse TMS results in preferential activation of distinct intracortical circuits in the M1 depending

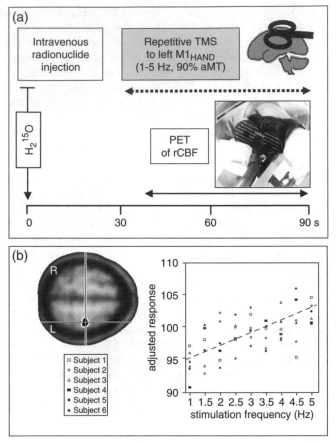

Fig. 35.3 Rate-dependent changes in the primary motor hand area ($M1_{HAND}$) during a continuous train of focal subthreshold repetitive (r)TMS. (a) The experiment used a parametric study design. $H_2{}^{15}O$ positron emission tomography (PET) measurements of regional cerebral blood flow (rCBF) were performed while a continuous train of rTMS was applied at an intensity of 90% of active motor threshold (aMT) and a rate of 1–5 Hz to the left $M1_{HAND}$. (b) Only the stimulated $M1_{HAND}$ showed a linear increase in rCBF with the frequency of rTMS ($P < 0.001$, uncorrected). There were no changes in rCBF in distant brain regions, presumably because TMS was given at a relatively low intensity. Reprinted with permission from Siebner *et al.* (2001).

on the interstimulus interval (ISI). Paired-pulse TMS offers interesting possibilities for combined TMS–PET studies because paired pulses can be used to preferentially excite distinct subsets of intracortical neurons with different anatomical and functional connections. Specific effects at the site of stimulation and in connected areas can then be quantified by mapping changes in rCBF or rCMRglc, providing a behavior-independent assay of connectivity in the human brain.

The feasibility of this approach was demonstrated in a PET study in which rCBF and motor-evoked potentials were recorded during single-pulse TMS and paired-pulse rTMS of the $M1_{HAND}$ (Strafella and Paus 2001). Using the conditioning–test stimulus paradigm introduced by Kujirai *et al.* (1993), paired-pulse TMS at ISIs of 3 and 12 ms were used to stimulate neuronal circuits subserving short-latency intracortical inhibition and facilitation, respectively. Relative changes in rCBF (i.e. rCBF during paired-pulse TMS minus rCBF during single-pulse TMS) were correlated with the degree of suppression and facilitation of EMG responses during paired-pulse TMS. Correlation analysis demonstrated that paired-pulse TMS was associated with different spatial patterns of rCBF changes depending on the ISI (Strafella and Paus 2001).

Using the same conditioning–test stimulus paradigm (Kujirai et al. 1993), Peller et al. (2005) gave paired-pulse TMS to the left PMd using ISIs of 2, 4, 6, 8, and 10 ms. Paired pulses were given at a rate of 0.75 Hz. Paired-pulse rTMS at an ISI of 4 ms caused a selective increase in rCBF compared with single-pulse TMS in distinct areas in left PMd, angular gyrus, and anterior cingulate cortex (Peller et al. 2005). These data indicate that on-line PET during paired-pulse TMS can be used to obtain an excitability profile and to map connectivity of nonmotor cortical areas. In another study, Peller et al. (unpublished data) combined paired-pulse rTMS at 0.75 Hz with PET to examine the connectivity of the left rostral PMd at very short intervals, ranging from 2.0 to 3.5 ms. The intensity of the first and second pulses of the stimulus pair was set at 95% of resting motor threshold.

A cluster in the right dorsolateral PFC showed a modulation of rCBF depending on the ISI (Figure 35.4). Regional rCBF was relatively increased at ISIs of 2.0 and 3.2 ms compared with the other ISIs, indicating a preferential stimulation of the cortico-cortical connections between left PMd and right PFC when ISIs of 2.0 and 3.2 ms were used for paired-pulse rTMS.

Off-line PET imaging of rTMS-induced regional plasticity

In the last decade, several rTMS protocols have been developed to induce lasting changes in excitability and neuronal processing in the stimulated cortex and connected areas, including continuous rTMS (Hallett 2000), intermittent 'burst' rTMS (Huang et al. 2005), or paired-associative stimulation (Stefan et al. 2000). The conditioning effects of rTMS on behavior, excitability, and synaptic activity have provided new insights into the plasticity of the intact human brain (Siebner and Rothwell 2003). Since the conditioning effects of rTMS can last

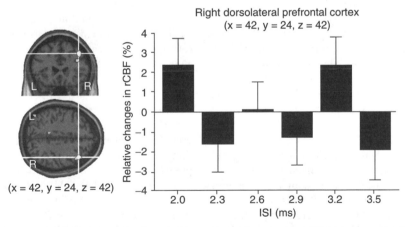

Fig. 35.4 Cortico-cortical connectivity revealed by paired-pulse repetitive (r)TMS to the rostral left dorsal premotor cortex (PMd). $H_2{}^{15}O$ PET was used to measure changes in rCBF during paired-pulse rTMS to left rostral PMd at six different interstimulus intervals (ISI). Paired-pulse rTMS was given at a frequency of 0.75 Hz and an intensity of 95% resting motor threshold. The right lateral prefrontal cortex (PFC) showed a modulation of rCBF depending on the ISI of paired-pulse rTMS (left panel; $P < 0.01$, uncorrected). The right panel shows that rCBF in this area showed relative increases in rCBF at ISIs of 2.0 and 3.2 ms. The rCBF profile suggests preferential activation of cortico-cortical connections between left PMd and right lateral PFC if paired pulses were delivered at an ISI of 2.0 and 3.2 ms.

up to several hours, it is possible to give rTMS outside the PET scanner and image the enduring effects of rTMS (i.e. off-line PET imaging). Because TMS and PET can be separated in space and time, activation patterns are not confounded by auditory and somatosensory stimulation.

If rTMS conditioning produces no changes in behavior this does not imply that the conditioning TMS protocol had no lasting effects on brain function. It is still possible that TMS has produced conditioning effects on neuronal activity, but that these changes are not expressed at a behavioral level. In fact, several PET studies have shown that rTMS produces substantial changes in synaptic activity in distributed areas without any change in behavior (Lee *et al.* 2003; Siebner *et al.* 2003a; Rounis *et al.* 2005). These changes may be a direct consequence of the conditioning effects of rTMS on synaptic activity, both in directly stimulated and anatomically connected areas. Alternatively, they may reflect a dynamic response of those parts of a network that have not been affected by rTMS, for instance to compensate for lasting disruptive effects of rTMS.

Off-line PET imaging of regional activity

In contrast to fMRI, off-line PET can be used to examine the spatial pattern, relative magnitude, and time course of changes in regional activity following rTMS (Figure 35.5). When given to the $M1_{HAND}$, a prolonged session of 'facilitatory' high-frequency rTMS at 90% of resting motor threshold raised the overall level of synaptic activity as indexed by an increase in normalized rCMRglc (Siebner *et al.* 2000) and rCBF (Rounis *et al.* 2005). Increased synaptic activity was observed in the stimulated left $M1_{HAND}$ and connected motor areas (Siebner *et al.* 2000; Rounis *et al.* 2005) and lasted for at least 50 min after the end of rTMS (Rounis *et al.* 2005). Similar increases in rCBF were found at the site of stimulation and in connected areas after low-frequency rTMS to the left $M1_{HAND}$ (Lee *et al.* 2003). These studies provide converging evidence that the conditioning effects of rTMS can spread from the site of stimulation to other areas within the stimulated network.

In a set of PET experiments, the same rTMS protocol was given to the left $M1_{HAND}$ or the left

"Conditioning-and-measure" approach

(a) Mapping the effects of rTMS conditioning on the "tonic" level of regional activity

> "Steady state" PET measurements (e.g. at rest)

(b) Mapping the effects of rTMS conditioning on "phasic" changes in regional activity

> PET measurements of task-related activity

(c) Mapping the effects of rTMS conditioning with TMS during PET

> PET measurements of TMS-induced activity changes

Fig. 35.5 Three complementary approaches using off-line positron emission tomography (PET) to evaluate long-term effects of rTMS on brain function.

rostral PMd in different groups of healthy volunteers. While rTMS to the left $M1_{HAND}$ led to an increase in rCBF (Lee *et al.* 2003; Rounis *et al.* 2005), rTMS given to the left PMd produced a consistent decrease in rCBF in the stimulated cortex and connected areas (Siebner *et al.* 2003a; Rounis *et al.* 2004). The similarity in local rCBF changes following 'inhibitory' 1 Hz rTMS and 'facilitatory' 5 Hz rTMS suggests that the cortical target area appears to define the direction of changes in rCBF whereas the rate of stimulation is less important. This does not contradict the finding that the intensity, temporal pattern, and duration of stimulation of rTMS determine the conditioning effects of rTMS on cortical excitability, but it serves to emphasize the fact that changes in rCBF cannot be readily interpreted as changes in cortical excitability.

As it is known that rTMS induces different effects on cortical excitability depending on the intensity, temporal pattern, and duration of stimulation (Siebner and Rothwell 2003), it is important to establish the relationship between the effects of changes in the stimulation parameters on local rCBF and changes in cortical excitability. A recent study showed that the effects of premotor 5 Hz rTMS differed substantially when the same number of rTMS trains were separated by an interval of 1 or 5 min

(Rounis *et al.* 2004). Exploring the site and dose dependency of rTMS-induced effects on regional synaptic activity will enable a closer understanding how rTMS modulates brain function in areas outside the primary motor, sensory, and visual cortices where external measures of excitability are unavailable.

One way to investigate the rCBF correlates of excitability changes caused by rTMS is to measure both. In separate experimental sessions, Chouinard *et al.* (2003) applied 1 Hz rTMS over the left PMd and M1$_{HAND}$, and examined the effects on corticomotor excitability and rCBF with single-pulse TMS over the left M1$_{HAND}$ and H$_2$^{15}O PET, respectively. Correlation analysis of changes in rCBF and changes in the amplitude of MEPs evoked from the M1$_{HAND}$ revealed that the inhibitory effect of 1 Hz rTMS to PMd and M1$_{HAND}$ on corticospinal excitability was associated with spatially distinct patterns of regional changes in synaptic activity.

Takano *et al.* (2004) used a similar approach to study the short-term modulation of regional excitability and rCBF in left M1$_{HAND}$ following a 30 s train of low-intensity 5 Hz rTMS (Takano *et al.* 2004). Single- and paired-pulse TMS was used to assess the conditioning effects of rTMS on motor cortical excitability at rest while changes in regional synaptic activity were measured with H$_2$^{15}O PET. Five-hertz rTMS caused an increase in rCBF in the stimulated M1$_{HAND}$ and a selective decrease in short-latency intracortical inhibition (SICI) without affecting short-latency intracortical facilitation or corticospinal excitability (Takano *et al.* 2004). The increase in rCBF and the decrease in SICI lasted for ~8 min. These results showed that, in the stimulated M1$_{HAND}$, the temporary attenuation of SICI was paralleled by an increase in synaptic activity, consistent with reduced efficacy of intracortical GABA$_A$-ergic synapses.

Changes in regional neuronal activity may not only occur at rest, but also during active states. Tamura *et al.* (2004) used SPECT to measure rCBF during a state of tonic pain induced by intradermal capsaicin injection (Tamura *et al.* 2004). One-hertz rTMS over M1$_{HAND}$ induced earlier recovery from acute pain compared with the sham or control conditions. SPECT measurements demonstrated an rCBF decrease in the right medial PFC and an increase in the caudal anterior cingulate cortex (ACC) after 1 Hz rTMS to the M1$_{HAND}$, suggesting that the beneficial effects on capsaicin-induced pain were caused by functional changes in the medial PFC and caudal ACC.

In various PET studies, changes in regional activity last from several minutes after short trains of rTMS (Takano *et al.* 2004) up to 1 h following 1800 stimuli of rTMS (Lee *et al.* 2003; Siebner *et al.* 2003a; Rounis *et al.* 2005). In humans, no PET study has examined the possibility that rTMS can produce effects on regional synaptic activity that last for more than 1 h. In anesthetized monkeys, repeated [^{18}F] deoxyglucose PET measurements were performed before, during, and up to 16 days after prolonged 5 Hz rTMS (2000 stimuli) over the right precentral gyrus. Decreases in rCMRglc were found in the motor/premotor cortices, while the anterior/ posterior cingulate and orbitofrontal cortices showed increases in rCMRglc. These changes lasted for at least 8 days, demonstrating that 5 Hz rTMS can have long-term effects on regional synaptic activity (Hayashi *et al.* 2004).

Metabolic PET and SPECT measurements can highlight regional changes in the overall level of synaptic activity but they provide no information about regional changes in neurotransmitter binding. Combining TMS with ligand–PET provides a means of assessing regional changes in neurotransmitter binding following focal rTMS. Strafella and colleagues used [^{11}C]raclopride PET to measure changes in extracellular dopamine concentration following high-frequency rTMS of the left dorsolateral PFC (Strafella *et al.* 2001) or the M1$_{HAND}$ (Strafella *et al.* 2003) in healthy volunteers (Figure 35.6). Focal rTMS to the dorsolateral PFC and M1$_{HAND}$ led to spatially restricted decreases in the [^{11}C]raclopride binding potential in the ipsilateral caudate nucleus and posterior putamen respectively. The areas showing a decrease in the [^{11}C]raclopride binding potential corresponded to the known corticostriatal projection zones of the stimulated cortices. The regionally specific decreases in [^{11}C]raclopride binding potential indicate that focal rTMS can induce a lasting increase in endogenous dopamine release in the corresponding striatal projection zone, presumably through repetitive stimulation of corticostriatal connections during rTMS.

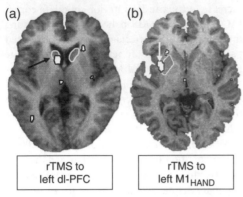

Fig. 35.6 Ligand positron emission tomography (PET) revealed that high-frequency rTMS can induce a lasting increase in endogenous dopamine release in the basal ganglia in healthy humans. (a) Focal high-frequency rTMS to left dorsolateral prefrontal cortex (dl-PFC) led to a reduction in [^{11}C]raclopride binding in the ipsilateral caudate nucleus, indicating a regional increase in endogenous dopaminergic neurotransmission. Reprinted with permission from Strafella *et al.* (2001). ©2001 Society of Neuroscience. (b) Focal high-frequency rTMS to left primary motor hand area (M1$_{HAND}$) caused a reduction in [^{11}C]raclopride binding in the left putamen. Reprinted with permission from Strafella *et al.* (2003). For both sites of stimulation the area of statistically significant change in binding closely corresponds to the known projection zone of corticostriatal efferents.

These studies provide the first evidence that focal rTMS to frontal cortex can influence neuro-transmitter release in distinct cortico-basal ganglia loops. These data in humans have been extended by a recent [^{11}C]raclopride PET in eight anesthetized monkeys (Ohnishi *et al.* 2004). Five-hertz rTMS over the right M1$_{HAND}$ induced a decreased [^{11}C]raclopride binding potential in the bilateral ventral striatum and an increased [^{11}C]raclopride binding potential in the right putamen.

Off-line PET imaging of rTMS-induced changes in task-related activity

In the previous section, we summarized PET studies which investigated how rTMS modulates neuronal activity in the brain when the behavioral context remains constant during PET scanning. Another pertinent question is whether rTMS can also produce lasting effects on the rapid modulation of regional activity and inter-regional coupling during a given task.

When a PET study is designed to detect task-related changes in activity, it should be recalled that the choice of the experimental task used during PET scanning will substantially influence the probability of detecting lasting effects of rTMS on the activation pattern. Clearly, it would be difficult to demonstrate task-related changes in activation after rTMS if the brain regions targeted by rTMS are not substantially engaged in the experimental task. Conversely, if the brain regions conditioned by rTMS are essential for the experimental task, task performance may be impaired after rTMS. This will introduce a performance confound which renders it difficult to assign changes in task-related activations to the effect of rTMS. Another point to consider is that there is no simple relationship between the effects of rTMS on steady-state synaptic activity and task-related activity. Large changes in steady-state synaptic activity are not necessarily paralleled by changes in task-related activity or task performance (Siebner *et al.* 2003a). In addition, changes in task-related activity may occur in different brain regions than steady-state changes in regional activity.

Using H$_2$^{15}O PET, Lee *et al.* (2003) measured rCBF at rest and during freely selected finger movements after 30 min of 1 Hz rTMS to the left M1$_{HAND}$ (Figure 35.7). Despite significant increases in synaptic activity in the stimulated rostral portion of the left M1$_{HAND}$, task performance and synaptic activity during finger movements were unaffected in the stimulated M1$_{HAND}$ (Lee *et al.* 2003). In contrast, a significant increase in movement-related activity was found in the right PMd and the inferomedial portion of the left M1$_{HAND}$, suggesting that maintenance of task performance involved activation of motor areas that were not directly stimulated by rTMS (Lee *et al.* 2003).

Analysis of effective connectivity suggested that the stimulated part of the left M1$_{HAND}$ became less responsive to inputs from premotor areas after 1 Hz rTMS, suggesting an impairment in the processing of premotor inputs at

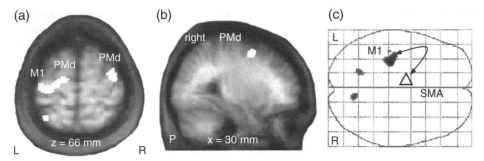

Fig. 35.7 Acute remapping within the motor system induced by subthreshold low-frequency repetitive (r)TMS to left primary motor hand area ($M1_{HAND}$). (a) Using $H_2^{15}O$ positron emission tomography (PET), regional synaptic activity was mapped at rest and during freely selected finger movements after 30 min of 1 Hz rTMS. Repetitive TMS increased synaptic activity in the stimulated left M1 and the dorsal premotor cortex, bilaterally. (b) rTMS induced an increase in movement-related activity in the dorsal premotor cortex (PMd) of the nonstimulated hemisphere. (c) Analyses of effective connectivity confirmed changes in coupling between frontal motor areas: rTMS led to an increase in coupling between the left caudal supplementary motor area (SMA) and an inferomedial portion of left $M1_{HAND}$. Reprinted with permission from Lee *et al.* 2003. ©2003 Society of Neuroscience.

the site of stimulation (Lee *et al.* 2003). Conversely, following rTMS there was increased coupling between an inferomedial portion of the left $M1_{HAND}$ and premotor areas during movement (Lee *et al.* 2003). This strengthening of functional coupling between premotor areas and nonstimulated parts of the left $M1_{HAND}$ may reflect rapid compensatory reorganization within the motor system to maintain functional integrity. These acute patterns of remapping may provide a neuronal substrate for compensatory plasticity of the motor system in response to a focal lesion, such as stroke.

Using the same experimental paradigm, a recent PET study extended the work of Lee *et al.* (2003) by providing evidence that the pattern of acute reorganization in the motor network following rTMS depends on the frequency of stimulation (Rounis *et al.* 2005). Rounis *et al.* (2005) used $H_2^{15}O$ PET to compare the effects of low-frequency (1 Hz) and high-frequency (5 Hz) rTMS on local activity and inter-regional connectivity within the motor system. Whereas 1 and 5 Hz rTMS induced a similar pattern of steady-state changes in rCBF at the site of stimulation and within areas of the motor network engaged by the task, rTMS had differential effects on movement-related increases in regional activity and changes in coupling between motor areas (Rounis *et al.* 2005).

These results suggest that rTMS-induced reorganization of the motor system differs according to rate-dependent modulation of excitability.

Lasting changes in inter-regional connectivity can also be tested by applying focal rTMS to a cortical area and subsequently examining the responsiveness of the stimulated network with on-line PET during focal TMS (Paus *et al.* 2001). Paus *et al.* (2001) conditioned the mid-dorsolateral PFC with 10 Hz rTMS. Repetitive TMS conditioning modulated the subsequent metabolic response of the fronto-cingulate circuit to TMS in healthy human volunteers. Similar changes were found in the rat cortex using electrical cortical stimulation and field-potential recordings (Paus *et al.* 2001).

Combining TMS and PET imaging in patients

Compared with studies on healthy individuals, a relatively small number of studies have used the combined TMS–PET approach in patients. This is surprising because the combined use of rTMS with PET can improve our understanding of the potential treatment effects of rTMS and reveal new insights into the pathophysiology of certain brain disorders. Several PET or SPECT studies have explored the effects of repeated sessions of prefrontal rTMS on rCBF and rCMRglc as a treatment for depression (Speer *et al.* 2000; Catafau *et al.* 2001; Mottaghy *et al.* 2002; Nadeau *et al.* 2002; Shajahan *et al.* 2002). These studies show that serial metabolic PET or SPECT studies provide important insights into the mechanism of action of rTMS and may help to predict antidepressant efficacy of different stimulation paradigms. In 10 patients with major depression, 10 daily treatments with 20 Hz rTMS applied over the left dorsolateral PFC led to bilateral increases in rCBF in the insula, basal ganglia, amygdala, uncus, hippocampus, parahippocampus, thalamus, cerebellum as well as the prefrontal and cingulate cortex (Speer *et al.* 2000). In contrast, 10 daily treatments with 1 Hz rTMS caused decreases in rCBF in the right PFC, left medial temporal cortex, left basal ganglia, and left amygdala (Speer *et al.* 2000). There was an inverse relationship between the mood changes following the two rTMS frequencies: individuals who improved with one frequency worsened with the other (Speer *et al.* 2000). Measurement of rCBF was also used to identify patterns of regional activity that may predict antidepressant efficacy of high-frequency rTMS to dorsolateral PFC. Before rTMS there was a significant left–right asymmetry, with more activity in the right hemisphere. Two weeks after treatment with rTMS this asymmetry was reversed. The rCBF at baseline in limbic structures was negatively correlated with the clinical outcome after rTMS, whereas rCBF in several neocortical areas showed a positive correlation.

In addition to studies of the antidepressant effects of prefrontal rTMS in depression, the combined TMS–PET approach represents a promising method with which to study the pathophysiology of neuropsychiatric disorders. For instance, PET imaging during rTMS can be employed to investigate regional changes in excitability and connectivity in patients suffering from epilepsy, schizophrenia, or stroke. Another interesting application is to image the patterns of acute or chronic functional reorganization induced by therapeutic interventions, such as neurorehabilitation or medication with centrally active drugs.

Off-line PET after rTMS can also be used to study differences in the modifiability of functional networks between patients and healthy

controls (Siebner *et al.* 2003a). Healthy individuals and patients with primary focal dystonia received 1800 stimuli of subthreshold 1 Hz rTMS (90% resting motor threshold) or sham stimulation to the left rostral PMd (Figure 35.8). Afterwards, rCBF was measured by PET at rest and during performance of freely selected finger movement in order to examine the pattern and time course of changes in rCBF produced by rTMS. In both groups, rTMS caused widespread bilateral decreases in synaptic activity in prefrontal, premotor, and primary motor cortex and in the left putamen. Patients showed significantly greater suppression of synaptic activity in lateral and medial premotor areas, putamen, and thalamus, indicating increased responsiveness to rTMS of the cortico-basal ganglia thalamic loop in focal arm dystonia (Siebner *et al.* 2003a). This finding agrees with other rTMS studies which have shown an abnormal pattern of rTMS-induced plasticity in focal hand dystonia (Quartarone *et al.* 2003, 2005).

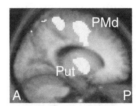
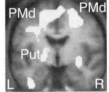
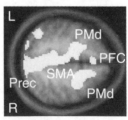

Fig. 35.8 Patients with focal arm dystonia have increased sensitivity to low-frequency repetitive TMS. The statistical parametric maps show regions in the brain where dystonic patients demonstrated a greater decrease in rCBF than healthy controls after premotor 1 Hz rTMS. PFC, prefrontal cortex; PMd, dorsal premotor cortex; Prec, precuneus; Put, putamen; SMA, supplementary motor area. Reprinted with permission from Siebner *et al.* (2003).

Conclusion and perspectives

The combined use of TMS and PET has considerably expanded the applications of TMS in basic neuroscience and clinical research. The existing data convincingly show that on-line PET during TMS provides a behavior-independent assay of cortical excitability and connectivity. However, little work has been done so far to explore how focal TMS interferes with task-related activity when given during the task. We anticipate that this approach may yield important insight into how focal TMS perturbs neuronal processing during a task, causing a 'virtual lesion'.

Off-line PET is an important method with which to investigate the effects of rTMS on synaptic activity in the intact human brain. In recent years, several new techniques and paradigms such as paired-associative stimulation (Stefan *et al.* 2000), theta-burst stimulation (Huang *et al.* 2005), or transcranial direct current stimulation (tDCS) (Paulus 2003) have been developed which can produce sustained changes in cortical excitability. Here, off-line PET imaging is of great relevance because it enables comparisons between the topographical and temporal profiles of changes in regional activity produced by various conditioning protocols. A recent study used $H_2^{15}O$ PET of rCBF at rest and during freely selected finger movements to map lasting changes in regional synaptic activity following 10 min of bipolar tDCS ($\pm$ 1 mA) given through electrodes that overlay left $M1_{HAND}$ and right frontopolar cortex (Lang *et al.* 2005). The results showed that tDCS provoked sustained and widespread changes in rCBF. Since the study used the same PET paradigm as previous off-line PET studies that have explored the after-effects of subthreshold 5 or 1 Hz rTMS to the left $M1_{HAND}$ (Lee *et al.* 2003; Rounis *et al.* 2005), it is possible to compare the effects induced by rTMS and tDCS. The pattern of changes in rCBF found after rTMS and tDCS were quite different, corroborating the notion that the conditioning effects of tDCS and rTMS are mediated through different neuronal mechanisms. We conclude that off-line PET imaging can make an important contribution to understanding the mechanisms of action of rTMS and has the potential to determine neural correlates of compensatory plasticity in both healthy subjects and disease states.

References

Cagnin A, Gerhard A, Banati RB (2002) The concept of in vivo imaging of neuroinflammation with [11C](R)-PK11195 PET. *Ernst Schering Research Foundation Workshop*, 179–191.

Catafau AM, Perez V, Gironell A, *et al.* (2001) SPECT mapping of cerebral activity changes induced by repetitive transcranial magnetic stimulation in depressed patients. A pilot study. *Psychiatry Research* 106, 151–160.

Chouinard PA, Van Der Werf YD, Leonard G, Paus T (2003) Modulating neural networks with transcranial magnetic stimulation applied over the dorsal premotor and primary motor cortices. *Journal of Neurophysiology* 90, 1071–1083.

Ferrarelli F, Haraldsson HM, Barnhart TE, *et al.* (2004) A [17F]-fluoromethane PET/TMS study of effective connectivity. *Brain Research Bulletin 64*, 103–113.

Fox P, Ingham R, George MS, *et al.* (1997) Imaging human intra-cerebral connectivity by PET during TMS. *Neuroreport 8*, 2787–2791.

Gilio F, Rizzo V, Siebner HR, Rothwell JC (2003) Effects on the right motor hand-area excitability produced by low-frequency rTMS over human contralateral homologous cortex. *Journal of Physiology 551*, 563–573.

Hallett M (2000) Transcranial magnetic stimulation and the human brain. *Nature 406*, 147–150.

Hayashi T, Ohnishi T, Okabe S, *et al.* (2004) Long-term effect of motor cortical repetitive transcranial magnetic stimulation [correction]. *Annals of Neurology 56*, 77–85.

Herholz K, Heiss WD (2004) Positron emission tomography in clinical neurology. *Molecular Imaging and Biology 6*, 239–269.

Huang YZ, Edwards MJ, Rounis E, Bhatia KP, Rothwell JC (2005) Theta burst stimulation of the human motor cortex. *Neuron 45*, 201–206.

Jahanshahi M, Rothwell J (2000) Transcranial magnetic stimulation studies of cognition: an emerging field. *Experimental Brain Research 131*, 1–9.

Kujirai T, Caramia MD, Rothwell JC, *et al.* (1993) Corticocortical inhibition in human motor cortex. *Journal of Physiology 471*, 501–519.

Lancaster JL, Narayana S, Wenzel D, Luckemeyer J, Roby J, Fox P (2004) Evaluation of an image-guided, robotically positioned transcranial magnetic stimulation system. *Human Brain Mapping 22*, 329–340.

Lang N, Siebner HR, Ward NS, *et al.* (2005) How does transcranial DC stimulation of the primary motor cortex alter regional neuronal activity in the human brain? *European Journal of Neuroscience 22*, 495–504.

Lee L, Siebner HR, Rowe JB, *et al.* (2003) Acute remapping within the motor system induced by low-frequency repetitive transcranial magnetic stimulation. *Journal of Neuroscience 23*, 5308–5318.

Logothetis NK (2002) The neural basis of the blood-oxygen-level-dependent functional magnetic resonance imaging signal. *Philosophical Transactions of the Royal Society of London B 357*, 1003–1037.

Logothetis NK, Wandell BA (2004) Interpreting the BOLD signal. *Annual Review of Physiology 66*, 735–769.

Mottaghy FM, Krause BJ, Kemna LJ, *et al.* (2000) Modulation of the neuronal circuitry subserving working memory in healthy human subjects by repetitive transcranial magnetic stimulation. *Neuroscience Letters 280*, 167–170.

Mottaghy FM, Keller CE, Gangitano M, *et al.* (2002) Correlation of cerebral blood flow and treatment effects of repetitive transcranial magnetic stimulation in depressed patients. *Psychiatry Research 115*, 1–14.

Mottaghy FM, Pascual-Leone A, Kemna LJ, *et al.* (2003) Modulation of a brain-behavior relationship in verbal working memory by rTMS. *Brain Research: Cognitive Brain Research 15*, 241–249.

Nadeau SE, McCoy KJ, Crucian GP, *et al.* (2002) Cerebral blood flow changes in depressed patients after treatment with repetitive transcranial magnetic stimulation: evidence of individual variability. *Neuropsychiatry, Neuropsychology and Behavioral Neurology 15*, 159–175.

Ohnishi T, Hayashi T, Okabe S, *et al.* (2004) Endogenous dopamine release induced by repetitive transcranial magnetic stimulation over the primary motor cortex: an [11C]raclopride positron emission tomography study in anesthetized macaque monkeys. *Biological Psychiatry 55*, 484–489.

Okabe S, Hanajima R, Ohnishi T, *et al.* (2003) Functional connectivity revealed by single-photon emission computed tomography (SPECT) during repetitive transcranial magnetic stimulation (rTMS) of the motor cortex. *Clinical Neurophysiology 114*, 450–457.

Pascual-Leone A, Bartres-Faz D, Keenan JP (1999) Transcranial magnetic stimulation: studying the brain-behaviour relationship by induction of 'virtual lesions'. *Philosophical Transactions of the Royal Society of London B 354*, 1229–1238.

Paulus W (2003) Transcranial direct current stimulation (tDCS). *Supplement to Clinical Neurophysiology 56*, 249–254.

Paus T, Wolforth M (1998) Transcranial magnetic stimulation during PET: reaching and verifying the target site. *Human Brain Mapping 6*, 399–402.

Paus T, Jech R, Thompson CJ, Comeau R, Peters T, Evans AC (1997) Transcranial magnetic stimulation during positron emission tomography: a new method for studying connectivity of the human cerebral cortex. *Journal of Neuroscience 17*, 3178–3184.

Paus T, Jech R, Thompson CJ, Comeau R, Peters T, Evans AC (1998) Dose-dependent reduction of cerebral blood flow during rapid-rate transcranial magnetic stimulation of the human sensorimotor cortex. *Journal of Neurophysiology 79*, 1102–1107.

Paus T, Castro-Alamancos MA, Petrides M (2001) Cortico-cortical connectivity of the human mid-dorsolateral frontal cortex and its modulation by repetitive transcranial magnetic stimulation. *European Journal of Neuroscience 14*, 1405–1411.

Peller M, Boehringer M, Drzezga A, Schwaiger M, Siebner HR (2004) Widespread suppression of regional synaptic activity in frontal areas during low-frequency rTMS of the left rostral dorsal premotor cortex. *NeuroImage 22*(Suppl. 1), WE49 (on-line).

Peller M, Boehringer M, Drzezga A, *et al.* (2005) Paired-pulse transcranial magnetic stimulation (TMS) can modulate the spread of TMS effects to connected cortical areas. *NeuroImage 26*(Suppl. 1), 506.

Piccini P, Whone A (2004) Functional brain imaging in the differential diagnosis of Parkinson's disease. *Lancet Neurology 3*, 284–290.

Quartarone A, Bagnato S, Rizzo V, *et al.* (2003) Abnormal associative plasticity of the human motor cortex in writer's cramp. *Brain 126*, 2586–2596.

Quartarone A, Rizzo V, Bagnato S, *et al.* (2005) Homeostatic-like plasticity of the primary motor hand area is impaired in focal hand dystonia. *Brain 128*, 1943–1950.

Rothwell JC (1999) Paired-pulse investigations of short-latency intracortical facilitation using TMS in humans. *Electroencephalography and Clinical Neurophysiology Supplement 51*, 113–119.

Rothwell JC, Hallett M, Berardelli A, Eisen A, Rossini P, Paulus W (1999) Magnetic stimulation: motor evoked potentials. The International Federation of Clinical Neurophysiology. *Electroencephalography and Clinical Neurophysiology Supplement 52*, 97–103.

Rounis E, Lee L, Siebner HR, *et al.* (2004) The effect of 5Hz-rTMS over the left dorsal premotor cortex on regional cerebral blood flow: a comparison between two patterns of stimulation. *NeuroImage 22*(Suppl. 1), WE57 (on-line).

Rounis E, Lee L, Siebner HR, *et al.* (2005) Frequency specific changes in regional cerebral blood flow and motor system connectivity following rTMS to the primary motor cortex. *NeuroImage 26*, 164–176.

Shajahan PM, Glabus MF, Steele JD, *et al.* (2002) Left dorso-lateral repetitive transcranial magnetic stimulation affects cortical excitability and functional connectivity, but does not impair cognition in major depression. *Progress in Neuro-psychopharmacology and Biological Psychiatry 26*, 945–954.

Siebner HR, Rothwell J (2003) Transcranial magnetic stimulation: new insights into representational cortical plasticity. *Experimental Brain Research 148*, 1–16.

Siebner HR, Willoch F, Peller M, *et al.* (1998) Imaging brain activation induced by long trains of repetitive transcranial magnetic stimulation. *Neuroreport 9*, 943–948.

Siebner HR, Auer C, Roeck R, Conrad B (1999a) Trigeminal sensory input elicited by electric or magnetic stimulation interferes with the central motor drive to the intrinsic hand muscles. *Clinical Neurophysiology 110*, 1090–1099.

Siebner HR, Peller M, Willoch F, *et al.* (1999b) Imaging functional activation of the auditory cortex during focal repetitive transcranial magnetic stimulation of the primary motor cortex in normal subjects. *Neuroscience Letters 270*, 37–40.

Siebner HR, Peller M, Willoch F, et al. (2000) Lasting cortical activation after repetitive TMS of the motor cortex: a glucose metabolic study. Neurology 54, 956–963.

Siebner H, Peller M, Bartenstein P, et al. (2001a) Activation of frontal premotor areas during suprathreshold transcranial magnetic stimulation of the left primary sensorimotor cortex: a glucose metabolic PET study. Human Brain Mapping 12, 157–167.

Siebner HR, Takano B, Peinemann A, Schwaiger M, Conrad B, Drzezga A (2001b) Continuous transcranial magnetic stimulation during positron emission tomography: a suitable tool for imaging regional excitability of the human cortex. NeuroImage 14, 883–890.

Siebner HR, Filipovic SR, Rowe JB, et al. (2003a) Patients with focal arm dystonia have increased sensitivity to slow-frequency repetitive TMS of the dorsal premotor cortex. Brain 126, 2710–2725.

Siebner HR, Peller M, Lee L (2003b) Applications of combined TMS–PET studies in clinical and basic research. Supplement to Clinical Neurophysiology 56, 63–72.

Siebner HR, Lang N, Rizzo V, et al. (2004) Preconditioning of low frequency repetitive transcranial magnetic stimulation with transcranial direct current stimulation: evidence for homeostatic plasticity in the human motor cortex. Journal of Neuroscience 24, 3379–3385.

Speer AM, Kimbrell TA, Wassermann EM, et al. (2000) Opposite effects of high and low frequency rTMS on regional brain activity in depressed patients. Biological Psychiatry 48, 1133–1141.

Speer AM, Repella JD, Figueras S, et al. (2001) Lack of adverse cognitive effects of 1 Hz and 20 Hz repetitive transcranial magnetic stimulation at 100% of motor threshold over left prefrontal cortex in depression. Journal of ECT 17, 259–263.

Speer AM, Willis MW, Herscovitch P, et al. (2003a) Intensity-dependent regional cerebral blood flow during 1-Hz repetitive transcranial magnetic stimulation (rTMS) in healthy volunteers studied with $H_2^{15}O$ positron emission tomography: I. Effects of primary motor cortex rTMS. Biological Psychiatry 54, 818–825.

Speer AM, Willis MW, Herscovitch P, et al. (2003b) Intensity-dependent regional cerebral blood flow during 1-Hz repetitive transcranial magnetic stimulation (rTMS) in healthy volunteers studied with H215O positron emission tomography: II. Effects of prefrontal cortex rTMS. Biological Psychiatry 54, 826–832.

Stefan K, Kunesch E, Cohen LG, Benecke R, Classen J (2000) Induction of plasticity in the human motor cortex by paired associative stimulation. Brain 123(Pt 3), 572–584.

Strafella AP, Paus T (2001) Cerebral blood-flow changes induced by paired-pulse transcranial magnetic stimulation of the primary motor cortex. Journal of Neurophysiology 85, 2624–2629.

Strafella AP, Paus T, Barrett J, Dagher A (2001) Repetitive transcranial magnetic stimulation of the human prefrontal cortex induces dopamine release in the caudate nucleus. Journal of Neuroscience 21, RC157.

Strafella AP, Paus T, Fraraccio M, Dagher A (2003) Striatal dopamine release induced by repetitive transcranial magnetic stimulation of the human motor cortex. Brain 126, 2609–2615.

Takano B, Drzezga A, Peller M, et al. (2004) Short-term modulation of regional excitability and blood flow in human motor cortex following rapid-rate transcranial magnetic stimulation. NeuroImage 23, 849–859.

Tamura Y, Okabe S, Ohnishi T, et al. (2004) Effects of 1-Hz repetitive transcranial magnetic stimulation on acute pain induced by capsaicin. Pain 107, 107–115.

Thompson CJ, Paus T, Clancy R (1998) Magnetic shielding requirements for PET detectors during transcranial magnetic stimulation. IEEE Transactions in Nuclear Science 45, 1303–1307.

Wittenberg GF, Werhahn KJ, Wassermann EM, Herscovitch P, Cohen LG (2004) Functional connectivity between somatosensory and visual cortex in early blind humans. European Journal of Neuroscience 20, 1923–1927.

Ziemann U (1999) Intracortical inhibition and facilitation in the conventional paired TMS paradigm. Electroencephalography and Clinical Neurophysiology Supplement 51, 127–136.

CHAPTER 36

Concurrent TMS and functional magnetic resonance imaging: methods and current advances

Sven Bestmann, Christian C. Ruff, Jon Driver, and Felix Blankenburg

Introduction

TMS is now routinely used for a wide range of applications in cognitive, clinical, and basic neuroscience. The precise physiological mechanisms by which TMS influences brain function are only partially understood and have been studied mainly for the motor system (Ziemann and Rothwell 2000; Di Lazzaro *et al.* 2003, 2004). For nonmotor cortical areas that do not lead to overt positive stimulation effects (such as observable motor-evoked potentials (MEPs)), the mechanisms underlying TMS effects may be more difficult to assess. Furthermore, some of the behavioral manifestations of TMS may not be due only to local effects at the stimulated site, but may involve TMS-induced modulation of remote activity in task-related brain regions connected with the stimulated area. Neuroimaging techniques can measure neuronal activity both directly underneath the coil and also in remote brain regions during application of TMS pulses (Paus *et al.* 1997). Combining TMS with functional magnetic resonance imaging (fMRI) thus promises to provide a more complete picture of the neural underpinnings of TMS effects. Although neuroimaging can in principle be implemented after repetitive TMS applications, to measure any enduring changes in brain activity (Paus 1999), here we focus in particular on combining fMRI with *concurrent* TMS, applied during scanning, as this may allow a more direct relation to be determined between TMS application and on-line neural activity.

Thus, we will focus on combining TMS with on-line fMRI concurrently, as a novel approach that allows stimulation inside the scanner, during the simultaneous recording of changes in brain activity evoked by TMS (Bohning *et al.* 1998). This can provide a noninvasive measure of both the local and remote effects of TMS stimulation, and whether any of these vary with psychological factors such as task-state (which might impact on 'effective connectivity' rather than purely fixed anatomical connections). Accordingly, we focus on dynamic and transient effects of TMS, rather than tonic and sustained effects. The latter are discussed in more detail by Siebner (Chapter 35, this volume; see also Paus 1999; Lee *et al.* 2006).

TMS has an excellent temporal resolution, on the order of milliseconds with a spatial resolution thought to be ~1 cm^2. Functional MRI provides a somewhat complementary approach, measuring on the order of seconds, but with a spatial resolution in the millimeter range that surpasses TMS and many other neuroimaging methods. The concurrent combination of both methods promises to overcome some of the inherent limitations of each method. Moreover, combining fMRI with TMS can bring a causal dimension into neuroimaging, by means of the TMS intervention.

Concurrent combination of TMS and fMRI imposes several major technical challenges. Accordingly we will first provide an overview of technical and methodological aspects to the approach. This overview seeks to sensitize the reader to potential problems and to provide recommendations for successful application of the combined method. The overview is necessarily rather general; several other reviews provide more in-depth coverage of the basic methods for MR Imaging (e.g. de Graaf 1998; Haacke et al. 1999; Stark and Bradley 1999).

In subsequent sections, we focus on recent applications of concurrent TMS–fMRI. Because of the wealth of information regarding TMS impacts on the motor system (due to directly observable MEPs, and the large literature on these), most TMS–fMRI studies to date have concerned the motor system. However, more recent studies demonstrate that concurrent TMS–fMRI can be applied outside the motor domain (for example, to test hypotheses concerning possible top-down influences upon the visual system). Other recent applications have interfered with task performance while concurrently measuring changes in brain activity both at the site of stimulation and in the wider brain network underlying task processing. This approach bridges the causal nature of TMS with the correlational inferences commonly afforded by fMRI in the context of behavioral or cognitive experiments.

Apart from its use in basic neuroscience, TMS is increasingly harnessed for clinical applications. The search for clinically effective applications of TMS provides a further rationale for seeking to understand the underlying neural mechanisms and affected networks. Concurrent TMS–fMRI can help to answer some pertinent questions in this respect also, including the following. Which (local and remote) brain areas are affected by TMS over a particular site? How does TMS over one brain region affect interconnected areas related to a particular clinical pathology or syndrome? While dealing with such general issues, we also consider their potential clinical relevance towards the end of this chapter.

Previous TMS studies without neuroimaging (using measures such as MEPs instead) have already implied that effects of TMS can depend on functional and effective connectivity, rather than merely fixed anatomical connectivity between brain regions. This may arise because the impact of a TMS pulse can depend on the excitability of connections (and/or the current level of activity) at the time a TMS pulse is applied. Presumably, the more excitable a given connection is at the time of stimulation, the more TMS will affect the corresponding circuit. For example, applying TMS during voluntary contraction versus rest affects the size and number of descending volleys evoked by TMS of motor cortex (Mazzocchio et al. 1994; Ridding et al. 1995; Fujiwara and Rothwell 2004), the amount of inferred interhemispheric inhibition from one motor cortex upon the other (Ferbert et al. 1992), and the coupling between frontal premotor areas and motor cortex (Strens et al. 2002). Further evidence comes from investigations of modulation by covert motor acts, such as motor imagery (Fadiga et al. 1999; Stinear and Byblow 2004) or action observation (Fadiga et al. 1995; Strafella and Paus 2000). These more cognitive manipulations have been shown to selectively modulate motor cortex excitability (as probed using TMS and measures of MEPs), depending on the type of imagined or observed movements, and are thought to reflect influences upon motor cortex from imagery processes involving also parietal and premotor regions.

Thus, there are several precedents for thinking that TMS influences may reflect not only strictly local effects, but also influences between brain regions, where functional or effective connectivity between such regions may vary with current motor or cognitive state. Combining TMS with fMRI concurrently may allow a new noninvasive probe of any such 'state-dependent' changes in cortical connectivity for the human brain.

Technical considerations

In their pioneering studies, Bohning *et al.* (1997, 1998) first demonstrated that TMS can be applied in the MR scanner. Although one might intuitively regard the combination of TMS and fMRI as next to impossible (due to the strong magnetic fields utilized by both methods and the susceptibility of fMRI to distortions of magnetic field homogeneity), Bohning *et al.* demonstrated that ferrous material can be removed from TMS coils, thus enabling safe insertion of the TMS probe into the MR scanner bore, when appropriate precautions are taken.

Coils and static imaging artifacts: practical considerations

Standard TMS coils are not appropriate for combined TMS and fMRI. Even after removal of ferromagnetic materials, the coil must be strengthened to prevent deterioration when used in the scanner. This can be achieved by appropriate casing of the TMS coil to withstand several thousand TMS pulses given inside the MR scanner. The handle on standard TMS coils severely limits the degrees of freedom for placement within the restricted MR environment. Therefore new MR-compatible coils have the handle removed. MR-compatible coil holders ensure stable and precise positioning of the TMS, providing several degrees of freedom for flexible placement. More recently, an automated and computer-operated coil-positioning device has been introduced with an estimated accuracy of approximately ±2 mm across all axes (Bohning *et al.* 2003a). This apparatus can provide accurate and consistent positioning of the TMS coil based on anatomical landmarks (Denslow *et al.* 2005a). However, stimulation of certain temporal or frontal structures may be quite difficult to achieve due to spatial limitations imposed by the MR scanner environment. As a further complication (which also has to be considered even in purely behavioral or cognitive studies), stimulation targeting those brain regions can evoke considerable face or neck muscle movements, or induce considerable discomfort due to stimulation of trigeminal nerve fibers, all of which may result in problematic head movement artifacts. A further practical

consideration is that experimental designs for concurrent TMS–fMRI (as also for behavioral or purely cognitive studies) have to take into account the 'click' sound emitted during typical TMS application, the possible somatosensory sensations on the scalp, and so on. While seemingly obvious, the need to consider such factors can become even more pressing when, for instance, combining TMS with whole-brain fMRI (such that auditory cortex activations may often be revealed due simply to the 'click' sound of the TMS; see below).

Turning to MR considerations, whereas T1-weighted MR images as commonly used for structural imaging show very little distortion in the presence of a TMS coil, echo-planar imaging (EPI) scans, as most commonly used for fMRI, can be prone to considerable geometric distortions (Baudewig *et al.* 2000). Image distortions occur when the homogeneity of the magnetic field of the MR scanner is disturbed, for instance by the presence of a TMS coil, even when distorting materials are removed from the coil. These image distortions, however, decrease with increasing distance from the coil. Given the scalp–cortex distance of ~1–2 cm, undistorted EPI images from the brain can generally be obtained (Figure 36.1) (Baudewig *et al.* 2000). Imperfect timing of MR gradients, eddy currents, nonlinearities in the receiver filters, and field inhomogeneities may also contribute to the image quality and are likely to vary between different scanner systems and TMS set-ups.

A shorter MR image data readout time can help to reduce slight geometric distortions caused by the TMS coil. Shortening the data readout time (also called echo time, TE) generally reduces signal loss due to magnetic field inhomogeneities because the data are acquired before transverse magnetization is significantly affected. Furthermore, oversampling EPI images in the phase-encoding direction (e.g. using a 64 × 96 image matrix rather than a 64 × 64 image matrix) can help to shift some image distortions, usually termed ghosting artifacts, outside the volume of interest.

Dynamic imaging artifacts

The magnetic fields generated by a TMS pulse constitute the dominant technical challenge to concurrent TMS–fMRI (Shastri *et al.* 1999) and may lead to what we will term 'dynamic'

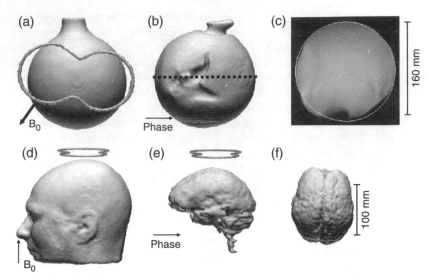

Fig. 36.1 (a) T1-weighted three-dimensional (3D) surface reconstruction of a spherical phantom in the presence of a TMS coil. (b) Reconstructed T2*-weighted echo-planar imaging (EPI) image with the same arrangement. The broken line refers to the cross-sectional image shown in (c). Image distortions in EPI images are clearly visible in direct vicinity of the TMS coil. (d) T1-weighted 3D brain surface reconstruction in the presence of a TMS coil placed over the vertex. (e, f) Reconstructed brain surface from T2*-weighted EPI with a slice orientation parallel to the TMS coil. These images illustrate the distance between TMS coil and cortical surface. No significant image distortions are visible in EPI sections even close to the TMS coil. Adapted from Baudewig *et al.* (2000).

imaging artifacts. EPI sequences used for fMRI can be divided into three principle phases. First, preparation pulses are applied, followed by a radiofrequency (RF) pulse exciting the desired slice (or volume of interest), and finally a period of data readout. This process is repeated for each slice of each acquired volume. Depending on the imaging system and the sequence being used, an EPI sequence typically lasts between ~50 and 120 ms, and is thus considerably longer than a single TMS pulse (~250 µs; Figure 36.2a). Even though a TMS pulse is very short, it can severely distort MR images (Shastri *et al.* 1999; Bestmann *et al.* 2003a).

Depending on the precise time of TMS pulse application, one can distinguish three principal categories of artifacts to deal with: (i) when TMS pulses occur prior to slice acquisition; (ii) TMS pulses during excitation pulses; and (iii) TMS pulses during slice acquisition (data readout).

TMS pulses applied before EPI

A single TMS pulse may affect a subsequent EPI section even when being applied up to 100 ms before slice acquisition onset (Shastri *et al.* 1999; Bestmann *et al.* 2003a). This value may vary considerably, depending on coil orientation, pulse intensity, and magnetic field strength (Figure 36.2b). These image artifacts are presumed to arise from residual currents in the TMS coil, as well as from currents induced by possible mechanical vibrations of the TMS coil following pulse application (Shastri *et al.* 1999). The resulting currents may induce local magnetic field distortions, which can be sufficient to affect nearby slices. Consequently, these EPI artifacts decrease at lower TMS intensities and larger distances between the imaging slice and the TMS coil because the distorting magnetic fields decrease with increasing distance from their source.

TMS pulses during slice excitation

TMS pulse application during MR image acquisition is a severe problem because the strong magnetic field induced by TMS pulses can act either as an efficient spoiling gradient that de-phases all transverse magnetization (i.e. signal of interest), or may interact unpredictably with slice excitation pulses. For example, water and lipids have different resonance frequencies, allowing selective excitation of one or the other by using frequency-specific excitation pulses to suppress signal from unwanted tissue components (such as lipids in the case of fMRI). Any TMS pulse applied at the same time as a frequency-selective RF pulse, however, will spoil the specificity of the required RF pulse. As a consequence, subsequent images of the same slice can be affected, or several spatially adjacent image slices may show strong signal fluctuations, rather than just the directly perturbed slice. One reason is that an unwanted suppression of water protons, i.e. the signal of interest, may occur. In summary, several slices (spatially adjacent slices and/or consecutively acquired slices) may experience signal changes not related to underlying physiological processes, when TMS pulses are applied during specific phases of slice acquisition.

Disturbance of slice-selective RF excitation by a simultaneous TMS pulse may cause persistent signal changes lasting over several acquisitions of a slice. The duration and magnitude of signal changes in successive images of the perturbed slice are dependent on several imaging sequence parameters, such as repetition time or the so-called 'flip angle' (Figure 36.3a). Importantly, these signal changes are not necessarily paralleled by geometric distortions of individual slices, so they may be overlooked unless the potential problem is appreciated. Moreover, as shown in Figure 36.3, the resulting signal changes can be on the order of true physiologically evoked BOLD response changes. Even after accounting for hemodynamic delays and discarding from the analysis the slice coinciding directly with TMS pulse application, these signal changes can thus potentially result in false-positive activations, including spatially adjacent slices (Figure 36.3b).

TMS pulses during data readout

Finally, the application of a TMS pulse during data readout, which follows slice excitation,

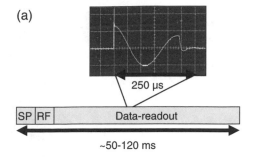

(a)

250 µs

SP | RF | Data-readout

~50–120 ms

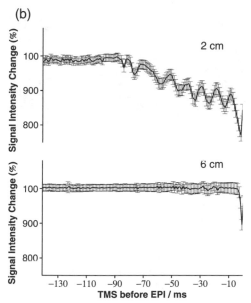

(b)

Fig. 36.2 (a) Schema of the temporal relationship between a TMS pulse and an echo-planar imaging (EPI) sequence. Whereas common EPI sequences last between 50 and 120 ms, a single biphasic TMS pulse is ~250 µs long. (b) EPI signal intensity changes plotted as a function of time (1 ms steps) between a single TMS pulse (100% of stimulator output) and the EPI slice excitation (RF) pulse. The signal was derived from central regions of interest in two transverse sections through a spherical phantom 2 and 6 cm away from a TMS coil. Signal fluctuations of up to 20% can be observed for TMS pulse–RF pulse intervals of ≤100 ms. This illustrates how a single TMS pulse, although several orders of magnitude shorter than an EPI sequence, can nevertheless perturb EPI images acquired even 100 ms later. RF, radiofrequency pulse; SP, suppression pulses. Adapted from Bestmann et al. (2003a).

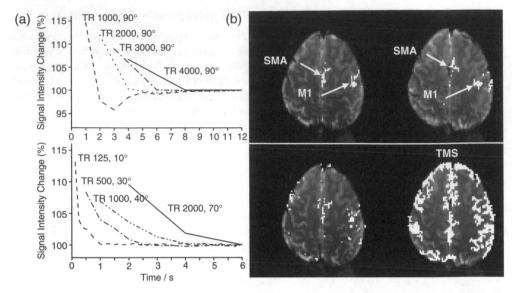

Fig. 36.3 (a) Echo-planar imaging (EPI) signal intensity changes in successive acquisitions following a single TMS pulse applied at 100% of stimulator output during the radiofrequency (RF) excitation pulse, shown as a function of repetition time (TR) (*upper graph*) or flip angle (*lower graph*) along x-axis. Signal intensity was derived from a central region of interest 6 cm away from the TMS coil. Depending on the degree of T1 saturation, the perturbation of steady-state longitudinal magnetization leads to signal increases of up to 15% in subsequent images. Note that several slice excitations (i.e. image acquisitions) are needed before steady state is regained. (b) Activation maps of a simple sequential finger-tapping task in the presence of a TMS probe but without TMS pulses (*upper row*). However, TMS pulses applied during RF excitation may translate into false-positive activations (*lower row*). While the directly perturbed slice (*lower right*) is massively affected, spatially adjacent slices may experience more subtle false-positive results (*lower left*). Adapted from Bestmann et al. (2003a). SMA, supplementary motor area; M1, primary motor cortex. (Plate 14)

causes an almost complete loss of MR signal due to an effective spoiling of all signal of interest remaining at the time of TMS pulse application (Figure 36.4). The later a TMS pulse is applied during data readout, the more image signal has been acquired so that the overall structure of the image can eventually be reconstructed. However, subtle image distortions can potentially remain even when a TMS pulse is applied towards the end of the data readout period, and may still translate into erroneous interpretation of functional MRI time series unless due caution is taken. Moreover, a TMS pulse applied very late during data readout will be close to the subsequent slice acquisition and may additionally distort the next image (see above).

Other potential artifacts

A more general problem is the transmission of RF noise into the scanner, especially via the antenna-like properties of the cable connecting the coil with the stimulator. This can increase the global noise level and therefore reduce sensitivity. The degree to which RF noise will interfere with fMRI may vary with several factors, e.g. resonance frequency of the MR system (~64 MHz at 1.5 T), particular TMS coils, TMS stimulator, and charging state of the TMS machine. For example, newer TMS machines typically have a larger number of built-in electronics that potentially increase RF noise transmission. An additional source of RF noise may occur when varying TMS intensities during scanning, as the

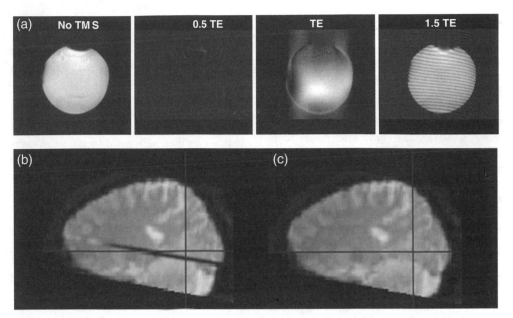

Fig. 36.4 (a) Effect of a single TMS pulse applied during echo-planar imaging (EPI) data readout in a slice 6 cm away from the TMS coil. Image perturbation occurs when the TMS pulse is applied during the early period of data readout; images gradually 'recover' when TMS pulses are applied at a later point of data readout. However, high-frequency distortions can still occur even when TMS is applied late during image acquisition and may then go undetected in human brain images. Adapted from Bestmann *et al.* (2003a). An EPI volume from a single subject with (b) one slice being perturbed by a single TMS pulse or (c) without TMS pulse application but with the TMS coil placed on the subject's head.

charging and recharging of the storage capacitors (thyristors) in itself may emit some RF noise. Appropriate timing of the intensity switching can alleviate this problem; for example, ensuring that intensity switching does not occur during RF slice excitation pulses. Moreover, leakage currents may be generated when switching stimulation intensities. This is because the excess energy in the storage capacitors of the TMS machine has to be released. In some machines, this release of excess energy may also induce small leakage currents in the TMS coil. While these currents are likely to be on the order of microvolts and are therefore completely irrelevant for standard (e.g. behavioral) use of TMS devices, they may be sufficient to distort MR images near the TMS coil. One solution to this is to prevent any currents from flowing between the TMS machine and the TMS coil between stimulation events by using appropriate remote-controlled switches. By completing the circuits via a switch only during the brief periods of TMS pulse discharge, no uncontrolled leakage currents can pass through to the TMS coil at other times.

Propagation of RF noise can be prevented with the use of special filters between the TMS machine and TMS coil, as well as ferrite sleeves around the TMS cable. Filters, however, may further reduce the TMS pulse intensity. This can be problematic as the resistive properties of the long cables needed to connect the TMS coil inside the scanner with the stimulator (typically ~8–10 m, depending on the scanner environment) also attenuate TMS output intensity. RF transmission into the scanner from other sources can be prevented, for example, by placing the entire stimulation unit in an RF-shielded cabinet inside the scanner room. In general, any noise transmission during scanning will compromise the signal-to-noise ratio (SNR) of functional imaging data, and thus may increase the number of false-negative observations.

TMS–fMRI pulse application strategies

In summary, both the insertion of a TMS coil into the scanner itself as well as the discharge of TMS pulses may compromise fMRI image quality. Taking appropriate precautions to address all of the potential issues mentioned above, however, can allow MR images to be obtained without severe losses of SNR, signal dropouts, or image disturbances. But this can require that neither fMRI nor TMS parameters will be applied in a way that would be optimal for either method alone. Depending on the desired TMS protocol, EPI imaging needs to be tailored accordingly. This may require limited volume coverage, prolonged acquisition time, or perturbation of specific slices. On the other hand, the restrictions of acquiring uncorrupted fMRI data may dictate that TMS cannot be applied at all possible frequencies and durations.

Initial fMRI studies of TMS primarily focused either on application of single TMS pulses, or on repetitive TMS applied at a frequency of 1 Hz (Bohning *et al.* 1999, 2000a,b; Nahas *et al.* 2001). Relatively sparse application of TMS pulses bypasses issues related to the synchronization of the TMS pulse with EPI imaging. Waiting periods need not prolong the time needed for a volume acquisition too prohibitively, while separation of TMS and image acquisition can then be more easily achieved. At 1 Hz, individual TMS pulses can be applied during temporal gaps between subsequent volume acquisitions of ~100 ms duration (Shastri *et al.* 1999; Bestmann *et al.* 2003a). But at higher stimulation frequencies or longer stimulation trains (or when single TMS pulses need to be presented at different times of an MR volume acquisition, i.e. being 'jittered'), achieving unperturbed imaging will require specially tailored imaging protocols.

Temporal gaps following volume acquisition

Introduction of temporal gaps between successive volume acquisitions allows for complete separation of TMS pulses and MR image acquisition. For example, Ruff *et al.* (2006) and Sack *et al.* (2007) applied TMS within waiting periods of ~500 ms following each volume acquisition (Figure 36.5a), thereby completely separating MR image acquisition and TMS pulse application. Although advantageous in several respects, this approach does slightly decrease temporal resolution, especially when longer pulse-trains and/or whole-brain coverage are desired. Furthermore, temporal jittering of TMS pulses then becomes difficult, when waiting periods are fixed with respect to the MR volume acquisition (i.e. after a volume is acquired).

Temporal gaps following slice acquisition

For longer pulse trains, TMS pulses and MR images can be interleaved by insertion of temporal gaps after each *slice*, rather than each *volume* (Bestmann *et al.* 2004). This requires waiting periods of sufficient length following each slice acquisition so that TMS pulses do not affect subsequent slices (Figure 36.5b). This can quickly increase the time required for volume acquisition unless brain coverage is limited by reducing the number of slices. Furthermore, the maximum stimulation frequency is limited by the time required for single-slice acquisition (usually in the range of 50–120 ms) plus the required waiting period. Another way to overcome this problem is to use sparse imaging sequences (as used, for example, in many auditory studies that present auditory stimuli during silent waiting periods) in which long temporal gaps (several seconds) between EPI volume acquisitions are introduced. The advantage of both these approaches is that pulse trains can then be applied at any desired length.

Perturbation of single slices by design

A further approach deliberately perturbs single slices (Figure 36.5c). As outlined above, TMS pulses must avoid direct RF-pulse and suppression-pulse interference. Slices corrupted by TMS pulses during data readout, however, can be identified and replaced. One strategy is to replace slices by interpolation between the preceding and subsequent acquisition of the same slice, although this introduces some degree of temporal filtering. Alternatively, affected slices can be included as covariates of no interest in a

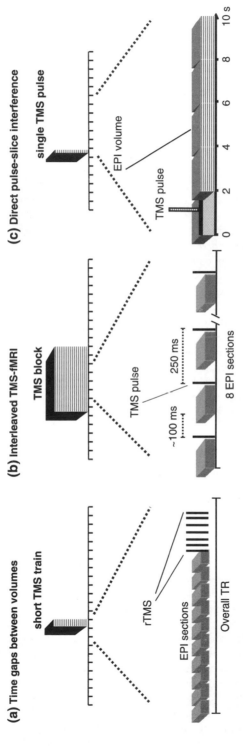

Fig. 36.5 Schematic of TMS–fMRI application strategies. (a) Insertion of temporal gaps following each volume acquisition provides waiting periods for TMS pulses, thereby separating TMS pulses and echo-planar imaging (EPI) (see Ruff *et al.* 2006; Sack *et al.* 2007). (b) This can also be achieved by insertion of temporal gaps following each slice acquisition. This dictates lower TMS stimulation frequencies but allows for long stimulation trains (see Bestmann *et al.* 2003a). (c) Direct TMS pulse interference with data readout destroys corresponding slices (gray region), but allows flexible jittering of pulses.

general linear model analysis. Using this strategy, TMS pulses can be flexibly applied and jittered with respect to imaging, which may be required for studies seeking to vary TMS on a trial-by-trial basis. Repeated perturbation of the same slice or slices should be avoided, however, as then the interpolation process could significantly reduce the overall variance in the affected slice. It is thus recommended to perturb all slices equally often. The introduction of spatial gaps (e.g. 50% of the slice thickness) between image slices can protect against the corruption of spatially adjacent slices, which might otherwise arise since slice excitation RF pulses also excite some small fraction of spatially adjacent tissue.

In summary, while adhering to all the safety guidelines that apply to TMS and MR in general, TMS and fMRI can be used in combination provided several further technical issues are addressed. This generally entails some restriction of flexibility in how both the TMS stimulation and the fMRI imaging parameters can be applied. Taking all of these points into account, a growing number of studies now show that combined TMS–fMRI is technically feasible. The remainder of this chapter will provide recent examples that illustrate how concurrent TMS–fMRI can address a variety of different research questions.

Investigating local and remote effects of TMS to motor-related structures on human brain activity with fMRI

As mentioned in the Introduction, the majority of studies using concurrent TMS–fMRI to date (but by no means all) have concerned the motor system. This focus on motor TMS accords with the extensive literature on motor TMS using overt measures, such as behavior or MEPs. Primary motor cortex is easily accessible to TMS stimulation and its consequences can readily be recorded from the muscles. Moreover, the anatomy and connectivity of the motor system is quite well characterized, and identification of targeted regions can proceed with some anatomical certainty, for example by identification of the

so-called hand-knob (or 'omega') in primary motor cortex (Dechent and Frahm 2003).

Changes in local activity at the targeted site

Several studies using single-pulse or 1 Hz TMS–fMRI to motor cortex have shown that activity increases with fMRI in the putative hand region of the primary motor cortex (Bohning *et al.* 1998, 1999, 2000a). These increases were spatially congruent with activity evoked by voluntary finger movements, suggesting that TMS to M1 can engage the same structures.

Early studies suggested a linear 'dose–response' relationship between evoked hemodynamic responses and TMS intensity (Bohning *et al.* 1999, 2003b) similar to that observed during voluntary finger movements (Kastrup *et al.* 2002). But a potential complicating issue when investigating TMS-evoked BOLD responses for the motor cortex is that afferent feedback will be generated by contralateral muscle responses for stimulation intensities at or above resting motor threshold. The intimate connectivity between primary somatosensory and primary motor cortex can make it difficult to dissociate hemodynamic changes related to direct TMS stimulation effects at the cortical level from their reafferent consequences. In other words, in the presence of induced muscle responses, one cannot readily determine whether increased responses in M1 following increased TMS intensity are a result of the direct TMS stimulation effect, the reafferent consequences of induced muscle twitches, or some interaction of both processes (see Bestmann *et al.* 2006, for discussion). While motor cortex has the advantage of being easily and precisely targeted, the consequences of its stimulation at higher intensities for fMRI activity are not always easy to interpret due to associated muscle twitches.

Applying TMS at intensities that do *not* evoke muscle responses can circumvent such interpretative problems. But significant changes in local activity in M1 have not always been reported with such subthreshold stimulation (Baudewig *et al.* 2001; Bestmann *et al.* 2003b, 2004). Only a few studies applied intensities known to act predominantly at the cortical level, with little or no descending activity. As shown in Figure 36.6,

110% RMT   **90% AMT**

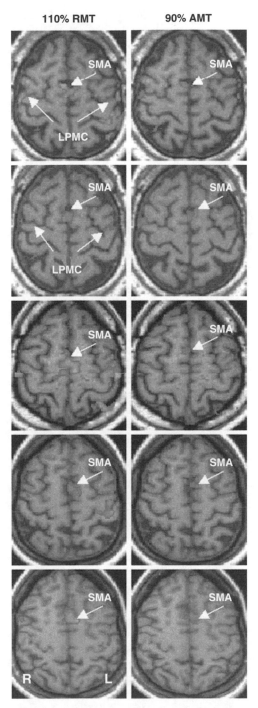

Fig. 36.6 Comparison of supra- and subthreshold TMS effects on local hemodynamics for TMS applied over primary motor cortex. The data from five individual subjects (rows) show localized activity increases in the putatively stimulated motor cortex hand regions during suprathreshold TMS (left panels), whereas no such increases are detected at subthreshold intensities (right panels), despite their known physiological impact on local activity. RMT, resting motor threshold; AMT, active motor threshold; SMA, supplementary motor area; LPMC, lateral premotor cortex; L, left hemisphere; R, right hemisphere. Adapted from Bestmann et al. (2004). (Plate 15)

trains of subthreshold intensity applied to M1 that did not evoke contralateral muscle activity also did not evoke significant changes in local activity (Bestmann et al. 2003b, 2004). Online recordings of electromyographic activity in contralateral target muscles confirmed the absence of overt muscle response. The absence of reliable M1 activations with subthreshold TMS may indicate that in those studies using suprathreshold TMS to M1 instead, reafferent signals from the induced twitch may have contributed significantly to observed changes in local BOLD signal. Stimulation over nonprimary motor areas can be used to sidestep this problem (Bestmann et al. 2005) as TMS pulses can then affect remote cortical and subcortical activity (see below) even at subthreshold intensities and without inducing twitches (Kujirai et al. 1993).

However, several fMRI studies did not report significant changes in local activity under the TMS probe when applying TMS to nonmotor regions, including sites over superior parietal (Kemna and Gembris 2003), premotor (Baudewig et al. 2001), or prefrontal cortex (Kemna and Gembris 2003; Li et al. 2004a; but also see Nahas et al. 2001). It has been reported that 10 s trains at 3 Hz to dorsal premotor cortex can lead to local increases in BOLD signal at the site of stimulation, at an intensity that would be above resting motor threshold for M1 stimulation (Bestmann et al. 2005). However, no such activity increases could be detected at subthreshold intensities known to affect premotor activity on other grounds (Civardi et al. 2001; Munchau et al. 2002). This may indicate that some internal threshold for TMS-evoked local activity needs to be passed in order to evoke significant hemodynamic changes. One explanation put forward involves a possible sigmoid relationship between local neural activity and changes in cerebral blood flow (Lauritzen and Gold 2003). This would suggest that relatively low levels of synaptic activity may not always be accompanied by concomitant BOLD signal changes, unless a certain threshold of activity is passed, and that the hemodynamic response may eventually saturate at very high activation levels. However, an in-depth parametric investigation spanning a large range of TMS stimulation intensities is not presently available, though would clearly be desirable. Moreover, the TMS input–BOLD signal output function may vary between different brain regions. Such issues concerning how neural activity relates to BOLD signals are a general concern for fMRI studies, and are not peculiar to concurrent TMS–fMRI only.

It may be worth noting that a clear match between TMS input and measured changes in local activity has not invariably been found by studies using other imaging modalities, such as positron emission tomography (PET) (Speer et al. 2003a,b) or single-photon emission computed tomography (Okabe et al. 2003). Here, both some decreases (Paus et al. 1998; Siebner et al. 2003a; Speer et al. 2003b) as well as some increases (Paus et al. 1997; Speer et al. 2003a) in regional cerebral blood flow or glucose utilization have been reported following TMS. Changes in physiological excitability may be paralleled by both positively (Siebner et al. 2000, 2001) and negatively correlated changes in regional blood flow (Paus et al. 1998; Chouinard et al. 2003). Moreover, several studies did not observe significant activity changes at the site of stimulation, yet still reported remote effects of TMS (Kimbrell et al. 2002; Okabe et al. 2003). A direct comparison between studies hitherto is difficult to conduct, given that a large range of stimulation protocols and stimulation sites have been used. Thus, the main conclusion that can be drawn from these studies so far is that local effects of TMS appear to vary considerably with the site of stimulation and the specific stimulation procedure. The absence of local TMS-evoked hemodynamic changes may not always indicate that the stimulation was ineffective. Indeed there may often be some remote effect on fMRI signals even when no significant local effect is found (see 'Changes in remote activity' below).

A detailed understanding of the factors contributing to any changes in local activity may become even more relevant when applying TMS to nonmotor regions, where physiological effects cannot be monitored by measures such as MEPs. Concurrent TMS–fMRI may then become even more important for quantifying the impact of TMS by measuring BOLD signal changes. Finally, we note also that measuring any change in local activity under the stimulated site may be important for assessing effective connectivity between that site and remote potentially interconnected sites.

Changes in remote activity

Some TMS studies (e.g. in the behavioral literature) have explicitly or implicitly assumed that TMS predominantly affects only the local site of stimulation. However, increasing evidence now shows that TMS pulses can activate both efferent and afferent pathways from the site of stimulation to interconnected areas. Studies on transcallosal inhibition (Ferbert et al. 1992) and inhibition of cerebello-cortical (Ugawa et al. 1991) or cortico-cortical (Civardi et al. 2001) connections in the motor domain provided some initial support for this notion. Many recent electrophysiological studies using double-coil approaches (to stimulate two different sites with a given temporal relation, often with M1 as one site to provide an MEP measure) lend further support to the notion that effects of TMS can propagate along anatomical connections to remote brain regions (Civardi et al. 2001; Munchau et al. 2002; Gilio et al. 2003; Rizzo et al. 2004). Such remote influences need to be demonstrated, as otherwise local influences may continue to provide the most parsimonious explanation for observed effects (see also Robertson et al. 2003). Nonetheless, recent TMS findings may add support to the general idea that 'almost nothing goes on internally in one area without this activity being transmitted to at least one other area' (Mumford 1992, p. 242).

It therefore becomes important to delineate which remote brain regions are activated by

TMS, and to characterize the appropriate stimulation parameters and conditions for this to arise. By combining TMS with concurrent neuroimaging, one may then chart a more complete picture of the neuronal underpinnings and dynamics of TMS effects, in principle across the whole brain. As outlined below, remote effects of TMS may turn out to be an asset rather than a limitation of the method, both for understanding causal interplay between interconnected brain regions, and for potential clinical applications. More generally, understanding these remote effects may in the future help to generate new hypotheses regarding the mechanisms by which TMS disrupts or improves task performance.

Several concurrent TMS–fMRI studies have now demonstrated changes in activity for interconnected brain regions even at low TMS intensities and short pulse train durations. These findings complement TMS–PET studies that analogously showed some activity changes in remote brain regions (Paus *et al.* 1997; Strafella *et al.* 2000, 2003; Paus 2002; Chouinard *et al.* 2003; see Chapter 35, this volume). However, the relatively high spatial resolution of fMRI (and its enhanced temporal resolution relative to PET, to allow event-related measures) may provide insights into the neuronal underpinnings of TMS that have not been accessible with other methods. Recent studies applying TMS to M1 have shown remote cortical and subcortical TMS-evoked BOLD signal changes with fMRI, including premotor cortex, supplementary motor area (SMA), thalamus, and basal ganglia (Bestmann *et al.* 2003b, 2004, 2005; Denslow *et al.* 2005b) (Figure 36.7). An important advantage of these studies over combined TMS–PET studies is that they can provide a more dynamic measure of brain responses evoked by short TMS stimulation epochs (Baudewig *et al.* 2001; Bestmann *et al.* 2004, 2005) or even single TMS pulses (Bohning *et al.* 2000b), rather than measuring more lasting accumulative effects of rTMS (Chapters 34 and 35, this olume).

More in-depth investigations of the effects of subthreshold TMS stimulation show that, even without evoked peripheral muscle responses, changes in BOLD response during TMS can arise in a complex network of interconnected brain regions. As demonstrated by Di Lazzaro *et al.* (2004), TMS at such subthreshold intensities predominantly targets the brain at the cortical level. However, when applied to primary motor cortex, BOLD signal changes are not only observed in dorsal premotor cortex and SMA (Bestmann *et al.* 2003b), but also in more ventral premotor regions and subcortical structures, such as the putative motor thalamus and the basal ganglia (Bestmann *et al.* 2004) (Figure 36.7). These changes cannot be explained by reafferent feedback since subthreshold stimulation does not evoke contralateral hand muscle activity. Similarly, short trains of TMS to dorsal premotor cortex evoke BOLD signal changes in a number of remote motor areas including SMA, contralateral dorsal and ventral premotor cortex, basal ganglia, cerebellum, and thalamus, even at TMS intensities below the resting motor threshold (Bestmann *et al.* 2005). These results agree with earlier TMS–PET findings showing activation of subcortical structures by rTMS to dorsal premotor cortex, albeit on a different time scale (Chouinard *et al.* 2003).

When applied to dorsolateral prefrontal cortex, relatively few pulses of 1 Hz TMS can evoke a similarly complex pattern of remote BOLD signal changes, including bilateral middle frontal cortex, contralateral orbitofrontal cortex, insula, and subcortical structures including hippocampus, thalamus, and putamen (Li *et al.* 2004a). The lack of overt peripheral effects (such as MEPs) during prefrontal TMS had hitherto limited the investigation of the physiological effects of TMS to such sites. The study by Li *et al.* reveals how combined TMS–fMRI can delineate the brain regions targeted by prefrontal TMS even in the absence of overt output associated with TMS application, and how these effects unfold on a relatively short temporal scale.

Studies of the motor system have acknowledged that a distinction between polysynaptic versus monosynaptic influences between two regions is not easily made with fMRI (Bestmann *et al.* 2003b, 2004). For example, TMS of M1 has been reported to evoke BOLD signal changes in bilateral SMA, bilateral ventrolateral thalamus, and bilaterally in the basal ganglia (Bestmann *et al.* 2004) (Figure 36.7). Rather

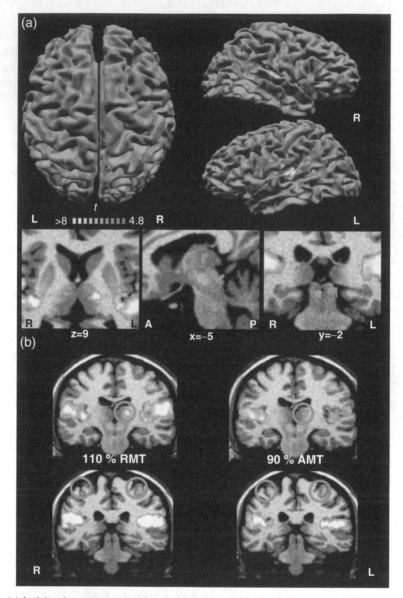

Fig. 36.7 (a) Activity changes to suprathreshold reptitive TMS (3 Hz) applied to left M1. Increased activity (red–yellow) is evoked in left (stimulated) M1/S1, dorsal premotor cortex, and supplementary motor area, but also in auditory cortex, post-central gyrus, and S2, presumably due to the 'click' sound of TMS plus the scalp somatosensory sensation it induces. Activity decreases (blue) can be seen in the contralateral M1/S1. At the subcortical level, activity increases can be detected in the ventrolateral thalamus and putamen. In addition, activity increases in subcortical parts of the auditory system. This illustrates the widespread activity changes induced by suprathreshold M1 TMS in several distinct motor regions, plus accompanying activity in the auditory and somatosensory system at both cortical and subcortical levels. (b) Both supra- and subthreshold stimulation trains evoke remote activity, even in subcortical structures such as the motor thalamus. However, no significant activity changes were observed in the directly stimulated regions at subthreshold intensities. In contrast to suprathreshold stimulation, subthreshold stimulation did not evoke significant electromyographic responses in the contralateral hand, thus confirming that remote activity during subthreshold stimulation was not due to afferent feedback from contralateral muscle activation. RMT, resting motor threshold; AMT, active motor threshold; L, left hemisphere; R, right hemisphere. Aadapted from Bestmann *et al.* (2004). (Plate 16)

than being evoked by direct monosynaptic long-range connections, some of these effects presumably originate from polysynaptic propagation of the effects of TMS, even across the corpus callosum (as might be tested in split-brain subjects). This illustrates that TMS may affect activity in an extended functional network underlying a specific brain state, rather than merely activating monosynaptic anatomical connections.

Extensions of the concurrent TMS–fMRI approach beyond the motor system

A recent study by Ruff et al. (2006) illustrates that the use of concurrent TMS–fMRI to study possible remote influences of TMS can go well beyond motor-related structures. Their study used the combined TMS–fMRI approach to test a hypothesis arising from the literature on visual attention, namely that anterior structures such as the frontal eye fields (FEFs) may causally modulate activity in posterior visual cortex (cf. Kastner and Ungerleider 2000; Moore et al. 2003). Short bursts of TMS were given at parametrically varied intensity either to the right FEF or to a vertex control site. Increased intensity of TMS to the right FEF led to a characteristic pattern of BOLD signal changes in retinotopic visual areas V1–V4 in posterior occipital cortex, with BOLD signal decreases for more foveal visual field representations, but BOLD signal increases in more peripheral visual field representations in the calcarine sulci of both hemispheres (Figure 36.8). This topographically specific activation pattern was consistently

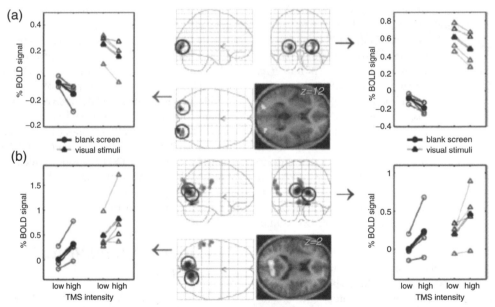

Fig. 36.8 Negative (a) and positive (b) correlations of BOLD signal changes with frontal eye field (FEF) TMS intensity. Graphs on either side display single-subject mean signal intensity (with group average in black) in the left-hemispheric (red circle) and right-hemispheric (blue circle) occipital regions for the two highest versus the two lowest TMS intensities. Activity in these visual regions was also higher with vs without visual stimulation (compare triangular with circular points in graphs), but the impact of high vs low intensity FEF stimulation was additive to this, and consistent across subjects. Further retinotopic mapping of occipital cortex (not shown here, for details see Ruff et al. (2006)) confirmed that visual areas V1–V4 were affected by FEF TMS intensity, with correspondingly increased activity for representations of the peripheral visual field but decreased activity for the central visual field in each of these areas. Adapted from Ruff et al. (2006). (Plate 17)

evoked in all subjects, was correlated with TMS intensity, and was specific to the FEF stimulation site (i.e. not present during vertex stimulation). Finally, these BOLD response changes were unaffected by the level of baseline activity in visual cortex, as manipulated by the presence or absence of visual input. This supported the hypothesis that the FEF stimulation effects on visual cortex reflect 'top-down' influences, consistent with microstimulation in monkeys (Moore and Armstrong 2003).

Furthermore, Ruff *et al.* (2006) were able to test a new behavioral prediction derived from their concurrent TMS–fMRI data, namely that FEF TMS should enhance peripheral vision relative to central, in line with the observed activity changes in retinotopic visual cortex. This new prediction was confirmed in a psychophysical test of the impact of FEF TMS (compared to the vertex control site again) upon judgments of relative contrast for Gabor patches (stimuli likely to activate only early visual cortex). This study thereby implies that the observed remote top-down influences of FEF TMS upon activity in visual cortex can have functional consequences for perception. The results of Ruff *et al.* (2006) thus illustrate that TMS–fMRI can be used to establish a link between functional and physiological consequences of TMS beyond the targeted site. Their study also implies that physiological or behavioral consequences of TMS may not always be attributable only to activity within the stimulated area itself, but may instead reflect remote influences on interconnected regions within a large-scale network.

All of the concurrent TMS–fMRI studies described above underline the importance of careful choice in control conditions. Irrespective of the stimulation site, it always has to be kept in mind that changes in BOLD response may occur due to the accompanying auditory and somatosensory stimulation evoked by TMS. These effects cannot be avoided easily: every TMS pulse (whether inside or outside the scanner (e.g. Nikouline *et al.* 1999)) evokes a sensory sensation on the scalp. In addition, auditory stimulation occurs due to the discharge noise of the TMS coil, while eye-blinks or pupil dilation may occur. All of these factors had to be taken into careful account in the Ruff *et al.* (2006) study, for instance, given its focus on any impact

of TMS upon activity in visual cortex. In that study, auditory cortex was affected similarly by the different TMS sites (FEF and vertex) whereas visual cortex was not. Moreover, blinks and pupil dilations were carefully assessed; these were found not to correlate with the applied TMS in that study, but nevertheless did produce their own effects on visual cortex which had to be separated from those due to TMS.

While the 'click' sound produced by TMS can be minimized using earplugs and headphones, the high sound pressure level ($\geq$90 dB) and bone conduction to the inner ear nevertheless renders auditory stimulation unavoidable. As shown in a number of studies, these 'secondary' effects of TMS often evoke bilateral BOLD response changes throughout the auditory and somatosensory system (Bohning *et al.* 2000b; Baudewig *et al.* 2001; Bestmann *et al.* 2003b, 2004, 2005) (Figure 36.7). Depending on the question and paradigm, such 'secondary' stimulation effects (for example, due to the sound and tactile sensation associated with TMS) may give rise to cross-modal interactions, which may influence cortical activity patterns and performance significantly (Paus 2000; Laurienti *et al.* 2002; Macaluso and Driver 2005). This highlights how one must control for possible influences of 'secondary' TMS effects, for example by directly comparing brain activity changes to TMS over different sites (such as FEF and vertex)(Ruff *et al.* 2006, Sack *et al.* 2007), by testing a wide range of TMS intensities to identify brain activity changes that covary with stimulation intensity *per se*, or by modelling eye-blinks and pupil diameter in the fMRI analyses (Ruff *et al.* 2006).

Concurrent TMS–fMRI studies of cognition

A further rationale for combining TMS and fMRI is that TMS manipulations can disrupt specific cognitive processes (as exemplified by many other contributions in this book), while it is also well-established that fMRI can localize neural substrates of particular cognitive processes. Combining the two approaches concurrently could thus bring the causal dimension introduced by TMS into the whole-brain localization dimensions of fMRI. One might thereby combine the correlative brain–behavior information provided by fMRI with the causal structure–function inference afforded by TMS. This might be achieved by delivering a controlled focal perturbation into a particular process with TMS, and at the same time visualizing its consequences for task-related activity, as well as for performance.

Based on several previous studies, Sack et al. (2007) reasoned that visuo-spatial processing may be particularly impaired by right parietal TMS due to an effect on several regions of the involved network, rather than on the site of stimulation alone. During scanning, they applied short bursts of TMS to the right or left intraparietal sulcus while participants performed a visuo-spatial or color discrimination task. TMS to the right site had previously been shown to lead to a significant behavioral impairment during a visuo-spatial but not a color discrimination task (Sack et al. 2002, 2005).

In the visuospatial task, participants had to judge whether the angle formed by the hands of a visually presented clock corresponded to a prespecified angle. In a control task, the same stimuli were presented but participants had to judge the color of the clock hands instead. Short bursts of TMS were applied at 13 Hz in 50% of trials for both tasks during the period of task processing. When trials with and without TMS were compared, right parietal TMS was found to prolong reaction times only for the visuo-spatial task. An analysis of the underlying BOLD response patterns revealed a decrease at the site of stimulation and ipsilateral medial frontal gyrus (Figure 36.9). Importantly, these TMS-related decreases in BOLD signal were correlated with task performance, suggesting a direct brain–behavior relation with respect to the causal TMS intervention. Note that this interrelation was not limited to the site of stimulation but was also observed in ipsilateral medial frontal gyrus. In other words, TMS to one putative control region for visuo-spatial processing, namely the right intraparietal sulcus, evoked a task-specific impairment in performance that was mirrored by an activity decrease not only at the site of stimulation, but also additional frontal regions conjointly activated during visuo-spatial processing.

Additionally, Sack et al. (2007) identified several brain regions exhibiting TMS-induced activity reduction in both tasks (clock angle and color). These included the SMA, FEF, and left parietal cortex, but here BOLD response decreases were not correlated with behavioral performance. This suggests that while TMS may influence several brain regions, not all of these regions contribute significantly to current task performance. Assessment of task performance in combined TMS–fMRI experiments may thus permit the delineation of task-relevant vs task-irrelevant remote stimulation effects, and correspondingly chart the causally relevant nodes of an extended brain network. These initial findings accord with the idea emerging in the clinical literature that visuo-spatial deficits following right parietal damage (e.g. after stroke) may reflect perturbed activity across a particular fronto-parietal network, rather than only at the lesioned parietal site (Corbetta et al. 2005).

Another approach is to use TMS to elicit specific perceptual phenomena that may or may not be reported on each individual trial, such as an induced sense of movement (SoM), while at the same time using fMRI to investigate the cortical correlates of such an evoked SoM. Bestmann et al. (2006) recently studied an amputee patient with some phantom-limb experiences for the missing lower arm and hand, in whom single TMS pulse applied to the putative former M1 hand area reliably elicited an SoM for the phantom hand. Concurrent fMRI was used to measure activity corresponding to this phenomenal perception. Importantly, TMS at high intensity could evoke SoM in this patient without eliciting overt responses in proximal arm muscles of the injured arm (as confirmed by concurrent EMG measurements during scanning) or in the face, thus ruling out reafferent feedback as a potential contribution to the TMS-induced phantom SoM for her. Moreover, the critical fMRI comparison was for trials with versus without a phantom SoM reported phenomenally, for the same intermediate TMS intensities. Note that this contrast completely factors out any nonspecific TMS effects upon brain activity, such as from auditory and somatosensory inputs due to TMS, to reveal neural correlates of just the induced phantom SoM itself. Areas activated more highly on trials where the patient did experience a TMS-induced phantom hand-movement included primary motor cortex,

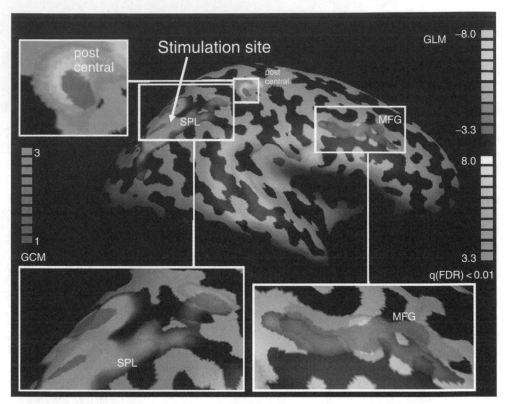

Fig. 36.9 The network of functional connectivity as revealed by Granger causality mapping (GCM) is superimposed on the general linear model (GLM) contrast for presence minus absence of right parietal TMS, in the clock-angle task. The GLM results are color coded in blue–orange, blue representing areas with a TMS-induced decrease of neural activity during angle task execution. The GCM is color coded in red, representing the brain areas showing functional connectivity during the angle task execution. Close-up windows are provided for the three regions of interest: right superior parietal lobe (SPL), right post-central gyrus, and right middle frontal gyrus (MFG). This shows that the task-specific TMS-induced activity modulations occur in the same brain areas that are functionally connected during the execution of specifically this visuo-spatial task. In contrast, the instantaneous GCM for the execution of the color task (on the same clock stimuli) did not reveal this task-dependent fronto-parietal network of functional connectivity, thereby mirroring the absence of color task-specific TMS-induced neural effects as revealed in the GLM analyses. Adapted from Sack *et al.* (2007). (Plate 18)

dorsal premotor cortex, anterior intraparietal sulcus, and caudal SMA, regions that are also involved in some hand-movement illusions and motor imagery in normals (Figure 36.10). These findings add support to proposals that a conscious sense of movement for the hand can be conveyed by activity within corresponding motor-related cortical structures, in the absence of reafferent feedback from hand muscles. This case report adds to previous attempts to chart the possible neural basis of conscious SoM, but now

includes the novel methodological approach of contrasting perceptual phenomena elicited by the identical TMS input, here in an amputee experiencing phantom movements. More generally, this demonstrates how combined TMS–fMRI can provide insight into the cortical underpinnings of perceptual phenomena which could not be investigated using conventional methodologies. It also shows that event-related TMS–fMRI, contrasting trials based on phenomenal report within a single individual, is now feasible.

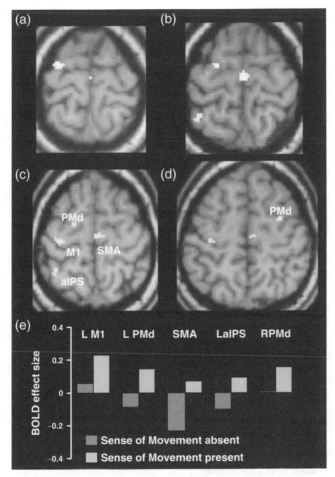

Fig. 36.10 Activity changes for the comparison of trials with versus without a phantom sense of movement (SoM) reported, at intermediate TMS intensities, in an amputee patient experiencing TMS-evoked phantom movements of her missing hand. When a conscious phantom SoM was perceived, activity increases were observed in several motor-related regions, including the left (stimulated) M1, left and right dorsal premotor cortex (PMd), left anterior intraparietal sulcus (aIPS), and caudal supplementary motor area (SMA). Importantly, TMS stimulation intensities were held constant for this comparison and were clearly below threshold for evoking peripheral muscle responses. The results are displayed on the patient's anatomical T1-weighted MRI. (a) Transverse section $z = 72$; (b) $z = 67$; (c) $z = 62$; (d) $z = 57$. (e) fMRI percentage signal change with respect to the session mean in peaks from these five motor-related regions (left M1, left and right PMd, SMA, left aIPS) for trials with or without evoked phantom SoM experienced. Adapted from Bestmann et al. (2006). (Plate 19)

The studies by Sack et al. (2007) and Bestmann et al. (2006) may provide pointers to future directions now made possible by concurrent TMS–fMRI, where causal interventions with TMS can be related not just to performance, and not just to local and remote brain activity, but to all of these aspects concurrently.

Clinical applications of combined TMS–fMRI

There is increasing interest in the clinical application of TMS (Cantello 2002; Curra et al. 2002; see Section VI, this volume), despite current limitations in knowledge on the exact local and remote brain influences that TMS can induce.

The concurrent combination of TMS with fMRI that we have described can be used to investigate how brain activity and connectivity are affected by acute and chronic injury to the brain. By understanding how TMS affects local and remote brain regions in health and disease, it may ultimately be possible to develop more effective clinical TMS interventions. Concurrent TMS–fMRI provides one way to investigate the basic physiology of TMS and also to characterize remote influences onto potentially disease-related brain regions. Its application to stroke, movement disorders, and psychiatric disorders such as major depression and schizophrenia may be of particular interest.

However, while an increasing number of studies have expressed some optimism that TMS can be applied to treat major disorders such as depression, this initial optimism has also been met with several reservations, since therapeutic effects from current TMS interventions have not always reached clinical significance. A general consensus seems to have emerged that dorsolateral prefrontal cortex may be a promising target site for interventions in depression. This was largely motivated by non-TMS neuroimaging studies related to depression, and by known anatomical connectivity of the prefrontal cortex with limbic structures. However, in the existing literature on TMS and depression, there is little consistency regarding the exact target location for TMS (left vs right, medial vs dorsolateral prefrontal cortex) or the effective stimulation parameters (with varied frequency, intensity, duration, and repetition of the procedure); (for an in-depth discussion of this topic, see Lisanby et al. (2003), Padberg and Moller (2003), Kozel and George (2004), and Paus and Barrett (2004), see also Chapters 40 and 41, this volume).

Against this background of clinical interest, effects of dorsolateral prefrontal 1 Hz rTMS have been investigated in healthy adults with fMRI (Nahas et al. 2001). Stimulation was only found to exert significant influences on BOLD signal in bilateral prefrontal regions at high stimulation intensities, indicating that TMS effects are dose-dependent. These BOLD signal changes were located in prefrontal sites connected to the stimulation site rather than directly under the coil, again suggesting that

effects of TMS (possibly including potential therapeutic effects) may be due to activity modulation in interconnected structures, rather than solely by changes at the site of stimulation itself.

Li et al. (2004a) have recently extended this approach in a study of 15 clinically depressed patients, showing that concurrent 1 Hz TMS–fMRI applied over the left dorsolateral prefrontal cortex increased BOLD signal near the site of stimulation as well as at a number of limbic regions including bilateral middle prefrontal cortex, right orbitofrontal cortex, insula, and left hippocampus. Additional BOLD signal increases were found at the level of the thalamus, putamen, and pulvinar. These remote effects were stronger and more widespread than in the study by Nahas et al. (2001). This may indicate that the 'reactivity' of some brain structures to TMS may not be comparable in health and disease (but note that no direct comparison was conducted), an observation that has frequently also been made for a number of movement disorders (Chapter 21, this volume).

There is little consensus as yet about the stimulation parameters most effective for therapeutic purposes, and a clinically valid characterization of the most effective TMS protocols would be of major importance. Nevertheless, the above-mentioned studies already show how TMS–fMRI can be used to chart the networks influenced by stimulation of putatively clinically relevant sites and to visualize the impact on subsequent activity changes in local and remote brain regions. For this, fMRI affords the high spatial resolution required for delineating relevant structures potentially influenced by TMS, including subcortical regions.

Combined TMS–fMRI may also be used to validate effects of neuroactive compounds by comparing the amount and topography of TMS-induced activity changes with and without drug administration. This approach is motivated by a wealth of studies showing that neuroactive drugs can change cortical excitability, and therefore the effectiveness of a given TMS pulse (see Chapter 13, this volume). Li et al. (2004b) compared in a randomized double-blind crossover study the TMS-evoked activity patterns revealed by fMRI before and after administration of lamotrigin (LTG), a use-dependent sodium-channel inhibitor or placebo. On different days, 1 Hz TMS–fMRI was applied to the left motor cortex or dorsolateral prefrontal cortex following a single oral dose of LTG or placebo. Under LTG, TMS-evoked BOLD signal changes in M1 were significantly lower. However, the reduced excitability of M1 following LTG administration might lead to diminished motor-evoked responses and therefore to reduced afferent feedback, potentially accounting for some of the observed relative activity decreases in M1. Subthreshold TMS might be used to address this.

Interestingly, the comparison of prefrontal TMS with and without LTG yielded TMS-evoked BOLD signal changes in limbic structures after LTG administration, suggesting a specific remote effect on these structures that could only be revealed by comparing TMS-induced activity patterns in different pharmacological states. It remains unknown if these results will generalize to a pathological population (for example, depressed individuals), and it is not fully clear which physiological mechanisms might have mediated the reported effects (for example, any possible role for changes in neurovascular coupling). Nevertheless, the study shows that combined TMS–fMRI holds promise for characterizing effects of neuroactive compounds throughout the human brain, and in particular any effects on causal interplay, or 'effective connectivity', between different brain regions in an extended network.

Outlook

The present chapter has presented an overview of recent developments in the combined use of TMS–fMRI. Now that concurrent TMS–fMRI is technically feasible, how can it be utilized so as to justify its technical challenge (Siebner *et al.* 2003b) and capitalize on the new opportunities afforded? Despite the methodological challenges, an increasing number of studies have used TMS–fMRI successfully to show how combining complementary research techniques can provide new perspectives on the underlying mechanisms of TMS effects, plus fresh insights into new questions regarding brain function (for example, concerning causal interplay between brain regions, and how these may vary with psychological and pharmacological state) that might otherwise not be approachable.

As we outlined above, seeking to understand the physiological bases of TMS effects may provide rationale enough for the new methods, and perhaps less is known about these physiological bases than commonly thought (e.g. as regards the importance of remote as well as local effects). Combined TMS–fMRI is by no means a panacea, but it can provide information that should complement other approaches (such as double-coil, MEPs, etc) to draw a more complete picture of the neural underpinnings of TMS and how it can affect cognition and behavior.

TMS can provide an independent well-controlled input or perturbation into a given cortical region, which in turn will connect with many other areas. It thereby provides a new opportunity to investigate 'effective connectivity' (and any changes in this with psychological, pharmacological or clinical state), by providing a causal link between the input given (the TMS pulse) and corresponding activity changes throughout the brain. Moreover, TMS–fMRI can now be used to compare such TMS-evoked effective connectivity in health and disease. In the context of task performance, TMS–fMRI can potentially be used to investigate connectivity changes during different states, with different degrees of involvement for interconnected brain regions during different tasks. Importantly, one can now visualize how entire networks, rather than just single brain regions, react to transient TMS input to a particular part of that network. Such applications, as well as the possible clinical value of TMS, illustrate that the remote effects of TMS may turn out to be an asset rather than a limitation for its use. One can now obtain a more complete picture of the induced activity changes throughout the brain, and use TMS–fMRI to systematically chart how these are related to behavioral, cognitive, and clinical changes.

Acknowledgments

We thank Jürgen Baudewig, Chris Chambers, Tomáš Paus, John Rothwell, Alexander Sack, and Nikolaus Weiskopf for helpful comments and ideas. The support of the Wellcome Trust and Medical Research Council (UK) is gratefully acknowledged.

References

Baudewig J, Paulus W, Frahm J (2000). Artefacts caused by transcranial magnetic stimulation coils and EEG electrodes in T(2)*-weighted echo-planar imaging. *Magnetic Resonance Imaging 18*, 479–484.

Baudewig J, Siebner HR, Bestmann S, *et al.* (2001). Functional MRI of cortical activations induced by transcranial magnetic stimulation (TMS). *Neuroreport 12*, 3543–3548.

Bestmann S, Baudewig J, Frahm J (2003a). On the synchronization of transcranial magnetic stimulation and functional echo-planar imaging. *Journal of Magnetic Resonance Imaging 17*, 309–316.

Bestmann S, Baudewig J, Siebner HR, Rothwell JC, Frahm J (2003b). Subthreshold high-frequency TMS of human primary motor cortex modulates interconnected frontal motor areas as detected by interleaved fMRI–TMS. *NeuroImage 20*, 1685–1696.

Bestmann S, Baudewig J, Siebner HR, Rothwell JC, Frahm J (2004). Functional MRI of the immediate impact of transcranial magnetic stimulation on cortical and subcortical motor circuits. *European Journal of Neuroscience 19*, 1950–1962.

Bestmann S, Baudewig J, Siebner HR, Rothwell JC, Frahm J (2005). BOLD MRI responses to repetitive TMS over human dorsal premotor cortex. *NeuroImage 28*, 22–29.

Bestmann S, Oliviero A, Voss M, *et al.* (2006). Concurrent TMS and fMRI in an amputee reveals cortical correlates of TMS-induced phantom hand movements. *Neuropsychologia 44*, 2959–2971.

Bohning DE, Pecheny AP, Epstein CM, *et al.* (1997). Mapping transcranial magnetic stimulation (TMS) fields in vivo with MRI. *Neuroreport 8*, 2535–2538.

Bohning DE, Shastri A, Nahas Z, *et al.* (1998). Echoplanar BOLD fMRI of brain activation induced by concurrent transcranial magnetic stimulation. *Investigations in Radiology 33*, 336–340.

Bohning DE, Shastri A, McConnell KA, *et al.* (1999). A combined TMS/fMRI study of intensity-dependent TMS over motor cortex. *Biological Psychiatry 45*, 385–394.

Bohning DE, Shastri A, McGavin L, *et al.* (2000a). Motor cortex brain activity induced by 1-Hz transcranial magnetic stimulation is similar in location and level to that for volitional movement. *Investigations in Radiology 35*, 676–683.

Bohning DE, Shastri A, Wassermann EM, *et al.* (2000b). BOLD-f MRI response to single-pulse transcranial magnetic stimulation (TMS). *Journal of Magnetic Resonance Imaging 11*, 569–574.

Bohning DE, Denslow S, Bohning PA, Walker JA, George MS (2003a). A TMS coil positioning/holding system for MR image-guided TMS interleaved with fMRI. *Clinical Neurophysiology 114*, 2210–2219.

Bohning DE, Shastri A, Lomarev MP, Lorberbaum JP, Nahas Z, George MS (2003b). BOLD-fMRI response vs. transcranial magnetic stimulation (TMS) pulse-train length: testing for linearity. *Journal of Magnetic Resonance Imaging 17*, 279–290.

Cantello R (2002). Applications of transcranial magnetic stimulation in movement disorders. *Journal of Clinical Neurophysiology 19*, 272–293.

Chouinard P A, Van Der Werf Y D, Leonard G, Paus T (2003). Modulating neural networks with transcranial magnetic stimulation applied over the dorsal premotor and primary motor cortices. *Journal of Neurophysiology 90*, 1071–1083.

Civardi C, Cantello R, Asselman P, Rothwell JC (2001). Transcranial magnetic stimulation can be used to test connections to primary motor areas from frontal and medial cortex in humans. *Neuroimage 14*, 1444–1453.

Corbetta M, Kincade MJ, Lewis C, Snyder AZ, Sapir A (2005). Neural basis and recovery of spatial attention deficits in spatial neglect. *Nature Neuroscience 8*, 1603–1610.

Curra A, Modugno N, Inghilleri M, Manfredi M, Hallett M, Berardelli A (2002). Transcranial magnetic stimulation techniques in clinical investigation. *Neurology 59*, 1851–1859.

Dechent P, Frahm J (2003). Functional somatotopy of finger representations in human primary motor cortex. *Human Brain Mapping 18*, 272–283.

de Graaf RA (1998). *In vivo NMR spectroscopy: principles and techniques.* Chichester: Wiley.

Denslow S, Bohning DE, Bohning PA, Lomarev MP, George MS (2005a). An increased precision comparison of TMS-induced motor cortex BOLD fMRI response for image-guided versus function-guided coil placement. *Cognitive and Behavioral Neurology 18*, 119–126.

Denslow S, Lomarev M, George MS, Bohning DE (2005b). Cortical and subcortical brain effects of transcranial magnetic stimulation (TMS)-induced movement: an interleaved TMS/functional magnetic resonance imaging study. *Biological Psychiatry 57*, 752–760.

Di Lazzaro V, Oliviero A, Pilato F, *et al.* (2003). Corticospinal volleys evoked by transcranial stimulation of the brain in conscious humans. *Neurological Research 25*, 143–150.

Di Lazzaro V, Oliviero A, Pilato F, *et al.* (2004). The physiological basis of transcranial motor cortex stimulation in conscious humans. *Clinical Neurophysiology 115*, 255–266.

Fadiga L, Fogassi L, Pavesi G, Rizzolatti G (1995). Motor facilitation during action observation: a magnetic stimulation study. *Journal of Neurophysiology 73*, 2608–2611.

Fadiga L, Buccino G, Craighero L, Fogassi L, Gallese V, Pavesi G (1999). Corticospinal excitability is specifically modulated by motor imagery: a magnetic stimulation study. *Neuropsychologia 37*, 147–158.

Ferbert A, Priori A, Rothwell JC, Day BL, Colebatch JG, Marsden CD (1992). Interhemispheric inhibition of the human motor cortex. *Journal of Physiology 453*, 525–546.

Fujiwara T, Rothwell JC (2004). The after effects of motor cortex rTMS depend on the state of contraction when rTMS is applied. *Clinical Neurophysiology 115*, 1514–1518.

Gilio F, Rizzo V, Siebner HR, Rothwell JC (2003). Effects on the right motor hand-area excitability produced by low-frequency rTMS over human contralateral homologous cortex. *Journal of Physiology 551*, 563–573.

Haacke EM, Lai S, Yablonskiy, DA, Lin WL (1995). In vivo validation of the BOLD mechanism – a review of signal changes in gardient echo functional MRI in the presence of flow. *International Journal of Imaging Systems and Technology 6*, 153–163.

Kastner S, Ungerleider LG (2000). Mechanisms of visual attention in the human cortex. *Annual Review of Neuroscience 23*, 315–341.

Kastrup A, Kruger G, Neumann-Haefelin T, Glover GH, Moseley ME (2002). Changes of cerebral blood flow, oxygenation, and oxidative metabolism during graded motor activation. *NeuroImage 15*, 74–82.

Kemna LJ, Gembris D (2003). Repetitive transcranial magnetic stimulation induces different responses in different cortical areas: a functional magnetic resonance study in humans. *Neuroscience Letters 336*, 85–88.

Kimbrell TA, Dunn RT, George MS, *et al.* (2002). Left prefrontal-repetitive transcranial magnetic stimulation (rTMS) and regional cerebral glucose metabolism in normal volunteers. *Psychiatry Research 115*, 101–113.

Kozel FA, George MS (2002). Meta-analysis of left prefrontal repetitive transcranial magnetic stimulation (rTMS) to treat depression. *Journal of Psychiatric Practice 8*, 270–275.

Kujirai T, Caramia MD, Rothwell JC *et al.* (1993). Corticocortical inhibition in human motor cortex. *Journal of Physiology 471*, 501–519.

Laurienti PJ, Burdette JH, Wallace MT, Yen YF, Field AS, Stein BE (2002). Deactivation of sensory-specific cortex by cross-modal stimuli. *Journal of Cognitive Neuroscience 14*, 420–429.

Lauritzen M, Gold L (2003). Brain function and neurophysiological correlates of signals used in functional neuroimaging. *Journal of Neuroscience 23*, 3972–3980.

Lee L, Siebner HR, Bestmann S (2006). Rapid modulation of distributed brain activity by transcranial magnetic stimulation of human motor cortex. *Behavioral Neurology 17*, 135–148.

Li X, Nahas Z, Kozel FA, Anderson B, Bohning DE, George MS (2004a). Acute left prefrontal transcranial magnetic stimulation in depressed patients is associated with immediately increased activity in prefrontal cortical as well as subcortical regions. *Biological Psychiatry 55*, 882–890.

Li X, Teneback CC, Nahas Z, *et al.* (2004b). Interleaved transcranial magnetic stimulation/functional MRI confirms that lamotrigine inhibits cortical excitability in healthy young men. *Neuro-psychopharmacology 29*, 1395–1407.

Lisanby SH, Morales O, Payne N, *et al.* (2003). New developments in electroconvulsive therapy and magnetic seizure therapy. *CNS Spectroscopy 8*, 529–536.

Macaluso E, Driver J (2005). Multisensory spatial interactions: a window onto functional integration in the human brain. *Trends in Neurosciences 28*, 264–271.

Mazzocchio R, Rothwell JC, Day BL, Thompson PD (1994). Effect of tonic voluntary activity on the excitability of human motor cortex. *Journal of Physiology 474*, 261–267.

Moore T, Armstrong KM (2003). Selective gating of visual signals by microstimulation of frontal cortex. *Nature 421*, 370–373.

Moore T, Armstrong KM, Fallah M (2003). Visuomotor origins of covert spatial attention. *Neuron 40*, 671–683.

Mumford D (1992). On the computational architecture of the neocortex. II. The role of cortico-cortical loops. *Biological Cybernetics 66*, 241–251.

Munchau A, Bloem BR, Irlbacher K, Trimble MR, Rothwell JC (2002). Functional connectivity of human premotor and motor cortex explored with repetitive transcranial magnetic stimulation. *Journal of Neuroscience 22*, 554–561.

Nahas Z, Lomarev M, Roberts DR, *et al.* (2001). Unilateral left prefrontal transcranial magnetic stimulation (TMS) produces intensity-dependent bilateral effects as measured by interleaved BOLD fMRI. *Biological Psychiatry 50*, 712–720.

Nikouline V, Ruohonen J, Ilmoniemi RJ (1999). The role of the coil click in TMS assessed with simultaneous EEG. *Clinical Neurophysiology 110*, 1325–1328.

Okabe S, Hanajima R, Ohnishi T, *et al.* (2003). Functional connectivity revealed by single-photon emission computed tomography (SPECT) during repetitive transcranial magnetic stimulation (rTMS) of the motor cortex. *Clinical Neurophysiology 114*, 450–457.

Padberg F, Moller HJ (2003). Repetitive transcranial magnetic stimulation: does it have potential in the treatment of depression? *CNS Drugs 17*, 383–403.

Paus T (1999). Imaging the brain before, during, and after transcranial magnetic stimulation. *Neuropsychologia 37*, 219–224.

Paus T (2000). Functional anatomy of arousal and attention systems in the human brain. *Progress in Brain Research 126*, 65–77.

Paus T (2002). Combination of transcranial magnetic stimulation with brain imaging. In: J Mazziotta, A Toga (eds), *Brain mapping: the methods*, 2nd edn, pp. 691–705. New York: Academic Press.

Paus T, Barrett J (2004). Transcranial magnetic stimulation (TMS) of the human frontal cortex: implications for repetitive TMS treatment of depression. *Journal of Psychiatry and Neuroscience 29*, 268–279.

Paus T, Jech R, Thompson CJ, Comeau R, Peters T, Evans AC (1997). Transcranial magnetic stimulation during positron emission tomography: a new method for studying connectivity of the human cerebral cortex. *Journal of Neuroscience 17*, 3178–3184.

Paus T, Jech R, Thompson CJ, Comeau R, Peters T, Evans AC (1998). Dose-dependent reduction of cerebral blood flow during rapid-rate transcranial magnetic stimulation of the human sensorimotor cortex. *Journal of Neurophysiology 79*, 1102–1107.

Ridding MC, Taylor JL, Rothwell JC (1995). The effect of voluntary contraction on cortico-cortical inhibition in human motor cortex. *Journal of Physiology 487*, 541–548.

Robertson EM, Theoret H, Pascual-Leone A (2003). Studies in cognition: the problems solved and created by transcranial magnetic stimulation. *Journal of Cognitive Neuroscience 15*, 948–960.

Rizzo V, Siebner HR, Modugno N, *et al.* (2004). Shaping the excitability of human motor cortex with premotor rTMS. *Journal of Physiology 554*, 483–495.

Ruff CC, Blankenburg F, Bioertomt O, *et al.* (2006). Concurrent TMS–fMRI and psychophysics reveal frontal influences on human retinotopic visual cortex. *Current Biology 16*, 1479–1488.

Sack AT, Sperling JM, Prvulovic D, *et al.* (2002). Tracking the mind's image in the brain II: transcranial magnetic stimulation reveals parietal asymmetry in visuospatial imagery. *Neuron 35*, 195–204.

Sack AT, Camprodon JA, Pascual-Leone A, Goebel R (2005). The dynamics of interhemispheric compensatory processes in mental imagery. *Science 308*, 702–704.

Sack AT, Kohler A, Bestmann S, *et al.* (2007). Imaging the brain activity changes underlying impaired visuospatial judgements: simultaneous fMRI, TMS, and behavioral studies. *Cerebral Cortex*, Epub.

Shastri A, George MS, Bohning DE (1999). Performance of a system for interleaving transcranial magnetic stimulation with steady-state magnetic resonance imaging. *Electroencephalography and Clinical Neurophysiology Supplement 51*, 55–64.

Siebner HR, Peller M, Willoch F, *et al.* (2000). Lasting cortical activation after repetitive TMS of the motor cortex: a glucose metabolic study. *Neurology 54*, 956–963.

Siebner HR, Takano B, Peinemann A, Schwaiger M, Conrad B, Drzezga A (2001). Continuous transcranial magnetic stimulation during positron emission tomography: a suitable tool for imaging regional excitability of the human cortex. *NeuroImage 14*, 883–890.

Siebner HR, Filipovic SR, Rowe JB, *et al.* (2003a). Patients with focal arm dystonia have increased sensitivity to slow-frequency repetitive TMS of the dorsal premotor cortex. *Brain 126*, 2710–2715.

Siebner HR, Lee L, Bestmann S (2003b). Interleaving TMS with functional MRI: now that it is technically feasible how should it be used? *Clinical Neurophysiology 114*, 1997–1999.

Speer AM, Willis MW, Herscovitch P, *et al.* (2003a). Intensity-dependent regional cerebral blood flow during 1-Hz repetitive transcranial magnetic stimulation (rTMS) in healthy volunteers studied with $H_2^{15}O$ positron emission tomography. I. Effects of primary motor cortex rTMS. *Biological Psychiatry 54*, 818–825.

Speer AM, Willis MW, Herscovitch P, *et al.* (2003b). Intensity-dependent regional cerebral blood flow during 1-Hz repetitive transcranial magnetic stimulation (rTMS) in healthy volunteers studied with $H_2^{15}O$ positron emission tomography. II. Effects of prefrontal cortex rTMS. *Biological Psychiatry 54*, 826–832.

Stark DD, Bradley W (1999). *Magnetic resonance imaging.* St Louis, MO: Mosby.

Stinear CM, Byblow WD (2004). Modulation of corticospinal excitability and intracortical inhibition during motor imagery is task-dependent. *Experimental Brain Research 157*, 351–358.

Strafella AP, Paus T (2000). Modulation of cortical excitability during action observation: a transcranial magnetic stimulation study. *Neuroreport 11*, 2289–2292.

Strafella AP, Paus T, Barrett J, Dagher A (2000). Repetitive transcranial magnetic stimulation of the human prefrontal cortex induces dopamine release in the caudate nucleus. *Journal of Neuroscience 21*, RC157.

Strafella AP, Paus T, Fraraccio M, Dagher A (2003). Striatal dopamine release induced by repetitive transcranial magnetic stimulation of the human motor cortex. *Brain 126*, 2609–2615.

Strens LH, Oliviero A, Bloem BR, Gerschlager W, Rothwell JC, Brown P (2002). The effects of subthreshold 1 Hz repetitive TMS on cortico-cortical and interhemispheric coherence. *Clinical Neurophysiology 113*, 1279–1285.

Ugawa Y, Day BL, Rothwell JC, Thompson PD, Merton PA, Marsden CD (1991). Modulation of motor cortical excitability by electrical stimulation over the cerebellum in man. *Journal of Physiology 441*, 57–72.

Ziemann U, Rothwell JC (2000). I-waves in motor cortex. *Journal of Clinical Neurophysiology 17*, 397–405.

CHAPTER 37

TMS and electroencephalography: methods and current advances

Risto J. Ilmoniemi and Jari Karhu

Introduction

Electroencephalography (EEG) combined with TMS provides detailed real-time information about the state of the cortex.[1] The measurement of the neuronal electrical activity elicited by TMS is a new modality for functional brain mapping. Many regions of the cortical mantle can be stimulated; in addition to the state of the stimulated area, the response informs us about the functional connectivity to other regions as well as about their state. By varying the TMS intensities, interstimulus intervals, induced current direction, and cortical targets, a rich spectrum of functional information can be obtained.

The new modality of TMS–EEG is straightforward and powerful; it allows one directly and noninvasively to:

- measure and map neuronal reactivity and the state of the cortex

- monitor how brain oscillatory activity is modulated by targeted stimulation
- measure functional connectivity between brain areas
- monitor the effects of repetitive (r)TMS during and after treatment
- monitor rTMS safety and alert user if epileptiform activity appears in the EEG.

In its simplest applications, EEG requires only two to four electrodes and can be a part of most TMS studies. In advanced connectivity and modulatory approaches where the whole brain should be monitored with high resolution, an array of 50–100 channels may be needed. In all cases, proper electrode and amplifier technology, careful experimental procedures, and suitable signal analysis methods must be used to eliminate the effect of the strong TMS pulse that may otherwise cause large readily visible disturbances in the EEG. Even more care is needed to avoid the subtle artifacts that are not visible in the unaveraged EEG but may cause misinterpretations of averaged evoked responses.

When used with magnetic resonance imaging (MRI) based targeting and conductor modelling, the TMS–EEG combination is a sophisticated

---

[1] By the state of cortex, we mean the configuration of cells, their membrane potentials, and the distribution of chemicals in both intra- and extracellular space. Any reaction of a patch of cortex is uniquely determined by its instantaneous state and by input from other neuronal areas or externally applied stimuli such as TMS.

brain-mapping tool. We can, for example, characterize multiple aspects of the local reactivity of the cortex and simultaneously determine how one area affects others. The precise targeting of TMS can improve the spatial resolution and reliability of EEG maps because the site of initial activation is accurately known.

A further unique strength of TMS-evoked responses is that they provide direct information about the timing of signals transmitted from the stimulated site to other areas. When this information is combined with modelling of neuronal sources, we obtain a measure of time-resolved functional connectivity and a method of studying the causality of activation in the observed neuronal network (Paus 2005).

History

The first TMS-evoked EEG measurements were performed by Cracco et al. (1989). They stimulated one hemisphere with a figure-8 coil in four subjects and observed transcallosal responses with an onset latency of 8.8–12.2 ms. These measurements were difficult because of the large artifacts produced by the strong TMS pulses. Somewhat later, the same group reported EEG responses to cerebellar TMS as recorded with scalp electrodes placed over the frontal and parietal regions (Amassian et al. 1992). From electrodes on the inter-aural line (for example, C3, Cz, C4), they observed onset latencies of 8.8–13.8 ms, while more frontal leads showed activity up to 3.5 ms later. The artifact in the latter study was reduced by differential measurement from neighboring electrodes and by a 20 cm^2 flexible metal strip that grounded the scalp between the coil and the electrodes.

The pioneering studies of Cracco, Amassian, and co-workers showed both the potential of EEG-evoked TMS studies and the need for new methods to overcome the large TMS-generated artifacts. The early measurements were limited to a few channels that were suitably oriented and distant with respect to the stimulation coil. Still, large artifacts were inevitable. It took almost a decade after the first studies to establish technologies that provide reliable and routine EEG measurements concurrently with TMS.

Techniques for recording TMS-evoked EEG

The main problem with TMS–EEG recording is the large electric field induced by TMS in the electrode leads. The voltage induced is $V = -d\Phi/dt$ where Φ is the magnetic flux threading the loop between active and reference leads, including the part of the loop that goes via the head. If the area enclosed by such loops is 10 cm^2 and if the average magnetic flux density threading the loop is 1 T and the rise time of the field 100 µs, the induced voltage is 10 V, about seven orders of magnitude larger than the signal features of ~1 µV we need to measure.

A more subtle and poorly understood problem is caused by electrode polarization. Even when the EEG amplifier input impedance is high, some current will pass via the electrode–electrolyte interface during the pulse, causing polarization that can take tens or hundreds of milliseconds to return to the baseline level.

If traditional large electrodes are used, strong currents are induced in the electrode material. These currents interact with the magnetic field, causing a force and thereby movement of the electrodes. It is well known that electrode movement can cause appreciable artifacts (see Virtanen et al. 1999).

Finally, electrode heating can be an issue, especially with large electrodes and when a large number of pulses are delivered. In extreme cases, the heating can cause pain or it may even burn the skin. The changing temperature might also produce artifacts in EEG, although it appears that these have neither been reported nor analyzed.

It should be pointed out that even properly designed EEG will detect any voltage that appears between the electrodes. Therefore, one must make sure that the measurements are performed in a location that is far away from strong electromagnetic disturbances such as power lines, large electric motors, etc. If such a space is not available, one may have to build an electrically shielded room (Faraday cage) to perform the studies.

In the following, we concentrate on TMS-specific requirements and describe methods that deal with the above-mentioned problems.

Electrodes and preparations for study

The purpose of the electrode is to form a good electrical contact from the skin to the amplifier input. The resistance of the electrode contact should be low compared to the input impedance of the amplifier and sufficiently low to avoid thermal noise in the contact resistance. Low contact resistance also improves the stability of recordings.

Magnetic stimulation poses additional requirements for electrodes. Roth *et al.* (1992) investigated the effect of eddy currents induced by the changing magnetic field. They found experimentally that heating is proportional to the square of TMS intensity and the square of electrode diameter, but independent of thickness. This is indeed what simple theoretical arguments will predict. In addition to heating, induced currents in the electrode give rise to a force which is proportional to the thickness of the electrode, the cube of its diameter, and the square of TMS intensity. Both inductive heating and forces are proportional to the conductivity of the electrode material.

Virtanen *et al.* (1999) found that a slit in an electrode annulus reduces heating by an order of magnitude. They also reported that it reduces the DC shift caused by TMS by an order of magnitude. With small Ag/AgCl pellet electrodes, this artifact nearly vanished; Ag/AgCl pellet electrodes are now used in the commercial TMS–EEG system that is based on the original design of Virtanen *et al.* (1999).

Ives *et al.* (2006) have developed conductive plastic electrodes in order to reduce eddy currents. However, high-quality EEG was not obtained until the plastic was covered by silver epoxy to create an Ag–AgCl surface.

In addition to the requirement that the electrodes have a suitable surface material and are either small or have low conductivity, their contact with the scalp is critical for high-quality TMS–EEG recordings. This requires careful preparation of the skin surface before making the electrode contacts. Usually wiping the scalp with alcohol and slight scraping of the skin directly underneath the recording electrode with electrode paste is sufficient.

Checking impedances of each individual electrode is necessary before each EEG recording session, and, if the recording takes longer than 1.5–2 h, within the session as well. Modern EEG systems usually have an automatic impedance checking system, which informs about problematic electrodes and allows quick correction of the contact without removing the cap either by re-cleaning the skin or by adding electrode paste. To reach the optimal quality of TMS-evoked EEG signals, it is useful to remember that the absolute value of individual impedances is less important than the balance in the array, i.e. similar impedances across the recording electrodes.

EEG amplifiers

A number of amplifiers have been developed to deal with electric or magnetic artifacts. Most of the 'TMS-compatible' amplifiers do not eliminate the artifact of the TMS pulse; instead, they remain operational and start producing clean EEG later, some 30–1000 ms after the pulse (Izumi *et al.* 1997; Iramina *et al.* 2002; Fuggetta *et al.* 2005). There are several methods that can help obtain high-quality EEG continuously during magnetic stimulation as will be explained below.

Virtanen *et al.* (1999) developed a 60-channel TMS-compatible EEG system comprising gain-control and sample-and-hold circuits to block the artifacts from induced voltages in the leads. The pre-amplifier was battery operated to avoid artifact coupling via electric leads; the signals were transmitted from the preamplifier to a light receiver via optical fibers.

There are two amplifier stages and two sample-and-hold circuits (Figure 37.1). The input to amplifier A1 is limited to ±9 V and the gain is reduced to unity for the duration of the pulse so as to keep A1 in the linear range during the pulse. A semiconductor switch (SW) is used to block the signal path to A2 during the pulse; the charge in capacitor C1 stays constant, but the input to A2 drops to zero. The sample-and-hold circuits S/H(A) and S/H(B), on the other hand, keep the voltage at input of filter FLT at a constant level during the pulse. Two S/H circuits with different timings are used to reduce the effect of the residual artifact passing S/H(A).

Although Virtanen *et al.* (1999) concluded that the new amplifier allows one to record artifact-free continuous EEG in the presence of TMS pulses, detectable artifacts remained,

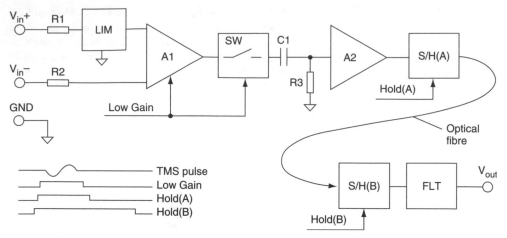

Fig. 37.1 Block diagram of the amplifier designed by Virtanen. Reproduced with permission from Virtanen *et al.* (1998).

especially in channels with electrodes directly under the coil. The artifact depended very much on the type of electrode, probably because of polarization or as the result of movement of the electrode; when an Ag/AgCl pellet electrode was used, the artifact was <1 μV. The basic design of Virtanen *et al.* has subsequently been developed and the remaining artifacts reduced. Examples of results are described later in this chapter.

An alternative way to deal with the TMS artifact has been used by Ives and co-workers (Ives *et al.* 1998, 2006; Thut *et al.* 2003a,b, 2005). They have developed a slew-rate-limited amplifier that prevents the electronics from saturating during the TMS pulse; the artifact is over within 30 ms in the worst case. The advantage of the system is its simplicity and that it can be inserted between any electrode and any commercial EEG system. On the other hand, the bandwidth is limited to <90 Hz; however, this is sufficient for most applications. When the effect of TMS on visually evoked potentials was studied, the remaining TMS-related contamination (physiological or instrumental) was eliminated by subtracting the averaged response to TMS presentation from the combined TMS–VEP trace.

Artifacts and how to deal with them

Although it is possible to record EEG without an artifact from the electromagnetic pulse itself, various other unwanted signals may contaminate

the responses. We have already discussed polarization of the electrolyte–electrode interface. Even if the problem appears to be solved, e.g. by using small Ag/AgCl pellet electrodes, subtle drifts with different time constants may cause artifacts that are difficult to distinguish from the physiological response.

Muscle artifacts

The electric field from a TMS coil is always stronger in the scalp than in the brain. Therefore, if the coil is held over excitable tissue such as muscle or nerves, these may be activated and an artifact in the EEG is created. The artifact is caused by the electrical activity, direct depolarization of the muscle itself, or scalp movements reflected in electrode contacts (Paus *et al.* 2001).

Muscle artifacts are most prominent when lateral aspects of the head or areas near the neck or forehead are stimulated. Temporal and frontal muscles, and, in some stimulating positions, masseter muscles are the most likely to be activated. These artifacts can be strong and may last up to 30 ms or more. The problem can be reduced by moving the coil to a more favorable location or orientation; reducing the TMS amplitude sometimes helps. TMS-evoked EEG activity does not always require suprathreshold stimulation (Komssi *et al.* 2004, 2007; Kähkönen *et al.* 2005). Even relatively minor reduction of the stimulation amplitude (on the order of 10% of individual motor threshold, MT) diminishes muscle artifacts significantly.

Eye-movement artifacts and blinks

TMS can cause eye movements or blinks by direct nerve stimulation, by activating brain circuits that control eye movements, or by a startle effect resulting from the coil click or scalp sensation. Additionally, subjects move their eyes and blink spontaneously. For high-quality recordings, it is advisable to use electrodes that measure both vertical and horizontal eye movements (e.g. Amassian *et al.* 1992). The simplest way to deal with the problem is then to discard data that overlap in time with eye movements or blinks.

Coil click and somatic sensation

The loud click (up to 120 dB) from the TMS coil obviously activates the subject's auditory system and gives rise to an auditory-evoked potential. This response, with maxima over central and parietal regions, may confound the TMS-evoked EEG and cause false interpretations (Nikouline *et al.* 1999; Tiitinen *et al.* 1999). Sometimes good hearing protection may be sufficient to deal with the coil click, but one has to be aware of the fact that a large part of the effect may be due to bone-conducted sound (Nikouline *et al.* 1999).

More complete elimination of the auditory click artifact can be obtained by masking the sound. Masking is preferably used in addition to hearing protection. Fuggetta *et al.* (2005) as well as Paus *et al.* (2001) played a 90 dB noise to subjects through insert earphones. Massimini *et al.* (2005) used a colored masking sound that had a spectrum similar to that of the coil click. These authors reported sufficient elimination of the click artifacts.

Since TMS also gives rise to sensory stimulation on the scalp; a somatosensory brain response can cause an artifact. However, this has been found to be either small or absent (Nikouline *et al.* 1999; Paus *et al.* 2001).

Mechanisms of TMS-evoked EEG generation

How does TMS evoke EEG activity?

When the stimulating pulse enters the brain, there is a complex interaction between the induced electric field and neuronal tissue.

The waveform and strength of the magnetic pulse, orientation and position of the coil relative to the head, properties of the intervening tissue, and distance between the coil and target neuronal tissue all have their effect on the locally induced electric field. As a first approximation, the effect of TMS is strongest where the induced electric field is strongest, i.e. in the superficial parts of the brain. A refinement to this approximation can be made if one takes into account the direction of the electric field with respect to the cortex. It has been observed that the TMS-induced electric field is most effective when the field is perpendicular to the cortex and directed into it, i.e. from the dendrites toward the soma of pyramidal cells (Fox *et al.* 2004).

EEG activity at low TMS intensities (typically, below the MT) probably has a different distribution than that at high intensities. Suprathreshold stimulation may lead to a combined activation of trans-synaptic pathways and the direct stimulation of the axonal pathways deeper in the gray matter or in the bending/tapering white matter structures in the border of the gray and white matter. With sufficiently strong intensities, it is possible to reach some subcortical structures such as the cerebellum. However, it is important to remember that with strong stimulation intensities the activating electric field remains strongest in the superficial cortex. Of particular importance for the genesis of EEG is the fact that both inhibitory and excitatory neurons are activated, most probably in different proportions depending on stimulus intensity and direction of induced current.

It is of interest to consider what type of responses would be obtained if the brain reacted to TMS in a linear fashion, with synaptically generated (primary) current densities directly proportional to the local induced electric field and with the same constant of proportionality everywhere. The induced field distribution is proportional to the lead field of the coil (Ilmoniemi *et al.* 1996). Because magnetic and electric lead fields in a spherically symmetric conductor or homogeneous conductor are orthogonal, linear and spherically symmetric brains would not generate any EEG in response to TMS. Thus, we may conclude that any TMS-evoked EEG responses reflect the brain's nonuniform structure and nonlinearity.

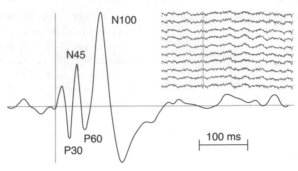

Fig. 37.2 A typical TMS-evoked response (TER) to the stimulation of left hand motor area in M1. The effect of a single pulse (vertical line in the insert) on spontaneous 'raw' EEG signal is practically invisible with dedicated EEG amplifiers.

What does TMS-evoked EEG reflect?

A typical scalp-recorded averaged TMS-evoked EEG signal is shown in Figure 37.2. It consists of several deflections, first in the form of rapid oscillations and then as lower-frequency waves. The responses depend on the instantaneous state of the cortex (Nikulin *et al.* 2003; Massimini *et al.* 2005) as well as on the exact location of the coil (Komssi *et al.* 2002). In non-REM (rapid eye movement) deep sleep state, Massimini *et al.* (2005) found that the EEG response is strong, but consists of fewer deflections than in the awake state (Figure 37.5). Nikulin *et al.* (2003), on the other hand, showed that the initiation of movement affects the excitability of the motor cortex, especially its N100 response.

The effect of a single pulse on spontaneous, 'raw' EEG signal is practically invisible even when the recording is technically immaculate (Figure 37.2, insert). As shown later in this chapter as well as in several articles (Paus *et al.* 2001; Fuggetta *et al.* 2005), the effects of TMS on spontaneous oscillatory EEG activity can be readily found and visualized by a multitude of analysis methods.

When the TMS-induced current causes sufficient membrane depolarization, action potentials are generated, which lead to synaptic transmission and excitatory or inhibitory postsynaptic potentials. It is believed that EEG is mainly generated by postsynaptic potentials instead of action potentials. The early part of the EEG response (up to 10 ms, cf. Figure 37.2) reflects the reactivity or functional state of the stimulated patch of cortex, while the later part also informs about functional connectivity from the stimulated area to other regions of the brain.

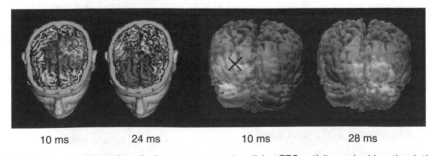

Fig. 37.3 *Left*: Spatiotemporal cortical current maps visualizing EEG activity evoked by stimulation of left sensorimotor hand area. First (10 ms after the stimulus) the electric activation is centered on the area of the stimulation, whereas later (at 24 ms) activation spreads to the contralateral hemisphere sensorimotor areas. *Right*: Ipsilateral (10 ms) and contralateral activation (28 ms) evoked by stimulation of occipital area. (Plate 20)

It should also be emphasized that functional connectivity is highly dependent on the global state of the brain (Kähkönen *et al.* 2001; Massimini *et al.* 2005).

As stated before, there is a complex interaction between the TMS-induced electric field and neuronal tissue. Likewise, in TMS-evoked EEG, neuronal excitation can spread via intra- and interhemispheric association fibers to other cortical areas and across projection fibers to deeper subcortical structures. A correct interpretation of TMS-evoked EEG must take into consideration the anatomical substrates for spreading of neuronal excitation. The recently introduced MRI technique, diffusion tensor imaging (DTI), allows one to perform tractography, which provides three-dimensional images of anatomical connections between brain areas. Integrating such anatomical information with TMS–EEG measurements provides direct information about individual anatomical CNS connections, which can be probed by targeted stimulation of one end of the tract and interrogation of EEG responses at the other end of multiple terminations.

What are the types of TMS-evoked activity?

TMS-evoked responses

In a pioneering set of experiments, motor and visual cortices of volunteers were stimulated (Ilmoniemi *et al.* 1997) (Figure 37.3). Stimulation of the left sensorimotor hand area elicited an immediate strong response at the stimulated site. Activation spread to adjacent ipsilateral motor areas within 5–10 ms and to homologous regions in the opposite hemisphere within 20 ms.

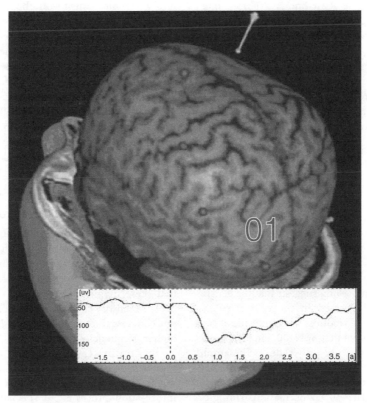

Fig. 37.4 After contralateral frontal stimulation at 100–120% of the motor threshold to nonmotor area (depression treatment area), EEG shows a clear stimulus-locked synchronization of activity around 10 Hz for about 1 s (wavelet analysis of O1-electrode EEG, centered at 10 Hz). (Plate 21)

Similar activation patterns were generated by magnetic stimulation of the visual cortex: after the immediate ipsilateral response, the contralateral response was observed at ~20 ms.

To locate the activity underlying the observed signals, various source-modelling techniques can be used. When only one area of the cortex is active, which may be a reasonable approximation or assumption in some cases, dipole modelling is appropriate. In more complicated situations, when the activation induced by TMS spreads in the stimulated neuronal network, continuous current estimates are more reliable in indicating the active areas. As in our example (Figure 37.3), minimum-norm estimation (Hämäläinen and Ilmoniemi 1994) can be used to observe the evolution of TMS-evoked brain activity, although original data and potential maps must always be carefully studied in order to evaluate the quality of the data and to detect possible artifacts.

Currently, the best-known TMS-evoked waveform elicited by single-pulse TMS of primary motor cortex (M1) consists of a fairly consistent series of scalp potentials, with positive waves peaking at around 30 and 60 ms and two negative waves peaking at 45 and 100 ms, respectively (Paus *et al.* 2001). A similar P30/N45/P60/N100 waveform has been observed by Tiitinen *et al.* (1999) during suprathreshold TMS delivered with a circular coil positioned over the vertex and, with slightly different latencies but a quite similar pattern, by Massimini *et al.* (2005) and Van Der Werf and Paus (2006) after stimulation of the premotor cortex.

Despite a similar pattern, these potentials do not seem to reflect a single process. For example, several characteristics of the N45 component distinguish it from the other components. The N45 amplitude correlates with intensity of TMS; dipole modelling suggests that the generator of N45 is located in the central sulcus; paired-pulse TMS reduces the amplitude of N45 (and P30) but not that of N100. Finally, paired-pulse stimulation not only reduces the amplitude of the P30 and N45 components but also decreases the amplitude of the time-locked TMS-induced 15–30 Hz oscillations riding on these averaged potentials (cf. 'TMS effects on synchronous oscillatory EEG activity at alpha and beta frequencies' below). Taken together,

it appears that different mechanisms may underlie the generation of each of these potentials (Paus *et al.* 2001).

Recently, Bender *et al.* (2005) observed that the N100 evoked by TMS at 105% MT exceeded 100 μV in 6–10-year-old children and that the response amplitude correlated positively with stimulus intensity and negatively with age. However, this can be partly explained by the fact that MT is higher in children than in adults; consequently, younger subjects were given stronger TMS pulses to reach MT. There was potentiation of the N100 amplitude during recording sessions: the average response amplitude was 125 μV during the first 10 trials and 139 μV during the second half of the trials. When the motor cortex was preactivated in a contingent negative variation paradigm, N100 was significantly suppressed.

This behavior of N100 supports the conclusion of Nikulin *et al.* (2003) that N100 reflects inhibitory mechanisms in the cortex. These authors proposed that the TMS-evoked N100 reflects inhibition and could be used to map the excitation state of the cortex. When measured from the motor cortex, N100 was found to be reduced markedly when the subject was about to perform a visually timed finger movement. However, TMS-evoked N100 reflects several overlapping processes in the brain, including a contribution from the coil click unless proper hearing protection and/or masking sounds are used.

TMS effects on synchronous oscillatory EEG activity at alpha and beta frequencies

The most common interpretation of alpha frequencies (8–13 Hz) in EEG/magneto-encephalography is that of idling state in 'normal' thalamo-cortical circuits, presuming that these frequencies represent the interplay between sensory cortices and subcortical (thalamic) circuits (Steriade and Llinas 1988). At the higher frequencies of beta (13–30 Hz) and gamma rhythm (30–60 Hz), the synchronization of neuronal activity within and across different cortical regions may reflect binding of information processed in specialized cortical modules even over long distances (Singer 1993).

A prominent change in the EEG activity is the blocking of oscillatory activity at alpha frequency, with a decrease in power, referred to as event-related desynchronization (ERD) (Pfurtscheller and Aranibar 1977, 1979; Pfurtscheller 1992). ERD has been associated with various sensory, motor and cognitive tasks (e.g. Klimesch 1996; Hari and Salmelin 1997; Pfurtscheller and Lopes da Silva 1999), is task and, to some extent, modality specific, and presumably reflects activation or anticipation breaking the 'idling state' in thalamo-cortical loops.

In sensorimotor cortex, in addition to the alpha frequency, the dominant frequency range is 15–30 Hz. The predominance of the beta rhythm over the precentral region was first reported by Jasper and Andrews (1938) with scalp EEG and subsequently confirmed with electrocortico-graphy (Jasper and Penfield 1949).

Paus *et al.* (2001) first documented that ongoing oscillatory EEG activity of sensorimotor cortex can be synchronized and modulated by magnetic stimulation delivered over the scalp. Single-pulse TMS induced a brief period of synchro-nized activity in the beta range (15–30 Hz) in the vicinity of the stimulation site. It was sug-gested that the TMS-induced oscillations likely reflected resetting of the cortical oscillators (Paus *et al.* 2001). The pulse may activate a population of 'idling' neurons that, owing to their membrane properties or intracortical con-nectivity, begin to oscillate, with the likelihood of activating these neurons being related to stimulus intensity.

In a follow-up study, low-frequency sub-threshold repetitive TMS was applied either over the primary motor cortex (local modulation) or over the dorsal premotor cortex (distal modula-tion) (Van Der Werf and Paus 2006). Modulation was evaluated with single suprathreshold test pulses of TMS applied over the primary motor cortex before and after the subthreshold low-frequency rTMS. Following rTMS applied over the primary motor cortex, but not the dorsal premotor cortex, the amplitude of the N45 in response to suprathreshold pulses tended to decrease (not significant), and subsequently increased (significant); neither type of repetitive TMS affected the amplitude of the beta oscilla-tion. The authors concluded that the beta oscil-lation is specific to stimulation of the primary motor cortex, but is not affected by modulation of either primary motor or premotor areas and

that the beta oscillatory response to pulses of TMS arises from resetting of ongoing oscilla-tions rather than their induction.

The same authors studied eight patients with Parkinson's disease who had undergone unilateral surgical lesioning of the ventrolateral nucleus of the thalamus and found that stimulation of the unoperated hemisphere (with thalamus) resulted in higher amplitudes of the single-trial-induced beta oscillations than in the operated hemisphere (with thalamotomy) (Van Der Werf *et al.* 2006). The beta oscillation obtained in response to pulses applied over the unoperated hemisphere was also higher than that obtained in healthy controls. The conclusion was that thalamotomy serves to reduce the abnormally high TMS-induced beta oscillations, and that the motor thalamus facilitates the cortically gen-erated oscillation through cortico-subcortico-cortical feedback loops. The concept of mod-ulating large-scale functional networks with stimulation and monitoring their oscillatory properties can be tested in all networks that produce strong enough synchronous activity to be measurable by scalp EEG.

TMS effects on synchronous oscillatory EEG activity in large-scale neuronal networks

Since the cortical motor areas operate in a network-like fashion (Gerloff *et al.* 1998), modulated oscillatory activity may reflect both the regional activity of individual areas and the degree of inter-regional communication (functional coupling). In EEG, event-related power thus reflects regional oscillatory activity of neural assemblies, while event-related coher-ence may reflect the functional coupling of oscillatory neural activity. For example, decreases in ERD and increases in coherence during motor task execution in humans are therefore parallel, representing both local and network properties of neuronal activation (Andrew and Pfurtscheller 1996; Classen *et al.* 1998; Gerloff *et al.* 1998; Manganotti *et al.* 1998).

It is tempting to suggest that the simple evalu-ation of oscillatory power would provide a measure of regional activation level of neuronal populations, and coherence/correlation meas-ures of EEG activity a measure of functional coupling of neuronal assemblies. However, it is unclear whether the reciprocal inhibitory/

excitatory effects across transcallosal or intra-hemispheric cortico-cortical connections are always the mechanism involved in synchronizing inter-regional coherence. An alternative explanation is provided by an enhancement of excitability of cortical pyramidal cells generating a more powerful and coherent feedback onto the thalamus that, as common pacemaker, synchronizes the cortical oscillations of thalamo-cortical pathways (Steriade and Amzica 1996; Destexhe *et al.* 1999). The 'third party' problem exists; there may be a node in the examined neuronal network that participates equally in the behavior of all neuronal assemblies showing regional coherence/correlation. Solving this problem requires the possibility of manipulating and simultaneously recording the behavior of other regions in the neuronal network. The combination of targeted excitatory/inhibitory pulses and simultaneous EEG provides this capacity and may open exciting new routes to investigation of causality and direct evaluation of effective connections in the human brain.

Indeed, in a study where primary motor cortex M1 was stimulated with several intensities adjusted according to the individual MT, sensorimotor oscillatory activity showed two distinct modes of behavior (Fuggetta *et al.* 2005). For the alpha rhythm, threshold TMS induced a small decrease in the amplitude of EEG oscillations over the stimulation site, while for both alpha and beta rhythms, a progressive synchronization was observed as the intensity of TMS was increased. The event-related coherence revealed that TMS enhanced the connectivity of both hemispheres, in particular within the first 500 ms following stimulation and only for the alpha rhythm. The increase in functional connectivity between cortical areas was minor for magnetic stimulation conditions compared with that for voluntary finger movements.

The concept can be extended to the stimulation of nonmotor areas. A similar synchronization of alpha range (~10 Hz) activity over occipital (visual) areas was found as a result of middle frontal gyrus stimulation, 5 cm anterior to the hand motor representation, an area commonly used in studies evaluating the use of rTMS for treatment of depression. The synchronous activation of neurons of both cortical and subcortical structures by artificial depolarization, which spreads through cortico-subcortico-cortical connections, is probably responsible for the short-lasting synchronization of the oscillatory activity. About 10 Hz spontaneous brain rhythms are mediated by thalamo-cortical neuronal loops and timed by discharges of pacemaking thalamic neurons; the intrinsic ion conductances of these neurons make them oscillate at 6–10 Hz (Steriade and Llinas 1988). The oscillating properties depend on the connectivity of different pacemakers and especially the modulating effect of the reticular system which is interconnected with all the thalamic nuclei (Steriade *et al.* 1990).

The state of vigilance and the amount of afferent input to oscillating networks, whether external or from intrinsic neuronal populations, can significantly change the oscillatory state of specific networks. It is well known from animal studies that stimulation of subcortical structures (e.g. reticular system) in mesencephalon and pons induces a regional synchronization of the cortical oscillatory activity (Moruzzi and Magoun 1949). Early human recordings showed that preparing for a movement or even imagining the movement, activating typically frontal neuronal populations, desynchronized and synchronized cortical 10 Hz activity in central and occipito-parietal areas (Gastaut *et al.* 1952; Chatrian *et al.* 1959). It is likely that this type of effect can also be seen after artificial depolarization of sufficiently large neuronal populations resulting in EEG changes in seemingly unconnected areas. The long-range EEG effects after magnetic stimulation may provide a substrate for exploring functional networks in the human brain.

In summary, the effects of stimulation intensity in all the cited studies (Paus *et al.* 2001; Fuggetta *et al.* 2005; Van Der Werf and Paus 2006; Van Der Werf *et al.* 2006) suggest the involvement of overlapping levels of modulation of oscillatory EEG activity. Low-intensity stimulation modifies the coupling of stimulated cortex to connected functional cortical areas through direct effects on cortical motor neurons and interneurons (cf. Kujirai *et al.* 1993; Ziemann *et al.* 1996a,b). High-intensity stimulation probably involves cortico-thalamic connections and subcortical structures, which may lead to distant EEG effects via indirect sub-cortical or transcallosal effective connections.

TMS-evoked EEG reflects local and global neuronal state

The TMS-evoked EEG response turns out to be a rich source of information about the state of the stimulated cortical area, and of the inter-connected local and large-scale networks, as well as the brain at large. The obvious fact is that nonfunctional or damaged areas do not respond well (Ilmoniemi et al. 1999) or that the res-ponses may be abnormal. Less obvious findings have also started to emerge.

Nikulin et al. (2003) discovered that the TMS-evoked N100 appears to reflect inhibition and could be used to map the excitation state of the cortex. The TMS-evoked motor-cortex N100 was markedly reduced when the subject was about to perform a finger movement. Komssi et al. (2004) deliberated on how the EEG amplitude depends on TMS intensity. They concluded that the measured TMS intensity–response amplitude curve could reveal significant information about the distribution of membrane potentials in the area under study.

The above findings immediately suggest that TMS–EEG allows one to directly probe the state of a cortical area during different experi-mental paradigms using external sensory stimuli. The data of Kicic et al. (2006) suggest that the EEG response elicited by frontally admin-istered TMS following sound presentation in a mismatch negativity paradigm depends on whether the stimulus is a standard or a deviant. Additionally, this would be an indication of the ability of TMS–EEG to reveal specific functional information about the cortical circuits outside the motor cortex.

Massimini et al. (2005) used the navigated brain stimulation (NBS) technique to target TMS to premotor area on subjects while awake and in non-REM sleep (Figure 37.5). They found that although TMS elicits strong synchronous activation during sleep, the neuronal activation remains very local. During wakefulness, on the other hand, TMS-triggered activation spreads to the contralateral hemi-sphere and to many associated ipsilateral areas.

The effective connectivity of cortical neuronal areas seems to break during deep sleep and gives room to a neuronal state optimized for purposes other than daytime sensory and motor tasks. Novel inroads to brain functions in sleep or during altered states of vigilance seem to be obtainable by TMS–EEG.

Monitoring EEG for safety

Contraindications for TMS are metal in the area of head/neck, recent trauma to the head area, and active epilepsy. Because of early case reports of seizures during single-pulse TMS performed on individuals with and without epilepsy (Homberg and Netz 1989; Hufnagel and Elger 1991), concern exists about the safety of performing such studies in individuals with epilepsy (Wassermann 1998). This concern has been extended to the studies of normal subjects and it has led to local, sometimes very stringent, practices. Indeed, in many cases optimal research and diagnostics of patients have been greatly hindered by the weakly justified fear of epileptic phenomena.

A recent study by Schrader et al. (2004) reviewed published data to determine a quanti-tative incidence of seizure in subjects with epilepsy undergoing single- and paired-pulse TMS and to explore conditions that may increase this risk. According to these authors, the crude risk of a TMS-associated seizure in patients with epilepsy ranges from 0.0 to 2.8% for single-pulse TMS and from 0.0 to 3.6% for paired-pulse TMS. Medically intractable epilepsy and lowering anti-epileptic drugs were associated with increased incidence. There was significant center-to-center variability that could not be explained by differences in patient population or by differences in reported stimulation parameters.

The risk of single- and paired-pulse TMS causing a seizure even in individuals with epilepsy appears small. The lowering of anti-epileptic drugs and the presence of medically intractable epilepsy increases the likelihood of atypical seizures during TMS. In all reported cases of a seizure during magnetic stimulation, the patients had their typical seizure followed by their typical recovery. In most cases, it was not clear if the seizure was actually induced by TMS

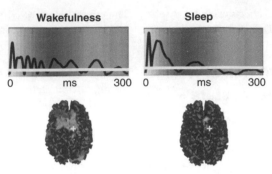

Fig. 37.5 Spatiotemporal cortical current maps evoked by cortical magnetic stimulation during wakefulness and deep (S3/S4 state) sleep. Black traces represent the global mean field power, and the horizontal yellow lines indicate significance levels. For each significant time sample, maximum current sources are plotted and color-coded according to their latency of activation (light blue, 0 ms; red, 300 ms). The yellow cross marks the TMS target on the cortical surface. (Plate 22)

or was merely a coincidence. Importantly, there have been no long-term adverse consequences in any individuals with epilepsy who have experienced a TMS-associated seizure.

A recent review of the literature reports that single- and paired-pulse TMS did not elicit adverse events in >850 children described in 28 published studies (Gilbert *et al.* 2004). Single- or paired-pulse TMS did not result in seizures, even in children with epilepsy or with conditions such as cerebral palsy that are associated with increased risk of seizures. The authors concluded that "there is 'no discernable evidence' that single- or paired-pulse TMS could cause harm".

The parameters of stimulation are certainly crucial for safety, and the guidelines for stimulation have been published by the International Federation of Clinical Neurophysiology (IFCN). For optimized safety, we suggest that a simultaneous monitoring of even a few channels of EEG allows early detection of any epileptiform activity, currently our greatest concern, and immediate stopping of stimulation. The EEG should be monitored specifically on patients at risk of epileptic phenomena, e.g. suffering from tumor, trauma, and epilepsy. Monitoring should also be part of diagnostics in children.

Novel avenues

In a number of studies, EEG has been used to measure effects of TMS or rTMS after the end of the TMS session(s). In some quite recent approaches EEG has provided timing information for delivery of TMS pulses.

Relationship between EEG and the efficacy of treatment by magnetic stimulation

When Schutter *et al.* (2001) applied 20 min of 1 Hz 130% MT rTMS (1200 pulses) over the right dorsolateral prefrontal cortex (F4) in healthy volunteers, left-hemisphere EEG theta activity was enhanced at 25–35 and 55–65 min after treatment. The authors considered this EEG change to be associated with the observed mood change (reduced anxiety). Jin *et al.* (2006) noted a 34% increase in frontal alpha amplitude following 2 weeks' daily treatment of schizophrenic patients with alpha-frequency rTMS to the dorsolateral prefrontal cortex bilaterally. Negative symptoms of schizophrenia were reduced by 30% with the individualized alpha-frequency rTMS but not with 3 or 20 Hz treatments.

On the other hand, Graf *et al.* (2001) reported no changes of EEG power spectrum topography after a total of 1600 rTMS pulses (in 2 s, 20 Hz trains delivered twice a minute) to left dorsolateral prefrontal cortex. Effects on subsequent sleep EEG were also small.

Enhancing cognitive performance with rTMS at human individual alpha frequency

Klimesch *et al.* (2003) applied rTMS at individual alpha frequency (IAF) to improve cognitive performance by influencing the dynamics of alpha desynchronization. Previous research had indicated that a large upper alpha power preceding a cognitive task was related to both large suppression of upper alpha power during the task and good performance. The hypothesis was that rTMS at individual upper alpha frequency (IAF + 1 Hz) would enhance alpha power, and thus improve task performance. Repetitive TMS was delivered to the mesial frontal (Fz) and right parietal (P6) cortex prior to a mental rotation task. TMS enhanced task performance and, concomitantly, the extent of task-related alpha.

Demonstrating cortical LTP in human by combined TMS and EEG

In animal models, a direct demonstration of long-term potentiation (LTP) is typically obtained by high-frequency electrical stimulation coupled with local field recordings of population responses. rTMS is increasingly being used to promote cortical reorganization, presuming that it can induce LTP. For example, rTMS of motor cortex can induce a potentiation of muscle motor-evoked potentials that outlasts the stimulation by several minutes.

Esser *et al.* (2006) recorded cortical EEG responses to single TMS pulses before and after applying rTMS to motor cortex (5 Hz, 1500 pulses). After rTMS, EEG responses at latencies of 15–55 ms were significantly potentiated. A topographic analysis revealed that this potentiation was significant at EEG electrodes located bilaterally over the premotor cortex, which is strongly interconnected to the site of stimulation. Thus, these findings can be taken as a direct demonstration of LTP induced by rTMS.

Cortical short-term plasticity in humans by combined TMS and EEG

Animal experiments suggest that cortical sensory representations may be remodelled as a consequence of changing synaptic efficacy by timing-dependent associative neuronal activity. Wolters *et al.* (2005) describe a timing-based associative form of plasticity in human somatosensory cortex. Median nerve stimulation was paired with TMS over the contralateral post-central region. Pairing increased exclusively the amplitude of the P25 component of the median nerve somatosensory-evoked potential (SEP), which is probably generated in the superficial

cortical layers of somatosensory area 3b. SEP components reflecting neuronal activity in deeper cortical layers (N20 component) or subcortical regions (P14 component) remained constant. Modulation of P25 amplitude was confined to a narrow range of interstimulus intervals (ISIs) between the median-nerve pulse and the TMS pulse; the sign of the modulation changed with ISIs differing by only 15 ms. These findings suggested a simple model of modulation of excitability in human primary somatosensory cortex, possibly by mechanisms related to the spike-timing-dependent plasticity of neuronal synapses located in upper cortical layers.

Studies of CNS-affecting drugs by combined TMS and EEG

TMS provides new possibilities for studying localized changes in the electrical properties of the human cortex (Kähkönen and Ilmoniemi 2004). For example, TMS combined with electromyography (EMG) has revealed that drugs blocking Na^+ or Ca^{2+} channels, such as phenytoin, lamotrigin, or carbamazepine, change the MT, whereas GABAergic agents vigabatrin, lorazepam, diazepam, and ethanol do not affect the MT, but increase intracortical inhibition and decrease facilitation. Dopamine receptor antagonists, such as haloperidol, decrease intracortical inhibition and increase intracortical facilitation.

Ethanol has been shown to modulate EEG responses evoked by motor-cortex TMS, the effects being largest at the right prefrontal cortex, suggesting that ethanol would have changed functional connectivity (Kähkönen *et al.* 2001). Furthermore, alcohol decreases amplitudes of EEG responses after the left prefrontal stimulation mainly in anterior parts of the cortex, which may be associated with the decrease of the prefrontal cortical excitability. Taken together, TMS combined with EEG provides a new insight into the actions of CNS drugs at the cortical level. Combining the well-known paired-pulse TMS techniques with simultaneous EEG may provide new routes to studies of CNS-affecting drugs.

Conclusion

TMS-induced peripheral measures give only indirect information about the excitability of the motor cortex; spinal mechanisms may contribute to the results. Cortical excitability and connectivity can be studied directly by combining TMS with EEG or other brain-imaging methods, not only in motor, but also nonmotor, areas.

Several groups have already demonstrated the ability of TMS–EEG to yield unique information on various aspects of reactivity and functional connectivity in the brain. Reactivity and connectivity are direct measures of the condition of the human neuronal system, and this new technique may be useful in both diagnostics and real-time monitoring during treatment.

Statement concerning conflict of interest

Prof. Risto Ilmoniemi is a founder, previous Chairman of the Board, previous Managing Director, present Chief Scientific Officer, and shareholder, and Prof. Jari Karhu is previous Managing Director, present Chief Medical Officer, and shareholder of Nexstim Ltd, a company that develops and markets Navigated Brain Stimulation systems that include TMS-compatible EEG.

References

Amassian VE, Cracco RQ, Maccabee PJ, Cracco JB (1992). Cerebello-frontal cortical projections in humans studied with the magnetic coil. *Electroencephalography and Clinical Neurophysiology 85*, 265–272.

Andrew C, Pfurtscheller G (1996). Event-related coherence as a tool for studying dynamic interaction of brain regions. *Electroencephalography and Clinical Neurophysiology 98*, 144–148.

Bender S, Basseler K, Sebastian I, *et al.* (2005). Transcranial magnetic stimulation evokes giant inhibitory potentials in children. *Annals of Neurology 58*, 58–67.

Chatrian GE, Petersen MC, Lazarte JA (1959). The blocking of the central wicket rhythm and some central changes related to movement. *Electroencephalography and Clinical Neurophysiology 11*, 497–510.

Classen J, Gerloff C, Honda M, Hallett M (1998). Integrative visuomotor behavior is associated with interregionally coherent oscillations in the human brain. *Journal of Neurophysiology 79*, 1567–1573.

Cracco RQ, Amassian VE, Maccabee PJ, Cracco JB (1989). Comparison of human transcallosal responses evoked by magnetic coil and electrical stimulation. *Electroencephalography and Clinical Neurophysiology 74*, 417–424.

Destexhe A, Contreras D, Steriade M (1999). Cortically-induced coherence of a thalamic-generated oscillation. *Neuroscience 92*, 427–443.

Esser SK, Huber R, Massimini M, Peterson MJ, Ferrarelli F, Tononi G (2006). A direct demonstration of cortical LTP in humans: A combined TMS/EEG study. *Brain Research Bulletin 69*, 86–94.

Fox PT, Narayana S, Tandon N, *et al.* (2004). Column-based model of electric field excitation of cerebral cortex. *Human Brain Mapping 22*, 1–14.

Fuggetta G, Fiaschi A, Manganotti P (2005). Modulation of cortical oscillatory activities induced by varying single-pulse transcranial magnetic stimulation intensity over the left primary motor area: a combined EEG and TMS study. *NeuroImage 27*, 896–908.

Gastaut MH (1952). Etude éléctrocorticographique de la réactivité des rhythms rolandiques. *Revue Neurologique 87*, 176–182.

Gerloff C, Richard J, Hadley J, Schulman AE, Honda M, Hallett M (1998). Functional coupling and regional activation of human cortical motor areas during simple, internally paced and externally paced finger movements. *Brain 121*, 1513–1531.

Gilbert DL, Garvey MA, Bansal AS, Lipps T, Zhang J, Wassermann EM (2004). Should transcranial magnetic stimulation research in children be considered minimal risk? *Clinical Neurophysiology 115*, 1730–1739.

Graf T, Engeler J, Achermann P, *et al.* (2001). High frequency repetitive transcranial magnetic stimulation (rTMS) of the left dorsolateral cortex: EEG topography during waking and subsequent sleep. *Psychiatry Research 107*, 1–9.

Hämäläinen MS, Ilmoniemi RJ (1994). Interpreting magnetic fields of the brain: minimum norm estimates. *Medical and Biological Engineering and Computing 32*, 35–42.

Hari R, Salmelin R (1997). Human cortical oscillations: a neuromagnetic view through the skull. *Trends in Neurosciences 20*, 44–49.

Homberg V, Netz J (1989). Generalised seizures induced by transcranial magnetic stimulation of motor cortex. *Lancet ii*, 1223.

Hufnagel A, Elger CE (1991). Induction of seizures by transcranial magnetic stimulation in epileptic patients. *Journal of Neurology 238*, 109–110.

Ilmoniemi RJ, Ruohonen J, Virtanen J (1996). Relationships between magnetic stimulation and MEG/EEG. In: J Nilsson, M Panizza, F Grandori (eds), *Advances in magnetic stimulation: mathematical modeling and clinical applications. Advances in occupational medicine and rehabilitation 2*, pp. 65–72. Pavia: Maugeri Foundation.

Ilmoniemi RJ, Virtanen J, Ruohonen J, *et al.* (1997). Neuronal responses to magnetic stimulation reveal cortical reactivity and connectivity. *Neuroreport 8*, 3537–3540.

Ilmoniemi RJ, Ruohonen J, Karhu J (1999a). Transcranial magnetic stimulation – a new tool for functional imaging of the brain. *Critical Reviews in Biomedical Engineering 27*, 241–284.

Ilmoniemi RJ, Ruohonen J, Virtanen J, Aronen HJ, Karhu J (1999b). EEG responses evoked by transcranial magnetic stimulation. *Electroencephalography and Clinical Neurophysiology 51*, 22–29.

Iramina K, Maeno T, Kowatari Y, Ueno S (2002). Effects of transcranial magnetic stimulation on EEG activity. *IEEE Transactions on Magnetics 38*, 3347–3349.

Ives JR, Pascual-Leone A, Chen Q, Schlaug G, Keenan J, Edelman RR (1998). Experience and early findings using transcranial magnetic stimulation (TMS) during functional magnetic resonance imaging (fMRI) in humans. *NeuroImage 7*, S33R.

Ives JR, Rotenberg A, Poma R, Thut G, Pascual-Leone A (2006). Electroencephalographic recording during transcranial magnetic stimulation in humans and animals. *Clinical Neurophysiology 117*, 1870–1875.

Izumi S, Takase M, Arita M, Masakado Y, Kimura A, Chino N (1997). Transcranial magnetic stimulation-induced changes in EEG and responses recorded from the scalp of healthy humans. *Electroencephalography and Clinical Neurophysiology 103*, 319–322.

Jasper HH, Andrews HL (1938). Electro-encephalography III Normal differentiation of occipital and precentral regions in man. *Archives of Neurology and Psychiatry 39*, 96–115.

Jasper H, Penfield W (1949). Electrocorticograms in man: effect of voluntary movement upon the electrical activity of the precentral gyrus. *European Archives of Psychiatry and Clinical Neuroscience 183*, 163–174.

Jin Y, Potkin SG, Kemp AS, *et al.* (2006). Therapeutic effects of individualized alpha frequency transcranial magnetic stimulation (alphaTMS) on the negative symptoms of schizophrenia. *Schizophrenia Bulletin 32*, 556–561.

Kähkönen S, Ilmoniemi RJ (2004). Transcranial magnetic stimulation: applications for neuropsychopharmacology. *Journal of Psychopharmacology 18*, 257–261.

Kähkönen S, Kesäniemi M, Nikouline VV, *et al.* (2001). Ethanol modulates cortical activity: direct evidence with combined TMS and EEG. *NeuroImage 14*, 322–328.

Kähkönen S, Komssi S, Wilenius J, Ilmoniemi RJ (2005). Prefrontal transcranial magnetic stimulation produces intensity-dependent EEG responses in humans. *NeuroImage 24*, 955–960.

Kicic D, Ilmoniemi R, Nikulin V (2006). Cortical excitability changes related to 'attention switching': a TMS-EEG demonstration. *Fourth Conference on Mismatch Negativity (MMN) and its Clinical and Scientific Applications*, Cambridge, UK, April 22–26, p. 92.

Klimesch W (1996). Memory processes, brain oscillations and EEG synchronization. *International Journal of Psychophysiology 24*, 61–100.

Klimesch W, Sauseng P, Gerloff C (2003). Enhancing cognitive performance with repetitive transcranial magnetic stimulation at human individual alpha frequency. *European Journal of Neuroscience 17*, 1129–1133.

Komssi S, Aronen HJ, Huttunen J, *et al.* (2002). Ipsi- and contralateral EEG reactions to transcranial magnetic stimulation. *Clinical Neurophysiology 113*, 175–184.

Komssi S, Kähkönen S, Ilmoniemi RJ (2004). The effect of stimulus intensity on brain responses evoked by transcranial magnetic stimulation. *Human Brain Mapping 21*, 154–164.

Komssi S, Savolainen P, Heiskala J, Kähkönen S (2007). Excitation threshold of the motor cortex estimated with transcranial magnetic stimulation electroencephalography. *Neuroreport 18*, 13–16.

Kujirai T, Caramia MD, Rothwell JC, *et al.* (1993). Corticocortical inhibition in human motor cortex. *Journal of Physiology 471*, 501–519.

Manganotti P, Gerloff C, Toro C, *et al.* (1998). Task-related coherence and task-related spectral power changes during sequential finger movements. *Electroencephalography and Clinical Neurophysiology 109*, 50–62.

Massimini M, Ferrarelli F, Huber R, Esser SK, Singh H, Tononi G (2005). Breakdown of cortical effective connectivity during sleep. *Science 309*, 2228–2232.

Moruzzi G, Magoun HW (1949). Brainstem reticular formation and activation of the EEG. *Electroencephalography and Clinical Neurophysiology 1*, 455–473.

Nikouline V, Ruohonen J, Ilmoniemi RJ (1999). The role of the coil click in TMS assessed with simultaneous EEG. *Clinical Neurophysiology 110*, 1325–1328.

Nikulin VV, Kicic D, Kähkonen S, Ilmoniemi RJ (2003). Modulation of electroencephalographic responses to transcranial magnetic stimulation: evidence for changes in cortical excitability related to movement. *European Journal of Neuroscience 18*, 1206–1212.

Paus T (2005). Inferring causality in brain images: a perturbation approach. *Philosophical Transactions of the Royal Society of London B 360*, 1109–1114.

Paus T, Sipila PK, Strafella APJ (2001). Synchronization of neuronal activity in the human primary motor cortex by transcranial magnetic stimulation: an EEG study. *Neurophysiology 86*, 1983–1990.

Pfurtscheller G (1992). Event-related synchronization (ERS): an electrophysiological correlate of cortical areas at rest. *Electroencephalography and Clinical Neurophysiology 83*, 62–69.

Pfurtscheller G, Aranibar A (1977). Event-related cortical desynchronization detected by power measurements of scalp EEG. *Electroencephalography and Clinical Neurophysiology 42*, 817–826.

Pfurtscheller G, Aranibar A (1979). Evaluation of event-related desynchronization (ERD) preceding and following voluntary self-paced movement. *Electroencephalography and Clinical Neurophysiology 46*, 138–146.

Pfurtscheller G, Lopes da Silva FH (1999). Event-related EEG/MEG synchronization and desynchronization: basic principles. *Clinical Neurophysiology 110*, 1842–1857.

Roth BJ, Pascual-Leone A, Cohen LG, Hallett M (1992). The heating of metal electrodes during rapid-rate magnetic stimulation: a possible safety hazard. *Electroencephalography and Clinical Neurophysiology 85*, 116–123.

Schrader LM, Stern JM, Koski L, Nuwer MR, Engel J Jr (2004). Seizure incidence during single- and paired-pulse transcranial magnetic stimulation (TMS) in individuals with epilepsy. *Clinical Neurophysiology* 115, 2728–2737.

Schutter DJ, van Honk J, d'Alfonso AA, Postma A, de Haan EH (2001). Effects of slow rTMS at the right dorsolateral prefrontal cortex on EEG asymmetry and mood. *Neuroreport 12*, 445–447.

Singer W (1993). Synchronization of cortical activity and its putative role in information processing and learning. *Annual Review of Physiology 55*, 349–374.

Steriade M, Amzica F (1996). Intracortical and corticothalamic coherency of fast spontaneous oscillations. *Proceedings of the National Academy of Sciences of the USA 93*, 2533–2538.

Steriade M, Llinas RR (1988). The functional states of the thalamus and the associated neuronal interplay. *Physiological Reviews 68*, 649–742.

Steriade M, Gloor P, Llinas RR, Lopes de Silva FH, Mesulam MM (1990). Report of IFCN Committee on Basic Mechanisms Basic mechanisms of cerebral rhythmic activities. *Electroencephalography and Clinical Neurophysiology 76*, 481–508.

Thut G, Northoff G, Ives JR, et al. (2003a). Effects of single-pulse transcranial magnetic stimulation (TMS) on functional brain activity: a combined event-related TMS and evoked potential study. *Clinical Neurophysiology 114*, 2071–2080.

Thut G, Theoret H, Pfennig A, et al. (2003b). Differential effects of low-frequency rTMS at the occipital pole on visual-induced alpha desynchronization and visual-evoked potentials. *NeuroImage 18*, 334–347.

Thut G, Ives JR, Kampmann F, Pastor MA, Pascual-Leone A (2005). A new device and protocol for combining TMS and online recordings of EEG and evoked potentials. *Journal of Neuroscience Methods 141*, 207–217.

Tiitinen H, Virtanen J, Ilmoniemi RJ, et al. (1999). Separation of contamination caused by coil clicks from responses elicited by transcranial magnetic stimulation. *Clinical Neurophysiology 110*, 982–985.

Van Der Werf YD, Paus T (2006). The neural response to transcranial magnetic stimulation of the human motor cortex. I. Intracortical and cortico-cortical contributions. *Experimental Brain Research 175*, 231–245.

Van Der Werf YD, Sadikot AF, Strafella AP, Paus T (2006). The neural response to transcranial magnetic stimulation of the human motor cortex. II. Thalamocortical contributions. *Experimental Brain Research 175*, 246–255.

Virtanen J, Ruohonen J, Näätänen R, Ilmoniemi RJ (1998). Instrumentation for the measurement of electric brain responses to transcranial magnetic stimulation. Report 1, Series B: Research Reports Helsinki University of Technology, Department of Electrical and Communications Engineering, Applied Electronics Laboratory.

Virtanen J, Ruohonen J, Näätänen R, Ilmoniemi RJ (1999). Instrumentation for the measurement of electric brain responses to transcranial magnetic stimulation. *Medical and Biological Engineering and Computing 37*, 322–326.

Wassermann EM (1998). Risk and safety of repetitive transcranial magnetic stimulation: report and suggested guidelines from the International Workshop on the Safety of Repetitive Transcranial Magnetic Stimulation, June 5–7, 1996. *Electroencephalography and Clinical Neurophysiology 108*, 1–16.

Wolters A, Schmidt A, Schramm A, et al. (2005). Timing-dependent plasticity in human primary somatosensory cortex. *Journal of Physiology 565*, 1039–1052.

Ziemann U, Lonnecker S, Steinhoff BJ, Paulus W (1996a). The effect of lorazepam on the motor cortical excitability in man. *Experimental Brain Research 109*, 127–135.

Ziemann U, Rothwell JC, Ridding MC (1996b). Interaction between intracortical inhibition and facilitation in human motor cortex. *Journal of Physiology 496*, 873–881.

SECTION VI

Therapeutic Applications of TMS

Sarah H. Lisanby

Focal noninvasive brain stimulation represents a previously unprecedented means of bridging the gap between the neuroscience of clinical disorders and the clinical science of their treatment. The ability to stimulate the brain noninvasively sparked early and sustained interest among clinical researchers regarding its therapeutic potential in psychiatry and neurology. Systemically administered pharmacological agents may be ineffective or cause treatment-interfering side-effects. TMS, on the other hand, avoids systemic side-effects, and stimulates the brain with a spatial and temporal specificity that cannot currently be achieved pharmacologically or via electroconvulsive therapy.

TMS has already attained regulatory approval in some countries for the treatment of depression, and a growing list of other disorders are under active study. Clearly the therapeutic potential of tools like TMS to expand treatment options for medication-resistant disorders merits close scrutiny; however, it is equally essential that enthusiasm not outpace evidence when judging the value of a novel therapeutic intervention. The literature on the clinical potential of TMS has matured to the stage where it is now

possible to compile a comprehensive review of the potential applications of TMS in a range of psychiatric and neurological disorders. This section on therapeutic applications of TMS provides a critical review of the state of the evidence for and against the therapeutic value of TMS.

While tools like TMS present great therapeutic promise, the realization of that promise demands an in-depth knowledge of pathophysiology of the illness in question, and of the mechanisms by which repeated activation of circuits can induce plastic changes in the functioning of those circuits. Since TMS is a focal intervention, its clinical utility will ultimately depend upon our knowledge of the circuitry of the underlying disorder that will guide the selection of the target cortical site for stimulation. Likewise, since TMS is administered in a phasic rather than tonic fashion, its appropriate use will hinge on knowledge of the dynamics by which repeated brain stimulation interacts with ongoing endogenous processes to induce lasting changes in behavior.

The ultimate therapeutic value of focal non-invasive brain stimulation may hinge not so much on the tool itself, but on how wisely we use it.

CHAPTER 38

Therapeutic potential of TMS-induced plasticity in the prefrontal cortex

Stanislav R. Vorel and Sarah H. Lisanby

Introduction

Transcranial magnetic stimulation is under active study in the treatment of a range of psychiatric and neurological disorders. The mechanisms by which intermittent focal brain stimulation with TMS could exert lasting effects in illnesses subserved by distributed neuronal networks are at present largely unknown. To be of lasting benefit beyond the period of stimulation, enduring changes in the functioning of the target pathways would need to be invoked. One proposed mechanism by which such lasting changes might come about is synaptic plasticity. As a noninvasive, relatively safe technique for fairly selective stimulation of cortical structures, TMS is a uniquely suitable research tool for the study of neural plasticity in humans.

The objective of this chapter is twofold. First, we discuss synaptic plasticity as a potential mechanism of enduring changes in function observed after relatively brief periods of repetitive (r)TMS. Second, we explore how principles of synaptic plasticity may be exploited in the rational design of future rTMS paradigms in psychiatric disorders, taking into account interactions between rTMS and pharmacological manipulations.

Given that prefrontal cortex (PFC) dysfunction is involved in a variety of neuropsychiatric disorders and that recovery of PFC function is associated with antidepressant treatment effects (Goldapple et al. 2004), we will focus on the therapeutic potential of rTMS-induced plasticity in the PFC. Implications for the interpretation of existing TMS literature and design of future interventions will be discussed.

Definition of plasticity

Plasticity may be broadly defined as a use-dependent enduring change in neural structure- and function. This definition would encompass such phenomena as acute amphetamine-induced dendritic sprouting of neurons in the nucleus accumbens and prefrontal cortex (Robinson and Kolb 2004), activity-dependent neurogenesis in the hippocampus (Deisseroth et al. 2004), and behavioral sensitization to psychostimulants (Ghasemzadeh et al. 2003). These examples illustrate a variety of neural plasticity effects with different induction protocols (e.g. drug administration, electrical stimulation) and outcome measures (e.g. locomotor activity, dopamine release, neurogenesis).

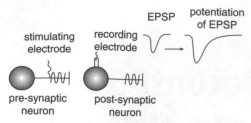

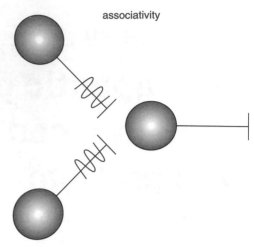

Fig. 38.1 Illustration of long-term potentiation of synaptic efficacy following repeated electrical stimulation of the presynaptic neuron as evidenced by an increase in excitatory postsynaptic potential (EPSP).

A more narrow definition of plasticity would be activity dependent-change in synaptic efficacy (Bliss and Collingridge 1993). Long-term potentiation (LTP) is the most widely studied version of synaptic plasticity and refers to enduring strengthening of synapses. LTP's counterpart is long-term depression (LTD), the enduring weakening of synaptic strength. Typically, 'synaptic' strength is measured as excitatory postsynaptic potentials (EPSPs) in response to electrical stimulation (Figure 38.1).

Characteristics of plasticity

Synaptic plasticity has several characteristics that can be described (Bliss and Collingridge 1993). Associativity refers to the concurrent stimulation of input neurons converging on the same output neuron, a condition that favors the induction of LTP (Figure 38.2). Coincidence refers to the simultaneous firing of pre- and postsynaptic neurons, a condition favoring LTP induction (Figure 38.3). Input specificity refers to the selective strengthening of active synapses, but not other synapses without input activity (Figure 38.4). Changes in synaptic efficacy are long-lasting and may occur in two directions. Synaptic strength can increase or potentiate up to a certain maximum at which 'saturation' occurs. The decrease in synaptic strength is known as depotentiation or depression.

Although LTP has predominantly been studied in the hippocampus, it has been demonstrated in other brain areas, including the PFC

Fig. 38.2 Illustration of associativity as a characteristic of long-term potentiation (LTP), whereby the concurrent stimulation of two input neurons converging on the same output neuron favors the induction of LTP.

(Floresco and Grace 2003; Maroun and Richter-Levin 2003). EPSPs were recorded in the PFC in response to electrical stimulations of afferent fibers originating in the ventral subiculum (VSUB), the outflow pathway of the hippocampus (Figure 38.5) (Gurden *et al.* 1999). After high-frequency stimulation of the VSUB, the EPSPs measured in the PFC in response to VSUB stimulation pulses were permanently enhanced. The enhancement was augmented after electrical stimulation of afferent dopamine fibers originating in the ventral tegmental area (VTA) concurrent with high-frequency stimulation of the VSUB. The experiment highlights associativity of active inputs from VTA and VSUB that converge on the same postsynaptic PFC neurons.

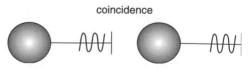

Fig. 38.3 Illustration of coincidence as a characteristic of long-term potentiation (LTP), whereby the simultaneous firing of pre- and postsynaptic neurons favors the induction of LTP.

specificity

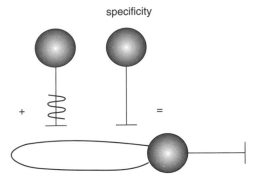

Fig. 38.4 Illustration of specificity as a characteristic of long-term potentiation, whereby active synapses are selectively strengthened (+), but those without input activity are not (=).

Neuroanatomical and neurochemical considerations

One possible mechanism of associativity is the cooperative action of postsynaptic receptors. For example, the excitatory neurotransmitter glutamate binds to multiple receptors. The ionotropic group of glutamate receptors includes alpha-amino-3-hydroxy-5-methyl-4-isoxazole propionic acid (AMPA) and N-methyl-D-aspartate (NMDA) receptors (Figure 38.6), which are cation channels activated by glutamate. During weak glutamate stimulation, only AMPA receptors permeable to sodium are active. During stronger glutamate stimulation, AMPA receptor activity will become sufficient to activate NMDA receptors. NMDA channels in their resting state are blocked by Mg^{2+}. After sufficient depolarization by sodium influx through AMPA channels the NMDA channels become unblocked and permeable to Ca^{2+} ions. Thus, the NMDA and AMPA receptors are able to integrate glutamate neurotransmission from convergent glutamatergic afferents (Bliss and Collingridge 1993).

The PFC receives convergent glutamate inputs from the VSUB, amygdala, and thalamus (Floresco and Grace 2003). In addition, there is convergence of glutamate- and dopamine-containing afferents on PFC neurons as mentioned

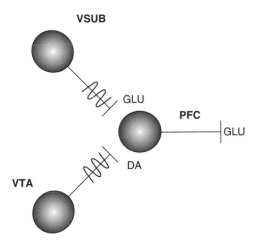

Fig. 38.5 Illustration of evidence for long-term potentiation in the prefrontal cortex (PFC). High-frequency stimulation of the ventral subiculum (VSUB) permanently enhanced the excitatory postsynaptic potentials measured in the PFC in response to VSUB stimulation. The enhancement was augmented after electrical stimulation of afferent dopamine (DA) fibers originating in the ventral tegmental area (VTA) concurrent with high-frequency stimulation of the VSUB, demonstrating associativity of active inputs from VTA and VSUB that converge on the same postsynaptic PFC neurons. GLU, glutamate. After Gurden et al. (1999).

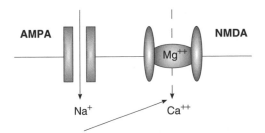

Fig. 38.6 Illustration of how AMPA and N-methyl-D-aspartate (NMDA) receptors integrate glutamate neurotransmission from convergent glutamatergic afferents. During weak glutamate stimulation (solid arrow), only AMPA receptors permeable to sodium are active and NMDA channels remain blocked by Mg^{2+}. During stronger glutamate stimulation and after sufficient depolarization by sodium influx through AMPA channels, the NMDA channels become unblocked and permeable to Ca^{2+} ions (dashed arrow).

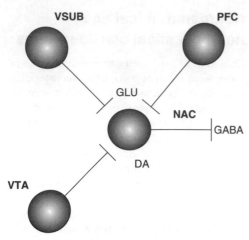

Fig. 38.7 Illustration of how the neuroanatomical convergence of glutamatergic (GLU) and dopaminergic (DA) inputs to the nucleus accumbens (NAC) sets up the possibility of plasticity. Repetitive TMS stimulation of prefrontal cortex (PFC) connections to the striatum could be coincident with inputs from ventral subiculum (VSUB) or ventral tegmental area (VTA) and provide a substrate for long-term potentiation.

above and in Figure 38.5 (Gurden *et al.* 1999). Similar convergence of glutamate and dopamine inputs is present in the nucleus accumbens (NAC) (Floresco *et al.* 1999, 2001). This neuroanatomical convergence sets up the possibility of associativity. Not surprisingly, then, LTP has been observed in the NAC and PFC. Coincidence-induced LTP could occur between inputs from amygdala or VSUB and outputs from PFC or NAC (Figure 38.7) (Floresco *et al.* 2001; Floresco and Grace 2003). The convergence in the NAC may be relevant for PFC rTMS-induced plasticity, since PFC rTMS induces dopamine release in the basal ganglia (Strafella *et al.* 2001) and since dysfunction of the basal ganglia (of which the NAC is part) is involved in several neuropsychiatric disorders.

Clinical significance of LTP

The hippocampus is critically involved in learning and memory. LTP in the hippocampus is therefore an attractive cellular model of memory storage in the hippocampus. Indeed, LTP

induction in the hippocampus accompanies the learning of new tasks (Rioult-Pedotti *et al.* 2000; Whitlock *et al.* 2006).

The PFC is more involved in working memory, rather than the storage of permanent memories (Levy and Goldman-Rakic 2000). This raises the question of what clinical significance LTP in the PFC might have. One possibility is that LTP could selectively strengthen specific inputs into the PFC, thereby filtering access of some inputs over other ones and thus representing a cellular mechanism of attention. Alternatively, LTP could strengthen specific neural circuits within the PFC or boost PFC metabolism and function overall, including executive, planning, and motor functions. Whereas LTP's possible role in PFC cognitive functions is of special interest with respect to psychiatric disorders, LTP in the PFC has been best studied in the motor cortex.

Repetitive TMS studies of LTP in the motor cortex

Repetitive TMS

TMS is a noninvasive, relatively safe technique for fairly selective stimulation of cortical structures in human and nonhuman subjects. TMS allows remote stimulation of superficial cortical structures up to 2 cm below the surface of the human skull (Bohning *et al.* 1997). TMS directly produces a large simultaneous activation of both excitatory and inhibitory neurons at the site of stimulation, resulting in inhibitory activity silencing local neural firing. More distant areas are activated through excitatory long-range pathways (Esser *et al.* 2005).

Motor cortex plasticity shares some features with LTP

It is no surprise that PFC LTP has been best characterized in the motor cortex because of several features that make it appealing for LTP experiments (Ilic and Ziemann 2005; Ziemann *et al.* 2006). Electrophysiological *outcome measures*, e.g. motor-evoked potentials (MEPs), are objective, well defined, and can be correlated with behavioral measures, e.g. acceleration or strength of movements.

The *coincidence* of pre- and postsynaptic neuron firing induces LTP in the motor cortex of primates (Jackson *et al.* 2006). Through an electronic chip implanted in the motor cortex, presynaptic inputs were triggered by neural firing recorded in postsynaptic outputs. Moreover, motor learning accompanied the LTP.

Convergent inputs from motor cortex and medial longitudinal nerve (MLN) address the *associativity* feature of LTP. Paired associative stimulation (PAS) of MLN and motor cortex results in LTP or LTD of MEPs, depending on the PAS interval. Fast repetitive thumb movements resulted in learning, as measured by an increase in the peak acceleration of practiced thumb movements. Movement practice prevented subsequent LTP induction, but fostered subsequent LTD, consistent with saturation of MEPs during motor learning (Ziemann *et al.* 2001, 2004). Thus, motor cortex plasticity possesses several characteristics of LTP.

LTP in humans

To date, plasticity studies in humans have been limited, since reported effects on synaptic plasticity are often weak, highly variable between individuals, and rarely last longer than 30 min. In comparison with the frequencies and train durations that have typically been used to induce LTP with direct electrical stimulation, parameters of stimulation for TMS protocols are relatively weak and limited by the risk of TMS side-effects, such as the risk of seizure. Recently investigators applied TMS in theta-burst fashion (50 Hz bursts, repeated at a 5 Hz frequency) to the motor cortex in humans (Huang *et al.* 2005). The protocol appeared safe and seems to hold promise as a powerful LTP inducer. Indeed, the authors observed long-lasting changes of up to 1 h in motor cortex electrophysiological measures. They observed accompanying behavioral potentiation as measured by the reaction time of the hand. Subsequent work has tested the clinical potential of theta burst stimulation (TBS) to the motor cortex to improve motor recovery following stroke, with some encouraging initial results in six chronic stroke patients (Talelli *et al.* 2007). Likewise, it would be of great interest to determine whether modulation of LTP via TBS

could have clinical significance in psychiatric disorders, specifically when applied to the dorsolateral prefrontal cortex (dlPFC). TBS has indeed been applied to brain regions outside of the primary motor cortex, with evidence of lasting inhibition demonstrated in the frontal eye fields (Nyffeler *et al.*, 2006) and the occipital cortex (Franca *et al.*, 2006). This work has suggested that the parameters of stimulation to effectively induce excitatory effects with intermittent TBS may differ for nonmotor cortical areas (Franca *et al.*, 2006).

While the evidence of lasting changes in the amplitude of corticospinal responses is consistent with LTP, it does not constitute proof that these changes come about via LTP. A direct measure of LTP in humans would be extremely useful to explain the mechanism of action of these lasting changes in function. A recent report describes one proposed measure which appears to offer supportive evidence of cortical LTP in humans (Esser *et al.* 2006). Cortical responses to single TMS pulses were measured through high-density electroencephalography (hdEEG) before and after applying rTMS (5 Hz, 1500 pulses) to the motor cortex. The authors observed significant potentiation of EEG responses bilaterally over the premotor cortex at latencies of 15–55 ms; magnetic artifacts prevented the measurement of earlier peaks. In addition, they observed potentiation of motor responses to TMS as measured by motor-evoked potentials (MEPs).

This study had several limitations. It is unclear how long the potentiation of EEG responses to TMS lasted, since responses were measured for only 10 min in quiet wakefulness. Hence, it is unclear if the study satisfied the longevity requirement of LTP. It is also unclear how MEPs could be potentiated while cortical activity was not. The behavioral significance of the MEPs (e.g. increased force or accelerated muscle contraction) was not assessed. Nevertheless, the hdEEG outcome measure is appealing, since it mimics EPSPs, the prototypical outcome measure in LTP experiments. It will be of note to see if this experiment can be replicated in other brain areas of psychiatric interest, e.g. the dlPFC, providing an intermediate outcome measure for the selection of TMS or TBS dosing parameters to maximize LTP- or LTD-like effects.

Pharmacological manipulation of motor cortex plasticity

Since many psychotropic medications affect neurotransmitter function, pharmacological studies of human motor cortex plasticity are relevant to consider (Boroojerdi 2002; Gu 2002; Ziemann *et al.* 2006). In general, while the inhibitory neurotransmitter gamma-aminobutyric acid (GABA) suppresses LTP, other neurotransmitter systems studied so far enhance LTP; they include glutamate, dopamine, norepinephrine, and acetylcholine. So far, serotonergic agents have failed to affect LTP induction. Of note, the dopamine receptor agonists cabergoline and methylphenidate enhance cortical LTP, in agreement with enhanced induction of PFC LTP in the rat (Gurden *et al.* 1999). It will be of interest to examine how norepinephrine, acetylcholine, serotonin, and other neurotransmitters affect PFC LTP in the rat. Another testable possibility is whether dopamine receptor agonists can potentiate the pharmacological effects of small doses of glutamate receptor agonists in humans, as the rat study suggests. The potential and need for translational neurobiological research is clear.

Conclusions: human motor cortex plasticity

Overall it has thus far been difficult, if not impossible, to link LTP, the cellular model of learning and memory, to behavior (Cooke and Bliss 2006). The difficulty is that studies attempting to establish such a link have essentially been correlative, so that a causal role of LTP in learning and memory cannot be definitively established. TMS experiments of plasticity in the human motor cortex have been limited by the intensity and frequency of TMS protocols. Effects on MEPs have been inconsistent and short-lasting. It remains to be demonstrated whether the observed potentiation of motor cortical output has to do with changes at the synaptic level. Nevertheless, the observation of robust ($\geq$50%) and enduring ($\geq$20 min) increases of MEPs by theta-burst stimulation is promising. It will be interesting to see if similar plasticity will apply to other brain areas of psychiatric interest, e.g. the dlPFC. Motor cortex

plasticity is of particular clinical interest to neurorehabilitation of motor function, e.g. after stroke (Kim *et al.* 2006; Talelli *et al.* 2007). The TMS experience with motor cortex plasticity should guide future studies of plasticity in other brain areas of psychiatric interest, e.g. the dlPFC.

Rational design of psychotropic interventions using rTMS

Clinical trials with rTMS in psychiatry to date have been encouraging, even despite the current lack of objective guideposts with which to select among the broad range of rTMS parameters of stimulation and to inform how rTMS would interact with pharmacotherapy. We propose that knowledge of the mechanisms by which repeated stimulation of dlPFC induces lasting changes in function would inform the rational design of rTMS interventions and the strategic selection of rTMS plus medication combination strategies to exert meaningful clinical benefit.

Furthermore, the neurobiological mechanisms of action of psychopharmacological agents are incompletely understood. Although they affect a variety of neurotransmitters, therapeutic onset of action is typically delayed in diverse psychiatric disorders. One medication, e.g. a selective serotonin reuptake inhibitor (SSRI), may be successful in a variety of psychiatric disorders, e.g. major depressive disorder (MDD) and obsessive–compulsive disorder (OCD). Conversely, one psychiatric disorder, e.g. MDD, may have a variety of effective medications, e.g. SSRI and psychostimulant. Taken together, these observations suggest that a variety of medications may affect common neurobiological mechanisms underlying a variety of psychiatric disorders. We propose that TMS of the dlPFC is a promising neurobiological tool to explore such common mechanisms.

Hypothesis

We propose that it may be clinically useful to design TMS approaches to psychiatric treatments based on the induction of long-lasting changes in cortical output using stimulation protocols similar to those that have been used to induce synaptic plasticity in animals and plasticity in the human motor cortex. Based on preclinical experiments we predict that concurrent administration of selected psychotropic medications and rTMS of the dlPFC might induce LTP accompanied by potentiated psychotropic effects, thereby amplifying the moderate effect sizes that have been reported to date.

Putative outcome measures for PFC plasticity in psychiatric disorders

LTP induction after dlPFC rTMS might affect a variety of outcome measures, including improvements in psychiatric symptoms measured by standard rating scales, long-lasting structural or functional changes (e.g. increased brain volume, increased metabolic activity in selective brain areas), or electrophysiological measures (hdEEG potentials, P50 wave, negative mismatch negativity (NMN)). For example, the dlPFC is involved in numerous neuropsychological functions that can be assessed by cognitive tasks (Bruder et al. 2004). Certain TMS paradigms have already been reported to enhance performance on selected cognitive tasks in human volunteers (Luber et al. 2006; Vanderhasselt et al. 2006). HdEEG can record electrophysiological responses to TMS. Some neurophysiological measures are abnormal in psychiatric disorders, such as the abnormalities in P50 wave suppression and NMN seen in schizophrenia. It will be interesting to see if dlPFC rTMS or rTMS-induced potentiation of psychopharmacological effects can restore P50 or NMN abnormalities concomitant with improvements in cognitive functioning. Dorsolateral PFC rTMS may also induce enduring changes in metabolic activity in selective brain regions as measured by functional magnetic resonance imaging (fMRI) or positron emission tomography (PET). For example, one study examined the effect of two frequencies of rTMS on regional brain metabolism and clinical effects in depressed patients (Speer et al. 2000). High-frequency rTMS of the PFC induced persistent increases in regional cerebral blood flow (rCBF) as measured by PET; low-frequency rTMS induced a persistent decrease. The mood changes following the two rTMS frequencies

were inversely related, so that individuals who improved with one frequency worsened with the other.

Major depressive disorder

Repetitive TMS has been mainly studied as an adjunctive treatment in medication-resistant depressive patients (George et al. 2000; Fitzgerald et al. 2006). This raises the question: What distinguishes resistant from nonresistant depression? Work is beginning to shed light on this important question. To take just one example, McMahon et al. (2006) found in a sample of 1953 depressed patients that variation in the gene encoding the serotonin 2A receptor predicted antidepressant response to citalopram. One possibility to consider within the framework of this chapter is that the brain of medication-resistant depressed patients may be less plastic and less sensitive to antidepressant psychotropic medications, perhaps as a consequence of altered receptor function. TMS measures of plasticity described above could be used to test this hypothesis, as has already been done in schizophrenia (see below). The answers to these questions may provide insights into the neurobiology of depression and medication resistance, but also suggest clinical diagnostic tools to predict whether depressed patients will respond to psychotropic medications or would benefit from more aggressive treatment early on in their depression.

Antidepressant medications have established efficacy in the treatment of depression. Medications have in common that antidepressant effects only occur after several weeks of treatment. The neurobiological explanation for this delay in onset is unknown, but is the target of ongoing psychopharmacological research. One possibility is that antidepressant action is weak and needs to accumulate over several weeks until therapeutic effects occur. Fitting the context of this chapter, perhaps concurrent rTMS could potentiate the pharmacological effects of antidepressant medications to accelerate therapeutic response via LTP-like mechanisms. Indeed, concurrent rTMS did hasten the response to antidepressant medications in inpatients with non-drug-resistant MDD (Rossini et al. 2005; Rumi et al. 2005). Unfortunately, the cited studies did not include a placebo control group to address the possibility of an inter-action or potentiation between rTMS and psychotropics. It will be interesting to see if this accelerated antidepressant response is accompa-nied by TMS plasticity measures of potentia-tion, including paired associative stimulation or rTMS-induced potentiation of hdEEG responses.

Concurrent electrical stimulation of the VTA and VSUB induces LTP in the PFC (Gurden *et al.* 1999). VTA electrical stimulation results in dopamine release, and it is well known that the dopamine releaser amphetamine has rapid but transient antidepressant efficacy (Tremblay *et al.* 2005). It will be interesting to capitalize on these observations and to determine whether concur-rent rTMS and amphetamine administration will induce fast, powerful, and enduring antide-pressant effects. Equally interesting will be to see whether concurrent rTMS and amphetamine induce LTP in the dlPFC.

Obsessive–compulsive disorder

Antidepressant medications that inhibit the serotonin reuptake site, including SSRIs, but also the tricyclic antidepressant clomipramine, are effective in the treatment of OCD. Higher dosages for longer periods of time are required in the treatment of OCD compared to depression. It will be interesting to see if the psychotropic effects can be accelerated and boosted by con-comitant rTMS. Indeed, as reviewed elsewhere in this section, rTMS of the lateral PFC has shown some activity in affecting symptoms of OCD (Greenberg *et al.* 1998) and it will be interesting to see if the therapeutic effect can be boosted by concomitant pharmacological treatment.

Schizophrenia

As reviewed elsewhere in this section, TMS has been successfully applied to the treatment of hallucinations in schizophrenic patients (Lee *et al.* 2005). Several investigators have found modest TMS-induced improvements in nega-tive symptoms, such as apathy, amotivation, and attention impairment (Cohen *et al.* 1999; Nahas *et al.* 1999; Jin *et al.* 2006). Negative symptoms are particularly refractory to psychotropic med-ications, with the possible exception of clozapine and dopamine receptor agonists. There is an

urgent need for novel treatment approaches. It will be useful to investigate whether TMS can potentiate psychotropic effects. Consistent neuro-physiological correlates of schizophrenia have included P50 and NMN. It will be interesting to see if dlPFC rTMS can reverse these abnormali-ties in schizophrenic patients who show cogni-tive improvements. There is also a need for translational studies of schizophrenia to gener-ate testable hypotheses (Floresco *et al.* 2005). For example, prepulse inhibition (PPI) is a widely accepted animal model of schizophrenia. If PFC LTP-induction can reverse PPI abnor-malities, neurotransmitters that play a role in animal PFC LTP may become promising targets for new psychopharmacological approaches.

Substance use disorders

Acute rTMS of the dlPFC results in reduced cig-arette smoking (Eichhammer *et al.* 2003). The acute effects, albeit modest, are promising for more protracted rTMS protocols in chronic nicotine dependence. Neuroimaging studies in chronic substance dependence suggest atrophy in the PFC from a variety of substances, includ-ing cocaine and alcohol (Franklin *et al.* 2002; Bjork *et al.* 2003). Although the clinical rele-vance of the atrophy (e.g. craving or cognitive distortions) is unclear, it will be interesting to see if rTMS of the dlPFC can compensate for the atrophy, perhaps through use of dependent plasticity mechansims.

Conclusions

Direct electrical brain stimulation has a long tradition in studying brain function and in psy-chiatric treatment (in the form of electrocon-vulsive therapy). TMS is a noninvasive, relatively safe technique for fairly selective remote induc-tion of electrical currents in cortical areas. TMS holds out the promise of new treatments based on the induction of long-lasting changes in cor-tical output using stimulation protocols similar to those that have been used to induce synaptic plasticity in animals and plasticity in the human motor cortex. Here we have discussed the potential value of using TMS to intentionally induce LTP-like changes by implementing established LTP characteristics as an unexplored yet conceptually sound approach to the research and treatment of neuropsychiatric disorders. Specifically, we argue that rationally designed rTMS paradigms might accelerate and potentiate psychopharmacological effects in a variety of psychiatric disorders. The realization of this promise will require rigorous translational neurobiological research to understand, and to exploit for therapeutic ends, the mechanisms by which repeated brain stimulation exerts lasting changes in brain function.

Acknowledgements

The authors thank Eve Vagg for her design of the figures. Dr Lisanby has received research support from Magstim Company, Neuronetics, and Cyberonics.

References

Bjork JM, Grant SJ, et al. (2003). Cross-sectional volumetric analysis of brain atrophy in alcohol dependence: effects of drinking history and comorbid substance use disorder. *American Journal of Psychiatry 160*, 2038–2045.

Bliss TV, Collingridge GL (1993). A synaptic model of memory: long-term potentiation in the hippocampus. *Nature 361*, 31–39.

Bohning DE, Pecheny AP, et al. (1997). Mapping transcranial magnetic stimulation (TMS) fields in vivo with MRI. *Neuroreport 8*, 2535–2538.

Boroojerdi B (2002). Pharmacologic influences on TMS effects. *Journal of Clinical Neurophysiology 19*, 255–271.

Bruder GE, Wexler BE, et al. (2004). Verbal memory in schizophrenia: additional evidence of subtypes having different cognitive deficits. *Schizophrenia Research 68*, 137–147.

Cohen E, Bernardo M, Masana J, et al. (1999). Repetitive transcranial magnetic stimulation in the treatment of chronic negative schizophrenia: a pilot study. *Journal of Neurology, Neurosergry and Psychiatry 67*, 129–130.

Cooke SF, Bliss TV (2006). Plasticity in the human central nervous system. *Brain 129*, 1659–73.

Deisseroth K, Singla S, et al. (2004). Excitation–neurogenesis coupling in adult neural stem/progenitor cells. *Neuron 42*, 535–552.

Eichhammer P, Johann M, et al. (2003). High-frequency repetitive transcranial magnetic stimulation decreases cigarette smoking. *Journal of Clinical Psychiatry 64*, 951–953.

Esser SK, Hill SL, et al. (2005). Modeling the effects of transcranial magnetic stimulation on cortical circuits. *Journal of Neurophysiology 94*, 622–639.

Esser SK, Huber R, et al. (2006). A direct demonstration of cortical LTP in humans: a combined TMS/EEG study. *Brain Research Bulletin 69*, 86–94.

Fitzgerald PB, Benitez J, et al. (2006). A randomized, controlled trial of sequential bilateral repetitive transcranial magnetic stimulation for treatment-resistant depression. *American Journal of Psychiatry 163*, 88–94.

Floresco SB, Braaksma DN, et al. (1999). Thalamic-cortical-striatal circuitry subserves working memory during delayed responding on a radial arm maze. *Journal of Neuroscience 19*, 11061–11071.

Floresco SB, Blaha CD, et al. (2001). Modulation of hippocampal and amygdalar-evoked activity of nucleus accumbens neurons by dopamine: cellular mechanisms of input selection. *Journal of Neuroscience 21*, 2851–2860.

Floresco SB, Grace AA (2003). Gating of hippocampal-evoked activity in prefrontal cortical neurons by inputs from the mediodorsal thalamus and ventral tegmental area. *Journal of Neuroscience 23*, 3930–3943.

Floresco SB, Geyer MA, et al. (2005). Developing predictive animal models and establishing a preclinical trials network for assessing treatment effects on cognition in schizophrenia. *Schizophrenia Bulletin 31*, 888–894.

Franca M, Koch G, Mochizuki H, Huang YZ, Rothwell JC (2006). Effects of theta burst stimulation protocols on phosphene threshold. *Clinical Neurophysiology 117*, 1808–1813.

Franklin TR, Acton PD, et al. (2002). Decreased gray matter concentration in the insular, orbitofrontal, cingulate, and temporal cortices of cocaine patients. *Biological Psychiatry 51*, 134–142.

George MS, Nahas Z, et al. (2000). A controlled trial of daily left prefrontal cortex TMS for treating depression. *Biological Psychiatry 48*, 962–970.

Ghasemzadeh MB, Permenter LK, et al. (2003). Nucleus accumbens Homer proteins regulate behavioral sensitization to cocaine. *Annals of the New York Academy of Sciences 1003*, 395–397.

Goldapple K, Segal Z, et al. (2004). Modulation of cortical-limbic pathways in major depression: treatment-specific effects of cognitive behavior therapy. *Archives of General Psychiatry 61*, 34–41.

Greenberg BD, Ziemann U, et al. (1998). Decreased neuronal inhibition in cerebral cortex in obsessive-compulsive disorder on transcranial magnetic stimulation. *Lancet 352*, 881–882.

Gu Q (2002). Neuromodulatory transmitter systems in the cortex and their role in cortical plasticity. *Neuroscience 111*, 815–835.

Gurden H, Tassin JP, et al. (1999). Integrity of the mesocortical dopaminergic system is necessary for complete expression of in vivo hippocampal–prefrontal cortex long-term potentiation. *Neuroscience 94*, 1019–1027.

Huang YZ, Edwards MJ, et al. (2005). Theta burst stimulation of the human motor cortex. *Neuron 45*, 201–206.

Ilic TV, Ziemann U (2005). Exploring motor cortical plasticity using transcranial magnetic stimulation in humans. *Annals of the New York Academy of Sciences 1048*, 175–184.

Jackson A, Mavoori J, et al. (2006). Long-term motor cortex plasticity induced by an electronic neural implant. *Nature 444*, 56–60.

Jin Y, Potkin SG, *et al.* (2006). Therapeutic effects of individualized alpha frequency transcranial magnetic stimulation (alphaTMS) on the negative symptoms of schizophrenia. *Schizophrenia Bulletin 32*, 556–561.

Kim YH, You SH, *et al.* (2006). Repetitive transcranial magnetic stimulation-induced corticomotor excitability and associated motor skill acquisition in chronic stroke. *Stroke 37*, 1471–1476.

Lee SH, Kim W, *et al.* (2005). A double blind study showing that two weeks of daily repetitive TMS over the left or right temporoparietal cortex reduces symptoms in patients with schizophrenia who are having treatment-refractory auditory hallucinations. *Neuroscience Letters 376*, 177–181.

Levy R, Goldman-Rakic PS (2000). Segregation of working memory functions within the dorsolateral prefrontal cortex. *Experimental Brain Research 133*, 23–32.

Luber B, Kinnunen LH, *et al.* (2006). Facilitation of performance in a working memory task with rTMS stimulation of the precuneus: frequency- and time-dependent effects. *Brain Research 1128*, 120–129.

McMahon FJ, Buervenich S, *et al.* (2006). Variation in the gene encoding the serotonin 2A receptor is associated with outcome of antidepressant treatment. *American Journal of Human Genetics 78*, 804–814.

Maroun M, Richter-Levin G (2003). Exposure to acute stress blocks the induction of long-term potentiation of the amygdala–prefrontal cortex pathway in vivo. *Journal of Neuroscience 23*, 4406–4409.

Nahas Z, McConnell K, Collins S, *et al.* (1999). Could left prefrontal rTMS modify negative symptoms and attention in Schizophrenia? *Biological Psychiatry 45*, 1s–147s.

Nyffeler T, Wurtz P, *et al.* (2006). Repetitive TMS over the human oculomotor cortex: comparison of 1-Hz and theta burst stimulation. *Neuroscience Letters 409*, 57–60.

Rioult-Pedotti MS, Friedman D, *et al.* (2000). Learning-induced LTP in neocortex. *Science 290*, 533–536.

Robinson TE, Kolb B (2004). Structural plasticity associated with exposure to drugs of abuse. *Neuropharmacology 47*(Suppl. 1), 33–46.

Rossini D, Magri L, *et al.* (2005). Does rTMS hasten the response to escitalopram, sertraline, or venlafaxine in patients with major depressive disorder? A double-blind, randomized, sham-controlled trial. *Journal of Clinical Psychiatry 66*, 1569–1575.

Rumi DO, Gattaz WF, *et al.* (2005). Transcranial magnetic stimulation accelerates the antidepressant effect of amitriptyline in severe depression: a double-blind placebo-controlled study. *Biological Psychiatry 57*, 162–166.

Speer AM, Kimbrell TA, *et al.* (2000). Opposite effects of high and low frequency rTMS on regional brain activity in depressed patients. *Biological Psychiatry 48*, 1133–1141.

Strafella AP, Paus T, *et al.* (2001). Repetitive transcranial magnetic stimulation of the human prefrontal cortex induces dopamine release in the caudate nucleus. *Journal of Neuroscience 21*, RC157.

Talelli P, Greenwood RJ, Rothwell JC (2007). Exploring theta burst stimulation as an intervention to improve motor recovery in chronic stroke. *Clinical Neurophysiology 118*, 333–342.

Tremblay LK, Naranjo CA, *et al.* (2005). Functional neuroanatomical substrates of altered reward processing in major depressive disorder revealed by a dopaminergic probe. *Archives of General Psychiatry 62*, 1228–1236.

Vanderhasselt MA, De Raedt R, *et al.* (2006). The influence of rTMS over the left dorsolateral prefrontal cortex on Stroop task performance. *Experimental Brain Research 169*, 279–282.

Whitlock JR, Heynen AJ, *et al.* (2006). Learning induces long-term potentiation in the hippocampus. *Science 313*, 1093–1097.

Ziemann U, Muellbacher W, *et al.* (2001). Modulation of practice-dependent plasticity in human motor cortex. *Brain 124*, 1171–1181.

Ziemann U, Ilic TV, *et al.* (2004). Learning modifies subsequent induction of long-term potentiation-like and long-term depression-like plasticity in human motor cortex. *Journal of Neuroscience 24*, 1666–1672.

Ziemann U, Meintzschel F, *et al.* (2006). Pharmacological modulation of plasticity in the human motor cortex. *Neurorehabilitation and Neural Repair 20*, 243–251.

CHAPTER 39

Methodological issues in clinical trial design for TMS

Mark A. Demitrack and Sarah H. Lisanby

Introduction

Insufficient efficacy, lack of tolerability, and concerns about unwanted adverse effects due to sustained systemic exposures to antidepressant pharmacotherapies are acknowledged clinical issues that have motivated the search for a new therapeutic platform using nonpharmacological options for the treatment of depression. TMS presents a number of notable differences from the therapeutic approaches that have dominated clinical psychiatry for the past few decades. First, while the presumed mechanism of action of TMS may involve some of the familiar biochemical targets known to be affected by pharmaceutical interventions, TMS may also possess biological mechanisms that may justify its being considered as a member of a unique emerging treatment platform distinct from established pharmacotherapy platforms. Second, the method by which TMS is administered raises important questions regarding skill acquisition and training needs for practitioners. Third, clinical research with TMS raises new challenges for clinical study design and implementation that must be understood for the practicing clinician to properly interpret reported findings. Fourth, the emergence of new device-based therapeutic

methods creates a paradigm shift in the way therapeutic planning occurs. This last point may be framed as a more practical question, namely, where might a modality like TMS fit in the therapeutic sequence relative to currently approved approaches? It is also worth noting that TMS differs in such striking ways from the current therapeutic context that new areas for discussion may be raised between the patient and clinician in treatment planning. In this chapter, we provide a perspective on these issues and discuss an overall framework for understanding the emergence of TMS as a new therapeutic approach and the implications of this technology for the study and treatment of neuropsychiatric disorders, with a focus on major depression.

Challenges posed by the search for effective treatments for major depression

Major depression is a common and potentially lethal illness that can be characterized by relapse, chronicity, and varying degrees of treatment resistance. Pharmacology offers effective, convenient, and time-efficient treatment for many patients. However, a substantial proportion of patients are not effectively treated with medications alone, may experience early illness relapse, or find the side-effects intolerable even in the face of effective symptom control. According to the National Comorbidity Survey Replication (NCS-R) study (Kessler et al. 2003), of patients diagnosed with major depression during a 12 month period who sought treatment for their illness (51.6% of all patients meeting diagnostic criteria), less than half (41.9% of the 51.6%, or 21.6% of the total population of patients with major depression) had an adequate treatment course. The Sequenced Treatment Alternatives to Relieve Depression (STAR*D) trial (Fava et al. 2006; McGrath et al. 2006; Nierenberg et al. 2006; Rush et al. 2006b; Trivedi et al. 2006a,b) used a semi-naturalistic treatment algorithm designed to model as closely as possible the sequence of treatment options most commonly applied in clinical practice for the treatment of major depression. Results from the STAR*D study showed that for patients who have no strong evidence of prior treatment *non*response, the likelihood of achieving remission of symptoms after either one (Level 1) or two (Level 2) sequential treatment trials is in excess of 50%. However, once clear prospective evidence of failure to achieve benefit has been demonstrated, the likelihood of good clinical outcome drops precipitously, and hovers at disturbingly low levels after three prospective treatments have failed. For example, the reported incidence of categorical remission in patients who were treated with the monoamine oxidase inhibitor tranylcypromine was 6.9% after patients had failed to receive benefit from any of the three preceding adequately administered antidepressant options.

Also illuminating are the data on patient attrition and the consequent implications for treatment tolerability. As patients proceeded through the sequential treatment levels in STAR*D, the discontinuation rate due to treatment intolerance or adverse events rose in an almost monotonic fashion: 8.6% at Level 1, 20.5% (range: 12.5–27.2%) at Level 2, 35.2% (range: 34.2–36.2%) at Level 3, and 32.1% (range: 21.6–41.4%) at Level 4. In other words, as the expectations of efficacy diminished with increasing evidence of resistance to prior treatment, the nonadherence to, and likely intolerability of, treatment options increased dramatically. Taken together with the observations regarding efficacy, these data paint a picture of measurable, but nevertheless limited, benefit afforded by the most commonly used antidepressant treatments.

The acute efficacy outcomes with ECT are superior to pharmacotherapy, making it the treatment of choice in more resistant forms of major depression. However, even with ECT, the likelihood of therapeutic success diminishes substantially when rigorous documentation of failure of prior adequate treatment is taken into account. For example, Prudic et al. (1996) described their experience with a sample of 100 patients with unipolar nonpsychotic major depression who achieved remission after an acute course of ECT. In their sample, the overall remission rate one week after the end of acute treatment was 57.0%. When the patient group was stratified based on their rigorously verified prior history of nonresponse to at least one adequately dosed antidepressant pharmacotherapy, the remission rate in this latter subgroup was considerably lower (7.7%). In a related study, these same investigators reported the remission rates observed in a naturalistic clinical sample of ECT as administered in seven community hospital settings (Prudic et al. 2004). The average remission rate across all sites ranged from 30.3% to 46.7% depending upon the stringency of the criteria used to define remission. Interestingly, the average number of adequately administered prior antidepressant treatments in this population was only 1.2. Both of these reports underscore the fact that even seemingly moderate levels of treatment resistance may reduce the likelihood of response even to our most effective treatments.

Also of concern is the durability of clinical response to antidepressant interventions. With increasing levels of prior treatment nonresponse, the likelihood that efficacy may be lost over time also increases, suggesting that the harder it is to achieve remission, the less stable that remission is. This point has been demonstrated in the STAR*D cohort (Rush et al. 2006b). The greater the number of treatment steps required for a patient to achieve an adequate clinical response, the greater their likelihood of experiencing a relapse of their illness during the follow-up interval. For those who achieved a meaningful clinical benefit after Level 1, 40.1% relapsed within 4.1 months of follow-up, on average. For those who did not achieve an adequate outcome until after reaching Level 4 of that algorithm, the average relapse rate during follow-up had risen to 71.1%, and this occurred on average within 3.3 months. The difficulty of maintaining the gains achieved during acute treatment in more difficult to treat patients has also been examined in the community-based ECT study from Prudic *et al.* (1996) discussed above. In that report, on average, symptom scores deteriorated at a rate of 4% per day during the immediate aftermath of an acute course of ECT. After 10 days of follow-up, nearly 40% of the benefit obtained from acute ECT treatment was lost.

These findings raise important issues to be taken into account for acute and long-term treatment planning in major depression. Overall, they indicate that clear and significant variations in the acute and long-term patterns of outcome in patients with major depression can be seen when the history of prior treatment response is examined. It is possible that failure of early response to one or two treatment exposures defines a clinically notable threshold, since the likelihood of good treatment outcome to later interventions seems substantially reduced once this threshold has been exceeded. Moreover, response is more difficult to achieve with each successive attempt at treatment. In addition, these results highlight an often unrecognized fact: beyond the successful outcome with first-line treatment interventions, tolerability and adherence are major clinical concerns in treatment planning. Indeed, the magnitude of the poor treatment adherence reported in the STAR*D dataset beyond the Level 1 intervention is especially notable. The reasons for this finding are unclear and may reflect problems intrinsic to the healthcare delivery context itself, clinical features or illness comorbidities present in the patients themselves that may amplify treatment intolerance, or undesired pharmacological consequences of the medications or medication combinations used in that study.

The points presented above underscore the fact that major depression represents a therapeutically heterogeneous condition. Furthermore, the STAR*D data regarding patient attrition in particular highlights the reality that tolerability and treatment adherence are important issues especially in these later stages of treatment complexity. It is therefore important to consider the benefit–risk profile of each treatment in a comprehensive sense, as the clinical condition becomes more enduring, complex, and poorly responsive to any intervention. Furthermore, measurement of clinical outcome in the context of a treatment study becomes a more daunting challenge. A nearly exclusive focus on symptom resolution is insufficient. Measures of functional outcome, tolerability, treatment adherence, patient acceptability of the treatment options, the health services and health economic implications of the treatment intervention, and a targeted assessment of safety should be core elements of any novel treatment development program.

Therapeutic neuromodulation: defining a new treatment platform

A growing body of research has begun to examine novel antidepressant interventions that are strikingly different from the familiar and dominant modality represented by the pharmacotherapies. One of these approaches is TMS. It has been proposed that TMS, along with other device-based therapies emerging in psychiatry, may define a potential new treatment platform, since they share some common features, and because of the strikingly different approach to treatment that they present in comparison with existing therapeutics for major depression. It therefore becomes important to consider the potential impact of these novel treatments on changing the clinical process of treatment of major depression for the patient and the clinician, and implications for clinical trial design.

Device-based therapies (e.g. TMS, vagus nerve stimulation, direct current depolarization) share a common process of delivering an electrical current to the CNS. The method and localization of the current delivery differs markedly with the different interventions. These approaches exploit the fact that neuronal tissue is an electrochemically active substrate, and also build on the emerging knowledge that the pathophysiology of psychiatric diseases like depression manifest themselves as functional disturbances in distributed neuronal networks in the brain (Mayberg 1997; Mayberg *et al.* 2005). These differing approaches share some similarity with the established treatment, ECT, but they differ in several important respects. Most notably, their therapeutic benefit does not rely on the induction of a seizure as is it does with ECT. Indeed, for TMS the method is applied entirely in an outpatient setting with no surgical intervention or anesthetic procedure. Moreover, the regional anatomic and spatial localization of therapeutic intervention is much greater with TMS than it is with ECT. Both of these features contribute to the markedly reduced medical morbidity associated with these newer approaches in comparison to ECT.

Another important way in which non-pharmacological somatic interventions differ from the pharmacotherapies is that they are applied episodically rather than continuously. Medications can be thought of as exerting a tonic sustained action on the brain, moving it from one neurochemical state to another. By comparison, therapeutic neuromodulation is more temporally and spatially localized, and therefore represents a dynamic interplay of the specific intervention with the adaptive response of the brain during the extended periods of absence of the therapeutic intervention itself. While the neurobiologic significance of this difference is not completely understood, the clinical implications of this difference could be considerable. For example, it is possibly that the episodic nature of their application may contribute to their tolerability. Episodic exposure could also reduce exposure-associated risks. Bearing in mind the striking degree of attrition observed in the later treatment levels in the STAR*D study, tolerability can have a substantial impact on the risk–benefit outcome.

Next, we summarize some features of the study design methods and implementation of a recently completed research development program for TMS. Understanding the challenges to the development of these newer therapeutic approaches will be increasingly important, since research in this arena will also ultimately involve the development of newer more complex study designs and more comprehensive methods of assessing clinical outcome. Given that these studies will also increasingly be performed in more difficult to treat segments of the depressive illness spectrum will also require the development of better methods of defining study populations for research trials.

Methodological considerations in the development of a novel therapeutic neuromodulation treatment: the example of TMS for the treatment of major depression

Critical issues considered in the study design

The development of TMS for the treatment of major depression was undertaken as a promising approach that may offer unique advantages over currently available treatments, especially in the area of clinical tolerability. Prior research provided supportive evidence of proof of concept (Burt *et al.* 2002); however, certain drawbacks were identified. A review of these issues highlights some of the critical differences in the approach to development of a procedure-based intervention such as TMS compared to established pharmaceutical interventions. For example, the selection of treatment parameters and site of stimulation are critical concepts in dose selection. Prior work explored a variety of treatment parameters, and appeared to provide the most credible support for high-frequency stimulation of the left prefrontal cortex. Accumulated evidence has also indicated that earlier work may have provided too few treatment sessions over too short a period of time (i.e. <2 weeks) to provide an optimal treatment effect, in other words the number of sessions and the duration of time over which these sessions were applied are both important variables to consider (Gershon *et al.* 2003). However, given the time-intensive nature of the treatment delivery, concern was also raised regarding whether patient attrition would be unacceptably high with study durations of >2 weeks. Clear definition of the study population was also critical, since interpretation of earlier studies has often been hampered by small samples of clinically heterogeneous populations. This heterogeneity is both diagnostic (e.g. inclusion of both unipolar and bipolar patients) and therapeutic in nature (i.e. allowance of concomitant psychotropic medication use). Methods of study blinding have also presented unique challenges for a procedure-based intervention; these challenges were addressed with the development of a novel sham or placebo treatment condition, and the additional use of an efficacy rating team that was distinct from the individuals involved in delivering the treatment sessions themselves. Finally, given the fact that currently available symptom outcome measures have been developed and refined in the context of the study of pharmaceutical antidepressants, the choice of a comprehensive and broad-based array of outcome measures with a pre-specified hierarchical and structured approach to hypothesis testing was used to ensure the optimal ability to interpret the results obtained.

Clinical development program for TMS in the treatment of major depression

The two main principles that guided the development of this program were (1) a study design team used all available information and expert consensus opinion regarding prior work in TMS to guide decisions regarding optimal administration of that treatment, and (2) standard approaches using accepted scientific and regulatory guidelines for the development of antidepressants were incorporated at all stages of study design and analysis.

A multicenter pivotal trial of TMS for the treatment of unipolar depression was designed which composed three separate clinical protocols that were related in temporal sequence to one another (see Figure 39.1). The first protocol (Study 101) was a randomized controlled clinical trial designed to examine the efficacy of TMS compared to a sham TMS treatment condition. Based on prior literature, treatment parameters and study duration (up to 6 weeks) were selected to represent a maximum feasible dose study design in order to test as exhaustively as possible the evidence for efficacy of TMS. Treatment parameters included stimulation at 120% of visual motor threshold, applied at 10 pulses per second, 4 s trains, with an inter-train interval of

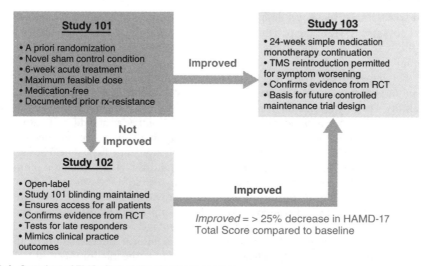

Fig. 39.1 Overview of TMS clinical development program.

26 s, for a total of 3000 pulses per session. At the time the study was designed, there was no prior work to guide estimates of patient adherence under study designs extending longer than 3 weeks; a consensus decision by the study development team was made to declare the primary efficacy outcome time point at 4 weeks to ensure that patient attrition was kept within an acceptable limit, while extending treatment for up to 6 weeks of acute treatment as a secondary endpoint to gain a better understanding of the benefit obtained from more extended treatment. The results of this randomized controlled trial have now been published (O'Reardon *et al.* 2007).

Following participation in the randomized study, patients had the option of entering a second open-label study (Study 102) which followed the same treatment sequence as the randomized controlled trial, and was available upon request for all patients (1) who had participated in the first study for at least 4 weeks of acute treatment, (2) who had received no clinical benefit from their randomized assignment, and (3) who were electing to discontinue study participation. The blinded treatment assignment received in Study 101 was not revealed upon entry to Study 102. In addition, an a priori criterion defining failure of clinical benefit in Study 101 was established; however, that specific criterion definition was concealed from the investigator in order to minimize likelihood of rating bias. The incorporation of an open-label extension study accompanying a randomized controlled trial is a relatively novel design, and has several advantages. From the patient's perspective, this ensures potential access to the experimental treatment at some point, and therefore can improve the recruitment and retention in the first study, thereby increasing its overall validity. From the research clinician's perspective, in the absence of a second confirmatory randomized trial, the outcome of this extended treatment provides the opportunity to examine patterns of clinical outcome that may be consistent with the randomized study results and therefore may lend further support to the conclusions from the controlled trial, or may help to interpret ambiguous results. Because the treatment blind from the randomized study was not broken upon entry into this study, outcomes can be examined in separate groups based upon prior treatment assignment in order to address questions of potential late responders to extended TMS exposure. Finally, the inclusion of an open-label treatment condition is particularly informative to the practicing clinician as these conditions are more similar to a naturalistic treatment setting than are those of a controlled clinical trial.

To determine whether the acute response to TMS could be maintained without abrupt loss of effect for a sufficient interval to allow continuation treatment on a known active antidepressant medication, a 3 week period of treatment transition, or Taper Phase, was included at the conclusion of the acute phase of either Study 101 or Study 102. Choice of medication in this phase was restricted to antidepressant monotherapy. This method of continuation therapy was chosen since it was felt that a specific regimen of maintenance TMS was premature to define with certainty prior to the availability of definitive acute treatment results, and because it was presumed that the proposed treatment-transition approach would, in any case, exist as a common choice of treatment continuation in clinical practice for a reasonable number of patients.

To examine persistence of benefit, a final open-label continuation trial (Study 103) was included for up to 24 weeks of clinical follow-up. During this study, all patients who had successfully established a meaningful level of clinical benefit from either Study 101 or Study 102, and who were able to transition to single-medication continuation treatment, were eligible for entry. As in Study 102, randomized treatment assignment from Study 101 was not unblinded upon entry into Study 103. If a patient experienced a protocol-defined level of symptomatic worsening, then TMS could be reintroduced at any time after entry, as an add-on treatment to the medication regimen, for a limited treatment interval on a defined schedule. The likelihood of TMS reintroduction, and the number and pattern of sessions during reintroduction, may provide clinically valuable information. For example, as in Study 102, these data provide an open-label replication of efficacy, and therefore can inform interpretation of the results observed in the randomized controlled trial. Furthermore, the timing, pattern, and number of TMS reintroduction sessions may serve as a meaningful descriptive starting point in the design of a controlled trial of maintenance treatment.

Characterization and definition of the study population

The inclusion criteria are one of the most critical aspects of the study design. Several considerations were taken into account in establishing these criteria. First, it was presumed that TMS might be sought as monotherapy for acute treatment; thus it would be important to the intrinsic efficacy of TMS in a medication-free patient population. Furthermore, to maximize the generalizability of the study results, we sought to examine the intervention in that part of the depression treatment continuum that most closely resembled the patients who were expected to commonly utilize this intervention in practice. Similarly, it is known that the most treatment-responsive population of patients with major depression are also the most placebo-responsive population. Therefore, studying a more treatment-resistant patient population would be expected to enhance the likelihood of detecting a true treatment signal by restraining the expected placebo response, and would likewise represent the population most likely to seek TMS as a clinical treatment. To address both of these issues, the use of a validated method to characterize the population in terms of prior treatment history was essential. The Antidepressant Treatment History Form (ATHF) was used to do this, as it is the only method that has been established as having prospective validity to predict subsequent treatment response (Sackeim 2001). Finally, diagnostic heterogeneity was constrained to establish the generalizability of the observed results to the larger population of patients with unipolar major depression. Patients with discrete depressive types that may be considered as clinically and perhaps biologically distinct groups, such as the depressed phase of bipolar illness or psychotic depression, were excluded from study.

Selection of outcome measures

Previous work has demonstrated that both the Hamilton Depression Rating Scale and the Montgomery–Asberg Depression Rating Scale are able to discriminate active TMS from a control condition (Fitzgerald et al. 2003; Avery et al. 2006). However, these studies have generally been small in size, and examined a relatively limited duration of treatment exposure. Few studies have determined whether the pattern and temporal course of treatment with TMS may be expected to parallel that observed with pharmaceutical treatments. In other words, are there modifications in measurement of outcome that are unique to these novel treatments and must be taken into account? It has been argued, for example, that the time course of action for vagus nerve stimulation accumulates with time, and may indeed be a relatively late-appearing phenomenon (Rush et al. 2006a). Hence, instruments designed to be sensitive to short-term change with pharmaceutical treatments may be reasonable, but not optimal, in studying the clinical effects of nonpharmacological interventions.

In the absence of a known theoretically superior rating method, we chose to employ a pre-specified ordering of hypothesis testing using established symptom-rating instruments, and to supplement these measures using additional examinations of functional status, quality of life, and global measures of clinical status. We expected that, while no one specific measure may be predicted to be a superior metric based on prior work, a coherent pattern of change across various methods of clinical assessment and several meaningful clinical domains of outcome should serve to lend strength to any conclusion of clinical efficacy. Further, we expected that the large sample size intended to be assembled in this multisite study would enhance future development of better assessment techniques unique to TMS and related technologies. The pre-specified list of test measures is presented in Table 39.1.

Safety assessment

Safety and tolerability are key issues not just for the risk:benefit ratio assessment of novel therapeutics, but also for their impact on patient adherence and compliance with a time-intensive treatment delivery paradigm. Patient adherence to the treatment protocol could be objectively verified by attendance at treatment sessions, as

Table 39.1 Pre-specified order of analysis of clinical outcomes

Primary outcome
1. MADRS total score change from baseline[a]
Secondary outcomes
1. HAM24 total score change from baseline[a]
2. HAM17 total score change from baseline[a]
3. Response rate (50% reduction: MADRS, HAMD24, HAMD17)[a]
4. MOSSF-36 and Q-LES-Q (functional status and quality-of-life outcomes)[b]
5. Remission rate (MADRS <10, HAMD24 <11, HAMD17 <8)[a]
6. HAMD scale factor scores change from baseline[a]
7. IDS-SR total score change from baseline[b]
8. CGI severity change from baseline[a]
9. PGI improvement change from baseline[b]

[a]Clinician-rated outcomes.
[b]Patient-rated outcomes.
MOSSF-36, Medical Outcomes Study Short Form 36; MADRS, Montgomery–Asberg Depression Rating Scale; HAMD, Hamilton Depression Rating Scale; Q-LES-Q, Quality of life Enjoyment and Satisfaction Questionnaire; IOS-SR, Inventory of Depressive Symptoms Self Report; CGI, Clinician Global Impressions; PGI, Patient Global Impressions.

well as the overall rate of discontinuation. It is well known that attrition in clinical trials of antidepressants is not uncommon, and can be high (e.g. >20–30%) (Leon *et al.* 2006). This considerably limits the generalizability of the conclusions drawn from such work, and can present significant statistical analytic problems in addressing the issue of missing data.

While the expected adverse events of acute short-term exposure to TMS are well documented in the literature, there are essentially no acute treatment trials that have examined continuous acute exposure in large samples for periods longer than 3 weeks, so knowledge of the time course of resolution of these common adverse events, or the potential appearance of late-appearing adverse events of medical consequence, is less well understood. In addition to general measures of safety reflected in spontaneously reported adverse events, there are specific safety concerns for TMS that require more targeted assessment, for example demonstration of a lack of effect on formally tested and appropriately sensitive measures of cognitive function. The TMS coil also produces an audible high-frequency click, which Loo

et al. (2001) have suggested may result in short-term changes in auditory threshold. Therefore, formal hearing testing was indicated in this trial involving chronic exposure over prolonged periods.

Methodology controls implemented in this study

Device-based therapies, which depend upon clinician delivery, present specific areas of interest with respect to study methodology controls. While some features of study methodology control, such as rater training, are a general concern of antidepressant clinical development, a number of areas of study quality control arise that are unique issues to the development of a novel therapeutic such as TMS. Because of the fact that this intervention is a clinical procedure, the question arises as to how the control condition will be established to conduct a properly blinded randomized controlled treatment study. Early research in TMS addressed this issue with a single-blind methodology. In this approach, only one active treatment coil is used, with the sham condition created by physically orienting the coil at a tangent or orthogonal to the surface of the head, so that the therapeutically actively magnetic field is oriented away from the brain. Several studies have shown that, even with this less than optimal condition, patients who are naïve to the TMS treatment setting, and hence have no prior knowledge of what constitutes a correct coil position, are unable to accurately predict the active from the sham treatment condition when asked directly either early or late in the treatment course. Nevertheless, others have studied these proposed sham treatment conditions, and have demonstrated that intracranial exposure to magnetic field may still be considerable with some coil orientations, therefore raising the possibility that certain sham conditions in some studies may have been partially active treatment conditions (Loo 2000). Sham coils were subsequently developed that contained a metal shield, effectively blocking the magnetic field from entering the brain. These coils looked more like active TMS from the patient perspective since they can be held in the same fashion as active TMS, but the investigator would still be unblinded by having to select which coil to use and in some cases the coils were nonidentical in appearance. Furthermore, the lack of scalp sensation from metal-shield sham coils could cue the patient.

A novel sham methodology was developed specifically for this study. All study sites were supplied with three identical treatment coils, identical in weight, external appearance, and acoustic properties during active pulsing. One coil was unblinded and known to be active, and was used for all open-label conditions and for treatment dose setting in all patients. The remaining two coils were blinded, and identified only by external labels by letter. The sham condition in one coil was achieved by the placement of an aluminum-based metal shield concealed in the coil housing that distorted the spatial orientation of the magnetic field such that <5% of the field generated by the sham coil reached the patient's cortical surface. Pre-specified randomization and coil assignment was indicated by electronic information previously recorded in flash memory embedded on a unique electronic treatment card assigned to each patient, and that was used to control the electronic console pro-

viding electrical power to the magnetic coils. When the patient's unique treatment card was inserted into the console, the operator was then prompted to attach the specific coil (labeled 'b' or 'c') defined by the randomized treatment assignment, and then to manually connect the appropriate coil prior to proceeding with treatment. This blinding method reduces the number of cues that could unblind the operator and patient.

It is now standard in the clinical development of new psychopharmacological agents to have specified training procedures for the efficacy ratings. Beyond those concerns about rater reliability, the study of novel therapeutic neuromodulation technology presents new issues with regard to standardization of training procedures for the operation of the device itself. In the case of TMS, this involves procedural training on the use of the device for the setting of the treatment parameters, a process referred to as motor threshold determination, as well as the method of locating the therapeutic target on the surface of the head. Satisfactory and reproducible methods to accomplish each of these steps were established in the engineering design of the device itself. Training and verification of competence of these techniques was established prior to first treatment, and then periodic on-site visual observation of treatment technique was conducted for the duration of this study. Reliability had to be demonstrated by each TMS operator on the determination of motor threshold, selection of treatment site, and delivery of the treatment session following a detailed study manual and TMS device operator's manual.

Study design limitations

While this study included a number of innovations presented above, there remain some limitations. For example, the possible impact of varying coil-to-cortex distance was not measured or taken into account. The risk of inadvertent unblinding upon crossover from the blinded to the open phase could not be avoided. Only a single-delivery paradigm (coil type, coil location, parameters of stimulation, treatment schedule, duration of treatment, and maintenance strategies) was tested; thus we could not examine the potential efficacy of other possible delivery paradigms.

Therapeutic neuromodulation: clinical considerations for treatment planning

Device-based approaches to therapeutic neuromodulation hold the promise of significant clinical advantages compared to existing treatments for major depression, but this promise needs to be supported with rigorous evidence in well-designed and properly blinded multicenter trials. Prominent among the potential advantages are their significantly improved tolerability profile, an especially critical issue in the more difficult to treat realm of major depression. Taking as an illustrative example the design and methods of a recently completed multisite TMS treatment study (O'Reardon *et al.* 2007), we have briefly reviewed several aspects of these new interventions that will require careful examination by the practitioner and researcher alike. First, for procedure-based therapies, training and skill acquisition for the administration of these therapies will be required. This represents a new field that may be unfamiliar to practicing psychiatrists who typically do not participate in clinical procedures unless they perform ECT (which defines a minority of practitioners). Second, these new treatment approaches demand the development of new approaches to clinical research, including innovations in study design and outcome measures. Understanding and properly interpreting the results of this new body of research will require close attention to strengths and limitations in trial design. Finally, these approaches create a different clinical context for the physician–patient interaction that remains to be explored. Considered together, these issues and challenges may herald a paradigm shift in our understanding of the potential for new therapeutic options in the treatment of major neurological and psychiatric illnesses. The current state of the evidence for the value of this approach to the treatment of psychiatric and neurological disorders will be carefully examined in the following chapters.

References

Avery DH, Holtzheimer PE, Fawaz W, *et al.* (2006). A controlled study of repetitive transcranial magnetic stimulation in medication-resistant major depression. *Biological Psychiatry 59*, 187–194.

Burt T, Lisanby SH, Sackeim HA (2002). Neuropsychiatric applications of transcranial magnetic stimulation: a meta analysis. *International Journal of Neuropsychopharmacology 5*, 73–103.

Fava, M, Rush, AJ, Wisniewski, SR, *et al.* (2006). A comparison of mirtazapine and nortriptyline follwing two consecutive failed medication treatments for depressed outpatients: a STAR*D Report. *American Journal of Psychiatry 163*, 1161–1172.

Fitzgerald PB, Brown TL, Marston NAU, *et al.* (2003). Transcranial magnetic stimulation in the treatment of depression: a double-blind, placebo-controlled trial. *Archives of General Psychiatry 60*, 1002–1008.

Gershon AA, Dannon PN, Grunhaus L (2003). Transcranial magnetic stimulation in the treatment of depression. *American Journal of Psychiatry 160*, 835–845.

Kessler RC, Berglund P, Demler O, *et al.* (2003). The epidemiology of major depressive disorder: results from the National Comorbidity Survey Replication (NCS-R). *Journal of the American Medical Association 289*, 3095–3105.

Leon AC, Mallinckrodt CH, Chuang-Stein C, *et al.* (2006). Attrition in randomized controlled clinical trials: methodological issues in psychopharmacology. *Biological Psychiatry 59*, 1001–1005.

Loo CK, Taylor JL, Gandevia SC, McDarmont BN, Mitchell PB, Sachdev PS (2000). Transcranial magnetic stimulation (TMS) in controlled treatment studies: are some "sham" forms active? *Biological Psychiatry 47*, 325–331.

Loo CK, Sachdev P, Elsayed H, *et al.* (2001). Effects of a 2–4 week course of repetitive transcranial magnetic stimulation (rTMS) on neuropsychological functioning, electroencephalogram, and auditory threshold in depressed patients. *Biological Psychiatry 49*, 615–623.

McGrath, PJ, Stewart, JW, Fava, M, *et al.* (2006). Tranylcypromine versus venlafaxine plus mirtazapine following three failed antidepressant medication trials for depression: a STAR*D Report. *American Journal of Psychiatry 163*, 1531–1541.

Mayberg HS (1997). Limbic-cortical dysregulation: a proposed model of depression. *Journal of Neuropsychiatry and Clinical Neuroscience 9*, 471–481.

Mayberg HS, Lozano AM, Voon V, *et al.* (2005). Deep brain stimulation for treatment-resistant depression. *Neuron 45*, 651–660.

Nierenberg, AA, Fava, M, Trivedi, MH, *et al.* (2006). A comparison of lithium and T3 augmentation following two failed medication treatments for depression: a STAR*D Report. *American Journal of Psychiatry 163*, 1519–1530.

O'Reardon JP, Solvason HB, Janicak PG, *et al.* (2007). Efficacy and safety of transcranial magnetic stimulation in the acute treatment of major depression: a multisite randomized controlled trial. *Biological Psychiatry*, in press.

Prudic J, Haskett RF, Mulsant B, *et al.* (1996). Resistance to antidepressant medications and short-term clinical response to ECT. *American Journal of Psychiatry 153*, 985–992.

Prudic J, Olfson M, Marcus SC, *et al.* (2004). Effectiveness of electroconvulsive therapy in community settings. *Biological Psychiatry 55*, 301–312.

Rush AJ, Sackeim HA, Marangell LB, *et al.* (2006a). Effects of 12 months of vagus nerve stimulation in treatment-resistant depression: a naturalistic study. *Biological Psychiatry 58*, 355–363.

Rush AJ, Trivedi MH, Wisniewski SR, *et al.* (2006b). Acute and longer-term outcomes in depressed outpatients requiring one or several treatment steps: a STAR*D report. *American Journal of Psychiatry 163*, 1905–1917.

Sackeim HA (2001). The definition and meaning of treatment resistant depression. *Journal of Clinical Psychiatry 62*(Suppl. 16), 10–17.

Trivedi MH, Rush AJ, Wisniewski SR, *et al.* (2006a). Evaluation of outcomes with citalopram for depression using measurement-based care in STAR*D: implications for clinical practice. *American Journal of Psychiatry 163*, 28–40.

Trivedi MH, Fava M, Wisniewski SR, *et al.* (2006b). Medication augmentation after the failure of SSRIs for depression. *New England Journal of Medicine 354*, 1243–1252.

CHAPTER 40

TMS in the treatment of major depressive disorder

Colleen Loo

Introduction

Several researchers using TMS as an investigative tool in neurology noted incidentally that a number of subjects reported mood changes after TMS stimulation (e.g. Bickford *et al.* 1987). As a result of these fortuitous observations, studies of TMS as a potential treatment for depression were initiated.

Evidence from preclinical studies on antidepressant effects of repetitive TMS

There is a growing body of literature on animal experiments (mainly involving rats) using behavioral paradigms or investigations of neurobiological mechanisms to assess the antidepressant effects of repetitive (r)TMS (for summary of studies, see Post and Keck 2001). Early studies reported results for rTMS comparable to those of antidepressants and electroconvulsive shock (ECS), i.e. reduced immobility time in the Forced Swim Test, increase in apomorphine-induced stereotypy, increased seizure threshold, downregulation of β-adrenergic receptors. However, these studies were also criticized methodologically in that stimulation was given to the whole rodent brain (due to mismatch between the size of the TMS coil and the rat head) rather than to frontal areas (as in clinical trials in depressed human subjects). Some later studies used specially designed coils to deliver localized frontal stimulation. Other methodological issues in rodent studies involve inadvertent effects arising from handling or the stress of restraint required during rTMS, and the multiplicity of investigative methods used (e.g. tissue homogenization vs sampling of extracellular concentrations via microdialysis, *in vivo* experiments vs brain slices). Nevertheless, the body of evidence is cautiously supportive of antidepressant effects for rTMS and findings are summarized below according to models of antidepressant mechanisms.

Models of stress coping

One of the most consistent findings is the reduction in immobility time in the Forced Swim Test in rats after acute (single session) or chronic (multiple session) rTMS (Post and Keck 2001), a finding that is consistent with the effects of antidepressants and electroconvulsive shock (ECS). This has been interpreted as an analogy of increased active coping under stress. As dopamine agonists reduce immobility in the Forced Swim Test, it is possible that this behavioral change is related to findings of increased dopaminergic transmission, particularly in the hippocampus and nucleus accumbens, after rTMS (Post and Keck 2001; Keck *et al.* 2002; Zangen and Hyodo 2002).

Regulation of the adrenocorticoid system

Other studies support a role for rTMS in normalizing the hypothalamic–pituitary–adrenal (HPA) system disinhibition associated with depression. As seen with antidepressants, rTMS in rats has been shown to reduce the output of corticotrophin, adrenocorticotrophic hormone (ACTH), and corticosterone in response to stress (Czeh et al. 2000, 2002; Keck et al. 2000a, 2001), possibly also by reducing hypothalamic vasopressin release (Keck et al. 2000b). In contrast, though, Hedges et al. (2003) found that rTMS led to increases in ACTH and corticosterone levels, despite controlling for experiment-related stress by including a control group which received sham rTMS. The finding of Ji et al. (1998) that rTMS induced activation of immediate–early gene expression in the paraventricular nucleus of the hypothalamus is consistent with the above observations that rTMS regulates neuroendocrine effects via the HPA axis.

Effects on neurotransmitters

Effects of rTMS on the intracerebral release of monoamines and amino acids have been studied, mainly in rats. While there are reports of increased dopamine transmission (see 'Models of stress coping' above) and dopamine transport (Ikeda et al. 2005), findings of effects on noradrenergic and serotonergic systems are less consistent between studies (for review see Post and Keck 2001), possibly because these changes are dependent on the exact brain areas stimulated and the stimulation paradigm used, factors which often differed between studies. Effects on other intracerebral neuroactive agents (e.g. taurine, serine, aspartate, glutamate, glutamine, arginine) and receptor systems (e.g. N-methyl-D-aspartate) have also been found after rTMS (Keck et al. 2000b; Lisanby and Belmaker 2000), though the role of these changes is unclear. Overall, the neurochemical effects of rTMS show both similarities and discrepancies with the effects of antidepressants and ECS.

Neuroprotective effects

Several studies have suggested that rTMS may have neuroprotective effects. Müller et al. (2000) reported increased brain-derived neurotrophic factor (BDNF) mRNA in the dentate gyrus and other hippocampal areas after rTMS in rats. Repetitive TMS has been reported to be neuroprotective against oxidative stress in mouse cells in vitro (Post et al. 1999) and to decrease the effects of neurotoxins on nigrostriatal neurons in rats (Funamizu et al. 2005). However, rTMS failed to reverse stress-induced suppression of hippocampal neurogenesis in rats (Czeh et al. 2002). The suggestion that rTMS may act by neuroprotective effects is in keeping with recent findings for antidepressants and ECS, though more work will be needed to clarify whether there are any effects of rTMS on neurogenesis.

Kindling and quenching

Several studies have reported lasting changes in neuronal excitability after acute or chronic rTMS, findings akin to the phenomena of kindling and quenching, and long-term potentiation and long-term depression. Jennum and Klitgaard (1996) reported kindling-like effects (reduced time to seizure onset in rats exposed to epileptogenic chemicals) after chronic rTMS (daily sessions for 30 days) at 50 Hz. Ebert and Ziemann (1999) found that acute 20 Hz rTMS did not affect the kindling process but raised the threshold for elicitation of after-discharges, i.e. epileptiform activity. Wang et al. (1996) reported variable induction of long-term potentiation or long-term depression in rodent auditory cortex with acute 8 Hz rTMS. Though the same rTMS paradigm was used in all cases, the polarity of the effects appeared to be specific for the animal studied and the exact placement of the recording electrodes. They also reported that the frequency (range 1–10 Hz) of rTMS influenced immediate effects on neuronal firing (spike rate), whereas others have found that the intensity of the stimulus also determines the induction and polarity of lasting changes (Ogiue-Ikeda et al. 2003).

While findings are not uniform across studies, the differences likely arise from differences in the areas of brain stimulated with rTMS, the exact location of electrodes used for evoking and recording neuronal responses, and variations in the parameters of rTMS given. Nevertheless, the above findings suggest that rTMS at certain stimulation parameters can induce lasting changes in neuronal function, in line with the effects of anticonvulsants (used as mood stabilizers) and electrical stimulation (cf. ECS), supporting a role for rTMS in the treatment of depression.

Evidence from clinical trials and meta-analyses

There are now a considerable number of open (i.e. non-sham-controlled) and sham-controlled trials of rTMS in depressed subjects. As placebo effects are likely to be associated with a new, novel, and high-technology treatment, and as the delivery of rTMS involves non-specific clinical effects (e.g. from daily attendance for rTMS and contact with research staff), sham-controlled studies are crucial to the assessment of the efficacy of rTMS and will be the main focus of this section.

Sham-controlled trials and meta-analyses

Given the relatively large number of trials involved and the small number of subjects in each trial (see Table 40.1 for summary of sham-controlled trials), an overview of the results is best approached by first considering the published meta-analyses (including a Cochrane review) which have attempted to combine the evidence from trials despite disparate study designs and stimulation parameters (McNamara et al. 2001; Holtzheimer et al. 2001; Burt et al. 2002; Kozel and George 2002; Martin et al. 2003; Couturier 2005).

Essentially, all but one of the meta-analyses found that rTMS was statistically superior to a sham control, and the main issue at hand concerns the clinical significance of the results, rather than statistical proof of the efficacy of rTMS. (The only meta-analysis to find a negative result applied selection criteria that excluded the majority of trials, leaving little power to detect any significant difference (Couturier 2005).)

Holtzheimer et al. (2001) conducted a separate analysis for trials with a parallel design. The latter distinction is important as the blinding of subjects in crossover studies is questionable, given the difference in scalp sensation with active and sham rTMS (for discussion of methodological issues see Chapter 39, this volume). The calculated weighted mean effect size for the seven trials (based on the difference in outcome between active and sham treatment and with studies weighted for sample size and variability) was 0.88. Though statistically convincing, none of the studies showed a mean reduction in Hamilton Rating Scale for Depression (HRSD) scores of ≥50% and only 22% of subjects showed this degree of improvement with active rTMS, compared with 7.9% of subjects after sham rTMS. (See Table 40.1 for summary of mean improvement in HRSD scores and number of responders for individual studies.)

A comprehensive review by Burt et al. (2002) reported on two separate meta-analyses, for nine open depression trials of rTMS and 16 controlled trials. Calculations yielded weighted mean effect sizes of 1.37 for the open studies and 0.67 for the controlled studies. These are moderate to large effect sizes, though the difference between the two values is interesting, suggesting that placebo effects contributed substantially to positive outcomes in the open studies. Again, though the results show strong statistical support for the efficacy of rTMS, the magnitude of the clinical response is relatively low, with subjects improving by 37% on average (as measured by HRSD and Montgomery–Asberg Depression Rating Scales (MADRS)) in the open studies, and 23.8% (active) and 7.3% (sham) in the controlled studies.

A Cochrane review by Martin et al. (2003) compared the results of 14 randomized controlled studies suitable for quantitative analysis. They found that high-frequency (>1 Hz) left prefrontal rTMS and low-frequency (<1 Hz) right prefrontal rTMS were statistically superior to a sham comparison, but only at one time point (immediately after the 2 weeks of treatment, with the difference not sustained 2 weeks later) and only for the HRSD. As the overall difference between active and sham treatment was not large, though significant (standardized mean difference of –0.35 for high-frequency left prefrontal rTMS), they concluded that there was not strong enough evidence at that stage to support the benefit of rTMS as an antidepressant treatment.

Studies published since these meta-analyses were undertaken (i.e. from 2003 onwards) have

Table 40.1 Summary of sham-controlled studies of repetitive TMS for the treatment of depression

Publication	Study design	Sham method	n	Medication resistant	Repetitive TMS treatment parameters	Mean % change, HRSD[a]	No. of responders[b]	Adverse events[c]
Pascual-Leone et al. 1996	Multiple crossover: L[d] active R active L sham R sham vertex active	Active coil at 45°[e]	17	Yes	10 Hz, 90%, 20 × 10 s, 5 days × 2000 stim[f]	L active: 48% Other: 4–12%	LDLPFC active: 4/17 Other: 0/17	Generalized seizure (1/17) Headache (7/17)
George et al. 1997	Crossover: L active L sham	Active coil at 45°	12	No	20 Hz, 80%, 20 × 2 s, 10 days × 800 stim	Active: 16% Sham: 13%	Active: 1/12 Sham: 0/12	Headache (4/12)
Klein et al. 1999	Parallel: R active R sham	Active coil off scalp, at 90°	70	No	1 Hz, 110%, 2 × 60 s, 10 days × 120 stim	Active: 47% Sham: 22%	Active: 17/36 Sham: 8/34	Discomfort/ muscle twitch (5/36) Headache (3/36)
Loo et al. 1999	Parallel: L active L sham	Active coil at 45°	18	Yes	10 Hz, 110%, 30 × 5 s, 10 days × 1500 stim	Active: 23% Sham: 25%	Active: 0/9 Sham: 1/9	Pain/discomfort (7/9) Mild headache (3/9)
Padberg et al. 1999	Parallel: L active 10 Hz L active 0.3 Hz L sham 10 Hz	Active coil, 1 wing at 90°	18	Yes	90%, 5 days × 250 stim 10 Hz: 5 × 5 s. 0.3 Hz: 75s	10 Hz: 6% 0.3 Hz: 19% Sham: 6%	All: 0/6	10 Hz: pain (3/6), headache (1/6) 0.3 Hz: pain (2/6), headache (1/6)
Berman et al. 2000	Parallel: L active L sham	Active coil at 45°	20	Yes	20 Hz, 80%, 20 × 2 s, 10 days × 800 stim	Active: 39% Sham: 0.5%	Active: 1/10 Sham: 0/10	Active: headache (6/10) Sham: headache (5/10)
Eschweiler et al. 2000	Crossover: L active L sham	Active coil at 90°	12	?[g]	10 Hz, 90%, 20 × 10 s, 5 days × 2000 stim	Active: 22% Sham: 7%	?	Headache (3/12)

Study	Design	N		Parameters	Response	Response	Side effects
George et al. 2000	Parallel: L active 20 Hz, L active 5 Hz, L sham (20/5 Hz)	Active coil, 1 wing at 45°	30 Yes	100%, 10 days × 1600 stim 20 Hz: 40 × 2s 5 Hz: 40 × 8 s	20 Hz: 26% 5 Hz: 48% Sham: 21%	20 Hz: 3/10 5 Hz: 6/10 Sham: 0/10	Headache (10/20) 5 Hz: pain (2/10)
Garcia-Toro et al. 2001	Add-on[h] to sertraline Parallel: L active L sham	Active coil at 90°	28 No	20 Hz, 30 × 2 s, 90%, 10 days × 1200 stim	Active: 38% Sham: 34%	Active: 4/11 Sham: 3/11	Muscle tension headache (3/11)
Lisanby et al. 2001	Add-on to sertraline Parallel: L active 10 Hz R active 1 Hz Sham	Active coil, 1 wing at 90°	36 Most	L 10 Hz: 20 × 8s R 1 Hz: 2 × 800s 110%, 1600 stim	L 10 Hz: 21% R 1 Hz: 20% Sham: 13%	?	?
Manes et al. 2001	Parallel: L active L sham	Active coil, handle on head	20 Yes	20 Hz, 20 × 2 s, 80%, 5 days × 800 stim	Active: 37% Sham: 32%	Active: 3/10 Sham: 3/10	Active: pain (1/10), headache (4/10), discomfort (4/10) Sham: discomfort (4/10), anxiety (1/10)
Dolberg et al. 2002	Parallel: active sham	?	20 ?	10 days	Active: 29% Sham: 17%	?	?
Padberg et al. 2002	Parallel: L active 100% L active 90% L sham	Active coil, 1 wing at 90°	30 Yes	10 Hz, 15 × 10s, 10 days × 1500 stim. 100%/90%/ sham	100%: 30% 90%: 15% Sham: 7%	100%: 3/10 90%: 2/10 Sham: 0/10	Mild headache/scalp numb (2/20) 100%: aversive sensation (2/10), unpleasant (2/10) 90%: aversive sensation (3/10), unpleasant (3/10)

Table 40.1 (*Cont.*) Summary of sham-controlled studies of repetitive TMS for the treatment of depression

Publication	Study design	Sham method	n	Medication resistant	Repetitive TMS treatment parameters	Mean % change, HRSD[a]	No. of responders[b]	Adverse events[c]
Fitzgerald et al. 2003	Parallel: L active 10 Hz R active 1 Hz Sham – L 10 Hz or R 1 Hz	Active coil, 1 wing at 45°	60	Yes	L 10 Hz: 20 × 5 s, 1000 stim R 1 Hz: 5 × 60 s, 300 stim 100%, 10 days	L 10 Hz: 13.5% R 1 Hz: 15% Sham: 0.76% (MADRS)	L 10 Hz: 8/20 R 1 Hz: 7/20 Sham: 2/20	Discomfort/pain (7/40, 10 Hz > 1 Hz) Headache (6/40) 10 Hz: dizziness (1/20) 1 Hz: dizziness (1/20)
Herwig et al. 2003	Add-on to antidepressants Parallel: L or R active (as per PET findings) Sham	Active coil to midline parieto-occipital area	25	Yes (9/25)	15 Hz, 100 × 2 s, 3000 stim × 10 days Active: 110% Sham: 90%	Active: 31% Sham: 2%	Active: 4/13 Sham: 0/12	Mild headache (n = 3)
Hoppner et al. 2003	Parallel: L active 20 Hz L sham 20 Hz R active 1 Hz	Active coil at 90°	30	No	L 20 Hz: 20 × 2 s, 90%, 800 stim R 1 Hz: 2 × 60 110%, 120 s, stim 10 days	L 20 Hz: 17% R 1 Hz: 10.5% Sham: 23%	L 20 Hz: 2/9 R 1 Hz: 1/10 Sham: 2/10	20 Hz: headache (1/9)
Loo et al. 2003a	Parallel: Bilateral active Bilateral sham	Inactive coil	19	Yes	15 Hz, 24 × 5 s, 90%, 15 days × 1800	Active: 24% Sham: 21%	Active: 2/9 Sham: 1/10	Pain (5/9), headache (3/9), anxious (3/9), tearful (1/9)
Nahas et al. 2003	Parallel: L active L sham	Active coil, 1 wing at 45°	23	No	5 Hz, 40 × 8 s, 110%, 10 days × 800 stim	Active: 25% Sham: 25%	Active: 4/11 Sham: 4/12	Nil reported
Holtzheimer et al. 2004	Parallel: L active L sham	Active coil at 45°	15	Yes	10 Hz × 5 s, 110%, 10 days × 1600 stim	Active: 32% Sham: 28%	Active: 2/7 Sham: 1/8	Mild pain (some subjects /7)

Study	Design	Coil orientation	N	Randomized	Parameters	Response	Adverse events (n)	Adverse events
Jorge et al. 2004	Parallel: L active L sham	Active coil, 1 wing at 90°	20	Yes	10 Hz, 110%, 20 × 5 s, 10 days × 1000 stim	Active: 38% Sham: 13%	Active: 3/10 Sham: 0/10	Active and sham: headache (6/20), discomfort (5/20, exacerbation of insomnia (1/20)
Kaufmann et al. 2004	Parallel: R active R sham	Active coil at 45°	12	Yes	1 Hz, 2 × 60 s, 110%, 10 days × 120 stim	Active: 48% Sham: 30%	Active: 4/7 Sham: 2/5	Nil reported
Koerselman et al. 2004	Parallel: L active L sham	Active coil at 45°	55	No	20 Hz, 20 × 2 s, 80%MT, 10 days × 800 stim	Active: 18.5% Sham 15.4%	?	Active and sham: suicidal ideation (2/55), dizziness (1/55)
Mosimann et al. 2004	Parallel: L active L sham	Active coil at 90°	24	Yes	20 Hz, 40 × 2 s, 100%, 10 days × 1600 stim	Active: 20% Sham: 17%	Active: 4/15 Sham: 0/9	Active: tearful (2/15), metallic taste (1/15), toothache (1/15), suicidal ideation (1/15), conjunctivitis (1/15) Sham: headache (2/9), dizziness (1/9), nausea (2/9)
Avery et al. 2006	Parallel: L active L sham	1 wing at 90°	68	Yes	10 Hz, 32 × 5 s, 110%, 15 days × 1600 stim	Active: 33% Sham: 16%	Active: 11/35 Sham: 2/33	Pain (14/35)

Table 40.1 (*Cont.*) Summary of sham-controlled studies of repetitive TMS for the treatment of depression

Publication	Study design	n	Medication resistant	Repetitive TMS treatment parameters	Mean % change, HRSD[a]	No. of responders[b]	Adverse events[c]
Christyakov et al. 2005	Parallel: L active 3 Hz + placebo L active 10 Hz + placebo R active 3 Hz + placebo R active 10 Hz + placebo Sham rTMS + clomipramine	Active coil at 90°	59	No	3 Hz: 5 × 30 s, 110%, 10 days × 450 stim 10 Hz: 10 × 5 s, 100%, 10 days × 500	?	L 3 Hz: 6/12 L 10 Hz: 1/10 R 3 Hz: 2/12 R 10 Hz: 2/9 Clomipramine: 2/16	Active (10 Hz > 3 Hz): scalp discomfort (1/43), pain and facial contraction (8/43) Sham: headache (1/16)
Miniussi et al. 2005	Crossover (8 week washout between): L active 17 Hz L sham 17 Hz coil and Crossover (8 week washout): L active 1 Hz L sham 1 Hz	Active coil, 60% MT, 25 cm wood between scalp	29 22	Yes Yes	17 Hz, 40 × 3 s, 110%, 5 days × 2040 stim 1 Hz, 40 × 10 s, 110%, 5 days × 2000 stim	17 Hz: 32% 1 Hz: 23% Sham (both groups): 17%	?	?
Rossini et al. 2005a	Parallel: L active 100% L active 80% L sham	Active coil at 90°	54	Yes	15 Hz, 20 × 2 s, 100%/80%/sham, 10 days × 600 stim	3 weeks post rTMS 100%: 69% 80%: 44% sham: 15%	3 weeks post rTMS: 100%: 11/18 80%: 5/18 sham: 1/16	Headache: 100% (2/12), 80% (2/15) Discomfort: 100% (3/12)

Study	Groups	Coil	N	Blind	Parameters	Response	Responders	Side effects
Rossini et al. 2005b	Add-on to escitalopram, sertraline or venlafaxine. Parallel: L active L sham	Active coil at 90°, 90% MT	99	No	15 Hz, 30 × 2 s, 100%, 10 days × 900 stim	Active: 51% Sham: 33%	Active: 25/49 Sham: 10/47	Active + venlafaxine: headache (1/17), cervical pain (1/17) Sham + escitalopram: agitation (1/17) Sham + venlafaxine: gastric symptoms (1/16)
Rumi et al. 2005	Add-on to amitryptiline Parallel: L active L sham	Placebo coil	46	No	5 Hz, 25 × 10s, 120%, 20 days × 1250 stim	Active: 57% Sham: 35%	Active:21/22 Sham:11/24	Active: headache (21/22), scalp pain (19/22), burning in the scalp (21/22), cervical pain (21/22) Sham: headache (22/24), scalp pain (17/24), burning in the scalp (15/24), cervical pain (18/24)
Su et al. 2005	Parallel: L active 20 Hz L active 5 Hz Sham	Active coil at 90°	30	Yes	100%, 10 days × 1600 stim 20 Hz: 40 × 2 s 5 Hz: 40 × 8 s	20 Hz: 58% 5 Hz: 54% Sham: 16%	20 Hz: 5/10 5 Hz: 6/10 Sham: 1/10	Active: pain – two dropouts (5 Hz), headache (20 Hz: 2/10; 5 Hz: 2/10), hypomania (1/20) Sham: headache (1/10)
Fitzgerald et al. 2006	Parallel Active: R 1 Hz + L 10 Hz Sham	Active coil at 45°	50	Yes	R 110%, 3 × 140 s, L 100%, 15 × 5 s, 30 days × 1170 stim	Active: 26% Sham: 1% (MADRS; sham up to 20 days)	Active: 11/25 Sham: 2/25	Active: headache (5/25), nausea (3/25) Sham: headache (2/25)

Table 40.1 (*Cont.*) Summary of sham-controlled studies of repetitive TMS for the treatment of depression

Publication	Study design	Sham method	n	Medication resistant	Repetitive TMS treatment parameters	Mean % change, HRSD[a]	No. of responders[b]	Adverse events[c]
Januel et al. 2006	Parallel: R active Sham	Placebo coil	27	No	1 Hz, 2 × 60 s, 90%, 10 days × 120 stim, then 6 days × 120 stim	Active: 54% Sham: 26%	Active: 7/11 Sham: 1/16	Active/sham: headache, 8% principally in the first session
Garcia-Toro et al. 2006	Parallel Active: R 1 Hz alternated with L 20 Hz Active (SPECT): 1 Hz + 20 Hz as per SPECT regions Sham	Active coil at 45°	30	Yes	R 30 × 60 s alternated with L 30 × 2 s, 110%, 10 days × 3000 stim	Active: 26% Active (SPECT): 28% Sham: 6%	Active: 2/10 Active (SPECT): 2/10 Sham: 0/10	Active: scalp discomfort, slight, self-limited and transitory muscle-tension headaches (7/20) Sham: ?
McDonald et al. 2006	Parallel Active L 10 Hz + R 1 Hz Active R 1 Hz + L 10 Hz Sham	Active coil at 90°	62	Yes	R 1 × 600 s, L 20 × 5 s, 110%, 10 days × 1600 stim	Active (L + R): 37% Active (R + L): 19% Sham: 29%	Active (L + R): 7/25 Active (R + L): 3/25 Sham: 1/12	Not reported
Herwig et al. 2006	Add-on to venlafaxine or mirtazapine. Parallel: L active L sham	Active coil at 45° over L temporal cortex, 90% MT	127	No	10 Hz, 110%, 15 days × 2000 stim	Active: 40% Sham: 37%	Active: 20/62 Sham: 20/65	Active: headache (3/62), scalp pain (1/62), nausea (1/62). Sham: headache (1/65), dizzy (1/65), scalp pain (2/65)

Study	Design	Coil arrangement	n	Medication	Treatment parameters	HRSD reduction[a]	Response/remission	Adverse events
Loo et al. (2007)	Parallel L active Sham	Inactive coil on head, active coil discharged 1 m away.	38	Yes	10 Hz, 30 × 5 s, 10 days × 2 sessions/day × 1500 stim	Active: 39% Sham: 26%	Active: 6/19 Sham: 3/19	Active: scalp pain (15/19), headache (8/19), tearfulness (4/19), brow twitching 3/19, nausea (1/19), anxiety (1/19), agitation (1/19) feeling high (1/19). Sham: nausea (1/19)
O'Reardon et al. (2007)	Parallel: L active L sham Antidepressants added during 3 week taper	Placebo coil	301	Yes	10 Hz, 75 × 4 s, 120%, 30 days × 3000 stim, then 6 days over 3 week taper	End 6 weeks: Active: 23% Sham: 13%	End 6 weeks: Active: 25% Sham: 14%	Active: headache (96/146), scalp pain/discomfort (77/146), muscle twitch (34/146), toothache (12/146). Sham: headache (87/155), scalp pain/discomfort (8/155), muscle twitch (5/155), toothache (1/155).

[a]Mean percentage reduction in Hamilton Rating Scale for Depression (HRSD) scores; increases indicated by ↑.

[b]Defined as ≥50% reduction in HRSD score from baseline. Denominator is number of completers in that treatment group. Based on ratings immediately after the end of rTMS except where otherwise stated. MADRS, Montgomery–Asberg Depression Rating Scale.

[c]Number of subjects per treatment condition who experienced adverse events. Refers to active treatment unless otherwise specified.

[d]L, left; R, right. Refers to dorsolateral prefrontal cortex except where stated otherwise.

[e]Arrangement used for sham TMS. 45° means that coil was tilted to contact the scalp at 45°. For '1 wing' only one side of the figure-8 coil was in contact with the scalp.

[f]Treatment parameters are 10 Hz, 90% resting motor threshold, 20 trains of 10 s duration, daily sessions for 5 days, total 2000 stimuli per stimulation session.

[g]?: data not available.

[h]Repetitive TMS was commenced simultaneously with new course of antidepressant medication.

generally supported the finding of antidepressant effects for rTMS, with several studies reporting that approximately half the sample receiving active rTMS achieved response (i.e. ≥50% reduction in depression scores from baseline) during the study period (see Table 40.1). The most definitive study to date has been a large multicenter trial of 301 patients with unipolar depression. Results supported statistically significant effects of high-frequency rTMS to the left prefrontal cortex (O'Reardon et al. 2007). Following 6 weeks of rTMS, 25% of patients met response criteria compared with 14% of sham-treated patients (P < 0.007).

Comparison of rTMS and ECT

Several studies randomized depressed subjects to receive rTMS or electroconvulsive therapy (ECT). Though often quoted as demonstrating comparable efficacy for rTMS and ECT, a careful inspection of the results actually suggests a superior efficacy for ECT (Grunhaus et al. 2000, 2003; Pridmore et al. 2000; Janicak et al. 2002; Eranti et al. 2007). In some of these studies the emerging trend in favor of ECT did not reach statistical significance (Grunhaus et al. 2000; Pridmore et al. 2000; Janicak et al. 2002), or was only significant on some measures (Pridmore et al. 2000), probably because of the small sample sizes involved. Others found ECT clearly superior (Eranti et al. 2007). Of interest, Grunhaus et al. (2000) found a clear difference between psychotic and nonpsychotic depression, with rTMS equivalent to ECT in efficacy only for the latter group.

It should be noted that the mean improvement with rTMS in the above studies (40–56% decrease in HRSD scores) is higher than in most depression treatment trials of rTMS. There was no sham comparison in these studies, making interpretation of this result difficult. Another important factor is that the outcomes of a 4 week course of rTMS were evaluated in these studies, in contrast to the 2 week treatment course in most sham-controlled studies. These studies have also been criticized for their modest ECT response (mean decrease in HRSD scores 48–66%). In comparison with the effect sizes quoted above for rTMS, Burt et al. (2002) calculated effect sizes of 2.26 for bilateral ECT

and 2.12 for high-dose right unilateral ECT, as given in research studies (Sackeim et al. 2000).

In other comparisons of rTMS and ECT, Dannon and Grunhaus (2001) reported that seven of 17 depressed subjects who had failed treatment with rTMS subsequently responded to ECT, and that relapse rates after treatment response to rTMS and ECT were similar at 6 month follow-up (20%, n = 41). In favor of rTMS is the lack of significant cognitive impairment found after a course of rTMS (see 'Predictors of response to rTMS' below).

Approaches to rTMS stimulation

There are many variations in the way rTMS can be given as a clinical treatment, involving choices over treatment site, stimulation parameters (stimulus frequency, intensity, waveform, number of stimuli per session), and treatment course (spacing and number of sessions). Studies to date have differed with respect to the above (see Table 40.1), with almost no two studies using identical rTMS parameters, except in deliberate attempts at replication. Thus, though the meta-analyses are useful for combining the results of multiple studies with small individual samples, it is necessary also to consider the results of individual studies which have used different rTMS approaches. As yet, there is little empirical evidence directly comparing the relative merits of these approaches.

Stimulation site

Most rTMS depression trials have given high-frequency rTMS to the left prefrontal cortex, encouraged by positive early results for this approach (George et al. 1995; Pascual-Leone et al. 1996). A few investigators have chosen to trial low-frequency (≤1 Hz) rTMS to the right prefrontal cortex. There are several theoretical reasons for this approach. One-hertz rTMS is relatively safe, with a much lower risk of inducing an accidental seizure (Wassermann 1998). It is more tolerable at a given stimulus intensity and does not require expensive TMS machines capable of high-frequency stimulation. Motor cortex studies suggest that high- and low- frequency rTMS have opposite effects on the excitability of neurons in brain cortex

(Pascual-Leone *et al.* 1994; Chen *et al.* 1997). There is considerable evidence from neuropsychological, lesion, and imaging studies that the left and right hemispheres have contrasting roles in mood regulation (Silberman and Weingartner 1987; Davidson 1995). Thus it might be expected that low-frequency rTMS to the right prefrontal cortex may be as likely to have antidepressant effects as high-frequency rTMS to the left prefrontal cortex.

Several open studies have reported on the treatment of depressed subjects with rTMS at <1 Hz to the right prefrontal cortex, or to a combination of right and left prefrontal cortices (e.g. Geller *et al.* 1997; Feinsod *et al.* 1998; Menkes *et al.* 1999; Kapiletti *et al.* 2001; Dragasevic *et al.* 2002). Overall results were positive, with mean changes in HRSD scores ranging from 26% to 42% over a 2 week period, though it is uncertain how much of this should be attributed to a placebo effect.

Several sham-controlled trials have examined the antidepressant effects of low-frequency right prefrontal rTMS in depressed subjects (Klein *et al.* 1999; Lisanby *et al.* 2001a; Fitzgerald *et al.* 2003; Hoppner *et al.* 2003; Kauffmann *et al.* 2004; Januel *et al.* 2006). In the first study Klein *et al.* (1999) reported promising results with 17 of 35 subjects improving by >50% in HRSD scores in the active group, compared with eight responders from the sham group of 32. However, subjects in this study were not medication resistant. Replication studies using the same (or almost identical) rTMS treatment parameters have reported both negative (Hoppner *et al.* 2003) and positive findings (Kauffmann *et al.* 2004; Januel *et al.* 2006). Other studies using a much higher number of TMS pulses either failed to find a significant difference between sham and active treatment (Lisanby *et al.* 2001a) or found significant differences, though the overall degree of improvement was small (Fitzgerald *et al.* 2003). Some of the above studies directly compared low-frequency right prefrontal rTMS and high-frequency left prefrontal rTMS but failed to find significant differences between the two approaches (Lisanby *et al.* 2001a; Fitzgerald *et al.* 2003; Hoppner *et al.* 2003). Taken together, the results of the above studies do not allow conclusions that low-frequency right prefrontal rTMS is more or less efficacious than high-frequency left prefrontal rTMS.

Another strategy has been the development of 'bilateral' rTMS, i.e. rTMS to both left and right prefrontal cortices. In the first study of simultaneous bilateral prefrontal rTMS, Loo *et al.* (2003a) tested the hypothesis that the superior efficacy of bilateral ECT over unilateral ECT arose from the stimulation of both frontal cortices. Subjects received simultaneous high-frequency rTMS to both prefrontal cortices or a sham control. Results were negative, with suggestions that high-frequency right prefrontal rTMS may have induced tearfulness in some subjects (Loo *et al.* 2003a). Other investigators have studied different combinations of low- and high-frequency rTMS to the left and right prefrontal cortices – these are considered in the section below on variations in site and frequency of stimulation.

Novel approaches involving rTMS to other cortical areas (e.g. parietal cortex, cerebellum) for the treatment of depression have also been proposed (Schutter and van Honk 2005), though there have been no published clinical trials of rTMS to these areas.

rTMS parameters – stimulus frequency

Apart from the low-frequency studies, most studies have given rTMS within the range of 5–20 Hz, as changes in cortical excitability have been demonstrated after rTMS in this range of frequency (Pascual-Leone *et al.* 1994). Preliminary results from rat studies suggest that higher stimulus frequencies may have greater antidepressant potency (Sachdev *et al.* 2002). Four sham-controlled studies have directly compared rTMS at different stimulus frequencies to the left prefrontal cortex (Padberg *et al.* 1999; George *et al.* 2000; Miniussi *et al.* 2005; Su *et al.* 2005). Padberg *et al.* (1999) stimulated at 10 and 0.3 Hz, George *et al.* (2000) and Su *et al.* (2005) (replication of George *et al.* 2000) used 20 and 5 Hz rTMS and Miniussi *et al.* (2005) compared 17 and 1 Hz rTMS. The total number of stimuli was kept constant as far as possible for the different groups within each study (see Table 40.1). While Su *et al.* found no difference between 20 and 5 Hz, the two earlier studies found that the lower frequency stimulation yielded superior outcomes, though a statistical difference between the two groups was not demonstrated. Miniussi *et al.* (2005) found better results with 17 than with 1 Hz rTMS, though the comparison was only over a 1 week treatment period and larger differences may have emerged over a longer treatment course.

Shajahan *et al.* (2002) attempted to investigate the differential effects of 5, 10, and 20 Hz rTMS to the left prefrontal cortex. Unfortunately the sample only included five subjects per group and no sham control group. Though on average subjects improved in depression scores by 40% and five subjects showed ≥60% improvement, no difference was evident between treatment groups. Overall, the sample sizes involved in the above studies and the lack of clear findings preclude any conclusions. Thus the optimum stimulus frequency for antidepressant efficacy remains unclear and the field awaits large adequately powered trials directly comparing the effects of a range of treatment frequencies in depressed subjects.

rTMS parameters – stimulus frequency and site of stimulation

Several studies have compared forms of rTMS involving a variety of combinations of stimulus frequencies and stimulation to the left and right prefrontal cortices. In a novel study, Conca *et al.* (2002) assigned 36 subjects to receive either 10 Hz rTMS to the left prefrontal cortex and 1 Hz rTMS to the right prefrontal cortex consecutively in each session (i.e. nonsimultaneous bilateral rTMS), 10 and 1 Hz rTMS (alternating trains) to the left prefrontal cortex, or 10 Hz to the left prefrontal cortex (i.e. a control group receiving more typical rTMS treatment). The total number of stimuli was the same for each group. All groups improved (23–31% reduction in HRSD scores) but no significant difference was found between them. It is possible that differences may have emerged with a much larger sample. Hausmann *et al.* (2004) compared 20 Hz left prefrontal rTMS alone, 20 Hz left prefrontal rTMS followed by 1 Hz right prefrontal rTMS, and a sham control. Disappointingly, no differences were found between the groups despite the use of relatively intense rTMS parameters (i.e. long stimulus trains, large number of stimuli). In the most comprehensive study of this type to date, Christyakov *et al.*

(2005) randomized depressed subjects to receive placebo medication and rTMS to the left or right prefrontal cortices, at 10 or 3 Hz, or to sham rTMS and clomipramine, over a 2 week period. Results clearly favored left prefrontal rTMS at 3 Hz, though the advantage over 10 Hz left prefrontal rTMS could also have resulted from the higher stimulation intensity used for the 3 Hz group (110% vs 100%).

rTMS parameters – stimulus intensity

Most rTMS trials have reported stimulus intensity relative to the subject's resting motor threshold, i.e. the lowest stimulus intensity necessary to produce a motor response in a relaxed contralateral muscle when TMS is given over the primary motor cortex. As reported in 'Kindling and quenching' above, animal studies have suggested that stimulus intensity may be crucial in determining the induction and polarity of lasting changes in neuronal excitability. Mathematical modelling has demonstrated that higher intensities of stimulation are associated with higher peak current intensities in underlying brain cortex (Padberg *et al.* 2002a).

In depressed subjects, two controlled studies directly compared different intensities of stimulation (see Table 40.1). Padberg *et al.* (2002a) gave rTMS at 100% and 90% of motor threshold, and used a sham condition mathematically modelled as corresponding to 40%. Rossini *et al.* (2005) compared rTMS at 100% and 80% of motor threshold, and sham treatment. In both studies, significantly greater clinical improvement was found for the group receiving the higher stimulation intensity, supporting intuitive expectations that a higher intensity of stimulation would be more effective.

The same phenomenon was described from a different perspective by Kozel *et al.* (2000) and Mosimann *et al.* (2002). The latter found that in older depressed subjects, a greater distance between the stimulating coil and brain cortex, i.e. greater prefrontal atrophy, was associated with a poorer antidepressant response to rTMS. Kozel *et al.* (2000) did not find this association in a younger sample. In these studies greater coil–cortex distance is a proxy for stimulation at reduced intensities as the intensity of the magnetic field in TMS diminishes rapidly with distance from the stimulating coil (Barker 1991).

Previous reviews of the rTMS literature have suggested that studies using higher stimulus intensities were more likely to have positive results (Gershon *et al.* 2003). However, this is not clear from an overview of sham-controlled studies published to date, with a large proportion of negative studies using stimulus intensities at the higher end of the range (110–120% motor threshold) (see Table 40.1).

Number of stimuli, length and spacing of treatment course

Likewise, an inspection of the published sham-controlled studies does not support the conclusion that treatment protocols using a higher total number of TMS stimuli per session had better therapeutic outcomes (see Table 40.1), though this may be possible. As mentioned above, it is difficult to compare studies directly due to the number of differences in treatment parameters and patient subtypes between studies. The few studies that reported a longer (3–4 week) course of rTMS have generally demonstrated continued improvement over the third and fourth weeks, though apart from Januel *et al.* (2006), Rumi *et al.* (2005), and Avery *et al.* (2006), these studies were either not sham-controlled (Grunhaus *et al.* 2000, 2003; Janicak *et al.* 2002) or provided continued rTMS in an open extension after an initial sham-controlled phase (Loo *et al.* 1999; Fitzgerald *et al.* 2003). Januel *et al.* (2006) found no continued improvement over the third and fourth weeks of rTMS but this may have been due to a reduction in the frequency of treatment sessions over these weeks.

Given that optimal efficacy is not achieved with other antidepressant treatments (e.g. medications) in a 2 week period (Stassen *et al.* 1993), it is to be expected that a treatment period of longer than 2 weeks would be necessary for optimal outcomes with rTMS. It is of interest that Rossini *et al.* (2005) found a substantial degree of further improvement over a 3 week follow-up period after completion of a 2 week course of rTMS. This suggests that the response to rTMS may have a delayed component and that other studies to date may have underestimated the antidepressant response to rTMS by evaluating outcomes immediately after the end of treatment (often necessitated to facilitate crossover of the sham group into active treatment). Arguing against this, though, is the finding of the Cochrane review (Martin *et al.* 2003) that no significant treatment effects were found 2 weeks after the completion of rTMS.

Further considerations

Apart from the above, there are a number of other factors in the way rTMS is administered that have not been adequately investigated.

Extrapolating from animal studies on the electrophysiological phenomena of long-term depression and long-term potentiation (Christie and Abraham 1994), it is likely that, apart from stimulus parameters such as frequency and intensity, the frequency of TMS sessions and prior stimulation conditioning may be important factors in the attempt to produce lasting effects on neuronal excitability. rTMS has been given each weekday in most studies so far but there is little information on the optimal frequency of treatment sessions – in practice this could range from multiple sessions per day to two to three sessions per week.

Recent TMS motor cortex studies support earlier observations from investigations in animals that prior conditioning of the cortex, perhaps by stimulation at different parameters, can enhance the effects of stimulation. For example, Iyer *et al.* (2003) demonstrated that the physiological depressant effect of 1 Hz rTMS on motor cortical excitability was considerably enhanced by prior conditioning with 6 Hz rTMS given at subthreshold stimulus intensity. Thus, optimal therapeutic effects with rTMS may require treatment with a particular sequence of stimulus parameters (varying in frequency, intensity, number of stimuli, length of stimulus trains, and/or intervals between trains), with the sequence extending across individual treatment sessions, over the treatment course, or possibly both. Furthermore, drawing parallels from the ECT literature, it may also be that adjustment of stimulus parameters over a treatment course, e.g. a gradual increase in stimulus intensity, is necessary for maximal therapeutic effects. Though researchers may be guided by findings from motor cortex and animal studies in understanding the likely implications of the above factors for the efficacy of rTMS, clinical trials involving large numbers of subjects will be necessary to test the relative outcomes of different stimulation paradigms.

The waveform of the electrical pulse used in TMS (and hence induced magnetic and electrical fields) varies between machines from different manufacturers. There is evidence from motor cortex studies to suggest that TMS with stimuli of different waveforms (e.g. biphasic versus monophasic) leads to discernibly different outcomes in terms of lasting changes in cortical excitability (Sommer *et al.* 2002; Taylor and Loo 2007). Thus, as in the development of ECT (Abrams 2002), refinement of the stimulus waveform may emerge as an important factor in optimizing therapeutic outcomes with TMS.

The most appropriate role for rTMS as a treatment in depression, whether it be a stand-alone treatment, an add-on to other treatments (antidepressants, ECT, etc.) for synergistic effects (Pridmore 2000; Garcia-Toro *et al.* 2001; Hausmann *et al.* 2004; Rumi *et al.* 2005), or an alternative means of inducing seizures (Lisanby *et al.* 2001b), is yet to be determined. The only controlled comparison of rTMS and an antidepressant (clomipramine) found a superior outcome for rTMS, though the 2 week treatment period studied was almost certainly suboptimal for both treatments and thus not an adequate evaluation of their relative antidepressant potentials (Christyakov *et al.* 2005).

Conclusion

Most of the sham-controlled data for the efficacy of rTMS in treating depression comes from studies with a 2 week sham-controlled period. The outcomes suggest clear statistical proof of superiority over placebo effects but not large clinical effects. However, the few studies which have reported results of longer periods of stimulation suggest that a longer course of rTMS is necessary for optimal therapeutic outcomes. Thus evidence for the efficacy of rTMS should be derived from sham-controlled studies of 4–6 weeks duration.

Predictors of response to rTMS

Noting the variability in subjects' response to rTMS treatment – both within and between studies – some investigators have attempted to predict those subjects likely to improve with

rTMS, or to identify subgroups which may benefit from specific types of rTMS.

There are early neuroimaging reports that cerebral hypoactivity and hyperactivity may predict response to high- and low-frequency rTMS respectively (Kimbrell *et al.* 1999; Speer *et al.* 1999), and that blood flow and metabolic levels in the anterior cingulate cortex may predict response to rTMS (Kimbrell *et al.* 1999; Mottaghy *et al.* 2002). However, further studies are required to confirm and extend these preliminary suggestions (see also 'Baseline cerebral blood flow and cerebral metabolic rate and subsequent response to rTMS' below).

Several observations regarding predictors of response have been made from clinical trials. rTMS has been reported to be less effective in psychotic depression (Grunhaus *et al.* 2000, 2003), older subjects (Figiel *et al.* 1998; Padberg *et al.* 1999; Su *et al.* 2005; Loo *et al.* 2007), and those with depressive episodes of longer duration (Holtzheimer *et al.* 2004). As mentioned in 'Approaches to rTMS stimulation' above, the finding of reduced response in older subjects may be attributable to prefrontal atrophy and hence a reduction in the effective stimulus intensity.

Conca *et al.* (2000a) analyzed a large number of subject factors in their trial of TMS as an add-on treatment and reported that a shorter episode of illness, relative frontal hypoactivity, faster activity in the baseline EEG, higher TSH levels, smaller doses of lorazepam, and higher trazodone levels were all correlated with response to TMS. However, the statistical significance of these findings is in doubt, as some differences were small in magnitude or only present on one of many occasions of measurement, and the analysis comprised a large number of comparisons in a small number of subjects without correction for the multiple comparisons.

Eschweiler *et al.* (2001) summarized data from 10 open and seven sham-controlled studies to derive lists of positive and negative predictors. Younger age, somatic signs of anxiety, and lack of cortical hyperactivity with 10 Hz treatment (and the converse for 1 Hz) were reported as positive predictors, with prefrontal atrophy, cognitive impairment, psychotic symptoms, and ECT failure listed as negative predictors. Given the many methodological differences between studies (discussed above), these lists are best understood as amalgamations of observations reported by different investigators, but the interaction between these factors and their relative importance are unclear.

Novel approaches to identifying predictors of response have been undertaken. Schiffer et al. (2002) reported that mood improvement with right rather than left lateral visual-field stimulation was associated with greater subsequent improvement with left prefrontal rTMS treatment. Right visual field stimulation is thought to activate the left hemisphere (Schiffer et al. 2002). These results are intriguing given the laterality hypothesis of mood regulation (Silberman and Weingartner 1986; Davidson 1995).

Padberg et al. (2002b) reported that depressed subjects who improved with sleep deprivation were less likely to improve with rTMS treatment and suggested that this may be related to the opposite effects of the two treatments on anterior cingulate cortex activity. However, the possibility that the prior sleep deprivation may have primed subjects against subsequent response to rTMS cannot be excluded, though the two interventions were separated by 5 days and HRSD scores had returned to baseline levels prior to rTMS. Nevertheless, these preliminary findings are interesting, suggesting the potential for subjects to be selected for rTMS treatment by undergoing a simple test. The results of both of these studies need to be replicated in many more subjects and tested with a sham control for rTMS before their clinical utility can be assessed.

Selecting patients who are likely to respond to rTMS treatment or tailoring aspects of rTMS treatment according to subject factors (e.g. higher stimulus intensities for older subjects) are approaches by which the efficacy of rTMS may be increased and merit further research in the development of rTMS as a treatment.

Safety

A few serious adverse effects have been reported with rTMS treatment in clinical trials. (For a review of adverse effects of rTMS in normal subjects and other experimental conditions, see Wassermann 1998; Loo et al., in press.) Seizures occurred in two depressed subjects receiving rTMS (Pascual-Leone et al. 1996; Conca et al. 2000b). In both cases stimulation parameters were relatively intense and were outside recommended safety guidelines (Wassermann 1998), and both patients had commenced medications which probably lowered their seizure thresholds (amitryptyline and haloperidol; venlafaxine) just prior to the incidents. Where rTMS is given within suggested parameter limits and subjects are carefully screening for seizure risk, the risk of seizure is very low. Repetitive TMS has also induced mania in a few depressed bipolar patients (Garcia-Toro 1999; Dolberg et al. 2002; Su et al. 2005) and hypomania in a subject with unipolar depression (George et al. 1995). Zwanzger et al. (2002) reported the unusual occurrence of rTMS precipitating delusions in a depressed nonpsychotic subject.

Apart from the above, the majority of depressed subjects receiving rTMS treatment have reported no adverse effects or minor side-effects (see Table 40.1). In most cases these are pain or discomfort during stimulation (due to stimulation of scalp nerves and muscles) and headache, often after rTMS. Repetitive TMS has been safely given to a pregnant depressed subject (Nahas et al. 1999) but evidence for its safety in pregnancy is only anecdotal at this stage.

Several studies administered comprehensive batteries of neuropsychological tests before and after a course of rTMS treatment, without finding any significant deficits after rTMS. As practice effects (leading to improved scoring on tests) may have obscured any detrimental effects of rTMS, it is reassuring that no differences were found between subjects receiving a course of active or sham rTMS in controlled studies (Padberg et al. 1999; Little et al. 2000; Loo et al. 2001, 2003a; Speer et al. 2001; Hoppner et al. 2003; Fitzgerald et al. 2003; Holtzheimer et al. 2004; Jorge et al. 2004; Mosimann et al. 2004; Januel et al. 2006; Avery et al. 2006).

Mechanisms of action of rTMS treatment

Apart from the animal studies reviewed above in 'Evidence from preclinical studies on antidepressant effects of rTMS', evidence for the mechanisms by which rTMS may exert an antidepressant effect comes mainly from neuroimaging studies of the effects of prefrontal rTMS on cerebral blood flow (CBF) and cerebral metabolic rate (CMR), or patterns of cerebral activity correlating with response to rTMS, and investigations of associated endocrine changes in clinical trials of rTMS in depressed subjects.

Baseline cerebral blood flow and cerebral metabolic rate and subsequent response to rTMS

A number of investigators have attempted to identify CBF and CMR patterns at baseline (i.e. prior to rTMS) which were associated with subsequent therapeutic response to a course of rTMS treatment.

Two studies giving subjects high (20 Hz) and low (1 Hz) frequency rTMS in crossover depression trials suggested that stimulation at these two frequencies had opposite cerebral effects, and that it may be possible by prior neuroimaging to predict subjects' differential responses to these two forms of rTMS. Speer et al. (1999) found that prefrontal hypoperfusion (as measured by ^{15}O PET) prior to rTMS treatment was associated with better response to 20 Hz rTMS and that prefrontal hyperperfusion was associated with better response to 1 Hz rTMS. Similarly, Kimbrell et al. (1999) reported that global hypometabolism (measured by [^{18}F]deoxyglucose PET) at baseline was associated with subsequent response to 20 Hz rTMS and that there was a trend for baseline global hypermetabolism to predict response to 1 Hz rTMS. Antidepressant response to 20 Hz rTMS was also correlated with baseline hypometabolism in several regions (cerebellar, temporal, anterior cingulate, occipital) (Kimbrell et al. 1999). In both studies, subjects who improved with rTMS at 20 Hz tended to worsen with rTMS at 1 Hz and vice versa. Furthermore, Speer et al. (2000) subsequently reported that 20 Hz rTMS led to increases in global and regional CBF whereas 1 Hz rTMS led to decreases in regional CBF (see 'Neuroimaging studies of effects of rTMS on cerebral activity' below).

There is remarkable concordance in the findings of these studies, possibly because of similarities in design, though different neuroimaging technology was used. Considered together, these results suggest that high- and low-frequency rTMS act therapeutically by specifically correcting regional and/or global cerebral hypo- and hyperactivity respectively, and that individuals are likely to respond to one treatment or the other (but not both), depending on their initial status.

However, the results of other studies comparing baseline neuroimaging findings and rTMS response do not fit neatly into the simple model proposed above. Teneback et al. (1999) found that responders to 20 or 5 Hz rTMS tended to have increased inferior frontal rCBF (measured with ^{99m}Tc bicisate single-photon emission computed tomography (SPECT)) at baseline compared to nonresponders, and this difference was further accentuated after treatment. Unfortunately, only six subjects in each group were available for this comparison and subjects receiving 20 or 5 Hz rTMS were pooled for the analysis.

In an ambitious study investigating associations between the effects of 0.2 Hz TMS on mood and a number of other measures (serum antidepressant levels, thyroid function, serum magnesium concentration, rCBF, rCMR, electroencephalogram), Conca et al. (2000a) sought to outline several factors predicting response. Among these, rCBF and rCMR (double-isotope SPECT–HMPAO (technetium-99m hexamethylpropylene amine oxime), [^{18}F]deoxyglucose) were assessed for the orbitofrontal, superior temporal, and basal ganglia regions prior to TMS. They reported that TMS responders ($n = 8$) only had hypoactivity in the frontal region, whereas nonresponders ($n = 4$) showed hypoactivity in all three areas at baseline. This could be construed as relative frontal hypoactivity predicting response to low-frequency TMS, again a finding conflicting with the first two reports above. However, limitations to the generalizability of these results include the small numbers of subjects (especially in the nonresponder cell), a unique use of TMS (given bilaterally to nine brain sites, as an add-on treatment), analysis of a large number of measures without correction for multiple comparisons, and confinement of the SPECT analysis to three brain regions.

Mottaghy et al. (2002) directly compared rCBF at baseline (^{99m}Tc bicisate SPECT) and percentage change in HRSD after a course of 10 Hz rTMS. They found that mood improvement was correlated with increased relative rCBF in neocortical areas (including the left inferior frontal region) and decreased relative rCBF in limbic areas. Again, these results do not concur with the observations by Speer et al. (1999) and Kimbrell et al. (1999).

The above studies reported associations between CBF and CMR findings at baseline and subsequent mood improvement after a course of rTMS treatment. However, the approach used cannot differentiate between improvement due to rTMS and improvement from other factors (e.g. spontaneous remission). Though some of the studies included control subjects who received sham rTMS (Kimbrell et al. 1999; Teneback et al. 1999), these subjects were not included in the neuroimaging/rTMS response analysis, possibly because of insufficient numbers.

Overall, despite the similarity of the findings of Speer et al. (1999, 2000) and Kimbrell et al. (1999), it is not possible to reconcile the above results into a single explanatory model, apart from the preliminary observation that abnormal rCBF and rCMR in the frontal regions appear to be associated with response to rTMS treatment.

In a logical development from the above, two studies have attempted to optimize rTMS treatment by using high- or low-frequency stimulation to correct hypo- or hypermetabolism in specific brain areas. Herwig et al. (2003) gave 15 Hz prefrontal rTMS to the hemisphere which showed ≥5% lesser glucose metabolism than the contralateral side. An rCMR difference of this magnitude could only be identified in five of 13 subjects. Though superior to sham stimulation, antidepressant effects in this study were not greater than those of other studies in which rTMS treatment was standardized, i.e. not individualized according to neuroimaging findings.

Only one study has directly compared the effects of rTMS tailored to baseline functional abnormalities and standard rTMS. Garcia-Toro et al. (2006) measured rCBF in four regions (left and right prefrontal, left and right temporo-parietal) in depressed subjects, and then gave 20 Hz rTMS to the region showing relative hyperperfusion and 1 Hz rTMS to the relatively

hypoperfused region. There was no difference in mood outcomes compared with a control group which received 20 Hz rTMS to the left prefrontal cortex and 1 Hz rTMS to the right prefrontal cortex, though both groups improved relative to a sham-control group.

Though the results of these two initial studies were not promising, the above approach may hold potential for the future optimization of rTMS and deserves further research. It is possible that more sophisticated analysis of baseline metabolism/perfusion is needed, e.g. patterns of activation in limbic networks, and that tailoring of rTMS to neuroimaging findings requires a paradigm other than high-frequency stimulation for hypometabolic areas and low-frequency stimulation for hypermetabolic areas.

Neuroimaging studies of effects of rTMS on cerebral activity

The literature on effects of prefrontal rTMS on CBF and CMR is complex as studies have varied in methodology in a number of areas: subjects (healthy or depressed, medication status); TMS parameters (e.g. frequency and intensity of TMS); scanning techniques – SPECT, positron emission tomography (PET), functional magnetic resonance imaging (fMRI); timing (imaging during or after TMS); and methods of analysis. These factors are likely to account for many of the differences in findings between studies, as reviewed elsewhere in this volume.

In keeping with findings of reduced rCBF and rCMR in the lateral prefrontal cortex in depressed subjects (Drevets 2000), most rTMS trials employ high-frequency rTMS to the left dorsolateral prefrontal cortex (DLPFC), with early reports that mood improvement in a depressed subject was accompanied by normalization of global hypometabolism (George et al. 1995).

Subsequently, several (but not all) studies have supported the concept that high-frequency left prefrontal rTMS leads to increased cerebral activity (rCBF, rCMR, BOLD response) at the stimulation site in depressed (Teneback et al. 1999; Speer et al. 2000; Catafau et al. 2001; Mottaghy et al. 2002; Loo et al. 2003b) and healthy subjects (Paus et al. 2001; Barrett et al. 2004).

Some studies have explored the simple hypothesis that high-frequency rTMS has an excitatory effect on brain functioning, reflected in rCBF or rCMR increases, with converse findings for rTMS at <1 Hz. Three studies compared the effects of high-frequency and 1 Hz left prefrontal rTMS within the same experimental paradigm. Speer *et al.* (2000) examined the effects of a 10 day course of left prefrontal stimulation in depressed subjects. They found only rCBF decreases after 1 Hz rTMS (right prefrontal cortex, left medial temporal cortex, left basal ganglia, left amygdala) and only rCBF increases after 20 Hz rTMS at the stimulation site, limbic and paralimbic areas (right prefrontal cortex, left cingulate gyrus, left amygdala, bilateral uncus, hippocampus, parahippocampus, thalamus, cerebellum). Of interest, they reported that rCBF changes after 1 Hz rTMS were of lesser area, magnitude, and significance than changes after 20 Hz rTMS. Disappointingly, the magnitude of rCBF changes was not correlated with the degree of clinical improvement.

Loo *et al.* (2003b) found that mean rCBF increased at the stimulation site during stimulation with 15 Hz left prefrontal rTMS and conversely slightly decreased (nonsignificant finding) during 1 Hz rTMS in depressed subjects, though the magnitude of the changes was small (~3%). Moreover, complex but differing networks of rCBF changes were found during 15 Hz (increases in the inferior frontal cortices, right dorsomedial frontal cortex, posterior cingulate, and parahippocampus; decreases in the right orbital corex, right subcallosal gyrus, and left uncus) and 1 Hz rTMS (increases in right anterior cingulate, bilateral parietal cortices, bilateral insula, and left cerebellum). Correlations between percentage rCBF changes at the stimulation site, thalamus, and basal ganglia were examined in an attempt to elucidate possible network connections. Positive correlations were found between the left DLPFC and left thalamus (15 Hz), and between left and right DLPFC (1 Hz), and a negative correlation between the left DLPFC and left basal ganglia (1 Hz).

Likewise, Barrett *et al.* (2004) reported that rCBF changes (overall mean increase) in the left mid-DLPFC during 10 Hz stimulation to this site in normal subjects were correlated with rCBF changes at remote cortical and subcortical sites (perigenual anterior cingulate gyrus, insula, thalamus, parahippocampus, caudate nucleus). No significant changes were found at the stimulation site during 1 Hz rTMS in this study.

Loo *et al.* (2003b) observed that far greater inter-individual variability in rCBF at the stimulation site was found with 1 Hz than with 15 Hz rTMS. This may explain why changes in cerebral activity at the stimulation site have been more consistently found with high-frequency rTMS (usually mean increases) than with 1 Hz rTMS.

Several studies have examined the immediate effects of 1 Hz rTMS to the left prefrontal cortex. Nahas *et al.* (2001) measured the BOLD (fMRI) response in normal subjects during 1 Hz rTMS. They found both local and distant changes (auditory cortex, bilateral prefrontal cortex) during rTMS, with greater activation after higher-intensity rTMS. The same group confirmed these findings in a later study in depressed subjects, reporting increases in the BOLD fMRI response in the left prefrontal cortex and associated limbic areas (bilateral midfrontal cortex, right orbitofrontal cortex, left hippocampus, mediodorsal thalamic nucleus, bilateral putamen, pulvinar, and insula) and decreases in the right ventromedial frontal cortex (Li *et al.* 2004). These findings appear at odds with the hypothesis that 1 Hz rTMS inhibits cerebral activity. However, the precise nature of the neuronal activity giving rise to the BOLD signal is unclear, including whether activation represents inhibitory or excitatory activity.

Kimbrell *et al.* (2002) investigated the effects of active and sham 1 Hz rTMS on the left prefrontal cortex in 14 normal subjects, using [^{18}F]deoxyglucose PET. Regional CMR effects at the stimulation site and distally were demonstrated during both active rTMS (decreases in right prefrontal cortex, bilateral anterior cingulate, bilateral basal ganglia, hypothalamus, midbrain, and cerebellum; increases in bilateral posterior temporal cortex, and bilateral occipital cortex) and sham rTMS (decreases in left dorsal anterior cingulate, left basal ganglia; increases in posterior association cortex, occipital cortex), including a global increase in CMR after sham stimulation. The authors postulated that the stress of TMS as a novel experience may have led to a global increase in CMR which was offset by the inhibitory effects of rTMS in the active stimulation group. Nevertheless, by directly comparing the changes with active and sham rTMS, this study found that rTMS led to a decrease in rCMR in the left superior frontal gyrus (near the stimulation site).

Speer et al. (2003) also found that the magnitude of rCBF changes at the stimulation site and remote sites during 1 Hz rTMS were inversely (right prefrontal cortex, left medial temporal lobe, bilateral parahippocampi, posterior middle temporal gyri) or positively (left anterior cingulate, cerebellum, right insula, auditory cortex, somatosensory cortex) correlated with the intensity of the stimulation in healthy subjects. This may have implications for therapeutic applications, suggesting that higher-intensity stimulation may be more efficacious.

Two studies attempted to differentiate rCBF changes in TMS responders and nonresponders. Unfortunately, as sample sizes were not large, only a small number of subjects in each cell were available for comparison. Teneback et al. (1999) reported on differences between six responders, six (partly matched) nonresponders, and nine subjects who received placebo rTMS. To increase the number of subjects in each cell, results of subjects receiving 5 or 20 Hz rTMS were pooled.

In an ambitious set of analyses, relationships between baseline and endpoint rCBF, change in rCBF, depression scores, and responder/ nonresponder/placebo status were explored (Teneback et al. 1999). In terms of rCBF changes in the regions of interest analyzed, responders showed increases in the cingulate, nonresponders had no significant changes, and subjects receiving placebo showed changes at a number of sites (increases in medial temporal and inferior frontal lobes, left DLPFC; decreases in other regions of medial temporal lobes). These results demonstrate the importance of a placebo control group, as rCBF changes appeared to accompany mood improvement, regardless of the TMS received. In a separate analysis comparing responders and nonresponders, the authors also reported that the former had higher rCBF in the anterior temporal lobes at baseline, with an increase in this difference after rTMS.

As absolute CBF measures are not possible with the SPECT technique, it is uncertain whether this represents a normalization of baseline hypoperfusion in responders, or an increase in hyperperfusion as a compensatory reaction to depression.

Nadeau et al. (2002) also studied the differences between TMS responders and nonresponders, with findings disparate from those above. Compared to nonresponders, reponders showed a reduction in orbitofrontal blood flow and/or a reduction in anterior cingulate blood flow after rTMS. Again, in the absence of absolute CBF measurements, it is difficult to know if rCBF decreased in these areas in responders or increased in nonresponders. However, there are a number of limitations to this study. The sample size only comprised eight subjects, of which seven received high-frequency left prefrontal rTMS at 80% motor threshold and one received low-frequency (0.3 Hz) right prefrontal rTMS at 120% motor threshold. Yet this important difference is apparently ignored with results pooled for analysis. Secondly a low threshold was chosen for the criterion of treatment response: ≥30% decrease in Beck Depression Inventory scores from baseline. By the more usual criterion (≥50% decrease in HRSD scores) only two subjects rather than six would have been classified as responders. In any case, the responder/non-responder analysis was based on two unequal groups, with only two subjects in one group.

In a different approach investigating rCBF changes after rTMS, Mottaghy et al. (2002) confined their analysis to the frontal lobes and compared relative left and right hemisphere values. In the group of 17 subjects scanned prior to rTMS, mean rCBF was less in the left frontal lobe than in the right. In the subset of nine subjects scanned after rTMS, no significant difference was found, suggesting that 10 sessions of rTMS had corrected this asymmetry. However, the change in the left:right ratio did not correlate with the change in HRSD scores.

Recently, other technologies have been used to examine the effects of high-frequency left prefrontal rTMS. Using magnetic resonance spectroscopy, Michael et al. (2003) measured the concentrations of several cerebral metabolites after one and five sessions of active or sham 20 Hz left prefrontal rTMS in healthy subjects. They reported a decrease in left prefrontal glutamate/glutamine levels after a single session of rTMS and an increase in left cingulate glutamate/glutamine levels after 5 days of rTMS. These findings are preliminary but suggest that prefrontal rTMS effects may be mediated via excitatory interconnections. Other studies emphasizing the role of prefrontal connections in the effects of rTMS have reported that high-frequency left prefrontal rTMS led to dopamine release in the striatum in healthy (Strafella *et al.* 2001) and depressed (Pogarell *et al.* 2007) subjects.

In a magnetoencephalographic study, Maihöfner *et al.* (2005) reported that eight depressed subjects had greater slow-wave (2–6 Hz) activity in the left prefrontal cortex compared with healthy controls prior to treatment, and that normalization of these levels was correlated with antidepressant response to a 10 day course of 10 Hz left prefrontal rTMS.

In summary, despite early forecasts and some findings that rTMS may treat depression by normalizing prefrontal hypoactivity, the body of evidence so far suggests that remote effects at limbic and paralimbic regions via prefrontal connections may be equally as important. Further studies examining correlations between cerebral activity patterns and response to rTMS are needed to clarify which areas within these networks have a major role in antidepressant effects. While neuroimaging findings support different mechanisms of action for high- and low-frequency rTMS to the prefrontal cortex, these are likely to be more complex than the simple hypothesis that high frequency activates, and low frequency depresses, cortical excitability. Different networks of brain regions appear to be activated by rTMS of different frequencies. There is some evidence from neuroimaging studies to suggest that the effects of 1 Hz rTMS on cerebral activity are less robust and less consistent between subjects than the effects of high frequency (≥5 Hz) rTMS. Higher stimulation intensities appear to lead to more widespread changes in cerebral activity, an observation that may have implications for the efficacy of rTMS.

Studies of neuroendocrine effects of rTMS

A few studies investigated the effects of prefrontal rTMS on hormonal changes in healthy subjects. Findings of changes in serum thyroid-stimulating hormone and cortisol levels were of small magnitude and not consistent between studies (George *et al.* 1996; Cohrs *et al.* 2001; Evers *et al.* 2001).

Szuba *et al.* (2001) measured TSH and mood immediately before and after a typical treatment session of high-frequency rTMS to the left prefrontal cortex in 14 depressed medication-free subjects. Active rTMS resulted in a significant increase in TSH and improvement in mood compared with sham stimulation. However, there was no correlation between mood and hormone changes. Nevertheless, it is interesting to note that three studies (of which two were sham-controlled and one controlled for stimulation at other brain sites) reported an increase in TSH after high-frequency rTMS to the prefrontal cortex (George *et al.* 1996; Cohrs *et al.* 2001; Szuba *et al.* 2001). This finding is in keeping with the evidence of thyroid function abnormalities in depression (Nemeroff and Evans 1989) and the correlation between TSH increases and mood elevation after sleep deprivation (Parekh *et al.* 1998).

Pridmore (1999) and Zwanzger *et al.* (2003) found that normalization of the dexamethasone response appeared to correlate with improvement in mood after a course of rTMS in depressed subjects. However, Zwanzger *et al.* (2003) also reported that excessive ACTH and cortisol release after corticotrophic-releasing hormone challenge in depressed subjects was not reduced after rTMS, even in responders, unlike findings after amitriptyline treatment, and probably implied a high risk for relapse.

Padberg *et al.* (2002c) measured neuroactive steroid levels (progesterones and dehydroepiandrosterone (DHEA)) after a 2 week treatment course of high-frequency rTMS to the left prefrontal cortex in 37 depressed unmedicated subjects. Though subjects improved significantly in mood, no significant hormonal changes or correlations between mood and hormone levels were found. This contrasts with changes accompanying successful antidepressant pharmacotherapy.

In summary, most investigators have found no or slight hormonal changes after prefrontal rTMS. It is important that some later studies compared the effects of active and sham stimulation, as hormonal effects may have been masked by physiological fluctuations, or changes found may have been falsely attributed to TMS rather than to other effects, e.g. stress from pain associated with stimulation. At this stage, there is no clear evidence from endocrine findings to explain the antidepressant effects of rTMS.

Conclusion

There is convincing statistical evidence for rTMS as an antidepressant treatment and several later studies have also reported clinically significant results, with a substantial proportion of subjects achieving a treatment response. Encouragingly, clinical trials to date have found rTMS to be safe when given within recommended parameter guidelines. Together, evidence from animal, neuroimaging, and neuroendocrine studies has provided several plausible hypotheses for the antidepressant effects of rTMS. There is wide scope for the manipulation of rTMS treatment parameters and the optimal combination of these for maximizing efficacy is yet to be established. Further research is needed to advance our understanding in the above areas, to identify the profiles of subjects most likely to benefit from rTMS, and to determine the role of rTMS with respect to other established antidepressant treatments.

Acknowledgements

'Evidence from clinical trials and meta-analyses' and Table 40.1 in this chapter were adapted from Loo C and Mitchell P, A review of the efficacy of transcranial magnetic stimulation (TMS) treatment for depression, and current and future strategies to optimize efficacy" *Journal of Affective Disorders* 88(3), 255–267, ©2005, with permission from Elsevier.

Dr Loo wishes to thank Kate Manollaras for assistance in preparing this chapter.

References

Abrams R (2002). *Electroconvulsive Therapy*, 4th edn. Oxford University Press, Oxford.

Avery DA, Holtzheimer III PE, Fawaz Q, *et al.* (2006). A controlled study of repetitive transcranial magnetic stimulation in medication-resistant major depression. *Biological Psychiatry, 59,* 187–194.

Barker AT (1991). An introduction to the basic principles of magnetic nerve stimulation. *Journal of Clinical Neurophysiology 8,* 26–37.

Barrett K, Della-Maggiore V, Chouinard PA, Paus T (2004). Mechanisms of action underlying the effect of repetitive transcranial magnetic stimulation on mood: behavioral and brain imaging studies. *Neuropsychopharmacology 29,* 1172–1189.

Berman RM, Narasimhan M, Sanacora G, *et al.* (2000). A randomized clinical trial of repetitive transcranial magnetic stimulation in the treatment of major depression. *Biological Psychiatry 47,* 332–337.

Bickford R, Guidi M, Fortesque P, Swenson M (1987). Magnetic stimulation of human peripheral nerve and brain: response enhancement by combined magnetoelectrical technique. *Neurosurgery 20,* 110–116.

Burt T, Lisanby H, Sackeim H (2002). Neuropsychiatric applications of transcranial magnetic stimulation: a meta-analysis. *International Journal of Neuropsychopharmacology 5,* 73–103.

Catafau AM, Perez V, Gironell A, *et al.* (2001). SPECT mapping of cerebral activity changes induced by repetitive transcranial magnetic stimulation in depressed patients. A pilot study. *Psychiatry Research 106,* 151–160.

Chen R, Classen J, Gerloff C, *et al.* (1997). Depression of motor cortex excitability by low-frequency transcranial magnetic stimulation. *Neurology 48,* 1398–1403.

Christie B, Abraham W (1994). Flip side of synaptic plasticity: long term depression mechanisms in the hippocampus. *Hippocampus 4,* 127–135.

Christyakov A, Kaplan B, Rubichek O, *et al.* (2005). Antidepressant effects of different schedules of repetitive transcranial magnetic stimulation vs. clomipramine in patients with major depression: relationship to changes in cortical excitability. *International Journal of Neuropsychopharmacology 8,* 223–233.

Cohrs S, Tergau F, Korn J, Becker W, Hajak G (2001). Suprathreshold repetitive transcranial magnetic stimulation elevates thyroid-stimulating hormone in healthy male subjects. *Journal of Nervous and Mental Disease, 189,* 393–397.

Conca A, Swoboda, E, Konig P, *et al.* (2000a). Clinical impacts of single transcranial magnetic stimulation (sTMS) as an add-on therapy in severely depressed patients under SSRI treatment. *Human Psychopharmacology 15,* 429–438.

Conca A, Konig P, Hausmann A (2000b). Transcranial magnetic stimulation induces 'pseudoabsence seizure'. *Acta Psychiatrica Scandinavica 101,* 246–248.

Conca A, Di Pauli J, Beraus W, et al. (2002). Combining high and low frequencies in rTMS antidepressive treatment: preliminary results. *Human Psychopharmacology 17*, 353–356.

Couturier JL (2005). Efficacy of rapid-rate repetitive transcranial magnetic stimulation in the treatment of depression: a systematic review and meta-analysis. *Journal of Psychiatry and Neuroscience 30*, 83–90.

Czeh B, Fischer AK, Fuchs E, et al. (2000). Chronic psychosocial stress in rats: effects of concomitant repetitive transcranial magnetic stimulation. *Society of Neuroscience Abstracts 867*, 15.

Czeh B, Welt T, Fischer AK, et al. (2002). Chronic psychosocial stress and concomitant repetitive transcranial magnetic stimulation: effects on stress hormone levels and adult hippocampal neurogenesis. *Biological Psychiatry 52*, 1057–1065.

Dannon PN, Grunhaus L (2001). Effect of electroconvulsive therapy in repetitive transcranial magnetic stimulation non-responder MDD patients: a preliminary study. *International Journal of Neuropsychopharmacology 4*, 265–268.

Davidson RJ (1995). Cerebral asymmetry, emotion, and affective style. In RJ Davidson and K Hugdahl, ed. *Brain asymmetry*, pp. 361–387, MA: Cambridge.

Dolberg O, Schreiber S, Grunhaus L (2001) Transcranial magnetic stimulation-induced switch into mania: a report of two cases. *Biological Psychiatry 49*, 468–470.

Dolberg T, Dannon PN, Schreiber S, Grunhaus L (2002). Transcranial magnetic stimulation in patients with bipolar depression: a double blind, controlled study. *Bipolar Disorders 4*, 94–95.

Dragasevic N, Potrebic A, Damjanovic A, Stefanova E, Kostic V (2002). Therapeutic efficacy of bilateral prefrontal slow repetitive transcranial magnetic stimulation in depressed patients with Parkinson's disease: an open study. *Movement Disorders 17*, 528–532.

Drevets, W (2000). Neuroimaging studies of mood disorders. *Biological Psychiatry 48*, 813–829.

Ebert U, Ziemann U (1999). Altered seizure susceptibility after high-frequency transcranial magnetic stimulation in rats. *Neuroscience Letters 273*, 155–158.

Eranti S, Mogg A, Pluck G et al. (2007). A randomized controlled trial with six-month follow-up of repetitive transcranial magnetic stimulation and electroconvulsive therapy for severe depression. *American Journal of Psychiatry 164*, 73–78.

Eschweiler G, Plewnia C, Batra A, Bartels M (2000). Does clinical response to repetitive prefrontal transcranial magnetic stimulation (rTMS) predict response to electroconvulsive therapy (ECT) in cases of major depression? *Canadian Journal of Psychiatry 46*, 845–846.

Eschweiler G, Plewnia C, Bartels M (2001). Which patients with major depression benefit from repetitive magnetic stimulation (rTMS)? *Fortschritte der Neurologie-Psychiatrie 69*, 402.

Evers S, Hengst K, Pecuch PW (2001). The impact of repetitive transcranial magnetic stimulation on pituitary hormone levels and cortisol in healthy subjects. *Journal of Affective Disorders 66*, 83–88.

Feinsod M, Kreinin B, Chistyakov A, Klein E (1998). Preliminary evidence for a beneficial effect of low-frequency, repetitive transcranial magnetic stimulation in patients with major depression and schizophrenia. *Depression and Anxiety 7*, 65–68.

Figiel GS, Epstein C, McDonald WM, et al. (1998). The use of rapid-rate transcranial magnetic stimulation (rTMS) in refractory depressed patients. *Journal of Neuropsychiatry and Clinical Neurosciences 10*, 20–25.

Fitzgerald PB, Brown TL, Marston NA (2003). Transcranial magnetic stimulation in the treatment of depression: a double-blind, placebo-controlled trial. *Archives of General Psychiatry 60*, 1002–1008.

Fitzgerald PB, Benitez J, de Castella A, et al. (2006) A randomized, controlled trial of sequential bilateral repetitive transcranial magnetic stimulation for treatment-resistant depression. *American Journal of Psychiatry 163*, 88–94.

Funamizu H, Ogiue-Ikeda M, Mukai H, Kawato S, Ueno S (2005). Acute repetitive transcranial magnetic stimulation reactivates dopaminergic system in lesion rats. *Neuroscience Letters 383*, 77–81.

Garcia-Toro M (1999). Acute manic symptomatology during repetitive transcranial magnetic stimulation in a patient with bipolar depression. *British Journal of Psychiatry 175*, 491.

Garcia-Toro M, Pascual-Leone A, Romera M, et al. (2001). Prefrontal repetitive transcranial magnetic stimulation as add on treatment in depression. *Journal of Neurology, Neurosurgery and Psychiatry 71*, 546–548.

Garcia-Toro M, Salva J, Daumal J, et al. (2006) High (20-Hz) and low (1-Hz) frequency transcranial magnetic stimulation as adjuvant treatment in medication-resistant depression. *Psychiatry Research 146*, 53–57.

Geller V, Grisaru N, Abarbanel JM, Lemberg T, Belmaker RH (1997). Slow magnetic stimulation of prefrontal cortex in depression and schizophrenia. *Progress in Neuro-Psychopharmacology and Biological Psychiatry 21*, 105–110.

George M, Wassermann EM, Williams W, et al. (1995). Daily repetitive transcranial magnetic stimulation (rTMS) improves mood in depression. *Neuroreport 6*, 1853–1856.

George M, Wassermann EM, Williams W, et al. (1996). Changes in mood and hormone levels after rapid-rate transcranial magnetic stimulation (rTMS) of the prefrontal cortex. *Journal of Neuropsychiatry and Clinical Neurosciences 8*, 172–180.

George MS, Wassermann EM, Kimbrell TA, et al. (1997). Mood improvement following daily left prefrontal repetitive transcranial magnetic stimulation in patients with depression: a placebo-controlled crossover trial. *American Journal of Psychiatry 154*, 1752–1756.

George MS, Nahas Z, Molloy M, et al. (2000). A controlled trial of daily left prefrontal cortex TMS for treating depression. *Biological Psychiatry 48*, 962–970.

Gershon A, Dannon P, Grunhaus L (2003). Transcranial magnetic stimulation in the treatment of depression. *American Journal of Psychiatry 160*, 835–845.

Grunhaus L, Dannon P, Schreiber S, *et al.* (2000). Repetitive transcranial magnetic stimulation is as effective as electroconvulsive therapy in the treatment of nondelusional major depressive disorder: an open study. *Biological Psychiatry 47*, 314–324.

Grunhaus L, Shcriber S, Dolberg O, Polak D, Dannon P (2003). A randomised controlled comparison of electroconvulsive therapy and repetitive transcranial magnetic stimulation in severe and resistant nonpsychotic major depression. *Biological Psychiatry 53*, 324–331.

Hausmann A, Kemmler G, Walpoth M, *et al.* (2004). No benefit derived from repetitive transcranial magnetic stimulation in depression: a prospective, single centre, randomised, double blind, sham controlled 'add on' trial. *Journal of Neurology, Neurosurgery and Psychiatry 75*, 320–322.

Hedges DW, Massari C, Salyer DL, *et al.* (2003). Duration of transcranial magnetic stimulation effects on the neuroendocrine stress response and coping behavior of adult male rats. *Progress in Neuro-Psychopharmacology and Biological Psychiatry, 27*, 633–638.

Herwig U, Lampe Y, Juengling F *et al.* (2003). Add-on rTMS for treatment of depression: a pilot study using stereotaxic coil-navigation according to PET data. *Journal of Psychiatric Research 37*, 267–275.

Herwig U, Spitzer M, Eschweiler G *et al.* (2006). Antidepressant transcranial magnetic stimulation – results from the first multi-center trial. *Biological Psychiatry 59*, 97S.

Holtzheimer PE, Russo J, Avery DH (2001). A meta-analysis of repetitive transcranial magnetic stimulation in the treatment of depression. *Psychopharmacology Bulletin 35*, 149–169.

Holtzheimer PE, Avery D, Schlaepfer TE (2004). Antidepressant effects of repetitive transcranial magnetic stimulation. *British Journal of Psychiatry 184*, 541–542.

Hoppner J, Schulz M, Irmisch G, Mau R, Schlafke D, Richter J (2003). Antidepressant efficacy of two different rTMS procedures. High frequency over left versus low frequency over right prefrontal cortex compared with sham stimulation. *European Archives of Psychiatry and Clinical Neuroscience 253*, 103–109.

Ikeda T, Kurosawa M, Uchikawa C, Kitayama S, Nukina N (2005). Modulation of monoamine transporter expression and function by repetitive transcranial magnetic stimulation. *Biochemical and Biophysical Research Communications 327*, 218–224.

Iyer M, Schleper N, Wassermann E (2003). Priming stimulation enhances the depressant effect of low-frequency repetitive transcranial magnetic stimulation. *Journal of Neuroscience 23*, 10867–10872.

Janicak PG, Dowd SM, Martis B, *et al.* (2002). Repetitive transcranial magnetic stimulation versus electroconvulsive therapy for major depression: preliminary results of a randomized trial. *Biological Psychiatry 51*, 659–667.

Januel D, Dumortier G, Verdon CM, *et al.* (2006) A double-blind sham controlled study of right prefrontal repetitive transcranial magnetic stimulation (rTMS): therapeutic and cognitive effect in medication free unipolar depression during 4 weeks. *Progress in Neuro-Psychopharmacology and Biological Psychiatry 30*, 126–130.

Jennum P and Klitgaard H (1996). Repetitive transcranial magnetic stimulations of the rat. Effect of acute and chronic stimulations on pentylenetetrazole-induced clonic seizures. *Epilepsy Research 23*, 115–122.

Ji RR, Schlaepfer TE, Aizenman CD, *et al.* (1998). Repetitive transcranial magnetic stimulation activates specific regions in rat brain. *Proceedings of the National Academy of Sciences of the USA 95*, 15635–15640.

Jorge RE, Robinson RG, Tateno A, *et al.* (2004). Repetitive transcranial magnetic stimulation as treatment of poststroke depression: a preliminary study. *Biological Psychiatry 1555*, 398–405.

Kapiletti S, Tsoukarzi S, Copolov E (2001). Efficacy pilot study of right prefrontal TMS in depression. *World Journal of Biological Psychiatry 2*(Suppl. 1), 1.

Kauffmann CD, Cheema MA, Miller BE (2004). Slow right prefrontal transcranial magnetic stimulation as a treatment for medication-resistant depression: a double-blind, placebo-controlled study. *Depression and Anxiety 19*, 59–62.

Keck ME, Engelmann M, Müller MB, *et al.* (2000a). Repetitive transcranial magnetic stimulation induces active coping strategies and attenuates the neuroendocrine stress response in rats. *Journal of Psychiatric Research 34*, 265–276.

Keck ME, Sillaber I, Ebner K, *et al.* (2000b). Acute transcranial magnetic stimulation of frontal brain regions selectively modulates the release of vasopressin, biogenic amines and amino acids in the rat brain. *European Journal of Neuroscience 12*, 3713–3720.

Keck ME, Welt T, Post A, *et al.* (2001). Neuroendocrine and behavioral effects of repetitive transcranial magnetic stimulation in a psychopathological animal model are suggestive of antidepressant-like effects. *Neuropsychopharmacology 24*, 337–349.

Keck ME, Welt T, Müller MB, *et al.* (2002). Repetitive transcranial magnetic stimulation increases the release of dopamine in the mesolimbic and mesostriatal system. *Neuropharmacology 43*, 101–109.

Kimbrell TA, Little JT, Dunn RT, *et al.* (1999). Frequency dependence of antidepressant response to left prefrontal repetitive transcranial magnetic stimulation (rTMS) as a function of baseline cerebral glucose metabolism. *Biological Psychiatry 46*, 1603–1613.

Kimbrell T, Dunn R, George M, *et al.* (2002). Left prefrontal repetitive transcranial magnetic stimulation (rTMS) and regional cerebral glucose metabolism in normal volunteers. *Psychiatry Research: Neuroimaging 115*, 101–113.

Klein E, Kreinin I, Chistyakov A, *et al.* (1999). Therapeutic efficacy of right prefrontal slow repetitive transcranial magnetic stimulation in major depression: a double-blind controlled study. *Archives of General Psychiatry 56*, 315–320.

Koerselman F, Laman DM, van Duijn H, van Duijin M, Willems M (2004). A 3-month, follow-up, randomized, placebo-controlled study of repetitive transcranial magnetic stimulation in depression. *Journal of Clinical Psychiatry* 65, 1323–1328.

Kozel FA, George M (2002). Meta-analysis of left prefrontal repetitive transcranial magnetic stimulation (rTMS) to treat depression. *Journal of Psychiatric Practice 8*, 270–275.

Kozel FA, Nahas Z, deBrux C, *et al.* (2000). How coil-cortex distance relates to age, motor threshold, and antidepressant response to transcranial magnetic stimulation. *Journal of Neuropsychiatry and Clinical Neuroscience 12*, 376–384.

Li X, Nahas Z, Kozel FA, Anderson B, Bohning DE, George MS (2004). Acute left prefrontal transcranial magnetic stimulation in depressed patients is associated with immediately increased activity in prefrontal cortical as well as subcortical regions. *Biological Psychiatry 55*, 882–890.

Lisanby SH, Belmaker RH (2000). Animal models of the mechanisms of action of repetitive transcranial magnetic stimulation (RTMS): comparisons with electroconvulsive shock (ECS). *Depression and Anxiety 12*, 178–187.

Lisanby SH, Pascual-Leone A, Sampson S, Boylan L, Burt T, Sackeim H (2001a). Augmentation of sertraline antidepressant treatment with transcranial magnetic stimulation. *Biological Psychiatry 49*, 81.

Lisanby SH, Schlaepfer TE, Fisch HU, Sackeim HA (2001b). Magnetic seizure therapy of major depression. *Archives of General Psychiatry 58*, 303–305.

Little JT, Kimbrell TA, Wassermann EM, *et al.* (2000). Cognitive effects of 1- and 20-hertz repetitive transcranial magnetic stimulation in depression: preliminary report. *Neuropsychiatry, Neuropsychology and Behavioral Neurology 13*, 119–124.

Loo C, Mitchell P, Sachdev P, McDarmont B, Parker G, Gandevia S (1999). Double-blind controlled investigation of transcranial magnetic stimulation for the treatment of resistant major depression. *American Journal of Psychiatry 156*, 946–948.

Loo C, Sachdev P, Elsayed H, *et al.* (2001). Effects of a 2- to 4-week course of repetitive transcranial magnetic stimulation (rTMS) on neuropsychologic functioning, electroencephalogram, and auditory threshold in depressed patients. *Biological Psychiatry 49*, 615–23.

Loo C, Mitchell P, Croker V, *et al.* (2003a). Double-blind controlled investigation of bilateral prefrontal transcranial magnetic stimulation for the treatment of resistant major depression. *Psychological Medicine 33*, 33–40.

Loo C, Sachdev P, Haindl W, *et al.* (2003b). High (15 Hz) and low (1 Hz) frequency transcranial magnetic stimulation have different acute effects on regional cerebral blood flow in depressed patients. *Psychological Medicine 33*, 997–1006.

Loo CK, Mitchell PB, McFarquhar T, *et al.* (2007). A sham-controlled trial of the efficacy and safety of twice-daily rTMS in major depression. *Psychological Medicine 37*, 314–349.

Loo CK, McFarquhar T, Mitchell P. A review of the safety of repetitive transcranial magnetic stimulation as a clinical treatement for depression. *International Journal of Neuropsychopharmacology*, in press.

Maihöfner C, Ropohl A, Reulbach U, *et al.* (2005). Effects of repetitive transcranial magnetic stimulation in depression: a magnetoencephalographic study. *Neuroreport 16*, 1839–1842.

Manes F, Jorge R, Morcuende M, Yamada T, Paradiso S, Robinson RG (2001). A controlled study of repetitive transcranial magnetic stimulation as a treatment of depression in the elderly. *International Psychogeriatrics 13*, 225–231.

Martin JLR, Barbanoj-Rodriguez M, Schlaepfer T, *et al.* (2003). Transcranial magnetic stimulation for treating depression (Cochrane Review). *The Cochrane Library Issue 1*, Update Software, Oxford.

McDonald WM, Easley K, Byrd EH, *et al.* (2006) Combination rapid transcranial magnetic stimulation in treatment refractory depression. *Neuropsychiatric Disease and Treatment 2*, 85–94.

McNamara B, Ray JL, Arthurs OJ, Boniface S (2001). Transcranial magnetic stimulation for depression and other psychiatric disorders. *Psychological Medicine 31*, 1141–1146.

Menkes DL, Bodnar P, Ballesteros RA, Swenson MR (1999). Right frontal lobe slow frequency repetitive transcranial magnetic stimulation (SF r-TMS) is an effective treatment for depression: a case–control pilot study of safety and efficacy. *Journal of Neurology, Neurosurgery and Psychiatry 67*, 113–115.

Michael N, Gösling M, Reutemann M, *et al.* (2003). Metabolic changes after repetitive transcranial magnetic stimulation (rTMS) of the left prefrontal cortex: a sham-controlled proton magnetic resonance spectroscopy (1H MRS) study of healthy brain. *European Journal of Neuroscience 17*, 2462–2468.

Miniussi C, Bonato C, Bignotti S, *et al.* (2005). Repetitive transcranial magnetic stimulation (rTMS) at high and low frequency: an efficacious therapy for major drug-resistant depression? *Clinical Neurophysiology 116*, 1062–1071.

Mosimann UP, Marre S, Werlen S, *et al.* (2002). Antidepressant effects of repetitive transcranial magnetic stimulation in the elderly: correlation between effect size and coil-cortex distance. *Archives of General Psychiatry 59*, 560–561.

Mosimann UP, Schmitt W, Greenberg BD, *et al.* (2004). Repetitive transcranial magnetic stimulation: a putative add-on treatment for major depression in elderly patients. *Psychiatry Research 30, 126*, 123–133.

Mottaghy FM, Keller CE, Gangitano M, *et al.* (2002). Correlation of cerebral blood flow and treatment effects of repetitive transcranial magnetic stimulation in depressed patients. *Psychiatry Research 115*, 1–14.

Müller MB, Toschi N, Kresse AE, Post A, Keck ME (2000). Long-term repetitive transcranial magnetic stimulation increases the expression of brain-derived neurotrophic factor and cholecystokinin mRNA, but not neuropeptide tyrosine mRNA in specific areas of rat brain. *Neuropsychopharmacology 23*, 205–215.

Nadeau SE, McCoy KJ, Crucian GP, et al. (2002). Cerebral blood flow changes in depressed patients after treatment with repetitive transcranial magnetic stimulation: evidence of individual variability. *Neuropsychiatry, Neuropsychology and Behavioral Neurology* 15, 159–75.

Nahas Z, Bohning D, Molloy M, Oustz J, Risch C, George M (1999). Safety and feasibility of repetitive transcranial magnetic stimulation in the treatment of anxious depression in pregnancy: a case report. *Journal of Clinical Psychiatry* 60, 50–52.

Nahas Z, Lomarev M, Roberts DR, et al. (2001). Unilateral left prefrontal transcranial magnetic stimulation (TMS) produces intensity-dependent bilateral effects as measured by interleaved BOLD fMRI. *Biological Psychiatry* 50, 712–720.

Nahas Z, Kozel A, Li X, Anderson B, George M (2003). Left prefrontal transcranial stimulation (TMS) treatment of depression in bipolar affective disorder: a pilot study of acute safety and efficacy. *Bipolar Disorders* 5, 40–47.

Nemeroff C, Evans D (1989). Thyrotropin releasing hormone (TRgH), the thyroid axis and affective disorders. *Annals of the New York Academy of Science* 533, 304–310.

Ogiue-Ikeda M, Kawato S, Ueno S (2003). The effect of repetitive transcranial magnetic stimulation on long-term potentiation in rat hippocampus depends on stimulus intensity. *Brain Research* 993, 222–226.

O'Reardon J, Solvason B, Janicak P, et al. (2007) Efficacy and safety of transcranial magnetic stimulation in the acute treatment of major depression: a multi-site randomized controlled trial. *Biological Psychiatry*, in press.

Padberg F, Zwanzger P, Thoma H, et al. (1999). Repetitive transcranial magnetic stimulation (rTMS) in pharmacotherapy-refractory major depression: comparative study of fast, slow and sham rTMS. *Psychiatry Research* 88, 163–171.

Padberg F, Zwanzger P, Keck ME, et al. (2002a). Repetitive transcranial magnetic stimulation (rTMS) in major depression: relation between efficacy and stimulation intensity. *Neuropsychopharmacology* 27, 638–645.

Padberg F, Schule C, Zwanzger P, et al. (2002b). Relation between responses to repetitive transcranial magnetic stimulation and partial sleep deprivation in major depression. *Journal of Psychiatric Research* 36, 131–135.

Padberg F, di Michele F, Zwanzger P, et al. (2002c). Plasma concentrations of neuroactive steroids before and after repetitive transcranial magnetic stimulation (rTMS) in major depression. *Neuropsychopharmacology* 27, 874–878.

Parekh P, Ketter T, Altshuler L, et al. (1998). Relationships between thyroid hormone and antidepressant responses to total sleep deprivation in mood disorder patients. *Biological Psychiatry* 43, 392–394.

Pascual-Leone A, Valls-Sole J, Wassermann EM, Hallett M (1994). Responses to rapid-rate transcranial magnetic stimulation of the human motor cortex. *Brain* 117, 847–858.

Pascual-Leone A, Rubio B, Pallardo F, Catala M (1996). Rapid-rate transcranial magnetic stimulation of left dorsolateral prefrontal cortex in drug-resistant depression. *Lancet* 348, 233–237.

Paus T, Castro-Alamancos M, Petrides M (2001). Cortico-cortical connectivity of the human mid-dorsolateral frontal cortex and its modulation by repetitive transcranial magnetic stimulation. *European Journal of Neuroscience* 14, 1405–1411.

Pogarell O, Koch W, Pöpperl G, et al. (2007). Striatal dopamine release after prefrontal repetitive transcranial magnetic stimulation in major depression: preliminary results of a dynamic [(123)I] IBZM SPECT study. *Journal of Psychiatric Research* 38, 74–77.

Post A, Keck ME (2001). Transcranial magnetic stimulation as a therapeutic tool in psychiatry: what do we know about the neurobiological mechanisms? *Journal of Psychiatric Research* 35, 193–215.

Post A, Müller MB, Engelmann M, Keck ME (1999). Repetitive transcranial magnetic stimulation in rats: evidence for a neuroprotective effect *in vitro* and *in vivo*. *European Journal of Neuroscience* 11, 3247–3254.

Pridmore S (1999). Rapid transcranial magnetic stimulation and normalization of the dexamethasone suppression test. *Psychiatry and Clinical Neurosciences* 53, 33–37.

Pridmore S, Bruno R, Turnier-Shea Y, Reid P, Rybak M (2000). Comparison of unlimited numbers of rapid transcranial magnetic stimulation (rTMS) and ECT treatment sessions in major depressive episode. *International Journal of Neuropsychopharmacology* 3, 129–134.

Rossini D, Lucca A, Zanardi R, Magri L, Smeraldi E (2005). Transcranial magnetic stimulation in treatment-resistant depressed patients: a double-blind, placebo-controlled trial. *Psychiatry Research* 137, 1–10.

Rumi DO, Gattaz WF, Rigonatti SP, et al. (2005). Transcranial magnetic stimulation accelerates the antidepressant effect of amitriptyline in severe depression: a double-blind placebo-controlled study. *Biological Psychiatry* 15, 162–166.

Sachdev P, McBride R, Loo C, Mitchell P, Malhi G, Croker V (2002). Effects of different frequencies of transcranial magnetic stimulation (TMS) on the forced swim test model of depression in rats. *Biological Psychiatry* 51, 474–479.

Sackeim H, Prudic J, Devanand D, et al. (2000). A prospective, randomised, double-blind comparison of bilateral and right unilateral ECT at different stimulus intensities. *Archives of General Psychiatry* 57, 425–434.

Schiffer F, Stinchfield Z, Pascual-Leone A (2002). Prediction of clinical response to transcranial magnetic stimulation for depression by baseline lateral visual-field stimulation. *Neuropsychiatry, Neuropsychology and Behavioral Neurology* 15, 18–27.

Schutter D, van Honk J (2005). A framework for targeting alternative brain regions with repetitive transcranial magnetic stimulation in the treatment of depression. *Journal of Psychiatry and Neuroscience* 30, 91–97.

Shajahan PM, Glabus MF, Steele JD, et al. (2002).Left dorso-lateral repetitive transcranial magnetic stimulation affects cortical excitability and functional connectivity, but does not impair cognition in major depression. *Progress in Neuro-Psychopharmacology and Biological Psychiatry* 26, 945–954.

Silberman E, Weingartner H (1986). Hemispheric lateralisation of functions related to emotion. *Brain and Cognition 5*, 322–353.

Sommer M, Lang N, Tergau F, Paulus W (2002). Neuronal tissue polarization induced by repetitive transcranial magnetic stimulation? *Neuroreport 13*, 809–811.

Speer AM, Kimbrell T, Wassermann E, Willis M, Post R (1999). Baseline absolute blood flow measured with oxygen-15 PET predicts differential antidepressants response to 1 Hz versus 20 Hz rTMS. *American Psychiatric Association Meeting, New Research*, abstract 191, p. 115.

Speer AM, Kimbrell TA, Wassermann EM, et al. (2000). Opposite effects of high and low frequency rTMS on regional brain activity in depressed patients. *Biological Psychiatry 48*, 1133–1141.

Speer AM, Repella JD, Figueras S, et al. (2001). Lack of adverse cognitive effects of 1 Hz and 20 Hz repetitive transcranial magnetic stimulation at 100% of motor threshold over left prefrontal cortex in depression. *Journal of ECT 17*, 259–263.

Speer AM, Willis MW, Herscovitch P, et al. (2003). Intensity-dependent regional cerebral blood flow during 1-Hz repetitive transcranial magnetic stimulation (rTMS) in healthy volunteers studied with $H_2^{15}O$ positron emission tomography: II. Effects of prefrontal cortex rTMS. *Biological Psychiatry 54*, 826–832.

Stassen H, Delini-Stula A, Angst J (1993). Time course of improvement under antidepressant treatment: a survival–analytical approach. *European Journal of Neuropsychopharmacology 3*, 127–135.

Strafella A, Paus T, Barrett J, Dagher A (2001). Repetitive transcranial magnetic stimulation of the human prefrontal cortex induces dopamine release in the caudate nucleus. *Journal of Neuroscience 21*, 157.

Su TP, Huang CC, Wei IH (2005). Add-on rTMS for medication-resistant depression: a randomized, double-blind, sham-controlled trial in Chinese patients. *Journal of Clinical Psychiatry 66*, 930–937.

Szuba MP, O'Reardon JP, Rai AS, et al. (2001). Acute mood and thyroid stimulating hormone effects of transcranial magnetic stimulation in major depression. *Biological Psychiatry 50*, 22–27.

Taylor JL, Loo CK (2007) Stimulus waveform influences the efficacy of repetitive transcranial magnetic stimulation. *Journal of Affective Disorders 97*, 271–276.

Teneback CC, Nahas Z, Speer AM, et al. (1999). Changes in prefrontal cortex and paralimbic activity in depression following two weeks of daily left prefrontal TMS. *Journal of Neuropsychiatry and Clinical Neurosciences 11*, 426–435.

Wang H, Wang X, Scheich H (1996). LTD and LTP induced by transcranial magnetic stimulation in auditory cortex. *Neuroreport 7*, 521–525.

Wassermann E (1998). Risk and safety of repetitive transcranial magnetic stimulation: report and suggested guidelines from the International Workshop on the Safety of Repetitive Transcranial Magnetic Stimulation, June 5–7, 1996. *Electroencephalography and Clinical Neurophysiology 108*, 1–16.

Zangen A, Hyodo K (2002). Transcranial magnetic stimulation induces increases in extracellular levels of dopamine and glutamate in the nucleus accumbens. *Neuroreport 13*, 2401–2405.

Zwanzger P, Ella R, Keck M, Rupprecht R, Padberg F (2002). Occurrence of delusions during repetitive transcranial magnetic stimulation (rTMS) in major depression. *Biological Psychiatry 51*, 602–603.

Zwanzger P, Baghai C, Padberg F, et al. (2003). The combined dexamethasone–CRH test before and after repetitive transcranial magnetic stimulation (rTMS) in major depression. *Psychoneuroendocrinology 28*, 376–385.

CHAPTER 41

TMS in bipolar disorder

Nimrod Grisaru, Bella Chudakov, Alex Kaptsan, Alona Shaldubina, Julia Applebaum, and R. H. Belmaker

Introduction

The relatively large number of existing studies of TMS in depression contrasts with the paucity of TMS studies in mania. Studies of any new treatment in mania are difficult to conduct because of problems of legal consent, involuntary hospitalization, and even cooperation with the TMS procedure. A search of the literature on the studies on TMS in the treatment of mania revealed only two controlled studies from our group (Grisaru et al. 1998b; Kaptsan et al. 2003) plus two open studies (Michael and Erfurth 2004; Saba et al. 2004). Despite the acknowledged difficulties of studying novel treatments in acute mania, the potential of focal stimulation with TMS to inform our understanding of the neurobiology of mania and of bipolar disorder in general should not be overlooked. This chapter reviews the existing animal and human literature on the clinical potential of TMS in mania and bipolar depression, and discusses potential future directions for this work.

Uncontrolled studies of TMS in the treatment of mania

Michael and Erfurth (2004) gave five TMS sessions during weeks 1 and 2, and three sessions during weeks 3 and 4, in an open design to manic patients. Nine bipolar inpatients diagnosed with mania were treated with right

prefrontal rapid TMS in an open and prospective study. Eight of the nine patients received TMS as add-on treatment to an insufficient or only partially effective drug therapy. During the 4 weeks of TMS treatment a sustained reduction of manic symptoms as measured by the Bech–Rafaelsen Mania Scale (BRMAS) was observed in all patients. Due to the open and add-on design of the study, a clear causal relationship between TMS treatment and reduction of manic symptoms could not be established. Saba et al. (2004) studied eight patients with TMS over right prefrontal dorsolateral cortex as an add-on treatment. While the results were positive after 14 days, the study was uncontrolled.

Right versus left prefrontal TMS in the treatment of mania

Because the studies of TMS in depression and normal volunteers suggested lateral specificity of TMS-induced mood effects, we designed a clinical trial to compare left vs right prefrontal TMS in mania (Grisaru et al. 1998b). The difficulties of drug-free studies of mania are well known (Licht et al. 1997), and we designed our study based on previous work in mania by our group (Biederman et al. 1979; Klein et al. 1999; Mishory et al. 2000) as an add-on study of left vs right prefrontal TMS to ongoing unrestricted drug treatment. Based on reports of rapid response of mania to electroconvulsive therapy (ECT), we hypothesized that the effect of TMS would be apparent early enough and strongly enough to be measurable even against the background of ongoing pharmacotherapy.

Patients admitted to the Beersheva Mental Health Center could enter the study if they met DSM-IV criteria for mania. No changes in clinical pharmacotherapy were made because of study participation. Patients were hospitalized for a mean of 8.6 days (range 1–38) before entering the study. Eighteen patients were enrolled. Two dropped out, one after four TMS treatments because of severe worsening and a positive drug urine screen, and the other before any TMS treatment because of change in diagnosis. Of the 16 completers, 12 were manic nonpsychotic and four were manic psychotic. Seven were male and nine were female; the average age was 36 years (range 20–52).

Concomitant drug therapy in the nine patients receiving left prefrontal TMS was lithium in six patients, carbamazepine in one patient, valproate in one patient, and no mood stabilizer in one patient; eight patients in this group also received neuroleptics (in chlorpromazine equivalents, mean 340 mg, range 150–600 mg).

Patients were assessed at four time points: 24 h before the first TMS (baseline), 3 and 7 days after the first treatment, and at the end of the study (day 14). Day 14 was usually 4 days after the final TMS. The following instruments were used: Clinical Global Impression (CGI) (Guy 1976), Mania Scale (Young et al. 1978), and Brief Psychiatric Rating Scale (BPRS).

A Cadwell high-speed magnetic stimulator with a 9 cm diameter circular coil was used. Motor threshold (Hallett and Cohen, 1989), defined as the lowest stimulation intensity over the motor cortex capable of inducing a finger movement at least five times out of 10, was assessed for each patient before the first treatment and 80% of individual patient motor threshold was then administered for all treatment days (George et al. 1995). Mean patient motor threshold was 67% for the left treatment group (range 50–80%) and 72% for the right treatment group (range 55–85%). Patients were given 10 daily consecutive sessions with 20 trains per session. Frequency was 20 Hz for 2 s per train; inter-train interval was 1 min. Each of the participants was given the stimuli over the right prefrontal cortex or the left prefrontal cortex, as randomized (R.H.B.).

The CGI improvement score at day 14 was significantly different for left- vs right-treated patients (median test, $P = 0.017$) (Figure 41.1). These results suggest that TMS stimulation in mania of the right prefrontal cortex has therapeutic effects. This is the opposite lobe to that reported to have antidepressant effects when stimulated at high frequency.

Table 41.1 Comparative patient demographics in the Kaptsan et al. (2003) and Grisaru et al. (1998b) studies

	Grisaru et al. 1998		Kaptsan et al. 2003	
	Right	Left	Right	Sham
n	7	9	11	8
Age (years)	35	32	44	40
Sex				
Male	2	5	5	4
Female	5	4	6	4
Nonpsychotic mania	7	5	2	1
Psychotic mania	0	4	9	7
CPZ equivalents	240	340	490	445
Motor threshold	72	67	63	65

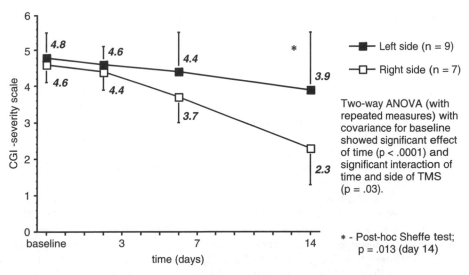

Fig. 41.1 The effect of right vs left prefrontal TMS on mania (Clinical Global Impression (CGI) severity scale; mean ± SD): ■, left side ($n = 9$); □, right side ($n = 7$). Two-way analysis of variance (with repeated measures) with covariance for baseline showed significant effect of time ($P < 0.0001$) and significant interaction of time and side of TMS ($P = 0.03$). *$P = 0.013$ (day 14; *post hoc* Sheffé test).

Interestingly, right unilateral ECT was not found to be effective in mania (Milstein *et al.* 1987) in a small group of patients. The effects of TMS in psychiatry may be complex, since certain stimulation patterns enhance neuronal activity and cause, for instance, a motor movement (Hallett and Cohen 1989), whereas other stimulation parameters can disrupt neuronal outflow and cause, for instance, speech arrest (Pascual-Leone *et al.* 1991). Thus further studies of frequency of the magnetic stimulus, its intensity, and its location will be necessary before the comparisons in hemisphere specificty with ECT can be made with confidence.

Sham-controlled trial of TMS in the treatment of mania

The results of Grisaru *et al.* (1998b) could have been due to worsening of mania by left TMS, as happens with monoamine reuptake inhibitor antidepressants (Wielosz 1983). Thus we designed a trial of right TMS vs sham TMS in mania (Kaptsan *et al.* 2003).

Patients admitted to the Beersheva Mental Health Center in Israel could enter the study if they met DSM-IV criteria for mania by consensus of two psychiatrists after clinical interview. No changes in clinical pharmacotherapy were made because of study participation. Patients with a history of epilepsy, neurosurgery, brain trauma, cardiac pacemaker implant, or drug abuse were excluded. The study was approved by our Helsinki Committee, and all patients gave written informed consent.

Twenty-five hospitalized patients were enrolled in the study. Importantly for the integrity of the blind, none of them had been a subject in a previous TMS study. Patients were hospitalized for a mean of 14.6 days (range 2–64) before entering the study, excluding one patient with 82 days of hospitalization as he had several phases in his hospitalization and thus the number of days before TMS was not a meaningful measure. The reason for the long pre-TMS hospitalization was the severity of the psychotic–manic state of most of the patients. In this period they were not able to sign informed consent or to cooperate with the study requirements.

Six of the patients dropped out, three after one right active TMS treatment because of manic uncooperativeness unrelated to TMS treatment, one after three right active treatments because of physical illness not due to TMS therapy, one during determination of motor threshold, reporting headache and unpleasant feelings, and one after five sessions of sham treatment because of severe worsening of mania and a positive urine drug screen. Of the 19 patients who completed the study, 16 had psychotic mania and three had nonpsychotic mania. Nine were men (five in the right TMS group) and 10 were women (six in the right TMS group). The patients' mean age was 41.6 (range 19–65); the mean age of the patients given right TMS was 43.8 years, and that of the patients given sham TMS was 39.6 years. Patients were randomized by one author (R.H.B.) unrelated to patient treatment, according to a prearranged random order.

Among the 11 patients receiving right active TMS, concomitant drug therapy consisted of valproate acid (mean dose 800 mg/day) for six patients, lithium (mean dose 1425 mg/day) for four patients, and no mood stabilizer for one patient. Ten of these patients also received neuroleptics (mean dose 490 mg/day in chlorpromazine equivalents, range 200–1100; one patient received olanzapine 15 mg/day).

Among the eight patients receiving sham TMS, four were receiving lithium, two were receiving valproate, one was receiving lithium plus valproate, and one was receiving lithium and carbamazapine. All of them also received neuroleptics (mean 445 mg/day in chlorpromazine equivalents, range 50–75). Table 41.1 shows the demographics of the patient population compared to that in the previous study (Grisaru et al. 1998b).

Patients were assessed at four time points: 24 h before the first TMS treatment session (baseline), 3 and 7 days after the first treatment, and at the end of the study (day 14). The following assessment instruments were used: CGI, Young Mania Rating Scale (YMRS), and BPRS. Rating scales were evaluated by a blind senior psychiatrist (Y.Y.) unrelated to the TMS treatment and the clinical treatment of the patients, and normally located on another clinical unit. The ward staff were not involved in the study

design and the TMS treatment was done in the human TMS laboratory, located away from the wards. Patients were blind to the hypothesis of the study.

A Cadwell high-speed magnetic stimulator with a 9 cm diameter circular coil was used. Motor threshold was determined for each patient only before the first treatment. Eighty percent of the individual patient motor threshold was then administered on all treatment days. Mean patient motor threshold was 63% for the right TMS group (range 52–75) and 65% for the sham group (range 40–100). In the subgroup receiving active anticonvulsant treatment, the mean patient motor threshold was 63.6% (range 60–67). For the subgroup not receiving active anticonvulsant treatment, mean patient motor threshold was 62.4% (range 52–75).

Patients were given 10 daily consecutive sessions of 20 trains per session. Frequency was 20 Hz for 2 s per train; inter-train interval was 1 min. Each of the participants was given the stimuli over the right prefrontal cortex or sham position (vertical position of the coil, 90° to the head but slightly off the scalp), randomly assigned by one of us (R.H.B.). The right prefrontal cortex was F3 on a rubber shower-cap with electroencephalography positions marked. Table 41.2 shows that there was no beneficial effect of right TMS compared with sham TMS. These results do not support a therapeutic effect of right TMS in mania. This study, when combined with our previous study of right TMS vs left TMS, could be interpreted to mean that left-sided TMS may have blocked the efficacy of concomitant antimanic medication. Furthermore, considering that high-frequency TMS to the left prefrontal cortex has been reported to exert antidepressant properties, then we might expect it to exacerbate mania. Indeed, cases of a mood switch to hypomania or mixed mania have been reported with left high-frequency TMS in the treatment of bipolar disorder (Garcia-Toro 1999; Dolberg et al. 2001; Huang et al. 2004). Interestingly, a switch into a mixed state has recently been reported in patients previously diagnosed with unipolar depression (cf. Rachid et al. 2006).

Several factors should be noted, however. Sixteen of the 19 patients in the present study

Table 41.2 Effect on mania of prefrontal TMS applied to the right side (n = 11) or sham TMS (n = 8)

Scales	Baseline		Day 3		Day 7		Day 14	
	Mean	SD	Mean	SD	Mean	SD	Mean	SD
BPRS								
Right TMS	35.4	9.9	33.3	7.1	28.5	5.9	25.6	5.9
Sham	36.5	6.2	35.1	5.7	32.4	3.8	30.4	10.9
YMRS								
Right TMS	27.4	8.6	26.2	8.0	18.9	8.1	15.7	9.4
Sham	30.5	7.9	26.9	6.9	21.4	7.0	14.1	7.3
CGI								
Right TMS	4.5	0.8	4.2	1.0	3.8	0.8	3.4	1.0
Sham	4.9	0.8	4.8	0.7	4.4	0.5	3.4	1.1

Modified from Kapstan et al. (2003).

BPRS, Brief Psychiatric Rating Scale; YMRS, Young Mania Rating Scale; CGI, Clinical Global Impression.

had psychotic mania. This group did more poorly in the previous study of TMS in mania (Grisaru et al. 1998b) as well as in studies of psychotic depression vs nonpsychotic depression with TMS (Grunhaus et al. 2000). The dose in chlorpromazine equivalents was 490 mg in this study for the right TMS group and 445 for the sham TMS group, whereas in the previous study it was 240 mg for right TMS and 340 for left TMS (see Table 41.1). The motor threshold was 72 for right TMS and 67 for left TMS for the previous study and in the present study 63 for right TMS and 65 for sham TMS. Thus lower TMS doses were given in the present study and this may have reduced TMS efficacy (see Table 41.1). Days of hospitalization before TMS were 8.6 in the previous study and 14.6 in the present study, suggesting inclusion of more resistant patients in the present study.

Some TMS researchers have expressed concern about the nature of sham TMS, and the possibility exists that our sham was active enough to obscure differences between the two treatment groups. It is also possible that a longer treatment period, a higher treatment intensity, or different parameters of location and frequency might be more therapeutic in mania. Testing a new treatment as an add-on to an effective treatment is a severe test. Most antidepressants in controlled studies do not show additive effects to another effective antidepressant (Nemets et al. 2001). Thus it will be important in the future to study

TMS as monotherapy in mild cases of mania where this will be ethically and practically possible. However, the data support the conclusion that right prefrontal TMS is safe in this patient population, and we cannot rule out the possibility that it could become an efficacious add-on treatment of bipolar mania if given at a more effective dosage.

TMS in animal models of mania

To explore the effects of rTMS in an animal model of mania, we used the amphetamine-induced hyperactivity model (Shaldivin et al. 2001). Amphetamine-induced hyperactivity is a well-investigated model of mania that includes reasonable face validity (increased activity, increase in secondary reinforcement value, increased aggression) and predictive validity as the effects are usually inhibited by lithium treatment (Robbins and Sahakian 1980; Robbins et al. 1983; Lyon 1990, 1991; Gessa et al. 1995) and were also reported to be reduced by carbamazepine (Maj et al. 1985) and valproate (Maitre et al. 1984).

Three experiments were performed to evaluate the effects of subacute (twice daily), daily chronic (7 days), and twice daily chronic (7 days) rTMS treatment in the amphetamine-induced hyperactivity model.

Male Sprague–Dawley rats (Harlan, Jerusalem; $n = 20$ for experiment 1; $n = 40$ for experiment 2; $n = 20$ for experiment 3), weighing 200–250 g at the beginning of experiment, were housed in an 'in-lab, rat only' colony room with 12 h light/dark cycle, constant temperature (22°C), and free access to food (standard rat chaw) and water. Rats were given a 1 week habituation period prior to the beginning of the experiment during which they were handled by the experimenter for 1 min every day. All experimental procedures were executed during the light phase of the light/dark cycle.

Repetitive TMS treatment was delivered with a Cadwell rapid stimulator and a 5 cm round coil. Each treatment session was 2 s long and used 25 Hz frequency at maximal machine capacity. During treatment, rats were held firmly, attached to a table, by one experimenter while another one applied rTMS or a sham control. As described previously (Fleischmann *et al.* 1995), the coil was held immediately above but not touching the rat's head, the pointer of the coil above the vertex of the skull, and the handle of the coil parallel to the rat's vertebral column. Control animals were held in a manner identical to TMS-treated animals, a coil held above their head, and they were exposed to the audible artifact of TMS given ~10 cm away. For experiment 1, rats were exposed to two treatment sessions, the first delivered ~24 h prior to amphetamine injection and the second delivered immediately after amphetamine injection, prior to placement in the automated activity monitors. For experiment 2, rats received seven daily rTMS sessions and were tested for amphetamine-induced hyperactivity ~24 h after the last session. For experiment 3, rats received 7 days of twice daily (morning and afternoon, ~9 h apart) rTMS treatments (totaling 14 sessions). The last session was administered immediately after amphetamine injection and prior to placement in the automated activity monitors. Control rats for each experiment received handling identical to that of the treatment group except that the rTMS apparatus was not activated and was replaced by sham audible artifact. The two- and seven-session schedules were chosen because they were previously demonstrated to be effective in a rat model of depression (Fleischmann *et al.* 1995). The twice daily schedule was added

because a similar schedule of ECS was reported to enhance amphetamine hyperactivity (Evans *et al.* 1976).

Immediately after amphetamine injection (for experiment 2) or immediately after the last rTMS session that followed the amphetamine injection (for experiments 1 and 3), rats were placed in automated activity monitors (Elvicom, Israel) measuring $38 \times 38 \times 35.5$ cm and left there for 30 min. The amount of locomotor activity, both horizontal and vertical, was recorded for each 10 min session and for the entire 30 min.

Subacute rTMS treatment produced a significant decrease in horizontal amphetamine-induced activity (analysis of variance (ANOVA), treatment effect: $F(1) = 6.62$, $P < 0.02$).

Daily rTMS treatment for 7 days significantly decreased amphetamine-induced horizontal hyperactivity (ANOVA, treatment effect: $F(1) = 5.46$, $P < 0.03$). The effect of twice daily treatment with rTMS was opposite to the effects of the previous treatment schedules. Twice daily rTMS treatment augmented horizontal amphetamine-induced hyperactivity (ANOVA, treatment effect: $F(1) = 5.36$, $P < 0.04$) (see Figure 41.2).

The results of Shaldivin *et al.* (2001) do not reveal a clear picture of TMS effects on amphetamine-induced hyperactivity. We hypothesized that TMS might reduce amphetamine-induced hyperactivity, as does lithium (Robbins and Sahakian 1980), and this effect was indeed apparent after two or after seven daily TMS treatments. However, TMS twice daily *enhanced* amphetamine-induced hyperactivity. The last effect may be similar to ECT effects since chronic twice daily ECS in rats was also demonstrated to increase amphetamine-induced hyperactivity (Evans *et al.* 1976), apparently in a manner similar to its enhancement of apomorphine-induced hyperactivity (Lerer and Belmaker 1982; Modigh *et al.* 1984).

Clark *et al.* (2000) studied TMS effects in human volunteers after amphetamine ingestion. TMS did not modify the psychostimulant effects of amphetamine. However, only a single TMS session was administered. Antidepressant treatments such as ECT or monoamine reuptake inhibitors are active after a single dose in rats but require a series of doses for mood effects in humans.

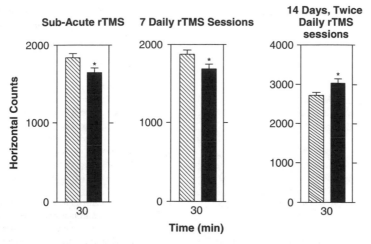

Fig. 41.2 Cumulative horizontal activity for groups treated with repetitive (r)TMS (filled bars) or sham (hatched bars) and with acute amphetamine injection. The y-axis represents activity counts. *Significant differences for a specific time period ($P < 0.05$). Modified from Shaldivin *et al.* (2001).

TMS in the treatment of bipolar depression

While the larger multicenter trials of TMS in the treatment of depression have generally excluded bipolar patients, several investigators have specifically studied rTMS in the treatment of bipolar depression (Dolberg *et al.* 2002; Nahas *et al.* 2003). The results show efficacy similar to that for unipolar depression. This in itself does not provide support for the concept of TMS as an anti-bipolar, or mood-stabilizing, treatment. A series of seven patients treated with TMS for bipolar depression and maintained on TMS for up to a year without mania (Li *et al.* 2004) suggests that TMS might be an anti-bipolar treatment (Grisaru *et al.* 1999). However, the utility of TMS as prophylaxis for subsequent manic or depressive episodes has not been reported in bipolar disorder.

While catatonia is usually considered in the depressive spectrum, it does not respond to antidepressant medications. ECT is a highly effective treatment of catatonia. Therefore, effectiveness of TMS in catatonia would strengthen the hypothesis that TMS has ECT-like, and therefore anti-bipolar, potential. We treated with open-label TMS one case of catatonia in a patient suffering from schizophrenia with moderate success (Grisaru *et al.* 1998a). Issues of consent and ethics of experimentation in this acute high-mortality syndrome make research in catatonia difficult.

Conclusions

Relative to the size of the literature on the potential value of TMS in depression, bipolar disorder has been less well studied with TMS to date. Human and animal studies on the potential value of TMS in mania have produced conflicting results. Given the focality of TMS as an intervention, there are opportunities to use this tool to determine whether there are prefrontal–laterality differences between the manic and depressed states that could inform the selection of TMS treatment paradigms, though the evidence for that is unclear at present. The potential value of TMS in the treatment of depression in the context of bipolar disorder has some support in the present literature, though more work is needed to clarify the risk of mood switch, and the potential of TMS as prophylaxis against future manic or depressive episodes. While the potential value of TMS in the treatment of mania is difficult to study and has produced conflicting results, the potential value of TMS in bipolar disorder should be a priority area for TMS research in the future given the severity of the syndrome and the side-effects of current treatments.

References

Biederman J, Lerner Y, Belmaker RH (1979). Combination of lithium carbonate and haloperidol in schizo-affective disorder: a controlled study. *Archives of General Psychiatry 36*, 327–333.

Clark L, McTavish SF, Harmer CJ, Mills KR, Cowen PJ, Goodwin GM (2000). Repetitive transcranial magnetic stimulation to right prefrontal cortex does not modulate the psychostimulant effects of amphetamine. *International Journal of Neuropsychopharmacology 3*, 297–302.

Dolberg OT, Schreiber S, Grunhaus L (2001). Transcranial magnetic stimulation-induced switch into mania: a report of two cases. *Biological Psychiatry 49*, 468–470.

Dolberg OT, Dannon PN, Schreiber S, Grunhaus L (2002). Transcranial magnetic stimulation in patients with bipolar depression: a double blind, controlled study. *Bipolar Disorder 4*(Suppl. 1), 94–95.

Evans JP, Grahame-Smith DG, Green AR, Tordoff AF (1976). Electroconvulsive shock increases the behavioural responses of rats to brain 5-hydroxytryptamine accumulation and central nervous system stimulant drugs. *British Journal of Pharmacology 56*, 193–199.

Fleischmann A, Prolov K, Abarbanel J, Belmaker RH (1995). The effect of transcranial magnetic stimulation of rat brain on behavioral models of depression. *Brain Research 699*, 130–132.

Garcia-Toro M (1999). Acute manic symptomatology during repetitive transcranial magnetic stimulation in a patient with bipolar depression. *British Journal of Psychiatry 175*, 491.

George MS, Wassermann EM, Williams WA, *et al.* (1995). Daily repetitive transcranial magnetic stimulation (rTMS) improves mood in depression. *Neuroreport 6*, 1853–1856.

Gessa GL, Pani L, Serra G, Fratta W (1995). Animal models of mania. In: GL Gessa, W Fratta, L Pani, G Serra (ed.), *Depression and mania from neurobiology to treatment*, pp. 43–66. New York: Raven Press.

Grisaru N, Chudakov B, Yaroslavsky Y, Belmaker RH (1998a). Catatonia treated with transcranial magnetic stimulation. *American Journal of Psychiatry 155*, 1630.

Grisaru N, Chudakov B, Yaroslavsky Y, Belmaker RH (1998b). Transcranial magnetic stimulation in mania: a controlled study. *American Journal of Psychiatry 155*, 1608–1610.

Grisaru N, Chudakov B, Yaroslavsky Y, Belmaker RH (1999). Is TMS therapeutic in mania as well as depression? In: W Paulus, M Hallett, PM Rossini and JC Rothwell (ed.), *Transcranial magnetic stimulation*, pp. 299–303. Amsterdam: Elsevier.

Grunhaus L, Dannon PN, Schreiber S, *et al.* (2000). Repetitive transcranial magnetic stimulation is as effective as electroconvulsive therapy in the treatment of nondelusional major depressive disorder: an open study. *Biological Psychiatry 47*, 314–324.

Guy W (1976). Clinical global impressions. *New clinical drug evaluation unit (ECDEU) assessment manual for psychopharmacology*, pp. 218–222.

Hallett M, Cohen LG (1989). Magnetism. A new method for stimulation of nerve and brain. *Journal of the American Medical Association 262*, 538–541.

Huang CC, Su TP, Shan IK (2004). A case report of repetitive transcranial magnetic stimulaiton-induced mania. *Bipolar Disorder 6*, 444–445.

Kaptsan A, Yaroslavsky Y, Applebaum J, Belmaker RH, Grisaru N (2003). Right prefrontal TMS versus sham treatment of mania: a controlled study. *Bipolar Disorder 5*, 36–39.

Klein E, Kreinin I, Chistyakov A, *et al.* (1999). Therapeutic efficacy of right prefrontal slow repetitive transcranial magnetic stimulation in major depression: a double-blind controlled study. *Archives of General Psychiatry 56*, 315–320.

Lerer B, Belmaker RH (1982). Receptors and the mechanism of action of ECT. *Biological Psychiatry 17*, 497–511.

Li X, Nahas Z, Anderson B, Kozel FA, George MS (2004). Can left prefrontal rTMS be used as a maintenance treatment for bipolar depression? *Depression and Anxiety 20*, 98–100.

Licht RW, Gouliaev G, Vestergaard P, Frydenberg M (1997). Generalisability of results from randomised drug trials. A trial on antimanic treatment. *British Journal of Psychiatry 170*, 264–267.

Lyon M (1990). Animal models of mania and schizophrenia. In: P Willner (ed.), *Behavioral models in psychopharmacology*, pp. 253–310. Cambridge: Cambridge University Press.

Lyon M (1991). Animal models for the symptoms of mania. In: AA Boulton, GB Baker, Martin- MD Iverson (ed.), *Animal models in psychiatry. I. Neuromethods*, Vol. 18, pp. 197–244. Clifton, NJ: Humana Press.

Maitre L, Baltzer V, Mondadoni C, Olpe HR, Baumann PA, Waldmeier PC (1984). Psychopharmacological and behavioral effects of anti-epileptic drugs in animals. In: HM Emrich, T Okuna, AA Miller (ed.), *Anticonvulsants in affective disorders*, pp. 3–13. Amsterdam: Elsevier.

Maj J, Chojnacka-Wojcik E, Lewandowska A, Tatarczynska E, Wiczynska B (1985). The central action of carbamazepine as a potential antidepressant drug. *Polish Journal of Pharmacology and Pharmacology 37*, 47–56.

Michael N, Erfurth A (2004). Treatment of bipolar mania with right prefrontal rapid transcranial magnetic stimulation. *Journal of Affective Disorders 78*, 253–257.

Milstein V, Small JG, Klapper MH, Small IF, Miller MJ, Kellams JJ (1987). Uni- versus bilateral ECT in the treatment of mania. *Convulsive Therapy 3*, 1–9.

Mishory A, Yaroslavsky Y, Bersudsky Y, Belmaker RH (2000). Phenytoin as an antimanic anticonvulsant: a controlled study. *American Journal of Psychiatry 157*, 463–465.

Modigh K, Balldin J, Eriksson E, Granerus AK, Walinder J (1984). Increased responsiveness of dopamine receptors after ECS: a review of experimental and clinical efidence. In: B Lerer, RD Weiner, RH Belmaker (ed.), *ECT: basic mechanisms*, pp. 18–27. London: Libbey.

Nahas Z, Kozel FA, Li X, Anderson B, George MS (2003). Left prefrontal transcranial magnetic stimulation (TMS) treatment of depression in bipolar affective disorder: a pilot study of acute safety and efficacy. *Bipolar Disorders 5*, 40–47.

Nemets B, Fux M, Levine J, Belmaker RH (2001). Combination of antidepressant drugs: the case of inositol. *Human Psychopharmacology 16*, 37–43.

Pascual-Leone A, Gates JR, Dhuna A (1991). Induction of speech arrest and counting errors with rapid-rate transcranial magnetic stimulation. *Neurology 41*, 697–702.

Rachid F, Golaz J, Bondolfi G, Bertschy G (2006). Induction of a mixed depressive episode during rTMS treatment in a patient with refractory major depression. *World Journal of Biological Psychiatry 7*, 261–264.

Robbins TW, Sahakian BJ (1980). Animal models of mania. In RH Belmaker, HM van Praag (ed.), *Mania: an evolving concept*, pp. 143–216. Lancaster: MTP Press.

Robbins TW, Watson BA, Gaskin M, Ennis C (1983). Contrasting interactions of pipradrol, d-amphetamine, cocaine, cocaine analogues, apomorphine and other drugs with conditioned reinforcement. *Psychopharmacology (Berlin) 80*, 113–119.

Saba G, Rocamora JF, Kalalou K, *et al.* (2004). Repetitive transcranial magnetic stimulation as an add-on therapy in the treatment of mania: a case series of eight patients. *Psychiatry Research 128*, 199–202.

Shaldivin A, Kaptsan A, Belmaker RH, Einat H, Grisaru N (2001). Transcranial magnetic stimulation in an amphetamine hyperactivity model of mania. *Bipolar Disorders 3*, 30–34.

Wielosz M (1983). Effects of electroconvulsive shock on monoaminergic systems in the rat brain. Thesis. *Polish Journal of Pharmacology and Pharmacology 35*, 127–130.

Young RC, Biggs JT, Ziegler VE, Meyer DA (1978). A rating scale for mania: reliability, validity and sensitivity. *British Journal of Psychiatry 133*, 429–435.

CHAPTER 42

TMS clinical trials involving patients with schizophrenia

Ralph E. Hoffman and Arielle D. Stanford

Introduction

Studies of the pathophysiological mechanisms underlying schizophrenia often make reference to abnormalities in cortical excitability and inhibition. This conceptual orientation is motivated by the fact that auditory hallucinations and delusions characteristic of schizophrenia appear to be 'activation' or 'breakthrough' symptoms, i.e. behaviors, thoughts, or perceptions that are inappropriate, intrusive, or out of place. Consistent with this view are neuroimaging data showing that auditory hallucinations in this patient group are accompanied by activation across a distributed network of cortical and subcortical regions (see, for instance, Shergill et al. 2000) and that delusions are associated with excessive activation of the left temporal cortex (Puri et al. 2001). Moreover, studies of early stages of sensory processing using pre-pulse inhibition and evoked potentials suggest that the cerebral cortex is less able to suppress responses to inputs in patients with schizophrenia (Alder et al. 1982; Swerdlow and Koob 1987), which again suggests enhanced cortical excitability and/or impaired inhibitory processes. In contrast, the so-called negative or deficit symptoms of schizophrenia,

which, by definition, reflect relative curtailment of ideation, affect, and actions, are associated with reduced activation, especially in frontal brain areas (see, for example, Wolkin et al. 1992; Chua and McKenna 1995; Schroder et al. 1996; Weinberger and Berman 1996; Vaiva et al. 2002).

Repetitive (r)TMS is increasingly being studied as an experimental intervention for patients with neuropsychiatric disorders. These approaches have been informed by animal studies of long-term potentiation (LTP) and long-term depression (LTD) showing that repeated stimulation of neural circuits can exert effects on synaptic efficacy in a fashion that persists, for varying amounts of time, beyond the period of stimulation. LTP and LTD are elicited by direct electrical stimulation of gray matter (for example, in the hippocampus, amygdala, and cerebral cortex) at high and low frequencies, respectively. The former tends to produce long-standing enhancements of synaptic efficacy, whereas the latter tends to produce reductions in synaptic efficacy, alterations which have been hypothesized to resemble neuroplastic changes that occur during mammalian brain development, learning, and memory (Scannevin and Huganir 2000; Jo et al. 2006). It has been proposed that rTMS may induce parallel shifts in neuroplasticity noninvasively in the human brain via excitation elicited in small regions of the cerebral cortex arising from magnetic pulses generated at the scalp. Analogous to LTP and LTD, higher-frequency rTMS (5 Hz or greater) tends to amplify brain excitability, whereas low-frequency rTMS (i.e. 1 Hz) tends to reduce brain excitability (Post *et al.* 1999). However, the molecular basis for frequency-dependent effects of rTMS on measures of cortical excitability has not been proven to be identical to those mechanisms underlying LTP and LTD.

The divergent effects of low- vs high-frequency rTMS are relevant to pathophysiological models of schizophrenia that assume excessive brain activation for positive symptoms and reduced brain activation for negative symptoms. This thinking has prompted a growing number of case series and clinical trials using rTMS, which are reviewed below.

Repetitive TMS studies targeting frontal regions in patients with schizophrenia

Initial reports utilizing rTMS in this patient group employed low-frequency stimulation and targeted frontal regions. The first such report involved 10 patients with schizophrenia and 10 patients with depression (Geller *et al.* 1997). The authors attempted to determine if mood changes could be induced and whether effects would be similar for the two patient groups. Very-low-frequency (once per 30 s) rTMS was administered to left and right prefrontal cortex, 15 pulses each. Two of 10 schizophrenic patients appeared to improve transiently. Feinsod *et al.* (1998) reported an open trial in which seven out of 10 patients with schizophrenia experienced decreased anxiety and restlessness in response to low-frequency frontal rTMS. A later double-blind study examining the effects of low-frequency rTMS to right prefrontal cortex in patients with schizophrenia did not find any improvement following active rTMS compared to sham stimulation (Klein *et al.* 1999a). This study was prompted by an earlier study demonstrating antidepressant effects using low-frequency right prefrontal rTMS in patients with major depression (Klein *et al.* 1999b).

The first study examining effects of higher frequency rTMS delivered to prefrontal cortex in schizophrenic patients was reported by Cohen *et al.* (1999). Twenty-hertz rTMS was delivered to left prefrontal cortex in patients in 2 s trains once per minute for 20 min each day for 10 days. Patients had chronic schizophrenia with predominantly negative symptoms. There was no comparison or control stimulation condition. Five of six patients demonstrated hypofrontality as determined by a single-photon emission computed tomography (SPECT) scan. Re-scanning after rTMS indicated no change in the hypofrontality. However, a within-subject comparison indicated that negative symptoms decreased significantly ($P < 0.02$). A trend towards improvement in neuropsychological test performance was detected although only performance in a delayed visual memory task improved to a statistically significant degree. This trend is of interest given studies associating neuropsychological deficits with negative symptoms (see, for instance, Gold *et al.* 1999).

Sachdev *et al.* (2005) studied four subjects with schizophrenia and stable negative symptoms using 15 Hz rTMS at 90% of motor threshold, 1800 pulses each session, daily for 20 sessions over 4 weeks over the left dorsolateral prefrontal cortex, the largest dose to date (total 36 000 pulses). Subjects showed a significant reduction in negative symptoms and improvement in function, with no change in positive symptoms. This improvement was maintained at the 1 month follow-up. Although the investigators excluded subjects with active depression, they also found a 33% reduction in the depression rating scores. There were no changes in cognitive measures (Digit Forwards and Backwards (Wechsler Memory Scale–III), Trails A and B, Symbol Digit Coding (Wechsler Adult Intelligence Scale–III), Verbal Fluency for letter and category, and the Wisconsin Card Sort Test).

Another open-label study by Jandl *et al.* (2005) provided additional support for this approach. Ten patients with schizophrenia and primarily negative symptoms were studied with 10 Hz rTMS over the left dorsolateral prefrontal cortex for 5 days. The Scale for the Assessment of Negative Symptoms (SANS) and electroencephalographic (EEG) recordings were obtained pre- and post-rTMS. SANS showed a modest improvement after rTMS (mean scores dropping from 49.0 to 44.7) with decreased in delta and beta and increased alpha-1 EEG activity in right fronto-temporal regions.

Another study investigated effects of 10 Hz rTMS administered to left prefrontal cortex (Yu et al. 2002). The main goal of this study was to determine effects on P300 abnormalities and elevated serum prolactin levels induced by antipsychotic drugs. Partial normalization of each of these abnormalities was detected although only five patients were studied. Given that elevated prolactin is likely due to dopamine blockade, partial normalization of prolactin levels suggests that high-frequency prefrontal rTMS may enhance dopaminergic function. This view is consistent with another recent TMS study in normal humans where high-frequency rTMS delivered to prefrontal cortex was shown to increase dopamine release using [^{11}C]raclopride positron emission tomography (Strafella et al. 2001) and with the hypothesis that reduced dopamine drive in prefrontal areas causes negative symptoms in patients with schizophrenia (Weinberger 1987).

Rollnik et al. (2000) reported the first randomized clinical trial comparing higher-frequency rTMS delivered to left prefrontal cortex and sham stimulation. Twelve schizophrenic patients with negative symptoms were studied. Two weeks of daily left prefrontal rTMS was administered using a double-blind crossover design. Each stimulation session consisted of twenty 2 s pulse trains at 20 Hz and 80% motor threshold. Comparing Brief Psychiatric Rating Scale (BPRS) values under active and sham treatment, there were no significant differences at baseline ($T = 1.99$, $P = 0.08$, with the active group tending to be somewhat greater than the sham group) but at follow-up BPRS values were significantly lower for active treatment (mean difference 5.80 ± 6.1; $t = -3.01$, $P = 0.015$). Interestingly, symptom changes associated with active rTMS did not appear to reflect predominantly negative symptoms.

In another randomized trial, 20 patients treated with 10 Hz rTMS at 110% motor threshold over 10 days demonstrated reductions in negative and depressive symptom ratings

compared to sham treatment (Hajak et al. 2004). This study used depressive scales validated in schizophrenic patients, designed to prevent confounding by extrapyramidal symptoms (EPS) and negative symptoms (Calgary Depression Scale for Schizophrenia). In a multiple regression analysis, depression ratings did not contribute to the response in negative symptoms, suggesting that negative symptoms arising from depression did not explain the rTMS response. Like the Cohen (1999) study, equivalent current dipole (ECD) SPECT scans were done before and after treatment course and demonstrated no change with active rTMS. Although their study revealed a trend for worsening of positive symptoms with active treatment, the clinical relevance of the change was questionable.

Recent randomized clinical trials by Holi et al. (2004) and Novak et al. (2006) cast doubt regarding this approach, however.

Holi et al. (2004) studied 22 chronically hospitalized patients with schizophrenia, who were randomly assigned to 10 sessions of real or sham rTMS. Twenty trains of 5 s, 10 Hz stimulation at 100% motor threshold were given 30 s apart. Effects on positive and negative symptoms were assessed. Although there was a significant improvement in both groups in most of the symptom measures, no differences were found between the groups. A decrease of >20% in the total Positive and Negative Syndrom Scale (PANSS) score was found in seven control subjects but only one subject in the real rTMS group showed this level of improvement. Specific analyses of negative symptoms were not provided.

Novak et al. (2006) studied 16 schizophrenia patients with predominantly negative symptoms on stable antipsychotic medication with 20 Hz rTMS (90% of motor threshold, 2000 stimuli per session) over 10 days using a double-blind sham-controlled parallel design. They failed to find significant effects of active rTMS compared to sham stimulation using PANSS, CGI, Montgomery–Asberg Depression Rating Scale (MADRS), and neuropsychological tests. Sham stimulation showed a trend for improvement over time on positive and negative subscales of PANSS and MADRS.

The different results of these controlled studies may be accounted for by somewhat different rTMS parameters. For instance, the Holi *et al.* study employed 10 Hz stimulation whereas the Rollnik *et al.* study employed 20 Hz stimulation. In the Hajak *et al.* study, subjects with schizoaffective disorder who might be more responsive to treatment were allowed in the study. Neither study assessed gender effects. Subjects in the Holi *et al.* study were mostly male, who tend to have a somewhat more severe course of schizophrenia. Perhaps a more important factor was that the Holi *et al.* study utilized chronically institutionalized patients whose total PANSS scores were >100. Patients such as these are likely to be especially resistant to any intervention.

An alternative innovative approach assessing effects of prefrontal rTMS on negative symptoms in schizophrenia based on EEG alpha wave abnormalities has been reported by Jin *et al.* (2006). Motivating their approach were studies indicating that (i) patients with schizophrenia have reduced alpha power and alpha coherence at rest (Stevens and Livermore 1982) and during cognitive processing relative to normal subjects (Colombo *et al.* 1989; Hoffman *et al.* 1991), (ii) level of negative symptoms in patients with schizophrenia was inversely correlated with alpha power, within-hemisphere alpha coherence, and between-hemisphere alpha coherence (Merrin and Floyd 1992, 1996), and (iii) improvement in negative symptoms following treatment with clozapine correlated with degree of normalization of photically driven alpha power (Jin *et al.* 1995, 1998). Jin *et al.* consequently hypothesized that setting the frequency of rTMS to the patient's individual peak alpha frequency (referred to as αTMS) would increase frontal alpha activity and reduce negative symptoms. Twenty-seven patients were studied with predominantly negative symptoms. Patients in the study received an unchanging dose of antipsychotic medication with stable symptoms for at least 30 days. Four rTMS interventions at three different frequencies and a sham condition were studied in a double-blind fashion. Each patient was randomized to one of two study groups and received two types of stimulation in a crossover design. The first group received real αTMS and sham αTMS. The second group received real 3 Hz and real 20 Hz rTMS. Treatment order for the two conditions

within groups was also randomized. Each condition consisted of 10 daily sessions during a 2 week period, with 2 weeks of no intervention between conditions. Stimulation was given 2 s/min for 20 consecutive minutes per session at 80% motor threshold. Individualized alpha frequency of rTMS was determined as peak EEG alpha averaged over frontal leads for each patient. Sham stimulation was given by applying an unplugged coil to the forehead while positioning an activated coil 2 feet away behind the patient. Using this four-cell randomization, 11 cases received αTMS, eight received sham stimulation, nine received 3 Hz rTMS, and nine received 20 Hz rTMS. αTMS produced statistically greater reductions in negative symptoms relative to the other three conditions. Moreover, the change in frontal alpha power induced by αTMS correlated with improvement in negative symptoms at a robust level ($R = 0.86$, $P = 0.01$).

One limitation of this study was that the sham condition may have been especially likely to cue subjects regarding whether they received real vs sham rTMS since the sham condition produced no vibration or scalp contractions, and was likely to have sounded more distant compared to active rTMS. However, similar differences were detected for other active rTMS interventions delivered at 3 and 20 Hz, suggesting that improvements associated with αTMS relative to sham were not due primarily to breakdowns in the blind. Most importantly, these data suggest that a benefit of rTMS in clinical populations could be unrelated to activation or inhibition *per se*, but instead could arise from 'tuning' and amplifying oscillatory cortical dynamics that impact on cognitive function and symptom generation.

One-hertz rTMS and auditory hallucinations

Another series of trials utilizing rTMS in patients with schizophrenia has focused on auditory hallucinations (AHs). This common symptom, occurring in 60–70% of patients with schizophrenia, often produces severe distress, disability, and behavioral dyscontrol. In ~25% of patients, AHs respond poorly or not at all to currently available antipsychotic medication (Shergill *et al.* 1998). One important feature of

AHs is that they are generally experienced as spoken speech with discernible loudness, timbre, and other 'percept-like' features. These characteristics suggest direct involvement of speech perception neurocircuitry. Support for this view derived from the observation that external noise can trigger AHs – a tendency that has been found to correlate with the level of cortical activation in temporal lobe speech-processing regions, during hallucination events (R.E. Hoffman, unpublished data).

An early ^{15}O positron emission tomography study found that activation in left temporo-parietal regions accompanied AHs (Silbersweig et al. 1995). These brain regions are adjacent to Wernicke's area and are also active during speech perception (Benson et al. 2001). We consequently predicted that 'suppressive' 1 Hz rTMS delivered to temporo-parietal cortex might reduce AHs. We targeted the left hemisphere TP3 site determined on the basis of the 10–20 International System of electrode placement (half way between T3 and P3) due to its proximity to left temporo-parietal cortex (Herwig et al. 2003), and because this area is readily accessible to scalp stimulation (Figure 42.1). We initially reported a study of 12 right-handed schizophrenic patients with medication-resistant auditory hallucinations comparing effects of 1 Hz active rTMS to sham stimulation to the TP3 site using a double-blind crossover design (Hoffman et al. 1999, 2000). Patients with a history of active drug or alcohol abuse, unstable medical condition, history of seizures or other neurological condition, subnormal IQ (i.e. estimated IQ <80), changes in antipsychotic medication within 30 days prior to initiation of the trial, or inability to provide an informed consent to participate in the study were excluded. Stimulation was administered at 80% of motor threshold. Sham stimulation was administered to the same location with coil tilted 45° off the scalp using the 'two-wing' method (i.e. where both 'wings' of the coil touched the scalp). Insofar as this was the first time that rTMS was administered in this brain area, we were cautious regarding patient safety/tolerability – the first day the patient received 4 min of stimulation, which was then increased by 4 min increments to 16 min on the final, fourth, day. Hallucination severity was

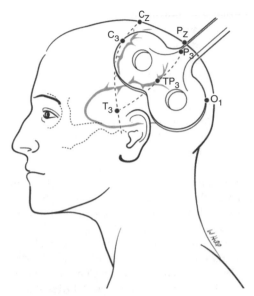

Fig. 42.1 Location of the stimulation for our repetitive TMS trials for auditory hallucinations. TP3 was defined as being midway between T3 and P3 according to the 10–20 International EEG electrode deployment system. This region falls near the posterior border of Wernicke's area at the junction of Brodmann's areas 39 and 40 of inferior parietal lobule (Herwig et al. 2003). C3 is the site used to elicit motor thresholds. Reproduced from Hoffman et al. (2003), p. 51. ©2003 American Medical Association. All rights reserved.

rated based on the Hallucination Change Score that was anchored to the patient's own narrative description of his hallucinations at baseline. This level of severity was assigned a score of 10. Subsequent assessments where no hallucinations occurred within the prior 24 h were assigned a score of 0 and hallucinations twice as severe were assigned a score of 20. No changes in antipsychotic or thymoleptic medication were made during the trial. Comparing endpoint data, statistically significant improvements in auditory hallucinations were detected for active rTMS relative to sham stimulation ($P < 0.01$). Therapeutic effects were brief, generally lasting less than 1 week. Concomitant anticonvulsant medication was found to curtail the symptom-reducing effects of rTMS, an effect that was highly statistically significant.

Insofar as anticonvulsant drugs limit trans-synaptic propagation of cortical activation (Applegate *et al.* 1997; Stefani *et al.* 1997), these data suggest that putative therapeutic effects of rTMS require propagation of activation. Other positive symptoms of schizophrenia were relatively unchanged by rTMS, suggesting that rTMS effects were relatively selective and specifically related to pathophysiology producing auditory hallucinations.

We consequently sought to determine if a more extended trial of rTMS administered to TP3 could produce more clinically significant sustained reductions in AHs. A sample of 50 patients was studied (Hoffman *et al.* 2003, 2005). All patients enrolled met DSM-IV criteria for schizophrenia or schizoaffective disorder based on the Structured Clinical Interview for DSM-IV (SCID), and reported AHs at least five times per day. Forty-two of these patients met criteria for medication resistance, defined as daily AHs persisting in spite of at least two adequate trials of antipsychotic medications that included at least one atypical antipsychotic medication. Ages ranged from 19 to 58 years. Exclusion criteria were the same as in our earlier protocol (Hoffman *et al.* 2000). All patients were naïve to rTMS (i.e. they had not been studied in our previous trial). Twenty-seven patients were randomly allocated to the active rTMS and 23 patients to the sham stimulation group. There were no statistically significant differences in age, gender, number of prior hospitalizations, duration of current hallucination episode (defined as the number of months since the patient last had a remission of AHs of 4 weeks or greater), number of patients with medication-resistant AHs, or prior treatment with ECT. Length of time of unremitting AHs was extended, with a mean of ~10 years in each group. No change in dose of antipsychotic or thymoleptic medication was made for 4 weeks prior to trial entry and during the trial itself. Study participants, clinical raters, and all personnel responsible for the clinical care of the participants remained blind to allocated condition. As in our first study, 1 Hz stimulation was used but at a modestly higher field strength (90% motor threshold vs 80% in our first study). Patients in this trial received 8 min of stimulation on day 1, 12 min of stimulation on

day 2, and 16 min of stimulation for the next 7 days (excluding weekends) for a total of 132 min of stimulation compared to 40 min for the first trial. Sham stimulation was administered at the same location, strength, and frequency, with the coil angled 45° away from the skull in a 'single-wing' tilt position. A neuropsychological battery was administered at baseline and at the end of each leg of the trial. Patients were assessed at baseline and after every third active/sham rTMS session using the Hallucination Change Score described above and the Auditory Hallucinations Rating Scale (AHRS) developed by our group (Hoffman *et al.* 2003, 2005), which assessed seven phenomenological components: hallucination frequency, loudness, 'realness', number of speaking voices, the typical length of individual hallucinations, degree that the patient could ignore hallucinations, and the level of distress produced. The PANSS and the Clinical Global Improvement Scale (CGI) were used to assess other symptoms. Two neuropsychological screening tasks – the Hopkins Verbal Memory Task (Benedict *et al.* 1998), and the Letter–Number Working Memory task (Gold *et al.* 1997) – were administered at the same intervals to assess evidence of cognitive decline that would prompt terminating the patient from the trial. A full neuropsychological test battery was administered at baseline and after each leg of the trial.

In terms of safety and tolerability, only headache (transient and responsive to acetaminophen) and lightheadedness (generally lasting <30 min) were expressed at a higher frequency for patients receiving active rTMS vs sham. Concentration complaints during active rTMS were no more frequent than for sham stimulation. Two patients in the open-label active group reported mild memory impairment. The research team later learned that one of these patients had just been started on benztropine by his outpatient psychiatrist, which could account for these memory difficulties. This patient's memory complaints lasted 3–4 days and improved when the patient was taken off benztropine. Memory complaints for the second case were only for 1 day. There were four dropouts. One patient was removed from the active trial and one patient from the sham trial due to drops in the Hopkins Verbal Learning Task beyond our 'stop' criteria. For the patient in the active trial, retesting demonstrated a return to baseline functioning. For the sham patient, worsening neuropsychological test performance was accompanied by worsening AHs, which may have produced these difficulties. One patient enrolled in the sham phase was removed from the study due to worsening psychotic symptoms. One patient dropped out after one session of active rTMS because he complained that his 'head felt weird'. Changes in performance for the full neuropsychological battery did not reveal noticeable trends toward declining function for patients following the double-masked active trial or differences in change scores comparing patients randomized to active vs sham stimulation.

In terms of clinical outcomes, the Hallucination Change Score, our primary outcome measure, was significantly lower for active compared with sham groups for the day 7 assessment ($t(44) -$ 2.53, $P = 0.02$) and the final (day 10) assessment ($t(43) = 2.70$, $P = 0.01$). The Hallucination Change Score dependence over time was characterized using a random-time model. The time effect ($F(1, 41.4) = 39.43$, $P < 0.0001$) and the group $\times$ time interaction ($F(1, 41.4) = 7.88$, $P = 0.0076$) were significant. The active group demonstrated a significant linear decrease in the hallucination change scores over time ($t(24.6) =$ -7.87, $P < 0.001$) as did sham group ($t(22.6) =$ -2.13, $P = 0.045$). Patients were classified as responders if hallucination severity was reduced by at least 50%. Using this criterion, 14/27 (51.9%) patients achieved responder status in the active group, compared with 4/23 (17.4%) of patients in the sham group, a difference that was statistically significant ($\chi^2 = 6.4$, $P = 0.01$). For CGI scores, mean $\pm$ SD = 2.84 $\pm$ 0.85 for the active group was reduced relative to the sham group at a robustly significant level (mean $\pm$ SD = 3.80 $\pm$ 0.88, $t(46) = -3.80$, $P = 0.0004$). Anticonvulsant drug treatment was not significantly associated with clinical outcome.

The one AHRS variable demonstrating significant treatment effects was frequency. The change of hallucination frequency over time was modelled by a random-intercept random-time model. There was a significant time effect ($F(1, 46.7) = 17.96$, $P = 0.0001$), and significant group $\times$ time interaction ($F(1, 46.7) = 11.47$,

$P = 0.0014$). The active group demonstrated a significant linear decrease in the frequency of hallucinations over time ($t(25.9) = -5.09$, $P < 0.0001$) whereas the sham group did not show a significant linear decrease ($t(20.7)$ $= -0.64$, $P = 0.53$).

Baseline AHRS phenomenological characteristics were assessed to determine if any variable appeared to be a moderator of rTMS effects as defined by Kraemer et al. (2002). One variable – hallucination frequency – had a significant interaction with treatment type (sham vs active ($F(1, 41) = 7.70$, $P = 0.008$) and therefore qualified as a candidate moderator of rTMS efficacy. Moderator effects were optimized if patients were dichotomized into high- and low-frequency hallucinations based on whether or not AHs occurred on average more than once every 10 min (Figure 42.2). Those patients with more frequent AHs demonstrated a greater differential effect when compared to sham stimulation, whereas patients with lower hallucination frequency demonstrated less robust differences between active and sham rTMS due to greater improvement following sham stimulation relative to active rTMS. These data suggest that active rTMS delivered to the left temporo-parietal cortex expressed efficacy in this study primarily via selective improvements in patients with more frequent AHs. Interestingly, these data suggest that less-frequent AHs are to some degree amenable to nonspecific interventions. In terms of the latter, most of our research patients had been hospitalized on a research unit, where they may have benefited from structure and the support provided by clinical staff as well as assurance of medication compliance.

Other variables emerged as 'nonspecific predictors', defined as variables demonstrating statistically significant correlations with endpoint Hallucination Change Scores but which failed to demonstrate statistically significant interactions with group allocation. Pooling data from the double-masked and nonmasked active rTMS (for those individuals randomized initially to the sham group) revealed that endpoint Hallucination Change Score was correlated with the number of acoustically distinct voices heard at baseline ($R = 0.45$, $P = 0.002$), California Verbal Learning Test (CVLT) short-delay recall

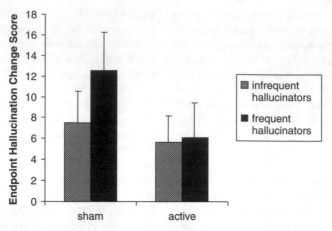

Fig. 42.2 A comparison of responses for sham vs active repetitive TMS with subjects divided into high and low baseline hallucination frequencies. Hatched bars, infrequent hallucinators; solid bars, frequent hallucinators. Modified with permission from Hoffmann *et al.* (2005), p. 101. ©2005 Society of Biological Psychiatry.

$(R = -0.46, P = 0.001)$, CVLT long-term recall $(R = -0.35, P = 0.016)$, and CVLT recognition discriminability $(R = -0.34, P = 0.022)$. In addition, hallucination frequency at baseline was correlated with endpoint hallucination severity $(R = 0.40, P = 0.003)$. CVLT findings are of interest given that verbal memory is known to rely on left temporo-parietal cortex (Ojemann 1978; Fiez *et al.* 1996), suggesting that greater pathophysiological involvement of this cortical area produces greater resistance to rTMS.

Other groups have now reported results of intervention studies using rTMS for AHs, which are summarized below.

D'Alfonso *et al.* (2002) described eight patients with persistent auditory hallucinations who were given a trial of 1 Hz rTMS with stimulation at 80% motor threshold and duration of 20 min of stimulation per day for 10 days. Statistically significant improvements in auditory hallucinations were detected relative to baseline, but improvements were modest. Reduced response may reflect at least three factors. First, one of the eight patients was left-handed and therefore had a 50/50 chance of being right-hemisphere dominant, which could reduce effects of rTMS administered to the left hemisphere. Second, another patient required a reduction in stimulation from 80% to 50% due to pain at the site of stimulation, which also may have reduced rTMS effects. Third, stimulation

in this study was not administered to left temporo-parietal cortex but instead to a more anterior left temporal region 2 cm above T3 per the 10–20 International EEG electrode placement system. Given that T3 often falls on the superior temporal gyrus (Homan *et al.* 1987), it is possible that the rostral displacement may have shifted the stimulation coil off the temporal lobe to a sensorimotor cortical area in some cases.

Poulet *et al.* (2005) described 10 dextral patients with schizophrenia and auditory hallucinations who were randomized to active vs sham rTMS to TP3 at 90% motor threshold using a crossover design. Each week of stimulation was separated by a 1 week washout period. The design was unique insofar as a total of 2000 pulses were administered each day in two separate sessions (1000 pulses each) for a total of 5 days. Thus the daily dose of rTMS given was roughly twice that of our later clinical trial (Hoffman *et al.* 2005) even though the number of stimulation sessions was much less. Improvements as robust as those detected in our trial were observed after 5 days in the Poulet *et al.* study, suggesting that higher daily 'dosages' of rTMS produced a more rapid response. In terms of the latter study, mean improvement of total AHRS scores was on average 56% with no improvement detected following the sham phase of the trial. Nonresponders were treated with medication having anticonvulsant properties, whereas responders were not treated with these medications, apparently in agreement with our first study finding that anticonvulsants curtailed rTMS clinical response (Hoffman *et al.* 2000). This group later reported data for 14 patients receiving active rTMS and 10 patients receiving sham rTMS using the same 5 day protocol of 2000 pulses each day (Brunelin *et al.* 2006). Robust improvements in total AHRS scores were again detected following active rTMS but not following sham rTMS. Moreover, source monitoring capacity was studied – where patients were tested in terms of their ability to recall the source of a list of words heard on headphones. This list included words previously read by the patient and words not previously seen by the patient. Source monitoring defects elicited by this task have been postulated to produce or contribute to the genesis of AHs (Keefe *et al.* 1999). Source monitoring performance improved following active rTMS but not sham rTMS. Reductions in AHRS scores correlated with source monitoring improvement at a trend level ($R = 0.37, P = 0.06$).

Chibbaro *et al.* (2005) studied 16 patients with schizophrenia and AHs. One-hertz rTMS was administered at 90% motor threshold during four sessions on successive days. Duration of each stimulation session was 15 min. Half the patients received active rTMS and half received sham stimulation. Both patient groups demonstrated a significant reduction in AHs as well as in positive symptoms at the end of the first week. However, at later time points up to and including 8 weeks following the trial, improvements in the sham group disappeared, while patients receiving active rTMS improvement retained improvement at a level that was statistically significant relative to the sham group.

Two studies compared effects of rTMS to TP3 vs the right homologous site (TP4) in patients with AHs. Lee *et al.* (2005) randomly allocated 39 patients with treatment-resistant AHs to active left vs right rTMS and sham stimulation. Symptoms were evaluated using the AHRS, the PANSS and the CGI. Active rTMS to both left and right temporo-parietal sites produced greater overall symptomatic improvements per CGI scores relative to sham stimulation ($P = 0.004$ and $P = 0.002$ respectively).

However, summed AHRS scores did not show significantly greater improvements for either active site vs sham stimulation. Right-sided rTMS did show reductions in the attentional salience subscale of the AHRS relative to sham stimulation at a trend level ($P = 0.07$). In contrast, Jandl *et al.* (2005) studied 14 right-handed hallucinating patients with schizophrenia and two left-handed hallucinating patients with schizophrenia. Each patient received five stimulation sessions to left and right sites respectively and an additional five sham stimulations, with each arm of the trial separated by 4 weeks. AH assessment utilized the hallucination subscale of the Psychotic Symptom Rating Scale (PSYRATS) (Haddock *et al.* 1999). Response was defined as at least a 50% reduction in the PSYRATS hallucination subscale scores, and partial response as a 30% reduction in these scores. There were five responders or partial responders in the group receiving left TP3 rTMS, with one partial responder following right-sided stimulation and no responders or partial responders following sham stimulation. These group differences were statistically significant based on χ^2-analysis ($P < 0.02$). However, there were no group differences for change in hallucination subscale PSYRATS scores. Of interest is that higher hallucination frequency was statistically associated with nonresponse to TP3 rTMS ($P < 0.05$) (M. Jandl, personal communication). This finding replicates prediction findings for our last trial (Hoffman *et al.* 2005), and suggests that high hallucination frequency is an indicator of more intractable pathophysiology.

There have also been two negative studies for rTMS applied to the TP3 site. The first, reported by McIntosh *et al.* (2004), utilized the lower-dosed 4 day protocol that we previously described (Hoffman *et al.* 2000), and found no significant improvement in AHs for active rTMS vs sham stimulation. Of note is that the stimulation was halted every minute for 15 s, which may have disrupted physiological effects of rTMS. The second was reported by Fitzgerald *et al.* (2005), who studied 33 patients with treatment-resistant AHs. rTMS was applied for 10 sessions for 15 min at 1 Hz and 90% motor threshold. Active treatment did not result in a greater therapeutic effect than sham on any measure except for the loudness of hallucinations where there was a significant reduction in the active vs the sham group over time. Of note is that the stimulation coil used in this study was not actively cooled and had to be switched once during each session. This interruption may have disrupted efficacy due either to delays in the stimulation or a failure to reposition rTMS at precisely the same site.

One additional study (Schönfeldt-Lecuona et al. 2004) tested the hypothesis that AHs are instances of inner speech that have been mislabelled as having a nonself origin (Frith and Done 1989; McGuire et al. 1996). Brain areas appearing to activate during inner speech based either on fMRI maps or using structural MRI were targeted. Active stimulation was applied over Broca's area and over the superior temporal gyrus corresponding to the primary auditory cortex (BA 22/42). rTMS did not lead to a significant reduction of hallucination severity for the patients overall. However, for the four patients where rTMS positioning was based on individualized fMRI maps, improvements following rTMS to the superior temporal site approached significance compared to sham stimulation ($P = 0.06$).

A recently completed study by our group also probed multiple cortical regions using 1 Hz rTMS based on individualized fMRI maps. Patients in this study reported especially severe hallucination syndromes. These data suggest that the site optimizing rTMS in this patient group is Wernicke's area itself (Hoffman et al. 2007).

Discussion

Although the number of studies using rTMS as a potential clinical intervention for schizophrenia is small, they show promise in terms of advancing our understanding of pathophysiological mechanisms and developing alternative interventions. Of the four randomized controlled studies utilizing higher-frequency rTMS studies in schizophrenia focusing on the prefrontal cortex with a sham control group, two studies were positive (Rollnik et al. 2000; Hajak et al. 2004) and two were negative (Holi et al. 2004; Novak et al. 2006). Of note is that patient populations may have influenced treatment efficacy: one study reflected an especially ill, chronically institutionalized group (Holi et al. 2004) that may have been more refractory

to interventions overall. The Jin et al. (2005) study suggests an alternative rTMS intervention strategy – namely 're-tuning' oscillatory frequencies associated with cognitive impairment or symptom expression. This approach may have applicability, not only for alpha wave abnormalities in schizophrenia, but also for other frequency bands reflecting functionally significant abnormalities in schizophrenia, such as deficient gamma resonances detected during performance of perceptual and cognitive tasks (Kwon et al. 1999; Spencer et al. 2003).

Low-frequency rTMS targeting speech processing areas has been studied more extensively, with many but not all studies showing greater efficacy for active rTMS compared to sham simulation in reducing AHs. A number of factors were likely to impact on reported results. In two of these studies concomitant anticonvulsant drugs reduced rTMS response (Hoffman et al. 2000; Poulet et al. 2005). Another factor likely to impact on efficacy was total 'dose' of rTMS. The most robustly positive studies (Hoffman et al. 2005; Poulet et al. 2005) delivered relatively large numbers of total pulses: 10 000 and 7920 pulses, respectively. In contrast, the negative study by McIntoch et al. (2004) delivered only 2400 pulses, and the borderline positive study reported by Jandl et al. (2006) utilized 4500 pulses. However, another borderline positive study reported by Lee et al. (2005) delivered a total of 12 000 pulses, and the Fitzgerald et al. (2005) study, which also reported a negative result, delivered a total of 7920 pulses.

Negative results are also likely to reflect reduced statistical power due to small sample sizes. Many of the studies described reported 10–15 patients receiving active stimulation. A recent meta-analysis, which accommodates small sample sizes of individual studies, found evidence of efficacy relative to the sham condition based on a combined total sample size of 232 (Hedge's $g = 0.51$, $P < 0.001$) (Sepehryl et al. 2006).

Another important factor in assessing and designing these studies is the likelihood of nonspecific improvements involving AHs due to study participation. These experiences are unstable at times and may be altered by a number of factors. One factor is emotional state, with negative emotions such as depression and anxiety tending to worsen AHs. These emotional factors are likely to be responsive to placebo effects arising from expectation of improvement from a novel experimental intervention. Assurance of medication compliance and abstinence from street drugs and alcohol due to more careful patient monitoring could also improve outcome. Social isolation significantly worsens AHs (Nayani and David 1996). Trial participation ensures regular daily contact with individuals carrying out the trial, which could provide a therapeutic benefit. These benefits are even more likely for the large percentage of patients enrolled in our trials who are admitted to an inpatient research unit where there is ready access to nursing staff, group meeting, etc. The importance of nonspecific improvement is highlighted by our last reported trial (Hoffman et al. 2005), which showed that hallucination frequency was a moderator of rTMS effects. For patients whose hallucination frequency was relatively low, differences between active and sham stimulation were modest – reflecting a high rate of improvement in the sham group (see Figure 42.2). Stated differently, the statistical weight of a treatment effect for rTMS was expressed primarily in the high-frequency hallucination subgroup, where nonspecific improvements (likely reflecting the factors outlined above) were much less pronounced. This observation is important because it suggests that a well-conducted study limited to patients with low hallucination frequencies would not have produced significant differences between active and sham groups using our protocol due to nonspecific improvements arising in the latter. Other factors that are likely to add 'noise' to outcome data (and thereby reduce effect size of clinical trial results) include the failure to ensure that patients are maintained on an unchanging dose of psychiatric medication for 4–6 weeks prior to trial initiation (such medication changes could generate delayed improvements that could elevate sham response), and inclusion of patients with unstable hallucination severity (where the likelihood of spontaneous worsening or improvement of AHs is great).

Future clinical trials are likely to benefit from strategies utilizing individualized neuroimaging to position rTMS. Adjustments for cortical atrophy could be critical given that effective power delivered to the cortex by rTMS falls off exponentially as the distance between skull and cortex increases (Nahas et al. 2004). Superior temporal lobe size has been found to negatively correlate with hallucination severity (Barta et al. 1990; Levitan et al. 1999) and reduced gray matter has been found to be associated with negative symptoms in patients with schizophrenia (Gur et al. 2000). rTMS trials targeting these symptom clusters may therefore benefit from adjusting stimulation strength according to skull–cortex distance.

Conclusions

These studies, considered together, suggest that rTMS holds promise as an intervention strategy for patients with schizophrenia. Rigorously designed trials with larger numbers of subjects are indicated in order to take into account nonspecific factors that could add noise to outcome data.

Acknowledgments

Studies described in this chapter were generously supported by two NARSAD Independent Investigator Awards, NIMH R21-MH63326 and 1R01MH06707 grants, grants from the Donaghue Medical Foundation and Dana Foundation, and NIH/NCRR/GCRC Program Grant RR00125.

References

Alder LE, Pachtman E, Franks RD, Pecevich M, Waldo MC, Freedman R (1982). Neurophysiological evidence for a defect in neural mechanisms involved in sensory gating in schizophrenia. Biological Psychiatry 17, 639–654.

Applegate CD, Samoriski GM, Ozduman K (1997). Effects of valproate, phenytoin, and MK-801 in a novel model of epileptogenesis. Epilepsia 38, 631–636.

Barta PE, Pearlson GD, Powers RE, Richards SS, Tune LE (1990). Auditory hallucinations and smaller superior temporal gyral volume in schizophrenia. American Journal of Psychiatry 147, 1457–62.

Benedict RHB, Schretlen D, Groninger L, Brandt J (1998). Hopkins Verbal Learning Test-Revised: Normative data and analysis of inter-form and test-retest reliability. Clinical Neuropsychologist 12, 43–55.

Benson RR, Whalen DH, Richardson M, et al. (2001). Parametrically dissociating speech and nonspeech perception in the brain using fMRI. Brain Language 78, 364–396.

Brunelin J, Poulet E, Bediou B, et al. (2006). Low frequency repetitive transcranial magnetic stimulation improves source monitoring deficit in hallucinating patients with schizophrenia. Schizophrenia Research 81, 41–45.

Chibbaro G, Daniele M, Alagona G, et al. (2005). Repetitive transcranial magnetic stimulation in schizophrenic patients reporting auditory hallucinations. Neuroscience Letters 383, 54–57.

Chua SE and McKenna PJ (1995). Schizophrenia–a brain disease? A critical review of structural and functional cerebral abnormality in the disorder. British Journal of Psychiatry 166, 563–582.

Cohen E, Bernardo M, Masana J, et al. (1999). Repetitive transcranial magnetic stimulation in the treatment of chronic negative schizophrenia: a pilot study (letter). Journal of Neurology, Neurosurgery and Psychiatry 67, 129–130.

Colombo C, Gambini O, Macciardi F, et al. (1989). Alpha reactivity in schizophrenia and in schizophrenic spectrum disorders: demographic, clinical and hemispheric assessment. International Journal of Psychophysiology 7, 47–54.

D'Alfonso AA, Aleman A, Kessels RP, et al. (2002). Transcranial magnetic stimulation of the left auditory cortex in patiens with schizophrenia: effects on hallucinations and neurocognition. Journal of Neuropsychiatry and Clinical Neurosciences 14, 77–79.

Feinsod M, Kreinin B, Chistyakov A, Klein E (1998). Preliminary evidence for a beneficial effect of low-frequency, repetitive transcranial magnetic stimulation in patients with major depression and schizophrenia. Depression and Anxiety 7, 65–68.

Fiez JA, Raichle ME, Balota DA, Tallal P, Petersen SE (1996). PET activation of posterior temporal regions during auditory word presentation and verb generation. Cerebral Cortex 6, 1–10.

Fitzgerald PB, Brown TL, Daskalakis ZJ, deCastella A, Kulkarni J (2002). A study of transcallosal inhibition in schizophrenia using transcranial magnetic stimulation. Schizophrenia Research 56, 99–209.

Fitzgerald PB, Benitez J, Daskalakis JZ, Brown TL, Marston NA, de Castella A (2005). A double-blind sham-controlled trial of repetitive transcranial magnetic stimulation in the treatment of refractory auditory hallucinations. Journal of Clinical Psychopharmacology 25, 358–362.

Frith CD and Done DJ (1989) Towards a neuropsychology of schizophrenia. British Journal of Psychiatry 153, 437–443.

Geller V, Grisaru N, Abarbanel JM, Lemberg T, Belmaker RH (1997). Slow magnetic stimulation of prefrontal cortex in depression and schizophrenia. Progress in Neuropsychopharmacology and Biological Psychiatry 21, 105–110.

Gold JM, Carpenter C, Randolph C, Goldberg TE, Weinberger DR (1997). Auditory working memory and Wisconsin Card Sorting Test performance in schizophrenia. Archives of General Psychiatry 54, 159–165.

Gold S, Arndt S, Nopoulos P, O'Leary DS, Andreasen NC (1999). Longitudinal study of cognitive function in first-episode and recent-onset schizophrenia. American Journal of Psychiatry 156, 1342–1348.

Gur RE, Cowell PE, Latshaw A, et al. (2000). Reduced dorsal and orbital prefrontal gray matter volumes in schizophrenia. Archives of General Psychiatry 57, 761–768.

Haddock G, McCarron J, Tarrier N, Faragher EB (1999). Scales to measure dimensions of hallucinations and delusions: the Psychotic Symptom Rating Scales (PSYRATS). Psychological Medicine 29, 879–889.

Hajak G, Marienhagen J, Langguth B, Werner S, Binder H, Eichhammer P (2004). High-frequency repetitive transcranial magnetic stimulation in schizophrenia: a combined treatment and neuroimaging study. Psychological Medicine 34, 1157–1163.

Herwig U, Satrapi P, Schoenfeldt-Lecuona C (2003). Using the international 10–20 EEG system for positioning of transcranial magnetic stimulation. Brain Topography 16, 95–99.

Hoffman RE, Buchsbaum MS, Escobar MD, Makuch RW, Nuechterlein KH, Guich SM (1991). EEG Coherence of prefrontal areas in normal and schizophrenic males during perceptual activation. Journal of Neuropsychiatry and Clinical Neuroscience 3, 169–175.

Hoffman RE, Boutros NN, Berman RM, Roessler E, Krystal JH, Charney DS (1999). Transcranial magnetic stimulation of left temporoparietal cortex in three patients reporting hallucinated 'voices'. Biological Psychiatry 46, 130–132.

Hoffman RE, Boutros NN, Hu S, Berman RM, Krystal JH, Charney DS (2000). Transcranial magnetic stimulation and auditory hallucinations in schizophrenia. Lancet 355, 1073–1075.

Hoffman RE, Hawkins KA, Gueorguieva R, et al. (2003). Transcranial magnetic stimulation of left temporoparietal cortex and medication-resistant auditory hallucinations. Archives of General Psychiatry 60, 49–56.

Hoffman RE, Gueorguieva R, Hawkins KA, et al. (2005). Temporoparietal transcranial magnetic stimulation for auditory hallucinations: safety, efficacy and predictors in a fifty patient sample. Biological Psychiatry 58, 97–104.

Hoffman RE, Hampson M, Wu K et al. (2007). Probing the pathophysiology of auditory hallucinations by combining functional magnetic resonance imaging and transcranial magnetic stimulation. Cerebral Cortex, in press.

Holi MM, Eronen M, Toivonen K, Toivonen P, Marttunen M, Naukkarinen H (2004). Left prefrontal repetitive transcranial magnetic stimulation in schizophrenia. Schizophrenia Bulletin 30, 429–434.

Homan RW, Herman J, Purdy P (1987). Cerebral location of the international 10–20 system electrode placement. Electroencephalography and Clinical Neurophsyiology 66, 376–382.

Jandl M, Bittner R, Sack A, et al. (2005). Changes in negative symptoms and EEG in schizophrenic patients after repetitive transcranial magnetic stimulation (rTMS): an open-label pilot study. Journal of Neural Transmission 112, 955–967.

Jandl M, Steyer J, Weber M, *et al.* (2006). Treating auditory hallucinations by transcranial magnetic stimulation: a randomized controlled crossover trial. *Neuropsychobiology 53*, 63–69.

Jin Y, Potkin SG, Sandman C (1995). Clozapine increases EEG photic driving in clinical responders. *Schizophrenia Bulletin 21*, 263–268.

Jin Y, Potkin SG, Sandman CA, Bunney WE Jr (1998). Topographic analysis of EEG photic driving in patients with schizophrenia following clozapine treatment. *Clinical Electroencephalogy 29*, 73–78.

Jin Y, Potkin SG, Kemp AS, *et al.* (2006). Therapeutic effects of individualized alpha frequency transcranial magnetic stimulation (αTMS) on the negative symptoms of schizophrenia. *Schizophrenia Bulletin 32*, 556–561.

Jo J, Ball SM, Seok H, *et al.* (2006). Experience-dependent modification of mechanisms of long-term depression. *Nature Neuroscience 9*, 170–172.

Keefe RS, Arnold MC, Bayen UJ, Harvey PD (1999). Source monitoring deficits in patients with schizophrenia: a multinomial modeling analysis. *Psychological Medicine 29*, 903–914.

Klein E, Kolsky Y, Puyerovsky M, Koren D, Chistyakov A, Feinsod M (1999a). Right prefrontal slow repetitive transcranial magnetic stimulation in schizophrenia: a double-blind sham-controlled pilot study. *Biological Psychiatry 46*, 1451–1454.

Klein E, Kreinin I, Chistyakov A, *et al.* (1999b). Therapeutic efficacy of right prefrontal slow repetitive transcranial magnetic stimulation in major depression: a double-blind controlled study. *Archives of General Psychiatry 56*, 315–320.

Kraemer HC, Wilson GT, Fairburn CG, Agras WS (2002). Mediators and moderators of treatment effects in randomized clinical trials. *Archives of General Psychiatry 59*, 877–883.

Kwon JS, O'Donnell BF, Wallenstein GV, *et al.* (1999). Gamma frequency-range abnormalities to auditory stimulation in schizophrenia. *Archives of General Psychiatry 56*, 1001–1005.

Lee S-H, Kim W, Chung Y-C, *et al.* (2005). A double-blind study showing that two weeks of daily repetitive TMS over the left or right temporoparietal cortex reduces symptoms in patients with schizophrenia who are having treatment-refractory auditory hallucinations. *Neuroscience Letters 376*, 177–181.

Levitan C, Ward PB, Catts S (1999). Superior temporal gyral volumes and laterality correlates of auditory hallucinations in schizophrenia. *Biological Psychiatry 46*, 955–962.

McGuire PK, Silbersweig DA, Wright I, Murray RM, Frrackowiak RSJ, Frith CD (1996). The neural correlates of inner speech and auditory verbal imagery in schizophrenia: Relationship to auditory hallucinations. *British Journal of Psychiatry 169*, 148–159.

McIntosh AM, Semple D, Tasker K, *et al.* (2004). Transcranial magnetic stimulation for auditory hallucinations in schizophrenia. *Psychiatry Research 127*, 9–17.

Merrin EL, Floyd TC (1992). Negative symptoms and EEG alpha activity in schizophrenic patients. *Schizophrenia Research 8*, 11–19.

Merrin EL, Floyd TC (1996). Negative symptoms and EEG alpha in schizophrenia: a replication. *Schizophrenia Research 19*, 151–161.

Nahas Z, Li X, Kozel FA, *et al.* (2004). Safety and benefits of distance-adjusted prefrontal transcranial magnetic stimulation in depressed patients 55–75 years of age: a pilot study. *Depression and Anxiety 19*, 249–256.

Nayani TH, David AS (1996). The auditory hallucination: a phenomenological survey. *Psychological Medicine 26*, 179–189.

Novak T, Horacek J, Mohr P, *et al.* (2006). The double-blind sham-controlled study of high-frequency rTMS (20Hz) for negative symptoms in schizophrenia: negative results. *Neuroendocrinology Letters 27*, 209–213.

Ojemann GA (1978). Organization of short-term verbal memory of human cortex: evidence from electrical stimulation. *Brain Language 5*, 331–340.

Post RM, Kimbrell TA, McCann UD, *et al.* (1999). Repetitive transcranial magnetic stimulation as a neuropsychiatric tool: present status and future potential. *Journal of ECT 15*, 39–59.

Poulet E, Brunelin J, Bediou B, *et al.* (2005). Slow transcranial magnetic stimulation can rapidly reduce resistant auditory hallucinations in schizophrenia. *Biological Psychiatry 57*, 188–191.

Puri BK, Lekh SK, Nijran KS, Bagary MS, Richardson AJ (2001). SPECT neuroimaging in schizophrenia with religious delusions. *International Journal of Psychophysiology 40*, 143–148.

Rollnik JD, Huber TJ, Mogk H, *et al.* (2000). High frequency repetitive transcranial magnetic stimulation (rTMS) of the dorsolateral prefrontal cortex in schizophrenic patients. *Neuroreport 11*, 4013–4015.

Sachdev P, Loo C, Mitchell P, Malhi G (2005). Transcranial magnetic stimulation for the deficit syndrome of schizophrenia: a pilot investigation. *Psychiatry and Clinical Neurosciences 59*, 354–357.

Scannevin RH, Huganir RL (2000). Postsynaptic organization and regulation of excitatory synapses. *Nature Reviews Neuroscience 1*, 133–141.

Schönfeldt-Lecuona C, Gron G, Walter H, *et al.* (2004). rTMS for the treatment of auditory hallucinations in schizophrenia. *Neuroreport 15*, 1669–1673.

Schroder J, Buchsbaum MS, Siegel BV, *et al.* (1996). Cerebral metabolic activity correlates of subsyndromes in chronic schizophrenia. *Schizophrenia Research 19*, 41–53.

Shergill SS, Murray RM, McGuire PK (1998). Auditory hallucinations: a review of psychological treatments. *Schizophrenia Research 32*, 137–150.

Shergill SS, Brammer MJ, Williams SCR, Murray RM, McGuire PK (2000). Mapping auditory hallucinations in schizophrenia using functional magnetic resonance imaging. *Archives of General Psychiatry 57*, 1033–1038.

Sepehryl AA, Tranulis C, Galinowski A, Stip E (2006). Repetitive transcranial magnetic stimulation treatment for auditory hallucinations in schizophrenia spectrum disorders: a meta-analysis. Poster presentation, Annual Meeting of the Society of Biological Psychiatry, Toronto, 2006.

Silbersweig DA, Stern E, Frith C, *et al.* (1995). A functional neuroanatomy of hallucinations in schizophrenia. *Nature 378*, 176–179.

Spencer KM, Nestor PG, Niznikiewicz MA, Salisbury DF, Shenton ME, McCarley RW (2003). Abnormal neural synchrony in schizophrenia. *Journal of Neuroscience 23*, 7407–7411.

Stefani A, Spadoni F, Bernardi G (1997). Voltage-activated calcium channels: targets of antiepileptic drug therapy? *Epilepsia 38*, 959–965.

Stevens JR, Livermore A (1982). Telemetered EEG in schizophrenia: spectral analysis during abnormal behavior episodes. *Journal of Neurology, Neurosurgery and Psychiatry 45*, 385–395.

Strafella AP, Paus T, Barrett J, Dagher A (2001). Repetitive transcranial magnetic stimulation of the human prefrontal cortex induces dopamine release in the caudate nucleus. *Journal of Neuroscience 21*, RC157.

Swerdlow NR, Koob GF (1987). Dopamine, schizophrenia, mania, depression: toward a unified hypothesis of cortico-striato-pallido-thalamic function. *Behavioral Brain Science 10*, 197–245.

Vaiva G, Cottencin O, Llorca PM, *et al.* (2002). Regional cerebral blood flow in deficit/nondeficit types of schizophrenia according to SDS criteria. *Progress in Neuropsychopharmacology and Biological Psychiatry 26*, 481–485.

Weinberger DR (1987). Implications of normal brain development for the pathogenesis of schizophrenia. *Archives of General Psychiatry 44*, 660–669.

Weinberger DR, Berman KF (1996). Prefrontal function in schizophrenia: confounds and controversies. *Philosophical Transactions of the Royal Society of London B 351*, 1495–1503.

Wolkin A, Sanfilipo M, Wolf AP, Angrist B, Brodie JD, Rotrosen J (1992). Negative symptoms and hypofrontality in chronic schizophrenia. *Archives of General Psychiatry 49*, 959–965.

Yu H-C, Liao K-K, Chang T-J, Tsai S-J (2002). Transcranial magnetic stimulation in schizophrenia (letter). *American Journal of Psychiatry 159*, 494–495.

CHAPTER 43

TMS in the study and treatment of anxiety disorders

Benjamin Greenberg and Sarah H. Lisanby

Introduction

As discussed elsewhere in this volume, TMS has been combined with a variety of other research methods to address many questions of interest in neuropsychiatry, including its promise as a therapy. Only a few exploratory studies of TMS as an anxiety disorder treatment have been reported. As with other disorders, in addition to direct therapeutic applications, TMS may also be a useful tool to study pathophysiology and the neurocircuitry and transmitter systems underlying psychiatric illness. Even in the case of treatment studies, the focal application of TMS in the treatment of anxiety disorders has been guided by present understanding of the neurocircuitry underlying these disorders. Furthermore, the potential to treat anxiety disorders with a focal brain intervention presents the opportunity to test these models for their functional relevance in patients, and to determine heterogeneity among patients in the neurobiological determinants of their illness. TMS may also be applied in the delineation of physiological endophenotypes of relevance to involvement of genetic factors or of particular neuroanatomical networks in anxiety disorders. TMS has also been used in intriguing studies of various behavioral dimensions of relevance to pathological anxiety. This work also remains in an early phase, but may be expected to develop further with the increasing emphasis on dimensional approaches to psychopathology and treatment, both within and across disorders that have traditionally been viewed as categorical entities. This chapter reviews the current state of the literature on these various uses of TMS in the study and treatment of anxiety disorders, and discusses the implications for our understanding of their patho-etiology.

Obsessive–compulsive disorder (OCD)

Models for the neurocircuitry of OCD

In OCD, intrusive and recurrent thoughts, images, or feelings lead to repetitive behaviors. While patients attempt to eliminate them, the obsessions persist, accompanied by marked anxiety. OCD is associated with functional impairment ranging from moderate to extreme. Functional neuroimaging studies beginning in the 1980s established a consistent association between activity in frontal–basal ganglia–thalamic circuits and OCD symptoms and in the response to treatment. Structural neuroimaging findings also generally support these neurocircuitry models, in which dysfunction in specific fronto-basal pathways was proposed to be central to OCD pathophysiology (Rauch *et al.* 1998; Saxena *et al.* 1998). The well-known clinical and familial associations between OCD and tic disorders are consistent with a key role for frontal cortex–basal ganglia–thalamic abnormalities in both illnesses. Although less familiar, a relationship between OCD and certain forms of dystonia (Cavallaro *et al.* 2002) further strengthens support for this model. In addition, OCD or OCD symptoms may appear or worsen markedly in response to brain lesions within this hypothesized circuitry. Moreover, neuropsychological studies have yielded an intriguing pattern of findings suggesting that cognitive processing in OCD reflects both corticobasal dysfunction and compensatory processes elsewhere (Savage and Rauch 2000).

Most relevant to TMS, OCD circuitry models have been used to generate hypotheses about how focal interventions targeting the brain systems implicated might have therapeutic effects. Dysfunction at any one of several places along the neuroanatomical circuits potentially involved in OCD could in theory account for its symptoms. One hypothesis is that a primary locus of abnormality is within orbitofrontal cortex (Insel 1992), the region most consistently implicated in OCD by imaging studies (Whiteside *et al.* 2004). Another widely held working hypothesis is that OCD symptoms arise as a response to defective filtering of cortical input by the basal ganglia (Rossi *et al.* 2005). One result of this abnormal processing would be dysregulated thalamic output to the cerebral cortex (termed defective 'thalamic gating'), leading to aberrant cortical and cortico-basal activity. Consistent with the hypothesis that striatal dysfunction contributes to symptomatology, structural neuroimaging has also found basal ganglia abnormalities in OCD (for review see Pujol *et al.* 2004). Of course, structural abnormalities, even if present, may be apparent on some measures but not others (e.g. Bartha *et al.* 1998; Russell *et al.* 2003). Heterogeneity within and across OCD patient samples also needs to be taken into account (see below).

Other evidence supporting overactive pre-frontal–subcortical function in OCD comes from neurosurgical treatment of patients with disabling OCD who prove extremely resistant to conventional pharmacological and behavioral treatments. Sterotactic lesion procedures, including anterior cingulotomy and anterior capsulotomy, have led to significant symptom improvement in prospective open-label treatment studies of patients selected according to strict criteria for treatment resistance and severity (for review see Greenberg *et al.* 2003). While based on empirical research, these procedures target neurocircuitry implicated in OCD by functional neuroimaging results. A newer nonablative procedure, deep brain stimulation (DBS) (Greenberg 2004), has also appeared promising in this OCD patient subgroup (Nuttin *et al.* 1999; Gabriels *et al.* 2003; Aouizerate *et al.* 2004; Greenberg *et al.* 2006). DBS targeting the anterior limb of the internal capsule and/or adjacent ventral striatum has also been shown to alter activity in the frontobasal networks implicated in OCD, either acutely (Rauch *et al.* 2006) or after chronic stimulation (Nuttin *et al.* 2003; Abelson *et al.* 2005). There is also the intriguing possibility that neuroimaging predictors of the response to DBS can be identified in OCD (Van Laere *et al.* 2006).

Although the therapeutic potential of TMS in OCD is unknown, the well-developed theoretical models of OCD can have great heuristic value in treatment development research. For example, combining TMS and imaging could help to answer questions central to the development of TMS as a potential OCD therapy. These include: (i) Can TMS modulate neural networks implicated in OCD?; (ii) Does TMS have acute behavioral effects consistent with such an effect on neural networks implicated in OCD?; (iii) Does TMS have lasting therapeutic effects in OCD?

TMS as an anatomical probe in OCD

Given the imaging results presented above, TMS studies targeting orbitofrontal cortex (OFC) in OCD would be of great interest. Unfortunately, OFC is not as accessible to TMS as other cortical regions due to its distance from the scalp. TMS most easily directly targets the lateral surface of the cortex, due to reduction in field intensity as a function of distance from the coil. While very intriguing studies have found physiological and information-processing effects consistent with effects of TMS on prefrontal cortex (van Honk *et al.* 2001; Schutter and van Honk 2006) the limited work in OCD to date has focused on more dorsal cortical areas.

The first TMS in OCD study (Greenberg *et al.* 1997) was not a treatment trial but rather used TMS as an anatomical probe to test the hypothesis that altering prefrontal cortex activity would affect OCD symptoms acutely. We administered single sessions of high-frequency (20 Hz) repetitive (r)TMS to left and right dorsolateral prefrontal cortex and to a parieto-occipital control site in a randomized controlled design. The prefrontal locations were defined as the site 5 cm anterior and 2 cm inferior to the hand area of primary motor cortex on each side. Twelve OCD patients were studied; they had on average moderately severe symptoms (a mean baseline Yale–Brown Obsessive–Compulsive Scale (Y-BOCS) score of about 20, but two of the 12 patients were severely affected). Eight patients were treated with serotonin reuptake inhibitors. Each site was stimulated, 2 days apart, with 20 Hz trains of 2 s each, once per minute for 20 min (800 pulses total per session) with a figure-8-shaped focal coil attached to a Cadwell high-speed magnetic stimulator. rTMS intensity was 80% of abductor pollicis brevis twitch threshold. We observed that right lateral prefrontal rTMS was followed by a significant reduction in compulsive urges, lasting at least 8 h. This effect was not seen after left prefrontal or parieto-occipital stimulation. These OCD patients, who were not clinically depressed at baseline as a group, also reported significant mood elevation for 30 min after right prefrontal stimulation. The results suggested that a single 20 Hz right prefrontal rTMS session may have produced effects on systems modulating compulsive urges for hours after stimulation ended, with a more transient effect on mood. This study had a number of limitations. The anatomical localization was relatively crude in comparison with today's standards. Importantly, there were no direct measures of cortical or subcortical function. With regard to placebo effects, which have been difficult to control for in TMS research

generally, the use of active stimulation to control sites used should have limited placebo responses.

These findings raised the intriguing possibility that right prefrontal rTMS interrupted ongoing cortical activity related to compulsive urges. Alternatively, stimulation may have indirectly enhanced activity in subcortical regions, which might have suppressed compulsions. These possibilities await testing in further studies combining rTMS with measures of regional brain activity in OCD. Such studies could include functional neuroimaging, cognitive tests, and electrophysiological methods. The unexpected laterality of the effects was intriguing. Converging evidence suggests that activity related to OCD symptom expression and also to the response to effective treatments may be lateralized. For example, there were opposite correlations between acute symptom provocation and orbitofrontal perfusion in the right and left hemispheres in one study (Rauch *et al.* 1994), disruption of abnormally correlated metabolic activity in brain regions specifically in the right hemisphere was associated with symptom improvement after cognitive-behavior therapy (Schwartz *et al.* 1996), and the location of anterior capsulotomy lesions in the right but not the left hemisphere appeared to be a key determinant of neurosurgical efficacy in OCD (Lippitz *et al.* 1997). Moreover, though there are exceptions, neuroimaging studies generally find changes in the right hemisphere associated with therapeutic improvement after medication treatment (e.g. Saxena *et al.* 2003).

Single- and paired-pulse TMS: physiological probes in OCD

Described elsewhere in this volume in great detail, paired-pulse TMS is well established as a probe of inhibitory and excitatory modulation in the primary motor cortex. This method is based on observations that TMS pulses too weak to produce motor-evoked potentials (MEPs) themselves can modulate the responses to stronger pulses that are above the threshold to produce MEPs. This happens when the weak pulses are presented milliseconds before the suprathreshold stimuli. Interest has focused on several phenomena, including the threshold excitability (the lowest intensity that produces MEPs, termed motor threshold or MT), and short-interval intracortical inhibition (SICI), the reduction in MEP amplitude when the weak and test pulses are separated by 2–5 ms. The pharmacological bases and associations of paired-pulse TMS with clinical diagnoses are discussed extensively elsewhere in this volume. Here, it is worth noting briefly that SICI has been thought to be mediated by activation of gamma-aminobutyric acid (GABA)-ergic cortical interneurons (e.g. Ziemann *et al.* 1996), and that abnormalities in this measure have been demonstrated in a number of neuropsychiatric disorders thought to involve abnormal basal ganglia function. Patients with Tourette's syndrome (TS) or focal dystonias, illnesses both thought to involve basal ganglia pathology, are related to OCD reduced paired-pulse inhibition compared to nonpatients (Ridding *et al.* 1995; Ziemann *et al.* 1997). Intrigued by these reports, we performed similar experiments in OCD. We also detected reduced inhibitory modulation with paired-pulse TMS (Greenberg *et al.* 1998, 2000). Although greatest in patients with comorbid OCD and tics, reduced SICI was observed even in OCD patients without tics.

A second, and unexpected, finding was that both resting and active motor thresholds were reduced in OCD patients compared with controls–evidence of increased cortical excitability (Greenberg *et al.* 2000). This could result from increased excitability intrinsic to motor cortex, or an increased excitatory drive from thalamo-cortical projections. Either phenomenon could be related to the consistent findings of increased cortical activity in OCD on functional neuro-imaging (see above). Moreover, either reduced intracortical inhibition or enhanced cortical excitability could both be due to an abnormality in basal ganglia function, consistent with the theories of OCD pathogenesis discussed above.

These single- and paired-pulse TMS findings provide additional evidence that OCD and TS, related to each other by phenomenology and heritability, may also have overlapping pathophysiological features. Further studies will be necessary to determine whether this phenomenon is preferentially observed in neuropsychiatric illnesses thought to be related to OCD and TS. Given the increasing interest in delineating subtypes of OCD, paired-pulse TMS measures may help to identify endophenotypes that are associated, in particular, with tic-related or even 'dystonia-related' OCD. A recent finding suggests, in contrast, that behavioral hyperactivity is strongly related to reduced SICI in TS patients (Gilbert *et al.* 2005), consistent with analyses of paired-pulse TMS from patients with TS, OCD, and attention deficit hyperactivity disorder, three disorders whose symptoms frequently overlap.

Additional work will be necessary to determine whether paired-pulse TMS measures can help screen new pharmacological treatment approaches in OCD, or could also usefully monitor therapeutic effects of agents that may act to decrease excitatory drive via reducing glutaminergic activity (Coric *et al.* 2003). There is empirical support for the idea that changes in glutaminergic function might be relevant to effects of successful pharmacological treatment of OCD (Rosenberg *et al.* 2000), which is unsurprising since glutamate is the predominant transmitter in the frontobasal circuitry implicated in OCD. Treatments for OCD could also act in part via enhancing inhibitory mechanisms. Those using GABA predominate in circuitry inhibited in OCD. In this regard, however, it is worth noting that the clinical utility of benzodiazepines and GABA-modulating anticonvulsants as OCD monotherapy has been mixed at best (e.g. Hollander *et al.* 2003).

Repetitive TMS: a possible OCD treatment?

Obsessive–compulsive disorder has been found to be among the top ten causes of disability in developed societies (Murray and Lopez 1997). New therapeutic modalities are needed, since available medication and behavioral treatments, though extremely beneficial for many patients, have limited efficacy or limited acceptance in a substantial group of affected individuals. Despite promising leads (Hollander *et al.* 2002), development of fundamentally new and effective medication treatment modalities has been largely lacking in recent years. Very few studies have explored the potential utility of rTMS as a treatment for OCD. A 2003 review found only three studies; only two of them were treatment studies (Martin et al. 2003). Since then, another pilot TMS treatment study has appeared (see below). Each of the three studies had a different design; no replication attempts have yet appeared.

In an open pilot trial, 12 patients with treatment-resistant OCD were randomly assigned to either right or left prefrontal rTMS daily for 2 weeks (Sachdev et al. 2001). Patients were assessed by an independent rater at 1 week, 2 weeks, and 1 month after TMS. There were significant group improvements in obsessions, compulsions, and total scores on the Y-BOCS after 2 weeks and at 1 month follow-up. Improvements remained significant for obsessions and tended toward significance for total Y-BOCS scores after controlling for post-rTMS changes in depression. There were no significant differences in outcome between the right- and left-sided rTMS groups. Two subjects (33%) in each group showed a clinically significant improvement that persisted at 1 month. However, without a sham treatment group, the possibility that these were placebo responses cannot be ruled out.

The only controlled trial published thus far (Alonso et al. 2001) used a very different technique. Eighteen patients were randomly assigned to real or sham TMS at 1 Hz, centered over right prefrontal cortex using a teardrop-shaped coil. Intensity was 110% of motor threshold for real rTMS and 20% for sham stimulation (which eight of the patients received). Each patient received 1200 pulses each of active or sham TMS per session in a total of 18 sessions. There were no group differences in OCD severity post-treatment. Two of 10 patients who received real rTMS, with checking compulsions, and one of eight receiving sham treatment, with sexual/religious obsessions, were considered responders. The techniques used would result in administration of far fewer pulses than after higher-frequency stimulation, and with a different magnetic field distribution than that produced by a figure-8 coil. The finding that low-frequency rTMS of the right prefrontal cortex failed to produce significant improvement in OCD is of interest in light of another report that short-term (2 days) rTMS at 1 Hz failed to affect obsessions, compulsions, or tics in patients with TS in a sham-controlled

crossover study (Munchau et al. 2002). A placebo-controlled crossover trial of rTMS was ineffective in TS. The 16 patients enrolled in the study received 2 days each of real and sham TMS. No improvements were found in motor or vocal tics or in obsessions and compulsions in the 12 patients who completed the protocol.

The most recent study (Mantovani et al. 2006) also used 1 Hz TMS, but a different target, the supplementary motor area (SMA), in patients with OCD, TS, or both disorders. The investigators hypothesized that using TMS over the SMA would improve symptoms by reducing the motor system hyperexcitability previously shown in OCD and TS (discussed above). A figure-8 coil was placed over the midline to stimulate the SMA bilaterally. Patients underwent 10 daily sessions of open rTMS, 1200 stimuli per session. Of the 10 patients participating, five had OCD and two had OCD plus TS. By the second week of treatment, statistically significant reductions were seen in the OCD and tic severity, and in measures of depressive and anxiety symptoms. Symptom improvements correlated with a significant increase of the right hemisphere resting motor threshold and were stable at 3 month follow-up. This intriguing study suggested that open rTMS to SMA resulted in both clinical improvement and in a normalization of right hemisphere hyperexcitability, thereby restoring hemispheric symmetry in motor thresholds. A sham-controlled study pursuing these results is presently underway. In this regard it is noteworthy that a single intravenous dose of clomipramine transiently reduced motor cortex excitability in depressed patients, seen as an increase in SICI, a decrease in ICF, and a small but significant increase in resting and active motor thresholds (Manganotti et al. 2001). The fact that the effects of 1 Hz TMS to SMA and intravenous clomipramine might have some parallels is interesting given the therapeutic efficacy of oral clomipramine in OCD. This finding suggests that TMS measures may be useful in research on the therapeutic effects of intravenous clomipramine in treatment-resistant OCD (Fallon et al. 1998; Koran et al. 2006).

In considering any therapeutic effects of TMS in OCD, it is worth noting that OCD symptoms are heterogeneous. Emerging data suggest that OCD symptom subtypes, identified within the categorical construct of OCD, may have different patterns of familiality, course, and response to treatment. Very recent functional imaging work suggests that the neuroanatomical underpinnings of these symptom subtypes might also be separable. It is reasonable to expect that attempts to probe brain mechanisms in OCD will be increasingly informed by advances in understanding of the phenotypes of the illness. Specifically, the hypothesis that different OCD symptom subtypes might be associated with abnormalities in different functional networks (Mataix-Cols *et al.* 2004) is being actively investigated in a number of centers and may have direct bearing on the designs of future work testing the therapeutic promise of TMS for OCD.

Post-traumatic stress disorder (PTSD)

Models for the neurocircuitry of PTSD

Cardinal symptoms of PTSD are the re-experiencing of events (such as visual flashbacks), avoidant behavior, and hyperarousal accompanied by marked anxiety. These symptoms, which by definition evolve after an extremely dangerous and frightening experience, cause significant interference in occupational and social functioning. Recent theories of PTSD pathogenesis suggest that mechanisms involved in normal threat assessment become dysregulated, so that fear responses associated with the original traumatic situation become overgeneralized and fail to extinguish (Rauch *et al.* 1998). This model invokes the common-sense proposal that brain regions associated with fear conditioning and extinction are important in PTSD. These areas include the amygdala, involved in threat assessment, in reallocation of resources in response to threat, and in fear conditioning itself, the hippocampus, believed to encode and access contextual information, and the medial prefrontal cortex, particularly the affective division of the anterior cingulate gyrus, believed to promote fear extinction via its descending influence on the amygdala. Dysfunction in any of these regions might therefore contribute to PTSD symptoms. Accumulating neuroimaging evidence supports this emerging conception of the neural circuitry

involved in PTSD (e.g. Kent and Rauch 2003). Here, it is most important to note that limbic and paralimbic activation appears associated with traumatic memory-related anxiety, and that prefrontal input could modulate PTSD-related subcortical activity.

rTMS: a possible PTSD treatment?

There is some preliminary research on therapeutic effects of TMS in PTSD. The possibility that prefrontal cortex stimulation might affect regional brain activity associated with PTSD symptoms was the focus of a small pilot study. Two PTSD patients received 1 Hz rTMS (80% motor threshold) delivered openly to right dorsolateral prefrontal cortex using a figure-8 coil (McCann *et al.* 1998). One patient received 17 daily sessions of 1200 pulses each over a 1 month period and reported selective improvement in PTSD symptoms without a change in global anxiety. The second patient had 1 Hz rTMS over the same region 30 times during a 6 week period, and also reported significant symptom improvement. In each case the apparent benefit persisted for less than 1 month after rTMS was discontinued. Baseline [^{18}F]deoxyglucose PET scans, obtained months before rTMS administration in both cases (an important caveat), showed that metabolism in right cerebral areas was higher than that in a reference healthy population. Repeat scans after TMS displayed a reduction in metabolism from the baseline images. This open pilot study in PTSD patients could not exclude placebo effects, or changes in severity due to the natural course of the illness, as explanations for the observed changes in clinical state.

Another preliminary study used a single session of even lower-frequency stimulation. Ten PTSD patients had a single session of 0.3 Hz stimulation. Intensity was 100% of the maximum output of a Magstim single-pulse stimulator. A total of 30 pulses were applied bilaterally over motor cortex (Grisaru *et al.* 1998). Both self and observer ratings of PTSD symptoms improved transiently, generally from 1 to 7 days after the procedure. Low-frequency TMS was well-tolerated. However, while there was some suggestion of a therapeutic effect on PTSD symptoms, this was transient.

A later study (Rosenberg *et al.* 2002) treated 12 patients with comorbid PTSD and depression openly, using rTMS over left frontal cortex as an adjunct to antidepressant medications. rTMS parameters were 90% of motor threshold, 1 or 5 Hz, 6000 stimuli over 10 days. Whereas 75% of the patients had an antidepressant response (sustained 2 months later in six patients) with improvement in anxiety, hostility, and insomnia, core PTSD symptoms improved only minimally.

In a recent controlled trial (Cohen *et al.* 2004), 24 patients with PTSD were randomly assigned to low-frequency (1 Hz), high-frequency (10 Hz), or sham rTMS in a double-blind design. Patients had 10 daily sessions over 2 weeks. Ten daily treatments of 10 Hz rTMS (at 80% motor threshold) over the right dorsolateral prefrontal cortex had notable therapeutic effects on the core PTSD symptoms of re-experiencing and avoidance. Anxiety was also reduced after a course of rTMS over the right dorsolateral prefrontal cortex.

Another controlled trial of right prefrontal rTMS for PTSD has been described in a preliminary report (Chae 2006). Eighteen patients with PTSD were randomly assigned to either 1 Hz or sham rTMS, delivered over 3 weeks. The Clinician Administered PTSD Scale (CAPS) was given at baseline and after 2, 4, and 8 weeks. TMS was well tolerated. While total CAPS scores reportedly decreased after TMS, scores on CAPS subscales changed differentially. There was a significant reduction in the re-experiencing subscale, and a trend toward a decrease in CAPS avoidance scores. CAPS hyperarousal scores were unchanged. As for OCD (discussed above), specific symptom dimensions of the categorical construct of PTSD may prove susceptible to TMS treatments, while others may not.

Taken together, the available data suggest that additional TMS work is warranted in PTSD, as it is for OCD, although clearly TMS work in either disorder remains early in development. It is intriguing that TMS may find particular application in investigating and possibly treating some symptoms of clinical disorders more than others, as in the example immediately above of symptom subtypes in PTSD. Such symptom subtypes, when represented as continuous scores on rating measures, might be described as 'pathological dimensions'.

Panic disorder

Relatively little work to date has examined the potential utility of TMS in the treatment of panic disorder. A recent study investigated the effects of pretreatment with active or sham 1 Hz rTMS to the right prefrontal cortex on panic attacks experimentally induced by cholecysto-kinin-tetrapeptide (CCK-4) in 11 normal volunteers in a randomized crossover protocol (Zwanzger *et al.* 2006). Active rTMS had no effect on CCK-4-induced panic as measured by the Acute Panic Inventory and the Panic Symptom Scale and measurements of heart rate, plasma adrenocorticotrophic hormone (ACTH), and cortisol. This finding is in contradistinction to the action of the anxiolytic alprazolam that blunts panic response to CCK-4.

Regarding therapeutic uses of TMS in panic, at present only case reports are available. Three cases with panic disorder were reported to have modest improvement with 1 Hz rTMS to the right prefrontal cortex (Garcia-Toro *et al.* 2002). In another case report, there was marked improvement in panic lasting 1 month and a reduction in CCK-4-induced panic following 1 Hz to the right prefrontal cortex (Zwanzger *et al.* 2002).

In an open trial of rTMS in six patients with panic disorder and comorbid depression, Mantovani *et al.* (personal communication) found that 1 Hz rTMS to the right prefrontal cortex produced a clinically significant and sustained improvement of panic symptoms in >80% of patients that appeared to be independent from antidepressant effects. It will be important to replicate this finding in a controlled setting to determine whether this approach will ultimately have value in treating panic disorder.

Animal models

Despite the difficulties in generalization from animal models to human illness, animal models remain important in the screening of novel therapeutic agents and may be useful for high-throughput studies to determine profiles of action across various behavioral models and determination of therapeutic range in dosage. The use of TMS in animal models is fraught by the difficulty in simulating in small animals the field distributions and strengths obtained in the human brain, and limitations on coil size due to material heating. Nevertheless, TMS has been attempted in animal models of anxiety. Hargreaves *et al.* (2005) found that while TMS had some activity on animal models of learned helplessness, there were no observable effects on the animal models of anxiety tested. This failure may be due to a number of etiologies, such as the technical limitations discussed above, the TMS dosages selected, or the relative lack of focality of TMS in rodent brains that may simultaneously stimulate circuits which may be anxiogenic. Nevertheless, it is clear that TMS as applied in rodents had no effects on the models that do show activity with effective anxiolytic medications. Better methods for applying TMS in animals should be developed. It is also possible that the animal models used to screen pharmacological agents may not necessarily reveal effects of TMS if TMS acts via mechanisms distinct from pharmacology.

Normal behavioral dimensions of relevance to anxiety disorders

TMS may be useful in investigating 'normal' behavioral dimensions, i.e. those continuously distributed in the general population. High or low levels of such traits can predispose individuals to anxiety disorders or other types of psychopathology. An example is the finding that motor cortex excitability on paired-pulse TMS correlated with neuroticism on the revised NEO personality inventory (NEO-PI-R) in 46 healthy volunteers (23 women, 23 men). The threshold and amplitude of MEPs were measured in response to single and paired (subthreshold–suprathreshold) TMS pulses at 3–15 ms intervals. The paired-pulse conditioned/unconditioned MEP amplitude ratios correlated with neuroticism, a stable measure of trait-level anxiety and other negative emotions, in the whole sample ($r = 0.48$; $P = 0.0006$) and in the men ($r = 0.63$; $P = 0.0009$) (Wassermann *et al.* 2001). This relationship appeared to reflect a factor that contributes to both personality and cortical regulation. It was not statistically significant in women, possibly because of confounding

hormonal influences on excitability (Smith *et al.* 1999). Decreased intracortical inhibition may be related more to trait anxiety and depression, which are high in OCD, than to OCD itself. However, the MEP threshold (significantly lowered in OCD) was unrelated to neuroticism.

TMS might, in combination with neuroimaging and personality psychology (research on individual differences) be used as one tool in translational research to understand and treat anxiety disorders. In a very intriguing recent report, brain morphometry was found related to extinction memory after fear conditioning. Specifically, healthy individuals with a thicker right medial orbitofrontal cortex displayed greater extinction memory, suggesting that the size of this brain region might explain individual differences in the ability to modulate fear, a trait central to the predisposition toward (or resilience to) anxiety disorders including PTSD (Milad *et al.* 2005). Moreover, the same research group found that cortical thickness at this medial orbitofrontal locus was positively correlated with extraversion and negatively correlated with neuroticism (Rauch *et al.* 2005). Thus, neuroimaging has identified brain structural markers that may predict an ability to recover from stressful life events and which are associated with personality traits known to predispose or protect individuals to symptoms of anxiety disorders.

TMS can be used to investigate brain mechanisms underlying how emotions are elicited in a situationally specific manner, as in anxiety disorders. For instance, rTMS research in healthy subjects suggests that the emotions anger and anxiety are lateralized in the prefrontal cortex. In an intriguing placebo-controlled study, 1 Hz rTMS at 130% of the individual motor threshold over the right prefrontal cortex reduced the vigilant emotional response to fearful faces in eight healthy subjects. These data provide further support for lateralization in processing anxiety and other emotions in the prefrontal cortex (van Honk *et al.* 2002). Such approaches, used in multidisciplinary studies with complementary measures of behavior, emotional traits, and cerebral activity, promise to further advance our understanding of relationships between clinically relevant behaviors and activity in specific neural networks.

Conclusions

TMS studies in anxiety disorders have the potential to make several unique contributions to the field. They hold promise in improving our understanding of physiological abnormalities and neuroanatomical networks mediating symptoms of this group of illnesses. TMS studies may elucidate pathogenesis while providing an impetus for pharmacological studies of new treatment approaches, such as agents that might normalize an excessive cortical excitability in OCD observed using TMS as a physiological probe. As emphasized above, investigation of possible therapeutic effects of rTMS in OCD, PTSD, or any anxiety disorder remains at a preliminary stage. There have been promising initial observations in OCD, which require systematic testing in controlled studies. Other research in progress might help guide trials of rTMS as a treatment. For example, DBS of subcortical targets in severe and extremely treatment-refractory OCD, termed 'intractable' illness, may, in combination with functional neuroimaging, may be particularly powerful in elucidating the relationship between activity within these networks and symptoms of illness. Coming full circle (or perhaps 'full circuit'), combined DBS–imaging studies could point to involvement of cortical regions that are more easily accessible to TMS. Investigation of all of these stimulation targets in treatment trials is to be encouraged.

References

Abelson JL, Curtis GC, Sagher O, et al. (2005). Deep brain stimulation for refractory obsessive–compulsive disorder. Biological Psychiatry 57, 510–516.

Alonso P, Pujol J, Cardoner N, et al. (2001). Right prefrontal repetitive transcranial magnetic stimulation in obsessive–compulsive disorder: a double-blind, placebo-controlled study. American Journal of Psychiatry 158, 1143–1145.

Aouizerate B, Cuny E, Martin-Guehl C, et al. (2004). Deep brain stimulation of the ventral caudate nucleus in the treatment of obsessive–compulsive disorder and major depression. Case report. Journal of Neurosurgery 101, 682–686.

Bartha R, Stein MB, Williamson PC, et al. (1998). A short echo 1H spectroscopy and volumetric MRI study of the corpus striatum in patients with obsessive–compulsive disorder and comparison subjects. American Journal of Psychiatry 155, 1584–1591.

Cavallaro R, Galardi G, Cavallini MC, et al. (2002). Obsessive compulsive disorder among idiopathic focal dystonia patients: an epidemiological and family study. Biological Psychiatry 52, 356–361.

Chae J-H (2006). Low frequency repetitive transcranial magnetic stimulation for the treatment of patients with posttraumatic stress disorder. International Anxiety Disorders Conference, Cape Town, South Africa.

Cohen H, Kaplan Z, Kotler M, Kouperman I, Moisa R, Grisaru N (2004). Repetitive transcranial magnetic stimulation of the right dorsolateral prefrontal cortex in posttraumatic stress disorder: a double-blind, placebo-controlled study. American Journal of Psychiatry 161, 515–524.

Coric V, Milanovic S, Wasylink S, Patel P, Malison R, Krystal JH (2003). Beneficial effects of the antiglutamatergic agent riluzole in a patient diagnosed with obsessive–compulsive disorder and major depressive disorder. Psychopharmacology (Berlin) 167, 219–20.

Fallon BA, Liebowitz MR, Campeas R, et al. (1998). Intravenous clomipramine for obsessive–compulsive disorder refractory to oral clomipramine: a placebo-controlled study. Archives of General Psychiatry 55, 918–924.

Gabriels L, Cosyns P, Nuttin B, Demeulemeester H, Gybels J (2003). Deep brain stimulation for treatment-refractory obsessive–compulsive disorder: psychopathological and neuropsychological outcome in three cases. Acta Psychiatrica Scandinavica 107, 275–282.

Gilbert DL, Sallee FR, Zhang J, Lipps TD, Wassermann EM (2005). Transcranial magnetic stimulation-evoked cortical inhibition: a consistent marker of attention-deficit/hyperactivity disorder scores in Tourette syndrome. Biological Psychiatry 57, 1597–1600.

Greenberg BD (2004). Deep brain stimulation in psychiatry. In SH Lisanby (ed.), Brain stimulation in psychiatric treatment, Vol. 23. Washington, DC: American Psychiatric Publishing.

Greenberg BD, George MS, Martin JD, et al. (1997). Effect of prefrontal repetitive transcranial magnetic stimulation in obsessive–compulsive disorder: a preliminary study. American Journal of Psychiatry 154, 867–869.

Greenberg BD, Ziemann U, Harmon A, Murphy DL, Wassermann EM (1998). Decreased neuronal inhibition in cerebral cortex in obsessive–compulsive disorder on transcranial magnetic stimulation. Lancet 352, 881–882.

Greenberg BD, Ziemann U, Cora-Locatelli G, et al. (2000). Altered cortical excitability in obsessive–compulsive disorder. Neurology 54, 142–147.

Greenberg BD, Price LH, Rauch SL, et al. (2003). Neurosurgery for intractable obsessive–compulsive disorder and depression: critical issues. Neurosurgery Clinics of North America 14, 199–212.

Greenberg BD, Malone DA, Friehs GM, et al. (2006). Three-year outcomes in deep brain stimulation for highly resistant obsessive–compulsive disorder. Neuropsychopharmacology 31, 2384–2393.

Grisaru N, Amir M, Cohen H, Kaplan Z (1998). Effect of transcranial magnetic stimulation in posttraumatic stress disorder: a preliminary study. *Biological Psychiatry* 44, 52–55.

Hollander E, Bienstock CA, Koran LM, *et al.* (2002). Refractory obsessive–compulsive disorder: state-of-the-art treatment. *Journal of Clinical Psychiatry* 63(Suppl. 6), 20–29.

Hollander E, Kaplan A, Stahl SM (2003). A double-blind, placebo-controlled trial of clonazepam in obsessive–compulsive disorder. *World Journal of Biological Psychiatry* 4, 30–34.

Insel TR (1992). Toward a neuroanatomy of obsessive–compulsive disorder. *Archives of General Psychiatry* 49, 739–744.

Kent JM, Rauch SL (2003). Neurocircuitry of anxiety disorders. *Current Psychiatry Reports* 5, 266–273.

Koran LM, Aboujaoude E, Ward H, *et al.* (2006). Pulse-loaded intravenous clomipramine in treatment-resistant obsessive–compulsive disorder. *Journal of Clinical Psychopharmacology* 26, 79–83.

Lippitz B, Mindus P, Meyerson BA, Kihlstrom L, Lindquist C (1997). Obsessive compulsive disorder and the right hemisphere: topographic analysis of lesions after anterior capsulotomy performed with thermocoagulation. *Acta Neurochirurgica Supplement* 68, 61–63.

McCann UD, Kimbrell TA, Morgan CM, *et al.* (1998). Repetitive transcranial magnetic stimulation for posttraumatic stress disorder. *Archives of General Psychiatry* 55, 276–279.

Manganotti P, Bortolomasi M, Zanette G, Pawelzik T, Giacopuzzi M, Fiaschi A (2001). Intravenous clomipramine decreases excitability of human motor cortex. A study with paired magnetic stimulation. *Journal of Neurological Sciences* 184, 27–32.

Mantovani A, Lisanby SH, Pieraccini F, Ulivelli M, Castrogiovanni P, Rossi S (2006). Repetitive transcranial magnetic stimulation (rTMS) in the treatment of obsessive–compulsive disorder (OCD) and Tourette's syndrome (TS). *International Journal of Neuropsychopharmacology* 9, 95–100.

Martin JL, Barbanoj MJ, Perez V, Sacristan M (2003). Transcranial magnetic stimulation for the treatment of obsessive–compulsive disorder. *Cochrane Database of Systematic Reviews*, CD003387.

Mataix-Cols D, Wooderson S, Lawrence N, Brammer MJ, Speckens A, Phillips ML (2004). Distinct neural correlates of washing, checking, and hoarding symptom dimensions in obsessive–compulsive disorder. *Archives of General Psychiatry* 61, 564–576.

Milad MR, Quinn BT, Pitman RK, Orr SP, Fischl B, Rauch SL (2005). Thickness of ventromedial prefrontal cortex in humans is correlated with extinction memory. *Proceedings of the National Academy of Sciences of the USA* 102, 10706–10711.

Munchau A, Bloem BR, Thilo KV, Trimble MR, Rothwell JC, Robertson MM (2002). Repetitive transcranial magnetic stimulation for Tourette syndrome. *Neurology* 59, 1789–1791.

Murray CJ, Lopez AD (1997). Global mortality, disability, and the contribution of risk factors: Global Burden of Disease Study. *Lancet* 349, 1436–1442.

Nuttin B, Cosyns P, Demeulemeester H, Gybels J, Meyerson B (1999). Electrical stimulation in anterior limbs of internal capsules in patients with obsessive–compulsive disorder. *Lancet* 354, 1526.

Nuttin BJ, Gabriels LA, Cosyns PR, *et al.* (2003). Long-term electrical capsular stimulation in patients with obsessive–compulsive disorder. *Neurosurgery* 52, 1263–1272; discussion 1272–1274.

Pujol J, Soriano-Mas C, Alonso P, *et al.* (2004). Mapping structural brain alterations in obsessive–compulsive disorder. *Archives of General Psychiatry* 61, 720–730.

Rauch SL, Jenike MA, Alpert NM, *et al.* (1994). Regional cerebral blood flow measured during symptom provocation in obsessive–compulsive disorder using oxygen 15-labeled carbon dioxide and positron emission tomography. *Archives of General Psychiatry* 51, 62–70.

Rauch SL, Whalen PJ, Dougherty DD, Jenike MA (1998). Neurobiological models of obsessive compulsive disorders. In: MA Jenike, L Baer, WE Minichiello (eds), *obsessive–compulsive disorders: practical management*, pp. 222–253. Boston, MA: Mosby.

Rauch SL, Milad MR, Orr SP, Quinn BT, Fischl B, Pitman RK (2005). Orbitofrontal thickness, retention of fear extinction, and extraversion. *Neuroreport* 16, 1909–1912.

Rauch SL, Dougherty DD, Malone D, *et al.* (2006). A functional neuroimaging investigation of deep brain stimulation in patients with obsessive–compulsive disorder. *Journal of Neurosurgery* 104, 558–565.

Ridding MC, Sheean G, Rothwell JC, Inzelberg R, Kujirai T (1995). Changes in the balance between motor cortical excitation and inhibition in focal, task specific dystonia. *Journal of Neurology, Neurosurgery and Psychiatry* 59, 493–498.

Rosenberg DR, MacMaster FP, Keshavan MS, Fitzgerald KD, Stewart CM, Moore GJ (2000). Decrease in caudate glutamatergic concentrations in pediatric obsessive–compulsive disorder patients taking paroxetine. *Journal of the American Academy of Child and Adolescent Psychiatry* 39, 1096–103.

Rosenberg PB, Mehndiratta RB, Mehndiratta YP, Wamer A, Rosse RB, Balish M (2002). Repetitive transcranial magnetic stimulation treatment of comorbid posttraumatic stress disorder and major depression. *Journal of Neuropsychiatry and Clinical Neuroscience* 14, 270–276.

Rossi S, Bartalini S, Ulivelli M, *et al.* (2005). Hypofunctioning of sensory gating mechanisms in patients with obsessive–compulsive disorder. *Biological Psychiatry* 57, 16–20.

Russell A, Cortese B, Lorch E, *et al.* (2003). Localized functional neurochemical marker abnormalities in dorsolateral prefrontal cortex in pediatric obsessive–compulsive disorder. *Journal of Child and Adolescent Psychopharmacology* 13(Suppl. 1), S31–38.

Sachdev PS, McBride R, Loo CK, Mitchell PB, Malhi GS, Croker VM (2001). Right versus left prefrontal transcranial magnetic stimulation for obsessive–compulsive disorder: a preliminary investigation. *Journal of Clinical Psychiatry 62*, 981–984.

Savage CR, Rauch SL (2000). Cognitive deficits in obsessive–compulsive disorder. *American Journal of Psychiatry 157*, 1182–1183.

Saxena S, Brody AL, Schwartz JM, Baxter LR (1998). Neuroimaging and frontal-subcortical circuitry in obsessive–compulsive disorder. *British Journal of Psychiatry Supplement*, 26–37.

Saxena S, Brody AL, Ho ML, Zohrabi N, Maidment KM, Baxter LR, Jr (2003). Differential brain metabolic predictors of response to paroxetine in obsessive–compulsive disorder versus major depression. *American Journal of Psychiatry 160*, 522–532.

Schutter DJ, van Honk J (2006). Increased positive emotional memory after repetitive transcranial magnetic stimulation over the orbitofrontal cortex. *Journal of Psychiatry and Neuroscience 31*, 101–104.

Schwartz JM, Stoessel PW, Baxter LR, Jr, Martin KM, Phelps ME (1996). Systematic changes in cerebral glucose metabolic rate after successful behavior modification treatment of obsessive–compulsive disorder. *Archives of General Psychiatry 53*, 109–113.

Smith MJ, Keel JC, Greenberg BD, *et al.* (1999). Menstrual cycle effects on cortical excitability. *Neurology 53*, 2069–2072.

van Honk J, Schutter DJ, d'Alfonso A, Kessels RP, Postma A, de Haan EH (2001). Repetitive transcranial magnetic stimulation at the frontopolar cortex reduces skin conductance but not heart rate: reduced gray matter excitability in orbitofrontal regions. *Archives of General Psychiatry 58*, 973–974.

van Honk J, Schutter DJ, d'Alfonso AA, Kessels RP, de Haan EH (2002). 1 hz rTMS over the right prefrontal cortex reduces vigilant attention to unmasked but not to masked fearful faces. *Biological Psychiatry 52*, 312–317.

Van Laere K, Nuttin B, Gabriels L, *et al.* (2006). Metabolic imaging of anterior capsular stimulation in refractory obsessive compulsive disorder: a key role for the subgenual anterior cingulate and ventral striatum. *Journal of Nuclear Medicine 47*, 740–747.

Wassermann EM, Greenberg BD, Nguyen MB, Murphy DL (2001). Motor cortex excitability correlates with an anxiety-related personality trait. *Biological Psychiatry 50*, 377–382.

Whiteside SP, Port JD, Abramowitz JS (2004). A meta-analysis of functional neuroimaging in obsessive–compulsive disorder. *Psychiatry Research 132*, 69–79.

Ziemann U, Lonnecker S, Steinhoff BJ, Paulus W (1996). Effects of antiepileptic drugs on motor cortex excitability in humans: a transcranial magnetic stimulation study. *Annals of Neurology 40*, 367–378.

Ziemann U, Paulus W, Rothenberger A (1997). Decreased motor inhibition in Tourette's disorder: evidence from transcranial magnetic stimulation. *American Journal of Psychiatry 154*, 1277–1284.

CHAPTER 44

Movement disorders

Mark Hallett and Alfredo Berardelli

Introduction

Cortical stimulation has proven valuable for studying human movement physiology, but its applications in the treatment of movement disorders are still developing. It is certainly possible that cortical stimulation could be clinically useful given the therapeutic success with deep brain stimulation in movement disorders. Of note, the optimal stimulation parameters for influencing the motor system are not yet established. Relevant parameters to investigate include site of stimulation, pattern and rate of stimulation, stimulation intensity, duration of stimulation, and periodicity at which stimulation should be delivered. Thus, systematic studies are required to definitively determine the therapeutic potential of repeated cortical stimulation in the treatment of movement disorders. Here we examine the current state of the evidence, and discuss the implications for future research directions in the treatment of movement disorders.

Tools for reversibly modulating cortical function

This chapter will focus on the potential therapeutic uses of TMS in movement disorders. However, there are a variety of other methods for superficial brain stimulation including transcranial direct current stimulation (tDCS), paired associative stimulation (PAS), deep brain stimulation (DBS), epidural electrical stimulation, and transcranial electrical stimulation (TES). It is useful to view the potential contribution of TMS within the context of other means of reversibly modulating cortical function with therapeutic potential.

The brain can be stimulated through the scalp with low levels of direct electrical current, called either tDCS or direct current polarization. This procedure is reviewed in more detailed elsewhere in this volume. tDCS is accomplished by passing 1 or 2 mA of current through large electrodes, typically widely spaced over the scalp. Recent modelling studies show that using stimulating currents of 2.0 mA, the magnitude of the current density in relevant regions of the brain is of the order of 0.1 A/m^2, corresponding to an electric field of 0.22 V/m (Miranda et al., 2006). Under the anode, the brain excitability is increased and under a cathode it is decreased. The precise physiology of how this happens is not yet clear.

Another mode of stimulation, PAS, takes advantage of the fact that an effective way to modulate synaptic efficacy is to deliver two inputs to the same cell at close to the same time. Depending on the precise timing of the two stimuli and the frequency of stimulation, synaptic strength can be increased or decreased. These two stimuli can be realized in humans with a peripheral stimulus paired with a TMS brain stimulus. A set of experimental paradigms using PAS has been developed by Classen and collaborators (Stefan et al. 2000; Wolters et al. 2003). If a median nerve stimulus at the wrist is paired with a TMS to the sensorimotor cortex at 25 ms, then the two stimuli arrive at about the same time and the motor-evoked potentials (MEPs) will be facilitated. If the interval is ~10 ms, the TMS comes ~15 ms before the median nerve volley arrives and the MEP will be depressed. The former has similarities with the phenomenon of long-term potentiation (LTP) and the latter with long-term depression (LTD).

The cerebral cortex can also be stimulated directly with epidural stimulation. While epidural stimulation is invasive, it is much less invasive than DBS which involves penetration into the brain. Direct cortical stimulation with short rapid trains of electrical stimuli has been used for many years for localization of function at the time of neurosurgical procedures. Now in a more continuous mode it can be examined for its potential as therapy in pain, movement disorders, and other applications, as reviewed elsewhere in this volume.

The mechanisms of action of DBS are incompletely understood, but operationally the function of targeted brain circuitry is altered in an acute rapidly reversible fashion. In the case of dystonia, the therapeutic efficacy appears to result mainly from a plastic change secondary to the stimulation since the clinical effect can be delayed by months, while onset of action in tremor can be nearly instantaneous.

Putative mechanisms of action of intermittent brain stimulation

While DBS and epidural cortical stimulation can be given continuously, other modes of stimulation that are applied only intermittently and for brief periods must rely on long-lasting rather than immediate effects to have a hope of exerting true clinical benefit that persists beyond the period of stimulation.

The putative mechanisms of action of intermittent brain stimulation are discussed at greater length in Chapter 38 of this volume. Briefly, long-lasting influences on brain function depend on changing synaptic strength or causing anatomical changes such as alterations in dendritic spines or sprouting. Since the anatomical changes may well be a secondary consequence of prolonged changes of synaptic strength, the basic physiological target of superficial brain stimulation is to change synaptic strength in a lasting fashion, such as via LTP and/or LTD. LTP is the basic mechanism of increasing synaptic strength, and LTD is the basic mechanism of decreasing synaptic strength. It is possible that increasing brain excitability for a period will lead to LTP in the

region, and decreasing brain excitability will lead to LTD, though this remains to be definitively tested. LTP and LTD can be induced by repetitive stimulation in the same synaptic pathway, called homosynaptic plasticity. On the other hand, methods that use the combinations of inputs, such as PAS, called heterosynaptic plasticity, may be more effective.

Repetitive TMS at slow rates, approximately between 0.2 and 1 Hz, can cause a measurable decrease in brain excitability (Chen *et al.* 1997); at rates of about 5 Hz or faster it can cause an increase in brain excitability (Pascual-Leone *et al.* 1994b; Berardelli *et al.* 1998). TMS can also be used repetitively in a mode where very short, very-high-frequency trains of stimuli are delivered at theta frequency (~5 Hz). This is called theta-burst stimulation (TBS) (Di Lazzaro *et al.* 2005; Huang *et al.* 2005). A typical paradigm would be three stimuli at 50 Hz, repeated at 5 Hz. If given intermittently, say 2 s of stimulation every 10 s, this leads to increased excitability. If given continuously over 40 s, this leads to decreased excitability. See Table 44.1 for a summary of the methods and their effects.

Potential therapeutic uses of TMS in movement disorders

General considerations

A challenge in applying intermittent cortical stimulation in the treatment of movement

Table 44.1 Summary of noninvasive methods for cortical excitation and inhibition

Method	Excitatory mode	Inhibitory mode
rTMS	High frequency, ≥5 Hz	Low frequency, 0.2–1 Hz
TBS	Intermittent	Continuous
tDCS	Anodal	Cathodal
PAS	Synchronous synaptic stimulation	Asynchronous synaptic stimulation

rTMS, repetitive transcranial magnetic stimulation; TBS, theta-burst stimulation; tDCS, transcranial direct current stimulation; PAS, paired associative stimulation.

disorders is that these disorders involve a distributed network of brain structures, some of which are deep and inaccessible to direct stimulation with rTMS. Comparable challenges are also presented by the treatment of the other disorders considered in this Section (e.g. depression, anxiety, schizophrenia). However, knowledge regarding the neural circuitry underlying movement disorders, cortical involvement in these disorders, and trans-synaptic effects of stimulation can guide the application of TMS in the hopes of maximizing therapeutic benefit. Nevertheless, therapeutic results with TMS have been mixed, as reviewed below.

Parkinson's disease

The observation that TMS, delivered at particular parameters of stimulation and timing relative to task performance (at low intensity at, or shortly after, the go-stimulus), can speed up the reaction time in patients with Parkinson's disease (PD) led to the idea that high-frequency rTMS might increase brain excitability and be able to be used for therapy in PD (Pascual-Leone et al. 1994a). In an initial trial with 5 Hz motor cortex stimulation at motor threshold, using the Purdue pegboard for assessment, this seemed to be the case (Pascual-Leone et al. 1994a), but the finding was not reproduced (Ghabra et al. 1999). On the other hand, other early studies suggested an improvement in pointing performance after 5 Hz rTMS (Siebner et al. 1999a), and an improvement on the Unified Parkinson's Disease Rating Scale (UPDRS) with rTMS (Siebner et al. 2000).

Repetitive TMS was performed on the left motor cortical area corresponding to the right hand in 12 off-drug patients with PD (Lefaucheur et al. 2004a). The effects of subthreshold rTMS applied at 0.5 Hz (600 pulses) or at 10 Hz (2000 pulses) using a real or a sham coil were compared to those obtained by a single dose of L-dopa. Real rTMS at 10 or 0.5 Hz, but not sham stimulation, improved motor performance. High-frequency rTMS decreased rigidity and bradykinesia in the upper limb contralateral to the stimulation, whereas low-frequency rTMS reduced upper limb rigidity bilaterally and improved walking.

Single sessions with TMS, however, have not proven to be very effective. That TMS repeated over long periods of time may be beneficial was first suggested by Shimamoto et al. (1999) although the stimulation frequency in this study was very slow (0.2 Hz). Subsequent studies with repeated sessions do appear to have more robust effects.

Thirty-six unmedicated PD patients were randomized to one of two groups: real rTMS (suprathreshold 5 Hz, 2000 pulses once a day for 10 consecutive days) and sham rTMS (Khedr et al., 2003). Total motor section of UPDRS, walking speed, and self-assessment scale were performed for each patient before rTMS and after the first, fifth, and 10th sessions, and then after 1 month. There were significant effects for the total motor UPDRS, walking speed, and self-assessment scale during the course of the study in the group of patients receiving real rTMS ($P = 0.0001$, 0.001, and 0.002), whereas no significant changes were observed in the group receiving sham rTMS except in self-assessment scale ($P = 0.019$).

In a double-blind placebo-controlled study, Lomarev et al. (2006) evaluated the effects of 25 Hz rTMS in 18 PD patients. Eight rTMS sessions were performed over a 4 week period. Four cortical targets (left and right motor and dorsolateral prefrontal cortex) were stimulated separately in each session, with 300 pulses each, at 100% of motor threshold intensity. During the four weeks, times for executing walking and complex hand movement tests gradually decreased (Figure 44.1). The therapeutic rTMS effect lasted for at least a month after treatment ended (Figure 44.2). Thus, not only do the effects continue to improve with repeated sessions, but the effects can be long-lasting. In this study, right-hand bradykinesia improvement correlated with increased MEP amplitude evoked by left motor cortex TMS after individual sessions, but improvement overall did not correlate with motor cortex excitability. For this reason, although short-term benefit may be due to MC excitability enhancement, the mechanism of cumulative benefit must have another explanation.

A recent meta-analysis includes 12 studies and concludes that the overall literature does show a positive effect of rTMS on Parkinson motor function (Fregni et al. 2005). On the other hand, TMS may also be detrimental as

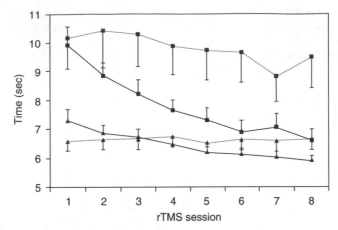

Fig. 44.1 Time of the 10 m walk and of the complex movement test during the course of repetitive (r)TMS treatment (mean ± SE). The figure shows the gradual decrease of time needed to execute the tests. Solid line, real rTMS group; dashed line, placebo group; squares, complex hand movement test; triangles, walk test. Measurements before and after each rTMS session were pooled. Times for left- and right-hand execution were pooled. From Lomarev *et al.* (2006) with permission.

demonstrated by the fact that rTMS over the supplementary motor area (SMA) worsened performance of spiral drawing and lengthened reaction time when tested 30 min after stimulation (Boylan *et al.* 2001). In this investigation, rTMS was given at up to 110% of patients' motor threshold (or highest tolerated) and at a frequency of 10 Hz. Forty 5 s trains were given over a period of 40 min for a total of 2000 pulses per session.

One study has investigated the effects of rTMS on vocal function in PD (Dias *et al.* 2006). Two different sets of rTMS parameters were investigated on 30 patients: active or sham 15 Hz rTMS of the left dorsolateral prefrontal cortex (LDLPFC) (110% of motor threshold (MT), 3000 pulses per session) and active 5 Hz rTMS of the primary motor cortex (M1)–mouth area (90% MT, 2250 pulses per session). rTMS of LDLPFC resulted in mood amelioration and subjective improvement of quality of life only (71.9% improvement, $P < 0.001$), but not in objective measures such as fundamental frequency ($P = 0.86$) and voice intensity ($P = 0.99$). rTMS of M1–mouth induced a significant improvement of the fundamental frequency (12.9% for men and 7.6% for women, $P < 0.0001$) and voice intensity (20.6%, $P < 0.0001$).

Using rTMS in a group of patients with advanced PD, Koch *et al.* (2005) investigated whether modulation of SMA excitability may result in a modification of a dyskinetic state induced by continuous apomorphine infusion. rTMS at 1 Hz was observed to markedly reduce drug-induced dyskinesias, whereas 5 Hz rTMS induced a slight but not significant increase.

There have been some early observations of the use of epidural cortical stimulation for PD with reports of some benefit (Canavero and Paolotti 2000; Canavero *et al.* 2002; Canavero and Bonicalzi, 2004), but the studies were not blinded and follow-up was short. This technique does not appear to help patients with multiple system atrophy and parkinsonism (Kleiner-Fisman *et al.* 2003), but at least one patient may have benefited (Canavero *et al.* 2003; Canavero and Bonicalzi 2004).

Treatment with tDCS has been performed in some open studies in Russia with some success, but these results need confirmation (Lomarev 1996).

Dystonia

Physiological findings in dystonia reveal a decrease in intracortical inhibition. Since rTMS delivered over the primary motor cortex at 1 Hz can induce an increase in inhibition, it was

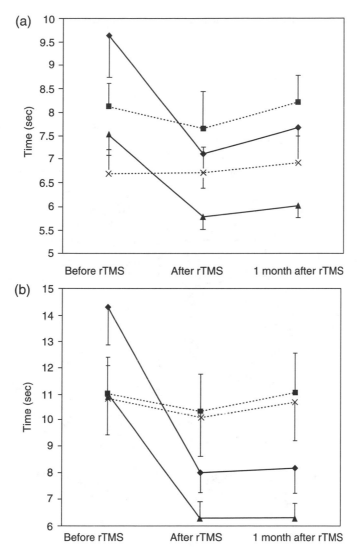

Fig. 44.2 Time of the 10 m walk (a) and complex movement test (b) before, after, and 1 month after the repetitive (r)TMS course (mean ± SE). The figure shows the decrease of time needed to execute the tests. Measurements for left and right hands were pooled. A solid line indicates the real rTMS group; the dashed line indicates the placebo group. Diamonds and squares indicate off-period of measurement; triangles and crosses indicate on-period of measurement. From Lomarev *et al.* (2006) with permission.

proposed that 1 Hz rTMS might ameliorate this deficit and be of clinical benefit in dystonia. A study with 1 Hz rTMS stimulation for 30 min at 10% less than motor threshold showed a normalization of the intracortical inhibition and some modest improvement in performance characterized mainly by a reduction in mean writing pressure (Siebner *et al.*, 1999b). Of 16 subjects, six patients noted a marked improvement of handwriting lasting more than 3 h, and in two of these patients the improvement persisted for several days.

Another therapeutic target in dystonia could be the premotor cortex since rTMS at 1 Hz can ameliorate the deficit in reciprocal inhibition in dystonia (Huang *et al.* 2004). Lefaucheur *et al.* (2004b) reported the effects of 1 Hz rTMS applied at 90% of resting motor threshold over the left premotor cortex in an open pilot study of three patients with severe generalized secondary dystonia including painful spasms in the proximal and axial musculature. A 20 min session of premotor rTMS was performed daily for five consecutive days. The rTMS sessions reduced the painful spasms for 3–8 days after the last session without any other significant beneficial effects. However, a slight reduction of the Movement score of the Burke, Fahn, and Marsden rating scale was observed for two patients, and of the Disability score for the third one. These were encouraging although uncontrolled results.

Murase *et al.* (2005) studied nine patients with writer's cramp and seven age-matched control subjects, using subthreshold 0.2 Hz rTMS with 250 stimuli applied to the motor cortex (MC), SMA, or premotor cortex (PMC). Stimulation of the PMC but not the MC significantly improved the rating of handwriting in

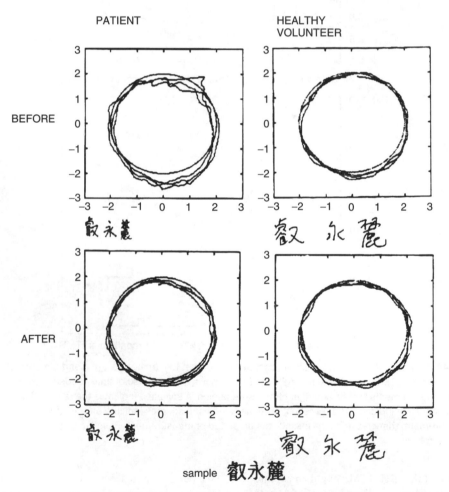

Fig. 44.3 Handwriting of patient 5 (a 41-year-old male) and a healthy control (a 47-year-old male) before and after repetitive (r)TMS over the premotor cortex. The mean tracking error from the target recorded by computer-assisted ratings of handwriting improved after rTMS in the patient. The speed of writing the three Chinese characters improved. It took 142 s for the patient to complete this task before rTMS, and he needed to touch the right wrist with his left hand (sensory trick). After rTMS, it took him 98 s and required no touching. On the contrary, it took 22 s for the normal subject to complete the task, and no clear change was observed after rTMS. From Murase *et al.* (2005) with permission.

the patient group (Figure 44.3). Improvement was seen in mean tracking error from the target ($P = 0.004$) and pen pressure ($P = 0.01$). This stimulation also produced physiological changes; the silent period was prolonged, indicating an increase in the amount of inhibition. rTMS over the other sites or using a sham coil in the patient group or trials in the control group revealed no clinical changes.

Other movement disorders

There have been a few studies of patients with Tourette's syndrome with mixed results. Two studies were quite favorable (Chae et al. 2004; Mantovani et al. 2006), whereas others were not (Munchau et al. 2002; Orth et al. 2005); the inconsistencies in outcomes may have arisen from different methods and patient populations. Further studies might be worthwhile with different parameters of stimulation to determine whether TMS might have benefit in this disorder.

There have been some reports of TMS being helpful in ataxia, although effects were weak (Shimizu et al. 1999; Ihara et al. 2005). One study has reported a brief mild beneficial effect in essential tremor (Gironell et al. 2002).

Conclusions

The availability of noninvasive tools to reversibly modulate cortical function enables previously unprecedented ability to study the clinical significance of motor networks in the pathophysiology of movement disorders. To date, clinical results with TMS in movement disorders have been mixed, and more work will be needed to clarify the potential clinical role of TMS. Work in PD and dystonia has shown promise, but well-controlled studies and systematic replications of early results are needed to make progress in this field.

References

Berardelli A, Inghilleri M, Rothwell JC, et al. (1998). Facilitation of muscle evoked responses after repetitive cortical stimulation in man. *Experimental Brain Research* 122, 79–84.

Boylan LS, Pullman SL, Lisanby SH, Spicknall KE, Sackeim HA (2001). Repetitive transcranial magnetic stimulation to SMA worsens complex movements in Parkinson's disease. *Clinical Neurophysiology* 112, 259–264.

Canavero S, Bonicalzi V (2004). Cortical stimulation for parkinsonism. *Archives of Neurology* 61, 606.

Canavero S, Paolotti R (2000). Extradural motor cortex stimulation for advanced Parkinson's disease: case report. *Movement Disorders* 15, 169–171.

Canavero S, Paolotti R, Bonicalzi V, et al. (2002). Extradural motor cortex stimulation for advanced Parkinson disease. Report of two cases. *Journal of Neurosurgery* 97, 1208–1211.

Canavero S, Bonicalzi V, Paolotti R, et al. (2003). Therapeutic extradural cortical stimulation for movement disorders: a review. *Neurological Research* 25, 118–122.

Chae JH, Nahas Z, Wassermann E, et al. (2004). A pilot safety study of repetitive transcranial magnetic stimulation (rTMS) in Tourette's syndrome. *Cognitive and Behavioral Neurology* 17, 109–117.

Chen R, Classen J, Gerloff C, et al. (1997). Depression of motor cortex excitability by low-frequency transcranial magnetic stimulation. *Neurology* 48, 1398–1403.

Dias AE, Barbosa ER, Coracini K (2006). Effects of repetitive transcranial magnetic stimulation on voice and speech in Parkinson's disease. *Acta Neurologica Scandinavica* 113, 92–99.

Di Lazzaro V, Pilato F, Saturno E, et al. (2005). Theta-burst repetitive transcranial magnetic stimulation suppresses specific excitatory circuits in the human motor cortex. *Journal of Physiology* 565, 945–950.

Fregni F, Simon DK, Wu A, Pascual-Leone A (2005). Non-invasive brain stimulation for Parkinson's disease: a systematic review and meta-analysis of the literature. *Journal of Neurology, Neurosurgery and Psychiatry* 76, 1614–1623.

Ghabra MB, Hallett M, Wassermann EM (1999). Simultaneous repetitive transcranial magnetic stimulation does not speed fine movement in PD. *Neurology* 52, 768–770.

Gironell A, Kulisevsky J, Lorenzo J (2002). Transcranial magnetic stimulation of the cerebellum in essential tremor: a controlled study. *Archives of Neurology* 59, 413–417.

Huang YZ, Edwards MJ, Bhatia KP, Rothwell JC (2004). One-Hz repetitive transcranial magnetic stimulation of the premotor cortex alters reciprocal inhibition in DYT1 dystonia. *Movement Disorders* 19, 54–59.

Huang YZ, Edwards MJ, Rounis E, Bhatia KP, Rothwell JC (2005). Theta burst stimulation of the human motor cortex. *Neuron* 45, 201–206.

Ihara Y, Takata H, Tanabe Y, Nobukuni K, Hayabara T (2005). Influence of repetitive transcranial magnetic stimulation on disease severity and oxidative stress markers in the cerebrospinal fluid of patients with spinocerebellar degeneration. *Neurological Research* 27, 310–313.

Khedr EM, Farweez HM, Islam H (2003). Therapeutic effect of repetitive transcranial magnetic stimulation on motor function in Parkinson's disease patients. *European Journal of Neurology* 10, 567–572.

Kleiner-Fisman G, Fisman DN, Kahn FI, Sime E, Lozano AM, Lang AE (2003). Motor cortical stimulation for parkinsonism in multiple system atrophy. *Archives of Neurology 60*, 1554–1558.

Koch G, Brusa L, Caltagirone C, *et al.* (2005). rTMS of supplementary motor area modulates therapy-induced dyskinesias in Parkinson disease. *Neurology 65*, 623–625.

Lefaucheur JP, Drouot X, Von Raison F, Menard-Lefaucheur I, Cesaro P, Nguyen JP (2004a). Improvement of motor performance and modulation of cortical excitability by repetitive transcranial magnetic stimulation of the motor cortex in Parkinson's disease. *Clinical Neurophysiology 115*, 2530–2541.

Lefaucheur JP, Fenelon G, Menard-Lefaucheur I, Wendling S, Nguyen JP (2004b). Low-frequency repetitive TMS of premotor cortex can reduce painful axial spasms in generalized secondary dystonia: a pilot study of three patients. *Neurophysiologie Clinique 34*, 141–145.

Lomarev MP (1996). [Effect of transcranial polarization on the state of supraspinal mechanisms of regulation of muscle tonus in patients with Parkinson's disease]. *Fiziologia Cheloveka 22*, 132–133.

Lomarev MP, Kanchana S, Bara-Jimenez W, Iyer M, Wassermann EM, Hallett M (2006). Placebo-controlled study of rTMS for the treatment of Parkinson's disease. *Movement Disorders 21*, 325–331.

Mantovani A, Lisanby SH, Pieraccini F, Ulivelli M, Castrogiovanni P, Rossi S (2006). Repetitive transcranial magnetic stimulation (rTMS) in the treatment of obsessive-compulsive disorder (OCD) and Tourette's syndrome (TS). *International Journal of Neuropsychopharmacology 9*, 95–100.

Miranda P, Lomarev M, Hallett M (2006). Modeling the current distribution during transcranial direct current stimulation. *Clinical Neurophysiology 117*, 1623–1629.

Munchau A, Bloem BR, Thilo KV, Trimble MR, Rothwell JC, Robertson MM (2002). Repetitive transcranial magnetic stimulation for Tourette syndrome. *Neurology 59*, 1789–1791.

Murase N, Rothwell JC, Kaji R, *et al.* (2005). Subthreshold low-frequency repetitive transcranial magnetic stimulation over the premotor cortex modulates writer's cramp. *Brain 128*, 104–115.

Orth M, Kirby R, Richardson MP, *et al.* (2005). Subthreshold rTMS over pre-motor cortex has no effect on tics in patients with Gilles de la Tourette syndrome. *Clinical Neurophysiology 116*, 764–768.

Pascual-Leone A, Valls-Solé J, Brasil-Neto J, Cammarota A, Grafman J, Hallett M (1994a). Akinesia in Parkinson's Disease. II. Effects of subthreshold repetitive transcranial motor cortex stimulation. *Neurology 44*, 892–898.

Pascual-Leone A, Valls-Solé J, Wassermann EM, Hallett M (1994b). Responses to rapid-rate transcranial magnetic stimulation of the human motor cortex. *Brain 117*, 847–858.

Shimamoto H, Morimitsu H, Sugita S, Nakahara K, Shigemori M (1999). Therapeutic effect of repetitive transcranial magnetic stimulation in Parkinson's disease. *Rinsho Shinkeigaku 39*, 1264–1267.

Shimizu H, Tsuda T, Shiga Y, *et al.* (1999). Therapeutic efficacy of transcranial magnetic stimulation for hereditary spinocerebellar degeneration. *Tohoku Journal of Experimental Medicine 189*, 203–211.

Siebner HR, Mentschel C, Auer C, Conrad B (1999a). Repetitive transcranial magnetic stimulation has a beneficial effect on bradykinesia in Parkinson's disease. *Neuroreport 10*, 589–594.

Siebner HR, Tormos JM, Ceballos-Baumann AO, *et al.* (1999b). Low-frequency repetitive transcranial magnetic stimulation of the motor cortex in writer's cramp. *Neurology 52*, 529–537.

Siebner HR, Rossmeier C, Mentschel C, Peinemann A, Conrad B (2000). Short-term motor improvement after sub-threshold 5-Hz repetitive transcranial magnetic stimulation of the primary motor hand area in Parkinson's disease. *Journal of Neurological Sciences 178*, 91–94.

Stefan K, Kunesch E, Cohen LG, Benecke R, Classen J (2000). Induction of plasticity in the human motor cortex by paired associative stimulation. *Brain 123*, 572–584.

Wolters A, Sandbrink F, Schlottmann A, *et al.* (2003). A temporally asymmetric Hebbian rule governing plasticity in the human motor cortex. *Journal of Neurophysiology 89*, 2339–2345.

CHAPTER 45

Brain stimulation in neurorehabilitation

Friedhelm C. Hummel and Leonardo G. Cohen

Introduction

The purpose of this chapter is to review the contributions of brain stimulation techniques to (i) the understanding of the mechanisms underlying recovery of motor function after stroke, and (ii) the development of therapeutic interventions designed to improve recovery of motor function in neurorehabilitation. The first part of the chapter will characterize some of the problems facing the field of neurorehabilitation in general. The second part will describe results of studies that contributed to our understanding of functional changes in the motor cortices of the intact and lesioned hemispheres after stroke. Finally, we will discuss some of the interventional approaches proposed to improve motor function.

The problem: motor dysfunction and disability following stroke

Stroke is the main cause of long-term disability among adults. Disability following stroke results in significant impairment of patients' quality of life (Dobkin 1995; Kolominsky-Rabas and Heuschmann 2002; Kolominsky-Rabas et al. 2006). Indeed, stroke has a major impact on the public healthcare system, responsible for 2–4% of healthcare expenses (Dobkin 1995; Kolominsky-Rabas et al. 2006). In Germany, for example, every year 200 000 to 250 000 patients suffer a stroke (Kolominsky-Rabas and Heuschmann 2002).

Only a very small proportion of survivors recover to the extent that they can continue their professional and private life as before the event (Kolominsky-Rabas et al. 2001; Kolominsky-Rabas and Heuschmann 2002; Lai et al. 2002). More than 60% of stroke survivors suffer from pronounced persistent deficits (Kolominsky-Rabas et al. 2001; Kolominsky-Rabas and Heuschmann 2002). Despite significant efforts dedicated to ameliorate this disability, complete recovery of motor function after stroke remains an elusive goal (Dobkin 2004, 2005; Hummel and Cohen 2006). Thus, the development of innovative and effective treatment strategies to enhance such recovery represents a significant purpose that could have a major impact on patients' quality of life and healthcare in general.

The tools: brain stimulation

In the last decade various therapeutic strategies have been proposed to address the aims of improving outcomes following stroke, such as physio- and ergotherapeutic strategies, constraint-induced movement therapy, bilateral arm training, or peripheral nerve stimulation (Whitall et al. 2000; Taub et al. 2002; Dobkin 2004, 2005; Hummel and Cohen 2006; Wolf et al. 2006). Here, we will focus on brain stimulation of cortical areas, proposed to enhance training effects in neurorehabilitative settings. Three main techniques will be discussed: transcranial magnetic stimulation (TMS), transcranial direct current stimulation (tDCS), and invasive electrical cortical stimulation (iCS).

Noninvasive brain stimulation methods including repetitive (r)TMS and tDCS are considered relatively safe if applied within published safety limits (McCreery and Agnew 1990; Wassermann 1998; Iyer et al. 2005). This is especially important for repetitive (r)TMS, as it can cause seizures at excessively high intensities or frequencies. However, guidelines for the use of rTMS and training protocols have made inadvertent seizures with rTMS rare. For tDCS, only brief tingling burning sensations occasionally accompanied by redness under the electrodes have been described; however, further studies are needed to define safety boundaries more precisely (Nitsche et al. 2003; Iyer et al. 2005; Gandiga et al. 2006). Transient headaches have been described for both. It is of note that both techniques have available sham or placebo stimulation, though challenges to true double-blind administration remain, especially for rTMS due to the scalp sensations produced by active stimulation (Gandiga et al. 2006).

Invasive cortical stimulation has also been used recently to attempt to ameliorate motor function after stroke. iCS involves implanting epidural electrodes on target cortical areas (so far the primary motor cortex has been the main one) and it has been applied with different degrees of success in animal models (Adkins-Muir and Jones 2003; Kleim et al. 2003; Plautz et al. 2003; Teskey et al. 2003) and humans (Brown et al. 2003, 2006). One potential advantage of invasive over noninvasive approaches is the focality of application; a clear disadvantage is their invasiveness. Clearly more work is required to assess safety and efficacy of these tools.

TMS has been utilized extensively as a diagnostic tool to evaluate, for example, central motor conductions in patients with brain lesions and neurological disorders such as multiple sclerosis (Claus 1990). TMS allows the measurement of different forms of corticospinal and intracortical excitability (Hallett 2000; Siebner and Rothwell 2003; Chen 2004), functional connectivity such as interhemispheric interactions (Ferbert et al. 1992), and interactions between primary and secondary motor areas

(Baumer et al. 2003). Additionally, TMS allows the study of the behavioral consequences of focal disruption of activity in specific cortical sites ('virtual lesion') (Hallett 2000; Pascual-Leone et al. 2000; Plewnia et al. 2003a). In combination with other neuroimaging and neurophysiological techniques, like functional magnetic resonance imaging (fMRI), electroencephalography (EEG), magnetoencephalography (MEG), or positron emission tomography (PET) (Arthurs and Boniface 2002; Hummel et al. 2002, 2004; Shmuel et al. 2006), TMS is the only noninvasive method to directly assess inhibitory activity within the motor cortex (Chen 2004). Using TMS to evaluate these functional properties of the motor cortex after lesions like stroke is of special interest in the field of neurorehabilitation, as it provides information about the intracortical changes taking place in motor cortical sites as the patients recover motor function. Understanding of these mechanisms has clear therapeutic implications.

One additional property of TMS, depending on the specific parameters of stimulation, is its ability to modulate excitability in focal cortical areas, at the site of stimulation, and in some cases in distributed networks. This particular property of noninvasive cortical stimulation has been used recently in neurorehabilitation with the purpose of enhancing or decreasing excitability of focal brain regions in patients with stroke (Hummel and Cohen 2006). Several 'proof-of-principle' studies, reviewed below, suggested that this approach could contribute to neurorehabilitative efforts.

The clues: intracortical processes associated with recovery of motor function after stroke

After ischemic stroke, the initial functional deficit, the degree and pattern of cortical reorganization, and the degree of functional recovery vary greatly among patients. These features are likely influenced by factors such as topography, type or size of the lesion, and the stage in the course of the rehabilitative process (acute–subacute–chronic). The relationship between these factors and their influence on the degree of impairment is far from being understood; nevertheless this understanding is the necessary basis to develop efficient therapeutic strategies and to make predictions in early stages of the disease about functional outcomes.

Functional brain imaging, such as fMRI, provides valuable information to address these open questions. For example, fMRI investigations have defined very clearly the possible alternative networks utilized by patients with chronic stroke to implement movements of the paretic hand (Calautti and Baron 2003; Rossini et al. 2003; Baron et al. 2004). This contribution has stimulated substantially the interest in imaging neurorehabilitative processes in humans. One limitation of activation studies with fMRI or PET is that they do not provide substantive information on the functional role of the activated regions. Does activation reflect excitatory or inhibitory processes? Is activity at one site directly influencing other cortical sites or does it represent an epiphenomenon? Noninvasive brain stimulation and particularly TMS provides a way to directly address some of these questions and in so doing it can provide information not available otherwise. In the motor domain, TMS can stimulate the primary motor cortex with high topographic and temporal resolution (Hallett 2000; Siebner and Rothwell 2003). Therefore, TMS in the framework of multimodal approaches provides unique information on the functional role of various cortical sites in neurorehabilitation.

Corticospinal and intracortical excitability after stroke

The functionality of the corticospinal tract can be evaluated by measuring latencies and recruitment curves (RCc) of the motor-evoked potentials (MEPs) to increasing TMS stimuli (Boroojerdi et al. 2001). RCs measure the increase in MEP amplitude with increasing intensity of single TMS pulses and convey information on the strength of corticospinal connections of the motor cortex (Chen 2000; Siebner and Rothwell 2003). Short-latency intracortical inhibition (SICI) and intracortical facilitation (ICF) are obtained by a technique previously described by Kujirai et al. (1993). In brief, application of a conditioning stimulus

preceding a test stimulus activates, depending on the interstimulus interval (ISI), intracortical inhibitory or facilitatory circuits. The response to a test shock is inhibited at ISIs of 1–5 ms and is facilitated at ISIs of 8–30 ms (Kujirai et al. 1993; Siebner and Rothwell, 2003). Gamma-aminobutyric acid (GABA)-ergic and glutamatergic agents influence SICI and ICF, respectively (Ziemann 1999, 2003, 2004) with a likely subcortical contribution to the latter (Di Lazzaro et al. 2006).

In several studies over the last decade, cortical excitability was evaluated during the acute, subacute, and chronic stages after stroke (Maertens de Noordhout et al. 1989; Cicinelli et al. 1997, 2003, 2006; Traversa et al. 1997; Liepert et al. 2000; Rossini 2000; Fraser et al. 2002; Bütefisch 2004; Bütefisch et al. 2003, 2005; Seitz et al. 2004; Di Lazzaro et al. 2006; Talelli et al. 2006). Motor thresholds, intracortical inhibition and facilitation, and the size of MEPs were evaluated in the affected and the intact hemisphere of stroke patients. Immediately after the stroke in the acute and subacute stage, the motor threshold of the affected hemisphere was increased and the MEP amplitudes significantly decreased as correlates of the damage to the motor system (Traversa et al. 2000; Delvaux et al. 2003). Additionally, intracortical inhibition in the affected hemisphere has been demonstrated in several studies to be reduced, compared to the intact hemisphere and age-matched controls (Traversa et al. 1997, 2000; Manganotti et al. 2002; Cicinelli et al. 2003). The changes in motor thresholds, MEP amplitudes, and intracortical inhibition progressively recovered from the acute to the chronic stage with follow-up until to 12 months (Traversa et al. 2000; Manganotti et al. 2002; Delvaux et al. 2003). For example, Delvaux et al. (2003) studied a group of 31 stroke patients who had suffered from an ischemic stroke in the middle cerebral artery territory with severe hand palsy at onset. They evaluated motor thresholds and MEP amplitudes from day 1 after stroke up to 1 year after the stroke. Initially after stroke, patients showed reduced MEP amplitudes which recovered progressively over a year, with most modifications found in the first 3 months. The changes in cortical excitability after a stroke are not restricted to the motor cortex of the affected hemisphere, but lead also to significant changes in the intact hemisphere with increased MEP amplitudes and motor cortical disinhibition (Manganotti *et al.* 2002). Manganotti *et al.* found that these changes in the unaffected hemisphere returned to baseline in well-recovered patients and stayed disinhibited in the group with lesser recovery, an issue that is not completely resolved (Liepert *et al.* 2000; Manganotti *et al.* 2002; Shimizu *et al.* 2002; Bütefisch *et al.* 2003).

Reduced activity in local inhibitory circuits, e.g. within the primary motor cortex leading to unmasking of latent intracortical connections (Buchkremer-Ratzmann and Witte 1997; Witte and Stoll 1997; Hickmott and Merzenich 2002), could potentially contribute to functional recovery. Thus, the presented results of reduced intracortical inhibition after stroke as well as changes in motor mapping could represent the correlate of compensatory mechanisms taking place in the brain to sustain recovery of motor function (Thompson *et al.* 1995; Traversa *et al.* 1997; Liepert *et al.* 1998; Rossini and Pauri 2000). Further support for this proposal comes from studies describing changes in motor mapping and excitability changes with rehabilitative treatments such as constraint-induced movement therapy (Liepert 2006; Liepert *et al.* 2006).

An additional area of substantial interest has been the search for specific parameters of motor excitability after stroke that could have value as predictors of functional recovery (Rapisarda *et al.* 1996; Pennisi *et al.* 1999, 2002; Delvaux *et al.* 2003; Hendricks *et al.* 2003). While so far inconclusive, it has been reported that the presence of MEPs could predict good outcome while its absence in the acute stage, if persistent over time, may be predictive of poorer functional outcome (Rapisarda *et al.* 1996; Pennisi *et al.* 1999, 2002; Delvaux *et al.* 2003; Hendricks *et al.* 2003). It has been proposed that the silent period following a single TMS pulse to the motor cortex might have predictive value on the degree of spasticity (van Kuijk *et al.* 2005).

Overall, these investigations from multiple groups shed some light on the mechanisms that could be operating during recovery of motor function after stroke.

Inter-regional interactions tested with TMS after stroke

In addition to the evaluation of changes in various measures of motor cortical excitability within the primary motor cortex, TMS has been used to probe the functional interactions among motor areas within the same hemisphere and between cerebral hemispheres. Some of these interactions recently studied after stroke provided novel information on the process of recovery of motor function after lesions of the CNS. Besides EEG, MEG, and fMRI coherence or effective connectivity analyses, TMS provides a method to study inter-areal interactions. For example, delivery of pairs of TMS stimuli applied to different sites at different ISIs allows the determination of the influence of a conditioning stimulus applied to one area on the test response to stimulation of a second area (Ferbert *et al.* 1992; Aglioti *et al.* 1993; Geffen *et al.* 1994; Meyer *et al.* 1995, 1998; Gerloff *et al.* 1998; Baumer *et al.* 2003; Murase *et al.* 2004; Duque *et al.* 2005; Koch *et al.* 2006). Examples are intrahemispheric premotor–primary motor cortex (M1) interactions (Baumer *et al.* 2003) and interhemispheric interactions (Ferbert *et al.* 1992; Murase *et al.* 2004; Duque *et al.* 2005; Koch *et al.* 2006).

There has recently been increasing interest in the evaluation of interhemispheric interactions in the process of functional recovery after stroke (Shimizu *et al.* 2002; Murase *et al.* 2004; Duque *et al.* 2005). Such interactions have been recognized as influential in other sensory domains for some time (Oliveri *et al.* 2001; Winhuisen *et al.* 2005; Heiss and Thiel 2006). More recently, the role of interhemispheric inhibition (IHI) in the motor domain has attracted attention. IHI tested with this paired-pulse technique is likely to reflect activity within glutamateric transcallosal connections that target pyramidal tract neurons in the opposite hemisphere through inhibitory $GABA_B$-ergic interneurons (Daskalakis *et al.* 2002; Chen 2004). Different forms of IHI have been described at ISIs of 10 and 40 ms using the paired-pulse technique and also the study of ipsilateral silent periods (Ziemann 1999; Chen 2004). On the other hand, there are also interhemispheric facilitatory processes that appear less stable (Ferbert *et al.* 1992).

Murase *et al.* (2004) reported that IHI (~10 ms) from the contralesional M1 to the ipsilesional M1 in preparation of movements by the paretic hand in a simple reaction-time paradigm is abnormally persistent in chronic stroke patients with predominantly subcortical lesions and substantial recovery (Murase *et al.* 2004; Duque *et al.* 2005). The finding that this abnormality was more prominent in patients with poorer function suggested a mechanistic link between abnormal IHI and poor motor performance (Murase *et al.* 2004). Testing the influence of the affected hemisphere onto the intact hemisphere in stroke patients showed no clear differences compared to controls in patients with subcortical lesions (Boroojerdi *et al.* 1996; Duque *et al.* 2005). Altogether, these findings suggested an imbalance of interhemispheric interactions between lesioned and intact hemispheres relative to healthy controls. It remains to be determined if particular intrahemispheric excitability changes other than MEP amplitudes and motor thresholds could themselves influence premovement IHI and therefore contribute to these findings. The most parsimonious explanation of these findings is that interhemispheric interactions may influence in the motor domain, as previously shown with other sensory processes, behavior after cortical lesions like stroke. A better understanding of these interactions, including the extent to which they are task dependent and their relative involvement in patients with different lesion locations or degree of impairment, could provide targets for the development of interventional strategies that promote functional recovery (for review see Ward and Cohen 2004).

TMS-induced 'virtual lesions' in the study of the functional relevance of reorganized cortical regions after stroke

TMS has been used to study the effects of transient disruption of different cortical regions after stroke during performance of particular tasks. This approach allowed the determination of the functional relevance of activation demonstrated in neuroimaging studies of stroke recovery. In these experiments, identification of TMS-induced disturbance of a particular motor behavior is often interpreted as indicative of a contributory role of the stimulated cortical site to that behavior. One caveat for the interpretation of these results is that this approach could reflect not only the particular role of the 'lesioned' site, but the extent to which the rest of the brain is able to compensate for that 'lesion' (Siebner and Rothwell 2003). A detailed description of this technique is provided in previous chapters of this volume.

This approach has been used to study the role of ipsilesional and contralesional M1 and dorsal premotor cortices (PMd) in functional recovery after stroke. These studies provided significant insight into the role of these areas in patients with subcortical stroke and marked initial paralysis but good subsequent recovery, Fridman *et al.* (2004) reported that transient disruption of activity in ipsilesional PMd elicited delays in reaction times in the recovered hand function. Interestingly, patients who recover to a lesser extent appear to engage the contralesional PMd more prominently and effectively (Johansen-Berg *et al.* 2002; Rushworth *et al.* 2003). These findings provided relevant insights for the formulation of possible interventional targets using brain stimulation in patients with stroke.

In terms of the role of the ipsi- and contralesional M1, various studies have been performed that shed some light on the role of these cortical sites in the process of functional recovery after stroke. In one study, application of TMS to the ipsilesional M1 disrupted performance of a simple reaction-time task and a motor-sequence test to a greater extent than stimulation of other control cortical sites (Werhahn *et al.* 2003), suggesting that the recovered functions relied predominantly on activity within the ipsilesional M1. Another important study showed that different cortical areas of the intact hemisphere such as the primary motor, the premotor, and the posterior parietal cortex might play a contributing role to functional recovery after stroke (Lotze *et al.* 2006).

Limitations to the interpretation of this type of study are that the inclusion criteria, motor tasks, and stimulating approaches often differ, making comparisons difficult. Previous work assessed the existence of fast-conducting corticospinal transmission from the intact M1 to muscles of the paretic limb (Caramia *et al.* 1996, 2000; Turton *et al.* 1996; Gerloff *et al.* 2006). For example, Turton *et al.* (1996) reported that it was possible to elicit MEPs in the paretic limb by stimulation of the contralesional M1 in patients with higher impairment (Turton *et al.* 1996).

A more recent study using a multimodal approach by combining PET, EEG coherence, and TMS in a well-defined group of subcortical stroke patients found no evidence for a recruitment of 'latent' uncrossed corticospinal connections from contralesional M1 to the muscles in the paretic limb (Gerloff *et al.* 2006). These results are consistent with the finding that downregulation of excitability in the contralesional M1 (for review see Hummel and Cohen 2006) does not adversely influence motor function in the paretic hand, but quite the opposite (Fregni *et al.* 2005; Mansur *et al.* 2005). These findings do not necessarily contradict the results of Johansen-Berg *et al.* (2002) or Lotze *et al.* (2006), for example, as the mechanisms underlying functional recovery are likely to differ, as previously proposed (Ward and Cohen 2004). Indeed, more work is necessary to clarify the relative role of ipsi- and contralesional areas in stroke recovery. In the following section, interventional strategies based on modulating activity in the affected or the intact hemisphere are discussed in detail.

The potential treatments: brain stimulation strategies to facilitate motor function after stroke

Previous studies with direct electrical stimulation of the cortex in animal stroke models (Adkins-Muir and Jones 2003; Kleim *et al.* 2003; Teskey *et al.* 2003) and findings from human studies of cortical excitability using TMS described above provided support for the testing of different interventional approaches to enhance functional recovery after stroke.

Enhancing excitability of the ipsilesional M1

One of the approaches proposed to influence motor function in the paretic hand after stroke

involves upregulating excitability within the ipsilesional motor cortices (Ward and Cohen 2004; Hummel and Cohen 2006). There is now direct evidence that cortical stimulation in human stroke patients can modulate cortical excitability (Di Lazzaro *et al.* 2006). Cortical stimulation has been applied to the ipsilesional M1 noninvasively using TMS and cortical polarization has been performed with tDCS (Fregni *et al.* 2005; Hummel and Cohen 2005; Hummel *et al.* 2005, 2006; Khedr *et al.* 2005; Kim *et al.* 2006). Cortical stimulation has also been performed invasively with direct epidural cortical stimulation (Brown *et al.* 2003, 2006). Hummel *et al.* showed that anodal tDCS applied over the ipsilesional M1 in patients with chronic subcortical stroke and moderate-to-good motor recovery transiently enhanced performance of skilled motor functions of the paretic hand that mimic activities of daily living (Hummel and Cohen 2005; Hummel *et al.* 2005). These behavioral gains were associated with enhanced corticospinal excitability and reduced intracortical inhibition within the motor cortex of the affected hemisphere, suggesting modulation of glutamatergic and GABAergic neurotransmission as possible underlying mechanisms (Hummel and Cohen 2005; Hummel *et al.* 2005). Using rTMS, Kim *et al.* (2006) described improvements in performance of movement sequences in patients with chronic stroke and relatively good recovery compared to sham stimulation. Performance of relatively simpler motor tasks was evaluated more recently. It was found that anodal tDCS over the ipsilesional M1 resulted in transient improvements of maximum pinch force and reaction times in a double-blinded placebo-controlled study design (Hummel *et al.* 2006). These proof-of-principle studies provided novel evidence that upregulation of excitability in the ipsilesional M1 may enhance transiently performance of a wide range of motor tasks, some of them skilled and complex (Hummel *et al.* 2005; Kim *et al.* 2006) and some relatively simpler and mediated predominantly by M1 function (Hummel *et al.* 2006).

One study by Khedr *et al.* (2005) showed that rTMS (3 Hz stimulation) applied daily in combination with customary rehabilitative treatment for 10 days within the first 2 weeks after the stroke event elicited improvements in motor performance that lasted for at least 10 days post-stimulation (Khedr *et al.* 2005). None of these studies reported adverse effects of the interventions; thus, when applied within the reported boundaries, tDCS and rTMS appear to be safe. One recent report describes preliminary safety data showing improvements in motor function with direct epidural stimulation to the ipsilesional M1 (Brown *et al.* 2006).

The results from these proof-of-principle studies underline the need to design and implement larger well-controlled randomized double-blind multicenter investigations to address the usefulness and safety of stimulation of ipsilesional M1 on functional recovery measures after stroke.

Downregulating excitability of the contralesional M1

An alternative strategy to correct abnormalities described in the intact hemisphere and also possible imbalances in IHI is downregulation of excitability in the contralesional M1 (Calford and Tweedale 1990; Murase *et al.* 2004; Duque *et al.* 2005). In healthy volunteers, downregulation of excitability in one motor cortex leads to enhanced cortical excitability in the opposite M1 (Plewnia *et al.* 2003b; Schambra *et al.* 2003) and to improved motor function in the ipsilateral hand (Kobayashi *et al.* 2004; Vines *et al.* 2006). These findings indicate that performance in the paretic hand could theoretically be influenced by purposeful modulation of activity in the intact hemisphere. Three recent studies evaluated this concept by applying low-frequency 1 Hz rTMS to downregulate excitability in the contralesional M1 (Mansur *et al.* 2005; Takeuchi *et al.* 2005; Boggio *et al.* 2006). Takeuchi *et al.* applied rTMS in one group of patients and sham stimulation in another group, whereas Mansur *et al.* used a sham-controlled crossover design. In both studies well-recovered chronic stroke patients were included in the protocols. The authors reported improved motor functions in the paretic hand, determined by reduced reaction times and enhanced pegboard task performance, and improved force acceleration compared with sham in the absence of effects on maximum force or finger tapping (Mansur *et al.* 2005; Takeuchi *et al.* 2005). Interestingly, a recent case report provided evidence for

functional improvements due to low-frequency rTMS to the intact hemisphere in a patient with more severe impairment levels (Boggio *et al.* 2006). Another possibility of downregulating excitability in the contralesional M1 is cathodal tDCS, which demonstrated comparable results to those achieved with rTMS (Fregni *et al.* 2005). The authors showed improvements in performance of motor tasks mimicking activities of daily living with the paretic hand. These improvements were documented for up to 25 min following the end of the stimulation period. One theoretical advantage of this approach over stimulation of the affected hemisphere is its application to healthy neural structures. The duration of these effects appears to lengthen with repetitive applications. Fregni *et al.* (2006) applied low-frequency rTMS to the intact motor cortex across 5 days. The authors showed improved performance of motor tasks after 5 days of rTMS and these improvements persisted for the tasks that were repeatedly performed on a daily basis but not for the nonpracticed tasks at follow-up 2 weeks later. Taking this into account, the progressive improvement of simple and choice reaction time tasks and the pegboard task is most likely due to the combination of rTMS and training of these tasks by repetitive performance and not solely due to repetitive application of rTMS.

Brain stimulation in nonmotor domains

The proposal of downregulating activity within the intact hemisphere to improve a specific function affected by a contralateral brain lesion has been implemented before in sensory domains. Based on general concepts of interhemispheric competition (Ledlow *et al.* 1978; Antonini *et al.* 1979; Calford and Tweedale 1990), downregulation of activity in the intact hemisphere has been experimentally used in trials of aphasia and neglect. In the language domain, one study evaluated the effects of low-frequency rTMS applied to Broca's homolog in the intact hemisphere over 10 repetitive daily sessions on picture naming in a group of four chronic stroke patients (Naeser *et al.* 2005). The authors described an immediate performance improvement which outlasted the intervention up to 8 months in three of the four patients. Comparable results were also reported in visuo-spatial neglect (Oliveri *et al.* 2001; Brighina *et al.* 2003; Shindo *et al.* 2006). Low-frequency rTMS applied to the intact hemisphere induced a persistent improvement in the viusospatial neglect 2–6 weeks after the end of the intervention. It will be important to replicate these results in larger samples and controlled conditions to determine the value of this approach.

Conclusions

Post-stroke recovery remains an important clinical focus. There are pressing needs for a greater understanding of the processes underlying neuronal injury, neural response to injury, predictors of outcome, and the mechanisms of recovery of function. Research using various tools for brain stimulation, particularly in the form of TMS, is enlarging the body of evidence regarding the neurophysiological intracortical processes associated with recovery of motor function after stroke. Such work may be informative regarding pathophysiological processes related to injury and recovery, which can be useful in guiding the development of novel therapeutic interventions. Various modalities of brain stimulation (e.g. TMS, tDCS, and invasive epidural cortical stimulation) have been proposed as novel strategies to enhance motor function when combined with conventional neurorehabilitative interventions after a stroke. Published proof-of-principle studies are promising, although the magnitude of improvements reported has been limited and the duration of the effects short-lasting. Replications and controlled trials are still few. Before these approaches can reach the clinic, a more thorough evaluation of safety and efficacy is required in large multicenter trials combining brain stimulation with training and conventional rehabilitative treatments.

References

Adkins-Muir DL, Jones TA (2003). Cortical electrical stimulation combined with rehabilitative training: enhanced functional recovery and dendritic plasticity following focal cortical ischemia in rats. *Neurological Research 25*, 780–788.

Aglioti S, Berlucchi G, Pallini R, Rossi GF, Tassinari G (1993). Hemispheric control of unilateral and bilateral responses to lateralized light stimuli after callosotomy and in callosal agenesis. *Experimental Brain Research 95*, 151–165.

Antonini A, Berlucchi G, Marzi CA, Sprague JM (1979). Importance of corpus callosum for visual receptive fields of single neurons in cat superior colliculus. *Journal of Neurophysiology 42*, 137–152.

Arthurs OJ, Boniface S (2002). How well do we understand the neural origins of the fMRI BOLD signal? *Trends in Neurosciences 25*, 27–31.

Baron JC, Cohen LG, Cramer SC, *et al.* (2004). Neuroimaging in stroke recovery: a position paper from the First International Workshop on Neuroimaging and Stroke Recovery. *Cerebrovascular Diseases 18*, 260–267.

Baumer T, Rothwell JC, Munchau A (2003). Functional connectivity of the human premotor and motor cortex explored with TMS. *Supplement to Clinical Neurophysiology 56*, 160–169.

Boggio PS, Alonso-Alonso M, Mansur CG, *et al.* (2006). Hand function improvement with low-frequency repetitive transcranial magnetic stimulation of the unaffected hemisphere in a severe case of stroke. *American Journal of Physical and Medical Rehabilitation 85*, 927–930.

Boroojerdi B, Diefenbach K, Ferbert A (1996). Transcallosal inhibition in cortical and subcortical cerebral vascular lesions. *Journal of Neurological Sciences 144*, 160–170.

Boroojerdi B, Battaglia F, Muellbacher W, Cohen LG (2001). Mechanisms influencing stimulus–response properties of the human corticospinal system. *Clinical Neurophysiology 112*, 931–937.

Brighina F, Bisiach E, Oliveri M, *et al.* (2003). 1 Hz repetitive transcranial magnetic stimulation of the unaffected hemisphere ameliorates contralesional visuospatial neglect in humans. *Neuroscience Letters 336*, 131–133.

Brown JA, Lutsep H, Cramer SC, Weinand M (2003). Motor cortex stimulation for enhancement of recovery after stroke: case report. *Neurological Research 25*, 815–818.

Brown JA, Lutsep HL, Weinand M, Cramer SC (2006). Motor cortex stimulation for the enhancement of recovery from stroke: a prospective, multicenter safety study. *Neurosurgery 58*, 464–473.

Buchkremer-Ratzmann I, Witte OW (1997). Extended brain disinhibition following small photothrombotic lesions in rat frontal cortex. *Neuroreport 8*, 519–522.

Bütefisch CM (2004). Plasticity in the human cerebral cortex: lessons from the normal brain and from stroke. *Neuroscientist 10*, 163–173.

Bütefisch CM, Netz J, Wessling M, Seitz RJ, Homberg V (2003). Remote changes in cortical excitability after stroke. *Brain 126*, 470–481.

Bütefisch CM, Kleiser R, Korber B, Muller K, Wittsack HJ, Homberg V, *et al.* (2005). Recruitment of contralesional motor cortex in stroke patients with recovery of hand function. *Neurology 64*, 1067–1069.

Calautti C, Baron JC (2003). Functional neuroimaging studies of motor recovery after stroke in adults: a review. *Stroke 34*, 1553–1566.

Calford MB, Tweedale R (1990). Interhemispheric transfer of plasticity in the cerebral cortex. *Science 249*, 805–807.

Caramia MD, Iani C, Bernardi G (1996). Cerebral plasticity after stroke as revealed by ipsilateral responses to magnetic stimulation. *Neuroreport 7*, 1756–1760.

Caramia MD, Palmieri MG, Giacomini P, Iani C, Dally L, Silvestrini M (2000). Ipsilateral activation of the unaffected motor cortex in patients with hemiparetic stroke. *Clinical Neurophysiology 111*, 1990–1996.

Chen R (2000). Studies of human motor physiology with transcranial magnetic stimulation. *Muscle and Nerve Supplement 9*, S26–32.

Chen R (2004). Interactions between inhibitory and excitatory circuits in the human motor cortex. *Experimental Brain Research 154*, 1–10.

Cicinelli P, Traversa R, Rossini PM (1997). Post-stroke reorganization of brain motor output to the hand: a 2–4 month follow-up with focal magnetic transcranial stimulation. *Electroencephalography and Clinical Neurophysiology 105*, 438–450.

Cicinelli P, Pasqualetti P, Zaccagnini M, Traversa R, Oliveri M, Rossini PM (2003). Interhemispheric asymmetries of motor cortex excitability in the postacute stroke stage: a paired-pulse transcranial magnetic stimulation study. *Stroke 34*, 2653–2658.

Cicinelli P, Marconi B, Zaccagnini M, Pasqualetti P, Filippi MM, Rossini PM (2006). Imagery-induced cortical excitability changes in stroke: a transcranial magnetic stimulation study. *Cerebral Cortex 16*, 247–253.

Claus D (1990). Central motor conduction: method and normal results. *Muscle and Nerve 13*, 1125–1132.

Daskalakis ZJ, Christensen BK, Fitzgerald PB, Roshan L, Chen R (2002). The mechanisms of interhemispheric inhibition in the human motor cortex. *Journal of Physiology 543*, 317–326.

Delvaux V, Alagona G, Gerard P, De Pasqua V, Pennisi G, de Noordhout AM (2003). Post-stroke reorganization of hand motor area: a 1-year prospective follow-up with focal transcranial magnetic stimulation. *Clinical Neurophysiology 114*, 1217–1225.

Di Lazzaro V, Pilato F, Oliviero A, *et al.* (2006). Origin of facilitation of motor-evoked potentials after paired magnetic stimulation: direct recording of epidural activity in conscious humans. *Journal of Neurophysiology 96*, 1765–1771.

Dobkin B (1995). The economic impact of stroke. *Neurology 45*, S6–9.

Dobkin BH (2004). Strategies for stroke rehabilitation. *Lancet Neurology 3*, 528–536.

Dobkin BH (2005). Clinical practice. Rehabilitation after stroke. *New England Journal of Medicine 352*, 1677–1684.

Duque J, Hummel F, Celnik P, Murase N, Mazzocchio R, Cohen LG (2005). Transcallosal inhibition in chronic subcortical stroke. *NeuroImage 28*, 940–946.

Ferbert A, Priori A, Rothwell JC, Day BL, Colebatch JG, Marsden CD (1992). Interhemispheric inhibition of the human motor cortex. *Journal of Physiology 453*, 525–546.

Fraser C, Power M, Hamdy S, *et al.* (2002). Driving plasticity in human adult motor cortex is associated with improved motor function after brain injury. *Neuron 34*, 831–840.

Fregni F, Boggio PS, Mansur CG, *et al.* (2005). Transcranial direct current stimulation of the unaffected hemisphere in stroke patients. *Neuroreport 16*, 1551–1555.

Fregni F, Boggio PS, Valle AC, *et al.* (2006). A sham-controlled trial of a 5-day course of repetitive transcranial magnetic stimulation of the unaffected hemisphere in stroke patients. *Stroke 37*, 2115–2122.

Fridmen EA, Hanakawa T, Chung M, Hummel F, Leiguarda RC, Cohen LG (2004). Reorganization of the human ipsilesional premotor cortex after stroke. *Brain 127*, 747–758.

Gandiga PC, Hummel FC, Cohen LG (2006). Transcranial DC stimulation (tDCS): A tool for double-blind sham-controlled clinical studies in brain stimulation. *Clinical Neurophysiology 117*, 845–850.

Geffen GM, Jones DL, Geffen LB (1994). Interhemispheric control of manual motor activity. *Behavioral Brain Research 64*, 131–140.

Gerloff C, Cohen LG, Floeter MK, Chen R, Corwell B, Hallett M (1998). Inhibitory influence of the ipsilateral motor cortex on responses to stimulation of the human cortex and pyramidal tract. *Journal of Physiology (London) 510*, 249–259.

Gerloff C, Bushara K, Sailer A, *et al.* (2006). Multimodal imaging of brain reorganization in motor areas of the contralesional hemisphere of well recovered patients after capsular stroke. *Brain 129*, 791–808.

Hallett M (2000). Transcranial magnetic stimulation and the human brain. *Nature 406*, 147–50.

Heiss WD, Thiel A (2006). A proposed regional hierarchy in recovery of post-stroke aphasia. *Brain and Language 98*, 118–123.

Hendricks HT, Pasman JW, Merx JL, van Limbeek J, Zwarts MJ (2003). Analysis of recovery processes after stroke by means of transcranial magnetic stimulation. *Journal of Clinical Neurophysiology 20*, 188–195.

Hickmott PW, Merzenich MM (2002). Local circuit properties underlying cortical reorganization. *Journal of Neurophysiology 88*, 1288–1301.

Hummel F, Cohen LG (2005). Improvement of motor function with noninvasive cortical stimulation in a patient with chronic stroke. *Neurorehabilitation and Neural Repair 19*, 14–19.

Hummel FC, Cohen LG (2006). Non-invasive brain stimulation: a new strategy to improve neurorehabilitation after stroke? *Lancet Neurology 5*, 708–712.

Hummel F, Andres F, Altenmuller E, Dichgans J, Gerloff C (2002). Inhibitory control of acquired motor programmes in the human brain. *Brain 125*, 404–420.

Hummel F, Saur R, Lasogga S, *et al.* (2004). To act or not to act. Neural correlates of executive control of learned motor behavior. *NeuroImage 23*, 1391–1401.

Hummel F, Celnik P, Giraux P, *et al.* (2005). Effects of non-invasive cortical stimulation on skilled motor function in chronic stroke. *Brain 128*, 490–499.

Hummel FC, Voller B, Celnik P, *et al.* (2006). Effects of brain polarization on reaction times and pinch force in chronic stroke. *BMC Neuroscience 7*, 73.

Iyer MB, Mattu U, Grafman J, Lomarev M, Sato S, Wassermann EM (2005). Safety and cognitive effect of frontal DC brain polarization in healthy individuals. *Neurology 64*, 872–875.

Johansen-Berg H, Dawes H, Guy C, Smith SM, Wade DT, Matthews PM (2002). Correlation between motor improvements and altered fMRI activity after rehabilitative therapy. *Brain 125*, 2731–2742.

Khedr EM, Ahmed MA, Fathy N, Rothwell JC (2005). Therapeutic trial of repetitive transcranial magnetic stimulation after acute ischemic stroke. *Neurology 65*, 466–468.

Kim YH, You SH, Ko MH, *et al.* (2006). Repetitive transcranial magnetic stimulation-induced corticomotor excitability and associated motor skill acquisition in chronic stroke. *Stroke 37*, 1471–1476.

Kleim JA, Bruneau R, VandenBerg P, MacDonald E, Mulrooney R, Pocock D (2003). Motor cortex stimulation enhances motor recovery and reduces peri-infarct dysfunction following ischemic insult. *Neurological Research 25*, 789–793.

Kobayashi M, Hutchinson S, Theoret H, Schlaug G, Pascual-Leone A (2004). Repetitive TMS of the motor cortex improves ipsilateral sequential simple finger movements. *Neurology 62*, 91–98.

Koch G, Franca M, Del Olmo MF, *et al.* (2006). Time course of functional connectivity between dorsal premotor and contralateral motor cortex during movement selection. *Journal of Neuroscience 26*, 7452–7459.

Kolominsky-Rabas PL, Heuschmann PU (2002). [Incidence, etiology and long-term prognosis of stroke]. *Fortschritte der Neurologie-Psychiatrie 70*, 657–662.

Kolominsky-Rabas PL, Weber M, Gefeller O, Neundoerfer B, Heuschmann PU (2001). Epidemiology of ischemic stroke subtypes according to TOAST criteria: incidence, recurrence, and long-term survival in ischemic stroke subtypes: a population-based study. *Stroke 32*, 2735–2740.

Kolominsky-Rabas PL, Heuschmann PU, Marschall D, *et al.* (2006). Lifetime cost of ischemic stroke in Germany: results and national projections from a population-based stroke registry: the Erlangen Stroke Project. *Stroke 37*, 1179–1183.

Kujirai T, Caramia MD, Rothwell JC, *et al.* (1993). Corticocortical inhibition in human motor cortex. *Journal of Physiology (London) 471*, 501–519.

Lai SM, Studenski S, Duncan PW, Perera S (2002). Persisting consequences of stroke measured by the Stroke Impact Scale. *Stroke 33*, 1840–184.

Ledlow A, Swanson JM, Kinsbourne M (1978). Differences in reaction times and average evoked potentials as a function of direct and indirect neural pathways. *Annals of Neurology 3*, 525–530.

Liepert J (2006). Motor cortex excitability in stroke before and after constraint-induced movement therapy. *Cognitive and Behavioral Neurology 19*, 41–47.

Liepert J, Miltner WH, Bauder H, *et al.* (1998). Motor cortex plasticity during constraint-induced movement therapy in stroke patients. *Neuroscience Letters 250*, 5–8.

Liepert J, Storch P, Fritsch A, Weiller C (2000). Motor cortex disinhibition in acute stroke. *Clinical Neurophysiology 111*, 671–676.

Liepert J, Haevernick K, Weiller C, Barzel A (2006). The surround inhibition determines therapy-induced cortical reorganization. *NeuroImage 32*, 1216–1220.

Lotze M, Markert J, Sauseng P, Hoppe J, Plewnia C, Gerloff C (2006). The role of multiple contralesional motor areas for complex hand movements after internal capsular lesion. *Journal of Neuroscience 26*, 6096–6102.

McCreery DB, Agnew WF (1990). Mechanisms of stimulation-induced damage and their relation to guidelines for safe stimulation. In: WF Agnew, DB McCreery (eds), *Neural prostheses: fundamental studies*, pp. 297–317. Englewood Cliffs, NJ: Prentice Hall.

Maertens de Noordhout A, Rothwell JC, *et al.* (1989). [Percutaneous electric and magnetic stimulation of the motor cortex in man. Physiological aspects and clinical applications.] *Revue Neurologique (Paris) 145*, 1–15.

Manganotti P, Patuzzo S, Cortese F, Palermo A, Smania N, Fiaschi A (2002). Motor disinhibition in affected and unaffected hemisphere in the early period of recovery after stroke. *Clinical Neurophysiology 113*, 936–943.

Mansur CG, Fregni F, Boggio PS, *et al.* (2005). A sham stimulation-controlled trial of rTMS of the unaffected hemisphere in stroke patients. *Neurology 64*, 1802–1804.

Meyer BU, Roericht S, Woiciechowsky C, Brandt SA (1995). Interhemispheric inhibition induced by transcranial magnetic stimulation: localisation of involved fibres studied in patients after partial callosotomy. *Journal of Physiology 487*, 68P.

Meyer BU, Roricht S, Woiciechowsky C (1998). Topography of fibers in the human corpus callosum mediating interhemispheric inhibition between the motor cortices. *Annals of Neurology 43*, 360–369.

Murase N, Duque J, Mazzocchio R, Cohen LG (2004). Influence of interhemispheric interactions on motor function in chronic stroke. *Annals of Neurology 55*, 400–409.

Naeser MA, Martin PI, Nicholas M, *et al.* (2005). Improved picture naming in chronic aphasia after TMS to part of right Broca's area: an open-protocol study. *Brain and Language 93*, 95–105.

Nitsche MA, Liebetanz D, Lang N, Antal A, Tergau F, Paulus W (2003). Safety criteria for transcranial direct current stimulation (tDCS) in humans. *Clinical Neurophysiology* 114, 2220–2222; author reply 2222–2223.

Oliveri M, Bisiach E, Brighina F, *et al.* (2001). rTMS of the unaffected hemisphere transiently reduces contralesional visuospatial hemineglect. *Neurology 57*, 1338–1340.

Pascual-Leone A, Walsh V, Rothwell J (2000). Transcranial magnetic stimulation in cognitive neuroscience–virtual lesion, chronometry, and functional connectivity. *Current Opinion in Neurobiology 10*, 232–237.

Pennisi G, Rapisarda G, Bella R, Calabrese V, Maertens De Noordhout A, Delwaide PJ (1999). Absence of response to early transcranial magnetic stimulation in ischemic stroke patients: prognostic value for hand motor recovery. *Stroke 30*, 2666–2670.

Pennisi G, Alagona G, Rapisarda G, *et al.* (2002). Transcranial magnetic stimulation after pure motor stroke. *Clinical Neurophysiology 113*, 1536–1543.

Plautz EJ, Barbay S, Frost SB, *et al.* (2003). Post-infarct cortical plasticity and behavioral recovery using concurrent cortical stimulation and rehabilitative training: a feasibility study in primates. *Neurological Research 25*, 801–810.

Plewnia C, Bartels M, Gerloff C (2003a). Transient suppression of tinnitus by transcranial magnetic stimulation. *Annals of Neurology 53*, 263–266.

Plewnia C, Lotze M, Gerloff C (2003b). Disinhibition of the contralateral motor cortex by low-frequency rTMS. *Neuroreport 14*, 609–612.

Rapisarda G, Bastings E, de Noordhout AM, Pennisi G, Delwaide PJ (1996). Can motor recovery in stroke patients be predicted by early transcranial magnetic stimulation? *Stroke 27*, 2191–2196.

Rossini PM (2000). Is transcranial magnetic stimulation of the motor cortex a prognostic tool for motor recovery after stroke? *Stroke 31*, 1463–1464.

Rossini PM, Pauri F (2000). Neuromagnetic integrated methods tracking human brain mechanisms of sensorimotor areas 'plastic' reorganisation. *Brain Research: Brain Research Reviews 33*, 131–154.

Rossini PM, Calautti C, Pauri F, Baron JC (2003). Post-stroke plastic reorganisation in the adult brain. *Lancet Neurology 2*, 493–502.

Rushworth MF, Johansen-Berg H, Gobel SM, Devlin JT (2003). The left parietal and premotor cortices: motor attention and selection. *NeuroImage 20*(Suppl. 1), S89–100.

Schambra HM, Sawaki L, Cohen LG (2003). Modulation of excitability of human motor cortex (M1) by 1 Hz transcranial magnetic stimulation of the contralateral M1. *Clinical Neurophysiology 114*, 130–133.

Seitz RJ, Bütefisch CM, Kleiser R, Homberg V (2004). Reorganisation of cerebral circuits in human ischemic brain disease. *Restorative Neurology and Neuroscience 22*, 207–229.

Shimizu T, Hosaki A, Hino T, *et al.* (2002). Motor cortical disinhibition in the unaffected hemisphere after unilateral cortical stroke. *Brain 125*, 1896–1907.

Shindo K, Sugiyama K, Huabao L, Nishijima K, Kondo T, Izumi S (2006). Long-term effect of low-frequency repetitive transcranial magnetic stimulation over the unaffected posterior parietal cortex in patients with unilateral spatial neglect. *Journal of Rehabilitative Medicine 38*, 65–67.

Shmuel A, Augath M, Oeltermann A, Logothetis NK (2006). Negative functional MRI response correlates with decreases in neuronal activity in monkey visual area V1. *Nature Neuroscience 9*, 569–577.

Siebner HR, Rothwell J (2003). Transcranial magnetic stimulation: new insights into representational cortical plasticity. *Experimental Brain Research 148*, 1–16.

Takeuchi N, Chuma T, Matsuo Y, Watanabe I, Ikoma K (2005). Repetitive transcranial magnetic stimulation of contralesional primary motor cortex improves hand function after stroke. *Stroke 36*, 2681–2686.

Talelli P, Greenwood RJ, Rothwell JC (2006). Arm function after stroke: neurophysiological correlates and recovery mechanisms assessed by transcranial magnetic stimulation. *Clinical Neurophysiology 117*, 1641–1659.

Taub E, Uswatte G, Elbert T (2002). New treatments in neurorehabilitation founded on basic research. *Nature Reviews Neuroscience 3*, 228–236.

Teskey GC, Flynn C, Goertzen CD, Monfils MH, Young NA (2003). Cortical stimulation improves skilled forelimb use following a focal ischemic infarct in the rat. *Neurological Research 25*, 794–800.

Thompson ML, Thickbroom GW, Laing B, Wilson S, Mastaglia FL (1995). Changes in the organisation of the corticomotor projection to the hand after subcortical stroke. *Electroencephalography and Clinical Neurophysiology 97*, S191.

Traversa R, Cicinelli P, Bassi A, Rossini PM, Bernardi G (1997). Mapping of motor cortical reorganization after stroke. A brain stimulation study with focal magnetic pulses. *Stroke 28*, 110–117.

Traversa R, Cicinelli P, Oliveri M, *et al.* (2000). Neurophysiological follow-up of motor cortical output in stroke patients. *Clinical Neurophysiology 111*, 1695–1703.

Turton A, Wroe S, Trepte N, Fraser C, Lemon RN (1996). Contralateral and ipsilateral EMG responses to transcranial magnetic stimulation during recovery of arm and hand function after stroke. *Electroencephalography and Clinical Neurophysiology 101*, 316–328.

van Kuijk AA, Pasman JW, Geurts AC, Hendricks HT (2005). How salient is the silent period? The role of the silent period in the prognosis of upper extremity motor recovery after severe stroke. *Journal of Clinical Neurophysiology 22*, 10–24.

Vines BW, Nair DG, Schlaug G (2006). Contralateral and ipsilateral motor effects after transcranial direct current stimulation. *Neuroreport 17*, 671–674.

Ward NS, Cohen LG (2004). Mechanisms underlying recovery of motor function after stroke. *Archives of Neurology 61*, 1844–1848.

Wassermann EM (1998). Risk and safety of repetitive transcranial magnetic stimulation: report and suggested guidelines from the International Workshop on the Safety of Repetitive Transcranial Magnetic Stimulation, June 5–7, 1996. *Electroencephalography and Clinical Neurophysiology 108*, 1–16.

Werhahn KJ, Conforto AB, Kadom N, Hallett M, Cohen LG (2003). Contribution of the ipsilateral motor cortex to recovery after chronic stroke. *Annals of Neurology 54*, 464–472.

Whitall J, McCombe Waller S, Silver KH, Macko RF (2000). Repetitive bilateral arm training with rhythmic auditory cueing improves motor function in chronic hemiparetic stroke. *Stroke 31*, 2390–2395.

Winhuisen L, Thiel A, Schumacher B, *et al.* (2005). Role of the contralateral inferior frontal gyrus in recovery of language function in poststroke aphasia: a combined repetitive transcranial magnetic stimulation and positron emission tomography study. *Stroke 36*, 1759–1763.

Witte OW, Stoll G (1997). Delayed and remote effects of focal cortical infarctions: secondary damage and reactive plasticity. *Advances in Neurology 73*, 207–227.

Wolf SL, Winstein CJ, Miller JP, *et al.* (2006). Effect of constraint-induced movement therapy on upper extremity function 3 to 9 months after stroke: the EXCITE randomized clinical trial. *Journal of the American Medical Association 296*, 2095–2104.

Ziemann U (1999). Intracortical inhibition and facilitation in the conventional paired TMS paradigm. *Electroencephalography and Clinical Neurophysiology Supplement 51*, 127–136.

Ziemann U (2003). Pharmacology of TMS. *Supplement to Clinical Neurophysiology 56*, 226–231.

Ziemann U (2004). TMS and drugs. *Clinical Neurophysiology 115*, 1717–1729.

CHAPTER 46

TMS and pain

Jean-Pascal Lefaucheur

Introduction

The idea that repeated stimulation of the brain with magnetic fields could produce analgesia has its origins in the profound clinical effects that electrical stimulation exerts on patients with chronic neuropathic pain. Chronic motor cortex stimulation (MCS) with surgically implanted epidural electrodes was first performed in the early 1990s and was found to provide significant pain relief in patients suffering from chronic drug-resistant central pain (Tsubokawa *et al.* 1991). Since this first report, numerous studies have confirmed the beneficial effects of the implanted MCS procedure in the treatment of chronic neuropathic pain of either peripheral or central origin (Meyerson *et al.* 1993; Tsubokawa *et al.* 1993; Nguyen *et al.* 1999; Carroll *et al.* 2000; Saitoh *et al.* 2000; Brown and Pilitsis 2005; Nuti *et al.* 2005; Rasche *et al.* 2006).

Given that electrical stimulation could exert such significant analgesic effects, there was considerable interest in whether similar effects could be achieved noninvasively via TMS. Testing that idea, Migita *et al.* (1995) reported that in one patient chronic neuropathic pain could be transiently relieved by the repeated application of a single pulse of TMS at low frequency (<0.3 Hz) over the motor cortex. This result was found predictive for the outcome of subsequent MCS implantation. When repetitive (r)TMS at frequencies closer to those used in MCS became available, it was tempting to determine whether rTMS could produce more significant and reliable analgesic effects. We first observed such effects by applying rTMS trains at 10 Hz over the motor cortex in a small series of patients

with chronic neuropathic pain (Lefaucheur *et al.* 1998). Since this preliminary report, a few studies have been performed to assess the value of rTMS to relieve neuropathic pain.

However, the relationships between TMS and pain are not restricted to the clinical observation of rTMS-induced analgesic effects. In the present chapter, we will also review the effects of pain on motor cortex excitability assessed by single- or paired-pulse TMS, the influence of rTMS on perception of innocuous or noxious peripheral stimuli, and the results obtained by applying peripheral magnetic stimulation to treat musculoskeletal pain.

Effects of pain on motor cortex excitability

Effects of acute phasic provoked pain

A few studies have examined the modulating effect of pain on motor cortex excitability. Painful electrical (Kaneko *et al.* 1998; Kofler *et al.* 1998; Classen *et al.* 2000; Urban *et al.* 2004) or laser (Valeriani *et al.* 1999, 2001) stimulation of the skin is able to reduce corticospinal motor output, as shown by a decrease in motor-evoked potential (MEP) amplitude.

It is still undetermined at which anatomical level a provoked pain can modulate the motor output resulting from cortical stimulation. The inhibitory influence of nociceptive afferents on motor responses could take place in various neural structures between the cortex and the spine, including the cingulate gyrus or the thalamus. Cortical interneurons were likely involved in laser experiments, because MEP inhibition began when the laser-evoked potentials (LEPs) peaked at the cortex (160 ms) (Valeriani *et al.* 1999, 2001). In addition, laser-induced MEP attenuation was observed in response to TMS, but not to high-voltage transcranial electrical anodal stimulation, which does not activate cortical interneurons, but directly activates the pyramidal tract, in contrast to TMS (Valeriani *et al.* 1999, 2001). In the studies based on painful electrical stimulation, spinal rather than cortical inhibition was evoked because MEP amplitude reduction takes place within a period of time corresponding to the cutaneous silent period (40–100 ms) (Kaneko *et al.* 1998; Kofler *et al.* 1998), which is of spinal origin. The discrepancy regarding the location of the interaction between nociceptive afferents and motor output could relate to the nature of the afferent pathways that were stimulated. In fact, laser pulses selectively stimulate small-diameter nociceptive afferents, while large-diameter proprioceptive and tactile fibers are stimulated concomitantly with nociceptive fibers in response to electrical stimulation.

Effects of prolonged tonic provoked pain

Similarly, tonic muscle pain induced by *in situ* injection of hypertonic saline (Le Pera *et al.* 2001; Svensson *et al.* 2003) or tonic cutaneous pain following topical application of capsaicin (Farina *et al.* 2001; Cheong *et al.* 2003) results in motor inhibition related to functional changes at cortical and/or spinal levels. In the case of muscle pain, MEP amplitude decreased 2–3 min after saline injection, preceding H-reflex reduction. Motor inhibition appeared to be of cortical origin in its early phase. In the case of cutaneous pain, MEP amplitude decreased 20–40 min after capsaicin application, i.e. at the onset of pain. With the same delay after capsaicin application, the silent period to cortical stimulation (cSP) was prolonged (Cheong *et al.* 2003). The enhanced motor inhibition was restricted to the territory of provoked pain. MEP amplitude recovered before the pain disappeared after capsaicin application, but remained reduced after pain recession in the case of intramuscular saline injection. This finding could reflect some

differences between these two experimental models regarding the type of nociceptors, nerve fibers, and pain pathways that were activated.

Physiological considerations

Despite one contradictory study (Romaniello *et al.* 2000), there is strong experimental support for the conclusion that either phasic or tonic provoked pain is able to reduce motor cortex excitability or corticospinal motor output. This could reveal a protective role of some spinal or intracortical processes from excessive motor responses to provoked pain.

High-frequency transcutaneous electrical nerve stimulation (TENS), usually applied to treat pain, was also shown to reduce MEP amplitude (Mima *et al.* 2004). However, this study enrolled healthy subjects without any pain syndrome. They felt tingling but not painful sensation in the stimulated area. It is likely that a wide range of noxious or innocuous peripheral sensory stimuli are able to impact on motor cortex excitability.

Effects of chronic pain

Whereas provoked pain tends to reduce motor cortex excitability in healthy subjects, a state of motor cortex disinhibition seems to characterize patients with chronic pain. For instance, while experimentally induced tonic muscle pain attenuates MEPs in healthy volunteers, MEP amplitude was found to be enhanced in muscle adjacent to chronic painful joint (On *et al.* 2004). This effect could represent a methodological confound for studies dosing rTMS intensities relative to motor threshold in pain patients. To better define the effects of chronic pain, comprehensive evaluation of motor cortex excitability, not restricted to MEP amplitude, but including resting motor threshold (RMT), cSP duration, intracortical inhibition (ICI), and facilitation (ICF), was performed in various series of patients suffering pain.

First, Salerno *et al.* (2000) showed increased RMT, normal MEP amplitude, shortened cSP, and reduced ICF and ICI (for long interstimuli intervals and suprathreshold intensities) in patients with fibromyalgia. Later, Schwenkreis *et al.* (2003) found normal RMT, MEP amplitude, and ICF, but reduced ICI for both hemispheres in patients suffering from complex regional pain syndrome (CRPS). Eisenberg et al. (2005) also reported ICI reduction in patients with CRPS, but only for the hemisphere corresponding to the painful side.

Recently, we assessed whether motor cortex excitability was modified in patients with chronic neuropathic pain of various origins (Lefaucheur et al. 2006a). These patients showed normal RMT and MEP amplitude, a tendency towards ICF reduction, but significant ICI reduction and cSP shortening, in correlation with pain level. Thus, chronic neuropathic pain, as CRPS, appeared to be associated with a state of motor cortex disinhibition. This might reveal some impairment in gamma-aminobutyric acid (GABA)-ergic mediation, related to some aspects of chronic pain. However, a variety of factors, in addition to chronic pain, could influence motor cortex excitability in this context, including the neurological lesion at the origin of pain, the severity of sensory or motor deficit, various cognitive parameters, or the effect of analgesic drugs.

Effects of cortical rTMS on perception of innocuous sensory stimuli

Normal subjects

Effects on sensory perception resulting from single-pulse TMS delivered to the motor cortex were first reported in healthy subjects (Cohen et al. 1991; Seyal et al. 1992; André-Obadia et al. 1999). In these studies, single-pulse TMS applied over hand motor cortical area was able to reduce or block the perception of nonpainful electrical sensory stimulation of the index finger, contralateral to the TMS pulse. This inhibition started when TMS was applied 300 ms before finger stimulation. Maximal attenuation was found when TMS occurred 20 ms after finger stimulation, at the time of afferent volley arrival in the primary somatosensory cortex. But TMS effect on sensory perception persisted when TMS was delivered up to 200 ms after finger stimulation, suggesting that cortico-subcortical processes, occurring after volley arrival in the primary somatosensory cortex, were essential for the conscious perception of

peripheral sensory stimuli. Finally, sensory perception was not altered if the TMS pulse occurred later than 200 ms after finger stimulation, confirming that late cortical responses to sensory stimulation (N200, P300) were post-perceptual responses. Sensory attenuation was maximal when TMS was applied at the site where the largest hand MEPs could be obtained. This attenuation occurred at a stimulus intensity that did not produce motor responses, and thereby could not result from contraction-induced sensory gating. Thus, sensory inputs can be blocked or attenuated within the lemniscal system due to intracortical or cortico-subcortical effects produced by the stimulation of the motor cortex. Not only sensory perception intensity but also stimulus location was found to be altered following motor cortex TMS (Seyal et al. 1997).

Paradoxically, the amplitude of somatosensory-evoked potentials (SEPs) increased for the intervals between TMS and peripheral stimulation that result in sensory perception attenuation (Kujirai et al. 1993; Seyal et al. 1993). The conditioning TMS pulse was thought to improve the synchronization of the afferent sensory volley in the primary somatosensory cortex, leading to SEP enhancement without concomitantly enhanced sensory perception, probably because sensory perception results from physiological processes more complex than SEPs. Tsuji and Rothwell (2002) applied TMS pulses paired with electrical stimulation of the motor point of a hand muscle (interstimulus interval of 25 ms) at 0.1 Hz for 30 min. Repeated single-pulse TMS increased the amplitude of the SEPs as well as of the MEPs recorded in the stimulated muscle for up to 10 min after the end of TMS application. This result implies that repeated conditioning TMS pulses may increase the responsiveness of the sensory cortex to synchronized afferent inputs.

In contrast to single-pulse TMS, low-frequency rTMS is able to reduce both sensory perception and SEP amplitude. Satow et al. (2003) reported that subthreshold rTMS delivered at 0.9 Hz for 15 min over the motor cortex transiently increased tactile threshold in the corresponding hand. At higher intensity (110% of RMT), 1 Hz rTMS over the right somatosensory impaired a tactile discrimination task performed with the left hand (Knecht *et al.* 2003). This effect lasted 2–8 min after 5–20 min of rTMS. Enomoto *et al.* (2001) also showed that the amplitude of the main components of the right median nerve SEPs (N20–P25 and P25–N33) decreased within 10 min after the application of 200 rTMS pulses delivered at 1 Hz and subthreshold intensity to the left motor cortex. This effect was thought to occur within the sensory cortex after the arrival of the thalamocortical volleys. SEP amplitude recovered 70–100 min after the end of the rTMS train. When the stimulation was applied over the premotor or the primary somatosensory cortex, SEPs remained unchanged. The absence of changes of the main sensory components of the SEPs was also observed after subthreshold stimulation of the sensorimotor cortex at 0.9 Hz (Satow *et al.* 2003) and of the motor or premotor cortex at 0.2 Hz (Urushihara *et al.* 2006). Nevertheless, in this latter study, the frontal N30 component of precentral origin was modified (increased in amplitude) after premotor cortex stimulation.

Seyal *et al.* (1995, 2005) assessed the interhemispheric influence of low-frequency rTMS (0.3 Hz) applied to one hemisphere onto the activity of the contralateral sensory cortex. These authors showed that the sensitivity to cutaneous stimulation was increased after ipsilateral stimulation of the parietal cortex but reduced after ipsilateral stimulation of the motor cortex. This latter result was associated with SEP amplitude decrease. Thus, the sensory cortex may be disinhibited by contralateral inhibition of the parietal cortex and inhibited by contralateral inhibition of the frontal cortex.

In contrast to low-frequency rTMS, high-frequency rTMS (5 Hz) applied over the primary sensory cortex improved tactile discrimination (Ragert *et al.* 2003). This was associated with a reduction of paired-pulse inhibition of the SEPs (Ragert *et al.* 2004). Moreover, after 25 trains of 50 pulses of 5 Hz rTMS applied over the cortical representation of the index finger within the primary somatosensory cortex, fMRI revealed an enlargement of this representation concomitant with the lowering of tactile discrimination thresholds of the finger (Tegenthoff *et al.* 2005). These changes recovered ~2 h after the termination of rTMS. Thus, high-frequency rTMS may

induce intracortical excitability enhancement in the sensory cortex, similarly as was shown in the motor cortex.

The effects of rTMS on innocuous thermal stimuli have been more rarely studied than those on tactile or nonpainful electrical stimuli. In 20 healthy individuals, cold detection thresholds were found to be significantly reduced after 500 rTMS pulses applied at low (1 Hz) or high (20 Hz) frequency over the motor cortex (Summers *et al.* 2004). An opposite result was observed in a series of 14 healthy volunteers, using a shorter train (50 stimuli) of rTMS applied at 5 Hz (Oliviero *et al.* 2005). Therefore, the way by which thermal sensitivity could be modified by rTMS in healthy subjects remains to be clarified.

In any case, there is strong evidence to conclude that the stimulation of motor cortical areas is able to modulate cutaneous sensitivity. Increase or decrease in sensory threshold that results from cortical stimulation depends on various parameters of stimulation, such as the frequency or the total number of pulses. Such influence of motor cortex activation on sensory discrimination is likely to be involved in the mechanisms of MCS-induced pain relief.

Patients with chronic pain

In a series of 46 patients with chronic neuropathic pain of various origins, we quantified the first perception thresholds for thermal (cold, warm) and mechanical (vibration, pressure) sensation in the painful zone and in the contralateral nonpainful homologous territory before and after a single session of rTMS applied for 20 min over the motor cortex, contralateral to the painful side (Lefaucheur *et al.*, unpublished data). Subthreshold rTMS delivered at 10 Hz significantly lowered thermal (cold, warm) thresholds in the painful zone, but not in the contralateral nonpainful territory. improvement in thermal sensory discrimination was associated with pain relief, in particular for patients with severe sensory deficit. In contrast to the effects of active 10 Hz rTMS, sensory thresholds did not change after sham 10 Hz rTMS or active 1 Hz rTMS.

A decrease in cold perception threshold was also observed in a series of 17 patients with chronic pain of non-neuropathic origin (low back pain) following high-frequency 20 Hz rTMS administered over the motor cortex (Johnson et al. 2006). A concomitant increase in warm perception threshold was reported in this study. This finding might reflect the differential influence of motor cortex rTMS on thermal stimulus perception with respect to Aδ- (cold) or C-fiber (warm) mediation. The influence of the nature of the involved sensory fibers will be discussed later regarding rTMS modulation of provoked pain in healthy subjects. Such a differential pattern of sensory changes in response to motor cortex rTMS might be lacking in the case of deafferentation pain.

In patients with neuropathic pain, it could be of value to assess the changes in sensory perception induced by motor cortex rTMS within the painful zone in order to predict the clinical outcome of subsequent MCS implantation. Indeed, we previously found that switching 'on' implanted MCS significantly lowered sensory detection thresholds in the painful zone, particularly in patients who were 'good responders' to the surgical procedure (Drouot et al. 2002). Thermal sensory relays are potentially hyperactive in patients with chronic neuropathic pain secondary to sensitization or deafferentation-induced disinhibition. By reducing neuronal hyperactivity in these structures, MCS might relieve pain and concomitantly improve innocuous thermal sensory discrimination. However, the exact mechanisms that lead to increased sensory perception, while blocking ongoing pain, remain to be determined.

Effects of cortical rTMS on perception of provoked pain

Normal subjects

The influence of rTMS on experimental pain was assessed in several studies (Table 46.1). Various neurophysiological techniques can be performed to induce pain in human subjects and to quantify the resulting nociceptive responses. One of the most elaborate methods is to apply a laser beam on the skin and to determine laser-induced pain thresholds or to record LEPs at the cortical level.

Two studies assessed the effects of TMS on laser-induced pain in healthy subjects. First,

Kanda et al. (2003) used a paired-pulse TMS procedure (50 ms interstimuli interval, suprathreshold intensity) delivered over various cortical targets at variable delays after a painful CO_2 laser stimulation of the dorsum of the hand. The laser pulse was perceived as more painful, compared to the control situation, when TMS was applied over the motor cortex 150–200 ms after the laser pulse, and was found less painful when TMS was applied over a medial frontal target 50–100 ms after the laser pulse. Second, Tamura et al. (2004a) showed that 1 Hz rTMS applied over the motor cortex transiently increased Tm:YAG-LEP amplitude (N2–P2 complex) in parallel with subjective pain ratings. Thus, when administered to the motor cortex, paired-pulse TMS and low-frequency rTMS appeared to enhance pain sensation evoked by CO_2 or Tm:YAG laser pulses and mediated by Aδ-fibers. Summers et al. (2004) also showed that high-frequency (20 Hz) rTMS over the motor cortex significantly lowered cold pain thresholds, leading to an increased susceptibility to cold pain, which is mainly mediated by Aδ-fibers. In contrast, Tamura et al. (2004b) showed that 1 Hz rTMS applied over the motor cortex suppressed tonic acute pain, which was provoked by intradermal capsaicin injection and probably mediated by C-fibers. These findings might indicate that in healthy subjects the effects of motor cortex rTMS on provoked pain depend on the type of fiber that mediates the pain.

The effects of 10 Hz rTMS applied to the motor cortex were also assessed on electrically induced pain, using alternating currents with sinusoid waveform (Neurometer device) (Yoo et al. 2006) or brief trains of five pulses at a frequency of 250 Hz (generating a nociceptive flexion reflex) (Mylius et al. 2007). The pain tolerance threshold to electrical stimulation was increased 30 min after the application of rTMS in the first study (together with an increase of the current perception threshold), whereas the scores of pain unpleasantness increased 20 min after rTMS in the second study. These results are difficult to interpret because the stimulus waveforms were different, as well as the type of assessment (e.g. pain tolerance vs pain unpleasantess) between the studies.

For cortical targets other than the motor cortex, conflicting results have been reported

Table 46.1 Effects of paired-pulse (pp) or repetitive (r)TMS on experimental pain

	Kanda et al. 2003	Töpper et al. 2003	Summers et al. 2004	Tamura et al. 2004a	Tamura et al. 2004b	Graff-Guerrero et al. 2005	Johnson et al. 2006	Mylius et al. 2006	Yoo et al. 2006	Mylius et al. 2007	Lefaucheur et al. unpublished data
Subjects/patients (n)	Normal subjects (9)	Normal subjects (4)	Normal subjects (40)	Normal subjects (13)	Normal subjects (7)	Normal subjects (180)	Patients with back pain (17)	Normal subjects (10)	Normal subjects (16)	Normal subjects (12)	Patients with neuropathic pain (32)
Pain stimulus	Laser stimuli	Cold water immersion	Heat pain, cold pain	Laser stimuli	Intradermal capsaicin injection	Cold water immersion, heat pain, pressure pain	Heat pain, cold pain	Painful electrical stimuli	Painful electrical stimuli	Painful electric stimuli	Laser stimuli
Type of coil	Figure-8	Figure-8	Figure-8	Figure-8	Figure-8	Figure-8	Figure-8	Figure-8	Figure-8	Figure-8	Figure-8
Cortical target	SM1, Occ, SII, MFC	F3, F4, Cz, P3, P4	M1	M1	M1	M1, DLPFC	M1	MFC	M1, MFC	M1	M1
TMS parameters	ppTMS (50 ms ISI)	rTMS, 15 Hz 1 train of 2 s	rTMS, 1/20 Hz 1 train of 8.3 min/ 12.5 trains of 2 s	rTMS, 1 Hz 1 train of 10 min	rTMS, 1 Hz 1 train of 5 min	rTMS, 1 Hz 1 train of 15 min	rTMS, 20 Hz 12.5 trains of 2 s	ppTMS (50 ms ISI)	rTMS, 10 Hz 50 trains of 1.8 s	rTMS, 10 Hz 20 trains of 5 s	rTMS, 10 Hz 20 trains of 10 s
Total pulse number	2	30	500	600	300	900	500	2	900	1000	2000
Intensity	120% RMT	110% RMT	95% MT	90% RMT	130% RMT	100% MT	95% MT	120–160% RMT	90% MT	80% RMT	90% RMT

Effects on provoked pain	SM1: increased pain Occ, SII: no change MFC: decreased pain	No change	Lowered cold pain threshold (only for 20 Hz rTMS)	Enhancement of laser-evoked potentials, correlated to pain relief	Reduced pain from 2 to 7 min after intra-dermal capsaicin injection	Increased tolerance to cold pain during rTMS (only for right DLPFC), but no change after rTMS	Lowered cold pain threshold, increased heat pain threshold	Increased pain	Pain tolerance threshold increased after M1 stimulation and decreased after MFC stimulation	Increased pain unpleasantness	Attenuation of laser evoked potentials, correlated to pain relief

M1, primary motor cortex; SM1, primary sensorimotor cortex; Occ, occipital cortex; SII, secondary sensory cortex; MFC, medial frontal cortex; DLPFC, dorsolateral prefrontal cortex; F3, F4, Cz, P3, P4, coordinates of the International 10–20 System; ISI, interstimulus interval; RMT, resting motor threshold.

about the value of TMS procedures to modulate provoked pain. As previously mentioned, Kanda *et al.* (2003) found that laser-induced pain could be significantly reduced in response to the stimulation of the medial frontal cortex. In contrast, Mylius *et al.* (2006) reported that a similar paired-pulse TMS protocol (50 ms interstimuli interval, suprathreshold intensity) applied over the medial frontal cortex increased the verbal pain report to a painful electrical stimulation that was able to elicit a nociceptive flexion reflex. In the same way, Yoo *et al.* (2006) showed that 10 Hz rTMS applied over the medial frontal cortex reduced pain tolerance threshold to electrical stimulation.

Amassian *et al.* (1997) found that ten pulses of rTMS applied at 20 Hz over the contralateral parietal cortex resulted in a significant attenuation of acute pain provoked by transient circulatory occlusion of the arm with a tourniquet. This effect was reversed by naloxone, suggesting an endorphin-mediated process. In contrast, Töpper *et al.* (2003) found no effect of rTMS administered to the parietal cortex on acute pain provoked by hand immersion in cold water (cold pressor test).

Finally, in a large series of healthy volunteers, Graff-Guerrero *et al.* (2005) studied the effect of low-frequency (1 Hz) rTMS on pain threshold and tolerance during the cold pressor test. The rTMS targets were the motor and the dorsolateral prefrontal cortices of the right and left hemispheres. Only the stimulation of the right dorsolateral prefrontal cortex induced a significant effect that was an increased tolerance to cold-induced pain. No change in cold, heat, or pressure pain was observed for the other sites of stimulation. This study suggested that inhibiting the right dorsolateral prefrontal cortex might enhance the tolerance to provoked pain.

Thus, painful stimulus perception is likely to be influenced by cortical stimulation, but this influence depends on various features regarding the nature of the recruited nociceptive afferents the parameters of stimulation, or the site of the cortical target. When applied to the motor cortex, paired-pulse TMS or 1 Hz rTMS appeared to increase phasic pain mediated by Aδ-fibers, whereas 1 Hz rTMS suppressed tonic pain mediated by C-fibers.

Patients with chronic pain

In patients with chronic neuropathic pain, we assessed the effects of high-frequency motor cortex rTMS on LEPs and laser-induced pain (Lefaucheur *et al.*, unpublished data). Thirty-two right-handed patients with unilateral hand pain of various neurological origins were enrolled in this study. The LEPs in response to Nd:YAG laser stimuli applied to painful or painless hand were recorded before and after a 20 min session of subthreshold rTMS applied at 10 Hz using an active or a sham coil over the motor cortex corresponding to the painful hand. Laser-induced pain was concomitantly scored on a 0–10 visual analog scale. At baseline, LEP amplitudes were reduced in response to painful hand compared to painless hand stimulation. This might result from the spinothalamic tract dysfunction accompanying neuropathic pain (Garcia-Larrea *et al.* 2002). After rTMS, LEP amplitude further decreased, whatever the type of rTMS (active or sham) and the side of laser stimulation (painful or painless hand), due to arousal factors that are known to greatly affect LEP amplitude (Lorenz and Garcia-Larrea 2003). However, active rTMS reduced painful hand LEP amplitude more profoundly than could be attributed to a retest effect. This decrease paralleled a reduction in subjective pain scores. Motor cortex rTMS affected more specifically the N2 component of the LEPs, suggesting a preferential influence on sensori-discriminative aspects of pain. The LEP parameters studied reflected Aδ-mediated pain. Another type of Aδ-mediated provoked pain, related to painful cold stimuli, was also found to be modified by motor cortex rTMS applied at 20 Hz in a series of 17 patients with chronic low back pain (Johnson *et al.* 2006). Following rTMS, cold pain threshold significantly decreased, while threshold to heat pain (mediated by C-fibers) increased, confirming the differential influence of motor cortex rTMS on pain according to the type of nociceptive afferents.

These studies showed that high-frequency motor cortex rTMS can reduce an acute pain that has been provoked in a cutaneous territory in which there already exists a chronic spontaneous pain. Then, the stimulation of the motor cortex is likely to interfere with various neural processes that are involved in various types of pain.

Effects of peripheral repetitive magnetic stimulation on muscle pain

Magnetic stimulation can be applied at a 'peripheral' level, not only on the scalp, to exert analgesic effects. A first sham-controlled study performed in 30 patients with musculoskeletal pain supported the efficacy of a single session of repetitive magnetic stimulation (rMS) applied at 20 Hz directly on a tender body region for 40 min (Pujol et al. 1998). A second study, with similar rMS parameters (40 trains of 5 s at 20 Hz, with 25 s pauses) but including repeated sessions (5 days a week for 2 consecutive weeks), confirmed the previous results in nine patients (Smania et al. 2003). Significant relief of the myofascial pain persisted at least 1 month. A subsequent comparative study showed that peripheral rMS was more effective than TENS at long term (3 months post-stimulation) for the treatment of myofascial pain (Smania et al. 2005). In contrast, rMS (30 trains at 20 Hz) was applied without success in cases of resistant tennis elbow (lateral epicondylitis) (Rollnik et al. 2003).

Thus, 'peripheral' rMS is emerging as a therapeutic tool for various musculoskeletal syndromes, but the lack of portability of rMS compared to TENS will probably limit its further development in clinical practice.

Effects of motor cortex rTMS on chronic pain

First reports and duration of the effects

In the last decade, several studies have been performed to test the ability of motor cortex rTMS to produce analgesic effects (Tables 46.2 and 46.3). First, Migita et al. (1995) delivered 200 TMS pulses at 0.2 Hz using a nonfocal circular coil centered over the motor cortex, contralateral to the painful side, in two patients with central pain. The first patient experienced 30% pain relief for 1 h, while TMS was ineffective for the second patient. TMS effects paralleled the outcome of subsequent MCS implantation.

Canavero et al. (2003) applied a similar protocol of repeated single-pulse TMS in a series of patients with chronic pain secondary to stroke or spinal cord lesion. The procedure consisted of two trains of 100 stimuli delivered at 0.2 Hz over the motor cortex using a figure-8 coil for arm stimulation or a double-cone coil for leg stimulation. From the nine patients enrolled in the study, one patient was relieved for allodynia and four patients for both spontaneous pain and allodynia. Pain relief lasted for 16 h in one case.

These two studies were based on very low frequency of stimulation with single-pulse TMS (0.2 Hz), compared with the frequencies used in chronic implanted MCS that range from 20 to 55 Hz (Nguyen et al. 2003). Frequency is considered as one of the most crucial parameters of stimulation, conditioning the functional result of rTMS despite high inter-individual variability (Maeda et al. 2000). High-frequency stimulation (>5 Hz) is able to excite the underlying cortex for a few minutes (Pascual-Leone et al. 1994), while low-frequency stimulation (<5 Hz) causes inhibition of motor responses (Chen et al. 1997). In our first placebo-controlled study, rTMS was applied to the motor cortex at high (10 Hz) or low (0.5 Hz) frequency, in a series of 18 patients with chronic pain secondary to thalamic stroke, brain-stem lesion, or brachial plexus lesion (Lefaucheur et al. 2001a). We found that rTMS administered at 10 Hz, but not at 0.5 Hz, resulted in pain relief, regardless of the side of the stimulated hemisphere (Lefaucheur et al. 2001a). This was the first demonstration of the ability of high-frequency motor cortex rTMS to relieve chronic neuropathic pain of peripheral or central origin. This finding was confirmed in subsequent studies (Lefaucheur et al. 2001b, 2004a,b, 2006a,b; Khedr et al. 2005; André-Obadia et al. 2006; Hirayama et al. 2006), even in the case of complex regional pain syndrome (CRPS) (Pleger et al. 2004).

Only two studies reported negative results in this domain (Rollnik et al. 2002; Irlbacher et al. 2006). Disappointingly, in one of these studies, more than one-third of the patients did not complete the full experimental design (Irlbacher et al. 2006). In addition, rTMS was applied at relatively low frequencies (1–5 Hz), and high-frequency rTMS (10–20 Hz) was shown to provide significantly more analgesia than low-frequency rTMS (0.5–1 Hz) (Lefaucheur et al. 2001a; André-Obadia et al. 2006). Concerning

Table 46.2 Effects of repetitive (r)TMS on chronic pain: primary motor cortex target

	Migita et al. 1995	Lefaucheur et al. 2001a	Lefaucheur et al. 2001b	Canavero et al. 2002	Rollnik et al. 2002	Lefaucheur et al. 2004a	Lefaucheur et al. 2004b	Pleger et al. 2004	Khedr et al. 2005	André-Obadia et al. 2006	Irlbacher et al. 2006	Lefaucheur et al. 2006a	Lefaucheur et al. 2006b	Lefaucheur et al. unpublished data
Total number of patients and pain origin	2 cerebral palsy + thalamotomy (1), putamen hemorrhage (1)	18 thalamic stroke (6), brain stem lesion (6), brachial plexus lesion (6)	14 thalamic stroke (7), trigeminal nerve lesion (7)	9 stroke (5), spinal cord lesion (4)	12 spinal cord lesion (2), osteomyelitis (1), peripheral nerve lesion (6), CRPS (2), phantom limb (1)	1 brachial plexus lesion (1)	60 thalamic stroke (12), brain stem lesion (12), spinal cord lesion (12), trigeminal nerve lesion (12), brachial plexus lesion (12)	10 CRPS (10)	48 stroke (24), trigeminal nerve lesion (24)	14 thalamic stroke (8), brain stem lesion (2), spinal cord, trigeminal nerve, brachial plexus, or nerve trunk lesion (4)	27 (13) thalamic stroke (3), brain stem lesion (7), spinal cord lesion (3), phantom limb (14)	22 thalamic stroke (10), spinal cord lesion (4), brachial plexus lesion (8)	36 thalamic stroke (5), brain stem lesion (4), spinal cord lesion (5), trigeminal nerve lesion (14), brachial plexus lesion (8)	46 thalamic stroke (13), spinal cord lesion (10), nerve lesion (13), brachial plexus lesion (10)
Type of coil	Circular	Figure-8	Figure-8	Figure-8, double-cone	Circular, double-cone	Figure-8	Figure-8	Figure-8	Figure-8	Figure-8	Figure-8	Figure-8	Figure-8	Figure-8
Target	M1	M1	M1	M1	M1	M1	M1	M1	M1	M1	M1	M1	M1	M1
Intensity	80% SO	80% RMT	80% RMT	100% SO	80% RMT	80% RMT	80% RMT	110% RMT	80% RMT	90% RMT	95% RMT	90% RMT	90% RMT	90% RMT

Frequency (Hz)	0.2	0.5/10	10	0.2	20	10	10	10	10	20	1/20	1/5 Hz	1/10	10	1/10
Train number duration	1 train of 16.7min	1 train of 20 min/20 trains of 5 s	20 trains 5 s	1 train of 16.7 min	20 trains of 2 s	20 trains of 5 s	20 trains of 5 s	10 trains of 1.2 s	10 trains of 10 s	10 trains of 10 s	1 train of 26 min/20 trains of 4 s	Unknown	1 train of 20 min/20 trains of 6 s	20 trains of 10 s	1 train of 20 min/20 trains of 6 s
Number of sessions	1	1	1	1	1	16 (16 months)	1	1	5 (1 week)	5 (1 week)	1	5 (1 week)	1	1	1
Number of pulses	200	600/1000	1000	200	800	1000 × 16	1000	120	2000 × 5		1600	500 × 5	1200	2000	1200
Sham control	None	Sham coil	Sham coil	Angled coil	Angled coil	Sham coil	Sham coil	Angled coil	Angled coil	Angled coil	Angled coil	Sham coil	Sham coil	None	Sham coil
Improved/unimproved	1/2	7/18 (for 10 only)	8/14	5/9	6/12	1/1	35/60	7/10	22/28 (real rTMS), 4/20 (sham rTMS)		5/14 (20 Hz), 1/14 (1 Hz), 4/14 (sham rTMS)	2/27 (real rTMS), 2/27 (sham rTMS)	Not determined	Not determined	Not determined
Pain relief percentage	30% and 0%	20% (10 Hz) (0.5 Hz sham: 4–7%)	30% (sham: no relief)	Unknown	4% (sham: 2%)	40% (sham: 15%)	21% (sham: 9%)	21% (sham: pain increase)	45% (sham: 5%)		11% (20 Hz) (sham: 8%)	6/5% (1/5 Hz) (sham: 10%)	33% (10 Hz) (sham: 11%)	27 /37% (face/hand pain)	24% (10 Hz) (1 Hz sham: 5–10%)
Comparison with placebo	Not applicable	P = 0.001 (10 Hz)	P = 0.013	Not done	P > 0.05	P = 0.0002	P = 0.0002	P = 0.02	P < 0.001		P = 0.04 (20 Hz vs 1 Hz and sham)	P = 0.08 (1 Hz), 0.06 (5 Hz)	P = 0.002 (10 Hz)	Not applicable	P < 0.0001 (10 Hz)
Pain relief maxima duration	1 h	Not determined	8 days	16 h	6 days	1 week in average after each session	Not determined	90 min at least	2 weeks at least after the last session		1 week	None	Not determined	1 week at least	Not determined

CRPS, complex regional pain syndrome; M1, primary motor cortex; S1, primary sensory cortex; PreM, premotor cortex; SMA, supplementary motor area; DLPFC: dorsolateral prefrontal cortex; PPC, posterior parietal cortex; SO, stimulator output; RMT, resting motor threshold.

Table 46.3 Effects of repetitive TMS on chronic pain: targets other than the primary motor cortex

	Reid et al. 2001	Töpper et al. 2003	Hirayama et al. 2005	Fregni et al. 2005	Sampson et al. 2006
Total number of patients and pain origin	1 teeth removal (1)	2 root avulsion (2)	20 thalamic stroke (7), brain stem lesion (5), spinal cord lesion (3), trigeminal nerve lesion (3), brachial plexus lesion (2)	5 visceral pain due to pancreatitis	4 Fibromyalgia
Type of coil Target	Figure-8 Left DLPFC	Figure-8 PPC	Figure-8 M1, S1, PreM, SMA	Figure-8 Right/left S2	Figure-8 Right DLPFC
Intensity	100% RMT	110% RMT	90% RMT	90% RMT	110% RMT
Frequency (Hz)	20	1/10	5	1/20	1
Train number and duration	30 trains of 2 s	1 train of 12 min/20 trains of 2 s	10 trains of 10 s	Unknown	2 trains of 800 s
Number of sessions	14 (3 weeks)	15 (3 weeks)	1	1	20 (4 weeks)
Number of pulses	1200 × 14	720/400 × 15	500	1600	1600 × 20
Sham control	None	None	Angled coil	Sham coil	None
Improved/ unimproved	1/1	2/2	10/20 (for M1 only)	Not determined	4/4
Pain relief percentage	42%	32/58% (1/10 Hz)	28% (M1)	62% (1 Hz, right S2)	82%
Comparison with placebo	Not applicable	Not applicable	$P < 0.01$ (M1)	$P = 0.037$ (1 Hz, right S2)	Not applicable
Pain relief maximal duration	4 weeks	11 min	3 days	Not determined	15–27 weeks

the other negative study, the cortical stimulation was not focal but performed with circular and double-cone coils, while the site and origin of pain were heterogeneous, including non-neuropathic pain syndromes (Rollnik et al. 2002). Nevertheless, in one patient of this latter study, pain relief was optimal 2 days after the rTMS session and lasted for 6 days. This observation was very similar to our own results. In a series of 14 patients with trigeminal neuralgia or thalamic pain, we found that pain level could be significantly reduced for 8 days by active and not sham 10 Hz rTMS, the maximal analgesic

effect being delayed by 2–4 days after the rTMS session (Lefaucheur et al. 2001b). This delay of action may be related to rTMS-induced plastic changes in cortical circuitry, and needs to be taken into account in the design of rTMS studies in the pain domain.

Clinical parameters and pattern of cortical activation

Most of the aforementioned studies support an effect of rTMS on neuropathic pain, but provide evidence of variability in response, with both responders and nonresponders described. As for MCS (Katayama *et al.* 1998), it is interesting to determine whether some neurological characteristics are associated with the extent of rTMS-induced analgesic effects. In a series of 60 patients with chronic neuropathic pain of various origins and located in the face or upper or lower limb, we studied the influence of the type of lesion at the origin of pain, the site of pain, and the degree of sensory loss within the painful zone on rTMS efficacy (Lefaucheur *et al.* 2004b). The most favorable condition was trigeminal lesion, facial pain, and absence of severe sensory loss within the painful zone. The worst condition was brain-stem stroke, limb pain, and severe sensory loss. However, it was not easy to delineate the respective influence of all these clinical variables on the final result of the rTMS procedure.

In a subsequent study, we showed that rTMS was more effective in relieving pain when the stimulation was applied to an area adjacent to the cortical representation of the painful zone rather than to the motor cortical area corresponding to the painful zone itself (Lefaucheur *et al.* 2006b). This finding contradicts the classical observation of a strict somatotopic efficacy of the implanted MCS procedure. This may result from differences in the pattern of the induced current flow between rTMS and epidural cortical stimulation. In pain studies, due to the postero-anterior direction of the figure-8 coil, motor cortex rTMS preferentially activates cortico-cortical interneuronal fibers, tangential to the surface of the cortex, evoking indirect I-waves with various delays in the corticospinal tract (Kaneko *et al.* 1996; Nakamura *et al.* 1996). Implanted epidural MCS could differ from rTMS concerning the recruited populations of fibers, evoking different I-waves, but also D-waves, due to the direct activation of the corticospinal tract, perpendicular to the surface of the cortex (Di Lazzaro *et al.* 2004).

Mechanisms and site of action

Whatever the differences in the pattern of motor cortex activation resulting from rTMS and epidural electrical stimulation, these methods probably share common mechanisms of action to produce analgesic effects.

We reported that rTMS applied at high frequency over the motor cortex improved sensory discrimination in association with pain relief (Lefaucheur *et al.*, unpublished data). The same result was observed in patients with neuropathic pain when switching 'on' implanted epidural MCS (Drouot *et al.* 2002). Similarly, 10 Hz motor cortex rTMS was found to act on the sensori-discriminative aspect of laser-induced pain in patients with chronic pain, as shown by preferential changes in the N2 component of LEPs (Lefaucheur *et al.*, unpublished data). Sensory discrimination improvement appeared to be specific for thermo-nociceptive signals conveyed by the spinothalamic tract. This precludes a mechanism of pain relief due to the reinforcement of the lemniscal 'gate control' over the nociceptive system. The functional integrity of the lemniscal system is essential to the efficacy of spinal cord stimulation (Sindou *et al.* 2003), but not of MCS (Garcia-Larrea *et al.* 1999).

Positron emission tomography showed that implanted MCSs lead to regional cerebral blood flow changes in thalamus, anterior cingulate/orbitofrontal area, anterior insula, and upper brain stem (Peyron *et al.* 1995; Garcia-Larrea *et al.* 1999). All of these structures are potentially involved in innocuous thermal perception processing (Casey *et al.* 1996; Davis *et al.* 1998), and thereby they could mediate the concomitant effects of MCS on spontaneous pain and thermal detection thresholds. Thus, MCS might reduce pain-related hyperactivity in thalamic relays or interfere with abnormal thalamothalamic or thalamocortical oscillations, via corticothalamic projections and connections between thalamic nuclei.

Increase in cerebral blood flow in the upper brain stem and modulation of nociceptive spinal reflexes (RIII) by switching 'on' MCS supports the role of descending controls triggered by the motor corticothalamic output (Peyron *et al.* 1995; Garcia-Larrea *et al.* 1999). These descending controls could take place in various brain-stem or spinal cord nuclei and be involved in the process of pain relief resulting from MCS. This hypothesis is reinforced by the low rate of efficacy observed in patients with brain-stem stroke or spinal cord lesion in response to motor cortex rTMS (Lefaucheur *et al.* 2004b) or implanted MCS (Tsubokawa *et al.* 1993; Fujii *et al.* 1997).

However, imaging studies suggested a preferential influence of implanted MCS on structures that participate in the motivational–affective aspect of pain, such as the cingulate/orbitofrontal cortex (Peyron *et al.* 1995; Garcia-Larrea *et al.* 1999). Tamura *et al.* (2004b) also showed by single-photon emission computed tomography that beneficial effects of motor cortex rTMS on capsaicin-induced acute pain correlated with a significant regional cerebral blood flow increase in the caudal part of the anterior cingulate cortex and decrease in the medial prefrontal cortex. This raises the possibility of using cortical targets involved in cognitive and emotional adaptation to chronic pain for therapeutic purposes.

As previously cited, Graff-Guerrero *et al.* (2005) showed in a large series of healthy volunteers that rTMS applied over the right dorsolateral prefrontal cortex increased tolerance to cold-induced pain. Reid and Pridmore (2001) reported the efficacy of repeated sessions of 20 Hz rTMS delivered to the left dorsolateral prefrontal cortex in a depressive patient with drug-resistant facial pain due to tooth removal. Pain decreased by 42% during the second week of stimulation and maintained 4 weeks after the end of treatment, unrelated to mood changes. Sampson *et al.* (2006) reported an open case series of right dorsolateral prefrontal cortex stimulation with 1 Hz rTMS reducing pain in four patients with fibromyalgia. Borckardt *et al.* (2006) stimulated the left dorsolateral prefrontal cortex with 10 Hz rTMS and found, in a sham-controlled trial of postoperative pain following gastric bypass surgery, that a single 20 min session reduced total morphine use by 40%, independent of effects on mood ratings. Finally, Kanda *et al.* (2003) found in normal subjects that paired-pulse TMS applied over the medial frontal cortex could reduce the perception of painful laser stimuli. The anatomical correspondence between this medial frontal target and the anterior cingulate cortex remains to be confirmed before considering this target for future rTMS trials in patients with chronic pain.

The value of various sensorimotor cortical regions as effective rTMS target in the attempt to relieve pain was assessed by Hirayama *et al.* (2006) with the help of a navigation system dedicated to rTMS practice. Compared to the primary motor cortex, neither the premotor cortex (including the supplementary motor area) nor the primary sensory cortex provides valuable targets for rTMS to produce neuropathic pain relief. In contrast, 1 Hz rTMS applied over the right secondary somatosensory cortex was found to reduce chronic visceral pain due to chronic pancreatitis (Fregni *et al.* 2005). In this latter study, the rTMS target was also defined by means of a navigation system.

MCS effects on chronic pain likely depend on the recruitment of fibers located within the motor cortex but projecting to remote structures, functionally connected with the motor cortex, and involved in pain and sensory processing. However, MCS may also impact on intracortical motor circuitry, as suggested by rTMS-induced changes in cortical excitability parameters (Lefaucheur *et al.* 2006a). We found that intracortical inhibition (ICI) assessed by the paired-pulse TMS paradigm was defective in the motor cortex corresponding to the painful hand, but improved in parallel with pain relief after active rTMS applied at 10 Hz over the motor cortex. This result was opposite to what induces high-frequency subthreshold rTMS in healthy subjects (Maeda *et al.* 2000; Peinemann *et al.* 2000). In healthy subjects, motor cortex inhibition is associated with the existence of 20 Hz cortical oscillations that are abolished in the presence of chronic or provoked pain (Juottonen *et al.* 2002; Raij *et al.* 2004). By restoring such oscillatory activity in the primary motor cortex, MCS could restore defective inhibitory mechanisms.

Conclusion: the place of rTMS in the management of chronic pain at present and in the future

The mechanisms underlying the analgesic effects of MCS are probably multifactorial, partly depending on pain origin and presentation, e.g. according to the presence of provoked pain symptoms or to the severity of sensory or motor deficits. The main interest of rTMS is to provide a noninvasive tool to study the efficacy and the role of the motor cortex or other cortical areas in the modulation of pain perception.

The analgesic effects resulting from a single session of rTMS are too short-lived and thereby incompatible with a durable control of chronic pain. Repeated sessions of rTMS on consecutive days are able to produce cumulative effects and to extend the effects of a single session, as shown by Khedr et al. (2005), but the best way to stimulate a targeted cortical area chronically for therapeutic purposes remains to implant electrodes.

Nevertheless, repeated daily rTMS sessions can be applied to control pain syndromes for a limited period, e.g. to help patients waiting for surgical implantation. We reported the case of a woman with drug-resistant chronic pain due to brachial plexus lesion, who experienced good pain relief for 16 months in response to repeated sessions of motor cortex rTMS until durable pain relief was obtained by the surgical implantation of a cortical stimulator (Lefaucheur et al. 2004a).

This case also revealed that, beyond the production of analgesic effects, rTMS could be used as a predictive tool to select patients for implantation. First, repeated single-pulse TMS over the motor cortex has been claimed to be predictive for the outcome of a subsequent chronic epidural electrical stimulation (Migita et al. 1995; Canavero et al. 2003). In a series of 40 patients with neuropathic pain, we found that the response to motor cortex rTMS was associated with a good surgical outcome in all cases, while the absence of response to rTMS trial did not predict the result of the implanted procedure, particularly for patients with lower limb pain (Lefaucheur et al., unpublished data). The positive prediction of the efficacy of MCS implantation according to the response to high-frequency rTMS was also reported by André-Obadia et al. (2006). However, various technical considerations are able to influence rTMS efficacy, and thereby the place of rTMS as a selection tool for surgical implantation remains to be precisely defined.

Any future therapeutic application of non-invasive transcranial cortical stimulation in chronic pain will require a substantial improvement in the methods of stimulation. The precentral cortical target was validated by the clinical results obtained with the implanted MCS procedure. However, better cortical targets for pain relief might exist, according to each type of neuropathic pain, located either within or outside the motor cortex. For instance, analgesic effects produced by the stimulation of prefrontal cortical areas are awaited, in particular on the cognitive and affective components of chronic pain. Navigation-guided rTMS will be required for such studies, in light of the recent study of Hirayama et al. (2006) assessing various sensorimotor cortical targets.

Another crucial issue is how to optimize the parameters of stimulation to raise the rate and duration of rTMS-induced pain relief at therapeutic level. At present, rTMS effects on chronic pain are quite low (ranging between 20% and 45%) and short-lived, while they appeared to correlate positively with the frequency of stimulation (better for 10–20 Hz than for lower frequencies). Various methodological parameters have never been questioned so far, such as the orientation of the coil and the waveform of the magnetic pulse. All relevant rTMS studies on pain were performed using a figure-8 coil with a postero-anterior orientation and delivering biphasic pulses. However, biphasic pulses were found more efficient when the current was induced with an antero-posterior orientation (Kammer et al. 2001). In addition, monophasic pulses were shown to provide stronger after-effects on cortical activity than biphasic pulses using rTMS (Sommer et al. 2002; Arai et al. 2005). Thus, rTMS efficacy might improve by changing coil orientation or pulse waveform, and also by delivering repetitive trains of paired rather than single pulses (Bestmann et al. 2004; Khedr et al. 2004).

Other exciting prospects to enhance intensity and duration of rTMS effects on chronic pain include the use of compound-frequency paradigms (theta burst) and pretreatment conditioning with transcranial direct current stimulation (tDCS). 'Theta-burst' stimulation consists of giving bursts of stimulation at gamma frequency (50 Hz), and repeating these bursts at a theta-range frequency (5 Hz) (Huang et al. 2005). tDCS utilizes weak anodal or cathodal constant direct currents applied on the scalp that cross the skull to induce prolonged changes in brain excitability and to alter the brain response to subsequently applied rTMS (Siebner et al. 2004; Lang et al. 2004).

The applications of such conditioning stimuli and compound frequency packages to cortical stimulation for pain relief are yet to be examined.

The technique of tDCS can also be used alone as an alternative technique to rTMS, and not only as a conditioning method. Anodal tDCS provides sustained cortical facilitation, whereas cathodal tDCS induces inhibition (Priori *et al.* 1998; Nitsche and Paulus 2000). The first application of tDCS to act on neuropathic pain was reported by Fregni *et al.* (2006a). They showed that anodal stimulation of the motor cortex for 20 min on five consecutive days produced significant pain relief in a series of patients with chronic pain due to spinal cord injury. Seven of 11 patients experienced pain reduction of 50% or more at the end of the 5 days of stimulation. The mean pain score was decreased by 58% at the end of the week of stimulation and remained decreased by 37% 2 weeks after the last day of stimulation. This rate of pain relief was higher than in any rTMS study. The analgesic effects were not confounded with effects on mood or anxiety. Anodal tDCS of the primary motor cortex (but not of the dorsolateral prefrontal cortex) was also shown to provide sustained pain relief in a series of 32 patients with fibromyalgia (Fregni *et al.* 2006b). This confirms that pain control is likely associated with a local increase in motor cortex excitability. Compared to rTMS, the tDCS technique seems easier to perform, may provide longer-lasting changes in cortical activity, and can be adapted to battery-driven portable stimulators. This technique opens new perspectives for therapeutic development of non-invasive transcranial cortical stimulation as an alternative method to surgical MCS to treat chronic neuropathic pain in the future.

Independent of its potential therapeutic application, the utility of TMS to study pain perception in healthy subjects and the pathophysiology of pain syndromes is great, as illustrated by the studies reviewed. Further work will be needed to define the ultimate clinical role of TMS in the management of pain.

References

Amassian VE, Vergara MS, Somasundaram M, Maccabee PJ, Cracco RQ (1997). Induced pain is relieved by repetitive stimulation (rTMS) of human parietal lobe through endorphin release. *Electroencephalography and Clinical Neurophysiology 103*, 179.

André-Obadia N, Garcia-Larrea L, Garassus P, Mauguière F (1999). Timing and characteristics of perceptual attenuation by transcranial stimulation: a study using magnetic cortical stimulation and somatosensory-evoked potentials. *Psychophysiology 36*, 476–83.

André-Obadia N, Peyron R, Mertens P, Mauguière F, Laurent B, Garcia-Larrea L (2006). Transcranial magnetic stimulation for pain control. Double-blind study of different frequencies against placebo, and correlation with motor cortex stimulation efficacy. *Clinical Neurophysiology 117*, 1536–44.

Arai N, Okabe S, Furubayashi T, Terao Y, Yuasa K, Ugawa Y (2005). Comparison between short train, monophasic and biphasic repetitive transcranial magnetic stimulation (rTMS) of the human motor cortex. *Clinical Neurophysiology 116*, 605–613.

Bestmann S, Siebner HR, Modugno N, Amassian VE, Rothwell JC (2004). Inhibitory interactions between pairs of subthreshold conditioning stimuli in the human motor cortex. *Clinical Neurophysiology 115*, 755–764.

Borckardt JJ, Weinstein M, Reeves ST, *et al.* (2006). Postoperative left prefrontal repetitive transcranial magnetic stimulation reduces patient-controlled analgesia use. *Anesthesiology 105*, 557–562.

Brown JA, Pilitsis JG (2005). Motor cortex stimulation for central and neuropathic facial pain: a prospective study of 10 patients and observations of enhanced sensory and motor function during stimulation. *Neurosurgery 56*, 290–297.

Canavero S, Bonicalzi V, Dotta M, Vighetti S, Asteggiano G (2003). Low-rate repetitive TMS allays central pain. *Neurological Research 25*, 151–152.

Carroll D, Joint C, Maartens N, Shlugman D, Stein J, Aziz TZ (2000). Motor cortex stimulation for chronic neuropathic pain: a preliminary study of 10 cases. *Pain 84*, 431–437.

Casey KL, Minoshima S, Morrow TJ, Koeppe RA (1996). Comparison of human cerebral activation pattern during cutaneous warmth, heat pain, and deep cold pain. *Journal of Neurophysiology 76*, 571–581.

Chen R, Classen J, Gerloff C, *et al.* (1997). Depression of motor cortex excitability by low-frequency transcranial magnetic stimulation. *Neurology 48*, 1398–1403.

Cheong JY, Yoon TS, Lee SJ (2003). Evaluation of inhibitory effect on the motor cortex by cutaneous pain via application of capsaicin. *Electromyography and Clinical Neurophysiology 43*, 203–210.

Classen J, Steinfelder B, Liepert J, *et al.* (2000). Cutaneomotor integration in humans is somatotopically organized at various levels of the nervous system and is task dependent. *Experimental Brain Research 130*, 48–59.

Cohen LG, Bandinelli S, Sato S, Kufta C, Hallett M (1991). Attenuation in detection of somatosensory stimuli by transcranial magnetic stimulation. *Electroencephalography and Clinical Neurophysiology 81*, 366–376.

Davis KD, Kwan CL, Crawley AP, Mikulis DJ (1998). Functional MRI study of thalamic and cortical activations evoked by cutaneous heat, cold, and tactile stimuli. *Journal of Neurophysiology 80*, 1533–1546.

Di Lazzaro V, Oliviero A, Pilato F, *et al.* (2004). Comparison of descending volleys evoked by transcranial and epidural motor cortex stimulation in a conscious patient with bulbar pain. *Clinical Neurophysiology 115*, 834–838.

Drouot X, Nguyen JP, Peschanski M, Lefaucheur JP (2002). The antalgic efficacy of chronic motor cortex stimulation is related to sensory changes in the painful zone. *Brain 125*, 1660–1664.

Eisenberg E, Chistyakov AV, Yudashkin M, Kaplan B, Hafner H, Feinsod M (2005). Evidence for cortical hyperexcitability of the affected limb representation area in CRPS: a psychophysical and transcranial magnetic stimulation study. *Pain 113*, 99–105.

Enomoto H, Ugawa Y, Hanajima R, *et al.* (2001). Decreased sensory cortical excitability after 1 Hz rTMS over the ipsilateral primary motor cortex. *Clinical Neurophysiology 112*, 2154–2158.

Farina S, Valeriani M, Rosso T, *et al.* (2001). Transient inhibition of the human motor cortex by capsaicin-induced pain. A study with transcranial magnetic stimulation. *Neuroscience Letters 314*, 97–101.

Fregni F, DaSilva D, Potvin K, *et al.* (2005). Treatment of chronic visceral pain with brain stimulation. *Annals of Neurology 58*, 971–972.

Fregni F, Boggio PS, Lima MC, *et al.* (2006a). A sham-controlled, phase II trial of transcranial direct current stimulation for the treatment of central pain in traumatic spinal cord injury. *Pain 122*, 197–209.

Fregni F, Gimenes R, Valle AC, *et al.* (2006b). A randomized sham-controlled proof-of-principle study of transcranial direct current stimulation for the treatment of pain in fibromyalgia. *Arthritis and Rheumatism 54*, 3988–3998.

Fujii M, Ohmoto Y, Kitahara T, *et al.* (1997). Motor cortex stimulation therapy in patients with thalamic pain. *Neurological Surgery 25*, 315–319.

Garcia-Larrea L, Peyron R, Mertens P, *et al.* (1999). Electrical stimulation of motor cortex for pain control: a combined PET-scan and electrophysiological study. *Pain 83*, 259–273.

Garcia-Larrea L, Convers P, Magnin M, *et al.* (2002). Laser-evoked potential abnormalities in central pain patients: the influence of spontaneous and provoked pain. *Brain 125*, 2766–2781.

Graff-Guerrero A, González-Olvera J, Fresán A, Gímez-Martín D, Méndez-Núñez JC, Pellicer F (2005). Repetitive transcranial magnetic stimulation of dorsolateral prefrontal cortex increases tolerance to human experimental pain. *Cognitive Brain Research 25*, 153–160.

Hirayama A, Saitoh Y, Kishima H, *et al.* (2006). Reduction of intractable deafferentation pain by navigation-guided repetitive transcranial magnetic stimulation (rTMS) of the primary motor cortex. *Pain 122*, 22–27.

Huang YZ, Edwards MJ, Rounis E, Bhatia KP, Rothwell JC (2005). Theta burst stimulation of the human motor cortex. *Neuron 45*, 201–206.

Irlbacher K, Kuhnert J, Röricht S, Meyer BU, Brandt SA (2006). Zentrale und periphere Deafferenzierungsschmerzen: Therapie mit der repetitiven transkraniellen Magnetstimulation? [Central and peripheral deafferent pain: Therapy with repetitive transcranial magnetic stimulation]. *Nervenarzt 77*, 1196–1203.

Johnson S, Summers J, Pridmore S (2006). Changes to somatosensory detection and pain thresholds following high frequency repetitive TMS of the motor cortex in individuals suffering from chronic pain. *Pain 123*, 187–192.

Juottonen K, Gockel M, Silen T, Hurri H, Hari R, Forss N (2002). Altered central sensorimotor processing in patients with complex regional pain syndrome. *Pain 98*, 315–323.

Kammer T, Beck S, Thielscher A, Laubis-Herrmann U, Topka H (2001). Motor thresholds in humans: a transcranial magnetic stimulation study comparing different pulse waveforms, current directions and stimulator types. *Clinical Neurophysiology 112*, 250–258.

Kanda M, Mima T, Oga T, *et al.* (2003). Transcranial magnetic stimulation (TMS) of the sensorimotor cortex and medial frontal cortex modifies human pain perception. *Clinical Neurophysiology 114*, 860–866.

Kaneko K, Kawai S, Fuchigami Y, Morita H, Ofuji A (1996). The effect of current direction induced by transcranial magnetic stimulation on the corticospinal excitability in human brain. *Electroencephalography and Clinical Neurophysiology 101*, 478–482.

Kaneko K, Kawai S, Taguchi T, Fuchigami Y, Yonemura H, Fujimoto H (1998). Cortical motor neuron excitability during cutaneous silent period. *Electroencephalography and Clinical Neurophysiology 109*, 364–368.

Katayama Y, Fukaya C, Yamamoto T (1998). Poststroke pain control by chronic motor cortex stimulation: neurological characteristics predicting a favorable outcome. *Journal of Neurosurgery 89*, 585–591.

Khedr EM, Gilio F, Rothwell J (2004). Effects of low frequency and low intensity repetitive paired pulse stimulation of the primary motor cortex. *Clinical Neurophysiology 115*, 1259–1263.

Khedr EM, Kotb H, Kamel NF, Ahmed MA, Sadek R, Rothwell JC (2005). Longlasting antalgic effects of daily sessions of repetitive transcranial magnetic stimulation in central and peripheral neuropathic pain. *Journal of Neurology, Neurosurgery and Psychiatry 76*, 833–838.

Knecht S, Ellger T, Breitenstein C, Ringelstein E, Henningsen H (2003) Changing cortical excitability with low-frequency transcranial magnetic stimulation can induce sustained disruption of tactile perception. *Biological Psychiatry 53*, 175–179.

Kofler M, Glocker FX, Leis AA, *et al.* (1998). Modulation of upper extremity motoneurone excitability following noxious finger tip stimulation in man: a study with transcranial magnetic stimulation. *Neuroscience Letters 246*, 97–100.

Kujirai T, Sato M, Rothwell JC, Cohen LG (1993). The effect of transcranial magnetic stimulation on median nerve somatosensory evoked potentials. *Electroencephalography and Clinical Neurophysiology 89*, 227–134.

Lang N, Siebner HR, Ernst D, *et al.* (2004). Preconditioning with transcranial direct current stimulation sensitizes the motor cortex to rapid-rate transcranial magnetic stimulation and controls the direction of after-effects. *Biological Psychiatry 56*, 634–639.

Le Pera D, Graven-Nielsen T, Valeriani M, *et al.* (2001). Inhibition of motor system excitability at cortical and spinal level by tonic muscle pain. *Clinical Neurophysiology 112*, 1633–1641.

Lefaucheur JP, Drouot X, Pollin B, Keravel Y, Nguyen JP (1998). Chronic pain treated by rTMS of motor cortex. *Electroencephalography and Clinical Neurophysiology 107*, 92.

Lefaucheur JP, Drouot X, Keravel Y, Nguyen JP (2001a). Pain relief induced by repetitive transcranial magnetic stimulation of precentral cortex. *Neuroreport 12*, 2963–2965.

Lefaucheur JP, Drouot X, Nguyen JP (2001b). Interventional neurophysiology for pain control: duration of pain relief following repetitive transcranial magnetic stimulation of the motor cortex. *Neurophysiologie Clinique 31*, 247–252.

Lefaucheur JP, Drouot X, Ménard-Lefaucheur I, Nguyen JP (2004a). Neuropathic pain controlled for more than a year by monthly sessions of repetitive transcranial magnetic cortical stimulation, *Neurophysiologie Clinique 34*, 91–95.

Lefaucheur JP, Drouot X, Ménard-Lefaucheur I, *et al.* (2004b). Neurogenic pain relief by repetitive transcranial magnetic cortical stimulation depends on the origin and the site of pain. *Journal of Neurology, Neurosurgery and Psychiatry 75*, 612–616.

Lefaucheur JP, Drouot X, Ménard-Lefaucheur I, Keravel Y, Nguyen JP (2006a). Motor cortex rTMS improves defective intracortical inhibition in patients with chronic neuropathic pain: correlation with pain relief. *Neurology 67*, 1568–1574.

Lefaucheur JP, Hatem S, Nineb A, Ménard-Lefaucheur I, Wendling S, Nguyen JP (2006b). Somatotopic organization of the analgesic effects of motor cortex rTMS in neuropathic pain. *Neurology 67*, 1998–2004.

Lorenz J, Garcia-Larrea L (2003). Contribution of attentional and cognitive factors to laser evoked brain potentials. *Neurophysiologie Clinique 33*, 293–301.

Maeda F, Keenan JP, Tormos JM, Topka H, Pascual-Leone A (2000). Modulation of corticospinal excitability by repetitive transcranial magnetic stimulation. *Clinical Neurophysiology 111*, 800–5.

Meyerson BA, Lindblom U, Linderoth B, Lind G, Herregodts P (1993). Motor cortex stimulation as treatment of trigeminal neuropathic pain. *Acta Neurochirurgica (Wien) Supplement 58*, 150–153.

Migita K, Uozumi T, Arita K, Moden S (1995). Transcranial magnetic coil stimulation of motor cortex in patients with central pain. *Neurosurgery 36*, 1037–1040.

Mima T, Oga T, Rothwell J, *et al.* (2004). Short-term high-frequency transcutaneous electrical nerve stimulation decreases human motor cortex excitability. *Neuroscience Letters 355*, 85–88.

Mylius V, Reis J, Kunz M, *et al.* (2006). Modulation of electrically induced pain by paired pulse transcranial magnetic stimulation of the medial frontal cortex. *Clinical Neurophysiology 117*, 1814–1820.

Mylius V, Reis J, Knaack A, *et al.* (2007). High frequency rTMS of the motor cortex does not influence the nociceptive flexion reflex but increases the unpleasantness of electrically-induced pain. *Neuroscience Letters 415*, 49–54.

Nakamura H, Kitagawa H, Kawagushi Y, Tsuji H (1996). Direct and indirect activation of human corticospinal neurons by transcranial magnetic and electric stimulation. *Neuroscience Letters 210*, 45–48.

Nguyen JP, Lefaucheur JP, Decq P, *et al.* (1999). Chronic motor cortex stimulation in the treatment of central and neuropathic pain. Correlations between clinical, electrophysiological and anatomical data. *Pain 82*, 245–251.

Nguyen JP, Lefaucheur JP, Keravel Y (2003). Motor cortex stimulation. In: BA Simpson (ed.), *Pain research and clinical management*, Vol. 15, *Electrical stimulation and the relief of pain*, pp. 197–209. Amsterdam: Elsevier.

Nitsche MA, Paulus W (2000). Excitability changes induced in the human motor cortex by weak transcranial direct current stimulation. *Journal of Physiology (London) 527*, 633–639.

Nuti C, Peyron R, Garcia-Larrea L, *et al.* (2005). Motor cortex stimulation for refractory neuropathic pain: four year outcome and predictors of efficacy. *Pain 118*, 43–52.

Oliviero A, Esteban MR, de la Cruz FS, Cabredo LF, Di Lazzaro V (2005). Short-lasting impairment of temperature perception by high frequency rTMS of the sensorimotor cortex. *Clinical Neurophysiology 116*, 1072–1076.

On AY, Uludag B, Taskiran E, Ertekin C (2004). Differential corticomotor control of a muscle adjacent to a painful joint. *Neurorehabilitation and Neural Repair 18*, 127–133.

Pascual-Leone A, Valls-Sole J, Wassermann E, Hallett M (1994). Response to rapid-rate transcranial magnetic stimulation of the human motor cortex. *Brain 117*, 847–858.

Peinemann A, Lehner C, Mentschel C, Munchau A, Conrad B, Siebner HR (2000). Subthreshold 5-Hz repetitive transcranial magnetic stimulation of the human primary motor cortex reduces intracortical paired-pulse inhibition. *Neuroscience Letters 296*, 21–24.

Peyron R, Garcia-Larrea L, Deiber MP, *et al.* (1995). Electrical stimulation of precentral cortical area in the treatment of central pain: electrophysiological and PET study. *Pain 62*, 275–286.

Pleger B, Janssen F, Schwenkreis P, Volker B, Maier C, Tegenthoff M (2004). Repetitive transcranial magnetic stimulation of the motor cortex attenuates pain perception in complex regional pain syndrome type I. *Neuroscience Letters 356*, 87–90.

Priori A, Berardelli A, Rona S, Accornero N, Manfredi M (1998). Polarization of the human motor cortex through the scalp. *Neuroreport 9*, 2257–2260.

Pujol J, Pascual-Leone A, Dolz C, Delgado E, Dolz JL, Aldoma J (1998). The effect of repetitive magnetic stimulation on localized musculoskeletal pain. *Neuroreport 9*, 1745–1748.

Ragert P, Dinse HR, Pleger B, *et al.* (2003). Combination of 5 Hz repetitive transcranial magnetic stimulation (rTMS) and tactile coactivation boosts tactile discrimination in humans. *Neuroscience Letters 348*, 105–108.

Ragert P, Becker M, Tegenthoff M, Pleger B, Dinse HR (2004). Sustained increase of somatosensory cortex excitability by 5 Hz repetitive transcranial magnetic stimulation studied by paired median nerve stimulation in humans. *Neuroscience Letters 356*, 91–94.

Raij TT, Forss N, Stancak A, Hari R (2004). Modulation of motor-cortex oscillatory activity by painful Aδ- and C-fiber stimuli. *NeuroImage 23*, 569–573.

Rasche D, Ruppolt M, Stippich C, Unterberg A, Tronnier VM (2006). Motor cortex stimulation for long-term relief of chronic neuropathic pain: a 10 year experience. *Pain 121*, 43–52.

Reid P, Pridmore S (2001). Improvement in chronic pain with transcranial magnetic stimulation. *Australian and New Zealand Journal of Psychiatry 35*, 252.

Rollnik JD, Wüstefeld S, Däuper J, *et al.* (2002). Repetitive transcranial magnetic stimulation for the treatment of chronic pain – a pilot study. *European Neurology 48*, 6–10.

Rollnik JD, Dauper J, Wustefeld S, *et al.* (2003). Repetitive magnetic stimulation for the treatment of chronic pain conditions. *Supplement to Clinical Neurophysiology 56*, 390–393.

Romaniello A, Cruccu G, McMillan AS, Arendt-Nielsen L, Svensson P (2000). Effect of experimental pain from trigeminal muscle and skin on motor cortex excitability in humans. *Brain Research 882*, 120–127.

Saitoh Y, Shibata M, Hirano S, Hirata M, Mashimo T, Yoshimine T (2000). Motor cortex stimulation for central and peripheral deafferentation pain. Report of eight cases. *Journal of Neurosurgery 92*, 150–155.

Salerno A, Thomas E, Olive P, Blotman F, Picot MC, Georgesco M (2000). Motor cortical dysfunction disclosed by single and double magnetic stimulation in patients with fibromyalgia. *Clinical Neurophysiology 111*, 994–1001.

Sampson S, Rome JD, Rummans TA (2006). Slow-frequency rTMS reduces fibromyalgia pain. *Pain Medicine 7*, 115–118.

Satow T, Mima T, Yamamoto J, *et al.* (2003). Short-lasting impairment of tactile perception by 0.9 Hz rTMS of the sensorimotor cortex. *Neurology 60*, 1045–1047.

Schwenkreis P, Janssen F, Rommel O, *et al.* (2003). Bilateral motor cortex disinhibition in complex regional pain

syndrome (CPRS) type I of the hand. *Neurology 61*, 515–519.

Seyal M, Masuoka LK, Browne JK (1992). Suppression of cutaneous perception by magnetic pulse stimulation of the human brain. *Electroencephalography and Clinical Neurophysiology 85*, 397–401.

Seyal M, Browne JK, Masuoka LK, Gabor AJ (1993). Enhancement of the amplitude of somatosensory evoked potentials following magnetic pulse stimulation of the human brain. *Electroencephalography and Clinical Neurophysiology 88*, 20–27.

Seyal M, Ro T, Rafal R (1995). Increased sensitivity to ipsilateral cutaneous stimuli following transcranial magnetic stimulation of the parietal lobe. *Annals of Neurology 38*, 264–267.

Seyal M, Siddiqui I, Hundal NS (1997). Suppression of spatial localization of a cutaneous stimulus following transcranial magnetic pulse stimulation of the sensorimotor cortex. *Electroencephalography and Clinical Neurophysiology 105*, 24–28.

Seyal M, Shatzel AJ, Richardson SP (2005). Crossed inhibition of sensory cortex by 0.3 Hz transcranial magnetic stimulation of motor cortex. *Journal of Clinical Neurophysiology 22*, 418–421.

Siebner HR, Lang N, Rizzo V, *et al.* (2004). Preconditioning of low-frequency repetitive transcranial magnetic stimulation with transcranial direct current stimulation: evidence for homeostatic plasticity in the human motor cortex. *Journal of Neuroscience 24*, 3379–3385.

Sindou MP, Mertens P, Bendavid U, Garcia-Larrea L, Mauguière F (2003). Predictive value of somatosensory evoked potentials for long-lasting pain relief after spinal cord stimulation: practical use for patient selection. *Neurosurgery 52*, 1374–1383.

Smania N, Corato E, Fiaschi A, Pietropoli P, Aglioti SM, Tinazzi M (2003). Therapeutic effects of peripheral repetitive magnetic stimulation on myofascial pain syndrome. *Clinical Neurophysiology 114*, 350–358.

Smania N, Corato E, Fiaschi A, Pietropoli P, Aglioti SM, Tinazzi M (2005). Repetitive magnetic stimulation: a novel therapeutic approach for myofascial pain syndrome. *Journal of Neurology 252*, 307–314.

Sommer M, Lang N, Tergau F, Paulus W (2002). Neuronal tissue polarization induced by repetitive transcranial magnetic stimulation? *Neuroreport 13*, 809–811.

Summers J, Johnson S, Pridmore S, Oberoi G (2004). Changes to cold detection and pain thresholds following low and high frequency transcranial magnetic stimulation of the motor cortex. *Neuroscience Letters 368*, 197–200.

Svensson P, Miles TS, McKay D, Ridding MC (2003). Suppression of motor evoked potentials in a hand muscle following prolonged painful stimulation. *European Journal of Pain 7*, 55–62.

Tamura Y, Hoshiyama M, Inui K, *et al.* (2004a). Facilitation of Aδ-fiber-mediated acute pain by repetitive transcranial magnetic stimulation. *Neurology 62*, 2176–2181.

Tamura Y, Okabe S, Ohnishi T, *et al.* (2004b). Effects of 1- Hz repetitive transcranial magnetic stimulation on acute pain induced by capsaicin. *Pain 107*, 107–115.

Tegenthoff M, Ragert P, Pleger B, *et al.* (2005). Improvement of tactile discrimination performance and enlargement of cortical somatosensory maps after 5 Hz rTMS. *PLoS Biology 3*, e362.

Töpper R, Foltys H, Meister IG, Sparing R, Boroojerdi B (2003). Repetitive transcranial magnetic stimulation of the parietal cortex transiently ameliorates phantom limb pain-like syndrome. *Clinical Neurophysiology 114*, 1521–1530.

Tsubokawa T, Katayama Y, Yamamoto T, Hirayama T, Koyama S (1991). Chronic motor cortex stimulation for the treatment of central pain. *Acta Neurochirurgica (Wien) Supplement 52*, 137–139.

Tsubokawa T, Katayama Y, Yamamoto T, Hirayama T, Koyama S (1993). Chronic motor cortex stimulation in patients with thalamic pain. *Journal of Neurosurgery 78*, 393–401.

Tsuji T and Rothwell JC (2002). Long lasting effects of rTMS and associated peripheral sensory input on MEPs, SEPs and transcortical reflex excitability in humans. *Journal of Physiology (London) 540*, 367–376.

Urban PP, Solinski M, Best C, Rolke R, Hopf HC, Dieterich M (2004). Different short-term modulation of cortical motor output to distal and proximal upper-limb muscles during painful sensory nerve stimulation. *Muscle and Nerve 29*, 663–669.

Urushihara R, Murase N, Rothwell JC, *et al.* (2006). Effect of repetitive transcranial magnetic stimulation applied over the premotor cortex on somatosensory-evoked potentials and regional cerebral blood flow. *NeuroImage 31*, 699–709.

Valeriani M, Restuccia D, Di Lazzaro V, *et al.* (1999). Inhibition of the human primary motor area by painful heat stimulation of the skin. *Clinical Neurophysiology 110*, 1475–80.

Valeriani M, Restuccia D, Di Lazzaro V, *et al.* (2001). Inhibition of biceps brachii muscle motor area by painful heat stimulation of the skin. *Experimental Brain Research 139*, 168–172.

Yoo WK, Kim YH, Doh WS, *et al.* (2006). Dissociable modulating effect of repetitive transcranial magnetic stimulation on sensory and pain perception. *Neuroreport 17*, 141–4.

Index

Note: Color plates are indexed **P1**, **P2** etc.